73rd World Foundry Congress (WFC 2018)

Creative Foundry

Scientific and Technical

Krakow, Poland
23 - 27 September 2018

ISBN: 978-1-5108-8041-2

Printed from e-media with permission by:

Curran Associates, Inc.
57 Morehouse Lane
Red Hook, NY 12571

Some format issues inherent in the e-media version may also appear in this print version.

Printed by Curran Associates, Inc. (2019)

For permission requests, please contact World Foundry Organization Ltd
at the address below.

World Foundry Organization Ltd
Winton House, Lyonshall
Kington, Herefordshire
HR5 3JP, United Kingdom

Phone: +44 (0) 1544 340332
Fax: +44 (0) 1544 340332

www.thewfo.com

Additional copies of this publication are available from:

Curran Associates, Inc.
57 Morehouse Lane
Red Hook, NY 12571 USA
Phone: 845-758-0400
Fax: 845-758-2633
Email: curran@proceedings.com
Web: www.proceedings.com

73rd World Foundry Congress

"Creative Foundry"

23rd–27th September 2018, Kraków, Poland

Congress Proceedings

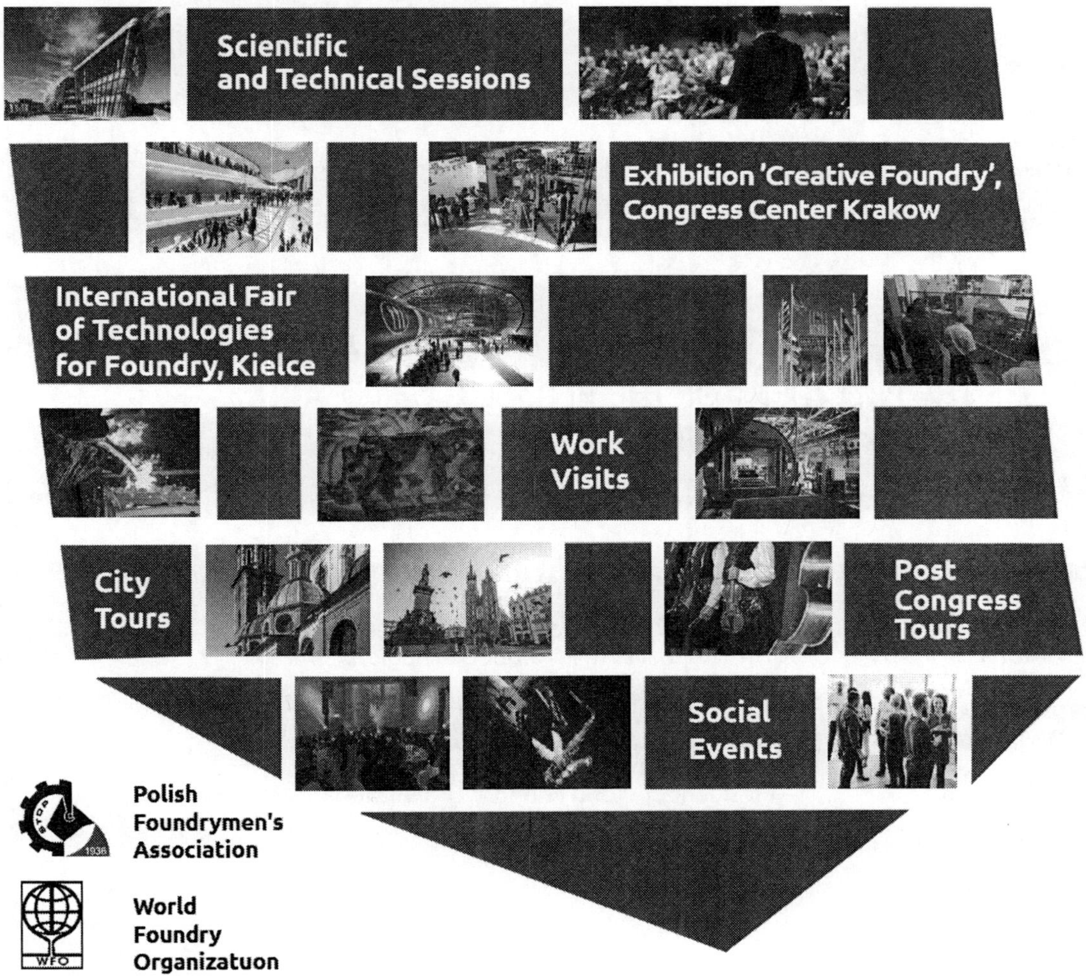

Polish Foundrymen's Association

World Foundry Organizatuon

Scientific and Technical

Organization

Executive of WFO

President	M. Fenyes (United Kingdom)
Vice President	U. Denizci (Turkey)
Immediate Past President	M.H. Kim (Korea)
Past President	X. Azpiri Gonzalez (Spain)
Traesurer	J.S. Suchy (Poland)

Members:
Y. Lou (China)
D. Dotson (United States of America)
U. Denizci (Turkey)
C. Kuhlgatz (Germany)
S. Panchal (India)
P.M. Cabbanne (France)
T. Torigoe (Japan)

General Secretary	A. Turner (United Kingdom)
Assistant Secretary	L. Postle (United Kingdom)
Media Manager	J. Call (USA)
Administration Secretary	Mr. J. J. Gozalez (Spain)

THE HONORARY PATRONAGE
OF THE PRESIDENT OF THE REPUBLIC OF POLAND
ANDRZEJ DUDA

Honorary Committee of the 73rd World Foundry Congress

Jarosław GOWIN	– Minister of Science and Higher Education
Jacek KRUPA	– The Marshal of the Małopolska Region
Jacek MAJCHROWSKI	– Major of the City of Krakow
Tadeusz SŁOMKA	– Rector of University of Science and Technology (AGH)
Ewa MAŃKIEWICZ-CUDNY	– President of NOT, The Polish Federation of Engineering Associatons
Mark FENYES	– President of World Foundry Organization

PUBLISHER/WYDAWCA:
STOWARZYSZENIE TECHNICZNE ODLEWNIKÓW POLSKICH/POLISH FOUNDRYMEN'S ASSOCIATION

Organizational Committee of the 73rd World Foundry Congress

Józef Szczepan SUCHY – Chairman
Tadeusz FRANASZEK – President of Polish Foundrymen's Association
Katarzyna LISZKA – General Secretary of Polish Foundrymen's Association
Andrew TURNER – General Secretary of World Foundry Organization

Michał BALICKI
Aldona GARBACZ-KLEMPKA
Aleksandra GRABARCZYK
Beata GRABOWSKA
Marcin GÓRNY
Joanna MADEJ
Jerzy Józef SOBCZAK
Michał SZUCKI
Emilia WILDHIRT
Dorota WILK-KOŁODZIEJCZYK
Jerzy ZYCH
Janusz LELITO
Magdalena JASIŃSKA
Magdalena PIETROŃ

Program Committee of the 73rd World Foundry Congress

Tadeusz FRANASZEK – President of Polish Foundrymen's Association
Józef Szczepan SUCHY – AGH University of Science and Technology
Rafał DAŃKO – Dean of Faculty of Foundry Engineering/AGH University of Science and Technology
Jerzy SOBCZAK – Foundry Research Institute, Krakow
Andrzej MOCHOŃ – Chairman of the Board/Kielce Trade Fairs
Przemysław TRAWA – Chairman of the Board/Poznań International Fair

Scientific Committee of the 73rd World Foundry Congress

Prof. Natalia Sobczak – Poland (Chairperson, Guest Editor)
Prof. Rajiv Asthana – USA (Vice-Chairman, Guest Editor)
Prof. Jan Szajnar – Poland (Vice-Chairman, Guest Editor)

Dr. Zdenka Zovko Brodarac – Croatia
Prof. Ludmil Drenchev – Bulgaria
Dr. Patricia Caballero – Spain
Dr. Rafał Cygan – Poland
Dr Donatella Giuranno – Poland/Italy
Prof. Edward Guzik – Poland
Prof. Milan Horacek – Czech Republic
Prof. Jolanta Janczak-Rusch – Switzerland
Prof. Jan Jezierski – Poland
Prof. Zenon Ignaszak – Poland
Prof. Savko Malinov – UK
Dr. Antton Meléndez – Spain
Prof. Sergey Mileiko – Russia
Prof. Primoz Mrvar – Slovenia
Prof. Anita Olszówka-Myalska – Poland
Prof. Hideo Nakae – Japan
Prof. Adel Nofal – Egipt
Prof. Tadeusz Pacyniak – Poland
Prof. Alberto Passerone – Italy
Prof. Marcin Perzyk – Poland
Prof. Yuriy Plevachuk – Ukraine
Dr. Wojciech Polkowski – Poland (Guest Editor)
T. Prucha – US (Guest Editor)
R.M. Purgert – US
Prof. Peter Schumacher - Austria
Dr. Mrityunjay Singh – USA
Prof. Augustin Sladek – Slovakia
Prof. Jerzy Józef Sobczak – Poland (Guest Editor)
Prof. Duru Stefanescu – US
Prof. Boris Straumal – Russia
Prof. Józef Szczepan Suchy – Poland (Guest Editor)
Dr. Lou Yanchun – China
Dr Vincent Vignal – France
Prof Gotthard Wolf – Germany
Prof. Pawel Zięba – Poland (Guest Editor)

Preface

Natalia Sobczak
Chairperson of Scientfic Committee of 73WFC

It is my honor and privilege to welcome you to the 73rd World Foundry Congress in Krakow, Poland on September 23–27, 2018.
The World Foundry Congresses (WFC), the most important events of the World Foundry Organization, are held bi-annually for the presentation of scientific and technical papers and providing an adequate networking opportunity for exchanging knowledge and experiences, the newest achievements and innovations in metal casting.

The 73rd WFC is organized under the motto "Creative Foundry", including broad thematic blocks:
- Materials,
- Technologies,
- Digitalization,
- Management,
- Ecology.

Industrial companies, universities, academia and research groups from 41 countries have submitted more than 390 contributions to present their achievements in research and research-implementation works. The 73rd WFC Program includes numerous Plenary Session, Technical Sessions, Poster Sessions and Young Researchers' Seminars.

This volume presents the 73WFC Proceedings that collect extended two-page abstracts of scientific or technical papers, including plenary lectures, oral presentations and posters. The 73rd WFC Proceedings is divided alike the Congress Program into different sections.
The Organizers are co-operating with following journals:
- Archives of Metallurgy and Materials – The Journal of Institute of Metallurgy and Materials Science and Committee on Metallurgy of Polish Academy of Science,
- Archives of Foundry Engineering,
- Foundry Trade Journal,
- Journal of Materials Engineering and Performance,
- Journal of Casting & Materials Engineering,
- International Journal of Metalcasting.

Participants can publish full-text papers after the Congress and after revieving procedure.
I would like to express my best gratitude to the authors of all contributions who shear their knowledge, expertise and best practices as the key factors contributing to Creative Foundry.

Organizing an event of this magnitude requires enormous work from a large group of individuals. Therefore, I would like to extend the gratitude to the members of Organizing Committee and Scientific Committee for their efforts in arranging this event.

Preface

Józef Szczepan Suchy
Chairman of the Organisational Committee of the 73rd WFC

Twenty seven years ago, Krakow was the host for foundry-men from all over the world. These years were especially important for Poland. Changes in political system and economy occurred. This was also a great challenge for people connected with the industry, also the foundry industry. Presently, the Polish foundry practice is the important industry sector of a great potential, open for investments and having a strong scientific and educational background. This, for sure, influenced the interests in the 73rd World Foundry Congress. We are very pleased that the participation in this event - one of the most important for our profession - is declared by persons from 42 countries. This interest is also confirmed by the record number of scientific and technical publications and - which is especially important - by the number of papers submitted for the session of young scientists and engineers.

The Congress organisers contributed a lot of effort to provide the congress participants – apart from numerous lecture and poster sessions and exhibitions – with the possibility of seeing the beauty of our country and Krakow. We also ensured – for the Authors of these interesting papers - the possibility of having them published in known scientific journals of international circulation, regardless of the Congress presentation and conference proceedings.

I would like to express my warm thanks to the Scientific Committee, to the team of the Organising Committee, especially to the Secretariat of the Congress main organiser, it means the Polish Foundry-men's Technical Association, efficiently supported by the Management and Secretariat of WFO. I would like to thank all institutions and persons, who were engaged in preparations of this great event, especially sponsors without which it would not be possible to ensure the proper standard.

The fact that entrusting to Poland the organisation of the 73rd WFC is of a great meaning is proofed by the honourable patronage of the President of the Republic of Poland as well as by the Government, City of Krakow and the whole technological environment.

We are hoping that our efforts will create an excellent atmosphere for expert meetings and discussions. Let it be also an occasion to discover our country and to make friends with Poland and with working here colleagues from the foundry sector.

Contents

Scientific

Materials:

Technical

Materials:

Scientific

MATERIALS

The 73rd World Foundry Congress
„Creative Foundry"
23 - 27 September 2018, Krakow, Poland

SOLIDIFICATION CHARACTERSITICS OF SILICON ALLOYED DUCTILE CAST IRONS

Stelian Stan, Iulian Riposan*, Mihai Chisamera, Iuliana Stan

POLITEHNICA University of Bucharest, Materials Science and Engineering Faculty, 313 Spl. Independentei, 060042 Bucharest, Romania

**Corresponding address: iulian.riposan@upb.ro*

Key words: casting, ductile iron, Si, Si-Mo, alloying, Mg-treatment, inoculation, thermal analysis, structure, graphite, ferrite

1. Silicon alloyed ductile cast irons

A review of Si / Si-Mo alloyed ductile iron data in the first part of the paper [1-6] shows that Si promotes ferrite, increases the strength and reduces the elongation and toughness by solid solution hardening (Fig. 1a).[1] The instability of a mixed ferrite-pearlite matrix [Si < 3%], could be replaced with more predictable and controllable ferritic grades [3.2 – 4.3%Si], with reduced hardness variation [± 4HB], increased cutting tool life, and consistently better mechanical properties [Rm=450-650MPa; $Rp_{0.2}$=350–500MPa; A=10-20%]. Mo additions in high Si ferritic grades (3.5–5.5%Si, 0.2–2.0%Mo) favours superior mechanical properties [Rm=450-550MPa; $Rp_{0.2}$=275 – 440MPa; A=4–10%] (Fig. 1b) and improved resistance to oxidation and corrosion at high temperatures.

Some important new knowledge was identified: Si segregation pattern and typical content to reach the maximum strength and ductility; graphite degeneration [Si and Al effects]; tolerance levels for other elements; increasing the strength and maximum working temperature up to 900-1000°C [Al, Ni, Co, Cr, V, Ti, Zr, etc supplementary alloying); carbides and phases characteristics; effects of Si and Al on transformation temperature and the oxide layer thickness; CGI–SiMoAl application; specific corrosion and fatigue behaviour in different media; crack growth mechanisms, etc.

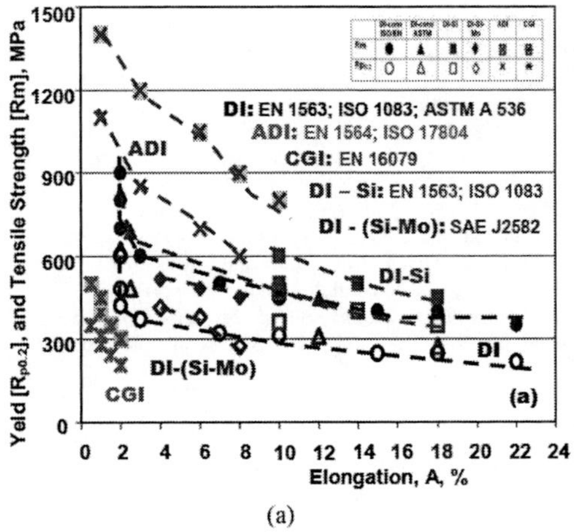

(a)

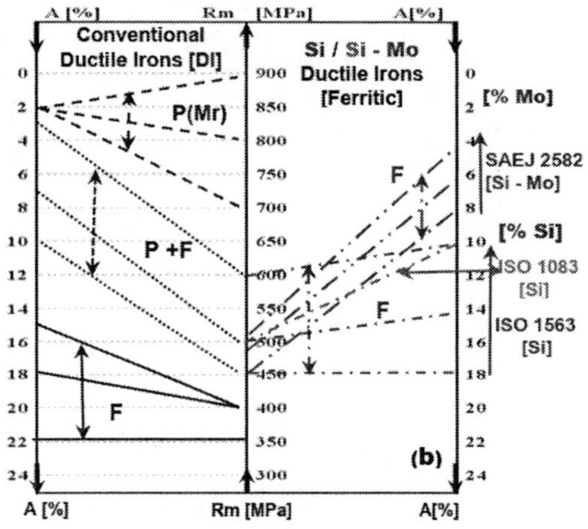

(b)

Fig. 1 Standardized Si [DI-Si] and Si–Mo [DI-(Si-Mo)] Ductile Irons [DI - Conventional (Si < 3%); ADI-Austempered Ductile Iron; CGI-Compacted Graphite Iron]

2. Experiments

Experiments studied the solidification pattern of three ductile iron compositions [2.5%Si; 4%Si and 4%Si-1.6%Mo (Si:Mo=2.5)], for 0.35–0.45%Mn and controlled on minor elements content. The cast irons tested were produced in a 10kg coreless induction furnace, 8000Hz frequency. The melt was superheated to 1550°C, held for 5 minutes and then tapped into a pre-heated tundish treatment ladle for magnesium treatment (2.0wt.% FeSiCaMg alloy). Inoculation in the ladle was applied, after Mg-treatment, with 0.3wt.% Ca, Ce, S, O-FeSi alloy; this alloy was selected as an effective inoculant to reduce chill and shrink tendency and to increase nodule count in ductile cast iron production. [6]

Conventional ceramic cups [0.75cm cooling modulus] were used for thermal analysis. Standard wedge samples [ASTM A 367] having a variable cooling modulus [0.11-0.35cm] and 20mm diameter bar were produced in resin sand moulds, for chill sensitiveness, macro-and-microstructure analyses.

The graphite characteristics were evaluated with Automatic Image Analysis [analySIS® FIVE Digital Imaging Solutions software].

3. Results

The influence of Si and Si-Mo content and inoculation on the representative temperatures and undercooling during the eutectic reaction and at the end of solidification was observed. Without inoculation a supplementary high Mo addition decreased the beneficial effect of Si on decreasing undercooling.

As expected, un-inoculated irons are characterized by low temperatures, at the beginning (TEU) and at the end (TER) of the eutectic reaction, and at the end of solidification (TES), while inoculation increases all of these temperatures. High Si [4%Si] ductile iron is characterized by higher values of ΔT_1 [TEU-Tmst] and ΔT_3 [TES-Tmst] [less negative] undercooling parameters, for both un-and-inoculated irons, compared to conventional material (2.5%Si). The combination of Si alloying with efficient inoculation led to positive values of the ΔT_3 parameter (Fig. 2).

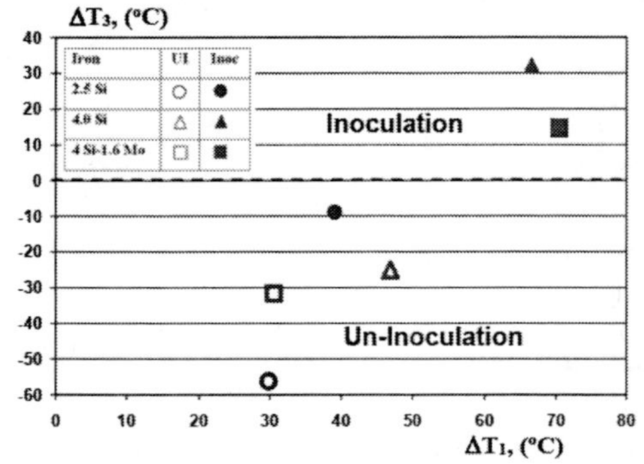

Fig. 2 Undercooling degrees (°C) referring to metastable eutectic temperature (Tmst), of un-inoculated (UI) and inoculated (Inoc) ductile irons, at the beginning of eutectic reaction (ΔT_1) and at the end of solidification (ΔT_3).

A different pattern was found for Si-Mo alloyed ductile irons, despite the same Si level and the same inoculation. Mo additions at higher levels in high Si ductile iron decreased the ΔTs eutectic interval and led to increased undercooling over the entire solidification period, in un-inoculated irons. Inoculation caused the greatest change for the eutectic reaction in Si-Mo ductile iron and compensates for the negative effect of Mo mainly during the eutectic reaction. It appears that Mo, at high levels, compromised the beneficial effect of Si alloying, while an efficient inoculation became mandatory to improve the solidification pattern up to the end of solidification.

The graphite phase and metal matrix characteristics are dependent on the solidification cooling rate, the contribution of inoculation and alloying system (Fig. 3). Despite that Si is known to favor chunky graphite formation, [6] effective inoculation decreased the sensitivity to form a dark colored porous region in the thermal center of castings. In completed solidification conditions the

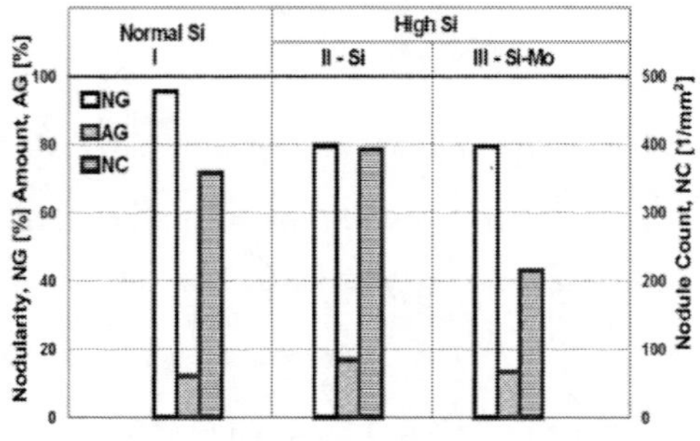

Fig. 3 Structure characteristics of inoculated cast irons

applied inoculation allowed good structure characteristics to be preserved, with the highest graphite amount and nodule count in Si-ductile iron, and the similar nodularity, medium graphite amount for the lowest nodule count, in Si / Mo = 2.5 ratio Si-Mo alloyed ductile iron.

Graphite nodularity values [NG-SSF] were also calculated for different levels of graphite sphericity shape factors [SSF], as influence of the alloying system. Normally, as SSF values increase, the corresponding NG-SSF values decrease. Using a minimum sphericity shape factor SSF value of 0.5, at least 90% nodularity was observed for all of the tested cast irons. If the minimum limit of the shape factor is increased, the nodularity appears to go down, especially in Si-alloyed irons.

It appears that a high Mo-content addition in 4%Si iron promotes pearlite formation, but a powerful inoculation avoided the formation of primary carbides, and favoured graphite particles morphologies. It was found that inoculation is also important for high-Si but particularly so for Si-Mo alloyed ductile cast irons, requiring a high efficiency inoculation procedure.

References

[1] S. Stan, I. Riposan, M. Chisamera, M. Barstow. (2018). Solidification Pattern of Silicon Alloyed Ductile Cast Irons. In 122nd AFS Congress, Fort Worth, TX, USA, P. 18-022.

[2] W. Stets, H. Loblich, G. Gassner, P. Schumacher. (2013). Solution Strengthened Ferritic Ductile Iron According DIN EN 1563:2012. In 2013 Keith Millis Symp. on Ductile Iron, Nashville, TN, USA, pp. 283-292.

[3] W. Menk. (2014). Development of cast iron for exhaust applications. In Europ. Cast Iron Meeting, Nancy, France.

[4] L. Dekker, B. Tonn. (2016). Occurrence and behavior of Mo containing precipitates in nodular cast iron at high temperatures. Int. J. Cast Met. Res., 29 (1-2), pp. 85-91.

[5] S. Xiang, C. Oberg, M. Ekstrom, B. Zhu, S. Jonson. (2017). High-temperature corrosion fatigue behavior of ductile cast irons for exhaust manifolds applications. In 11th Int. Symp. on the SPCI, Jonkoping, Sweden, Paper no. 16.

[6] I. Riposan, T. Skaland. (2017). Modification and Inoculation of Cast Iron. In D.M. Stefanescu, Volume Editor, Cast Iron Science and Technology Handbook (pp.160-176), American Society of Materials, ISBN 978–62708-133-7.

The 73rd World Foundry Congress
„Creative Foundry"
23 - 27 September 2018, Krakow, Poland

EFFECT OF INTENSIVE COOLING OF ALLOY ALSI7MG WITH ALLOY ADDITIONS ON THE MICROSTRUCTURE AND MECHANICAL PROPERTIES

Bogusław Pisarek [1]*, Cezary Rapiejko [1], Tadeusz Pacyniak [1]

[1]Lodz University of Technology/ Department of Materials Engineering and Production Systems, 1/15 Stefanowskiego Street, 90-924 Lodz, Poland
**Corresponding address: boguslaw.pisarek@p.lodz.pl*

Key words: Solidification Process, Aluminum Alloy, Alloy Additions, Intensive Cooling, Mechanical Properties

1. Introduction

The Department of Materials Engineering and Production Systems, Lodz University of Technology (KTMiSP PŁ), has realized, within the Applied Studies Program, the project „Elaboration of innovative technologies of producing complex construction high quality precisions casts from light metal alloys". The project was implemented within the frames of a consortium, in which KTMiSP PŁ was one of its members. The practical aim of the project was to develop innovative technologies of producing precision casts in the lost wax casting technology from the newly-elaborated aluminum alloys [1]. The technology has to be an alternative for the pressure casting processes for products of high technical and economical effectiveness in the small and medium lot production. One of the project tasks realized at KTMiSP PŁ was to examine the effect of the physico-chemical state of the liquid metal on the microstructure and physico-chemical properties of the experimental casts. The task includes the application of intensive cooling of the mold and its effect on the microstructure and mechanical properties of the elaborated aluminum alloys.

2. Work methodology

The investigations of the effect of intensive cooling of the ceramic mold on the mechanical properties of the alloys were performed for the aluminum alloy AlSi7Mg0.3 (AK7) and AlSi7Mg+4%Cu+2.5%Ni+0.1% of Cr, V, Mo, W each (AK7+) [1]. The alloys were melted in an induction furnace in a AC20-type crucible from silicon carbide. The alloy was modified with strontium (AlSr10) and refined with nitrogen (N_2). The casting temperature was 750 °C. For the construction of the gating system, models of the following, used at the Labour Co-operative ARMATURA (Lodz, Poland), were applied: the pouring basin, the cross-gate and the ingates WD25 and WD45. The ceramic mold for the TDA tests and for the production of strength samples – produced at the Labour Cooperative ARMATURA – consists of seven layers. The molds were prepared by means of the investment casting method [2]. The quartz sand granularities of each coating varied between 0.1 mm and 1 mm. After applying the seven layers on the wax model, composed of a mixture of paraffin, stearic acid, synthetic wax and copolymer, the molds were subjected to drying, and next, the molding sand was melted from them at 150 °C. The following step in the mold preparation process was annealing at 800 °C in a tunnel furnace, and then the mold's cooling to 180 °C. The mold prepared in this way was filled with the liquid alloy and next submerged in a water solution of Polihartenol-E8 (Fig. 1).

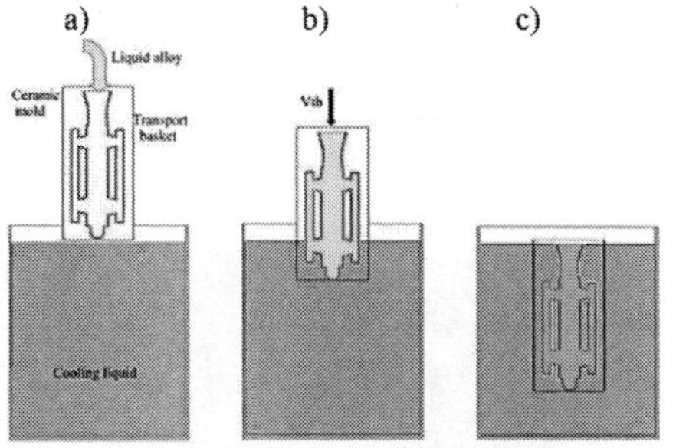

Fig. 1. Steps of alloy pouring(a) and cooling of the ceramic mold intensively cooled (b, c); Vtb=0.07 m/s

In this way, the mold for the casts assigned for the strength tests as well as the tester ATD10C-PŁ were prepared. The liquid metal was poured into the previously prepared tester ATD10C-PŁ, which, differently from the conventional method, was also submerged in the water solution of Polihartenol-E8 (Fig. 2).

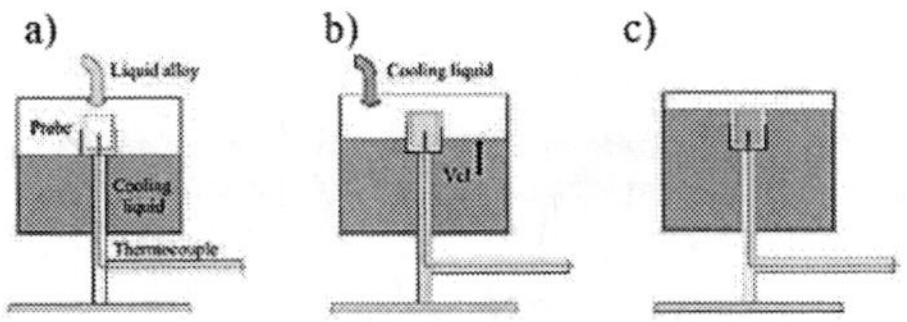

Fig. 2. Steps of alloy pouring (a) and cooling of the TDA probe intensively cooled (b, c); Vcl=0.07 m/s

The tester was used to perform a thermal and derivative analysis (TDA) of the solidifying alloy. In order to reveal the phases in the alloy's microstructure, the samples were etched in a 4% water solution of hydrofluoric acid. The metallographic tests were carried out by means of the optical microscope Nikon Eclipse MA200. The tests of the mechanical properties were conducted on the machine Instron 4485. The HB hardness measurements were performed by the Brinell method with a 2.5 mm diameter steel ball, on the hardness tester KB Prüftechnik. The simulation of the process of casting into the ceramic mold (before its submersion in the cooling liquid) as well as cooling of the examined alloys was performed by means of the MAGMA5.3 program.

3. Description of achieved results of own researches

A simulation of the solidification process of the alloy in the ceramic mold (Fig. 3) was used to determine the time, calculated from the moment of filling the mold, after which the filled mold could be submerged in the cooling liquid without the risk of the cast being gasified by the water vapour.

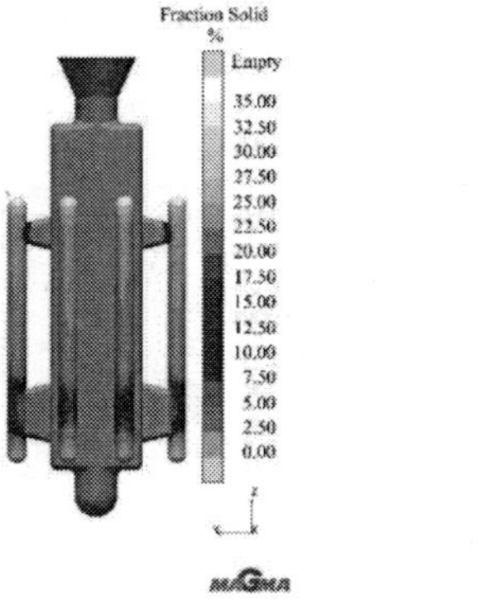

Fig. 3. Fraction of solid after 19.6 s (AlSi7Mg)

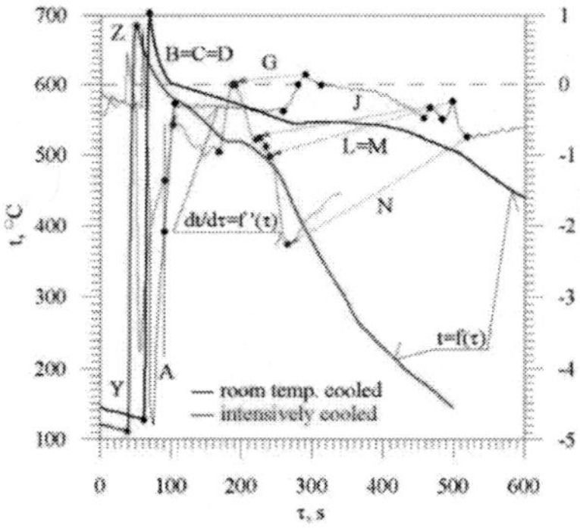

Fig. 4. Comparison of TDA characteristics of AK7+ alloy in room temperature cooled and intensively cooled AK7+s

Intensive cooling of the cast, both AK7 and AK7+, in the ceramic mold (Fig. 4) increases the kinetics of the thermal processes of the alloy crystallization dt/dτ=f '(τ), lowers the temperature of phase transformations t=f(τ) and shortens the time of alloy crystallization $\tau_{cryst}=\tau_N-\tau_A$, which leads to a refinement of the alloys' microstructure. The refinement of the microstructure of intensively cooled alloys and the introduction of thermal stresses caused by the cooling process is the main reason for the increase of UTS and HB of the examined alloys (Fig. 5).

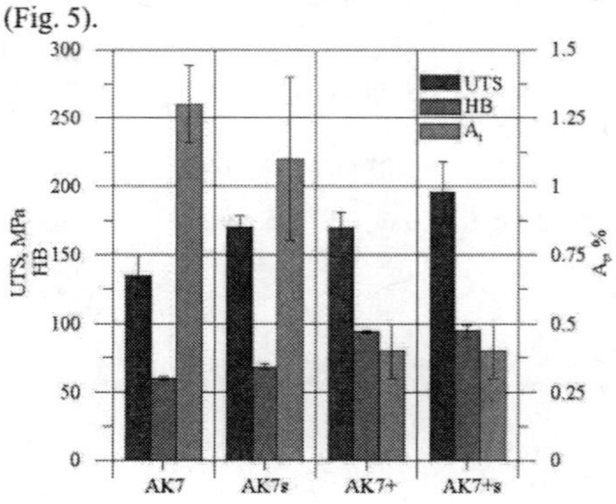

Fig. 5. Mechanical properties of alloys AK7 and Ak7+ in room temperature cooled and intensively cooled AK7s and AK7+s

4. Conclusions

The following conclusions can be drawn from the studies of intensive cooling of alloys without additions: AK7, as well as with Cu, Ni, Cr, Mo, V and W additions, AK7+, in a water solution of Polihartenol-E8:

- in reference to the microstructure:
  - for the examined alloys, both AK7 and AK7+, a microstructure refinement was observed mostly of the primary phase α and eutectic precipitations of phase β(Si),
- in reference to the mechanical properties:
  - a tensile strength UTS increase was observed (for alloy AK7 by 26 %, for AK7+ by 15 %),
  - the elongation At of the examined alloys was reduced (for AK7 by 15 %, for AK7+ the elongation did not change),
  - the hardness of alloy AK7 increased by 15 %, and of alloy AK7+ by 0.7 %.

Acknowledgments

The research was realized within the project PBS I, financed by the National Centre for Research and Development. Project No.: 178739. Project realized in 2013-2015.

References

[1] Pisarek, B.P., Rapiejko, C., Szymczak, T. & Pacyniak, T. (2017). Effect of alloy additions on the structure and mechanical properties of the AlSi7Mg0.3 alloy. Archives of Foundry Engineering. Volume 17 (1), 137-142

[2] Rapiejko, C., Pisarek, B.P. & Pacyniak, T. (2017) Effect of Intensive Cooling of Alloy AM60 with Chromium and Vanadium Additions on Cast Microstructure and Mechanical Properties. Arch. Metall. Mater. Volume 62 (1), 319-324

The 73rd World Foundry Congress
„Creative Foundry"
23 - 27 September 2018, Krakow, Poland

THE INFLUENCE OF WALL THICKNESS ON THE FATIGUE LIMIT OF V-NOTCHED BARS MADE OF AS-CAST OR HEAT TREATED DUCTILE IRON

Stefano Masaggia [1]*, Giovanni Meneghetti [2]

[1]R&D Department, Zanardi Fonderie S.p.A., via Nazionale, 3 – 37046 Minerbe (VR), Italy
[2] Department of Industrial Engineering, University of Padova, via Venezia, 1 – 35131 Padova, Italy
**Corresponding address: mst@zanardifonderie.com*

Key words: Ductile Irons, Fatigue, Fracture Mechanics, Stress Intensity Factor, Mode I loading

A fatigue model, previously published in the technical literature [1,2], has been applied to estimate the fatigue limit of ductile irons and heat treated ductile irons in presence of V-shaped notches. When the notch tip radius approaches reduced values, the model is based on the mode I Notch-Stress Intensity Factor (N-SIF). Conversely, when the notch tip radius is large, the classical approach based on the linear elastic stress concentration factor is matched. This model contains the well known Kitagawa-Takahashi [3] as a special case.

1. Introduction

In the recent past, Atzori, Lazzarin and Meneghetti proposed a fatigue model valid for components that include sharp as well as rounded V-shaped notches of any size [2]. This model is the generalization of a previous one proposed by Atzori and Lazzarin [4]. The equation to estimate the fatigue limit of sharp V-shaped notches is the following [2]:

$$\Delta\sigma_{g,th} = \frac{\Delta K_{I,th}^{V}}{\sqrt{\pi}\left(a_{eff} + a_0^V\right)^{\gamma}} \quad (1)$$

where $\Delta K_{I,th}^V$ is the threshold value of Notch-Stress Intensity Factor, a_{eff} is the effective notch depth and a_0^V is a material and notch geometry-dependent parameter. Appropriate expressions to evaluate all previous material and stress parameters are reported in [2]. It should be noted that when the notch opening angle reduces to zero (i.e. γ=0.5), Eq. 1 exactly match the Linear Elastic Fracture Mechanics (LEFM). The model reported in following Fig. 1 is useful to assess the fatigue limit of any kind of notch (long and short cracks, blunt and crack-like notches with parallel flanks and blunt and sharp notches with an opening angle different from zero), including surface defects, by means of a unified approach; an application of this model is reported in [5,6]. Further an application of such engineering design tool to V-notched bars obtained from as-cast as well as heat treated cast iron Y-blocks of different wall thickness is performed in the present work.

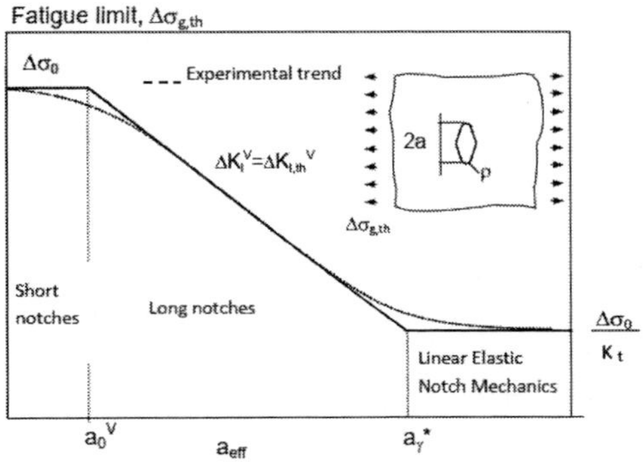

***Fig. 1.** Fatigue limit estimation of a notched component subject to mode I loading*

2. Fatigue characterisation

The proposed fatigue design tool has been applied to estimate the rotating bending fatigue limit of round specimens machined from separately cast test bars made of as-cast or heat treated ductile iron, having relevant wall thickness 25 mm, 50 mm and 75 mm respectively; specimens were characterised by smooth surface (6.5 mm diameter) or by the presence of a V-notch having a 45° notch opening angle (8-1-0.1 and 10-2-0.1: the code indicates gross diameter, notch depth of and notch tip radius respectively). The singularity degree of V-notch characterized by this opening angle is close to the value relevant to a crack 0°: then all the V-notches have been treated as cracks having the same depth of the notches from the stress analysis point of view. A short stair-case procedure was adopted to determine the fatigue limits at 50% probability of survival; run out was assumed at $5 \cdot 10^6$ cycles. Table 1 reports the stair-case results of plain specimens in terms of stress amplitude $\sigma_{Ag50\%}$ (defined as half the difference between the maximum and the minimum stress) referred to the gross section and, furthermore, the threshold range ΔK_{th} which was calculated by fitting the data relevant to the V-notched specimens. Results referred to relevant wall thickness 25 mm have also been represented as Kitagawa-Takahashi diagram in Figs. 2-5.

Table 1. Fatigue tests results

Material	w. t.[mm] (HBW)	$\sigma_{Ag50\%}$ [MPa]	ΔK_{th} [MPa·√m]
DI ferritic	25 (156)	278.6	12.66
	50 (145)	277.5	14.34
	75 (161)	265.0	16.54
DI pearlitic	25 (250)	350.0	17.71
	50 (239)	315.0	20.14
	75 (244)	326.7	19.76
IDI® [7]	25 (255)	390.0	15.96
	50 (249)	321.4	18.75
	75 (242)	314.0	19.53
ADI 1050	25 (346)	441.3	20.25
	50 (349)	362.2	24.69
	75 (354)	355.7	23.13

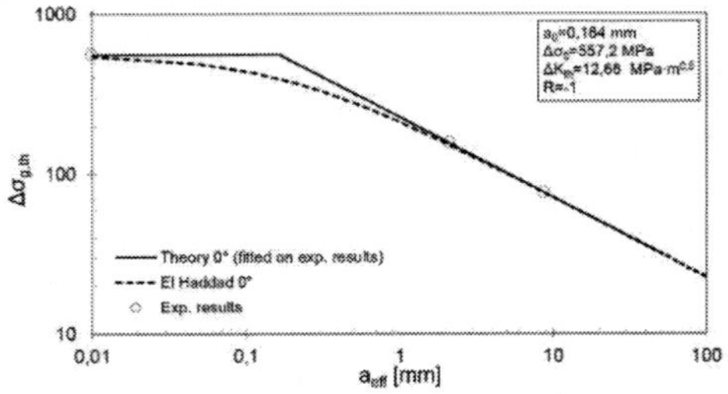

Fig. 2. Kitagawa diagram of DI ferritic (w.t. 25mm)

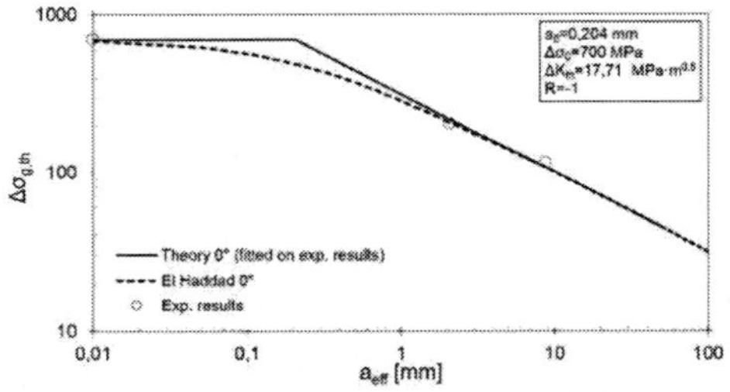

Fig. 3. Kitagawa diagram of DI pearlitic (w.t. 25 mm)

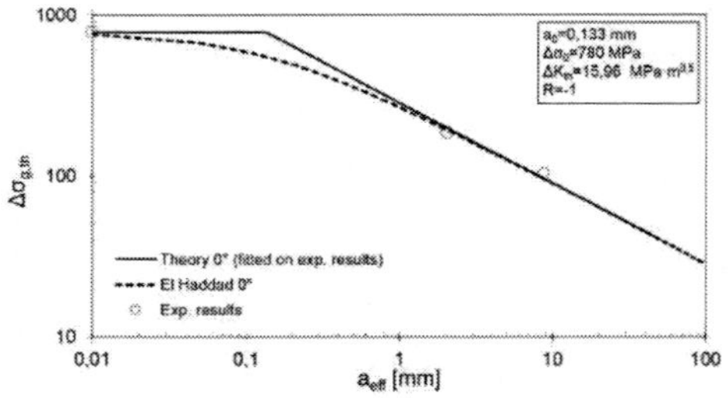

Fig. 4. Kitagawa diagram of IDI® (w.t. 25 mm)

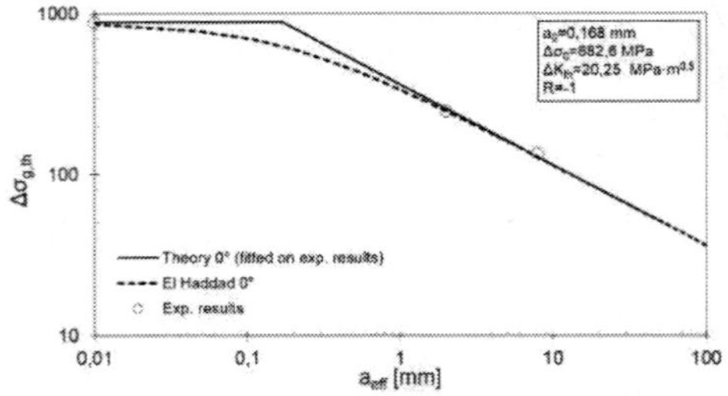

Fig. 5. Kitagawa diagram of ADI 1050 (w.t. 25 mm)

3. Conclusion

The paper illustrated the application of a fatigue testing procedure in presence of notch effects. The material fatigue behavior has been characterized experimentally by means of smooth specimens, in order to derive the plain fatigue limit, and by means of sharply notched specimens, in order to found the threshold value of the stress intensity factor. All theoretical estimations indicate a sharp behavior of the notched specimens, i.e. the 0.1 mm notch tip radius is ineffective on the fatigue strength, which can be assimilated to that of notch having a zero notch tip radius. A complete set of data to be used in fatigue design was provided. Many efforts in terms of investigation at microscale are now continuing to better understand the microstructure response vs wall thickness under severely notched conditions, since the threshold value ΔK_{th} remains consistent.

References

[1] Atzori, B., Lazzarin, P., Meneghetti, G., "Fracture Mechanics and Notch Sensitivity", Fatigue & Fracture of Engineering Materials and Structures, 26, pp.257-267 (2003), Blackwell.

[2] Atzori, B., Lazzarin, P., Meneghetti, G., "A unified treatment of the mode I fatigue limit of components containing notches and defects", International Journal of Fracture, 133, pp. 61-87 (2005).

[3] Kitagawa, H., Takahashi, S., "Applicability of fracture mechanics to very small cracks in the early stage", Proceedings of the 2nd International Conference on Mechanical Behavior of Materials, pp. 627–631 (1976).

[4] Atzori, B., Lazzarin, P., "Notch sensitivity and defect sensitivity under fatigue loading: two sides of the same medal", International Journal of Fracture 107, L3-L8 (2001).

[5] Masaggia, S., "The development of ADI and IDI in Italy", Proceedings of the 10th International Fatigue Conference, 2, pp. 1459-1476 (2010), Elsevier.

[6] Meneghetti, G., Masaggia, S., "Estimation of the fatigue limit of components made of Austempered Ductile Iron weakened by V-shaped notches", Proceedings of the 70th World Foundry Congress, pp. 387-395 (2012), Curran.

[7] Bronzato, M., Ilibasic, Z., Zanardi, F., Minerbe (Italy), Patent n° WO2008/003395 (A2) (2008), World Intellectual Property Organization - International Bureau.

The 73rd World Foundry Congress
„Creative Foundry"
23 - 27 September 2018, Krakow, Poland

DEVELOPMENT OF A SQUEEZE SEMISOLID HIGH PRESSURE DIE CASTING PROCESS FOR MAGNESIUM STRUCTRURAL PARTS

I. Vicario [1]*, I. Crespo [1], D. Val [2], U. Weiss [3], D. Cao [2], I. Martinez de la Pera [4]

[1]Tecnalia Research & Innovation, C/Geldo. Edificio 700. E-48160. Derio – Bizkaia, Spain
[2] Grupo Antolin Ingeniería S.A., Ctra. Madrid-Irún, Km. 244,8. E09007. Burgos, Spain
[3] Ford Research & Innovation Center Aachen, Süsterfeldstr. 200. D-52072. Aachen, Germany
[4] Aurrenak, S. Coop. C/ Vitorialanda, 15. E-01010. Vitoria – Gasteiz, Spain
**Corresponding address: iban.vicario@tecnalia.com*

Key words: Structural parts, magnesium alloys, High Pressure Die Casting, Semi-solids, Vacuum HPDC

1. Introduction

Between the different casting procedures, High Pressure Die Casting (HPDC) is the most common method used to produce aluminium and magnesium castings, due to the excellent balance of cost and properties for high production volumes. Nevertheless, conventional HPDC process presents certain limitations in terms of final mechanical properties, due to the intrinsic porosity, so different processes have been developed in order to overcome this problem, including semi-solid processing and vacuum HPDC casting. However, the vacuum and the semi-solid processes increases the complexity and cost of the process. Trying to tackle this, EFEVE project was born to develop a different production process able to produce high performance magnesium cast structural parts.

The developed new process is based on the injection of the alloy at low speeds, in a range slightly over the standard semi-solid parameters with a modified die with thick gates and at high temperatures. In despite of working with speeds that are not in the lamellar flow, the obtained parts present very low porosity, what allows to use thermal treatments to increase the ductility without blistering. The demonstration has been performed in a magnesium body joint, using standard HPDC, Vacuum HPDC and the new squeeze casting process.

2. Experimental conditions

A New Generic Body Joint part has been designed, with similarities to the demonstrator used in NADIA European project to compare the results of the new developed process. In Fig. 1 we can observe the definition of the new geometry of the Body Joint.

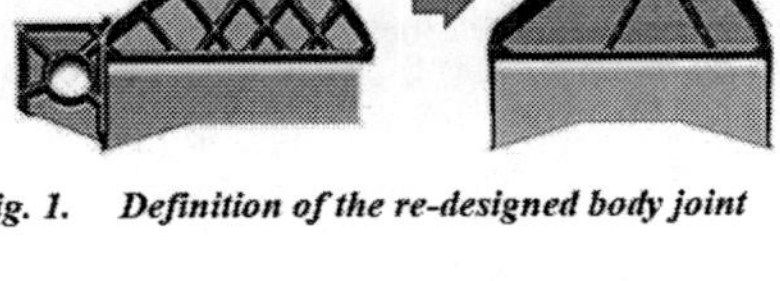

Fig. 1. Definition of the re-designed body joint

For the re-design of the Body Joint the process feasibility for both SCMg and HPDC production, a peak force reduction to reduce passenger body loads, an homogeneous energy absorption and an increased part symmetry to prevent bifurcation were taken into account. A 3 stage procedure was established to optimize the body joint, as shown in Fig. 2.

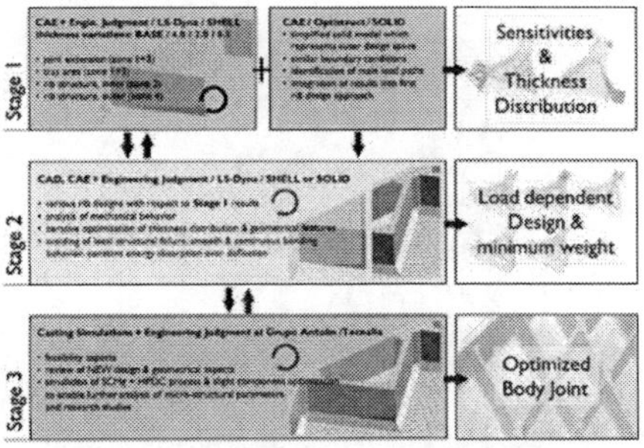

Fig. 2. Definition of the re-designed body joint

In order to obtain squeeze speeds a valve incorporated in the cellule permits to obtain second injection speeds between 0.5 and 1 m/s, the cellule has been equipped with a high vacuum device and advanced data controllers have been installed in the HPDC cellule and die in order to control the parameters of the process as shown in Fig. 3.

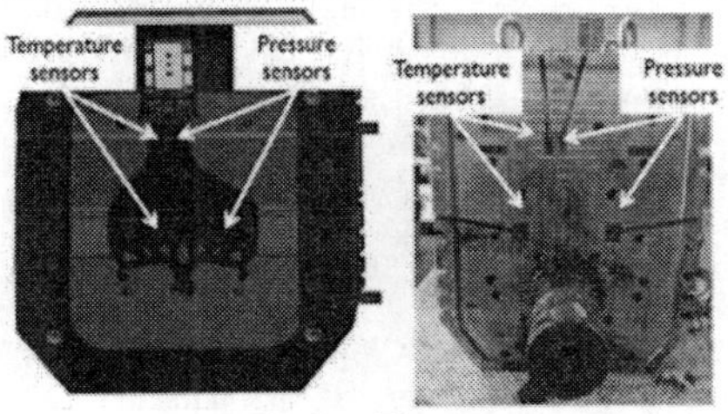

Fig. 3. Die sensors

In order to determinate the position of the overflows and the correct design and process parameters, Magma and Procast simulations were taken into account.

3. Results and discussion

The laboratory test defines a series of parameters suitable for the new Mg squeeze casting process. In Table 1 there are resumed the main parameters.

Table 1. Main parameter comparison

Parameters	Mg Squeeze	Mg HPDC
Mold Temperature	300ºC	200-250ºC
Alloy Temperature	690ºC	680ºC
Specific pressure	650-1200 bar	<700 bar
Cycle time	57 s	60 s
2nd plunger speed	0.5-2 m/s	6-10 m/s

X-ray analysis where used in order to determinate the internal porosity of obtained samples. In Fig. 4 from green to red color the correspondent areas with and without acceptable porosity and in Fig. 5 the situation of every check point in the part.

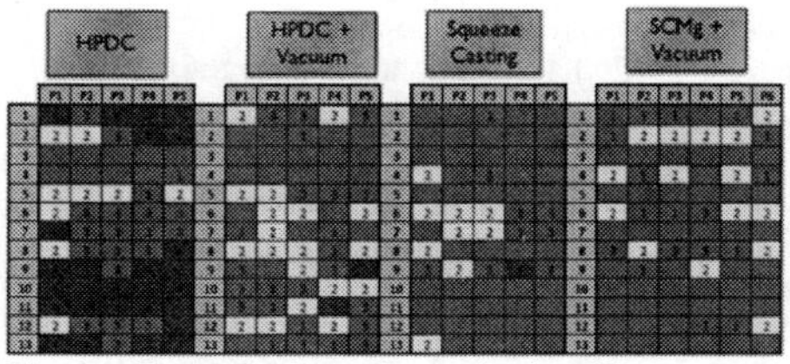

Fig. 4. Determination of porosity in different parts and areas

Fig. 5. Definition of RX studied areas

As we can observe, the Vacuum Squeeze casting almost eliminate the porosity and allows thermal treatments. Tensile tests were performed in order to determinate mechanical properties. SCMg material without heat treatment shows improved mechanical properties compared to HPDC material.

A significant improvement is being introduced by T4 heat treatment of SCMg parts: Elongation to fracture practically is doubled and hardening behaviour is improved, enabling more widespread and homogeneous plastic deformation in crash load cases, as shown in Fig. 6.

Penetration test results show a clear advantage for the new SCMg process with T4 thermal treatment, delivering a 40 % increase of the average penetration force level and 30 % improved energy absorption as shown in Fig. 7.

In the axial tests the new SCMg process with a T4 thermal treatment demonstrated the potential to highly increase the absorption energy (up to 90 %), as shown in Fig. 8.

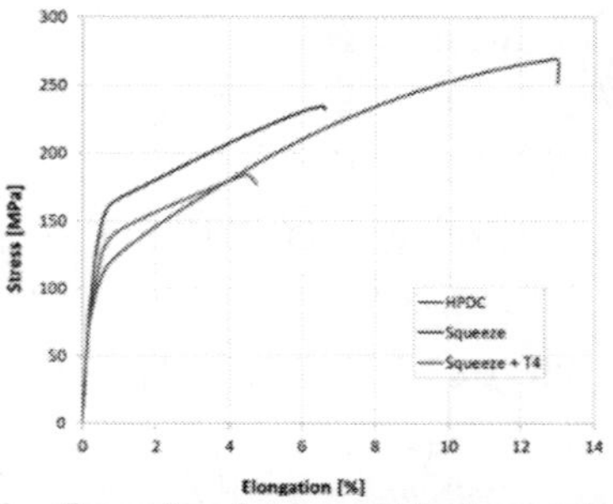

Fig. 6. Comparison between obtained properties for HPDC, SC and SC+T4 in AM60B alloy

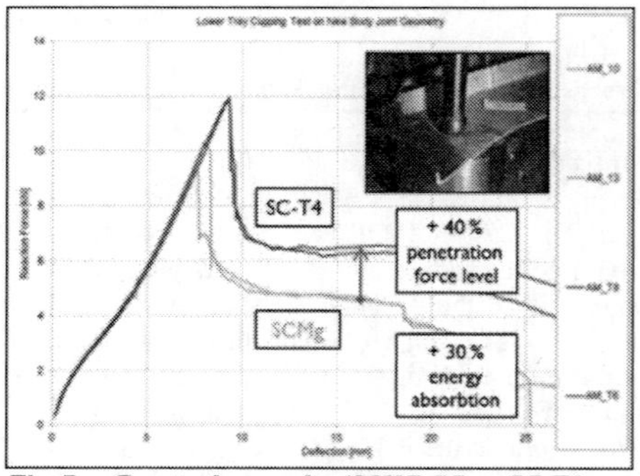

Fig. 7. Penetration results AM60B SC vs. SC-T4

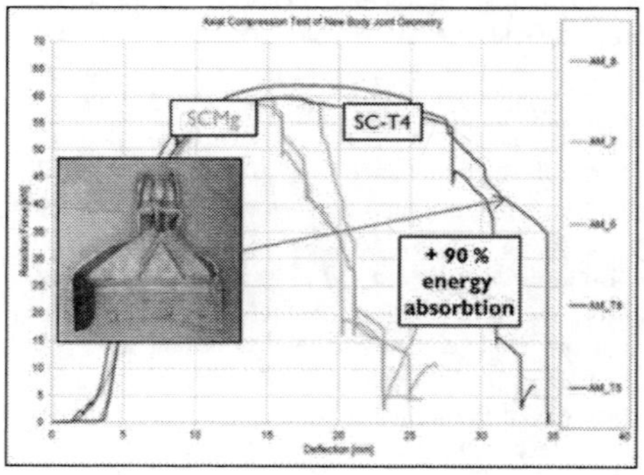

Fig. 8. Example of obtained results with a SC Mg with and without T4 thermal treatment

4. Conclusions

Comparing the state-of-the-art HPDC process for AM60 material with the new SCMg process with T4 heat treatment, the new SCMg-T4 process successfully has shown a unique performance potential (with heat treatment), unrivalled low porosity (even compared to state-of-the-art vacuum HPDC), high strength + high ductility and an attractive cost-benefit ratio.

Acknowledgments

This work has been carried out under the framework of the EFEVE European project funded by the European Union through its Seventh Framework Programme for research, technological development and demonstration under grant agreement No 314582.

ON QUANTITATIVE METALLOGRAPHY VACUUM CASTED NI-BASED SUPERALLOY USED IN AERO JET ENGINE CONSTRUCTION – TURBINE BLADES SEM STRUCTURE ANALYSIS AFTER VARIOUS WORKING HOURS

Juraj Belan [1]*, Alan Vaško [1], Lenka Kuchariková [1], Eva Tillová [1], Mária Chalupová [1]

[1]University of Žilina, Department of Materials Engineering, Faculty of Mechanical Engineering, Univerzitná 8215/1, 01026 Žilina, Slovakia
**Corresponding address: juraj.belan@fstroj.uniza.sk*

Key words: cast Ni-based superalloy, vacuum casting, turbine blades, SEM γ'-phase analysis, microstructure changes

1. Introduction

As the aero jet engines become more powerful and sophisticated peace of engineering craftworks also demand for advanced materials used at their construction became more important. Materials used at aero jet engine construction can be divided into following groups; i) materials for aero jet engine body construction (e. i. high temperature steels, composite materials), ii) materials used in low pressure and low temperature section (fans and compressors materials such Al alloys, Ti alloys, and high temperature steels), iii) materials used in hot section and high pressures of jet engine (for European development Ni-based superalloys were more important)

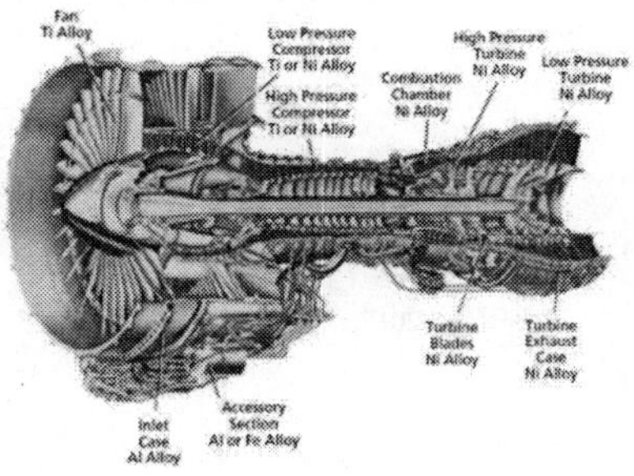

Fig. 1. Material distribution in aero jet engine [1].

Superalloys are heat resistant alloys of nickel, iron–nickel and cobalt that frequently operate at temperatures exceeding 537.7°C (1000° F). However, some superalloys are capable of being used in load bearing applications in excess of 85% of their incipient melting temperatures. They are required to exhibit combinations of high strength, good fatigue and creep resistance, good corrosion resistance, and the ability to operate at elevated temperatures for extended periods of time (i.e., metallurgical stability) [2]. Their combination of elevated temperature strength and resistance to surface degradation is unmatched by other metallic materials. Superalloys are the primary materials used in the hot portions of jet turbine engines, such as the blades, vanes and combustion chambers, constituting over 50% of the engine weight. For creep resistant increasing are superalloys produced as cast; with polyedric (equiaxed), direction solidified or single crystal structures; because grain boundaries decreasing the creep rupture life. That is the reason for using of vacuum induction melting (VIM) method (used casting tree and vax models), Fig. 2, for turbine blades creating polyedric coarse microstructure of austenitic FCC γ matrix with primary MC carbides inside of grains and secondary carbides $M_{23}C_6$ or M_6C as well at grain boundaries.

Fig. 2. An example of cast 'tree' (turned upside down) with the mold shell partly removed [3].

2. Experimental material and methods

The cast Ni–base superalloy ZhS6K was used as an experimental material. Alloy ZhS6K is a former USSR superalloy which was used in the DV-2 jet engine. It is used for turbine rotor blades and whole-cast small-sized rotors with a working temperature of up to 800÷1050°C. The alloy is made in vacuum furnaces. Parts are made by the method of precise casting. The temperature of the liquid at casting in a vacuum to form is 1500÷1600°C, depending on the part's shape and its quantity. The cast ability of this alloy is very good, with only 2÷2.5% of shrinkage.

This alloy was evaluated at the starting stage, the stage with normal heat treatment after 600, 1000, 1500 and 2000 hours of regular working (for these evaluations, real ZhS6K turbine blades with a protective alitize layer were used as an experimental material. The chemical composition received from SPECROMAXx in wt % is presented in Table 1. For evaluation of structural characteristics the following quantitative metallography methods were used, secondary dendrite arm spacing measurement, for number of γ' - phase particles (N) coherent testing grid with 9 area probe of square shape was used, and for volume of γ′ - phase particles (V) coherent testing grid with 50 dot probes made of backslash crossing were used [4].

Table 1. Experimental alloy's chemical composition.

C	Ni	Co	Ti	Cr	Al	W	Mo	Fe
0.202	64.7	4.76	3.05	12.44	5.22	5.28	3.48	0.192

3. Results

A typical microstructure of ZhS6K Ni-base superalloy as-cast is showed on Fig. 3. Microstructure of as-cast superalloy consists of significant dendritic segregation caused by chemical heterogeneity (Fig. 3a) and particles of primary MC and secondary $M_{23}C_6$ carbides. Primary carbides (Ti, Mo, and W)C are presented as an block shape particles mainly inside of grains. Secondary carbides are presented as a "Chinese" script shape particles on grain boundaries.

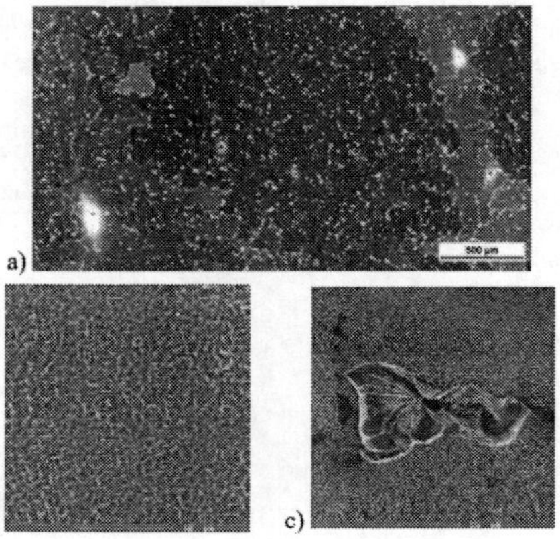

Fig. 3. Microstructure of as-cast Ni - base superalloy ZhS6K: a) dendritic segregation, b) matrix and γ′ phase, c) γ/γ′ eutectic, etch. Beraha III.

However, microstructure also contains solid solution of elements in base nickel matrix - so called γ phase (Ni(Cr, Co, and Fe)) and strengthening phase, which is product of artificial age – hardening and has significant influence on mechanical properties and creep rupture life - so called γ′ phase (gamma prime, Ni_3(Al, and Ti)), Fig. 3b. Of course, both of these phases, γ (gamma) and γ′ (gamma prime) are creating an eutectic γ/γ′, Fig. 3c. The first characteristic what has been evaluated is dendrite arm spacing. The evaluation were provided on ZhS6K turbine blades used in DV-2 (LPT - Low Pressure Turbine and HPT - High Pressure Turbine) aero jet engine at starting stage (basic heat treatment) and after in engine exposition (at real working temperatures) for 600, 1000, 1500, and 2000 hours. Results are in Table 2. The number and volume of γ′ - phase have significant influence on mechanical properties of this alloy, especially on creep rupture life. Average satisfactory size of γ′ - phase (u) is about 0.35-0.45 μm and also carbide size should not exceed size of 5 μm - because of fatigue crack initiation [5]. Another risk of using high temperature loading or annealing is creation of TCP phases, such σ - phase or Laves phase, in range of temperature 750°C - 800°C. The evaluation of γ′ phase was provided on heat treated turbine blades of DV-2 aero jet engine after various time of work. Obtained results are showed also in Table 2 [4, 6].

Table 2. Results for secondary dendrite arm spacing and γ′-phase size and volume in ZhS6K turbine blade

Type of blade	Secondary dendrite arm spacing - D [μm]
1°LPT blade - 0 hrs.	24.38
HPT blade - 600 hrs.	24.78
HPT blade - 1000 hrs.	27.98
HPT blade - 1500 hrs.	48.73
HPT blade - 2000 hrs.	66.66

Work [hrs.]	N [μm^{-2}]	V [%]	u [μm^2]
0	0.98542	67.2	0.6819
600	1.1242	67.6	0.60131
1000	1.1004	59	0.53615
1500	0.81938	57.4	0.7005
2000	0.6968	40.6	0.5826

4. Conclusions

For the turbine blades, which have been worked at normal loading and for various durations (600, 1000, 1500 and 2000 hours), the following results were achieved:

- The medium distance of secondary dendrite arms "D" grows in dependence on the time of work, caused by changes of the grain size of the γ-matrix.
- The inter-metallic phase-γ′ was evaluated with the methods of quantitative metallography; this evaluation shows gradual morphology changes of the γ′-phase - coarsening, spheroidisation and reprecipitation.

Acknowledgments

The project presented in this article is supported by Scientific Grant Agency of Ministry of Education of The Slovak Republic and the Slovak Academy of Sciences, No. 1/0533/15, No. 049ŽU-4/2017.

References

[1] Campbell F.C. (2006). Manufacturing Technology for Aerospace Structural Materials. Amsterdam: Elsevier Ltd.

[2] Davis J.R. (2001). Superalloys. In (Eds. Davis J.R.), Alloying: Understanding The Basics, (pp. 290-307), Ohio: ASM International.

[3] Choudhury I.A., El-Baradie M.A. (1998). Machinability of Nickel-Base Super Alloys: A General Review. J. Mater. Process. Technol., Vol. 77, pp. 278-284.

[4] Belan J.(2012). Study of Advanced Materials for Aircraft Jet Engines Using Quantitative Metallography. In (Eds. Agarwal R.K.), Recent Advances in Aircraft Technology, (pp. 49-74), Split: IN-TECH.

[5] Donachie M.J., Donachie, S.J. (2002). Superalloys – A technical Guide (2nd edition). Ohio: ASM International.

[6] Kopas P., Vaško M., Handrik M. (2014). Computational modeling of the microplasticization state in the nodular cast iron. Applied Mechanics and Materials, Vol. 474, pp. 285-290.

The 73rd World Foundry Congress
„Creative Foundry"
23 - 27 September 2018, Krakow, Poland

QUALITY ASSESSMENT OF Al CASTINGS PRODUCED IN SAND MOULDS USING IMAGE AND CT ANALYSIS

Lenka Kucharíková [1]*, Eva Tillová [1], Michaela Smardžiová [2], Milan Uhríčik[1], Juraj Belan [1], Ivana Švecová [1]

[1]*University of Žilina, Faculty of Mechanical Engineering, Department of Materials Engineering, Univerzitná 8215/1, 010 26 Žilina, Slovakia*
[2] *Slovak University of Technology in Bratislava, Advanced Technologies Research Institute, Faculty of Materials Science and Technology in Trnava, Jána Bottu 2781/25, 917 24 Trnava, Slovakia*
**Corresponding address: lenka.kucharikova@fstroj.uniza.sk*

Key words: sand castings, aluminum cast alloys, image analysis, computed tomography, properties of aluminum cast alloys

1. Introduction

The automotive and aerospace industries put a lot of attention toward aluminum alloys instead of heavy steel or cast iron structure [1]. The properties of aluminum cast alloys such as: good corrosion resistance, high strength to weight ratio, good formability, good machinability and scrap reuse etc. are a great advantages for use such material for industry applications [1-4].

Scrap reuse for production secondary aluminium alloys comes with many benefits: environmental, economic and low energy consumption. Aluminium alloys are considered to be 100 % recyclable. Recycling saves almost 95 % of the energy needed to produce prime aluminum from ore and creates only about 5 % as much CO_2 as by primary production [5-7].

Aluminium alloys usually contain a certain amount of Si, Cu, Fe, Mn, Mg and Zn that are present either undeliberate, or they are added deliberately to provide special material properties [2, 8]. The Iron belong to common impurity in aluminum cast alloys. Iron is highly soluble in liquid aluminum and its alloys and has very little solubility in the solid (max. 0.05 wt %, 0.025 atom. %). Therefore it tends to combine with the other elements to form intermetallic phase particles of various types [9]. The usage of scrap for production aluminium cast led to increasing amount of iron in material. This fact lead to formation higher amount of iron intermetallic phases which morphology influence the properties of castings. A lot of scientific work shows some theories about the relationship between microstructure and mechanical properties for aluminium alloys. Therefore, it is necessary to pay attention to examine mechanical properties considerably affected from microstructure. The aim of this study is to investigate the quality assessment of sand castings from aluminium alloys.

2. Experimental procedure

The AlSi7Mg0.3 aluminum cast alloy bars were produced by gravity casting into the sand mould (Fig. 1) in company Uneko spol.s.r.o. The experimental material were made from aluminium scrap and therefore were casted three series of bars with different content of Fe. The series A of experimental bars have basic chemical composition for this type of materials; the series B were samples with higher content of 0.454 wt. % Fe and the series C were samples with Fe content of 0.655 wt %.

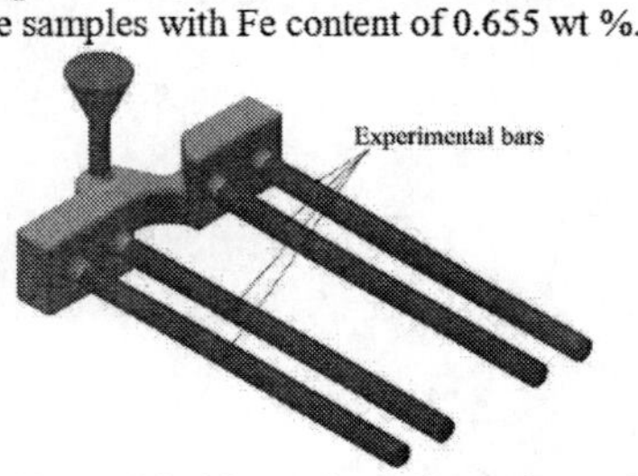

Fig. 1 The model of the sand casted experimental bars

The dimension of experimental bars were 300 x 20 mm (Fib. 1). From these bars were made samples for mechanical test with turning and milling operations. The chemical composition of these experimental materials was checked out using arc spark spectroscopy and is shown in Table 1.

Table 1 The chemical composition of experimental material, wt%

Element	A	B	C
Al	92.253	91.673	91.486
Si	7.028	7.340	7.315
Mg	0.354	0.302	0.292
Fe	0.123	0.454	0.655
Cu	0.013	0.021	0.030
Mn	0.009	0.009	0.010
Zn	0.036	0.020	0028
Ti	0.123	0.118	0.120
Other	balance	balance	balance

The experimental materials were not modified, grain refined or heat treated. This type of material belongs to the group of aluminium alloys using especially for casting in mechanical, electrical engineering, automotive and aerospace industries for shipbuilding, and so on [10]. The high quality of castings from such materials is needed.

The main influence of higher amount of iron is expected on to porosity formation and decreasing mechanical properties thanks morphology of these phases. Assessment of porosity changes was done using image analysis software NIS Elements and computed tomography (CT) with ZEISS METROTOM. The results represent assessment of each samples used for mechanical test. Mechanical properties – tensile strength (UTS) changes were measured on INSTRON Model 5985 according to the standard ISO 6892-1:2009. Test rates and control are set according to Method A recommended ranges. Template is intended for specimens that produce a clearly defined linear elastic region and homogenous deformation. The calculated results include UTS and Ductility (A). The evaluated UTS and Ductility reflect average values of at least six separate specimens.

3. Experimental results and discussion

The microstructure of experimental materials consist of matrix - dendrites α-phase, eutectic (mechanical mixture of eutectic Si and α-phase), intermetallic Fe- and Mg-rich phases and porosity (Fig. 2).

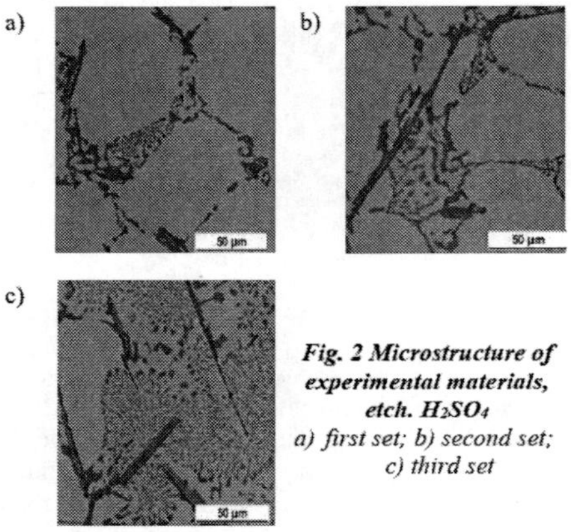

Fig. 2 Microstructure of experimental materials, etch. H_2SO_4
a) first set; b) second set; c) third set

The higher amount of iron in series B and C of experimental materials demonstrated higher amount of iron intermetallic phases group β (AlFeSi-needles form) as phases group α (Al(FeMn)Si – skeleton like from). Image analysis assessment also confirms that the material series A has a very small average length of Fe-rich needles (28.19 μm^2) in comparison to the other series of the same material (B = 53.37 μm^2, C = 43.55 μm^2).

To examine the effect of the Fe content on the porosity formation in experimental materials, were samples subject to quantitative analysis of area size and volume fraction of porosity. The results confirms that with increasing amount of Fe in material decrease volume of porosity (1.5 %, 1.0 %, and 0.9 %) and increase maximum area size of porosity (12 5723 μm^2, 125066 μm^2, and 219627 μm^2). Therefore this assessment decelerate that the higher influence on porosity formation have technology of casting. Because, the CT assessment shows that samples made form ones bars have different amount and distribution of porosity. This fact have a great influence on the properties.

The results of mechanical properties shows increasing of UTS and A with increasing amount of iron. The highest properties were on samples with 0.454 % of Fe (UTS = 150 MPa and A = 1.91 %). The samples series A have the lowest properties (UTS = 141 MPa and A = 1.45 %).

4. Conclusions

Quality assessment of sand castings microstructure confirmed that higher Fe content leads to mostly formation of needles Fe intermetallic phases with growing length. The increasing content of Fe leads to decreasing of content of porosity in experimental material, but to creating of larger pores. Mechanical properties dependent on area from which were made experimental samples moor as on the content of iron.

Acknowledgments

The authors acknowledge the VEGA N°1/0533/15, KEGA 049ŽU-4/2017 for the financial support of this work.

References

[1] Shriwas K.A. & Kale C. V. (2016). Impact of aluminum alloys and microstructures on engineering properties – review. Journal of Mechanical and Civil Engineering, Vol. 13 (3), 16-22.

[2] Grosselle F., Timelli G. & Bonollo, F. (2010). Doe applied to microstructural and mechanical properties of Al-Si-Cu-Mg casting alloys for automotive applications. Materials Science and Engineering A 527, 3536 – 3545

[3] Li R. X., Li R. D., He L. Z., Li C. X., Gruan H. R. & Huet Z. Q. 2004. Age hardening behavior of cast Al–Si base alloy, Materials Letters, 58, 2096-2101

[4] Irizalp S.G. & Saklakoglu N. (2014). Effect of Fe-rich intermetallic on the microstructure and mechanical properties of thixoformed A380 aluminum alloy. Engineering Science and Technology, an International Journal, 17(2), 58-62.

[5] Slagsvold M. (2011). Effect of Fe and Si content in Aluminium Alloys as a result of increased recycling. PhD Thesis. Norwegian University of Science and Technology.

[6] Das S. K. (2006). Designing Aluminum Alloys for a Recycling Friendly World. Materials Science Forum Vol. 519-521, 1239 - 1244.

[7] Senčáková L. & Virčíková E. (2007). Life cycle assessment of primary aluminum production. Acta Metallurgica Slovaca, Vol. 13 (3), 412-419.

[8] Rios C.T., Caram R., Bolfarini C., Botta F. W. J. & Kiminami C. S. 2003. Intermetallic compounds in the Al-Si-Cu system. Acta microscopia, 12, 77-82

[9] Taylor J.A. 2012. Iron-containing intermetallic phases in Al-Si based casting alloys. Procedia Materials Science. Vol. 1, 19-33.

[10] Bangyikhan K. (2005). Effects of Oxide film, Fe-rich phase, Porosity and their Interactions on Tensile Properties of Cast Al-Si-Mg Alloys. Ph.D. thesis, University of Birmingham.

The 73rd World Foundry Congress
„Creative Foundry"
23 - 27 September 2018, Krakow, Poland

MECHANICAL AND FATIGUE PROPERTIES OF NODULAR CAST IRONS

Alan Vaško [1]*, Juraj Belan [1], Lenka Kuchariková [1], Eva Tillová [1]

[1] *University of Žilina, Faculty of Mechanical Engineering, Department of Materials Engineering, Univerzitná 8215/1, 010 26 Žilina, Slovakia*
**Corresponding address: alan.vasko@fstroj.uniza.sk*

Key words: nodular cast iron, mechanical tests, fatigue tests, low frequency fatigue, high frequency fatigue

1. Introduction

The majority of service interruptions and cracks at the present time is caused by fatigue processes, degradation of materials properties in long time service and influence of external environment. The most significant of them is the failure caused by mechanical fatigue. The importance of fatigue results especially from the economical point of view (costs for development of new materials, testing equipments, methods of testing of prototypes and final products etc.) and safety of people whose life also depends on working reliability of devices [1,2].
Nodular cast iron is a group of cast structural materials with a wide application in engineering practice (especially in the automotive industry). It combines high tensile strength and plasticity with high fatigue strength [3]. The paper presents the results of low and high frequency fatigue tests carried out on nodular cast iron. The specimens of synthetic nodular cast irons from three different melts were studied in the high cycle fatigue region (from 10^5 to 10^8 cycles) using fatigue experimental equipments for low and high frequency cyclic loading. The ralationship $\sigma_a = f(N)$ and fatigue strengths were determined experimentally; mechanical properties and microstructures were also investigated.

2. Experimental material and methods

Three different melts of nodular cast iron were chosen for experiments and the specimens from them were used for metallographic analysis, mechanical tests and fatigue tests. Similar chemical composition of the melts was achieved by different charge composition (Table 1). The basic charge of individual melts consisted of pig iron, steel scrap and some additives for the regulation of chemical composition (carburizer and ferrosilicon FeSi75 or metallurgical silicon carbide SiC90). The content of these additives was chosen to achieve similar resultant chemical composition of the melts (eutectic degree $S_C \approx 1.2$).

Table 1. Charge composition of experimental melts

Melt number	pig iron (%)	steel scrap (%)	additives
1	40	60	C + SiC90
2	0	100	C + SiC90
3	40	60	C + FeSi75

note: C – carburizer

The metallographic analysis of specimens from experimental melts was made by the light metallographic microscope Neophot 32. The microstructure of specimens was evaluated according to STN EN ISO 945 and by automatical image analysis [4].
The tensile test was carried out according to STN EN ISO 6892-1 by means of the testing equipment ZDM 30 with a loading range F = 0 to 50 kN. The impact bending test was carried out according to STN EN ISO 148-1 by means of the Charpy hammer PSW 300 with a nominal energy of 300 J. The Brinell hardness test was carried out according to STN EN ISO 6506-1 by means of the testing equipment CV-3000 LDB with a hardmetal ball of diameter D = 10 mm forced into specimens under the load F = 29 430 N [5].
The fatigue tests were carried out according to STN 42 0362 at low and high frequency sinusoidal cyclic push-pull loading (stress ratio R = –1) at ambient temperature (T = 20 ± 5 °C). Low frequency fatigue tests were carried out at frequency f ≈ 120 Hz using the fatigue experimental machine Zwick/Roell Amsler 150HFP 5100 (Fig. 1a). High frequency fatigue tests were carried out at frequency f ≈ 20 kHz using the ultrasonic fatigue testing device KAUP-ZU (Fig. 1b) [6].

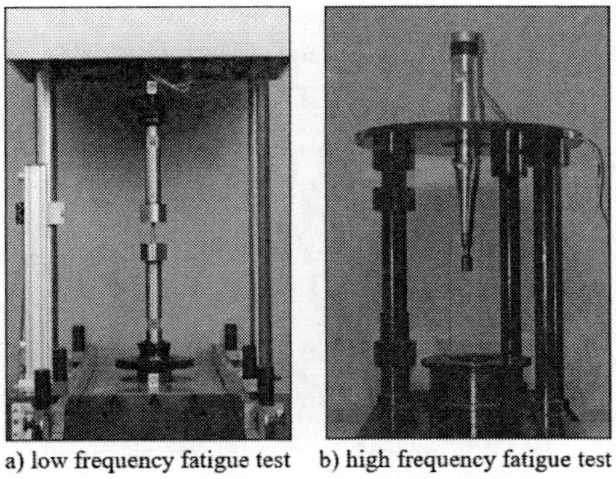

a) low frequency fatigue test b) high frequency fatigue test

Fig. 1. Experimental device for fatigue tests

3. Experimental results and discussion

All the specimens from experimental melts are ferrite-pearlitic nodular cast irons with similar content of ferrite and pearlite in the matrix, similar shape, size and count of graphitic nodules (Table 2).

Table 2. Microstructure of the specimens

Melt number	STN EN ISO 945	Count of graphitic nodules (mm^{-2})	Content of ferrite (%)
1	80%VI5/6 + 20%V6	199.8	74.0
2	70%VI5/6 + 30%V6	179.8	78.0
3	70%VI5/6 + 30%V6	151.0	65.2

Mechanical properties (tensile strength R_m, elongation A, absorbed energy K0 and Brinell hardness HBW) of the specimens from experimental melts are summarised in Table 3. The best mechanical properties were reached in the melt 1, formed by 40% of pig iron and 60% of steel scrap with SiC additive, which has the highest ratio of perfectly-nodular graphite, the highest count of graphitic nodules and almost the highest content of ferrite.

Table 3. Mechanical properties of the specimens

Melt number	R_m (MPa)	A (%)	K0 (J)	HBW 10/3000
1	539.0	4.0	30.6	192.3
2	515.7	3.7	17.2	182.3
3	462.6	2.7	24.0	181.3

The results of fatigue tests (relationship between stress amplitude σ_a and number of cycles to failure N_f) obtained at low frequency cyclic loading (f ≈ 120 Hz) and high frequency cyclic loading (f ≈ 20 kHz) are shown in Fig. 2.

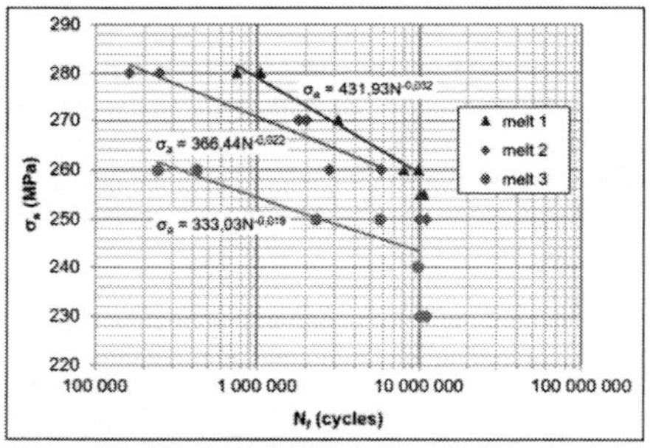

a) low frequency cyclic loading (f ≈ 120 Hz)

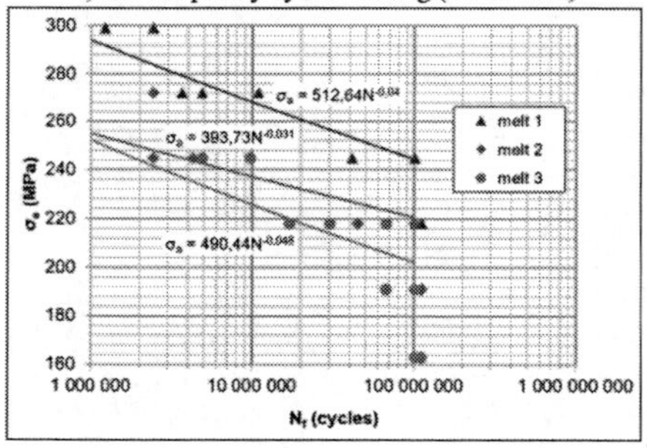

b) high frequency cyclic loading (f ≈ 20 kHz)

Fig. 2. ***Wöhler curves*** $\sigma_a = f(N)$

The values of fatigue strength σ_c determined for N = 10^7 cycles (for LFCL) or N = 10^8 cycles (for HFCL) in comparison with tensile strength R_m are given in Table 4. The fatigue strength in analysed specimens of nodular cast iron increases with an increasing tensile strength. The highest fatigue strength for LFCL as well as for HFCL was reached in the melt 1, formed by 40% of pig iron and 60% of steel scrap with SiC additive, which also has the best mechanical properties.

Table 4. Fatigue strength σ_c for low and high frequency cyclic loading

Loading/ frequency		LFCL f ≈ 120 Hz	HFCL f ≈ 20 kHz
Melt number	R_m (MPa)	σ_c (MPa) for N = 10^7 cycles	σ_c (MPa) for N = 10^8 cycles
1	539.0	255	218
2	515.7	250	191
3	462.6	230	163

4. Conclusions

The experimental results can be summarized to the following points:

- The best mechanical properties (tensile strength, elongation, absorbed energy and Brinell hardness) as well as fatigue properties (fatigue strength for low and high frequency cyclic loading) were reached in the specimens from the melt 1, formed by 40 % of pig iron and 60 % of steel scrap with SiC additive;
- The values of fatigue strength obtained at high frequency cyclic loading are in a good agreement with the results obtained at low frequency cyclic loading. Moreover, application of high frequency cyclic loading is characteristic with the significant time, energy and work saving.

Acknowledgments

This work has been supported by the grant projects VEGA No. 1/0533/15 and KEGA No. 049ŽU-4/2017.

References

[1] L. Várkoly, J. Zuidema, B. Várkolyová, M. Chalupová. (1998). Fatigue failure of materials – theory and solved examples. Žilina: EDIS.

[2] O. Bokůvka, G. Nicoletto, M. Guagliano, L. Kunz, P. Palček, F. Nový, M. Chalupová. (2014). Fatigue of materials at low and high frequency loading. Žilina: EDIS.

[3] R. Konečná, L. Bubenko, G. Nicoletto. (2012). Microstructure vs. near-threshold fatigue crack growth behavior of an heat-treated ductile iron. Medziagotyra. 18 (1), 23-27.

[4] P. Skočovský, A. Vaško. (2007). Quantitative evaluation of structure of cast irons. Žilina: EDIS.

[5] A. Vaško, P. Skočovský. (2014). Properties and using of metal materials. Žilina: EDIS.

[6] F. Nový, R. Ulewicz, O. Bokůvka, L. Trško, J. Lago. (2016). Reliability and safety of structural elements in the gigacycle region of loading. Communications. 18 (3), 83-87.

The 73rd World Foundry Congress
„Creative Foundry"
23 - 27 September 2018, Krakow, Poland

EFFECTS OF MANGANESE CONTENT AND COOLING RATE ON FATIGUE IN LIMIT HEAVY SECTIONAL SPHEROIDAL GRAPHITE CAST IRON

Naoto Shiraki[1]*, Tsubasa Hirose[2] and Ryosuke Fujimoto[3]

[1] *Tokyo City University, Faculty of Engineering, Setagaya, Tokyo, Japan*
[2] *Tokyo City University, Graduate school of Engineering, Setagaya, Tokyo, Japan*
[3] *Toshiba Machine Co.,LTD, Numazu, Shizuoka, Japan*
* *Corresponding address: nshiraki@tcu.ac.jp*

Key words: Spheroidal graphite cast iron, Manganese, Cooling rate, Axial loading fatigue test, Fatigue limit, Defect size

1. Introduction

The purpose of this study is to investigate that effect of manganese (Mn) content and cooling rate to characteristic of fatigue limit in spheroidal graphite cast iron where Mn content and cooling rate are changed. As specimens, spheroidal graphite cast irons with 0.3, 0.6 and 0.9 mass% Mn content were produced respectively. Cooling rate was changed to 0.22 K/s and 0.08 K/s by ingot size. The pearlite area ratios increase as Mn content increases and cooling rate is high.

The graphite spheroidization rate decreases as Mn content increases and cooling rate is low. Tensile test was carried out in air at room temperature. Pearlite area and graphite spheroidization rate ratios decided by Mn content and cooling rate have influence on hardness and tensile strength. The axial loading fatigue test was also carried out using an electric-hydraulic fatigue tester with a capacity of ±98 kN. The test was conducted using a sinusoidal load control system with a load repetition frequency of 17Hz, and stress ratio, R=-1 and 0.05. The number of repetitions for the fatigue test was 1×10^7 cycles. The test was carried out at room temperature. The specimen used was the 1 type of 8.00mm in diameter. All the fracture surfaces were observed with a scanning electron microscope. The relationship between the characteristics of fatigue limit and the size of fracture origins (defect size) was investigated.

The fatigue limit increases as Mn content increases and cooling rate is high, and the fatigue limit decreases when it becomes the stress ratio, R=0.05. Fracture origins were micro-shrinkage, aggregate graphite and unspheroidizing graphite. In cooling rate 0.22 K/s fracture origins were micro-shrinkage and aggregate graphite, but in cooling rate 0.08 K/s were micro-shrinkage and unspheroidizing graphite. The reason of this phenomenon was to provide abnormal graphite by cooling rate was slow. It is known that the fatigue limit of a material with a defect can be estimated using the 4-parameter model. In this study, the fatigue limit estimated using the 4-parameter method obtained good result.

Recently, the percentage of the high tensile steel in the iron scrap is also increased with usage rate increase of the high tensile steel, which had weight saving of a car for its object. Because high tensile steel includes manganese (Mn) much compared with normal steel, Mn content in the cast iron is increased, and it's said that they have an influence on the microstructure of ferritic spheroidal graphite cast iron [1]. Ferritic spheroidal graphite cast iron is used for large casting parts. There is the influence on the microstructure of matrix and graphitic shape because it's slow in the cooling rate until coagulation. Therefore influence of the Mn content and the cooling rate exerted on the fatigue limit of heavy sectional spheroidal graphite cast iron can't be ignored. In this study, Mn content and the size of heavy sectional spheroidal graphite cast iron were changed, and influence of the Mn content and the cooling rate exerted on the fatigue limit was investigated.

2. Sample materials and experimental procedure

Sample materials targeted chemical components shown in Table 1 and rare-earth elements in spheroidizing agent was performed casting by adding so as to be 200 ppm in the melt. Specimen for tensile test and axial load fatigue test changed the Mn content with 0.3, 0.6 and 0.9 mass%. The specimens were taken from six kinds of h100×w100×t100mm and h500×w500×t500mm size of heavy sectional ingot. The cooling rate until the eutectic temperature from 1523K was 0.08K/s by h500×w500×t500 mm and 0.22K/s by h100×w100×t100mm respectively. Specimen code is indicated with Mn content and the cooling rate (for example, FCD3-0.22).

Table 2 shows microstructure of each sample, the ratio of graphite spheroidization, the mean diameter of graphite, the number of graphite in per the unit area and the ration of perlite area ratio. The ratio of perlite area in microstructure increases in with increase of the Mn content in each cooling rate.

Table 1 Chemmical compositions of experimental materials. (mass%)

Material	Ingot size,mm	C	Si	Mn	P	Mg	S	RE
FCD3-0.22	100	3.4~ 3.5	2.5	0.3	≦0.03	0.05	≦0.01	0.002
FCD3-0.08	500			0.3				
FCD6-0.22	100			0.6				
FCD6-0.08	500			0.6				
FCD9-0.22	100			0.9				
FCD9-0.08	500			0.9				

Table 2 Comparison of microstructures, ratios of graphite spheroidization, mean diameter of spheroidal graphite, and number of graphite in each material.

Material	FCD3-0.22	FCD3-0.08	FCD6-0.22	FCD6-0.08	FCD9-0.22	FCD9-0.08
Microstructure 200μm						
Mean diameter of graphite, μm	46	60	51	59	51	67
Number of graphite in microstructure, μm	51	31	459	31	42	20
Ratio of graphite spheroidization, %	83	73	84	75	79	66
Pearlite area, %	16	10	29	11	49	34

Especially, it increased in the ratio of perlite area large in 0.22 K/s with the fast cooling rate. When the cooling rate became slow, diameter of graphite was larger, and the graphite spheroidization was decreased. For this reason, graphite size becomes large, and graphite couldn't maintain any more spheroidal form. Especially, spheroidization indicated the low by a sample of Mn content 0.9 mass%.

In the axial fatigue test, electro-hydraulic servo pulsar was used with a loading frequency of 17 Hz and stress ratio R=-1 and 0.05 at room temperature in the atmosphere. The number of cycle to discontinue test was set to 1×10^7 cycles. The fracture origins were observed by scanning electron microscope (SEM). The shape of the axial load fatigue test specimen is 1-type and 8mm diameter at parallel position followed by JIS Z2273 [2]

3. Experimental results and discussion

Figures 1a and 1b show the relation between the fatigue limit compared with each stress ratio, the Mn content and the cooling rate. In the case of same cooling rate, fatigue limit increases with increase of the Mn content. Fatigue limit was decreased to 50-60 MPa depend on the difference in stress ratio in every Mn content. Fatigue limit of R=0.05 was decreased to 50-60 MPa compared with R=-1 depend on the difference Mn content and the cooling rate.

The observation using SEM (Figs. 2a-c) showed that all fracture origins observed in this study were either a micro-shrinkage cavity (Fig. 2a), aggregate graphite (Fig. 2b), or unspheroidal graphite (Fig. 2c). There are no difference in fracture origin by the Mn content and compared with the stress ratio for the kind of fracture origin of each material. Main fracture origins were a micro-shrinkage cavity and aggregate graphite in cooling rate 0.22K/s, but in cooling rate 0.08K/s, fracture origin were a micro-shrinkage cavity and unspheroidal graphite. The plotting the results of defect size as same sample on gumbel paper, the cooling rate became slow, defect

(a) Micro-shrinkage (b) Aggregate graphite (c) Unspheroidizing graphite

Fig.2 Example of fracture origins observed by SEM, (a) Micro-shorinkage, (b) Aggregate graphite (c) Unspheroidized grapite

size was larger. This results agreed with well the results of fatigue limit.

4. Conclusion

Axial load fatigue test was carried out using the spheroidal graphite cast iron which changed the Mn content and the cooling rate, following conclusions were obtained.

(1) The cooling rate became slow in each Mn content, defect size of fracture origin was to become large, and the fatigue limit was decreasing. In the case of same cooling rate, the fatigue limit was increased with increase of the Mn content.

(2) Main fracture origins were a micro-shrinkage cavity and aggregate graphite in cooling rate 0.22K/s, but in cooling rate 0.08K/s, fracture origin were a micro-shrinkage cavity and unspheroidal graphite respectively. A Microsoft Word file (.doc) is required. A total size should not exceed 8 MB.

References

[1] R. Fujimoto, S. Hiratsuka, H. Horie, T. Hareyama, and S. Honma, J.JFS (in Japanese), Vol.85 (2013) pp.651-658.
[2] JIS Z 2273 : General Rules for Fatigue Testing of Materials, (1978).

INFLUENCE OF DEFECTS ON HPDC STRENGTH AND EFFECTIVENESSE OF GAS POROSITY DISPERSION BY ATOMIZED FLOW

Eitaro Koya *, Masahiko Nakagawa

Honda R&D Co., Ltd., Motor Cycle R&D Center, 3-15-1 Senzui, Asaka-Shi, Saitama, Japan
**Corresponding address: eitaro.koya@mail.a.rd.honda.co.jp*

Key words: HPDC, Strength, Gas porosity, J Factor, Atomization

1. Introduction

High-pressure diecasting (HPDC) is a method of rapidly filling molten metal into a die by means of an injection mechanism. The process is susceptible to certain defects— oxide inclusion, cold flak, gas porosity and shrinkage porosity —which can affect the strength of the resulting casting. Here, we first examine relations between defect formation and casting strength by means of a statistical analysis of the locational distribution of strength within cast parts. Second, and following on recent research on the effect of differing "J Factor"[1] (J Factor: an indicator of the degree of the metal atomization upon passing through the ingate), we reveal (a) differences in strength distributions under differing J values and (b) characteristics of defects formed under high J values.

2. Locational distribution of strength within cast parts

With regards to molten aluminum, a J Factor of 1,064 or above is considered indicative of atomized flow [2]. Here, the authors investigated whether parts cast at equivalent gate speeds but differing J values exhibit any differences in their resulting strength distributions. Here, we examined two castings, Part A and Part B. Manufacturing conditions and number of sampling are shown below in Table 1.

Table 1. Manufacturing conditions and number of sampling

	Gate Speed (m/s)	J Value	Weight (kg)	Number of Sampling
Part A	38	1800	3.8	20
Part B	34	750	4.3	10

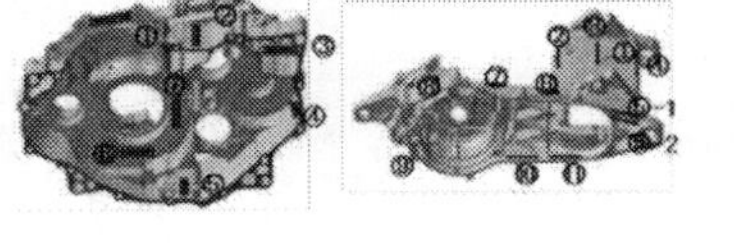

Fig. 1. Castings and measurement locations

Both parts were cast from JIS ADC12 aluminum diecasting alloy under no depressurization with a casting machine having a clamping force of 800 t. Fig. 2 shows tensile test results as plotted on normal probability paper. The measured values fall into three distinct groups, each of the same slope. Of the two parts, Part A showed a clear grouping into high- and low-strength distributions.

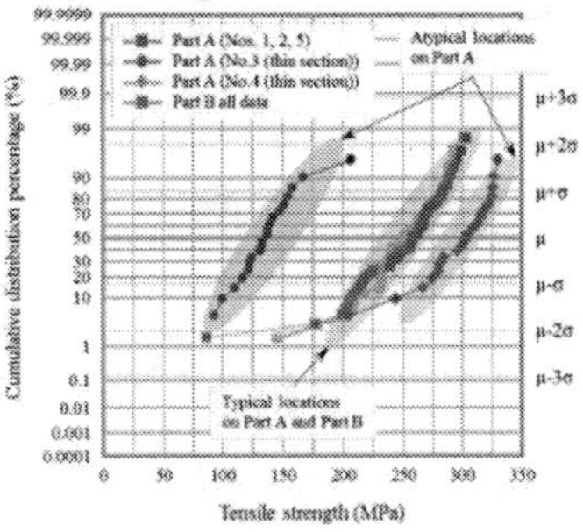

Fig. 2. Distribution of tensile strength measurements

3. Relation between defects and strength

We first employed a micro focus x-ray CT system to examine relations between defects and strength. We found, however, that oxide inclusion could not be readily detected by micro-focus x-ray; and so instead we used (a) SEM observation of tensile-test fracture surfaces to determine defect type/region followed by (b) image analysis to determine the surface area of each type of defect. Fig. 3 presents typical defect areas (%) observed within the three strength groups.

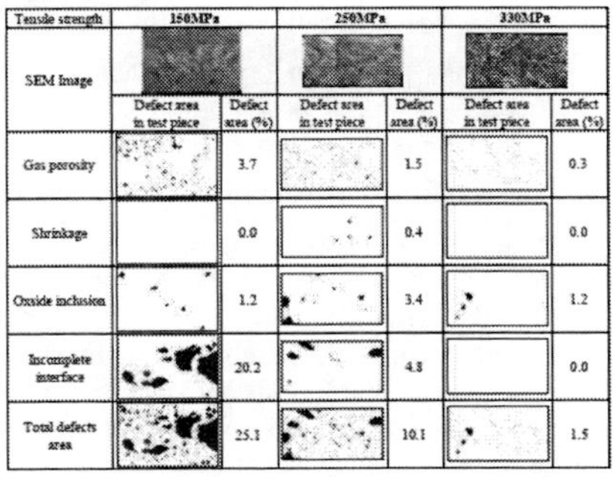

Fig. 3. Strength and defect areas: observations

Fig. 4 is a plot of relations between total defect area ratio (Sa (%)) and gas porosity area ratio (Sg (%)). Middle-strength locations, the most commonly observed type, have an Sg/Sa ratio ranging from roughly 4 to 8. As for the low-strength locations, however, we see that the relationship with gas porosity diminishes. At the high-strength locations, on the other hand, we do note a higher proportion of gas porosity.

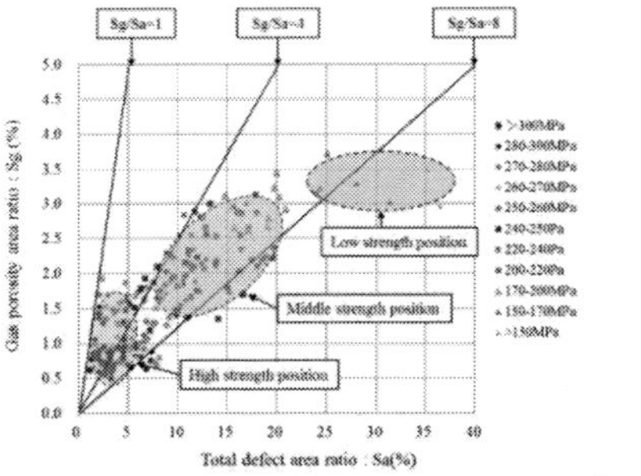

Fig. 4. Relation between total defect area ratio (Sa (%)) and gas porosity area ratio (Sg (%))

4. Observation of low strength positions

SEM observation of low-strength locations reveals considerable oxide inclusion and extensive areas of comparatively smoothness that are distinct from cold flakes. During tensile testing, cracks tended to occur along boundaries once deformation enters the plastic region. Fig. 5 shows a micrographic comparison of a boundary observed within this specimen and a cold flakes. Cold flakes are the result of a transport into the die of flakes formed upon rapid cooling within the injection sleeve. Along the boundary apparent here, however, we note no structural differences. This leads us to deem such boundaries to be "incomplete interface areas." The proportion of such incomplete interface areas is higher within low strength positions.

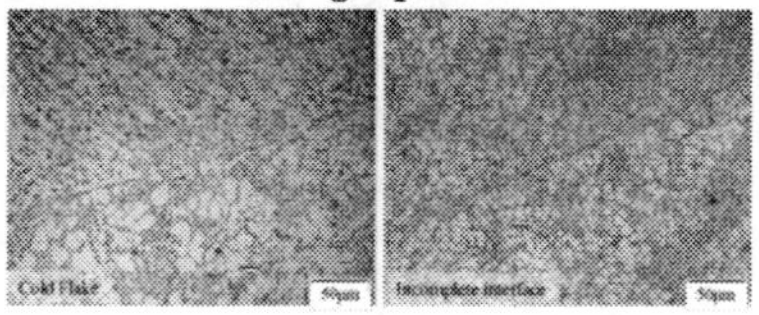

Fig. 5. Microstructure of incomplete interface and cold flakes

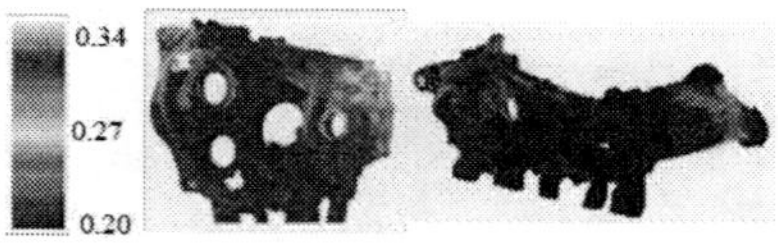

Fig. 6. Solidification ratios upon completion of filling

Fig. 6 compares the results of a simulation of solid phase ratios within the two castings upon the completion of the metal filling.

In castings with a high proportion of incomplete interface areas, low strength positions correspond to confluences of the metal flows having high solid-phase ratios.

5. Observation of high strength positions

ADC 12 aluminum alloy is generally produced from recycled aluminum and contains many impurities. Thus, we normally cannot anticipate tensile strength in excess of 300 MPa. One characteristic of high-strength locations as revealed by SEM observation is that the gas porosity within them is finely dispersed. High-strength positions have a higher proportion of gas porosity defects to total defects. Fig. 7 shows the frequency distribution of gas porosity by short diameter within high-strength positions.

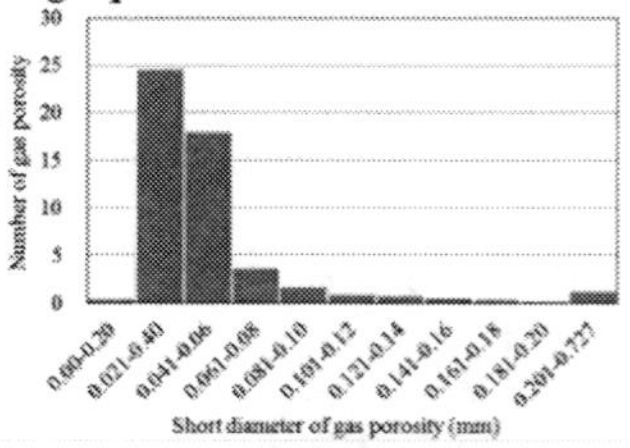

Fig. 7. Frequency distribution of gas porosity short diameter within high strength positions

Within the high strength positions, gas porosity of diameters of up to 100 μm are finely dispersed. Fig. 8 shows fatigue test results for the high-, middle- and low-strength positions within Part A. The high-strength locations exhibit a higher fatigue strength, just as they exhibit a higher tensile strength.

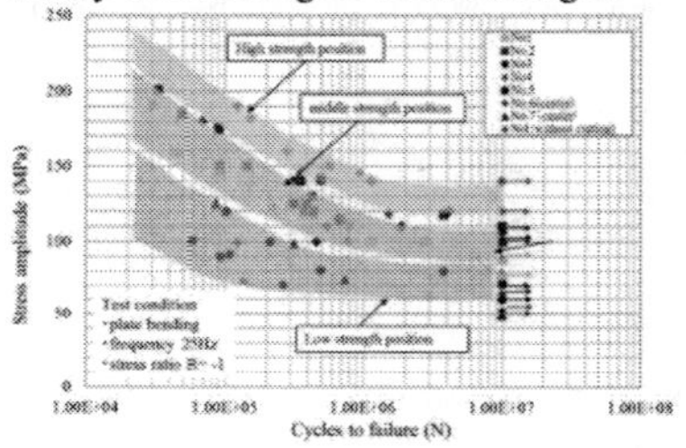

Fig. 8. Plane-bending fatigue test results by position

From these results, we find that HPDC parts can be cast to high strength by finely dispersing gas porosity within them. Such fine dispersions have only been observed within parts cast under atomized flow as indicated by a J value of 1,064 or above. From this, we conclude that atomization of the metal contributes to a fine dispersion of gas porosity.

6. Conclusions

We have identified incomplete interface areas, seen at confluences of the metal flows with high solid phase ratios, as a factor acting to substantially lower strength within HPDC castings. Also, finely dispersing gas porosity by means of atomized flow is thought to contribute to the obtainment of high strength within castings.

References

[1] Kaneuchi, T., Nakano, H., "Effect of J-Value for Internal Quality of HPDC Products at Various Casting Alloys.", Japan Die Casting Congress Transactions 131-136, 2008.

[2] North America Die Casting Association (NADCA), "GATING MANUAL", 35-36, Publication #5.

STUDY OF AlSi7MgCu ALLOY WITH IMPROVED PROPERTIES IN AS-CAST STATE

Zdenka Zovko Brodarac [1]*, Davor Stanić [2], Letian Li [3]

[1] *University of Zagreb Faculty of Metallurgy, Aleja narodnih heroja 3, 44103 Sisak, Croatia*
[2] *CIMOS-P.P.C. Buzet, d.o.o., Most 24, 52420 Buzet, Croatia; Croatia Polytechnic Pula - College of Applied Sciences, Riva 6, 52100 Pula, Croatia*
[3] *FEI Netherlands, Achtseweg Noord 5, Bldg, Eindhoven, Netherlands*
**Corresponding address: zovko@simet.hr*

Key words: AlSi7MgCu alloy, copper, microstructure, solidification sequence, mechanical properties

Designing of new chemical composition of AlSi7MgCu alloy with extra addition of Cu represents a challenge in order to achieve advanced mechanical properties already in as-cast state. A wide range of complex reactions and intermetallic phases occurs due to numerous alloying (Si, Mg, Cu) and trace elements (Fe, Mn) interaction. Exact determination of alloy behaviour was performed by modelling of equilibrium phase diagram and metallographic investigations. Wide spectra of favourable intermetallic phases enable an overview of theirs strong influence comprehending to the mechanical properties development already in as-cast state.

1. Introduction

Excellent castability and favourable relation between strength and weight, especially in the heat-treated state, indicate conventional AlSi7Mg alloy as a frequent choice for complex geometry castings with high properties demand. The solidification sequence of hypoeutectic AlSi7Mg alloys begins with development of primary aluminum dendrites α_{Al} and formation of dendritic network, followed by eutectic reaction (α_{Al} + β_{Si}) on the primary grains α_{Al} or independently on present nucleus and/or other impurities with different crystallographic orientation [1]. The way of eutectic occurs determines the amount and morphology of eutectic phase, and also the porosity ratio in the microstructure. The primary and most important alloying element in conventional AlSi7Mg alloy is Si which is characterized by high fluidity and reduction in shrinkage, followed by Mg responsible for strength increase [2].

Magnesium addition up to a 0.7 wt. % has a strengthening effect due to precipitation of α_{Al} -Mg_2Si eutectic phase and/or Mg-rich intermetallic with other alloying elements mostly due to transformation of the deleterious Al_5FeSi platelets into a π-phase with a stoichiometry $Al_8Mg_3FeSi_6$ [3].

Also, Cu is commonly used as an alloying element in order to increase the strength of foundry alloys, especially when heat treatment is applied. In Al-Si alloys, Cu is usually added in levels between 1.5 – 3.5 wt.% and forms the secondary eutectic phase Al_2Cu [4]. Copper decreases significantly the melting point and eutectic temperature of the alloy. Therefore, the copper increases the solidification range of an alloy, and facilitates the condition of porosity formation [5,6].

Innovated chemical composition with extra Cu addition and determined solidification sequence of AlSi7MgCu as well as developed microstructure comprehend to significant increase of mechanical properties already in as-cast state.

2. Experimental

Characterization of newly designed AlSi7MgCu alloy has been performed in as-cast state. Investigation covered the development of equilibrium phase diagram, microstructural, and mechanical properties analysis.

Modelling of equilibrium phase diagram has been performed using ThermoCalc (TCW 5.0) software. Newly designed AlSi7MgCu alloy melt was prepared in induction furnace. Chemical composition analysis was performed using optical emission spectrometer ARL-3460. Metallographic investigations comprehends metallographic investigations (Olympus GX 51) and microstructural investigations (scanning electron microscopes Tescan Vega TS 5136 MM and Helios NanoLab G4 DualBeamTM). Mechanical tensile properties investigations were performed on testing machine MTS 810 [7].

3. Results and discussion

New chemical composition of AlSi7MgCu alloy with extra addition of copper has been designed and compared with conventional AlSi7Mg alloy, as shown in Table 1.

Table 1. The chemical compositions of AlSi7Mg/ AlSi7MgCu alloys

Element, wt.%	Si	Fe	Cu	Mn	Mg
AlSi7Mg	7.008	0.101	0.130	0.010	0.320
AlSi7MgCu	7.527	0.235	1.435	0.076	0.348

The newly designed alloy complies with the EN 42000 AC standard for AlSi7Mg alloy in relation to the content of the base alloying elements (Si and Mg), and trace elements such as Fe and Mn. Deviation has been implemented with significant increase in Cu content.

Modelling of newly designed AlSi7MgCu alloy by ThermoCalc (TCW 5.0) program resulted with calculated equilibrium phase diagram. Interaction of alloying and trace elements reveals wide range of intermetallic phases, followed with calculation of solidification sequence of AlSi7MgCu.

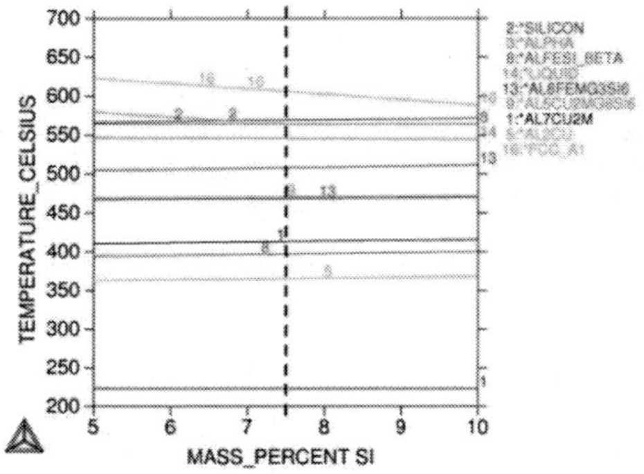

Fig. 1. Modelling of solidification sequence of AlSi7MgCu alloy

Microstructure of AlSi7MgCu alloy was investigated using light (LM) and scanning (transmission) electron microscopy (SEM/STEM), as shown in Figure 2.

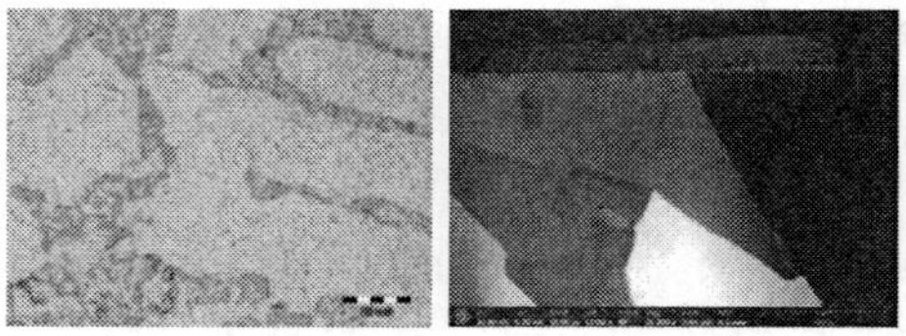

Fig. 2. Micrographs of AlSi7MgCu alloy obtained using a) LM and b) SEM/STEM

Detail metallographic analysis of SEM/STEM images using EDS enabled recognition of following phases: dendrite network development (α_{Al}), intermetallic Al_5SiFe and modified Chinese script phase in $Al_{15}(Fe,Mn,Cu)_3Si_2$, main eutectic ($\alpha_{Al} + \beta_{Si}$), compact complex intermetallic $Al_5Cu_2Mg_8Si_6$ and $Al_8FeMg_3Si_6$ phases and secondary eutectic phases' Mg_2Si and Al_2Cu.

Mechanical properties of AlSi7MgCu alloy have been also investigated. Comparison of those obtained for new AlSi7MgCu and conventional AlSi7Mg alloy is shown in Table 2.

Table 2. Tensile mechanical properties of AlSi7MgCu alloy

Alloy	$R_{p0,2}$, MPa	R_m, MPa	A_{50}, %
AlSi7Mg	120	165	6,3
AlSi7MgCu	143	234	3,1

Significant increase in yield and tensile strength of innovative AlSi7MgCu alloy in as-cast state has been noticed. It can be attributed to the complex interactions of $Al_{15}(Fe,Mn,Cu)_3Si_2$ phase modified with Cu addition as well as to the formation of compacted $Al_5Cu_2Mg_8Si_6$ and $Al_8Mg_3(Fe,Mn)Si_6$ phases'. Elongation has been significantly decreased.

4. Conclusions

Designing of innovative AlSi7MgCu alloy reveals a wide range of complex reactions and possible intermetallic phases due to the interaction of alloying and trace elements. Evolution of microstructure and determination of solidification sequence enables characterization of new alloy and its properties. Determination of microconstituents' interactions in solidification process was performed by modelling of equilibrium phase diagram compared with simultaneous thermal analysis results, microstructural and mechanical properties investigations. An extra addition of Cu (up to 1,435 wt.%) as a secondary alloying element initiates additional interaction with Fe, Mn and Mg. Determination of AlSi7MgCu alloy solidification sequence and correlation with revealed morphology and distribution of microstructure constituents enables prediction of its behaviour.
Wide spectra of favourable intermetallic phases enable an overview of strong connections comprehending to the improved mechanical properties development already in as-cast state.

Acknowledgments

Investigations have been performed in collaboration with companies CIMOS-P.P.C. Buzet, d.o.o., Buzet, Croatia. Investigations were performed within the research topic "Design and Characterization of Innovative Engineering Alloys", Code: TP167 funded by the University of Zagreb within the Framework of Financial Support of Research.

References

[1] A.K. Dahle, J. Hjelen, L. Arnberg, Formation of hypoeutectic Al-Si allyos, In Conference Proceedings of the 4th International Conference on Solidification Processing, Sheffield, 1997, 527-530

[2] L. Bäckerund, G. Chai, J. Tamminen, Solidification Characteristics of Aluminium Alloys: Foundry Alloys Vol. 2, Stockhlom: AFS/Skanaluminium; 1999.

[3] N.A. Belov, D.G. Eskin, A.A. Aksenov, Multicomponent phase diagrams, Applications for commercial aluminum alloys, Elsevier, London, 47-52, 2005

[4] Z. Zovko Brodarac, N. Dolić, F. Unkić, Influence of copper content on microstructure development of AlSi9Cu3 alloy, Journal of mining and metallurgy Section: B, Metallurgy. 50 (2014) 1; 53-60

[5] F.H. Samuel, A.M. Samuel, H.W. Doty, Factors controlling the type and morphology of Cu-containing phases in 319 Al alloy, AFS Trans., 104 (1996) 893–901.

[6] C.H. Cacers, M.B. Djurdjevic, T.J. Stockwell, J. H. Sokolowski, The effect of Cu content on the level of microporosity in Al–Si–Cu–Mg casting alloys, Scripta Mater. 40 (5) (1999) 631–637

[7] EN 10002-1:1998 Metallic materials – Tensile testing – Part 1: Method of test (at ambient temperature)

The 73rd World Foundry Congress
„Creative Foundry"
23 - 27 September 2018, Krakow, Poland

DATA MINING METHODS FOR PROPERTIES PREDICTION WITH TDA CURVES OF THE HYPOEUTECTIC Al-Si ALLOYS

Zenon Pirowski [1], Krzysztof Jaśkowiec [1], Stanisława Kluska-Nawarecka [1], Dorota Wilk-Kołodziejczyk [1,2,*], Krzysztof Regulski [2], Tomasz Szymczak [3], Grzegorz Gumienny [3]

[1] Foundry Research Institute, Krakow, Zakopiańska 73, Poland,
[2] AGH University of Science and Technology, Krakow, Poland,
[2] Lodz University of Technology, Lodz, Poland
**Corresponding address: dorota.wilk@iod.krakow.pl*

Key words: Al-Si alloy, crystallization, TDA curves, mechanical properties, data mining

1. Introduction

Al-Si alloys are commonly used in the foundry of aluminum-silicon alloys, which may also contain alloying additives. These alloys are characterized by good castability, corrosion resistance, machinability, heat resistance, electrical and thermal conductivity, low density (2.7 g/cm^2), thermal expansion and low shrinkage. Al-Si alloys are also characterized by high strength properties in the group of low density alloys. Thanks to these properties, Al-Si alloys have found a wide range of applications, especially in the automotive, aerospace and electrical machinery industries as well as in the production of household appliances.

Current research, aimed at improving the properties of Al-Si alloys, mainly relate to the use, as modifiers, elements from the group of lanthanides and transition metals. From the group of the lanthanides are tested include: La [1,2], Ce [2], Sm [3] Eu [4] Gd [5] Er [6,7] or Yb [8]. The mentioned elements or their mixtures used in appropriate amount reduce the size of precipitations of silicon eutectic. A similar effect on eutectics in Al-Si alloys is exerted by transition metals such as: Nb [9,10], Zr [11], Sc [11,12] or Mo [13]. Other tests carried out over recent years to increase the properties of Al-Si alloys consist in the reduction of hydrogen and non-metallic pollutants [14], development of modern heat treatment technology [15] and the use of specific casting conditions, eg high temperature overheating of the external polymers [16]. Crystallization of silumin determines the fragmentation of the microstructure, which causes a significant increase in their strength properties. The danger of separation of intermetallic phases in Al-Si alloys containing Cr, Mo, V and W increases with decreasing the heat dissipation rate from the crystallizing casting. Therefore, these additives are not suitable for use in Al-Si alloys cast into sand and ceramic molds. The high rate of heat removal from the casting to the pressure mold gives the possibility of supersaturation of Al-Si alloys with these additives. This should lead to solidification of Al-Si alloys and, as a consequence, improvement of its strength properties. Accordingly, the addition of Cr, Mo, V and W to Al-Si alloys intended for pressure casting seems to be the most advantageous. The preliminary tests carried out confirmed the possibility of obtaining higher strength properties of Al-Si alloys cast in this technology as a result of the introduction of the discussed high-melting elements [17-19].

Crystallization subjected to modifications through the influence of alloying additives can be described by means of ATD curves. Thermal and derivative analysis (ATD) is currently the best method of controlling the production of alloys. On the derivative curve, there are thermal effects from the crystallization of all phases emitted from the liquid and solid alloys. The mechanical properties of the alloy are a function of the microstructure. Therefore, there are specific statistical relationships between the characteristic values of ATD curves and the chemical composition of the alloy, metallurgical operations, microstructure and mechanical properties. An important advantage of the ATD method is the short time of microstructure control and the resulting mechanical properties.

2. Applied methods and results

ATD properties and curves were developed on the basis of experimental studies on Al-Si alloy of the original EN-AC 46000 grade (EN-AC AlSi9Cu3 (Fe)). This is a typical Al-Si alloy hypoeutectic pressure die casting. The additions of Cr, Mo, V and W in the form of mortars AlCr15, AlMo8, AlV10 and AlW8 were introduced into the Al-Si initial. Castings were made from the obtained alloy.

In order to determine the relationship between characteristic points of ATD curves and mechanical properties, data mining techniques were used. The aim of the task is to analyze data aimed at acquiring and formalizing knowledge about occurring phenomena. These phenomena can be described by triggering changes in mechanical properties. The collection of data mining methods includes the most popular tools such as induction of decision trees, artificial neural networks (ANN - Artificial Neural Networks), as well as modern ones such as the support vector machine (SVM - Support Vector Machine) or MARSplines. Multivariate Adaptive Regression Splines (MARSplines) is an implementation of a generalization of the technique, the purpose of which is to find values of output (dependent) variables based on input (predictive) variables. MARSplines is a nonparametric procedure that does not need any assumptions about the functional dependence of variables dependent on independent variables. The dependence is built from coefficients and so-called base functions, fully determined by data. The general mechanism of action of MARSplines can be imagined as a multiple, linear linear regression. Section limits (determined on the basis of data) determine the "applicability ranges" of individual linear equations. MARSplines is adapted to tasks in which there are both quantitative and qualitative predictive variables. The entrance space is divided into areas in which separate regression or classification functions are defined. This approach makes MARSplines particularly useful with a larger number of dimensions on the input (more than two variables), when in the case of other techniques, the problem of dimensionality begins to threaten. In all of these algorithms, learning proceeds according to a certain pattern: on the basis of a part of data (about 70% of cases), a regression model is built. This model can take a different form: a set of rules, a matrix of distances, weights on connections, or a matrix of vectors. This model is tested in the learning process with approx. 15% of cases, and the final form of the model is validated (verified) using the last separate group of data - approx. 15%. The model can be evaluated using several parameters, including coefficient of determination (matching). The final operation of the model can be tested for consistency of matching results with the input data of the dependent variable. The process of machine learning or discovering knowledge can be symbolically represented in three steps: (1) discovering knowledge in data; (2) prediction - creation of prognostic models; (3) creation of knowledge bases - codification of results for future use. The developed models will be able to continue autonomous learning during industrial processes. The development of models of dependencies between ATD curves and mechanical properties using data mining methods can improve the quality control process to a degree more perfect than traditional multiple regression models known from the literature [20]. Methods of data mining are not limited to linear models, which often better reflect the physical nature of phenomena than statistical models.

References

[1] Yu-Chou Tsai, Cheng-Yu Chou, Sheng-Long Lee, Chih-Kuang Lin, Jing-Chie Lin, S.W. Lim. (2009). Effect of trace La addition on the microstructures and mechanical properties of A356 (Al–7Si–0.35Mg) aluminum alloys. Journal of Alloys and Compounds, 487, 157–162.

[2] E. Aguirre-De la Torre, R. Pérez-Bustamante, J. Camarillo-Cisneros, C.D. Gómez-Esparza, H.M. Medrano-Prieto, R. Martínez-Sánchez. (2013). Mechanical properties of the A356 aluminum alloy modified with La/Ce. Journal of Rare Earths, 31 (8), 811-816.

[3] Qiu Hongxu, Yan Hong, Hu Zhi. (2013). Effect of samarium (Sm) addition on the microstructures and mechanical properties of Al–7Si–0.7Mg alloys. Journal of Alloys and Compounds, 567, 77–81

[4] J.H. Li, X.D. Wang, T.H. Ludwig, Y. Tsunekawa, L. Arnberg, J.Z. Jiangb and P. Schumacher. (2015). Modification of eutectic Si in Al–Si alloys with Eu addition. Acta Materialia, 84, 153–163

[5] Shi Zhiming, Wang Qiang, Shi Yuting, Zhao Ge, Zhang Ruiying. (2015). Microstructure and mechanical properties of Gd-modified A356 aluminum alloys. Journal of Rare Earths, 33 (9), 1004–1009.

[6] Z.M. Shin, Q. Wang, G. Zhao, R.Y. Zhang. (2015). Effects of erbium modification on the microstructure and mechanical properties of A356 aluminum alloys. Materials Science & Engineering A, 626, 102–107

[7] P. Pandee , U. Patakham, C. Limmaneevichitr. (2017). Microstructural evolution and mechanical properties of Al-7Si-0.3Mg alloys with erbium additions. Journal of Alloys and Compounds, 728, 844–853

[8] J.H. Li, S. Suetsugu, Y. Tsunekawa, P. Schumacher. (2013). Refinement of Eutectic Si Phase in Al-5Si Alloys with Yb Additions. Metallurgical and Materials Transactions A, 44A, 669–681.

[9] L. Bolzoni, M. Nowak, N. Hari Babu. (2015). Grain refinement of Al-Si alloys by Nb-B inoculation. Part I: Concept development and effect on binary alloys, Materials and Design, 66, 366–375.

[10] L. Bolzoni, M. Nowak, N. Hari Babu. (2015). Grain refinement of Al–Si alloys by Nb–B inoculation. Part II: Application to commercial alloys. Materials and Design, 66, 376–383.

[11] Cong Xu , Wenlong Xiao, Ruixiao Zheng, Shuji Hanada, Hiroshi Yamagata, Chaoli Ma. (2015). The synergic effects of Sc and Zr on the microstructure and mechanical properties of Al–Si–Mg alloy. Materials and Design, 88, 485–492.

[12] Cong Xu, Fang Wang, Hussain Mudassar, Chengyuan Wang, Shuji Hanada, Wenlong Xiao, and Chaoli Ma. (2017). Effect of Sc and Sr on the Eutectic Si Morphology and Tensile Properties of Al-Si-Mg Alloy. Journal of Materials Engineering and Performance, 26 (4), 1605–1613.

[13] A.R. Farkoosh, X. Grant Chen , M. Pekguleryuz. (2015). Dispersoid strengthening of a high temperature Al–Si–Cu–Mg alloy via Mo addition. Materials Science & Engineering A, 620, 181–189.

[14] O. Majidi, S.G. Shabestari, M.R. Aboutalebi. (2007). Study of fluxing temperature in molten aluminum refining process. Journal of Materials Processing Technology, 182, 450–455

[15] J. Pezda. (2014). The Effect of the T6 Heat Treatment on Hardness and Microstructure of the EN AC-AlSi12CuNiMg Alloy. Metalurgija 53 (1), 63-66.

[16] J. Piątkowski, B. Gajdzik, T. Matuła. (2012). Crystallization and structure of cast A390.0 alloy with melt overheating temperature. Metalurgija 51 (3), 321-324.

[17] T. Szymczak, G. Gumienny, T. Pacyniak. (2016). Effect of Cr and W on the Crystallization Process, the Microstructure and Properties of Hypoeutectic Silumin to Pressure Die Casting. Archives of Foundry Engineering, 16 (3), 109-114.

[18] T. Szymczak, G. Gumienny, T. Pacyniak. (2016). Effect Effect of Vanadium and Molybdenum on the Crystallization, Microstructure and Properties of Hypoeutectic Silumin. Archives of Foundry Engineering, 15 (4), 81-86.

[19] T .Szymczak, G. Gumienny, K. Walas, T. Pacyniak. (2015). Effect of Tungsten and Molybdenum on the Crystallization, Microstructure and Properties of Silumin 226. Archives of Foundry Engineering, 15 (3), 61-66.

[20] S. Pietrowski, G. Gumienny, B. Pisarek, R. Władysiak. (2004). Production control of advanced casting alloys with TDA method. Archiwum Technologii Maszyn i Automatyzacji 24 (3), 131-144.

The 73rd World Foundry Congress
„Creative Foundry"
23 - 27 September 2018, Krakow, Poland

EFFECT OF COMPOSITION AND MICROSTRUCTURE ON THE FATIGUE LIFE OF QUATERNARY SnZnAgCu LEAD-FREE ALLOY

Krystyna Pietrzak [1]*, Adam Klasik [2], Maria Maj [3]*, Natalia Sobczak [1,4]

[1] Institute of Precision Mechanics 3 Duchnicka Str., 01-796 Warsaw, Poland
[2] Motor Transport Institute, 80 Jagiellonska Str., 03-301 Warsaw, Poland
[3] AGH University Science and Technology, Faculty of Foundry Engineering, 23 Reymonta Str., 30-059 Cracow, Poland
[4] Foundry Research Institute, 73 Zakopianska Str., 30-418 Cracow, Poland
**Corresponding address: krystyna.pietrzak@imp.edu.pl, mmaj@agh.edu.pl*

Key words: Lead-free alloy, microstructure, properties

In the article the results of studies of quaternary SnZn4.5Ag1Cu1 lead-free alloy were presented. The microstructure was examined by means both light microscopy (LM) and scanning electron microscopy (SEM). The identification of chemical composition of alloy in micro-areas was performed by means of SEM/EDS method. The qualitative as well as quantitative assessment by means of combinatorial method was carried out in the light microscopy range. Comprehensive characterization of mechanical behavior especially focusing on alloy fatigue life was performed by means of the original modified low cycle fatigue method (MLCF). The article presents the microstructural conditions of fatigue life of the SnZn4.5Ag1Cu1 alloy.

1. Introduction

In accordance with the European Union recommendations (ROHS 2011/65/UE), lead as a hazardous metal should be withdrawn as soon as possible from the composition of all materials applied in various design solutions for the electrical and electronic equipment. The results of the studies are based, among others, on continuously prepared material bases, which are an important aid in the development of new lead-free alloys with different chemical compositions. For example, the NIST database [1], last updated in 2016, is one of the most widespread databases focused mainly on lead-free solders. It contains a series of data on physico-chemical properties, including also the wettability, as well as functional properties such as mechanical characteristics resulting from static tensile test, creep behaviour, resistance to thermal shocks and shear resistance (the latter one referring, for example, to solder joints). At the same time, searching for literature data describing the properties of lead-free alloys, many other publications can be found in addition to the above mentioned database [1]. Their authors analyze in detail the effect of the proposed chemical composition and microstructure on various functional properties [2-7] depending on the technical use of alloys, and although some of these publications are of a typical review nature [2,3], there is no literature data on the microstructure and numerous mechanical parameters based on measurements performed each time on a single specimen only.

2. Test materials

The quaternary SnZn4.5Cu1Ag1 lead-free alloy was examined in detail in terms of its chemical composition, microstructure and mechanical characteristics. The alloy was prepared from metals of 99.9% purity. The metals were melted in an Ar atmosphere and then cast at a temperature exceeding by about 50°C the point of liquidus. Casting was performed in such a way as to obtain ready-made samples for mechanical tests and microstructure examinations. text - Times New Roman, size 10

3. Microstructure

The microstructure was studied by both light microscopy (LM) and scanning electron microscopy (SEM/EDS).

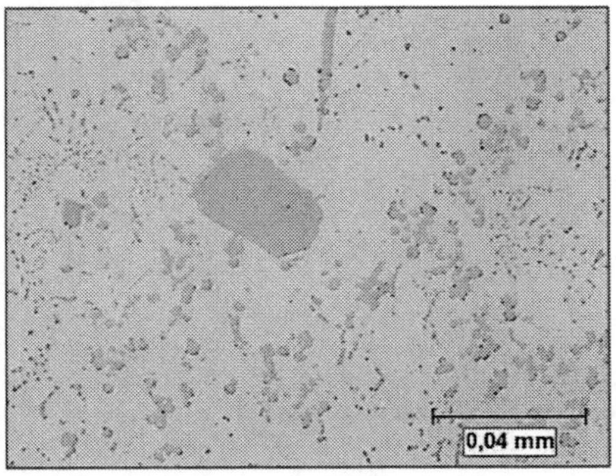

Fig. 1. Typical microstructure of the quaternary SnZnCu1Ag1 lead-free alloy

The typical microstructure of the examined alloy is shown in Figure 1. The occurrence of intermetallic compounds of $AgZn_3$ and Cu_5Zn_8 in the microstructure of the SnZn4.5Cu1Ag1 alloy was confirmed by SEM/EDS.
Geometric parameters of the microstructure were determined by a combinatorial method based on the phase quanta theory [8].

4. Mechanical characteristics

Comprehensive characterization of the mechanical behaviour, focusing in particular on the alloy fatigue life, was performed using the modified low-cycle fatigue test (MLCF) [9]. The following mechanical parameters were measured: UTS -ultimate tensile strength, $R_{0.05}$ – elastic limit, $R_{0.2}$ – yield point, Z_{go} - assessed fatigue strength, K' – stress coefficient under cyclically varying loads, R_A – accommodation limit, E – Young's modulus and also fatigue parameters: b – Basquin's coefficient, c – fatigue ductility exponent, ε_{max} – maximum allowable strain; n' –strain hardening exponent under cyclically varying loads (Fig. 2).

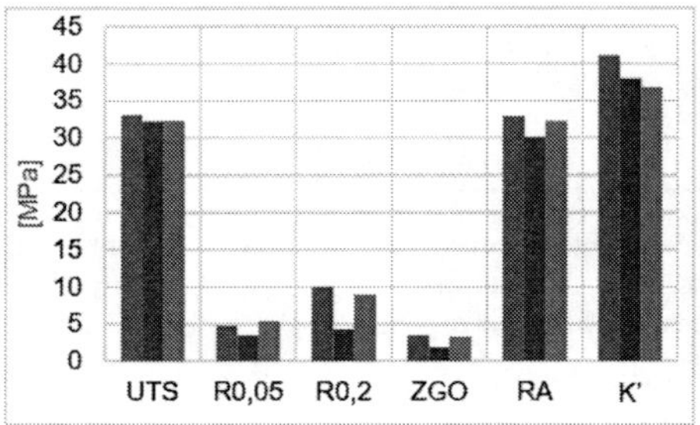

a

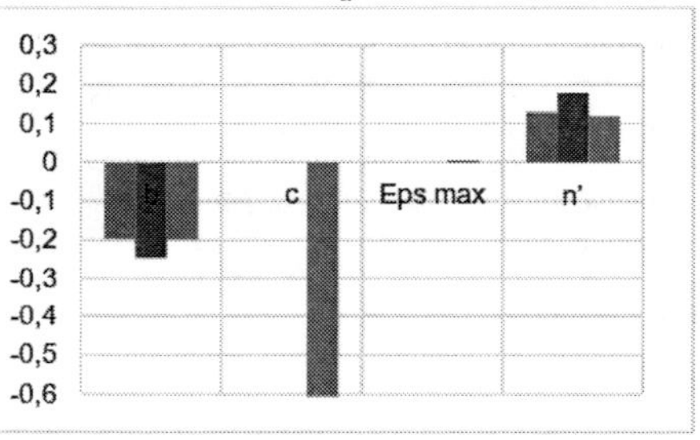

b

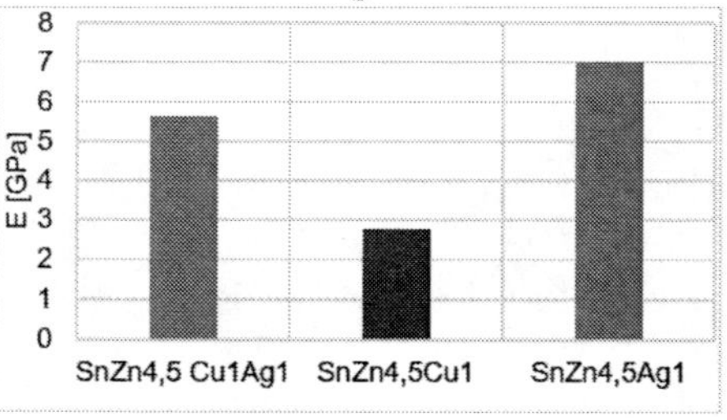

c

Fig. 2. Mechanical parameters of the quaternary SnZnCu1Ag1 lead-free alloy ▇ and of the ternary SnZnCu1▇ [10] and SnZnAg1 ▇ [11] lead-free alloys as determined by the modified low-cycle fatigue test (MLCF).

Figure 2 also shows the comparison of results obtained for the quaternary SnZn4.5Cu1Ag1 alloy with the results obtained previously for the ternary alloys of SnZn4.5Cu1 [10] and SnZn4.5Ag1 [11]. From the comparison it follows that the quaternary SnZn4.5Cu1Ag1 alloy is characterized by some mechanical parameters superior to those obtained in the ternary SnZn4.5Cu1 and SnZn4.5Ag1 alloys.

5. Conclusions

Based on the results obtained, it can be concluded that from the point of view of mechanical characteristics, fatigue life in particular, the addition of Ag and Cu to the binary SnZn4.5 alloy has a beneficial effect on this alloy. Moreover, the advantages of both test methods, i.e. the combinatorial method in the quantitative assessment of the microstructure and the MLCF method in characterizing the mechanical behaviour of the alloys tested, were also demonstrated.

Acknowledgments

The article was in part based on research carried out in the framework of the EU COST ACTION MP0602 with financial support from the Ministry of Science and Higher Education of Poland.

References

[1] www.nist.gov/programs-projects/lead-free-solder, 2013, updated 2016.

[2] Yi. Li, C.P. Wong (2006). Recent advances of conductive adhesives as a lead-free alternative in electronic packaging: Materials, processing, reliability and applications. Materials Science and Engineering 51, 1–35.

[3] L. Zhang, Ch. He, Y. Guo, J. Han, Y. Zhang, X. Wang (2012). Development of SnAg-based lead free solders in electronics packaging. Review. Microelectronics Reliability 52, 559–578.

[4] W.R. Osório, L.C Peixoto, L.R. Garcia, N. Mangelinck-Noël, A. Garcia. (2013). Microstructure and mechanical properties of Sn–Bi, Sn–Ag and Sn–Zn lead-free alloys. Journal of Alloys and Compounds. 572, 97–106.

[5] H.R. Kotadia, P.D. Howes, S.H. Mannan (2014). A review: On the development of low melting temperature Pb-free solders. Microelectronics Reliability. 54, 1253–1273.

[6] J. Shen, Y. Pu, H. Yin, D . Luo, J. Chen (2014). Effects of minor Cu and Zn additions on the thermal, microstructure properties of Sn–Bi-based solder alloys. Journal of Alloys and Compounds. 614, 63–70.

[7] L. Ratke, S. Brück, R. Mathiesen, A. Ludwig, M. Gruber-Pretzler, B. Tonn, K. Gzovskyy, L. Gránásy, G. Tegze, J. Agren, L. Höglund, L. Arnberg, E. Gust, G. Anger, M. Lauer, R. Garen, & B. Reifenhäuser (2007). Lead-free Bearing Alloys for Engine Applications. Results of the ESA-MAP Project MONOPHAS. Trans. Indian Inst. Met. 60, 103-111.

[8] B.K. Kęsy (1990). Microstructure as arrangement of unitary phase parts and stereological parameters. In Stermat'90: the third International Conference on Stereology in Materials Science, October 1-4 1990 (226 -231). Krakow-Katowice Poland: Publisher is not specified in proceedings.

[9] M. Maj, A. Klasik, K. Pietrzak, D. Rudnik (2015). Modified low-cycle fatigue (LCF) test. Metalurgija = Metallurgy 54, 207–210.

[10] K. Pietrzak, A. Klasik, M. Maj, A. Wojciechowski, N. Sobczak (2017). Comparative Studies of Microstructure and Fatigue Life of Selected Lead-free Alloys. Archives of Foundry Engineering 17, 111-116.

[11] K. Pietrzak, A. Klasik, M. Maj, N. Sobczak, A. Wojciechowski (2018). Microstructural aspects of fatigue life of Sn-Zn lead-free solders with 1% of Ag addition, prepared for printing in Archives of Foundry Engineering.

The 73rd World Foundry Congress
„Creative Foundry"
23 - 27 September 2018, Krakow, Poland

HETEROEPITAXIAL GROWTH OF PASSIVATING LAYERS ON RUTILE IN CONTACT WITH MOLTEN ALUMINIUM AND MOLTEN A356 ALUMINIUM ALLOY

Anton Salomon *, Christiane Ullrich and David Rafaja

TU Bergakademie Freiberg, Institute of Materials Science, Gustav-Zeuner-Straße 5, 09599 Freiberg
**Corresponding address: anton.salomon@ww.tu-freiberg.de*

Key words: Filtering, Spark Plasma Sintering, electron microscopy, functional coating, interfacial reactions

1. Introduction

Rutile-coated ceramic foam filters were recently shown to exhibit high melt flow rate and moderate filtration efficiency for different inclusion types in A356 alloy [1]. These properties were mainly attributed to a specific wetting behaviour [1]. The most important products of the reaction between rutile (tetragonal TiO_2) and molten aluminium or between rutile and molten aluminium alloy A356 (AlSi7Mg0.6) were identified by a combination of phase analysis and thermodynamic calculation to be corundum (α-Al_2O_3) and magnesium titanate ($MgTiO_3$), respectively [2].

However, the previous experiments [2] were performed on model systems approaching the thermodynamic equilibrium. In this contribution, it is discussed, in which way the functional rutile coatings covering ceramic foam filters react with molten aluminium and molten aluminium alloys in a short time. For this study, rutile was applied as a functional coating [3] on corundum-based ceramic foam filters. The use of an SPS apparatus for melting the aluminium alloy impedes the macroscopic flow of the melt and excludes mechanically induced wear on the metal-ceramic interface.

2. Sample preparation

The filter material was prepared in form of small moulds from high-purity corundum (Martoxide MR70, Albemarle Corp., USA) that was sintered at 1600°C, coated with rutile (R320 from Sachtleben, Germany) via slip casting and finally sintered at 1300°C [1,4].

The aluminium and aluminium alloy melts were produced within the moulds from the respective powder in a Spark Plasma Sintering (SPS) device (HPD 25, FCT Systeme GmbH, Germany). The purity of the Al powder (Goodfellow, Germany) was 99.9 %. The aluminium alloy AlSi7Mg0.6 (TLS Technik, Germany) contained 7.1 mass% Si and 0.6 mass% Mg with 0.1 mass% Fe as the main impurity.

The powders were melted under a working gas (residual air) pressure between 0.5 and 5 Pa with the heating rates of about 1500 $K \cdot min^{-1}$. The dwell times at 750°C were up to 300 min. After the respective reaction time, the samples were rapidly cooled with about 300 $K \cdot min^{-1}$.

3. Sample characterisation

The products of the reaction between the aluminium or aluminium alloy melt and TiO_2 were analysed using scanning electron microscopy with energy dispersive X-ray spectroscopy (SEM/EDX) and electron backscatter diffraction (SEM/EBSD, SEM LEO-1530, Carl Zeiss AG, Germany), and electron probe microanalysis with wavelength dispersive X-ray spectroscopy (EPMA/WDX, JXA8900 RL, Jeol Ltd., Japan).

4. Rutile in contact with pure Al

After 60 min at 750°C, a newly formed corundum layer with a thickness of about 1 µm was found at the TiO_2/Al interface (Fig. 1). After 300 min of continuous contact between molten Al and TiO_2 at 750 °C, the corundum layer thickness had only slightly increased to about 1.5 to 2 µm. The stoichiometry of the layer (α-Al_2O_3) was confirmed by the EPMA/WDX analysis.

Figure 1 shows the results of EBSD combined with EDX that was performed on the interface between Al and TiO_2 after contact for 60 min at 750°C. The local grain orientation maps with corresponding inverse pole figures (Figs. 1b,c) were used to identify an

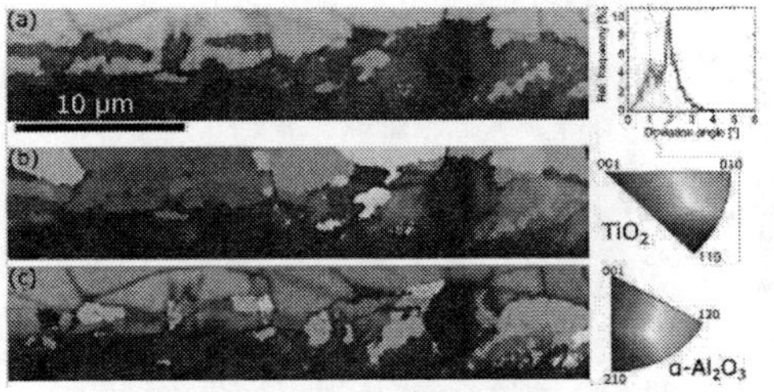

***Fig. 1.** Results of the EBSD analysis performed on a rutile coating after it reacted with molten Al (for 60 min at 750 °C). (a) Phase map showing the distribution of rutile (yellow) and corundum (red) grains. Black areas within the colourised region are pores, other non-indexed regions below represent solidified aluminium. The green lines mark the interfaces between rutile and corundum crystallites having the orientation relationship $(001)_{rutile} \parallel (100)_{corundum}$ and $[010]_{rutile} \parallel [001]_{corundum}$. A histogram of the local deviations from this orientation relationship with the corresponding colour code is shown in (a). The orientation relationship and the local deviation from it are based on the orientation maps with inverse pole figures (colour coding respective to upward growth direction) for rutile (b) and for corundum (c), respectively.*

orientation relationship between the newly formed α-Al_2O_3 crystals growing into the TiO_2 matrix. The corundum grows on rutile with the orientation relationship $(001)_{rutile} \parallel (100)_{corundum}$ and $[010]_{rutile} \parallel [001]_{corundum}$ with less than 3 ° deviation from a perfect match as displayed in the frequency distribution along with a colour comparison on the micrograph in Figure 1a. Initial α-Al_2O_3 formation is believed to consume oxygen stemming from oxide skins on the aluminium powder particles thereby minimising the amount of oxygen solved in the melt.

5. Rutile in contact with molten AlSi7Mg0.6

In contact with molten AlSiMg0.6 alloy at 750 °C, an $MgTiO_3$ interlayer was formed on the surface of the functionalized rutile coating (Fig. 2). Again, EPMA/WDX analysis was applied to confirm the $MgTiO_3$ stoichiometry, and the crystal structure and orientation was verified with EBSD.

The thickness of the reaction layer was approximately 5 ± 1.5 μm after 60 min. It increased slightly to about 7 ± 2 μm after 300 min. The large errors are attributed to the needle-like shapes of the $MgTiO_3$ crystallites at the $MgTiO_3/TiO_2$ interface. The well-defined orientation relationship $(001)_{rutile} \parallel (100)_{titanate}$ and $[010]_{rutile} \parallel [001]_{titanate}$ between neighbouring TiO_2 and $MgTiO_3$ crystallites was found using the EBSD grain orientation analysis (Figs. 2a-c). It is identical with the orientation of corundum growing on rutile discussed above. Along the whole $MgTiO_3/TiO_2$ interface, this relationship is met with a deviation of less than 4 ° (Fig. 2a). The heteroepitaxy between newly formed $MgTiO_3$ and TiO_2 guarantees a good adhesion.

Most probably, $MgTiO_3$ results from the reaction of TiO_2 with MgO. Magnesium stems from the molten alloy. The main source of oxygen needed for the $MgTiO_3$ formation is the oxide skin covering the AlSiMg0.6 powder particles. Consequently, the metal melt filter that was functionalised by the rutile coating decreases the oxygen concentration in the melt. However, the formation of the $MgTiO_3$ layer decreases also the Mg concentration in the AlSiMg0.6 alloy. These effects were confirmed by WDX, which did not reveal any Mg and O concentration about the detection limit of approx. 0.1 to 0.5 at.% at the filter surface. The limited supply of Mg and O in our system is regarded to be the main reason for the low growth rate observed for longest dwell times.

Fig. 2. Results of the EBSD analysis performed on a rutile coating that was in contact with molten AlSi7Mg0.6 alloy for 60 min at 750 °C. (a) Phase map showing the distribution of rutile (yellow) and MgTiO3 (blue). The areas reproduced in black are pores/voids, the grey region at the bottom of the micrograph is the solidified alloy AlSi7Mg0.6. The green spots and lines mark the interfaces between the TiO2 and MgTiO3 grains having the orientation relationship (001)rutile || (100)titanate and [010]rutile ‖ [001]titanate. A histogram of the local deviations from this orientation relationship with the corresponding colour code is shown. This analysis is based on the orientation maps of rutile (b) and MgTiO3 (c) with the colour coding of the crystallographic directions (inverse pole figures) related to the macroscopic layer growth direction (upward).

6. Conclusions

The reaction of rutile with molten pure Al and with molten AlSi7Mg0.6 alloy at 750 °C was studied using scanning electron microscopy, electron backscatter diffraction and X-ray spectroscopy. The exposure of TiO_2 to molten Al led to the formation of a thin layer of thermodynamically stable corundum exhibiting the orientation relationship to rutile, which was described as $(001)_{rutile} \parallel (100)_{corundum}$ and $[010]_{rutile} \parallel [001]_{corundum}$. The exposure of TiO_2 to molten AlSi7Mg0.6 led to the formation of $MgTiO_3$ with the following orientation to rutile: $(001)_{rutile} \parallel (100)_{titanate}$ and $[010]_{rutile} \parallel [001]_{titanate}$.

These orientation relationships, which are driven by a heteroepitaxy of the counterparts, are considered as the reason for the excellent adhesion between the newly formed corundum or $MgTiO_3$ layer and the TiO_2 functional coating. Furthermore, the formation of α-Al_2O_3 and/or $MgTiO_3$ was shown to consume oxygen dissolved in both melts (Al and AlSi7Mg0.6), which confirms the effectiveness of TiO_2 as a reactive filter. A slight drawback is a concurrent Mg depletion in the aluminium alloy.

Acknowledgments

This work was financially supported by the German Research Foundation (DFG) in the frame of the Collaborative Research Centre SFB 920. The authors would like to thank Ms. Claudia Voigt for providing the filter ceramic and Dr. Dietrich Heger for the EPMA/WDX analyses.

References

[1] C. Voigt, E. Jäckel, F. Taina, T. Zienert, A. Salomon, G. Wolf, C. Aneziris & P. LeBrun. (2016). Filtration Efficiency of Functionalized Ceramic Foam Filters for Aluminum Melt Filtration. Metal. Mater. Trans. B. 48, pp. 497-505.

[2] A. Salomon, C. Voigt, O. Fabrichnaya, C. G. Aneziris & D. Rafaja. (2017). Formation of Corundum, Magnesium Titanate, and Titanium(III) Oxide at the Interface between Rutile and Molten Al or AlSi7Mg0.6 Alloy. Adv. Eng. Mat.. 19 (9), 1700106.

[3] C. Aneziris, S. Dudczig, M. Emmel, H. Berek, G. Schmidt & J. Hubalkova. (2013). Reactive filters for steel melt filtration. Adv. Eng. Mat.. 15, pp. 46-59.

[4] C. Voigt, T. Zienert, P. Schubert, C. Aneziris & J. Hubalkova. (2014). Reticulated porous foam ceramics with different surface chemistries. J. Am. Cer. Soc.. 97, pp. 2046-2053.

The 73rd World Foundry Congress
„Creative Foundry"
23 - 27 September 2018, Krakow, Poland

NON-METALLIC INCLUSIONS AND THEIR INFLUENCE ON THE MECHANICAL PROPERTIES OF 18CrNiMo7-6 STEEL TREATED IN DIFFERENT CRUCIBLES

Mikhail Seleznev[1*] Anja Weidner[1], Johannes Gleinig[1], Sebastian Henschel[1], Jens Fruhstorfer[2], Steffen Dudczig[2], Lutz Krüger[1], Christos G. Aneziris[2], Horst Biermann[1]

[1] *TU Bergakademie Freiberg, Institute of Materials Engineering, Gustav-Zeuner-Str. 5, 09599 Freiberg, Germany*
[2] *TU Bergakademie Freiberg, Institute of Ceramic, Glass and Construction Materials, Agricolastr. 17, 09599 Freiberg, Germany*
**Corresponding address: Mikhail.seleznev@iwt.tu-freiberg.de*

Key words: Non-metallic inclusions, fatigue lifetime, very high cycle fatigue, energy dissipation, fracture toughness

1. Introduction

Non-metallic inclusions (NMI) occur in steels after the casting process either as exogenous inclusions caused by e.g. refractory linings or as endogenous inclusions formed after the melt treatment (e.g. deoxidation) during the solidification of the material. Majority of individual endogenous NMI are usually in the size of 2-5 µm. However, they have the tendency to sinter or agglomerate to larger NMI. The endogenous NMI can be removed e.g. by means of metal melt filtration using ceramic foam filters with active or reactive coatings [1]. Thus, Emmel et al. [1] studied the effect of filter surface chemistry on the entrapment of non-metallic inclusions at the filter surface. NMI have a detrimental effect on the mechanical properties of steels (e.g. [2]), since they act as stress concentrators during both cyclic and static loading. Thus, cyclic loading of materials is a very promising technique to detect the most critical non-filtered inclusion within the tested specimen volume. NMI cause even at small stress amplitudes fatigue failure at very high number of cycles (> 10^7 cycles) – the so-called very high cycle fatigue (VHCF) regime. The fatigue failure in this regime is mainly caused by internal defects. Here, the size, the location and the chemical composition have, in particular, an influence on the fatigue life [3,4]. In addition, the fracture toughness is influenced by NMI. Here, the size distribution and the location of NMI affect the scatter of fracture toughness in the transition temperature range, see in particular [3,5,6].

The present contribution focuses on the effect of different carbon-bonded crucible materials for ingot casting on the size distribution of NMI in the steel 18CrNiMo7-6. Furthermore, the fatigue behavior, the strength and deformation behavior at quasi-static loading rates and the dynamic fracture toughness were investigated. The inclusion distributions were quantified at metallographic sections by means of optical microscopy (OM) and/or scanning electron microscopy (SEM). Fracture surfaces were analyzed applying SEM with energy dispersive X-ray spectroscopy (EDS).

2. Materials and Methods

The case hardening steel 18CrNiMo7-6 was treated in a metal casting induction furnace under Argon atmosphere (see e.g. Ref. [7]). Three different crucibles were applied: (i) carbon-bonded alumina (A-C), (ii) carbon-bonded alumina-zirconia-titania (AZT-C), and (iii) AZT-C with carbon nanotubes (AZT-C-n). The steel was molten in the crucible, held at 1,587°C for 1h, and controlled cooled to room temperature (RT) (see ref. [8]). The obtained cylindrical steel blocks had a diameter of 200 mm and a height of 95 mm. Small shrinkage cavities were removed by hot-isostatic pressing. Specimens for mechanical testing and for metallographic analysis were sampled in vertical direction from the cylinder. Since high strength levels were necessary for studying the influence of inclusion on the fatigue live (usually above 1200 MPa), the fatigue specimens were austenitized and quenched. Tempering was omitted in order to keep the strength level of the steel matrix as high as possible. The specimens for the static loading and fracture toughness testing were heat treated including tempering (see [8]).

The fatigue behavior was studied on the steel treated with the three different crucible materials. The fatigue tests were performed with an ultrasonic fatigue testing machine, which allows up to 20 kHz of testing frequency. The tests were performed at symmetric push-pull loading conditions ($R = -1$) and ambient temperature. The run-out limit was 10^9 cycles. Further details of the experimental set-up and the used specimen geometry are given in [9].

Quasi-static tensile tests at different temperatures were performed on the steel treated in the A-C and AZT-C crucibles. The dynamic fracture toughness was evaluated on pre-cracked SENB specimens ($a_0/W = 0.5$). Hence, the behavior in the ductile-to-brittle transition regime, and the reference temperature was evaluated. Details of used instrumented pendulum impact tester and the determination of the displacement can be found elsewhere (see ref. [10].

Metallographic sections were analyzed regarding the size distribution of NMI both by OM and also at selected sections by an automated SEM for particle analysis (FEI-ASPEX PSEM eXpress) enabling the analysis of NMI in terms of size, location and chemical composition.

3. Results and Discussion

The different crucible materials had an influence on the chemical composition of the treated steel, especially on the elements carbon, aluminum and oxygen. It was found that the amount of aluminum was reduced below the detection limit by all applied crucible materials. In addition, the AZT-C and the AZT-C-n crucible materials significantly reduced the oxygen content.

In addition, an influence of the crucible material on the frequency distribution of the NMI in the treated steel was observed. Thus, the steel treated in the A-C crucible contained the highest amount of NMI at all and, in particular, with a maximum of the size distribution at around 2-5 µm. In contrast, the AZT-C and AZT-C-crucibles resulted in total in a smaller amount of NMI. Hereby, a bimodal size distribution was observed with a maximum at even smaller inclusions (around 1 µm) and a second maximum at larger inclusions above 15-20 µm. The microstructural investigations on the NMI in the differently treated steels revealed in general two main types of inclusions: (i) plain manganese sulphides and (ii) so-called duplex inclusions consisting of an oxide core surrounded by manganese sulphide shell. The size distribution of these two types of NMI revealed also an influence of the different crucible materials.

The fatigue behavior of the steel treated with different crucibles was significantly influenced by the NMI. Here, not the total amount of inclusions was crucial. However, the different size distributions had to be taken into account for the interpretation of the fatigue strength, which is significantly different for the three crucible materials. Thus, the steel treated in the A-C crucible showed the highest fatigue strength. The treatment in the AZT-C-n crucible resulted in the lowest fatigue strength. The differences in the fatigue strength are well correlated to the size of NMI found on the fracture surfaces. The critical size of NMI causing fatigue failure is with 15 µm the smallest for the A-C treated steel, whereas the size of detrimental NMI is with 100 to 200 µm the highest for the AZT-C+n treated variant. Since the fatigue strength is mainly affected by the largest inclusion in the tested volume, the high number of small inclusions as obtained by the A-C treatment had no negative effect in terms of fatigue strength.

This is in contrast to the results obtained from the quasi-static tensile tests, which revealed a higher strength but a lower ductility for the steel treated with the A-C crucible. Furthermore, the energy dissipation was much lower for this steel. The lower deformability and, hence, the lower energy dissipation until fracture for the A-C treated steel was attributed to the higher inclusion content in this steel.

4. Summary and Conclusions

The content and the size distribution of NMI has to be taken into account for the interpretation of the mechanical behavior of the steel. Moreover, whereas the total amount of NMI is crucial for the static properties and the energy absorption, the size of inclusions above 10 µm is crucial for the fatigue behavior.

Acknowledgment

The authors thank the German Research Foundation (DFG) for the financial support of the investigations in the Collaborative Research Center 920, subprojects C04, C05, and C01. We express our special thanks to Tilo Zienert for fruitful discussions. We acknowledge the support of Dr. Thilo Kreschel for the chemical analysis, Birgit Witschel for quantitative optical microscopic analysis, Tim Lippmann for performing fatigue tests, and Patrick Schlenz for specimen preparation.

References

[1] Emmel, M., Aneziris, C.G., Schmidt, G., Krewerth, D. & Biermann, H. (2013). Influence of the chemistry of surface functionalized ceramic foam filters on the filtration of alumina inclusions in steel melts. Adv. Eng. Mater. 15 (12), 1188–1196.

[2] C. Klinger, D. Bettge, Axle fracture of an ICE 3 high speed train. Eng. Fail. Anal. 2013, 35A, 66.

[3] Henschel, S., Krewerth, D., Ballani, F., Weidner, A., Krüger, L., Biermann, H., Emmel, M. & Aneziris, C.G. (2013). Effect of filter coating on the quasi-static and cyclic mechanical properties of a G42CrMo4 casting. Adv. Eng. Mater. 15 (12), 1216–1223.

[4] Krewerth, D., Lippmann, T., Weidner, A. & Biermann, H. (2016). Influence of non-metallic inclusions on fatigue life in the very high cycle fatigue regime. Int. J. Fatigue 84, 40-52.

[5] Wallin, K., Saario, T. & Törrönen, K. (1984). Statistical model for carbide induced brittle fracture in steel. Met. Sci. 18 (1), 13–16.

[6] Zhang, L. & Thomas, B. G. (2003). Inclusions in continuous casting of steel. In: XXIV Steelmaking National Symposium. Morelia, Mich, Mexico, 138–183.

[7] Dudczig, S., Aneziris, C.G., Dopita, M., Rafaja, D. (2013) Application of oxide coatings for improved steel filtration with the aid of a metal casting simulator. Adv. Eng. Mater. 15 (12), 1177-1187.

[8] Henschel, S., Gleinig, J., Lippmann, T., Dudczig, S., Aneziris, C.G., Biermann, H., Krüger, L., Weidner, A. (2017). Effect of crucible material for ingot casting on detrimental non-metallic inclusions and the resulting mechanical properties of 18CrNiMo7-6 steel. Adv. Eng. Mater. 19 (9), 1700199.

[9] Krewerth, D., Lippmann, T., Weidner, A. & Biermann, H. (2015). Application of full-surface view in situ thermography measurements during ultrasonic fatigue of cast steel G42CrMo4. Int. J. Fatigue. 80, 459–467.

[10] Henschel, S. & Krüger, L. (2015). Deflection measurement in low-blow Charpy impact tests. Mater. Test. 57 (10), 837–842.

The 73rd World Foundry Congress
„Creative Foundry"
23 - 27 September 2018, Krakow, Poland

Profiling the Quality of Bentonite Clay with Dilatometry

J. Thiel[1]*, S. Ravi[1], V. Lafay[2]

[1]University of Northern Iowa, Metal Casting Center, Cedar Falls, Iowa, USA
[2] Imerys, S&B Industrial Materials, Cincinnati, OH, USA
**Corresponding address: e-mail: Thiel@uni.edu*

Key words: Bentonite, Foundry Clays, Green Sand Binder, Sand Additives, Molding Binders

1. Abstract

Bentonite clay (Montmorillonite) is one of the basic components of a green sand system. The bentonite clay serves as a sand binder that is both cost effective and abundant. The green sand process is currently used in the production of more castings than any other molding method. The quality of the clays used has been historically evaluated using standard tests including CEQ, (Cation Exchange Quotient), chemical analysis, swelling index, and test casting results.

The University of Northern Iowa Metal Casting Center has developed a method of determining clay quality using the thermal expansion profile determined by dilatometry. This method can quickly predict not only the quality of the castings the clay will produce but also the durability of the clay in a green sand system over several casting cycles. The method can also be used to determine how sand additives will affect the performance of the clay in the system.

2. Introduction

Trials were conducted to evaluate three different green sand premixes for physical and thermal properties over four cycles. Phase I of the testing consisted of baseline testing of the three green sand mixtures. A clay content of 9% and compactability in the range of 42-45 was used to prepare the test specimen required. The tests conducted for each mixture included:

- Green compression strength
- Green shear strength
- Compactability
- Moisture
- Wet tensile strength
- Methylene blue clay
- Permeability
- Loss on Ignition

Additionally, thermal expansion tests were conducted using the University's high temperature aggregate dilatometer to measure the expansion characteristics, surface viscosity and sinter temperature of the three green sand mixtures.

Phase II of the testing consisted of four cycles of performance testing by pouring grey iron step-block test castings. After each cycle, a methylene blue clay test was run on the shakeout sand to keep the clay content consistent at 9% clay. The green sand mixtures were then characterized for physical and thermal properties using the tests described in phase I.

Results obtained from the baseline and cycle tests were evaluated to compare the three different premixes. The test castings were sectioned, sand blasted and photographed to evaluate the performance of the green sand mixtures. Additionally, the castings were scanned using a Romer arm laser scanner to obtain color maps. The castings were compared numerically using the color maps.

3. Thermal Expansion

Thermal expansion tests were run on the green sand samples utilizing the university's high temperature aggregate dilatometer, shown in figure 1. The cylindrical expansion cores had a height of approximately 1.6 inches (4.06 cm) and a diameter of 1.1 inches (2.8 cm). The samples were heated to 1650 C (3002 F) at heating rates of 5 C per minute from room temperature to 600 C (1112 F) and 10 C per minute from 600 C (1112 F) to 1650 C (3002 F). The resulting deformation was recorded.

Results of the thermal expansion tests showed what was recognized as the alfa/beta quartz transformation at 573C as expected. Most foundry sands are composed of alpha-quartz. These sands expand slowly to around 573C (1063F) where quartz undergoes a phase change to beta-quartz[2]. This phase change is associated with a rapid volume change 4.74% and a linear expansion of 1.56% (.015 mm/mm). This rapid expansion is what is usually associated with veining defects and thermal fracture of sand grains. Table 1 illustrates the phase changes associated with temperature increases in quartz. Beta quartz transforms to beta-tridymite at 870C (1598F) in the presence of a liquid. This transformation is accompanied by a 12.44% change in volume which is substantially greater than the original alpha to beta-quartz transformation.

Table 1. Phase transformation temperatures of quartz

Phase Transition	Temperature	Density Change (Volume)	Linear Change
alpha-quartz to beta-quartz	573C (1063F)	2.65 to 2.53 (+4.74%)	+1.56% (0.0156 in/in)
beta-quartz to beta-tridymite	870C (1598F)	2.53 to 2.25 (+12.44%)	+3.99% (0.0399 in/in)
beta-tridymite to beta-cristobalite	1470C (2678F)	2.25 to 2.20 (+2.27%)	+0.75% (0.0075 in/in)
beta-quartz to beta-cristobalite	1470C (2678F)	2.53 to 2.20 (+14.71%)	+4.74% (0.0474 in/in)

We see this transformation when engineered sand additives (ESA) are used or when sodium is present in the sand as in bentonite bonded green sands. Sodium and aluminum are both known to stabilize tridymite transformation.

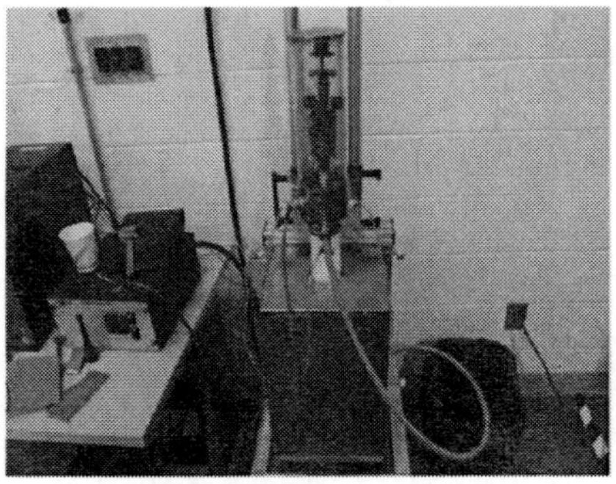

Fig. 1. High temperature aggregate dilatometer

Figure 2. shows a typical green system sand including 8% of a commercial sodium bentonite with a 1.6% seacoal addition. We can see drastic increases in volume at both the alfa/beta transformation as well as the tridimite transformation temperatures. When the sand is cycled through the casting process, the peak

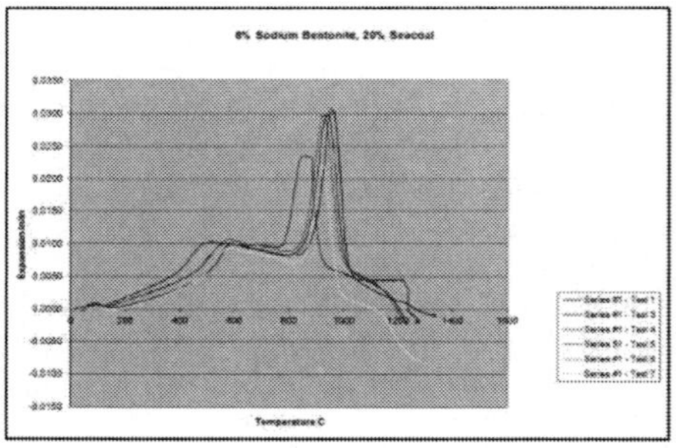

Fig. 2. Linear expansion through several casting cycles

expansion level of the tridimite transformation is only slightly affected declining after 3 to four casting cycles. Figure 3. Illustrates the relationship between the wet tensile and the peak tridimite transformation. We see by the graph that the wet tensile of the mulled sand is directly proportional to the peak tridimite transformation. It is well established that the wet tensile properties of green system sand are an important characteristic reflecting on resistance to both expansion and erosion defects.

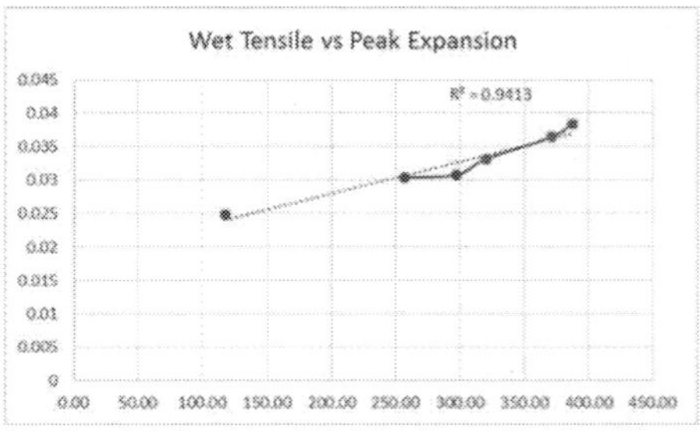

Fig. 3. Relationship of Wet Tensile results with tridimite expansion

Figure 4. Illustrates a green system sand containing a second sodium bentonite clay with a 1.6% seacoal addition. We can see by the tridimite expansion that although the results are initially high, the results decrease quickly with successive casting cycles. This indicates that the tested bentonite will provide good initial results but will most likely fade quickly with time requiring additional amounts of bentonite to maintain performance. This affect was not only seen in bentonites but also in the use of additives.

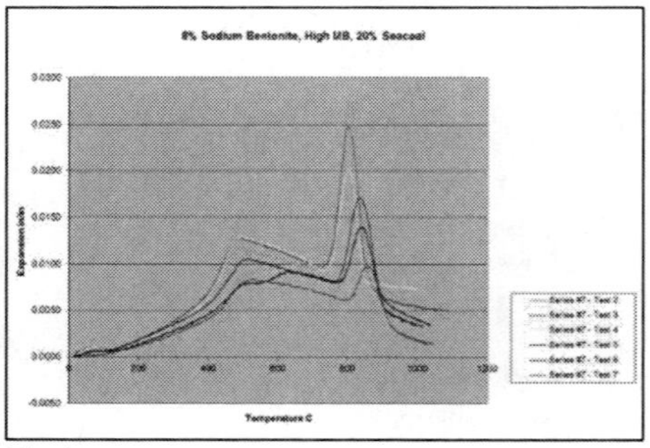

Fig. 4.

4. Acknowledgments

The authors would like to acknowledge the support of Imerys Minerals for both technical information and materials.

The 73rd World Foundry Congress
„Creative Foundry"
23 - 27 September 2018, Krakow, Poland

FATIGUE BEHAVIOR ON HEAVY SECTION DUCTILE IRON CASTING MANUFACTURED IN GRADE EN-GJS-500-14.

M.A. Altuna [1*], J. Uranga[1], G. Pedrero[1], U. Uribe [2]

[1] *Grupo WEC, Polígono Industrial N°38, 20829 Itziar-Deba, Spain*
[2] *IK4-AZTERLAN, Aliendalde Etxetaldea, 6, 48200 Durango, Spain*
**Corresponding address: a.altuna@grupowec.com*

Key words: Heavy section ductile iron, EN-GJS-500-14, microstructural evolution, fatigue behavior, number of cycles

1. Introduction

In the past decade, in order to improve efficiency in wind energy, the size of turbines has been increasing and consequently the weight of the castings has become a great limitation. In the next few years, the challenge will be to improve castings by increasing the strength of material without increasing the weight. To achieve that, the spheroidal iron commonly used in wind cast components (EN-GJS-400-18-LT) will be replaced by new materials, known as nodular iron hardened by solid solution. Nowadays, the relation between microstructure and mechanical properties is very poor in these new grades in heavy castings and there is a lack of information about fatigue behavior. In heavy section nodular cast iron parts, anomalous microstructures can be formed, especially at the core, because of low cooling rates. Furthermore, these grades are more prone to presenting porosity defects. All this causes an important decrease in mechanical properties.

In this paper, the relation between the microstructure and fatigue behavior will be presented for a heavy casting manufactured in grade EN-GJS-500-14.

2. Design of experiments

A 15 Tn casting with thicknesses between 150-250 mm was manufactured in grade EN-GJS-500-14. In order to increase solidifications rates, in the critical areas chillers were places from outside. The section with chillers will be denominated "Fast Cooling" (FC) and the section without chillers "Slow Cooling" (SC).

Metallographic analyses were carried out for FC and SC sections to verify the matrix microstructure, the graphite morphology and the nodule count (the number of graphite particles per mm^2 of area). The samples were taken from different solidification rates, estimated by Magmas software. Tensile test specimen, with a diameter of 14 mm, were performed at room temperature, according to the EN ISO 6892-1. Fatigue specimens of 9 mm diameter were machined according to ASTM E466. The axial fatigue tests were carried out in a hydraulic dynamic test machine, with maximum load capacity of ±100 KN and the maximum test frequency of 15 Hz. The tests were performed at a maximum tension of 550 MPa with R=-1.

3. Results and Discussion

3.1 Microstructure

With the aim of understanding the microstructure evolution along the entire wall, a detailed microstructure analysis was done for FC and SC sections for different positions: 5 mm from outer surface (A), 50 mm from outer surface (B), 50 mm from inside surface (C) and 5 mm from inside surface (D). As seen in Fig. 1, the solidification time increases dramatically as it moves away from the outside surface and it varies from 125 min to 330 min and from 3 min to 160 min for SC and FC sections, respectively.

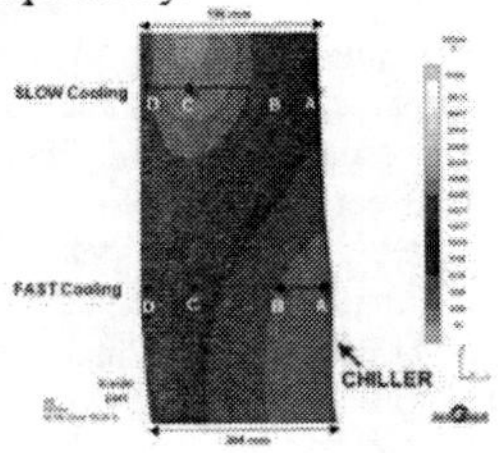

Fig. 1. ***Cooling time for FC section (with chillers) and SC section (without chillers).***

In Fig. 2 the evolution of the microstructure along the wall thickness can be observed, where the differences between both sections are very clear, mainly in the outer side (A and B). The microstructure is totally ferritic but significant differences are observed in the morphology and number of the nodules for different positions. For both FC and SC sections, samples closer to the surface have higher nodule counts, smaller nodules sizes and a higher degree of nodularity than those samples further away from the surface. Moreover, this effect is much more pronounced for FC section. It is important to mention that spiky traces appear from 50 mm for the inner side (C): 3% and 5% of spiky graphite for FC and SC sections, respectively. In ductile irons, in order to avoid the formation of chunky and spiky graphite an optimum balance of antinodularising elements (Sb, Sn, Bi or Pb) and nodularising elements (rare earths) is needed.

Increase of antinodularising elements favors the formation of spiky graphite, whereas the excess of nodularising elements favors the chunky graphite.
In heavy castings manufactured in grade EN-GJS-400-18, the optimum balance for Sb/Ce is near 1.2-1.8. However, in nodular iron hardened by solid solution, as the silicon content is higher than in EN-GJS-400-18 the formation of chunky graphite is favored, so higher Sb/Ce ratios are needed. In this trial, the ratio Sb/Ce was 2.3 and the obtained microstructure demonstrates that this ratio was too high for this grade.

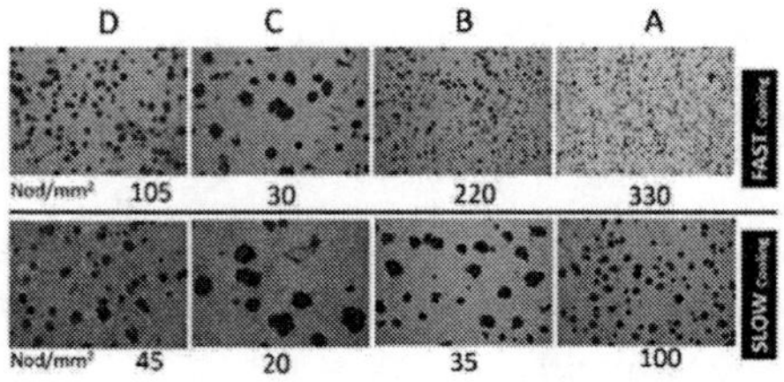

Fig. 2. Microstructural evolution along the wall thickness. x100

3.2 Mechanical Properties

The samples for tensile and fatigue tests were taken from area B (50 mm away from the outer surface). The Yield and Tensile Strength are slightly lower for SC sections (405 MPa versus 411 MPa in Yield Strength and 486 MPa versus 527 MPa in Tensile Strength). However, a significant difference is observed in the elongation, which is less than half for the SC section (11.4% versus 4.4%).
Regarding the fatigue tests, five tests were performed taking samples from area B, where the number of nodules per mm^2 are 220 and 35 for FC and SC sections, respectively. The number of cycles for the FC section range from 454,414-666,214, the average being 580,000 cycles. For the SC section, 15,886-28,840 cycles were achieved, the average being 21,300 cycles, almost 30 times less than in the FC sections.

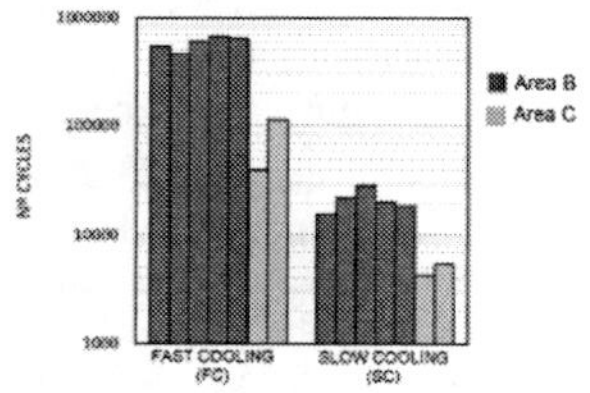

Fig. 3. Number of cycles for FC and SC sections.

Furthermore, in order to study the effect of degenerated graphite in the number of the cycles, two tests were carried out in area C. If area B and C are compared, for both FC and SC sections, a reduction in the number of the cycles occurred: from 580,000 cycles to 77,000 cycles in the case of FC section, and from 21.300 cycles to 4.800 cycles in SC section. In both cases, in addition to the presence of spiky graphite, there is an important reduction in the number of nodules, and therefore it is not easy to estimate what the real impact of each parameter is.
The cycles for area B of the SC sections are 21,300, whereas the cycles for area C in the FC sections are 77,000. In both areas, the number of nodules is very similar (25-30 nod/mm^2), whereas area C in the FC sections contains spiky graphite and area B in the SC sections not. However, even the presence of spiky graphite, the number of cycles is higher, which is a surprising result. Consequently, in addition to the number of nodules and spiky graphite, other parameters affect the number of cycles. In order to clarify this issue the fracture surfaces were analyzed by electron microscopy. In all the cases, the origin of the crack was a microshrinkage, except in one case (area C of the FC section) where the fatigue nucleation was originated at spiky graphites. Therefore, it can be concluded that the effect of microshrinkage is more critical than the presence of spiky graphite, at least when the spiky graphite is less than 5%.

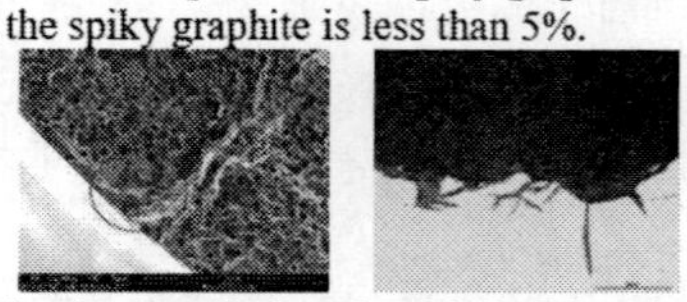

Fig. 4. Detail of crack origin in FC section: microshrinkage in area B and Spiky graphite in area C.

It is well known that the presence of shrinkages has a negative effect on the mechanical performance of a casting, resulting in a significant decrease in the service life. Comparing the grade EN-GJS-500-14 with EN-GJS-400-18, as the content of carbon is lower in the former case, there is a growing tendency of shrinkage formation, and moreover this effect is more pronounced in heavy castings. This considerably hinders the production of castings for wind turbines in grade EN-GJS-500-14 and therefore the use of chillers is needed to increase the solidification rate, favoring a finer microstructure and reducing the microshrinkage formation.

4. Conclusions

1. The placement of chillers has a big effect on the number of graphite nodules: near 10 times more nodules are formed on the surface than in the core.
2. The average number of cycles is 30 times less in “Slow Cooling” sections than in “Fast Cooling” section.
3. Fatigue tests combined with fractographic analyses show that the effect of shrinkage on the number of cycles is more critical than spiky graphite, at least when spiky graphite is less than 5%.
4. In heavy castings manufactured in grade EN-GJS-500-14, the use of chillers is needed to increase the solidification rate, favoring a finer microstructure and reducing the microshrinkage formation.

The 73rd World Foundry Congress
„Creative Foundry"
23 - 27 September 2018, Krakow, Poland

ADI -The Material Revolution and its Applications at CMRDI

ADEL NOFAL [1]*

[1]Central Metallurgical R&D Institute, Egypt, Cairo,
**Corresponding address: e-mail: adelnofal@hotmail.com*

Key words: Austempered Ductile Iron, Production Technologies, Applications

In the last three decades, the revolutionary material; the austempered ductile iron (ADI) with its unique combination of mechanical properties, has been offering the design engineer alternatives to conventional material/process combinations [1-2]. The excellent properties of this material have opened new horizons for cast iron to replace steel castings and forgings in many engineering applications with considerable cost benefits. Moreover, the sustained efforts worldwide of the automotive industry to use lightweight materials have eroded the market for the heavier iron castings. Currently, ADI with its super strength can successfully compete with the lightweight alloys, a point which has yet to be fully understood by many design engineers.

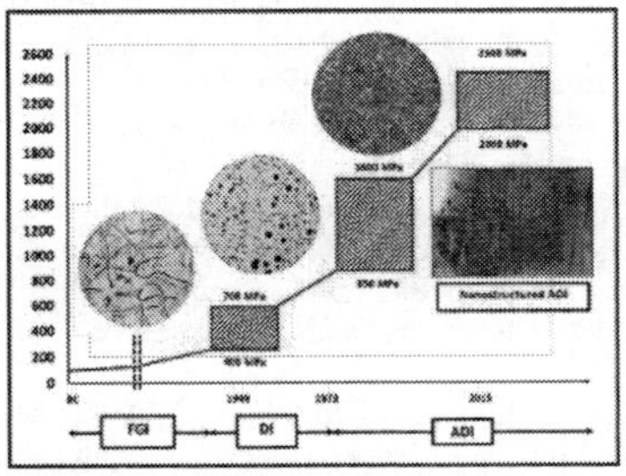

Fig. 1. Strength development of cast iron alloys with time.

This review is an attempt to compile the results of the worldwide explosion of research and development that followed the announcement of the first production of this material, meanwhile, reference is made to the work carried out at Central Metallurgical R&D Institute (CMRDI) over the past decade. It is not intended to provide an in-depth investigation of any specific technique, but to present a macro-analysis of the current state of metallurgy, processing and applications of ADI.

Better understanding of the strengthening mechanisms of ADI has led to the development of new techniques that contribute to enhance strength and toughness of the alloy, some examples will be discussed in details such as:-

Ausformed ADI; [3-6] where mechanical processing component was added to the conventional heat treatment as a driving force to accelerate the rate of stage I austempering (see Fig. 2.).

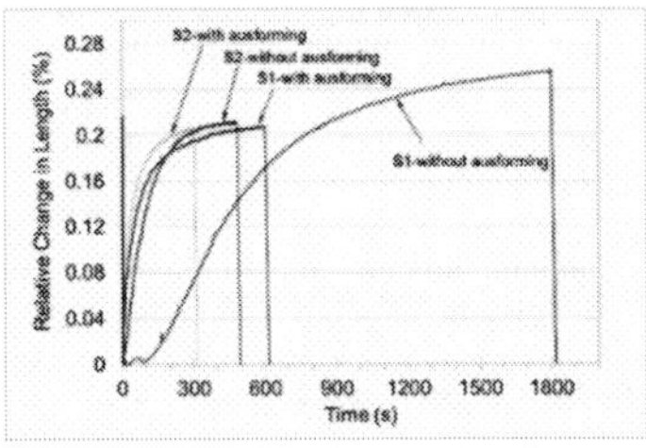

Fig. 2. Transformation Kinetics of ausferrite as related to ausforming and intercritical austenitizing.

Squeeze cast ADI; where superior quality ADI castings were produced through squeeze casting of molten iron in a permanent mold, followed by in-situ heat treatment of the hot knocked-out castings in the austenite range followed by normal austempering in a salt bath.

Two step austempering [7] to achieve finer ausferrite at higher undercooling during austempering treatment followed by austempering at higher temperature where higher austenitic carbon is promoted due to the enhanced diffusion rate of carbon into austenite.

Fig. 3. Microstructure of two-step austempered iron.

A recently published patent shows that the combination of a very high yield strength, fatigue strength and fracture toughness, that cannot be achieved in structural materials could be obtained in nano-structured ADI. The exceptional combination of mechanical and physical properties obtained is comparable to maraging steel without excessive alloying and costly processing.

The excellent abrasion resistance of ADI could still be remarkably increased through the development of:

- Carbidic ADI-ductile iron containing carbides subsequently austempered to form ausferritic matrix with an engineered amounts of carbides. Methods used to introduce carbides in the microstructure of ADI include alloying with carbide stabilizers, higher cooling rate during solidification, using ductile iron with lower CE as well as surface chilling. Carbides may also be mechanically introduced, where crushed Mx Cy carbides are strategically placed in the mold cavity at desired locations.

- Bainitic/martensitic (B/M) ADI containing less expensive alloying elements such as Si and Mn in the range of 2.5 - 3.0%.
- Selective surface treatment, where parts of the casting subjected to excessive wear may be locally hardened, either by induction heating and then austempered or by surface laser processing.

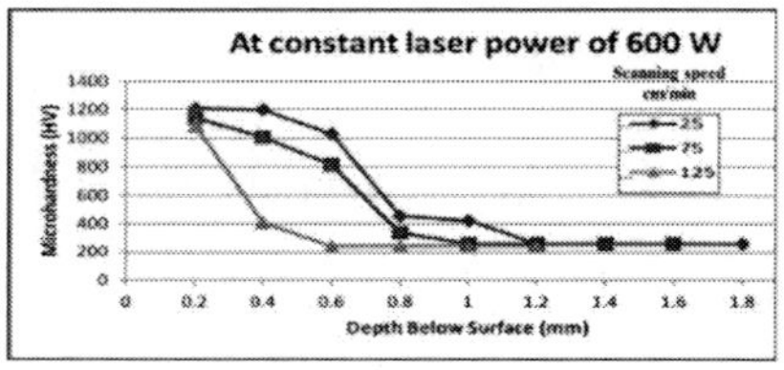

Fig. 4. Hardness gradient of laser surface treated ADI.

The review analyses the key features of important processing techniques of ADI such as:

- Cold rolling, where as thin as 3 mm sheets have been successfully produced at CMRDI with enhanced strength and hardness properties [8].
- Welding, whether used as repair welding for the parent ductile iron castings before austempering or to weld already austempered parts to each other or to other materials such as steel and ductile iron.

Some of the machining difficulties, that appeared with the first use of ADI as engineering material still persist today, which are mainly related to the work hardening and deformation induced martensitic formation from retained austenite [10]. Recent work at CMRDI clearly indicates that cutting force during machining of ADI is closely related to martensitic transformation, which in turn, is a function of cutting depth and speed of cutting.

A novel development of ADI is the intercritically austempered ductile iron; an exciting engineering material with a favorable combination of good strength and ductility, excellent fatigue strength and good machinability resulting from a microstructure containing colonies of proeutectoid ferrite together with isolated islands of austenite. In contrast to the conventionally austempered ADI, where the austempering temperature and time play the dominant role in determining the material properties, the austenitizing time and temperature strongly affect the mechanical properties since this determines the α & γ ratio and how much carbon and alloying elements are put into solid solution in γ before quenching to the austempering temperature. The performance of different ADI types under very high strain rate dynamic deformation were jointly assessed by CMRDI and Fraunhofer Institute in Germany using a servo hydraulic testing machine at crash-like strain rate of 50 sec-1. Higher deformations are recommended for suspensions parts, such as automotive steering knuckles, in particular for legal reasons in context of accidents. Different ADI grades proved to be candidate materials for steering knuckles production.

Different applications of ADI cover automotive e.g. gears, crankshafts, connecting rods, camshafts, transmission as well as suspension components and steering knuckles as well as earthmoving, defense and agricultural consumables such as plough blades. Case studies will be shown with emphasis on the experience of CMRDI in production of gears of different types and agricultural parts in its experimental foundry. ADI and steel properties related to gear performance will be compared such as structural integrity, abrasion resistance, bending fatigue, teeth conformation, noise and vibration reduction, manufacturing cost, weight reduction and gear power loss, the comparison seems to be in favor of ADI. The CMRDI experience with the use of ADI in agricultural applications will be discussed. Production of thin wall ADI components may offer potentials for new applications. Thin wall ADI castings are capable to build complex thin wall parts of high strength.

References

[1] A. Nofal (2013). Advances in the metallurgy applications of ADI, Journal of Metallurgical Engineering (ME), Vil. 2, issue 1, pp 1-18.

[2] F. Zanardi, F. Bonollo, G. Angella, N. Bonora, G. Iannitti, A. Ruggiero (2017); a contribution to new material standards for ductile irons and austempered ductile irons, International Journal of metalcasting, Vol. 11 issue 1, pp. 136-147.

[3] H. Nasr El-Din, A.A. NOFAL, K.M. Ibrahim and A.A. Ramadan (2006); Ausforming of austempered ductile iron alloyed with nickel, The International J. of Cast Metals Research, Vol. 19, No. 3, pp. 137-150.

[4] M. Soliman, A. NOFAL, and H. Palkowski (2015); Alloy and process design of thermo-mechanically processed multiphase ductile iron, Materials & Design, Vol. 87, PP. 450-465.

[5] M. Soliman, H. Palkowski, and A. NOFAL (2011): Thermo-mechanically processed multi-phase ductile iron, Microstructure development, Key Engineering Materials, 457, pp. 199-204.

[6] M. Soliman, H. Ibrahim, ***A. NOFAL***, and H. Palkowski (2015): Effect of thermomechanical processing parameters on phase transformation and hardness of dual matrix ductile iron, International Journal of Cast Metals Research, Advance article.

[7] J. Yang, S. Putatunda (2004), Improvement in strength and toughness of austempered ductile cast iron by a novel two-step austempering process; Materials and Design, Vol. 25, pp. 219-230.

[8] H.N. El-Din, ***A.A. NOFAL***, A. Ramadan and M. Ibrahim (2004): Structure characteristics and mechanical properties of cold-rolled austempered ductile iron alloyed with nickel, Canadian Metallurgical Quaterly, vol. 43, No. 2, pp. 203-218.

The 73rd World Foundry Congress
„Creative Foundry"
23 - 27 September 2018, Krakow, Poland

INTERACTION OF AlSi7Mg WITH OXIDIC FILTER MATERIALS

Beate Fankhänel [1]*, Michael Stelter [1], Claudia Voigt [2], Christos G. Aneziris [2]

[1]TU Bergakademie Freiberg, Institute for Nonferrous Metallurgy and Purest Materials, Leipziger Straße 34, 09599 Freiberg, Germany
[2] TU Bergakademie Freiberg, Institute of Ceramic, Glass and Construction Materials, Agricolastraße 17, 09599 Freiberg, Germany
**Corresponding address: beate.fankhaenel@inemet.tu-freiberg.de*

Key words: sessile drop method, interface, reactive coating, AlSi-alloy, oxide ceramic

The development of new active filters for the melt filtration in aluminum industry requires investigations of the interaction of aluminum and its alloys with novel filter materials. The requested filter effects require information about the wettability and potential chemical reactions between the filter material and the metal. Wetting experiments (sessile drop technique at 950 °C under vacuum) of an AlSi7Mg alloy on oxide ceramics (Al_2O_3, $MgAl_2O_4$, $3Al_2O_3{\cdot}2SiO_2$, TiO_2, $LiAl(Si_2O_6)$) were carried out. The results show that all substrates but the pure Al_2O_3 form reactive systems in presence of the AlSi7Mg alloy.

1. Introduction

The filtration of aluminum melts during casting process is an already established method to purify aluminum melts. The removal of oxidic and other nonmetallic inclusions in aluminum melts is essential to avoid defects in aluminum goods and to produce high quality products. Thereby ceramic foam filters made of alumina are common filter materials which were chosen because of their good thermo-mechanical properties but not with regards to best filtration efficiency. During the last years new filtering techniques and new active filter materials have been developed to expand the range of application and to improve filter efficiency [1-4]. Hereby several authors supposed an influence of the wetting behavior on the filtration efficiency [4-6]. The aim of the present work is to compare the wettability and the interaction of an aluminum alloy on/with different oxide ceramics ($MgAl_2O_4$, $3Al_2O_3{\cdot}2SiO_2$, TiO_2) as material for new active/reactive filter systems. The wettability was determined in comparison to pure alumina. Investigations of the interfaces and the reaction zones completed the examinations.

2. Experimental

A commercial AlSi7Mg alloy was used for all sessile drop measurements. All substrates were bulk sintered tablets. The substrates made of Al_2O_3 (alumina), $MgAl_2O_4$ (spinel), $3Al_2O_3{\cdot}2SiO_2$ (mullite) and TiO_2 (rutile) consisted 100 % of the particular filter material. The raw powders were pressed and sintered with the same parameters as for filters used and described in [7]. All substrates for the sessile drop experiments featured the same chemistry and particle size distributions as the equivalent filter surfaces. All sessile drop experiments were carried out following the same heating procedure which was described in detail in [8]. The maximum temperature was 950°C and a vacuum of $7.4{\cdot}10^{-6}{\pm}0.4{\cdot}10^{-6}$ mbar on average was measured within the furnace after deoxidation. The contact angle (θ) was calculated from the measured height and diameter of the drop taken from the images of the droplets during the experiments.

3. Results

Figure 1 shows the typical development of the contact angle for the AlSi7Mg alloy on the different ceramic substrates during the sessile drop experiments. After the droplet's formation the initial value of the contact angle can be measured. These values are affected by the oxide layer around the metal melt and do not represent the wettability of the aluminum alloy on the substrates. The initial values of the contact angle stayed almost constant until the process of deoxidation was nearly finished and the melt came into direct contact with the substrate's surface. Hence the resulting wetting behavior leveled out and could be detected. [9]

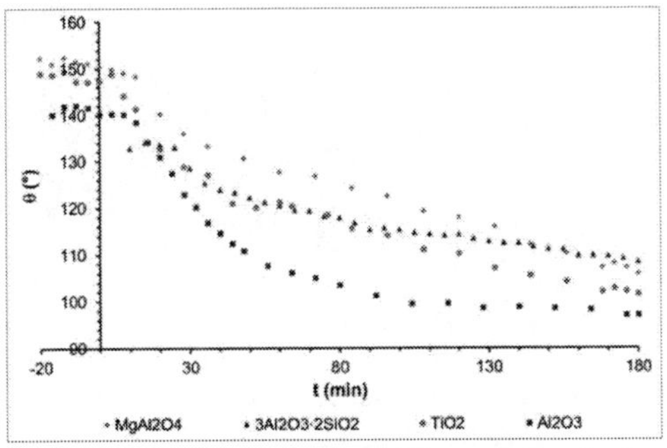

Fig. 1. ***Typical development of the contact angle of AlSi7Mg alloy droplets on different oxide ceramics under vacuum. The time was set to be zero when the maximum temperature (950°C) was reached.***

A non-wetting phenomenon was observed for all ceramics. The development of the contact angle in the system $AlSi7Mg/Al_2O_3$ is characteristic for non-reactive systems and acts as reference here. After deoxidation the contact angle decreased rapidly. Now the pure metal was in direct contact to the ceramic at the entire area of the interface and the contact angle leveled out to an almost constant value.

In case of the other ceramic substrates a different wetting behavior was determined. Though the decrease of the contact angle started right after deoxidation the velocity of this process differed comparing to the pure alumina substrate.

In terms of the spinel substrate an almost constant decrease of the contact angle was detected during the whole measuring period. The reason for this can be a reaction between the alloy and the spinel which will lead to a change in surface chemistry. Actually, it seems as if at the beginning of the process the decrease of the contact angle run fast until deoxidation is completed and afterwards the velocity of the decrease was determined by a reaction rate. The $AlSi7Mg/MgAl_2O_4$ interface showed a magnesium-free reaction zone with some parts of pure aluminum (Fig. 2), which can be an indication that an alumina interface was created and the contact angle may approach the value measured for the system $AlSi7Mg/Al_2O_3$.

Fig. 2. Cross section perpendicular to the $AlSi7Mg/MgAl_2O_4$ interface and EDX mapping of Al and Mg

In case of TiO_2 the contact angle first decreased quickly as it was observed with the alumina substrate but leveled out later at a slightly higher value. Investigations of the $AlSi7Mg/TiO_2$ interface showed an approximately 1 µm thick reaction zone between the AlSi7Mg droplet and the TiO_2 substrate (Fig. 3, left). EDX measurements showed that this reaction zone consists of aluminum and oxygen with some parts of silicon and titanium. This means a thin alumina layer has been developed and the final contact angle probably represents the system $AlSi7Mg/Al_2O_3$, too.

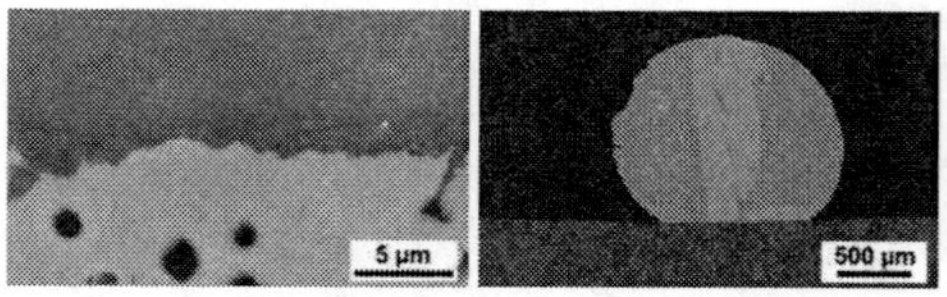

Fig. 3. Cross section perpendicular to the $AlSi7Mg/TiO_2$ interface (left) and cross section of the solidified sessile drop on a $3Al_2O_3{\cdot}2SiO_2$ substrate (right)

Regarding the mullite substrate a continuous decrease of the contact angle could be noticed, too. A reaction zone has been developed around the droplets on the mullite substrates, which can be the reason for the mass loss of the droplets. Silicon crystals could be detected within the droplets after the experiments (Fig. 3, right). Such large crystals are obviously the result of a reduction of the SiO_2 in the mullite by aluminum. Laurent et al. observed a comparable behavior of the contact angle in case of aluminum on SiO_2 at 800°C and 900°C [10]. They supposed that the SiO_2 reduction by aluminum starts already before the spreading of the droplet takes place. The reason for this assumption was seen in the extended steady state of the contact angle at the beginning of the sessile drop experiments and the slower decrease to its final value in comparison to a pure alumina substrate. That indicates that the formation of new Al_2O_3 by the reduction of silica causes a delay in wetting. This effect could be observed in our experiments as well. In case of the system $AlSi7Mg/3Al_2O_3{\cdot}2SiO_2$ the visible formation of the droplet and the decrease of the contact angle has been delayed comparing to the other systems.

4. Summary

The AlSi7Mg alloy forms a reactive system not only in contact with the mullite but also in presence of the spinel and the rutile. An alumina rich interface was created as reaction product in all cases.

Acknowledgments

The authors would like to thank the German Research Foundation (DFG) for supporting the investigations in terms of the Collaborative Research Centre SFB 920 "Multi-Functional Filters for Metal Melt Filtration - A Contribution towards Zero Defect Materials", subprojects C06 and A02.

References

[1] L.N.W. Damoah & L. Zhang. (2011). AlF_3 reactive Al_2O_3 foam filter for the removal of dissolved impurities from molten aluminum: Preliminary results. Acta Mater. 59(3), 896-913.

[2] H. Ni, Z. Yu, H. Mingyu, G. Weibiao & B. Sun. (2006). Purifying effects and mechanism of a new composite filter. Mater. Sci. Eng. A 426(1-2), 53-58.

[3] M. Zhou, D. Shu, K. Li, W.Y. Zhang, H.J. Ni, B.D. Sun & J. Wang. (2003). Deep filtration of molten aluminum using ceramic foam filters and ceramic particles with active coatings. Metall. Mater. Trans. A 34(5), 1183-1191.

[4] C. Voigt, B. Fankhänel, E. Jäckel, C.G. Aneziris, M. Stelter & J. Hubálková. (2015). Effect of the filter surface chemistry on the filtration of aluminum. Metall. Mater. Trans. B 46(2), 1066-1072.

[5] S. Bao, M. Syvertsen, A. Kvithyld & T.A. Engh. (2014). Wetting behavior of aluminium and filtration with Al_2O_3 and SiC ceramic foam filters. T. Nonferr. Metal. Soc. 24(12), 3922-3928.

[6] M. Syvertsen & S. Bao. (2015). Performance Evaluation of Two Different Industrial Foam Filters with LiMCA II Data. Metall. Mater. Trans. B 46(2), 1058-1065.

[7] C. Voigt, T. Zienert, P. Schubert, C.G. Aneziris & J. Hubálková. (2014). Reticulated Porous Foam Ceramics with Different Surface Chemistries. J. Am. Ceram. Soc. 97(7), 2046-2053.

[8] B. Fankhänel, M. Stelter, C. Voigt & C.G. Aneziris. (2017). Interaction of AlSi7Mg with Oxide Ceramics. Adv. Eng. Mater. 19(9), 170084.

[9] V. Laurent, D. Chatain, C. Chatillon & N. Eustathopoulos. (1988). Wettability of monocrystalline alumina by aluminium between its melting point and 1273 K. Acta Metall. 36(7), 1797-1803.

[10] V. Laurent, D. Chatain & N. Eustathopoulos. (1991). Wettability of SiO_2 and oxidized SiC by aluminium. Mater. Sci. Eng. A135, 89-94.

The 73rd World Foundry Congress
„Creative Foundry"
23 - 27 September 2018, Krakow, Poland

THE MODIFIED LOW-CYCLE FATIGUE TEST AS A QUICK AND ECONOMIC CRITERION OF THE QUALITY OF DUCTILE IRON AFTER NORMALIZING ANNEALING

Maria Maj [1]*, Krystyna Pietrzak[2], Adam Klasik[3]

[1.] *AGH University of Science and Technology, Faculty of Foundry Engineering, 23 Reymonta Str., 30-059 Cracow, Poland*
[2.] *Institute of Precision Mechanics 3 Duchnicka Str. 01-796 Warsaw, Poland*
[3] *Motor Transport Institute 80 Jagiellonska Str. 03-301 Warsaw, Poland*
**Corresponding address: mmaj@agh.edu.pl*

Keywords: cast iron, mechanical properties, microstructure

The results of the EN-GJS-600-3 cast iron testing by a modified low-cycle fatigue method (MLCF) are presented along with the results of metallographic examinations carried out by the light microscopy. Based on the results of the conducted mechanical and structural tests, it has been found that the MLCF method can be used for a quick and economically justified assessment of the quality of ductile iron after normalizing annealing.

1. Low cycle fatigue test (LCF) vs modified low cycle fatigue test (MLCF)

When searching for literature data on fatigue strength, of ductile iron in particular, a lot of information can be found [1-8]. The authors describe various mechanisms governing the fatigue phenomenon itself and its microstructural conditions [1-3] as well as the initiation and propagation of fatigue cracks depending on the microstructure of cast iron metal matrix and morphological characteristics of graphite precipitates [4-6]. Some of the data also include the description of theoretical models developed to predict the fatigue life at different temperatures [7-8]. In most cases, however, the experimental studies of fatigue behaviour performed by the low-cycle fatigue test (LCF method - [9]) require testing of at least 10 samples because the measurement data obtained from a single sample give only one point on the fatigue curve. Thus, the accuracy of the method increases with the increasing number of samples. The situation becomes even more complicated if the material tested is characterized by microstructural heterogeneities, as then the number of samples necessary for a reliable analysis must increase even more.

Taking into account all these drawbacks, a modified version of the low cycle fatigue test (MLCF) was developed and proposed for use. The MLCF method allows obtaining a lot of static and dynamic mechanical parameters based on the measurements taken in each case on one specimen only [10]. The fatigue strength (Zgo) necessary to calculate the test parameters is determined from a test curve developed for the different types of materials [11].

2. Test materials and methods

The EN-GJS-600-3 ductile iron after normalizing annealing was subjected to mechanical tests and to qualitative and quantitative analysis of microstructure made by the light microscopy.

3. Mechanical properties

The mechanical properties of EN-GJS-600-3 cast iron were determined from the results of fatigue tests carried out according to the MLCF [10, 11] procedure. The results of the measurements are shown in Tables 1 and 2.

Table 1. Mechanical parameters of EN-GJS-600-3 cast iron - MLCF method

No.	E	UTS	$R_{0.02}$	$R_{0.2}$	Zgo	Ra
	GPa	MPa				
1	198	569	339	479	189	534
2	194	557	341	469	186	533
3	179	533	343	487	191	492
4	199	601	348	486	192	573
5	193	493	308	484	183	492
6	180	533	344	488	192	492
7	180	492	431	446	182	452
8	190	573	411	489	226	532
Avg	189	544	358	479	193	513
S	8	39	41	15	14	38
CV [%]	4	7	11	3	7	7

Table 2. Mechanical parameters of EN-GJS-600-3 cast iron - MLCF method

No.	b	c	n'	K' MPa	$\varepsilon_{max}\cdot 10^6$
1	-0.096	-0.446	0.106	933	1069
2	-0.095	-0.535	0.091	849	899
3	-0.089	-0.343	0.171	1461	1156
4	-0,099	-0,525	0,104	939	1001
5	-0,086	-0,364	0,15	1267	1005
6	-0,089	-0,333	0,155	1345	1116
7	-0,087	-0,360	0,16	1318	977
8	-0,081	-0,569	0,092	873	1141
Avg	-0.090	-0.434	0.129	1123	1046
S	0.006	0.097	0.033	248	90
CV [%]	7	22	26	22	9

4. Microstructure

The microstructure of ductile iron as examined by light microscopy is shown in Figure 1. This is the microstructure typical of the normalized state.

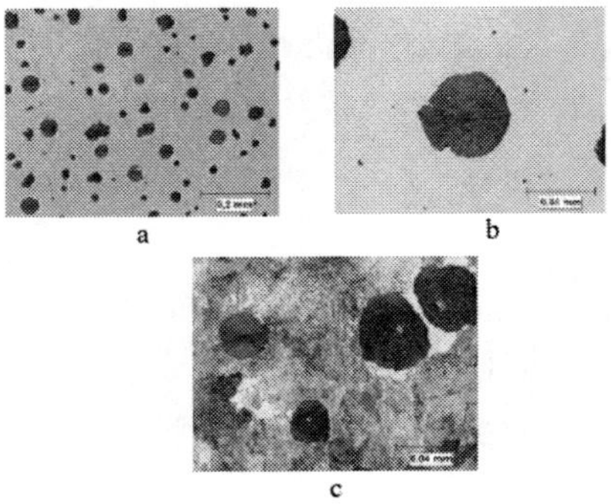

Fig. 1. Microstructure of the EN-GJS-600-3 cast iron: a), b) unetched state, 100x, 500x, c) etched state, 500x

The microstructure is composed of graphite precipitates visible in the unetched state (Fig. 2a, b) and of a pearlitic matrix visible in the etched state with very small quantities of ferrite precipitates distributed mainly around the graphite precipitates (Fig. 2c). Due to small differences observed in the cast iron matrix, quantitative studies of cast iron microstructure were limited to the graphite precipitates only. The following geometrical parameters were determined: $A_A = V_V$ [%] - the volume fraction, N_L - estimator of the relative area, Ω- coefficient of microstructural anisotropy, l – chord of graphite precipitates, F- Feret diameter of graphite precipitates (Table 3).

Table 3. Geometrical parameters of graphite precipitates in EN-GJS-600-3 cast iron

No.	A_A	N_L	N_A	Ω	F	l
	%	1/mm	1/mm^2		µm	µm
1	10	65	682	1	9	16
2	10	61	687	1	9	17
3	10	67	829	1	8	15
4	10	64	670	1	9	16
5	11	65	845	1	9	15
6	10	61	695	1	9	17
7	10	68	820	1	8	15
8	11	62	636	1	9	14
Avg	10	64	733	1	9	16
S	1	3	84	0	1	1
CV [%]	5	5	11	0	11	6

They prove that most of the measured geometric parameters of the spheroidal graphite show almost no changes, because the coefficients of variation are less than 10%. Only the coefficients of variation of the geometric parameters N_A and F take higher values equal to 11% (Table 3). At the same time, for the determined mechanical parameters: $R_{0.02}$, c, n ', K (Table 2), the coefficients of variation also take values greater than 10% (11%, 22%, 26% and 22% respectively). Therefore, it can be assumed that these parameters are much more structurally sensitive than others.

5. Conclusions

1. Based on the tests and analyses carried out, it was found that most of the determined mechanical parameters and geometric microstructure parameters show only small variations, which is evidenced by the values of the specified coefficients of variation.
2. From the point of view of static and dynamic mechanical behaviour, four mechanical parameters, i.e. $R_{0.02}$ c, n' and K, have proved to be most structurally-sensitive.
3. The higher variability of the $R_{0.02}$, c, n 'and K parameters results from the occurrence of identified variations in the size of graphite precipitates and their number per unit area (parameters F and N_A respectively).
4. The MLCF method can be used for quick and economically justified determination of the quality of cast iron in normalized state from the point of view of its mechanical and microstructural characteristics.
5. The obtained results allow predicting the similar results for other materials.

References

[1] V. Di Cocco, F. Iacoviello. (2015). Pearlitic ductile cast irons: fatigue initiation micromechanism. Procedia Engineering. 109, 465-472.
[2] F. Iacoviello, V. Di Cocco, A. Rossi. (2013). Pearlitic ductile cast iron: damaging micromechanisms at crack tip. Frattura ed Integrità Strutturale. 25, 102-108.
[3] H. Germann, P. Starke, E. Kerscher, D. Eifler. (2010). Fatigue behaviour and lifetime calculation of the cast irons EN-GJL-250, EN-GJS-600 and EN-GJV-400. Procedia Engineering. 2, 1087-1094.
[4] L. Bubenko, R. Konečná, G. Nicoletto. (2009). Observation of fatigue crack paths in nodular cast iron and ADI microstructures. Materials Engineering. 16, 13-18.
[5] P. Čanžar, Z. Tonković, J.Kodvanj. (2012). Microstructure influence on fatigue behaviour of nodular cast iron. Materials Science & Engineering. A 556, 88–99.
[6] M. Cavallini, O. Di Bartolomeo, F. Iacoviello. (2008). Fatigue crack propagation damaging micromechanisms in ductile cast irons. Engineering Fracture Mechanics. 75, 694–704.
[7] Y. Ma, X.Wang, I. An (2012). Fatigue Life Prediction of Ductile Iron Based on DE-SVM Algorithm. Physics Procedia. 33, 1309 – 1315.
[8] X. WU, G. Quan, R. Macneil, Z. Zhang and C. Sloss. (2014). Failure Mechanisms and Damage Model of Ductile Cast Iron Under Low-Cycle Fatigue Conditions. Metallurgical and Materials Transactions. A, 2014-5097.
[9] S. Kocańda, A. Kocańda. (1989). Low cycle fatigue strength of metals. Warsaw: PWN.
[10] M. Maj, (2012).Fatigue life of selected casting alloys. Katowice-Gliwice: Archives of Foundry Engineering
[11] M. Maj, W. Moćko, K. Pietrzak, A. Klasik. (2009). Methodological conditionings of a modified low cycle fatigue method of tempered 41Cr4 steel in comparison to some other materials. Archives of Foundry Engineering. 9, 129-134.

The 73rd World Foundry Congress
„Creative Foundry"
23 - 27 September 2018, Krakow, Poland

INFLUENCE OF INVESTMENT CASTING PARAMETERS ON CREEP RESISTANCE OF NI-BASED SUPERALLOY

Łukasz Rakoczy [1]*, Rafał Cygan [2], and Małgorzata Grudzień [3]

[1]*AGH University of Science and Technology, Faculty of Metals Engineering and Industrial Computer Science, av. Mickiewicza 30, 30-059 Krakow, Poland*
[2] *Consolidated Precision Products, ul. Hetmanska 120, 35-078 Rzeszow, Poland*
[3] *Foundry Research Institute, ul. Zakopianska 73, 30-418 Krakow, Poland*
**Corresponding address: lrakoczy@agh.edu.pl*

Key words: creep, rafting, IN713C, grain refinement, stereological parameters

In this paper the high temperature creep properties of polycrystalline IN713C Ni-based superalloy have been investigated taking into consideration different casting parameters: melt pouring temperature and composition of primary coating of shell mould.
The size of the grains in the castings can be controlled by thermal methods and the introduction of seeds for crystallization, namely surface modification.
The selection of the optimal grain size in precision castings depends on the final position of the component in the aircraft engine, which determines the mechanism of microstructural degradation during service. In order to meets the requirements of industrial standards time to rupture must be longer than 25 hours with elongation till aforementioned rupture min. 5% independently from the size of grain.

The aim of this investigation was to prepare the correlation between casting parameters with microstructure and creep resistance of IN713C Ni-based superalloy.

1. Introduction

Ni-based superalloys are used for the manufacturing of several hot section components of jet engines, such as turbine disks, blades, and vanes. Superalloys provide structural integrity during extreme service conditions, resistance to creep deformation and hot corrosion at high homologous temperatures [1]. The preferable high-temperature performance, in particular, the increase in yield strength with temperature results from the $L1_2$ ordered γ' phase, which precipitates within the face-centered cubic Ni-rich matrix.
The microstructural changes during high-temperature exposure and the deformation mechanisms (Fig. 1) are heavily coupled with the strain and stress distributions at the γ/γ' interfaces and also with the loss of coherency between the γ matrix and γ' precipitates [2-3].

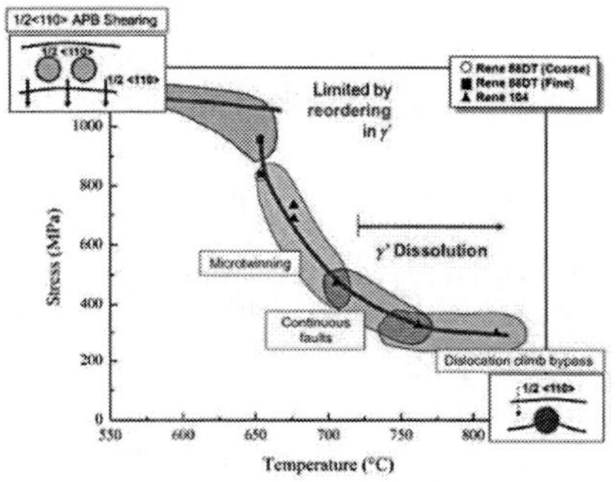

Fig. 1. ***Deformation map for the Rene88DT and Rene104 Ni-based superalloys. Dependence of the deformation mechanism on the temperature and stress [4]***

2. Experimental procedure

Ten investment castings with different pouring temperatures and composition of prime coat of shell mould were produced. The chemical composition (wt%) of the used alloy was measured as C: 0.11%, Al: 5.77%, Cr: 14.08%, Mo: 4.25%, Ti: 0.91%, Nb: 2.33%, S: 0.002%, P: 0.004%, Co: 0.04%, Fe: 0.04%, B: 0.01%, Zr: 0.06%, and balance Ni. A master heat charges were melted in a zirconia crucibles inside vacuum Balzers furnace (VIM IC) and poured into preheated ceramic moulds covered with alumina silicate Fiberfrax® insulation. The five melt pouring temperature from range 1400°C-1540°C were selected. The difference of prime coating composition originated from addition of cobalt aluminate inoculant, namely 0 and 5 wt.%. The cast bar blanks were machined into cylindrical specimens for creep test with a geometry presented in Fig. 2.

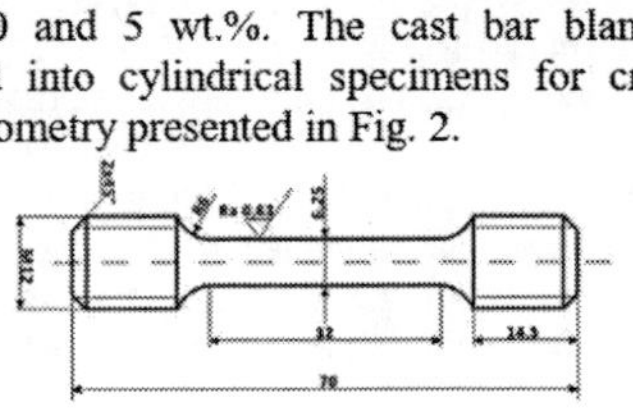

Fig. 2. ***Geometry of creep test sample***

Creep test was carried out in accordance with the requirements of ASTM E139 standard under

982°C/152 MPA. Microstructure of IN713C in as cast condition and after creep test was investigated using light and scanning electron microscopy equipped with EDS detector. Captured images were subjected to further image analysis using ImageJ software in order to calculate several stereological parameters. Each of the samples received a unique signature which is presented in Table 1.

Table 1. Description of investigated variants

Symbol	Pouring temperature	Content of cobalt aluminate [wt%]
A1	1400	
A2	1450	
A3	1480	0
A4	1520	
A5	1540	
B1	1400	
B2	1450	
B3	1480	5
B4	1520	
B5	1540	

3. Microstructure in as cast condition

Regardless of the pouring temperature and the inoculant content, inside different size grains a dendritic structure with substantially large local inhomogeneities was observed. The γ' phase surrounded by matrix fulfill dendrite cores (Fig. 3a), whereas enrichment of interdendritic spaces with alloying elements additionally caused formation other phases (Fig. 3b). Based on scanning electron microscopy and EDS analysis it was stated that within interdendritic spaces occur also γ' as eutectic island γ/γ', Mo-rich borides M_3B_2, Nb-rich MC-type carbides and compound of Ni with Zr, namely Ni_7Zr_2.

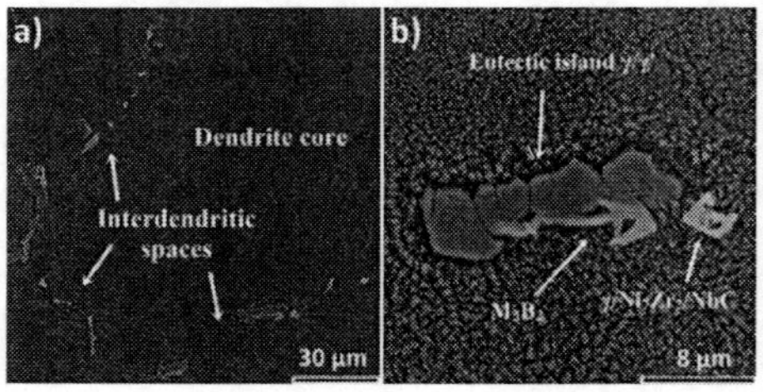

Fig. 3. Microstructure in as cast condition: a) dendrite core and interdendritic space; b) constituents in interdendritic region

4. Creep tests results

Creep test was carried out at a homologous temperature 0.82 and so Nabarro-Herring creep dominates and takes place by migrating of vacancies within grains from the stretched to the compressed regions and by the volume diffusion in the opposite course. Below at a homologous temperature of 0.4-0.7 Coble creep that occurs through diffusion along grain boundaries prevails. Creep rate through Nabarro-Herring mechanism is $1/d^2$, whereas via Coble is $1/d^3$.

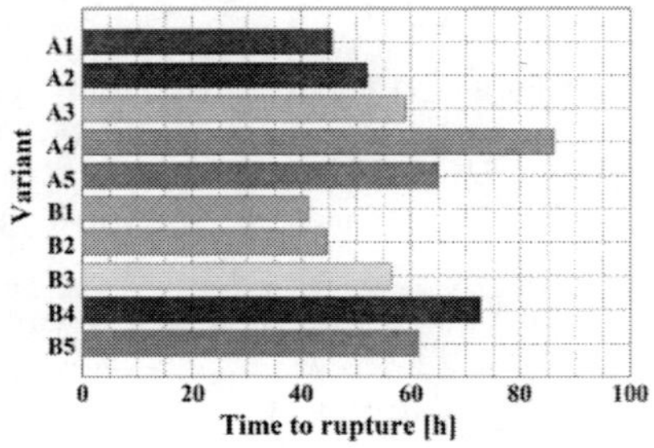

Fig. 4. Influence of selected variant on time to rupture during creep testing

Results of creep test presented in Fig.4 reveal that minimum time to rupture have been met by all specimen. The shortness time to rupture was registered for the sample poured at 1400°C with a 5% of $CoAl_2O_4$ in prime coat. The highest creep resistance, namely above 85 h to rupture was noted for sample poured at 1520°C. Depending on the temperature and stress, there are three ranges in which creep mechanisms predominate. At high temperature/low stress (982°C/152 MPa) creep involves rafting with a short primary stage and steady state represented by a plateau and pronounced strain appearing at the tertiary stage.

5. Summary

The present paper present two variants of grain size control, namely pouring temperatures from 1400°C to even 1540°C and also the using of additional crystallization nucleants originated from $CoAl_2O_4$ in prime coat. Comprehensive study carried out by authors reveal how to select optimal investment casting parameters in order to obtain microstructure with favorable stereological parameters.

Acknowledgments
This research was supported by National Centre for Research and Development, Grant No. LIDER/227/L-6/14/NCBR/2015.

References

[1] Roger Reed. 2006. The Superalloys: Fundamentals and Applications. Cambridge. Cambridge University Press

[2] Łukasz Rakoczy, Małgorzata Grudzień, Anna Zielińska-Lipiec. 2018. The contribution of microstructural constituents on hot cracking of MAR-M247 nickel based superallloy. Archives of Metallurgy and Materials. 63 (1). 181-189.

[3] Luis Thebaud et.al. 2018. Is there an optimal grain size for creep resistance in Ni-based disk superalloy?. Materials Science&Engineering A. 716. 274-283.

[4] Roger Reed, Catherine Rae. 2014. 22-Physical Metallurgy of the Nickel-Based Superalloy, In Physical Metallurgy (Fifth Edition), edited by David Laughlin and Kazuhiro Hono. Oxford. Elsevier

The 73rd World Foundry Congress
„Creative Foundry"
23 - 27 September 2018, Krakow, Poland

EFFECT OF MELT POURING TEMPERATURE AND COMPOSITION OF PRIMARY COATING OF SHELL MOULD ON TENSILE DEFORMATION BEHAVIOR OF IN713C SUPERALLOY

Łukasz Rakoczy [1]*, Małgorzata Grudzień [2], and Rafał Cygan [3]

[1] AGH University of Science and Technology, Faculty of Metals Engineering and Industrial Computer Science, av. Mickiewicza 30, 30-059 Krakow, Poland
[2] Foundry Research Institute, ul. Zakopianska 73, 30-418 Krakow, Poland
[3] Consolidated Precision Products, ul. Hetmanska 120, 35-078 Rzeszow, Poland
**Corresponding address: lrakoczy@agh.edu.pl*

Key words: Tensile, antiphase grain boundary, ordering, gamma prime, IN713C

In the present study investment casting of IN713C Ni-based superalloy poured in different temperature into two types of shell mould was carried out. The investigation concerned mainly on the change of grain size depending on the selected pouring variant, strength at ambient temperature and also effect of deformation on the morphology of microstructural constituents. As a result of grain refinement through surface modification using cobalt aluminate pronounced increasing of yield strength was observed.
The main aim of this investigation was to find the best combination of melt pouring temperature and composition of shell mould in order to obtain high strength and relatively fine grain structure desired in the intermediate temperature range service.

1. Introduction

The strength of Ni-based superalloys depend strongly upon the initial condition of microstructure, which, in turn, is controlled by the alloying elements and the investment casting parameters. This is particularly for yield stress, which is a sensitive function of the distribution of the $L1_2$ ordered phase. The presence of γ' precipitates in matrix hinders the dislocation movement, despite the coherency between γ and γ' phases [1]. The planar fault widely known as the antiphase boundary is of profound significance in superalloys metallurgy. This is because the γ' phase deforms very differently from the disordered γ matrix in which it exists. In matrix, slip deformation takes place due to dislocation slide on the a/2< $1\bar{1}0$ >{111} system. Lattice vector equal a/2< $1\bar{1}0$ > is the shortest one accessible and is present in the close-packed plane. In the ordered γ' phase the close-packed planes are also {111}. Lattice vector a<100> do not cover with them. The preferred slip system should be a<110>{001}, a/2< $1\bar{1}0$ >{001} or a/2< $1\bar{1}0$ >{111}. At relatively low temperature takes place the last one slip system [2, 3].

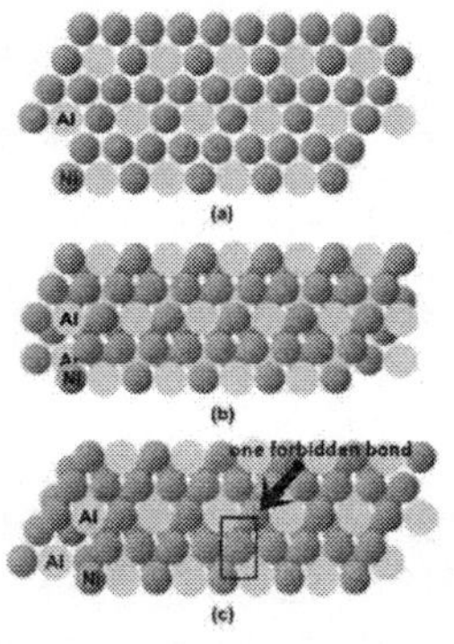

***Fig. 1.** {111} plane of the ordered γ' (a); Two planes stacked in accordance to the structure of crystal (b); location of atoms after displacing of the second plane by a/2< $1\bar{1}0$ > on {111} [2]*

2. Experimental procedure

Base metal used in this investigation was Inconel 713C melted in a zirconia crucible inside vacuum Balzers furnace and poured into preheated shell moulds covered with alumina silicate insulation. The chemical composition (wt.%) of the alloy was measured using Optical Emission Spectroscopy as C: 0.11%, Al: 5.77%, Cr: 14.08%, Mo: 4.25%, Ti: 0.91%, Nb: 2.33%, S: 0.002%, P: 0.004%, Co: 0.04%, Fe: 0.04%, B: 0.01%, Zr: 0.06%, and balance Ni. The melt pouring temperature from 1400°C to 1540°C were selected. The prime coatings from five shell moulds contain 5 wt.%. of $CoAl_2O_4$ inoculant. Detail of prepared castings with a unique signature are presented in Table 1. The cast bar blanks were machined into cylindrical specimens M12x70mm. The microstructure of IN713C directly after investment casting and after tensile testing was investigated using light and electron microscopy. Captured images were also analyzed via ImageJ software in order to calculate stereological parameters of the microstructure.

Table 1. Description of prepared samples

Symbol	Melt pouring temperature	Content of $CoAl_2O_4$ [wt%]
C1	1400	
C2	1450	
C3	1480	0
C4	1520	
C5	1540	
D1	1400	
D2	1450	
D3	1480	5
D4	1520	
D5	1540	

3. Microstructure of IN713C in as cast condition

The morphology of two strengthening phases in IN713C are presented in Fig. 2. Segregation of alloying elements during crystallization of castings led to creation of Nb-rich carbides, eutectic islands γ/γ', Mo-rich borides and intermetallic phase Ni_7Zr_2 in interdendritic spaces. In dendrite cores only fine cubic shaped γ' precipitates divided by γ channels are observed. Total content of γ'-formers Al+Ti in alloy, namely 7.68 wt% allow to obtain high volume fraction of $Ni_3(Al, Ti)$ phase .

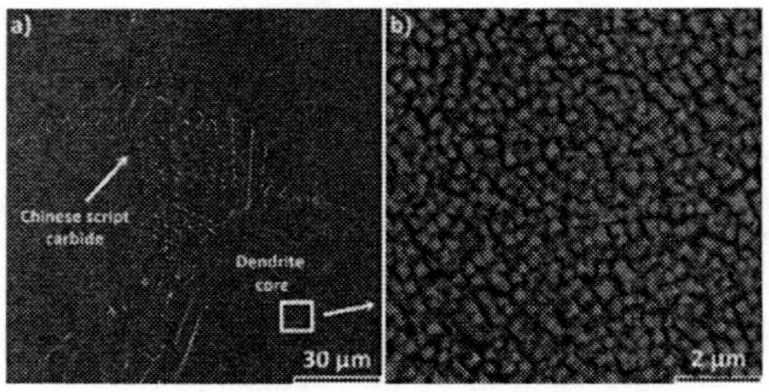

Fig. 2. The main strengthening phase in IN713C: a) MC-type carbides; b) γ' precipitates

4. Tensile testing results

In order to compare strength of castings, the yield strength at 0.2% offset depending on the variant is presented in Fig. 3. Independently from melt pouring temperature the higher value of yield strength are measured for variants "D". Surface modification led to grain refinement during crystallization of castings.

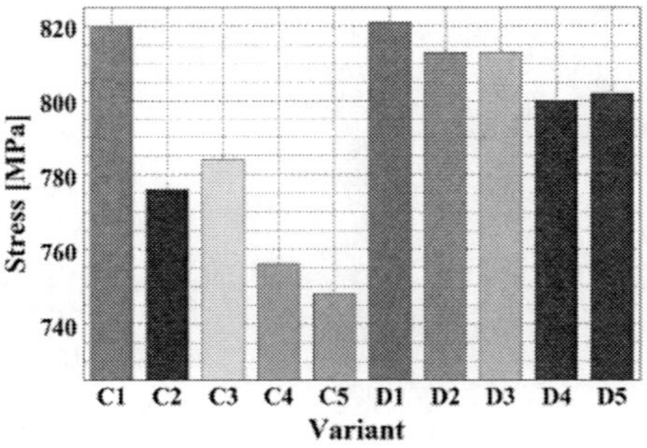

Fig. 3. Influence of selected variant on yield strength of IN713C

Difference between the lowest and higher values is above 70 MPa. The most favourable result was observed for sample D1 poured from 1400°C. The mean size of the grains in IN713C influences the strength since neighboring grains usually have various crystallographic orientations. There are exists two reasons why the grain boundaries act as a barrier to dislocation slip. Firstly, the dislocation has to change its direction when passing from one into the second grain with a different orientation. Moreover, disordering of atoms at grain boundary space causes a discontinuity of slip planes from one into the second grain.

5. Summary

The mean size of the grain is changed with nucleation rate and velocity of crystal growth. The grains of castings were refined by the addition of the cobalt aluminate to the prime coat of moulds. Fine grain structure is desired for components exposed to low cycle fatigue, like aerospace vanes clusters. Appropriate selection of investment casting parameters led to increasing of yield point by 10% up to 821 MPa. Analysis of fracture revealed that cracking propagates through equiaxed grains along the interdendritic areas.

Acknowledgments

This research was supported by National Centre for Research and Development, Grant No. LIDER/227/L-6/14/NCBR/2015.

References

[1] Tresa Pollock, Sammy Tin. 2006. Nickel-Based Superalloys for Advanced Turbine Engines: Chemistry, Microstructure, and Properties. Journal of Propulsion and Power. 22(2). 361-374

[2] Roger Reed. 2006. The Superalloys: Fundamentals and Applications. Cambridge. Cambridge University Press

[3] Roger Reed, Catherine Rae. 2014. 22-Physical Metallurgy of the Nickel-Based Superalloy, In Physical Metallurgy (Fifth Edition), edited by David Laughlin and Kazuhiro Hono. Oxford. Elsevier

The 73rd World Foundry Congress
„Creative Foundry"
23 - 27 September 2018, Krakow, Poland

PREDICTION OF MICROSTRUCTURE OF GREY CAST IRONS BY ELECTRICAL RESISTIVITY MEASUREMENTS

Mitja Petrič [1], Primož Mrvar [1], Sebastjan Kastelic [1]

[1] University of Ljubljana, Faculty of Natural Sciences and Engineering, Aškerčeva 12, Ljubljana, Slovenia

Key words: Solidification, grey cast iron, electrical resistivity, graphite shape, microstructure

Simple thermal analysis is well established method for solidification path monitoring of metals and alloys especially in aluminium and cast iron foundry industry. The results give us information of solidified microstructure in terms of phase fractions, shapes of microstructural constituents, presence of carbides, inclusions, etc. The measurement is based on heat release during solidification which is exothermal process. The heat release can be connected to individual phase or eutectic and in this way the microstructure can be predicted.

Electrical resistivity is a property of materials which is changed by temperature and it depends also on microstructure [1, 2].

In cast irons the microstructure usually first consists from austenite dendrites and eutectic graphite which can be lamellar, nodular etc. Austenite is later transformed to pearlite and/or ferrite. The involved phases and amounts of phases and also the shape of phases have big influence on electrical resistivity of material. It is reported that shape and size of present phases has important influence on the electrical resistivity of material. If the phases are large and acicular or lamellar the electrical resistivity is higher because there are more interfaces – barriers for electrons to overcome. On the other hand if phases are smaller and more rounded the electrical resistivity is lower [3, 4].

There is no difference at cast irons. Very few reports have been found in literature. Stefanescu et al. [5] reported that electrical resistivity of lamellar grey cast irons is higher than the one of nodular cast irons. There is also stated that the matrix also influences the electrical resistivity and it is higher if it is pearlitic than ferritic [6].

Electrical resistivity measurements were performed on different lamellar and nodular cast iron melts during the range of solidification and further cooling. The "in situ" measurement was performed in a sand mould made through the Croning process. The shape of a casting is a 210 mm long bar with square cross-section of 330 mm^2. Electrical resistivity was measured using four probe technique and the temperature was measured simultaneously. The samples were metallographically investigated by light microscopy in order to evaluate the microstructure.

Results of measurements are given in Figs. 1 and 2. It is seen that the electrical resistivity of lamellar cast iron is decreasing in liquid but when the solidification starts it is suddenly increased since first lamellas of graphite appear. When solidification is finished the resistivity reaches a plateau after which eutectoid transformation takes place and after it decreases.

At nodular cast iron the shape of the electrical resistivity curve is similar to those of pure metals since graphite nodules do not much affects the conduction electrons. After the solidification starts the curve decreases faster and after solidification it slows down. At eutectoid reaction at app 300 – 400 s there is a small plateau and then decreases continuously with temperature.

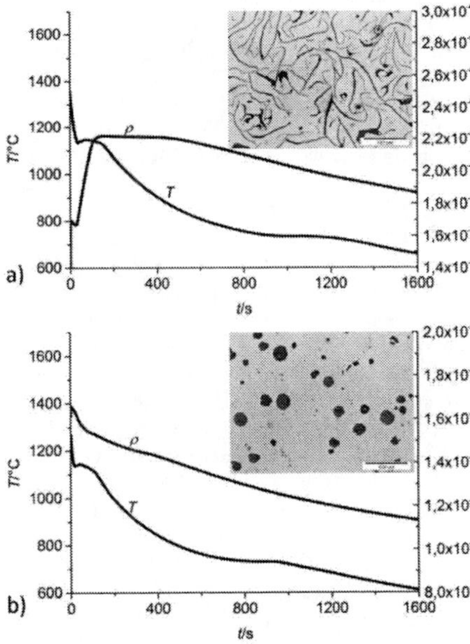

Fig. 1. Measured cooling curves and electrical resistivity curves with corresponding microstructures: lamellar grey cast iron (a), nodular cast iron (b)

Fig. 2 is presenting comparison of electrical resistivity curves versus temperature where the difference between lamellar and nodular graphite is explicitly seen. Also the drops in electrical resistivity at eutectoid transformation are detected.

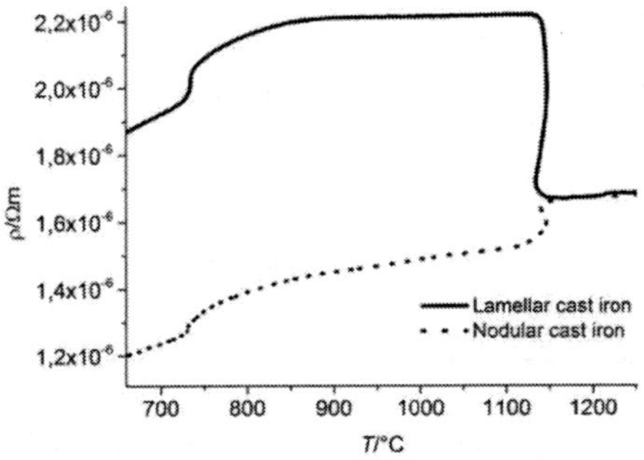

Fig. 2. Comparison of electrical resistivity curves vs. temperature

From obtained results it can be concluded that such measurements can provide information about graphite shape already in early stage of a measurement. After eutectoid transformation is finished also the matrix can be comparatively determined.

References

[1] A.R. Sinigoj. 1999. Basics of Electromagnetics. Ljubljana: Publisher of Faculty of Electrical engineering and Faculty for Computer and Information Science. (in Slovenian)

[2] T.E. Faber. 1972. An introduction to the theory of liquid metals. New York: Cambridge University Press

[3] L. Weber, C. Fischer & A. Mortensen. (2003). On the influence of the shape of randomly oriented, non-conducting inclusions in a conducting matrix on the effective electrical conductivity. Acta Materialia, Volume 51, 495-505

[4] L. Weber, J. Dorn & A. Mortensen. 2003. On the electrical conductivity of metal matrix composites containing high volume fractions of non-conducting inclusions. Acta Materialia, Volume 51, 3199-3211

[5] D.M. Stefanescu & S.T. Craciun. (1973). Bestimmung der graphitform im gusseisen durch messung des spzifischen elektrischen widerstandes waehrend der erstarrung. Giesserei-praxis, Volume 11, 197-200

[6] D.M. Stefanescu & H. Weingert. (1971). Einfluss des graphits und der metallischen grundmasse auf den spezifischen elektrischen widerstand von gusseisen mit lamellen- bzw. kugelgraphit. Giesserei-praxis, Volume 14, 256-261

The 73rd World Foundry Congress
„Creative Foundry"
23 - 27 September 2018, Krakow, Poland

RESEARCH AND ANALYSIS OF THE FOUNDRY COATINGS INFLUENCE ON THE MOULD FILLING PROCESS

Yevgeny Marukovich [1*], Yuri Nikolaichik [2]

[1] Institute of Technology of Metals, National Academy of Sciences of Belarus, Byalynitsky-Biruli Str., 11, 212030, Mogilev, Belarus

[2] Belorussian National Technical University, Mechanical and Technological faculty, Nezavisimosty Ave., 65, 220013, Minsk, Belarus

*Corresponding address: foundry@bntu.by

Key words: foundry coating, mould, filling process, grey iron

It is well known that foundry, mould or refractory coatings is one of the most effective tools for prevention of castings surface defects. The use of foundry coatings for moulds and cores during casting is quite necessary as a means of achieving high quality surface finish of castings more especially in complex internal channels created by use of cores [1]. The foundry coatings are also used for protection of castings from metal penetration, sand burn-on (as the result of chemical reaction on the border between melt and foundry mould, mould and core erosion (as the result of sand expansion defects). Foundry coatings protect bonded sand cores and moulds from producing defective castings during the casting process by providing a barrier between the core and the liquid metal [2].

In the present paper the foundry coatings influence on the mould filling process was studied. For this investigation we used a technological test of a spiral form (Kerry spiral, GOST 16438-70). Kerry spiral consists of a bowl, a half-mould of the top and a half-mould of the bottom, with a spiral cavity. The construction of the gate system ensures filling the mould with a constant speed. The area of the trapezoidal spiral canal is 48 mm^2. In the Figure 1 you can see spiral moulds with foundry coating based on aluminosilicate refractory filler with nanostructured refractory additives (Fig. 1a) and based on aluminosilicate refractory filler only (Fig. 1b). In Also the experiment we also used a mould without foundry coating (Fig. 1c). In Fig. 2 you can see spiral castings from grey iron wich were made in the moulds according Figure 1.

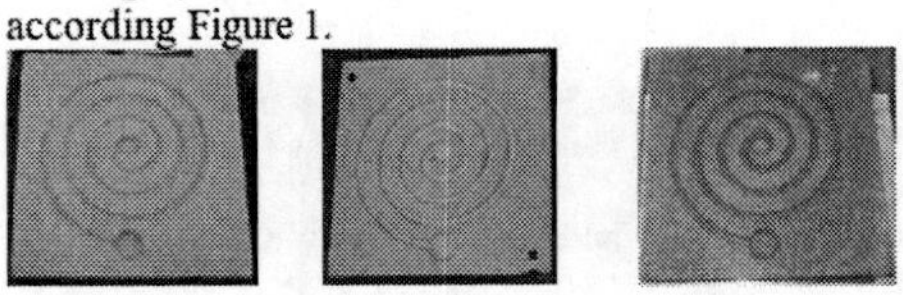

a *b* *c*

Fig. 1. Lower half-moulds with spiral chanel

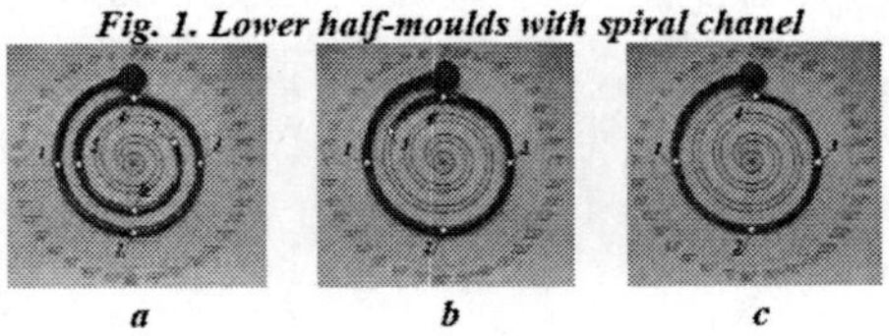

a *b* *c*

Fig. 2. Spiral castings from grey iron

The lenght of the Kerry spiral castings, which were received during the experiments, are given below in Table 1. The measurement error did not exceed 5%.

Table 1. Results of the experiments

Mould	Length of castings, L, mm
With foundry coating based on aluminosilicate refractory filler and nanostructured refracrory additives	851
With foundry coating based on aluminosilicate refractory	661
Without foundry coatings	524

Due to computer simulation of the hydrodynamic processes, the main factors which affected the experiment's results were identified. It was made with initial and boundary conditions corresponding to the real experiment (Fig. 3). The results of the experiments show that the foundry coatings reduce the rate of drop of the melt temperature when it is moving along a spiral. Thus, residence time of the melt in the liquid state increases.

a *b* *c*

Fig. 3. Results of the filling process computer simulation

Firstly, acording to Table 1, a foundry coating with nanostructured additives makes it possible to obtain a spiral length 190 mm (28%) longer than with a foundry coating without nanostructured additives. Secondly, it makes it possible to obtain a spiral length 327 mm (60%) longer than in the mould without foundry coating. Analysis of the temperature change dynamics in the mould without a foundry coating (control points 1, 2, 3, 4 (Fig. 2c) with the coordinates 90° (L = 154mm), 180° (L = 284mm), 270° (L = 797mm), 360° (L = 529 mm)) shows (Fig. 4) that the melt is moving along a spiral by an uneven pulsating flow. During the mould filling process the amplitude of temperature fluctuations on its surface is of a shock character (reaches high values in short time intervals). Two distinct peaks (curve 1, Fig. 4) could be seen at control point 1. This characterizes the oscillatory nature of the filling process.

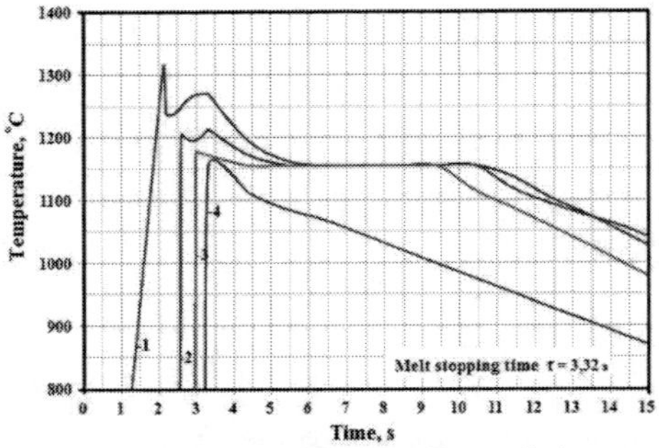

1, 2, 3, 4 – control points on Kerry spiral

Fig. 4. Temperature changes on the surface of the mould without a foundry coating

Fig. 4 gives information about temperature changes on the surface of the mould without a foundry coating. At the begining, its increase is close to pouring temperature. Then the temperature practically instantaneously (0.2 s) decreases by 81°C and after that increases again by 38°C within 0.69 s. Similar temperature fluctuations can be observed at control point 2 (curve 2, Fig. 4) but their amplitude is less than in the first case (temperature decreases by 16°C and temperature increases by 25°C). At point 3 the temperature reaches 1180 °C, which is close to the liquidus temperature of the alloy (T_{lik} = 1188°C for grey iron). Then we can see no repeated oscillations. The movement of the melt along the spiral continues to zero fluid temperature (T = 1154°C) (point 4). After this point temperatures pulsating stoppes and it decreases monotonically.

The resuslts of the real experiment and computer simulation allow us to describe a mechanism of the Kerry spiral mould filling (without foundry coatings). At the first time (from 0 to 2.1 s), overheated melt comes from the gate system to the mould cavity and rapidly moves forward to maximum length. During this stage the head part of the melt flow contacts with cold surface of the mould (20-25°C) and it intensively cooled. At this time on the melt meniscus begins to form thin films of crystallizing melt, which are broken by the moving melt flow and captured by it. Therefore in this period (from 2.1 to 2.3 s) the temperature decreases abruptly due to growth of the melt viscosity and decline of the melt flow velocity. This leads to intensification of crystallization processes at the melt flow meniscus. Such a sharp slowing of the melt flow leads to rapid increase of the pressure and causes a micro hydraulic shock. As a result, a second jump of temperature takes place again within the period from 2.3 s to 2.98 s. The crust of solid metal, which has arisen during the shortest stoping, is broken again by the accelerating flow coming from the gate system. And at the end (in the time interval from 2.98 to 3.2 s), the flow velocity is significantly slowed down and when the necessary amount of solid phase reaches the meniscus the metl flow stops completely.

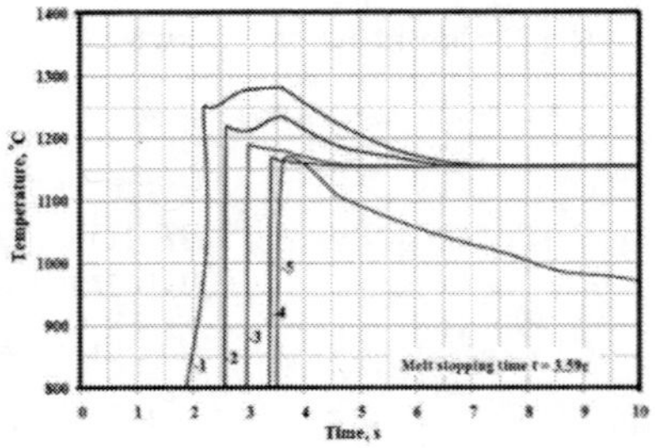

a

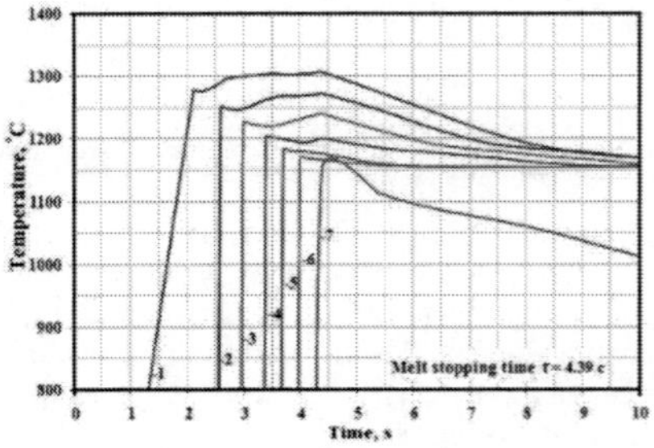

b

1, 2, 3, 4, 5, 6, 7 – control points on Kerry spiral; a – mould with foundry coating based on aluminosilicate refractory filler; b - mould with foundry coating based on aluminosilicate refractory filler and with nanostructured refractory additives

Fig. 5. Temperature changes on the surface of the mould without foundry coatings

As we can see from Figure 5, the foundry coating changes the time of melt liquidity and increases it by 0.39 s (by 1.1 times) for the mould with foundry coating (Fig. 5a) and by 1.19 s (by 1.3 times) for the mould with foundry coating based on aluminosilicate refractory filler and nanostructured refractory additives (Fig. 5b). This allows provides better results of the casting spiral length. The characteristic peaks of the temperature pulsation are smoothed out and leveled in time through the foundry coatings cause thermal resistance on the border between melt and foundry mould. And for a foundry coating with nanostructured refractory additives its thermal resistance is higher than for a foundry coating without this additives.

The results of this study proved that the foundry coating is important to produce castings not only with excellent surface finish but it contributes to production of thin-walled castings without defects.

References

[1] U.C. Nwaogu, N.S. Tiedje. 2011. Foundry Coating Technology: A Review. Materials Sciences and Application. Volume 2, p.1143-1160.

[2] U.C. Nwaogu, T. Poulsenb, R.K. Stageb, C. Bischoffb, N.S. Tiedjea. 2011, New sol–gel refractory coatings on chemically-bonded sand cores for foundry applications to improve casting surface quality. Surface and Coatings Technology, Volume 205, Issue 16, p. 4035-4044.

The 73rd World Foundry Congress
„Creative Foundry"
23 - 27 September 2018, Krakow, Poland

INFLUENCE OF FILTER SURFACE CHEMISTRY ON THE FILTRATION OF ALUMINUM MELT

Claudia Voigt [1]*, Fabio Taina [2], Pierre Le Brun [2] Christos Aneziris[1]

[1] *Technische Universität Bergakademie Freiberg, Institute of Ceramic, Glass and Construction Materials, Agricolastr. 17,09599 Freiberg, Germany*

[2] *Constellium Technology Center, Parc Economique Centr'Alp, CS10027, 38341 Voreppe Cédex, France*

**Corresponding address: claudia.voigt@ikgb.tu-freiberg.de*

Key words: filtering, ceramic foam filter, LiMCA, aluminum, non-metallic inclusions

1. Introduction

Melt cleanliness is essential for high quality aluminum production [1]. The main cleanliness criteria are dissolved hydrogen (potentially resulting in porosity) and non-metallic inclusions.

For cast parts, non-metallic inclusions may impair the castability, the mechanical properties, and the machinability of the components. In thin rolled products (e.g. aluminum cans and foil) inclusions can create pinholes, while in thick rolled products (used e.g. in transportation), inclusions may restrict fatigue life. These effects of inclusions illustrate the fact that the continuous improvement of melt cleanliness is essential to meet the ever-increasing quality requirements of today's final products. Inclusions also have different origins, e.g. from scrap melting, oxidation of the melt surface, and erosion of the refractory material. In aluminum castings, oxides (Al_2O_3, MgO, $MgAl_2O_4$, $3Al_2O_3{\cdot}2SiO_2$, SiO_2), carbides, and nitrides are the most common types of inclusion. A simple, relatively cheap, and efficient way to meet the increasing requirements of aluminum castings relies on ceramic filters during the casting process. The ceramic foam filter (CFF) is the most common filter type for aluminum. A significant amount of work has been done to assess the efficiency of filters [2].

The filter material is selected based on processing, thermo-mechanical, and economic criteria but not according to the filtration efficiency because there is only limited information, on the influence of the filter surface material on the filtration efficiency of liquid aluminum.

Syvertsen et al. [3] compare the filtration of aluminum of Al_2O_3 and SiC foam filters, and find for the SiC filters. A possible explanation for the better filtration efficiency of SiC could be the better wetting of the SiC. The measured wetting angle at 1373 K (1100°C) for the SiC filter material is 39° and for the Al_2O_3 filters material 84° [3].

In order to contribute to the understanding of the impact of surface chemistry on aluminum filtration, four different filter surfaces were produced: Al_2O_3 (alumina), $MgAl_2O_4$ (spinel), $3Al_2O_3{\cdot}2SiO_2$ (mullite), and TiO_2 (rutile). Experimental casts of AlSi7Mg0.3 were performed under controlled test conditions. The casts were monitored by LiMCA, and the spent filters were characterized by metallographic methods.

2. Materials and Methods

The pilot filtration line used for the trials is shown in Figure 1 and consists of a 750 kg resistance tilting furnace, a launder system, a filter box, and a ladle to collect the filtered material. A metallic grid was put at the exit of the tilting furnace in order to retain the oxide films and prevent premature filter clogging.

The ingots (AlSi7Mg0.3, EN AC-42100) were melted in the furnace at 1023 K (750 °C). A rotary degassing treatment was applied in the furnace (rotor with 6 blades, Ar 500 Nl/h, 10% Cl_2, 20 min), and the metal was left to settle.

In order to guarantee the reproducibility of the trials, the casting line was equipped with thermocouples and a metal level indicator. The filter box was preheated and once a set point was reached the filter used for the trial was set in and preheated up to 923 K (650°C) with a hot air burner. After the casting, the metal solidified in the filter.

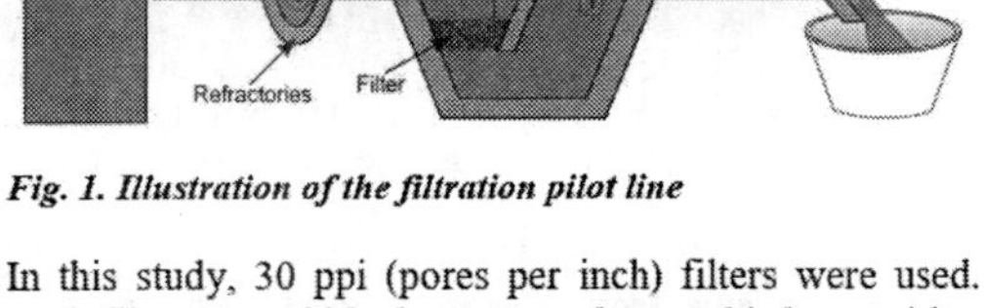

Fig. 1. Illustration of the filtration pilot line

In this study, 30 ppi (pores per inch) filters were used. Each filter pyramid had a truncated pyramid shape with a square section of 90 mm x 90 mm on the small side (melt outlet), 115 mm x 115 mm on the large section (melt entry) and a thickness of 50 mm. The preparation of the filters is described at Voigt et al. [4]. The described process generates filters with comparable pore size and varying surface properties.

In order to assess the inclusion content of the melt, two LiMCA II units (Liquid Metal Cleanliness Analyzer) (ABB Inc., Quebec, Canada) were installed on the casting line, one before and one after the filter. The LiMCA measurements are expressed in k/kg (thousands of inclusions per kg of molten aluminum). N20 and N40 are often used, representing the total amount of inclusions larger than 20 and 40 µm, respectively, expressed in k/kg. These measurements also enabled the determination of the filtration efficiency E of the filters.

3. Results

The morphology of the experimental filters was assessed before being used for casting trials. The porosity, the pore, and struts sizes are comparable. The filters exhibited very similar structural properties. Considering that they were manufactured from the same substrate, this proves the reproducibility and comparability of the coating thickness independently of the viscosities and chemical composition of the slurries.

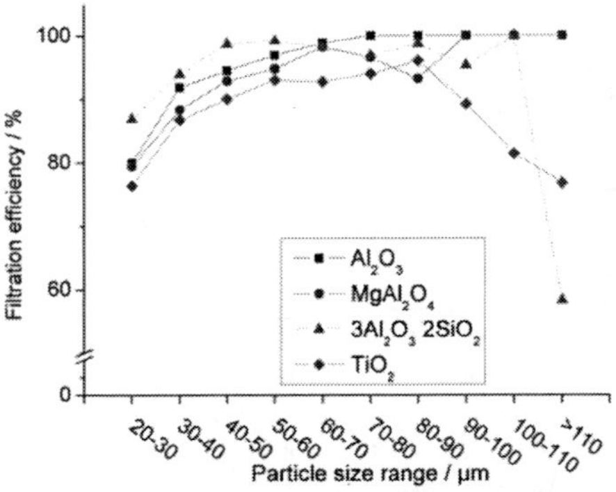

Fig. 2. Filtration efficiency of the four different coatings as a function of the particle size range (The typical error on the efficiency is around 5%)

The inclusion removal efficiency was calculated from the LiMCA data.

The overall efficiency was calculated from the counts of the whole cast. The efficiency was higher than 80% for all casts based on the N20 index, and higher than 90% for all casts for larger inclusions (N40 index).

A more detailed analysis can be done by focusing on size classes. Following this approach, the mean inclusion count of each size class (width 10 µm) was calculated before and after the filter, providing the mean efficiency for this specific class during casting. The results are presented in Figure 2. Differences were observed for the large inclusion sizes: while the spinel and alumina coatings have increasing efficiencies for larger inclusions, mullite and rutile coatings exhibited lower efficiencies for larger inclusions. This behavior was already observed in a previous work by Voigt et al. [5].

Before analyzing the inclusion capture mechanism on the filter surface, the stability of the coating had to be checked in order to identify any filter degradation during casting. This was done by combined SEM, EDX and EBSD.

The SEM analysis show that no damage had occurred to the coatings.

4. Discussion

The methodology used in this study allowed a rigorous comparison of the behavior of four filter surfaces with different chemistries (Al_2O_3, $MgAl_2O_4$, $3Al_2O_3{\cdot}2SiO_2$ and TiO_2). The characterization of the initial filters showed that the porous structure was similar in each filter (in terms of mean porosity, strut thickness). During each test about 730 kg were cast in 40 to 75 min, which corresponds to filtration velocities ranging from 3 to 7 mm/s (calculated based on the ratio of the mean metal flow to the empty filter cross section surface and the metal density). These values are in the range of velocities recommended in the literature for ceramic foam filters (CFFs) for aluminum filtration [2]. Observation of the spent filters showed that the filters had a suitable stability for aluminum filtration: the filters were not damaged after removal (for instance, they were not cracked) and microscopic observations showed there was no deterioration of the coatings. In addition, there was no impact on the chemical composition of the melt when the functionalized filters were used.

The efficiency of the filtration was characterized in real time by LiMCA measurements. The assessed total filtration efficiencies (calculated on the basis of the N20 data) were above 75% for all four filters. These values are considered to be very high compared to industrial data, which would range typically from 40 to 60% [2]. However, it should be noted that there was no refiner used in the reported lab-scale trials. It has been shown that the use of refiner (TiB_2, TiC) prevents the formation of bridges inside the filter pores, resulting in a reduction in inclusion capture [6]. As already reported in the LiMCA results presentation, two filter surfaces appeared to be less efficient in inclusion capture: mullite and rutile. This was true in particular for large inclusion sizes (see Figure 2). Additional research is required in this area to confirm the initial findings of this work.

In a next step the measured filtration efficiencies were correlated with measured contact angles, see presentation WETTABILITY MEASUREMENT OF AlSi7Mg ON Al_2O_3, $MgAl_2O_4$, $3Al_2O_3{\cdot}2SiO_2$ and TiO_2 AT 730°C.

Acknowledgements

The authors would like to thank the German Research Foundation (DFG) for supporting these investigations in terms of the Collaborative Research Centre 920 "Multi-Functional Filters for Metal Melt Filtration – A Contribution towards Zero Defect Materials", Subprojects A02, S03, A03 and A06. The authors also would like to acknowledge the support of G. Schmidt and J. Hubálková (S01).

References

[1] Le Brun P.(2008) Melt Treatment — Evolution and Perspectives. Light Metals, 621- 626.

[2] Laé E, Duval H, Rivière C, Le Brun P, Guillot JB (2006) Experimental and Numerical Study of Ceramic Foam Filtration. Light Metals 753-758.

[3] Syversen M, Kvithyld A, Bao S, Nordmark A,Johansson A (2014) Parallel Laboratory and Industrial Scale Aluminium Filtration Tests with Al_2O_3 and SiC Based CFF Filters. Light Metals 1041 – 1046.

[4] Voigt C, Zienert T, Schubert P, Aneziris CG, Hubálková J (2014) Reticulated Porous Foam Ceramics with Different Surface Chemistries. J. Am. Ceram. Soc. 97 (7): 2046-2053.

[5] Voigt C, Fankhänel B, Jäckel E, Aneziris C.G, Stelter M, Hubálková J (2015) Effect of the Filter Surface Chemistry on the Filtration of Aluminum. Metall. Mater. Trans. B 46: 1066-1072.

[6] Towsey N, Schneider W, Krug HP, Hardman A, Keegan NJ (2001) The influence of grain refiners on the efficiency of ceramic foam filters. Light Metals 291-295.

The 73rd World Foundry Congress
„Creative Foundry"
23 - 27 September 2018, Krakow, Poland

EFFECT OF TRACE ELEMENTS ON MICROSTRUCTURE AND MATERIAL PROPERTIES OF AN ALUMINIUM ALLOY

Thomas Pabel [1], Tose Petkov [1], Hartmuth Schroettner [2,3], Mihaela Albu [3], Anita Rossmann-Perner [3], Peter Schumacher [1,4], Jiehua Li [4]

[1]Austrian Foundry Research Institute, Parkstrasse 21, 8700 Leoben, Austria
[2]Institute of Electron Microscopy and Nanoanalysis, Graz University of Technology, Steyrergasse 17, 8010 Graz, Austria
[3]Graz Centre for Electron Microscopy, Steyrergasse 17, 8010 Graz, Austria
[4]Chair of Casting Research, University of Leoben, Franz-Josef-Strasse 18, 8700 Leoben, Austria
e-mail: office@ogi.com

Key words: Aluminium, SEM, TEM, ThermoCalc Simulation, intermetallic phases

1. Introduction

The aim of the project is to determine the fundamental effects of impurities and individual micro-alloying elements (e.g. as vanadium V, titanium Ti, calcium Ca, zirconium Zr and phosphorus P) or combinations of these trace elements in aluminium alloys and their impact on the quality of aluminium castings.

As a result, limiting values and tolerances for individual impurities (trace elements) should be defined and determined.

Additional practical methods were developed to support a reliable series production of high-quality alloys and castings as well as the procurement of aluminium alloys by the foundries.

2. Investigations

The casting trials in industry-related standards were performed based on a high purity alloy Al Si7Mg0.3 with systematic addition (30 ppm, 300 ppm, 3000 ppm) of micro-alloying elements, controlled by chemical analysis with an emission spectrometer.

Technological properties as flowability, hot crack susceptibility and shrinkage cavity formation were evaluated. Static and dynamic material testing, like tensile tests, hardness measurements and Woehler curves was implemented.

The thermo-physical properties as specific heat capacity, thermal expansion, temperature conductivity, density and heat conductivity were measured.

Phase calculations by ThermoCalc-Software, to simulate the formation of intermetallic phases and their impact on the microstructure are compared to classical metallography and additional SEM/TEM-investigations.

SEM/EDXS/EBSD-measurements for the micro-characterization of the structure and the composition of the intermetallic phases are made on cross sections (Figure 1).

TEM-lamellas are prepared from phases of interest using the FIB-technique (Figures 2-3) and transferred to TEM for using EDXS/EELS/EFTEM and diffraction methods for the nano-characterization of these phases.

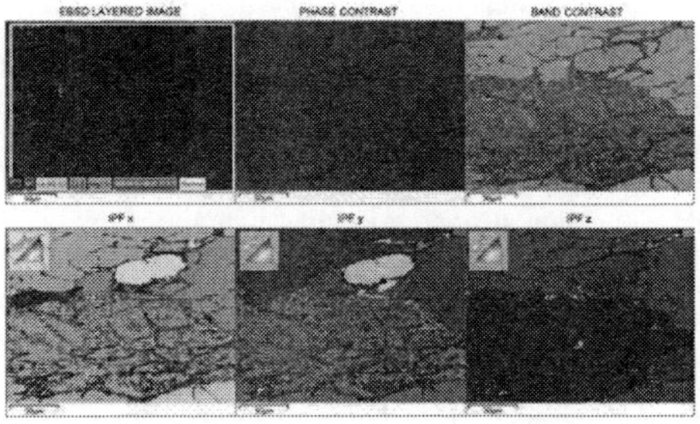

Fig. 1. EBSD investigation of AlSi7Mg0.3 doped with 30 ppm calcium and 30 ppm phosphorus. Inverse pole figure (IPF) colouring with reference directions X, Y, Z.

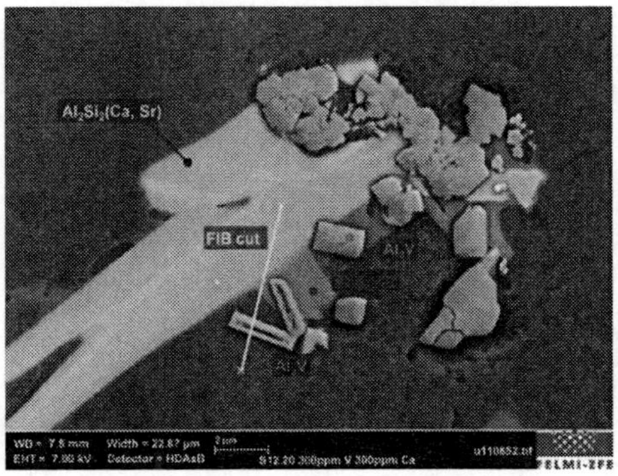

Fig. 2. Backscattered Electron (BSE) image of AlSi7Mg0.3 doped with 300 ppm vanadium and 300 pm calcium.

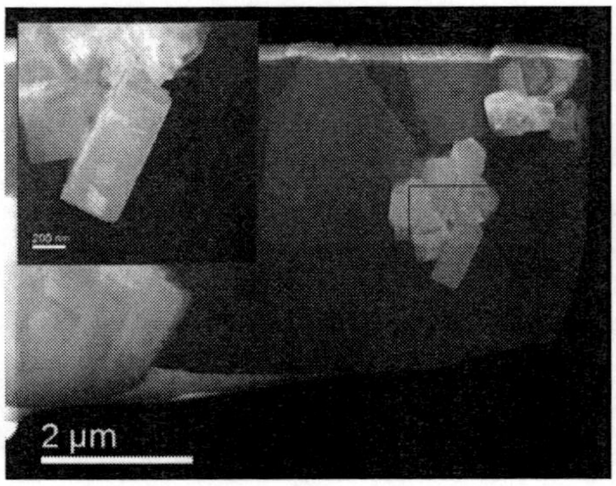

Fig. 3. HAADF-image of the Al_3V-phase together with Al_2Mg_2Si-phases and eutectic Si.

3. Conclusion

The examinations show a significant correlation between the thermodynamic calculations (ThermoCalc-Software) and the actual casting trials close to industrial conditions.

Already minor traces of calcium lead to the formation of intermetallic Al_2Si_2Ca precipitations.

Titanium, vanadium and zircon tend to precipitate in the intermetallic Al_3M-phases when the solubility limit is exceeded.

The casting characteristics worsen significantly with an increase of trace elements leading to primary phases in the liquid.

The thermos-physical properties, tensile strength and elongation at fracture decline with the increase of trace elements. Especially element combinations with calcium are critical.

Acknowledgments

The authors want to thank the Austrian Research Promotion Agency (FFG) for financial support (PN845411).

The 73rd World Foundry Congress
„Creative Foundry"
23 - 27 September 2018, Krakow, Poland

PREPARATION OF MAGNESIUM BORATE FIBERS BY ELECTROSPINNING

Enrico Storti [1]*, Paolo Colombo [2], Christos G. Aneziris [1]

[1] *TU Bergakademie Freiberg, Institute for Ceramics, Glass and Construction Materials, Agricolastraße 17, 09599 Freiberg, Germany*
[2] *Università degli Studi di Padova, Department of Industrial Engineering, Via Marzolo 9 35131 Padova, Italy*
**Corresponding address: enrico.storti@ikgb.tu-freiberg.de*

Key words: Electrospinning, Magnesium borate, Sol-gel, Filtering

1. Introduction

Electrospinning represents an interesting bottom-up approach for the development of 1D nanoscaled materials. It is a relatively simple and inexpensive to obtain continuous nanofibers for a broad range of applications[1]. Ceramic nanofibers synthesized via e-spinning were only introduced in 2002 to be first used in components such as gas sensors and catalysts. In this work, magnesium borate fibers were produced by electrospinning from sol-gel precursors and annealed under different conditions. $Mg_2B_2O_5$ offers excellent mechanical properties combined with good heat and corrosion resistance. It is used as anti-wear material but can also work as catalyst, luminescent material, or fused cast refractory in steel-making environments[2]. In view of a possible application of electrospun fibers on ceramic filters for metal filtration, the same treatment as for carbon-bonded alumina filter substrates was applied on the fibers.

2. Experimental

As precursors for the magnesium borate phase, high- purity boron ethoxide and magnesium ethoxide were used. PVP with 1,300,000 molecular weight was required to generate the fibrous structure. First a 40 wt% precursor solution was prepared in 2-Methoxyethanol using a magnetic stirrer. A 1:1 molar ratio between $C_4H_{10}MgO_2$ and $C_6H_{15}BO_3$ was chosen in order to obtain $Mg_2B_2O_5$ as the primary phase after heat treatment in air. Separately, a 25 wt% solution of PVP was prepared in N,N-Dimethylformamide, again by means of magnetic stirring. Since 2-Methoxyethanol and DMF are miscible in one another, the e-spinning solution was obtained by mixing the precursor and the PVP solutions in a 2:1 weight ratio until homogenization. The production of the fibers was performed in a NE100 unit provided by Inovenso (Istanbul, Turkey). The prepared e-spinning solution was loaded into a 10 ml syringe connected to a 0.8 mm diameter conductive brass needle and fed at 0.5 ml/h with a syringe pump. e-spinning was carried out at room conditions (T~21°C, R.H.~40%) and at a voltage of 15 kV using a DC power supply. An aluminum plate covered with aluminum foil was used as the collector, and the distance between nozzle and collector was 10-12 cm.

After the e-spinning process the as-spun fiber mat was removed from the aluminum foil. Samples were then heat treated at 900°C in air with three different heating rates of 1, 5, and 10 °C/min and a holding time of 1 hour. The resulting samples were called, respectively, MB_1, MB_5 and MB_10. Separately, another fiber mat was treated at 800°C for 3 h in a retort filled with calcined petcoke, exactly as carbon-bonded filters for steel melt filtration[3]. This sample was called MB_C. The microstructure evaluation before and after annealing/coking of the samples was carried out by means of optical microscopy and scanning electron microscopy in combination with energy dispersive X-ray spectroscopy (EDS). Differential scanning calorimetry (DSC) coupled with simultaneous thermogravimetric (TG) analysis was performed on a green fiber sample up to 1000°C in air, with a heating rate of 10 °C/min[4]. X-ray diffraction analysis (XRD) was conducted after heat treatment to determine the phase composition of the samples. The specific surface area of the samples was determined by N_2 adsorption-desorption measurements at 77 K. Samples were first degassed at 90 °C for 30 min and then at 350°C for 5 h under vacuum. The specific surface areas were calculated from the linear form of the BET equation taking the cross-sectional area of the nitrogen molecule to be $16.2 \cdot 10^{-20}$ m^2 (linearity region between 0.05 and 0.35 p/p_0).

3. Results and discussion

The green sample consisted of continuous fibers with random orientation, which is usually the case for a basic laboratory equipment equipped with a plate collector. The fiber diameters were mainly in the 500-700 nm range. Defects as droplets or beads were not observed in the investigated sample. SEM micrographs of the MB_5 sample after calcination in air at 900°C is presented in Fig. 1. The fibers consisted of small particles joined together to form thin filaments. This structure is known as "necklace" and is typical for fibers produced by e-spinning[5]. The average diameter was slightly smaller than in the original mat and located in the 300-500 nm range. This shrinkage was a result of the remarkable weight loss (~70%), related to the decomposition of the PVP and the reaction of alkoxides, during the heat treatment. The heat treatment in reducing conditions delivered fibers with comparable diameters, but smaller particles and no necklace structure (see 0).

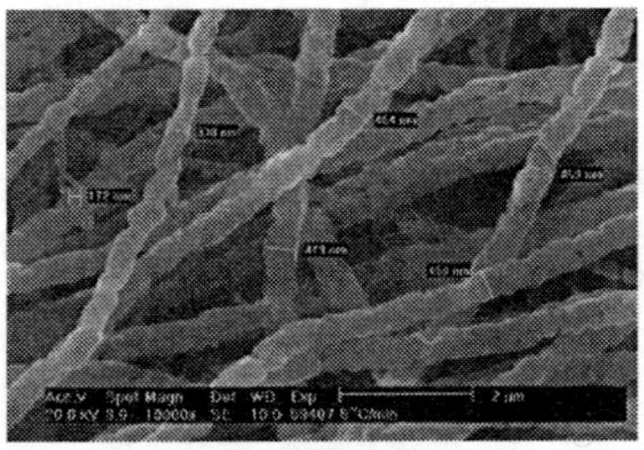

Fig. 1. MB_5 fiber mat.

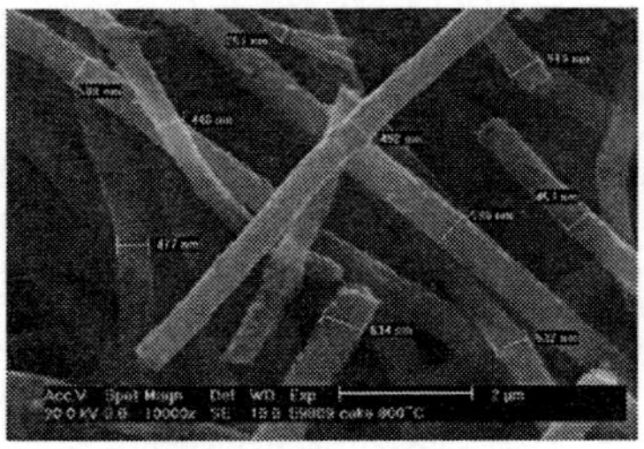

Fig. 2. MB_C fiber mat.

The XRD patterns were typical for highly crystalline materials and the main phase detected in all samples calcined in air was $Mg_2B_2O_5$. In addition, magnesium borate $Mg_3B_2O_6$ was also detected as a secondary phase. A higher intensity of the $Mg_3B_2O_6$ peaks was observed for the MB_1 sample, while the other two samples showed almost identical patterns. We assumed that a slower treatment promoted the loss of boron through evaporation: the volatilization is a significant phenomenon for some glass melt components (such as alkali oxides, lead, boron and phosphorus) which have high vapor pressures at high temperature[6]. $Mg_3B_2O_6$ was found as the main phase in the mat treated in reducing atmosphere (MB_C), followed again by $Mg_2B_2O_5$. In addition, broad peaks and a strong background noise compared to the other XRD patterns suggested the presence of an amorphous phase and smaller crystals. It was reported that, if MgO is present, the T_g of vitreous boric oxide can reach values above 900°C[7]. It is likely that a remarkable amount of boron was retained in this glass phase, hence the different crystalline phase composition.

The magnesium borate fibers synthesized in air were quite dense (only ~4 m^2/g BET surface), as revealed by adsorption-desorption analysis. Their isotherms, were of type II (common for disperse, non-porous or macroporous materials) according to the classification by Brunauer et al.[8]. The use of sol-gel precursors allowed to obtain well crystallized and dense fibers. The fibers treated in reducing atmosphere (MB_C) presented an isotherm of type IV (typical for porous materials) and a much stronger adsorption. The plateau at high relative pressures suggested a complete filling of pores, characteristic of a material with a limited range of pore sizes. The calculated BET specific surface for this sample was ~57 m^2/g. The decomposition of the organic part possibly generated a lot of pores, and the lower temperature hindered the sintering of the material.

4. Conclusions

The electrospinning technique for the synthesis of magnesium borate fibers was investigated. A solution of alkoxide precursors and polyvinyl-pyrrolidone was used. Fibers with average diameters in the 500-700 nm range were detected in the green fiber mats. Heat treatments at 900°C in air formed crystalline and dense fibers with slightly smaller dimensions (300-500 nm). The main phase detected by XRD analysis was $Mg_2B_2O_5$, while $Mg_3B_2O_6$ was also present as a secondary phase. The heat treatment at 800°C in reducing atmosphere also delivered thin fibers with higher specific surface area: in this case part of the boron from the precursors was retained in a glassy phase, hence the main compound detected by X-ray diffraction was $Mg_3B_2O_6$.

Acknowledgments

The studies were carried out with financial support from the German Research Foundation (DFG) within the framework of the Collaborative Research Center SFB 920.

References

[1] Sigmund W, Yuh J, Park H, Maneeratana V, Pyrgiotakis G, Daga A, Taylor J & Nino JC. (2006). Processing and Structure Relationships in Electrospinning of Ceramic Fiber Systems. J Am Ceram Soc. 89 (2), 395–407.

[2] Hu ZS, Lai R, Lou F, Wang LG, Chen ZL, Chen GX & Dong JX. (2002). Preparation and tribological properties of nanometer magnesium borate as lubricating oil additive. Wear. 252 (5–6), 370–4.

[3] Emmel M, Aneziris CG. (2012). Development of novel carbon bonded filter compositions for steel melt filtration. Ceram Int. 38 (6), 5165–73.

[4] Storti E, Roso M, Modesti M, Aneziris CG & Colombo P. (2016). Preparation and morphology of magnesium borate fibers via electrospinning. J Eur Ceram Soc. 36, 2593–2599.

[5] Lu X, Zhang D, Zhao Q, Wang C, Zhang W & Wei Y. (2006) Large-scale synthesis of necklace-like single-crystalline $PbTiO_3$ nanowires. Macromol Rapid Commun. 27 (1), 76–80.

[6] Shelby JE. (2005) Introduction to Glass Science and Technology. Cambridge: The Royal Society of Chemistry.

[7] Yiannopoulos YD, Chryssikos GD & Kamitsos EI. (2001) Structure and properties of alkaline earth borate glasses. Phys Chem Glas. 42 (3), 164–72.

[8] Brunauer S, Emmett PH & Teller E. Adsorption of Gases in Multimolecular Layers. (1938) J Am Chem Soc. 60 (2), 309–19.

The 73rd World Foundry Congress
„Creative Foundry"
23 - 27 September 2018, Krakow, Poland

CERAMIC FILTERS WITH COATINGS BASED ON NANO-MATERIALS OR CALCIUM ALUMINATES WITH CARBON FOR STEEL MELT FILTRATION

Enrico Storti [1]*, Anne Schmidt [1], Steffen Dudczig [1], Christos G. Aneziris [1], Christoph Wohrmeyer [2], Chris Parr [3]

[1]*TU Bergakademie Freiberg, Institute for Ceramics, Glass and Construction Materials, Agricolastraße 17, 09599 Freiberg, Germany,*

[2] *Kerneos S GmbH, Centroallee 275 46047 Oberhausen, Germany*

[3] *Kerneos, 8 rue de Graviers 92521 Neuilly sur Seine Cedex, France*

**Corresponding address: enrico.storti@ikgb.tu-freiberg.de*

Key words: Ceramic foam filters, Carbon nanotubes, Alumina nanosheets, Calcium aluminates, Filtering

1. Introduction

It is known that non-metallic inclusions in metal parts can severely degrade their properties[1]. Ceramic filters (in particular foams) after the ladle treatment have been widely used for years by foundries to improve the purity of ferrous and non-ferrous alloys[2]. To improve the performance of commercial filters, coatings based on carbon nanotubes and alumina nanosheets, or calcium di- and hexaluminate were developed and tested in a steel melt containing artificially-generated alumina inclusions. After the tests, the inclusions remaining in the steel were analyzed and classified by size and chemistry by means of an ASPEX scanning electron microscope.

2. Experimental

Filter production

Prismatic carbon-bonded alumina filters (10 pores per inch, 125 x 22 x 22 mm³) were produced via the Schwartzwalder process using a two-steps approach, according to the parameters described by Emmel et al.[3]. After drying, the filters were heat treated at 800 °C in retorts filled with calcined petcoke (Müco, Germany), to approach reducing conditions.

The raw materials used for the preparation of the nano-based coatings were MWCNTs, Al_2O_3-nanosheets and modified coal tar pitch powder (used as a binder). The additives were Xanthan powder (dispersing agent and thickening agent), carboxylic acid preparation (dispersing agent) ammonium-ligninsulfonate and alkylpolyalkyleneglycolether. Xanthan was first coupled with pulsed ultrasonication to disperse the nanotubes in water[4]. The alumina nanosheets were dispersed separately in water with the carboxylic acid, stabilized with xanthan and then mixed with the CNTs dispersion in a 3:1 ratio (CNTs:ANSs). Finally, the coal tar pitch powder was added from another dispersion prepared by horizontal ball milling. The compositions with calcium aluminates are protected by patent (application number EP 16305289.7; extension number PCT/EP2017/056296). CA2 and CA6 were both supplied by Kerneos. A modified coal tar pitch powder was used as a binder. The additives were ligninsulfonate, modified polycarboxylate ether and alkylpolyalkylene-glycolether in the same concentrations as for the filters. For each composition the spraying slurry was prepared by means of mechanical stirring. All coatings were applied by cold spraying. After application and drying of the coating, the filters were thermally treated again under reducing conditions (see above) in order to achieve the bonding and close some residual porosity and cracks from the first cycle.

Melting experiment

The filters were tested in a metal casting simulator (Systec, Germany) located in Freiberg. For each sample, ≈40 kg of commercially available 42CrMo4 steel (AISI 4142) were melted under argon in a special alumina crucible. Before the immersion of the filter, defined alumina impurities were created in the steel melt as described by Dudczig[5]. Once the desired temperature was reached, endogenous alumina inclusions were generated by adding 0.5 wt.% of iron oxide and then 0.05 wt% of aluminum (related to the steel mass) to the melt.

Finally, the ceramic foam samples were dipped into the steel (at a temperature of ≈1650°C) and rotated with 30 revolutions per minute for the desired time. At the end all samples were removed from the melt and cooled down in a chamber under argon in order to prevent oxidation of the carbon fraction.

Inclusion analysis

After solidification, steel samples were cut from the blocks at a standard position (i.e. at 2/3 height, 1/3 diameter from the edge). In the case of calcium aluminate coatings, steel samples were taken directly from the melt during the experiment. Hence, the results cannot be directly compared. The metallographic specimens were polished and analyzed by means of an ASPEX SEM (FEI, USA): areas of approximately 110 mm² were scanned for about 8 h. Inclusions >0.6 µm were detected using a BSE (back-scattered electron) detector. For each found particle the position, geometry, orientation and chemical composition were analyzed, this last one by EDS (energy-dispersive X-ray spectroscopy). Finally, the inclusions were classified by composition according to our classification rule.

3. Results and discussion

No macroscopic damage was observed on the filters after the test and only the areas which were in direct contact with the melt showed a color change from black to gray, related to decarburization of the surface and formation of a new phase. Optical and scanning electron microscopy confirmed the presence of new layers on the surface. Alumina particles in various shapes were found and the clogging effect increased with increasing immersion time. The CA2-C filter showed a higher roughness than the CA6-C and a lot of pores infiltrated with steel (white spots). The different Ca content (reactivity) and wettability of the materials may explain this morphology. In addition, many clusters of nano-sized inclusions with dendritic shape were found on the outer layer (Fig. 1). We believe that these particles are the actual endogenous inclusions produced by our treatment before the immersion test, while the underlying layers should derive from the reaction of the filter material at the interface with the melt. Nano-materials were not observed and were likely involved in these interactions.

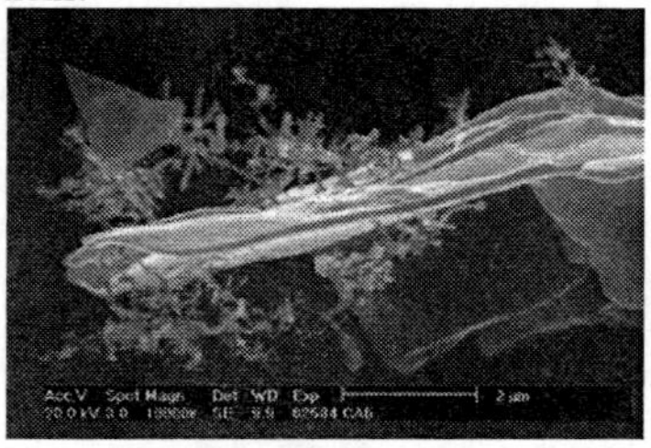

Fig. 1. ***Detail of the CA6-C filter after immersion, with collected endogenous inclusions.***

The majority of inclusions in the reference melt (not filtered) consisted of Al_2O_3, as expected due to the oxide/Al treatment. Regarding the inclusion size, most of the detected particles in the steel samples had a surface area <20 μm^2, while only few inclusions had >50 μm^2. The use of a filter always had a positive impact on the steel purity (high efficiency η), especially for a 10 s immersion. It was observed that the nano-coated filter produced silicates in the steel as a side effect (Table 1), consistent with the observations by Storti et al.[6] Alumina-coated filters produced instead silica inclusions, likely due to impurities in the raw material[7]. CA-coated filters had promising performance (Table 2) and seemed to have no such drawback, but the steel blocks need to be also analyzed to compare the results. We believe that nano-coated filters may be well suited for small castings.

4. Conclusions

Coatings based on nano-sized materials, or calcium aluminates were explored. Fine endogenous inclusions were efficiently collected during an immersion test. Nano-sized additives produced silicates in the melt as a side effect. Both CA-coated compositions may deliver excellent filtration efficiencies.

Table 1. Number of inclusions found in the steel samples (nano-coated filters) by ASPEX, classified by chemistry. Values for alumina-coated filters taken from the work of Schmidt et al.[7]

	Inclusions per cm²				
Chemical class	No filter	Al_2O_3-coated 10 s	CNTs-ANSs 10s	Al_2O_3-coated 60 s	CNTs-ANSs 60 s
Al_2O_3	1248	371	56	439	446
Al-Mn-Mg-Fe-Ca-silicate	0	0	2	0	234
SiO_2	5	9	1	18	3
Other	122	9	1	16	82
$\eta(Al_2O_3)$	-	70%	95%	65%	64%

Table 2. Number of inclusions found in the steel samples (CA-coated filters) by ASPEX, classified by chemistry.

	Inclusions per cm²			
Chemical class	After treatment, pre filter	After CA2-C 10 s	After treatment, pre filter	After CA6-C 10 s
Al_2O_3	132	4	408	1
Al-Mn-Mg-Fe-Ca-silicate	11	7	50	2
SiO_2	1	2	1	0
Other	61	34	144	29
$\eta(Al_2O_3)$	-	97%	-	100%

Acknowledgments

The studies were carried out with financial support from the German Research Foundation (DFG) within the framework of the Collaborative Research Center SFB 920. Thanks for Kerneos for providing the calcium aluminate materials.

References

[1] Zhang L. (2013). Nucleation, growth, transport, and entrapment of inclusions during steel casting. Jom. 65 (9), 1138–44.

[2] Apelian D, Mutharasan R & Ali S. (1985). Removal of inclusions from steel melts by filtration. J Mater Sci. 20, 3501–14.

[3] Emmel M & Aneziris CG. (2012). Development of novel carbon bonded filter compositions for steel melt filtration. Ceram Int. 38 (6), 5165–73.

[4] Storti E, Emmel M, Dudczig S, Colombo P & Aneziris CG. (2015). Development of multi-walled carbon nanotubes-based coatings on carbon-bonded alumina filters for steel melt filtration. J Eur Ceram Soc. 35 (5), 1569–80.

[5] Dudczig S, Aneziris CG, Emmel M, Schmidt G, Hubalkova J & Berek H. (2014). Characterization of carbon-bonded alumina filters with active or reactive coatings in a steel casting simulator. Ceram Int. 40 (10), 16727–42.

[6] Storti E, Dudczig S, Hubálková J, Gleinig J, Weidner A, Biermann H & Aneziris CG. (2017). Impact of nanoengineered surfaces of carbon-bonded alumina filters on steel cleanliness. Adv Eng Mater.

[7] Schmidt A, Salomon A, Dudczig S, Berek H, Rafaja D & Aneziris CG. (2017). Functionalized Carbon-Bonded Filters with an Open Porous Alumina Coating: Impact of Time on Interactions and Steel Cleanliness. Adv Eng Mater. 1700170.

The 73rd World Foundry Congress
„Creative Foundry"
23 - 27 September 2018, Krakow, Poland

THE EFFECT OF THE SUBSTITUTION OF SILICON BY ALUMINUM ON THE MECHANICAL PROPERTIES OF GRAY IRON

E. Aguado [1*], M. Ferrer [1], P. Larrañaga [1], D. M. Stefanescu [2], R. Suárez [1,3]

[1]IK4-Azterlan, Aliendalde 6, E-48200, Durango, Spain
[2]The Ohio State University, Columbus, OH and The University of Alabama, Tuscaloosa, Alabama, USA
[3]Veigalan Estudio 2010, Askatasun Etorbidea 16, 2I, E-48200, Durango, Spain
* Corresponding address: eaguado@azterlan.es

Key words: aluminum alloyed gray iron, high tensile strength gray iron, Fe-C-Al alloy.

1. Background

Gray iron, a Fe-C-Si alloy, is one of the most common ferrous alloys used in the metal casting industry. Silicon is required to avoid metastable solidification (cementite). While small additions of Al may promote the pinholes or a greater amount of slag, several investigators have examined the influence of Al as a potential substitute for Si in gray iron [1,2] and in compacted graphite iron [3-6].

It was documented that higher percentages of Al alter the microstructure of gray cast iron, producing some potential benefits, including increased strength, reduced chill and lower hardness.

This article investigates the effect of aluminum additions as a substitute for silicon, focusing on producing gray iron with high tensile properties.

2. Experimental procedure

Experimental heats were produced in a 100kg medium frequency induction furnace from low-Si pig iron.

Typical base iron compositions included 3.08-3.19% C, 0.1-0.16% Si, 0.53-0.63% Mn, 0.28-0.36% Cr, 0.069-0.089% Sn.

The aluminum levels ranged from 0.5 to 5.5%. From each heat thermal analysis (TA) cups, resin-bonded Y-shaped casting type II (UNE-EN-1563), and plates of 2.5, 3.5 and 6 mm thickness were poured from non-inoculated and inoculated irons. The inoculation was made with a commercial Zr and Mn based inoculant.

3. Results and discussion

The effect of aluminum on the tensile strength is shown in Fig. 1. It is seen that in the range 2 to 3.5% Al tensile strength exceeds 450 MPa in four samples. Strengths as high as 466 MPa with a moderate hardness of 250 HB were obtained at 3.15% Al levels, for an iron with 3.08% C and 0.08% Si.

The graphite area measured from the microstructure inspection decreases with increasing Al for the same C and Si contents. This statement is supported through the graph in Fig. 2 and the microstructures in Fig. 3. This behavior is expected, as Al increases the solubility of carbon in the austenite, which implies less precipitated graphite.

The lower amount of graphite and higher amount of austenite for the same amount of carbon is the main reason for the increase in strength with higher aluminum.

Higher Al content produces an increase in the liquidus and eutectic temperatures, as shown in Fig. 4. The data in the figure were obtained from calculations carried out with the thermodynamic software Thermocalc. Because the $T_L - T_{Gr}$ interval decreases with higher Al less coarsening of primary austenite occurs. The temperature of ferrite formation also increases.

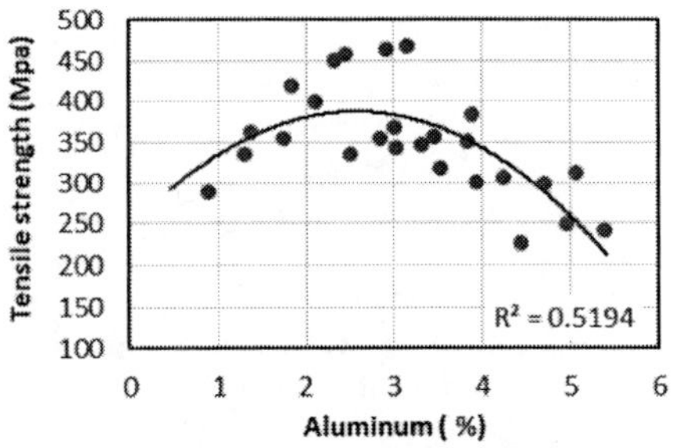

Fig. 1. Effect of aluminum content on the tensile strength of inoculated samples in Y-block Type II

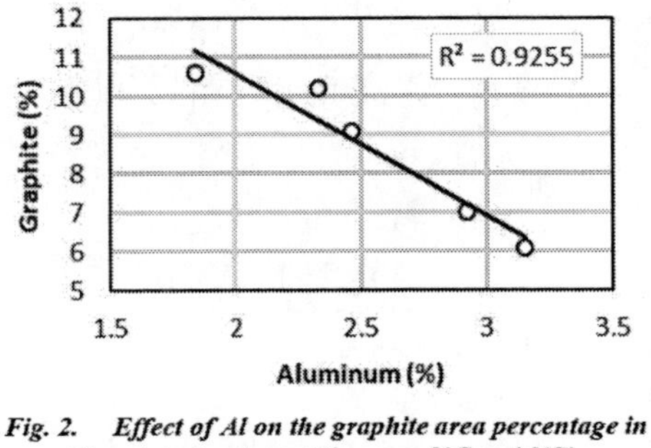

Fig. 2. Effect of Al on the graphite area percentage in the microstructure at the same %C and %Si.

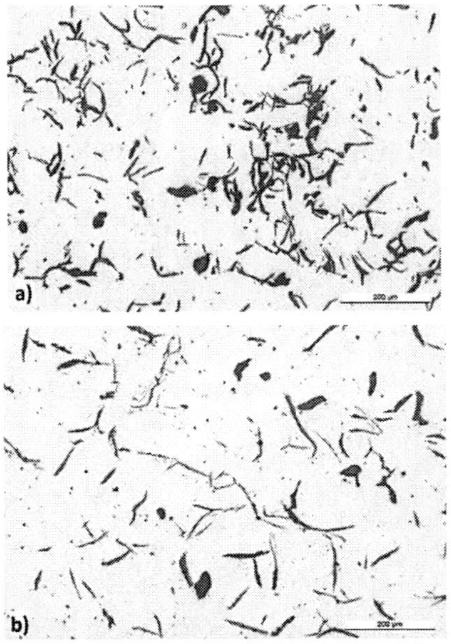

Fig. 3. Effect of Al on the graphite shape and size of inoculated samples: a) 1.84% Al; b) 3.15% Al

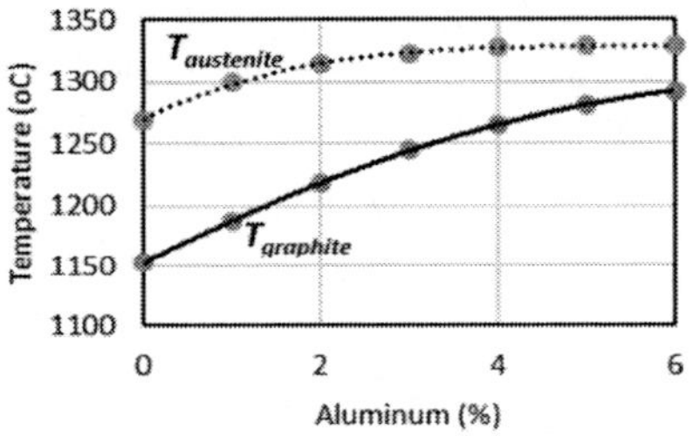

Fig. 4. Effect of Al on the temperature of formation of the primary austenite and graphite; Thermocalc calculation.

The typical matrix of a high-strength Fe-C-Al cast iron consists of fine pearlite and very few patches of ferrite in the vicinity of the graphite (Fig. 5).

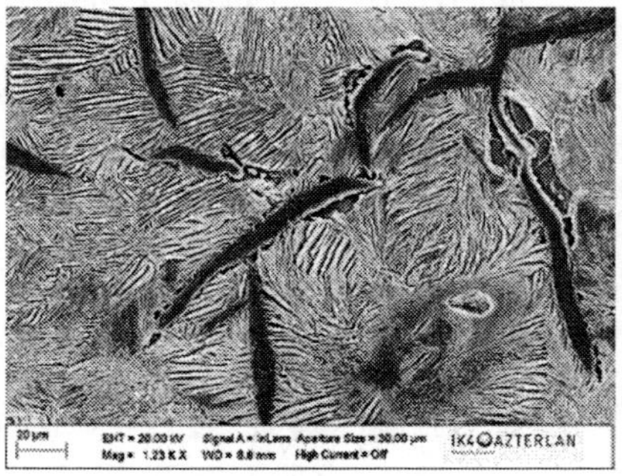

Fig. 5. SEM image of an Fe-C-Al iron with 3.15% Al; etched

One of the main advantages of aluminum cast iron is the low propensity for carbide formation. In non-inoculated irons, carbides were only found in irons with less than 1.5% Al. Thus, plates as thin as 2.5 mm were cast free of carbides from irons with more than 2% Al, which makes this material very desirable for thin wall castings.

SEM-EDAX analysis was performed on selected samples to document the nature of the graphite nuclei in the Fe-C-Al irons. It was concluded that the main nuclei are Mn sulfides, as in regular Fe-C-Si irons. However, the MnS does not appear to nucleate on oxides but rather on aluminum nitrides.

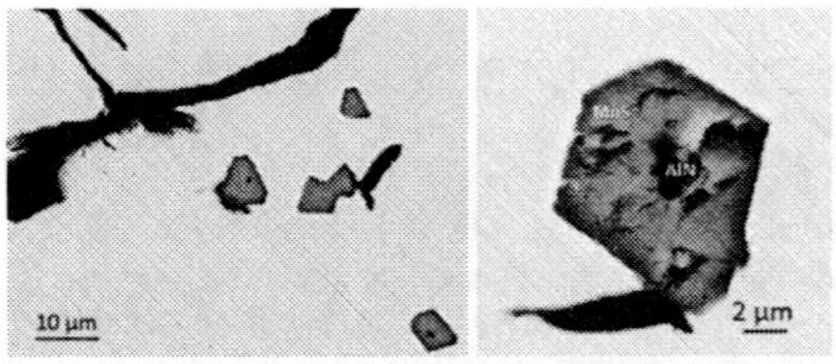

Fig. 6. MnS nuclei for graphite in a 2.87% C, 3.87% Al iron

4. Conclusions

Experimental laboratory trials on standard Y-blocks and on thin plates have demonstrated that pearlitic Fe-C-Al irons can be produced through regular foundry techniques. These irons can have tensile strengths exceeding 450 MPa on standard 25 mm Y-blocks, with average hardness of 250 HB. They can also be used for thin-wall castings (2.5 mm) with no risk of carbide formation.

Acknowledgments

The authors would like to acknowledge Diputacion Foral de Bizkaia for supporting this research.

References

[1] C. Defrancq, J. Van Eeghem, A. DeSy. (1969). Study of the Inoculation of Gray Cast Irons from the Fe-C-Al System; Development of a New Flake Graphite Cast Iron with Very High Strength. In: 36th Int. Foundry Congress. Beograd: CIATF

[2] C. Defrancq, J. Van Eeghem, A. DeSy. (1973). Further Development of Aluminum Cast Iron Inoculated with High Amounts of Calcium. In: 40th Int. Foundry Congress. Moscow: CIATF

[3] D.M. Stefanescu, F. Martinez. (1982). Compacted/ Vermicular Graphite Cast Irons in the Fe-C-Al System. AFS Trans. Vol. 90 pp 39-46

[4] F. Martinez, D.M. Stefanescu. (1983). Properties of Compacted/ Vermicular Graphite Cast Irons in the Fe-C-Al System Produced by Ladle and In-Mold Treatment. AFS Trans. Vol. 91 pp 593-606

[5] D.M. Stefanescu, F. Martinez, I.G. Chen. (1983). Solidification Behavior of Hypoeutectic and Eutectic Compacted Graphite Cast Irons, Chilling Tendency and Eutectic Cells. AFS Trans. Vol. 91 pp 205-216.

[6] D.M. Stefanescu, F. Martinez. (1985). Compacted Graphite Cast Irons in the Iron-Carbon-Aluminum System", U.S. Patent no. 4,501,612.

PREDICTION OF CHUNKY GRAPHITE ON THE BASE OF NUMERICAL SIMULATION AND EXPERIMENTAL DATA

Branko Bauer [1]*, Ivana Mihalic Pokopec [1], Mitja Petrič [2], Primož Mrvar [2]

[1] *University of Zagreb, Faculty of mechanical engineering and naval architecture, Ivana Lučića 5, Zagreb, Croatia*
[2] *University of Ljubljana, Faculty of natural sciences and engineering, Aškerčeva cesta 12., Ljubljana, Slovenia*
**Corresponding address: branko.bauer@fsb.hr*

Key words: chunky graphite, spheroidal graphite cast iron, numerical simulation, cooling curve, microstructure

Over the last fifteen years, the need for heavy section ductile iron castings have increased in different industry sectors. This refers primarily to the manufacture of castings for wind turbines, manufacture of containers for permanent storage of nuclear waste and industry of heavy-duty vehicles [1,2]. Chunky graphite (CHG) is a graphite degeneration which normally appears in the thermal center of large castings and decreases the mechanical properties [3,4]. On macro-scale, it is optically visible on cut or machined surfaces as a black spot. Microscopic observation shows that CHG consists of large cells of highly branched and interconnected graphite strings. The main causes of a CHG occurrence are low cooling rate and chemical composition of the melt [1-3]. A great number of studies have been conducted to describe the CHG formation, but a clear understanding of its appearance and a safe mastering of the melt preparation to avoid CHG are not yet available.

In the present work the cooling curves were recorded in the cone blocks and standard TA cups. According to measured data from the cone block, melt characteristics and heat transfer coefficient between casting and mould were adjusted in the ProCAST® simulation software. Correlation of the simulation results with the experimentally obtained CHG area was discussed.

The experimental study was performed on large cone blocks Φ 300 mm x 350 mm, cast in sodium silicate bonded sand molds. Near-eutectic spheroidal graphite cast iron EN-GJS-400-18 with 0.7 wt% Ni was poured through KALPUR feeder, Figure 1 a). Metallic charge and the melt treatment for spheroidization and inoculation are described in previous work [5]. Two molds were equipped with three K-type thermocouples, two in the mold cavity, Figure 1 b) and one in the mold. Pouring temperature for both moulds was 1380 °C and pouring time 27 s. Just before pouring the melt in the mould, the chilled coupon was analysed by optical emission spectrometer (ARL 3460). C and Si contents were determined from the results of thermal analysis in TA cup using the ATAS® system. Chemical composition is listed in Table 1.

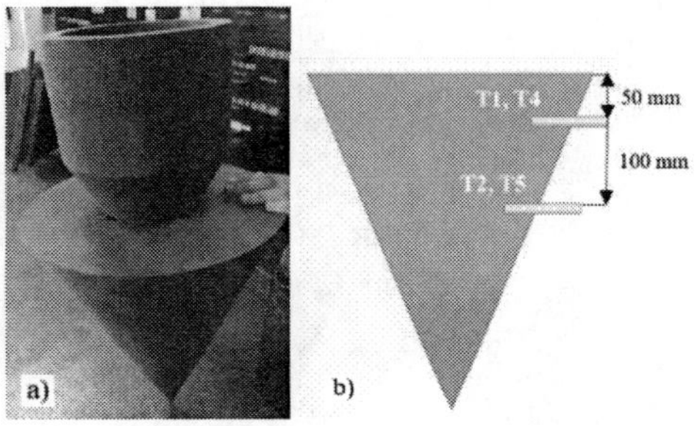

Fig. 1. ***Cone pattern with feeder and location of thermocouples in the mould cavity.***

Table 1. Chemical composition of the melt

Mass fraction, wt%						
C	Si	Mn	S	P	Mg	Co
3.25	1.75	0.16	0.011	0.032	0.036	0.017
Ni	Cr	Cu	Sn	Mo	Ti	Al
0.72	0.03	0.016	0.004	0.003	0.007	0.0095

The recorded cooling curve in TA cup of the spheroidized melt shows that the solidification takes place in two phases. Characteristic temperatures are listed in Table 2. Using this curve, the nucleation of the melt can be estimated and it has also been observed in work [6] a relationship between the recalescence measured on TA cups and the volume of the blocks affected by CHG.

Table 2. Characteristic temperatures (°C) evaluated on TA cup.

T_L	T_{EU}	T_{ER}	T_S
1155	1139	1141	1096

Cooling curve obtained on cast cone block with thermocouple T4 is shown in Figure 2. Characteristic temperatures obtained from first derivative of the cooling curve for spheroidized melt with in stream inoculation are listed in Table 3. Simulated cooling curve adjusted according measured data from the cone block is also shown in Figure 2. and it is in good agreement with experimental cooling curve.

Cone block was afterwards sectioned along the vertical symmetry plane for evaluation of the zone affected by CHG, Figure 3a).

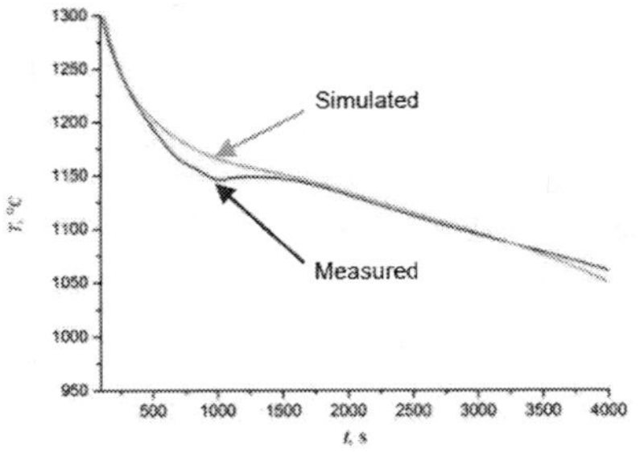

Fig. 2. Cooling curve recorded on the cone block and simulated cooling curve.

Table 3. Characteristic temperatures (°C) evaluated on cone block.

T_N	T_L	T_E	T_{EU}	T_{ER}	T_S
1168.3	1160	1152.3	1145.4	1148.7	1132.2

According to the results of solidification software with melt adjusted with experimentally obtained parameters, it could be observed that thermal modulus Figure 3b) is a very good predictor of affected CHG area in the real casting, Figure 3a).

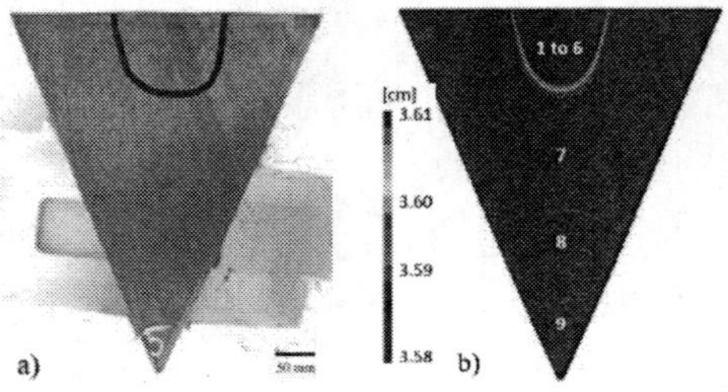

Fig. 3. Cross-section of the cone block and thermal modulus (cm).

Simulated cooling curves from 1 to 6 are related to the area where the CHG was observed while for the cooling curves 7 to 9 there is no danger of CHG occurrence, Figures 3 and 4.

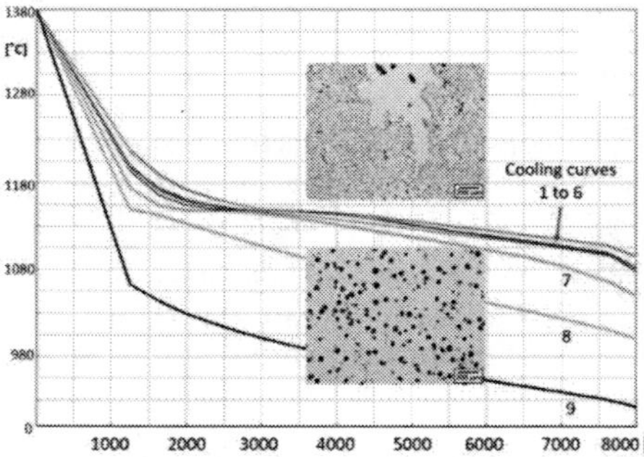

Fig. 4. Simulated cooling curves of the cone block.

According to the results of simulated cooling rates on Figure 5, low cooling rate area below 0.022 °C/s does not match with the area affected by CHG in the real castings.

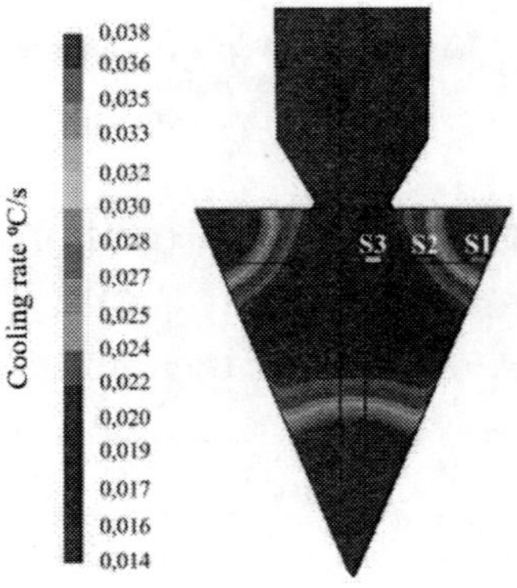

Fig. 5. Cooling rates in the cone block, S1=0.036 °C/s, S2=0.022 °C/s, S3=0.016 °C/s.

For a given melt a relationship between the area of the cone block affected by CHG and the thermal modulus calculated by simulation software has been observed. If the chemical composition of the melt is known, experimental cooling curves data can be used in the simulation to calculate critical areas which are susceptible to the formation of CHG. This can be used in designing complex geometry of heavy-sectioned castings in order to avoid CHG.

Acknowledgments

This work is supported by the foundry MIV d.d. Varaždin.

References

[1] J. Lacaze, L. Magnusson Aberg & J. Sertucha. (2013). Review of microstructural features of chunky graphite in ductile cast irons. In Keith Millis Symp. on Ductile Cast Iron, October 15-17 (360-368). Nashville: AFS.

[2] I. Riposan, M. Chisamera & S. Stan. (2010). Performance of heavy ductile iron castings for windmills. China foundry. 7 (2), 163-170.

[3] R. Källbom, K. Hamberg & L.E. Björkegren. (2005). Chunky Graphite - Formation and Influence on Mechanical Properties in Ductile Cast Iron. In Proc. of the Gjutdesign 2005, June 13-14 (63-86). Espoo: VTT.

[4] I. Mihalic Pokopec, P. Mrvar & B. Bauer. (2017). Effect of inoculation on the formation of chunky graphite in ductile-iron castings. Materials and technology. 51 (2), 275-281.

[5] I. Mihalic Pokopec, P. Mrvar & B. Bauer. (2018). Influence of chunky graphite on latent heat. Materials and technology. 52 (1), 59-65.

[6] J. Sertucha, R. Suarez, I. Asenjo, P. Larranaga, J. Lacaze, I. Ferrer & S. Armandariz. (2009). Thermal analysis of the formation of chunky graphite during solidification of heavy-section spheroidal graphite iron parts. ISIJ International. 49 (2), 220-228.

The 73rd World Foundry Congress
„Creative Foundry"
23 - 27 September 2018, Krakow, Poland

PRELIMINAR MECHANICAL PROPERTIES AND MICROSTRUCTURES OBTAINED AFTER CASTING A MODIFIED AL-ZN-MG WROUGHT ALLOY

Raquel Hidalgo [1]*, Iñaki Hurtado [1], Nuria Herrero-Dorca [1], Jon Ander Esnaola[1], Edurne Ochoa de Zabalegui[2], Patxi Rodríguez[2], A. Kortabarria[2], A. Rios[3], I. Cia[4]

[1] Mondragon University, Loramendi 4, 20500, Arrasate-Mondragón, Guipúzcoa, Spain
[2] EDERTEK Technology Center of FAGOR EDERLAN Group, Isasi 6, 20500 Arrasate-Mondragón, Guipúzcoa, Spain
[3] FAGOR EDERLAN S.COOP, Torrebaso 7, 20540 Eskoriatza, Guipuzcoa, Spain
[4] MAPSA S.COOP, Ctra Echauri 11, 31160 Orcoien, Navarra, Spain
**Corresponding address: rhidalgo@mondragon.edu*

Key words: 7xxx aluminium alloy, microalloying, A356 alloy, casting, mechanical properties

Nowadays, the automotive industry has to comply with increasingly strict fuel efficiency standards and emission regulations due to the overall concern for environmental protection. However, at the same time, the weight of the vehicles is increasing because of safety, luxury and performance features. For this reason, the use of aluminium in cars is growing and this trend is expected to continue in future years.

Casting parts represent 64% of the total of aluminium in vehicles [1] and A356 (Al-Si-Mg) alloy is the most widely used due to its good specific strength, excellent castability and high corrosion resistance. However, wrought aluminium alloys are known to exhibit superior mechanical properties compared to casting aluminium alloys. Where possible, therefore, development way would be using wrought alloy compositions as casting alloys if their castability could be improved. Consequently, in order to be able to cast wrought alloys, efforts have been oriented towards special casting technologies and developing novel aluminum alloys [2, 3]. Despite some good results, these techniques are difficult to transfer to the automotive industry due to their complexity and higher costs.

Therefore, from an industrial point of view, composition modification of existing commercial alloys to adapt them to traditional casting processes would be a better alternative.

Among commercial aluminium alloys, 7xxx series (Al-Zn-Mg-Cu) are considered as excellent candidate materials for high-duty structural automotive applications because of their high strength-to-weight ratio and good ductility. However, copper in aluminium alloys can produce corrosion problems. Avoiding copper, within the 7xxx series, there are two candidates: 7005 and 7072 series. Since 7005 alloy presents better properties, it will be the basis for the alloy to be developed in this study. Zn and Mg are the main alloying elements in this alloy and $MgZn_2$ (η and η´ phases) the principal strengthening precipitate.

Zr and Ti may also be added in order to improve the mechanical properties. The addition of Zr up to 0.19 wt% can increase the strength and fracture properties of the alloys due to the formation of Al_3Zr phase [4]. Ti is well known for its effect on grain refinement. In an Al-5Mg-1.5Si-0.6Mn alloy, addition of 0.20 wt% increase the elongation in the as-cast condition from 11% to 18% , the yield strength from 136 MPa to 157 MPa and the ultimate tensile strength from 296 MPa to 308 MPa. The improved mechanical performance can be attributed to the reduced tendency for hot tearing due to Ti addition [5].

In the present work, an Al-5Zn-1.5Mg alloy has been cast in a permanent mould with Zr, TiBAl and Ti additions in order to study their effects on the final mechanical properties, both in as cast and heat treated conditions. Afterwards, those properties were compared with the ones obtained in an A365 alloy cast in the same conditions.

1. Experimental procedure

The Al-Zn-Mg alloy was prepared from 99 wt.% pure Al, adding Al-50 wt%. Zn master alloy and pure Mg. Four alloys were cast and their compositions are shown in Table 1.

Table 1. Chemical composition (wt%) of the alloys.

	Zn	Mg	Zr	Ti	Si	Al
Base Alloy	4.96	1.47	0.01	0.01	0.13	bal
Base Alloy +Zr	4.99	1.46	0.1	0.01	0.13	bal
Base Alloy +Zr+ TiBAl	5.05	1.5	0.09	0.02	0.14	bal
Base Alloy +Zr+ TiBAl+Ti	5.03	1.45	0.1	0.09	0.14	bal

The composition of each alloy was chosen with the purpose of studying the effect of each alloying element separately.

Tensile bars were cast for each composition. It is worth mentioning that during the casting, no feeding problem was detected and no cracks were visible in the specimens even under X-Ray inspection (Fig. 1).

The specimens were subjected to a T6 heat treatment consisting on solution treating at 480°C for 1 hour followed by quenching and ageing at 150°C for 3 hours.

2. Microstructure

As cast

The intermetallic phases identified with the SEM-EDS analysis are shown in Fig. 2. There are four different phases in the alloys: T (AlMgZn(Cu)), Mg_2Zn, Mg_2Si and $Al_3Fe(Mn)$. There are no visible phases containing Zr or Ti.

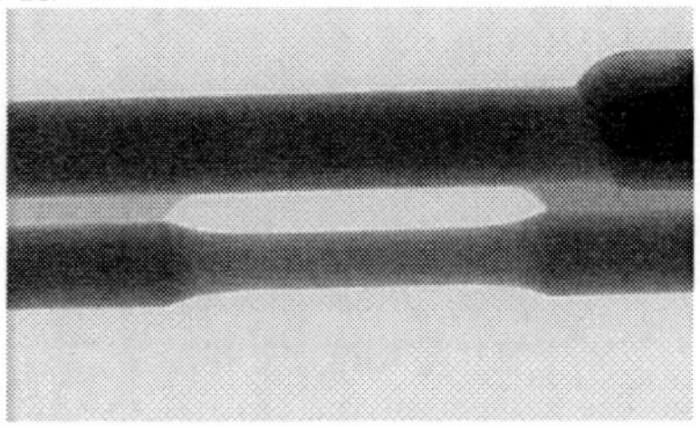

Fig. 1. Foto rayos X

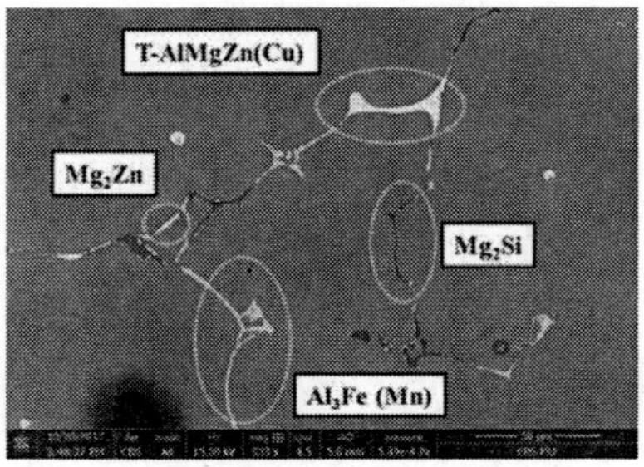

Fig. 2. Phase identified in as cast alloys.

After heat treatment

After heat treatment (480°C for 1 hour followed by quenching and ageing at 150°C for 3 hours), the only visible phases in SEM are Mg_2Si and Al_3Fe (Fig. 3). The Zn-containing phases have been dissolved and precipitated again with a size smaller than the SEM resolution.

3. Hardness

Fig. 4 shows the Brinell hardness of the alloys. As can be observed, after heat treatment, alloys with Zr, TiBAl and Ti additions present higher Brinell hardness than the A356 aluminium alloy after T6.

4. Conclusions

- A base Al-Zn-Mg alloy with different Zr, TiBAl and Ti additions were cast with no feeding problems and no visible cracks in the specimens.
- The phases identified in the as cast microstructure are T (AlMgZn(Cu)), Mg_2Zn, Mg_2Si and $Al_3Fe(Mn)$.
- After a heat treatment consisting of solution at 480°C for 1 hour followed by quenching and ageing at 150°C for 3 hours, T (AlMgZn(Cu)) and Mg_2Zn were dissolved.
- After heat treatment, alloys with Zr, TiBAl and Ti additions present higher Brinell hardness than an A356 aluminium alloy after T6.

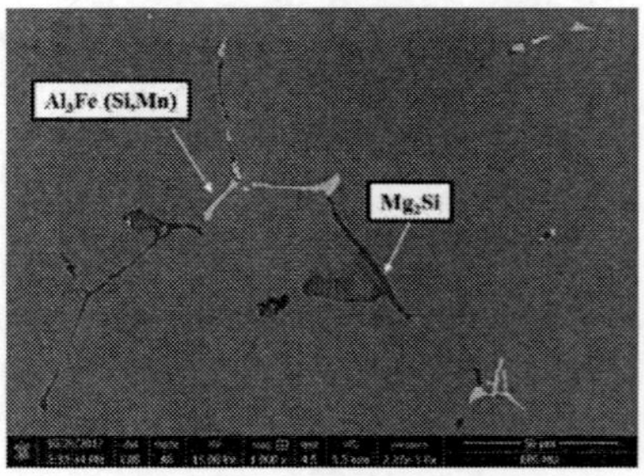

Fig. 3. Phase identified after heat treatment.

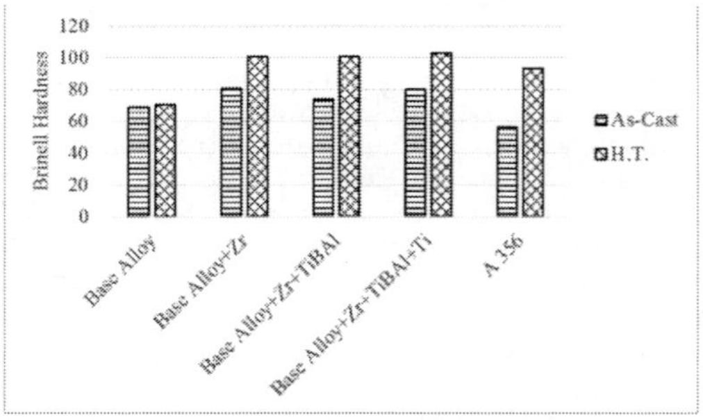

Fig. 4. Brinell Hardness of the alloys.

References

[1] D. Worldwide, "Aluminum Content in Cars," 2016.

[2] R. Ghiaasiaan, X. Zeng, and S. Shankar, "Controlled Diffusion Solidification (CDS) of Al-Zn-Mg-Cu (7050): Microstructure, heat treatment and mechanical properties," *Mater. Sci. Eng. A*, vol. 594, pp. 260–277, 2014.

[3] H. Yu, M. Wang, Y. Jia, Z. Xiao, C. Chen, Q. Lei, Z. Li, W. Chen, H. Zhang, Y. Wang, and C. Cai, "High strength and large ductility in spray-deposited Al – Zn – Mg – Cu alloys," *J. Alloys Compd.*, vol. 601, pp. 120–125, 2014.

[4] J. Zhang, D. Ding, W. Zhang, and S. Kang, "Effect of Zr addition on microstructure and properties of Al–Mn–Si–Zn-based alloy," *Trans. Nonferrous Met. Soc. China*, vol. 24, no. 12, pp. 3872–3878, 2014.

[5] S. Ji, D. Watson, and Y. Wang, "Effect of Ti Addition on Mechanical Properties of High Pressure Die Cast Al-Mg-Si Alloys," *Mater. Sci. Forum*, vol. 765, pp. 23–27, 2013.

[6] J. Shin, T. Kim, D. Kim, D. Kim, and K. Kim, "Castability and mechanical properties of new 7xxx aluminum alloys for automotive chassis / body applications," *J. Alloys Compd.*, vol. 698, pp. 577–590, 2017.

The 73rd World Foundry Congress
„Creative Foundry"
23 - 27 September 2018, Krakow, Poland

EFFECT OF SINGLE AND DOUBLE SOLUTION TREATMENT ON SECOND PHASE DISSOLUTION OF AN AL-8.5SI-1.5CU-0.4MG ALLOY WITH DIFFERENT INITIAL COOLING RATES

Raquel Hidalgo [1*], Iñaki Hurtado [1], Nuria Herrero-Dorca [1], Jon Ander Esnaola[1], Edurne Ochoa de Zabalegui[2], Patxi Rodríguez[2], A. Kortabarria[2], A. Rios[3], I.Cia[4]

[1] *Mondragon University, Loramendi 4, 20500, Arrasate-Mondragón, Guipúzcoa, Spain*
[2] *EDERTEK Technology Center of FAGOR EDERLAN Group, Isasi 6, 20500 Arrasate-Mondragón, Guipúzcoa, Spain*
[3] *FAGOR EDERLAN S.COOP, Torrebaso 7, 20540 Eskoriatza, Guipuzcoa, Spain*
[4] *MAPSA S.COOP, Ctra Echauri 11, 31160 Orcoien, Navarra, Spain*
**Corresponding address: rhidalgo@mondragon.edu*

Key words: solution treatment, cooling rate, aluminum alloy, dissolution, microstructure.

Global trends toward CO2 reduction are making aluminium alloys one of the most attractive material for this purpose. Within the casting aluminium alloys, A356 (Al-7Si-0.3Mg) is the most widely used in automotive and aeronautical industries due to its high ductility. However, A319 (Al-7Si-4Cu) has greater tensile strength. If both properties are required, a mix between the above mentioned properties would be necessary, which could be obtained with the combination of both in the Al-Si-Cu-Mg alloy system. Nevertheless, in order to get good properties, composition is not the only important parameter to take into account; heat treatment is another variable that has an effect on the final behavior of the alloys.

With a view to improving the mechanical properties, solution heat treatment is routinely carried out to maximize the concentration of the age-hardening constituent in solid solution, together with the homogenization of the casting, and the modification of the morphology of the eutectic silicon particles. The control of the solution-treatment temperature is very critical because, if the melting point is exceeded, there is localized melting at the grain boundaries and, thus, the mechanical properties are reduced. In an Al-Si-Cu-Mg alloy, the Cu containing phases are expected to melt at 507 °C. That is the reason why most of the recommended heat treatments of this alloys restrict the solution temperature below the final solidification point so as to avoid the melting of copper-containing phases [1].

A two-step solution treatment suggested by Sokolowski *et al* [2] is reported to significantly reduce and refine the amount of the as-cast copper-rich phase in 319 alloys, giving rise to better homogenization prior to ageing, and thus improving their mechanical properties [3]. Their results show that a two-step solution treatment of 495°C/2h followed by 515°C/4h produce the optimum combination of strength and ductility compared to the traditional single-step solution treatment of 495°C/8h [4].

Another key role in the solution step is the size of the intermetallic phases. The larger they are, the more difficult it would be to dissolve them. This size is related to the solidification rate of the alloys. A higher cooling rate will produce smaller intermetallic particles that would be easier to dissolve, even when the quantity could be higher.

In this study, the effect of the microstructure and single and double solution heat treatments on the phase dissolution are studied for an Al-8.5Si-1.5Cu-0.4Mg alloy for two different initial cooling rates.

1. Experimental procedure

The Al-Si-Cu-Mg alloy was prepared from an A356 alloy, adding Al-50 wt% Cu master alloy and pure Mg and Si. The alloy was modified with 200 ppm Sr using an Al-10 wt% Sr master alloy and TiBAl was added to get an appropriate grain refinement. The chemical composition of the alloy is shown in Table 1.

Table 1. Chemical composition (wt.%) of the alloy.

Si	Cu	Mg	Fe	Mn	Ti	Sr	Al
8.5	1.6	0.34	0.12	0.005	0.11	0.025	bal

Two types of samples were obtained, one with SDAS=35 μm and another with SDAS= 16 μm.

In this study, two types of solution heat-treatment were carried out; Single Solution Treatment (SST) at 505°C for 4, 6 and 8h and Double Solution Treatment (DST) consisting in 505°C for 6 h. followed by a second step at 520°C for 2h (Fig. 1).

2. Thermodynamic calculations

Thermodynamic FactSage software (CALPHAD method) was used to predict the phases formed during the solidification of the alloy. As can be seen in Fig. 2, based on an equilibrium cooling condition; the solidification of this alloy would finish at 530°C. The phases predicted are θ (Al_2Cu), Q phase ($Al_3Mg_9Si_7Cu_2$), π phase ($Al_8FeMg_3Si_5$) and β-Fe phase (Al_5FeSi).

3. Phase identification with SEM

The morphology and chemical composition of intermetallic phases formed during solidification were investigated by means of SEM-EDS analysis.

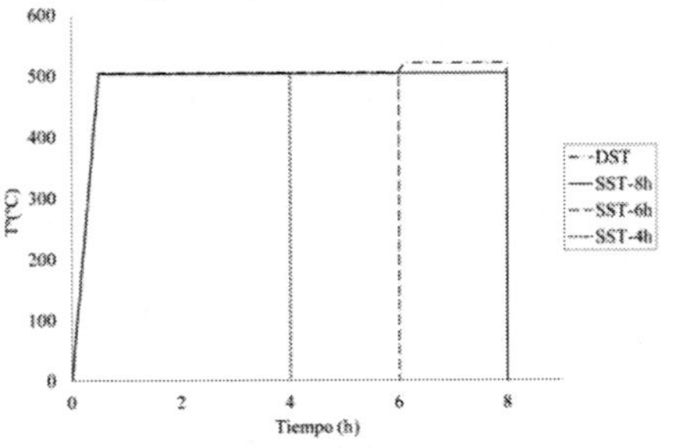

Fig. 1. Diagram showing the solution treatments applied to the alloy.

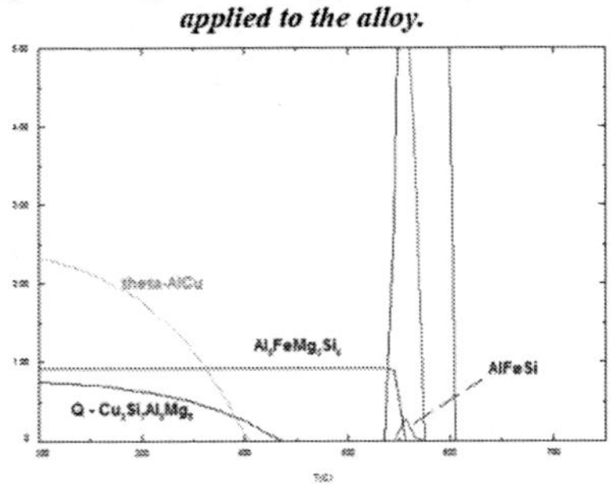

Fig. 2. Phase equilibria calculated with FactSage for Al-8.5Si-1.5Cu-0.4Mg-0.1Fe

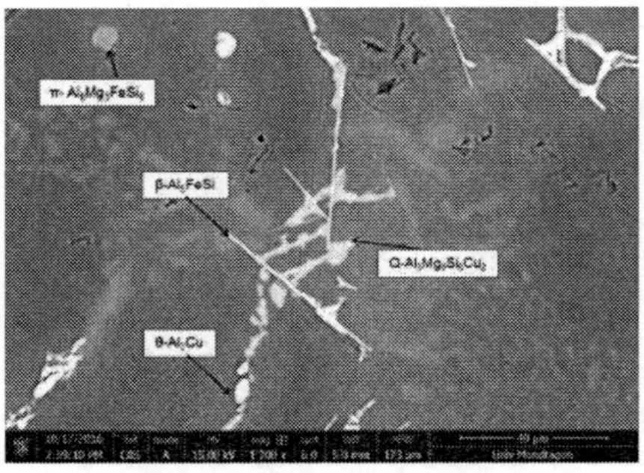

Fig. 3. SEM micrograph showing the as-cast microstructure of the developed alloy SDAS=35 μm.

In Fig. 3 the blocky bright white particles and the gray intermetallics with Chinese script morphology are identified as the θ (Al_2Cu) and Q ($Al_3Mg_9Si_7Cu_2$) phases respectively. The π phase ($Al_8FeMg_3Si_5$) appears as dark gray intermetallics and β (Al_5FeSi), which is the other Fe-containing phase, has a plate-like shape but it is seen as needle like form in the metallographic section.

It is worth mentioning that all phases identified in the microstructure were well predicted by the thermodynamic calculation in equilibrium (Fig. 2).

4. Phase dissolution

As was mentioned before, two types of solution treatment were chosen (Fig. 1). Table 2 shows the different intermetallic phases present after each heat treatment and for each cooling condition.

In the alloy with SDAS=16 μm, a solution treatment consisting in 505°C for 6 hours is enough for dissolve θ phase whereas it is necessary 8 hours for SDAS=35 μm.

The use of a two step solution treatment (505°C for 6 h followed by 520°C for 2h) allows to dissolve Q phase in alloy with SDAS=16 μm. However, it remains stable in alloy with SDAS=35 μm, primarily in zones around microshrinkage.

Table 2. Effect of different solution treatment (Single Solution Treatment SST for different times, Double Solution Treatment) on the intermetallic phase dissolution for two cooling rates.

	SST 4h	SST 6h	SST 8h	DST
SDAS 16 μm	θ Q β	Q β	Q β	β
SDAS 35 μm	θ Q β	θ Q β	θ Q β	Q β

5. Conclusions

- The phases predicted by the thermodynamic calculation are θ (Al_2Cu), Q phase ($Al_3Mg_9Si_7Cu_2$), π phase ($Al_8FeMg_3Si_5$) and β-Fe phase (Al_5FeSi) which are in good agreement with the SEM observations.
- In the alloy with 16 μm SDAS, a SST at 505°C for 6 hours is enough to dissolve θ phase while a DST (505°C for 6 h followed by a second step at 520°C for 2h) is necessary to dissolve the Q phase. However, in the alloy with 35 μm SDAS θ is dissolve for a DST and Q phase remains stable primarily in zones around microshrinkage.

References

[1] F.H. Samuel, "Incipient melting of Al5Mg8Si6Cu2 and Al2Cu intermetallics in unmodified and strontium-modified Al – Si – Cu – Mg (319) alloys during solution heat treatment," *J. Mater. Sci.*, vol. 33, pp. 2283–2297, 1998.

[2] J.H. Sokolowski, X.C. Sun, G. Byczynski, D.O. Northwood, D.E. Penrod, R. Thomas, and A. Esseltine, "The removal of copper-phase segregation and the subsequent improvement in mechanical properties of cast 319 aluminium alloys by a two-stage solution heat treatment," *J. Mater. Process. Tech.*, vol. 53, no. 1–2, pp. 385–392, 1995.

[3] A.M. Samuel, H.W. Doty, S. Valtierra, and F.H. Samuel, "Defects related to incipient melting in Al – Si – Cu – Mg alloys," *Mater. Des.*, vol. 52, pp. 947–956, 2013.

[4] A.M.A. Mohamed and F.H. Samuel, *Heat Treatment-Conventional and Novel Applications*. 2012.

The 73rd World Foundry Congress
„Creative Foundry"
23 - 27 September 2018, Krakow, Poland

EFFECT OF MICRO-SCALE GAS BUBBLES ON STEEL FILTRATION: A NUMERICAL STUDY

Amjad Asad *, Rüdiger Schwarze

Technische Universität Bergakademie Freiberg, Institute of Mechanic and Fluid Dynamics, Lampadiusstraße 4, 09599 Freiberg, Germany
**Corresponding address: Amjad.Asad@imfd.tu-freiberg.de*

Key words: Filtering, Non-metallic inclusions, Simulation, Induction furnace, Open FOAM

1. Introduction

Non-metallic inclusions arise in the steel melt during casting operations. This may reduce the strength and the machinability of the steel product. Therefore, inclusion removal plays a significant role in order to enhance the quality of steel. Using a ceramic filter is an innovative way to remove the non-metallic inclusions before they are trapped by the solidifying shell. The capability and the efficiency of the ceramic filter are studied at Technische Universität Berakademie Freiberg in terms of the Collaborative Research Center (CRC 920) by operating a real steel casting simulator [1-2].

In the present work, the numerical model adopted in Asad et al. [3, 4] is extended to account for the possible formation of small bubbles at the filter walls and at the inclusions surface. For this purpose, two numerical models are created in order to consider the contribution of the CO-bubbles in the inclusion removal. The results show that considering the formation of CO-bubbles at the inclusions' surface leads to an increase in the number of removed inclusions after a short immersion time of the filter (10 seconds).

2. Numerical model

In both models, it is assumed that the volume of the bubble V_b increases as a linear function of the time t due to different reactions in melt, e.g. diffusion of carbon (C). $V_b = c\ t$, where c is the growing rate of the volume of arising CO-bubbles.

2.1 Bubbles arising on inclusions surface (model 1)

In model 1, the CO-bubbles arise directly on the rough surface of an inclusion, which can grow up and bring the non-metallic inclusions to the free surface of the melt. This assumption is sketched in Fig. 1. In this model, it is assumed that the initial volume of the bubble is equal to zero and then it increases linearly until it reaches the maximum prescribed volume. The agglomerate of inclusion and the CO-bubble are represented by an equivalent particle that has an effective mass m_{eff}, effective volume V_{eff} and effective density ρ_{eff}.

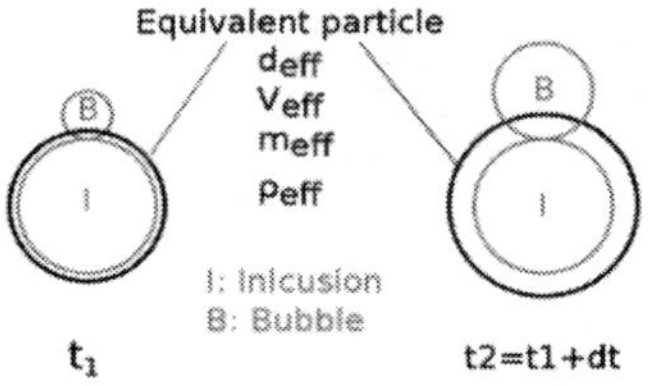

Fig. 1. Sketch to clarify the numerical model 1.

2.2 Bubbles arising at filter walls (model 2)

The model 2 is based on the assumption that the CO-bubbles arise on the filter wall because of the high roughness of the filter. Then, these bubbles leave the filter when their buoyancy is high enough to overcome the high surface tension of the steel melt. The bubbles catch non-metallic inclusions and bring them to the free surface of the melt. In this model, the inclusion is assumed to have been removed from the melt and deleted from the simulation, as they touch the CO-bubbles.

3. Results and Discussion

Four different cases are investigated for different growth rates c of the volume of the CO-bubble (see Table 1). Here, $d_{b,max}$ stands for the maximal diameter of the CO-bubble that can be reached at the end of the simulation (after 10 seconds).

Table 1. The cases investigated in the present article

	c [$\mu m^3\ s^{-1}$] x 10^3	$d_{b,max}$ [μm]
Case 1	52	100
Case 2	177	150
Case 3	144	200
Case 4	818	250

Fig. 2 shows the distribution of the equivalent particles in the melt after 5 seconds of starting the CO-bubbles formation on the surface inclusions for two cases. It is evident from Fig. 2 that the equivalent particles can reach the top surface of the melt and leave the lower part of the melt faster in Case 4 compared to Case 1. The ratio of the number of removed inclusions to the initial number of the inclusion, inserted at the beginning of the simulation, $\eta = n_f / n_t$ depends considerably on the constant c.

Fig. 2. The distribution of the equivalent particles after 5 s in case of model 1 for two cases (see Table 1): case 1 (top), case 4 (bottom)

In Fig. 3, it can be seen that the increase of η depends strongly on c. In Case 4, the number of removed inclusions increases considerably faster than in other cases. This can be attributed to the fast increase in the effective volume of the equivalent particles and the corresponding buoyancy force. That means that the inclusions can reach the top surface faster. The increase of η, observed by adopting model 1, the formation of CO bubbles in direct contact with the inclusions, may explain the huge difference between the experiment conducted by Storti et al. [5] and the simulation performed by Asad et al. [3-4].

In the simulation setup with model 2, CO-bubbles with a diameter of 2 μm are injected at the walls of the filter from 32 points with a rate of 50 bubble/second. As can be seen in Table 2, the values of η obtained at the end of the simulation (after 10 seconds), are much smaller compared to those obtained in model 1. The reason for this fact is that the larger CO-bubbles rise rapidly to the top surface of the melt and do not follow the melt flow. Therefore, they are not able to catch many inclusions on the way to the top surface. Although the smaller CO-bubbles follow the melt flow better, they cannot catch too many inclusions because the contact probability between the bubbles and the inclusions is rather low. Moreover, it must be mentioned here that the distribution of the bubbles in the flow fields depends strongly on the flow conditions as the bubbles are injected.

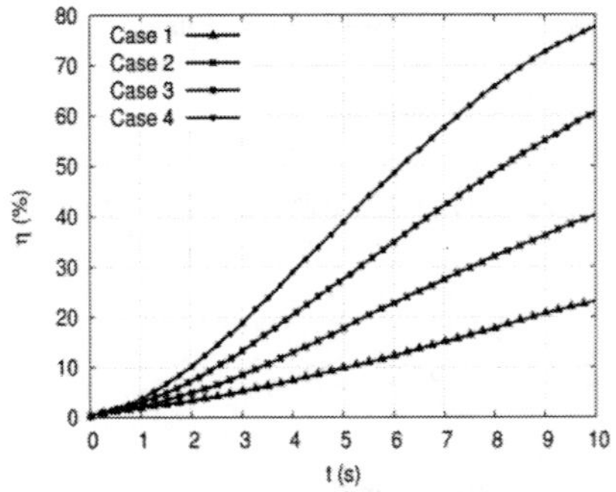

Fig. 3. The behavior of the ratio of the number of the removed inclusions to the total initial number of the inclusions η (see Table 1).

Table 2 The ratio of the number of removed inclusions to the initial number of the inserted inclusions η for the studied cases (see Table 1).

	η [%]	
	Model 1	Model 2
Case 1	23	0.05
Case 2	40	0.12
Case 3	61	0.2
Case 4	78	0.32

Acknowledgments

The authors would like to thank the German Research Foundation for financially supporting the subproject B06.

References

[1] C. G. Aneziris, S. Dudczig, J. Hubálková, M. Emmel & G. Schmidt. (2013). Alumina coatings on carbon bonded alumina nozzles for active filtration of steel melts. Ceramics international. Volume 39 (3), 2835-2843.

[2] S. Dudczig, C. G. Aneziris, M. Dopita & D. Rafaja. (2013). Application of Oxide Coatings for Improved Steel Filtration with the Aid of a Metal Casting Simulator. Advanced Engineering Materials. Volume 12 (15). 1177- 1187.

[3] A. Asad, C. Kratzsch, S. Dudczig, C. G. Aneziris, R. Schwarze, Numerical study of particle filtration in an induction crucible furnace. (2016). International Journal of Heat and Fluid Flow. Volume 62. 299-312.

[4] A. Asad, E. Werzner, C. Demuth, S. Dudczig, A. Schmidt, S. Ray, C. G. Aneziris & R. Schwarze. (2017). Numerical Modeling of Flow Conditions during Steel Filtration Experiments. Advanced Engineering Materials. Volume 19 (9).

[5] E. Storti, S. Dudczig, A. Schmidt, G. Schmidt, & C. G. Aneziris. (2017). Filter Functionalization with Carbon Nanotubes and alumina Nanosheets for Advanced Steel Filtration. Volume 88 (10).

The 73rd World Foundry Congress
„Creative Foundry"
23 - 27 September 2018, Krakow, Poland

3-DIMENSIONAL IMAGING OF SPHEROIDAL GRAPHITE BY ULTRA-HIGH VOLTAGE ELECTRON MICROSCOPY

Hidefumi Maeda [1*], Kanako Inoue [2], Hidehiro Yasuda [2], Akira Sugiyama [3]

[1]*Ryukoku University, Faculty of Science & Technology, 1-5 Yokotani, Ooe-cyo, Seta, Ootsu, Shiga, Japan*
[2] *Osaka University, Research Center for Ultra-High Voltage Electron Microscopy, 7-1 Mihogaoka, Ibaraki, Osaka, Japan*
[3] *Osaka Sangyo University, Graduate School of Engineering, 3-3-1 Nakakakiuchi, Daito, Osaka, Japan*

**Corresponding address: hidefumi@rins.ryukoku.ac.jp*

Key words: UHVEM, Three-dimensional imaging, Internal structure of spheroidal graphite, Spheroidal graphite cast iron

1. Introduction

The morphology and the distribution of the spheroidal graphite in spheroidal graphite cast iron is important to improve its mechanical property. The spheroidal graphite cast iron which supports bases of our life at various places, was produced by effective and minute control during spheroidal graphite formation. A lot of scientific report about the formation theory of spheroidal graphite has been published, but the theory is not yet uniformed [1][2]. To understand the formation of spheroidal graphite correctly, it is significant to visualize an internal structure of the whole spheroidal graphite 3-dimensionally and to explore the trace of the formation processes.

By using the ultra-high voltage electron microscopy (UHVEM), the transmission images can be taken from spheroidal graphite samples with over 10 micro meters thickness. There are expectations that these specific visual images will provide an understanding of crystal growth in spheroidal graphite formation.

2. Experimental Procedure

The ingots of spheroidal graphite cast iron was prepared by using a 50 kg high-speed high-frequency induction furnace. Regarding a reactivity of Mg, two spheroidizing agent were used, one was 3% and the other was 8% in a Mg composition. The mold was designed according to Y-block test pieces of JIS G5502. In order to change the cooling rate, the leg width of the test pieces was changed in 8, 16, 25, 40 mm. The observation samples were finally cut out about $25 \times 20 \times 10$ μm^3 (w×h×t) from the central region of arbitrary spheroidal graphite by FIB.

The internal structure of spheroidal graphite was observed by a dark field image with the UHVEM (Hitachi H-3000 with a maximum acceleration voltage of 3.0 MeV in Osaka University, Japan). This UHVEM enables high angle inclined observation and it is possible to observe with the tomographic method to analyze the internal structure three-dimensionally. The data was captured in the range of ± 65 degrees every 2 degrees continuously. Under 2.0 MeV and ×3000 magnification condition, internal contrast was retained even at an incident angle of 65°. Also, no change in the internal contrast was observed by long-time beam irradiation while data capture.

3. Results and Discussion

The Fig. 1 was focused on the central region in spheroidal graphite. In the electron microscope images, the contrasts derived from diffraction phenomena and from scattering-absorption phenomena are mainly observed. In this UHVEM, it is possible to separate them by serial inclined observation. For example, inside all the spheroidal graphite, there were weak contrasts looked like a part of annual rings with radial boundaries, as shown in Fig. 1 (arrow 1). These contrasts changed in an intensity depending on the angle of incident beam, so these contrasts were thought to be due to diffraction phenomena. At the same time, there are the radially extending dark contrasts seen on the outer region of spheroidal graphite (arrow 2). These dark contrasts did not blink, so that these were considered to be due to the absorption-scattering phenomena.

The region which have the weak diffraction contrasts (mother phase of graphite sample) has been filled by graphite crystals with nearly circular annual ring-like contrasts of growth trace, and the center of them was considered the crystallographic center of crystal growth. According to the thin film diffraction pattern, it was found that the diffraction spots close to the c-axis of graphite was appeared in the direction of stacking the layer.

The radially extending dark contrasts occurred due to existing inclusions in spheroidal graphite in a cast iron. As a result of analysis by FE-SEM/EDX, main component of the inclusions was iron. It is inferred that these radially extended inclusions formed by excluding carbon from the cast iron components. These inclusions brought the "double structure" inside the spheroidal graphite depending on the condition of casting. It was observed that the spheroidal graphite consisted of two regions with the radial inclusions or without ones. This boundary looks like a shell had a very well spherical shape. The radial inclusions were not existed inside of this spherical shell. In the comparison with various samples made by different casting conditions, it was confirmed that the faster the solidification rate was, the smaller this spherical shell became, and then finally the "double structure" disappeared. The spherical shell of this "double structure" was considered to be an important event for the spheroidization of graphite. These results might be suggested that a solidification morphology had a nearly spherical form when molten metal reached at the eutectic point, and that iron begun to crystallize afterwards. This "double structure" causes the material characteristic difference between the internal and the external region of the spherical shell [3][4].

Furthermore, as shown in Fig. 1, the circular annual ring like contrasts (arrow 1) has an eccentric center on the dark grain contrast (arrow 3) and has the white and short contrasts (arrow 4) between the stacking layers. Even in case of presence of a center nuclear material, it was found that only one limited point on the nuclear material could be a certain starting point of graphite crystal growth. The white contrasts were considered as empty gaps.

4. Conclusion

The UHVEM observation of the internal structure of spheroidal graphite supported an imaging concretely by chasing a trace of a three-dimensional crystal growth. These data provided a number of new cases to be verified in the formation of a spheroidal graphite.

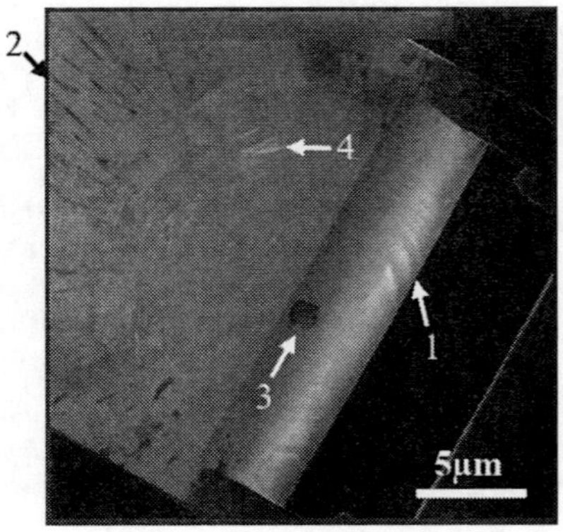

Fig. 1. UHVEM image of a spheroidal graphite. (The line across the sample is a trace of cutting the sample bottom.)

Acknowledgments

We wish to thank the all staff of Sankyo Alloy Foundry Co., Ltd. (Japan) for supplying a lot of experimental samples.

This work was supported by "Advanced Characterization Nanotechnology Platform, Nanotechnology Platform Program of the MEXT, Japan" at the Research Center for Ultra-High Voltage Electron Microscopy (Nanotechnology Open Facilities) in Osaka University.

References

[1] Doru M. Stefanescu. (2005) Solidification and modeling of cast iron-A short history of the defining moments. Mat. Sci. Eng. A. 413-414. 322~333

[2] Jung Sanghoon, et al. (2007). From Invention of Spheroidal Graphite Cast Iron to Today. J. JFS, Vol. 79 (No 10), pp.605~615.

[3] Hideki Masuda, et al. (2015) Cross-sectional observation of ferrum casting ductile using multi-mode SPM. Proc. the 71rd Annual Meeting of JSM, May13 (pp236 PHOTO-9) ICC Kyoto.

[4] KoshiroYamane, et al. (2014). Influence of Mg Content on Graphite Growth in Hypereutectic Cast Iron and Crystallographic Feature of Graphite. J. JFS, Vol.86(No.6). pp.461~470

The 73rd World Foundry Congress
„Creative Foundry"
23 - 27 September 2018, Krakow, Poland

MECHANICAL BEHAVIOUR OF NODULAR CAST IRONS AFTER PROLONGED HIGH TEMPERATURE EXPOSURE

A. Morri [1]*, L. Ceschini [2], S. Masaggia[3]

[1]University of Bologna, Department of Industrial Engineering, V.le Risorgimento 4 40136 Bologna, Italy
[2]University of Bologna, Department of Civil, Chemical, Environmental, and Material Engineering, V.le Risorgimento 2 40136 Bologna, Italy
[3]R & D Department, Zanardi Fonderie S.p.A., Via Nazionale 3, 37046, Minerbe, Italy
*Corresponding address: alessandro.morri4@unibo.it

Key words: Nodular Cast Irons, Microstructure, High Temperature Exposure, Mechanical Properties, ADI

Ductile cast irons (DI) are widely used in industrial applications for their good castability, low cost and high range of mechanical properties that can be tailored for specific applications by changing matrix microstructure. However their use is limited to components working at room temperature, since prolonged exposure at high temperature can lead to decomposition of both ausferrite and perlite, with a consequent strength reduction [1-3]. To date most of the research activity on this topic was mainly focused on investigating microstructural evolution of DIs, rather than establishing relationships among time and temperature of exposure, and final mechanical properties.

The aim of the present paper was to correlate the effects of prolonged high temperature exposure on microstructure and residual strength of ductile cast irons with different matrix microstructures, after an isothermal high temperature exposure (degradation).

1. Materials and Methods

The following materials were investigated: the ferritic ductile iron EN GJS/400-18 (FDI), the pearlitic ductile iron EN GJS/800-2 (PDI), the austempered ductile iron ISO 17804/JS/1050-5 (ADI), the perferritic isothermed ductile iron - Zanardi internal Standard 101:2007 (IDI). In order to evaluate the thermal stability of these cast irons, high temperature exposure tests were carried out by isothermal soaking at temperatures between 200 and 600°C and time up to 300 hours, in a muffle furnace.. Microstructural characterization was carried out with optical microscope (OM). Mechanical properties of degraded samples were evaluated by means of Brinell hardness measurements and tensile tests, both at room and 500°C, according to ASTM E8.

2. Results

Microstructural analyses confirm that the prolonged high temperature exposure leads to the decomposition of the ausferrite and pearlite, while it has no appreciable effects on ferrite. The microstructure of PDI is stable up to 500°C, regardless of time(?), while at higher temperatures pearlite spheroidization firstly occurs (Fig. 1) and then, with prolonged exposure, it decomposes into ferrite and graphite.

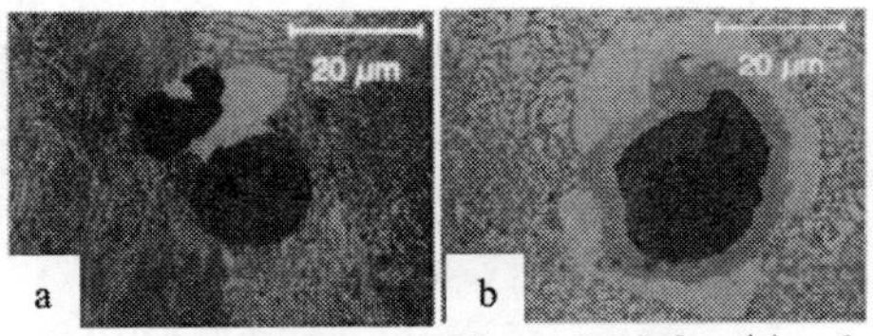

Fig. 1. Microstructure of PDI samples. before (a) and after (b) thermal exposure at 550°C for 120 hours

A similar evolution of the pearlite present in the perferritic microstructure of IDI cast irons has been observed. However they are stable only up to 400°C, since perferritic microstructure has a finer pearlite respect to PDI [4] and is therefore more prone to spheroidization.

In ADI cast irons, decomposition of ausferrite is not appreciable up to 400°C and only samples exposed at 500°C clearly show the decomposition of the carbon-rich austenite into carbides and ferrite (Fig. 2). Mantaining the ADI at high temperature the complete decomposition of the ausferrite leads to a fully ferritic microstructure.

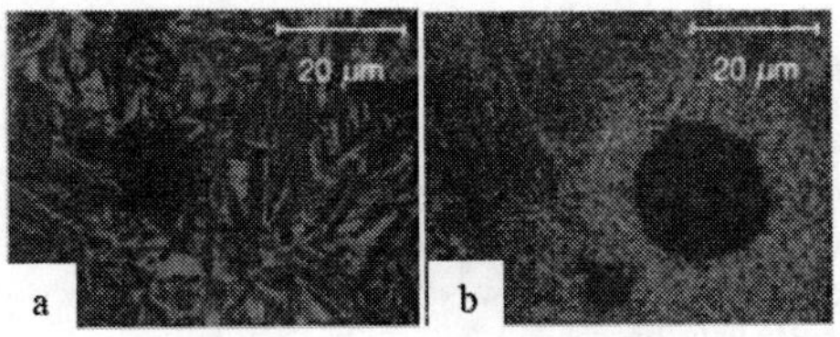

Fig. 2. Microstructure of ADI samples, before (a) and after (b) thermal exposure at 500°C for 72h.

Decomposition of both pearlite and ausferrite is a diffusion controlled process, clearly dependent on both time and temperature exposure.

In order to relate microstructural evolution to mechanical properties after thermal exposure, several samples of the studied ductile cast irons were subjected to prolonged high temperature soaking, and the corresponding hardness variations were evaluated. The results are reported in the hardness-time-temperature curves (HTT) of Fig. 3.

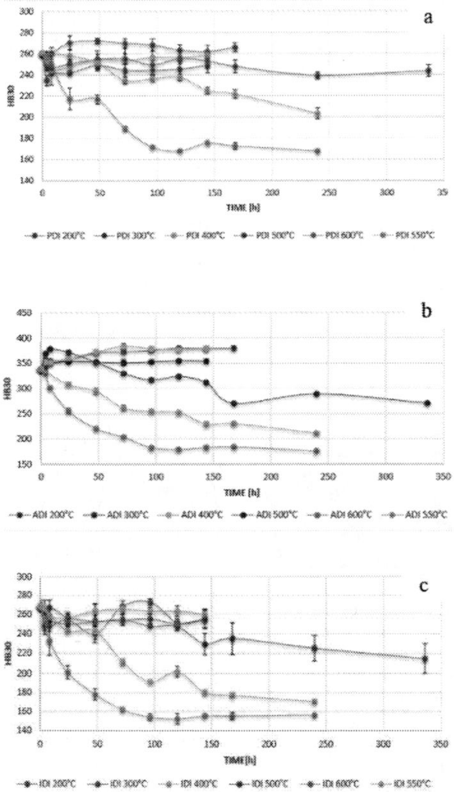

Fig. 3. Hardness-time-temperature curves (HTT) of PDI (a), ADI (b) and IDI (c)

Figure 3 clearly shows that while PDI hardness is substantially unaffected by high temperature exposure up to 500°C, IDI and ADI are stable only up to 400°C and at 500°C, after 240 hours, their hardness decrease is of about 15%.

Results of the tensile tests carried out at both room and 500°C, after thermal exposure at 500°C up to 240h (Fig. 4), highlight a reduction of both yield (YS) and ultimate tensile strength (UTS) at 500°C of about 60% respect to UTS and YS at room temperature, regardless of the matrix microstructure and of the high temperature exposition time the samples have been subjected.

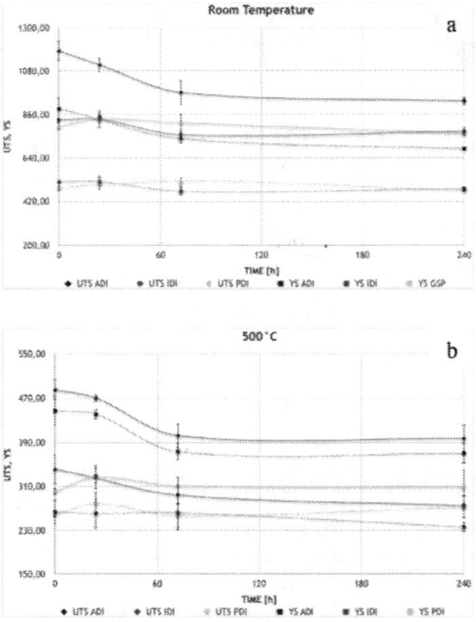

Fig. 4. Results of tensile tests carried out on DIs exposed at 500°C for time up to 240h. Tests at room temperature (a), tests at 500°C (b).

3. Conclusions

Several ductile cast irons with different matrix microstructures (perlitic, ausferritic.....) were compared in terms of their microstructural evolution, residual hardness and tensile properties, after prolonged high temperature exposure.

PDI The highest thermal stability among the tested ductile cast irons was observed in the PDI

Even if ADI displayed the most evident strength reduction after exposure at 500°C, they always showed UTS and YS higher than the others DIs.

References

[1] A. Gazda. (2010). Analysis of decomposition processes of ausferrite in copper–nickel austempered ductile iron. J. Therm Anal Calorim. 102, 923–930.

[2] S. Ghodrat, Leo A. I. Kestens. (2016). Microstructural Dependence of Tensile and Fatigue Properties of Compacted Graphite Iron in Diesel Engine Components. Steel research int. 87 (6), 772-779.

[3] M. Ajabshiri, S. Sharafi, A Moeini. (2012) Comparison of High-Temperature Properties and Thermal Shock Resistance of Austempered Ductile Irons (ADI) with Those of Pearlitic Ductile Cast Irons. Met. Mat Trans, 43A, 56-60

[4] S. Masaggia. (2010). The development of ADI and IDI in Italy. Procedia Engineering 2, 1459–1476.

The 73rd World Foundry Congress
„Creative Foundry"
23 - 27 September 2018, Krakow, Poland

INFLUENCE OF SHRINKAGE POROSITY ON FRACTURE UNDER TENSILE STRESS IN DUCTILE CAST IRON

Marcos López*, Diego Fernandino, Roberto Boeri, Juan Massone

[1] *Metallurgy Division, INTEMA – CONICET, National University of Mar del Plata, 10850 Colón Av., Argentina.*
**Corresponding address: marcosl@fi.mdp.edu.ar*

Key words: SGI, Shrinkage porosity, classification, influence on damage evolution

1. Introduction

One of the major problems affecting the quality of cast parts are shrinkage defects that occur during solidification. When high strength pieces are produced, it becomes critical to ensure that they are structurally sound. Typical classification of shrinkage defects [1] includes three different types of shrinkage cavities: a) the concentrated shrinkage; b) the dispersed macroshrinkage, and c) the dispersed microshrinkage. Concentrated shrinkage forms because of metal contraction during cooling of the liquid and the phase change. It can be open or closed to the atmosphere. The dispersed macroshrinkage consists of cavities of millimeters size resulting from the lack of feeding during solidification. It appears in heavy sections or at hot spots of the casting and can be prevented by an optimization of the riser design. The microshrinkage appears usually dispersed along the whole volume of the casting. It consists of micron-size cavities that are formed between the eutectic grains or dendrite arms. The presence of these micro cavities could affect the mechanical answer of the part.

The macroscopic deformation of ferritic SGI is generally considered to induce a homogeneous local deformation at the microscopic level; however, the experimental evidence shows that the mechanical properties and deformation behavior at microscopic level are heterogeneous. [2-3]. This is caused by the heterogeneity in the mechanical properties of the microconstituents. However, a lack of knowledge about the effect of the microsegregated zones of the metallic matrix and the dispersed microshrinkage on the occurrence and sequence of damage mechanisms is evident.

This work focuses on the study of shrinkage porosity in SGI, identifying the morphology and location of porosity in cast samples especially developed, and assessing the influence of dispersed microshrinkage cavities as a preferential site for cracks initiation and propagation, by means of in-situ tensile tests and Digital Image Correlation (DIC) analysis.

2. Experimental procedure

A test casting was especially designed to develop large and small shrinkage cavities, employing dimensions and geometric relationships that violate the basic design rules for sound castings [4]. Then, a system containing the test casting and feeders was simulated employing the software MagmaSoft®, and modifications in the geometry were introduced with the following objectives:

- To obtain macro and microporosity in the pieces.
- To allow hot shakeout, necessary to apply the DAAS technique [5].
- To allow a correct filling for the whole piece, ensuring a non-turbulent flow.

The designed model is shown in Fig. 1.

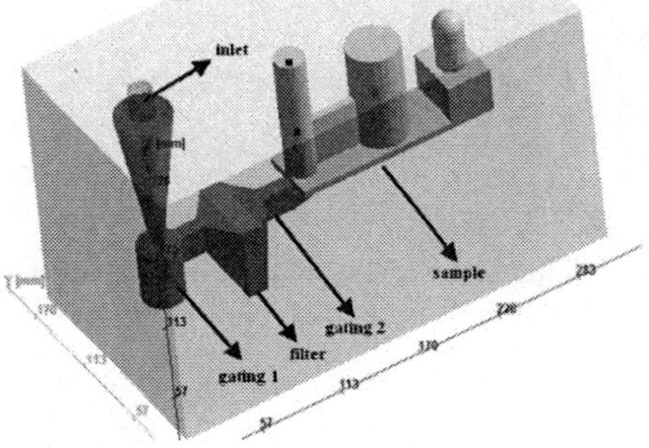

Fig. 1. 3D view of the model.

Melt was prepared using a 50 kg capacity medium frequency induction furnace. Pig iron, steel scrap, ferroalloys and carbon raiser were used as raw materials in order to obtain ductile iron with near-eutectic chemical composition. Molds were made using resin bonded silica sand. Table 1 lists the chemical composition used for the melt.

Table 1: Chemical composition.

%C	%Si	%Mn	%S	%P	%Mg	%Cu	%Ni
3.21	2.76	0.13	0.02	0.03	0.053	0.85	0.53

Castings were sectioned, grinded and polished, looking for shrinkage cavities at the different volumes. From those regions where microshrinkage was revealed, small test samples were machined preserving the defect close to the central part of the test sample. Fig. 2 shows a sketch of the test sample.

The sample was tested under uniaxial loading and the microscopic observations were performed "in-situ" by adapting a Digital microscope camera Leyca MC120HD to a tensile test equipment specially designed for this purpose. Fig. 2 shows the experimental "in situ" tensile test setup.

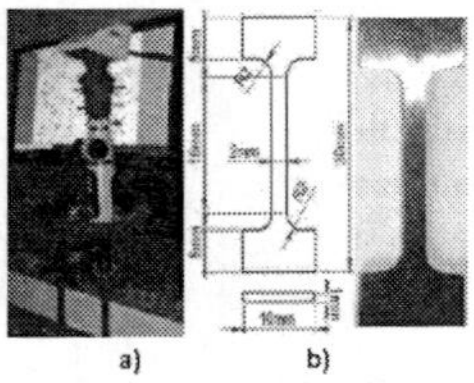

Fig. 2. In situ tensile test. a) Experimental setup, b) Test sample dimensions.

Prior to the tensile test, the sample was etched using a color etchant reagent sensitive to the segregation of Si in order to identify the microsegregated zones. DIC analysis was performed by using the Ncorr open source 2D DIC program presented by Blaber et al (2015) [6]. A pit etching procedure was also applied to the tensile test sample with the objective of improving the resolution of the local strain mapping.

3. Results and discussion

Fig. 3 shows the polished surface of one of the samples and the MagmaSoft simulation of the same section. Defective (D) and other apparently sound (S) areas are recognized.

Fig. 3. Different zones in a specimen and correlation with MagmaSoft simulation.

Zone S, from where test samples were machined, is located near the sample surface and is apparently free from cavities. Nevertheless, when careful microscopic examination is carried out, very small and disperse shrinkage (microshrinkage) cavities can be identified at the last to freeze zones (LTF), as shown in Fig. 4 after color etching.

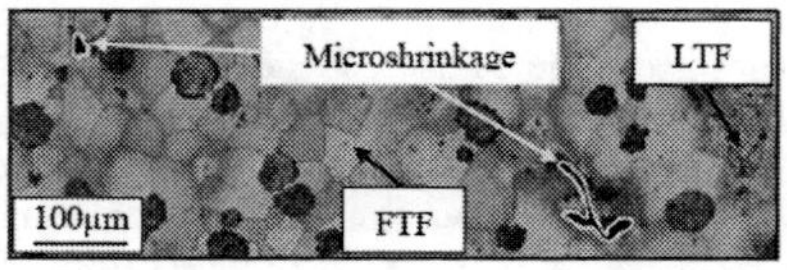

Fig. 4. Optical micrograph after color etching.

Fig. 5 shows the microstructural damage evolution at two different macroscopic strain values (ε). The shrinkage porosity is pointed by yellow arrows. Note the random speckle pattern on the samples used for DIC analysis (e.g. the small dark dots in the square of Fig. 5a). It is possible to observe that the microshrinkage cavities do not act as a main crack initiation site (see Fig. 5a) and that they are not in the main crack propagation path as it is shown in Fig. 5b. From damage evolution and DIC analysis it is possible to support that microshrinkage cavities, in spite of acting as a stress raiser at the beginning of the deformation stage, are not a predominant site for crack initiation and propagation. On the other hand, this has been found to be true for the matrix-nodule interface debonded zones, where the plastic deformation bands, which act as predominant sites for crack initiation and propagation, tend to be localized. Thus, when the strain continues to increase, the cracks propagate across the internodular ligaments.

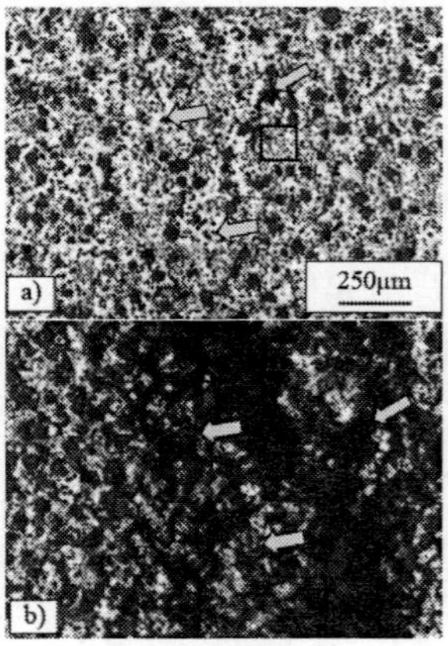

Fig. 5. Optical micrograph of damage evolution. Yellow arrows indicate the shrinkage porosity location a) ε=0%, b) ε=19.8%

4. Conclusions

- It was possible to identify microshrinkage cavities even in zones of the casting apparently sound.
- Microshrinkage cavities are always located at last to freeze zones.
- When regions with microshrinkage are loaded in tension, neither crack initiation nor crack propagation are in relationship with the micro cavities.

References

[1] D.M. Stefanescu, H.Q. Qiu, C.H. Chen. (1995) Effects of Selected Metal and Mold Variables on Dispersed Shrinkage in SG Cast Iron. AFS Transactions. 57, 189-197.

[2] D.O. Fernandino, A.P. Cisilino, S. Toro, P.J. Sánchez. (2017). Multi-scale analysis of the early damage mechanics of ferritic ductile iron. Int. J. Fract. 207, 1–26.

[3] K.A. Kasvayee, K. Salomonsson, E. Ghassemali, A. E.W. Jarfors. (2016). Microstructural strain distribution in ductile iron; comparison between finite element simulation and digital image correlation measurements. Mater Sci Eng. 655, 27–35.

[4] J. Campbell. (2003). Casting. Butterworth-Heinemann, Second Edition.

[5] G. Rivera, R. Boeri, J. Sikora. (2002). Revealing and characterising solidification structure of ductile cast iron. Materials Science and Technology. 18, 691-697.

[6] J. Blaber, B. Adair, A. Antoniou. (2015). Ncorr: Open-Source 2D Digital Image Correlation Matlab Software. Experimental Mechanics.

The 73rd World Foundry Congress
„Creative Foundry"
23 - 27 September 2018, Krakow, Poland

Metallography for cast Al-alloys using FIB-SEM

Tonica Bončina [1*], **Franc Zupanič** [1]

[1]*University of Maribor, Faculty of Mechanical Engineering, Smetanova 17, SI-2000 Maribor, Slovenia*
**Corresponding address: tonica.boncina@um.si*

Key words: aluminium alloy, casting, focussed ion beam, characterization, microstructure

Metallography includes a number of research methods. The list of these methods was systematically described by Exner in 1999 in Practical Metallography [1]. The methods were divided into the metallographic preparation of samples, optical, quantitative, chemical, surface, dynamic and visual metallography, as well as special metallographic methods. Since then, the development of methods, research equipment and knowledge of the characterization of materials has been enormous.

One of the new techniques, which has emerged, is the focused ion beam (FIB), initially used in the microelectronics industry. A milestone in the use of focused ion beam was the installation of the ion column to a scanning electron microscope. Since then, an electron microscope has not been anymore just a tool for visualization of surfaces and for the determination of microchemical composition but also a tool for material processing.

We would like to present the use of a dual-system (FIB/SEM) in metallography of cast Al-alloys for the everyday practical use for analysing, characterising and controlling the quality of aluminium alloys and its applications for the scientific research. The FIB can find its application in the sample preparation and as an imaging tool.
Microstructures of cast aluminium alloys consist of numerous micro- and nanosized microstructural constituents. Several standard methods have been used for metallographic investigations: light microscopy (LM), scanning (SEM) and transmission (TEM) electron microscopy combined with several analytical techniques, and X-ray diffraction (XRD). Additional information regarding the microstructure can be obtained by dual beam microscopy: focussed ion beam (FIB) and electron beam (SEM) [2, 3].

The FIB/SEM system enables cleaning and etching the sample surface, determination of microstructural constituents (size, shape and distribution) at and below the surface, determination of microchemical composition of the microstructural components at and under the surface, determination of corrosion products and corrosion mechanisms, 3D-imaging, failure analysis, preparation of TEM lamellas, nanosamples for the atom probe tomography (APT) and different nanotips for other methods.
There are some characteristics (advantages and disadvantages) of FIB analysis which have to be taken into account. The most important are the analysis of small areas or volumes at a precise location, expensive and time-consuming analysis, non-destructive method (in case of small samples that can be put it into a microscope chamber), 3D-analysis and possibility of contamination of samples with Ga+ ions.

Ion etching and 3D-microscopy

A rough metallographic preparation of samples by grinding and polishing causes the formation of the deformation layer and the change of the microstructure. With the ions, it is possible to remove surface layers at selected places and revealed the microstructure; the process is called ion etching.

With the ion milling, when higher ion current is applied, we can produce a deep hole with one surface perpendicular to the initial mechanically polished surface. This surface is indicated by "FIB cross section" in Fig. 1a, and the initial mechanically surface is denoted by "ion etched" in Fig. 1b. This, so-called, 3D-microscopy reveals 3D-shapes of phases and also indicates a spatial phase distribution. Fig. 1 shows the same cross-section imaged by secondary (Fig. 1a) and backscattered (Fig. 1b) electrons and by secondary electrons induced by ions (Fig. 1c). Each image carries different information. The silicon phase (β-Si) is hardly seen in SE- and BSE-images, while it is easily recognised in FIB-SE-image.

Figure 2a shows a quasicrystalline dendrite of an *i*-phase extracted from the Al-Mn-Be-Cu alloy covered with small particles. The particles on the quasicrystal's surface can be seen, but there are no indications how far they extend inside the particle. Using a combination of particle extraction techniques, FIB cross-sectioning and EDS- analysis we can simultaneously see the morphology of phases, their internal structure (Fig. 2b) and microchemical compositions of the phases. The process is much faster than making cross-sections in bulk material since much less material has to be removed [4]. This method can also be upgraded by 3D-reconstruction and 3D-mapping.

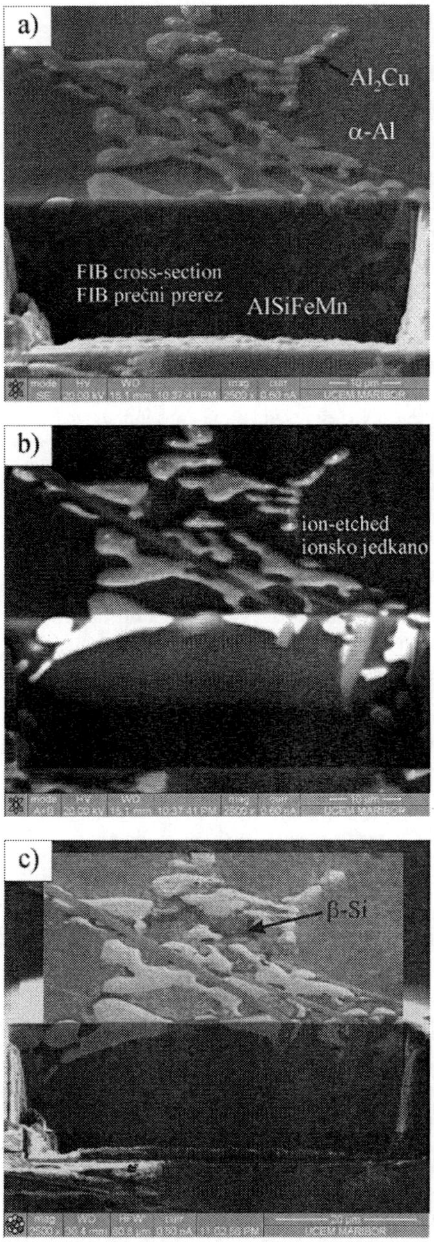

Fig. 1: FIB cross-section of an interdendritic region in the alloy A383: a) SE-image, b) BSE-image and c) FIB-SE-image.

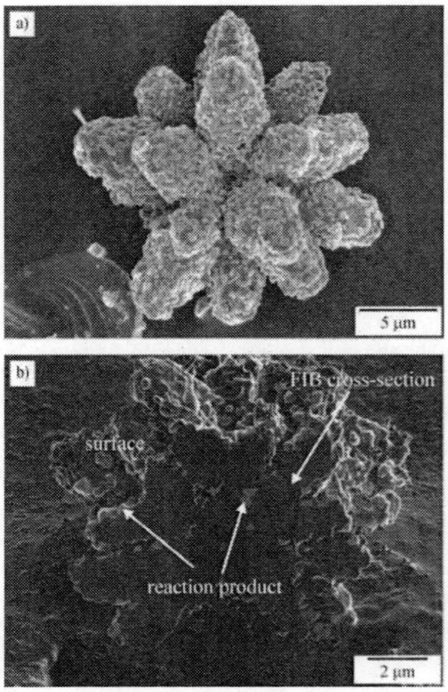

Fig. 2: (a) SEM micrograph of an extracted quasicrystalline particle (alloy Al—Mn—Be—Cu after annealing at 500 °C for 5 hours) covered by the reaction products, and (b) FIB cross-section of the extracted quasicrystalline particle [4].

Acknowledgement

The work was carried out in the framework of the Slovenian smart specialisation, programme Materials and Technologies for New Applications, MARTINA, No. OP20.00369 and Infrastructure programme UM I0-0029 financed by the National Research Agency ARRS.

References

[1] H.E. Exner, Metallography: Technique of metallurgical investigation or the science of describing metals?, PRAKT METAL, 36(3), 1999, pp. 115-137

[2] M. Voncina, J. Medved, T. Boncina, F. Zupanic, Effect of Ce on morphology of alpha(Al)-Al2Cu eutectic in Al-Si-Cu alloy, Trans. Nonferrous Met. Soc. China, 24 (2014) 36-41.

[3] F. Zupanič, B. Markoli, I. Naglič, T. Weingartner, A. Meden, T. Bončina T. Phases in the Al-corner of the Al-Mn-Be system. Microscopy and microanalysis, ISSN 1431-9276. Oct. 2013, vol. 19, pp. 1308-1316.

[4] T. Bončina, B. Markoli, I. Anžel, f. Zupanič. Metallographic techniques for the characterization of quasicrystalline phases in aluminium alloys. V: *ICQ10 Proceedings. Part 1*, 10th International Conference on Quasicrystals, July 6-11, 2008, Zurich, Switzerland, (Zeitschrift für Kristallographie, ISSN 0044-2968, Vol. 223, iss. 11/12, 2008). Zürich: Oldenburg. 2008, pp. 747-750

The 73rd World Foundry Congress
„Creative Foundry"
23 - 27 September 2018, Krakow, Poland

STRENGTHENING OF Al-CASTING ALLOYS BY QUASICRYSTALLINE PRECIPITATES

Franc Zupanič [1]*, Christian Gspan [2], Tonica Bončina [1]

[1] *University of Maribor, Faculty of Mechanical Engineering, Smetanova 17, SI-2000 Maribor, Slovenia*
[2] *Institute of Electron Microscopy and Nanoanalysis, 8010 Graz, Steyrergasse 17/III, Austria*
**Corresponding address: franc.zupanic@um.si*

Keywords: aluminium, quasicrystal, precipitation hardening, hardness, microscopy

1. Introduction

Aluminium quasicrystalline alloys belong to a novel class of high-strength alloys. Initially, they have been. by rapid solidification [1, 2]. Recently, they were also manufactured by several conventional casting methods, provided that the solidification rates exceed few hundred K/s [3]. In such cases, an icosahedral quasicrystalline phase (IQC) forms during solidification, either as a primary phase or as a constituent of a binary eutectic (Al_{ss} + IQC), where Al_{ss} stands for the aluminium-rich solid solution.

The strengthening effect arises from the supersaturated solid-solution and from the presence of hard and relatively uniformly distributed quasicrystalline particles, which attain sizes from few hundred nanometres up to few micrometres. It has been found that IQC-phase can also be formed during heat treatment. In this case, it precipitates from the Al_{ss} with the shape of spherical precipitates with sizes in the range between 10 and 50 nm in the Al-Mn-Cu-Be alloys [3]. In this case, an additional hardening is obtained by precipitation hardening.

In the present investigation, the effects of heat treatment conditions, and pre-deformation of the alloys on the precipitation behaviour have been studied. A detailed analytical transmission electron microscopy study has been carried out using different imaging modes, and several nanoanalytical techniques.

2. Experimental

The investigated alloys contained 3.64 wt% Mn, 0.46 wt% Be, 3.50 wt% Cu and 92.40 wt% Al ($Al_{95.5}Mn_{1.8}Be_{1.2}Cu_{1.5}$). The chemical composition was determined using AES-ICP (Atomic Emission Spectroscopy – Inductively Coupled Plasma). The alloy was prepared using vacuum induction melting in an alumina crucible and cast into a steel mould. Afterwards, the alloy was cast into a copper mould to achieve cooling rates around 1000 K/s.

In the first series, the as-cast samples were heat-treated by artificial ageing at 200°C, 300°C, 400°C and 500°C for 24 hours in the air (T5 treatment). Afterward, they were heat-treated at 300°C and 400°C for different periods of time to obtain the ageing curves.

The lamellae from the as-cast and heat-treated specimens for transmission electron microscopy (TEM) were either prepared using focused ion beam (FIB by FEI, Helios) or by precise ion polishing (PIPS I, GATAN). High-resolution TEM (HRTEM), energy-filtered TEM (EFTEM), scanning TEM (STEM), electron energy-loss spectroscopy (EELS) and energy-dispersive X-ray spectroscopy (EDXS) were carried out in an FEI Titan 60-300 image corrected electron microscope. The indentation hardness tests were carried out using a Fisherscope H100C testing machine (Fisher Technology). The loading and unloading times were 10 seconds each. The maximum load applied was 1 N.

3. Results

After the gravitational casting of the alloy into the copper mould, the microstructure consisted of dendritic or cellular Al-rich solid solution with (α-Al + IQC) and (α_{Al} + Θ-Al_2Cu) eutectics in the interdendritic regions.

In this abstract we only give the main results of heat treatment of the investigated alloy after heat treatment at 400 °C. The hardness of the alloy in the as-cast condition was around 130 HV. During the heat treatment, the hardness first dropped, but after 10 minutes started to increase, and attained approximately 125 HV after one h. Further annealing up to 24 h did not decrease the hardness substantially. The as-cast hardness probably arises from the distribution of quasicrystalline phase in the interdendritic region, and from the solution hardening.

During the heat treatment, the solution hardening started to decrease, and after some period precipitates began to form, increasing the hardness. This result shows that the hardness of the alloy does not change significantly. It can be stated that the hardness is not increased in comparison to the initial as-cast state. On the other hand, it is clear, that the alloy possesses excellent heat resistance. Namely, the hardness of many aluminium alloys drops significantly when heated to 400°C. This is true for both wrought and precipitation hardening aluminium alloys.

It seems that these alloys have an excellent potential for high-temperature application.
The measuring of other mechanical properties at elevated, such as compression and creep tests, is ongoing.

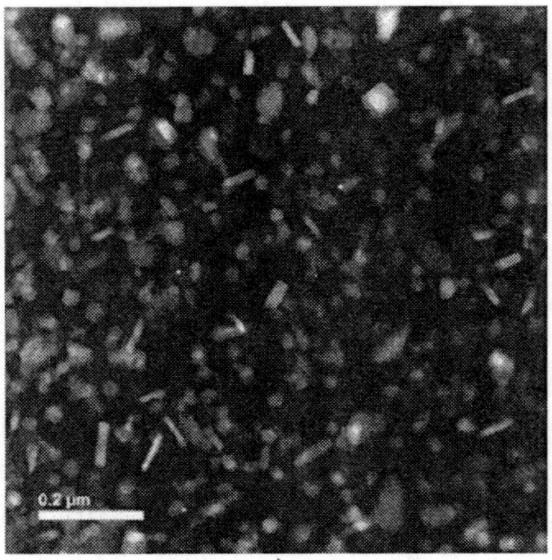

a)

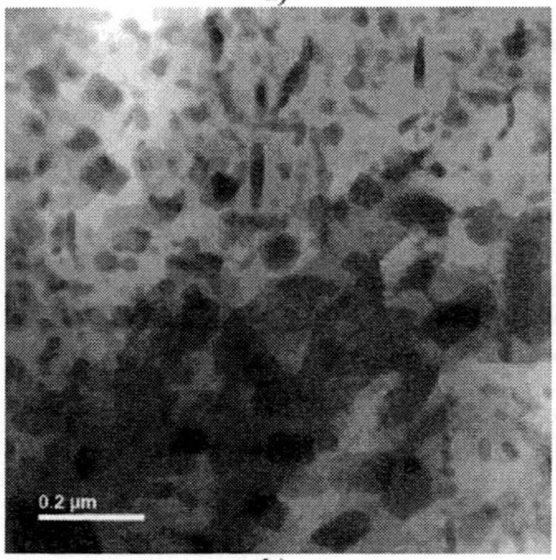

b)

Fig. 1. Transmission electron micrographs of the gravitationally cast sample of the alloy $Al_{95.5}Mn_{1.8}Be_{1.2}Cu_{1.5}$ after one h heat treatment at 400°C a) without pre-deformation, b) with pre-deformation

Fig. 1a shows the transmission electron micrograph after one h of heat treatment at 400 °C. After this heat treatment, the highest hardness was attained in the gravitationally cast alloys. The icosahedral phase formed during solidification was in the intercellular spaces. In the centres of dendrites, there were plenty of precipitates. The elongated precipitates were identified as T-$Al_{20}Mn_3Cu_2$, and the spherical precipitates have a structure of the icosahedral quasicrystal. The number density of precipitates was rather high, and they were slightly coarser than in the melt-spun ribbons.

The pre-deformation caused the formation of slightly coarser precipitates (Fig. 1b).

4. Conclusions

The peak hardness of the gravitationally cast cylindrical castings of the alloy $Al_{95.5}Mn_{1.8}Be_{1.2}Cu_{1.5}$ occurred after one h of ageing at 400 °C.

Two types of precipitates in the microstructure: T-phase ($Al_{20}Mn_3Cu_2$) and IQC. The precipitates were slightly finer when the alloy was directly aged after casting (T5 treatment). When the alloy was plastically deformed after casting, the same types of precipitates appeared but were coarser.

The first results showed that the alloy Al-Mn-Be-Cu with icosahedral phase during solidification and icosahedral precipitates indicates excellent potential for high-temperature applications at considerably higher temperatures than conventional wrought and precipitation hardening alloys.

References

[1] H.J. Kang, X.Z. Li, T.M. Wang, D.M. Liu, Y.Q. Su, Z.L. Hu, J.J. Guo & H.Z. Fu, Crystal-quasicrystal transition depending on cooling rates in directionally solidified Al-3Mn-7Be (at.%) alloy, Intermetallics, 44 (2014) 101-105.

[2] T. Oz, E. Karakose & M. Keskin, Impact of beryllium additions on thermal and mechanical properties of conventionally solidified and melt-spun Al-4.5 wt.%Mn-x wt.%Be (x=0, 1, 3, 5) alloys, Mater. Des., 50 (2013) 399-412.

[3] F. Zupanic, D. Wang, C. Gspan & T. Boncina, Precipitates in a quasicrystal-strengthened Al-Mn-Be-Cu alloy, Mater. Charact., 106 (2015) 93-99.

The 73rd World Foundry Congress
„Creative Foundry"
23 - 27 September 2018, Krakow, Poland

CHARACTERIZATION OF COMPOSITE CERAMIC MATERIALS USED DURING INVESTMENT CASTING OF AIRCRAFT ENGINES COMPONENTS

Rafał Cygan [1*]

[1] *Consolidated Precision Products, ul. Hetmanska 120, 35-078 Rzeszow, Poland,*

**Corresponding address: rafal.cygan@cppcorp.com*

Key words: investment casting, ceramic composite, Ni-based superalloy, casting defects

The article presents the results of research on new composite ceramic materials used in the manufacturing of hot section components of aircraft engines. In the manufacturing of aircraft engines components, investment casting technology in multi-layer ceramic moulds is widely used. It enables producing metal replicas geometrically complex, spatial shapes of wax sculptures.

The main of the work was to prepare and test in industrial conditions newly-developed composite ceramic moulds to ensure optimization of the investment casting process and reduction of final costs of components.

1. Introduction

The development of metallurgy processes, as well as precision casting techniques (lost-wax casting), allowed to obtain components characterized by high mechanical properties, dimensional accuracy, and low content of harmful elements. Precision casting has been used for thousands of years in a wide range of applications, from artistic castings to the most responsible components for the army and aerospace [1-7]. Manufacturing process of turbines blades step by step is presented in Fig. 1.

Fig. 1. Investment casting process

2. Experimental procedure

In order to develop new ceramic materials, the Consolidated Precision Products Corp. Rzeszow cooperates with the Laboratory for Aerospace Materials at the Rzeszow University of Technology. One of the examples of such cooperation is the research project funding by National Centre for Research and Development Poland "New technology for investment casting manufacturing critical engine components with a new generation ceramic materials". One of its goals is to develop a chemical composition and apply in industrial condition modern casting binders that are friendly to human health and the environment. As a consequence of the research a polymer acrylic emulsion, containing 48-50 wt % of the solid phase and 1-10 wt% glass fibers was developed. The use of composite glass fibers increased the gases permeability and mechanical properties of the moulds and also allowed to reduce the number of applied layers, thus the thickness of the shell moulds. The light-transmittance fibers with a diameter of 5-10 μm were used (Fig. 2).

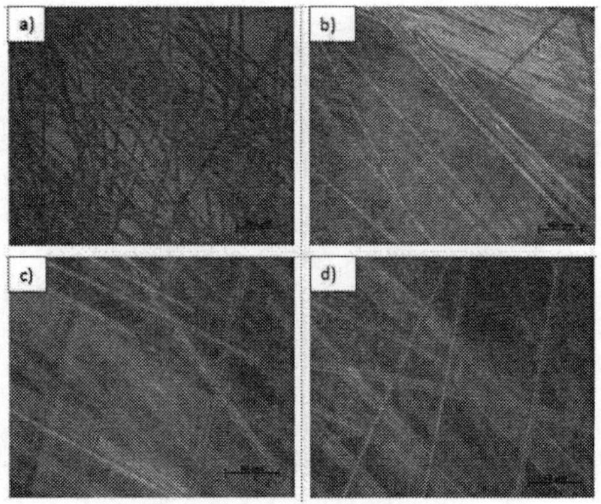

Fig. 2. Glass fibers used to make binders

Technological trials were conducted in casting moulds made of developed ceramic mixtures. Wax models of turbine vanes and gating systems were combined into complete wax models, and then were degreased in the washing agent. The wax models prepared in this way were dipped in a ceramic mass and then covered with ceramic powder. A total of six ceramic layers were applied to obtain a wall thickness of approx. 8 mm (Fig. 3a). Subsequently, the ceramic moulds were placed in an autoclave and subjected to a wax smelting process (temperature 165°C, pressure 7 bar). Wax residues were fired from ceramic moulds in an electric chamber furnace at 600°C.

The next step was washing in water to remove residues of ceramic masses and powders from the mould cavity and then drying at 80°C for 4 hours. The molds were insulated with 2 layers of Fiberfrax alumino-silicate fibres. Ni-based superalloy IN713C was poured at 1550°C under vacuum ($2x10^{-3}$ Tr) into ceramic moulds. The turbine vanes segments (Fig. 5b) were subjected to machining and final control operations

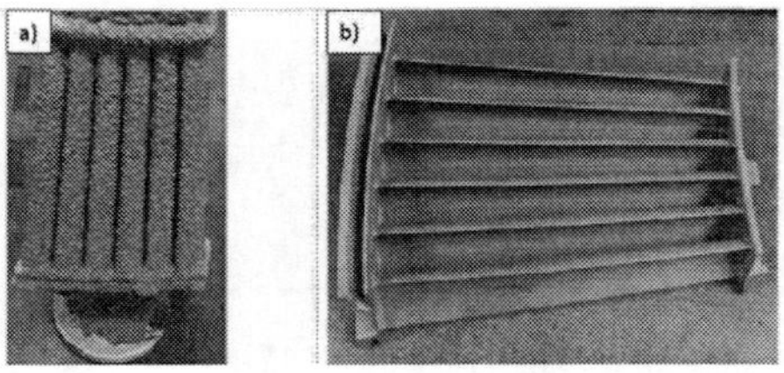

Fig. 3. Ceramic shell mould (a), turbine vane cluster (b)

3. *Summary*

Complicated geometry and increasing quality requirements for castings force the looking for new ceramic materials for building moulds, as well as technological solutions ensuring reduction of manufacturing costs, in particular for reducing the number of defective castings.

The newly-developed ceramic moulds presented in the paper are in line with the latest trends in the aircraft industry. The use of water binders, acrylic emulsions and glass fibers as fillers will help to reduce environmental pollution and increase the efficiency of the investment casting process. The turbine vane clusters made using new ceramic shell moulds, meet all quality requirements (dimensions, acceptable casting defects, surface roughness, porosity) set by the final recipients.

The developed composite ceramic mixture containing glass fibers allowed to reduce the number of layers from 8 to 6, which has a direct impact on the increase in savings resulting from shortening the production time of moulds by about 25% and reducing their weight by 15%. The ceramic moulds are made at CPP Poland using robotic station and the reduction in the number of shells contributes directly to the increase of the production line capacity. In addition, these moulds are characterized by higher mechanical properties and higher gas permeability, which also increases the area of their application.

Acknowledgments

This research was supported by National Centre for Research and Development, Grant No. LIDER/227/L-6/14/NCBR/2015.

References

[1] P. Wasilewski. 2001. Odlewnictwo jako jedna z najstarszych metod wytwarzania, Archiwum Odlewnictwa, 1 (1). 9-38.

[2] K. Stölzel. 1978. Giesserei über Jahrtausende, VEB Deutscher Verlag, Leigzig

[3] M. Gromada, A. Świeca, M. Kostecki, A. Olszyna, R. Cygan. 2015. Ceramic cores for turbine blades via injection moulding. Journal of Materials Processing Technology Volume 220. 107-112

[4] R. Cygan, P. Zywicki,T. Szczech, M. Kwiatkowski. Revert alloy impact on mechanical properties of nickel based superalloys used in production of aircraft engine critical components. 2016. 72nd World Foundry Congress. 225-226

[5] J. Baker. 1997 Five Thousand Years of Investment Castings, BICTA Bulletin 26. 6–8.

[6] D. Szeliga, K. Kubiak, W. Ziaja, R. Cygan. 2014. Influence of silicon carbide chills on solidification process and shrinkage porosity of castings made of nickel based superalloys. International Journal of Cast Metals Research Volume 27, Issue 3. 146-160

[7] P. Rokicki, G. Budzik, K. Kubiak, J. Bernaczek, T. Dziubek, M. Magniszewski, A. Nowotnik, J. Sieniawski, H. Matysiak, R. Cygan, A. Trojan. 2014. Rapid prototyping in manufacturing of core models of aircraft engine blades. Aircraft Engineering and Aerospace Technology Volume 86, Issue 4. 323-327

The 73rd World Foundry Congress
„Creative Foundry"
23 - 27 September 2018, Krakow, Poland

INFLUENCE OF THE SELECTED SUPERALLOY AND CERAMIC CRUCIBLE ON THE MELTING AND MELT-POURING PROCESS DURING THE INVESTMENT CASTING OF THE AIRCRAFT COMPONENTS.

Rafał Cygan [1]*, Łukasz Rakoczy [2], Rafał Nowak [3], and Natalia Sobczak [3]

[1]Consolidated Precision Products, ul. Hetmanska 120, 35-078 Rzeszow, Poland
[2]AGH University of Science and Technology, Faculty of Metals Engineering and Industrial Computer Science, av. Mickiewicza 30, 30-059 Krakow, Poland
[3]Foundry Research Institute, ul. Zakopianska 73, 30-418 Krakow, Poland
**Corresponding address: rafal.cygan@cppcorp.com*

Key words: investment casting, crucible, vacuum melting, superalloys,

1. Introduction

The paper presents the results of research on ceramic crucibles used in the production of critical aircraft engines components. In the manufacturing of these components, investment casting technology in multi-layer ceramic moulds is widely used. It enables producing metal replicas geometrically complex, spatial shapes of wax sculptures. One of the key parameters is the way of superalloys melting under reduced pressure using induction furnaces.

The main aim of the research was to determine the physicochemical phenomena occurring at the liquid metal-ceramic crucible boundary during the melting and melt-pouring processes.

Two different crucibles and superalloys were used:

- Al_2O_3
- ZrO_2
- Co-based MAR M509
- Ni-based IN713C

2. Experimental procedure

2.1. Wettability measurements

Investigation of wetting kinetics was carried out using the drop method according to a contact heating (CH) procedure. The CH procedure is based on preheating of the set alloy - ceramic crucible to the assumed temperature, annealing at this temperature, and then cooling. The superalloy and ceramic crucible are in constant contact with each other during the whole cycle. During the tests, images of the tested object were recorded - a drop of alloy placed on a smooth and flat surface (Fig. 1). Until the melting point of the melt was reached (preheating), continuous image recording was carried out at a rate of 1 frame per second. After reaching the melting point, the image was recorded at a rate of 10 frames per second using a high-resolution Microtron 1310 digital camera.

Fig. 1. Preparation of sample for wettability measurements

2.2 Investigation of residual gases

The gasification of ceramic crucibles was determined based on a continuous recording of the residual gases using a Pfeifer QMS-200 spectrometer (Fig. 2) from PFEIF (at 1200°C and 30 minutes).

Fig. 2. Apparatus set for a comprehensive investigation of liquid metals and alloys at high homologous temperature

3. Summary

The performed research allowed observation of the behavior of the system liquid metal - ceramic crucible. The shape change of the observed drop liquid (Fig. 3) allowed to determine the contact angle of liquid metal. This parameter is extremely important during the melting process using induction vacuum furnaces. The research allowed to optimize the process depending on the melted superalloy (MAR M509 or IN713C) and allowed to choose the optimal ceramic material of the crucible.

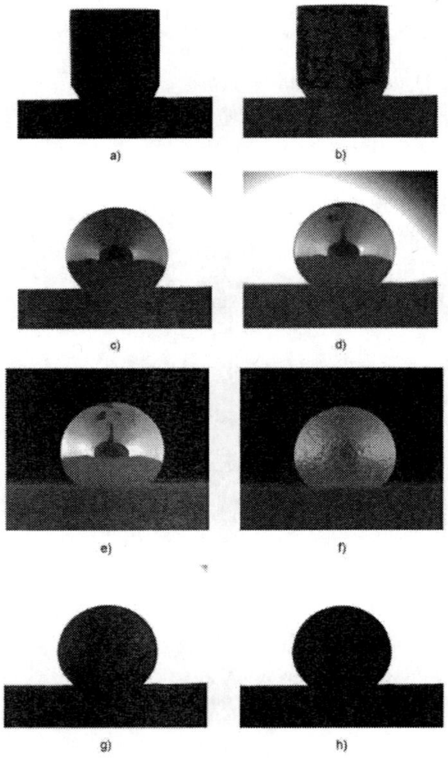

Fig. 3. Wettability investigation of set MAR M509/ZrO_2 *crucible: a) T=21°C; b) start of melting T=1359°C; c) creation of spherical drop T=1381°C; d) T=1499°C t=0min; e) T=1501°C t=5 min ; f) solidification of drop T=1281°C; g) T=979°C, lack of delamination; h) T=255°C*

The selection of the appropriate ceramic crucible and the optimization of technological parameters have resulted in the following benefits:

- shortening the melting time,
- limiting the contact time of liquid metal with the outer layer of the crucible,
- limiting the erosion of ceramic material,
- reducing the amount of non-metallic inclusions.

These benefits directly affect the improvement of efficiency and competitiveness of Consolidated Precision Products Company Poland.

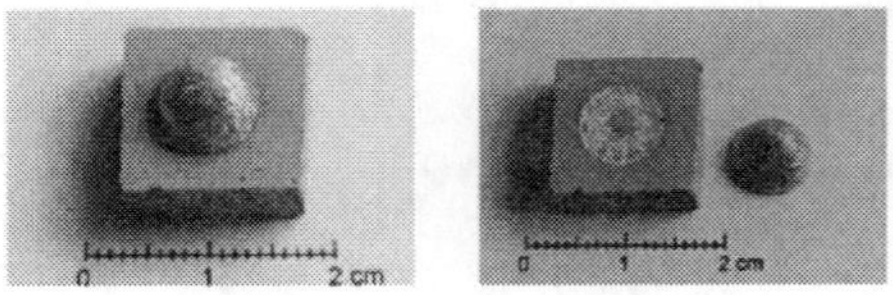

Fig. 4. Set IN713C/ crucible Al_2O_3 *set after investigation (5 min at 1500°C)*

Acknowledgments

This research was supported by National Centre for Research and Development, Grant No. LIDER/227/L-6/14/NCBR/2015.

References

[1] N. Sobczak, M. Singh, R. Asthana. 2005. High-temperature wettability measurements in metal/ceramic systems – Some methodological issues. Current Opinion in Solid State and Materials Science 9. 241-253

[2] N. Sobczak, R. Nowak, W. Radzwill, J. Budzioch, A. Glenz. 2008. Experimental complex for investigations of high temperature capillarity phenomena, Material Science and Engineering: A 495. 43-49

[3] M. Gromada, A. Świeca, M. Kostecki, A. Olszyna, R. Cygan. 2015. Ceramic cores for turbine blades via injection moulding" Journal of Materials Processing Technology Volume 220. 107-112

[4] R. Cygan, P. Zywicki, T. Szczech, M. Kwiatkowski. 2016. Revert alloy impact on mechanical properties of nickel based superalloys used in production of aircraft engine critical components. 72nd World Foundry Congress. 225-226

[6] D. Szeliga, K. Kubiak, W. Ziaja, R. Cygan. 2014. Influence of silicon carbide chills on solidification process and shrinkage porosity of castings made of nickel based superalloys. International Journal of Cast Metals Research Volume 27 Issue 3. 146-160

[5] L. Liggieri, A. Passerone. 1989. Automatic technique for measuring the surface tension of liquid metals. High temperature technology 7. 82-86

The 73rd World Foundry Congress
„Creative Foundry"
23 - 27 September 2018, Krakow, Poland

MORPHOLOGICAL EVOLUTION OF SEMISOLID Mg-Al-La-Ca MAGNESIUM ALLOY PRODUCED BY MECHANICAL STIRRING PROCESS (MSP)

Sérgio Bartex[1]*, Lírio Schaeffer[1], Vinicius de Barcellos[1]

[1]*Federal University of Rio Grande do Sul, Engineering School, Bento Gonçalves Av., 9500, ZIP Code 91501-970, Porto Alegre, RS, Brazil*
**Corresponding address: bartex@ufrgs.br*

Key words: Solidification, Rheocasting, semisolid slurry, microstructural evolution.

1. Introduction

As the lightest structural metallic materials, magnesium alloys have exceptional characteristics including low density, good fusibility, high specific strength and good recyclability. All these great characteristics bring significant application value in industrial fields such as the aerospace and automobile industry and telecommunications [1].
Ferry et al. [2] studied the influence of both calcium (1.0wt%Ca) and lanthanum (3.0wt%La) addition to a magnesium alloy system (Mg-4.0wt%Al, 0.5wt%Zn). They proposed the substitution of Rare Earths elements by lanthanum (La) to increase creep resistance. However, the structure has shown dendritic matrix and very coarse compounds which decrease other mechanical properties behavior.
A good way to reduce the microstructural heterogeneity is processing magnesium alloy by Mechanical Stirring Process (MSP). This method consists on applying shear stresses in magnesium alloy during the semisolid state. The shear stresses may break the dendritic structures and may create a globular morphology improving mechanical properties [3]. There are some different ways to obtain the slurry by magnesium alloys: ultrasonic vibration [4], mechanical stirring [5], gas bubble stirring [6], frequency electro-magnetic stirring [7].
In this work, mechanical stirring process was used to brake dendritic structures in order to evaluate the influence of mechanical stirring time on the magnesium alloy Mg-6Al-3La-1Ca morphology.

2. Materials and Methods

The experiments were performed with magnesium alloy ingots, nominal composition Mg-6wt%Al-3wt%La-1wt%Ca, produced by a Brazilian company, which chemical composition is presented in Table 1.
A resistance-type electric closed furnace with a protective gas atmosphere (commercial argon gas, 99.9%) was designed to prevent any reaction between oxygen and molten magnesium to avoid ignition [2]. A schematic representation of the assembly used in solidification experiments is presented in Fig. 1.

Table 1. Chemical composition of magnesium alloy performed in this work (wt%).

Element (wt %)					
Mg	Balance	**Mn**	0.300	**Ni**	0.001
Al	5.900	**Fe**	0.005	**Ca**	0.988
Zn	0.030	**Cu**	0.002	**La**	2.770

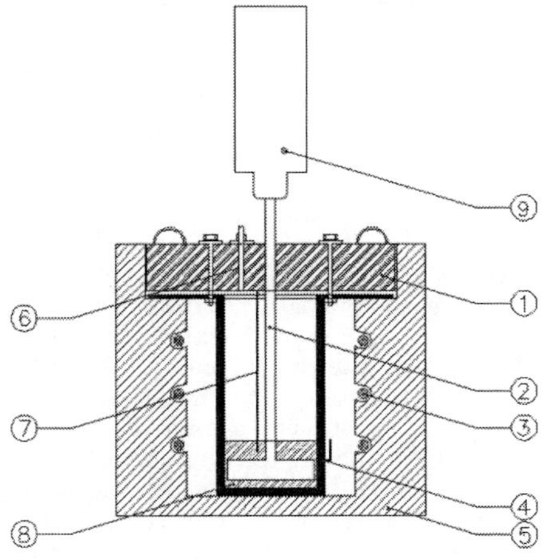

Fig. 1. Schematic representation of experimental casting arrangement: (1) cover; (2) mechanical stirrer; (3) electrical resistances; (4) external thermocouple; (5) thermal insulation; (6) argon inlet; (7) internal thermocouple; (8) magnesium slurry; (9) stirring engine (950 rpm).

A magnesium alloy ingot of 1 kg was melted inside the furnace at 650°C and after the temperature was reduced to 605°C to obtain 0.29 solid fraction. At the latter temperature, mechanical stirring was applied with different stirring times: 0, 1, 2, 4, 8 and 10 minutes. After each defined time, a cylindrical sample 70 x 3 mm was extracted from the semisolid magnesium slurry by suction and immediately cooled in water 25°C [8]. Cross sections of the samples were cut to be analyzed by metallography techniques. The specimens were etched by a Nital 3% solution (3 mL HNO_3 + 97mL ethanol) and investigated using an optical microscope.

3. Results and Discussion

The Fig. 2 shows the morphological evolution in function of processing time.

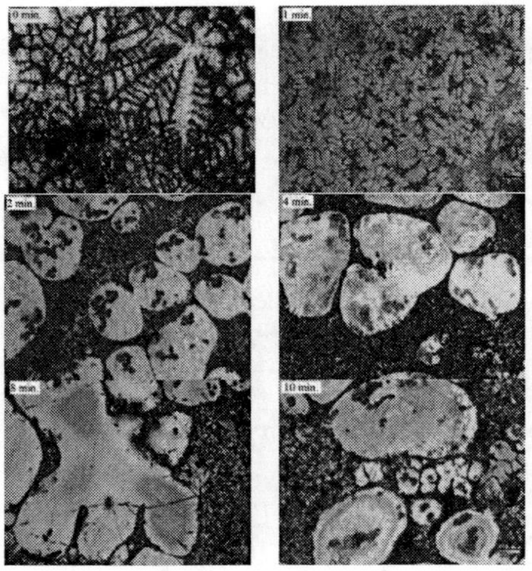

***Fig. 2.** Morphological evolution in function of time in Mg-6Al-3La-1Ca alloy processed by MSP. All specimens stirred at 605°C, 950 rpm, time t (t= 0, 1, 2, 4, 8, 10 minutes).*

Fig. 2 shows the dendritic structure formed by magnesium (α-Mg) and other phases in the interdendritic region (0-minute stirring time – as cast) [2]. 1-minute stirring time is not effective to create globules and can only break the dendritic structures creating a refinement process. However, stirring time equal or superior to 2 minutes can both break the dendritic structure and create globules into the matrix. These globules are formed by magnesium (α-Mg) and interglobular region is formed by other phases. 2-minute stirring time showed a refined globular structure in comparison to the superior stirring times. At 4-minute and 8-minute stirring time, the globules started to grow and became coarser. This behavior is in accordance with the Ostwald Ripening phenomena which is associated with the system free energy decrease, increasing the area of the globules [9]. On the other hand, at 10-minute stirring time the dendritic nucleation began in the remaining liquid creating, in addition to the globules, small dendritic structures in the interglobular zone.

4. Conclusions

The morphological evolution investigation in a semisolid magnesium alloy (Mg-6Al-3La-1Ca) was done. The Mechanical Stirring Process (MSP) can effectively change this alloy microstructure. After two minutes stirring time the structure is composed by globular forms. The globules size increased over stirring time due to coarsening phenomena. So, the most suitable time, between all stirring times evaluated, is two minutes.

Acknowledgments

The authors acknowledge financial support provided by CNPq (The Brazilian Research Council) and FAPERGS (The Scientific Research Foundation of the State of Rio Grande do Sul).

References

[1] X. Fang, S. Lü, L. Zhao, J. Wang, L. Liu, S. Wu. (2016). Microstructure and mechanical properties of a novel Mg–RE–Zn–Y alloy fabricated by rheo-squeeze casting. JMADE. 94, 353–359.

[2] T. V. Ferri, A.P. Figueiredo, C.R.F. Ferreira, W. Hormaza, C.A. Santos, J.A. Spim. (2010). Mechanical properties as a function of microstructure in the new Mg-Al-Ca-La alloy solidified under different conditions. Materials Science Engineering A. 527, 4624–4632.

[3] M. Esmaily, M. Shahabi-Navida, N. Mortazavi, J.E. Svensson, M. Halvarsson, M. Wessén, A.E.W. Jarfors, L.G. Johansson. (2014). Microstructural characterization of the Mg–Al alloyAM50 produced by a newly developed rheo-casting process. Materials Characterization. 9 5, 5 0 – 6 4.

[4] R. Guan, L. Chen, F. Cao, Z. Zhao, and Y Ren. (2011). Semisolid die forging process, microstructures and properties of AZ31 magnesium alloy mobile telephone shells. International Journal of Minerals, Metallurgy and Materials Volume 18, N° 6, Page 665-671.

[5] C.D. Yim, K.S. Shin. (2005). Changes in microstructure and hardness of rheocast AZ91HP magnesium alloy with stirring conditions. Materials Science and Engineering A. 395, 226–232.

[6] R. Canyook, J. Wannasin, S. Wisuthmethangkul, M.C. Flemings. (2012). Characterization of the microstructure evolution of a semi-solid metal slurry during the early stages. Acta Materialia. 60, 3501–3510.

[7] C. Wang, A. Chen, L. Zhang, W. Liu, G. Wu,W. Ding. (2015). Preparation of an Mg–Gd–Zn alloy semisolid slurry by low frequency electro-magnetic stirring. Materials and Design. 84, 53–63.

[8] X. Fang, S. Lü, L. Zhao, J. Wang, L. Liu, S. Wu. (2016). Microstructure and mechanical properties of a novel Mg–RE–Zn–Y alloy fabricated by rheo-squeeze casting. Materials and Design 94, 353–359.

[9] Y. Chen, L. Zhang, W. Liu, G. Wu,W. Ding. (2016). Preparation of Mg–Nd–Zn–(Zr) alloys semisolid slurry by electromagnetic stirring. Materials and Design. 95, 398–409.

The 73rd World Foundry Congress
„Creative Foundry"
23 - 27 September 2018, Krakow, Poland

DETERMINATION OF GRAY CAST IRON AGE STRENGTHENING BY NON-DESTRUCTIVE METHODS: EFFECT OF ALLOYING ELEMENTS

Vaucheret Alexis[1,2,*], Rossi Frédéric[1], Quesada Jean[1], Albertalli Eric[3]

[1] *LaBoMaP, Arts et Métiers ParisTech, 71250 Cluny, France*
[2] *Université de Lyon, ECAM Lyon, INSA-Lyon, LabECAM, F-69005 Lyon, France*
[3] *Renault ACI, 15 Avenue Pierre Piffault, 72000 Le Mans, France*
**Corresponding address: alexis.vaucheret@ecam.fr*

Key words: age strengthening, cast iron, hardness, inert frequency, nitrogen

1. Introduction

This study was led in collaboration with Renault Le Mans in order to better understand gray cast iron age strengthening. It causes a modification of cast iron machinability and mechanical properties in a short time, until they reach their metallurgical stability.

2. Literature review

A few studies concern gray cast iron age strengthening and they were principally led by Richards and Nicola [1, 2]. As shown in the Fig. *1*, they have proved that the Ultimate Tensile Test (UTS) increases with aging time. Vickers micro-hardness of the cast iron's matrix follows the same evolution. At the same time, the machinability of the parts and their surface roughness improves with age strengthening and so the wearing tool decreases [1].

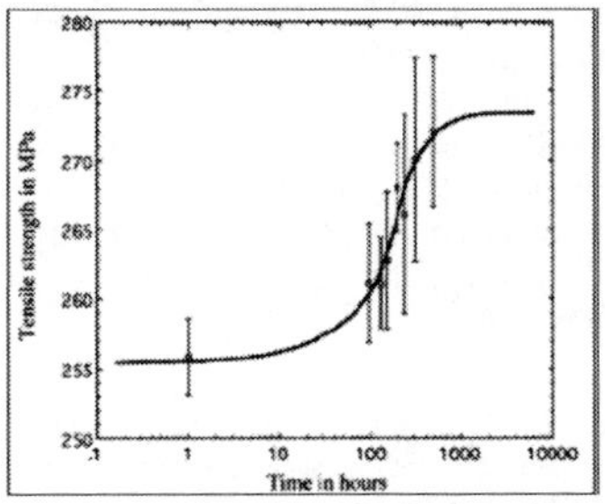

Fig. 1. Ultimate Tensile Stress depending on aging time maturation

These studies showed that age strengthening was caused by nitrogen in solid solution in ferrite of cast iron matrix. Nitrogen could form Guinier-Preston zones which would change into another type of precipitate structure as aging progresses. These precipitates are probably stable iron nitrides Fe_4N which formation led to matrix enhancement by dislocation locking mechanism that could occur with a Fe-N or Fe-C-N precipitate. So the influence of cast iron alloying elements is decisive on the aging time and intensity, as shown by Richards and Anish [2]. Alloying elements will affect aging by reducing the amount of nitrogen in solubility able to react with iron, by forming nitrides of titanium, boron aluminium and so on. All the elements where nitride formation enthalpy on Ellingham diagram is lower than iron one will react with nitrogen before iron. A prediction model of the gain of UTS during a 20-day period of aging based on the influence of alloy elements in cast iron was proposed [2]. However, the model presents some restriction because it doesn't take into account the influence of carbon which is mainly responsible for the amount of nitrogen in solution in cast iron [3].

3. Experimental method

Based on these studies, this work wants to monitor cast iron age strengthening by non-destructive mechanical tests and observe aging in-situ on casted parts.

a. Rockwell hardness

As shown on past research, micro-hardness of cast iron ferritic matrix increases with aging [1]. This study wants to correlate the aging with macro-hardness of the parts. The first thing was to choose the type of hardness to use. For this, the heterogeneity of the material between the matrix and the flake graphite makes it necessary to use a spherical penetrator. A non-normalized hardness was selected using an indentor's diameter of 1/8 inch and a load of 250 daN. The measurement of this particular Rockwell hardness was able to follow the aging by giving hardness value with enough allowance to observe the gain of hardness and by minimizing the measuring error.

b. Inherent frequencies

Richards and Nicolas studies have shown that age strengthening increases mechanical properties [1, 2]. Regarding these results, Young modulus of the material also should rise. To verify this hypothesis, it has been decided to follow the evolution of Young modulus by measuring the inherent frequencies during aging with a free system. The natural frequencies, measured in a free system, are directly correlated with Young modulus:

$$E_t = E_0\left(\frac{\omega_{nt}^2}{\omega_{n0}^2}\right) \quad (1)$$

with 0: non-aging state and t: t-day aging state

c. Monitoring of the cast iron age strengthening

The following step of the study was the experimental validation of age strengthening monitoring. The aim was to observe and quantify cast iron aging's intensity and time with Rockwell hardness and inherent frequencies. In agreement with Renault, the cast iron parts were poured in laboratory to minimize the time between the monitoring of the aging and the casting. For this validation, measures of hardness and natural frequency were done daily to monitor the aging at room temperature. The Fig. *2* showed that. The evolution of hardness and inherent frequencies is the same during the aging. Moreover, it seems to be in accordance with Richard and Nicola results [1]. The aging shows an increase of the hardness and an evolution of the value of each inherent frequency followed by a stabilization phase. Finally, the monitoring of the aging using classical and non-destructive mechanical tests seems to be possible.

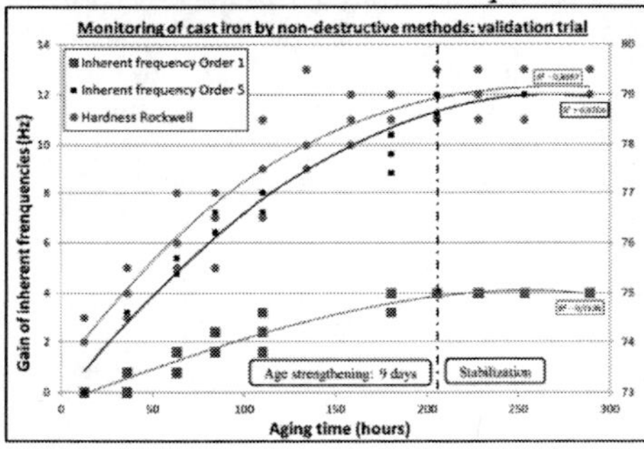

Fig. 2. Correlation of the different methods of monitoring

4. Influence of alloying elements on age strengthening

As said by Richards [1,2] and Anish [5], alloying elements have an influence on age strengthening especially nitrogen and all the elements able to form nitrides before iron. The aim of this study is to show the effects of these elements. To do that, four different cast irons (EN GJL 200) with variable free nitrogen content were poured with the same cooling rate. Free nitrogen is defined as nitrogen in excess and able to react with iron and could be calculated as:

$$N_{Free} = N_{Total} - \frac{33}{100}Ti \quad (2)$$

Figures 3 and 4 show the results of the age strengthening monitoring of each sample. It possible to see that the aging time increases with the amount of free nitrogen. In the same way, the gain of mechanical properties, hardness and Young Modulus, follows the same evolution. This evolution is in accordance with the literature review [1].

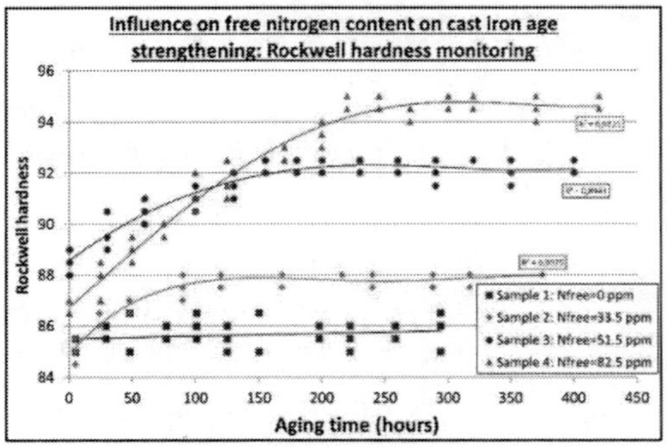

Fig. 3. Influence of free nitrogen on age strengthening: hardness monitoring

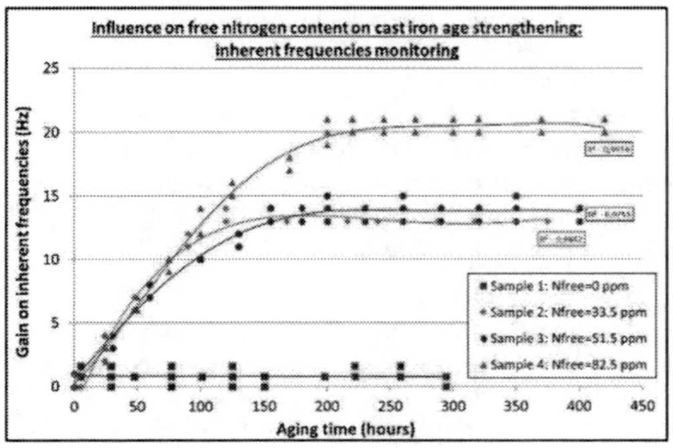

Fig. 4. Influence of free nitrogen on age strengthening: inherent frequencies monitoring

5. Conclusion

The study shows the possibility to monitor cast iron age strengthening by non-destructive methods. This allows following this phenomenon directly on industrial parts. This type of monitoring could be useful to guarantee the metallurgical stability of parts before machining. Moreover, the effects of alloying elements on gray cast iron age strengthening, especially free nitrogen, has been confirmed by this study.

References

[1] VL. Richards, W. Nicola,(2003) Age strengthening of gray cast iron phase III, International Journal of Cast Metals Research, 16:1-3, 275-280.

[2] VL. Richards, TV. Anish, S. Lekakh, DC. Van Aken, W Nicola, (2006) Composition effects on age strengthening of gray iron, University of Missouri-Rolla,

[3] Casting Plant and Technology International, (2009) Improving the mechanical properties of grey cast iron by alloying with nitrogen, N°1

[4] C. Gatellier, PV. Riboud, A. Rist et MF. Ancey-Moret, (1974) Equilibres thermodynamiques en sidérurgie, Techniques de l'ingénieur, Dossier M 1730

[5] TV. Anish, (2007) Age strengthening of gray cast iron: alloying effects and kinetics study, Masters Theses, 4554

The 73rd World Foundry Congress
„Creative Foundry"
23 - 27 September 2018, Krakow, Poland

THE EFFECT OF Sr AND IMPURITIES INTERACTION ON THE Al-7%Si-0.3%Mg ALLOYS

Teru Fukuhara[1]*, Hisao Kazuta [1],Arisa Osugi[1] Susumu IKENO[2], Seiji Saikawa [3]

[1]Shoko Co, LTD. Metal Ceramics Division Light Metals Dept. Raw Materials Group ,Tokyo105-8432 , Japan
[2]University of Toyama, Graduate School of Science and Engineering for Education, Toyama 930-8555, Japan
[3]University of Toyama, Graduate School of Science and Engineering for Research, Toyama 930-8555, Japan
**Corresponding address: teru-fukuhara@shoko.co.jp*

Key words: Al-Si alloy, casting, Sr, modifying, impurities

1. Introduction

The Al-Si alloy system is one of the most frequently used aluminium foundry alloys due to their excellent castability, strength-to-weight ratio, resistance to corrosive attack. In addition, small amount of Mg content occurred of good enhance of strength by help of solution treatment and aging treatment. Therefore the morphology of eutectic Si will to a large extent affect the alloys final properties.

In commercial Al-Si alloys such as A356 or A360 which are used auto wheel the eutectic Si will be coarse and flake-like. This morphology has been claimed to be the result of higher fractions of impurity particles acting as heterogeneous nucleation. Strontium are the most widely used and a very effective element for modifying the morphology of eutectic silicon. It was found that the eutectic nucleated at very low Sr contents and grew from the primary phase .

Increasing the Sr level to between 70 and 110ppm resulted in nucleation of independent eutectic grains with no relations to the primary dentrites[1)]. Small amount of Sr addition improves its mechanical properties. On the other hand, Na and Ca are also modifying elements used commercially to treat hypoeutectic Al-Si foundry alloys in order to refine and modify the eutectic Si from the coarse acicular to fibrous or lamellar morphology. However the influence of Ca on modification is more smaller than Sr and Na, and it is treated as impurities because it is easy to mix into the secondary alloys. An influence of impurities on the casting structure and the characteristics are still unresolved problem. Phosphorus additions are known to be an efficient means of nucleating primary silicon in hypereutectic alloys, and AlP-particles are proposed to be the main nucleant.

On the other hand, Na has been used for a long time as an improved material for Al - Si alloys in the aluminum casting industry, therefore many casting foundry companies still use Na as a refining treatment agent. The influence of combined additions of Na and other improved elements such as Sr ,Ca, and P on casting quality tained in the Al - Si based alloy was investigated.

However in resent almost casting products are cast from not only virgin material but also return scraps. Therefore casting products contain of inevitable impurity such as Na, Ca, P, Sr.

In spite of the situation, the influence on the mechanical properties and solidification structure in Al-7Si-0.3Mg (wt%) alloys are not clear.

In this study it indicates combined additions Sr grain refiner and impurities on mechanical properties. Cooling curves were obtained using tissue analysis of a Sr-refined Al - Si - Mg alloy and a quasi - equilibrium apparatus, using 2N-Al at commercial level and 5N-Al with low impurity.

2. Experiment procedure

The chemical compositions of the alloys and alloy codes are shown in Table 1.

Five kinds of alloys contained 0-200ppm Sr (0Sr, 10Sr, 25Sr, 50Sr, 100Sr and 200Sr), two kind of Al based alloy (2N, 5N) were cast in this study.

Before melting process, compositions of the raw material were adjusted with weighing of 99.8%Al, 99.99%Si, 99.9%Mg and Al-10%Sr to achieve the target compositions shown in Table1. Each alloy with a total weight about 5kg was melted in an alumina crucible using an electrical furnace. When the temperature was heated to 720-730°C molten metal was cast into permanent mold which controlled at about 200- 230°C after BN spraying to the cavity of Y block.

The apparatus which the molten metal could be slowly cooled at a cooling rate close to equilibrium state (CR≈0.05°C/s) and held at any temperature, in addition, the microstructure could be rapidly freezed by water-quenching (quasi-equilibrium solidification apparatus) was used to obtain the microstructures freezed at any temperature during solidification.

Schematic drawing of quasi-equilibrium solidification apparatus is shown in Fig. 1. About 3.5g of specimens were cut from the sampling portion of the Y-block shaped castings and enclosed into SUS304 capsule with K-type thermocouple. The enclosed capsule was hung in the furnace by the wire. After the alloy was melted at 700°C and held for 30min, the molten metal was slowly cooled in the furnace by cutting off the power. The lower insulator was removed and the wire and thermocouple was cut immediately at the predetermined temperature during slowly cooling, the microstructures were rapidly freezed by water-quenching the enclosed capsule. Consequently, the microstructures during and after solidification were obtained. The temperature changes were measured by the thermocouple connected to recorder at the same time with slowly cooling, the cooling curves were obtained.

The specimens taken out from the capsule were polished with waterproof sandpaper, buffed with 1μm diamond paste and etched in a 0.5% HF solution. The microstructures were observed by optical microscope (OM) and field emission scanning electron microscope (FE-SEM).

Table1. The target composition of the sample and the analysis results

Purity of aluminum	Alloy	Chemical composition					Code
		Si	Mg	Fe	Sr	Al	
2N	Al-7Si-0.3Mg	7.1	0.32	0.09	1	Bal.	2N0Sr
	Al-7Si-0.3Mg-100Sr	7.1	0.32	0.09	120	Bal.	2N100Sr
5N	Al-7Si-0.3Mg	7.1	0.31	<0.01	1	Bal.	5N0Sr
	Al-7Si-0.3Mg-10Sr	6.99	0.31	<0.01	13	Bal.	5N10Sr
	Al-7Si-0.3Mg-25Sr	7.1	0.31	<0.01	34	Bal.	5N25Sr
	Al-7Si-0.3Mg-50Sr	7.2	0.32	<0.01	60	Bal.	5N50Sr
	Al-7Si-0.3Mg-100Sr	7.3	0.35	<0.01	106	Bal.	5N100Sr
	Al-7Si-0.3Mg-200Sr	7.2	0.33	<0.01	218	Bal.	5N200Sr

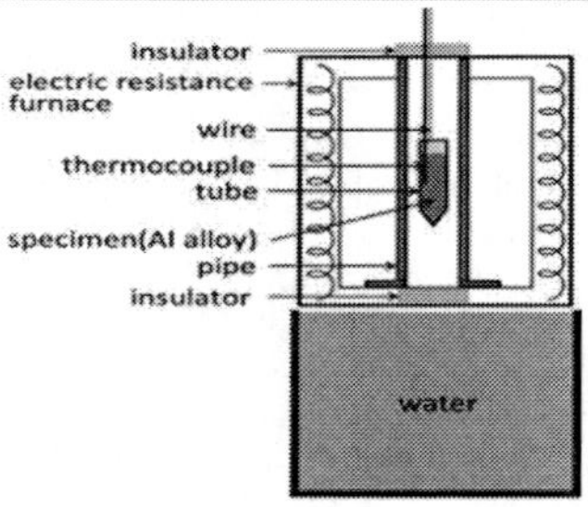

Fig. 1. Schematic diagram of pseudo-equilibrium device

3. Result and discussion

The 2N0Sr and 5N0Sr cooling curves are shown in Fig.2. At 5N0Sr, undercooling at about 3 ° C was observed. On the other hand, almost no undercooling was observed at 2N0Sr.

Figure 3a, b, c and d show the microstructures of 2N alloys and 5N alloys, respectively. In all of the alloys, the primary crystallized α-Al phase and the eutectic phase (α-Al phase and Si phase) which is the secondly crystallized phase were observed. Comparing the microstructure of 2N 0Sr and 5N 0Sr was not containing Sr, the eutectic structure was finer in 5N 0Sr than 2N 0Sr. Both alloys of the 2N and the 5N had a rod-like eutectic Si phase.

Figure 4 shows a eutectic cell formed epitaxially (arrow A) from a fine plate-like Si phase with AlP as heterogeneous nucleation site. It is considered that the eutectic Si phase of 2N0Sr became coarse due to heterogeneous nucleation by AlP. In addition, it is considered that undercooling only occurred because the content of P was extremely small at 5 N 0 Sr and heterogeneous nucleation by AlP did not occur. Comparing 2N100Sr and 5N100Sr with 100 ppm Sr added to alloys using commercial purity 2N and high purity 5N, the refinement effect of eutectic structure by Sr is almost equivalent.

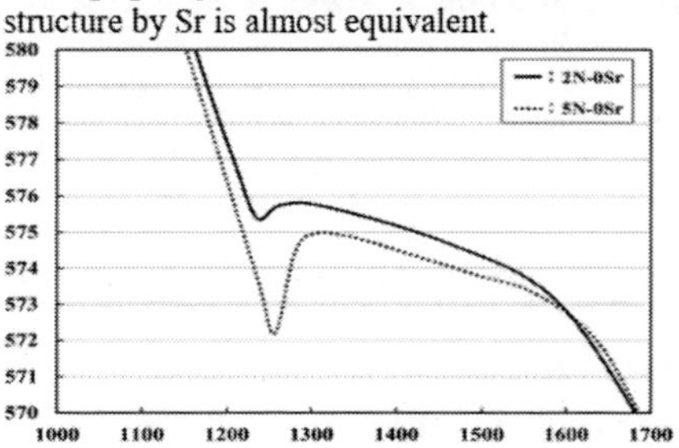

Fig. 2. The obtained cooling curve

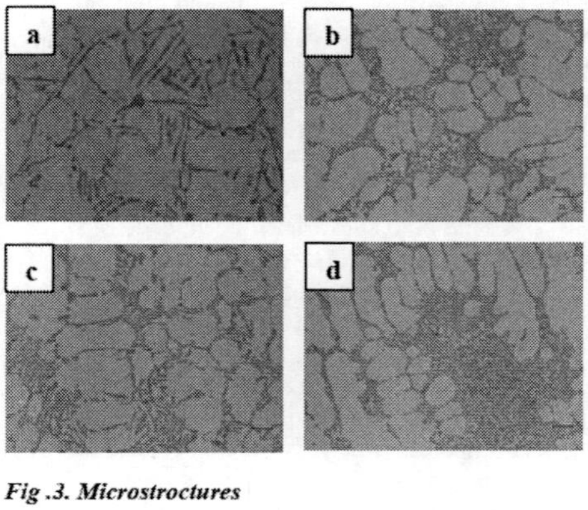

Fig .3. Microstroctures

a) 2N-0Sr, b)2N-100Sr , c)5n-0Sr , d)5N-100Sr

Fig. 4. Microstructure of 2N 0 Sr

References

[1] L. Lu, K. Nogita, A,K. Dahle, Material Science Engineering, A399 (2005) 244-253

The 73rd World Foundry Congress
„Creative Foundry"
23 - 27 September 2018, Krakow, Poland

DEVELOPMENT OF *IN SITU* FABRICATION PROCESS OF CLAD MATERIALS BY USING TANDEM TWIN-ROLL CASTING

Shinji Kumai [1]*, Yusuke Takayama [2]

[1]*Tokyo Institute of Technology, Department of Materials Science and Engineering, 2-12-1 O-okayama, Midori-ku, Tokyo 152-8552, Japan*
[2]*Tokyo Institute of Technology, Department of Metallurgy and Ceramics Science, 2-12-1 O-okayama, Midori-ku, Tokyo 152-8552, Japan*
**Corresponding address: kumai.s.aa@m.titech.ac.jp*

Key words: twin-roll casting, energy-saving, clad sheets fabrication, microstructure, mechanical properties

1. Introduction

Twin-roll casting is a direct strip production method from the molten alloy. The vertical-type high-speed twin-roll casting method possesses an extremely high productivity and an excellent heat extraction ability compared to the conventional horizontal-type twin-roll casting. The rapid cooling effect provides significant microstructure refinement and mechanical properties improvement in various kinds of cast aluminum alloy products [1,2]. Both great reduction in energy consumption and manufacturing processes are also attained. In the present paper, the potential of the casting method is introduced by using the application of the energy-saving single-step production of aluminum alloy clad sheets as an example.

2. Vertical-type high-speed twin-roll caster

Figure 1 shows a schematic illustration of the laboratory-scale vertical-type high-speed twin-roll caster. The diameter and width of the roll were 300 mm and 10 mm, respectively. In the caster, a large feeding nozzle is used to maintain a high melt pool level during the strip casting. The feeding nozzle is mounted on the rolls, so that the bottom-end of the nozzle has direct contact with the roll surface. A pair of side dams is set on the side-edges of the nozzle and the casting roll. The caster is equipped with a pair of pure copper rolls (made of tough-pitch copper). No lubricant is applied to the roll surface. The casting rolls are internally cooled with running water during the casting. One of the rolls is loosely fixed by being supported by a series of springs to control the roll separating force, while the other one is fixed firmly. The roll separation force is controlled by using springs connection in parallel. The initial roll separating force is obtained by setting their length to the pre-determined values. During strip casting, the pre-set force is applied on the strip near the roll gap by enlarging the roll gap. The molten metal in the crucible is poured into the nozzle from the fixed roll side toward the movable roll side. The pouring is conducted as fast as possible, but very carefully so that the maximum height of the melt pool in the nozzle is kept constant. Roll rotating speed is controlled in a range from 30 to 90 m/min. The initial roll gap is controlled in a range from 0.5 to 2 mm. This procedure allows us to obtain a long strip. The length depends on the volume of the charged molten metal.

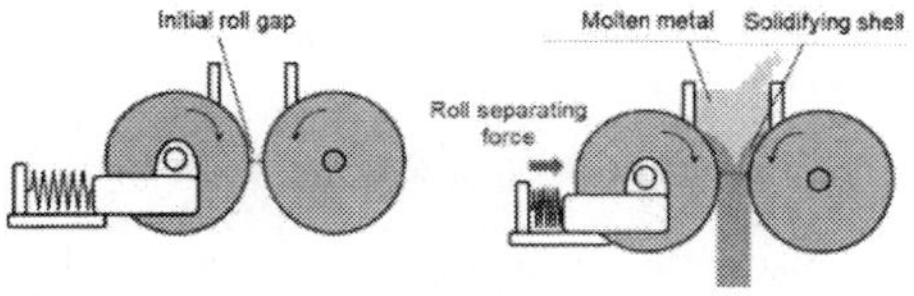

Fig. 1. Schematic illustration of the vertical-type high-speed twin-roll caster. Before casting (left) and during casting (right).

3. Tandem twin-roll caster for *in situ* fabrication of clad materials

Development of *in situ* fabrication process of aluminum alloy clad materials is a representative example of advanced application of the vertical-type high-speed twin-roll casting. Two sets of roll casters are arranged vertically. The core strip fabricated at the first (top) caster is introduced to the second caster, and in that place, the skin metal is put on the core strip. The clad materials can be produced from molten alloys in a single step. Production of multi-layered clad materials and controlling of the clad ratio can be made by using multiple sets of roll casters, as shown in Fig. 2.

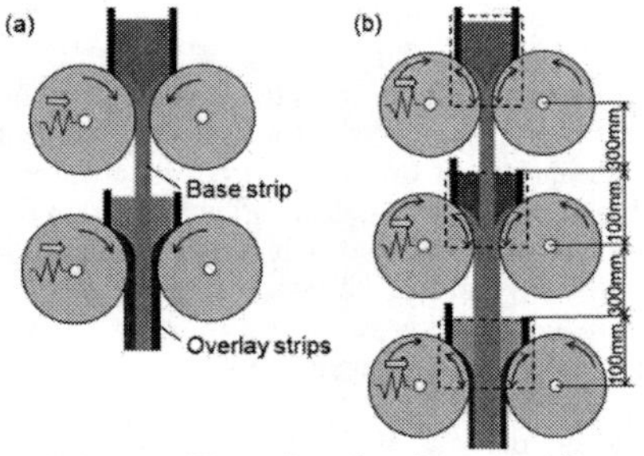

Fig. 2. Schematic illustration of tandem twin-roll caster. (a) Double-type (b) Triple-type

This novel process can reduce a number of process steps, production energy compared to the conventional hot roll bonding process, as shown in Fig. 3. The original microstructure of clad strip is fine because of the high cooling rates of the twin-roll casting. In addition, chemical reactions and solute diffusion at the clad interface are prevented, and coarsening of secondary particles is suppressed. A number of hot rolling and annealing processes is inevitable for thickness reduction in the conventional hot roll bonding. Therefore, this is clearly a strong point of the present method.

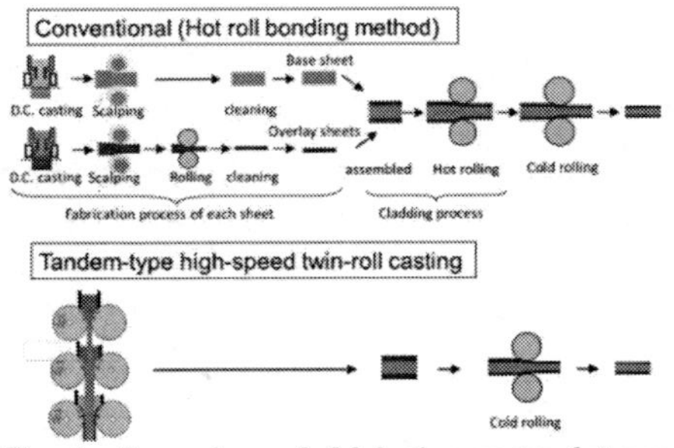

Fig. 3. Comparison of fabrication routes between conventional method and the tandem-type high-speed twin-roll casting.

4. Fabrication of 4045/3003/4045 clad material

A 4045/3003/4045 clad material used for brazing sheet of automobile heat exchanger was fabricated by this method. The microstructure and mechanical properties were compared with those of the conventional hot roll bonded clad materials. The clad sheets having the same thickness (6.0 mm) and clad ratio (12%) were fabricated both by the hot roll bonding method and the vertical-type high-speed tandem twin-roll casting. The clad sheets were cold rolled to 0.17 mm in thickness, annealed at 400 °C for 2 h, cold rolled to 0.1 mm in thickness, and finally subjected to heat treatment for brazing at 600 °C for 3 min. Microstructures and mechanical properties of the clad sheets were investigated at each stage. Al-Mn-base dispersed particles in the core (3003) and Si particles in the overlay (4045) of the cold rolled sheet with 0.17 mm thickness originating from the twin-roll cast sheets were smaller than those originating from the hot roll bonded sheets, as shown in Fig. 4. A fully recrystallized structure was observed after annealing in the sheet originating from the hot roll bonding, while the fibrous structure still remained in the sheet originating from the twin-roll casting. Yield strength, tensile strength, and elongation of the annealed sheet originating from the twin-roll casting were higher than those originating from the hot roll bonding, as shown in Fig. 5. After heat treatment for brazing, the tensile strength was the same in both clad sheets. Consequently, it was demonstrated that the present method can produce clad materials with good mechanical properties equivalent to those of the hot roll bonded clad sheet. It was also effective with respect to the large reduction in a number of process steps, production energy compared to the conventional hot roll bonding process.

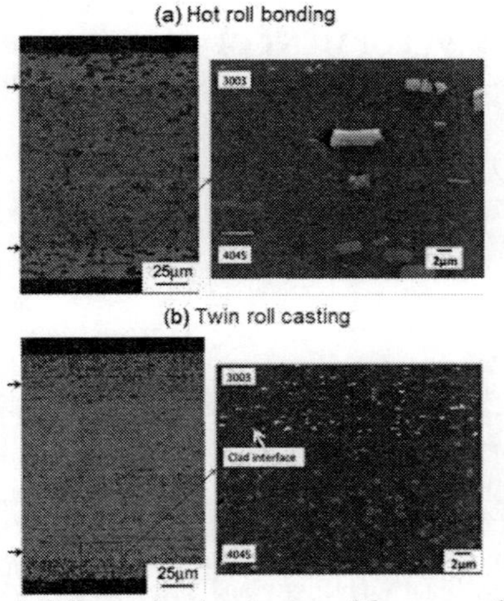

Fig. 4. Difference in microstructure of the cross section of the clad sheet fabricated by hot roll bonding and twin roll casting.

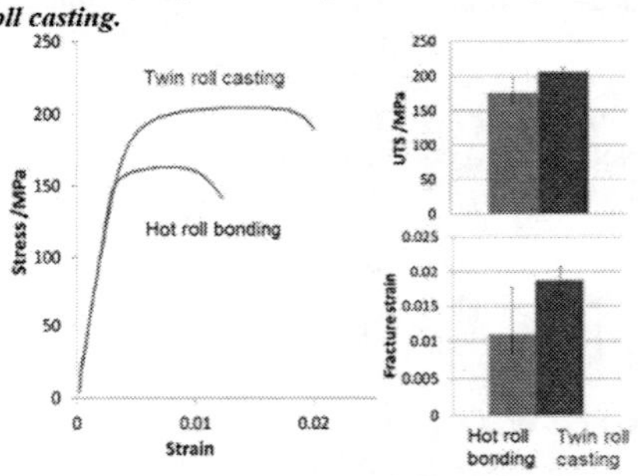

Fig. 5. Comparison of mechanical properties of the clad sheet fabricated by hot roll bonding and twin-roll casting.

5. Summary

In order to make the best use of the vertical-type high-speed twin-roll casting, the *in situ* fabrication of 4045/3003/4045 clad material was performed. The clad strip with no interface reaction layer between core and skin materials was successfully obtained in a single step. The clad sheets produced by using the clad strip exhibited superior mechanical properties compared to those fabricated by the conventional hot roll bonding process.

References

[1] Shinji Kumai, (2017), Fundamentals and Applications of Vertical-type High-speed Twin-roll Casting, USB Proceedings of the 14th Asian Foundry Congress, Incheon, Korea, Korea Foundry Society.

[2]Shinji Kumai, Yusuke Takayama, Ryoji Nakamura, Daisuke Shimosaka, Yohei Harada and Min-Seok Kim, (2017), Application of Vertical-type High-Speed Twin-Roll Casting for Up-Grade Recycling and Clad Sheets Fabrication of Aluminum Alloys, Materials Science Forum, Vol. 877, pp 56-61, Trans Tech Publications, Switzerland.

The 73rd World Foundry Congress
„Creative Foundry"
23 - 27 September 2018, Krakow, Poland

AN INVESTIGATION ON MICROSTRUCTURAL AND MECHANICAL PROPERTIES OF CERAMIC MOULDS APPLIED IN THE INVESTMENT CASTING OF CRITICAL PARTS OF AIRCRAFT ENGINES

Marzanna Ksiazek[1]*, Adam Tchorz[1], Izabela Krzak[1], Krzysztof Zaba[2], P. Kurdziel[3], A. Dydak[3], R. Sitek[4], J. Mizera[4]

[1] *Foundry Research Institute, 73 Zakopianska Str., 30-418 Cracow, Poland*
[2] *AGH University of Science and Technology, Department of Non-Ferrous Metals, 30 Mickiewicza Ave, 30-059 Cracow, Poland*
[3] *The Specodlew Enterprise of Foundry Innovation Ltd., 3 Rotmistrza Witolda Pileckiego Str., 32-050 Skawina, Poland*
[4] *Warsaw University of Technology, Department of Materials Science, 141 Wolska Str., 02-507 Warsaw, Poland*
**Corresponding address: marzanna.ksiazek@iod.krakow.pl*

Key words: ceramic moulds, investment casting, porosity, microstructure, bending strength

The present study is concerned with the investigation of microstructure, mechanical properties and porosity of two types of ceramic moulds designed for precise casting of parts of aviation turbines, the method by lost wax casting, i.e. moulds based on quartz with an alcohol and water binder. The bending strength was determined for model ceramic moulds depending on the annealing temperature. In order to verify the technological quality of casting ceramic moulds, created on the basis of quartz and two different binders, SEM microstructure tests were carried out and the average porosity of casting ceramic moulds was determined using X-ray computed tomography (CT). The results obtained prove that the application in moulding mixtures of water-dispersive binders, it increases the mechanical properties of moulds with definitely better their permeability.

1. Introduction

In the process of manufacturing of the critical parts of aircraft engines, there is commonly used the method of precision casting in multilayer ceramic moulds (CM) referred to as the method of investment patterns. In ceramic moulds, there are distinguished layers that differ considerably one from another and have entirely different destination. The first layer (at the model) is designed to reproduce the original shape of the model and also to ensure that there is no chemical interaction with a model material during melting inside the autoclave, and to prevent the reaction between the liquid metal and the ceramic mould during casting of the liquid metal. Subsequent (constructional) layers are responsible for the mechanical strength of the mould, which should be high enough to prevent the mould from being damaged in the casting process and sufficiently elastic to avoid cracks in castings caused by too high stresses during crystallization of the casting. [1-3].

The basic components forming CM are: base material (e.g. sand grains or quartz flour) and binding material (mostly hydrolysed ethyl silicate – $(C_2H_5O)Si$). Presently, due to environmental regulations, more and more often is used more expensive binding material on the base of colloidal silica – sols of orthosilicic acid. Hence, the proposed method of X-ray computed tomography supplemented with a structural and mechanical analysis of the technological construction of a multilayer ceramic mould is a way of controlling its (technical) quality at all stages of its production.

2. Experimental procedure

In this work, there were made 7-layer ceramic moulds for precision castings of critical parts of aircraft engines. The model moulds were made by immersing the wax model set in suitable ceramic mixtures, which in turn were sprinkled with coarse grain ceramic powder. In order to prepare the moulds there were provided 2 ceramic mixtures. The selection of powders and binding materials was made on the basis of recognition of materials used in the industry. There were used materials commercially available in the offer of companies: Wacker Chemical Corporation (ethyl silicate), Grace Davision (Ludox), Remet UK (Remasol), Quartz Sand Mine „Grudzien Las" (flour and quartz sand). The composition of mixtures in the individual layers of CM is listed in Table 1.

Model samples of ceramic moulds (CM) selected for tests and designated as **K** (matrix of quartz sand and ethyl silicate binder and **R** (matrix of quartz sand and Remasol binder). The moulds are based on melted models, basing on the mixture of ceramics and the binder. The first layer, so-called prime coat, is made of quartz flour with the grain size of 75 μm as well as Ludox and and Remasol Premium Plus binders. Six structural layers were made of quartz sand and the binder - hydrated ethyl silicate and Remasol Plus.

The moulds were dried after applying each layer in a forced-air drying at ambient temperature and humidity for 12 h.

Table 1: Composition of ceramic mixtures for individual layer of the ceramic mould (CM)

Mould	1-st layer	Constructional layers (2-7)	Sprinkling
K	Base material quartz flour (75 μm) binder: Ludox	Base material quartz sand (0.1-0.3 mm) binder: ethyl silicate	quartz sand on the layer 7 coarse grain sand (0.5-1 mm)
R	Base material quartz flour (75 μm) binder: Remasol Premium Plus	Base material quartz sand (0.1-0.3 mm) binder: Remasol Plus	quartz sand on the layer 7 coarse grain sand (0.5-1 mm)

The mechanical strength of model samples of ceramic moulds was determined in a static bending test. The test was performed under conditions of 4-point bending of square-section beams. Samples were loaded to the breaking point at a rate 1mm/min, recording the value of breaking strength. The bending strength of the ceramic material was determined using the formula:

$$\sigma_f = \frac{3}{2}\frac{F_f(L-l)}{d \cdot h^2} \quad (1)$$

where: σ_f- breaking strength on bending [MPa]; F_f - breaking load [N]; L- spacing of supports: L=25 [mm]; l-load distribution [mm]; d- sample width [mm]; h- sample height [mm].

Microstructures of the casting ceramic moulds were also observed in FEI's Scios scanning microscope. In addition, using X-ray technique of computed tomography (CT), the porosity of casting ceramic moulds made from the above mentioned ceramic mixtures was estimated.

3. Result and discussion

Fig. 1 shows the maximum bending strengths for K and R moulds in the "raw state" and after the annealing process at temperatures: 850, 900 and 950°C. The highest bending strength is characteristic to R type CM after annealing at the temperature 950°C for 1 h. This is due to the fact that at this temperature takes place the process of sintering of the ceramic mould material, which significantly increases the mechanical strength. These moulds have also better surface quality; Ra parameter equals to 2.10μm. This means that CM, based on water binding materials, are characterized by greater resistance to failure during the process of wax model melting and pouring the liquid alloy.

The results of microstructural examinations of cross sections of casting ceramic moulds indicate a more developed surface of SiO_2 particles and their differentiation in size with alcoholic binding material (Fig. 2). Compounds with colloidal silica have a more compact structure, SiO_2 particles are sharp-edged and irregular and smaller in size, which is likely to favour better coverage of the wax model surface (ensure a better compaction in the model compound). Particles of smaller sizes have slower sedimentation,

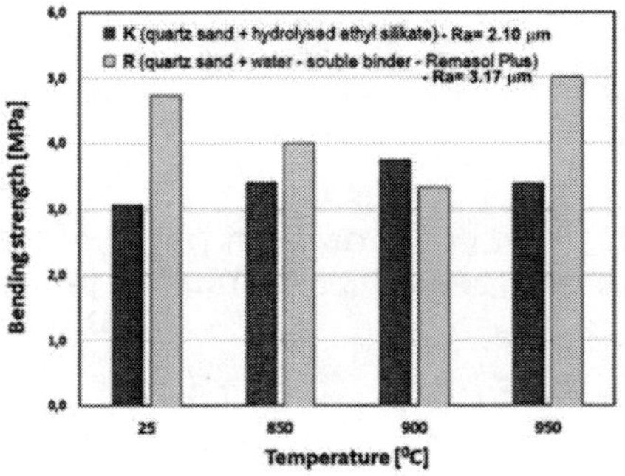

Fig. 1. The bending strength of CM samples at temperatures of: 25, 850, 900 and 950° C.

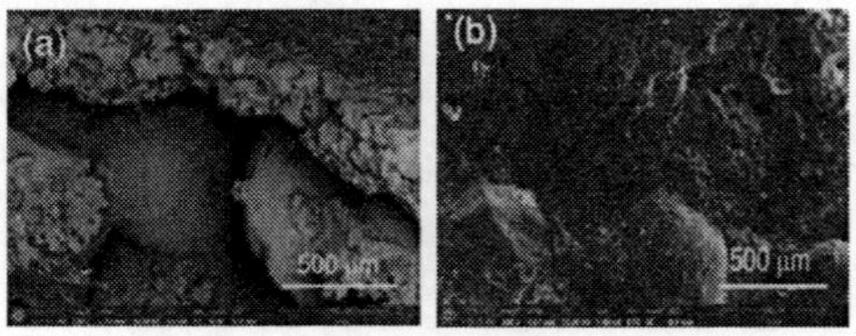

Fig. 2. Scanning micrograph of cross-sections of samples of ceramic moulds: (a) type K and (b) type R

providing a uniform consistency of the poured compound in the mould making process, which as a consequence ensures good strength properties of CM. Porosity of ceramic casting moulds (type K and R) is similar and is included in the range of 2.6 - 2.9%. The distribution of pores is uniform on the layers cross-sections, which guarantees a sufficient permeability of moulds.

4. Summary

It has been demonstrated, that quartz based multi-layered ceramic moulds with water-dilutable and alcohol based binding materials for precision casting of critical parts of aircraft engines meet the basic technological criteria applied in the precision foundries: (i) the porosity of ceramic moulds is comparable to the both applied binding materials, and (ii) as a result of the introduction of „water dilutable" binding material in the place of ethyl silicate, there was improved the mechanical durability and the quality of ceramic moulds.

Acknowledgments

This work has supported by The national Centre for research and Development under the project No PBS3/A5/54

References

[1] Chakrabarti B.K. (2002). Drying conditions and their effect on ceramic shell investment casting proces. Materials Science and Technology. 18 , pp. 935-940

[2] Karwiński, A., Sobczak, N., Leśniewski, W., Wieliczko, P. (2010). The Technology of Ceramic Mould Manufacture for Investment Casting of Titanium Alloys. Proceedings of the 69th World Foundry Congress

[3] Matysiak H. (2011). Porosity and durability of ceramic moulds used in Bridgman precision casting process. Materials Engineering. 1, pp. 17-21 (in Polish)

The 73rd World Foundry Congress
„Creative Foundry"
23 - 27 September 2018, Krakow, Poland

EFFECT OF TEMPERATURE ON THE STRUCTURE AND MAGNETIC PROPERTIES OF ZINC FERRITE NANOPARTICLES

Angelika Kmita[1]*, Jan Żukrowski[1], Juliusz Kuciakowski[1,2], Antoni Żywczak[1], Marcin Sikora[1]

[1]AGH University of Science and Technology, Academic Centre for Materials and Nanotechnology, A. Mickiewicza 30 Ave., 30-059 Krakow, Poland
[2] AGH University of Science and Technology, Faculty of Physics and Applied Computer Science, A. Mickiewicza 30 Ave., 30-059 Krakow, Poland
**Corresponding address: akmita@agh.edu.pl*

Key words: thermal decomposition, zinc ferrite nanoparticles, magnetic properties, Mössbauer spectroscopy

1. Introduction

Nanomaterials, due to their special properties, have attracted interest of foundry industry [1] e.g. in:

a) *investment casting:*
- liquid ceramic slurry with colloidal silica
- modification of wax pattern surfaces by colloidal solutions of nanoparticles.

b) *moulding sands technology:*
- adhesive nanocomposite (binder) e.g.: water glass modified by nanoparticles,
- protective coatings with nanoparticles,

c) *in alloys:*
-modern nanocrystalline materials, such as e.g.: Ni-Al alloys etc. [2].

Among the nanomaterials available in large quantities, zinc ferrite nanoparticles might be attractive for emerging modifications in casting and moulding sands technologies. Zinc ferrite $ZnFe_2O_4$ belongs to a group of ferrites, i.e. iron oxides of the general formula $(Me_{1-x}^{2+}\ Fe_x^{3+})[Me_x^{2+}\ Fe_{2-x}^{3+}]O_4$, where ( ) denotes tetrahedral sites and [] denotes octahedral sites.

The oxygen atoms build a network of close-packed face-centered cubic lattice/system of atoms. At x=0 divalent cations reside in tetrahedral sites (A) while Fe^{3+} cations reside in octahedral sites [B] of normal spinel structure. At x = 1 an inverted spinel structure is created, while at the intermediate x, the mixed spinel is formed [3,4]. Zinc ferrite $ZnFe_2O_4$ is widely used in various fields as e.g.: magnetic storage, biomedicine, ferrofluids [4,5], catalyst in the hydrogen production from the decomposition of water, catalysts in methanol decomposition on CO and hydrogen [4], or as decomposition catalyst greenhouse gases (CO_2) [4], for electromagnetic wave absorber applications [5], nanofiller in rubber composites [8] or in casting technology [3,6]. The studies of thermal decomposition of zinc ferrite material carried out up to date comprised of micro-sized and larger materials, and aimed at the reclaiming zinc from waste material containing zinc ferrite or the production of hydrogen from water splitting under the action of concentrated solar energy [7].

Stoichiometric zinc ferrite has nominally a normal spinel structure. However, upon increasing Fe/Zn molar ratio in nonstoichiometric zinc ferrite an increasing occupation of tetrahedral site by Fe ions is expected due to charge transfer from tetrahedral to octahedral Fe sites that are responsible for occurrence of net magnetic moment. In the case of the nanoparticles, the situation is even more complicated due to interplay of the inversion of the cations distribution, size effects, and non-stoichiometry [3]. Properties of the nanoscale ferrite materials depend mainly on the chemical composition, size and morphology (porous nanorods), hollow microspheres etc. [4].

Reducing the particle size of the magnetic substances including ferrites from micron to nanometer size results in a significant change in their properties, e.g. physico-chemical, magnetic. Many of these properties, e.g. magnetic coercivity (H_c), magnetic susceptibility (X), depends not only on the size of the nanoparticles, but also on their shape and morphology. It comes from the fact that with decreasing particle size, the magnetic energy decreases, which at a particle size is comparable to the thermal energy. As a result, there are thermal fluctuations of the magnetic moments of the particles, which have a strong influence on their properties. Zinc ferrite of micron and larger size, at room temperature is paramagnetic, and nano-sized zinc ferrite at the same temperature is superparamagnetic or ferrimagnetic. Thus, it is reasonable to question whether nanocrystalline ferrites undergo thermal decomposition in the same way as their mesoscale counterparts.

In this article we discuss the decomposition processes of non-stoichiometric zinc ferrite nanoparticles as a functionalizing agent for foundry binders. The process can be used in casting or in the technology where active oxygen is needed, e.g. water splitting into H_2 (obtaining clean energy) or decomposition of hazardous compounds from the BTEX group [4].

2. Materials and methods

The starting material for this study was zinc ferrite nanoparticles: **sample A** ***($Zn_xFe_yO_4$, were molar ratio Zn:Fe ≅ 0.5)*** particles having an average grain size of 50 - 100 nm (Fig 1a) and **sample B** ***($Zn_xFe_yO_4$, were molar ratio Zn:Fe ≠ 0.5)*** (Fig. 1b) particles having an average grain size of 5 nm. The morphology of the "as prepared" material and upon thermal decomposition was performed using a transmission electron microscope (TEM).

a) b)

Fig. 1. TEM image of zinc ferrite nanoparticles: a) sample A; b) sample B

The chemical composition of nanoparticles was analysed with inductively coupled plasma optical emission spectroscopy (ICP-OES) using a Varian Vista-Pro CCD simultaneous ICP-OES (Agilent 5100). Transition temperatures and specific heat of phase transformation in the process of decomposition were conducted using the thermal analysis/differential scanning calorimetry (TGA/DSC, TA INSTRUMENTS SDT Q600). The experiment was conducted in Ar (99.9999%) atmosphere, carrier gas flow of 100 mL/min with heating rate ß =10 °C/min. The heating temperature range was from 50 to 1400 °C. Determination of the level of defects in the output material (sample A and B) and the degree of decomposition of there were made by means of Mössbauer spectroscopy (MS-4 Renon spectrometer) in the temperature range of 80-300 K.

3. Results

Thermogravimetric analysis of zinc ferrite revealed the existence of characteristic temperature ranges, in which there are a significant losses in the mass of the samples. An exemplary TG/DTG plot for sample A is visible in Fig. 2.

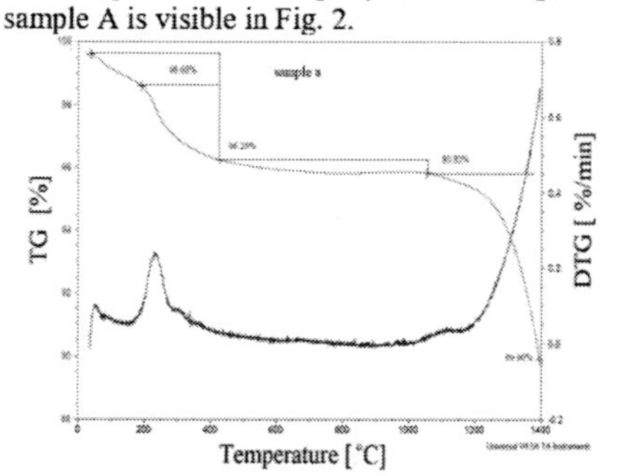

Fig. 2. TG/DTG curve of the zinc ferrite nanoparticles for sample A

Mössbauer spectra were collected for nanoparticles of zinc ferrite and for the final products of decomposition for sample A and for sample B. An exemplary Mössbauer spectra for sample A are shown in Fig. 3. They reveal a significant modification of iron containing phases.

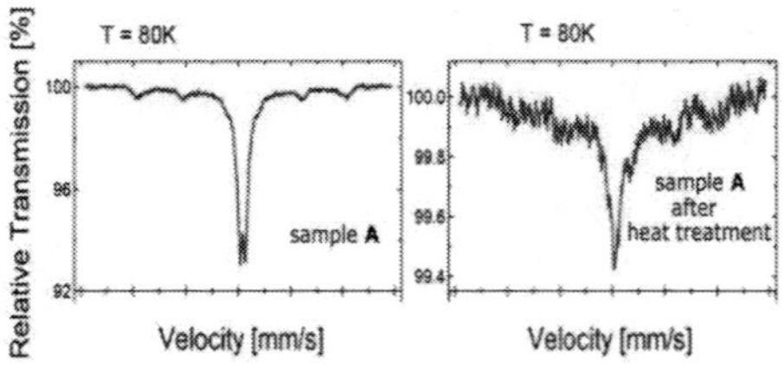

Fig. 3. ^{57}Fe Mössbauer spectra measured at 80 K for nanoparticles (sample A) and products of their thermal decomposition

4. Summary

As a result of thermal decomposition:

- products obtained with particle size at the nano and mesoscale level of the order of 200 - 500 nm,
- TG /DTG analysis showed a loss weight at 1400 °C in an amount of: 7.91% for sample A and 12 % for sample B,
- Mössbauer spectroscopy have shown that the decomposition products were rich in Fe^{2+} (not present in the starting zinc ferrite) and Fe^{3+}.

Acknowledgments

This work was supported by the National Science Centre, Poland (grant number UMO-2016/23/D/ST8/00013)

References

[1] A. Kmita, R. Dańko, M. Holtzer, B. Hutera, & B. Stypuła. (2014). Utilisation of nanoparticles in technologies of producing castings for the needs of power engineering. Nanomaterials and Energy. 3, 53-60.

[2] J. Sobczak. (2003). Selected aspects of nanotechnology and nanomaterials. Composites 3/8, 387-391 (in Polish).

[3] A. Kmita, A. Pribulova, M Holtzer, P. Futas & A. Roczniak. (2016). Use of specific properties of zinc ferrite in inovative technologies. Arch. Metall. Mater. 61 4, 2141-2146.

[4] M. Qin, Q. Shuai, G. Wu, B. Zheng, Z. Wang & H. Wu. (2017). Zinc ferrite composite material with controllable morphology and its applications. Materials Science & Engineering. B 224 125-138.

[5] S. J. Yaghmour, M. Hafez, K. Ali & W. Elshirbeeny. (2012). Polymer Composites, 4, 1672-1677.

[6] A. Kmita, J. Żukrowski, K. Hodor, H. Smogór & M. Sikora. (2017). Zinc ferrite nanoparticles as perspective functional materials for applications in casting technologies. Metalurgija. 56, 1-2, 29-32.

[7] B. Janković, S. Stopić, A. Guven & B. Fiedrich. (2014). Kinetic modeling of thermal decomposition of zinc ferrite from neutral leach residues based on stochastic geometric model. Journal of Magnetism and Magnetic Materials. 358/359, 105-118.

The 73rd World Foundry Congress
„Creative Foundry"
23 - 27 September 2018, Krakow, Poland

INFLUENCE OF THE HEATING RATE ON THE THERMAL DECOMPOSITION KINETICS OF THE CHEMICAL CURED BINDER

Mariusz Holtzer[1], Grzegorz Piwowarski[1], Agnieszka Roczniak[1], Angelika Kmita[2*]

[1]AGH University of Science and Technology, Faculty of Foundry Engineering, A. Mickiewicza 30 Ave , 30-059 Krakow, Poland
[2]AGH University of Science and Technology, Academic Centre for Materials and Nanotechnology, A. Mickiewicza 30 Ave., 30-059 Krakow, Poland
**Corresponding address: akmita@agh.edu.pl*

Key words: thermal decomposition, chemical cured binder, kinetic, activation energy.

1. Introduction

The article discusses thermal decomposition of chemical cured binder in oxygen conditions. This binder is used in the moulding sand technology_for the production of metal castings. The process was studied using a convectional dynamic thermogravimetric analysis TG/DTG and DSC at different heating rates ß: 5 - 30 (K min^{-1}) under synthetic air purge. The parameters of thermal degradation of chemical cured binder kinetics was investigated using the two different kinetics models: (Kissinger) and (Ozawa-Flynn-Wall) methods.

There are many reasons why the mechanism and kinetics of thermal decomposition of synthetic resins are very important [1-6]. From a practical point of view, the most important is that the rate and the temperature of thermal decomposition, as well as its thermal hazard potential.

Knowledge of these parameters is especially important in case of applying synthetic resins as binders for moulding and core sands in the metal casting industry, where they are subjected to the high temperature influence of the order: 1000-1500 °C (liquid alloy temperature), it means significantly above the thermal decomposition of resins (500-600 °C). Under the influence of the heat originated from the solidifying casting the binder thermal destruction occurs and several substances - often hazardous for the environment and for humans - are generated. The kind and amount of these compounds depends on resin applied as the binder base, on the temperature and the time of this temperature influencing the moulding sand and also on the character of the atmosphere in the mould (inert, oxidising, reducing). Investigations were performed in the synthetic air atmosphere.

The objective of this work was to calculate the kinetic parameters (activation energy) using TG/DTG and DSC curves obtained by thermogravimetry analysis of the commercial

2. Materials and methods

The subject of the study was commercial binder (the ester - cured alkaline phenolic resin) used in the sand moulding process.

TG/DTG/DSC methods

Thermal degradation of cured resin were performed using Thermogravimetric Analysis /Differential Scanning Calorimetry (TGA/DSC). The system is equipped with an apparatus for simultaneous thermogravimetric and differential scanning calorimetry analyses (TA INSTRUMENTS SDT Q600). Sample masses ranging about 10 mg heated from 20 °C to 700 °C, at heating rates ß: 5 - 30 (K min^{-1}). Synthetic air was used as carrier gas at a flow rate of 100 mL/min.

3. Results

The DSC curves recorded during the 50 up to 750 °C of the cured resin for different heating rate are shown in Figure 1 and Figure 2.

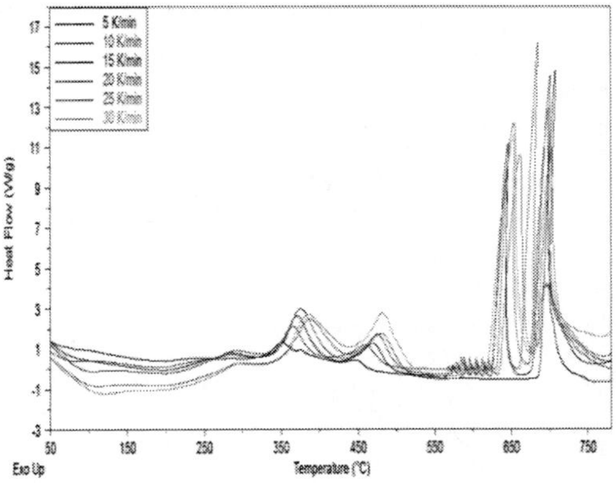

Fig. 1 DSC curves of the cured binder, under a synthetic air (50 mL/min) with a different heating rate

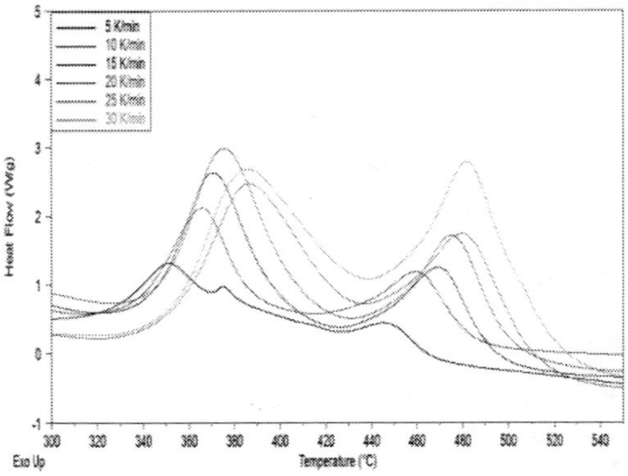

Fig. 2 DSC curves of the cured binder, under a synthetic air (50 mL/min) with a different heating rate

4. Kinetic analysis

Generally, the thermal degradation of the solid materials may be assumed that occurs after following equation:

$$A \rightarrow B_{liquid\ or\ solid} + C_{gas}$$

The equation which describes the rate of the heterogeneous thermal degradation reaction of a solid, at constant heating rate, has the following form:

$$d\alpha/dt = k(T)f(\alpha)$$

where: α - conversion degree of the reaction at constant temperature:

Expressing the dependence of the reaction rate constant k on the temperature by the Arrhenius equation:

$$k(T) = A \exp(-E/RT)$$

Activation energy of thermal degradation reaction for resin can be calculated by:

- **Kissinger equation expression as:**

$\ln(ß/T^2_p) = -E_a/RT_p + \ln(AR/E_a)$

A and E_a were obtained by the plotting of - ln $(ß/T_p^2)$ versus $1/T_p$ at different heating rates ß.

Calculate an value for activation energy E_a as follows:

$E_a = -2.19 \times 8.314\ J\ mol^{-1}K^{-1}\ [d\ lg(ß/T^2_p)]/\ d\ (1/T)]$

- **Ozawa-Flynn-Wall equation expression as:**

$\log ß = \log(AE/R) - 2{,}315 - 0{,}4567\ (E/RT_p) - \log f(\alpha)$

A and E were obtained by the plotting of log ß versus $1/T_\alpha$ at different heating rates.

Calculate an value for activation energy E as follows:

$E = -2.19 \times 8.314\ J\ mol^{-1}K^{-1}\ [d \log ß / d(1/T)]$

4. Summary

These methods are based on the principle according to which the reaction rate at a constant conversion is only a function of temperature. The apparent activation energy (AAE) is depend on the destruction degree of binder. This process occurs in a few stages of different activation energy. It was shown that the mechanism of thermal destruction of the tested binder under oxidative conditions is diametrically different than in an inert atmosphere [7]. Since the atmosphere in the mould changes during the casting process the next step of our investigations will constitute measurements in the reducing atmosphere.

Acknowledgments

This work was supported by the National Science Centre, Poland (grant number UMO-2016/23/D/ST8/00013).

References

[1] Y. Chen, Z. Chen, S. Xiao & H. Liu. (2008). A novel thermal degradation mechanism of phenol-formaldehyde type resins. Thermochim. Acta 476, 39-43.

[2] M.V. Alonso, M. Oliet, J.C. Domínguez, E. Rojo & F. Rodríguez. (2010). Thermal degradation of lignin–phenol–formaldehyde and phenol–formaldehyde resol resins. J. Therm. Anal. Calorim. 105, 349-359.

[3] V.S. Sereblo, B.A. Rits & M. Srgeeva.(1974). Curing and thermal degradation of phenol-formaldehyde binders. Russ. Cast. 2 55-56.

[4] B. Howell, C. Yumin & B. Duane. (2003). Assessment of the thermal degradation characteristics of isomeric poly(styrene)s using TG/MS and TG/GC/MS. Thermochim. Acta 396, 167-177.

[5] Authors' own research unpublished.

[6] J.J. Sunol, J. Saurina, F. Carrillo & X. Colon. (2003). Comparison of the thermal behavior of three cellulosen fibres mercerized or submitted to solar degradation. J. Therm. Anal. Calorim. 72, 753-758.

[7] A. Kmita, G. Piwowarski, A. Roczniak & M. Holtzer. The activation energy of the thermal degradation of the cured alkaline resole resin. (in press).

The 73rd World Foundry Congress
„Creative Foundry"
23 - 27 September 2018, Krakow, Poland

REVEALING OF MICROSTRUCTURAL FEATURES IN HAYNES® 282® SUPERALLOY SUBJECTED TO VARIOUS HEAT TREATMENTS

Adelajda Polkowska[1]*, Wojciech Polkowski[1], Małgorzata Warmuzek[1], Grzegorz Włoch[2] Robert M. Purgert[3]

[1] Foundry Research Institute, Zakopiańska 73 Str., 30-418 Krakow, Poland
[2] AGH University of Science and Technology, Faculty of Non-Ferrous Metals, Mickiewicza 30 Av., 30-059 Krakow, Poland
[3] Energy Industries of Ohio, 6100 Oak Tree Blvd, Cleveland, OH 44131, USA
**Corresponding address: adelajda.polkowska@iod.krakow.pl*

Key words: nickel superalloy, microstructure, chemical etching, metallography, microscopy

1. Introduction

Despite of an observed development of many "new and advanced" materials (e.g. metal matrix composites, carbon fiber-composites, bulk metallic glasses or intermetallic based alloys), recent market analyses and forecasts clearly point toward the irreplaceability of nickel superalloys in the most economy and ecology related crucial areas, including transportation, energy production, manufacturing and other applications where, high-performance and reliable materials are required [1]. Due to an outstanding combination of very good creep resistance and corrosion resistance at high temperature and chemically aggressive environments, superalloys are widely applied in devices working under extreme conditions. In fact, actually superalloys represents up to 70 % of the weight of new gas-turbine engines (that is the fundamental application field of these materials) [2]. Moreover, superalloys are key players in the global efforts devoted to a development of new highly efficient power generators e.g. clean coal-fired ultra-supercritical steam plants [3].

In the view of a well-established positive dependence between raising the working temperature and an increase of engines and power plants efficiency (also in terms of a more economic management of non-renewable fuels), a continuous works on an improvement of superalloys material and technology design are carried out.

Haynes® 282® alloy is newly developed age hardened, wrought nickel-based superalloy designed for high temperature applications. This wrought alloy may successfully substitute actually applied superalloys (e.g. Inconel 718, Waspaloy, R-41) in many gas turbine engine applications by substantially exceeding their working temperature limitations. However, due to a strategic importance of Haynes® 282® alloy, so far a very limited information may be found about the effect of applied processing on structural changes taking place upon the proposed annealing treatment (also regarding a precipitation kinetics and distribution of "second phase" particles).

As recommended by the producer of Haynes® 282® alloy, its typical heat treatment should include following consecutive stages [4]: (i) a solution treatment carried out in in the range of 1050 to 1200°C with air cooling or water quenching followed by a two-step aging treatment. The first step of aging is applied for the sake of initiation of γ' precipitation and carbide stabilization, while the second for completing the precipitation of γ' phase. Nevertheless, since the selection of heat treatment parameters strongly alter the morphology of precipitated phases and thereby affect the properties of the alloy, a detailed microstructural characterization seems to be a necessary supplementation to the alloy treatments.

In this work, we have applied various metallographic and microscopic techniques in order to reveal microstructural features in Haynes® 282® nickel based superalloy after each particular step of its heat treatment.

2. Materials and methods

The material investigated was Haynes® 282® nickel superalloy in the form of sheets with dimensions of 100×100×1.60 mm^3. The nominal composition of alloy is listed in Table 1.

Table 1. The nominal chemical composition of Haynes 282 alloy (in weight %).

Cr	Co	Mo	Ti	Al	Fe	Mn	Si	C	B	Ni
20	10	8.5	2.1	1.5	1.5	0.3	0.15	0.06	0.005	bal.

The material provided by the Haynes Company in a solution annealed state was subjected to the heat treatment including three steps as follows: (i) the solutionizing at 1150°C/1h/air cooling; (ii) the first step aging at 1010°C/2h/air cooling and (iii) the second step aging at 788°C/8h/air cooling.

After that, small pieces were cut off from the heat treated sheets, and then subjected to microstructural characterization. The cross-sectioned specimens were mounted and subjected to mechanical grinding and polishing. Subsequently, the polished surfaces were chemically etched in order to reveal specific microstructural constituents. The following etchants were applied: (i) Beraha: 33 ml HCl+67 ml destilled water+1 g $K_2S_2O_5$; (ii) Marble: 10 g $CuSO_4$+100 ml HCl+50 ml destilled water; (iii) Aqua regia – 20 ml HNO_3+60 ml HCl; (iv) 50 ml HCl+1 ml destilled water.

The microstructural characterization was performed by means of light microscopy (LM) (Carl Zeiss Axio Observer ZM10) and scanning electron microscopy coupled with energy dispersive X-ray spectroscopy (FEI Scios FEG SEM/EDS) working under various imaging modes.

3. Results and conclusions

The Haynes® 282® alloy in the as-received state (in as solution annealed condition) was characterized by an average grain size of matrix (the γ solid solution) of around 150 µm. Additionally, beside of primary Mo- and Ti- rich MC-type carbides preferentially located inside grain volumes (Fig. 1), some Ti-based nitrides were also found (having orange color as observed on light microscope – Fig. 2).

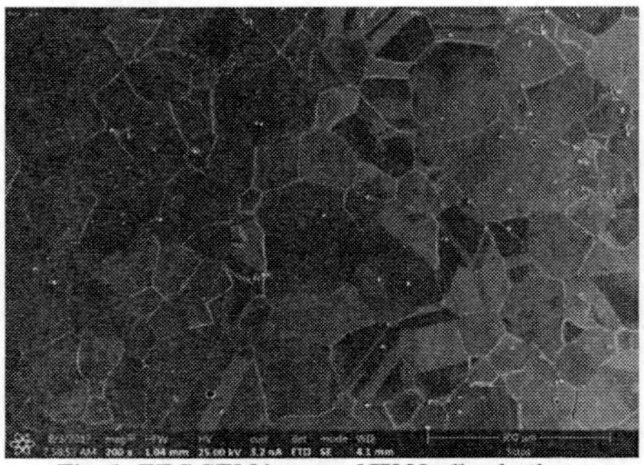

Fig. 1. FEG SEM image of H282 alloy in the as received state (as-solution annealed condition). Chemical etching in aqua regia.

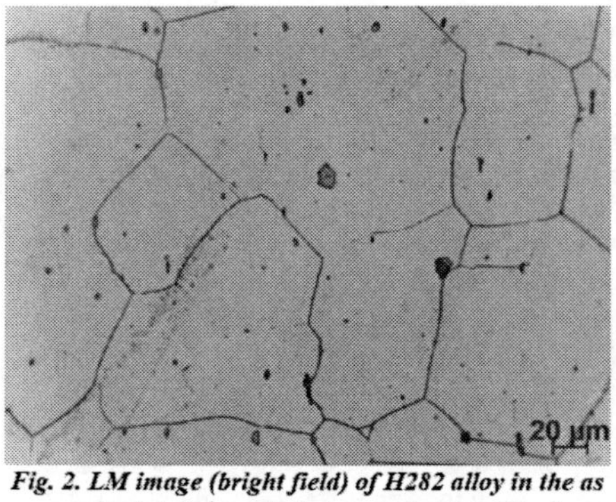

Fig. 2. LM image (bright field) of H282 alloy in the as received state (as-solution annealed condition). Chemical etching in aqua regia.

The first step of aging treatment (at 1010°C/2h/air cooling) resulted in a precipitation of Cr-enriched $M_{23}C_6$ carbides distributed along grain boundaries (Fig. 3). Since the application of second step aging at 788°C/8h gave the grain boundary formation of undesirable needle-like σ phase (Fig. 4) that exhibits detrimental effect on mechanical properties, some further optimization of heat treatment parameters, is needed.

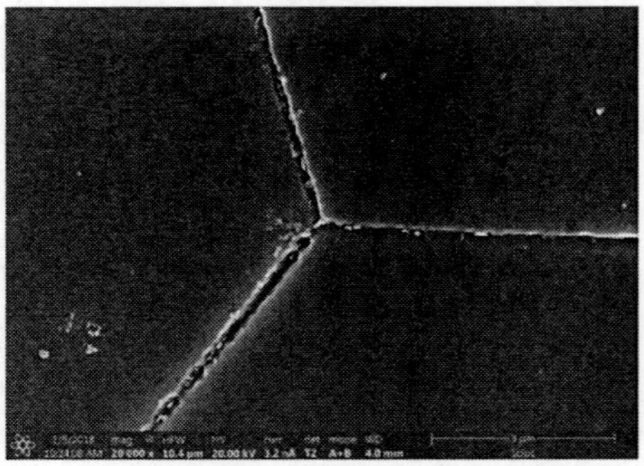

Fig. 3. FEG-SEM image (T2 in-lens detector) showing Cr-enriched $M_{23}C_6$ carbides distributed at grain boundaries after the first step of aging treatment. Chemical etching in the Marble reagent.

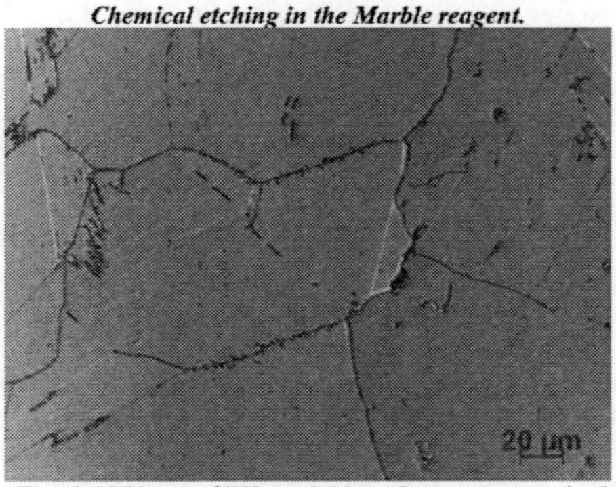

Fig. 4. LM image (differential interference contrast) of H282 alloy showing precipitation of needle like σ phase after the 2nd aging step. Chemical etching in Marble reagent.

Acknowledgments

A financial support from the Polish National Science Centre under Grant no. UMO-2016/23/D/ST8/01269 is gratefully acknowledged.

References

[1] J. Bedder, R. Baylis. (2013). Into the melting pot: the superalloy market and its impact on minor metals. Minor Metals Conference. 28 February - 1 March 2013, London

[2] U.S. International Trade Commission. (1986). Competitive assessment of the U.S. forging industry: report to the President on investigation no. 332-216 under section 332 of the Trade Act of 1930, as amended.

[3] R. Viswanathan, J.F. Henry, J. Tanzosh, et al.. (2005) U.S. program on materials technology for ultra-supercritical coal power plants, Journal of Materials Engineering and Performance 14, 281-292

[4] HAYNES®282®ALLOY:http://www.haynes.ch/doc/haynes/282_h3173.pdf

The 73rd World Foundry Congress
„Creative Foundry"
23 - 27 September 2018, Krakow, Poland

MICROSTRUCTURE EFFECTS OF Y_2O_3 ADDITION TO A356 ALLOY

Mohamed E. Moussa[1], Shimaa El-Hadad[1,*], Waleed Khalifa[2]

[1]Central Metallurgical Research & Development Institute, P.O. Box 87, Helwan-Egypt

[2] Cairo University, Faculty of Engineering, 12613, Giza-Egypt

*Corresponding address: shimaam@yahoo.com

Key words: Please provide 5 key words divided with commas - Times New Roman, size 9 (use style Key words)

Aluminum alloys are the major lightweight alloys used in the automotive industry. [9]. Improving the mechanical properties of these alloys via microstructure control has been the target of many research works. In the foundry industry, microstructure is controlled mainly through modification and grain refinement processes.
The present work investigates the effect of dispersing Y_2O_3 powder in the molten 356 cast alloy on the grain size of α-aluminum. A series of sand casting experiments were carried out with Y_2O_3 addition ranges from 0.0 to 2.5 wt%. The microstructure was investigated and the hardness was evaluated and related to the microstructure of the cast samples. It was observed that addition of can partially refine α-Al grains; however, the amount of addition should be optimized to avoid the possible adverse effects on the microstructure.

1. Introduction

Aluminum and its alloys are the ideal candidates for use in pistons, cylinder heads, intake manifolds, differential housings and the other automotive parts, etc. [1]. Among these alloys, Al-Si alloys are the best candidate for automotive applications because of their unique combination of excellent castability, low density and good mechanical properties. The mechanical properties of these alloys are strongly influenced by their microstructure features. The microstructure is controlled mainly by modification and grain refining processes. Modification has been a well-known foundry practice employed for refining the needle like eutectic silicon to fine particles [2]. Grain refinement of Al-Si alloys is done by addition of grain refiners into the melt before pouring in order to decrease the size of α-Al grains, increase homogeneity and reduce the hot cracking susceptibility. Therefore, many investigations worked on improving the grain refinement process in order to enhance the mechanical properties [3]. The most commonly used chemical grain refiners are Al-Ti-B and Al-Ti-C master alloys. The efficiency of these grain refiners can be decreased in the presence of other elements as Zr and V as shown in Fig. 1 [4]. Some limited works [5,6] recently tried the oxide dispersion as a replacement of the traditional chemical refiners and could obtain promising results. Following the recent concept of grain refining by oxide dispersion, the current work was aiming at evaluating the microstructure effects of Y_2O_3 addition on the sand cast hypoeutectic A356 alloy.

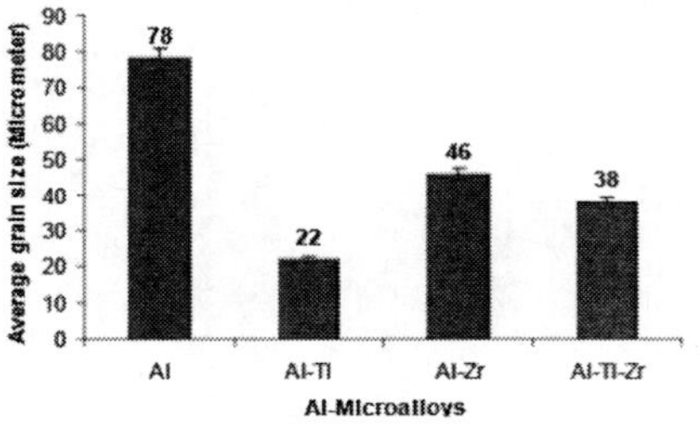

Fig. 1 Effect of Zr addition on Al-alloy refined by Ti addition [4].

2. Experimental work

Alloy A356 alloy was prepared by melting in (100 Kg) induction furnace. The Y_2O_3 was added as master powder mixture of (Al-30 wt%Y_2O_3). The addition levels ranged from 0.0-2.5 wt%. The samples were cut at 2 cm from the bottom of the mould and then prepared for microstructure investigation. Image analysis measurements were done to evaluate the grain size of α-Al. The chemical composition of the alloy excluding Y_2O_3 was shown in Table 1. The hardness of the prepared A356 without and with alloy Y_2O_3 additions was also measured to evaluate the effect of grain refining by Y_2O_3 on the mechanical properties of the samples.

Table 1 Chemical composition of the Al-Si alloys used in the current study in wt%

Al	Si	Fe	Cu	Mg	Mn	Zn
Bal.	7.68	0.016	0.017	0.54	0.01	0.33

4. Results and discussion

The microstructure of the as-cast investigated A356 alloy without Y_2O_3 addition was shown in Fig. 2. The microstructure consists of α-Al grains with eutectic Si as secondary phase. The structure of this sample was fully coarse dendritic since no grain refiners were used. The average grain size of this sample without Y_2O_3 addition was about 220 μm. Figure 4 shows the microstructure of the investigated A356 alloy with different additions of Y_2O_3.

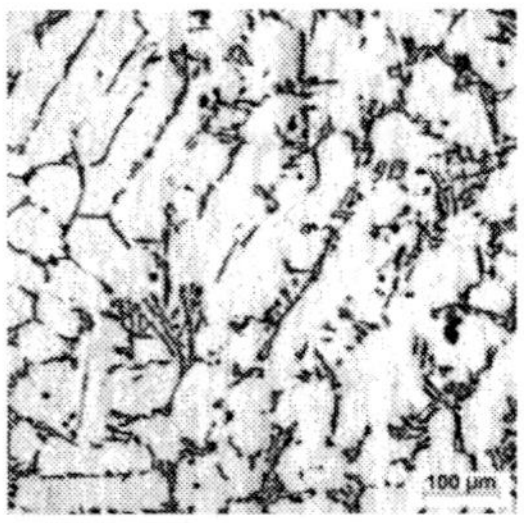

Fig. 3 Microstructure of the A356 sample with 0.0 wt% Y_2O_3

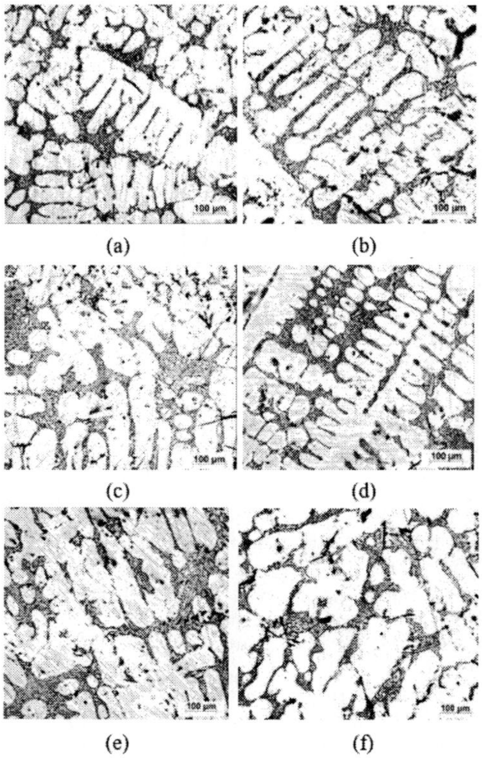

Fig. 4 Microstructures of the cast samples with addition of a) 0.3, b) 0.6, c) 1.0, d) 1.5, e) 2 and f) 2.5 Y_2O_3 in wt%

Comparing microstructure in Fig. 3 to those in Fig. 4, it can be observed that addition of Y_2O_3 altered the size of α-Al grains. In the microstructure of the alloy with addition of Y_2O_3 there are obvious changes in the size of primary α-Al grains.

In Fig. 4(a), with 0.3 wt% Y_2O_3 addition, most of primary α-Al are still dendrites, but their average size is reduced to about 150 μm. With gradual increase in the amount of Y_2O_3, the grain size decreased further. When the Y_2O_3 addition reached 1.5 wt%, most of primary α-Al become rosette-like shape as clear from Fig. 4(d) and their average size was significantly reduced to about 80 μm. Moreover, the dentertic-like phase decreased. However, when the Y_2O_3 content was further increased to 2.5 wt% (Fig. 4e), it was found that the average size of primary α-Al slightly increased again to about 100 μm. Therefore, the optimal addition of Y_2O_3 for obtaining fine grain size of primary α-Al in the investigated alloy was decided to be 1.5 wt%.

Usually, rare elements have strong surface chemical activity, which can be absorbed on the grain interface, hindering the growth of grains, resulting in grain refinement, and thus playing the role of refiners [7]. It was shown that Y can be considered an effective refiner for α-Al in semisolid Al-Si alloy [8]. The refining mechanism of Y_2O_3 comes mainly from its impact on the growth of the primary phase by segregation of Y_2O_3 at the liquid-solid interface which restricts the growth of primary α-Al grains during solidification and reduce the grain size.

5. Conclusions

Y_2O_3 showed a considerable refining influence on the investigated A356 alloy. The morphology of primary α-Al changed from dendritic into rosete like grains after adding Y_2O_3, and its dimension also decreased. The addition of 1.5 wt.% was found to be the optimum amount to have this refining effect.

Acknowledgments

The authors would like to acknowledge Eng. Said Abdeltwab and Eng. Moustafa Shehata for their help during the experimental work.

References

[1] Miller WS, Zhuang L, Bottema J, Wittebrood AJ, De Sme P, Haszler A, Vieregge A. Recent development in aluminum alloys for the automotive industry [J]. Materials Science and Engineering A, 2000, 280, 37-49.

[2] Davis JR. ASM Specialty Handbook: Aluminum and Aluminum Alloys [M]. 2nd ed. ASM International, 1993.

[3] Backerud, L., Chai, G., Tamminen, J., 1990. Solidification Characteristics of Aluminum Alloys. Foundry Alloys. AFS/Skanaluminium, Stokholm, Vol. 2.

[4] Y. Birol, 2006.Grain refining efficiency of Al-Ti-C alloys"*Journal of Alloys and Compounds*, Vol. 422, 128-131

[5] Zheng Liu, Xiao-mei Liu, Yong-mei Hu, 2011. Microstructure refinement by Y_2O_3 in Ssemisolid A356 Alloy. Avanced Materials Research,Vol. 8, 666-70.

[6] Gwang-Ho Kim, Sung-Mo Hong, Min-Ku Lee, Soon-Ho Kim, Ikuo Ioka, Byoung-Suhk Kim, Ick-Soo Kim, 2010. Effect of oxide dispersion on dendritic grain growth characteristics of cast aluminum alloy. Materials Transactions, Vol. 51, 1951-1957.

[7] Y.G. Liu, P.M. Zhang, G. Chen: Nonferrous Metals Vol. 57(2005), p. 15.

[8] Z. Liu, Y.M. Hu, X.M. Liu: Acta Metallurgica Sinica Vol.23(2010), p. 277

The 73rd World Foundry Congress
„Creative Foundry"
23 - 27 September 2018, Krakow, Poland

FABRICATION OF ALUMINUM BASED FUNCTIONALLY GRADED MATERIALS BY CENTRIFUGAL CASTING AND THEIR APPLICATION OF GRINDING WHEELS

Yoshimi Watanabe [1*], Takayasu Sugiura [2], Tadachika Chiba [1], Hisashi Sato [1]

Hideaki Tsuge [3]

[1]Nagoya Institute of Technology, Department of Physical Science and Engineering, Gokiso-cho, Showa-ku, Nagoya 466-8555, Japan
[2]Nagoya Institute of Technology, Department of Engineering Physical, Electronics and Mechanics, Gokiso-cho, Showa-ku, Nagoya 466-8555, Japan
[3]Industrial Research Institute of Gifu Prefecture, Oze 1288, Seki 501-3265, Japan

**Corresponding address: yoshimi@nitech.ac.jp*

Key words: Functionally graded materials (FGMs), Centrifugal casting, Carbon fiber reinforced plastic (CFRP), Diamond, Cubic boron nitride (cBN)

1. Introduction

In functionally graded materials (FGMs), the composition and/or microstructure gradually changes over the volume, resulting in corresponding changes in the properties of the materials. A number of FGMs fabrication methods have been proposed, and one of them is centrifugal method which is application of centrifugal casting. In the method, a centrifugal force applied to a homogeneous molten composite assists the formation of the desired gradation. The composition gradient is then achieved primarily due to the difference in the centrifugal force produced by the difference in density between the molten metal and solid particles. However, since the velocity of a solid particle in a molten metal is dependent on the square of the particle diameter, it is quite difficult to control graded distributions of dispersion nanoparticles in FGMs in the case of the conventional centrifugal casting.

Alternatively, as a new processing technique for FGMs, the centrifugal mixed-powder method has been proposed for overcoming these problems. At first, a predesigned mixed powder consists of metal-matrix particles and dispersion particles is prepared. Basically, the melting point of dispersion particles should be higher than that of metal-matrix particles. The mixed powder is inserted into a spinning mold. After that, a metal-matrix ingot is melted in a crucible. This molten metal matrix is poured into the spinning mold. The poured molten metal matrix penetrates into the space between the particles due to the applied centrifugal force. The heat from the poured molten matrix melts the metal-matrix particles. Finally, ring- or disc-shaped FGMs can be obtained. In this study, aluminum based FGMs with diamond or cubic boron nitride (cBN) particles are fabricated by the centrifugal mixed-powder method. Recently, as novel drilling machine for the carbon fiber reinforced plastic (CFRP), gyro-driving drilling machine has been proposed by Takekoshi, where the gyro-driving drilling machine makes a hole by gyro-movement of grinding wheel. In this study, the fabricated FGMs is used as the grinding wheels.

2. Experimental Procedures

Pure Al particles of 99.9% purity having sizes in the range of 75 - 150 μm and abrasive grains (cBN particles or diamond particles) having sizes in the range of 125 - 150 μm were used as primary materials. The volume fraction of abrasive grains in the powder mixture was 25 vol%. The prepared mixed-powder was set in a compact vacuum centrifugal caster. The mold was prepared by a lost-wax process, and the sample size was approximately ϕ 20 mm x 39 mm excluding sprue and runner. The hollow section of the cast was made by use of core. Then, the pure Al of 99.99% purity (30g) was placed in the melting crucible surrounded by the induction coil. After the Al melt reached to the desired temperature, the mold and the crucible were rotated for 99 s and the molten Al was poured into the mold cavity, where applied centrifugal force was 80 *G*.

The microstructural observations of the fabricated FGMs were carried out with an optical microscope (OM) and scanning electron microscope (SEM) after the polishing. Composition of each of the FGMs was identified by SEM with energy-dispersive X-ray spectroscopy (EDS). The defect population in each of the FGMs were analyzed by X-ray computed tomography (CT).

After the polishing, the FGMs for bonding force test were deeply etched by 10% hydrofluoric acid for 200 sec to 250 sec. This selective etching results in the exposure of the diamond or cBN particles from Al matrix. The bonding force between diamond or cBN particles and the Al matrix was tested at room temperature using a bonding force measurement system. In the bonding force test, 50 diamond or cBN particles in each FGMs sample were picked to be sheared.

Drilling experiments were carried out by using gyro-driving grinding wheel system. The thermosetting CFRP laminate having a size of 100×100×5 mm^3 was clamped to the fixture mounted on a force sensor in the gyro-driving grinding wheel system. Drilling tests were carried out until 50 holes with the condition: a feed rate of 5 mm/min, a wheel rotation speed of 7000 rpm, a spindle rotation speed (tool rotation speed) of 2800 rpm, and dry machining.

3. Results and Discussion

Figures 1(a) and 1(b) show the SEM secondary electron images of Al-diamond and Al-cBN FGMs, respectively. As can be seen from these figures, Al particles in the mixed-powder were fully melted by heat from molten Al. Subsequently, the diamond and cBN particles were successfully embedded in the Al matrix. Moreover, these figures reveal angular edges of Diamond and cBN particles and a clear interface between Al matrix and particles.

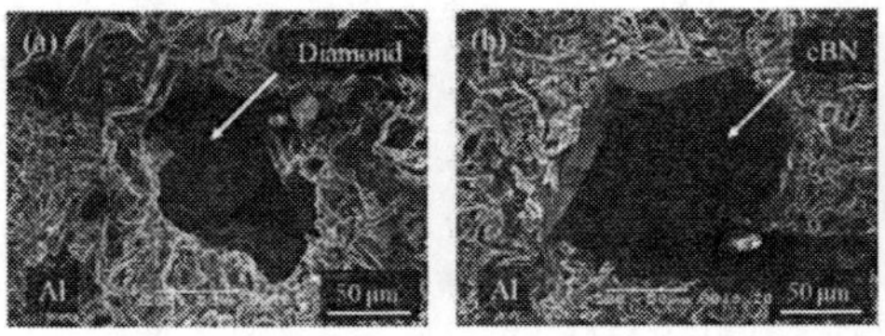

Fig. 1. SEM secondary electron images of Al-Diamond and Al-cBN FGMs

A defect analysis was performed on the X-ray CT scan of the Al-Diamond and Al-cBN FGMs, and results are shown in Figs. 2(a) and 2(b), respectively, where the 3D image made semi-transparent to view the location, shape and sizes of the defects. Although some defects could be observed in both of the FGMs, the size is small.

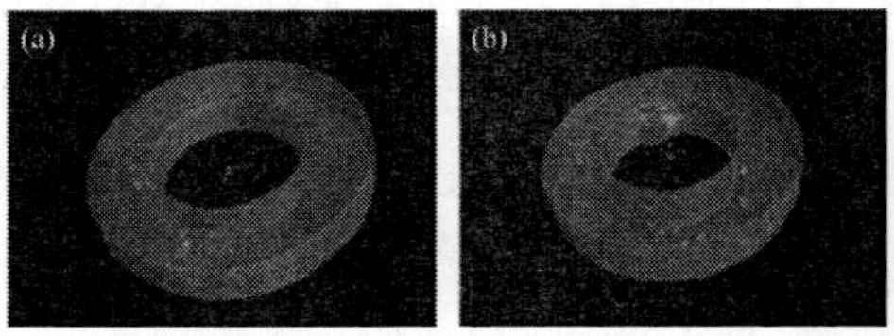

Fig. 2. Defect analysis on X-ray CT of (a) Al-Diamond and (b) Al-cBN FGMs.

To study the bonding force between the particles and the Al matrix, bonding tests were carried out. By the bonding tests, three kinds of the fracture and the detachment-types are mainly observed. When the probe touches the particle and started to push it, Al matrix is shear deformed without the fracture and the detachment of the particle. This behavior is defined as type A. On the other hand, when the bonding tests are carried out, the particles are sometimes fragmented by shear deformation. This fracture behavior is denoted as type B. In case of type B, the force is dropped down without plastic deformation, since brittle fracture of the particles occurs. The behavior of type C is the detachment of the particle from the matrix. The shear behavior of type A presents the highest bonding force among type A, type B and type C.

Figures 3(a) and 3(b) are appearance frequencies of type A, type B and type C by bonding tests for Al-Diamond and Al-cBN FGMs, respectively. As shown in Fig. 3(b), in the Al-cBN FGMs, shear behavior of type A occurs most frequently while fracture under type C is not observed. In case of the Al-Diamond FGMs, all of shear behavior types occur, although occurrence frequency of type C is the lowest among three types. Therefore, it is expected that the fabricated Al-cBN FGMs as well as Al-Diamond FGMs can be used as the grinding wheels.

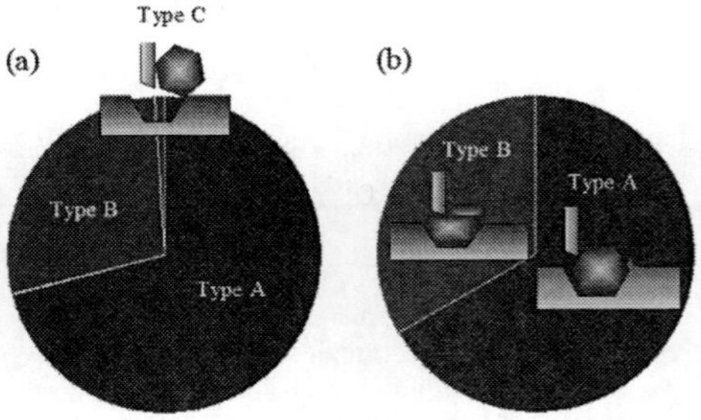

Fig. 3. Appearance frequency of type A, type B and type C by bonding tests for (a) Al-Diamond and (b) Al-cBN FGMs.

Figures 4(a) and 4(b) show the CFRP plate after making 30 holes with Al-Diamond FGMs and Al-cBN FGMs, respectively. Good hole-quality without burring and delamination were obtained by drilling using Al-Diamond FGMs. Therefore, Al-Diamond FGMs fabricated by the centrifugal mixed-powder method can be used for grinding wheels.

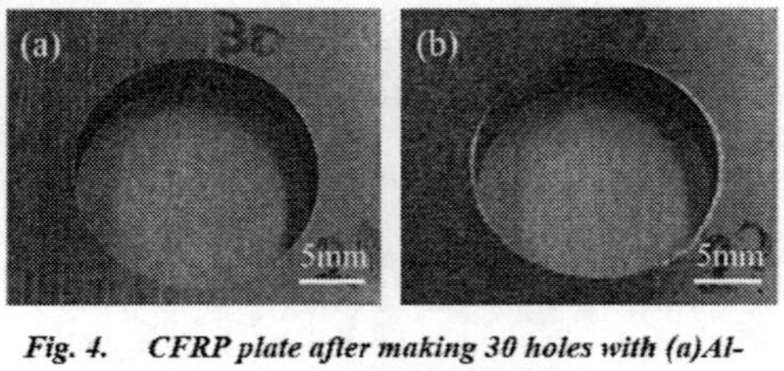

Fig. 4. CFRP plate after making 30 holes with (a)Al-Diamond and (b) Al-cBN FGMs.

4. Summary

In this study, as an attempt to improve the tool life of grinding wheels, Al-Diamond and Al-cBN FGMs have been fabricated by the centrifugal mixed-powder method. It is found that the fabricated Al-based FGMs would be suitable for the gyro-driving grinding wheel system with dual-axis-driven mechanism.

The 73rd World Foundry Congress
„Creative Foundry"
23 - 27 September 2018, Krakow, Poland

INTERFACIAL PHENOMENA IN NI ALLOYS-OXIDES SYSTEMS OF INTEREST FOR INVESTMENT CASTING

Fabrizio Valenza[1]*, Gabriele Cacciamani[2], Maria Luigia Muolo[1], Alberto Passerone[1], Rafal Nowak[3], Jerzy Sobczak[3], Natalia Sobczak[3,4]

[1] *CNR-ICMATE - National Research Council – Institute of Condensed Matter Chemistry and Technologies for Energy, 6 de Marini, 16149, Genoa, Italy*
[2] *UNIGE – DCCI – University of Genova – Department of Chemistry and Industrial Chemistry, 31 Dodecaneso, 16146, Genoa, Italy*
[3] *Foundry Research Institute, 73 Zakopianska, 30-418, Krakow, Poland*
[4] *Institute of Precision Mechanics, Duchnicka 6, 01-796 Warsaw, Poland*

**Corresponding address: fabrizio.valenza@ge.icmate.cnr.it*

Key words: Ni-alloys, wetting, casting, interfaces

When dealing with investment casting of innovative superalloys, issues related to high-temperature metal-ceramic interactions are of great importance. High-performance Ni- and Co-based alloys have very high melting temperatures (up to 1500°C) and contain highly reactive elements (e.g. Al, Ti, Hf, Y) that, even when present at very low concentration, form, in contact with the ceramic mould, non-metallic inclusions as internal and surface casting defects [1-3]. Therefore, acquiring knowledge on interfacial phenomena may assist the choice of the right alloy-refractory combination in relation to the processing parameters as surface finishing, temperature, heating-cooling rates, atmosphere conditions, etc.

In order to deepen the knowledge about the role of active elements in the interfacial behaviour of superalloys, the wetting of "simple" Ni alloys containing other elements (e.g. Cr, Hf) in contact with oxides chosen among those used in contact with superalloys (e.g. Al_2O_3, mullite) was studied. Preliminary results were obtained on sapphire by the sessile drop method at 1500°C.

After wettability tests, the metal/ceramic couples were let cooling down for further analyses (SEM-EDS, confocal microscope, etc.).

Moreover, for some samples, a special technique was adopted: as shown in Figure1, after measurements and still at the testing temperature, the liquid drop was pushed out the substrate by a special alumina manipulator in order to reveal the interface and analyse the modifications occurred on the ceramic substrate.

Fig. 1. Example of sessile drop and sequence of the drop removal

As shown in Table 1, low percentages of Hf added to Ni (up to 3 at.%) have a negligible effect on wetting; increasing the Hf content to 8.5 at.% led to decrease of a contact angle drop to 90°.

Table 1. Contact angles of Ni-Cr-Hf alloys on sapphire

Alloy	Contact angle, deg
Pure Ni	113
Ni-Hf 0.5 at.%	113
Ni-Hf 1.2 at.%	116
Ni-Hf 3 at.%	108
Ni-Hf 7.7 at.%	95
Ni-Hf 8.5 at.%	90
Ni-Cr 22 at.%	91
Ni-Cr 21-Hf 3.8 at.%	89

However, the presence of Hf in the liquid drop, even at low percentages as those found in many industrial superalloys, caused a strong interaction with the sapphire and the formation of interfacial HfO_2. Figure 2 shows the substrates after test with pure nickel or with Ni-0.5%Hf; in the case of pure nickel, the surface is mostly intact while in the case of Ni-0.5% Hf alloy, the zone of contact during test and the zone touched by melt during drop removal have been strongly modified during test. As shown in SEM pictures of Figure 3 and 4a, a continuous layer of HfO_2, ~ 1 μm thick, formed at the metal/ceramic interface. As observed in samples where the liquid was removed during the high-temperature tests, the formation of interfacial HfO_2 layer occurred during the contact between the liquid alloy and the substrate and not during cooling. Moreover, for higher Hf-containing alloys (Figure 4b) big HfO_2 inclusions were found in the bulk of the solidified drops.

Figure 5a shows the morphology of a sapphire surface after contact with Ni-Hf 22 at.% melt; in this sample, the drop was mechanically removed after solidification and the HfO_2 layer remained adhered to the drop leaving the sapphire etched. The same surface was analysed by confocal microscope (Figure 5b) that revealed that about 0.5 µm of the sapphire substrate has been consumed by the interfacial reaction. Moreover, further EDS analysis have shown that Al is present in the solidified drops with concentrations increasing with the Hf content. These evidences demonstrate that HfO_2 was formed by reaction of Hf inside the melt with Al_2O_3 following the simple redox reaction:

$$Hf + 2/3\ Al_2O_3 \leftrightarrow HfO_2 + 4/3\ Al$$

having a ΔG_r° = -57 kJ/mol calculated at T = 1500°C (thermodynamic data from [4]).

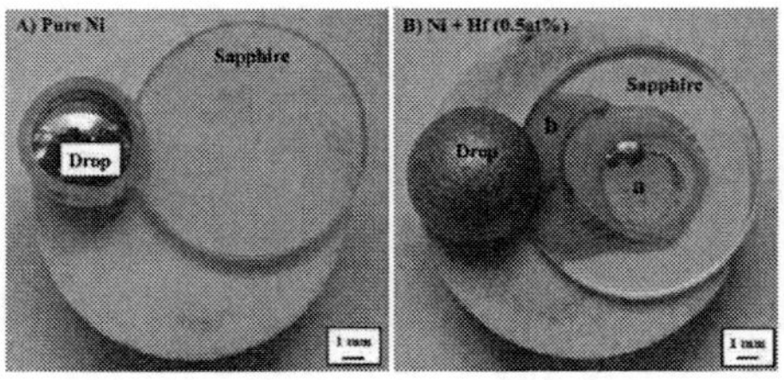

Fig. 2. A) sapphire after contact with liquid Ni. B) sapphire after contact with liquid Ni-Hf; a: contact zone during test, b: trail of the liquid moved out from the substrate.

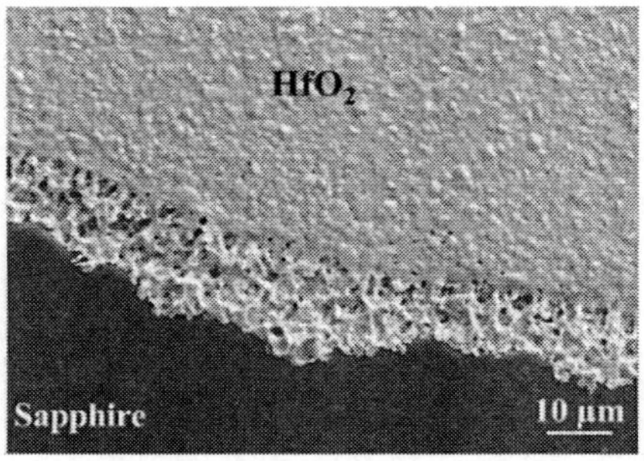

Fig. 3. SEM image of a sapphire substrate after test with Ni-Hf 1.2 at.% and removal of the drop in the liquid state. A HfO_2 layer covers the liquid contact zone.

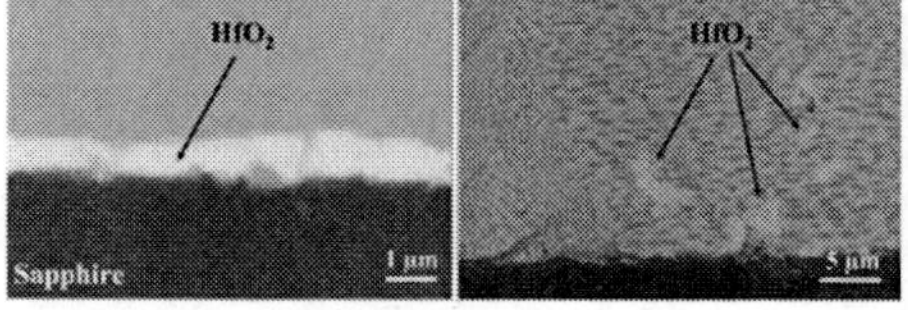

Fig. 4. Cross-sectioned metal-ceramic couples. a) formation of interfacial HfO_2 layer for low Hf content; b) formation of big HfO_2 inclusions for high Hf content.

For Ni-Cr 22 at.% alloys a contact angle of 91° was obtained with sapphire. Despite this strong effect on wetting, Cr has not macroscopic effects on the sapphire substrate and no reaction product were found in the solidified drop. Adding Hf to Ni-Cr alloy has a negligible effect on wetting and the same interfacial reactivity, i.e. formation of HfO_2, observed for Ni-Hf alloys.

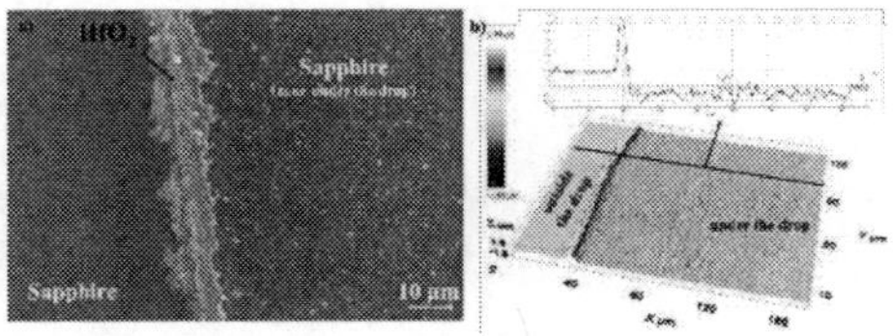

Fig. 5. a) sapphire substrate after mechanical drop (Ni-Hf 22 at.%) removal showing that the sapphire was etched by the formation of HfO_2 (SEM image); b) 3D profile and 2D profile over a line. The surface height is indicated by different colors.

To summarize, it can be said that when Hf is present in Ni-based melts in contact with Al_2O_3 ceramics, strong interfacial phenomena are expected leading to formation of HfO_2 layers or inclusions. On the other hand, Cr has a strong effect from the point of view of interfacial adhesion with ceramics.

More studies are ongoing to interpret and predict the phenomena occurring during industrial high-temperature liquid-assisted processes by the determination of the relevant thermodynamic data (phase diagrams, activities, energy quantities) obtained by CALPHAD.

Acknowledgments

This work is part of the Project INSURFCAST (Innovative Surfaces for Superalloys Casting Processes) financed within the ERA-LEARN 2020 support action funded by EU-H2020, M-ERA.NET Joint Call 2016.

INSURFCAST

References

[1] Valenza F., Muolo M.L. & Passerone A. (2010) Wetting and interactions of Ni-and Co-based superalloys with different ceramic materials. Journal of Materials Science. 45(8), 2071-2079.

[2] Valenza F., Nowak R., Sobczak N., Passerone A., Di Foggia M., Muolo M.L. (2010). Advances in Science and Technology. 70, 130-135.

[3] Sobczak N, Nowak R, Asthana R, Purgert R. (2010) Wetting in high-temperature materials processing: the case of Ni/MgO and NiW10/MgO. Scripta Materialia. 62(12), 949-954.

[4] Barin I., Knacke O. (1973) Thermochemical properties of inorganic substances, Berlin: Springer.

The 73rd World Foundry Congress
„Creative Foundry"
23 - 27 September 2018, Krakow, Poland

NOVEL METHOD OF THERMAL CONDUCTIVITY MEASUREMENT USING STEFAN–BOLTZMANN LAW

Piotr Wieliczko [1]*, Wojciech Leśniewski [1], Edward Czekaj [1], Marek Wawrylak [1],
Foundry Research Institute, Zakopiańska 73, 30-418, Kraków, Poland
**Corresponding address: piotr.wieliczko@iod.krakow.pl*

Key words: thermal conductivity, thermal radiation, ceramic shell molds, aluminium castings

1. Introduction

The article presents a novel method that allows measurement of thermal conductivity that uses Stefan–Boltzmann law. The developed method presented in this work can be used to determine thermal conductivity when ceramic shell molds are produced.

Knowing the values of heat capacity and thermal conductivity in the function of temperature it is possible to carry out computer simulation of the cooling and solidification process of a liquid metal in the mold.

2. Thermal conductivity measurements

Heat transport through the layer of porous ceramic material takes place by conduction and radiation through pores. It is necessary to know the thermal conductivity of ceramic molds for aluminium based alloy castings in temperature range of solidification and cooling, i.e. 600 K to 1000 K. These temperatures are rather low in foundry industry, hence considerable technical problems when determining the thermal conductivity can be found.

The amount of energy transferred through a unit of area in a unit of time is described by the Fourier differential equation as shown below:

$$\frac{\partial Q}{\partial t} = -k \oint_S \nabla T \cdot dS, \text{W} \qquad (1)$$

where:
Q – amount of heat, transferred through sample in time t, J,
k – thermal conductivity, W/(m·K),
∇T – temperature gradient, K/m,
dS – surface area element, m^2.

For a steady heat flow through a flat sample with parallel walls, after integration, the Fourier equation can be written in the following form:

$$\frac{Q}{t} = \left(\frac{k}{d}\right) A(T_b - T_t), \text{W} \qquad (2)$$

where:
d – thickness, m,
A – surface area, m^2,
T_b, T_t – temperature of the bottom and the top of the sample, K.

In laboratory based research, a plate apparatus were used to measure the thermal conductivity that allows measurements of the materials characterized by low heat conductivity (ceramics). The principle of the performed measurement was to determine thermal conductivity by the amount of heat conducted by the flat-parallel test sample, heated on one side with a flat heater, on the other side, cooled by a flat water cooler. The thermal conductivity then can be calculated from the formula shown below:

$$k = \frac{(Q/t)\, d}{A(T_b - T_t)}, \text{W/(m·K)} \qquad (3)$$

3. Novel idea of thermal conductivity measurements

Ceramic shell molds are characterized by quite different properties of both surfaces. The inner surface preferentially is designed to contact with a liquid metal possesses low roughness in order to obtain the best quality of casting. The outer surface of the mold shows the surface corresponding to the grain size of the ceramic material used for the last mold layer. The thickness of the mold is strictly related to the number of ceramic layers, and the grain size of the ceramic material that was used, however due to variable sizes of the ceramic materials the thermal parameters of such materials vary significantly. Variable graining of the external surface precludes the use plate apparatus for measurements. The grinding of the surface required in this case significantly changes the way the heat is discharged.

The design of the new equipment by the authors of this work allowing determination of thermal conductivity of shell ceramic samples taking into account the fact that when a liquid metal is pouring into ceramic mold, its internal surface provides very good contact with a liquid metal, while its external surface discharges heat into the space mainly through the emission of thermal radiation. In this wok it was assumed that when the heat flowing through the ceramic plate is radiated to the half-space by thermal radiation, then the value of Q/t can be calculated from the well-known Stefan–Boltzman equation. The equation defines the power radiated by the unit surface to the half-space:

$$\left(\frac{Q}{t}\right) = n^2\sigma T^4, \mathrm{W} \quad (4)$$

where:
n – index of refraction,
σ – Stefan-Boltzmann constant, $5.670\cdot10^{-8}$ W/(m^2·K^4),
T – temperature, K.

The calculations assume that the refractive index, n, equals 1 and that the calculated power is related to the unit surface.

$$k = \frac{\sigma T^4 d}{A(T_b - T_t)}, \mathrm{W/(m{\cdot}K)} \quad (5)$$

The above presented formula allows calculation of thermal conductivity can be considered in correct way if assumption that the energy sent from the sample can be determined by the Stefan–Boltzman equation is postulated, i.e. there is no convective air movement over the sample and that the emissivity is equal to 1. These assumptions are sufficiently accurate because above 573 K thermal energy radiated by a measuring object, which is proportional to T^4, reaches values exceeding the energy occurring in other forms of heat transport. The surface of ceramics in foundries has usually high emissivity. In this case, the surface is further developed by a significant graining of the surface, which is why the calculation assumes that the emissivity is equal to 1. The designed construction of the new device is shown in Fig. 1.

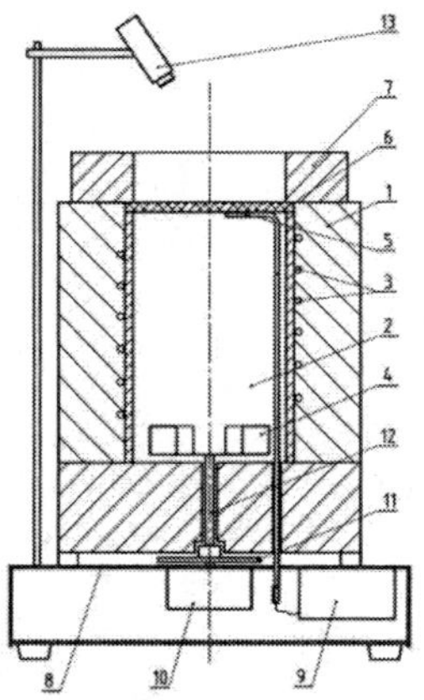

Fig. 1. Construction of apparatus for determining thermal conductivity of ceramic shell samples

In order to measure thermal conductivity, the heating chamber (2) was closed with the sample (6). The sample (6) was constantly in contact with a thermocouple (5). The heating power of the measuring device was adjusted by the control system (9); the temperature of the bottom surface of the sample was obtained. When the thermal conditions are stable, the temperature of the bottom surface is measured by a thermocouple (5), the temperature of the top surface of the sample is measured by pyrometer 13 [1].

Measurements of thermal conductivity of ceramic mold samples were carried out until a maximum value of the lower surface sample reached 1223 K. The temperature of upper surface of samples, measured by a pyrometer was obviously lower and its value allowed defining the radiated thermal energy and finally thermal conductivity.

The results of thermal conductivity of ceramic shell material obtained in this study from non-stabilized (monoclinic) zirconium oxide in the form of flour (325 mesh) and sand (0.1-0.3 mm and 0,5-1 mm) bound by colloidal silica are shown in Table 1. The tested material was highly porous therefore the values obtained in this work were lower than the values reported elsewhere for yttrium-stabilized zirconium oxide which are close to 2-3 W/(m·K) in this temperature range [2].

Table 1. The results of thermal conductivity measurement for a sample with a thickness of 6 mm made from zirconium dioxide

T, K	629	710	790	868	945	1021
k, W/mK	0.45	0.48	0.51	0.55	0.59	0.63

4. Summary

The novel method for thermal conductivity measurements allows determination of the values in the inhomogeneous ceramic shell molds materials. The research conducted in this work indicates that the structure of the mold material plays a significant role in heat conduction. In this work, thermal conductivity measurements were carried out in temperature range of 623 K to 1023 K. The obtained data can be used further in computer simulations, to predict cooling and solidification processes of aluminum based alloys in ceramic shell molds.

References

[1] W. Leśniewski, E. Czekaj. 2012, Patent no: 210610. Poland. Patent Office of the Republic of Poland.
[2] K.W. Schlichting, N.P. Padture, P.G. Klemens. (2001). Thermal conductivity of dense and porous yttria-stabilized zirconia. Journal of Materials Science 36, 3003 –3010.

The 73rd World Foundry Congress
„Creative Foundry"
23 - 27 September 2018, Krakow, Poland

PHYSICO-CHEMICAL, STRUCTURAL AND DERIVATOGRAPHIC STUDIES OF BENTONITE CLAYS FROM NATIONAL DEPOSITS

Jadwiga Kamińska [1]*, Michał Angrecki [1], Jarosław Jakubski [2], Dariusz Drożyński [2]

[1]Foundry Research Institute, Department of Technology, Zakopianska 73, 30-418 Cracow, Poland
[2] AGH University of Science and Technology, Faculty of Foundry Engineering, Mickiewicza 30, 30-059 Cracow, Poland
**Corresponding address: jadwiga.kaminska@iod.krakow.pl*

Key words: Moulding sand, bentonite clay, bentonite, structural studies, derivatographic analysis

1. Introduction

One of the indispensable conditions to make the casting process successful is that the casting mould cavity retains its original shape undeformed when poured with molten metal. To achieve this goal it is necessary to use binding materials that confer to the sand mixture proper strength. The basic bonding materials are different types of moulding clays and clay minerals [1, 2].

In the foundry industry, it is possible to use argillaceous rocks rich in clay-containing minerals, such as: kaolinite, halloysite, hydromica, montmorillonite, palygorskite, vermiculite and allophane. Due to their binding capacity, montmorillonites are the most interesting material in the family of plastic clay rocks. The basic clay rock containing montmorillonite, used as a common binding material for classical foundry sand mixtures, is bentonite [3].

In Poland, the bentonite deposits are found in trace amounts. The exploitation of bentonites is carried out only in the Krzeniów deposit, where they constitute a mineral accompanying the extraction of basalt. Domestic demand for bentonite raw materials is almost entirely satisfied by imports from Slovakia, Turkey, Italy, Germany and the Czech Republic, in order of import volume [4].

The research conducted at the Department of Technology is aimed at the foundry use of domestic bentonite clay deposits, which constitute a mineral accompanying the extraction of other raw materials, to obtain a modern hybrid bentonite with high physico-chemical and technological properties.

2. Materials

In this article, the basic properties of bentonite clay called beidellite, originating from overburden deposits of the Turoszów Mine, are presented. As a reference material, foundry bentonite from one of the Slovak deposits was used. Both materials were subjected to the process of activation, pre-drying and milling to obtain an appropriate granulation consistent with the requirements of PN-85/H-11003 [5].

3. Methods

Materials prepared in this way were subjected to physical and chemical tests (montmorillonite content, swelling index, loss on ignition and pH), their chemical composition was determined and qualitative phase analysis was performed [6]. Derivatographic tests were also carried out, allowing simultaneous execution of thermal differential analysis and thermogravimetric analysis. Changes occurring in the samples during heating were registered in the form of two curves:

- TG - graph showing changes in sample temperature during heating,
- DTG - differential plot of sample weight loss during heating.

The TG and DTG curves were used to determine the temperature of the beginning of the process running in parallel with the sample loss of weight (T_p), as well as the temperature at which the rate of the sample decomposition process assumed the highest values (T_{max}).

4. Results

Figure 1 compares the derivative curves of the tested clays and bentonites after the initial drying and grinding process in non-activated condition.

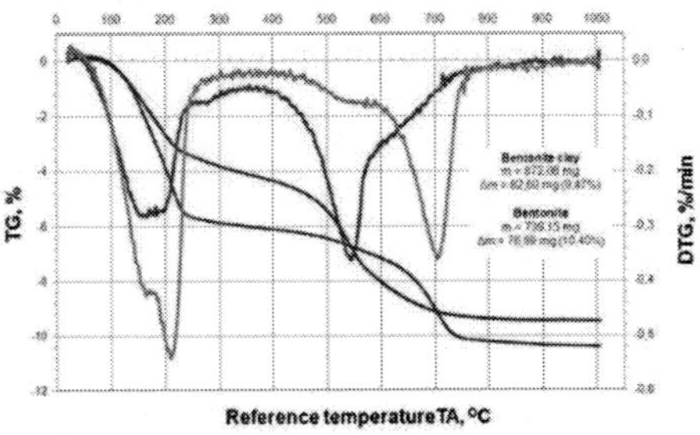

Fig. 1 Heat resistance of clays and bentonites as determined from the thermogravimetric analysis (TG) and thermal differential analysis (DTG)

Acknowledgments
The article was prepared as part of the project No. LIDER / 21/0125 / L-8/16 / NCBR / 2017.

References

[1] Lewandowski J. L. (1997). Tworzywa na formy odlewnicze. Wydawnictwo Akapit, Kraków.

[2] Pezarski F., Palma A., Izdebska-Szanda I. (2011). Innowacyjna technologia produkcji bentonitu odlewniczego z wykorzystaniem wyselekcjonowanych frakcji powstających przy wytwarzaniu sorbentów. MOTROL, 13, s. 269–280.

[3] Wyszomirski P., Lewicka E. (2005). Bentonity jako uniwersalny surowiec wielu dziedzin przemysłu. Gospodarka surowcami mineralnymi, 21/3, s. 5-19.

[4] Biuletyn Państwowego Instytutu Geologicznego (2009). Bentonity i iły bentonitowe.

[5] Norma PN-85/H-11003: Odlewnicze materiały formierskie - Gliny formierskie.

[6] Kościówko H., Wyrwickiego R. (1996). Metodyka badań kopalin ilastych - Państwowy Instytut Geologiczny.

The 73rd World Foundry Congress
„Creative Foundry"
23 - 27 September 2018, Krakow, Poland

WETTABILITY IMPROVEMENT OF WOVEN FABRIC FOR ALUMINIUM CASTING REINFORCEMENT

Sylvia Cruz*, Daniel Sánchez, Mª Eugenia Rodríguez and Manel da Silva

[1]*Eurecat, Centre Tecnològic de Catalunya. Avda. Universitat Autonoma 23, Cerdanyola del Vallès, 08290, Spain*
**Corresponding address: sylvia.cruz@eurecat.org*

Key words: Aluminium metal matrix composites, carbon and glass fabrics, electroless nickel deposition

Aluminium EN AC 47100 reinforced with glass and carbon fabrics were obtained by gravity permanent mould casting process. To increase the wettability of the woven fabrics with aluminium alloy, both glass fabrics and carbon fabrics, were coated with Nickel with an electroless treatment.
Nickel coating improved the adherence between aluminium and fabrics respect the fabric without coating.

1. Introduction

Aluminium metal matrix composites (AlMMC) are potential materials for various applications including aerospace, automobiles and aircraft, due to their good physical and mechanical properties [1]. The most common reinforcements used are carbides (SiC, B_4C, TiC), oxides (Al_2O_3, SiO_2) and fly ash, which increases the hardness, density and tensile strength of aluminium alloys [2].
Use of carbon fibers or carbon nanotubes as reinforcement, provide tribological and mechanical properties enhancement to the aluminium alloys [3]. However, carbon fibers and fabrics exhibit very poor wettability by liquid aluminium alloys, to improve it, a proper pre-heating process can lead almost complete infiltration of the fiber according with Kachold et al [4], non-coated carbon fibers starts to oxidize when heated above 600°C in air. But, according to different studies the reaction between carbon and aluminium in the interface produce Al_3C, influencing the mechanical properties of the final composite. A possible approach to control the aluminium carbides formation is the use of coatings on the reinforcement, the coatings are also employed to increase the fabrics wettability [5].
Many authors have reported the use of A356 as aluminium matrix alloy to produce AlMMC, using different method and reinforcements [6] but EN AC 47100 have been less studied, being one of the most employed alloy in the High Pressure Die Casting process (HPDC).
In this report, the gravity die casting method was employed to fabricate EN AC 47100 reinforced with carbon and glass fabrics. Employing both, pre heating and coating process to increase the wettability. The fibers, before and after the electroless treatment, were analyzed with a SEM, using EDX technique to assess the deposition of Ni on the fabrics. The interface was also analyzed using optical microscopy, to stablish the difference between the fabrics and the coating in the wettability with aluminium alloy.

2. Experimental Procedure

For composites sample fabrication was employed Aluminium cast alloy EN AC 47100 (Table 1 show the chemical composition), carbon and glass fabrics 0/90° with a grammage of 600 and 500 g/m^2 respectively.

Table 1. Chemical composition (%wt) of EN AC 47100

Cu	Mg	Fe	Pb	Sn	Ni	Zn	Mn	Si	Sb	Ti	others
1.11	0.098	0.722	0.085	0.013	0.044	0.569	0.207	12.574	0.1238	0.040	0.034

The woven fabrics were previously coated with Nickel. Firstly, was cleaned in acetone during 20 min. Surface was sensitized and activated immersing the fabrics in 5g $SnCl_2$/20ml HCl balanced 1L and then 0,10g $PdCl_2$/10 drops HCl balanced 1L also at room temperature during 20 minutes each one. The Nickel solution content: 5g $NiSO_4$ $6H_2O$, 5g NaOH, 5ml Formaldehyde, 25 g potassium sodium tartrate tetrahydrate and balanced with water in 1 l of capacity. The immersion time was 30 minutes.
Aluminium is molten and infiltrate on the mould that content the woven fabrics. Both carbon and glass fabrics are heating in air using IR at 260°C during 2 minutes for enhancing their wettability. According with the sequence shown in Fig. 1.

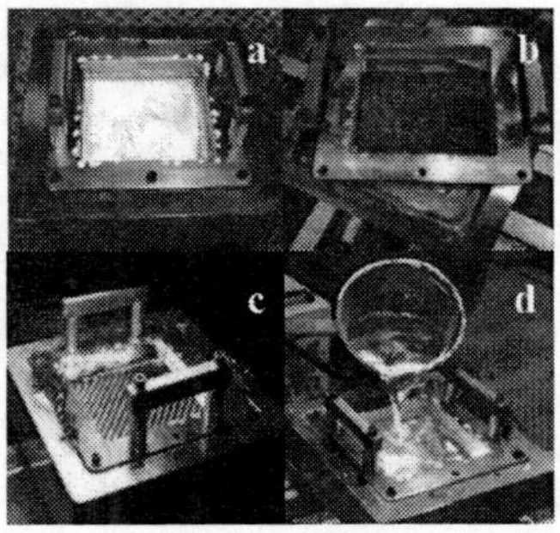

Fig. 1. Sequence of the process. a) Die, b) steel frame to carbon fabric in the die, c) pre-heated IR, d) casting aluminium alloy on the die

3. Results and Discussion

SEM images (Fig. 2) show the presence of Nickel particles in the carbon fabric after electroless process

(Fig. 2b). The selected area in the carbon fabrics (Fig. 2c) and EDX (Fig. 2e) verified the presence of C and O, in the case of the carbon fabric after electroless process the EDX (Fig. 2f) confirmed that the brilliant particles are enrichment in Nickel. That result confirms the efficiency of electroless process to coat the fabrics.

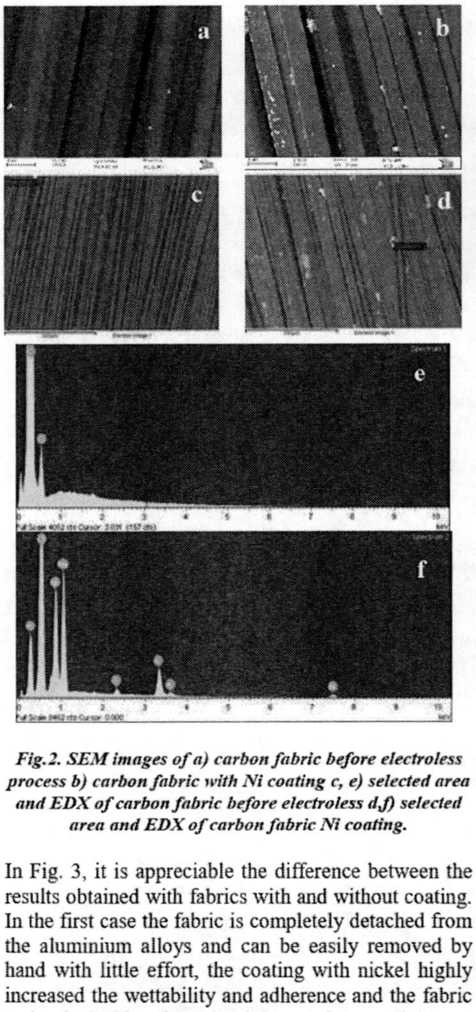

Fig.2. SEM images of a) carbon fabric before electroless process b) carbon fabric with Ni coating c, e) selected area and EDX of carbon fabric before electroless d,f) selected area and EDX of carbon fabric Ni coating.

In Fig. 3, it is appreciable the difference between the results obtained with fabrics with and without coating. In the first case the fabric is completely detached from the aluminium alloys and can be easily removed by hand with little effort, the coating with nickel highly increased the wettability and adherence and the fabric resisted inside the aluminium after applying a considerable tension.

Optical microscopy of the transversal section (Fig. 4a) confirm the beneficial effect of the coating in the adherence and wettability, being this improvement more accused in the glass coating compared to the carbon fabric, where the aluminium tries to flow through the dense fabric leaving a smaller (Fig. 4b) gap.

Fig. 3. a) Composite with glass fabric without coating b) glass and carbon fabrics with Ni coating.

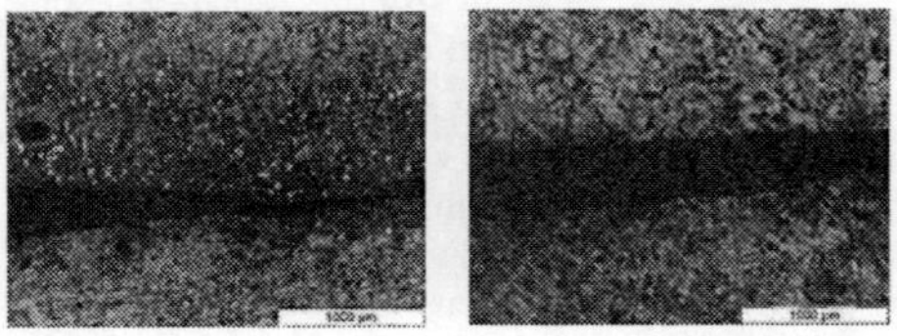

Fig. 4. Optical microscopy of transversal section of a) glass fabric with coating and b) carbon fabric with coating.

4. Conclusions

The SEM-EDX analysis, verified the presence of Nickel particles in the carbon fabrics and the efficiency of electroless process to produce an effective coating. The glass fabric presented better wettability with aluminium EN AC-47100 alloy, than carbon fabric. In the case of glass fabrics the gap observed in subsequent metallographic inspection was smaller.

References

[1] Hasan Ali Alhashmy, Michel Nganbe. (2015). Laminate squeeze casting of carbon fiber reinforced aluminum matrix Composites Materials and Design 67. 154–158.

[2] B. Vijaya Ramnath, C. Elanchezhian1, RM. Annamalai, S. Aravind, T. Sri Ananda Atreya, V. Vignesh and C. Subramanian. (2014). Aluminium metal matrix composites. Rev. Adv. Mater.Sci 38. 55-60.

[3] M. Dhanashekara, V.S. Senthil Kumarb. (2014). Squeeze Casting of Aluminium Metal Matrix Composites- An Overview Procedia Engineering 97. 412 – 420.

[4] Franziska S. Kachold, Rafal Kozera, Robert F. Singer, and Anna Boczkowska. (2016). Mechanical Properties, Surface Structure, and Morphology of Carbon Fibers Pre-heated for Liquid Aluminum Infiltration. Journal of Materials Engineering and Performance 25(4). 1502-1507.

[5] Ziyuan Shi, Xuezhi Wang, Zhimin Ding. (1999).The study of electroless deposition of nickel on graphite fibers. Applied Surface Science 140. 106–110.

[6] Pu-yun DONG, Hai-dong ZHAO, Fei-fan CHEN, Junwen LI. (2013). Microstructures and properties of A356-10%SiC particle composite castings at different solidification pressures. Trans. Nonferrous Met. Soc. China 23. 2222-2228.

The 73rd World Foundry Congress
„Creative Foundry"
23 - 27 September 2018, Krakow, Poland

RESEARCH ON COATINGS AND INFILTRATION TO STRENGTHEN CERAMIC LOST CORES USED IN HIGH PRESSURE DIE CASTING PROCESSES

Mikel Merchán *, Maider García de Cortázar, Haize Galarraga, Jorge Bárcena, Antxon Artola

Fundación TECNALIA R&I, Paseo Mikeletegi 2, 20009 Donostia-San Sebastián (Guipúzcoa), Spain
**Corresponding address: mikel.merchan@tecnalia.com*

Key words: HPDC, lost core, aluminium, coating, infiltration

Abstract

Lost cores used to manufacture complex aluminium components through High Pressure Die Casting (HPDC) processes need to withstand very high pressures. Conventional sand cores used in other casting processes, such as sand casting or low pressure die casting, do not support these pressures, so advanced ceramic cores must be used. These cores must be strong enough to withstand the casting process and, at the same time, have enough porosity to be easily removed to obtain the finished part. Due to this porosity, aluminium penetrates the core surface during the casting process. In this work, two research lines have been followed to address this problem: application of coatings to seal the surface and infiltration with silica.

1. Introduction

HPDC is the most widely used casting process to manufacture aluminium components [1], thanks to its versatility, high production rate and the possibility of producing thin-walled components, thus reducing the weight of the casting [2]. The weight reduction of components is mandatory in an industry such as the automotive who is continuously searching for a reduction in the fuel consumption of future cars. This is the way of complying with the EU standards, that are continuously tightening the emission limits of cars [3].

From the last 25 years, aluminium has substituted many iron-based automotive components thanks to its good strength-to-weight ratio and lower density. However, some of the components continue being manufactured in iron because they are very difficult to be produced in aluminium through HPDC due to their complexity. Components having undercuts, for example, require the use of lost cores which are later removed from the casting and create internal cavities.

The cores used in the High Pressure Die Casting-Lost Core (HPDC-LC) process need to withstand pressures around 800 bars, so they need to be strong enough not to get broken during the injection process. Sand cores used in sand casting and low pressure die casting processes are not suitable since they cannot cope with the process specific parameters. Some researchers

[44-5] have worked in the development of salt cores based on water-soluble inorganic salts that are easily removed from the casting. Bending strength values up to 20 MPa have been reached in some formulations [5]. However, the high corrosivity of these salts and the penetration of aluminium during the casting process represent a problem.

In this work, ceramic cores have been used to produce aluminium castings. The problem in this case is the high porosity that the core must have to be later removed from the casting. Due to this porosity, aluminium penetrates the core surface during the injection process, creating an interface between the aluminium part and making the posterior removal of the core from the part impossible.

In this work, the solution to this problem has been addressed in two ways. On the one hand, different foundry coatings have been studied as a mean of sealing the surface and avoiding the penetration of aluminium. On the other hand, the strengthening effect of infiltrating highly porous ceramic cores with silica has been studied.

2. Experimental

For the coating trials, six commercial refractory paints from different suppliers were traded-off. Preliminary trials were performed first, using ceramic bars and coating one of their faces. These bars were dried and inserted in a sand mould with the coated face exposed. Aluminium was melted in an induction furnace and poured over the different coated samples at 700°C to study the reaction between the coating and the aluminium.

Three paints were selected to continue with the study in HPDC trials. A simplified geometry was defined for these trials. Ceramic cores were painted with two different coating methods (brushing and spraying) to evaluate the effect of the painting process on the coating behaviour.

For the HPDC trials, a 900-ton industrial die casting machine was used. First, cores were inserted in the cavity and the die was closed. Then, aluminium was injected and, when solidified, the aluminium part was extracted. Finally, the core was removed from the casting. From each part, a sample was cut and a metallographic specimen was prepared.

An optical microscope was used to analyse whether the aluminium had penetrated the core or not. The microscope was also used to measure the thickness density levels were used. A colloidal dispersion of silicon (IV) oxide in water (30%) from Alfa Aesar was employed for these trials. Previously sintered bars were infiltrated in air and in vacuum during infiltration periods of 1, 2 and 4 hours. The effect of the infiltration media and the infiltration time in the porosity reduction and in the bending strength was studied. Bending tests were performed following the UNE EN 60672-2 standard using the three-point bending test method.

3. Results

In the preliminary trials no penetration of aluminium in the ceramic was seen in any of the samples due to the absence of applied pressure. However, it was noticed that the coating reacted with the aluminium in the uncoated sample as well as in three of the coated samples. With the other three coatings, no signs of reaction were visible. So, in the HPDC trials only the three paints with better results (zirconium silicate-based paint, zirconium oxide-based paint and boron nitride-based paint) were studied.

From the coating methods comparison, it was concluded that with the spraying method the thickness of the applied coating was more uniform, so the ceramic cores for the HPDC trials were sprayed. Five cores were painted with each of the three refractory paints. And five more cores were left uncoated.

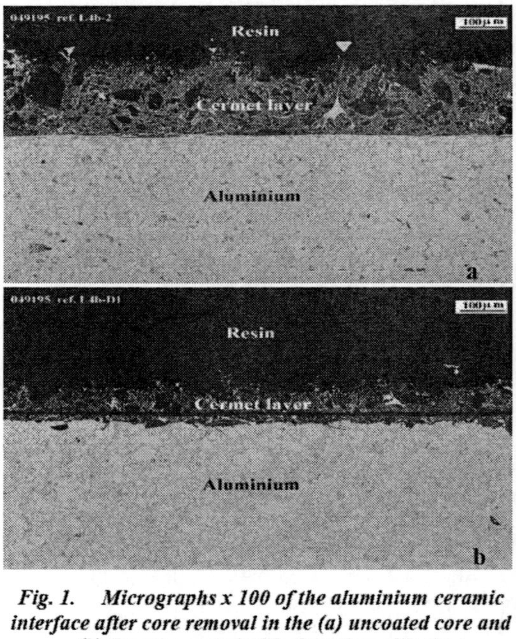

Fig. 1. Micrographs x 100 of the aluminium ceramic interface after core removal in the (a) uncoated core and (b) the core coated with zirconium silicate

The microstructural analysis revealed that in the uncoated ceramic core a cermet layer of around 200 μm thickness was formed as a result of the penetration of aluminium during the casting (Fig. 1a). In the core coated with zirconium silicate there is also some nitride, no penetration was observed, and they could be completely removed from the casting.

Fig. 2 shows the bending strength increase of the bars infiltrated with silica in air and vacuum conditions at different infiltrating periods. The bending strength of the non-infiltrated bar of 1.7 g/cm^3 density was taken as baseline value. Increases up to 135% were obtained infiltrating bars in air during two hours with porosity reductions up to 6%.

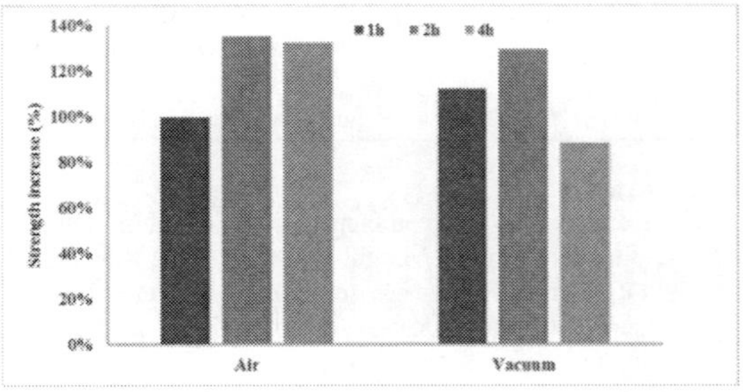

Fig. 2. Effect of the infiltration time and medium in the strength increase of the bars with 1.7 g/cm^3 density

4. Conclusions

- Coating ceramic cores with zirconium oxide or boron nitride avoids aluminium penetration during the HPDC-LC process and permits removing the core from the casting.
- Infiltrating ceramic bars with silica for two hours either in air or vacuum conditions, increases bending strength more than 130%. Future work will be carried out in HPDC trials of infiltrated cores.

Acknowledgements

This work has been carried out under the framework of the CORE 4.0 European project funded by the European Union through its Horizon 2020 Framework Programme under grant agreement No 701197.

References

[1] The Aluminium Association. http://www.aluminum.org

[2] J. Collot. (2007). Review of new process technologies in the aluminum die-casting industry. Materials and Manufacturing Processes Vol. 16 (5), 595-617.

[3] European Commission. https://ec.europa.eu/clima/policies/strategies/2020_en

[4] P. Jelínek & E. Adámková. (2014). Lost Core for High-Pressure Die Casting. Archives of Foundry Engineering. Vol.14 (2), 101-104.

[5] J. Yaokawa, D. Miura, K. Anzai, Y. Yamada & H. Yoshii. (2007). Strength of Salt Core Composed of Alkali Carbonate and Alkali Chloride Mixtures Made by Casting Technique. Materials Transactions. Vol.48 (5), 1034-1041.

MECHANICAL AND STRUCTURAL CHARACTERIZATION OF CAST IRON USING SYNCHROTRON LIGHT

Lennart Elmquist [1]*, Stephen Hall [2], Torsten Sjögren [3], Erik Dartfeldt [3], Peter Skoglund [4], Jessica Elfsberg [4], Marta Majkut [5]

[1] Swerea SWECAST, Research and Business Development, P.O Box 2033, 550 02 Jönköping, Sweden
[2] Lund University, Division of Solid Mechanics, Lund, Sweden
[3] Research Institutes of Sweden, RISE Safety and Transport/Mechanics Research, Borås, Sweden
[4] Scania CV AB, 151 87, Södertälje, Sweden
[5] European Synchrotron Radiation Facility, Grenoble, France
**Corresponding address: lennart.elmquist@swerea.se*

Key words: SiMo ductile iron, synchrotron light, DVC, graphite, crack propagation

The paper describes work done using synchrotron light to investigate the microstructure and how it behaves in 3D when a load is applied. Two different cast iron materials with different matrix structures and graphite morphologies were investigated; SiMo51, which is basically a spheroidal iron (SGI) alloyed with Si and Mo, and a lamellar graphite iron (LGI). The tensile test specimens were loaded in steps, at which x-ray tomography as well as 3DXRD measurements were made to characterize the microstructure. The result shows how the crack propagates and which path it takes through the materials. DVC was applied to analyze the strain fields. This work also shows how useful synchrotron experiments can be in the study of the mechanical behavior of cast iron.

1. Introduction

The development of highly sophisticated combustion engines is driven by the desire to improve efficiency and power density as well as to reduce emissions. Such developments will inevitably lead to an increase of the combustion pressure and temperature, which has the consequence of an increased loading on the materials. Today, spheroidal graphite iron (SGI) alloyed with silicon and molybdenum, SiMo-alloys, are commonly used in components such as manifolds in exhaust systems, where materials that can withstand the harsh conditions are needed. The mechanical behaviors of these materials have been studied by others [1-4]. However, the microstructural analysis used to correlate the mechanical behavior to microstructural properties is performed on polished 2D surfaces taken from the position of final fracture of the specimen. Instead, for a detailed understanding of the micromechanics of the material an in-situ 3D analysis approach is needed. Lamellar graphite iron (LGI) is another cast iron material used in components for combustion engines. Also in this case a better understanding of understanding and models that can predict how the material will behave under load. In this work x-ray tomography, together with Digital Volume Correlation (DVC) and 3D x-ray diffraction (3DXRD) during in-situ loading has been used to identify and characterize the deformation mechanisms and corresponding phase interactions. The results will provide a better understanding of the influence of the different microstructural length scales on the macroscopic mechanical response and explain the observed crack propagation paths.

2. Experimental procedure

The experiment was performed at the European Synchrotron Radiation Facility in Grenoble, France, beamline ID11. Before the experiment, tensile test samples were machined from each material. A tensile test stage developed for the specific setup was used to apply and control the load steps from outside the beamline. One sample from each material was monotonically loaded in steps between which x-ray tomography as well as 3DXRD measurements were made. Figure 1 shows the sample geometry and the setup inside the beamline.

Fig. 1. Setup positioned in the beam line. The small picture in the left corner shows the tensile test specimen.

Besides, also 2D microstructure investigations were used to correlate and verify the different phases procedures and studied using optical microscopy and SEM with EDS. In this part of the work tomography and DVC results were correlated with the 2D microstructures.

3. Results and discussion

In the case of LGI it is very clear how the graphite lamellas act as stress concentrations. It is also very clear that cracks appear very early and that the propagation finds the easiest path by connecting to flakes with shortest distance through the matrix. Figure 2 shows results obtained by DVC for load step 4, (nominal stress 127 MPa).

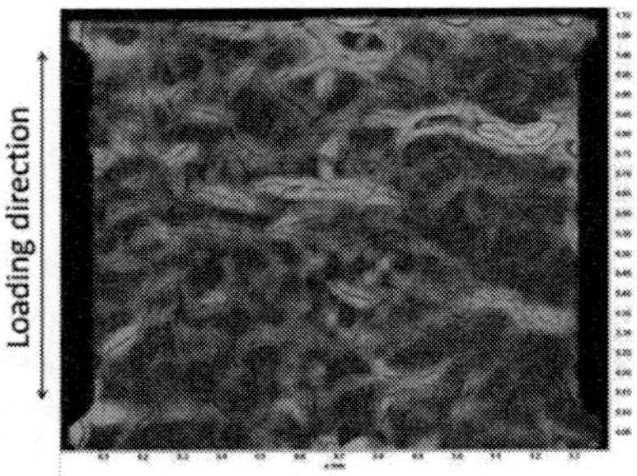

Fig. 2. DVC results for LGI sample at load step 4.

Due to the high levels of Si and Mo, the microstructure of the SiMo materials contains molybdenum carbides and also spheroidized pearlite [5]. These structures are found together with shrinkage defects in regions corresponding to the last to freeze areas. Figure 3 shows the microstructure. It is clear how certain areas are enclosed, containing more than single nodules.

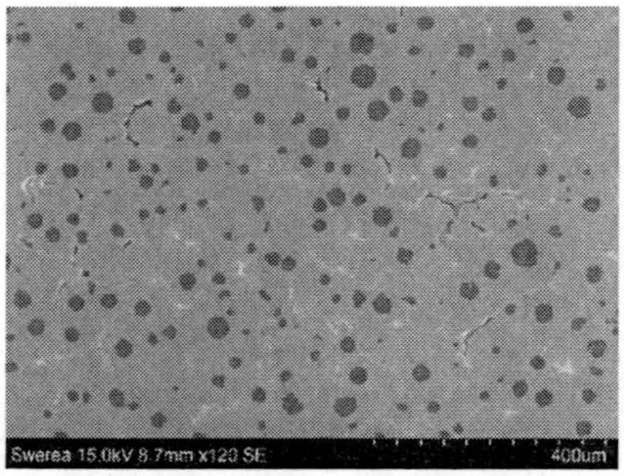

Fig. 3. SEM picture of the SiMo material.

A similar behavior is found when comparing with tomography pictures, see Figure 4. In this picture another important observation was made, the presence of compacted graphite. Figure 5 shows DVC results for the SiMo material. The strain distribution reveals a shear behavior that depends on the size and dimension of the sample. It also shows how cracks are formed when the graphite delaminates from the matrix material.

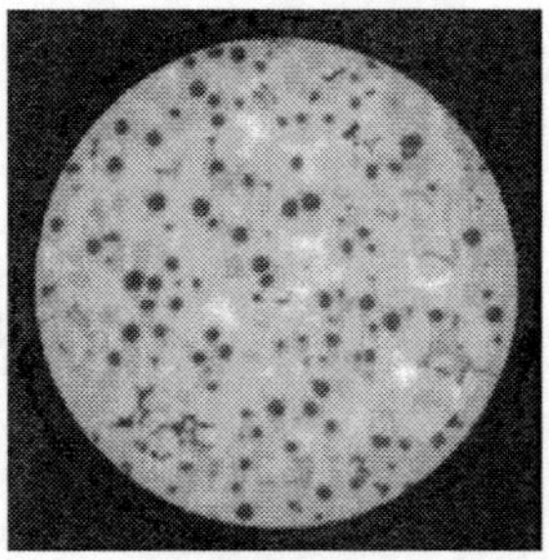

Fig. 4. X-ray tomography picture of SiMo sample.

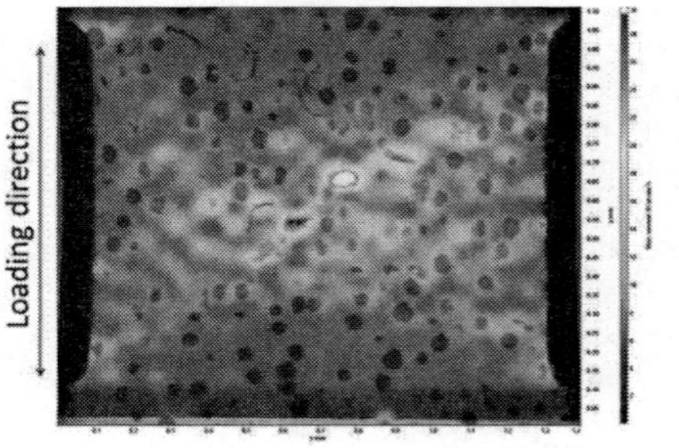

Fig. 5. DVC analysis of SiMo. Strain distribution and crack initiation (nominal stress 537 MPa).

4. Conclusion

The overall conclusion is that synchrotron light is very useful to study the influence of microstructures and defects on the mechanical behavior. It is shown how cracks nucleate and propagate and DVC gives some very helpful information to understand the strain distribution during loading. Based on the results it is suggested to redesign the sample geometry and due to the flake size in the LGI relative to sample size SGI samples will be more useful for future experiments.

Acknowledgments

The authors acknowledge European Synchrotron Radiation Facility (ESRF) in Grenoble, France, for hosting the experiment.

References

[1] Norman, V., Skoglund. P., Moverare J. (2016), Damage evolution in compacted graphite iron during thermomechanical fatigue testing, International journal of Cast Metals Research, 29 (1-2), 26-33.

[2] Norman, V., Skoglund, P., Leidermark, D., Moverare, J. (2016), The effect of superimposed high-cycle fatigue on thermo-mechanical fatigue in cast iron, International Journal of Fatigue, 88, 121-131.

[3] Nadot, Y., Mendez, J., Ranganathan, N., (2004), Influence of casting defects on the fatigue limit of nodular cast iron, Int. J. of Fatigue, 26 (3), 311-319.

[4] Hervas, I., Thuault, A., Hug, E., (2015), Damage Analysis of a Ferritic SiMo Ductile Cast Iron Submitted to Tension and Compression Loadings in Temperature, Metals, 5 (4), 2351-2369.

[5] Ekström, M., 2013, Development of a ferritic ductile cast iron for increased life in exhaust applications, Licentiate thesis, KTH, Stockholm, Sweden.

The 73rd World Foundry Congress
„Creative Foundry"
23 - 27 September 2018, Krakow, Poland

THERMAL STABILITY OF A RESIN BINDER USED IN MOULDING SAND TECHNOLOGY

Mariusz Holtzer[2], Angelika Kmita[1]*, Agnieszka Roczniak[2], Aleksandra Benko[3],

[1]*AGH University of Science and Technology, Academic Centre for Materials and Nanotechnology, A. Mickiewicza 30 Ave., 30-059 Krakow, Poland*
[2] *AGH University of Science and Technology, Faculty of Foundry Engineering, A. Mickiewicza 30 Ave., 30-059 Krakow, Poland*
[3] *AGH University of Science and Technology, Faculty of Materials Science and Ceramics, A. Mickiewicza 30 Ave., 30-059 Krakow, Poland*

**Corresponding address: holtzer@agh.edu.pl*

Key words: "slow" and "flash" pyrolisis, thermal stability, resin binder, Py-GC/MS, TGA/DSC/FTIR

1. Introduction

As part of the research program conducted at the AGH University of Science and Technology, regarding the thermal stability of materials used in the moulding and core sand technology, a group of binder resins based on phenol-formaldehyde resins - from different manufacturers - were tested. In our previous works [1-3], we had presented the investigation of the pyrolysis of some of the commercially available binders and the hereby article is a continuation of these studies presenting the analysis of resin used in the same technology but obtained from a different manufacturer. As found in the preliminary studies, the binder shows some differences in thermal resistance and the compounds released, as compared to previously evaluated products. Knowledge of the thermal destruction products as well as the temperature ranges of their release is important both from the point of view of the technology and environment protection. [4-9].

2. Materials and methods

The subject of the study were two types of cured phenol-formaldehyde resol resins: sample **A** and **B**, both being commercial binders used in the core and moulding sands technology. In the study, two analytical techniques were applied to investigate the thermal decomposition of the resins: Thermogravimetric Analysis/Differential Scanning Calorimetry coupled with Fourier Transform Infrared Spectroscopy (TGA /DSC/FTIR) and Pyrolysis Gas Chromatography coupled with Mass Spectrometry (Py/GC/MS). The first method simulated the "slow" pyrolysis while the second – the "flash" pyrolysis.

TGA/DSC/FTIR method

For the TGA/DSC, the STA 449 F3 Jupiter (NETZSCH) thermal analyzer was used, while gases were investigated by the Tensor 27 spectrometer (Bruker Optics). Samples were heated with a heating rate of 10°C/min, from 40°C to 1200°C, under a protective flow of pure nitrogen (50 mL/min. A total of 16 scans, recorded with a spectral resolution of 4 cm^{-1}, were collected and averaged for each temperature point. The spectra were recorded in the medium infrared spectral range (600–4500 cm^{-1}). Based on the appearance of different bands in the spectrum, at selected temperatures/time points, the spectra were extracted and analyzed for their composition using a Bruker NIST/EPA Vapor Phase Library.

Py/GC/MS method

Simulation of "flash" pyrolysis was carried out using Py/GC/MS. The final temperature of pyrolysis was: 500, 700, 900 and 1100°C. The pyrolysis products were separated on a capillary column and then identified based on the mass spectral library NIST MS Search 2.0 Libera

3. Results and discussion

The TGA and DSC curves recorded during the "slow" pyrolysis of the cured resins are shown in Figure 1.

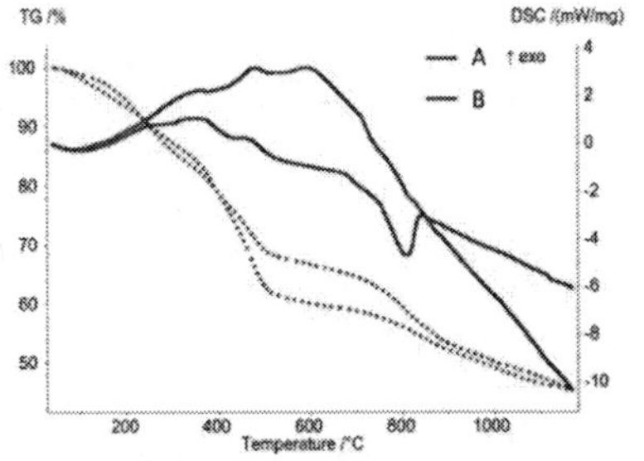

Fig. 1. TG (dotted line) and DSC (solid line) curves of the cured binders: sample A (green line) and sample B (red line).

For both resins, mass reduction curves are roughly the same, with slightly faster mass reduction for the

A resin in the temperature range between 595 and 850°C.

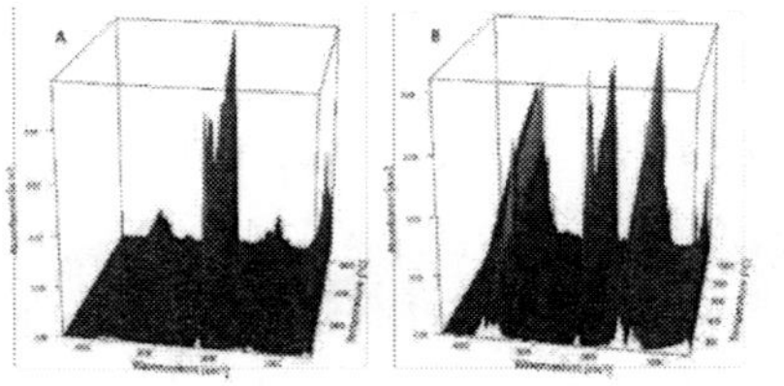

Fig. 2. FTIR spectra of the gaseous products of the thermal decomposition of the cured resins (3D chromatographic plot representing changes in absorbance as a function of increasing temperatures and wave numbers).

As it can be seen from the DSC curves, in this region of resin B there is an ongoing endothermic reaction (most likely a decomposition) being a possible cause for differences in the TGA curve. In both cases, characteristic inflections of the TGA curve, when the moulding sand reduction rate was decreased, can either suggest ongoing of additional chemical reactions, (for example: synthesis and/or oxidation between the released gaseous products and/or remnants of the sample) or that the loss of compounds with the boiling point around that temperature has ended [9]. Figure 2 presents the FTIR chromatograms of the gaseous products evolved during the thermal decomposition of the cured resins. It can be seen that types of products emitted are roughly the same. However, for resin B, proportions of compounds change, with decreased amount of CO_2 and higher amounts of water, methane, ammonia and phenol with its derivatives [3]. When combined with the overall amount of heat released from resin B (observed in the DSC curve) these results indicate less complete combustion of resin B, causing an increased potential harmfulness for the foundry workers during the processes of metal casting, when compared to resin A.

In the „flash" pyrolysis numerous volatiles, including HAPs are produced. Most likely, these compounds may become trapped in the mould and released during the casting knocking out. This indicates that the process is potentially harmful for both the environment and the foundry workers. Figure 3 shows the example of a chromatogram obtained for sample A at 700 °C.

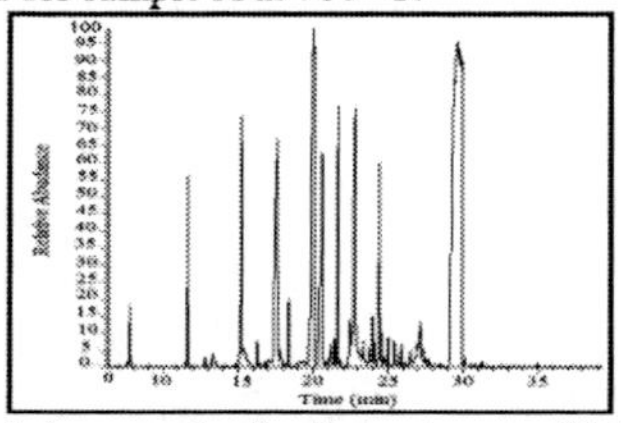

Figure 3 An example of a chromatogram obtained for sample A at 700 °C

4. Summary

The pyrolysis products depend on the temperature of the pyrolysis.

In case of the "slow" pyrolysis, evaluated using TGA / DSC / FTIR technique compounds: CO_2, NH_3 phenol, propaic acid and acetic acid were identified.

In contrast, during the "flash" pyrolysis, the released products were: CO, CO_2, phenol and its derivatives, benzene and its derivatives, etc.

Ratios of the decomposition products of the cured phenol-formaldehyde resol resins can vary, causing either decreased or increased potential harmfulness, what can be seen in the case of tests carried out on 2 types of cured resins from 2 different manufacturers.

Acknowledgments

This work was supported by the Dean's Grant no 15.11.170.599 realized in 2018 at AGH.

References

[1] A. Kmita, C. Fischer, K. Hodor, M. Holtzer & A. Roczniak. (2016). Thermal decomposition of foundry resins: A determination of organic products by thermogravimetry/gas chromatography/ mass spectrometry (TG/GCMS). Arabian Journal of Chemistry, doi.org/10.1016/ j.arabjc.2016.11.003

[2]A. Roczniak, M. Holtzer & A. Kmita. (2017). The use of Py-GC/MS (pyrolysis/gas chromatography/mass spectroscopy) in the analysis of foundry binders: alkaline phenolic resol resin. Transactions of Foundry Research Institute. 57, 19-28.

[3] A. Kmita, A. Benko, A. Roczniak, A. Frączek-Szczypta & M. Holtzer. (2017). Pyrolysis of organic ester cured alkaline phenolic resin: identification of products. Journal of Analytical and Applied Pyrolysis, doi.org/10.1016/j.jaap.2017.12.014 (available online).

[4] Y. Zhao, N. Yan and M.W. Feng. (2013). Thermal degradation characteristics of phenol–formaldehyde resins derived from beetle infested pine barks, Thermochim. Acta. 555, 46-52.

[5] A. Gardziella, L.A. Pilato & A. Knop. (2000). in: Phenolic resins: chemistry, applications, standardization, safety and ecology. Springer Berlin Heidelberg, Berlin, Heidelberg, 488-513.

[6] Y. Wang, F.S. Cannon, M. Salama, J. Goudzwaard & J.C. Furness. (2007). Characterization of hydrocarbon emissions from green sand foundry core binders by analytical pyrolysis, Environ. Sci. Technol. 41, 7922-7927.

[7] Y. Wang, Y. Zhang, L. Su, X. Li, L. Duan, C. Wang & T. Huang. (2010). Hazardous air pollutant formation from pyrolysis of typical Chinese casting materials, Environ. Sci. Technol. 45, 6539-6544.

[8] N. Tiedje, R. Crepaz, T. Eggert & N. Bey. (2010). Emission of organic compounds from mould and core binders used for casting iron, aluminium and bronze in sand moulds. J. Environ. Sci. Health A Tox. Hazard. Subst. Environ. Eng. 45, 1866-1876.

[9] T. Lis, N. Korzec, W. Frohs, J. Tomala, A. Frączek-Szczypta & S. Błażewicz. (2016). Wood-derived tar as a carbon binder precursor for carbon and graphite technology. J. Wood Chem. Technol. 36, 393-400.

The 73rd World Foundry Congress
„Creative Foundry"
23 - 27 September 2018, Krakow, Poland

DEFORMATION BEHAVIOR OF PURE COPPER CASTINGS WITH AS-CAST SURFACES FOR ELECTRICAL PARTS

Ikuzo Goto [1]*, Setsuo Aso [1], Ken-ichi Ohguchi [1], Kengo Kurosawa [2], Hiroyuki Suzuki [3], Hiroyuki Hayashi [3], Jun-ichi Shionoya [3]

[1]Akita University, Graduate School of Engineering Science, 1-1 Tegatagakuen-machi, Akita 010-8502, Japan
[2] Akita Industrial Technology Center, 4-11Arayamachi-aza-sanuki, Akita 010-1623, Japan
[3] Sanwa Tekki Corporation, 6-4-6 Minami-shinagawa, Shinagawa-ku, Tokyo 140-8669, Japan
**Corresponding address: goto@gipc.akita-u.ac.jp*

Key words: pure copper, as-cast surface, elongation, surface pinholes, directional solidification

1. Introduction

Crimp terminals and branched sleeves are used for electrical wire connections in applications such as electrical transmission and substation equipment. Because these products require high electrical conductivity, they are made from pure copper and are manufactured by sand-mold casting to achieve the near-net-shape and high-mix low-volume production. The terminals and the sleeves are connected to wires by crimping them using a specialized hydraulic tool. To avoid cracking during the crimping process, these electrical parts should be ductile. In the previous work, we investigated the deformation behaviors of pure copper castings using machined specimens (without as-cast surfaces) [1-4]. However, the crimping sections of the electrical parts principally have as-cast (unmachined) exterior surfaces. In the present study, to clarify this effect, we prepared pure copper specimens with as-cast surfaces, and subsequently conducted tensile tests using the specimens.

2. Experimental procedure

No. 1 copper wire (JIS H 2109, high purity scrap, ≥99.9 mass%) and electrolytic copper were melted under atmospheric conditions using a high-frequency induction furnace and a graphite crucible. The melt was sequentially treated with a $(C_2F_4)_n$-based commercial degasifying (dehydrogenating) agent and phosphor copper deoxidizer (Cu-15 mass%P, 0.4mass% of the raw materials). The deoxidized melt was poured into molds under a superheat of 50°C. Figure 1 shows the casting design. The sand mold was made from silica sand and water glass by the CO_2 process. For the casting design shown in Fig. 1(b), a water-cooled copper chill was used for directional solidification along the longitudinal direction, where the sand mold was preheated to approximately 200°C.

The tensile properties were evaluated using tensile tests. JIS Z 2241 No. 8B specimens that have as-cast surfaces were used in the tests; these specimens were prepared by sand strip using a brass brush. The gauge length was 10 mm. The elongation was measured in real time using a vision sensor. The crosshead speed was set to 0.1 mm/s. The corresponding actual strain

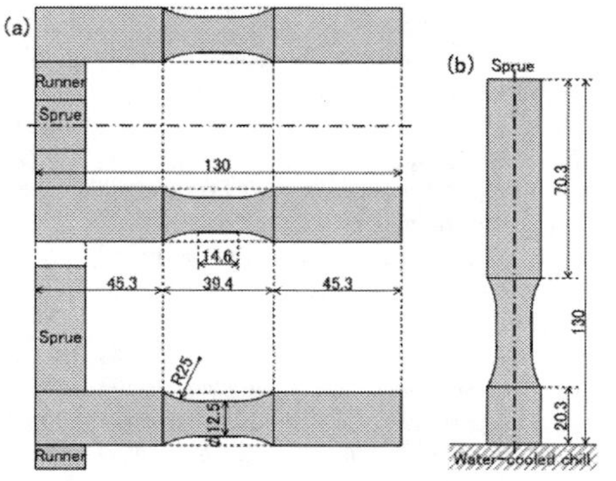

Fig. 1. Schematic of the Casting design (unit: mm.)

rates ranged from 0.10 to 0.24% s^{-1}. The vertical cross-section of the specimens was also observed by optical microscopy and scanning electron microscopy (SEM). The samples of the cross-section were electropolished.

3. Results and discussion

Figure 2 shows the relationship between uniform and rupture elongation when the Cu-Cu_2O eutectic phase is absent. The previous results using machined specimens under the same casting and test conditions are also shown in Fig. 2 [2,4]. Irrespective of the presence or absence of the as-cast surfaces, the tensile strengths stably ranged from 165 to 195 MPa.

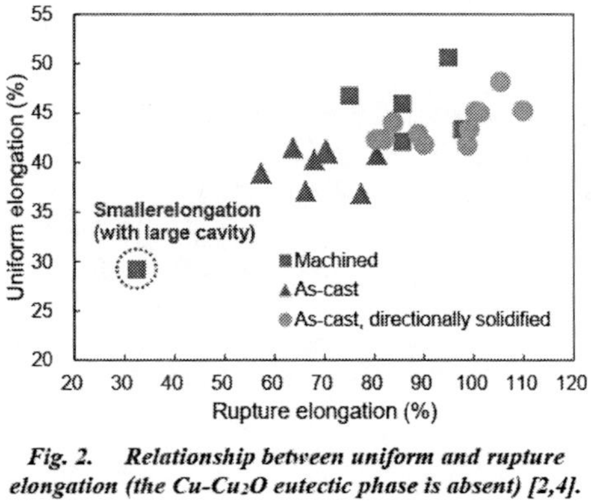

Fig. 2. Relationship between uniform and rupture elongation (the Cu-Cu_2O eutectic phase is absent) [2,4].

Although a machined specimen exhibited substantially smaller uniform and rupture elongations shown in Fig. 2, this was because of a happenstantial large cavity [4]. As shown in Fig. 2, as-cast specimens (non-directionally solidified) exhibited slightly smaller elongations than the sound machined specimens. Meanwhile, directionally solidified as-cast specimens exhibited similar or larger elongations than the machined sound specimens.

Figures 3 and 4 show the cross-sectional macro- and microstructure, respectively, of the gauge sections. As shown in Fig. 3(a), pinholes were observed in the vicinity of the as-cast surfaces. These surface pinholes are generally caused by the dissolved gases and/or gas-generating reaction with the mold [5,6], which were not prevented by the directional solidification. Although rounded defects such as the cavity and the pinholes have less stress concentration, these reduce the cross-sectional area [1,3,4]. As a result, the surface pinholes might promote necking and fracture of the as-cast specimens, possibly leading slightly smaller elongations [3,4]. In addition, as shown in Figs. 3(a) and 3(b), the columnar crystal grain structures along the solidification directions were observed inside the both as-cast specimens.

As shown in Fig. 4(a), arrays of microporosity along the grain boundary were also observed inside the as-cast specimens, which were similarly caused in the machined specimens [4]. On the other hand, as shown in Fig. 4(b), almost no microporosity was observed inside the directionally solidified as-cast specimens. This is because rapid cooling using the water-cooled chill leads to less microporosity along with the supersaturation of the hydrogen and/or greater temperature gradient [2-4,7]. The above results suggest that the presence of not only the surface pinholes but also the microporosity affects the both elongations, and that the use of the chill counteracts the adverse effect of the pinholes.

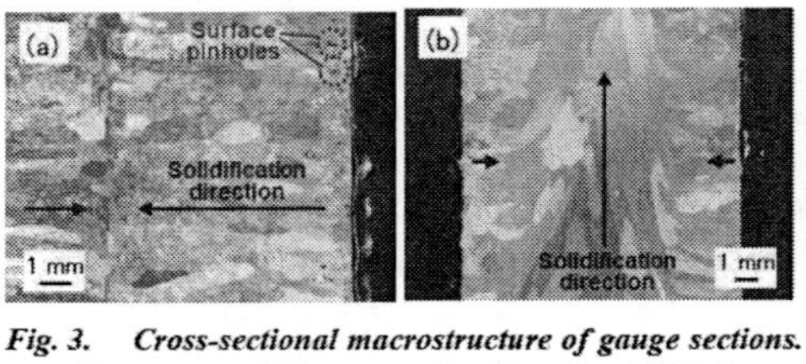

Fig. 3. Cross-sectional macrostructure of gauge sections. (a) as-cast and (b) as-cast, directionally solidified.

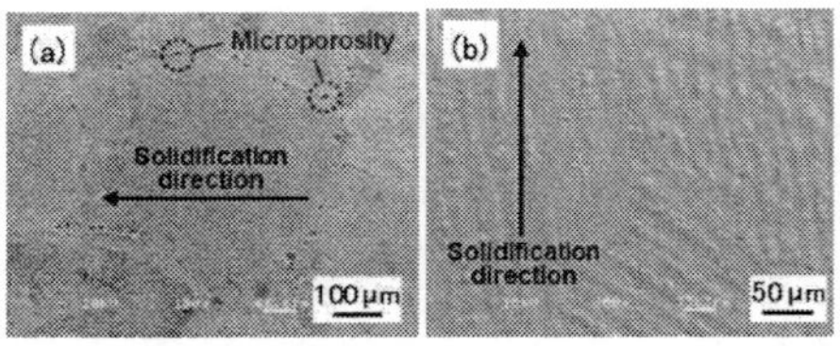

Fig. 4. Cross-sectional SEM micrographs of gauge sections. (a) as-cast and (b) as-cast, directionally solidified.

Table 1 shows the summary of the results from this study [2,4]. As shown in Table 1, casting soundness including both macro- and microscopic levels is a dominant factor influencing the elongations [2-4], which implies that another factor such as surface roughness has less effect on the both elongations. On the basis of this study, larger elongations than those shown in Fig. 2 may be achieved when both the surface pinholes and the microporosity are absent.

Table 1. Summary of the results from this study [2,4].

Specimen	Large cavity	Surface pinholes	Micro-porosity	Elonga-tions
Machined	Absent	Absent	Present	Large
	Present	Absent	Present	Small
As-cast	Absent	Present	Present	Medium
Directionally solidified	Absent	Present	Almost absent	Large

4. Conclusions

(1) Although as-cast specimens exhibit smaller uniform and rupture elongations than machined specimens, directionally solidified as-cast specimens exhibit similar elongations to the machined specimens.

(2) Surface pinholes are observed in the vicinity of the as-cast surfaces, which are not prevented by the directional solidification.

(3) Although arrays of microporosity along the grain boundary are caused inside the as-cast specimens, almost no microporosity is caused inside the directionally solidified as-cast specimens.

(4) The presence of not only the surface pinholes but also the microporosity affects the elongations.

Acknowledgments

This work was supported in part by the Japan Institute of Copper.

References

[1] I. Goto, K. Kurosawa, K. Ohguchi, S. Aso, H. Suzuki, H. Hayashi & J. Shionoya. (2015). Factors Influencing Deformation Behavior of Pure Copper Castings for Electrical Parts. J. Japan Inst. Met. Mater. 79 (10), 511-517.

[2] I. Goto, S. Aso, K. Ohguchi, K. Kurosawa, H. Suzuki, H. Hayashi & J. Shionoya. (2016). Effect of Solidification Conditions on Deformation Behavior of Pure Copper Castings for Electrical Parts. Proc. 6th Korea-Japan Conf. for YFE, 25-27 Aug. (124-127). Seoul: KFS.

[3] I. Goto, S. Aso, K. Ohguchi, H. Oguri, K. Kurosawa, H. Suzuki, H. Hayashi & J. Shionoya. (2017). Factors Influencing Deformation Behavior and Electrical Conductivity of Pure Copper Castings Fabricated by Industrial Process. J. Japan Inst. Met. Mater. 81(3), 133-142.

[4] I. Goto, S. Aso, K. Ohguchi, K. Kurosawa, H. Suzuki, H. Hayashi & J. Shionoya. (2019). Effect of Solidification Conditions on the Deformation Behavior of Pure Copper Castings for Electrical Parts. Mater. Trans. submitted.

[5] Dogokin-imono-no-seisangijutsu-kaiteiban-henshuiinkai. (2014). Dogokin-imono-no-seisangijutsu. (264). Tokyo: Materials Process Technology Center.

[6] JFS. (2016). Casting Defects and Countermeasures. (48, 212). Tokyo: JFS.

[7] E. Niyama. (2001). Chuzo-dennetsu-kogaku. (97-98). Tokyo: Agne Gijutsu Center.

The 73rd World Foundry Congress
„Creative Foundry"
23 - 27 September 2018, Krakow, Poland

AMPLITUDE DEPENDENCE OF INTERNAL DAMPING OF MAGNESIUM ALLOYS BEFORE AND AFTER PLASTIC DEFORMATION

Milan Uhríčik *, Zuzana Dresslerová , Peter Palček , Tatiana Oršulová

University of Žilina, Faculty of Mechanical Engineering, Department of Materials Engineering, Univerzitná 8215/1010 26 Žilina, Slovak Republic
**Corresponding address: milan.uhricik@fstroj.uniza.sk*

Key words: internal damping, magnesium alloy, plastic deformation, ultrasonic resonance, FEM analysis.

1. Introduction

Engineering magnesium alloys are the lightest commercial alloys being developed, very attractive for automotive and electrochemical industry. Magnesium alloys are used especially for production of light weight-walled casting and except many covers and lockers in automotive engines are used for the mass production of parts of mobile phones, cameras or notebooks, so components, where low weight is ultimate. Magnesium alloys have also good damping properties (belong to the group of the HIDAMETS materials – High Damping METals) [1].

2. Experimental material

For experimental measurements of amplitude dependence of internal damping were used magnesium alloys of type AZ31, AZ61 and AZ91. Result of the spectrometer chemical analysis is shown in the Tab. 1, which shows only the most constituent elements. Magnesium alloys were delivered without heat treatment and were produced by the squeeze-casting method. Microstructures are shown in Figure1.

Table 1. Chemical composition of magnesium alloys in wt.%

Elements	Al	Zn	Mn	Si	Mg
AZ31	2.98	0.655	0.202	0.067	balance
AZ61	6.88	1.200	0.229	0.079	balance
AZ91	7.81	0.645	0.195	0.053	balance

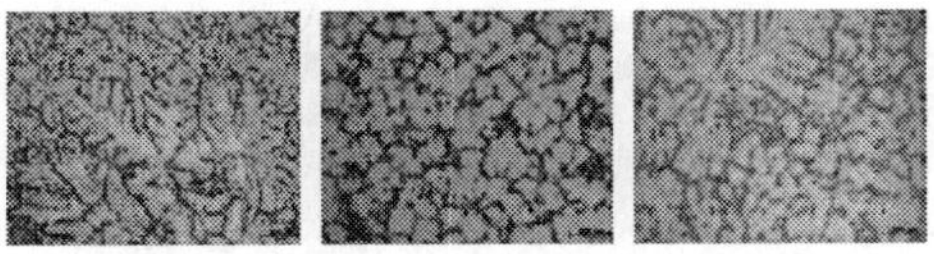

a) b) c)

Fig. 1. Microstructures of a) AZ31 b) AZ61 c) AZ91

3. Experimental equipment

The experimental equipment consists from electronic and mechanical part (Fig.2). The mechanical part (Fig.3) consists from three basic parts. From the transducer which serves like the source and at the same time like the detector of ultrasonic waves. Of the rod which is made of titanium and has a cylindrical shape with a diameter of 12 mm. The last part of device is the specimen, which is designed to fulfill the resonance condition that means the natural frequency must be approximately the same like the frequency of the test equipment.

One of the electronic parts is a generator which produces a sine wave. The electric signal is then amplified and transformed into a mechanical wave by using the piezo-ceramic transducer. The ultrasonic wave is amplified in aluminum horn and spreads into the specimen by the titanium rod. After passing through the specimen the wave is reflected at the free end and spreads back through the entire device [2].

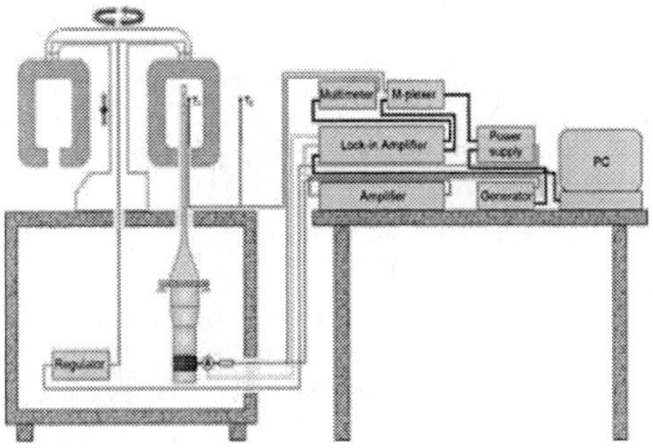

Fig. 2. Ultrasonic resonance device for measurement of internal damping

Fig. 3. Mechanical part of device with the specimen at the end

4. Results of experimental measurements

Before measuring of internal damping, it was necessary to determine the dependence of the amplitude of displacement of the test rod from the current input to the ultrasonic transducer using the MTI-2100 measuring device. After that it was also necessary to determine the amplitude of deformation from the amplitude of displacement, which was done in the ANSYS program. Results of the FEM analysis (Fig.4) point to the place where the greatest deformation is.

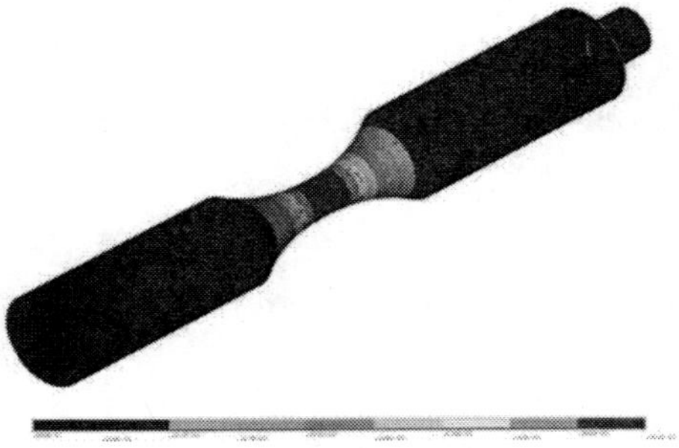

Fig. 4. Result of FEM analysis

Measurement of internal damping was performed on heat treated test rods (solution treatment at 390°C with a residence time of 22 hours and rapid cooling in water) to ensure a homogeneous structure. The measurement was carried out several times, increasing the maximum amplitude of the vibration voltage up to 7 volts. Figures 5 and 6 show the measurement rates of internal damping of magnesium alloys depending on the amplitude of deformation at driving voltages of up to 6 volts and 7 volts.

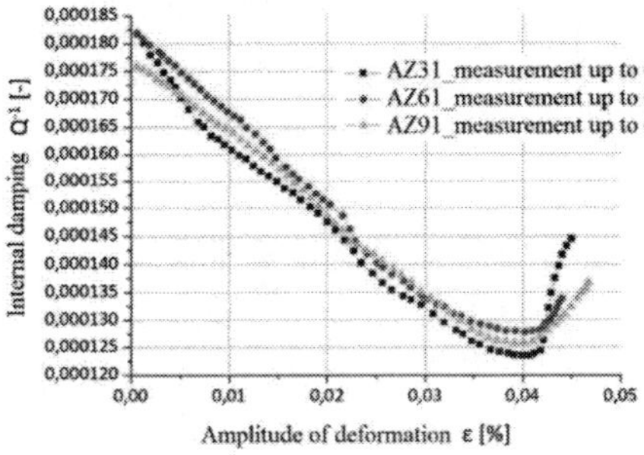

Fig. 5. Comparison of amplitude dependencies of the internal damping of experimental alloys measured to 6V.

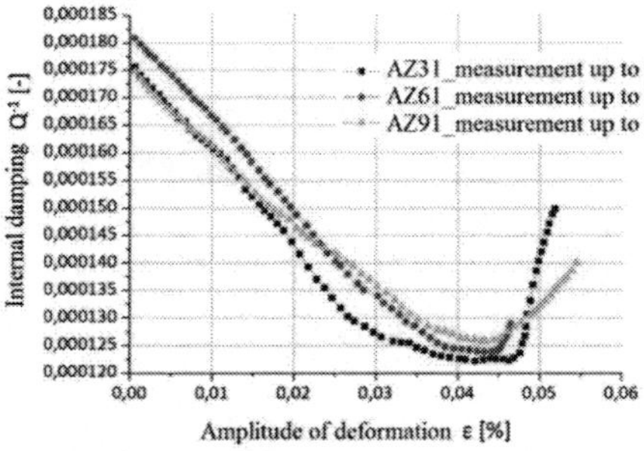

Fig. 6. Comparison of amplitude dependencies of the internal damping of experimental alloys measured to 7V.

Other amplitude-dependent measurements of internal damping were performed on specimens after plastic deformation. It was an uniaxial deformation when specimens subjected to pressure deformation. Several loads were performed and after each of them the plastic strain was calculated and one internal damping course measured. Figure 7 shows the measurement rates of internal damping of magnesium alloys depending on the amplitude of deformation, at various percentages of deformation.

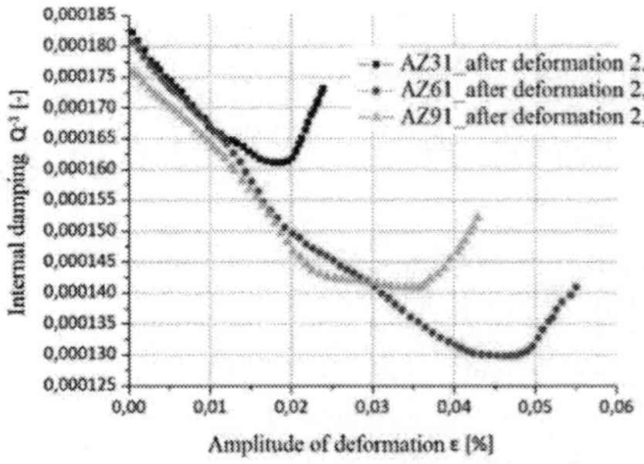

Fig. 7. Comparison of amplitude dependences of internal damping of experimental alloys after plastic deformation

Comparing the amplitude dependences of the internal damping of the individual alloys after the plastic deformation, it can be seen that ε_{cr} (critical amplitude of deformation) does not appear at the same amplitude of the deformation even at the same value of the internal damping. Whereas on test rods that were not plastically deformed, ε_{cr} occurred approximately at the same amplitude of deformation

This effect is probably related to strength or hardness of material after plastic deformation. The stronger the material, the higher the amplitude of deformation is necessary to apply, in order to break dislocations from weakly anchoring points.

Acknowledgments

This work has been supported by Scientific Grant Agency of Ministry of Education of Slovak Republic and Slovak Academy of Sciences, No.1/0683/15.

References

[1] Watanabe, H., Sasakura, Y., Ikeo, N. & Mukai, T. (2015). Effect of deformation twins on damping capacity in extruded pure magnesium. Journal of Alloys and Compounds. Volume 626, pp. 60-64.

[2] Dresslerová, Z., Palček, P. & Uhríčik, M. (2015). Influence of homogenization annealing on internal damping depending on the vibration amplitude measured on specimens AZ31 an AZ91. Manufacturing Technology. Volume 15, Issue 4, pp. 1213-2489.

The 73rd World Foundry Congress
„Creative Foundry"
23 - 27 September 2018, Krakow, Poland

INFLUENCE OF MISCHMETAL ON IMPACT TOUGHNESS AND MORPHOLOGY OF G20Mn5 CAST STEEL FRACTURES

Justyna Kasińska*

Kielce University of Technology, Department of Metal Science and Manufacturing Processes, Faculty of Mechatronics and Machine Engineering, Al. Tysiąclecia Państwa Polskiego 7, 25-314 Kielce, Poland
**Corresponding address: kasinska@tu.kielce.pl*

Key words: modification, mischmetal, rare earth elements, G20Mn5, impact toughness

Low-alloyed cast steels are vastly applied because of relatively low production costs, however in order to increase their properties such micro additives as Nb, V or Ti have started to be used [1]. The interesting alternative are rare earth elements having been examined by researchers since the half of the 20th century. For the last decade rare earth elements have been becoming extremely popular in world-wide markets thanks to their practical applications in several industrial fields, like electronics or nanotechnologies [2-5].

1. Material and experiment

G20Mn5 manganese cast steel was smelted in an industrial induction furnace with volume of 300 kg. For modification purpose mischmetal of the following chemical composition was used: 49.8% Ce, 21.8% La, 17.1% Nd, 5.5% Pr and 5.35% of REMs. 2 series of smelts of the chemical composition specified in Table 1 were performed. After cutting flashes off, they were heat-treated by hardening (920 C/water) and tempering (720 C/air). Added mischmetal caused significant grain fragmentation (Fig .1) and changes in morphology and dimensions of non-metallic inclusions (Fig. 2).

Table 1. Chemical composition of cast steel being smelted

	Cast steel without REMs	Cast steel with REMs
Element	Mass %	
C	0.19	
Mn	1.14	
Si	0.41	
P	0.02	
S	0.018	
Cr	0.15	
Mo	0.15	
Fe	97.80	
Ce	<0.00500	0.0288
La	<0.00100	0.0176

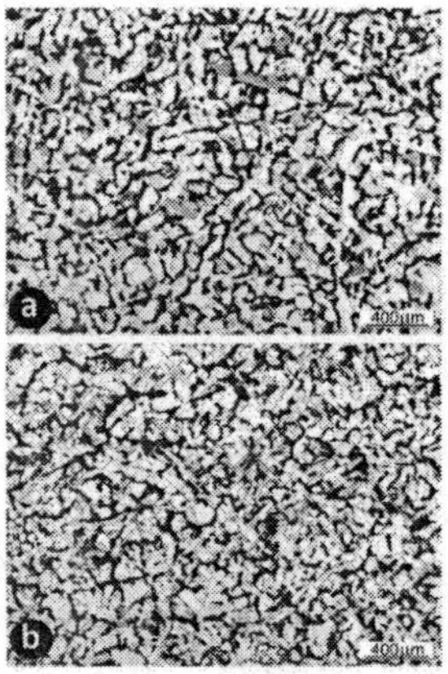

Fig. 1. G20Mn5 cast steel microstructure without heat treatment, a – non-modified cast steel, b – cast steel with added REMs

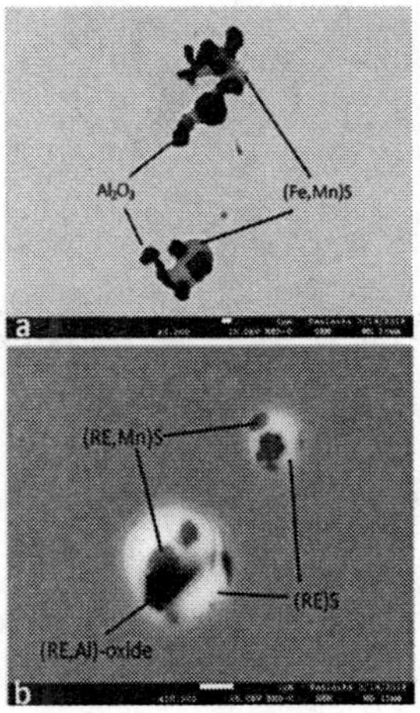

Fig. 2. SEM micrographs in BSE modes, a – non-modified cast steel, b – cast steel with added REMs

Adding of mischmetal to the cast steel did not result in changed yield limit and yield strength that achieved the average values of approximately 348 MPa and ca. 550 Mpa for both variants, whereas impact toughness of the cast steels being examined did change (Table 2).

Table 2. Impact toughness of modified and non-modified cast steel

	Without REMs	With added REMs
	Impact toughness (J)	
+20 C	113, 113, 123	140, 151, 134
-30 C	58, 67, 66	62, 53, 102
-40 C	48, 48,54	76, 60, 61

2. Morphology of impact-test fractures

Impact-test fractures for samples fractured in the ambient temperature were ductile in both cases. Differences could be seen with regard to dimensions of non-metallic inclusions observed within fracture areas and thus in dimples surrounding them (Fig. 3). As the temperature was decreasing, fractures became brittle and strips of ductile cracks could be locally observed. In the modified cast steel these strips were more frequent covering the greater part of the fracture surface than in the non-modified cast steel. The cracking process was also influenced by grain fragmentation that could be seen in multiple faults subsequently accumulating in the form of river patterns.

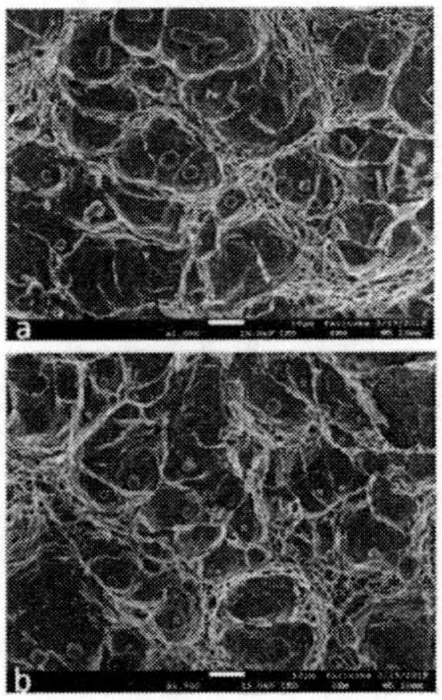

Fig. 3. SEM micrographs of the fracture surface of the G20Mn5, test temperature +20°C, a – non-modified cast steel, b – cast steel with added REMs

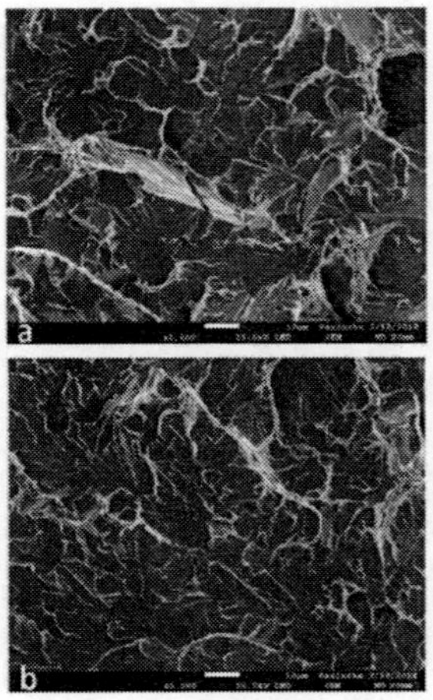

Fig. 4. SEM micrographs of the fracture surface of the G20Mn5, test temperature -40°C, a – non-modified steel, b – cast steel with added REMs

3. Conclusions

G20Mn5 cast steel can be successfully modified in industrial conditions. After adding rare earth metals impact toughness turned out to be increased, when comparing to non-modified cast steel. Different grain dimensions were observed, as well as different shapes, dimensions and dispersion of non-metallic inclusions, having subsequent influence on the cracking process.

Acknowledgments

The author gratefully acknowledge the financial support provided by the Ministry of Science and Higher Educations for the project number 01.0.10.00/2.01.01.01.0040 MNSP.MKMT.18.003.

The author would like to thank for the effective cooperation with HARDTOP Centrum Odlewnictwa Sp. z o.o., Charsznica, Poland.

References

[1] Kalandyk B., Sierant Z., Sobula S. ((2009) Improvement of the microstructure, yield stress and impact toughness of medium carbon steel by vanadium additions. Przegląd Odlewnictwa , 58 (3), 108-113.

[2] Xiang CH., Yanxiang L. (2007). Fracture toughness improvement of austempered high silicon steel by titanium, vanadium and rare earth elements modification, Materials Science and Engineering A, vol. 444, 298 – 305.

[3] Longmei Wang, Qin Lin, Jingwen Ji, Denian Lan. (2005). New study concerning development of application of rare earth metals, Journal of Alloys and Compounds, vol. 408 – 412, 384 – 386.

[4] Gambino R.J., McGuire T.R. (1986), Enhanced magneto-optic properties of light rare-earth-transition metal amorphous alloys, vol. 54-57 (3), 1365-1370.

[5] Mitsunori Yada, Masami Mihara, Shingi Mouri, Masako Kuroki, Tsuyoshi Kijima. (2002). Rare Earth (Er, Tm, Yb, Lu) Oxide Nanotubes Templated by Dodecylsulfate Assemblies, Advanced Materials, vol. 14 (4), 309 - 313.

The 73rd World Foundry Congress
„Creative Foundry"
23 - 27 September 2018, Krakow, Poland

TECHNOLOGY OF ALLOY LAYERS ON SURFACE OF CASTINGS

Tomasz Wróbel*, Natalia Przyszlak, Agnieszka Dulska

Silesian University of Technology, Department of Foundry Engineering, Towarowa St. 7, 44-100 Gliwice, Poland
**Corresponding address: tomasz.wrobel@polsl.pl*

Key words: cast iron, cast steel, stainless steel, layer, insert

1. Introduction

In mechanical engineering industry is observed a growth of demand on casting with special properties, as corrosion or abrassion resistance. High usable properties are often required only on surface layer of casting, wherefore, application of expensive materials for the whole casting is unprofitable and unjustified.

Multi-layered casting technology lets obtain good usable properties on working surface layer of the casting, while the base fulfils a purpose of supporting structure, which is not exposed on direct influence of external factors. The technology of forming such materials in foundry process is one of the most economical methods of improving working layer of casting. Manufacturing of layer elements has place directly in casting process. It is possible to control thickness of the layer. In addition this technology lets avoid cracks in heat affected zone, what is a common phenomenon when surface layer is manufactured in welding method.

The basic technology of forming layered castings in "liquid-solid" configuration is mould cavity preparation method. It consist of preparing the mould by fixing solid insert on surface of mould cavity, what is followed by pouring molten metal. The components (insert and base material) of such material vary in chemical and physical properties what enable to simultaneously use all their advantages i.e. insert - hardness, high mechanical, abrasion or corrosion resistance and base material - ductility, machinability. Correct preparation of the mould relies on placing the insert in such surface of the mould where composite surface layer, characterizing by different properties compared to the basic cast, is expected to be formed.

The insert can take the form of granular particles, monolithic plate or spacial shape.

The quality of the joint depends on temperature and pressure of liquid metal. In general, lower temperature difference between the insert and base material aids to obtain more stable joint, however, the mutual diffusion of the basic components and components of metals depends on their mutual solubility in the liquid and solid state [1-5].

The aim of the study was to manufacture layered casting and obtain stable diffusion joint between:

- monolithic inserts in a form of steel plate made of X2CrNi18-8 and grey cast iron,
- monolithic inserts in a form of steel plate made of X2CrNi18-9 and cast steel,
- granular inserts in a form of particles of high carbon ferrochromium and cast steel,
- spacial insert made of Ti6Al4V alloy and grey cast iron (Fig. 1).

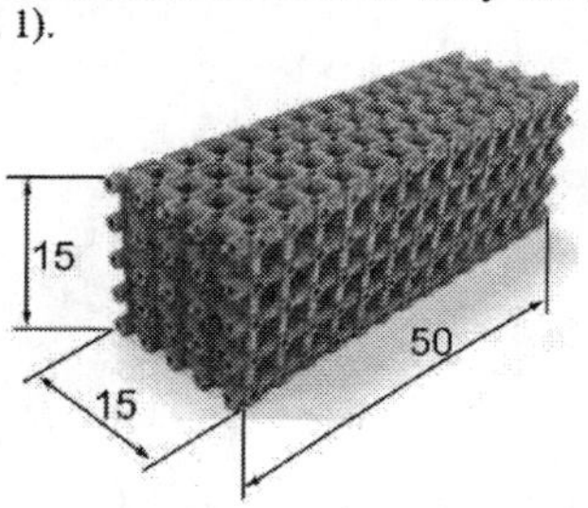

Fig. 1. Shape of the insert used to manufacture layered casting, insert made of Ti6Al4 using 3D printing.

2. Range of studies

In the range of studies were made layered castings consisting of two basic parts: base part and working part (layer). Base part was made of typical casting material (i.e. grey cast iron with flake graphite, ferritic-pearlitic cast steel) and it is characterised by lower properties. However, working part has i.e. better corrosion and abrasive resistance and it is characterised by high hardness. The layer is manufactured with the insert (i.e. monolithic, granular, 3D printed). Chemical composition of materials is presented in Table 1 and 2.

On the basis of numerous studies i.e. structure (fig. 2), usable properties, temperature measurement, was affirmed that producing layered casting according presented method, with monolithic, granular or spacial insert, lets obtain high quality layers, which are characterised by good durability, hardness and abrasion resistance.

Table 1. Chemical composition of materials used as base parts of layered castings.

Grey cast iron, mass contents in %								
C	Mn	Cr	Si	Cu	Al	P	S	Fe
3,1	0,6	0,8	1	0,2	0,2	0,03	0,03	rest
Carbon cast steel, mass contents in %								
0,6	0,6	0,09	0,18	0,28	-	0,03	0,02	rest

Table 2. Chemical composition of materials used as working part of layered castings; insert: - granular FeCrC insert(granularity 0.32-0.64 mm, thickness 5 mm); -monolithic insert made of X2CrNi18-8, thickness 1mm; - monolithic insert made of X2CrNi18-9; spacial insert made of Ti6Al4, printed using SLM method.

Alloy steel X2CrNi18-8, mass contents in %								
C	Cr	Ni	Mn	Si	Mo	P	S	Fe
0,06	20,3	7,91	0,98	0,59	0,08	0,04	0,01	rest
Alloy steel X2CrNi18-9, mass contents in %								
C	Cr	Ni	Mn	Si	Mo	P	S	Fe
0,03	19,2	9,8	1,45	0,9	0,2	0,03	0,01	rest
Ferrochromium, mass contents in %								
C	Cr	Fe	Si	P	S			
7,92	62,5	28,8	0,75	0,026	0,02			
Alloy titanium Ti6Al4, mass contents in %								
Al	V	C	Fe	O	N	H	Ti	
5,5	3,5	0,08	0,3	0,2	0,05	0,015	bal.	

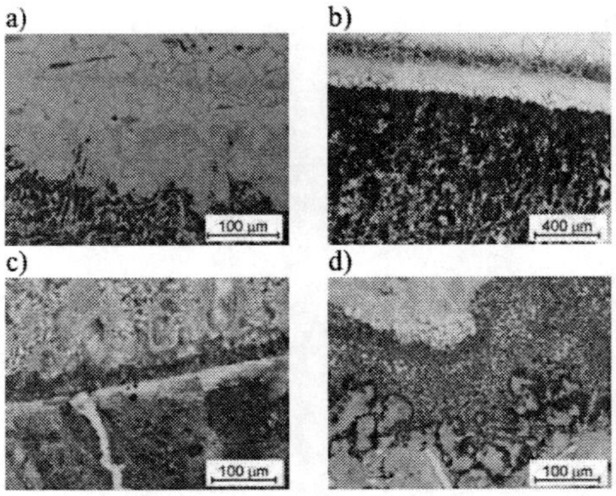

Fig. 2. Microstructure of layered casting, working part (insert) made of: a) X2CrNi18-8; b) X2CrNi18-9; c) FeCrC 800; d) Ti6Al4

3. Summary

Durable diffusion connection between base parts of layered casting - where working part is made of alloy steel and base part is made of cast steel or grey cast iron - is a result of carbon transport toward lower concentration, by extension, from base part to working layer. In presented technology, base part material is poured into mould cavity, where working part material, in the form of plate, is placed. Therefore, in the first step, carbon transport is a result of diffusion in liquid-solid system and after solidification process termination, in solid state.

Discussed diffusion processes are thermally activated, so carbon transport is a succession of heat transport in the analogues direction - from „hot" metal to „cold" monolithic layer.

Production of working part in this case proceeds in following steps:

- humidification of monolithic insert by liquid metal,
- carbon diffusion from liquid metal towards monolithic insert,
- melting of low thickness layer on the surface of monolithic insert,
- carbon diffusion in solid state.

However, production of layered casting with granular ferrochromium and cast steel proceeds in following steps:

- pouring the liquid cast steel into mould cavity, where granular insert was previously placed,
- heat transport from cast steel to insert,
- formation of thin, solidified layer of cast steel,
- diffusion in solid state in insert layer adjacent to solidified layer of cast steel
- decrease of Cr and C concentration, increase of Fe in above-mentioned insert area,
- the transition to liquid state of part of the insert adjacent to cast steel,
- diffusion in liquid state,
- reverse carbon diffusion from transition zone and cast steel,
- crystallization and formation of surface alloy layer.

Formation of working surface layer for the castings consisting of cast iron and spacial insert made of titanium alloy Ti6Al4 proceeds as below:

- pouring the mould with spacial insert with liquid cast iron,
- heating and dissolving of insert components,
- elements (mainly C and Ti) diffusion,
- crystallization and formation of surface layer rich in titanium carbides.

References

[1] Wróbel, T., Wiedermann, J. & Skupień, P. (2015). Bimetallic castings in a chromium–nickel stainless steel working surface layer configuration with a grey cast iron base. Transactions of the Indian Institute of Metals. 68 (4), 571-580.

[2] Dulska, A., Baron, C., & Szajnar, J. (2016). The analysis of the effects of heat and mass movement during alloy layer forming process on steel cast. METAL 2016. 25th Anniversary International Conference on Metallurgy and Materials, Brno, Czech Republic, May 25th - 27th, 2016 (pp. 110-115). Conference proceedings. Ostrava : Tanger, 2016.

[3] Przyszlak, N., Dulska, A., Wróbel, T. & Szajnar, J. (2018). Grey Cast Iron Locally Reinforced Using 3D Printing Scaffold Insert. Archive s of Foundry Engineering. 18 (1), 99-102.

[4] Górny, Z. (1987). Multi-layers and reinforced castings. Solidification of Metals and Alloys. 11, 135-164.

[5] Zhi, X., Han, Y., Liu, J., Zhao, M. & Ma, S. (2014). Casting process optimization of a bimetal wear-resistant block using liquid-solid processing. International Journal of Materials Research. 105(10), 953-960.

The 73rd World Foundry Congress
„Creative Foundry"
23 - 27 September 2018, Krakow, Poland

STRENGTH PROPERTIES OF CERAMIC MOULDS CONTAINING WASTE MOULDING SAND AFTER INITIAL RECLAMATION AS A SUBSTITUTE FOR BASE SAND

Michał Angrecki [1]*, Jadwiga Kamińska [1], Jarosław Jakubski [2], Piotr Wieliczko [1]

[1] *Foundry Research Institute, Department of Technology, Zakopianska 73, 30-418 Cracow, Poland*
[2] *AGH University of Science and Technology, Faculty of Foundry Engineering, Mickiewicza 30, 30-059 Cracow, Poland*
**Corresponding address: michal.angrecki@iod.krakow.pl*

Key words: Ceramic shell molds, waste molding sand, reclamation sand, lost wax method, strength properties

According to the Act on Waste, the basic principle of proceeding is to prevent the occurrence of the waste, and if this is not possible, to provide a useful reclaim compatible with the principles of environmental protection.

The foundry industry generates large amounts of waste. In the production cycle of one tonne of sound castings, approximately one tonne of waste is generated, and it must be either recycled or stored in appropriate landfills. The waste is mainly composed of used moulding and core sands. Increasing prices of raw materials and waste disposal costs force foundries to intensively search for ways to reduce the volume of the used materials, on the one hand, and to effectively re-use the waste generated in the production process, on the other hand.

Sand moulds with binders are single-use moulds, but moulding sand from which the casting mould is made can be re-used in the mould making process if the foundry has proper installation for moulding sand reclamation. If the foundry can not reclaim the waste sand, the waste is disposed to a landfill. Reclamation of moulding sands does not always ensure full re-use. If the foundry produces small-size castings, the degree of burnout of the binder keeping the sand grains together will be low. Then the reclamation process will not ensure adequate cleaning of the surface of the base sand grains and re-use of the reclaimed moulding sand will give rise to surface defects in castings. Therefore, in foundry practice, only about 80% of the reclaimed moulding sand is used as a base material for re-moulding. The missing amount ismade up with fresh silica sand. The conclusion is that even if the foundry reclaims its moulding sand, about 20% of the total sand volume is anyway a waste material and as such is disposed to a landfill.

Depositing waste in landfills means, in addition to environmental pollution, also a large financial burden for the foundry in the form of fees for landfilling.

In this article, the attention was focused on the results of studies of the possibility to re-use this fraction of moulding sand which is directed to landfills as a useless waste.

Mechanical properties of new ceramic material based on the waste moulding sand were tested. For the tests, waste moulding sands made by various foundry technologies were selected, including sand with furfuryl resin, sand with phenol-formaldehyde resin and sand with water glass. The waste moulding sand was first subjected to preliminary crushing. Next, the granular fractions useful for the manufacture of individual layers of the ceramic material were separated. Ceramic specimens, in which the waste moulding sand was a substitute for fresh silica sand in subsequent layers of the aggregate, were prepared. As a reference, similar specimens were used but in those specimens all layers were made from the fresh ceramic materials. On thus prepared ceramic specimens, the bending strength of the ceramic material was determined in the temperature range from 20 to 900°C. The ceramic specimens containing the waste moulding sand were reported to be entirely free from any surface defects of the crack type. Based on the obtained values of the bending strength, it was concluded that the waste moulding sand had no adverse effect on the strength of the ceramic material.

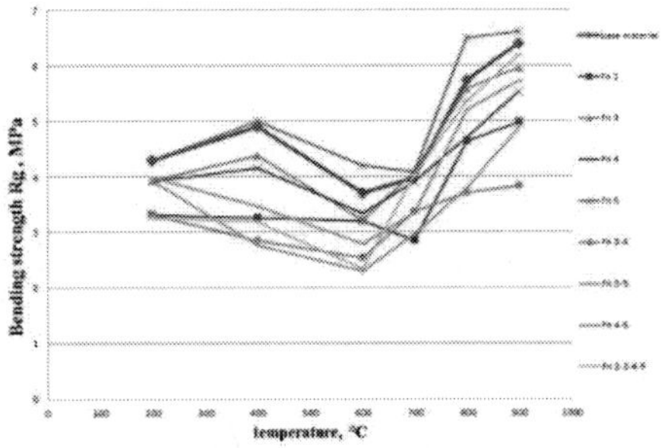

Fig. 1. Bending strength of ceramic specimens as a function of temperature.

Acknowledgments
The article was prepared as part of the project No. LIDER / 21/0125 / L-8/16 / NCBR / 2017.

References

[1] Izdebska-Szanda I., Angrecki M., Palma A. (2013). Recycling of waste moulding sands with new binders. Archives of Foundry Engineering 13 (2), s. 43-48.
[2] Dańko J., Holtzer M., Dańko R. (2010). Problems of scientific and development research concerning the reclamation of used foundry sands. Archives of Foundry Engineering 10, s. 29-34.
[3] Skrzyński M., Dańko R., Kamińska J. (2013). Reclamation of mixtures of spent sands of inorganic and organic type. 13/4, s. 93-96.
[4] Angrecki M., Izdebska-Szanda I., Palma A. (2015) Prace Instytutu Odlewnictwa 55, s. 15-22.
[5] Stefański Z., Izdebska-Szanda I., Angrecki M.(2014). Application of a New Innovative Ceramic Material for Investment Casting Technology. Archives of Foundry Engineering 14, s. 51-56.

The 73rd World Foundry Congress
„Creative Foundry"
23 - 27 September 2018, Krakow, Poland

EXPERIMENTAL AND SIMULATION STUDIES OF CORES MAKING PROCESS WITH BLOWING METHODS

Rafał Dańko [1]*, Michał Szucki [1], Mateusz Skrzyński [1]

[1]*AGH University of Science and Technology, Faculty of Foundry Engineering, Reymonta 23, 30-059 Krakow, Poland*
**Corresponding address: e-mail: rd@agh.edu.pl*

Key words: Sand core, core shooting, core blowing, core sand, computer simulation

The article presents the results of experimental research of making casting cores using the blow method. The tests included various parameters of the blowing process (eg. blow pressure and blow hole diameter) as well as various shapes of the core box. Another variable parameter was the venting degree of the core box. During the tests the flow of sand-air stream from the blowing chamber to the core box was analyzed and recorded with the use of PHANTOM V210 high-speed camera. Another element of the research was using the geometric parameters of the core box and core blower as well as the parameters of the blow process used in the experimental studies to carry out the numerical simulation. Two different numerical models were used. Then obtained results were compared with experimental data.

1. Introduction

Optimization of cores' making process with blowing methods covers a number of issues. The most important are [1]:

- selection of the appropriate core sand, characterized by fluidity that allows filling the core box,
- selection of the parameters of the blow, taking into account the rate of pressure increase in the blowing chamber, which results from the construction of the blow valve,
- proper construction of the core box, especially the appropriately selected venting system.

In the presented studies, all of the listed above topics were considered.

In order to optimize the blowing process, numerical methods that allow to assess the nature of the process, the correctness of the construction of the core boxes and a number of other parameters are more and more often used [2-5].

2. Experimental stand

The experimental stand includes the following elements:

- experimental core blower,
- experimental core boxes,
- measuring and control system.

The schematic diagram of the experimental stand is shown in Figure 1.

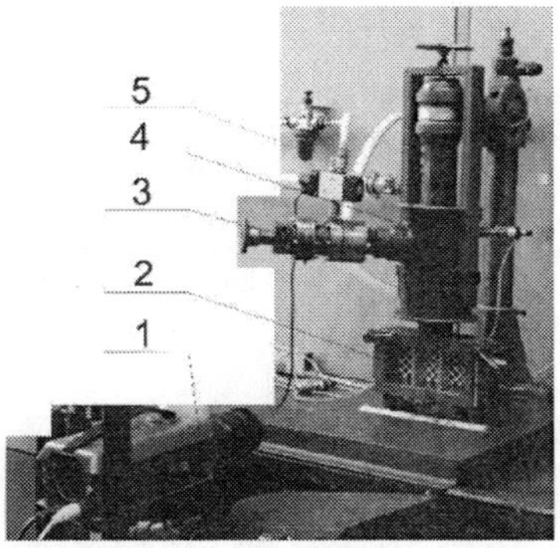

Fig. 1. Diagram of the experimental system: 1-PHANTOM fast camera, 2-Experimental core box, 3-Hansberg valve, 4-Shooting chamber, 5-Piezoelectric pressure converter (sensor)

The test core blower used in the tests is characterized by the perforated blowing chamber with a volume of 3 dm^3. The perforations in the chamber are designed in such a way that their active surface area is about 4%. Moreover the design of the blowing chamber allows the use of exchangeable shapes reproducting the blow hole (various blow hole diameters). The presented tests used blow pressure of 0.4 MPa, 0.5 MPa, 0.6 MPa and blow hole diameters of 10mm and 20mm respectively.

The studies were performed on cuboid core box (Vc = 795 cm^3) with and without internal shapes changing the nature of the sand-air stream flow, shown in figure 2. In each type of used system it was possible to adopt a different number of vents. In the present work the total area of active vents for all analysed cases was 4.2 cm^2.

The core sand for the classical cold-box technology was used in the study. In earlier works it was shown that this type of core sand is characterized by very good fluidity and can be used as the reference material [6].

Fig. 2. Core box with and without internal shapes

The measurement-control system, based on a PLC controller, allowed for the measurement of pressure in the blowing chamber and in the core box. Performed experiments showed that the rate of pressure increase in blowing chamber for the applied pressure values was in the range of 5.5 to 7.6 MPa / s.

3. Numerical simulations

The obtained experimental result (Fig. 3a) were compared with simulation data. Numerical tests included the analysis of the core blowing process in the Flow-3D software. This program uses so-called Drift-Flux model for tracking core sand-air interface (Fig. 3b).

a)

b)

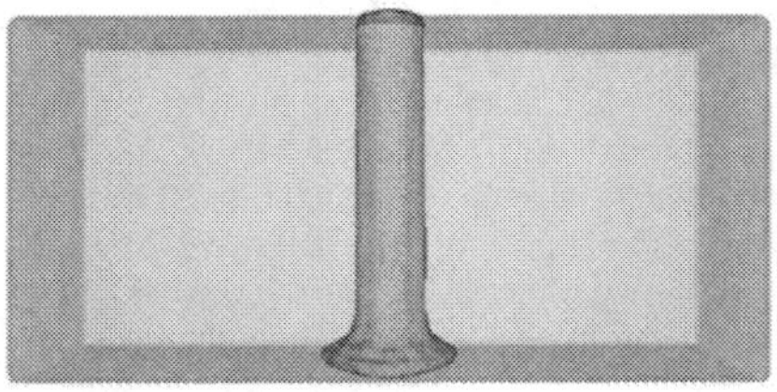

c)

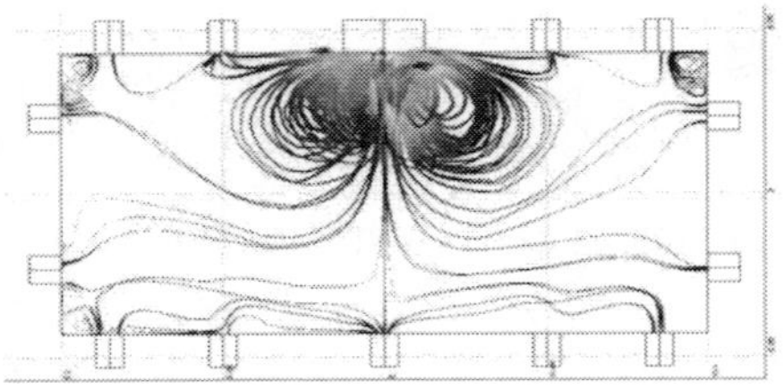

Fig. 3. Blow process (pressure 0.4MPa and d=20 mm); a) image recorded during experiment, b) simulation result from Drift-Flux model, c) streamlines in gas phase (Mixture model)

Because in the Drift-Flux model the air is usually treated as adiabatic "gas bubble" where the variation of pressure are neglected, this solution does not allow the analysis of the air evacuation process. To solve this problem additional computations were made with so-called mixture model implemented in COMSOL Multiphysics software (Fig. 3c).

4. Summary

The obtained simulation results are in good correlation with experimental data. With the use of the mixture model it is possible to analyse the local pressure/velocity changes in gas phase. The air evacuation process is characterized by highly turbulent flows and strongly influence the final core properties.

Acknowledgments (optional)

This work was supported by the Polish NCN project UMO-2014/15/B/ST8/00206.

References

[1] Boenisch D., Knauf M. (1991). Kernschiessen - Untersuchungen mit neuartigen Prüfkörpern und verschiedenen Kernbindern . Giesserei 18, 640-646.

[2] Dańko J. (1992). Teoretyczna i doświadczalna weryfikacja modelu Aksjonowa obliczania maszyn dmuchowych. Archiwum Technologii Budowy Maszyn. 10, 7-15.

[3] Zhang Yi-zhong, Lu Gao-chun, Ni Chang-jiang, Jing Tao, Yang Lin-long, Wu Qin-fang (2017). GPU based numerical simulation of core shooting process. China Foundry. 14(5), 392-397.

[4] Winartomo B, Vroomen U, Hrigpolaczek A B, et al. . (2013). Multiphase modelling of core shooting process. International Journal of Cast Metals Research. 18 (1), 13-20.

[5] Ni Chang-jiang, Lu Gao-chun, Zhang Qing-dong, et al. (2016). Influence of core box vents distribution on flow dynamics of core shooting process based on experiment and numerical simulation. China Foundry. 13(1), 22-29.

[6] Fedoryszyn A., Dańko R. (2013). Examinations of parameters influencing the outflow of two phase air-sand stream from machine chamber and core box filling in shooting process. Archives of Metallurgy and Materials. 58(3), 903-906.

The 73rd World Foundry Congress
„Creative Foundry"
23 - 27 September 2018, Krakow, Poland

A STUDY OF Mg-Cu INTERREACTION IN COPPER-ALLOYED ADI

Marcin Górny[1]*, Edward Tyrała[1], Magdalena Kawalec[1], Gabriela Sikora[1]

[1]*AGH University of Science and Technology, Faculty of Foundry Engineering, Reymonta 23, 30-059 Krakow, Poland*
**Corresponding address: mgorny@agh.edu.pl*

Key words: ADI, casting, copper, magnesium, fracture, spheroidal graphite cast iron

Abstract

In the present work, the issue of precipitates from the Mg-Cu system in copper-alloyed ADI (Austempered Ductile Iron) is introduced. The research work was carried out on spheroidal graphite cast iron that was austenitized at 900°C, followed by austempering at 380°C. The research is also focused on spheroidal graphite cast iron followed by oil quenching. The microstructures and fractures were investigated using optical and scanning electron microscopy. It was revealed that, in castings made of copper alloyed ADI, highly dispersed particles from Mg-Cu system are created. It has been revealed that their number and size strongly depends on austenitizing process. Precipitations from Mg-Cu system influence impact properties of ADI castings.

1. Introduction

Copper is one of the most-common alloying elements used in cast iron [1-4], especially in high strength spheroidal graphite cast iron (SGI) and austempered ductile iron (ADI) [5]. Usually copper is added in amount up to 1.5 wt.%. in SGI - Cu promotes the pearlite formation, stabilizes the austenite, increases the graphite nodule count, increases: the tensile strength, yield strength and hardness, decreases: the chilling tendency, ductility and impact resistance and raises the ductile to brittle transition temperatures [4-5]. In ADI - Cu increases the wall thickness that can be successfully austempered, increases the amount of retained austenite in the microstructure [6] and restrains carbide formations in the ausferrite [5]. It segregates towards graphite nodules and makes the shape of the graphite nodule more rounded. From the published literature, there is little information on the formation of highly dispersive copper-magnesium precipitates in the microstructure of ADI [7]. This study investigated the effect of the addition of copper at the level of 1.0 wt.% on the formation of the Mg-Cu precipitates that are formed during ADI production.

2. Experimental

The experimental melts were carried out in a 2.1 dm^3 capacity crucible using an electrical induction furnace. After melting and preheating to a temperature of 1490°C, the metal was held for 2 minutes followed by spheroidization and inoculation presses (bell method). For spheroidization and inoculation master alloy (Fe-Si-Mg-6% Mg) and inoculant (73-78% Si, 0.75-1.25% Ca, 0.75-1.25% Ba, 0.75-1.25% Al, Fe: balance) were used, respectively. Y-shaped ingots blocks with 13 mm in its bottom part, following the ASTM A 536-84 Standard were used. Two melts were prepared: without copper (Alloy A), with copper (Alloy B). The chemical composition of investigated alloys is shown in Table 1.

Table 1. Chemical composition of the investigated ADI, wt. %

Alloy	C%	Si%	Mn%	Mg%	Cu%
A	3.60	2.66	0.38	0.042	0.01
B	3.63	2.63	0.35	0.041	1.02

The heat treatments procedure consisted in:

Series I

(a) Austenitizing in a silite furnace at 900±2°C for 40 min, (b) austempering in a salt bath of $NaNO_2$-KNO_3 at 380±2°C for up to 120 min, and (c) air cooling at room temperature.

Series II

(a) Austenitizing in a silite furnace at 900±2°C for 20, 60 and 180 min, (b) quenched in oil bath.

The samples were examined using an optical microscope (Leica MEF 4M) and scanning electron microscope (Jeol JSM 5500LV) equipped with an energy-dispersive spectroscopy (EDS) detector.

3. Results

The "as-cast" microstructure of the Alloy A (copper free SGI) is composed of graphite nodules and ferritic-pearlitic metallic matrix, with pearlite fraction at a level of 10%. The addition of Cu decreased ferrite fraction to a level of 5 %.

A metallographic analysis of ADI revealed that the addition of Cu increased the austenite fraction by 6 % (using line scan method). The effect of Cu alloyed ADI on the fracture morphology is illustrated in Fig. 1. Fractures of the specimens (Alloy A and Alloy B) exhibit a mixture of ductile and cleavage fracture modes.

In alloy B fracture (Fig. 1) revealed the presence of ellipsoidal-shaped precipitations from Mg-Cu system identified as Mg_2Cu [7].

According to Zho [8] a thin Cu film on the spheroidal graphite surface is formed.

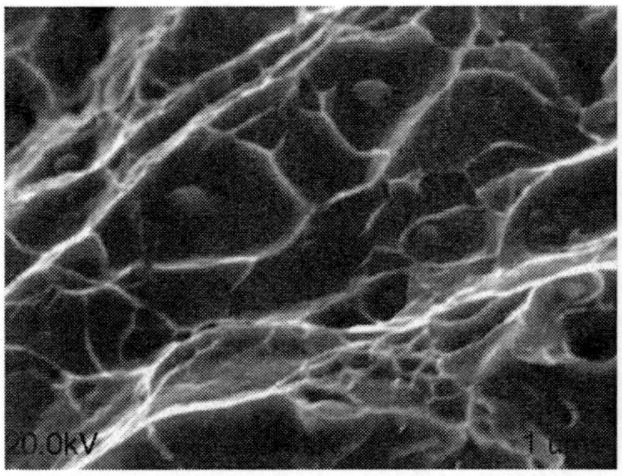

Fig. 1. Fracture morphology of Cu-alloyed ADI after impact Charpy test (Series I, Alloy B)

In turn according to Dierickx [9] during austenitizing of SGI an increase of the Mg content at the graphite-matrix interface occurs. This makes the possible reaction between Mg and Cu. Schematic of interaction between Mg and Cu during austenitization stage of ADI production is shown in Fig. 2.

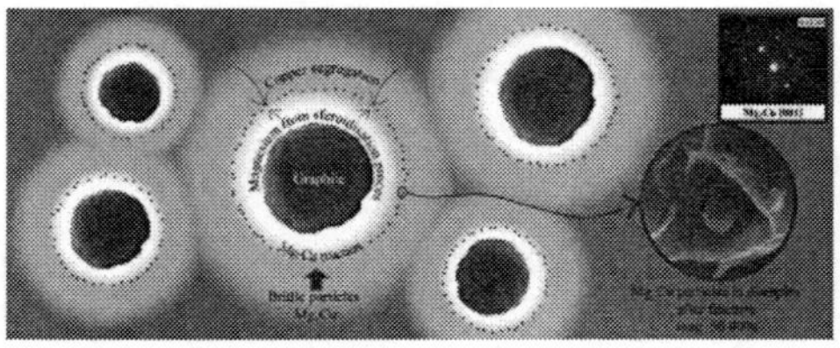

Fig. 2. Schematic of interaction between Mg and Cu during austenitization stage of ADI production

From second series of tests (austenitization process) it follows that as holding time increases the size of Mg-Cu precipitates increases as well. The maximum size of these precipitates was as high as 3.3 µm at the time of 3 hour of austenitization. In addition, Mg-Cu precipitates (Series II, alloy B) are located on both edges and cleavage surfaces (Fig. 3).
The impact properties of ADI for both series are listed in Tables 2.

Table 2. Impact strength (J) of ADI as a function of austempering time (Series I) [7]

Time, min	Alloy A	Alloy B
40	169	90
60	174	129
80	160	144
120	160	130

Impact strength of copper alloyed ADI is lower in comparison to unalloyed ADI (Table 2). The change in properties is generally related to graphite shape and number, austenite fraction, ferrite platelets, and the presence of other phases like those identified Mg-Cu precipitates.

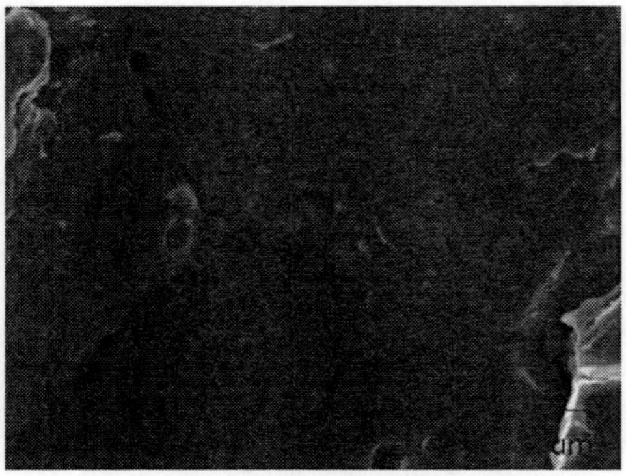

Fig. 3. Fracture morphology of Cu-alloyed ADI after impact Charpy test (Series II, Alloy B)

4. Conclusions

1. The interaction between Mg and Cu during austenitization process of heat treatment can produce Mg-Cu precipitates. The number and the size of these precipitates strongly depends on austenitizing time.
2. Mg-Cu precipitates in copper alloyed ADI (Alloy B), can have a negative effect on the Charpy impact strength of ADI.

References

[1]. K. Osamura, H. Okuda, S. Ochiai, M. Takashima, K. Asano, M. Furusaka, K. Kishida, and F. Kurosawa. (1994). Precipitation Hardening in Fe-Cu Binary and Quaternary Alloys. ISIJ Int. Vol. 34, pp 359–65.

[2]. M. Górny, E. Tyrała, and H.F. Lopez. (2014). Effect of Copper and Nickel on the Transformation Kinetics of Austempered Ductile Iron. J. Mater. Eng. Perform. Vol. 23 pp. 3505–3510.

[3]. U. Batra, S. Ray, and S.R. Prabhakar. (2004). The influence of nickel and copper on the austempering of ductile iron. J. Mater. Eng. Perform. Vol. 13, pp. 64–68.

[4]. C. Labrecque and M. Gagne. (1998). Review ductile iron: 50 years of continuous development. Can. Metall. Q. vol. 37, pp. 343-378.

[5]. M. Bamberger (2016). Austempered Ductile Cast Iron. Encycl. Iron, Steel, Their Alloy. 5th ed., Taylor and Francis, New York, USA. pp.196-216.

[6]. A.P. Druschitz, R.E. Aristizabal, E. Druschitz, C.R. Hubbard, T.R. Watkins, L. Walker, and M. Ostrander. (2012). In situ studies of intercritically austempered ductile iron using neutron diffraction. Metall. Mater. Trans. A 43, pp. 1468-1476.

[7]. M. Górny, E. Tyrała, G. Sikora, and Ł. Rogal. (2018). Identification of Mg2Cu Particles in Cu Alloyed Austempered Ductile Iron. Met. Mater. Int. Vol. 24, pp. 95-100.

[8]. Y. Zou, M. Ogawa, H. Nakae. (2012). Interaction of Boron with Copper and Its Influence on Matrix of Spheroidal Graphite Cast Iron. ISIJ International. Vol. 52, No. 3, pp. 505–509.

[9]. P. Dierickx, C. Verdu, A. Reynaud, and R. Fougeres. (1996). A Study Of Physico-Chemical Mechanisms Responsible For Damage Of Heat Treated And As-Cast Ferritic Spheroidal Graphite Cast Irons. Scripta Mater. Vol. 34, pp. 261-268.

The 73rd World Foundry Congress
„Creative Foundry"
23 - 27 September 2018, Krakow, Poland

ENVIRONMENTALLY FRIENDLY FOUNDRY MOLDING AND CORE SANDS

Katarzyna Major-Gabryś *

AGH University of Science and Technology, Faculty of Foundry Engineering, Department of Molding Materials, Mold Technology and Foundry of Non-ferrous Metals, Mickiewicza 30, 30-059 Krakow, Poland
**Corresponding address: katmg@agh.edu.pl*

Key words: innovative technologies, environmental protection, foundry engineering, molding and core sands, development trends

1. Introduction

Contemporary economy requires the production of products that meet the highest quality criteria. Such products include the production of the most responsible castings used in the automotive, construction, metallurgical and cement industries as well as the production of castings for the art and jewelry industry. In the context of the Polish economy, the production of castings is a significant branch. One of the basic factors affecting the quality of finished castings, mainly in terms of dimensional accuracy and surface smoothness, is the quality of molds and foundry cores. These in turn depend on the type and quality of the applied molding and core sands.

This paper is devoted to the above issues. It discusses the currently used technologies as well as development trends in the foundry and molding sands. These trends must take into account not only the technological quality and economics of their production process, but also the increasingly stringent requirements related to environmental protection [1].

The paper describes the current state of knowledge in the field of changes that have occurred in the field of foundry molding sands due to the increasing requirements regarding environmental protection. The possible development directions in this field were indicated.

Technological defects occurring in environmentally friendly processes, such as molding sands with hydrated sodium silicate, have been described. The paper also covers the harmfulness to the environment caused by various resins widely used in the foundry technology practice, detailing the changes that are introduced in them. Such processes include, for example, the cold-box technology or the method of lost wax casting. New - environmentally friendly technologies, which for example include molding sands with geopolymers, as well as molding sands bound with biotechnological factors have also been described [1].

The paper presents the results of author's own research devoted to selected technologies of molds and core production.

2. Development directions of molding and core sand

Due to the variety of foundry alloys used, casting technologies, their various mass and quality requirements, it is necessary to use a variety of molding and core sands. From the point of view of the size of the cast, the greater its mass is, the greater the effect of the casting alloy on the molding or core sand is. The type of cast and its pouring temperature make it necessary to select the right molding sand due to the chemical reactions on the alloy - mold interface. Additionally, the cast quality class required by the recipient makes it necessary to select the appropriate molding and core sand.

The trends in modern foundry engineering regarding molding sands and mold technology result from three main requirements:

- ✓ ensuring appropriate technological properties of the mold and core,
- ✓ high economy of the process,
- ✓ low harmfulness to the environment [1].

Providing adequate technological properties of the mold and core as well as high process economy are very important conditions for the development of molding sands. High economic efficiency of the process consists of its efficiency, including the speed of making molds and cores (speed of molding sand binding process), material prices, manufacturing costs, costs of casts' knocking-out, finishing and cleaning, as well as employment costs of employees, including qualified staff. However, in recent years, the dominant factor in the development of molding and core sands technology is the need to comply with high environmental protection requirements [1]. A new instrument in the protection of the environment as a whole (protection of air, water and soil) are the so-called integrated permits. They cover emissions to the environment: dust, gases, energy, electromagnetic fields, noise, heat, sewage and waste, and savings of resources and materials. Integrated permits were introduced into the European Union legislation by so-called The IPPC Directive of 1996 [2].

The development of legal regulations related to environmental protection in the countries of the European Union perfectly illustrates the number of new legal provisions in this area and the dynamics of their formation, as shown in Figure 1 [3].

The data shows that:

- ✓ In the years 1865 – 1970 → 20 acts related to environmental protection were introduced,
- ✓ In the years 1971 – 1985 → 94 acts related to environmental protection were introduced,
- ✓ In the years 1986 – 2014 → 800 acts related to environmental protection were introduced [3].

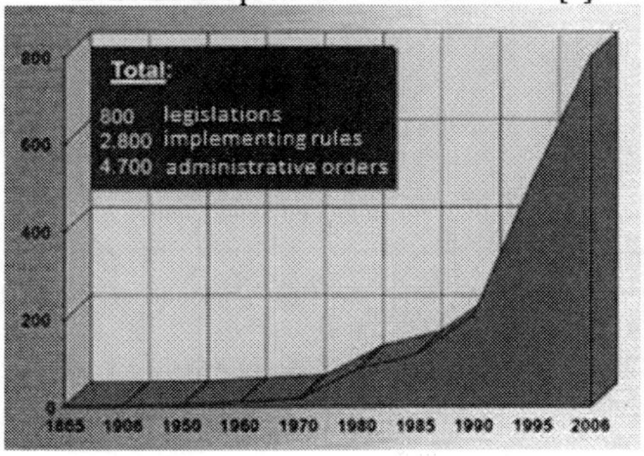

Fig. 1. Development of regulations related to environmental protection in the EU [3]

Trends related to the need to comply with high environmental protection requirements are intensifying to such an extent that even technologies that have been working well for decades in production processes must be replaced with more ecological solutions. Examples of such activities can be:

→ coal dust

In terms of limiting emissions of harmful substances into the environment from molding sands with bentonite, various technologies are used, such as coal dust substitutes or introduction to molding sand substances that neutralize - in a large extent - harmful compounds [4-5].

→ flammable protective coatings

Protective coatings have been used in foundry engineering for a long time. The so far used flammable protective coatings were mainly produced on the basis of organic solvents, mainly isopropyl alcohol. The requirements of the European Union, however, require the elimination of the use of organic solvents in the industry. That forces coating manufacturers to develop solutions based on water-based solvents, i.e. water soluble coatings with low water content, which do not require drying in heating chambers [6].

→ molding sands with ethyl silicate

Due to the process of hydrolysis of ethyl silicate used for the preparation of these molding sands with the use of organic solvents, precise casting was forced to search for new binders. The effect of research is the implementation of molding sands with colloidal silica [7].

→ molding sands with furfuryl resins

On 1 December 2010, entered into force a Regulation of the European Parliament and of the Council (EC) (No 1272/ 2008 of 16 December 2008 on classification, labeling and packaging of substances and mixtures), which classifies furfuryl resins containing more than 25 % of the free furfuryl alcohol as toxic. Furfuryl resins, which contain a lower level of free furfuryl alcohol, are regarded as harmful [8].

→ cold-box technology

Solvents based on PAHs used in cold-box technology (ensuring molding sands' bending strength R_g^u= 5-6 MPa) have been replaced with mineral solvents (ensuring molding sands' bending strength R_g^u= 3-3.5 MPa). The tendencies of replacement of the organic solvents with ecological solvents are also noticed in cold-box technology, in which carbon atoms are replaced with silicon atoms [9].

3. Conclusions

The results of the analysis show there is a need to develop new systems for bonding molding and core compounds that guarantee low harm to the environment and at the same time ensure high technological requirements. This paper presents modern systems for bonding molding and core compounds that provide low harm to the environment, but at the same time ensures high technological properties.

Acknowledgments

The research was funded as a part of AGH Research Project no. 11.11.170.318.

References

[1] Major-Gabryś K. (2016). Odlewnicze masy formierskie I rdzeniowe przyjazne dla środowiska. Katowice-Gliwice. Archives of Foundry Engineering.

[2] Forsight Technologiczny Odlewnictwa Polskiego, pod redakcją J. Tybulczuka (2009). Instytut Odlewnictwa, ISBN 978-83-88-770-42-5, Kraków.

[3] Bacanu I.A. (2014). Nowoczesne technologie Hüttenes-Albertus: Spojrzenie na przyszłość technologii w światowym odlewnictwie. In IV Conference Hüttenes-Albertus Poland'2014. Iława, Poland.

[4] Głowacki C.R., Crandell G.R., Cannon F.S., Clobes J.K., Yoight R.C., Furness J.C., McComb B.A., Knight S.M. (2003). Emission Studies at a test Foundry using an Advanced Oxidation – Clear Water System. AFS Transactions, 111, 579-598.

[5] Wang Y., Cannon F.S., Neill D., Crawford K., Yoight R.C. (2004). Effects of Advanced Oxidation Treatment on Green Sand Proprieties and Emissions. AFS Transactions, 112, 635-648.

[6] Seeger K. (2012). Zastosowanie pokryć wodnych przy formowaniu ręcznym. In III Conference Hüttenes-Albertus Poland'2012 (pp. 9-18), Zakopane, Poland.

[7] Karwiński A. (1999). Ekosil – "wodne" spoiwo dla odlewnictwa precyzyjnego, Biuletyn Instytutu Odlewnictwa, 5, 3-15.

[8] Benz N., Fourberg C. (2012). Przyjazne dla środowiska żywice furanowe z zawartością wolnego alkoholu furfurylowego mniejszą niż 25%. In III Conference Hüttenes-Albertus Poland'2012 (pp. 129-137), Zakopane, Poland.

[9] Serghini A. (2010). Krzemianowe systemy cold-box – czy mogą osiągnąć szczyt? In II Conference Hüttenes-Albertus Poland'2010 (pp. 1-10), Kazimierz Dolny, Poland.

The 73rd World Foundry Congress
„Creative Foundry"
23 - 27 September 2018, Krakow, Poland

CAST IRON "VARI – MORPH" (VM) WITH GRAPHITE OF SEVERAL FORMS – MATERIAL FOR CASTINGS OF SPECIAL DESTINATIONS

Jerzy Zych[1], Janusz Postuła[2]

[1]AGH University of Science and Technology, Faculty of Foundry Engineering, Mickiewicza 30 Av., 30-059 Krakow, Poland
[2]FANSULD Cast Iron Foundry; Zielona 22 Str., 26-700 Końskie, Poland
**Corresponding address: jzych@agh.edu.pl*

Key words: cast iron, compact graphite, index quality, thermal fatigue, castings of special destinations

1. Introduction

Cast iron is a material, which mechanical, physical, technological and functional properties are formed by graphite (its shape, distribution and size) and by a metallic matrix, which the most often is ferrite, pearlite or their mixture. Generally, three grades of cast iron are produced: grey cast iron (flake graphite), vermicular cast iron (compact graphite) and spheroidal cast iron (nodular graphite). It is assumed that in the casting structure of each cast iron grade occurs graphite, homogenous in respect of its shape, and the homogeneity degree is not smaller than 90 - 95%. Grey and spheroidal cast irons dominate in the world industrial practice.

The technological standards do not predict producing cast irons with a mixed graphite. Cast iron with a mixed graphite forms, with controlled fractions of individual shapes, which proposed name is VM cast iron ('Vari - Morph'), can become an interesting material for castings of a special requirements. This VM cast iron is characterised by physical and mechanical properties, which can not be achieved when graphite is homogeneous in respect of its shape.

A control of the graphite form can be realised in many ways, influencing by small additions of certain elements not its composition but the surface tension. The shape of graphite precipitates depends on the surface tension value. The influence of the magnesium content on the graphite precipitates shapes is presented in Figure 1. VM cast iron is between cast iron with flake (F) graphite and vermicular (V) graphite and in between vermicular (V) and spheroidal (S) graphite.

Cast irons with graphite F+V will be characterised by a good thermal conductivity and better plasticity and strength than grey cast iron. Especially interesting is cast iron with a mixed form: graphite (S) + graphite (V). Such cast iron will have high strength and plasticity, tightness, relatively good ability of damping vibrations and good thermal conductance. It was shown in papers [1-5], that such cast iron is characterised also by a high heat fatigue resistance higher than vermicular cast iron.

In contemporary methods of mechanical working, with the application of CNC machines, decreasing of HB at maintaining high R_m is attempted. Cast iron grades of a high index quality (IQ) - which is defined as the ratio R_m/HB - are looked for. The effect of strength increase can be achieved by substituting, at least partially, flake graphite by compact graphite while remaining the ferritic matrix. This will allow to improve the IQ value. The strength R_m in grey cast iron can be increased by the limitation of the graphite (C) amount and the ferrite substitution by pearlite. However, these traditional operations are decreasing IQ and cause worsening of the cast iron machining

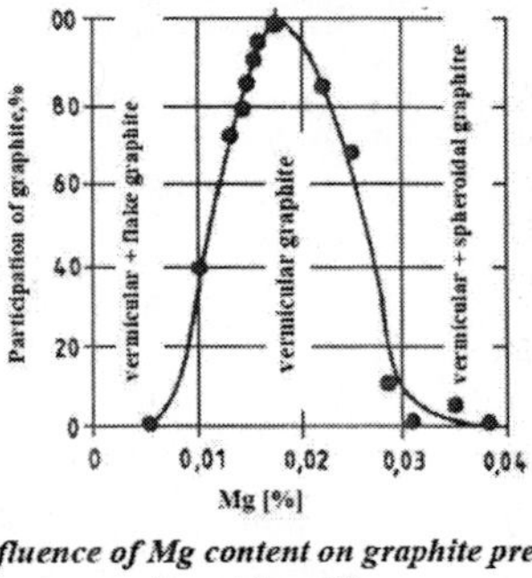

Fig. 1. Influence of Mg content on graphite precipitates in cast iron [6].

The results of investigations of the selected group of physical (thermal conductivity), mechanical (R_m and A_5) and functional properties (heat fatigue resistance) of „MV" cast iron is presented in the hereby paper. The ultrasound technique was applied for assessing the graphite compactness degree.

2. Materials and methods

The group of cast iron grades, which chemical composition oscillated near the eutectic value, was subjected to investigations (C=3.3÷3.6%, Si =2.6÷2.95%, Mn < 0.1; P < 0.02%; S < 0.01%; Mg =0.005÷0.030%). Spheroidisation was performed by the bell method using FeSi75Mg6 alloys. The graphite shape classification was based on the indicator shape of graphite ξ, which was defined by formula (1):

$$\xi = \frac{A_i}{P_i^2} \qquad (1)$$

where: A_i , P_i - surface (m^2) and diameter (m), respectively, of a single precipitate, i - number of graphite precipitates.

Indicator ξ takes the following values: 0-0.03 (ξ<0.03) for flake graphite, 0.035-0.065 for vermicular graphite, 0.065-0.08 (ξ>0.065) for nodular graphite.

The thermal fatigue was estimated by the L.F. Coffin method, considering that MV cast iron should be used for constructions subjected to a cyclic heating. The influence of the indicator shape of graphite on the thermal fatigue is presented in Figure 2. The more compact graphite form, the higher heat fatigue resistance. The dependence between the graphite form shown in the microstructure and the heat fatigue resistance of the cast iron is presented in Figure 3.

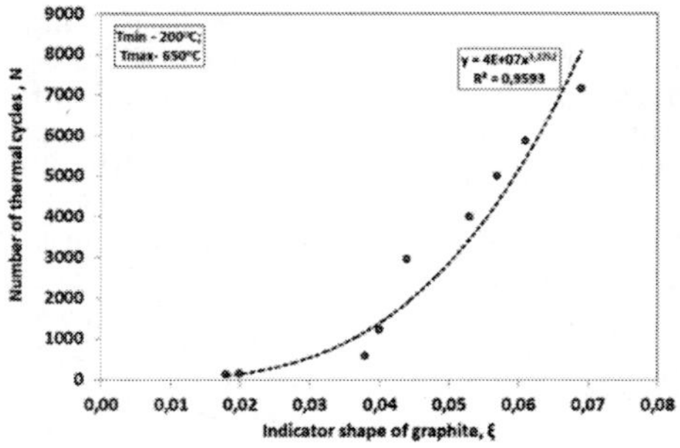

Fig. 2. Influence of the indicator shape of graphite on the thermal fatigue.

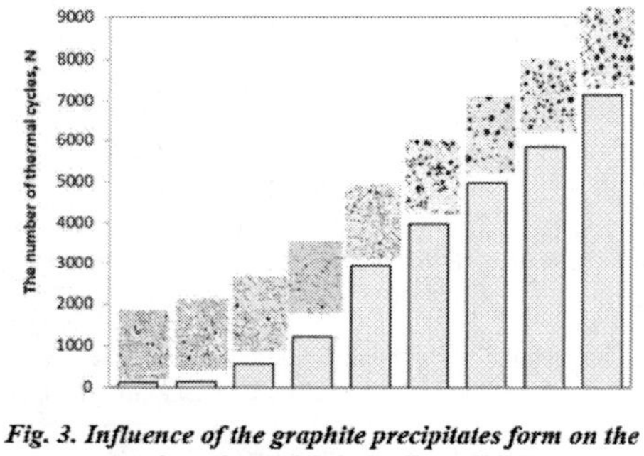

Fig. 3. Influence of the graphite precipitates form on the cast iron thermal fatigue; T_{max}=650°C.

The ultrasound wave velocity, which obtains the higher values the more compact are graphite precipitates, can be the good indicator of the compactness degree of graphite precipitates.

The dependence of the longitudinal wave velocity (2 MHz) on the heat fatigue resistance is shown in Figure 4. There is a possibility of predicting the heat fatigue resistance on the bases of ultrasound investigations.

The casting which was previously made of grey cast iron EU GJL250, and after the technology change will be made of MV cast iron is shown in Figure 5. This casting operates under heat fatigue conditions.

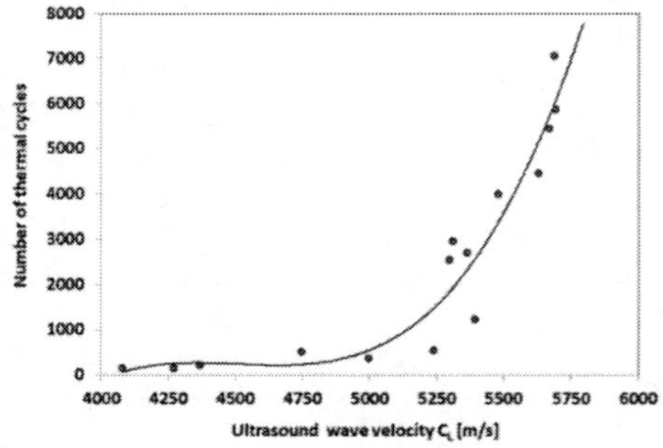

Fig. 4. Dependence between the cast iron ability to propagate the ultrasound wave (C_L) and the heat fatigue resistance.

Fig. 5. Element of the heating furnace - casting of MV cast iron operating under conditions of cycling heating

3. Conclusions

The performed investigations confirmed that considering cast irons with several graphite forms (VM) as not suitable material, is erroneous. This is the material for operations under conditions of thermal fatigue, at demanding a high tightness, etc. Its production technology is easier than e.g. of 'pure' vermicular cast iron.

Acknowledgments

This research received funding from NCBiR No. POIR 01.01.01-00-0042/17-00.

References

[1] J. Zych: The role of graphite morphology in forming of thermal fatigue resistance of cast iron. Inżynieria Materiałowa; ISSN 0208-6247 (2015) Nr 5, p. 343–347.

[2]J. Zych: Cast iron thermal fatigue resistance – influence of graphite shape. ISSN 0860-097X; (Seria Mechanika). S. 51–64.

[3] K. Kudala: Thesis - Department of Foundry AGH, Kraków-Poland (2015).

[4] J. Postuła: Thesis - Department of Foundry, AGH Kraków-Poland (2017).

[5] Iron Castings Handbook. Covering data on Gray, Malleable, Ductile, White, Alloy and Compacted Graphite Irons. (1981).

[6] E. Nechtelberger, H. Puhr et al.; Stand der Entwicklung von Gusseisen mit Vermiculagrahit – Herstellung, Eingenschaften und Anwendung. Giesserei-Praxis., Nr 22, S. 359-372; Berlin (1982).

The 73rd World Foundry Congress
„Creative Foundry"
23 - 27 September 2018, Krakow, Poland

LEAD-FREE CASTING BRASSES. ANALYSIS OF MICROSTRUCTURE AND PROPERTIES COMBINED WITH THE CASTING TECHNOLOGY

Janusz Kozana*, Aldona Garbacz-Klempka, Marcin Piękoś

AGH-University of Science and Technology, Faculty of Foundry Engineering,
Reymonta 23, 30-059 Krakow, Poland
**Corresponding address: jkozana@agh.edu.pl*

Key words: lead-free brass, microstructure, mechanical properties, alloy additions, valve alloys

1. Introduction

Among casting alloys, widely described in the scientific literature [1-2], casting bronzes constitute one of the important groups, used for many years and still developed, since they create certain technological problems. Copper and zinc based various alloys developed and assessed under laboratory and industrial conditions indicate several advantageous properties. They also indicate different, both mechanical and technological properties and corrosion resistance [3,4].

However, contemporary more and more emphasis - in the aspect of forming alloys properties - is attached to the ecological factor. Harmful elements from the group of heavy metals should be, in a maximal possible way, eliminated from alloys i.e. from the environment of humans. Ordinary casting brasses - lead-free [5,6], containing 10 to 40% of zinc as well as arsenic brasses of the increased resistance to dezincification are well suited for the so-called eco-brasses.

A wide group of Cu-Zn alloys, mainly low-zinc, is applied for producing semi-products and elements of machines and devices in a form of sections - extruded and drawn - of various dimensions and cross-sections. Ordinary brasses are also casting brasses. Several brasses rated to the ecological group occur in standards e.g. acc. to DIN the alloy marked by MS-60, acc. to EN12164 e.g. CW511L.

Apart from copper, zinc and additions of tin, arsenic, and aluminium, these alloys contain also contaminations but within the standard limits. They are produced mostly in three forms: as extruded or drawn sections and as casting materials, the so-called pigs obtained by the gravitational casting and with applying the crystallizer and semi-continuous casting and then either drawn or extruded.

The form/quality of the charge material for melting, more precisely the way of its production is very important for the quality of produced castings. By analogy to the similar way of producing casting alloys from the group of AlSi alloys (pigs and extruded ingots) slightly better properties could be expected in these alloys. This is related to the quality of precipitates occurring in the microstructure, which can essentially influence the final quality of castings.

The results of investigations indicating differences in properties of produced castings in dependence of the charge quality, mould technology and applications of alloying additions, are presented in the hereby paper.

2. Methodology of investigations

The performed investigations contained the total research process:

- preparation of the induction melting furnace and melting the metal charge in chamotte-graphite crucible,
- heating the metal bath up to a temperature of 1100 °C, bath purification, introduction of alloying additions,
- casting alloys to metal and sand moulds,
- testing the kinetics of crystallisation and solidification.

The samples for investigating the chemical composition as well as macro and microstructure, using optical and scanning microscopy, were taken from the obtained castings. The specimens for assessing mechanical properties were also prepared. Investigations were performed in the Laboratory of Non-ferrous Metals Casting in the Faculty of Foundry Engineering of AGH University of Science and Technology in Krakow.

3. Selected results of the performed investigations

Examples of the obtained images of microstructures are shown in Figs. 1-2.

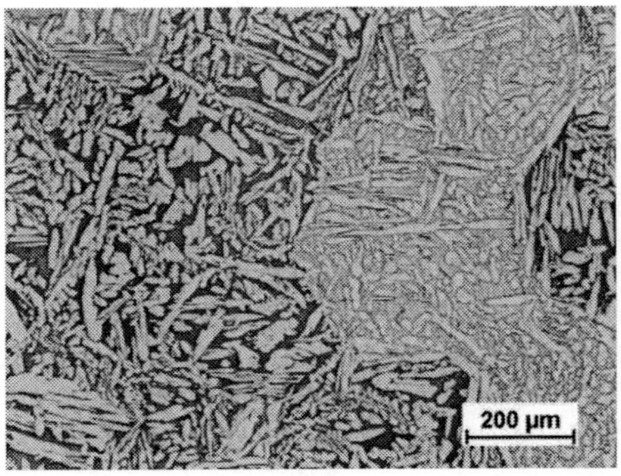

Fig. 1. Microstructure of MS-60 alloy, casting from the sand mould, obtained from the charge in the form of extruded ingots

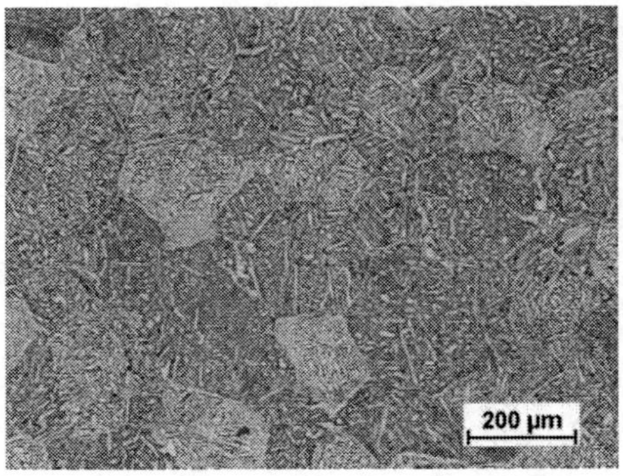

Fig. 2. Microstructure of MS-60 alloy, casting from the metal mould, obtained from the charge in the form of ingots cast by gravity into the ingot mould.

4. Results of testing mechanical properties

The selected results of testing mechanical properties of alloys from the eco-brasses group are presented in Figs. 3 and 4. The obtained results are summarized in conclusions

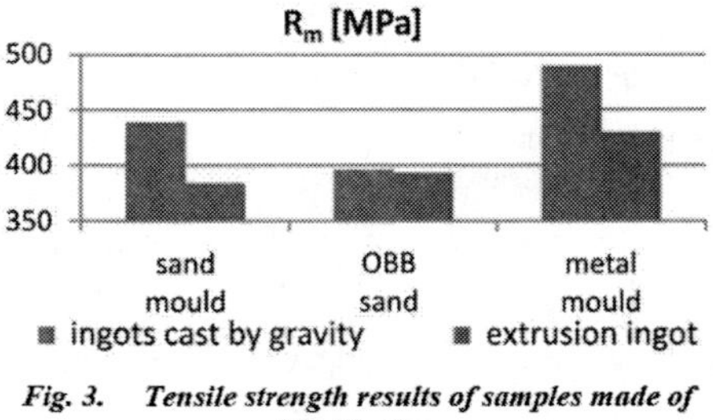

Fig. 3. Tensile strength results of samples made of MS-60 alloy.

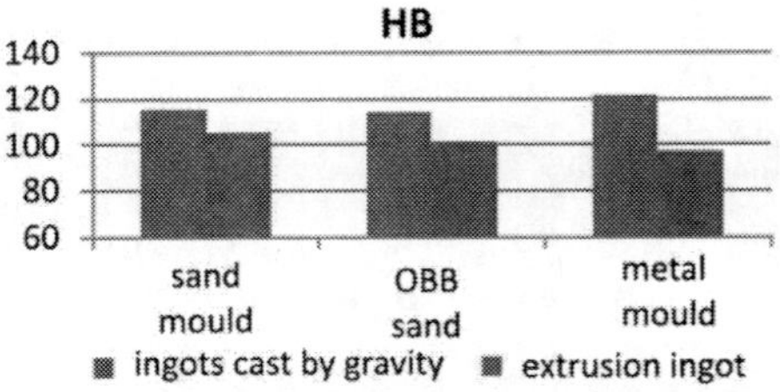

Fig. 4. Results of hardness tests of samples made of MS-60 alloy.

5. Conclusions

The presented results indicate the possibility of obtaining various mechanical properties from the same material (CuZn39), but produced in a different way. The divergences in the values of indices: UTS, YS and HB are mainly caused by differences in microstructures and grains sizes in the obtained alloys. In the case of these alloys the grain size is decisive. Regardless that the chemical composition (in conformity with standards) of both charges was nearly identical, the tensile strength results were significantly different. Castings of pigs produced by the most often applied method, it means by the gravitational casting into ingot moulds and free solidifying, have tensile strength 20% higher. On the other hand, the influence of alloying additions of various elements, especially arsenic, even at its negligible fraction in the alloy, significantly inhibits dezincification process of the analysed alloys from the Cu-Zn group.

Acknowledgments

The financial support of the State Committee for Scientific Research of Poland under the grant numbers 11.11.170.318 – 11, is acknowledged.

References

[1] Rzadkosz, S., Kranc, M., Garbacz-Klempka, A., Kozana, J. & Piękoś, M. (2015). Refining processes in the copper casting technology. Metallurgy. 54(1), 259-262.

[2] Rzadkosz, S., Garbacz-Klempka, A., Kozana, J., Piękoś, M. & Kranc, M. (2014). Structure and properties research of casts made with copper alloys matrix. Archives of Metallurgy and Materials. 59(2), 775-778.

[3] Rzadkosz, S., Kozana, J., Garbacz-Klempka, A. & Piekoś, M. (2014). Modelling the structure and properties of chosenbrasses. In A. Garbacz-Klempka, J. Kozana, M. Piękoś (Eds.), Nauka i technologia. Monograph (99-106), Kraków: Akapit.

[4] Romankiewicz, F, Romankiewicz, R., Michalski, M. & Ro mankiewicz, A. (2006). Study of intermetallic phases in leaded brass. In Polska Metalurgia w latach 2002-2006 (717-720). Kraków: Akapit.

[5] Rzadkosz, S., Kozana, J., Garbacz-Klempka, A., & Piekoś M. (2015). Structure analysis and properties of unleaded brasses. Archives of Metallurgy and Materials. Vol. 60(1), , s. 323-328. DOI: 10.1515/amm-2015-0053.

[6] Kondracki, M., Gawroński, J. & Szajnar, J. (2005). Lead elimination from fixture copper-zinc alloys. 26(1), 47-55.

The 73rd World Foundry Congress
„Creative Foundry"
23 - 27 September 2018, Krakow, Poland

INFLUENCE OF STRUCTURAL DISCONTINUITIES ON THE FATIGUE LIFE OF ALUMINIUM ALLOYS OF 4XXO SERIES

Jerzy Zych, Jarosław Piekło, Maria Maj, Aldona Garbacz-Klempka*, Marcin Piękoś

AGH University of Science and Technology, Faculty of Foundry Engineering, Reymonta 23 Str., 30-059 Kraków, Poland
**Corresponding address: agarbacz@agh.edu.pl*

Key words: low-cycle fatigue, fracture surface, fracture mechanics, aluminium alloys, FEM analysis

Fatigue behaviour investigations of two aluminium alloys 4XXO, acc. PN-EN 1706 (within the range below 10^4 cycles at the cycle asymmetry coefficient R= -1) were performed in the hereby study. The obtained results are presented as diagrams: stress amplitude σ_a – number of load cycles to the specimen failure $2N_f$. On the basis of the microscopic images of sample crack surfaces the influence of the noticed casting defects on the cycle number decrease, were analysed at the given stress amplitude level. For each of defects the coefficient of stress intensity $K_{I(sz)}$, $K_{I(POW)}$ was determined on the bases of images and dimensions, by means of analytical solutions. Their numerical models were also made and the stress intensity coefficients $K_{I(MES)}$ were determined by means of the finite element method.

Liquid aluminium has a tendency for hydrogen adsorption and oxidation, thus being the reason of an unavoidable presence of gaseous porosities and oxide intrusions in aluminium castings. When the casting is not properly supplied its fatigue resistance can be additionally decreased by the shrinkage porosity. A gaseous porosity is usually of a round shape while shrinkage porosity forms irregular three-dimensional voids [1]. Both kinds of porosity can occur with aluminium oxide films. Structural discontinuities - such as voids caused by porosities - can have various sizes. Thus, it is possible to single out a microporosity when the diameter or the largest dimension of the void does not exceed 100 μm and a porosity when the void size is larger than 100 μm [2]. Porosities and macroporosities have a high influence on decreasing the fatigue properties and alloy strength, while the microporosity presence in the casting lowers only its fatigue strength within the range of a large number of cycles. The microporosity influence should be examined jointly with the alloy microstructure features since the microcrack initiation placement depends on the mutual ratio of dimensions and distances of voids and microstructure phases. There is the opinion that - in a similar fashion as alloy plastic properties - its fatigue depends on distances between arms of secondary dendrites (SDAS) and on sizes of silicate eutectic precipitates [3].

Macro and microscopic analyses were performed by means of the optical and scanning microscopy. Investigations of the chemical composition and fatigue were also carried out. Numerical models of crack initiations within the defect areas were designed and FEM calculations performed.

Fatigue tests were performed on round samples (ϕ = 8 mm) of the measuring base l = 20 mm. Samples were cut out from plates of a thickness of 10 mm and subjected to the heat treatment T6 of the following parameters: hyperquenching T_p = 540°C for 4h, cooling in water of a temperature T_{chl} = 55-60°C and aging at a temperature T_{starz} = 180°C for 12h. The same heat treatment was applied for both tested alloys.

The dependence of the number of cycles on the stress level is presented by the diagram in Fig. 1. A relatively high scatter of the results of tests - performed at the same stress amplitude - is seen.

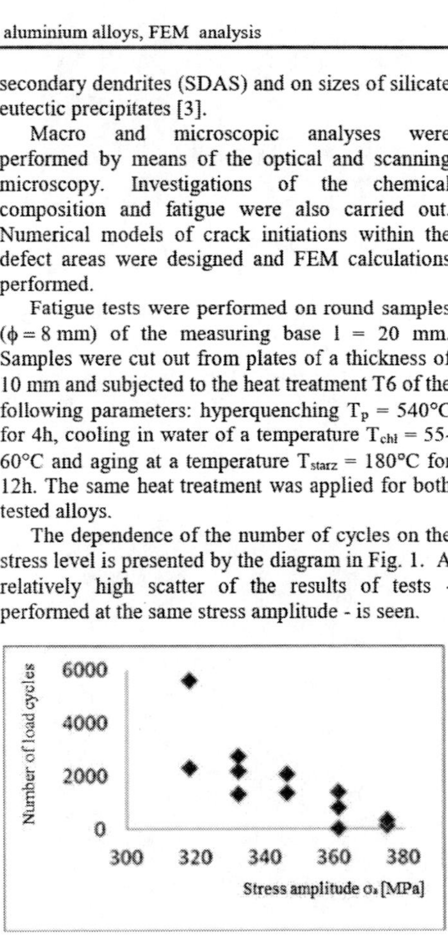

Fig.1. Dependence of the stress amplitude on the number of load cycles to a specimen failure - determined for samples of PR1 alloy

In a few cases the presence of casting defects within the sample fracture zone was the cause of a significant decrease of the number of load cycles to a specimen failure. This justifies undertaking investigations on the influence of structural discontinuities on the alloys fatigue lives.

Crack surfaces of samples, after fatigue tests, were analysed in order to find eventual non-metallic inclusions, porosities or oxide films, and to determine their influence on a fatigue life decrease as compared with the remaining samples, in which this type of defects were not found. The typical fatigue crack surface on which defects were not found is presented in Fig. 2.

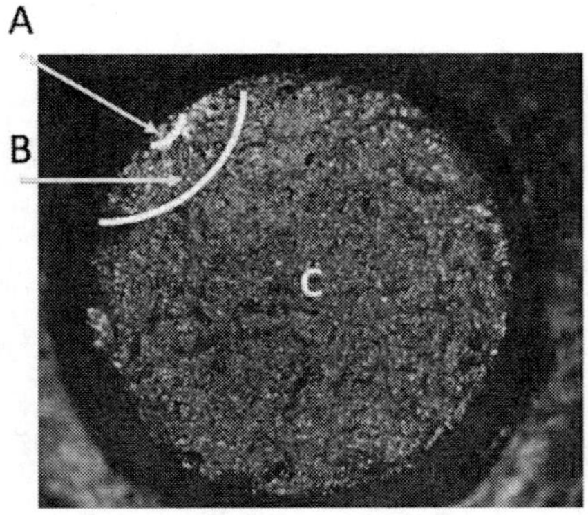

Fig. 2. Fatigue crack surface: with a single place of the crack initiation - PR1 alloy, stress amplitude 318 MPa, number of cycles to specimen failure $2N_f = 5600$ A - crack initiation, B - stable crack growth, C - process zone

When the stress amplitude was higher than 318 MPa it was possible to see a few places of the crack initiation, which were developing simultaneously. The example of such crack surface is presented in Fig. 3.

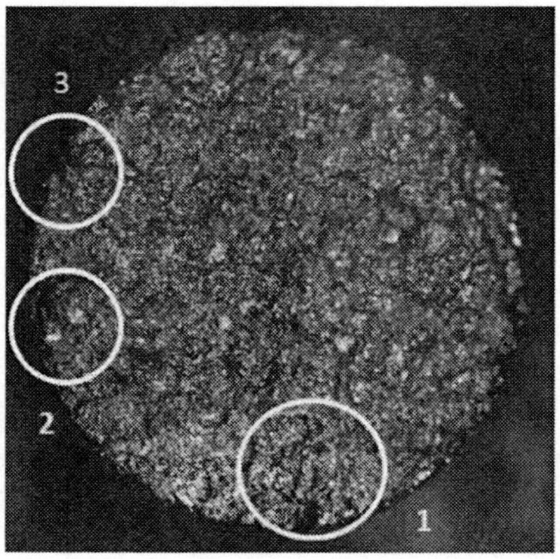

Fig. 3. Fatigue crack surface with three marked places of the crack initiation - PR1 alloy, stress amplitude 361 MPa, number of load cycles to specimen failure $2N_f = 1386$.

When in the specimen crack surface occurred defects such as porosity or non-metallic precipitates out of which cracks were initiated the numerical models of defects were made. The FEM calculations were performed to determine the stress intensity coefficient K_I in selected places of the outline encompassing the surface of the defect. The example of the model of one defect, found in the crack surface, is shown in Fig. 4.

Fig. 4. Geometrical model and the image of the local stress increase in the vicinity of defect - according to the HMH criterion

The performed investigations indicate a significant influence of casting defects on fatigue properties of the tested alloys, from 4XXO series. The obtained results indicate large differences in the number of cycles to a specimen failure between samples loaded by the same stress amplitude, in dependence of the size and location of the defect. The number of sources of the crack initiation, identified on the fracture surface, had also an influence on the number of load cycles leading to the specimen failure.

Acknowledgments

Research was performed within the project: INNOLOT/I/10/NCBiR/2014-Innogear and grant number 11.11.170.318–10 funded by the State Committee for Scientific Research of Poland.

References

[1] X.B. Cao, J. Zhao, J.H. Fan, M.H. Zhang, G.J. Shao & Q. Hua (2014). Influence of casting defects on fatigue behavior of A356 alluminium alloy. International Journal of Cast Metals Research. Vol. 27(6), pp. 362-368.

[2] K. Gal, N. Yang, M. Horstemayer, D.L. McDowell, J. Fan (1999). The debonding and fracture of Si particles during the fatigue of a cast Al-Si alloy. Metallurgical and Materials Transactions. Vol. 30A (December), pp.3079-3088.

[3] Q.G. Wang, D. Apelian, D.A. Lados (2001). Fatigue behavior of A356-T6 aluminium cast alloys, part II. Effect of microstructural constituents. Journal of Light Metals 1, pp. 85-97

The 73rd World Foundry Congress
„Creative Foundry"
23 - 27 September 2018, Krakow, Poland

EFFECT OF ALUMINUM MELT TREATMENT BY NANOSECOND ELECTROMAGNETIC PULSES ON STRUCTURE AND PROPERTIES OF CASTINGS

Vladislav Deev [1]*, Ernst Ri [2], Evgeny Prusov [3]

[1] *National University of Science & Technology (MISIS), Department of Foundry Technology, 119049, Moscow, Leninsky Pr. 4, Russian Federation*
[2] *Pacific National University, Department of Foundry Engineering and Metal Technology, 680042, Khabarovsk, Tihookeanskaya Str. 136, Russian Federation*
[3] *Vladimir State University named after Alexander and Nikolay Stoletovs, Department of Functional and Constructional Materials Technology, 600000, Vladimir, Gorky Str. 87, Russian Federation*
**Corresponding address: deev.vb@mail.ru*

Key words: aluminum alloys, physical melt treatment, nanosecond electromagnetic pulses, microstructure, properties

1. Introduction

Aluminum casting alloys are currently one of the most common materials used to produce shaped castings for various purposes. Their structure and a complex of technological, mechanical and operational properties determine the quality of castings. Largely, the structure and properties of a particular alloy depend, besides the position on the phase diagram, on the initial charge, the technological parameters of the melting and the methods of the melt treatment, on the conditions of crystallization and the methods of casting [1]. Under constant technological conditions for the castings production, the methods for efficient melt treatment, which lead to a refined cast structure, are the simplest tool for influencing the properties of the obtained alloys. In recent years, significant attention is paid to the methods of melt treatment during melting and crystallization, which do not lead to a change in its chemical composition, that is, do not include impurity modification by any elements. These are physical methods – such as thermal-temporal and thermal-rate treatment [2, 3], various electromagnetic effects [4, 5] and nanosecond electromagnetic pulses (NEPs) [6]. Despite certain successes in the development of these technologies for processing the melts, many questions remain concerning the effect of processing methods on the properties of castings.

This work is devoted to study of the influence of technological parameters of melt treatment by NEPs on structure and properties of the A356 alloy.

2. Materials and Methods

Melting was carried out in alundum crucible in electric resistance furnace. As the charge materials, the A356 ingot alloy was used. The processing of the melts was carried out using a special installation. The source of the radiation energy was the GNP generator with the following characteristics: pulse duration 0.5 ns, amplitude 8 kV, pulse repetition rate 1 kHz. The coaxial type transducer was made in the form of steel rods with a diameter of 4 mm and a length of 300 mm. The effect of NEPs on the liquid phase was accomplished by immersing the transducer into the melt, isolating it from contact with the liquid metal by using a quartz tip transparent to nanosecond electromagnetic waves.

In the experiments, the following parameters for the melt treatment by NEPs were used: heating to a temperature of 900°C and subsequent treatment of liquid aluminum after a five-minute holding at this temperature with NEPs for 5, 10, 15, 20 and 25 min. The pouring temperature was 720-730 °C. The structure of alloy was studied using a Planar Micro-200 microscope. Wear resistance was investigated according to the standard procedure of GOST 23.208. Hardness was studied on the TK-2 tester. Heat resistance tests were performed on a Q-1000 derivatograph in the temperature range 200-900°C.

3. Results and Discussion

As can be seen in Fig. 1, an increase of the NEPs treatment time of the melt up to 15 min leads to refinement the dendrites of the α-solid solution, siliceous eutectic and excess phases. This time, apparently, is optimal from the point of view of the decay of the microinhomogeneities and clusters in the melt. As a result, the melt becomes more homogeneous, which leads during crystallization to an increased number of nucleation centers and a uniform fine-grained structure. With further increasing of the NEPs treatment time up to 20-25 min, the structural constituents are coarsened again. It can be assumed that for a given treatment time the energy conditions of the system change and coagulation of dispersed particles takes place, with the melt becoming more heterogeneous, including large complexes of clusters and different particles. Supposedly, this leads to formation a coarser structure during crystallization.

The results of the microstructure study are well combined with the results of studying the alloy properties. The wear resistance and hardness of the A356 alloy varies with the NEPs treatment time. Thus, the maximum value of these properties is observed at 15 min of treatment.

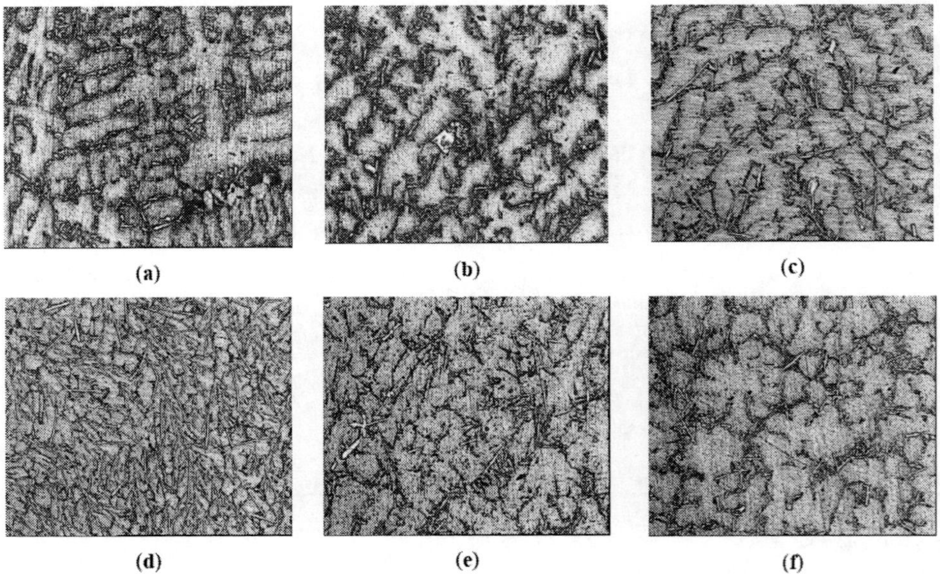

Figure 1. Microstructure of A356 alloy, treated by NEPs with different duration (×100): without treatment (a), 5 min (b), 10 min (c), 15 min (d), 20 min (e), 25 min (f)

With an increase in the NEPs treatment time up to 15 min, the high-temperature oxidation resistance of A356 alloy decreases sharply by a factor of 2-3 in the case of temperature tests of 200-500°C. With prolonged irradiation of the melt by NEPs (25 min), the alloy high-temperature oxidation resistance was increased in comparison with the untreated alloy.

4. Conclusion

It was shown on the example of alloy A356 that the expediency of using NEPs for melt treatment should be justified by the specific purpose of the obtained casting. Thus, for achievement of increased mechanical properties of casting, the necessary NEPs treatment time is up to 15 minutes, which will result in the refinement of the structural constituents of the alloy.

It is necessary to continue research on the influence of NEPs on the structure and properties of aluminum alloys. It is advisable to use NEPs and other physical methods of the melt treatment in technologies of industrial alloys and cast composite materials production.

Acknowledgments

The work has been carried out on the equipment of Center "Applied Material Science" at Pacific National University with financial support from the Ministry of Education and Science of Russia in accordance of the state order (State registration 11.3014.2017/4.6).

References

[1] Campbell J. (2015). Complete Casting Handbook. 2nd ed. Oxford: Butterworth Heinemann.

[2] Deev V.B., Prusov E.S., Kutsenko A.I. (2018) Theoretical and experimental evaluation of the effectiveness of aluminum melt treatment by physical methods. La Metallurgia Italiana. 2, 16-24.

[3] Deev V.B., Ponomareva K.V., Prikhodko O.G., Smetanyuk S.V. (2017) Influence of temperatures of melt overheating and pouring on the quality of aluminum alloy lost foam castings. Russian Journal of Non-Ferrous Metals. 58(4), 373-377.

[4] Chen H., Jie J., Fu Y., Ma H., Li T. (2014) Grain refinement of pure aluminum by direct current pulsed magnetic field and inoculation. Transactions of Nonferrous Metals Society of China. 24, 1295-1300.

[5] Haghayeghi R., de Paula L.C., Zoqui E.J. (2017) Comparison of Si refinement efficiency of electromagnetic stirring and ultrasonic treatment for a hypereutectic Al-Si alloy. Journal of Materials Engineering and Performance, 26, 1900-1907.

[6] Krymsky V.V., Shaburova N.A., Litvinova E.V. (2016) Microstructure and properties of cast metal treated with electromagnetic pulses while in molten state. Materials Science Forum. 843, 106-110.

The 73rd World Foundry Congress
„Creative Foundry"
23 - 27 September 2018, Krakow, Poland

INVESTIGATIONS OF THE INFLUENCE ZONE OF CHILLS - ON THE CASTING OF THE PLATE MADE OF AlSi7Mg ALLOY, AFTER THE HEAT TREATMENT T6

Marcin Piękoś *, Jerzy Zych

AGH University of Science and Technology, Faculty of Foundry Engineering,
Reymonta 23 Str., 30-059 Kraków, Poland
**Corresponding address: mpiekos@agh.edu.pl*

Key words: AlSi7Mg alloy, chills, thermal characteristics, cooling rate, mechanical properties

The influence of the chill on the AlSi7Mg alloy properties after the heat treatment T6, was realised in the system of the horizontally cast plate of dimensions 160x240 mm and thickness of 10 mm. The cooling course in individual casting zones was recorded, which allowed to determine the solidification rate in individual crystallisation stages. Castings were subjected to the heat treatment T6 process. Several properties of the alloy such as: hardness HB, density, tensile strength R_m, elongation A_5 were determined. The microstructure images were presented and the structural SDAS parameter determined.

This study constitutes the continuation of investigations of the solidification rate influence on the AlSi7Mg alloy properties [1].

The ability of a fast heat abstraction by the casting mould in crucial places of the solidifying casting allows to minimise unfavourable results of the crystallisation process. In the traditional casting based mainly on sand moulds technology, we are often dealing with the occurrence of micro-shrinkages, shrinkage cavities and similar defects which decrease properties or even cause scrap castings. Oxide contaminations non-metallic inclusions porosities related to alloy gasification should be eliminated by means of refining at a preparation stage of the liquid casting alloy. The correctly designed casting, properly placed in the mould, correctly selected gating system and eventual feeders should warrant obtaining healthy castings. Increasing the solidification rate in the given zone, it means the so-called local acceleration of the crystallization process by the chills application allows to achieve castings of the needed microstructure parameters and free of defects.

Placement of chills in sand moulds is aimed at increasing cooling rates of casting walls causing structure refinements and in effect improving casting properties. The chill directly influencing the casting solidification changes all features and properties of the material (casting walls). Thus, the operation range of the chill can be determined by detecting one or a few features (e.g. density) or one property (e.g. R_m) in samples taken from places in various degrees distant from the chill edge. Such method of assessing the operation zone of chills applied in the casting production of Al–Si alloys was applied in the hereby study [2, 3, 4].

The plate of dimensions 160×240×10 mm was horizontally cast in two variants: without a chill and with the chill placed above the plate. Technical copper was the material for the chill of dimensions 50×50×50 mm. Copper as well as its alloys are characterised by the highest thermal conductivity.

To determine the solidification rate of the plate, which was cast in the sand mould with water glass (5%) as a binder, thermocouples were placed inside the plate in such way that the temperature was measured in the half of the plate thickness. Schematic presentation of the plate is shown in Figure 1.

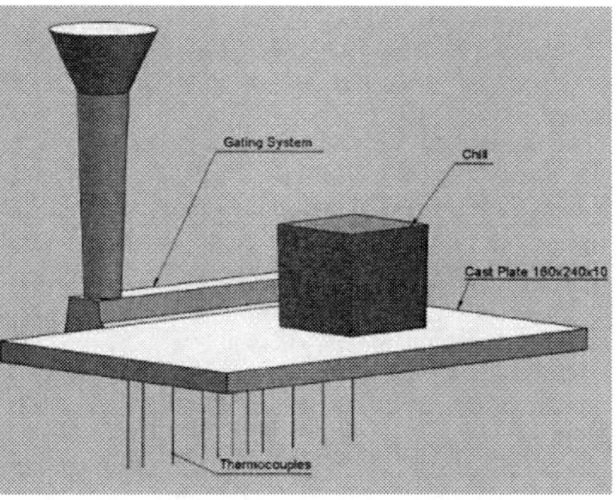

***Fig. 1.** Schematic presentation of the system of the horizontal plate casting, of dimensions 160×240×10 mm*

The performed measurements of the cooling course in individual plate points allowed to determine temperatures characteristic for transitions occurring during the casting solidification. The example of thermal analysis ADT with the determined characteristic points is presented in Figure 2. Stages on the cooling curve, used in this study at determining solidification rates coefficients are segments: "P_k - C" - crystallisation of solution α, "C - G" - crystallisation of eutectic α+β (together with multicomponent eutectics and intermetallic phases), "P_k - G" - total crystallisation period of the casting [5, 6, 7].

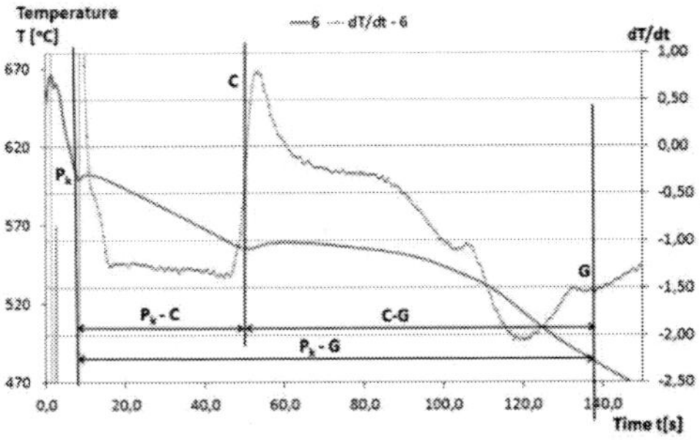

Fig. 2. Example of the cooling course and the determined characteristic transitions

The thermal analysis allowed to determine the cooling and solidification rates of individual stages of the solidification of the plate casting without the chill (BO) and with the chill (OCH). The results are presented in Fig. 3.

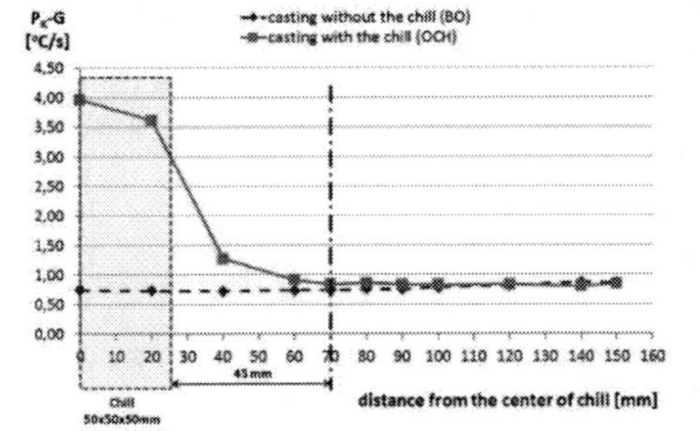

Fig. 3. Graphical presentation of the solidification rates P_k-G of the plate casting without the chill (BO) and with the chill (OCH)

The zone of the chill placement is marked by the green dashed line. Differences in solidification rates of both castings are significant. The maximum crystallisation rate for the whole crystallisation process, stage P_k-G, was found under the chill: 3.96 °C/s. The crystallisation rate of the plate without the chill, in the same place was equal 0.74 °C/s.

The results of measuring the alloy hardness HB in individual points of the plate casting are shown in Figure 4.

Thus, in determining the influence zone of the chill all obtained test results should be taken into account:

- on even page: top - 2 cm, bottom - 2 cm, right side - 2 cm, left side - 3 cm;
- crystallisation rate of the casting P_k-G,
- tensile strength R_m,
- alloy elongation A_5,
- alloy hardness HB,
- alloy density ρ,
- structural parameter *SDAS*.

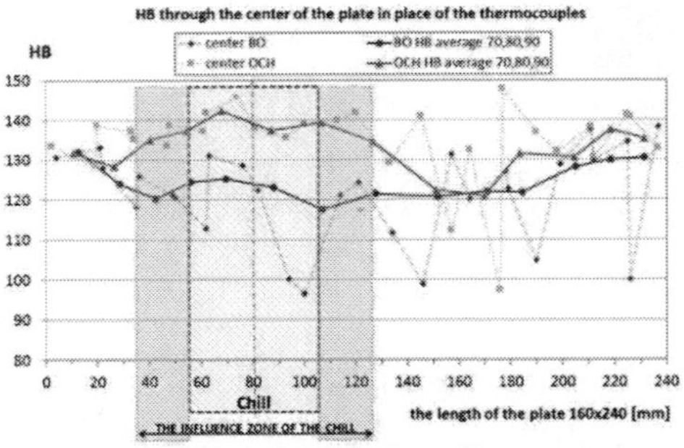

Fig. 4. Hardness values HB in selected places of the tested alloy

The performed investigations as well as the analysis of the results allowed to determine the influence zone of the chill. For the casting of the plate of dimensions 160×240×10 mm, made of AlSi7Mg alloy, this zone reached from 20 to 35 mm from the chill wall. However, it should be noticed that the efficient influence of the chill, i.e. the place up to which the chill efficiently influences the solidifying casting of the plate of dimensions 160mm x 240mm and thickness 10 mm was 20 mm measured from the chill wall. It is easy to notice that there is a certain dependence between the thickness of the casting wall and the influence zone of the chill, being not less than 2g, where g is the casting wall thickness.

The next aim of successive investigations will be finding the confirmation that there is the dependence between the casting wall thickness and the influence zone of the chill for other thicknesses of walls. We would like to prove that this principle is of a universal character.

Acknowledgments

Research was performed within the project: INNOLOT /I/10/NCBiR/2014– Innogear.

References

[1] J. Zych, M. Piękoś. (2016). Wykorzystanie szybkości stygnięcia do sterowania strukturą i właściwościami stopu AlSi7Mg po obróbce cieplnej T6. Nauka i technologia: odlewnictwo metali nieżelaznych, Kraków: Wydawnictwo Akapit.

[2] Z. Poniewierski. (1989). Modyfikacja siluminów. Warszawa: WNT.

[3] M. Tokarski. (1985). Metaloznawstwo metali i stopów nieżelaznych w zarysie. Wydawnictwo Śląsk

[4] S. Pietrowski. (1999). Krystalizacja, struktura i właściwości siluminów tłokowych. Wydawnictwo Politechniki Łódzkiej.

[5] Ł. Bernat, J. Hajkowski, M. Hajkowski. (2006). Struktura i porowatość odlewów ze stopu aluminium a właściwości mechaniczne. Archiwum Odlewnictwa. Rocznik 6(22).

[6] P. Wasilewski. (2003). Porównanie metod badania krzepnięcia i krystalizacji stopów metali. Archiwum Odlewnictwa. Rocznik 3(10).

[7] L.A. Dobrzański, R. Maniara, J. Sokolowski. (2006) Kinetyka krystalizacji i struktura odlewniczego stopu AC AlSi7Cu4. Archiwum Odlewnictwa. Rocznik 6, Nr 21(1/2).

The 73rd World Foundry Congress
„Creative Foundry"
23 - 27 September 2018, Krakow, Poland

MICROALLOYING OF Al-5%Cu ALUMINUM ALLOY WITH NICKEL ALUMINIDES AND RARE-EARTH METALS

Ernst Ri [1]*, Khosen Ri[1], Vladislav Deev [2], Evgeny Prusov [3]

[1] *Pacific National University, Department of Foundry Engineering and Metal Technology, 680042, Khabarovsk, Tihookeanskaya Str. 136, Russian Federation*
[2] *National University of Science & Technology (MISIS), Department of Foundry Technology, 119049, Moscow, Leninsky Pr. 4, Russian Federation*
[3] *Vladimir State University named after Alexander and Nikolay Stoletovs, Department of Functional and Constructional Materials Technology, 600000, Vladimir, Gorky Str. 87, Russian Federation*
Corresponding address: erikri999@mail.ru

Key words: aluminum alloys, master alloy, rare-earth metals, eutectic, α-solid solution, solubility, microhardness

1. Introduction

Prospects for the creation of aircraft details operating under conditions of increased loads and temperatures, as well as the extended duration of their impact, determine tasks to improve the performance of structural materials, develop new alloy compositions and production technologies, including methods of alloying and modification. Modern aluminum-based materials can acquire high performance due to microalloying. Currently, one of the most effective alloying elements is scandium due to the formation of an anomalously supersaturated solid solution during crystallization and its subsequent disintegration with the precipitation of dispersed secondary intermetallides Al_3Sc [1]. Also discussed the creation and development of new master alloy types – complex master alloys containing nickel and rare-earth metals (REM) [2]. Intermetallic compounds such as nickel and REM (Ce, La) aluminides are proposed to be used as master alloys components for increasing the physical, mechanical and operational properties (heat resistance, corrosion resistance, high temperature oxidation resistance, wear resistance) of cast Al-Cu alloys.

This work is aimed at investigation the effect of microalloying by the synthesized ligature with nickel and REM (Ce, La) aluminides on the structure formation, elements distribution and microhardness of the structural constituents of Al-5% Cu aluminum alloy.

2. Materials and Methods

Al-Cu ingot alloy (5.0 wt% Cu) was used as an initial charge. To obtain the master alloy, aluminum grade A7 (>99.7 wt% Al) and complex modifier Al-Ce (20 wt.% Al, 3 wt% Ca, 25 wt.% REM, Ni bal.) were used. It consists of nickel aluminides (Al_3Ni) and REM ($Al_{11}Ce_3$, $Al_{11}La_3$) and eutectic Al + Al_3Ni + $Al_{11}REM_3$. The Al-5% Cu melt was superheated to a temperature of 900 °C. After a melt holding during 5 min, a master alloy of 0.05 to 0.3 wt% was added through the variation range of the addition of 0.05 wt%. The melt was cooled at a rate of 20 °C/min. Microhardness of the alloy structural constituents was determined on the PMT-3 tester. X-ray microanalysis was performed on the Hitachi SU-70 Analytical Field Emission SEM with the extensions Thermo Scientific UltraDry and Thermo Scientific MagnaRay.

3. Results and Discussion

Microhardness of the α-solid solution varies according to the extreme dependence with its maximum at 0.15% of the master alloy, increases from 825 MPa for the initial alloy to 950 MPa (Fig. 1a). In this case, microhardness of the eutectic also changes according to the extreme dependence with the maximum at 0.15% of the master alloy, followed by a slight decrease at 0.3% of the master alloy.

From Fig. 1b it follows that the maximum solubility of Cu in the α-solid solution (3.5% Cu) is observed at the addition of 0.15% of the master alloy, and sharply decreases at 0.3% of the master alloy. The Al content varies inversely with its minimum in α-solid solution (95% Al). The maximum solubility of Cu in the α-solid solution promotes a sharp increase in microhardness. According to the Al-Cu phase diagram, Cu at room temperature dissolves in an amount of 0.2% in the α-solid solution, but under conditions of nonequilibrium crystallization at casting with a copper content in alloy more than 0.2%, the α-solid solution is supersaturated. In such solid solution, natural aging can occur (formation of the Ginier-Preston zone GP-1) and the microhardness of α-solid solution is additionally increased up to 0.15% of the master alloy.

At further increase of the master alloy addition, the content of Cu in the α-solid solution decreases and the Cu concentration in the initial alloy becomes lower; microhardness is also reduced due to the weakening of natural aging effect.

Fig. 1c shows the distribution curves of Al and Cu in the eutectic. In it, copper is in a bound state in the form of $CuAl_2$ compound. It was found that the eutectic is not homogeneous in content of Cu:

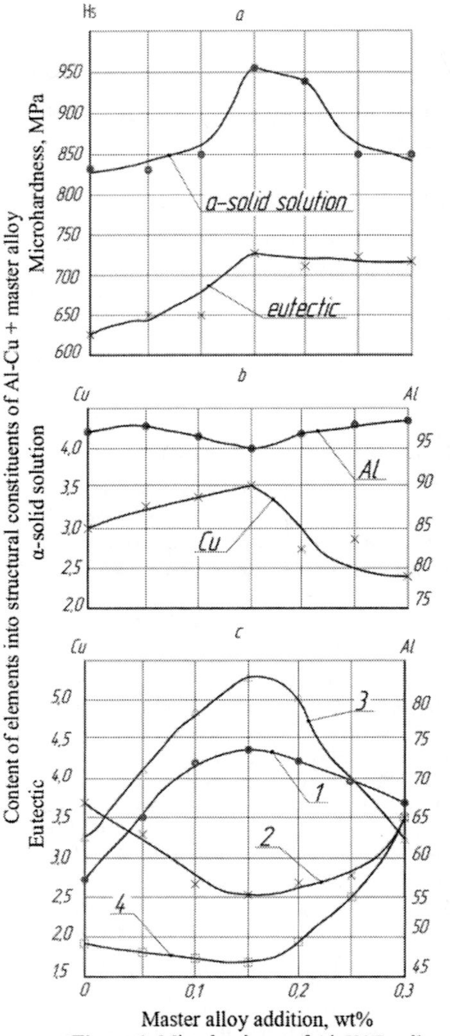

Figure 1. Microhardness of Al-5%Cu alloy structural constituents

- eutectic enriched with Cu (>20%), corresponds to curve 2 (Cu) and curve 1 (Al);
- eutectic depleted with Cu (10-20% Cu), corresponds to curve 4 (Cu) and curve 3 (Al).

In these zones, the minimum copper content is observed at the 0.15% of master alloy addition.

At further increase of the master alloy addition, an increase in the Cu concentration is observed in these two zones and at the 0.3% of the master alloy, the Cu content in the eutectics is leveled. Accordingly, the Al content in the eutectic changes in the reverse order (curves 1 and 3 in Fig. 1c). Consequently, simultaneously with a decrease in the Cu concentration in eutectic at 0.15% of the master alloy, the amount of $CuAl_2$ intermetallide decreases. With an increase of the additive up to 0.3%, due to an increase in the Cu content in the eutectic of the two types, the proportion of $CuAl_2$ intermetallics with increased microhardness in the eutectic composition increases, but it is also a fragile component of the eutectic, which contributes to some decrease in the microhardness of the eutectic.

An increase in the Cu content in the α-solid solution (Fig. 1b) leads to a decrease in the content of the eutectic down to 0.15% of the master alloy (curves 2 and 4 in Fig. 1c). This means that the fraction of the intermetallide $CuAl_2$, which has a microhardness commensurate with the α-solid solution (530 and 570 MPa, respectively), decreases in the eutectic composition. It can be assumed that the main reason for the increase in the microhardness of the eutectic is a change in the structural components of the eutectic at 0.15% of the master alloy and their microalloying with nickel, and also a decrease in the iron content which promotes formation of brittle phases ($CuAl_2$, $Al_xSi_yFe_z$).

4. Conclusion

On the example of microalloying of the Al-5% Cu alloy with a master alloy containing Ni and REM (Ce, La) aluminides it was shown that a redistribution of Al and Cu occurs in α-solid solution and eutectic. This is reflected in the refinement of α-solid solution and eutectic at 0.15% of the master alloy addition and, accordingly, increases the microhardness of α-solid solution by 100 MPa and eutectic by 125 MPa. At the next stage of this research, the results of the microalloying influence on the nature of distribution of Ni and Fe in the structural constituents, physical-mechanical and operational properties (heat resistance, corrosion resistance, high temperature oxidation resistance, wear resistance) of the Al-5% Cu cast alloy will be obtained.

Acknowledgments

Studies were carried with financial support from the Ministry of Education and Science of Russia in accordance of the state order (State registration 11.3014.2017/4.6).

References

[1] Zhang Y., Gao K., Wen S., Huang H., Wang W., Zhu Z., Nie Z., Zhou D. (2014) Determination of Er and Yb solvuses and trialuminide nucleation in Al–Er and Al–Yb alloys. Journal of Alloys and Compounds. 590, 526-534.

[2] Gostishchev V.V., Kim E.D., Ri E.Kh., Khimukhin S.N. (2017) Obtaining of complex-alloyed nickel aluminides and complex ligatures by metal oxides' metallothermy. Tsvetnye Metally. 10, 79-84.

The 73rd World Foundry Congress
„Creative Foundry"
23 - 27 September 2018, Krakow, Poland

FABRICATION METHOD OF SILICON CARBIDE BY INFILTRATION OF MOLTEN Fe-Si ALLOY THROUGH TWO-STEP REACTION SINTERING

Yoshitsugu Hanada[1]*, Yang Xiao[1], Akio Sonoda[1], Hyo-Gyoung Kang[1], Hideaki Nagayoshi[1], Tatsuya Tokunaga[2], Hidenori Era[2]

[1]*FUJICO Co., Ltd., Research and Development Center, 4-31 MakiyamaShin-machi, Tobata-ku, Kitakyushu-shi, Fukuoka, Japan*
[2]*Kyushu Institute of Technology, Graduate School of Engineering, Department of Material Science, 1-1 Sensui-cho, Tobata-ku, Kitakyushu-shi, Fukuoka, Japan*
**Corresponding address: y-hanada.fujico@kfjc.co.jp*

Key words: Silicon carbide, molten metal infiltration, reaction sintering, iron silicide, composite material

1. Introduction

Silicon carbide (SiC) is a commonly used material with the excellent properties at high temperature such as low density, high hardness, high wear resistance, and structural stability [1]. In addition, SiC composes of silicon (Si) and carbon (C), so that their materials are immune to dry up for the time being in the earth. In future, SiC can be supplied stable to product, and to provide equipment for guaranteeing the stable and secure operations for a long time. However, SiC ceramics is too hard to machine.

The purpose of this study is to promote providing with new application of SiC for diverse industrial plant, and we engage in developing to improve machinability of SiC. Therefore, we have dealt here with the fabrication method of SiC by two-step reaction sintering [2]. In this paper, the molten material of Fe-Si alloy [3] to infiltrate spontaneously into the preformed body is used. It will be expected that the developed SiC enabled improvement of electrical conductivity further than the conventional SiC. Further, it could be applied for the precision electrical machining of SiC. Not much is available on fabrication of SiC by the two-step reaction sintering with Fe-Si alloy. Here we report the introduction of the fabrication method and research the micro-structure of sintered SiC with Fe-Si alloy.

2. Experimental

The two-step reaction sintering is processed as first and second sintering. Raw material powders were SiC, Si and C. Also, phenol resin was used to bind their powders. The infiltration materials were Si metal and Fe-Si alloy (75mass%Si). Graphite crucible was used to sinter in a furnace. The outline of the fabrication method is shown in Fig.1. Raw material powders were weighed by specified quantities and dry mixed, press shaped, first sintered, and second sintered to obtain a SiC body. The compounding ratio composed of 73mass% SiC, 2mass% Si, 9mass% C and 16mass% resin. The first sintering was carried out at temperature of 973K in an inert gas atmosphere. The second sintering was at temperature of 1693K in a vacuum. The molten metal was spontaneously infiltrated into the sintered body. The Si component of the molten alloy was reacted with C of the body. The inner gap was filled by molten alloy. The sintered body solidificated together around surplus molten alloy.

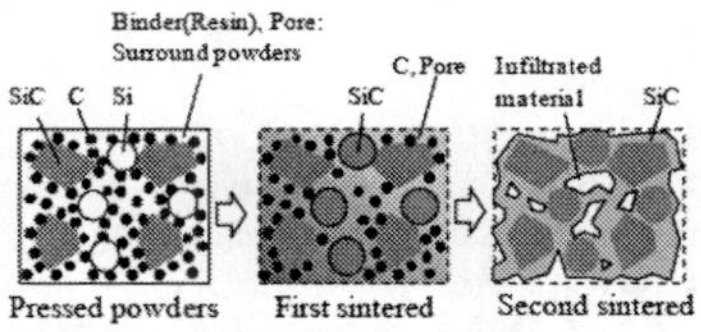

Fig. 1. ***Schematic illustration of two-step reaction sintering method.***

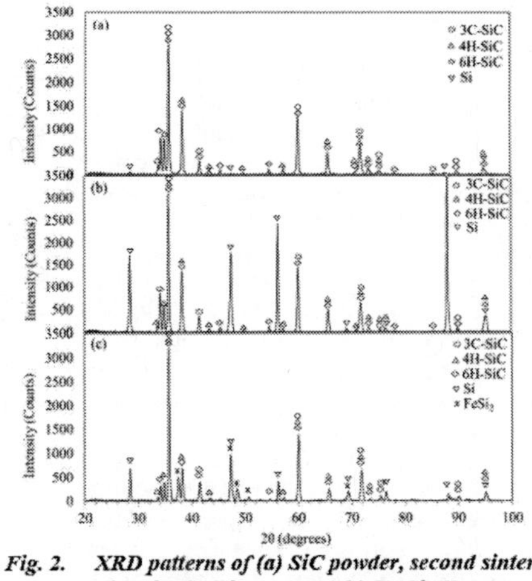

Fig. 2. ***XRD patterns of (a) SiC powder, second sintered SiC with (b) Si metal and (c) Fe-Si alloy.***

Analyzed terms were a crystal structure using X-ray diffraction (XRD) measurement, secondary electron images and element mappings using Electron Probe Micro Analyzer (EPMA), hardness, bending strength and fracture toughness by indentation fracture method.

3. Results and discussion

The X-ray diffraction pattern of SiC for the raw material is shown in Fig. 2(a). The pattern indicated that SiC and Si ware existed, the SiC had three types

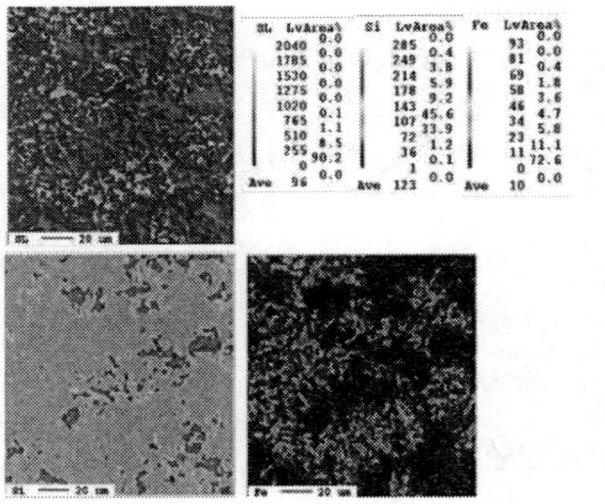

Fig. 3. SEM image and element mappings of Si and Fe of second sintered SiC with Fe-Si alloy.

of crystal structure 3C, 4H and 6H. The amount of Si was small. About 200 different polytypes structure of SiC has been reported so far, the most common SiC polytypes are 3C, 4H, 6H and 15R [4]. Next, X-ray diffraction patterns of sintered SiC with Si metal and Fe-Si alloy are shown in Fig. 2(b) and 2(c). These were similar to Fig. 2(a) and also SiC structure types of 3C, 4H and 6H can be confirmed. Reaction types of SiC had no differentiation by sintering materials of Si metal and Fe-Si alloy. However, Si peaks became strong in Fig. 2(b) and new peak of $FeSi_2$ appears in Fig. 2(c). From the binary phase diagram, it is cleared that the structure of Fe-75mass%Si alloy consists Si metal and $FeSi_2$ intermetallic compound. And also, there was no the reacted product of Fe.

Fig. 3 shows SEM image and element mapping of Si and Fe of second sintered SiC with Fe-Si alloy. From the qualitative analysis result, they could be identified that the dark gray grains are SiC, the light gray parts are Si, the white parts are $FeSi_2$. The structure was completely separated Si and $FeSi_2$ in the body, also many SiC grains were grown and combined. Even if the structure was such status, porosity was extremely low. The Fe-Si alloy had a sufficiently excellent performance of infiltration into the preformed body.

Table 1 shows properties of reaction sintered SiC with Si metal and Fe-Si alloy. The density of sintered SiC with Si metal was 2.75 g/cm^3, that with Fe-Si alloy was 3.25 g/cm^3. It is cleared that sintered SiC with Fe-Si alloy has a bigger component of Fe as compared to one with Si metal. Vickers hardness of each sintering were 1.9 GPa and 2. 0GPa. Bending strengths were 299 MPa and 290 MPa. These values were generally average at best. The high hardness of the material is due to SiC structures. Therefore, hardness was unchanged independently of the sintered matrix. And also, Si metal and Fe-Si alloy are both hard brittle material, and matrix sizes of sintered SiC was unchanged when the porosity of preformed body was same, so that the strength between Si metal and Fe-Si alloy was scarcely generated. The fracture toughness of sintered SiC with Fe-Si alloy was higher

Table 1. Comparison with properties of sintered SiC.

Properties	Sintering materials	
	Si metal	Fe-Si alloy
Compounding ratio of Si (mass%)	100	75
Density ($10^3 kg/m^3$)	2.75	3.25
Vickers hardness (GPa)	1.9	2.0
Bending Strength (MPa)	299	290
Fracture toughness ($MPa/m^{0.5}$)	3.0	4.1
Electrical resistivity (Ωm)	$1x10^{-1}$	$9x10^{-3}$

than Si metal. The matrix of sintered SiC is brittle in same way the bending strength, however, the fracture toughness was improved of about 30%. For the result of observation tested indentation, it was presumed that the matrix boundary of $FeSi_2$ is effective to suppress propagation of a crack. The electrical resistivity of sintered SiC with Fe-Si alloy was greatly lower than Si metal. It can be easily explained the result, because of the electrical conductivity of Fe-Si alloy was lower than Si metal, and that of reaction sintering was lower than normal sintering. As a result, the sintering with Fe-Si alloy was found the same behavior of Si sintering, namely the molten Fe-Si alloy can be infiltrate into the deep part of the preformed body.

4. Conclusions

The Fe-Si alloy had the excellent infiltration performance into sintered body and was able to react sintering same as Si metal, so that the structure of sintered SiC with Fe-Si alloy had very low porosity. There were no reaction products of Fe, however $FeSi_2$ was existed in the matrix after sintering. Considering the results, influence affecting to react formation of SiC by molten Fe-Si alloy was small. Therefore, it is can be used in same way as a composite material of SiC/$FeSi_2$. In this experimental condition, the electrical resistivity of the SiC was increased by reaction sintering with Fe-Si alloy without decreasing of the mechanical performance compared to Si metal. It will be expected that the fabrication method greatly improves the property of sintering SiC for controlling raw and infiltration materials in future.

References

[1] H. Tanaka. (2002). Sintering of Silicon Carbide and Theory of Sintering. J. Ceram. Soc. Japan. 110 (1286), 877-883.

[2] J. N. Ness & T.F. Page. (1986). Microstructural evolution in reaction-bonded silicon carbide. J. Mater. Sci. 21, 1377-1397.

[3] A. Ciftja, T.A. Engh & M Tangstad. (2010). Wetting properties of molten silicon with graphite materials. Metall. Mater. Trans. A. 41A, 3183-3195.

[4] L.K. Frevel, D.R. Petersen & C.K. Saha. (1992). Polytype distribution in silicon carbide. J. Mater. Sci. 27, 1913-1925.

[5] Ch. Kloc, E. Arushanov & M. Wendl. et al. (1995). Preparation and properties of FeSi, α-$FeSi_2$, β-$FeSi_2$ single crystals. J. Alloys Compd. 219, 93-96.

The 73rd World Foundry Congress
„Creative Foundry"
23 - 27 September 2018, Krakow, Poland

MICROSTRUCTURE AND TRIBOLOGICAL PROPERTIES OF Co-Cr ALLOYS USED FOR METAL ELEMENTS IN PROSTHETIC TECHNIQUE

Joanna Augustyn-Nadzieja *, Łukasz Frocisz , Piotr Matusiewicz

AGH University of Science and Technology Faculty of Metals Engineering and Industrial Computer Science,
Mickiewicza Av. 30, 30-059 Krakow, Poland
**Corresponding address: jap@agh.edu.pl*

Key words: Co-Cr alloys, wear, microstructure, dendrite crystallization, dental prosthetic

1. Introduction

The human teeth are a very characteristic "system", which to a large extent is exposed to various processes that can cause its destruction. Due to this reason, materials used in dental prosthetic should been characterized by good wear resistance correspond to high biocompatibility and mechanical properties. In addition, abrasion wear, cracking and pits formation should be as small as possible for such purpose [1, 2].

Alloys used in dentistry in particular, should be characterized by [3]:

- high yield strength, especially due to the possibility of exposure to large forces,
- adequate ductility to prevent breakage during the bending procedure,
- applicable hardness, making it easier for the dentist technician to mechanical treatment (final) of the finished element.

Tribological wear is a process of destroying and removing material from a surface of solids as a result of friction. It is manifested by a continuous change of dimensions and shapes of corresponding elements. Among the elementary processes of tribological wear, the following are distinguished: abrasive, abrasive and corrosion, adhesive, fatigue [4].

Detailed information about the tribological properties of commercially used Co-Cr alloys is not well recognized. Therefore, the aim of presented work is an abrasive wear of Co-Cr alloys. Mostly used as: skeletal dentures, retention elements or metal frameworks for ceramics, eg crowns or bridges.

2. Subject of study

Samples for testing were made of Co-Cr-Mo and Co-Cr-Mo-W alloy with the trade name respectively Wironit LA and Colado CC. Tested samples were cast by means of the lost wax method. The first stage of the process consisted in the preparation of wax models for each alloy, of a specified size: 20x4x4 mm (dimensions of the samples required to perform the dry abrasion tests at T-05 tribotester. The following stage included placing the wax models in a ring, into which a phosphate-based refractory was cast. When the refractory had bounded, the formed crucible was placed in the furnace and the process of heating. The end of the process took place at 950°C. Then, the metal casting process was started. After casting, samples were drawn from the furnace and cooled in the open air. The final stage of preparation of the samples was removed mechanically runner, sandblasting and oxidation.

3. Research methodology

The microstructure analysis of the tested alloys with the cobalt matrix was performed by microscopic observations using light microscopy and scanning electron microscopy (SEM) with the EDS analysis. A significant part of the research consisted of abrasion resistance tests which were carried out using a block on ring T-05 tester. The tests of tribological wear of investigated alloys were made in metal-metal dry contact under the load 100 N for two friction paths 500 and 1000 meters. The measure values was amount of material removed in the friction process. Tribological tests were supplemented by SEM observations of samples surfaces, in addition profilometric tests were performed to obtain information about the detrition depth.

4. The results of research

As-crystallized structure of the tested Wironit LA (Co-Cr-Mo) and Colado CC (Co-Cr-Mo-W) alloys was characterized by dendritic crystallization. Typical for materials prepared in such way. (Fig. 1). It was composed of the large branches of dendrite arms embedded in the matrix, which was a solid solution of chromium, molybdenum, and carbon in the A1 β-Co phase. In the interdendritic spaces were located, among others, the precipitates of $M_{23}C_6$ carbides, arranged in characteristic bands [5]. The greatest tendency to segregation in the as-cast state was exhibited by presence of $M_{23}C_6$, MC and M_6C carbides typical for chromium, molybdenum and tungsten. Carbides precipitated in the interdendritic spaces has testified the chemical heterogeneity observed within the entire volume of the alloy.

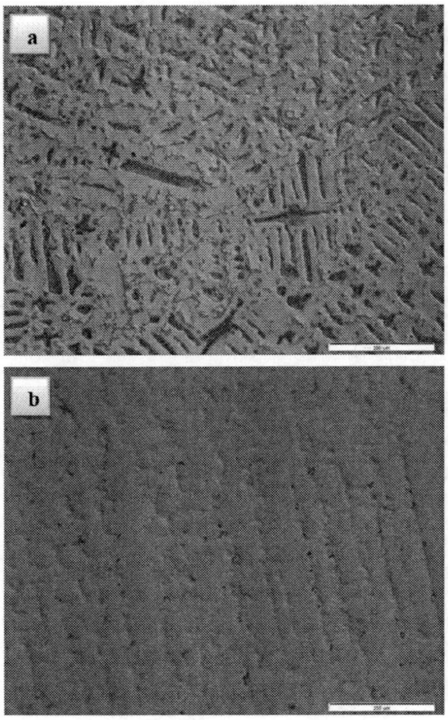

Fig. 1. Microstructures of investigated alloys: Co-Cr-Mo (a), Co-Cr-Mo-W (b)

The abrasion tests performed by the use of dry abrasion T-05 type tester, revealed that the Wironit LA alloy (Co-Cr-Mo) was characterized by intensive wear, the weight loss of the sample was 0.00357 g for the friction path of 500 m, and respectively for the Colado CC alloy (Co-Cr- Mo-W) was 0.00329 g. For the friction path of 1000 m, observed tribological relationships between the tested Co-Cr alloys were reversed, i.e. the Colado CC alloy had a mass loss of 0.00828 g, whereas for the Wironit LA alloy this value was 0.00633 g. For both tested alloys, the mechanism of wear based on SEM tests can be defined as microplowing and microcutting (Fig. 2).

5. Summary

The studied Co-Cr-Mo (Wironit LA) and Co-Cr-W-Mo alloys (Colado CC) were characterized by clearly outlined areas of dendrites and interdendent spaces enriched with alloying elements. Increasing the friction path from 500 m to 1000 m resulted in a change in the weight loss of the samples. The surface wear mechanism for both tested Co-Cr alloys was the same.

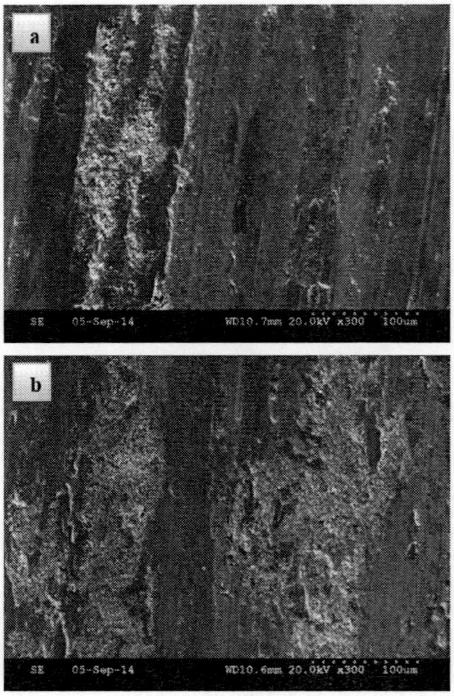

Fig. 2. SEM images of investigated alloys samples surface after abrasion: Co-Cr-Mo (a), Co-Cr-Mo-W (b)

Acknowledgments

The work has been implemented within the framework of statutory research of AGH University of Science and Technology, contract No 11.11.110. 299 AGH.

References

[1] Marciniak J., Kaczmarek M., Ziębowicz A. (2008), Biomaterials in dentistry. Gliwice: Publisher of the Silesian University of Technology (in Polish).

[2] Augustyn-Pieniążek J., Kurtyka P., Sulima I., Stopka J. (2015). Selected properties and tribological wear alloys Co-Cr-Mo and Co-Cr-Mo-W used in dental prosthetics. Archives of Metallurgy and Materials, Vol. 60 (3), pp. 1569-1574.

[3] Combe E.C. (1997). Introduction to dental material science, Warszawa: Publishing house Sonmedica (in Polish).

[4] Frocisz Ł., Krawczyk J., Bała P., Madej M. (2015). The relationship between tribological properties and microstructure of 90CrV6 steel. Ostrava: TANGER, Metal 2015, pp. 585-590.

[5] Giachci J.V., Morando C.N., Fornaro O., Palacio H.A. (2011). Microstructural characterization of as - cast biocompatible Co-Cr-Mo alloys. Materials Characterization 62, pp. 53-61.

The 73rd World Foundry Congress
„Creative Foundry"
23 - 27 September 2018, Krakow, Poland

OPTIMIZATION OF MOULD PREHEATING PROCESS TO REDUCE FORMATION OF GASES DURING INVESTMENT CASTING OF Ni-BASED ALLOYS

Małgorzata Grudzień [1]*, Rafał Nowak [1], Zenon Pirowski [1], Rafał Cygan [2], Łukasz Rakoczy[3]

[1] *Foundry Research Institute, Zakopianska 73 Str., 30-418 Krakow, Poland,*
[2] *Consolidated Precision Products, Hetmanska 120 Str., 35-078 Rzeszow, Poland*
[3]*AGH University of Science and Technology, Faculty of Metals Engineering and Industrial Computer Science, Mickiewicza 30 Av., 30-059 Krakow, Poland*
**Corresponding address: malgorzata.grudzien@iod.krakow.pl*

Key words: gasification, residual gas analysis, mould, nickel superalloys

1. Introduction

The investment casting widely known as lost-wax casting is the most popular commercially used technique for manufacturing of jet engines components including the critical parts from hot section. Investment casting is an extremely important technique to the manufacturing of the components from superalloys because it enables the development of parts with complex geometries, dimensional accuracy, and high surface quality. The molten metal is cast into ceramic shell moulds fabricated through the 'dip and stucco' technique.

Fig. 1. Wax model of casting [1]

The ceramic moulds are consists of several different shells:

a) primary coat - this layer accurately reproduces the shape of the wax pattern (Fig. 1) and counteract the reaction of liquid metal with ceramic mould,

b) second coat - transition layer between the primary coat and mould backup

c) several-coat mould backup - ensuring mechanical strength of the mould, which should be strong enough not to break down when pouring liquid metal (Fig. 2).

The basic components forming the ceramic mould are: matrix material (sand or flour: SiO_2, $ZrSiO_2$, $FeCr_2O_4$, Al_2O_3) and binder. One of the most popular technologies for making ceramic moulds uses hydrolyzed ethyl silicate as a binder. Due to harmful effects on the environment is often replaced by ecological binders. In precision foundries, the hydrolyzed ethyl silicate is replaced with a binder based on colloidal silica (sol of an orthosilicic acid). However, such a binder has many disadvantages. The ceramic mass with colloidal silica binder cause to improper wetting of the wax model, contributing to its wrong reproduction. In order to improve the adhesion of the ceramic mass to the wax model, substances improving wettability are used. Then, compounds are added to prevent aeration and foaming of the ceramic mass during mixing.

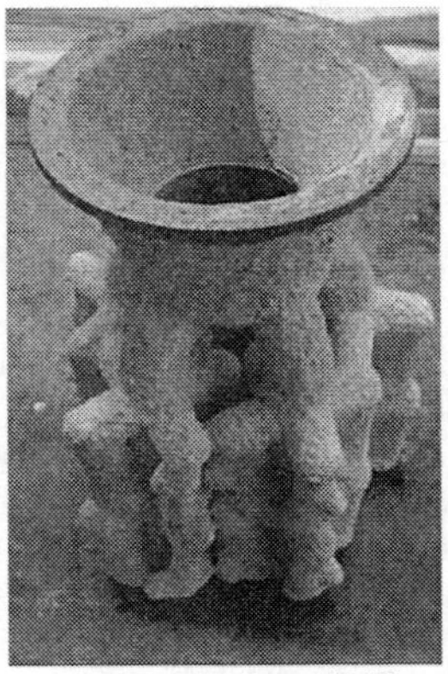

Fig. 2. Shell mould [1]

2. Experimental procedure

To obtain the temperature regime for preheating, raw moulds were perform tests of gasification at high temperature equipment for liquid metal investigation [2]. The tested moulds were produced through 'dip and stucco' technique. The tests were perform under high vacuum condition which allows to continuously acquisition the chemical composition of gasses in the testing chamber by quadrupole mass spectrometer. The temperature regime for preheating raw moulds were with two isothermal steps: 1) at 500 °C by 1 hour and 2) at 700 °C by 30 minutes. The heating rate were 8 °C/min and cooling rate 20 °C/min.

The raw sample of moulds were heated in air at temperature 220 °C by 3 hours to observe change of the on surface.

3. Residual gas analysis

The residual gas spectra shows strong gasification of raw moulds up to 300 °C and it is connected with gasification of water and carbohydrate from organic like waxes and alcohols resolvents. In Fig. 3 is shown the 3D graph of change the chemical composition of atmosphere in testing chamber during heating, isothermal steps and cooling of mould.

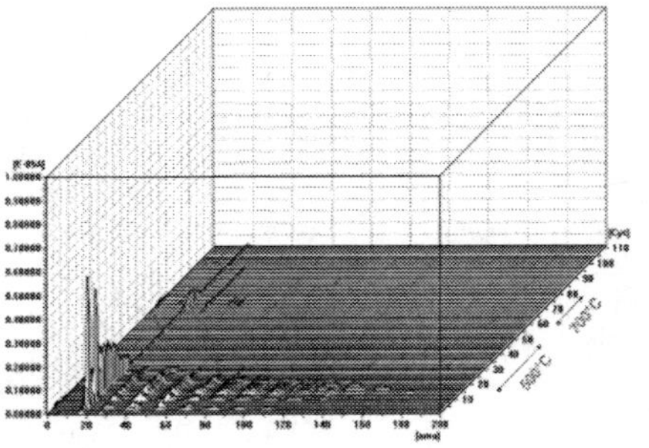

Fig. 3. The chemical composition of gases in testing chamber during heating mould 3.

The strongest gasification was observed in mould 2 and 4 which contain tetraethyl orthosilicate. For this moulds were necessary to add one more isothermal step in heating regime at ~350 °C for 45 min and 70 min, respectively. For all testing moulds, during the isothermal step at 500 °C the vacuum level increase what originated from reducing the organic substances. Mostly reduced were intensity of peaks with atomic mass above 60 amu, which is correlated with high-carbon carbohydrate. At 700 °C were observed peaks corresponding to gasification of water, CO^{++} and CO_2^{+} and this type of peaks showed at high temperature is connected with decomposition of low-carbon carbohydrate from deeper sand layer of mould. The strong gasification at 300°C suggest that the wax used to produce moulds were not completely removed and it is still inside the first layers. The samples of raw moulds were heated on heating plate on Al_2O_3 supports with final layer down to support. To melt wax and not allow it to completely burn/decompose the low temperature (220 °C) were chosen. The moulds were preheated by 3 hours and Al_2O_3 supports were used to observe any change of color of it. Fig. 4 shows the surface of moulds after heating. It is visible change of colors in final layer and light brown marks.

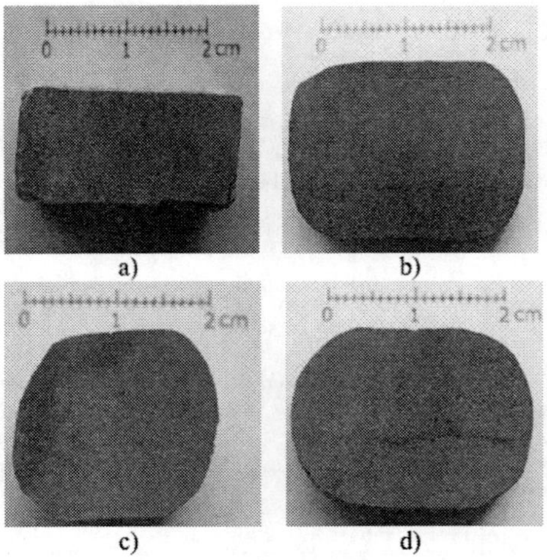

Fig. 4. Moulds samples after heating 220 °C by 3 hours on air: a) mould 1, b) mould 2, c) mould 3, d) mould 4

Acknowledgments

This research was supported by National Centre for Research and Development, Grant No. TECHMATSTRATEG1/348072/2/NCBR/2017.

References

[1] R. Cygan, Ph.D. thesis, The reasons for the formation of imperfections in super-thin-walled parts cast from nickel superalloys (in Polish), AGH Krakow, 2010

[2] Sobczak N., Schmidt J., Kazakov A.: Device for examinations of surface properties of liquid metals, Poland, patent no. 166953 (date: 26.07.1991)

The 73rd World Foundry Congress
„Creative Foundry"
23 - 27 September 2018, Krakow, Poland

CHARACTERIZATION OF PRIMARY MICROSTRUCTURE OF THIN-WALLED NI-BASED SUPERALLOY CASTING

Małgorzata Grudzień [1]*, Zenon Pirowski [1], Krzysztof Jaśkowiec [1], Adam Bitka [1], Rafał Cygan [2], Łukasz Rakoczy [3]

[1] Foundry Research Institute, Zakopianska 73 Str., 30-418 Krakow, Poland
[2] Consolidated Precision Products, Hetmanska 120 Str., 35-078 Rzeszow, Poland
[3]AGH University of Science and Technology, Faculty of Metals Engineering and Industrial Computer Science, Mickiewicza 30 Av., 30-059 Krakow, Poland

**Corresponding address: malgorzata.grudzien@iod.krakow.pl*

Key words: investment casting, inconel, heating rate, temperature field, pouring

In this paper the influence of selected investment casting parameters on the microstructure of Inconel 713C nickel-based superalloy have been investigated. The temperature inside the shell mould was precisely measured during the investment casting of thin-walled elements. The primary microstructure was characterized by relatively large local inhomogeneities inside interdendritic spaces relative to dendrite cores. As a consequence of alloying elements segregation during solidification, constituents like eutectic γ/γ' islands, phases rich in molybdenum, zirconium, and niobium were observed.

The main goal of the research was to show the correlation between selected investment casting parameters and the primary microstructure of nickel-based superalloy.

1. Introduction

Important technological problem occurring in the production of nickel superalloy is the developing of the conditions of the casting process and crystallization of castings. The primary microstructure is largely influenced by the temperature of preheating of shell mould and melt pouring temperature. The number of layers of the shell mold results from the provision of sufficiently high mechanical properties to prevent deformation and cracking of the mould. The problem of the occurrence of defects in lost-wax castings despite the use advanced technology is not yet fully solved. This is mainly due to the lack of a comprehensive development and evaluation of the change in the temperature field on the cross-section of the ceramic mould during the preheating, pouring and solidification of the liquid superalloy [1, 2].

2. Experimental procedure

The investment casting mould with eleven shells was prepared (Fig. 1). The chemical composition (wt %) of the Inconel 713C was measured as C: 0.10, Al: 5.74, Cr: 14.09, Mo: 4.21, Ti: 0.91, Nb: 2.35, S: 0.002, P: 0.004, Co: 0.04, Fe: 0.04, B: 0.01, Zr: 0.06, and balance Ni. A master heat charge was melted in a zirconia crucible inside vacuum Balzers furnace and poured into preheated ceramic mould. The designed preheating and pouring temperatures equaled 1250°C and 1470°C respectively. Shell mould was made through the 'dip and stucco' method. Alcohol-based coating with $ZrSiO_4$ as a matrix was used to construct the mould. The composition of a layer was chosen in order to prevent ceramic-metal interaction and ensure good thermal conductivity. Temperature measurement in shell mould was carried out based on signals registered by four thermocouples; a) the first- between 2nd and 3th layer; the second- between 3rd and 4th layer; the third- between 5th and 6th layer; the fourth- between 7th and 8th layer.

Fig. 1. Model of shell mould

3. Temperature distribution in shell mould

The temperature registered by thermocouples mounted in shell mould is presented in Fig. 3. Pouring of liquid through the filter into shell mould was performed after 111 min of preheating. Temperatures after 50, 100 of preheating, maximum registered during pouring, and during cooling after 150 min of process are shown in Table 1.

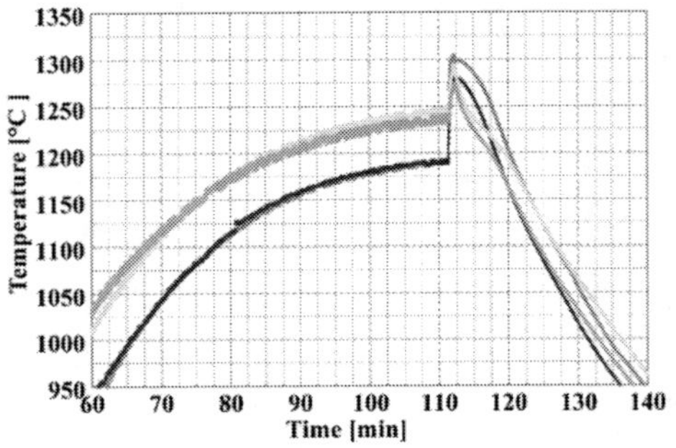

Fig. 2. Temperature registered by thermocouple: #1green; #2red; #3yellow; #4blue

Table 1. Temperature in shell mould during process

Time [min]	Temperature [°C] #1	#2	#3	#4
50	713	749	845	897
100	1178	1180	1232	1221
maximum registered during pouring	1305	1280	1293	1273
150	852	822	871	840

4. Primary microstructure

Microstructure of Inconel 713C in as cast condition is characterized by a dendritic structure with substantially large local inhomogeneities (Fig. 3a). Dendrite cores consist of cubic-shaped γ' precipitates divided by γ phase channels.

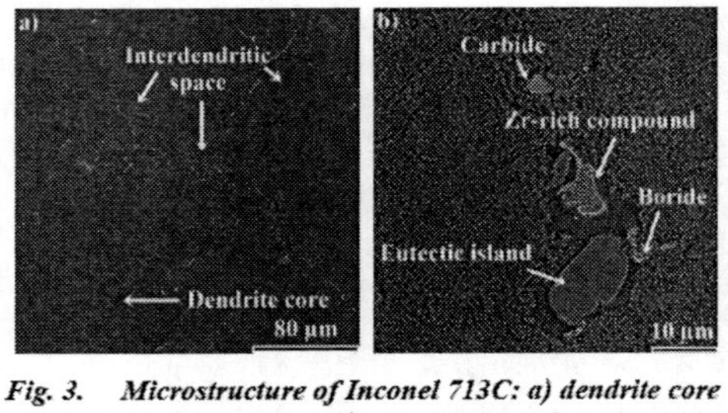

Fig. 3. Microstructure of Inconel 713C: a) dendrite core and interdendritic spaces; b) constituents in interdendritic spaces

Segregation of alloying elements during crystallization led to the creation of eutectic γ/γ' and carbides, while the very low solubility of Zr and B in γ and γ' phases to borides and also intermetallic compound (Fig. 3b). EDS analysis carried out in dendrite cores and interdendritic spaces is expressed as k coefficient which is define concentration in dendrite core divided by concentration in interdendritic space (Fig. 4).

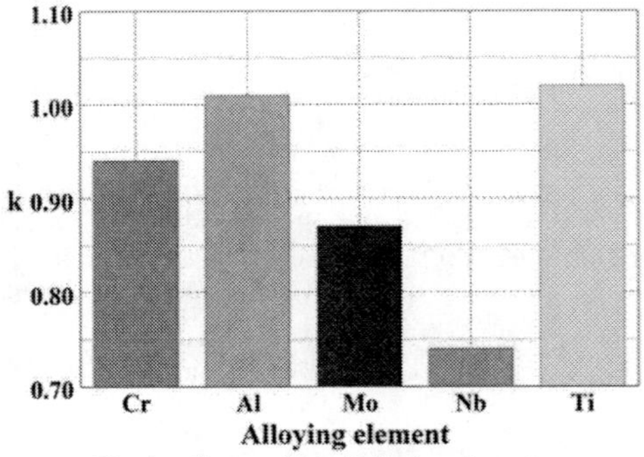

Fig. 4. Segregation of alloying elements

5. Summary

The paper presents a detailed analysis of the temperature distribution during the investment casting process in a multi-layer shell mould. Adequate selection of the preheating and pouring temperature as well as the preheating time affects the quality of the castings and the price of its manufacturing. The obtained microstructure is characterized by the high volume fraction of the main strengthening phases, namely γ'. Heterogeneity of the cores, and thus the presence of borides, eutectic γ/γ' and Zr-rich phase can be reduced by heat treatment.

Acknowledgments

This research was supported by National Centre for Research and Development, Grant No. TECHMATSTRATEG1/348072/2/NCBR/2017.

References

[1] D. Szeliga et.al. 2017. Investigation of casting–ceramic shell mold interface thermal resistance during solidification process of nickel based superalloy. Experimental Thermal and Fluid Science. 87. 149-160

[2] H. Long et.al. 2018. Microstructural and compositional design of Ni-based single crystalline superalloy - A review. 743. 203-220

STRUCTURE AND MECHANICAL PROPERTIES OF AUSTEMPERED GREY IRON (AGI)

Krzysztof Jaśkowiec [1]*, Zenon Pirowski [1], Adam Kowalski [1], Adam Bitka [1], Małgorzata Grudzień [1]

[1] Foundry Research Institute, Zakopianska 73 Str., 30-418 Krakow, Poland
**Corresponding address: e-mail: krzysztof.jaskowiec@iod.krakow.pl*

Key words: austempering, AGI, grey iron, heat treatment, microstructure

Ausferritic gray cast iron in comparison with spheroidal cast iron shows a number of favorable features. Is attractive for materials engineers due to high abrasion resistance and thermal conductivity, very good machinability, excellent vibration and noise damping ability [1-2]. At the Foundry Research Institute in Krakow, the manufacturing technology of this material was developed and then the most important mechanical properties were determined.

The main aim of the work was to determine the effect of selected chemical composition of grey iron on mechanical properties.

1. Designed variants

Table 1. Temperature in shell mould during process

Variant	C	Si	Mn	Cu	Mo	Cr
AG1	3.65	2.07	0.56	0.75	0.25	-
AG2	3.10	1.75	1.23	-	-	0.40
AG3	3.20	1.75	1.18	0.48	-	0.45
AG4	3.20	1.75	1.17	0.51	0.20	0.38

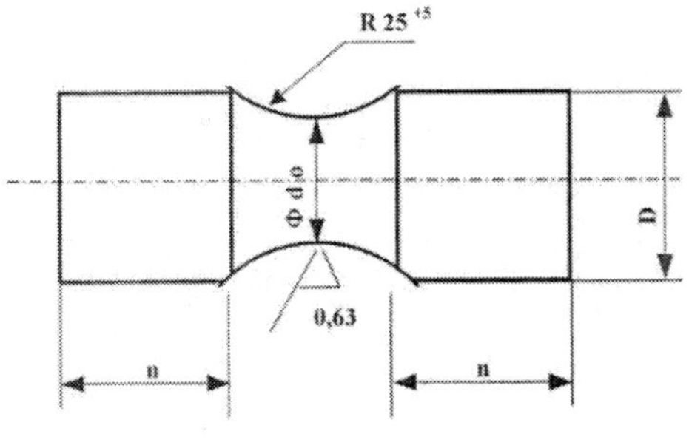

Fig. 1. Tensile test specimen d_0 = 18 mm. D = 25 mm, n = 50 mm

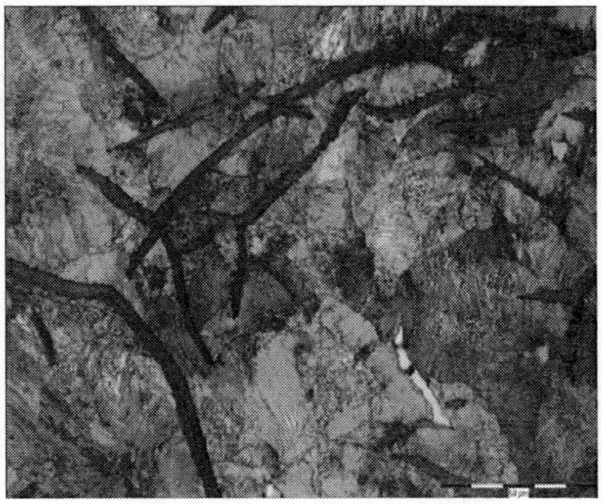

Fig. 2. Microstructure of variant no. AG2. Perlite, graphite, caribides

The conducted dilatometric studies of cast irons involved the registration of heating and cooling curves. The tests were carried out by Leitz dilatometer in the temperature range from 0÷1200˚C. Based on the test results, it was found that the austenitizing and austempered range with isothermal transformation was 880÷950°C and 270÷360°C respectively. The bars were quenched with isothermal transformation according to cycles A and B is shown in Table 3.

Table 2. Mechanical properties of designed variants in as cast condition

Variant	Strength R_m [MPa]	Hardness HB
AG1	255	210
AG2	326	252
AG3	281	195
AG4	287	198

2. Influence of heat treatment on mechanical properties

Table 3. Heat treatment parameters

Variant	Cycle	Austempering		Isothermal quenching	
		Temp. [°C]	Time [min]	Temp. [°C]	Time [min]
AG1,AG2	A	920	120	270	120
AG3,AG4	B	920	120	300	120

Table 4. Mechanical properties after heat treatment

Variant	Strength R_m, [MPa]	Hardness HBW	Elongation A_5,%
Cycle A			
AG 1	420	315	1,2
AG 2	578	366	0,9
AG 3	425	335	1,6
AG 4	477	342	1,4
Cycle B			
AG 1	398	315	-
AG 2	545	345	1,0
AG 3	422	302	1.7
AG 4	421	308	1,8

Fig. 3. Microstructure of AG2 cast iron, hardening with isothermal transformation at 270°C

The most favorable mechanical properties were obtained to variant AG2. The microstructure of this cast iron is shown in Fig. 3.

The cast iron AG2 was subjected to additional tests. The content of austenite in the matrix was determined after hardening at temperatures: 270, 330 and 360°C at the hardening times: 2, 5, 15, 30, 120 and 300 min. The content of austenite in the matrix after austempering for 120 min was: at 270°C - 18.6%, at 330°C - 39.5%, at 360°C - 41.1%.

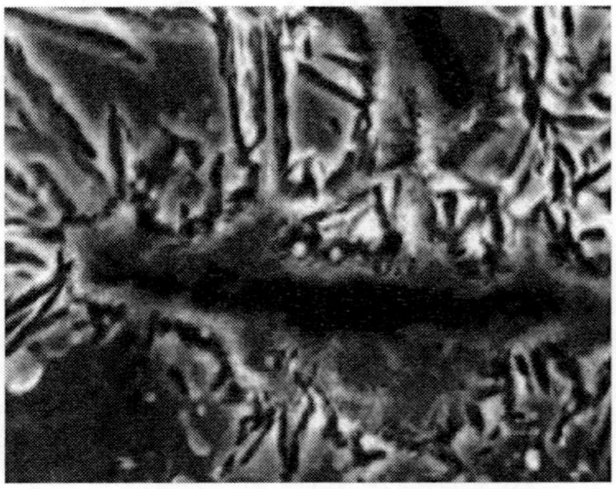

Fig. 4. Segregation of alloying elements in AG2 cast iron. Ferrite, austenite and graphite flake

3. Summary

The highest tensile strength and hardness were obtained in cast iron austempered at 270°C. The most preferred variant is AG2, containing 3.10% C, 1.75% Si, 1.23% Mn and 0.40% Cr. The tensile strength was 587 MPa, 1% elongation at hardness of 366 HB. Noteworthy are also properties of other variants. The AGI cast iron described in this paper can fill the gap between gray cast iron and cast iron with vermicular graphite.

Acknowledgments

This research was supported by National Centre for Research and Development, Grant No. TECHMATSTRATEG1/348072/2/NCBR/2017

References

[1] A. Kowalski, A. Pytel, J. Turzyński: Properties and application of new types of ausferritic cast iron. Conference Proceedings. Slovenian Fundrymen Society. Faculty of Natural Sciences and Engineering. University of Liubliana. Portoroż 12-14 September 2007

[2] B. Kovacs, J. Keough.1994. Bainitisches usseisen mit lamellengraphit (AGI) – hohefestigkeit und Dämpfungsfähigkeit. Konstruieren + Giessen. no 3, p. 16

The 73rd World Foundry Congress
„Creative Foundry"
23 - 27 September 2018, Krakow, Poland

DETERMINING THE EFFECT OF AUSTEMPERING TEMPERATURE ON THE MORPHOLOGY OF AUSFERRITE IN ADI USING COMPUTER IMAGE ANALYSIS AND X-RAY DIFFRACTOMETRY

Krzysztof Jaśkowiec[1]*, Zenon Pirowski[1], Józef Turzyński[1], Adam Kowalski[1], Barbara Mrzygłód[2], Małgorzata Grudzień[1]

[1] *Foundry Research Institute, Ferrous Alloys Department, 30-043 Krakow, Zakopianska 73 Str., Poland*
[2] *AGH University of Science and Technology, Faculty of Metals Engineering and Industrial Computer Science, 30-059 Kraków, Mickiewicza 30 Av., Poland*
**Corresponding address: krzysztof.jaskowiec@iod.krakow.pl*

Key words: ADI, chemical composition, heat treatment, austenite ,carbon in austenite

The microstructure of spheroidal graphite cast iron matrix subjected to austempering is called ausferrite, which is composed of ferrite, similar to bainitic ferrite in steels, and retained austenite with high carbon levels (1.8 - 2.2% C). Cast iron with this microstructure is called ADI (Austempered Ductile Iron). Chemical composition of cast iron used in this research is shown in table 1.

Table 1. Chemical composition of spheroidal graphite cast iron

C	Si	Mn	Mg	Ni	Cu	Mo
3,50	2,70	0,35	0,07	0,70	0,70	0,18

The melt was poured into keel blocks YII type according to EN-PN 1653. Samples cut from the bottom part of the keel blocks were heat treated under parameters as shown in table 2.

Table 2. Heat treatment parameters

Austenitizing		Austempering	
Temperature, °C	Time, min	Temperature, °C	Time, min
900	120	280, 320, 360	2, 5, 15, 30, 60, 120, 180

The aim of this research was to analyse the ausferrite morphology in ADI. The austenite lattice parameters, austenite carbon content and austenite volume fraction in the matrix were determined using HZG4 X-ray diffractometer (Table 3). Microstructural analysis of cast iron was made. Microstructure observations were conducted using AXIO OBSERVER Z1M metallographic microscope. Hardness measurements were also made using a standard Rockwell machine.

Figure 1 shows a plot of austempering time vs. austenite content in the matrix for different austempering temperature.

Table 3. Ausfferite characteristics – melt A1

Austempering tempera. [° C]	Time, [min]	Carbon content in austenite, [%]	volume fraction of austenite, [%]
280	2	1.174	17.39
	15	1.232	24.64
	30	1.414	32.11
	60	1.668	34.64
	120	1.809	30.83
	180	1.881	28.91
320	2	1.270	23.30
	15	1.504	37.81
	30	1.696	40.48
	60	1.791	40.82
	120	1.863	39.16
	180	1.905	39.13
360	2	1.320	34.93
	15	1.660	39.97
	30	1.784	42.60
	60	1.823	43.30
	120	1.830	41.84
	180	1.828	38.66

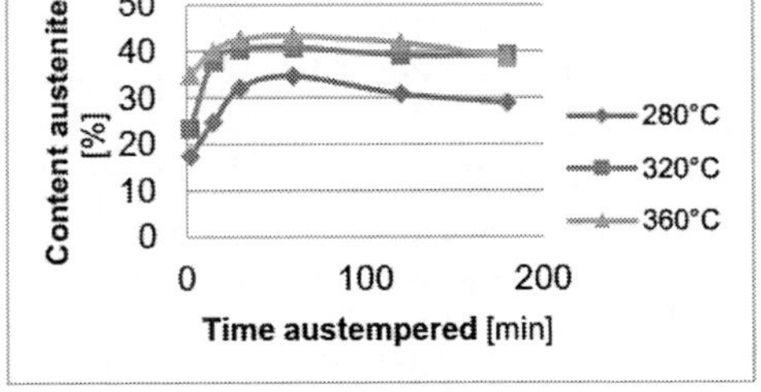

Fig. 1. Effect of austempering time on austenite content with different austempering temperature

The austenite content in ADI matrix increases with both austempering reaction time and temperature. The maximum austenite content was obtained for austempering at 360°C for 60 min.

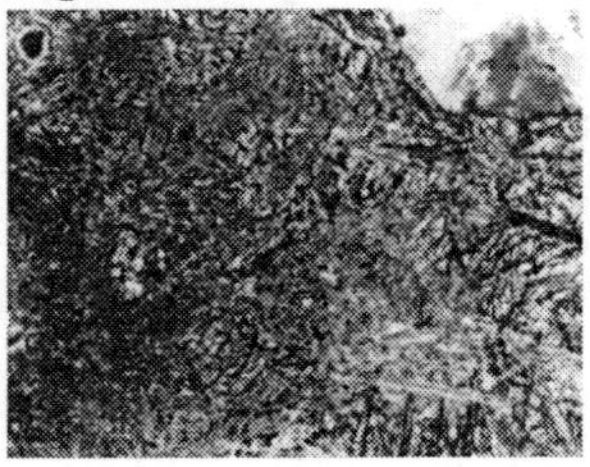

Fig. 2. ADI microstructure after austempering at 280°C/300 min, 1000x, spheroidal graphite, ferrite, austenite, etched in nital

Table 4. Hardness of ADI cast iron – melt Al

austempering temp., °C	austempering time, min.	Hardness, HRC
280	2	58,3
	5	55,8
	15	48,1
	30	42,7
	120	41
	180	40,3
	300	41
320	2	57,7
	5	48,2
	15	40,2
	30	35,5
	120	33,5
	180	33,7
	300	30,8
360	2	50,5
	5	47,7
	15	38,8
	30	34
	120	31
	180	31
	300	30,3

Figure 4 shows the influence of heat treatment parameters on cast iron hardness.

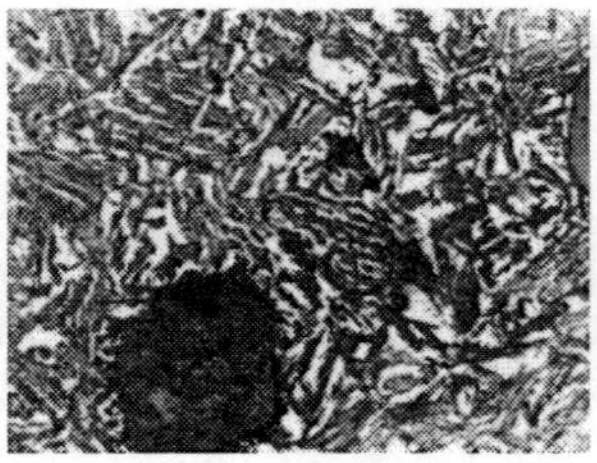

Fig. 3. ADI microstructure after austempering at 360°C/300 min, 1000×, spheroidal graphite, ferrite, austenite, etched in nital

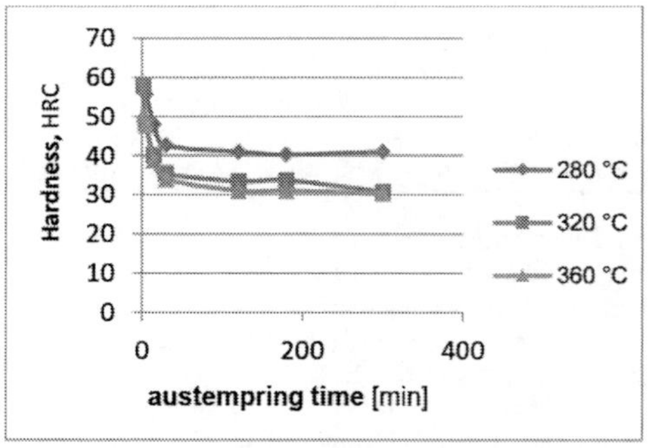

Fig. 4. Influence of austempering time on hardness

Results obtained in this research show strong influence of austempering time on austenite carbon content and austenite volume fraction in ADI matrix, which increase up to 120 min of austempering.

The highest austenite carbon content of 1.905% was obtained in sample austempered at 320°C for 180 min. Maximum austenite volume fraction in ADI matrix of 43.3% was achieved after austempering at 360°C for 60 min.

Acknowledgments

This research was supported by National Centre for Research and Development, Grant No. TECHMATSTRATEG1/348072/2/NCBR/2017

Literatura

[1] Voigt R.C., Loper C.R. Jr.: Austempered ductile irons - influence of microstructure control. 3rd International Symposium on the Physical Metallurgy of Cast Iron. August 1984, p.377, Stockholm, Sweden

[2] Kovacs B.V.: On the terminology and structure of ADI. Trans. AFS, 1994, 83, p. 417.

[3] Kowalski A., Rączka J.: Przebieg procesu kształtowania struktury i własności mechanicznych żeliwa sferoidalnego z podwyższoną zawartością Mn i dodatkami Ni, Cu i Mo. I Krajowa Konferencja Naukowa "Materiałoznawstwo - Odlewnictwo - Jakość". Politechnika Krakowska. Kraków 20 - 22.02.1997

The 73rd World Foundry Congress
„Creative Foundry"
23 - 27 September 2018, Krakow, Poland

USE OF THE ATND METHOD TO ASSESSMENT OF EN AC-AlSi9Mg ALLOY HARDNESS MOULDED IN METAL MOULDS

Jacek Pezda *

University of Bielsko-Biała, Department of Production Engineering and Automation, ul. Willowa 2, 43- 309 Bielsko-Biała, Poland,
**Corresponding address: jpezda@ath.bielsko.pl*

Key words: aluminum alloys, crystallization, hardness, regression analysis, ATND

1. Introduction

Al-Si casting alloys are characterized by soft matrix and hard precipitations of silicon. Except silicon, these alloys can consist of small additions of other elements affecting mechanical properties of alloys by change of structure and course of crystallization process [1,2]. Mechanical properties of the Al-Si alloys, directly determined by shape, size and distribution of silicon crystals, belong to basic factors influencing application field of the Al-Si alloys [1,3]. Improvement of mechanical properties of the Al-Si alloys in area of standard methods is connected mainly with change of morphology of Si precipitations through: usage of modification of alloy, maintaining correct temperature of superheating and pouring into moulds, and possible heat treatment. Modification is the most effective method enabling obtainment of optimal structure of castings. Quality of the modification depends on correct dosing of inoculants, temperature of metal and time elapsing from modification to solidification of the alloy [1].

In case of hypoeutectic and near-eutectic alloys the modification results in change of precipitations morphology of silicon crystals from flakes to fibrous and spherical, also interfacial space in α(Al)+β(Si) eutectic mixture changes [4]. One from the most common methods serving to assessment and estimation of grain size, extend of the modification of alloy and temperatures of phase transitions of alloy is thermal analysis, consisting in recording of temperature in course of solidification process of the alloy. Very often, however, simple thermal analysis is not sensitive enough to assess thermal processes being in course, and in such situations it is more helpful to simultaneously draw a curve depicting derivative $dt/d\tau$, which emphasizes less clear changes present on cooling curve $t= f(\tau)$ (method of thermal-derivative analysis - ATD). Apart from these two methods there also exist another methods based on thermal effects (DTA, DSC) [5], thermal-electric (ATD-AED) [6] and thermal-voltage (ATND) [7]. In course of measurements with use of the ATND method (thermal-voltage-derivative method) are recorded values of generated voltage and temperature of investigated test piece. Value of generated voltage is within range of ±5 mV. Measuring probes measure differences in potential within volume of the test piece. Combination of thermal analysis with measurement of generated voltage results in more clear picture of changes occurring during crystallization and solidification of alloy. Course of crystallization process is presented in form of diagram created in time of solidification of alloy; the diagram reflects changes in the structure during solidification of alloy (formation of phases and eutectic mixtures). Thermal curves (t and $dt/d\tau$) in the ATND method are supplemented with voltage curves (U and $dU/d\tau$).

In the presented study it has been implemented the author's methodology of thermal-derivative-voltage analysis (ATND), and regression analysis aimed at obtainment of mathematical dependencies enabling prediction of HB hardness of the EN AC-AlSi9Mg silumin at stage of its preparation.

2. Methodology of the research

To the investigations the EN AC-43300 (AlSi9Mg) alloy was used, which can be characterized by very good casting properties, good machinability and workability, and corrosion resistance. To prepare test pieces, the alloys were melted in electric resistance furnace at temperature of about 760°C. The next treatment consisted in refining of the investigated alloys. The Rafal 1 preparation in quantity of 0.4% mass of charge has been used to the refining.

Refined alloys, after removal of oxides and slag from metal-level, were modified with AlSr10 master alloy in quantity of 0.4% mass of charge (0.04% Sr) at temperature 780 °C.

The investigated alloy was poured into metallic mould destined to production of standardized test pieces; the mould was adapted to recording of crystallization processes using the ATND method [7]. The metal mould was heated to temperature of 250°C. Measurement of Brinell hardness was performed according to the PN-EN ISO 6506-1:2008 standard with use of the Brinell hardness tester of PRL 82 type, with steel ball having 10 mm diameter, under load of 9800N sustained for 30 seconds. On the base of obtained values of characteristic points from the ATND method (independent variables) and HB 10/1000/30 hardness (dependent variable) it has been created a files with data for regression analysis, performed with use of the software StatSoft Statistica ver. 10.

First order polynomial was taken as function of tested object (1).

$$z = b_0 + b_1x_1 + b_2x_2 + b_3x_3 + ... + b_nx_n \pm \varepsilon \quad (1)$$

where:

z – dependent variable HB 10/1000/30,

$x_1, x_2, ..., x_n$ – independent variables ($t_1 \div t_4$, $U_1 \div U_4$),

$b_0, b_1, ..., b_n$ - regression estimators,

ε - standard error of estimation.

3. Description of obtained results

Course of crystallization process of the investigation alloy recorded with use of the ATND method, together with marked characteristic points for the thermal curve (t_1 - t_4) and voltage curve (U_1 - U_4) are presented in the Figure 2

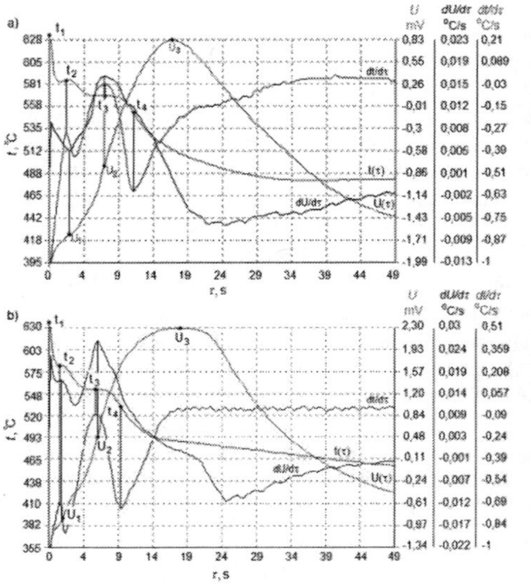

Fig. 1. Curves of the ATND method with marked characteristic points for the investigation alloy: a) refined, b) modified

HB10/1000/30 hardness of the alloy was included within range of 61-64.5 for the refined alloy and 59.5-63 for the refined and modified alloy.

In result of the performed regression analysis it has been obtained complete linear model for the dependent variable HB10/1000/30, described by the equation (2):

$$HB\,10/1000/30 = -113{,}5 + 0{,}07t_1 - 0{,}007t_2 + 0{,}09t_3 + 0{,}16t_4 + 1{,}8U_1 - 0{,}03U_2 - 1{,}26U_3 + 2{,}74U_4 \quad \pm 0{,}89 \qquad (2)$$

In result of performed elimination of the irrelevant variables it has been obtained dependency (3), for which regression coefficient amounts to R=0.84, deformation coefficient amounts to R^2=0.71. Statistical significance of the effect of influence of independent variables on the dependent variable HB is fulfilled at the level of p=0.00018<a=0.05 ($F_{obl} > F_{(\alpha;df1;df2)}$) ($F_{obl}$ = 9.13 > $F_{(0,05;5;18)}$ = 2.77).

$$HB\,10/1000/30 = -68{,}1 + 0{,}07t_1 - 0{,}16t_4 + 1{,}75U_1 - 1{,}23U_2 + 2{,}45U_4 \quad \pm 0{,}81 \qquad (3)$$

The Fig. 2 illustrates how anticipated and real (recorded) values of *HB10/1000/30* hardness are shaped for the refined and modified alloy.

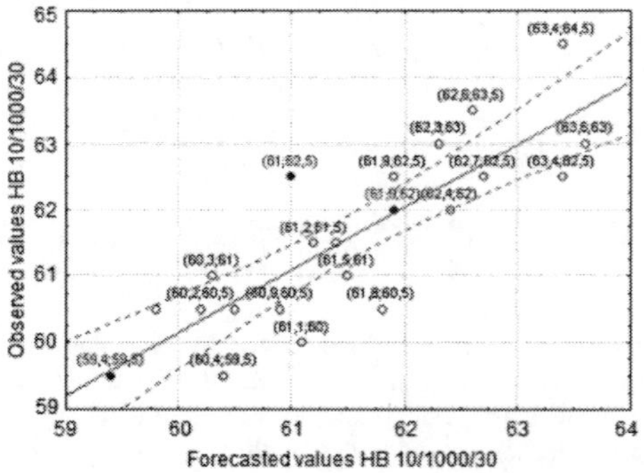

Fig. 2. Anticipated and real values of HB10/1000/30 hardness for the EN AC-AlSi9Mg alloy

4. Conclusion

The ATND method reflects crystallization process of the investigated alloy. Presence of correlation between values of temperature and voltage of characteristic points from the ATND method and HB 10/100/30 hardness of the EN AC-AlSi9Mg alloy has been confirmed on the base of the performed investigations. Obtained dependency enables in laboratory conditions prediction of hardness of investigated alloy on stage of its preparation (melting) with significance level of α = 0.05. There exists a necessity to verify obtained results in industrial conditions.

References

[1] S. Pietrowski. (2001). Siluminy, Wydawnictwo Politechniki Łódzkiej, Łódź.

[2] S. Rzadkosz, L. Staszczak. (2007). Effect of selected microadditives on mechanical properties of aluminum alloys. Archives of Foundry Engineering. 7(1), 85-88.

[3] J.X. Chen et al. (2017). Study on Eutectic Microstructure and Modification Mechanism of Al-Si Alloys. Materials Science Forum, 877, 97-103.

[4] A.K. Dahle, et al. (2005). Eutectic modification and microstructure development in Al–Si Alloys. Materials Science and Engineering A. 413-414, 243-248.

[5] M. Sikora, J. Piątkowski. (2007). Zastosowanie analizy termicznej ATD i skaningowej DSC w badaniach topnienia i krzepnięcia stopów Al-Si. Rudy i Metale Nieżelazne. 52(6), 317-321.

[6] J. Pezda, M. Dudyk, T. Ciucka & P. Wasilewski. (1998). Wielomianowe modele właściwości, mechanicznych stopów aluminium. Krzepnięcie Metali i Stopów. 38, 131-136.

[7] J. Pezda. (2016). Prediction mechanical properties of AlSi13Cu2Fe alloy using the ATND methods. Materials Research. 16(1), 252-257.

The 73rd World Foundry Congress
„Creative Foundry"
23 - 27 September 2018, Krakow, Poland

WEAR PROPERTIES OF MILLED CARBON FIBER-REINFORCED ALUMINUM ALLOY COMPOSITES

Kazunori Asano*, Muhammad Faiz Bin Zainuddin

Department of Mechanical Engineering, Faculty of Science and Engineering, Kindai University, Higashiosaka 577–8502, Japan
**Corresponding address: asano@mech.kindai.ac.jp*

Key words: carbon fiber, aluminum composite, thermal conductivity, mechanical property, sliding wear

1. Introduction

Reinforcing with non-metallic fibers have been proposed to improve the properties of aluminum alloy. Carbon fibers are widely-used reinforcements due to their low density, high strength, high rigidity, self-lubricating properties and low thermal expansion. Carbon fibers are roughly divided into the PAN-based one and the pitch-based one. Although several investigations have been carried out on the wear properties of the PAN fiber-reinforced aluminum composites fabricated by the casting process [1, 2], there have been few reports on the properties of the pitch fiber composite. In the present study, the aluminum alloy composites, which are reinforced with these carbon fibers, were fabricated by the squeeze casting process. The effects of the fiber reinforcement on the sliding wear behavior of the alloy were investigated.

2. Experimental procedure

The JIS-AC8A aluminum alloy (Al-12Si-1Mg-1Cu-1Ni) was used as the matrix metal, and the milled PAN and pitch-based carbon fibers were used as reinforcements. Table 1 lists the properties of the fibers. Carbon fibers were mixed with water, then pressed in the mold to form a cylindrical fibrous preform, followed by drying. The fiber volume fraction in the preform was 30 vol%.

The composite was fabricated by squeeze casting. The mold was preheated with the preform in the mold at 673 K, the molten alloy (973 K) was then poured into the mold and quickly pressed by a plunger (40 MPa) for infiltrating into the preform.

The disk test pieces of 10-mm diameter and 1.5-mm thickness for the thermal conductivity measurements were machined from the composite. After measuring the thermal diffusivity using the laser flash technique, the thermal conductivity was calculated using the values of thermal diffusivity, density (measured using the Archimedean principle) and specific heat (by DSC). Compressive tests were performed at ambient temperature. The crosshead speed was 2.5×10^{-5} m/s. The composites were subjected to a wear test using the pin-on-disk method. Aluminum and the composites were cut into the pin specimens. High-carbon chromium bearing steel was processed into a disk shape for the counterpart. The wear tests were carried out under contact loads (P) of 17, 27 and 37 N and at the sliding speeds (v) of 0.3, 1.5 and 3.0 m/s. Wear mass loss, coefficient of friction (μ) during the wear test, temperature of the pin specimen during the wear test were measured. The worn surfaces of the specimens were examined using an SEM equipped with an EPMA.

Table 1. Properties of the carbon fibers.

	Length (μm)	Diameter (μm)	Thermal Conductivity (W/(m·K))	CTE ($\times10^{-6}K^{-1}$)	Elastic Modulus (GPa)
PAN	130	7	10	-0.4	230
Pitch	150	9.7	900	-1.2~0	940

3. Results and discussion

Fig.1 shows the optical micrographs of the composites, showing that the fibers were in a random arrangement in the composites.

Table 2 lists the properties of the unreinforced AC8A alloy and composites. The reinforcement with the PAN fiber decreased the thermal conductivity but increased the 0.2% proof stress of the alloy. In contrast, the reinforcement with the pitch fiber increased the thermal conductivity but decreased the proof stress of the alloy.

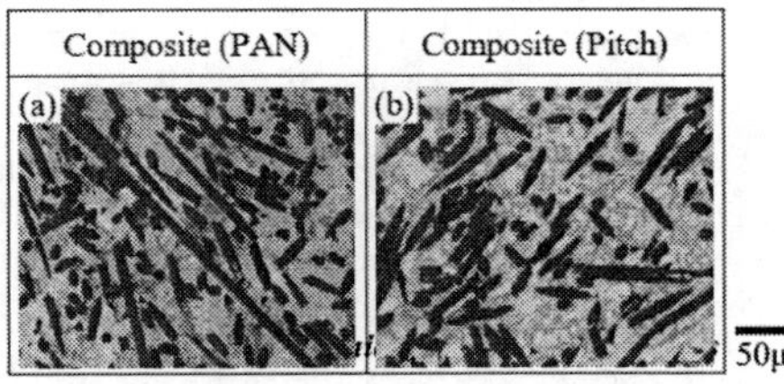

Table 2. Properties of unreinforced AC8A alloy and composites.

	λ (W/(m·K))*	$\sigma_{0.2}$ (MPa)**
Unreinforced AC8A	125	184
Composite (PAN)	52	260
Composite (Pitch)	184	137

* Thermal conductivity measured at ambient temperature.
** Compressive 0.2% proof stress measured at ambient temperature.

Fig. 2 represents the effects of v on the wear rate (WR) of the unreinforced alloy and composites. The

WR of the unreinforced alloy drastically increased when the υ exceeded 1.5 m/s under a load of 37 N. The fiber-reinforcement drastically reduced the *WR*, especially under a high load. The reinforcement with the pitch-based fiber has a substantial effect on decreasing the *WR*.

The μ during the wear test is shown in Fig. 3. The error bars represent the range between the maximum and minimum values (scatter). The large range represents a greater change in the μ during the wear test. The scatter was independent of the υ. The scatter slightly decreased by the fiber-reinforcement under every υ, while the reinforcement did not have a significant role in changing the average values of μ. Although the difference in the average values between the two composites was low, the scatter for the pitch fiber composite was slightly lower than that for the PAN fiber one.

Fig. 4 shows the surface roughness values (R_a) of the specimens after the wear test. Although the change in the average values was not apparent by the fiber reinforcement at every υ, the scatter of the values due to the measured points decreased by the reinforcement.

Under the severe wear conditions, such as υ and P were 3.0 m/s and 37 N, respectively, the temperature of the PAN composite sharply rose after the wear started, while that of the pitch composite only slightly increased. Results of the EPMA map analysis revealed that the worn surfaces of the counterpart combined with the unreinforced alloy mainly consisted of aluminum. In contrast, iron, carbon and aluminum were detected on the surfaces of the counterparts combined with the composites, suggesting that the fiber has an effect on suppressing the seizure of the aluminum alloy with the counterpart. Wide area of the surface was covered with the crumbled fibers when combined with the pitch composite, while the covered area was limited when combined with the PAN composite. Near the worn surfaces, plastic flow was hardly observed for the pitch composite, while a slight plastic flow was observed for the PAN composite, even though the proof stress of the pitch composite was smaller than that of the PAN composite.

These results indicate that the effect of the thermal conductivity is greater than that of the proof stress on the wear properties of the composites. The high thermal conductivity can improve the radiation of heat which was generated by the friction, leading to the suppression of seizure between the aluminum alloy matrix and the counterpart. The decrease in the seizure would lead to a decrease in the wear, and the scatter in the coefficient of friction value. The pitch fibers in the composite would act as a heat radiator near the worn surface during the wear, leading to the suppression of the frictional heat and seizure, and subsequent decrease in the wear loss.

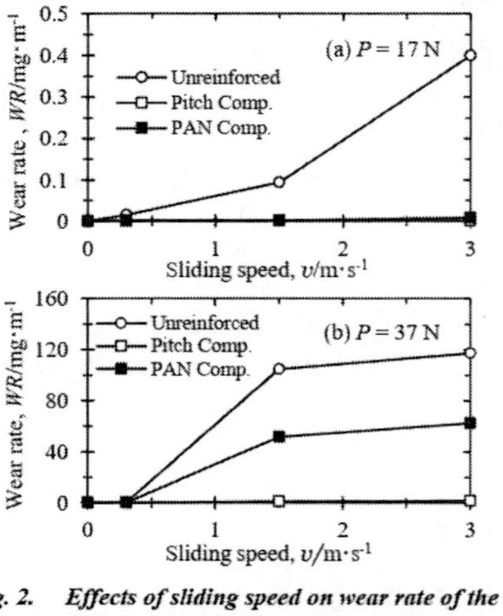

Fig. 2.* *Effects of sliding speed on wear rate of the unreinforced AC8A alloy and composites.

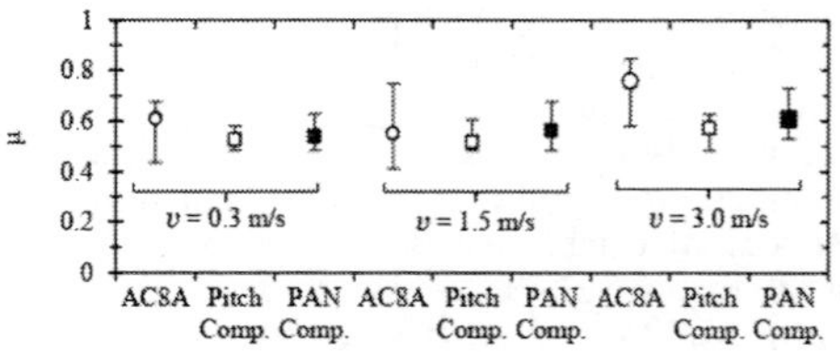

Fig. 3* *Coefficient of friction (μ) of the specimens during the wear test (P=27 N). Error bars in the figure represent the range between the maximum and minimum values.

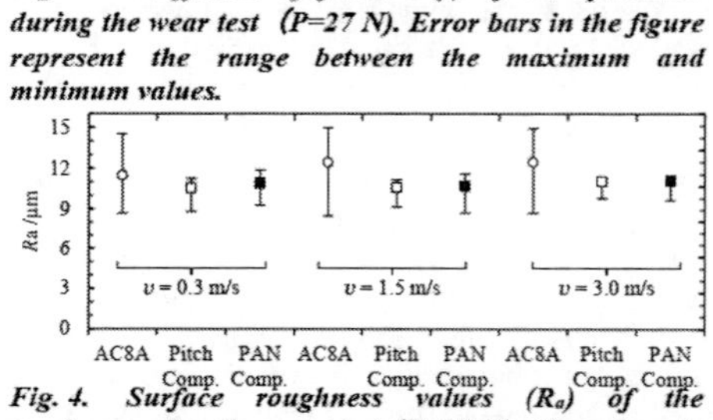

Fig. 4.* *Surface roughness values (R_a) of the specimens after the wear test (P=37 N). Error bars in the figure represent the scatter of the values due to the measured points.

4. Conclusion

The wear loss of the pitch composite was smaller than that of the PAN composite under severe wear conditions. It was mainly attributed to its higher thermal conductivity, rather than its proof stress.

References

[1] C.S. Ramesh, H. Adarsha, S. Pramod & Z. Khan. (2013). Tribological Characteristics of Innovative Al6061-Carbon Fibre Rod Metal Matrix Composites. Mater. Design 50, 597-605.

[2] L. Liu, W. Li, Y. Tang, B. Shen & W. Hu. (2009). Friction and Wear Properties of Short Carbon Fiber Reinforced Aluminium Matrix Composites. Wear 266, 733-738.

The 73rd World Foundry Congress
„Creative Foundry"
23 - 27 September 2018, Krakow, Poland

THE DECREASING MICROSTRUCTURE DEGRADATION OF FIBER BASE FEEDER SLEEVE WITH VARIOUS REFRACTORY COATINGS

Haydar Kahraman[1, 2]*, Ibrahim Hayri Keser [2,3], Ozge Aksın Artok [1], Umit Cöcen [3]

[1]Cukurova Kimya Endüstrisi A.Ş., Organize Sanayi Bölgesi, 50. Yıl Cad. No: 10, Manisa, TURKEY
[2] Dokuz Eylül University, Graduate School of Natural and Applied Science, Tınaztepe Campus Buca, Izmir, TURKEY
[3] Dokuz Eylül University, Faculty of Engineering, Department of Metallutgical and Materials Engineering, Tınaztepe Campus Buca, Izmir, TURKEY
**Corresponding address: h.kahraman@cukurovakimya.com.tr*

Key words: Feeder sleeves, refractory coating, ductile iron, micro structure, chunky graphite

Feeder sleeves are consisted of insulation, exothermic and binder components. Fiber base feeder sleeves are commonly used because of low cost, high yield, too low weight and easy moldability. However, intreaction with molten metal occurs a bit on long time depend of porosity structure. This phenomena can cause contamining same ingeradient to molten metal. Especially for ductile iron alloys, when spherodial graphites are occurred, metalic aluminum in sleeve can cause degradation. So the mechanical properties of casting part can get worse. In this study, experimental design and analysis procedure are determined for explaining this phenomena and inhibiting this degradation. Rich and poor metalic aluminyum included feeder sleeves are prepared. Then some of them are coated with different refractory coating. All sleeves are used with same ductile iron alloy. The metal sample in sleeves are examined with optical and scanning electro microscope for microstructure analysis. The chemical composition change in structure is determine with EDS analysis. Finally, the hardness testing were applied to determine mechanical differences between zones.

1. Introduction

Exothermic feeder sleeves are used for increasing the casting yield especially in more shrinkage shown metal at solidification [1]. Exothermic reaction is occurred between metallic aluminum and metal oxide agent. After pouring, some metallic aluminum particles can be mixed to casting metal [1,5,6].

Ductile iron consists of ferritic, perlitic or austenite main phase and spheroidal graphite. The distribution and dimension of spheroidal graphite effects the mechanical properties of casting.

The graphite has come together as a flake shape during solidification. For a theory if some inoculant agents are used like Mg, Ce, the graphite transverse spheroidal shape, because of oxygen bonding with these agents. So, the surface energy increase and graphite flake get rolling by themselves. Some easily oxidable element like aluminum can prevent these phenomena and spheroidal graphite can be decreased.

2. Experimental

Three types of feeder sleeves were tested, these are aluminum rich sleeve, aluminum poor sleeve and aluminum poor sleeve with refractory coating. EN-GJS-400 ductile iron alloy is used for casting trial. Feeder sleeve shape and sample zone are shown in Figure 1. The sleeves are full with ductile iron alloys that surrounded with green sand. The chemical composition of ductile iron is given in Table 1. The pouring temperature is 1450°C.

All samples are grinded and polished for optical microscope analysis. Microstructure photo taken from different area of sample with Nikon ECLIPSE ME 600 optical microscope. And then the morphology of selected samples is analyzed with Jeol 650 SEM. The EDS analysis is done inner and outer sections of samples.

The hardness test was applied with Brinell Hardness Test Method defined in ASTM E10. The indenter diameter is 5 mm, measurement time is 15 seconds and force is 250 Kgf.

3. Results and Discussion

4.1 Microstructure Analyses with Optical Microscope

As seen in all micro structure images, in generally graphite degradation and carbide phase occur at metal that closes to sleeve. On the other hand, high graphitization and sherardization can be seen at the inner zone (Figs. 2-4).

Microstructure studies show that both graphitization and sherardization are very well at aluminum poor and coated sleeve. The degradation is very few and spheroidal graphite percentage is pretty much at inner coated (Figs. 3,4).

4.2 Surface Morphology With SEM

SEM images are taken from the regions where the graphite structure is degraded. Flake graphite can be seen in SEM images. (Figs. 5,6). The spheroidal graphite structures are completely or partially deteriorated.

Graphite degradation in the structure of poor and rich aluminum sleeves outer metals is clearly seen as in microstructure images (Fig. 5). However, graphite in the microstructure of the coated feeder metal is more spherical (Fig. 5). Therefore, EDS analysis to determine the elements affecting graphite degradation.

4.3 Chemical Analysis with EDS

In order to demonstrate the elemental differences in microstructure, EDS measurements are taken at different regions of the sample. Microstructure images are compared and analyzed from the deteriorated and spherical regions.

Chemical changes is detected by EDS analysis in the areas where construction is deteriorated. According to the results of the analysis, it is seen that the amount of aluminum increased 6.7 times compared to the areas which is deteriorated in the regions where the graphite structure is not deteriorated (Table 1). The increases of aluminum elements in these parts neutralizes the elements that enable the spheroidal of graphite such as cerium (Ce) and magnesium (Mg) [2].

Table 1. EDS result of different region in Figure 7

Region in Figure 7	EDS Result			
Region 1	Component	Mole Conc.	Conc.	
	C	29.549	8.820	wt.%
	Fe	61.725	85.662	wt.%
	Si	4.704	3.283	wt.%
	Ce	0.088	0.308	wt.%
	Al	0.182	0.122	wt.%
	Mg	0.257	0.155	wt.%
		100.000	100.000	wt.%
Region 2	Component	Mole Conc.	Conc.	
	C	29.999	8.910	wt.%
	Fe	62.709	86.596	wt.%
	Si	4.178	2.902	wt.%
	Ce	0.031	0.107	wt.%
	Al	0.027	0.018	wt.%
	Mg	0.240	0.144	wt.%
		100.000	100.000	wt.%

Aufderheide and Showman have found that the amount of aluminum measured from the top of the feeder is higher than the amounts of aluminum measured in the middle and bottom of the feeder. As a result of this, the amount of aluminum increased and graphite formation was observed in spheroidal cast iron microstructure in the microstructure images taken from the top and bottom of the feeder [3].

Lute et al. investigated the microstructure effect of the amount of aluminum in the feeder sleeve composition in ductile iron applications. Lute et al. found that graphite degradation in the microstructure of feeder metal due to difference the weight ratio of aluminum in feeder sleeve composition. As a result, Lute et al. have shown that increasing the amount of aluminum in the direction of the progress of the heat in the casting stage, the formation of graphite in spheroid cast iron structure is degraded [4].

4.3 Hardness Test Result

The carbide phase increases with graphite degradation, so the hardness increases. If the lamellar or spheroidal graphite can be occurred, hardness of the test zone decreases. As shown in Table 2, poor aluminum and coated sample, the graphitization was very well in, has low hardness values. However hardness values are high in rich aluminum sleeve that has degraded graphite.

Table 2. Hardness result of metal of different feeder sleeve

	Hardness(HB)			
Feeder sleeve	Outer	Inner	Center	Avarage
Aluminium rich sleeve	156	138	156	159
	165	129	213	
Aluminum poor sleeve	159	148	141	148
	148	136	159	
Aluminum poor with coated sleeve	124	156	140	144
	123	159	163	
	156	154	167	

4. Conclusion

In this study compared with literature studies results are consistent in terms of aluminum element effect on graphite degradation. At the same time this study show that the refractory coating of the inner part of the feeder sleeve is reduced graphite degradation in spheroidal cast iron.

References

[1] Twardowska, H. & Aufderheide, R.C. (2002). Patent No. US 006360808B1. Dublin Ohio.

[2] Alagarsamy, A. & Aufderheide, R. (2005). Casting Defect: Doughnut Defect. The Ductile Iron News. Issue 12.

[3] Aufderheide, R.C. & Showman R.E. (2001). Casting Problems with New Sleeve Technology. The Ductile Iron News. Issue 3.38-43.

[4] Lute, C.A., Aufderheide R.C. & Showman R.E. (2001). Exothermic Riser Sleeves Can Cause Flake Graphite in Ductile Iron. AFS Casting Congress 105th. Dallas. Vol 109. Paper No 01-086. 1115-1120.

[5] Kahraman, H., Keser, I.H., Tozan, M.K. & Cocen, U. (2015). The Effects of Auxiliary Mold Components on Casting Metals. 13th International Foundry Trade Fair With Technical Forum. 16-20 June. Düsseldorf. Poster Presentation.

[6] Thomas, J.W., Hardin, R.A. & Beckermann, C. (2014). Thermophysical Properties for ASK Chemical and Exochem Riser Sleeves for Steel Castings. Proceedings of the 68th SFSA Technical and Operating Conference, Paper No. 4.6, Steel Founders' Society of America, Chicago, IL.

The 73rd World Foundry Congress
„Creative Foundry"
23 - 27 September 2018, Krakow, Poland

EFFECT OF MOLTEN MATEL TEMPERATURE ON MOLD FILLING IN EVAPORATIVE PATTERN CASTING

Toru Maruyama [1]*, Mitsuyoshi Tamaki[2], Gou Nakamura[2], Keisuke Nakamura[2]

[1] Kansai Univ., Dept. of Chemistry and Mater. Engineering, 3-3-35Yamate-cho, Suita, Japan
[2] Graduate school of Kansai Univ., Dept. of Chemistry and Mater. Engineering, 3-3-35Yamate-cho, Suita, Japan
**Corresponding address: tmaru@kansai-u.ac.jp*

Key words: evaporative pattern casting, mold filling, molten metal temperature, polystyrene foam, in-situ observation

Abstract

Mold filling rate in evaporative pattern casting is measured on wide range of molten metal temperatures from 700°C to 1400°C. A polystyrene columnar foam was used as foamed patterns, and bottom pouring method was applied to a casting design. A tin alloy, copper alloys and a cast iron were used as molten metals for changing liquids temperature. The mold filling rate increases with increasing molten metal temperature in relatively lower temperature. In contrast, the mold filling rate decreases with increasing molten metal temperature in high temperature. These phenomena was verified by in-site observation.

1. Introduction

The evaporative pattern casting (EPC) process is one of a metal casting process by using a polymer foam as a casting pattern [1]. The polymer foam is decomposed thermally by heat transfer from a molten metal, and the molten metal is filled into a cavity formed as a result of the thermal decomposition of the foamed pattern. Polystyrene is conventional foamed material rather than methyl methacrylate polymer. However, polystyrene forms residue of thermal decomposition products during casting, and the residue may cause casting defects [2]. For avoiding casting defects, sufficient decomposition of foamed patterns and the exhaust of thermal decomposition products from cavity to outside are required. It is important that decomposition heat is given to foamed pattern. The mold filling rate is expected to increase with increasing molten metal temperature because sufficient thermal decomposition of the foamed pattern is expected with higher molten metal temperature. In contrast, mold filling rate may be decreased with increasing molten metal temperature because a large amount of thermal decomposition gas can be formed. In this study, mold filling rate was measured on changing molten metal temperature, and for verifying mold filling, in-situ observation was carried out.

2. Experimental Procedures

The casting design of bottom pouring method was applied to avoid turbulence flow of molten metals. A density of the polystyrene foamed pattern was 15.4 kg m^{-3}. Dimension of a columnar pattern was 40 mm in outer diameter and 250 mm in height. Tough sensors for molten metal and gas pressure ports were inserted into the foamed pattern. The sensor points were set at 5mm, 65mm, 125mm, 185mm and 245mm form bottom of the pattern. A Sn-Cu alloy, Cu Alloy and cast iron (CE: 4.3) were cast into the mold from 700°C to 1400°C. The vacuum was not applied in casting.

Similar type of casting design was applied for in-situ observation. Figure 1 shows the casting design for in-situ observation. A density of the foamed pattern was 11.4 kg m^{-3}. A silica glass was set on one side of the foamed pattern. Both top and bottom end of the silica glass was fixed with sand mold as shown in Fig. 1. Grooves of which a direction to length direction of the foamed pattern was vertical were cut into the foamed pattern surface contacted with the silica glass.

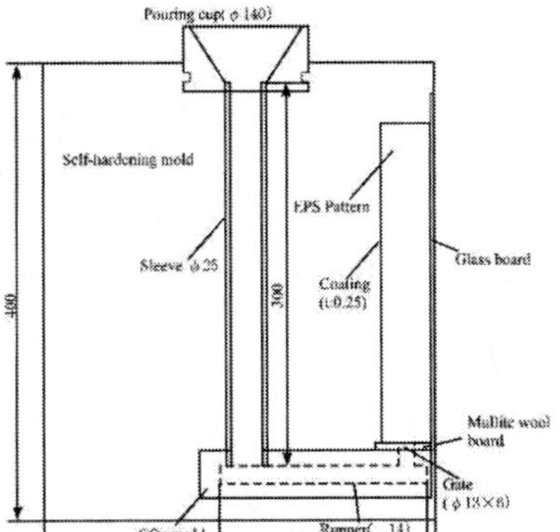

Fig. 1. ***Casting design of EPC for in-situ observation by using a silica glass plate as an observation window***

3. Results and discussion

Figure 2 shows mold filling progress during casting on various molten metal temperatures. Temperatures shown in the Fig. 2 show molten metal temperatures during mold filling measured with thermocouples in a cavity. The plots show locations of tip of molten metal and arrival times at touch sensor point. The slopes of line connected with plots mean mold filling rate. The mold filling rate increased with increasing molten metal temperature when the temperature under 900°C. The mold filling rate in the first half of mold filling at 1000°C was smaller than that at 900°C although the mold filling in the latter half at 1000°C was larger than that at 900°C. In addition, the mold filling rate at 1200°C was slower than that at 600°C. These results shows the mold filling rate is not simply increased with increasing molten metal temperature.

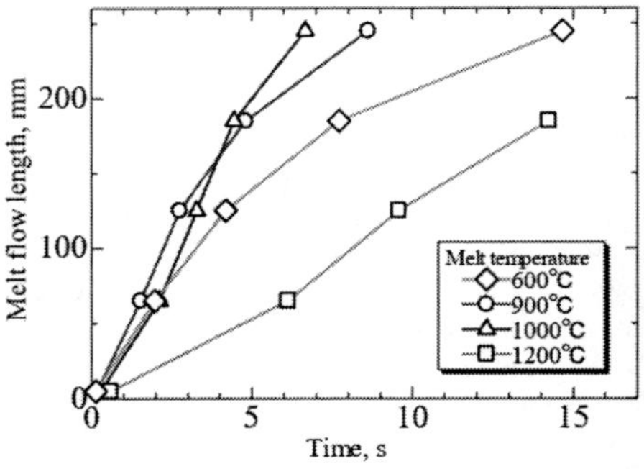

Fig. 2. Mold filling during evaporative pattern casting on various molten metal temperatures

Figure 3 shows influence of the maximum gas pressures in the gas gap on molten metal temperature. The gas pressure decreased with progress of mold filling due to decrement of back pressure. The maximum pressures were almost same even the molten metal temperature was changed. It is considered that the gas pressure in the gas gap depends on the back pressure, and the gas pressure is not main factor to decrease the mold filling rate.

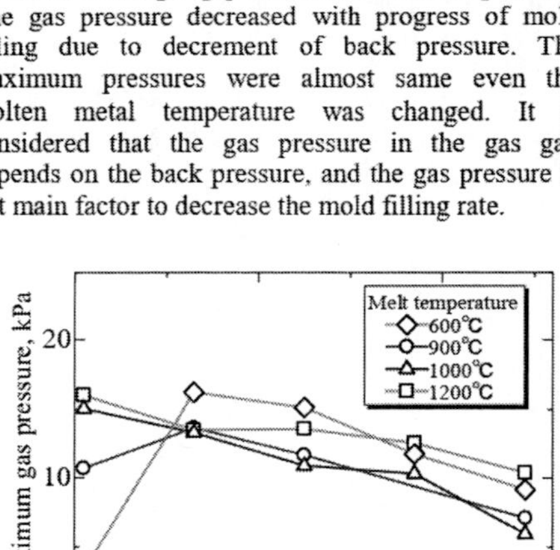

Fig. 3. Influence of maximum gas pressure in gas gap on molten metal temperature

According the result of Fig. 2, it is indicated that the mold filling rate in the initial stage of mold filling at high temperature of molten metals is slow. For verifying the phenomenon, in-situ observation was carried out. Figure 4 shows appearances of the thermal decomposition of the foamed polystyrene pattern and the molten metal flow. From just after casting to 1.2s after casting, the foamed pattern was decomposed and a cavity formed, and the molten metal flowed into the cavity. From 1.2s to 2s, two third of the foamed pattern was decomposed like as black fog, even the molten metal flow was almost stopped in the position at 1.2s. After 2s, the molten metal was started to flow up again. The molten metal was filled into the cavity which was formed by thermal decomposition of the foamed pattern. However, the molten metal was not filled to the top end of the cavity. The result of Fig. 4 supports the result of Fig. 2. It is considered that too high temperature of molten metal causes the decrement of mold filling rate in the initial stage of mold filling.

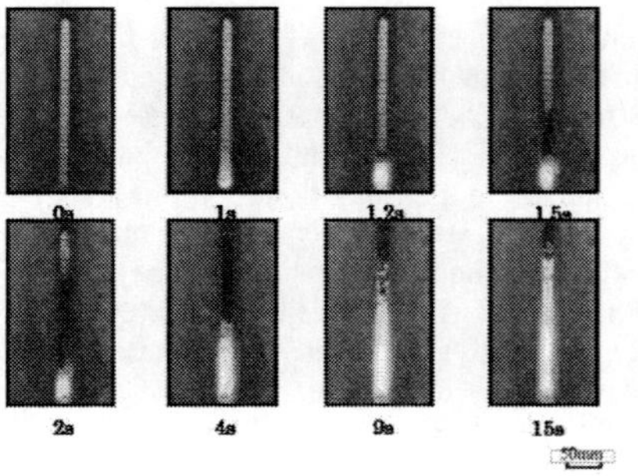

Fig. 4. Appearances of thermal decomposition of formed polystyrene and mold filling

4. Conclusions

The mold filling rates at various molten metal temperatures melt were measured during evaporative pattern casting (polystyrene foamed pattern) and the results was verified by in-situ observation. The following results and conclusions were obtained:

(1) The mold filling rate increases with increasing molten metal temperature under about 900°C. In contrast, the mold filling rate decreases with increasing temperature over 900°C, especially, in the initial stage of mold filling.

(2) The gas pressure in the gas gap decreased with mold filling progress, and the mold filling rate did not depend on the gas pressure without vacuum.

(3) As the result of in-situ observation, the mold filling rate in a period of initial stage of mold filling was very slow even the polystyrene foam was decomposed continuously.

References

[1] F. Sonnenberg (1992), Lost Foam Casting Made Simple, Schaumburg, American Foundry Society.

[2] Kobayashi T. and Kasuya Y. (1992), Behavior of Produced Gas and Mist with Thermal Decomposition in the Evaporative Pattern, IMONO, 64, 318-324

The 73rd World Foundry Congress
„Creative Foundry"
23 - 27 September 2018, Krakow, Poland

INFLUENCE OF HEAT TREATMENT ON THE MICROSTRUCTURE AND CORROSION RESISTANCE OF AUSTEMPERED DUCTILE IRON

Halina Krawiec[1], Vincent Vignal[2]*, Janusz Lelito[1], Edward Tyrała[1]

[1] *AGH-University of Science and Technology, Faculty of Foundry Engineering, Reymonta 23 Street., 30-059 Krakow, Poland*
[2] *ICB, UMR 6303 CNRS - Université Bourgogne - Franche Comté, BP 47870, 21078 Dijon Cedex, France*
**Corresponding address: vincent.vignal@u-bourgogne.fr*

Key words: Austempered Ductile Iron (ADI), microstructure, heat treatment, corrosion

1. Introduction

Austempered ductile iron (ADI) because of combination of high strength and ductility, good wear resistance and high fatigue strength is widely used in automotive, agriculture, energy and other industries [1-4]. The excellent mechanical properties of ADI are the results of microstructure called "ausferrite" that consists of ferrite (α) and high carbon austenite (γ_{HC}). Moreover, ADI has graphite spheres. Some oxide inclusions can also be found. Such complex microstructure has significant influence on the mechanical properties and corrosion resistance of ADI. In order to obtain "ausferrite" microstructure, the austempering heat treatment must be performed. Fig.1 shows a schematic diagram of the austempering process. The heat treatment consists of three steps: 1) ductile iron (DI) is heated to a high temperature (about 900°C - matrix is fully austenitic), 2) the metal is cooled down to a temperature in the range of 250-450°C and held at that temperature for the required duration, 3) the casting is air cooled to room temperature. During austempering, two reactions proceed:

1) austenite (γ) decomposes into ferrite (α) and high carbon austenite (γ_{HC})

2) if the casting is held at the austempering temperature too long decomposition of the high carbon austenite into ferrite and ε-carbide takes place.

The heat treatment parameters such as austenitizing and austempering times and temperatures have significant influence on the evolution of ADI's microstructure and mechanical properties [3, 5, 6].

The microstructure of cast iron has essential influence on its corrosion resistance. Gray iron and cast iron undergo graphitic corrosion which causes the selective leaching of iron from the graphite network. This type of corrosion takes place because graphite is cathodic with respect to iron (matrix). Therefore, the galvanic cell graphite/iron is formed. It can be mentioned that ductile iron is less prone to graphitic corrosion than the grey iron [7].

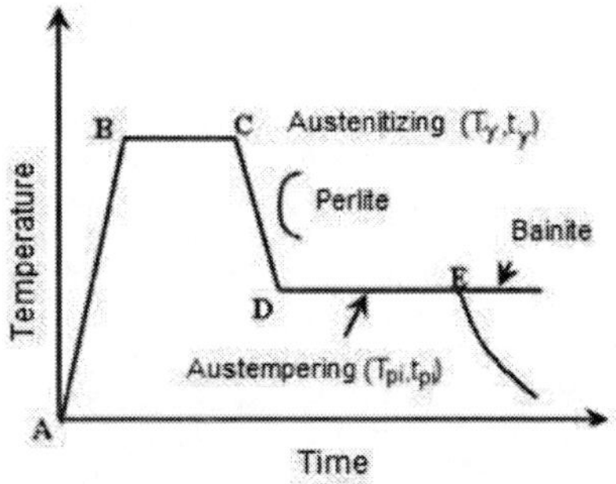

Fig. 1. Schematic diagram of the austempering process

The previous studies have revealed that the ADI's specimens with upper ausferritic microstructure are less prone to corrosion in sodium chloride solution [8]. Such behaviour is related with greater coarsening of the austenite grains and broad ferrite needles. In this case there is less austenite/ferrite grain boundaries. The interfaces are very often weak places where corrosion can start. It has been revealed that corrosion of ADI starts at graphite/matrix and inclusions/matrix interfaces [9]. Moreover, the chemical composition of ADI has influence on its corrosion behaviour. It was revealed, that the presence of silicon hinders anodic dissolution of ADI [8, 10].

2. Material and experimental

The chemical composition of ductile iron used in these experiments was C: 3.48 wt.%, Mn: 0.63%, Si: 2.58%, P: 0.03%, S: 0.017%, Cr: 0.06%, Ni: 0.04%, Cu: 0.32%, Mg: 0.054%, Mo: 0.07%, Ti: 0.015%, Sn: 0.006%, Pb: 0.001%, V: 0.009%, W: 0.038%, Zn: 0.003. The austempered holding temperature of 280°C was selected. The heat treating procedures were as follows: (1) austenitizing at 900°C for 2 hours, (2) austempering at 280°C for 10 minutes (sample 1.1) and for 150 minutes (sample 7.1), (3) water cooling down to room temperature of 25°C. The heat treatment was performed in a salt bath furnace. Before corrosion tests, ADI specimens were mechanically ground with silicon carbide (SiC) emery papers down to 4000 grit, polished with diamond pastes down to 1 μm and then ultrasonically rinsed in ethanol. In order to reveal the microstructure, both specimens were chemically etched in Nital solution.

Electrochemical corrosion tests were performed by means of the Potentiostat Autolab PGSTAT302N. Polarization curves (LSV), open circuit potential (OCP) vs. time plots and electrochemical impedance spectroscopy diagrams (EIS) were determined. Moreover, corrosion tests were also performed in a salt fog chamber.

3. Results

Fig. 2. shows the microstructure of ADI austempering at 280°C for 10 and 150 minutes. These FE-SEM images show the morphology of ferrite plates in the ADI cast iron structure as a function of the holding time. Notable are the ferrite plate clusters in the form of packages. Both the amount and thickness of the ferrite plates increase with the holding time at the isothermal transformation temperature. FE-SEM images also allow showing preferential places for ferrite nucleation, which are areas around graphite balls.

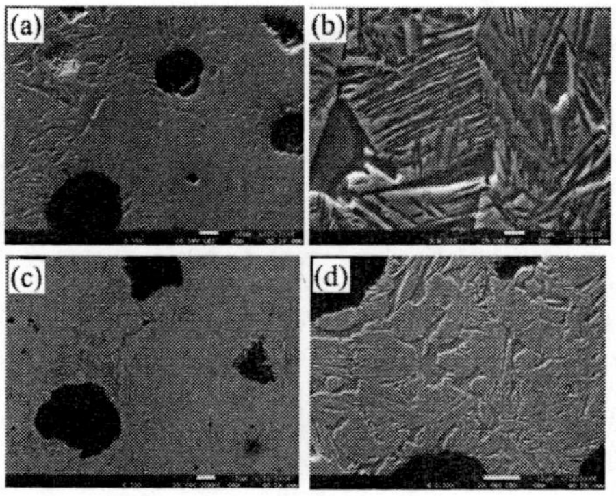

Fig. 2. FE-SEM images after etching of (a-b) sample 1.1 and (c-d) sample 7.1

The corrosion behavior of these samples was also studied by means of LSV, OCP and EIS measurements in NaCl based media (neutral pH). The influence of the microstructure was discussed and the nature of weak sites (graphite / matrix interfaces) was identified.

The corrosion resistance of these two samples was also investigated using salt spray and condensation water (relative humidity of 90%) tests at 35°C for 7 days. A NaCl based solution was used to generate the atmosphere. These two types of experiments were performed in a salt spray corrosion test chamber Ascott S120iP. Results were discussed considering the microstructure of specimens.

References

[1] Cheng-Hsun Hsu, Jung-Kai, Rung-Jie Tsai, Material Science and Engineering A 398, (2005), 282-290.
[2] A.R. Ghaderi, M. Nili Ahmadabadi, H.M. Ghsemi, Wear 255, (2003), 410-416.
[3] P. Prasad Rao, Susil K. Putatunda, Material Science and Engineering A 349, (2003), 136-149.
[4] Jianghuai Yang, Susil K. Putatunda, Material Science and Engineering A 406, (2005), 217-228.
[5] Olivera Eric, Milan Jovanovic, Leposava Sidanin, Dragan Rajnovic, Slavica Zec, Material and Design 27, (2006), 617-622.
[6] J. Zimba, D.J. Simbi, E. Navara, Cement & Concrete Composites 25, (2003), 643-649.
[7] Corrosion understanding the basics, ed. by J.R. Davis ASTM International 2000, 162-164.
[8] H. Krawiec, J. Lelito, E. Tyrała, J. Banaś, Journal of Solid State Electrochemistry, 13 (2009), 935-942
[9] H. Krawiec, V. Vignal, J. Banaś, Journal of The Electrochemical Society, 153 (2006) 231-237
[10] H. Krawiec, B. Stypuła, J. Stoch, M. Mikołajczyk, Corrosion Science, 48 (2006), 595-607

The 73rd World Foundry Congress
„Creative Foundry"
23 - 27 September 2018, Krakow, Poland

MECHANICAL PROPERTIES OF PHENOLIC URETHANIC SAND AFTER THERMAL REGENERATION AT 930°C

Tulio Nascimento[2]*, Willian Pasini[1], Leonardo Pereira[1], Matheus Bellé[1], Regis Amaral[1] Vinicius Karlinski [1], Cleber Lessa[2]

[1]*Federal University of Rio Grande do Sul, School of Engineering, Foundry Laboratory, Bento Gonçalves Ave. 9500, Postal code: 15.021, Porto Alegre, Brazil*
[2]*Federal Institute of Science, Technoloy and Education of Rio Grande do Sul, Avelino Antônio de Souza Street 1730 Caxias do Sul. Brazil*
**Corresponding address: tulio.nascimento@caxias.ifrs.edu.br*

Key words: Sand; phenolic urethane resin; thermal regeneration, mechanical properties, grain size

1. Introduction

The foundry industry generates great volumes of solid residue, such as mould sand and general dust [1]. Mould sand with resin addition difficults its regeneration and reuse, generating a great amount of disposal to industrial landfills, increasing the production cost. Therefore, it is sought to regenerate the used sand, assembling again its required properties to mould confection, thereby avoiding the use of fresh (new) sand, reducing environmental damage.

The regeneration of foundry sand implies on a superficial cleaning of its grains and removal of the resulting particles, aiming to redirect the used sand with a similar condition of fresh sand, allowing its reuse on the process without detriment of mould or core quality. To perform the grain cleaning step, the following treatments have been employed more frequently: mechanical, chemical, heat and humid treatments [2,3].

The performed researches indicate thermal regeneration of moulding sands system provides regenerated sands of good quality, and therefore may completely replace mechanical regeneration. This simplifies the technological line of operations related to the regeneration process. [4]

In the present work it was aimed the thermal regeneration of a phenolic urethane sand at 930°C after 5 cycles of regeneration followed by preparation and moulding, evaluating the following properties: tensile, shear and flection strength, besides the acid demand value (ADV) and grain morphological characterization.

2. Materials and Methods

The specimens were prepared in a sand mixer using 8kg of fresh silica sand and resin (Part 1: phenol formaldehyde 0.5%, Part 2: methyl-diphenyl-isocyanate 0.5%, amine catalyst 0.05%). The fresh sand presented the following qualification results of raw material according to Table 1.

Table 1. Silica-based sand properties.

Humidity	0.01%
Module	48.13 AFS
Fines	1.16%
AFS Clay	0%

The tensile tests followed the ABIFA CEMP 162, the flection tests followed the ABIFA CEMP 181, the shear tests in accordance with a laboratory specific method, using the universal test machine and specimens with dimensions of 50.8 mm diameter and 10.0 mm thick; the acid demand value was analyzed following the ABIFA CEMP 101 and grain morphological analysis was performed by optical microscopy.

For each cycle (sand mix, moulding and mechanical tests), fragmentation of sand lumps was performed during 30 minutes in a sand mixer and, afterwards, calcination was performed in an electric resistance furnace at 930°C for 3 hours, at last, the sand was sieved in a #140 mesh.

3. Results and Discussion

The tests results indicate a decreasing trend on the tensile and shear strength values after the third thermal regeneration cycle and the flection strength after the fourth cycle (Fig. 1).

The behavior of the tensile strength and the shear strength decreased during the thermal regeneration cycles at 930°C. This effect occurs due to the polymeric film disruption which reduces the grain cohesion. In a real production condition, the effect mentioned above causes mould breaking, which generates sand inclusion defects.

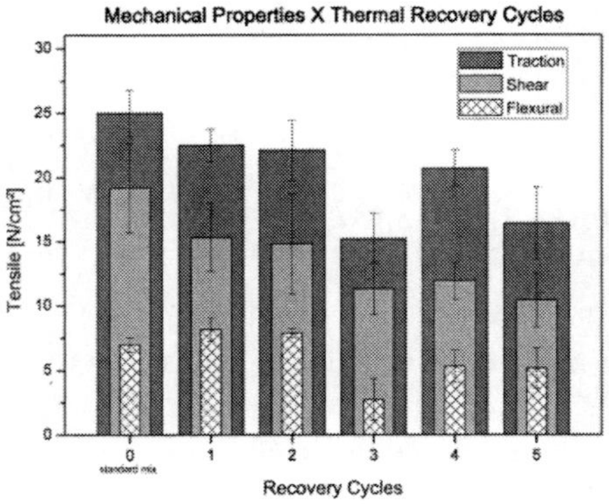

Fig. 1. Graphic of the tests after each step of calcination.

Cycle 3 showed an inadequate result for the evaluated properties due to the non-withdrawal of the fines on the #140 mesh.

The acid demand value indicates a decrease on the acidity due to basic contaminants remaining on the sand grains after each regeneration cycle. The sand grain presented sub angular characteristics and after the regeneration there was a grain breakage, however maintaining the sub angular morphology.

The probable causes for the persistent decrease may be connected to the alteration on the acid demand value of the sand, from neutral or close to neutral to basic, which would result in a greater reactivity of the system, producing more friable specimens (Table 2).

Table 2. Acid Demand Value (ADV).

Cycles	ADV (ml)
Standard Mix	48.7
Cycle 1	49.1
Cycle 2	49.0
Cycle 3	44.6
Cycle 4	44.5
Cycle 5	44.1

The breakage of sub angular grains would lead to an increase on the resin content, diverging from the research scope, which is the use of a same proportion of binders through the cycles. The presence of some basic contaminants is responsible for the slight decrease on the mechanical properties. Fig. 2 clearly shows the presence of contaminants after 5 cycles of regeneration on the left side of the picture, different from the right side composed by fresh sand.

Fig. 2. Sand grain morphology comparing the sand after 5 cycles and the fresh sand.

4. Conclusion

Thermal regeneration, on static condition, was an efficient regeneration process for foundry sand. The results showed a decrease on the sand mechanical properties after each cycle of regeneration. However, these results still allow the fabrication of moulds and cores for cast production without casting defects.

Acknowledgments

The authors acknowledge financial support provided by CNPq (The Brazilian Research Council) and FAPERGS (The Scientific Research Foundation of the State of Rio Grande do Sul).

References

[1] Dantas, J.M. (2003) Montagem, Comissionamento e Operação de um Sistema de Recuperação de Areia de Fundição: Regenerador Térmico. Plano de Trabalho da Fase II - Instituto de Pesquisas Energéticas e Nucleares – São Paulo.

[2]Comissão De Meio Ambiente Da Abifa. (1999). Manual de Regeneração e Reuso de Areias de Fundição. Associação Brasileira de Fundição. Revista ABIFA (p.1-49). Sao Paulo: ABIFA.

[3] Mariotto, C.L. (2000). Regeneração de Areias: Uma Tentativa de Discussão Sistemática. Revista Fundição & Matérias-Primas. Caderno Técnico, n°.42, v.33, p.A - T.

[4] Łucarz M. (2015). Thermal reclamation of used moulding sands. Metalurgija. v.54 (1), 109-112.

The 73rd World Foundry Congress
„Creative Foundry"
23 - 27 September 2018, Krakow, Poland

DEVELOPMENT OF NUMERICAL MODEL FOR CORE GAS GENERATION BY THE REACTION WITH BINDER IN SAND CORES AND TRANSPORT IN MOLTEN METAL

Sung-Bin Kim*, Dong-Hoon Roh and Eung-Su Kweon

AnyCasting Software Co., Ltd.,
583 Yangchun-ro, Gangseo-gu, Seoul, South Korea
**Corresponding address: ksb@anycasting.com*

Key words: core gas, porous media, decomposition rate of binder, transport of gas in melt, core gas defect

Abstract

A numerical model for prediction of core gas in aluminum and iron castings has been developed. It describes the movement of gas in the porous core using the Darcy's equation, and the gas concentration in core due to a binder pyrolysis using the mass conservation equation. The decomposition rate of solid binders was described using the Arrhenius relationship, and the movement of gas in melt has been estimated using the Basset-Boussinesq-Oseen equation based upon the Eulerian formulation. The model was applied to the prediction of core gas defects in a gray iron casting. Comparison between the model predictions and experimental results indicated that the model can be applied not only to the prediction of core gas defects, but also to the gating design improvement including the vents and runners for the controlling of core gas defects.

1. Introduction

Once the sand core is covered by liquid metal, the escape of the core gases is limited to the area of the core prints, furthermore, the rate of heating of the core is often greater than that of the mold because it is usually surrounded on all sides by hot metal, and the volume of the core is much less. All these factors contribute to the internal pressure within the core rising rapidly to high values [1]. High local gas pressures produced by pyrolysis of organic materials and the evolution of water as steam drive the gases into the molten metal. This gas produces blow defects and allows hydrogen absorption which causes pinholes when the hydrogen is rejected during solidification. The gas evolved is usually entrapped at or near the cope surface and under cores [2]. If the gas pressure in the core exceeds the metal head pressure, gas will bubble from the core into the metal, and during this time, gas bubbles through the metal to produce dross, oxide folds, and slag defects [3]. Several numerical approaches have been tried for the bubbling and floating-up of core gas which are simulated under the assumption that if core surface element pressure is higher than melt pressure acting on the element, gas bubbles are formed, followed by floating up with the buoyancy force [4].

2. Core Gas Model Description

2.1 Mathematical formulation

The movement of gas in the porous core can be understood from the Darcy's equation as Eq. 1, where $\vec{V}_g$ is the velocity of core gas, k is the instrinsic permeability of sand core, μ_g and P_g are the viscosity and pressure of core gas, respectively.

$$\vec{V}_g = -\frac{k}{\mu_g}\nabla P_g \tag{1}$$

The core gas concentration is calculated from the mass conservation equation as Eq. 2, where c_g and c_b are the concentrations of core gas and binder, respectively, and ε is the porosity of core [5].

$$\varepsilon\frac{\partial c_g}{\partial t} = -\nabla\cdot\left(c_g\vec{V}_g\right) - (1-\varepsilon)\frac{dc_b}{dt} \tag{2}$$

The decomposition rate of solid binder can be described using the Arrhenius relationship by the pyrolysis study [6] as Eq. 3, where A is the empirical reaction rate constant, E_b is a component binding energy, R is the universal gas constant, and T_c is the temperature of core.

$$\frac{dc_b}{dt} = -A\cdot\exp\left(-\frac{E_b}{RT_c}\right)c_b \tag{3}$$

The movement of gas in melt is calculated from the Basset-Boussinesq-Oseen equation in the Eulerian formulation assuming the shape of gas is sphere in melt as Eq. 4, where the subscripts g and f stand for gas and local melt flow, V, ρ, D and μ are velocity, density, diffusivity and viscosity, respectively, and $\vec{g}$ is the gravity force[7].

$$\vec{V}_g = \vec{V}_f + \tau\left(\vec{g} - \frac{d\vec{V}_f}{dt}\right), \tau = \frac{(\rho_p - \rho_f)D_g^2}{18\mu_f} \tag{4}$$

2.2 Boundary conditions around core

Melt/core interface pressure gradient: The core gas is allowed to pass out of the core when its pressure is greater than that of the melt, but no melt is allowed to enter the core as shown in Fig. 1 [8].

$$\nabla P_g^{surface} = \begin{cases} 0 & for\ P_g \le P_f \\ P_f - P_g & for\ P_g > P_f \end{cases} \tag{5}$$

Air/core interface pressure gradient: The gas can flow across the boundary of cavity and vent depending on the pressure gradient.

$$\nabla P_g^{air} = P_{air} - P_g \leq 0 \qquad (6)$$

Metal mold/core interface pressure gradient: The gas does not flow into the metal mold.

$$\nabla P_g^{m\_mold} = 0 \qquad (7)$$

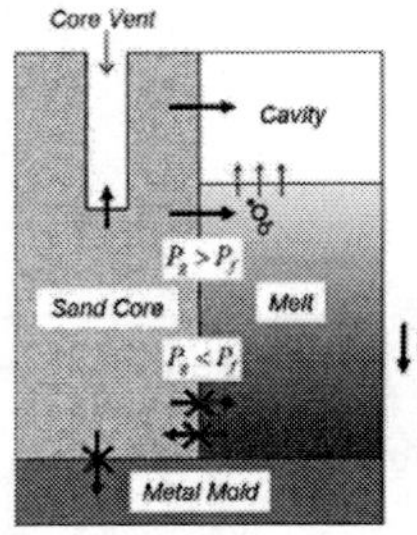

Fig. 1. Boundary conditions around sand core

3. Model Verification

The model predictions were evaluated for actual applications of ferrous alloy, such as bed plate castings made of gray cast iron. The solid models of the casting and two different gating system designs are presented in Fig. 2. The experimental casting conditions were as follows: material gray cast iron GC250 which is similar to ISO250 and FC25, green sand mold, shell sand core, core and mold temperature 40°C, pouring temperature 1380°C.

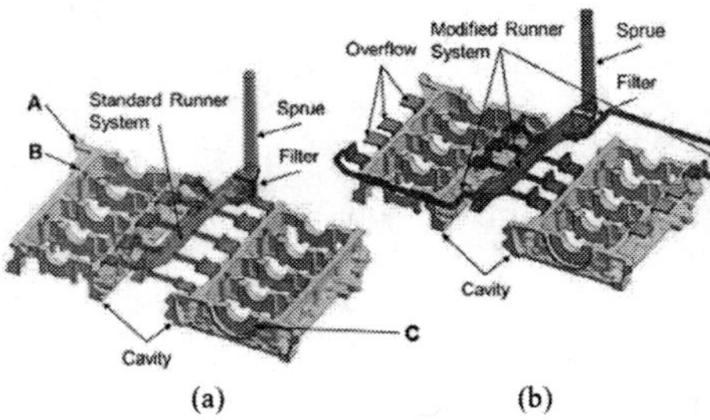

Fig. 2. Improvement of gating design for decreasing core gas defects

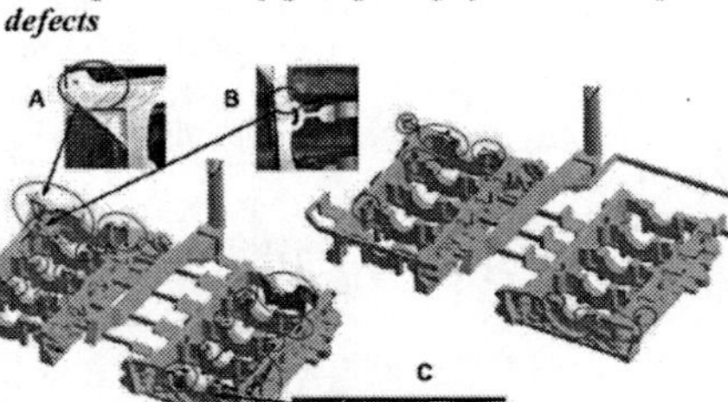

Fig. 3. Verification of core gas model with the casting experiments and the control of core gas defects

A comparison between the predictions of the core gas model with the results of core gas defects in the experimental castings is displayed in Fig. 3 (a) as the photo of A, B and C. It is seen that the model predicts core gas defects of these areas are similar to those of the experimental results. In order to reduce these defects, the gating design was improved based on the core gas predictions as shown in Fig. 2 (b), such as adding three overflows to each cavity, extending the length of one runner and changing its in-gate position to control the defects at the end of product as in C area of Fig. 3 (a). Although the core gas defects still remain, the defects in this gating design are much smaller than Fig. 3 (a) as shown in Fig. 3 (b). When the amount of core gas generated over time is observed, two peaks appear; The first peak is due to the rapid rise of the surface temperature of core, while the central temperature is not yet high. The second peak is due to the fact that as the temperature in the center of core increases over time, a large amount of gas is generated and then it is gradually decreased.

4. Summary

A numerical model for predicting core gas defects in aluminum and iron castings based on the binder pyrolysis within sand cores and the transport of core gas in melt during and after filling of mold. The model is intended to assist aluminum and iron foundries in minimizing gas defects in castings by accurately simulating the formation of core gas related defects.

Acknowledgments

This work was supported by the Institute for Information & Communications Technology Promotion (IITP) through a grant funded by the Korea government (MSIT) (No. 2016-0-00072, Development of Integrated Design Software and Optimization of Manufacturing Process in Automotive and Transport Equipment).

References

[1] Campbell, J. (2015) Complete Casting Handbook (2nd ed.), pp.564-569, Butterworth-Heinemann.

[2] Winardi, L. Littleton, H.E. and Bates, C.E. (2005) New Technique for Measuring Permeability of Cores Made from Various Sand, Binders, Additives, and Coatings, Trans. AFS, vol. 113, pp. 393-406.

[3] Foundry Management & Technology (May 2007) Core and Mold Gas Evolution, p.17.

[4] Ohnaka, I., Zhu, J., Sako, N., et al. (2010) What should we do in computer simulation of casting ?, MCSP-VIII, pp. 3-8.

[5] Starobin, A., Hirt, C.W. & Goettsch, D. (2009) A Model for Binder Gas Generation and Transport in Sand Cores and Molds, MCWASP-XII, pp.345-352, TMS.

[6] Font, R., Fullana, A., Caballero, J.A., et al. (2001) Pyrolysis Study of Polyurethane, J. Anal. Appl. Pyrolysis, vo.58, pp.63-77.

[7] Hryb, D., Cardozo, M., Ferro, S., et al. (2009) Particle Transport in Turbulent Flow using both Lagrangian and Eulerian Formulations, Int. Cummun. Heat & Mass Transfer, vol.36, pp.451-457.

[8] AnyCasting (2017) User Manual, AnyCasting Software

The 73rd World Foundry Congress
„Creative Foundry"
23 - 27 September 2018, Krakow, Poland

GRAIN BOUNDARY WETTING IN THE TUNGSTEN–NICKEL ALLOYS

Boris Straumal[1-3]*, Andrey Mazilkin[1,3], Svetlana Protasova[1], Alexandr Straumal[1,2], Brigitte Baretzky[3]

[1] *Institute of Solid State Physics RAS, Ac. Ossipyan str. 2, 142432 Chernogolovka, Russia*
[2] *National University of Science and Technology «MISIS», Leninskii prosp. 4, 119049 Moscow, Russia*
[3] *Karlsruhe Institute of Technology (KIT), Institute of Nanotechnology, Hermann-von-Helmholtz-Platz 1, 76344 Eggenstein-Leopoldshafen, Germany*
**Corresponding address: straumal@issp.ac.ru*

Key words: grain boundaries, wetting, tungsten alloys, nickel, phase transitions

Pure tungsten is expensive and difficult to machine and manufacture. Therefore, the heavy tungsten-based alloys were developed which maintain some of the useful characteristics of tungsten, such as density and X-ray shielding capabilities, but which would be easier to machine and less expensive. The distribution of Ni, Fe or Cu-rich binder in the W/W grain boundaries (GBs) is formed during the liquid-phase or activated sintering as well as post-sintering processing and is governed by the GB wetting phenomena. One distinguishes the complete and incomplete (or partial) GB wetting. In first case the liquid phase (in W-alloys it is Ni, Fe or Cu-rich melt) forms the zero contact (dihedral) angle $\theta=0$ with a W/W GB. As a result, the melt has to completely substitute a W/W GB, it forms rather thick layers and separate tungsten grain from each other. If the specific area of GBs is high and melt is in a deficit, the liquid phase cannot form thick films between all grains and form the few-nm thin "pancackes" instead (or so-called intergranular films – IGFs).

In case of incomplete (or partial) GB wetting, the liquid phase forms the non-zero contact (dihedral) angle $\theta>0$ with a W/W GB. In this case the GBs between lenticular liquid particles or GB triple junctions (TJs) filled by melt remain "dry". If to change temperature, pressure etc, one can observe the transition between complete and partial GB wetting. In the majority of systems, the direct transition between complete and partial GB wetting takes place and respective, additional tie-lines appear in the two-phase "solid solution+melt" fields of the bulk phase diagrams. The wetting tie-line should continue in a single-phase area of a phase diagram as a line of so-called prewetting phase transition.

However, in certain conditions the so-called pseudopartial (or pseudoincomplete, or frustrated complete) GB wetting appears in a phase diagram between complete and partial wetting. In case of pseudopartial GB wetting, the thin GB layer can coexist with large droplets (or particles) of the wetting phase with a non-zero dihedral (contact) angle $\theta>0$. Thus, such IGFs can be observed in the two-phase (or multiphase) fields of bulk phase diagrams, in the broad intervals of concentration, temperature and/or pressure. The IGFs driven by the pseudopartial GB wetting can drastically modify the properties of polycrystals, in particular those of technologically important Fe–Nd–B-based hard magnetic alloys, W–C–Co cemented carbides and Al-based light alloys. The W–Ni alloys also can contain the few nm thin Ni-rich IGFs.

Fig. 1a shows the microstructure of W–Ni alloy annealed at 1050°C. Two GB triple junctions and five GBs are visible. In TJ2 (left-hand side) all three GBs form the non-zero dihedral (contact) angles with the channel of Ni-rich phase in TJ (namely, θ = 81°, 57° and 34°). In TJ1 (right-hand side) the GB1 forms the non-zero contact angle θ = 39° and other two GBs form zero contact angle with the channel of Ni-rich phase in TJ1. Arrows in Fig. 1a indicate direction across the GBs for which the elemental profiles have been obtained. Result of the EDS elemental profiling is shown in Fig 1b (GB1) and Fig. 1c (GB2). According to these plots, GB layer of (Ni) phase is observed. The thickness of this layer is ~12 nm for GB1 (θ = 0° in TJ1) and ~3 nm for GB2 (θ = 39° in TJ1 and θ = 57° in TJ2).

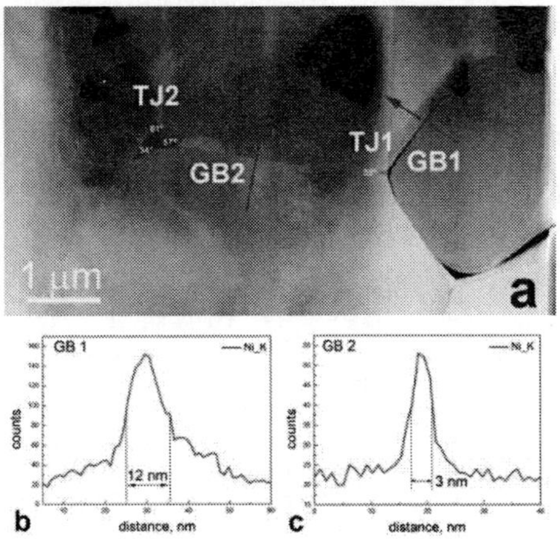

Fig. 1. STEM HAADF image of the W–Ni alloy annealed at 1050°C. Arrows denote the directions of the elemental profiles (b and c) across the grain boundaries.

HRTEM image of GB2 also reveals the presence of thin (Ni) phase layer along GB2 (Fig. 2). It has a uniform thickness of ~3 nm along all the length of GB2.

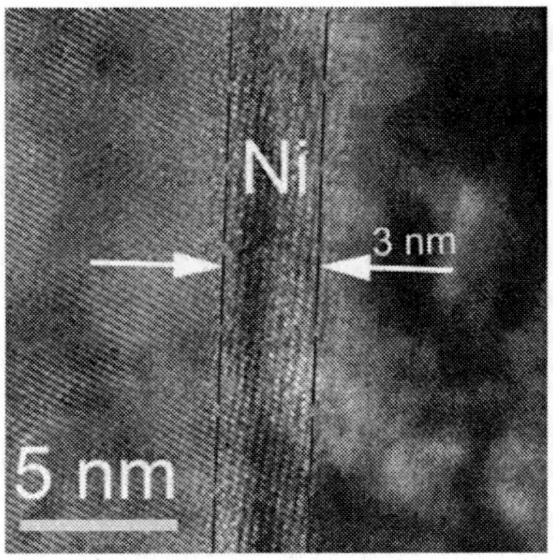

Fig. 2. HRTEM image of the GB2 (see Fig. 1a); 3 nm thick Ni layer is visible between two tungsten grains.

In order to understand the special features of observed W–Ni microstructure, let us consider the droplet of a liquid phase between two solid grains in a partially melted two- or multicomponent polycrystal. If a liquid droplet partially wets (PW) the boundary between two solid grains, then $\sigma_{gb} = 2\,\sigma_{sl}\cos\theta$, where σ_{gb} is the free energy of a grain boundary, σ_{sl} is the free energy of a solid/liquid interface, and θ is the contact (dihedral) angle. The GB between liquid droplets (or liquid channels in TJ) remains dry and contains only the adsorbed atoms with coverage below one monolayer. In this case the GB can exist in the equilibrium contact with the liquid phase (left-hand side GBs in TJ2 in Fig. 1a). In case of complete wetting (CW), $\sigma_{gb} > 2\,\sigma_{sl}$, the contact angle $\theta = 0°$, and liquid spreads between grains. In this case the GB is completely substituted by the liquid phase (GB1 in Fig. 1a). The transition from incomplete to complete GB wetting proceeds at a certain T_w if the energy of two solid-liquid interfaces $2\,\sigma_{SL}$ becomes lower than the GB energy $\sigma_{GB} > 2\,\sigma_{SL}$. The GB wetting temperatures T_w, depend both on GB energy and solid-liquid interfacial energy which, in turn, depend on the crystallography of these interfaces.

In case of complete wetting the liquid spreads between grains. What happens, if the amount of liquid is small and GB area is large? In this case the liquid spreads until two abutting solid grains begin to interact through the liquid layer. The liquid forms a kind of "pancake" having a thickness e_s of ~ 2-5 nm. $e_s = (A/4\pi S)^{1/2}$, where $S = \sigma_{gb} - 2\sigma_{sl}$ being the spreading coefficient on a strictly "dry" GB and A is the Hamaker constant. In case of complete wetting $A > 0$ and $S > 0$. Such "pancake" between the grains is formed by the deficit of a wetting phase, i.e. in the α+L two-phase area of a phase diagram, but very close to the solidus line. The thin "pancake" is very similar to that observed in our W/W GB and presented in Fig. 2. However, in our case the amount of Ni-rich binder is high and one cannot speak about deficit of the liquid phase. What is the reason for the coexistence of thin Ni-rich GB layer between tungsten grains and high contact (dihedral) angle θ?

In the majority of cases the direct transition occurs from partial into complete wetting, for example by increasing temperature. However, in some cases the state of pseudopartial wetting (PPW) occurs between partial (PW) and complete (CW) wetting. In this case the contact (dihedral) angle $\theta > 0$, the liquid droplet does not spread between the grains, but a thin (few nm) precursor film forms around the droplet and separates both abutting grains. Such a precursor film is very similar for the liquid "pancake" in case of complete wetting and deficit of the liquid phase. The case with precursor is known as "pseudopartial (or pseudoincomplete, or frustrated complete) wetting" and becomes possible when $A < 0$ and $S > 0$. In case of pseudoincomplete wetting the precursor film coexists together with liquid droplets, and in case of complete wetting the droplets disappear forming the "pancake".

In a classic case, wetting or prewetting IGFs only exist close to the solidus line of a bulk phase diagram. On the contrary, the IGFs connected with pseudopartial GB wetting can be observed in a broader range of phase fields in binary and muticomponent phase diagrams, including two- and multiphase regions, coexisting with the pseudopartial-wetting phases that form non-zero contact (dihedral) contacts where they meet the pseudopartial-wetting complexions. Therefore, such IGFs can be used as a fine instrument for the so-called grain boundary engineering. As we saw, the thin IGFs can improve the magnetic properties (Fe–Nd–B-based hard magnetic alloys) [1], fracture toughness and strength (W–C–Co cemented carbides) [2] or formability (Al-based light alloys) [3, 4]. We observed in this work that such IGFs, driven by the pseudoincomplete GB wetting, exist also in the W–Ni alloys. They can explain, for example, the peculiarities of GB diffusion in W-based alloys.

Acknowledgments

Authors thank the Russian Foundation for Basic Research for the partial financial support.

References

[1] B.B. Straumal, A.A. Mazilkin, S.G. Protasova, G. Schütz, A.B. Straumal, B. Baretzky. (2016) Observation of pseudopartial grain boundary wetting in the NdFeB-based alloy. J. Mater. Eng. Perform. 25, 3303–3309.

[2] B.B. Straumal, I. Konyashin, B. Ries, A.B. Straumal, A.A. Mazilkin, K.I. Kolesnikova, A.M. Gusak, B. Baretzky. (2015) Pseudopartial wetting of WC/WC grain boundaries in cemented carbides. Mater. Lett. 147, 105–108.

[3] B.B. Straumal, X. Sauvage, B. Baretzky, A.A. Mazilkin, R.Z. Valiev. (2014) Grain boundary films in Al–Zn alloys after high pressure torsion. Scripta Mater. 70, 59–62.

[4] B.B. Straumal, A.A. Mazilkin, X. Sauvage, R.Z. Valiev, A.B. Straumal, A.M. Gusak. (2015) Pseudopartial wetting of grain boundaries in severely deformed Al–Zn alloys. Russ. J. Non-Ferrous Met. 56, 44–51.

GRAIN BOUNDARY WETTING PHENOMENA IN THE NdFeb-BASED COMMERCIAL ALLOYS

Svetlana Protasova[1], Alexandr Straumal[1,2], Andrey Mazilkin[1,3], Boris Straumal[1-3]*, Brigitte Baretzky[3]

[1] *Institute of Solid State Physics RAS, Ac. Ossipyan str. 2, 142432 Chernogolovka, Russia*
[2] *National University of Science and Technology «MISIS», Leninskii prosp. 4, 119049 Moscow, Russia*
[3] *Karlsruhe Institute of Technology (KIT), Institute of Nanotechnology, Hermann-von-Helmholtz-Platz 1, 76344 Eggenstein-Leopoldshafen, Germany*
**Corresponding address: straumal@issp.ac.ru*

Key words: grain boundaries, wetting, NdFeB-alloys, magnets, phase transitions

The developments of last decade show that properties of fine-grained and nanograined materials are critically controlled by the behaviour of grain boundaries (GBs) and triple junctions (TJs). Moreover, the most advanced experimental methods like high-resolution electron microscopy (HREM) and atom probe microscopy allowed observing that GBs and TJs are frequently not atomically thin and smooth but contain the few nm thick layers or so-called complexions. These layers can appear in equilibrium, non-equilibrium (transient) or steady-state structures. Most interesting, from our point of view, is the phenomenon of the so-called pseudopartial (or pseudoincomplete) GB wetting (marked as PPW in the generic phase diagram Fig. 1g). It is intermediate between complete (CW, Fig. 1g) and partial (PW, Fig. 1g) GB wetting.

Let us consider a partially melted two- or multicomponent polycrystal. It corresponds to the condition that temperature is between the solidus temperature T_S and the liquidus temperature T_L. Consider the droplet of a liquid phase on the surface of a solid phase or between two solid grains. Usually one distinguishes partial and complete wetting of surfaces or interfaces. If the liquid droplet partially wets a solid surface (Fig. 1a) then $\sigma_{sg} - \sigma_{sl} = \sigma_{lg} \cos\theta$, where σ_{sg} is the free energy of solid/gas interface, σ_{sl} is the free energy of solid/liquid interface, σ_{lg} is that of liquid/gas interface and θ is the contact angle. If the liquid droplet partially wets the boundary between two solid grains (Fig. 1b), then $\sigma_{gb} = 2\,\sigma_{sl} \cos\theta$, where σ_{gb} is the free energy of a grain boundary (GB). The free surface or GB which is not covered by the liquid droplets remains dry and contains only the adsorbed atoms with coverage below one monolayer. In this case the GB can exist in the equilibrium contact with the liquid phase. In the case of complete wetting (Figs. 1c,d) $\sigma_{sg} > \sigma_{lg} + \sigma_{sl}$ or $\sigma_{gb} > 2\,\sigma_{sl}$, the contact angle is zero, and liquid spreads over the free surface or between grains. In this case the GB separating the grains is completely substituted by the liquid phase.

In the majority of cases the direct transition occurs from partial wetting into complete wetting, for example by increasing temperature or decreasing pressure. However, in some cases the state of pseudopartial wetting occurs (PPW in Fig. 1g) between partial and complete wetting. In this case the contact angle $\theta > 0$, the liquid droplet does not spread over the substrate, but the thin (few nm) precursor

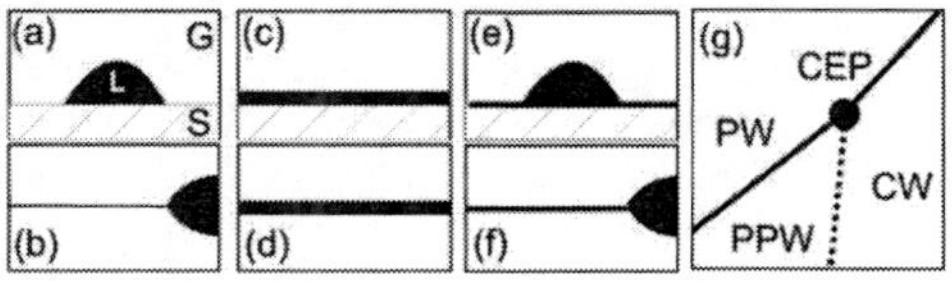

Fig. 1. The schemes for the wetting of free surfaces and GBs. (a) partial surface wetting, L – liquid phase, S – solid phase, G – gas phase; (b) partial GB wetting; (c) complete surface wetting; (d) complete GB wetting; (e) pseudopartial surface wetting; (f) pseudopartial GB wetting; (g) generic wetting phase diagram, PW – partial wetting, CW – complete wetting, PPW – pseudopartial wetting, CEP – critical end point, thick lines mark the discontinuous (first order) wetting transition, thin line mark the continuous (second order) wetting transition.

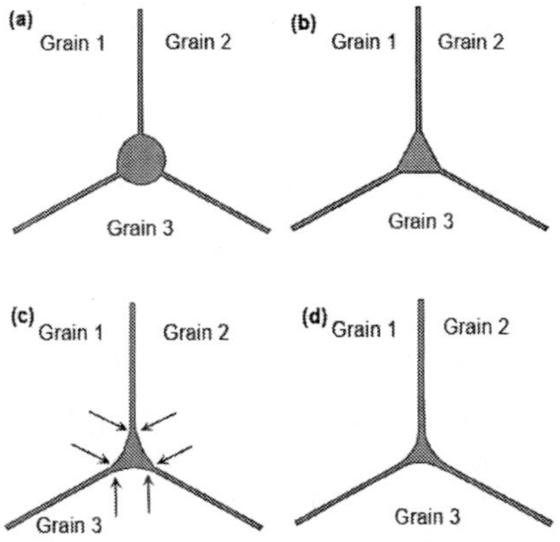

Fig. 2. Different configurations of liquid phase in the GB triple junction and thin quasi-liquid thin layers in the GB. (a) Pseudopartial GB wetting, θ > 60°. (b) Pseudopartial GB wetting, θ = 60°. (c) Pseudopartial GB wetting, θ < 60°. The contact points between liquid phase in TJ and quasi-liquid layers in the GB are shown by arrows. (d) Complete GB wetting, θ = 0°.

film exists around the droplet and separates substrate and gas (Fig. 1e). Such precursor film is very similar for the liquid "pancake" which forms in case of complete wetting and deficit of the liquid phase. This case is called pseudopartial wetting, it is possible when $A < 0$ and $S > 0$. In case of pseudopartial wetting the precursor exists together with liquid droplets, and in case of complete wetting the droplets disappear forming the "pancake".

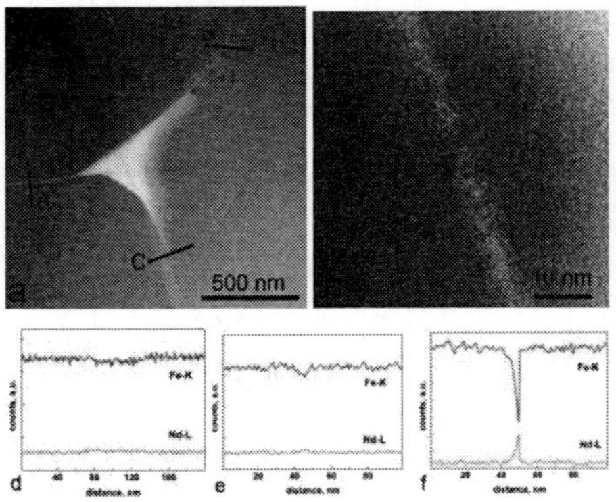

Fig. 3. (a) STEM micrograph of a triple joint between three $Nd_2Fe_{14}B$ grains filled by the Nd-rich phase. The positions of concentration profiles are shown (A, B and C). (c) HREM micrograph of a GB A containing the uniformly thin layer of a Nd-rich phase. (d, e, f) Fe and Nd concentration profiles in the locations C, B and A, respectively

In Fig. 3a the STEM micrograph of a triple joint between is shown. The chemical composition of these three grains measured by the EDXS in TEM corresponds to that of $Nd_2Fe_{14}B$ hard magnetic phase. The triple joint shown in Fig. 3a is filled by the Nd-rich phase which was liquid during the liquid phase sintering and annealing at 900°C. The Figs 3d, 3e, and 3f show the Fe and Nd concentration profiles across all three $Nd_2Fe_{14}B/Nd_2Fe_{14}B$ GBs in the locations C, B and A (Fig. 3a), respectively. The first two profiles do not contain any Fe and/or Nd maxima or minima. It means that the respective GBs C and B remain "dry" and are not enriched (depleted) by the Fe and/or Nd. The GBs C and B have the non-zero contact angle with Nd-rich phase in the triple junction. In other words, the GBs C and B are incompletely (partially) wetted by the Nd-rich melt. This situation corresponds to the scheme shown in Fig. 1b.

The GB A is different. The concentration profile in Fig. 3f shows that the GB A is enriched by Nd and depleted by Fe. The thickness of the Nd maximum and Fe miniumum is about 5 nm. The uniformly thin light-grey layer of a Nd-rich phase in GB A is clearly visible also in Fig. 3a. Fig. 3b shows the conventional TEM micrograph of this GB and Fig. 3c contains the micrograph of the same GB with a thin layer of a Nd-rich phase. Both TEM and HREM also witness that the Nd-rich GB layer is uniformly thin and has a thickness of about 5 nm. The GB A, similar to the GBs C and B, also has the non-zero contact angle with Nd-rich phase in the triple junction. Since GB A is not "dry" and contains the thin Nd-rich layer, this situation corresponds to the scheme shown in Figs 1f and 2c. In other words, the GB C is pseudo-incompletely (or pseudo-partially) wetted by the Nd-rich melt.

If one observes the thin GB layers of a constant thickness, it is not easy to distinguish, whether one has the case of (1) prewetting/prewetting in the one-phase area of a bulk phase diagram; (2) thin GB "pancake" due to the deficit of wetting phase or (3) pseudopartial wetting by a liquid or solid phase. The big problem is that most frequently the bulk liquid or solid phase can be found only in the GB triple junction (TJ) "pockets". The pseudopartial wetting can be clearly identified only if the contact angle $\theta \geq 60°$ and the solid/liquid interface is convex (Figs. 2a,b). If the contact angle $\theta < 60°$ and the solid/liquid interface is concave (Figs. 2c,d), the difference becomes very fine. If the solid/liquid interface has a discontinuity (two tangentials) between TJ pocket and GB layer (Fig. 2c), the pseudopartial wetting takes place. If the solid/liquid interface is continuous (one tangential) between TJ pocket and GB layer (Fig. 2d), the complete wetting with the deficit of wetting phase takes place.

Can the $Nd_2Fe_{14}B/Nd_2Fe_{14}B$ GBs contain the few nanometer thick Nd-rich layers like the prewetting GB layers in metals or oxides? Already in early 1990-ies such few nm thin Nd-rich GB layers were indeed observed in the NdFeB-based alloys. Later the Nd-rich uniformly thick layers between $Nd_2Fe_{14}B$ grains with thickness below 5 nm were observed in various alloys in many experimental works. What is the physical reason for the formation of such uniformly thick GB layers of few nanometer thickness? Two possibilities exist for the explanation of this phenomenon. First one is the complete GB wetting by a liquid phase in case of a deficit of a melt. In this case the liquid pockets in the GB triple junctions (TJs) could still completely wet the TJs and form the continuous network along TJs. Second possible explanation for the existence of the uniformly thin GB layers is the pseudopartial GB wetting (like it is shown in the scheme in Figs. 2a,b,c). In this case there should be a shape discontinuity (break of the first shape derivative) in the contact point between liquid TJ pocket and thin GB layer.

The results obtained in our work show without any doubt that the NdFeB-based alloys, together with few completely wetted GBs, can contain GBs either incompletely or pseudoincompletely wetted by the Nd-rich melt.

Acknowledgments

Authors thank the Russian Foundation for Basic Research (contract 18-03-00284) as well as The Allianz Industrie Forschung (Germany) for the partial financial support.

The 73rd World Foundry Congress
„Creative Foundry"
23 - 27 September 2018, Krakow, Poland

EFFECT OF NICKLE CONTENT ON MICROSTRUCTURAL EVOLUTION IN AUSTEMPERED SOLUTION STRENGTHENED FERRITIC DUCTILE CAST IRON

Tatsuya Tokunaga [1]*, Yeong-Jin Kim [2], Hidenori Era [1]

[1]Department of Materials Science and Engineering, Kyushu Institute of Technology, 1-1, Sensui-cho, Tobata-ku, Kitakyushu 804-8550, Japan
[2] Graduate School of Engineering, Kyushu Institute of Technology, Now at Sharp Corporation, Sakai, Osaka 590-8522, Japan
**Corresponding address: tokunaga@post.matsc.kyutech.ac.jp*

Key words: solution strengthened ferritic ductile cast iron, austempering, bainite, retained austenite, nickel addition

1. Introduction

Austempered ductile cast iron (ADI) has recently attracted attention as an important engineering material due to its excellent properties such as high strength with good ductility, good fracture toughness, good wear resistance, and so forth [1]. These superior properties of ADI are attributed to its microstructure that consists of bainitic ferrite and stable carbon-enriched retained austenite with graphite nodules. ADI can be obtained from heat treatment that consists of austenitising at around 950 °C followed by a quenching and isothermal holding in the bainitic transformation temperature range. Further improvement of the mechanical properties of ADI has been carried out by addition of alloying elements and optimization of heat treatment conditions [2-5]. Recently, solution-strengthened ferritic ductile cast iron (SSFDI) has been developed by increasing the Si content in the matrix [6]. In this study, the effects of Ni content and austempering temperature on the microstructural evolution during austempering in SSFDI with different Ni contents were investigated.

2. Experimental

The materials used in this study were SSFDIs with different Ni content, namely 0 mass%, 0.1 mass%, 0.57 mass%, and 1.06 mass%, hereafter referred as to SSFDI, 0.1Ni, 0.6Ni, and 1.1Ni, respectively. The starting materials were melted by high frequency induction heating, and then, the melts were cast into Y-block sand mold with a cross section of 25 mm × 40 mm which was prepared in accordance with Japanese Industrial Standards (JIS) Code G5502, after spheroidization and inoculation treatments. The chemical composition of the materials is shown in Table 1. Disk-shaped samples with a diameter of about 4 mm and a thickness of about 1 mm were used in the experiment. The temperature profile of the austenitising and austempering heat treatments was as follows. First, the sample was heated to 900 °C at a rate of 50 °C/s using an infrared gold image furnace and kept at this temperature for a period of 30 min under flowing argon gas. The sample was then quenched by a helium-gas blow to temperatures between 375°C and 325°C and held at these temperatures for a period of 2 h, after which the sample was allowed to air-cool to room temperature. The cooling rate during the helium-gas quenching was approximately 30 °C/s in the region between 800°C and 600°C. The microstructure was observed with optical microscopy, X-ray diffraction (XRD) measurement and electron back-scattered diffraction pattern (EBSP) analysis. The Vickers hardness of the heat-treated samples was measured under a load of 0.9807 N.

Table 1. Chemical composition (mass%) of the materials used.

Samples	C	Si	Mn	P	S	Cu	Mg	Ni
SSFDI	3.2	3.8	0.3	0.02	0.01	0.02	0.04	0
0.1Ni	3.02	3.66	0.32	0.023	0.009	0.025	0.038	0.1
0.6Ni	2.95	3.83	0.32	0.022	0.011	0.028	0.047	0.57
1.1Ni	2.99	3.66	0.32	0.023	0.008	0.026	0.048	1.06

3. Results and discussions

Optical micrographs of the samples with different Ni content austempered at 350°C are shown in Fig. 1. The phase identification by means of XRD patterns and EBSP analysis revealed that all the samples were composed of ferrite and austenite, as well as graphite. The matrix of the sample exhibited a coarser bainitic ferrite microstructure with increasing Ni content.

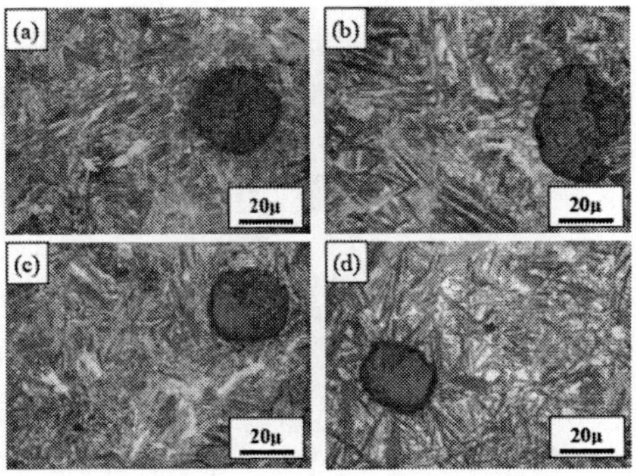

Fig. 1. Optical micrographs of the samples austempered at 350°C: (a) SSFDI, (b) 0.1Ni, (c) 0.6Ni, and (d) 1.1Ni samples.

Similar microstructural characteristics were also observed from the samples austempered at 325 and 375°C. XRD analysis, optical microscopy and EBSP analysis revealed that the amount of retained austenite in the austempered sample increased with increasing Ni content and austempering temperature, and the quantitative result for samples austempered at 350°C obtained from the EBSP analysis is shown in Fig. 2. It is well known that Ni is an austenite stabilizing element, and this may lead to the stabilization of retained austenite in the sample with higher Ni content. In addition, our preliminary experiment [7] indicated that the progress of isothermal austenite transformation for SSFDI increased with increasing Ni content, and the sample with higher Ni content may contain large amounts of C due to the diffusion from the graphite nodules [8], resulting in the stabilization of the retained austenite during austempering.

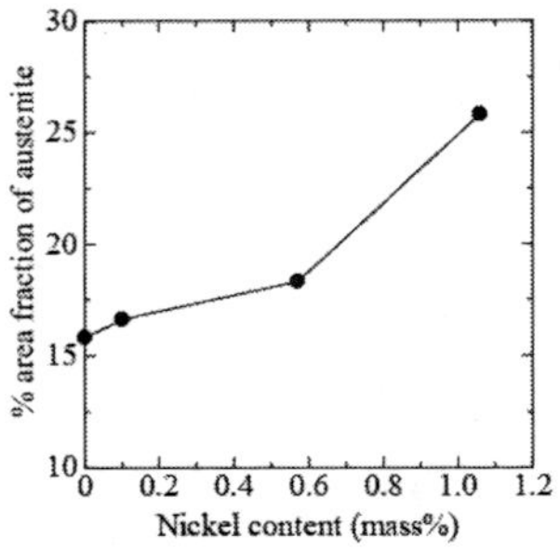

Fig. 2. Effect of Ni content on area fraction of retained austenite in SSFDI samples austempered at 350°C.

Figure 3 shows the change in the Vickers hardness value of the matrix of the samples austempered at various temperatures. According to the measurements, the Vickers hardness of the matrix for each sample decreased with the increase in austempering temperature, and this change may be attributed to the increase in the retained austenite in the matrix. However, although the samples with higher Ni content were found to contain more retained austenite in the matrix as shown in Fig. 2, the hardness of the samples tended to increase with increasing Ni content at the same austempering temperature. The reason for this may be due to the higher C content in the samples with higher Ni content.

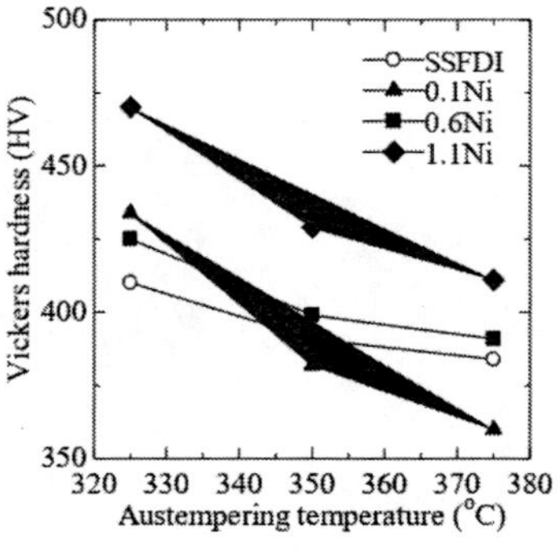

Fig. 3. Change in the Vickers hardness value of the matrix of the samples austempered at various temperatures.

4. Summary

The effects of Ni content and austempering temperature on the microstructural evolution during austempering in SSFDI with different Ni contents were investigated. The results obtained are summarized as follows:

(1) The bainitic ferrite microstructure for SSFDI was found to be much coarser with increasing Ni content.

(2) The amounts of retained austenite for the samples after austempering increased with increasing Ni content and austempering temperature.

(3) The Vickers hardness of the matrix for each sample decreased with the increase in austempering temperature, in contrast, increased with increasing Ni content.

Acknowledgment

The authors are grateful to Hinode, Ltd., Japan for providing the materials used in this study.

References

[1] Y. Tanaka & H. Kage. (1992). Development and Application of Austempered Spheroidal Graphite Cast Iron. Mater. Trans, JIM. 33 (6), 543-557.

[2] T. Kobayashi & H. Yamamoto. (1988). Development of High Toughness in Austempered Type Ductile Cast Iron and Evaluation of Its Properties. Metall. Trans. A. 19A (2), 319-327.

[3] Y. Mi. (1995). Effect of Cu, Mo, Si on the Content of Retained Austenite of Austempered Ductile Iron. Scr. Metall. Mater. 32 (9), 1313-1317.

[4] J. Yang & S.K. Putatunda. (2004). Improvement in Strength and Toughness of Austempered Ductile Cast Iron by a Novel Two-Step Austempering Process. Mater. Des. 25 (3), 219-230.

[5] A.S. Benam. (2015). Effect of Alloying Elements on Austempered Ductile Iron (ADI) Properties and Its Process: Review. China Foundry. 12 (1), 54-70.

[6] R. Larker. (2009). Solution Strengthened Ferritic Ductile Iron ISO 1083/JS/500-10 Provides Superior Consistent Properties in Hydraulic Rotators. China Foundry. 6 (11), 343-351.

[7] T. Tokunaga, T. Furukawa & H. Era. (2015). Effect of Ni Content on the Ferrite-to-Austenite Transformation during Austenitising in Spheroidal Graphite Cast Iron. Reports of the 167th JFS Meeting. 63 (in Japanese).

[8] K. Ogi, A. Sawamoto & J. Hiwatashi. (1988). Transformation Process and Behavior of Carbon in Austempering Heat Treatment of Spheroidal Graphite Cast Iron. J. JFS. 60 (4), 233-239 (in Japanese).

The 73rd World Foundry Congress
„Creative Foundry"
23 - 27 September 2018, Krakow, Poland

DAMAGE MICROMECHANISM IN SPHEROIDAL CAST IRON AS AFFECTED BY GRAPHITE/MATRIX INTERFACE INTERACTIONS

Małgorzata Warmuzek*, Adelajda Polkowska, Andrzej Gazda

Foundry Research Institute, Materials Testing Laboratory, 30-418 Kraków, Zakopiańska 73, Poland
**Corresponding address: malgorzata.warmuzek@iod.krakow.pl*

Key words: cast iron, microstructure, spheroidal graphite, fracture, SEM

Abstract

In this work, the observations of the fracture surface after standard tensile test of several kinds of spheroidal cast irons, ferritic and ADIs, have been carried out by means of scanning electron microscope. The local crack path in the area of grahite/matrix (G/M) interface has been analyzed as affected by matrix phase composition and austempering treatment parameters. The G-M debonding effects at G/M interface have been revealed. The obtained results allowed to identify some determination factors for debonding mode at G/M interface and its role in final damage mechanism.

Introduction

The microstructural determination of damage in spheroidal cast iron has been a subject of many works [1, 2] taking into account the microstructural heterogeneity introduced by graphite nodules. The microscope observations of G/M interface topography, exposed after final crack, have revealed many microstructural details, concerning graphite nodule cracks mode and G/M interface debonding course. They seemed to play an important role in sequence of local decohesion effects preceding the final crack of specimen. Thus, in the recently considered models of the cast iron damage, the microstructural features, such as G/M interface morphology, internal structure of spheroids and anisotropy of graphite crystal lattice have been recognized important factors, although their particular impact has not been identified unequivocally until now. In this work the systematic observations of fracture surface have been focused on revealing the details of graphite spheroids structure and G/M interface morphology as involved in a cracks course and damage mechanism determination in several kinds of spheroidal cast irons both ferritic and ADIs.

Experimental

Materials for examination were specimens of spheroidal cast iron, of different phase composition of matrix: of matrix ferritic (SNi9, elongation 17.8%) and ferritic-pearlitic (SNi11, elongation 12%), in as-cast state, and ADIs after isothermal heat treatment, both one-step (ADI390, elongation 6.9%) and cyclic (ADIcyclic, elongation 8.4%%). Fracture surface topography after standard tensile test has been observed by means of scanning electron microscope SCIOS FEG equipped with in-lense SE detector.

Results and discussion

The microscope observations of fracture surface of the examined specimens revealed some microstructural details in the microregions composed of graphite and matrix at G/M interface as determined by macroscopic material ductility, metal matrix phase composition and internal sectorial structure of nodule. The G-M debonding mode at G/M interface seemed to be subordinated to local strain rate, as reported previously [2]. The pure pole G-M debonding, without significant nodule destruction, was observed mainly in the most ductile, ferritic cast iron (SNi9), the high strain rate promoting this mechanism, could be assumed in. Thus, G-M debonding in ferritic cast iron took place due to break some interatomic connections between spheroid and matrix. In the ADIs specimens of polyphase matrix, and of less ductility, the observed microstructural effects have reflected some interactions at G/M interface (Figs. 1, 2). The metal particles embedded in graphite spheroid surface, exposed after specimen crack, could be simultaneously recognized in the bottom of the complementary dimple. On both surfaces, curved graphene layers have been identified. Thus, one can say, that graphite nodule was removed from matrix after separating graphene layers in its external shield, composed of less ordered and more dispersed graphene blocks than those in the conical sectors in its core [3-5]. The debonding between two zones of nodule, that primary one formed in as-cast state and that secondary formed due to isothermal annealing [4], has been also recognized. Nevertheless, nodules destroyed by diagonal cracks have been also identified on the fracture surface in all the examined specimens. These cracks seemed to pass through intersectorial boundaries in the nodule core (Fig. 3a). However, the presence of either curved (Fig. 1b) or twinned (Fig. 3b) graphene layers on the debonded interfaces allows

a)

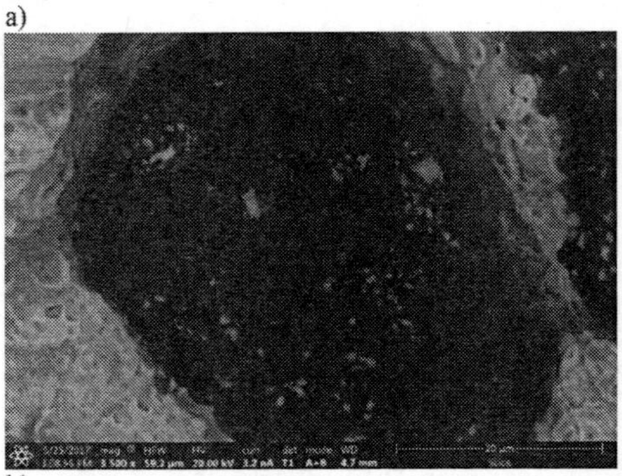

b)

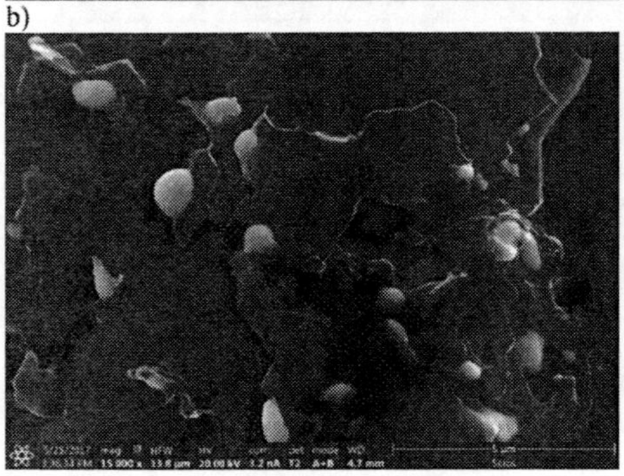

Fig. 1. Fracture surface, ADI after cyclic austempering, SEM, a) graphite nodule embedded partially in alloy matrix, b) graphite nodule surface exposed after crack, delaminated graphene layers with embedded metal particles

Fig. 2. Fracture surface, bottom of dimple exposed after removing graphite nodule from alloy matrix specimen ADI after cyclic austempering, SEM

an assumption that internal cracks between sectorial blocks of nodule core [4] have developed due to break of the weak van der Waals bonds between graphite basal planes, while strong covalent bonds, connecting carbon atoms in a basal plane have been sustained longer. Thus, the fracture surface observations, carried out in this work, showed that G-M debonding mechanism

a)

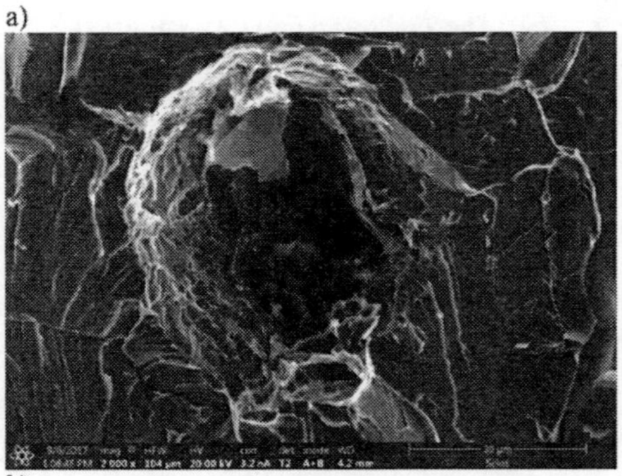

b)

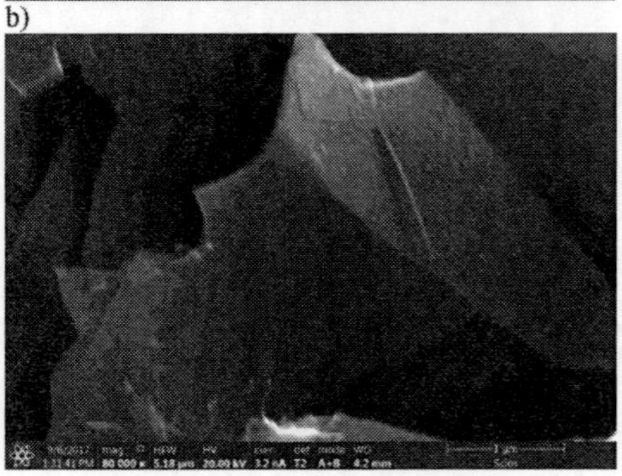

Fig. 3. Fracture surface, specimen SNi11, as cast, SEM, a) internal crack in graphite nodule partially embedded in alloy matrix, b) visible graphene layers with growth steps and twins

at G/M interface has been controlled by alloy chemical composition and matrix microstructure state, while graphite nodule destruction has been mainly determined by its internal structure and graphite crystal lattice anisotropy.

4. References

[1] N. Bonora, A. Ruggiero. (2005) Micromechanical modeling of ductile cast iron incorporating damage. part I: Int. J. Solids. Struct. 42 1401-1424.

[2] V. DiCocco, F. Iacoviello, A. Rossi, D. Iacoviello. (2014). Macro-and microscopical approach to the damaging micromechanisms analysis in a ferritic ductile cast iron. Theor . Appl. Fract.Mech, 42, 6926-33

[3] J.P. Monchoux, C. Verdu, G. Thollet, R. Fougères, A. Reynaud. (2001) Morphological changes of graphite spheroids during heat treatment of ductile cast irons, Acta Mat., 49, 4355–62

[4] K. Theuwissen, J. Lacaze, L. Laffont. (2016) Structure of graphite precipitates in cast iron. Carbon, 96, 1120-128

[5] J. Qing, V.L. Richards, D.C. Van Aken. Growth stages and hexagonal-rhombohedral structural arrangements in spheroidal graphite observed in ductile iron. (2017). Carbon, 116, 456-469

The 73rd World Foundry Congress
„Creative Foundry"
23 - 27 September 2018, Krakow, Poland

GRAIN SIZE PREDICTION MODEL IN ALUMINIUM CASTINGS MANUFACTURED BY LOW PRESSURE TECHNOLOGY

I. Lizarralde[1], E. Sal[1], P. Rodríguez[2], E. Ochoa de Zabalegui[2], I. Cia[3], A. Rios[4], A.I. Fernández-Calvo[1]

[1] *IK4-AZTERLAN, Aliendalde Auzunea 6, 48200-Durango, Spain*
[2] *EDERTEK Technology Center of FAGOR EDERLAN Group, Isasi 6, 20500-Mondragón, Spain*
[3] *MAPSA S.COOP., Ctra Echauri 11, 31160-Orcoien, Spain*
[4] *EDERLAN S. Coop., Torrebaso 7, 20540-Eskoriatza, Spain*
* *Corresponding address: afernandez@azterlan.es*

Key words: grain refinement, AlSi7Mg alloy, thermal modulus, cooling rate, solidification evolution.

The grain refinement in a real casting manufactured by Low Pressure Die Casting, LPDC, such as a wheels and steering knuckles depend on the grain refinement potential of the metal and the geometry/process parameters.

For this study, the effect of cooling rate on AlSi7Mg alloy with different metal qualities in terms of grain refinement was tested. The grain size has been metallographically evaluated in cylindrical test pieces and in real wheels manufactured at MAPSA. Thermolan® -Al system has been used to evaluate the nucleation potential in term of grain size on the standard cup. The grain growth evolution has been modeled taking into account undercooling measurements located in the center of the cylindrical test parts for refined and unrefined metal. Different grain growth coefficient has been obtained depending on the melt quality and local solidification rate. The grain size measured in a real casting, a wheel, was used to calibrate the model for a real part in LPDC for different grain size potential.

1. Introduction

Aluminum-silicon alloys are the most widely used alloys to produce aluminium castings because they have very good foundry properties and may be adapted to span a large range of mechanical properties. Grain size refinement is one of the melt treatments commonly applied to these foundry alloys. The grain size has a very clear effect on the shrinkage behavior: well refined alloys generally tend to display less and more finely dispersed shrinkage. Thus, in those parts which comprise zones difficult to feed adequately a good grain refinement can decisively reduce the porosity [1-2].

The quality of the melt preparation in terms of grain refinement has been extensively studied by thermal analysis techniques for standard cups in AlSi7Mg alloy [3-12]. However, obtaining a proper grain refinement in standard cup does not assure that grain refinement is correct in real parts. The objective of the present work is to predict the grain size in real part by measuring the grain size potential by Thermolan®-Al system [7,8] and considering the local cooling rate in the different areas. A real wheel part has been used for validation of the model.

2. Experimental results and discussion

Alloy AlSi7Mg0.3 with Ti content varying between 0.08 and 0.14 wt. % was studied at different grain refinement condition obtained by addition of different quantities of Ti5B1 grain refinement. The grain size, GS_{TH-Al}, evaluated by Thermolan®-Al system additions ranging from 0.4 to 1.0 mm.
The grain size, GS; was evaluated by metallographic analysis in two different types of castings:
- a test piece with 5 cylinders was casted in permanent mould
- a real wheel casted in low pressure die casting.

The cylinders have a height equal to their diameter varying from 24 mm to 72 mm. Their cooling rate at 615 °C range between −32.0 °C/s to −8.5 °C/s. A thermocouple located at the centre of each cylinder records the temperature evolution during cooling and solidification. An example is shown in Fig. 1.

Fig. 1. ***Cooling curves of the cylinders of different thermal modulus***

The grain growth evolution, r_i, has been modelled using Eq. (1), considering the instantaneous nucleation and grain growth depending on the undercooling, ΔT; and grain growth coefficient, μ, [13]. The urdercooling was obtained from the difference between 615°C and solidification temperature measure in the centre of each cylinder. Solid fraction evolution, f_s; was calculated considering spherical grains. When the end of solidification is reached the grains come in contact one with another and an impingement correction must be considered ($1-f_s$) in Eq. 2, where N is the number of nucleation points [13].

$$r_i = r_{i-1} + [\mu\,(\Delta T_i)^2]\cdot\Delta t \qquad (1)$$

$$(df_s/dt) = 4N\pi r^2(\mu\,\Delta T^2)\cdot(1-f_s) \qquad (2)$$

The comparison of modelled data with experimental data allow to obtain the grain growth coefficient, which varies depending on the liquidus cooling rate, as it is shown in Fig. 2.

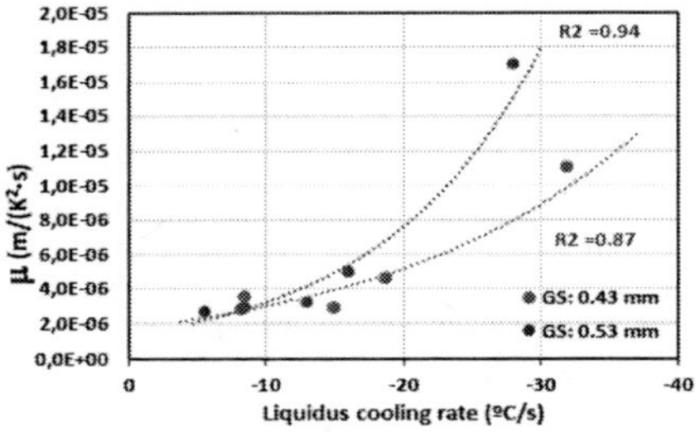

Fig. 2. Grain growth evolution vs liquidus cooling rate

To validate the model in a complex real casting, an automotive wheel has been selected and casted in Low Pressure Die Casting, LPDC, with different grain size potential. The local solidification time predicted by simulation software PROCAST is shown in Fig. 3. The final thermal modulus depends strongly on the geometry and the die thermal management. Six different locations where selected to evaluate grain size metallographically. An example of the grain size variation in the wheel for different solidification time for a coarse grain size of 0.79 mm evaluated by Thermolan®-Al as is shown in Fig. 3.

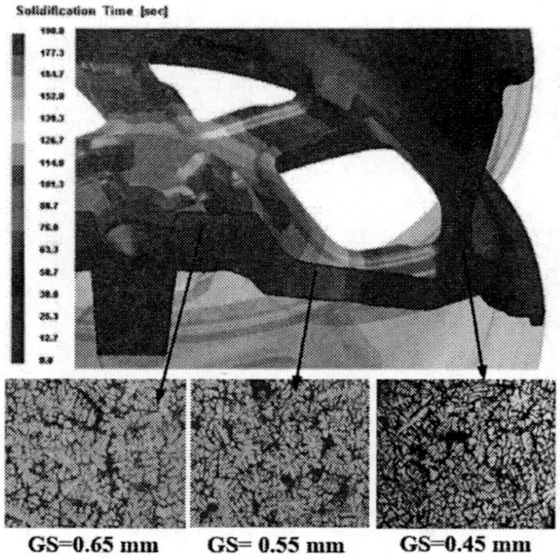

Fig. 3. Solidification time predicted by Procast and grain size measured metallographically in a real wheel

3. Conclusions

The grain refinement in a real casting manufactured by low pressure die casting, LPDC, depend on the grain refinement potential of the metal, casting geometry and the die thermal management.

For this study, the effect of cooling rate on AlSi7Mg alloy with different metal qualities in terms of grain refinement was tested. The grain size has been metallographically evaluated in cylindrical test pieces and in real wheels manufactured at MAPSA. Thermolan-Al® system has been used to evaluate the nucleation potential in term of grain size on the standard thermal analysis cup.

The grain growth evolution has been modeled taking into account undercooling measurements located in the center of the cylindrical test parts for refined and unrefined metal. Different grain growth coefficients have been obtained depending on the melt quality and thermal modulus. The grain size measured in a real casting, a wheel, was used to calibrate the model for a real part in LPDC for different grain size potential.

Thus, thermal analysis of the melt prior to casting allows to predict if the desired grain size will be obtained at different locations of the part or if corrective actions must be taken to improve the grain refinement potential of the melt.

Acknowledgments

This research was financially supported by Spanish Ministerio de Economía y Competitividad, RETOS programme, No. RTC-2015-3822-4.

Acknowledgment to Laurentiu Nastac for the help in microstructural modelling.

References

[1] ASM Int. (1992). ASM Metals HandBook Vol. 15, Castings. 9th Ed. Materials Park, Ohio, USA.

[2] J. Gilbert Kaufman, E.L. Rooy. (2005) Aluminum alloy castings: Properties, Processes and Applications, AFS, ASM Int., Materials Park, Ohio, USA.

[3] J.E. Gruzleski, B.M Closset. (1990). The treatment of liquid aluminum-silicon alloys. AFS. Illinois, USA.

[4] R.I. MacKay, M. Djurdjevic, J.H. Sokolowski, W.F. Evans. (2000). AFS Transaction, Vol. 108, p. 511.

[5] A. Apelian, G.K. Sigworth, K.R. Whaler. (1984). AFS Transaction, Vol. 92, p. 297.

[6] A. Niklas, U. Abaunza, A.I. Fernández-Calvo, J. Lacaze, R. Suárez, (2011). China Foundry, vol. 8 (1), p. 89

[7] I. Lizarralde, A. Niklas, A.I. Fernández-Calvo, J. Lacaze, (2013). Light Metals, pp. 327-331

[8] I. Lizarralde, S. Eguskiza, F. Santos, A. Niklas, A.I. Fernández-Calvo. (2014). 71st WFC. Bilbao, Spain.

[9] F. Bekaert, E. Wettinck. (1996). Aluminium 72, pp. 442-47

[10] G.K. Sigworth, T.A. Kuhn. (2007). International Journal of Metalcasting, pp. 31-40.

[11] J.A. Spittle. (2008). FTJ, pp. 308-314

[12] D.C. Ibarra. (1999). Control of grain refinement of Al-Si-alloys by thermal analysis", PhD thesis, McGill University of Montreal, Canada.

[13] D.M. Stefanescu, (2015). Science and Engineering of Casting Solidification. 3rd Ed. Springer, Switzerland.

The 73rd World Foundry Congress
„Creative Foundry"
23 - 27 September 2018, Krakow, Poland

THE ROLE OF ALUMINIUM IN THE CAST IRON SPHEROIDIZING PROCESS

Marek Soiński[1]*

[1]The Jacob of Paradies University, The Department of Technology, 66-400 Gorzów Wielkopolski, ul. Teatralna 25, Poland
*Corresponding address: marek.soinski@gmail.com

Key words: alloyed cast iron, aluminium, spheroidization, crystallization, graphite precipitates

Abstract
The work discusses the influence of aluminium on the spheroidization of cast iron treated with cerium mixture and graphitized with 75% ferrosilicon. It was proved that the impeding influence of aluminium on the crystallization of nodular graphite, often reported in professional literature, is related to the increase of the maximum eutectic temperature TS caused by this element.

1. Introduction

Aluminium is used as an alloying element in cast iron for a range of reasons, but mainly to increase its fire resistance; this property of the cast iron is also favourably influenced by the occurrence of separate, preferably nodular, graphite precipitates. These precipitates are detached from one another, so that the graphite 'skeleton', characteristic for the cast iron containing flake graphite, does not occur in the material. By the same the danger of its burning-out and forming a net of micro-channels, which would facilitate penetration of oxygen into the walls and accelerate the destruction of casting, is avoided. However, the production of fully nodular aluminium cast iron is not easy, because the presence of aluminium significantly impedes crystallization of nodular graphite.

2. Author's investigations

The production of fully nodular cast iron with aluminium addition is extremely difficult. Spheroidization of low-aluminium silicon-rich cast iron with cerium mixture and 75% ferrosilicon gives usually an alloy containing both nodular and vermicular graphite precipitates (Fig. 1).

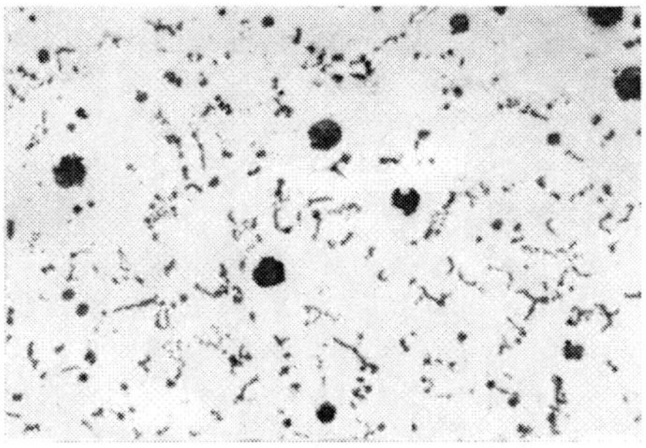

Fig. 1. ***Graphite in cast iron containing 3.13% Al, 3.43% C, 4.21% Si, 0.102% Mn, 0.034% P, and 0.002% S after spheroidization with cerium mixture and 75% ferrosilicon. Specimen cut out of Ø20 mm rod, non-etched, magn. 100×***

This material can exhibit the ferritic matrix in the as-cast state, which is an advantageous feature as far as its fire resistance is considered (Fig. 2).

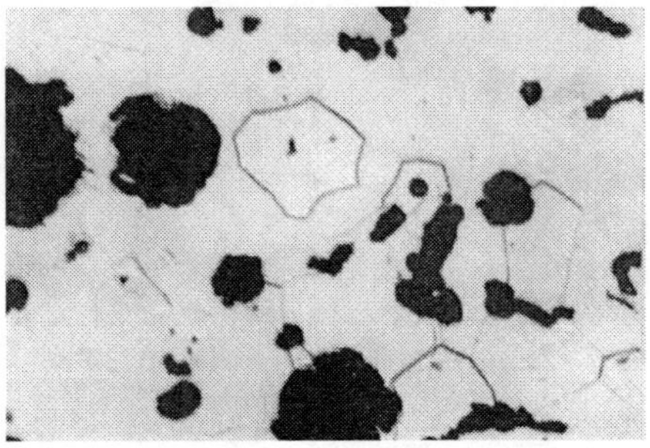

Fig. 2. ***Microstructure of cast iron of chemical composition as in Fig. 1. Specimen cut out of Ø20 mm rod, etched with Nital, magn. 300×***

However, the practical significance of obtaining the fully nodular fire resistant material cannot be overestimated. Therefore it is reasonable to investigate the role of aluminium in the process of cast iron spheroidization. A series of experimental researches with respect to the cast iron containing about 3% of Al, and about 4% of Si, was performed in order to explain what the hampering influence of aluminium on graphite spheroidization relies on. The experiments included:

– determination of maximum (from the process effectiveness point of view) quantity of nodulizer (cerium mixture) and graphitizing modifier (75% ferrosilicon);

– determination of characteristic temperature points during the cooling process for the investigated alloy (using the equipment for derivational thermal analysis);

– assessment of the course of graphite crystallization in the spheroidized cast iron by the method of 'freezing' material samples;

– determination of distribution of elements in the cast iron structure, and primarily in graphite precipitates (by means of scanning electron microscope and X-ray microanalyser.

The detailed results of experiments allow to state that:

– the degree of graphite spheroidization in the examined cast iron increases with an increase of the eutectic saturation degree of the alloy;

– nodular and close-to-nodular graphite crystallizes during the initial stage of cast iron solidification (Fig. 3), however most of graphite phase crystallizes as a part of eutectics;

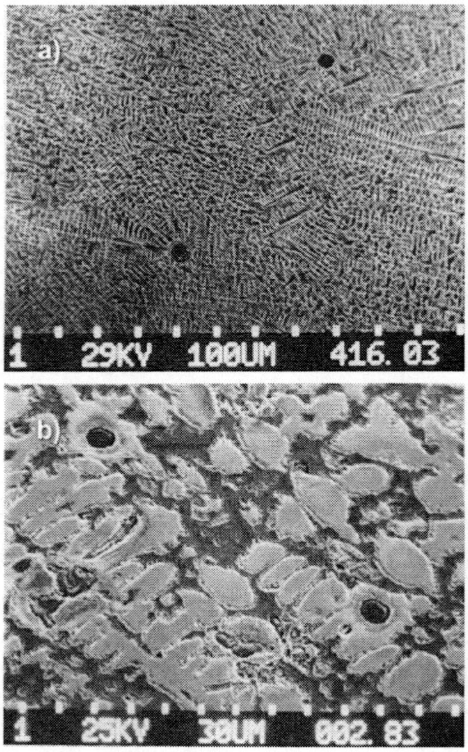

***Fig. 3.** Structures of cast iron specimens (2.73% Al, 2.93% C, 4.33% Si, 0.120% Mn, 0.077% P, and less than 0.01% S) 'frozen' after 7 seconds (a) and 35 seconds (b) from the moment when the temperature of the solidifying alloy dropped down to the liquidus temperature. Deep etched with Nital, scanning electron microscope*

– aluminium can be contained within the nodular graphite precipitates (it is being enveloped inside during graphite crystallization) – thus its merely presence is not an obstacle for arising of precipitates of such a shape;

– the presence of about 3.2% Al and about 3.7% Si in cast iron leads to the increase in the temperature of eutectic crystallization up to about 1200°C.

A series of detailed data concerning investigations with respect both to low-aluminium cast iron solidification and to its properties one can find in References [1], [2], and [3].

3. Summary

An addition of about 3% Al in cast iron leads to the increase of TS temperature by about 40°C, as compared with cast iron of similar chemical composition, but without aluminium. The supercooling of eutectics to the range of 1130-1135°C, which provides the nodular shape of graphite precipitates, becomes impossible. Graphite precipitating in higher temperature cannot take the nodular shape; vermicular graphite occurs in the structure.

The impeding influence of aluminium on the graphite spheroidizing process is therefore related to the increase of maximum temperature TS of eutectic crystallization.

References

[1] Soiński M. S. (2001). Spheroidizing of low-aluminium silicon-containing cast iron with cerium mischmetal. Częstochowa: Ed. of the Faculty of Process and Material Engineering and Applied Physics of Częstochowa University of Technology.

[2] Soiński M. S. (2012). Low-aluminium cast iron. Częstochowa: Ed. of the Faculty of Process and Material Engineering and Applied Physics of Częstochowa University of Technology.

[3] Soiński M. S., Schierbaum K.-D., Nadler E. & Göpel W. (1998) Rasterelektronenmikroskopische und röntgen-analytische Gefügeuntersuchungen von Al- und Si-legiertem Gußeisen mit Kugel- und Vermiculargraphit. Zeitschrift für Metallkunde, vol. 89 (1), 32-36.

GENERAL CONCEPT OF CAST METAL MATRIX COMPOSITES DESIGN

Evgeny Prusov *, Vladimir Kechin

Vladimir State University named after Alexander and Nikolay Stoletovs, Department of Functional and Constructional Materials Technology, 600000, Vladimir, Gorky Str. 87, Russian Federation
**Corresponding address: eprusov@mail.ru*

Key words: design methodology, cast metal matrix composites, selection strategies, targeted properties, materials discovery

1. Introduction

Innovations in the field of creating new materials throughout the world history laid the foundation of scientific and technological progress. However, at the present stage, traditional approaches based on intuitive scientific search and "trial and error" methods no longer satisfy the intensively growing needs of high-tech industries in new materials corresponding to ever-hardening operating conditions of details [1-3]. Such approaches are characterized by high labor-consuming, resource-consuming and in most cases do not lead to positive results when trying to create materials with the required level of properties and the necessary set of operational characteristics. In this connection, a worldwide trend in this direction is the search for methods to accelerate the development of new materials with targeted properties.

Realizing the critical importance of intensifying the creation and implementation of modern functional and structural materials to achieve global economic and military superiority, in 2011 the U.S. Government launched a national infrastructure project for the development of the "material genome" (Materials Genome Initiative, MGI). The main objective of the MGI is to ensure the development of new materials based on the creation of databases on the properties of substances and computer tools to predict the properties and behavior of materials, as well as the development of breakthrough methods of modeling and data analysis [4]. Similar projects have been initiated in other countries [5], particularly in Japan (Elements Strategy Initiative, 2012), India (Mapping the Materials Genome, 2012), China (Key Technology and Supporting Platform for Materials Gene Engineering, 2016). In the European Union, in order to accelerate the development of materials with targeted functional properties, the NoMaD (Novel Materials Discovery) program was launched in 2015 within the framework of the HORIZON 2020 project. Thus, the trends of recent years in the development of new materials are characterized by increased attention to the development of computational methods with their subsequent application for accelerated searching of the components for designing new materials.

Modern metal matrix composites can contain dozens of different alloying and reinforcing components, each of which can have a significant influence on the mechanical and operational properties of designed materials. The complex mechanism of reinforcement of the metal matrix composites and the large number of components used in the alloying and reinforcing complexes do not allow the use of traditional empirical methods of selecting the components. In this regard, the expediency of using for this purpose new approaches developed within the framework of the Materials Genome Initiative and the basic principles of a system for the design of alloys in combination with methods of physical-chemical and technical-economic analysis, as well as mathematical design of the experiment with the construction of functional dependencies "composition-property." Such a concept will allow laying scientific and methodological foundations for the criterial choice of groups of alloying and reinforcing components of cast metal matrix composites, ensuring the solution of the task of creating new functional and constructional materials taking into account the specified operating conditions.

This work is aimed at development of the general concept of metal matrix composites design for specified exploitation conditions.

2. Materials and Methods

The basis for developing a general concept for the metal matrix composites design was the method of system analysis, implemented in the classical methodology "design of alloys" [6]. The term "design of alloys" means the creation of new alloys with the targeted combination of mechanical and operational properties. The process of alloys design in traditional form is represented by the following stages: objective statement for alloy development; alloy basis selection; alloying elements selection and harmful impurities definition; alloying complexes selection; final alloy composition selection.

In this work, we have proposed to consider cast metal matrix composites as foundry alloys. It allows to use the basic principles of alloys design theory in their development, that is based on the physic-chemical analysis, Mendeleev's Periodic Table of Elements and technical-economic analysis. Unlike conventional alloys, metal matrix composites have multiple levels of matrix reinforcing, including major alloying elements of the matrix alloy, as well as exogenous and endogenous reinforcing particles. Within a system approach to optimization of cast composites, selection of alloying and reinforcing components and their classification have been carried out.

3. Results and Discussion

In generalized view the proposed stage-by-stage system of metal matrix composites design can be represented as follows.

At the first stage, the technical specification for the development of metal matrix composite has been formulated and the main requirements determined by operational conditions of details have been appointed. Selection of a basis of composite material has been carried out, as a rule, according to the data of technical-economic analysis. It is worth to mention that most low-cost metal won't always provide the greatest economic benefit. For example, it's often advantageous to use more expensive base metal for parts of friction units of ground transportation and aviation technics if it has a lower density. Parts weight reduction will reduce fuel consumption and increase the mass of transporting freights, which makes cost-effective use of composite alloys based on aluminum and magnesium.

Selection of series of alloying elements and reinforcing components is based on the use of the physical-chemical analysis method for the predictive assessment of nature and mechanism of the components interaction with each other and with the matrix alloy. Criterion approach, based on the results of metal-chemical and technical-economic analysis, allows to allocate groups of reinforcing components for matrix alloys by indicating the possibility of heterogeneous structure formation, thermodynamic stability, minimum solubility in matrix metal at temperatures of manufacturing and exploitation, lack of phase transformations at a temperature range of parts working and some other criteria. One of the main criteria for alloying components selection is supposed to consider their influence on the nature of interphase interaction between matrix alloys and exogenous reinforcing particles.

Application of the stated approach allows to choose components of metal matrix composite alloys with various schemes of reinforcing reasonably, including exogenously reinforced, endogenously reinforced and complex reinforced alloys. It is necessary to use the method of experiments design for relevant mathematical models derivation at the stage of the optimal composition of metal matrix composite alloy selection, including alloying elements concentration ranges and reinforcing components volume fractions.

According to the above scientific provisions, the authors have carried out research on the creation of new complex-reinforced composite alloys.

4. Conclusions

A criterial approach based on the results of physical-chemical and technical-economic analyzes has been developed, which makes it possible to identify groups of potential alloying and reinforcing components for matrix alloys at metal matrix composites design. The proposed concept can form the basis for the development of technological principles for the cast metal matrix composites production and practical recommendations for the selection of their optimal compositions depending on the set operating conditions of the products, which will reduce the cost of conducting expensive experimental studies.

References

[1] Xue D. et al. (2016) Accelerated search for materials with targeted properties by adaptive design. Nature Communications. 7, 11241.

[2] Srinivasan S., Broderick S.R., Zhang R., Mishra A., Sinnott S.B., Saxena S.K., LeBeau J.M., Rajan K. (2015) Mapping chemical selection pathways for designing multicomponent alloys: an informatics framework for materials design. Scientific Reports. 5, 17960.

[3] Vojtech D. (2010) Challenges for research and development of new aluminum alloys. Metalurgija. 49, 181-185.

[4] De Pablo J.J., Jones B., Kovacs C.L., Ozolins V., Ramirez A.P. (2014) The Materials Genome Initiative, the interplay of experiment, theory and computation. Current Opinion in Solid State and Materials Science. 18, 99-117.

[5] Lin H.Q. (2016) Boosting computational capabilities. Nature Materials. 15, 693-694.

[6] Gulyaev B.B., Zaplatkin Yu.Yu. (1992) The synthesis of multicomponent casting alloys. Litejnoe Proizvodstvo. 6, 4-6.

The 73rd World Foundry Congress
„Creative Foundry"
23 - 27 September 2018, Krakow, Poland

INVESTIGATION OF THE EFFICIENCY OF GRAIN REFINERS ON THE HOT TEARING IN Al8Si3Cu ALLOY

Muhammet Uludağ [1]*, Remzi Çetin [2] and Derya Dispinar [3]

[1] Bursa Technical University, Metallurgical and Materials Engineering, Bursa-16310, Turkey
[2] KTO Karatay University, Materials Science and Nanotechnology Engineering, Konya-42020, Turkey
[3] Istanbul University, Metallurgical and Materials Engineering, Istanbul-34320, Turkey
**Corresponding address: dr.uludagm@gmail.com*

Key words: Degassing, Grain refinement, Bifilms, Hot tearing

1. INTRODUCTION

Surface of most liquid metals is covered by a film that forms as a result of the reaction of atmosphere with the metal. These films are harmless as long as they remain on the surface. However, they can be entrained into the bulk of the melt when the surface is disturbed, such as with additions to the melt, and/or when new surfaces are created, such as during pouring. These entrained films are called bifilms because of their double, film-on-film nature [1]. To produce high quality aluminum castings, it is essential to start with a melt that is substantially free from bifilms and bifilm-opening agents such as hydrogen dissolved in liquid aluminum and intermetallics such as Fe-containing phases [2].

If the initial melt quality is too low that a good casting cannot be made from it, it is fruitless to use the best practices in filling and feeding of castings. Conversely high quality melts can be damaged significantly when surface oxide films are entrained into the casting during melt transfers and in poorly-designed filling systems. In aluminum alloys, Al-Ti-B grain refiners are added to alter the coarse α-Al to finer dendrites.

Anyalebechi [3] reported that grain refinement with Al-5Ti-0.2B additions increased pore sizes and N in 2024 alloy products. Similar results were reported by Fakhraei et al. [4] who observed that as Ti amount increased in the melt, porosity increased in an Al-20%Mg alloy. As an alternative to Ti grain refinement, Al-3B master alloy has been used which produces AlB_2 and B in the melt to act as heterogeneous nucleates for grains [5].

Dispinar et al. [6] investigated the differences in porosity distribution between Ti-B and B grain refinement additions. More localized porosity was observed with B grain refinement compared to the evenly distributed pores with Ti-B grain refinement.

Moreover, Schaffer and Dahle [7] and Dispinar et al. [6] observed that Ti tended to sink to the bottom due to its higher density, which is known as fading. With B grain refinement, no fading effect was observed.

2. EXPERIMENTAL PROCEDURE

Chemical composition of A380.1 alloy that was used in this study is given **Table 1**.

Table 1. Chemical composition (in wt%) of the alloy used in this study.

Si	Fe	Cu	Mn	Mg	Zn	Ti	B	Al
8.14	0.64	3.12	0.44	0.22	0.49	0.02	0.001	rem.

Ingots were cut to be charged into the crucible. A resistance furnace that has 22 kg capacity was used to melt the alloy.

In non-degassed melts, grain refinement and modification additions were made once the melt temperature reached 1013K (740°C).

For degassing process, a T-type graphite lance was used [8]. Degassing process was started when melt temperature reached 1013K (740°C) and was carried out for 20 minutes with argon. Gas flow rate was held at 2 L/min. Master alloy additions were made when the degassing was completed and after the surface of the melt was skimmed.

3. RESULTS AND DISCUSSION

Casting quality and hot tearing sensitivity were investigated. The results of bifilm index, diameter of bifilm index, hot tearing degree and changes in HTS are given in Figure 1, 2, 3 and 4 respectively.

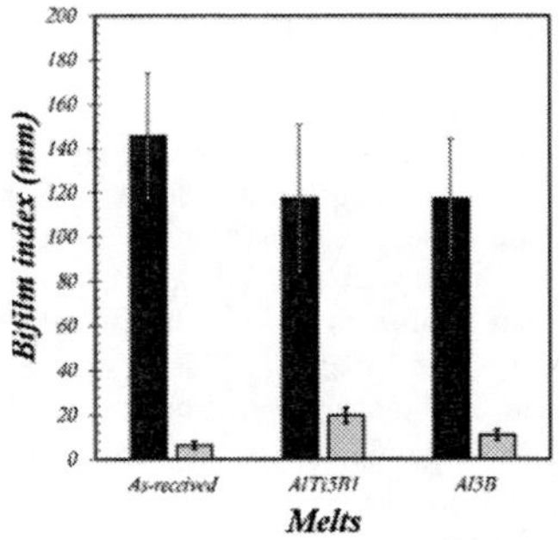

Figure 1. Results of bifilm index of the melts depending on degassing. Black bars are for no degassing condition and grey bars are for degassed condition

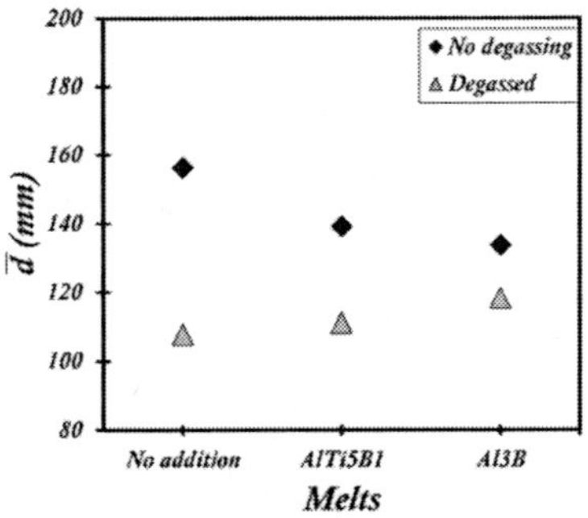

Figure 2. Changes of pore diameters on RPT samples

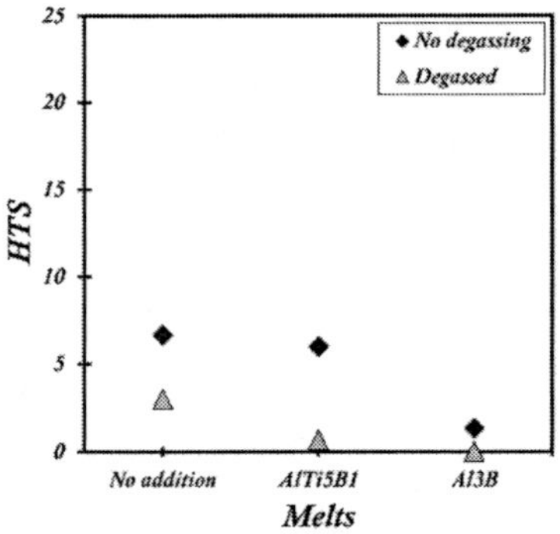

Figure 3. Hot tearing degree of the casting trials

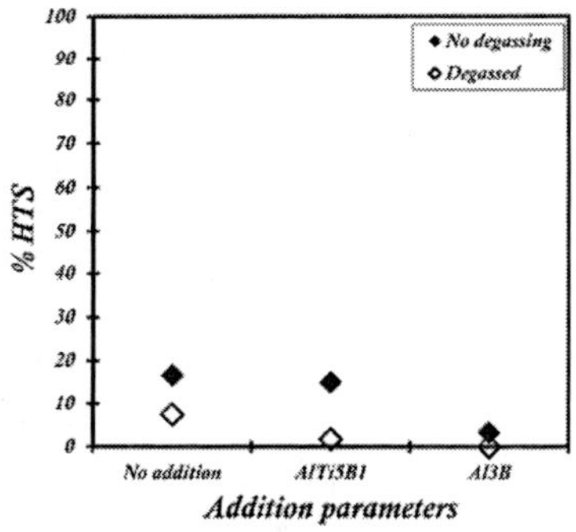

Figure 4. HTS % of the casting trials

Differences between degassed and non-degassed samples is clear for all conditions. While varying levels of porosity is evident in non-degassed castings, the number of pores is significantly reduced after degassing. This is an indication that rising bubbles during degassing had cleared the old bifilms in the melt. Degassing reduces pore sizes and consequently pore size distribution is shifted to lower sizes. This shift can be at least partially attributed to reduction in H level in the melt that would be expected after degassing with argon. With lower level of dissolved hydrogen in the melt, entrained bifilms would inflate less under reduced pressure, resulting in lower pore sizes.

4. CONCLUSIONS

Best HTS is obtained in degassed melts with Al3B addition which was zero. The extrapolation of data indicates that when number density of pores are decreased in Al3B addition, HTS reaches a negative value which shows that Al3B is the best means of eliminating hot tearing tendency in A380 alloy.

Best quality melt is obtained after when A380 was grain refined by Al3B even when melt was not degassed.

REFERENCES

1.Campbell, J., *Entrainment defects*. Materials Science and Technology, 2006. **22**(2): p. 127-145.

2.Campbell, J., *Complete Casting Handbook: Metal Casting Processes, Metallurgy, Techniques and Design*. 2015: Elsevier Science.

3.Anyalebechi, P.N., *Hydrogen-induced gas porosity formation in Al–4.5 wt% Cu–1.4 wt% Mg alloy*. Journal of Materials Science, 2013. **48**(15): p. 5342-5353.

4.Fakhraei, O., M. Emamy, and H. Farhangi, *The effect of Al–5Ti–1B grain refiner on the structure and tensile properties of Al–20% Mg alloy*. Materials Science and Engineering: A, 2013. **560**: p. 148-153.

5.Yuan, L., D. Chao, and Y.-X. Li, *Grain refining mechanism of Al-3B master alloy on hypoeutectic Al-Si alloys*. Transactions of Nonferrous Metals Society of China, 2011. **21**(7): p. 1435-1440.

6.Dispinar, D., et al., *Degassing, hydrogen and porosity phenomena in A356*. Materials Science and Engineering: A, 2010. **527**(16): p. 3719-3725.

7.Schaffer, P.L. and A.K. Dahle, *Settling behaviour of different grain refiners in aluminium*. Materials Science and Engineering: A, 2005. **413**: p. 373-378.

8.Dispinar, D. and J. Campbell, *Porosity, hydrogen and bifilm content in Al alloy castings*. Materials Science and Engineering: A, 2011. **528**(10): p. 3860-3865.

The 73rd World Foundry Congress
„Creative Foundry"
23 - 27 September 2018, Krakow, Poland

THE EFFECT OF Sr MODIFICATION AND Ti GRAIN REFINEMENT ON THE MECHANICAL PROPERTIES OF A356

Muhammet Uludağ [1]* and Derya Dispinar [2]

[1] *Bursa Technical University, Metallurgical and Materials Engineering, Bursa-16310, Turkey*
[2] *Istanbul University, Metallurgical and Materials Engineering, Istanbul-34320, Turkey*
**Corresponding address: dr.uludagm@gmail.com*

Key words: A356 Alloy, Mechanical properties, Sr modification, Grain refinement, Quality index

1. INTRODUCTION

A356 alloy is used more in automotive industrial as automotive parts such as cylinder block, wheel etc. some features that are low density, high corrosion resistance, high strength etc. are reasons to be preferred for this alloy. It has a good mechanical properties when the heat treatment is done and casting process is carried out without any casting defects [1].

Mechanical properties can be affected easily by casting defects such as porosity [2]. Porosity is one of the most important defects to decrease properties of the alloy and it is formed by oxide films that are claimed by Dispinar and Campbell [3].

Degassing process is mostly used to clean liquid metal in order to produce cast part that has no any casting defect. On the other hand, mechanical properties can be arranged with microstructure of the alloy.

Some element additions are done to modify Si paticles or refined grains. While Sr addition is preffred to modify Si particles, AlTixBy is preffered to rafine α-Al denrites. Because coarse Si particles and large α-Al denrites affect mechanicak properties negatively. Coarse Si particles behave like a crack.

When the a-Al dendrites are in large, porosity can be occurred easily and gotten bigger. Besides their effects of microstructure, these master alloys add to the liquid metal to prevent casting defects. It was claimed that Sr addition has a negative effect on porosity formation that decreases mechanical properties. On the other hand, Grain refiners has a positive effect on porosity formation. AlTixBy has an effect on decreasing porosity less than AlxB (Ti-free) [1, 4-6].

In the present work, AlTi5B1 and AlSr15 master alloys added to A356 alloy to modify microstructure. Trials were done both before degassing and after degassing. Effect of modification and grain refinement on mechanical properties depending on degassing process was investigated.

2. EXPERIMENTAL PROCEDURE

A356 alloy was provided from Eti Aluminyum company in Turkey and used in this study as primer. Chemical composition of the alloy used in this study is given in Table 1. An electrical resistance furnace and SiC crucible were used to melt ingots.

Table 1. Chemical composition of A356 alloy used in this study

Si	Fe	Cu	Mn	Mg	Zn	Ti	Al
7,00	0,20	0,02	0,03	0,35	0,04	0,09	Rem.

Degassing process was carried out with Ar along 20 minutes. AlSr15 was added in amount of 30 ppm and AlTi5B1 was added in amount of 10 ppm. After addition of Sr at 740°C liquid temperature, it was waited for 10 minutes and started to casting. The waiting time for Ti addition was 5 minutes. A tensile test mold that has siz cylindirical bars in diameter 13 mm and in length 150 mm was used to produce samples. Mechanical properties were tested and microstructural analysis were done on tensile test samples.

3. RESULTS AND DISCUSSION

Tensile test was done for all addition parameters depending on degassing process and results of UTS, YS and e % are given in Figures 1, 2 and 3, respectively.

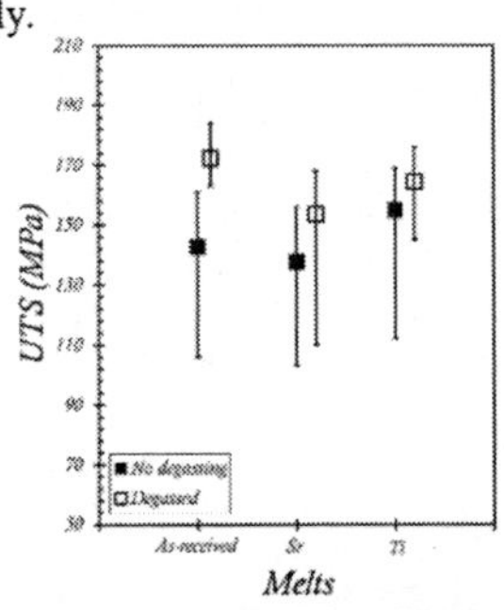

Figure 1. Changes on UTS of the A356 depending on addition parameters for no degassing and degassed conditions.

Effect of addition parameters on UTS of A356 alloy can be seen clearly on the Figure 1. It can be said that while Sr has a negative effect on UTS, Ti has a positive effect on UTS for no degassed condition. However, Sr and Ti have negative effect on UTS fo degassed condition. It can be concluded that from this results Sr increased casting defects that are cause for decreasing of UTS for all degassing process. On the other hand, Ti increased defects for only degassed condition. However, Ti help to clean defects in the liquid for no degassing condition. Same results with UTS can be presented for elongation (Figure 3). On the other hand, Sr and Ti have positive effect on YS for no degassing and degassed condition (Figure 2). It can been seen that YS was increased with Sr and Ti additions. Increasing with Ti addition on YS is more than Sr addition.

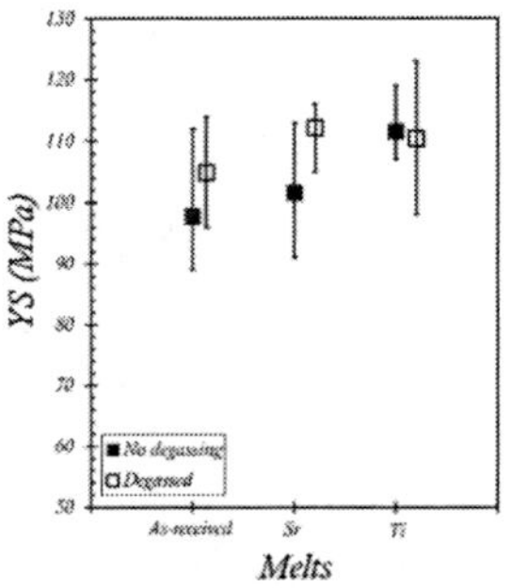

Figure 2. Changes on YS of the A356 depending on addition parameters for no degassing and degassed conditions.

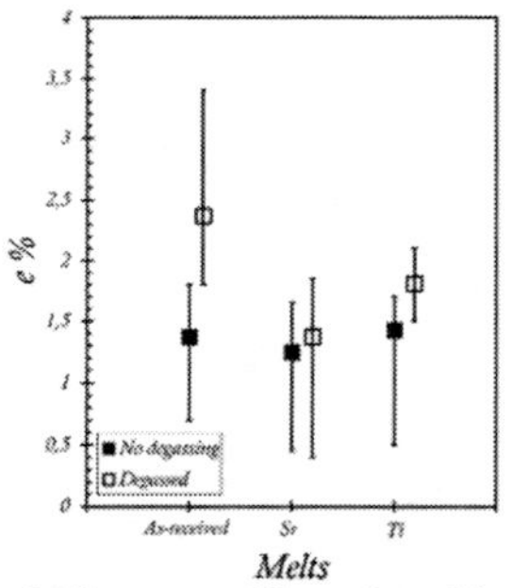

Figure 3. Changes on Elongation of the A356 depending on addition parameters for no degassing and degassed conditions.

Tiryakoğlu and Campbell [7] ranked mechanical properties of the Al alloys with a quality index (Q_T) that is claimed by them. Q_T was adapted for the mechanical results of this study and changes of Q_T values of the addition parameters are given in Figure 4. It can be seen that on the Figure 4, Q_T values have same tendency with UTS and elongation values depending on addition parameters and degassing process. It was concluded that Sr should never add to liquid alloy to increase mechanical properties. However, Tİ can be added to liquid alloy only when degassing process is not applied.

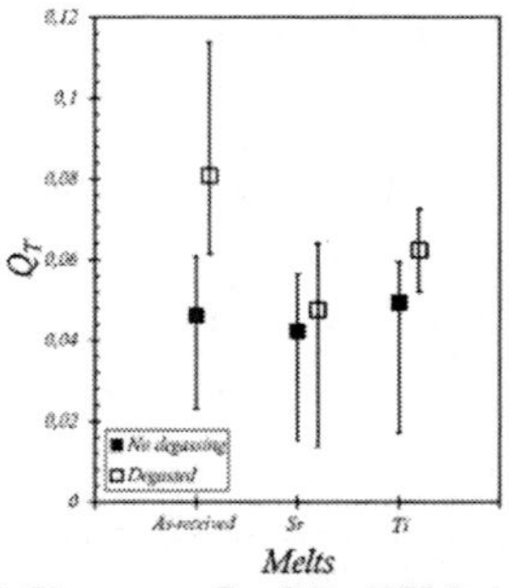

Figure 4. Changes on Q_T of the A356 depending on addition parameters for no degassing and degassed conditions.

4. CONCLUSIONS

Effect of Sr and Ti on mechanical properties depending on degassing process was investigated in this study. All results of the tensile tests show that Sr plays an important role on the mechanical properties to decrease them. However, Ti has a positive effect on mechanical properties when degassing process is not applied.

REFERENCES

[1].Campbell, J., 2015, *Complete casting handbook: metal casting processes, metallurgy, techniques and design*. Butterworth-Heinemann.

[2].Campbell, J., 2003, *Castings*. Butterworth-Heinemann.

[3].Dispinar, D. and J. Campbell, 2011, *Porosity, hydrogen and bifilm content in Al alloy castings.* Materials Science and Engineering: A, **528**(10): p. 3860-3865.

[4].Haro-Rodríguez, S., et al., 2011, *On influence of Ti and Sr on microstructure, mechanical properties and quality index of cast eutectic Al–Si–Mg alloy*. Materials & Design, **32**(4): p. 1865-1871.

[5].Mostafaei, M., et al., 2016, *Evaluation of the Effects of Rotary Degassing Process Variables on the Quality of A357 Aluminum Alloy Castings.* Metallurgical and Materials Transactions B, **47**(6): p. 3469-3475.

[6].Wenwen, D., et al., 2003, *Microstructure and mechanical properties of Mg–Al based alloy with calcium and rare earth additions.* Materials Science and Engineering: A, **356**(1): p. 1-7.

[7].Tiryakioğlu, M. and J. Campbell, 2014, *Quality index for aluminum alloy castings.* International Journal of Metalcasting, **8**(3): p. 39-42.

The 73rd World Foundry Congress
„Creative Foundry"
23 - 27 September 2018, Krakow, Poland

INFLUENCE OF T6 HEAT TREATMENT ON SECONDARY AlSi9Cu3(Fe) ALLOY PRODUCED BY SEMI-SOLID SEED PROCESS

Alberto Fabrizi*, Giulio Timelli, Alessandro De Mori

University of Padova, Department of Management and Engineering, Stradella San Nicola 3, 36100, Italy
**Corresponding address: fabrizi@gest.unipd.it*

Key words: Al–Si alloy, Semi-solid, SEED, T6 heat treatment, microstructure.

1. Introduction

The typical thermal treatment for Al-Si based alloy castings is T6 temper which consists in a solution heat treatment followed by quenching and then age hardening.

It is widely known that the T6 heat treatment for high-pressure die-cast (HPDC) components is still a limited practice in industry. Internal or sub-surface pores can usually appear in the HPDC castings during the solutioning treatment at elevated temperatures (470-500°C), and unacceptable blisters can occur on the component surface [1].

Semi-solid metal (SSM) processes have been considered a valid method for producing sound light-weight parts with satisfactory response to heat treatments.

Few studies have payed attention on the age hardening response of a secondary Al-Si alloy produced via SSM process although secondary Al foundry alloys are the most used Al diecasting alloys in the automotive industry.

In this work, the microstructure and the precipitation hardening response of a SSM secondary AlSi9Cu3(Fe) alloy used for automotive components were investigated as function of different T6 temper parameters.

2. Experimental procedure

In this study, automotive components were manufactured with a secondary AlSi9Cu3(Fe) alloy (EN AC-46000) by a SSM process. The chemical composition of the alloy, measured on separately poured samples, is shown in Table 1.

Table 1. Chemical composition of the experimental AlSi9Cu3(Fe) alloy (wt%).

Si	Fe	Cu	Mn	Mg	Zn	Cr	Al
8.82	0.69	2.1	0.2	0.32	0.9	0.05	bal.

The preparation of the semi-solid slurry was carried out by means of the rheocasting Swirled Enthalpy Equilibration Device (SEED) process; more details are reported in [2].

The specimens were drawn from the as-rheocast components and then T6 heat treated according to the following conditions: solution treatment inside a muffle furnace, water quenching at room temperature and then artificial ageing. The solution treatments were performed at 450, 470 and 490°C for 1, 3 and 6 hours. The artificial ageing was carried out respectively at 160, 180 and 220°C for ageing times ranging between 1 and 30 hours.

Microstructural investiagations were perfomed at differtent scales by using optical microscope (OM), scanning and trasmission electron microsocopes (SEM and TEM). Vickers hardness mesuraments were performed to evaluate the strengthening response of the alloy according to the solution and ageing treatment conditions.

3. Results and conclusions

The hardness values of the AlSi9Cu3(Fe) alloy as function of the solution treatment conditions is reported in Fig. 1.

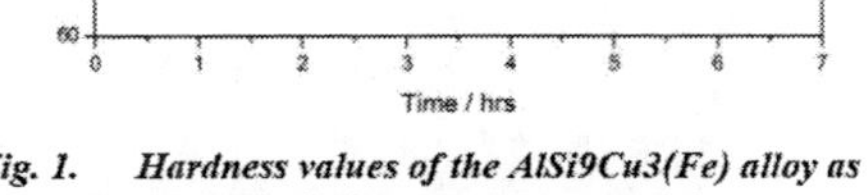

Fig. 1. Hardness values of the AlSi9Cu3(Fe) alloy as function of solution temperature and time.

The hardness of the alloy increases by increasing the solution temperature and time. However, as the alloy solutioned at 490°C for 3 and 6 h showed evident blistering phenomena on the specimen surface, the solution condition considered for the following ageing treatment was 470°C for 6 h.

After the solution heat treatment, the eutectic Si particles fragment and show a more spheroidal shape (Fig. 2) due to the reduction of the total Al/Si interfacial energy during heat treatment; moreover, smaller eutectic Si particles dissolve and favour the coarsening of bigger ones and fine dispersoids precipitate within the α-Al phase. The age hardening behaviour at 160, 180 and 220°C of the 470°C-solutioned alloy is shown in Fig. 3.

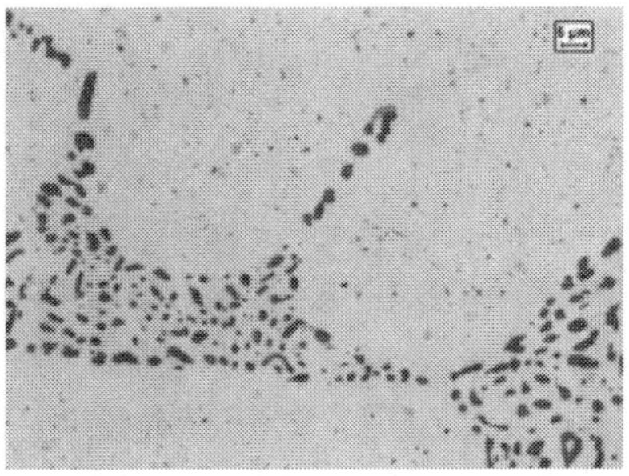

Fig. 2. OM micrograph of the AlSi9Cu3(Fe) alloy after solution heat treatment at 470°C for 6 h.

The results indicate that the maximum hardness value (peak ageing) is achieved at 160°C for 24 h. Upon increasing the ageing temperature, the peak hardness takes place at shorter times as a consequence of faster hardening kinetics; however, the peak value progressively decreases as the temperature increases.

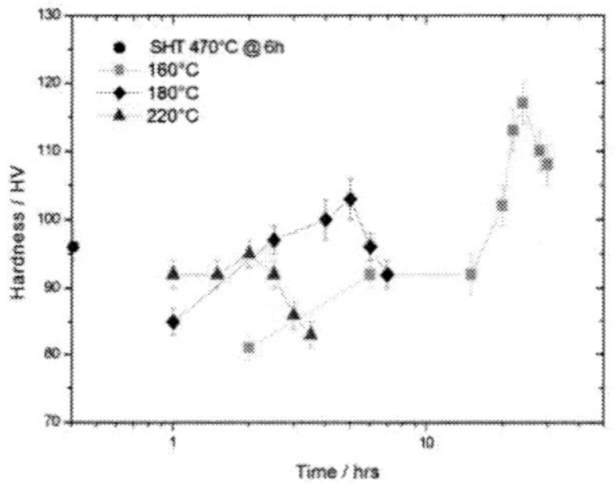

Fig. 3. Age hardening curves of the AlSi9Cu3(Fe) alloy after solution treatment at 470°C for 6 h.

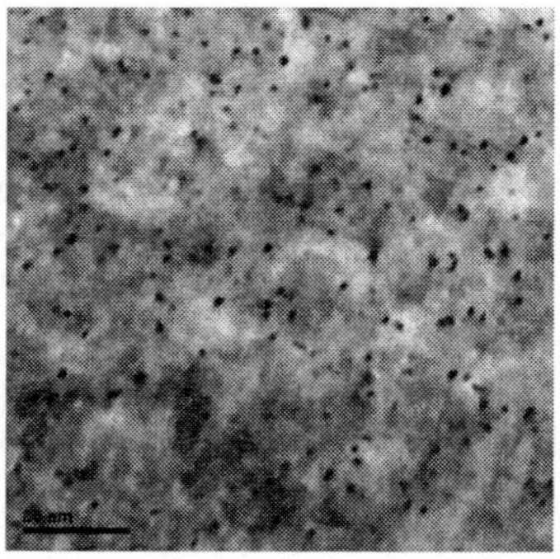

Fig. 4. TEM micrographs of the AlSi9Cu3(Fe) alloy after peak ageing treatment at 160°C.

In Fig. 4, the TEM micrograph of the peak aged condition at 160°C for 24 h reveals a large amount of fine black dots with a square-like shape homogeneously distributed in the α-Al matrix. According to the literature [3] and to the experiemental TEM analysis, these precipitates can be identified as the metastable coherent β" phase. After ageing at 220°C for 2 h (peak ageing), large platelet-shape precipitates apper in the α-Al matrix (Fig. 5). As documented in literature [4,5], θ' phase forms fine platelets lying on {100}Al planes.

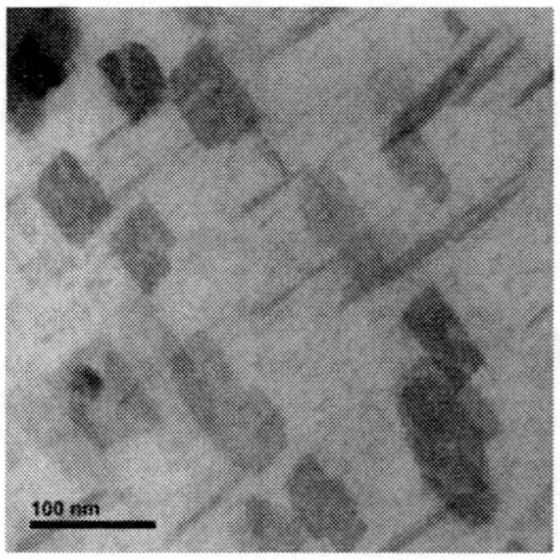

Fig. 5. TEM micrographs of the AlSi9Cu3(Fe) alloy after peak ageing treatment at 220°C.

4. Conclusions

In conclusion, the best strengthening is achieved after artificial ageing at 160°C for 24 h by needle-shaped β" and eventually Q' precipitation; with a further increment of ageing temperature, the formation of large θ' platelets explains the incremental softening of the alloy.

References

[1] R.N. Lumley, R.G. Odonnell, D.R. Gunasegaram & M. Givor. (2007). Heat Treatment of High-Pressure Die Castings. Metallurgical and Materials Transactions A. 38, 2564-2574.

[2] S. Capuzzi, S. Ferraro & G. Timelli. (2016). Development of Heat Treatments for Automotive Components die-cast with Secondary Aluminium Alloy at Semi-Solid state. Trade Foundry Journal. 11, 270-274.

[3] Y.J. Li, S. Brusethaug & A. Olsen. (2006). Influence of Cu on the mechanical properties and precipitation behavior of AlSi7Mg0.5 alloy during aging treatment. Scripta Materialia. 54, 99–103.

[4] R.X. Li, R.D. Li, Y.H. Zhao, L.Z. He, C.X. Li, H.R. Guan & Z.Q. Hu. (2004). Age-hardening behavior of cast Al–Si base alloy. Materials Letters. 58, 2096–2101.

[5] D. Ovono Ovono, I. Guillot & D. Massinon. (2007). Determination of the activation energy in a cast aluminium alloy by TEM and DSC. Journal of Alloys and Compounds. 432, 241–246.

The 73rd World Foundry Congress
„Creative Foundry"
23 - 27 September 2018, Krakow, Poland

SHRINKAGE POROSITY FORMATION IN CAST IRON COMPONENTS

Attila Diószegi [1]*, Jessica Elfsberg [2], Zoltán Diószegi [3] and André Koerich Ramos [4,5]

[1] *Jönköping University, Department of Materials and Manufacturing, Box 1026, SE-55111 Jönköping, Sweden*
[2] *Scania CV AB, SE-151 87 Södertälje, Sweden*
[3] *Volvo Group Trucks Operation, SE-541 87 Skövde, Sweden*
[4] *TUPY S.A., 3400 – 89227-901 - Joinville – SC – Brazil*
[5] *UDESC - Joinville – SC – Brazil*

**Corresponding e-mail: attila.dioszegi@ju.se*

Key words: cast iron, shrinkage porosity, density variation, volume change

Shrinkage porosity formation mechanisms at production of cast iron components is particularly complex and became more and more difficult to understand since the shape casting complexity have been increased more and more in advanced component applications. The present work intends to summarize the state of the art in understanding the shrinkage porosity formation mechanism of cast iron based on a series of key publications, MSc and PhD theses performed in collaboration between cast iron component users, producers and researchers.

1. Introduction

Beside beneficial properties, such tensile properties, thermal conductivity, machinability defect formations tendencies are detracting the advances of producing cast iron components. The critical mechanism at shrinkage porosity formation is related to the sum of the densities of the precipitated phases formed during solidification. The metallic phases precipitates to a denser crystallographic aggregate also known as austenite in relation to the original liquid iron while the carbon forms low density graphite crystals [1]. In connection to the solidification process it has been demonstrated also the role of the solved gaseous elements as hydrogen and nitrogen facilitating the shrinkage porosity formation [2].

2. Experimental observations

Advances in understanding the solidification mechanisms and shrinkage porosity formation were possible due to a series of advances in the investigation methods introduced during the last two decades. The main solidification structure, the austenite grains were observed first after introduction of the heat treatment method named Direct Austempering After Solidification (DAAS) aimed to preserve the crystallographic orientation of the austenite grains instead of the eutectoid transformation [3]. The austenite grains formed on the mould surface classified also as columnar grains creates a container like vessel.

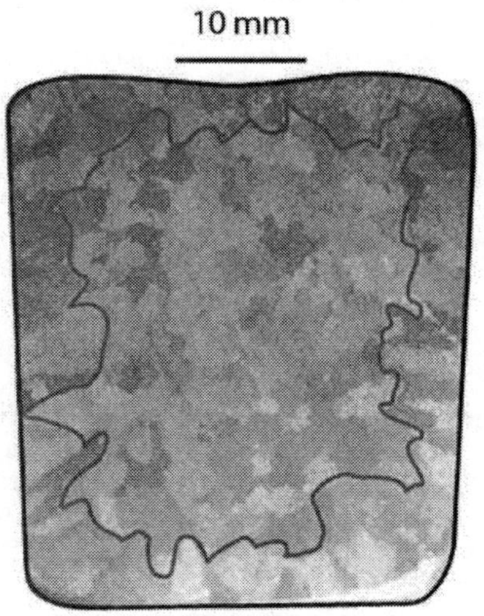

Fig. 1. Columnar and equiaxed grains separated by the columnar to equiaxed transition line in a lamellar graphite iron.

Despite the balance between the high density primary austenite and the low-density graphite phase precipitation a local liquid depression was observed in the extra-dendritic liquid. The driving force of the liquid depression were attributed to the columnar zone expansion caused by the graphite precipitation and volume increase of the vessel [4] Complex shaped casting components were observed to form either a closed vessel separating the equiaxed austenite grains and the extra-dendritic liquid from the surrounding or a discontinuous columnar zone where the equiaxed grains and the extra-dendritic liquid comes in direct contact with the moulding material. A discontinuous columnar zone was found to be the main reason for shrinkage porosity formation at solidification of lamellar graphite iron (LGI) when the surrounding gaseous environment were sucked in between the equiaxed grains.

The evidence for the in-moving external gaseous environment is found on the internal surface of the shrinkage porosity strongly oxidized by the environmental oxygen [5].

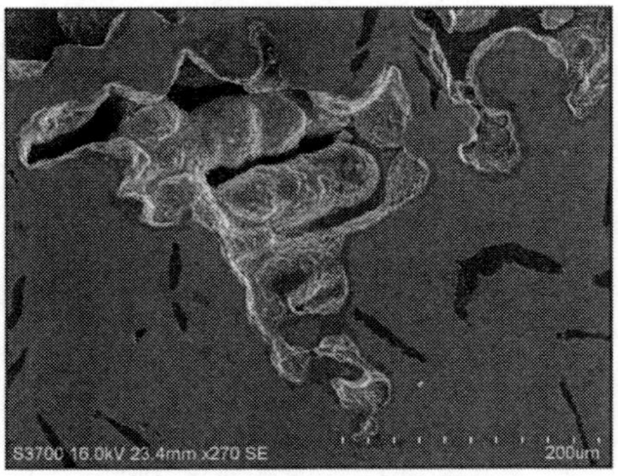

Fig. 2. Shrinkage porosity in lamellar graphite iron with oxidized dendritic surface [5].

Compacted and nodular graphite iron components (CGI, SGI) are produced mainly from near eutectic compositions and less complex geometrical shapes where the columnar grains form dominantly closed vessels. Consequently, the observed shrinkage porosities in CGI and SGI components nucleates in the extra-dendritic zone without any trace of oxidization [6].

Investigated shrinkage cavities in CGI were found to contain strong coarsened dendrite arms, mainly primary arms missing secondary or higher degree arms. The coarsening phenomena in CGI was demonstrated in separate isothermal coarsening experiments where similar coarsened dendritic structure was found as inside the investigated shrinkage cavities [7].

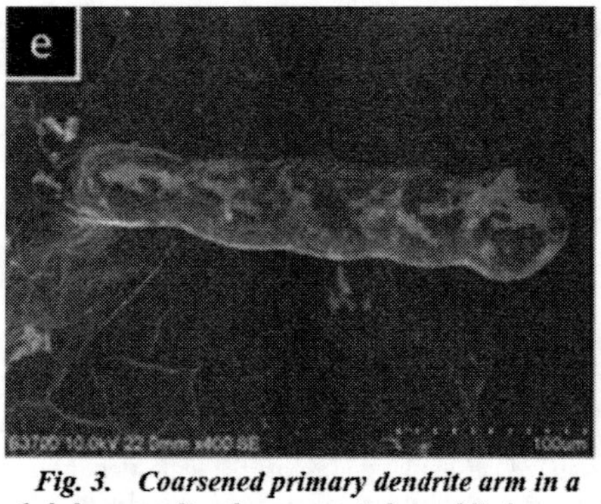

Fig. 3. Coarsened primary dendrite arm in a shrinkage cavity of a compacted graphite iron cast component [6].

3. Conclusion

Advances of methods for investigation the solidification of cast iron promoted the understanding of the shrinkage porosity formation at casting of cast iron components. It has been demonstrated that the macrostructure formation in combination with the casting geometry plays an important role for the shrinkage formation tendency. The extra dendritic zone is the place where the shrinkage porosity forms either by intake of the outer gaseous atmosphere in the case of LGI or by local nucleation driven by the local liquid depression in CGI and SGI. Further advances in understanding the solidification and volume change phenomena are required to understand the influence of metallurgical treatment, chemical composition and moulding material stability with respect to shrinkage porosity formation mechanisms.

Acknowledgments

The present work was realised as a part of the project: Appointment of Professorship in Foundry Technology at Jönköping University financed by Swedish Knowledge Foundation, Scania CV AB, Volvo Group Trucks Operations and Jönköping University. Tupy S.A and UDESC are strategic research partners of Jönköping University. All participating personnel from these companies are greatly acknowledged.

References

[1] K. Hellström: Density variations during solidification of lamellar graphite iron. Licentiate theses at Jönköping University.2017.

[2] A. Diószegi, J. Elfsberg, Z. Diószegi: Solubility of Hydrogen and Nitrogen in liquid cast iron during melting and mold filling. Proceedings of the 71st World Foundry Congress, 21-25th May 2016, Nagoya, Japan.

[3] Rivera, G.; Calvillo, P.; Boeri, R.; Houbaert, Y.; Sikora, J. Examination of the solidification macrostructure of spheroidal and flake graphite cast irons using DAAS and ESBD. Mater. Charact. 2008, 59, 1342–1348.

[4] P. Svidró: On the Mechanisms of Volume Change Related Casting Defects Formation in Lamellar Graphite Iron. PhD Theses at Royal Institute of Technology, Stockholm, 2018.

[5] L. Elmquist: Defect Formation during Solidification in Gray Iron Castings. PhD Theses Chalmers University of Technology. 2009.

[6] S. Vazehrad: A Study on Factors Influencing the Microstructure and Shrinkage Porosity Formation in Compacted Graphite Iron. Licentiate thesis at Royal Institute of Technology, Stockholm, 2014.

[7] J.C. Hernando: Morphological Characterization of Primary Austenite in Cast Iron. Licentiate theses at Jönköping University. 2017.

The 73rd World Foundry Congress
„Creative Foundry"
23 - 27 September 2018, Krakow, Poland

INFLUENCE OF CARBON ON THE FORMATION OF MICROSTRUCTURE AND MECHANICAL PROPERTIES OF Co-Cr-Mo AND Co-Cr-W-Mo CAST ALLOYS

Joanna Augustyn-Nadzieja [1]*, Piotr Matusiewicz [1], Sławomir Kąc [1], Janusz Krawczyk [1]

[1]AGH University of Science and Technology / Faculty of Metals Engineering and Industrial Computer Science, A. Mickiewicza Av. 30, 30-059 Krakow, Poland,
**Corresponding address: jap@agh.edu.pl*

Key words: cast cobalt alloys, dendritic microstructure, metallography, nanohardness, dental alloys

1. Introduction

Co-Cr-Mo and Co-Cr-W-Mo alloys are currently the most common used in non-precious metal prosthetics. The influence of the alloy microstructure on its mechanical as well as functional properties is very significant. The fundamental strengthening phase for Co-Cr alloys are the carbides, which are responsible for increasing the mechanical properties of the alloy, such as hardness or abrasion resistance [1], and on the practical side, bending of clamps in a prosthesis. The precipitation of $M_{23}C_6$ carbides in the alloy may favor the microcracks nucleation, which reduces plasticity, impact resistance and resistance to cracking of these alloys and may initiate of prosthetic fractures. $M_{23}C_6$ carbides present at the cobalt matrix shows a wide range of the acceptable content of constituent elements. If the content of molybdenum in the cobalt alloy reaches the level of 6÷8 mass %, MC or M_2C carbides may be formatted in microstructure [2, 3, 6, 7]. Carbon as a main alloying element have a significant influence at the alloy mechanical properties. Especially on the plasticity of the Co-Cr alloys. The carbon content of 0.3-1.0% allows to obtain high strength and creep resistance. Strengthening of the cobalt alloys is associated with alloy additives such as molybdenum, tungsten, tantalum or niobium. The role of alloy additives is the formation and precipitation in the alloy of carbides, located in the interdendritic spaces of the alloy. In addition, alloying elements added to the cobalt solid solution reduce the energy of the stacking faults [3].

2. Subject of study

For the study Co-Cr-Mo alloy were tested (trade name - Biosil F) and Co-Cr-W-Mo alloy (trade name - Gialloy CB-H). The chemical composition of tested alloys is given in Table 1.

Table 1. Chemical composition of investigated Co-Cr-Mo and Co-Cr-W-Mo cast alloys (mass. %)

Co	Cr	Mo	W	C	Mn	Nb	V	Si
64.8	28.5	5.3	-	0.4	0.5	-	-	-
58	25	1	9	<0.04	-	2	2	1

3. Research methodology

The microstructure was observed with a light microscope Leica DM 4000, scanning electron microscopy Hitachi S-3500N with the EDS Noran 986B-1SPS and the Hitachi SU-70 analyzer. Nanohardness studies were performed by the use of CSM Instruments 50-183 NHT. The measurements were performed at a load of 20 mN using a Berkovich indenter [4]. Measurement time was 10 s with increasing force equal 40 mN / min.

4. The results of research

The primary microstructure of the tested alloys, after crystallization under conditions typical for prosthetic castings, consists of very large crystallites (Fig. 1), which are a solution of chromium, molybdenum and carbon (for the Biosil F alloy) and additionally tungsten for the Gialloy CB-H alloy in the β-Co, and carbide precipitations of the type $M_{23}C_6$, which cause a significant strengthening of the alloy. In the cast cobalt alloys carbides MC, M_2C and M_6C and M_7C_3 may also occur [5-7].

EDS analysis revealed that for both: dendrites and the interdendritic spaces, segregation of the Cr and Mo compared have occurred. Addition of Mo in Co-Cr alloys improves the ductility of the alloy, increases the range of solidification and changes the morphology of the precipitated material by segregation producing. Additionaly amounts of eutectic carbides have increased [6]. In the case of the alloy Gialloy CB-H (Co-Cr-W-Mo) to the areas of dendrites "migrates" tungsten and niobium.

The increasing of tungsten concentration reduce the stacking fault density and increases the fraction of the intermetallic compound σ Co-Cr-W-Mo [7]. Nanohardness measurements (HV) shows that alloy Biosil F exhibit value about 694 HV - for dendritic matrix and 846 HV in the interdendritic areas.

For the Gialloy CB-H Dendritic matrix nanohardness was about 571 HV and for interdendritic areas 696 HV.

The results obtained for nanohardness Co-Cr alloys allow us to conclude that the interdendritic spaces have a higher hardness compared to dendritic matrix. The specified property depends on the content of mass % carbon , thus the amount of carbides, which are formed in the core and interdendritic areas.

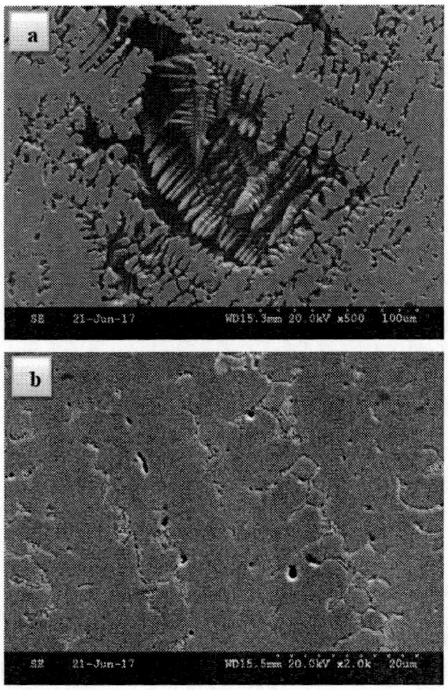

Fig. 1. SEM images of investigated alloys Co-Cr-Mo Biosil F (a), Co-Cr-Mo-W Gialloy CB-H (b)

5. Summary

- The investigated Co-Cr alloys were characterized by heterogeneous microstructure with clearly outlined dendrites.
- The microstructure observed on the scanning electron microscope possesses artifacts and gas microbubbles (pores, shrinkage porosity). In the areas of dendritic cores as well as interdendritic spaces were enriched in Mo, Co and Cr, additionally in W and Nb in the case of the Gialloy CB-H alloy. The austenitic matrix was a chromium in cobalt solid solution.
- The higher nanohardness in inter-dendritic spaces is caused by alloying elements segregation to these areas and by formation of the carbides.

Acknowledgments

The work has been implemented within the framework of statutory research of AGH University of Science and Technology, contract No 11.11.110. 299 AGH.

References

[1] Augustyn-Pieniążek J., Kurtyka P., Sulima I., Stopka J. (2015). Selected properties and tribological wear alloys Co-Cr-Mo and Co-Cr-Mo-W used in dental prosthetics. Archives of Metallurgy and Materials, Vol. 60 (3), pp. 1569-1574.

[2] Górny Z. (2008). Cobalt casting alloys. Institute of Foundry, Krakow (in Polish).

[3] Bojar Z. (2010). Analysis of the structure's impact on corrosion resistance and cracking mechanism of Vitalium type cobalt alloys. WAT, Warsaw (in Polish).

[4] Oliver W.C., Pharr G.M. (1992). An improved technique for determining hardness and elastic modulus using load and displacement sensing indentation experiments. Journal Materials Research Vol. 7, Issue 6, pp.1564-1583.

[5] Karpiński R., Walczak M., Śliwa J. (2015). Tribological research of cobalt alloys used as biomaterials. Journal of Technology and Exploitation in Mechanical Engineering, Vol. 1, Issue 1-2, pp. 17-32.

[6] Lee S-H., Takahashi E., Nomura N., Chiba A. (2006). Effect of Carbon Addition on Microstructure and Mechanical Properties of a Wrought Co-Cr-Mo Implant Alloy. Materials Transactions, Vol. 47, No. 2, pp. 287-290.

[7] Podrez-Radziszewska M., Haimann K.,. Dudziński W, Morawska-Sołtysik M. (2010). Characteristic of intermetallic phases in cast dental CoCrMo alloy. Archives of Foundry Engineering Vol. 10, Issue 3, pp. 51-56.

The 73rd World Foundry Congress
„Creative Foundry"
23 - 27 September 2018, Krakow, Poland

INFLUENCE OF THE LOW Ti ADDITION ON GRAY CAST IRON PROPERTIES PRODUCED WITH INCREASED STEEL SCRAP IN THE CHARGE

Peter Futas 1*, Alena Pribulova 2, Marcela Pokusova1, Jozef Petrik2, David Mahut2

2 Slovak university of technology in Bratislava, Vazovova 5, 81243 Bratislava, Slovakia,
1 Technical university of Kosice, Letna 9, 04011 Kosice, Slovakia,
**Corresponding address: peter.futas@tuke.sk*

Key words: steel scrap, cast iron, mechanical properties, synthetic cast iron

The use of steel scrap in the charge of cast iron production is very effective for economic reasons. In the long time, the negative impact of steel in the batch is known to have multiple gray cast iron properties, respectively also ductile iron, in particular to increase the hardness of HB and tensile strength Rm, respectively deterioration in RG maturity and relative hardness. These features increase the cost of mechanical machining of castings. The increased brittleness of the castings was also confirmed.

The cause of these negative mechanical and physical properties is most often referred to as the increased nitrogen content of the steel and hence also of the cast iron produced.

The most effective procedure for eliminating the adverse impact of steel on the quality of gray cast iron is the addition of titanium (0.2-0.3%). In more detail, however, this addition was not used in practice, so we verified its effect under operating conditions. We have also researched change the other properties of gray cast iron, in particular a change in hardness from the wall thickness of the casting, liability to chillout, to shrinkages and residual tensions.

The statistically evaluated quality in foundry (122 melts) with quality of synthetic gray cast iron (100% steel scrap in charge) with quality of gray cast iron alloyed with 0,183 % Ti with portion of 35% steel scrap in charge was compared which is the common practice. Tthe nitrogen content was measured, which had a significant effect on the resulting properties.

A new knowledge is the pronounced overflow of the cast iron in the induction furnace, depending on the melting speed and in particular on the surface area of the charge surface. Also, the high carburization of small-scale steel scrap with a large surface (steel sheets) even after an initial low nitrogen content of about 30 to 40 ppm at 120 to 140 ppm. This knowledge is verified in practice [1,2,3]).

On the basis of the initial statistical results (122 melts [4,5,6]) three experimental melts of gray cast iron by standard EN-GJL 250 with constant degree of saturation (Sc) were realized:

1. The most commonly produced (current production) halfsynthetic gray iron
2. Same the A and was alloying with titanium
3. Fully synthetic gray cast iron (100% steel scrap)

This melts were realized in two medium frequency induction furnace „Siemens" with following parameters:

- the nominal content of the furnace 6 tons
- melting furnace power 5350kg/h
- nominal frequency 500 Hz

The cast iron was by the bar qauge treated FeSi75.

Next samples were a casted from the melts:

- samples for chemical analysis
- stairs test (R-Block with a thickness of 100, 50 and 20 mm – standard thickness of the walls castings, fig. 1), to determine the sensitivity difference of the thickness the walls of the castings hardness (cooling rate)
- test bars with diameter of 30 mm (tensile strength - Rm)
- wedge-shaped test bar, fig. 2 [7]
- samples to determine the tendencies to shrinkages create (cylinder with diameter ø95 mm and height of 150 mm)
- samples to determination of nitrogen content in gray cast iron

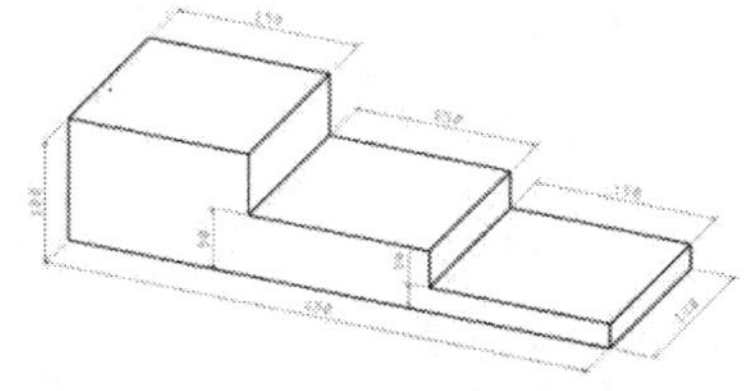

Fig. 1. Stairs-test (R-block)

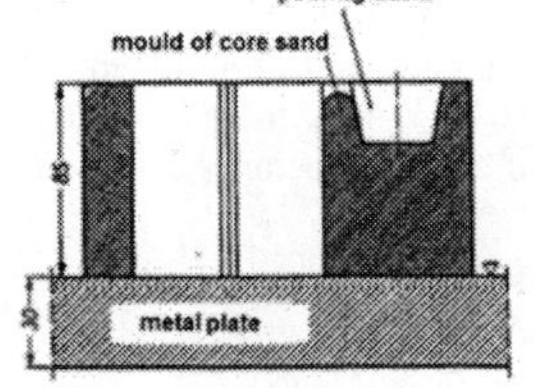

Fig. 2. Wedge-shaped test bar

Table 1. The results of the chemical composition, contents of N2, mechanical properties, depth of chillout and tensile stresses

Melt	Portion of steel scrap [%]	wt [%]			Sc	CE [%]	Rm [MPa]	HB	Depth of chill wedge [mm]
		C	Si	N_2					
1	33	3,23	1,612	0,0113	0,848	3,721	297	185 *(205)	5
2	35,5	3,32	1,486	0,0091	0,863	3,773	268	217 *(224)	1
3	97,8	3,28	1,69	0,0175	0,869	3,805	257	243 *(312)	10

Note: * hardness HB measured on thin wall of casting (20 mm)

The Samples for metalografical analysis were taken from the test bars and prepared by a standard manner.

R-Block was cuted in each of the cross sectional areas. On every section the Brinell hardness was measured. The hardness was measured with the durometer HPO 3000 (terms: 10/3000/10), i.e. the diameter of ball 10 mm, strength duration 3000 N and period of load 10 sec.

Tensile strength was measured on a test-bars (Ø30 mm) on the universal crackle machine ZWICK.

Table 1 shows the mechanical properties, degree of saturation Sc, depth of chill wedge and tensile stresses.

The tensile strength Rm and hardness HB in melts 1. – 3. shows fig. 3. All melts were undereutectic cast iron with degree of saturation Sc<1 and carbon equivalent CE< 4,23. Chemical composition and mechanical properties of melts according to EN standard for cast iron GJL-250. All melts were innoculated on furnace channel the same procedure (approx. 0.2 % wt. "INOCAST") or SiC into the furnace (approx. 0,3 % wt.).

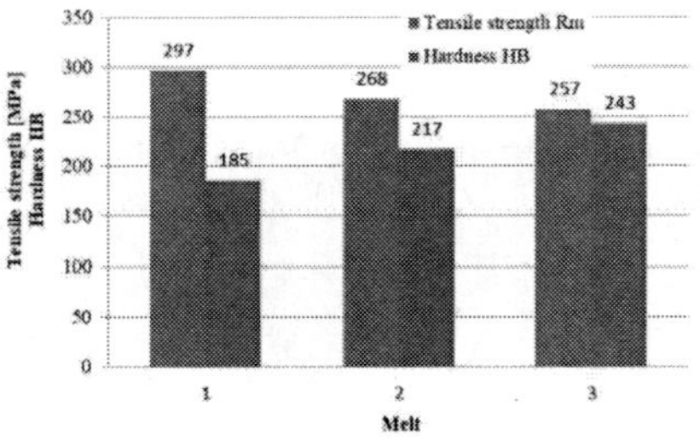

Fig. 3 Tensile strength Rm and Hardness HB in melts

The chemical composition of melted cast iron responded to requirements of identical Sc 0,848 - 0,869. The nitrogen content in melts was lowest at melt "2" (N2 = 0,0091%), which was alloyed with titanium. The highest nitrogen content was in melt "3" (N2 = 0,0175). This melt was fully synthetic gray cast iron and steel scrap was the main source of nitrogen.

Tensile strength Rm in all melts complied with the EN standards. In melt "3", i.e. synthetic gray iron there were the lower value (257 MPa). Melt "5" require higher overheating (over 1450 °C). The expected impact of the high content of N=0,0175% on the tensile strength was not showed.

Hardness HB was in the range of 185 to 243 in the thickness of 50 mm which corresponding with standard STN and EN-GJL-HB255 (185-200 HB). It was markedly higher in melt "3" (fullsynthetic gray cast iron).

Acknowledgments (optional)

A This work was supported by the Scientific Grant Agency of The Ministry of Education of the Slovak republic No. VEGA1/0073/17, VEGA 1/0703/16, and APVV-16-0485.

References

[1] Kagawa, A., Okamoto, T.: Partition of alloying elements in freezing cast irons and its effect on graphitization and nitrogen blowhole formation, Osaka, 2011, The institute of scintific and Industrial research Osaka University

[2] Wilberfors, F., Ingvar, Svensson, I.: The effect of nitrogen and inoculation on the tensile properties and microstructure of cast iron lamellar, Södertälje, Jönköpings university, 2010

[3] Qujie, Zhai, Hanqi, H.: Effect of nitrogen on matrix structure of gray cast iron, Acta Metallurgica Sinica, series A, Vol. 6, No. 6, pp. 370 372, 1993

[4] Futáš, P., Pribulová, A.: Quality criteria as implement for advisement of metallurgical quality of grey iron, Evaluation of people and products features, Celje 2014, p. 99 – 118

[5] Futáš, P., Jelč, I., Bartošová, M.: Usage of Quality Criteria for Advertisement of Metalurgical Quality of Gray Cast Iron, Quo Vadis Zlievarenstvo, 4. ročník medzinárodnej konferencie, Tále, 10. - 12.10.2012, Košice, TU, 2012, s. 58-64, ISBN 978-80-553-1127-2

[6] Futáš, P., Jelč, I., Vasková, I. a kol.: The Gist of Thermal Stresses of Cast Iron Castings, Manufacturing technology, June 2013, Vol 13, No. 2, p. 173 – 177

[7] Roučka, J.: Metalurgie litin, Brno, 1998

The 73rd World Foundry Congress
„Creative Foundry"
23 - 27 September 2018, Krakow, Poland

THE WEAR MECHANISM OF MILL BEATERS FOR COAL GRINDING MADE-UP FROM HIGH MANGANESE CAST STEEL

Janusz Krawczyk [1]*, Piotr Matusiewicz[1], Łukasz Frocisz[1], Joanna Augustyn-Nadzieja[1], Sławomir Parzych[2]

[1]AGH University of Science and Technology / Faculty of Metals Engineering and Industrial Computer Science, A. Mickiewicza Av. 30, 30-059 Krakow, Poland
[2]Cracow University of Technology / Faculty of Mechanical Engineering, Warszawska St. 24, 31-155 Krakow, Poland
**Corresponding address: jkrawcz@agh.edu.pl*

Key words: cast steel, Hadfield, wear, coal grinding

1. Introduction

In Poland the main energy source is an energy obtain in coal power plants. The mostly common used power plants fuel are hard and brown coal. Relevant issue is a fuel preparation to obtain the most efficient incineration. The high fuel incineration efficiency is strictly related to the efficiency of whole power plant installation and energy production. For this purpose coal is milled to the possible the smallest fraction. The finest grinding increases the incineration rate and decreases losses caused by incomplete fuel combustion. Contemporaneously with the decreasing of carbon fraction more intense is wear of the mill elements. It leads to the decreasing the mill efficiency in a time unit. Materials used in mills should be characterized by the high wear resistance [1].

2. Material for investigation

Material for the investigation was 400x120x40 mm and 400x80x40 mm plates. Such plates are used as a beaters in beater mill for coal grinding. By the producer material was classified as L120G13 high manganese cast steel (Hadfield cast steel). Chemical composition analysis shows that for the investigated material the carbon content is much lower to classified it as Hadfield cast steel (Table 1).

Table 1. Chemical composition of investigated cast steel (mass %)

C	Mn	Si	P	Cr	Ni	Mo	Al	Co	Fe
0.43	12.20	0.75	0.060	0.42	0.12	0.03	0.03	0.13	Bal.

Investigated plates ware characterized by the typical for crystallization microstructure. The frozen, columnar and equiaxed grain zones ware observed. Dendritic character of crystallization was also observed (Fig. 1).

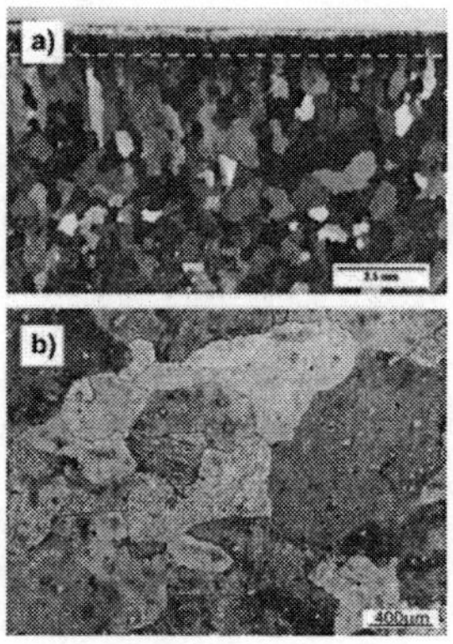

***Fig. 1.** Macrostructure (a) and microstructure (b) of investigated material with the dendritic character of crystallization*

Analyzed cast steel due to the high alloying elements content and heat treatment have a high tendency to deformation strengthening. Investigations of material deformation strengthening was performed by the hardness measurements of the cast interior before and after deformation Hardness measurement for investigated material shows value 190 HBW. Deformation strengthening obtain by the indenter delving should increase the material hardness to 500 HBW. During the investigation 450 HBW was obtain.

3. Investigations results

As a result of exploitation investigated plates ware highly worn. The main mechanism of wear was erosion, moreover the chipping of the former austenite grains was observed. Significant role in the chipping of former austenite process have a manganese cementite. Documentation of the wear stages is presented at Figure 2.

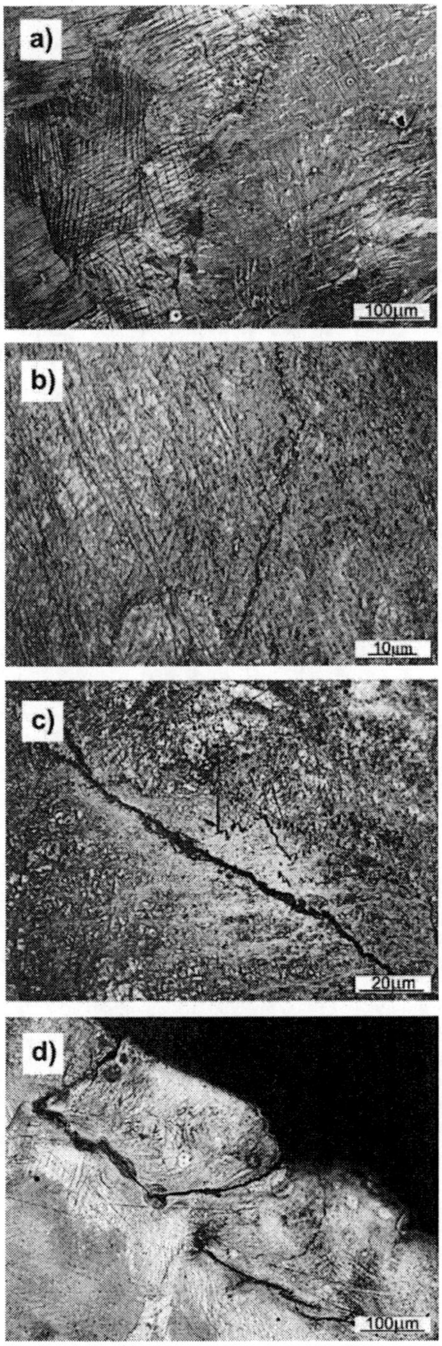

Fig. 2. Microstructures of investigated materials in different wear stages: plastic deformation (a), deformation concentration at the area of manganese cementite (b), grain boundary dekohesion (c), chipping of the former austenite grains (d)

Analysis of the hardness profile from material surface have shown that exploitation caused the deformation strengthening of investigated material (Fig. 3).

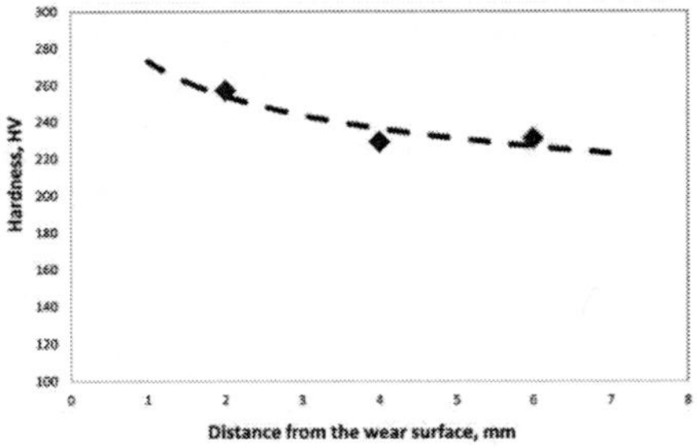

Fig. 3. Material strengthening due to the exploitation

4. Conclusions

1. Decrease of the alloying elements content, especially the elements affect on deformation strengthening - such as carbon (about 0,7% decrease), resulted in high intense wear.
2. The presence of carbides in microstructure, caused the cracks formation along the area of carbides precipitations. It indicates that material was not fully supersaturated.

Acknowledgments

The work has been implemented within the framework of statutory research of AGH University of Science and Technology, contract No 11.11.110. 299 AGH. The Authors would like to thank Adrian Burza for help in the realization of the presented investigations.

References

[1] Z. Stradomski. (2010). The microstructure in problems of wear-resistant steel castings. Technical University of Czestochowa. Poland (in Polish).

The 73rd World Foundry Congress
„Creative Foundry"
23 - 27 September 2018, Krakow, Poland

ROLE OF CHEMICAL COMPOSITION ON SECONDARY CEMENTITE MORPHOLOGY IN ALLOYED CAST STEELS

Janusz Krawczyk [1]*, Piotr Matusiewicz [1], Łukasz Frocisz [1], Joanna Augustyn-Nadzieja [1]

[1] *AGH University of Science and Technology / Faculty of Metals Engineering and Industrial Computer Science, A. Mickiewicza Av. 30, 30-059 Krakow, Poland*

**Corresponding address: jkrawcz@agh.edu.pl*

Key words: Adamite type cast steel, ledeburite cementite, secondary cementite, mill rolls, chemical composition

1. Introduction

Hypereutectoid and hypoeutectic cast steel has a carbon content from 1.2% to 2.1% about the 1% mass Cr, and as a main additions: Mn, Ni, Mo and Si. One of the most common usage of such kind of materials are mill rolls, called adamite mill rolls. [1]. Chemical composition of these materials influent on the volume fraction of microconstituents in microstructure [2].

Secondary cementite volume fraction in ledeburite in a significant way influent on the fracture toughness and tribological properties of such materials [1-3].

It a relevant to design the chemical composition of alloyed cast iron to obtain the proper volume fraction and morphology of secondary cementite precipitation in a former austenite grains.

2. Material and methodology of investigations

Material for the research was sampled from the seven mill rolls made-up from alloyed cast steel. Chemical composition of investigated samples are presented in Table 1.

Microstructure of investigated materials are presented in Figure 1. On the basis of microstructural investigation the volume fraction of each microconstituents was analyzed.

Table 1. Chemical composition of investigated cast steels (mass %)

No.	C	Mn	Si	Cr	Ni	Mo
1	2.10	0.50	0.70	1.20	1.20	0.20
2	2.00	0.70	1.12	0.84	0.51	0.33
3	2.00	1.10	1.20	1.20	0.30	0.10
4	1.99	0.80	1.30	1.24	1.81	0.35
5	1.83	0.64	0.56	1.30	0.47	0.33
6	1.34	0.65	1.12	0.83	0.60	0.19
7	1.22	0.73	0.51	0.99	0.47	0.42

3. Results

Carbon content increase equaled the increase of ledeburitic cementite and graphite volume fraction. As a consequence the volume fraction of secondary cementite has decreased. When the hypoeutectic composition of the materials was obtain, by the increase of the carbon content, higher tendency to Widmansttäten secondary cementite formation was observed (Fig. 2).

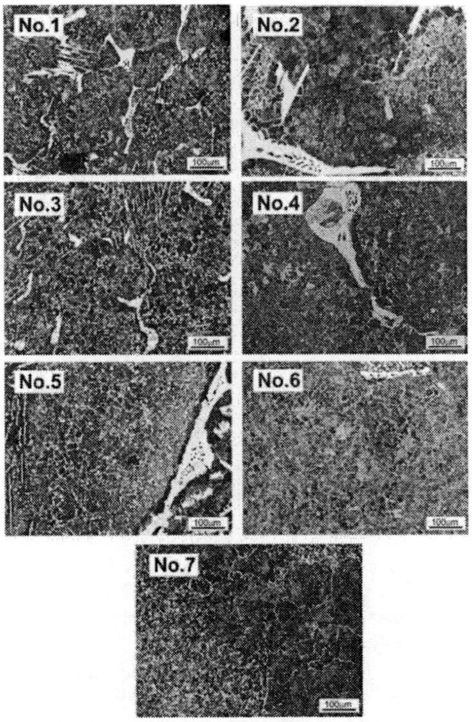

Fig. 1. Microstructures of investigated cast steels (Table 1)

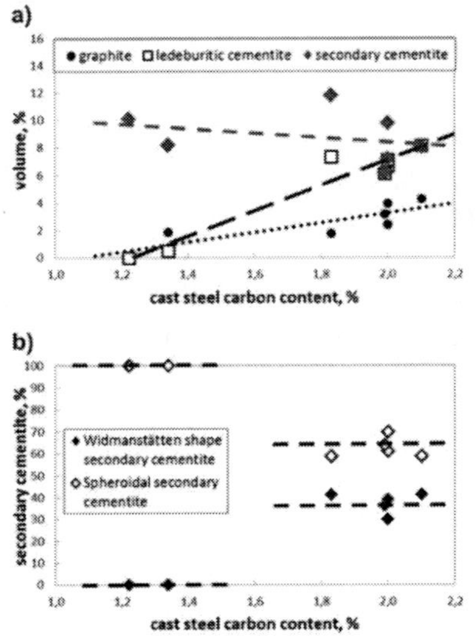

Fig. 2. Influent of carbon content on: a) volume fraction of microconstituents, b) volume fraction and morphology of secondary cementite, in investigated materials

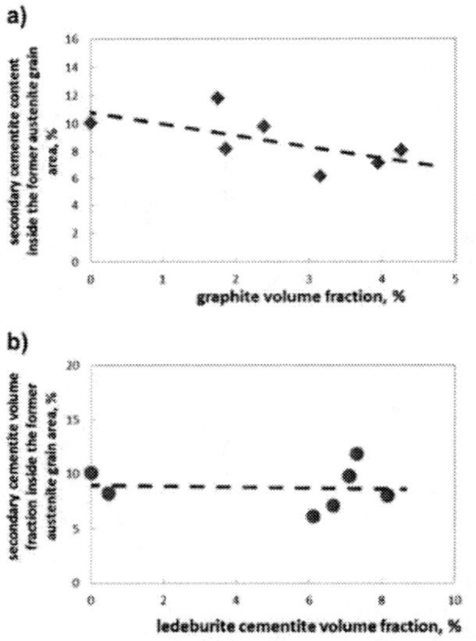

Fig. 3. Relation of volume fractions of: graphite (a), ledeburite cementite (b), to secondary cementite

Secondary cementite volume fraction decrease is related to the graphitization process. Also part of the secondary cementite was integrating into the ledeburite cementite (Fig. 3).

In the case of the cast steels with the 2 mass % of carbon, increase of other alloying elements content resulted in the decrease of secondary cementite volume fraction (Fig. 4).

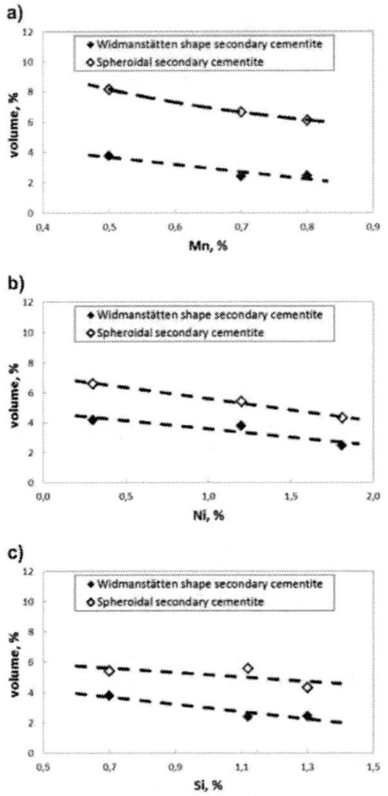

Fig. 4. Influence of: a) Mn, b) Ni, c) Si, content on volume fraction of secondary cementite

Acknowledgments

The work has been implemented within the framework of statutory research of AGH University of Science and Technology, contract No 11.11.110. 299 AGH.

The authors would like to thank Sławomir Kruczała and Izabela Orłowska for help in the realization of the presented investigations.

References

[1] J. Krawczyk. (2007). The dependence between of tribological wear of adamite cast steel from the microstructure. Tribologia. 212, 38(2), 325–337 (in Polish).

[2] J. Krawczyk. (2011). The effect of cast steel chemical composition on its microstructure and tribological properties in mill rolls. Journal of Achievements in Materials and Manufacturing Engineering. 47(2), 137–148.

[3] J. Krawczyk, E. Rożniata & J. Pacyna. (2005). The influence of hypereutectoid cementite morphology upon fracture toughness of chromium–nickel–molybdenum cast steel of ledeburite class. Journal of Materials Processing Technology. 162–163, 336–341.

The 73rd World Foundry Congress
„Creative Foundry"
23 - 27 September 2018, Krakow, Poland

DETERMINATION OF CONTENT OF PHENOL IN FOUNDRY RESINS BY PYROLYSIS GAS CHROMATOGRAPHY-MASS SPECTROMETRY METHOD

Sylwia Żymankowska-Kumon [1]*

[1]AGH University of Science and Technology, Faculty of Foundry Engineering,
Reymonta 23 Street, 30-059 Kraków, Poland
**Corresponding address: szk@agh.edu.pl*

Key words: gas chromatography, mass spectrometry, pyrolysis, resin, phenol

1. Introduction

Foundry engineering belongs to an industrial branch of an increased professional risk. During the production process workers are exposed to dangerous and harmful agents, among which the emission of hazardous substances is essential [1,2].

One of the sources of hazardous substances are moulding sands containing resins, which at high temperatures undergo thermal decompositions. Depending on the kind of the applied binder, under such conditions, the following substances can be formed and emitted to the environment: phenol, formaldehyde, furfuryl alcohol, aromatic hydrocarbons from BTEX group (benzene, toluene, ethylbenzene, xylenes), PAHs (polycyclic aromatic hydrocarbons) etc. [3-5].

Phenolic resins are a group of the most versatile polymers. Although they came into existence at the very start of the age of polymers, they continued to be developed into more and more applications. Usually phenolic resins are divided into two different types, novolacs and resoles. Both have high temperature stability up to about 350° C, high water and chemical stability [3,6]. The phenol is the one of main compounds of foundry resins used in the cold-box technology, but also for resin-coated sands used for forming hot core blocks and shell moulds.

Phenol is not classifiable as to its carcinogenicity to humans but its vapors are corrosive to the eyes, the skin, and the respiratory tract. Its corrosive effect on skin and mucous membranes is due to a protein-degenerating effect. The substance may cause harmful effects on the central nervous system and heart, resulting in dysrhythmia, seizures, and coma [6].

2. Research and methodology

The aim of the research was to use the pyrolysis gas chromatography-mass spectrometry method (Py-GC/MS) to assess the phenol emission from resin-coated sands (based on phenolic resin) used in the production of foundry hot cores and shell moulds (called S1, S2 and S3). Gas chromatography is one of the useful techniques, which permits to determine the percentage composition of mixtures of chemical compounds. The good point of this method is, among others, its high sensitivity due to which very small samples of the analyzed substances can be used (approximately 1μl or 1 mg) [7,8]. Identification of phenol was carried out by means of the pyrolysis gas chromatography-mass spectrometry method. The process parameters are shown in Figure 1.

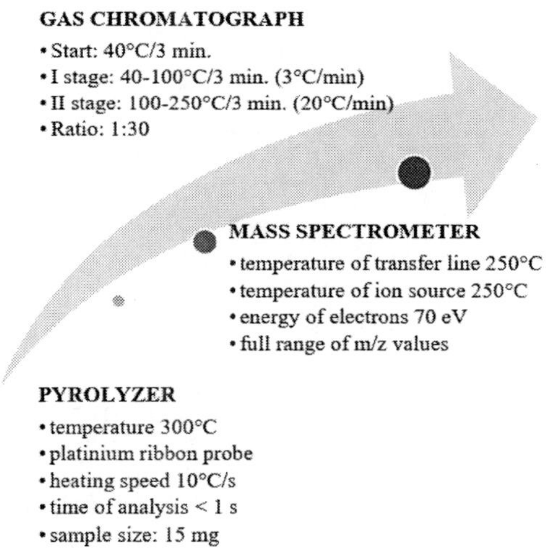

Fig. 1. The parameters of the Py-GC/MS process

3. Results

Figures 2-4 present the results of research in the form of chromatograms for the tested materials. The chromatograms contain the main signals come from carbon dioxide and phenol, and in the case of samples "S3" - small content of benzene was identified. The area for a retention time of more than 30 minutes contains signals related to the column coating (stationary phase: 1,4-bis(dimethylsiloxy)phenylene dimethyl polysiloxane) and is not analyzed.

The tested materials contain in their chemical composition mainly: phenol, salicylic acid, hexamethylenetriamine and calcium stearate (base for all) and in the case of sample S3 additionally bisphenol A (increased carbon dioxide emission in the chromatographic spectrum – Fig. 4).

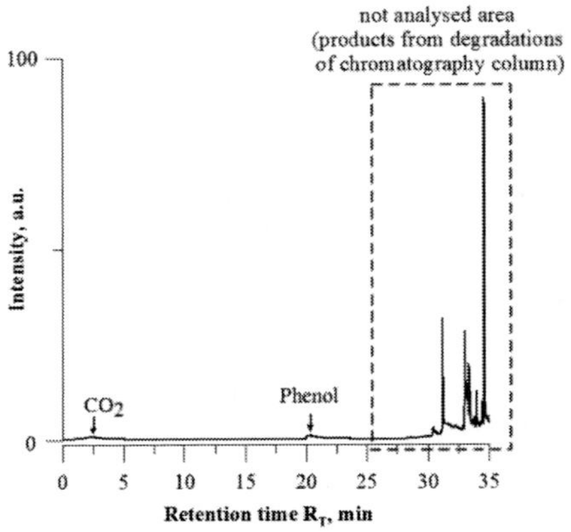

Fig. 2. Chromatogram of sample S1

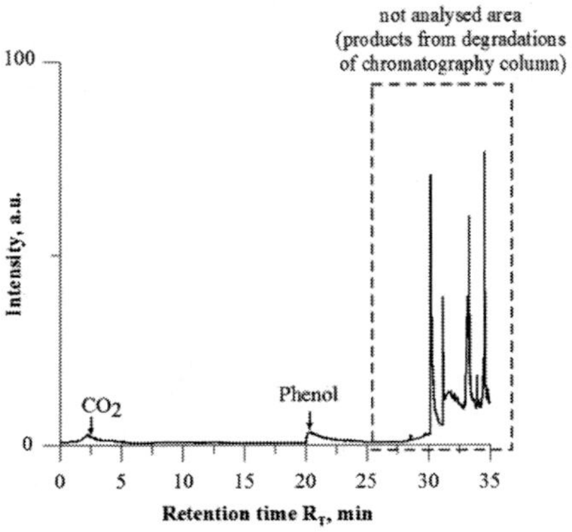

Fig. 3. Chromatogram of sample S2

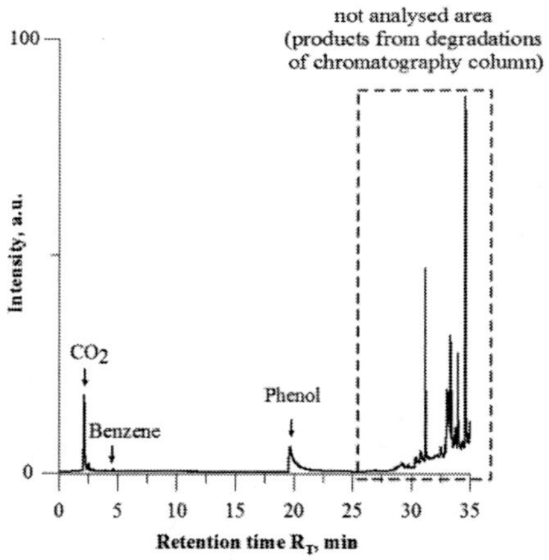

Fig. 4. Chromatogram of sample S3

The phenol emission is the highest for sample S2 and S3, the smallest for sample S1 (Table 1). Probably this is due to the higher proportion of resin in sample S3 or too little hardener (no total polymerization of the binder).

Table 1. Emission of phenol in tested samples

Sample	Phenol emission, μg/mg	Relation to the volume of the whole sample	
		%	mg
S1	34,86	3,49	0,523
S2	83,63	8,36	1,254
S3	94,22	9,42	1,413

4. Conclusions

The research shows that the identification of phenol by pyrolysis gas chromatography-mass spectrometry method (Py-GC/MS) is possible and reliable, especially based on the internal standards of the compounds. In the samples of tested resin-coated sands, the phenol emission does not exceed the environmental standards, however, it is the highest for samples S2 and S3. After taking into account the user's properties, the chemical composition needs to be adjusted in order to achieve the lowest possible phenol emission.

Acknowledgments

The study has been realized within Dean's grant no. 15.11.170.596.

References

[1] Gerber A.H. (2000). Patent US No.US006046252A. United States of America.

[2] Karbhari V.M. (2013). Non-Destructive Evaluation (NDE) of Polymer Matrix Composites. United States of America. Woodhead Publishing.

[3] Asahi Yukizai Corporation. What Are Phenol Resins? 28.03.2018, http://www.asahi-yukizai.co.jp.

[4] Biedermann M., Grob K. (2006). Phenolic resins for can coatings: I. Phenol-based resole analysed by GC–MS, GC×GC, NPLC–GC and SEC. LWT – Food Science and Technology, 39 (6), 633-646.

[5] Li C., Ma Z., Zhang X. et al (2016). Silicone-modified phenolic resin: Relationships between molecular structure and curing behavior. Thermochimica Acta. 639, 53-65.

[6] Pilato L. (2013). Phenolic resins: 100 Years and still going strong. Reactive and Functional Polymers. 73 (2), 270-277.

[7] Żymankowska-Kumon S., Kolczyk J (2016). The chromatographic analysis of selected products of thermal decomposition from moulding sands performed in cold-box technology. Transactions of the Foundry Research Institute. 16 (4), 369-378.

[8] Grabowska B., Kaczmarska K. et al (2017). TG-DTG-DSC, FTIR, DRIFT, and Py-GC-MS studies of thermal decomposition for poly(sodium acrylate)/ dextrin (PAANa/D). Journal of Casting & Materials Engineering. 1 (1), 27-32.

The 73rd World Foundry Congress
„Creative Foundry"
23 - 27 September 2018, Krakow, Poland

INFLUENCING CRYSTALLIZATION OF AUSTENITIC MANGANESE STEEL BY MODIFICATION WITH COMPLEX ALLOYING ELEMENTS

Štefan Eperješi[1], Alena Pribulová[1], Peter Futáš[1], Marianna Bartošová[1]

[1]Technical Universityof Kosice, Faculty of Materials, Metallurgy and Recycling, Košice
Letná 9 Str., 042 00 Košice, Slovak Republic

Corresponding address: stefan.eperjesi.2@tuke.sk

Key words: crystallization, Hadfield´s steel, modification, complex alloying element, structural components

In special demanding practical conditions, where in addition to mechanical strain also other factors can operate (such as environment aggressiveness, abrasive or erosive influence of particles of working environment), high alloyed steels are used.

Most commonly used is steel alloyed by 12 to 14% of manganese. Structure of this steel is made of austenite in normal temperatures after heat treatment, thanks to what it is known as austenitic manganese steel (also Hadfield´s steel). This steel has a highly important place among other high alloyed steels, mainly because of production of special castings highly resistant to depreciation in conditions of abrasion environment in current influence of dynamic stress.

Most of researches [1-4] dealing with study of Hadfield's steel is using pseudo-binary phase diagram Fe – C with 13% of Mn (Fig. 1) for description of its basic structure.

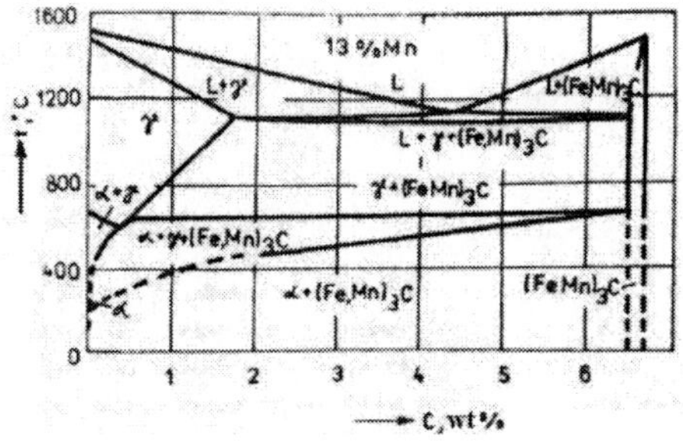

Fig. 1. Phase diagram Fe – C with 13% Mn [2]

This diagram described phase composition of Hadfield´s steel in as cast state only partially, because also influence of other elements in basic chemical composition of steel is not negligible. Basic structure of austenitic manganese steel depends greatly also on conditions of solidification of the casting, which are in given case different from equilibrium conditions.

Structure of the casting is after undercooling made mostly by austenite, which co-exists with eutectics, secondary carbides, fine pearlitic structure called troostite and inclusions (Fig. 2).

Fig. 2. As cast structure of Hadfield´s steel, mag. 200x, Nital 2% etch [5]

By many observations [5-8] it was found, that strength and especially plasticity of casted steel is significantly influenced by character of structure created in the process of primary crystallisation and purity of basic austenitic structure.

Under the name structure of primary crystallisation, it is understood the size of crystals, their form and orientation, redistribution of chemical composition, amount and dimensions of pores, cavities, cracks etc.

Investigation of macrostructure and fractures of castings from austenitic manganese steel showed, that this steel exhibits increased tendency to create column crystallisation (trans-crystallisation) (Fig.3).

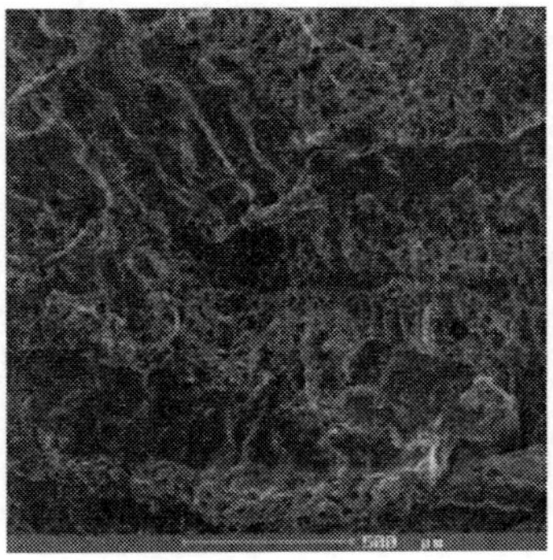

Fig. 3. Trans-crystalline ductile failure with dimple morphology, dendritic fracture character [9]

As a result of continuous distribution of chemical compounds during primary crystallisation and creation of column structure, chemical heterogeneity between casting surface and its thermal axis is obvious as well as between composition of dendrite and dendritic phase, especially in amount of carbon, which causes non-desirable consequences – instability of austenite and phase changes. Intensive growth of column crystals continues until growing compounds run into some obstacle (until thermal gradient is sufficiently high) and when spontaneous creation of centres of crystallisation in molten metal does not occur due to the homogeneous temperature decrease of liquid phase in the casting.

Change of character of primary crystallisation of manganese austenitic steel is influenced by even small amounts of alloying elements, modification elements or grain refining elements. Influence of modification and grain refining elements is based on physical-chemical processes during crystallisation. Modification elements for austenitic manganese steel can be boron, titanium, nitrogen, cerium and others. During modification of this steel, formation of fine grained equiaxed structure occurs in wide temperature range. Main objective of this work is influencing and observation of primary crystallisation of austenitic manganese steel after modification of complex modification elements on base of cerium, magnesium and lanthanum.

Acknowledgements
This work was supported by the Scientific Grant Agency of The Ministry of Education of the Slovak republic No. VEGA1/0073/17, VEGA 1/0703/16.

References

[1] Fremunt, Pr. et al 1985. Austenitic manganese steel. Pages 64-78. Slevarenska rocenka.
[2] Vlasov V.I., Komolova E.F: 1963. Cast high manganese steel. Masgiz.Moskva
[3] Sulte Ju.A., Kurbatov M.I.: 1961. The influence of melting factors on the properties of high manganese steel. Metalurgizdat.
[4] Janssen K., Jellinghaus W.: 1956. Perlit- und Martensitbildung in Manganhartstahl. Archivfür das Eisenhüttenwesen, č.9. Page 573
[5] Vilcko J.:1978. Effect of melting on the properties of Hadfield steel used on castings. Technical University. Kosice.
[6] SUBRAMANYAM, D.K. 2005. Specialty Steels and Heat-Resistant Alloys, Austenitic Manganese Steels. ASM Handbook: Properties and Selection: Irons, Steels, and High Performance Alloys. 10th ed. Materials Park, OH: ASM International. Volume 1. ISBN 161503827223. Pages1274 - 1302.
[7] KUYUCAK, S. 2004. Austenitic Manganese Steel Castings, Metallography and Microstructures. Vol 9, ASM Handbook, ASM International. Pages 701–708
[8] Akeel, D. and Omar, A. 2007. Phase Transformations of Hadfield Manganese Steels. Eng. & Technology. vol. 25, no. 26. Pages 227-289.
[9] Stefan Eperjesi: 1990. The influence of endogenous and exogenous gases on the occurrence of foundry errors in austenitic manganese steel castings. Technical University. Kosice.

The 73rd World Foundry Congress
„Creative Foundry"
23 - 27 September 2018, Krakow, Poland

DETERMINATION OF TRANSFORMATION TEMPERATURES AND SOLIDIFICATION SEQUENCE OF THE Mg-Al-La-Ca ALLOY

Sérgio Bartex[1]*, Lírio Schaeffer[1], Vinicius de Barcellos[1]

[1]*Federal University of Rio Grande do Sul, Engineering School, Foundry Laboratory, Bento Gonçalves Av., 9500, ZIP Code 91501-970, Porto Alegre, RS, Brazil*
**Corresponding address: vinicius.karlinski@ufrgs.br*

Key words: solidification, thermal analysis, microstructural evolution, magnesium alloy, rare earth element.

1. Introduction

Because of the excellent specific strength, magnesium alloys could be utilized in automobile, aerospace and electronic devices to reduce the overall weight and save energy [1].

The most commonly used cast magnesium alloys are Mg-Al system, which constitute more than 90% of magnesium used in structural applications. However, drastic losses of strength and creep resistance at moderate temperatures have restricted the use of Mg–Al alloys to temperatures lower than 125°C [2]. Both, Calcium and Lanthanum are effective alloying elements for Mg–Al and similar alloys in order to improve their high-temperature mechanical properties and creep resistance [3-4].

Many researches have worked with Ca and La additions in magnesium alloys [2-4]. The phases formed during the solidification of these alloys have already been identified but the continuous solidification sequence remains unclear.

For this propose, this work aim to determine the transformation temperatures and solidification sequence in Mg-6Al-3La-1Ca system. Another objective is to evaluate the solidification temperature range verifying if this alloy has good properties to Rheocasting processing.

2. Materials and Methods

The experiments were performed with magnesium alloy ingots. The alloy chemical composition is presented in Table 1.

Table 1. Chemical composition of magnesium alloy performed in this work (wt%).

Element (wt %)					
Mg	Balance	**Mn**	0.300	**Ni**	0.001
Al	5.900	**Fe**	0.005	**Ca**	0.988
Zn	0.030	**Cu**	0.002	**La**	2.770

A resistance-type electric closed furnace with a protective gas atmosphere (commercial argon gas, 99.9%) was designed to prevent any oxygen coming into contact with casted magnesium and, in this way, avoiding the ignition. A schematic representation of the assembly used in solidification experiments is presented in Fig. 1.

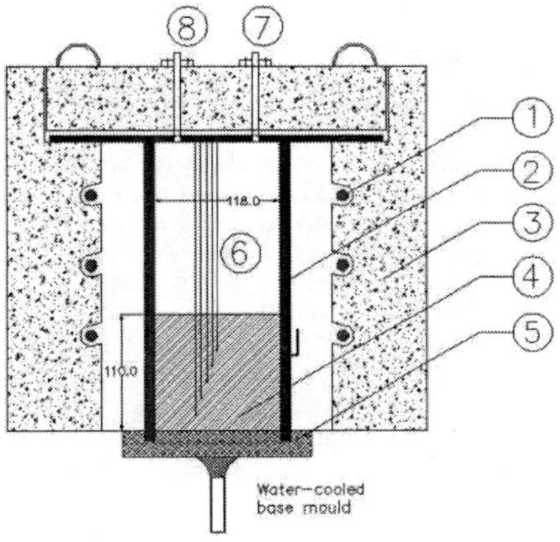

Fig 1. Schematic representation of experimental casting arrangement: (1) Electric resistance; (2) Steel crucible (diameter 118 mm); (3) Thermal insulation; (4) Molten magnesium; (5) Water-cooled base mould; (6) Thermocouples; (7) Argon inlet; (8) Argon outlet. Dimensions in millimeter (mm).

A magnesium alloy ingot of 2 kg was melted inside the furnace at 740°C. The slow cooling condition was obtained by powering-off the furnace when the metal reached the desired temperature. The cooling conditions were as close as possible to thermodynamic equilibrium.

Temperatures in the ingots were measured during solidification by a bank of type K thermocouples (1.6 mm diameter, AISI/SAE 446 chromium-steel) accurately positioned concerning the metal/mould interface. 5 thermocouples were used and they were positioned according to the following distances: 15, 30, 45, 60 and 75 mm away from the base as indicated in Fig. 1. The temperature data were used to estimate solidification parameters and to determine the cooling rates before the ingots solidification.

The samples for metallography were analyzed in a cross-section of the mentioned positions (15, 30, 45 and 75 mm). The polished samples were etched by a Nital 3% solution (3 mL HNO3 + 97mL ethanol) and initially investigated using an optical microscope for metallographic analysis. The samples were analyzed in Differential Scanning Calorimeter (DSC). The cooling rate applied in both Thermal Analysis and DSC was 0.03 °C/s.

3. Results and Discussion

Fig. 2 shows the cooling curve at a distance of 15mm from the base and first derivate curve.

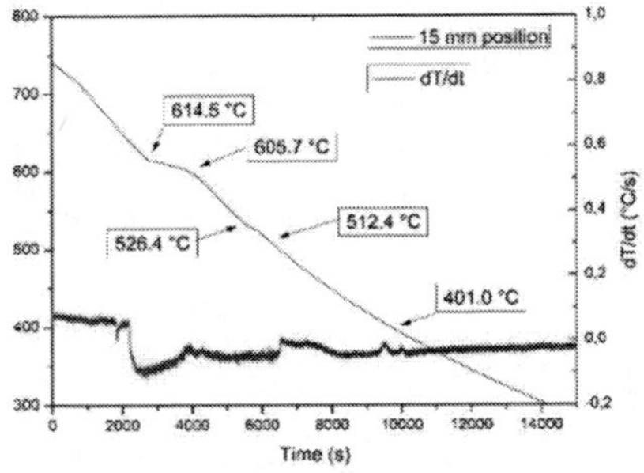

Fig 2. Thermal analyses at a distance of 15 mm from the base showing the temperatures transformation and phases precipitation: 614.5°C - matrix α-Mg; 605.7°C - acicular phase $Al_{11}La_3$; 526.4°C eutectics - Al_2Ca; 512.4°C eutectics-Mg_2Ca. 401.0°C - solid transformation.

To increase the accuracy of experimental data, DSC was performed in a sample extracted from the ingots and results are shown in Fig. 3

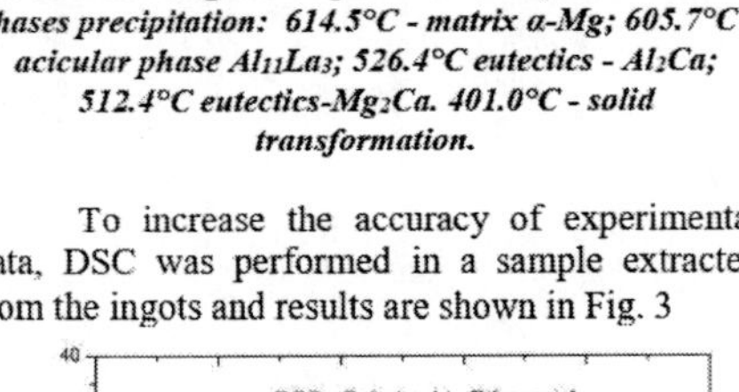

Fig 3. DSC results showing the transformation temperatures for Mg-6Al-3La-1Ca alloy.

The results from Fig. 2 and Fig. 3 are very close and show the solidification sequence in Mg-6Al-3La-1Ca alloy. The *liquidus* temperature is 614°C and *solidus* temperature is 512°C, approximately. The microstructure is presented in Fig. 4. It can be seen in Fig. 4 very coarse structures precipitated along to the matrix as described in images.

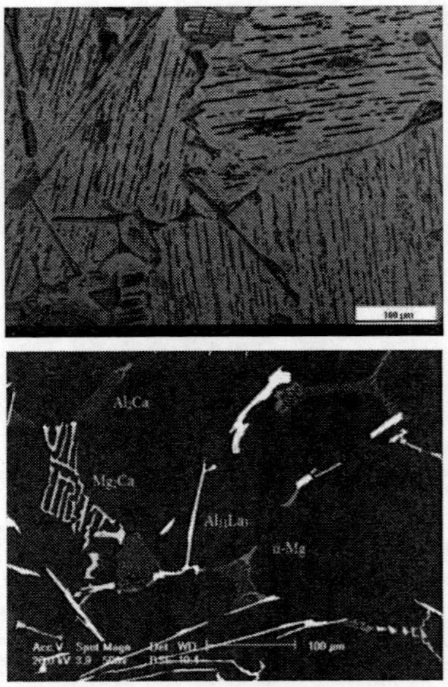

Fig 4. Optical and MEV analysis showing the following phases: matrix α-Mg; acicular phase $Al_{11}La_3$; eutectics Al_2Ca and Mg_2Ca.

4. Conclusions

The thermal analysis shows the follow solidification paths: 614.5°C matrix α-Mg; 605.7°C acicular phase $Al_{11}La_3$; 526.4°C eutectics Al_2Ca; 512.4°C eutectics Mg_2Ca. Besides that, the *liquidus* and *solidus* temperature were, respectively, 614°C and 512°C, meaning wide temperature range is appropriated to Rheocasting process.

Acknowledgments

The authors acknowledge financial support provided by CNPq (The Brazilian Research Council) and FAPERGS (The Scientific Research Foundation of the State of Rio Grande do Sul).

References

[1] D. Lu, Q. Zhang, W. Wang, F. Guan, X. Ma, L. Yang, X, Wang, Y. Huang, B. Hou. (2017). Effect of cooling rate and the original matrix on the thermal diffusion alloyed intermetallic layer on magnesium alloys. Materials and Design. 120, 75–82.

[2] B. Kondori, R. Mahmudi. (2017) Effect of Ca additions on the microstructure and creep properties of a cast Mg–Al–Mn magnesium alloy. Materials Science & Engineering A. 700, 438–447.

[3] T. V. Ferri, A.P. Figueiredo, C.R.F. Ferreira, W. Hormaza, C.A. Santos, J.A. Spim. (2010). Mechanical properties as a function of microstructure in the new Mg-Al-Ca-La alloy solidified under different conditions. Materials Science Engeneering A. 527, 4624–4632.

[4] W. Zhang, W. Xiao, F. Wang, C. Ma. (2016). Development of heat resistant Mg-Zn-Al-based magnesium alloys by addition of La and Ca: Microstructure and tensile properties. Journal of Alloys and Compounds. 684, 8-14.

The 73rd World Foundry Congress
„Creative Foundry"
23 - 27 September 2018, Krakow, Poland

PREPARATION, PROPERTIES AND APPLICATIONS OF THE NOVEL POLYMER BINDERS BioCo

Beata Grabowska*, Karolina Kaczmarska, Artur Bobrowski

AGH University of Science and Technology, Faculty of Foundry Engineering, Reymonta 23, 30-059 Krakow, Poland
**Corresponding address: beata.grabowska@agh.edu.pl*

Key words: polymer binder, cross-linking, thermal degradation, moulding sand, foundry

In foundry engineering, not only resins but also synthetic or natural polymers are used as binding materials for moulding and core sands. In the group of well-known and commonly used polymers there are e.g. polyurethanes, polystyrene and polyacrylates, which can be used as binders or their components. Currently, many research teams are focused in their studies on the development of biopolymers or biomaterials in moulding sands technology. Among the biopolymers there is mainly interest in starch and its modified derivatives [1-5]. At the Faculty of Foundry Engineering (AGH University of Science and Technology) are also conducting the researches on the polymers application in the foundry technology. The studies are focused on novel binders BioCo polymeric materials designed for application in the moulding sands engineering. In this paper polymeric compositions consisting of acrylic derivatives and modified natural polymers are described [6-11]. Polymeric binders BioCo are aqueous polymer solutions with the participation of synthetic polymers and modified biopolymers (Table 1).

Table 1. Characteristics of polymer binders BioCo

Polymer binders	Constitution
BioCo1	aqueous composition PAA/CMS of poly(acrylic acid) (PAA, BASF) and carboxymethyl starch (CMS, Xenon)
BioCo2	aqueous composition PAA/D of poly(acrylic acid) (PAA, BASF) and dextrin (D, Fluka)
BioCo3	aqueous composition PAANa/D of poly(sodium acrylate) (PAANa, BASF) and dextrin (D, Fluka)

Crosslinkink of the novel polymeric binder BioCo can be done either physically by the exposure to radiation (UV or microwaves) or by chemical methods using $Ca(OH)_2$ and CO_2 or glutaraldehyde [6] (Fig. 1). Based on spectroscopic analysis (FT-IR, FT-Raman) was noticed [6], that the type of used agent and the mechanism of reaction were found to be decisive factors influencing the mechanical properties of moulding sands. The microwave radiation occurred to be the most effective means for binding in the polymer binder-sand grains system (Fig. 2). It has been proven that in the field of microwave radiation under via adsorption goes the transient formation of networks of hydrogen bonds. These bonds are mainly responsible for upholding crosslinked structures in the binder-sand grains system [6,7].

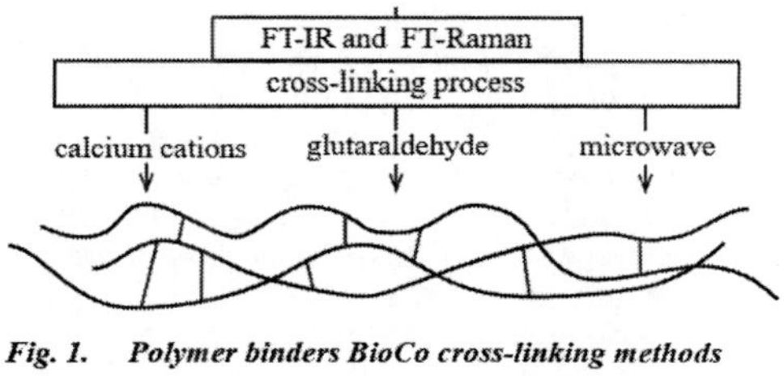

Fig. 1. Polymer binders BioCo cross-linking methods

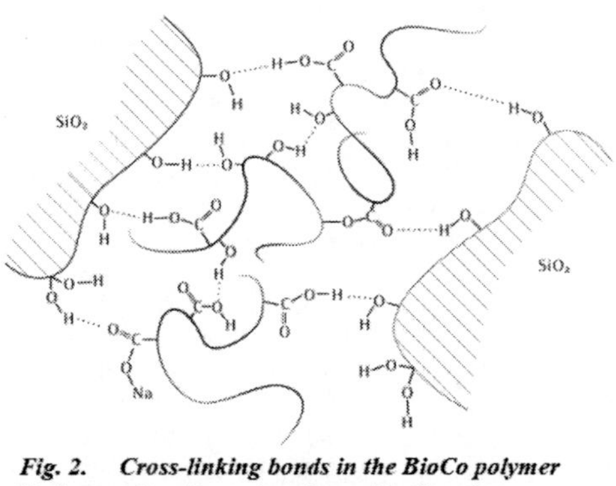

Fig. 2. Cross-linking bonds in the BioCo polymer binder-sand grains system formed in the microwave field [7]

The polymer binders BioCo were submitted to thermal decomposition to determine the conditions impact of this process and importance the structure of the polymers on the progress of material destruction in high temperature. The research conditions were considered as similar to the phenomena occurred between liquid metal and moulding sand material.

Results of comprehensive thermal analysis (TG-DTG-DSC, TG-MS, TG-IR) allowed to determine the range of temperature below thermal decomposition - in this range binders preserving their binding properties [8]. The results were analyzed in three temperature ranges:

I range: 20-300°C – physical transformations and reversible reactions,

II range: 300-400°C – thermal degradation and decomposition process,

III range: above 400°C – progressive destruction.

The obtained results of the thermal analysis exerted a decisive influence on planning the optimum conditions for crosslinking by microwave radiation, as well as by conventional heating in regard to processes going on in the system casting-mould [8]. Moulding sands containing BioCo binders after hardening in the microwave field have an appropriate mechanical strength. It has been shown that moulding sands containing polymer binders are characterised by compression strength ca. 2 MPa and bending strength to 1 MPa – after 1 hour of sample storage after curing. Additional the permeability (determined after 1 h and 24 h storage) of microwave-cured moulding sand with polymer binders BioCo was about 340×10^{-8} $m^2/Pa{\cdot}s$. The measurements carried out after 1 h and 24 hours of the storage showed friability of the cured moulding sands being about 15 (±1)%. It has been ascertained that usage of the novel binders BioCo in form of the investigated polymer compositions designed for foundry moulds, is also adequate for iron castings full fitting all requirements necessary in that case. The removal of castings from the moulds and core-boxes is quite easy. The iron castings obtained in this technology do not have of defects like pittings, deformation of surfaces or gaseous porosities. The determination of the mechanical strength of the obtained casting samples confirmed properties typical for the spheroidal cast iron [9].

Additionally binders BioCo should be regarded as environmentally friendly materials [10, 11]. The amount of BTEX compounds emitted from moulding sands was held on a low level during pouring a liquid iron (Py-GC-MS investigations).

Research is currently under way to determine the effect of silicate and carbon additives on the thermal properties of BioCo3 binder, including on the course of thermal destruction, and identification of degradation products. It will be also important to determine the effect of inorganic additives on the bonding ability of polymer binders BioCo to the selected technological properties [12].

References

[1] Manas C., Salil R. (2008). Industrial Polymers, Specialty Polymers and Their Applications. CRC Press Taylor&Francis Group.

[2] Patterson M.A., Thiel J. (2011). Replacement of Phenol-Formaldehyde in Urethane Binders. Transactions of the American Foundry Society. 119 (11–117), 331–338.

[3] Fox J.T., Cannon F.S., Brown N.R., Huang H., Furness J.C. (2012). Comparison of a new, green foundry binder with conventional foundry binders. International Journal of Adhesion and Adhesives. 34, 38–45. doi:10.1016/j.ijadhadh.2011.11.011

[4] Nwaogu U.C., Tiedje N.S. (2011). Foundry Coating Technology: A Review. Materials Sciences and Applications. 2 (8), 1143-1160. DOI:10.4236/msa.2011.28155.

[5] Zhou, X., Yang, J., Qian, F., Qu, G. (2010). Synthesis and application of modified starch as a shell-core main adhesive in a foundry. Journal of Applied Polymer Science. 216 (5), 2893-2900. DOI:10.1002/app.31781.

[6] Grabowska B., Sitarz M., Olejnik E., Kaczmarska K. (2015). FT-IR and FT-Raman studies of cross-linking processes with Ca^{2+} ions, glutaraldehyde and microwave radiation for polymer composition of poly(acrylic acid)/sodium salt of carboxymethyl starch – Part I. Spectrochimica Acta Part A: Molecular and Biomolecular Spectroscopy.135,529–535doi:10.1016/j.saa.2014.07.031

[7] Grabowska B., Sitarz M., Olejnik E., Kaczmarska K., Tyliszczak B. (2015). FT-IR and FT-Raman studies of cross-linking processes with Ca^{2+} ions, glutaraldehyde and microwave radiation for polymer composition of poly(acrylic acid)/sodium salt of carboxymethyl starch – In moulding sands, Part II. Spectrochimica Acta Part A: Molecular and Biomolecular Spectroscopy. 151, 27–33. doi:10.1016/j.saa.2015.06.084

[8] Grabowska B., Hodor K., Kaczmarska K., Bobrowski B., Kurleto-Kozioł Ż., Fischer C. (2016) Thermal analysis in foundry technology. Pt. 2, TG-DTG-DSC, TG-MS and TG-IR study of the new class of polymer binders BioCo. Journal of Thermal Analysis and Calorimetry, 130 (1), 301–309.

[9] Grabowska B., Holtzer M., Dańko R. Górny M., Bobrowski A., Olejnik E. (2013). New BioCo binders containing biopolymers for foundry industry. Metalurgija. 52(1), 47–50.

[10] Grabowska B., Bulwan M., Zapotoczny G., Grabowski G. (2012). Biodegradation of new polymer foundry binders composition of poly(acrylic acid)/dextrin. Polimery. 57 (7–8), 529–534.

[11] Grabowska B., Kaczmarska K., Bobrowski A., Żymankowska-Kumon S. (2017). TG-DTG-DSC, FTIR, DRIFT, and Py-GC-MS studies of thermal decomposition for poly(sodium acrylate)/dextrin (PAANa/D) – new binder BioCo3. Journal of Casting & Materials Engineering. 1(1), 27-32.

[12] Grabowska B., Kaczmarska K., Bobrowski A., Drożyński D., Żymankowska-Kumon S., Cukrowicz S. (2017). Polymer Binder BioCo3 with Silicates and its Application to Microwave-Cured Moulding Sand. Archives of Foundry Engineering. 17 (4), 51–60, DOI: https://doi.org/10.1515/afe-2017-0130.

The 73rd World Foundry Congress
„Creative Foundry"
23 - 27 September 2018, Krakow, Poland

INFLUENCE OF THE MASTER ALLOY Ti-Fe ON THE MICROSTRUCTURE AND SELECTED PROPERTIES OF COPPER AND COPPER ALLOYS

Marcin Piękoś [1]*, Janusz Kozana [1], Aldona Garbacz-Klempka [1]

[1]*AGH University of Science and Technology, Faculty of Foundry Engineering, Reymonta 23 Str., 30-059 Kraków, Poland*
**Corresponding address: mpiekos@agh.edu.pl*

Key words: copper alloys, bronze, alloying additions, titanium, nickel

1. Introduction

Copper is one of more expensive and in short supply metals. This element is characterised by several valuable properties, due to which is widely applied in technique. Striving for the development of copper alloys, by improving their properties, is undertaken all the time. This is related to increasing their tensile and bending strength, improving plasticity and even their thermal resistance. Attempts to improve the corrosion resistance and to hold the high electrical conductivity are also undertaken. This type of improvements can be achieved in many ways, e.g. by the heat treatment, by the modification and also by the introduction of the proper alloying additions [1-3].

2. Investigation methodology

The plan contained performing investigations with using the binary foundry alloy Ti-Fe, in order to determine the titanium (as the main alloy reagent) influence when it is introduced into copper and into the copper with silica alloy. In addition, at the introduction of the selected foundry alloy Ti-Fe the ferrous fraction in the analysed copper and silicon bronze was increasing. The presence of ferrous is sometimes determined as contaminations, while sometimes it constitutes the alloying addition causing grains refinement in Cu alloys.

As the first stage of investigations the series of melts - at various contents of the Ti73Fe master alloy, introduced in amounts of 5%, 15% and 25%, in relation to the metal charge - were performed.

In the second stage, the silicon addition of approximately 4 wt%, was introduced to copper followed by 0.5% and 1% of the Ti73Fe master alloy.

The melt was performed in the induction furnace with the chamotte-graphite crucible. The addition of the Ti73Fe master alloy was introduced at the bath temperature being 1100 - 1200°C. The finished alloys were cast into metal moulds. Samples were prepared from the obtained alloys and investigations of their microstructure performed. The ultimate tensile strength, elongation A_5, reduction of area as well as hardness HB were assessed.

During the research the mentioned below features were analysed: the chemical composition of the alloys obtained according to the designed experiments, using the Spectro Midex energy dispersion X-ray fluorescence spectrometer, macro- and microstructures were investigated by the NIKON SMZ 745Z and Eclipse LV 150 microscopes (OM). Samples were ruptured by means of the INSTRON device, model 1115.

3. Results of the selected tests

The examples of the characteristic images of microstructures from the first research stage, i.e. assessments of the influence of various additions of Ti-Fe master alloy to electrolytic copper M1E, are shown in Figs. 1 and 2. This stage provided information and allowed to verify the available reference knowledge.

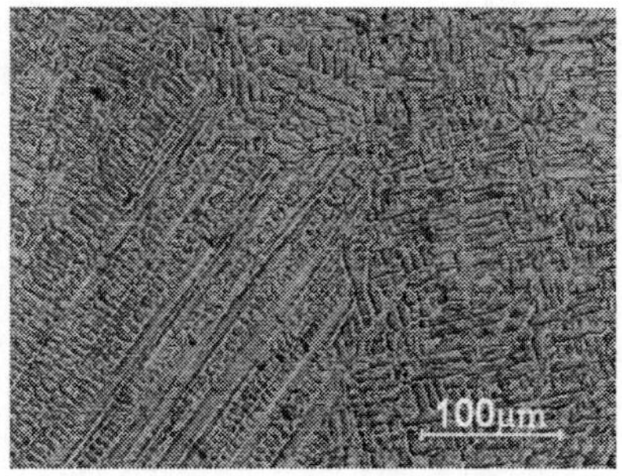

Fig. 1. Microstructure image of Cu+5%Ti73Fe, 200x

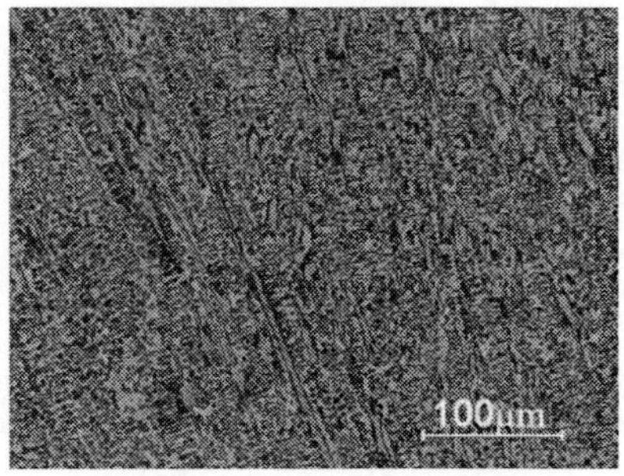

Fig. 2. Microstructure image of Cu+10%Ti73Fe, 200x

In the next stage, the master alloy Ti-Fe in various amounts was introduced to binary silicon bronze CuSi4.

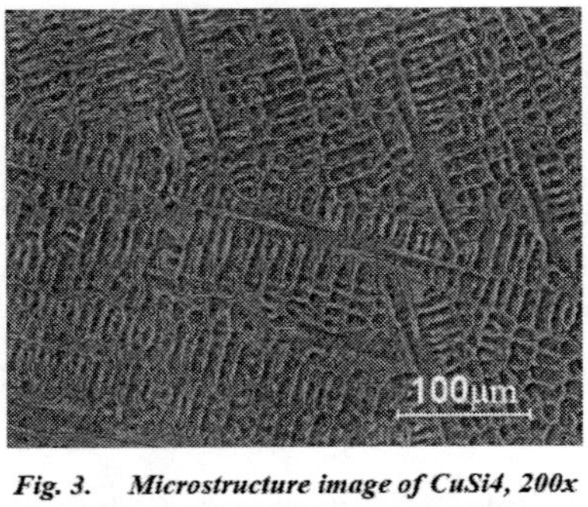

Fig. 3. Microstructure image of CuSi4, 200x

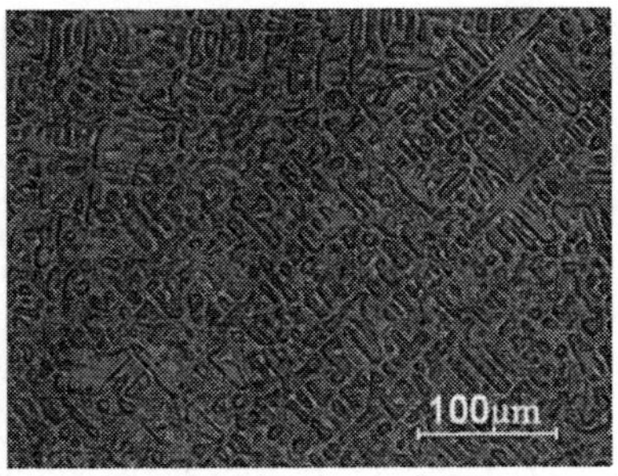

Fig. 4. Microstructure image of CuSi4+0.5%Ti73Fe, 200x

The obtained images of microstructures from the first and second stage indicate - after the preliminary analysis - a significant influence of the applied addition both into pure copper and into the silicon bronze. Changes are mainly related to the crystallization process of grains, however the morphology of dendrites is also changing.

Significant changes in the microstructure usually cause changes in mechanical properties. The selected results of these investigations are presented in Figs. 5 and 6.

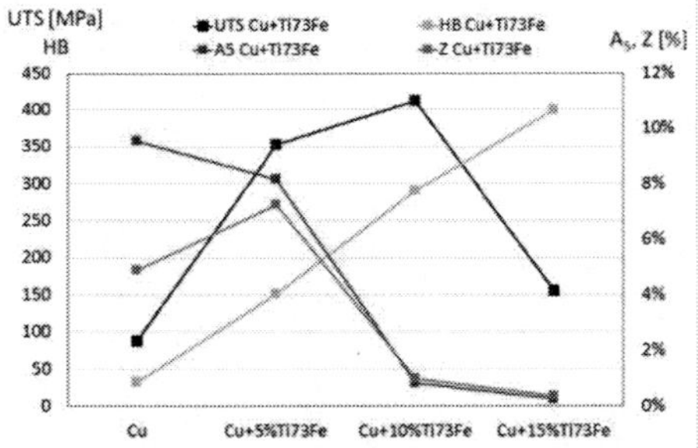

Fig. 5. Graphical listing of the mechanical properties results of samples from individual experiments Cu+Ti73Fe

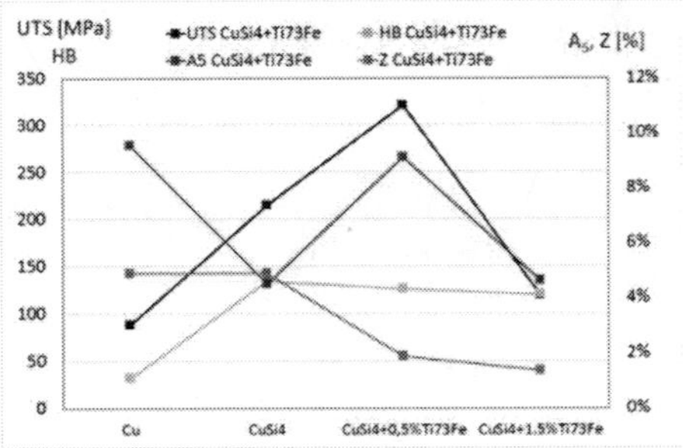

Fig. 6. Graphical listing of the mechanical properties results of samples from individual experiments CuSi4+Ti73Fe

4. Conclusions

The performed metallographic tests and the obtained results of the selected mechanical properties of the analysed alloys revealed the justification of the undertaken research. Microstructure changes were obtained, mainly within grains but also in developed dendrites of the solid solution. Changes are occurring when titanium together with iron is introduced into copper as well as into binary silicon bronze. The maximum copper strengthening occurs when approximately 12 wt% of titanium is added. Simultaneously the plasticity - expressed by parameter A_5 - is decreased. Similar situation was observed at the master alloy Ti-Fe introduction into bronze CuSi4, however at much lower fraction of titanium in the alloy, i.e. 0.3 wt%. In this case we are dealing with the material in which the silica addition is decisive in forming the analysed properties. The applied master alloy was favourably acting both in the aspect of solving the alloy as well as in the improvement of strength properties.

Acknowledgments

The financial support of the State Committee for Scientific Research of Poland under the grant numbers 11.11.170.318 – 11 is acknowledged.

References

[1] Eze, A.A, Jamiru, T., Sadiku, E.R., Durowoju, M.O., Kupolati, W.K., Ibrahim, I.D., Obadele, B.A., Olubambi, P.A. & Diouf, S. (2018) Effect of titanium addition on the microstructure, electrical conductivity and mechanical properties of copper by using SPS for the preparation of Cu-Ti alloys. Journal of Alloys and Compounds. 736, 163-171. from doi:10.1016/j.jallcom.2017.11.129

[2] Takahashi, T., Hashimoto, Y. & Koyama, K. (1989). Effect of Ti and Zr Additions on the Internal Oxidation of Cu–Si Alloy. Materials Transactions, JIM. 30(2), 127-136. from doi:10.2320/matertrans1989.30.127

[3] Polok-Rubiniec, M., Konieczny, J., Labisz, K. & Włodarczyk-Fligier A., (2016). Microstructure and Service Properties of Copper Alloys. Archives of Metallurgy and Materials, 61(3), pp. 1277-1282. from doi:10.1515/amm-2016-0211

The 73rd World Foundry Congress
„Creative Foundry"
23 - 27 September 2018, Krakow, Poland

INVESTIGATIONS OF THE TITANIUM INFLUENCE ON THE STRUCTURE AND SELECTED PROPERTIES OF TIN BRONZES

Janusz Kozana[1], Aldona Garbacz-Klempka[1]*, Marcin Piękoś[1], Małgorzata Perek-Nowak[2], Paweł Pałka[2]

[1]*AGH-University of Science and Technology, Faculty of Foundry Engineering, Reymonta 23, 30-059 Krakow, Poland*
[2]*AGH-University of Science and Technology, Faculty of Non-Ferrous Metals, Mickiewicza 30, 30-059 Krakow, Poland*
**Corresponding address: jkozana@agh.edu.pl*

Key words: Tin bronze, Titanium, Nickel, Microstructure, Mechanical properties

1. Introduction

Tin bronzes, known and applied already in the antiquity, still constitute the important and irreplaceable group of alloys used for production of machines and devices parts. Good points of these alloys are, among others: good fluidity, strength, plasticity, high corrosion resistance and also abrasion resistance at high loads [1]. However, due to higher and higher requirements in front of modern materials and high prices of tin and its being in short supply, substitutes of this group of alloys are looked for and applied [2-3]. This is realised by developing of new alloys based on alloying additions, purification, modification [4] and the way of casting.

Regardless of the fact that several multi-component bronzes based on Cu-Sn matrix were developed and described in the scientific literature, looking for innovatory products of special properties by means of modern research methods, innovative approaches to alloying and deep analysis of results, is fully justified. This allows to modify alloys from the Cu-Sn system and to widen their application range. The main aim is thus reaching the improvement of the castings quality [5-6].

The results of the influence of titanium on the CuSn8 bronze are presented in the hereby paper. Investigations were performed within the macro- and microstructure with using the optical and scanning microscopy SEM-ESD and X-ray diffraction (XRD). Based on thermal analysis diagrams the crystallisation process was analysed. Within this research stage such parameters as UTS, A_5 and HB of alloys cast to metal moulds were measured.

2. Methodology of investigations

The investigations concerning changes caused by titanium additions to tin bronzes were realised in the Laboratory of Non-Ferrous Metals Casting of the Faculty of Foundry Engineering of AGH – University of Science and Technology, Krakow. The influence of titanium additions was analysed with using two base alloys: CuSn8 and CuSn8Ni3.

Melts were carried out in the inductive furnace of a medium frequency, in the chamotte-graphite crucible. The Ti addition was introduced into a metal bath at a temperature of 1150-1200°C. Ready alloys, in dependence of the planned investigations, were cast into metal or ceramic moulds. On the bases of the recorded crystallisation diagrams the investigations of the alloys crystallisation were also made.

Investigations of the chemical composition, phase composition and macro- and microstructure observations, were also performed. The tensile strength parameters UTS, elongation A_5 and hardness HB, were determined.

3. Investigation results

Investigations of the alloys crystallisation allowed to observe and determine changes during solidifying of the tested alloys (Figs. 1-2). This allowed to determine the different character of solidifying the CuSn8 and CuSn8Ni3 alloys with titanium additions.

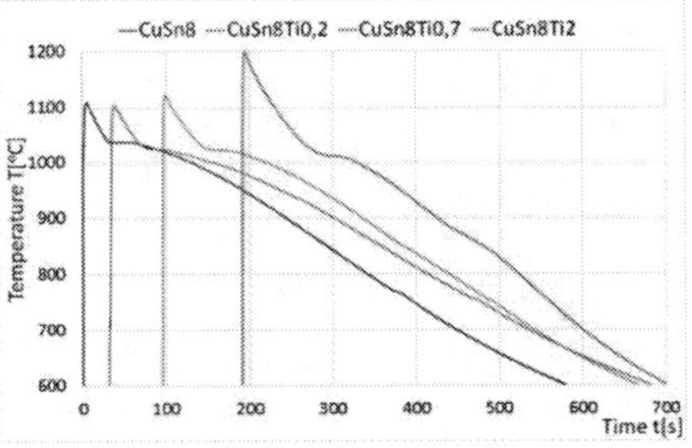

Fig. 1. The cooling characteristics of the CuSn8 alloy with the Ti addition

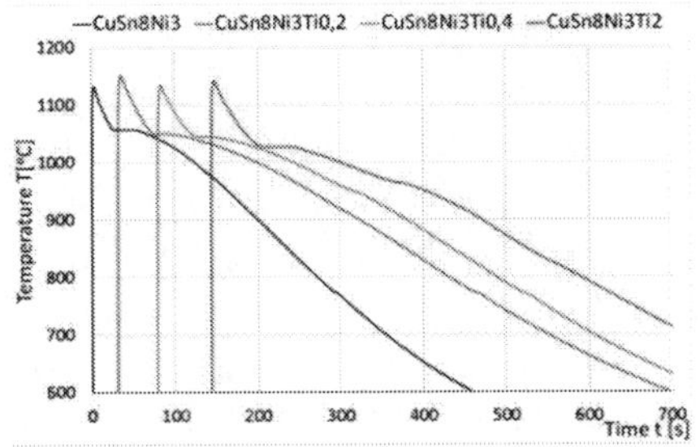

Fig. 2. The cooling characteristics of the CuSn8Ni3 alloy with the Ti addition

The macrostructure images indicate also the variable character of solidifying CuSn8 alloys (Figs. 3-4).

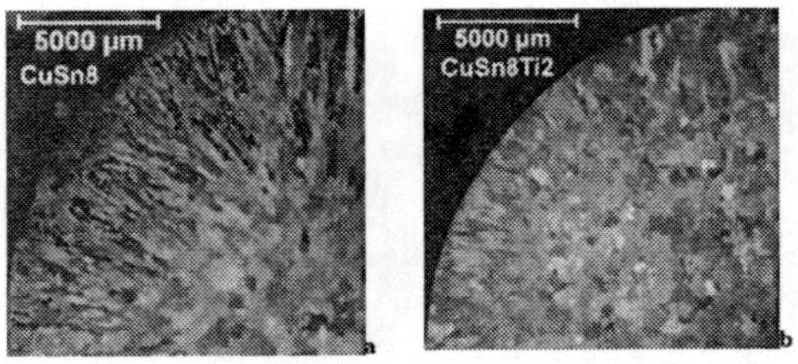

Fig. 3. ***Macrostructure of the CuSn8 alloy (a) and CuSn8Ti2 alloy (b), metal mould***

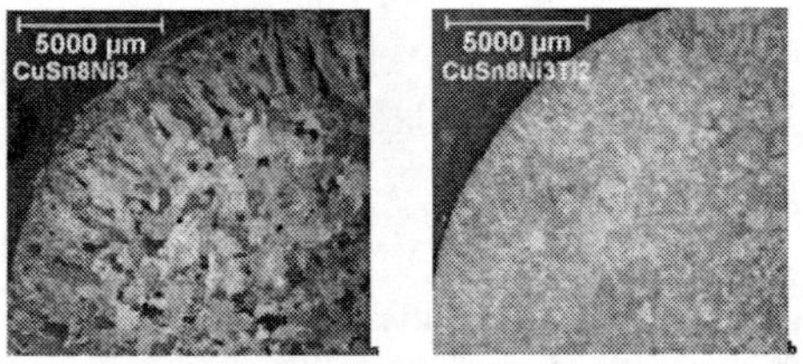

Fig. 4. ***Macrostructure of the CuSn8Ni3 alloy (a) and CuSn8NiTi2 alloy (b), metal mould***

The grains shape and their fractions on the cross-section of the sample-ingot are changing. Under the influence of the introduced titanium directional crystals, related to the crystallisation from the metal mould walls are vanishing. Also the area of occurrence of equiaxial crystals in the middle part of the sample-ingot is disappearing.
The titanium addition causes in alloys: CuSn8 and CuSn8Ni3, narrowing the edge zone (directional crystallisation) and a significant refinement of equiaxial grains in the central part of the sample. The influence of titanium addition causes also changes in the microstructure images (Figs. 5-6).

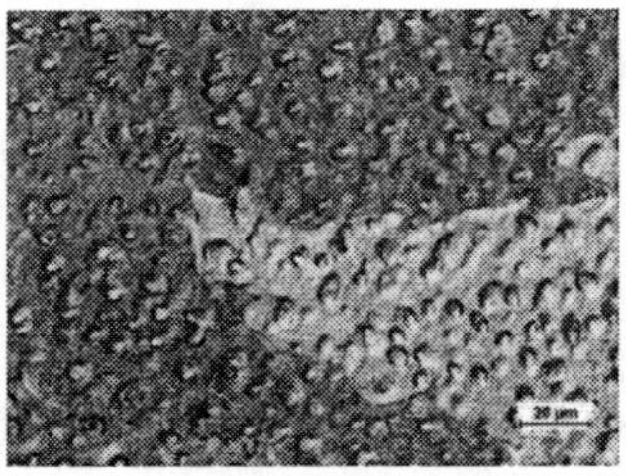

Fig. 5. ***Microstructure of the CuSn8, metal mould***

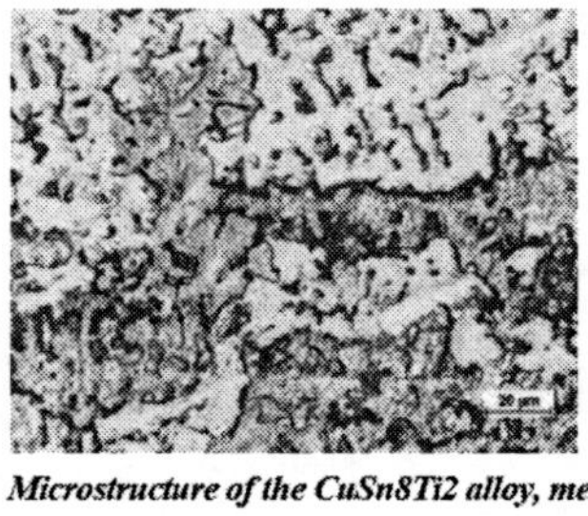

Fig. 6. ***Microstructure of the CuSn8Ti2 alloy, metal mould***

The microstructure images indicate, first of all, the dendrite character of solidifying, characteristic for tin bronzes with changes in the solid solution of Sn in Cu and with precipitates of the secondary crystallisation, related to the presence of alloying additions of nickel and titanium.
The obtained results of mechanical properties (Fig. 7) indicate the increase of tensile strength and hardness at quite intensive lowering of plasticity of the tested alloys.

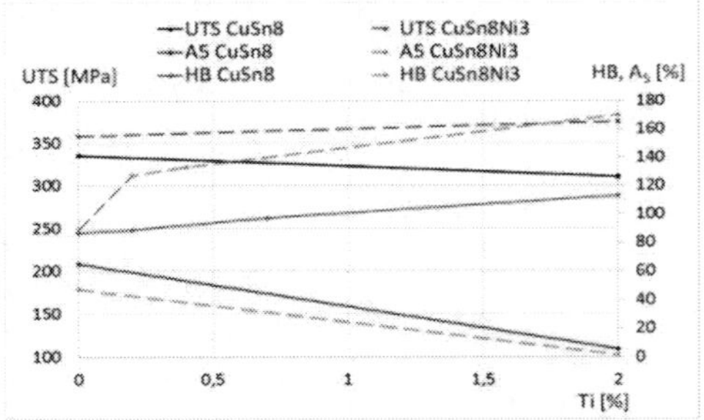

Fig. 7. ***Graphical presentation of the mechanical properties results of samples from individual experiments on CuSn8Ti and CuSn8Ni3Ti***

All this indicates the usefulness of undertaken investigations leading towards improvements of properties of copper and tin based bronzes, with additions of nickel and titanium. The analysed CuSnNiTi alloys can constitute cheaper (in consideration of the decreased tin content) and simultaneously better alloys, in relation to the presently applied tin bronzes, including also multi-component bronzes with additions of phosphor, zinc and lead.
Further investigations of technological-casting properties and also mechanical wear and corrosion resistances, are aimed at the development of alloys of high mechanical and technological properties.

Acknowledgments
The financial support of the State Committee for Scientific Research of Poland under the grant numbers 11.11.170.318 – 11 is acknowledged.

References

[1] Rzadkosz, S., Zych, J., Garbacz-Klempka, A., Kranc, M., Kozana, J., Piękoś, M., Kolczyk, J., Jamrozowicz, Ł. & Stolarczyk, T. (2015). Copper alloys in investment casting technology. Metalurgija Metallurgy 54(1), 293–296.

[2] Rzadkosz, S., Kranc, M., Garbacz-Klempka, A., Kozana, J., & Piękoś, M. (2015).. Metalurgija Metallurgy 54(1),259-262.

[3] Rzadkosz, S., Garbacz-Klempka, A., Kozana, J., Piękoś, M. & Kranc M. (2014). Structure and properties research of casts made with copper alloys matrix. Archives of Metallurgy and Materials 59(2), 775-778.

[4] Szajnar, J., Kondracki, M. & Stawarz, M. (2003). Modification of CuSn8 tin bronze and its influence on tin segregation. Archives of Foundry 3(10), 315-322.

[5] Ma, B. & Lian, F. (2013). Study on the use of CuSnTi brazing alloy for induction brazing of diamond grits surface-treated by direct current plasma chemical vapor deposition, International Journal of Refractory Metals and Hard Materials 41, 339-344. doi.org/10.1016/j.ijrmhm.2013.05.006

[6] Weitzer F., Perring L., Shibayanagi T., Naka M. & Schuster J. C. (2000) Determination of the crystal structure of CuSnTi by full profile Rietveld analysis. Powder Diffraction 15(2):91-93, DOI 10.1017/S088571560001

The 73rd World Foundry Congress
„Creative Foundry"
23 - 27 September 2018, Krakow, Poland

COMPARISON OF THE PROPERTIES OF THE ALKALINE-PHENOLIC BINDERS FOR THE MOULDING SAND FOR THE STEEL CASTINGS

Mariusz Łucarz[1,*], Dariusz Drożyński[1], Jan Jezierski[2,3], Rafał Kania[3]
Andrzej Kaczmarczyk[3]

[1] *AGH University of Science and Technology, Faculty of Foundry Engineering, 23 Reymonta Str., 30-059 Kraków, Poland*
[2] *Silesian University of Technology, Department of Foundry Engineering, 7 Towarowa Str., 44-100 Gliwice, Poland*
[3] *Pioma-Odlewnia Sp. z o.o., 38 Romana Dmowskiego Str., 97-300 Piotrków Trybunalski, Poland*
**Corresponding address: eumar@agh.edu.pl*

Key words: alkaline-phenolic binder, sieve analysis, molding sand properties, thermal analysis

1. Introduction

The results of the examinations of the moulding sand with the alkaline-phenolic resin have been presented in the article. The four different binders have been examined for the same resin and hardener ratio for two different sand matrix. The aim was to find the binder which can be the best replacement for the furfuryl resin used as a binder at the foundry at the moment. The alkali-phenolic binding system is a two-reagent process with the water reactor and liquid ester as a reaction partner. It can be used for every alloy. The lack of nitrogen and sulphur in the resin is very beneficial during the steel and ductile iron casting. The main feature of the alkali-phenolic process is that during the binding reaction the two-stage setting is going on. At the room temperature the initial (partial) binding (binder polymerisation) giving the moulding or core sand enough strength for handling and causes that it is thermoplastic in the average temperature. The thermoplasticity of the binder compensates thermal sand expansion, what practically eliminates moulding sand cracking and the leak of the liquid alloy. The heat going out the liquid alloy (casting) finishes the resin cross-linking process and causes that this technology characterizes high dimensional accuracy and liquid alloy erosion [1-2].

2. Methodology and the results

The experiments on the moulding sand with the alkali-phenolic resin have been carried out on the binders labelled: A1, A2, A3, A4 and four hardeners: 1, 2, 3, 4. The sand matrix was always the silica sand of two grades. The first was SIBELCO (S) type of the following properties: sieve analysis – medium sand (0,20/0,32/0,16), with the average grain size $d_L = 0,25$ mm, uniform (J85) with the specific surface area 8,75 m^2/kg. The second was (P) type – medium sand (0,40/0,20/0,32), with the average grain size $d_L = 0,33$ mm, uniform (J91) with the specific surface area 22 m^2/kg. All sands were mixed in the laboratory roll mixer type LM-1. To the weighed portion of the sand matrix the hardener has been dosed and mixed for 1,5 minute, then the resin was dozed and next 1,5 minute mixing were applied. The sand was then loaded into the dies of standardized shape to produce the samples for the particular sand property tests. The dies were mounted on the vibration compaction device LUZ-1 by WADAP Wadowice, which was switched on for 15 seconds with the maximum vibrations amplitude of 2 mm. After the visual check the samples were stored for hardening under the ambient conditions. The 8-like shape samples were produced for tensile strength R_m^u, longitudinal for the bending strength R_g^u and round ones for the permeability P^u and ścieralności S. The composition of the moulding sand was presented in Table 1.

Table 1. Composition of the moulding sand with the alkali-phenolic resin

Sand label	Resin	Hardener	Resin [wt.] Hardener [% to resin, wt.]
A1	Estrofen	PR	A1-1,2 1-25
A2	Sinotherm 8255	Akivator J 120	A2-1,2 2-25
A3	Fenotec 280ES	Fenotec HC Esters (HC30)	A3-1,2 3-25
A4	Permabind 440	Permabind Plus 7 Hardener	A4-1,2 4-25

In the figure 1 the results of the binders' thermal analysis were presented. As it can be seen their decomposition based on the degradation and thermal destruction is similar. The temperature of 150°C, 450°C and over 700°C, in which the visible mass changes of the examined binders appear. In comparison to the furfuryl binder the following degradation and destruction steps are visible what gives the alkali-phenolic resin some specific features of compensation of the sand matrix thermal expansion, with keeping the proper mechanical properties [3-4].

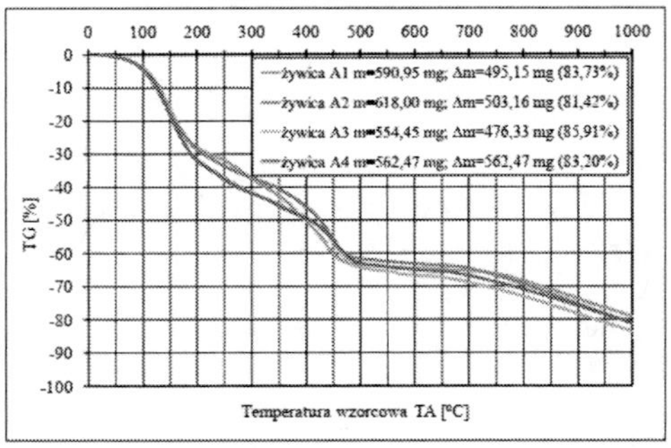

***Fig. 1.** Thermal analysis of the alkaline-phenolic resins*

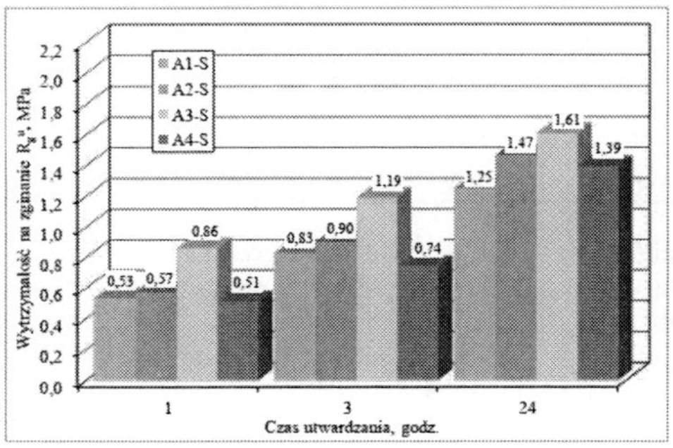

***Fig. 2.** Bending strength R_g^u of the moulding sand on the base of SIBELCO (S) sand matrix*

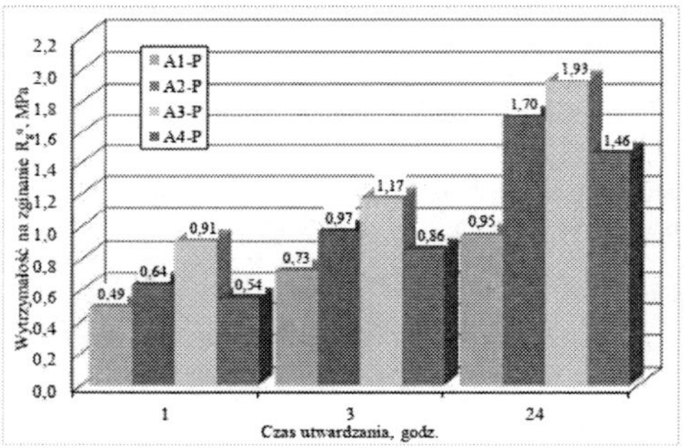

***Fig. 3.** Bending strength R_g^u of the molding sand on the base of the (P) sand matrix*

The selected results of the moulding sand properties examination have been presented in figures 2-5. Among them the most interesting is bending strength R_g^u and abrasibility S. After the research it has been shown that the best technological properties have been achieved for the resins labelled A2 and A3. It is justified by the exemplary bending strength R_g^u and abrasibility S results for the two sand matrix of different average grain size d_L.

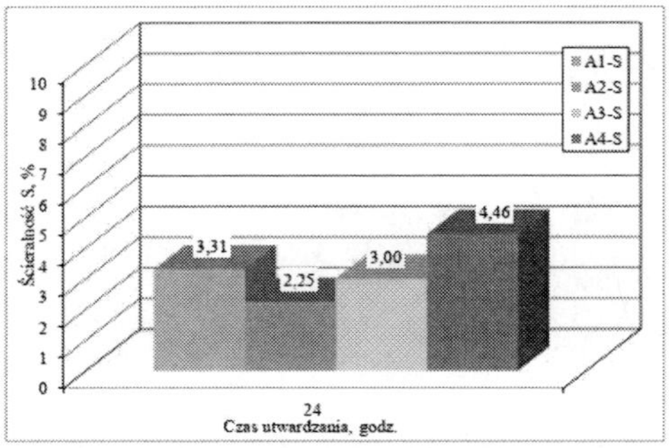

***Fig. 4.** Abrasibility S of the moulding sand on the SIBELCO (S) sand matrix*

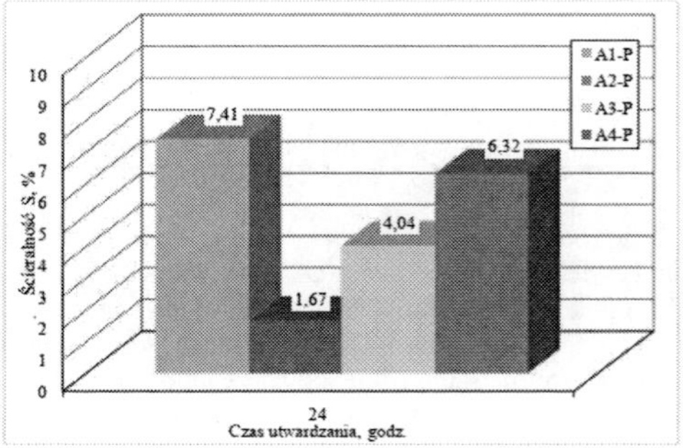

***Fig. 5.** Abrasibility S of the moulding sand on the (P) sand matrix*

Acknowledgments

The study was conducted under the project "Demonstration of the technology of manufacturing innovative cast steel castings to be used in extreme operating conditions", co-financed by the National Centre for Research and Development, under Operational Programme Intelligent Development 2014-2020. Action 1.1 R&D projects of enterprises, Sub-action 1.1.2. R&D works related to manufacturing pilot/demo system, contract POIR.01.01.02-00-0148/16-00 dated May 30, 2017

References

[1] Holtzer M., Bobrowski A., Drożyński D., Makselon J., Isendorf B. (2013). Investigations of Properties of Moulding Sands with Resins Applied in the Alphaset Technology. Archives of Foundry Engineering. 13(1), 31-37.

[2] Holtzer M., Żymankowska-Kumon S., Bobrowski A., Kmita A., Dańko R. (2015). Influence of the Reclaim Addition to the Moulding Sand Matrix Obtained in the ALPHASET Technology on the Emission of Gases - Comparison with Moulding Sand with Furfuryl Resin. Archives of Foundry Engineering. 13(1), 31-37. 15(1), 121-125.

[3] Łucarz M., Grabowska B., Grabowski G. (2014). Determination of parameters of the moulding sand reclamation process, on the thermal analysis bases. Archives of Metallurgy and Materials. 59(3), 1023-1027.

[4] Łucarz M. (2015). Setting temperature for thermal reclamation of used moulding sands on the basis of thermal analysis. Metallurgy. 54(2), 319-322.

The 73rd World Foundry Congress
„Creative Foundry"
23 - 27 September 2018, Krakow, Poland

AMBIENT TEMPERATURE INFLUENCE ON THE PROPERTIES OF THE MOULDING SAND WITH THE ALKALINE-PHENOLIC BINDER

Mariusz Łucarz[1*], Dariusz Drożyński[1], Jan Jezierski[2,3], Arnold Bul[3], Andrzej Kaczmarczyk[3]

[1] *AGH University of Science and Technology, Faculty of Foundry Engineering, 23 Reymonta Str., 30-059 Kraków, Poland*
[2] *Silesian University of Technology, Department of Foundry Engineering, 7 Towarowa Str., 44-100 Gliwice, Poland*
[3] *Pioma-Odlewnia Sp. z o.o., 38 Romana Dmowskiego Str., 97-300 Piotrków Trybunalski, Poland*
**Corresponding address: eumar@agh.edu.pl*

Key words: alkaline-phenolic binder, sieve analysis, moulding sand properties, thermal analysis.

1. Introduction

One of the important issues when the binder is being selected for the moulding sand is its stability with the ambient temperature changes. The paper presents the results of the technological properties examination of the moulding sand prepared on the base of four alkaline-phenolic resins under the various temperature conditions. The aim was to check the temperature influence on the moulding sand and to find the binder with the technological properties least affected by the temperature changes in the foundry shop floor. The alkaline-phenolic resins have been examined because they are more and more commonly used due to the ecological issues causes the binder with the furfuryl alcohol must be replaced. The tensile strength of the moulding sand prepared on the base of fresh sand matrix in the alkaline-phenolic process is generally lower than for the organic resins as phenolic urethanes or resins hardened by acids. Low strength is achieved for the fine sand grains, too. But for the most applications the tensile strength is satisfying for the moulds and cores making together with the better surface quality of the castings. These aspects encourage to fully characterize the moulding and core sand with the alkaline-phenolic binder. What is more, this binding system was accepted in England as an industrial standard for the steel castings manufacturing. All of these reasons cause that such technology is nowadays the most convenient and the most ecological technology for foundry industry [1-2].

2. Methodology and results

The experiments on the moulding sand with the alkali-phenolic resin have been carried out for the binders labelled: A1, A2, A3, A4 and four hardeners: 1, 2, 3, 4 respectively. As a sand matrix the typical silica sand was selected for all mixtures – medium size (0,40/0,20/0,32), average grain size d_L = 0,33 mm, uniform (J91) with the specific surface area 6,22 m^2/kg. All sands were prepared in the laboratory roll mixer type LM-1. To the weighed portion of the sand the hardener has been dosed and mixed for 1,5 minute, then the resin was dozed and next 1,5 minute mixing was applied. The sand was then loaded into the dies of standardized shape to produce the samples for the particular sand property tests. The dies were mounted on the vibration compaction device LUZ-1 by WADAP Wadowice, which was switched on for 15 seconds with the maximum vibrations amplitude of 2 mm. After the visual check the samples were stored for hardening under the ambient conditions. The 8-like shape samples were produced for tensile strength R_m^u, longitudinal ones for the bending strength R_g^u and round ones for the permeability P^u and abrasibility S. The composition of the moulding sand was presented in Table 1.

Table 1. Composition of the moulding sand with the alkali-phenolic resin

Sand label	Resin	Hardener	Resin [wt.] Hardener [% to resin, wt.]
A1	Estrofen	PR	A1-1,2 1-25
A2	Sinotherm 8255	Akivator J 120	A2-1,2 2-25
A3	Fenotec 280ES	Fenotec HC Esters (HC30)	A3-1,2 3-25
A4	Permabind 440	Permabind Plus 7 Hardener	A4-1,2 4-25

Moulding sand with the composition given in Table 1 was prepared under different temperature conditions:

a) in the laboratory: ambient temperature tot ≈ 19,6°C, relative humidity W_w≈34%, labelled with additional T20,
b) in the shop floor: ambient temperature tot ≈ 9°C, relative humidity W_w≈50%, labelled with additional T10,
c) in the shop floor: ambient temperature tot ≈ 0°C, relative humidity W_w≈30%, labelled with additional T0.

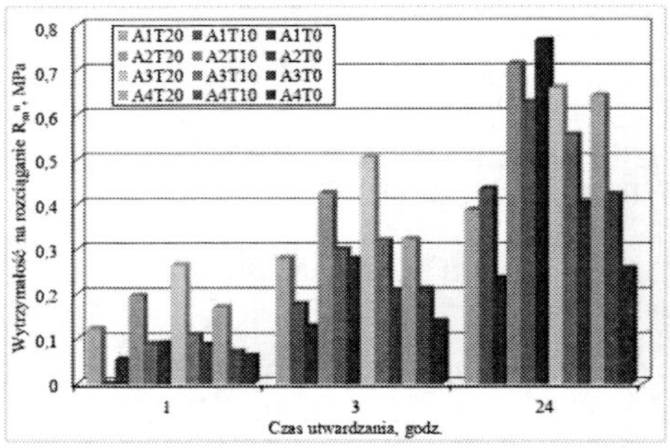

***Fig. 1.** Influence of the hardening time on the tensile strength versus ambient temperature*

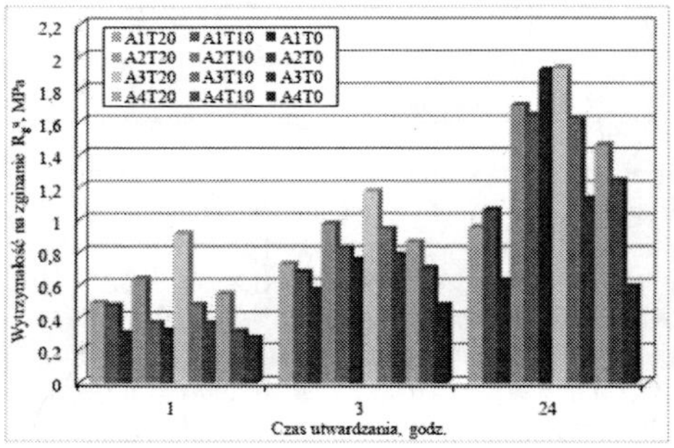

***Fig. 2.** Influence of the hardening time on the bending strength versus ambient temperature*

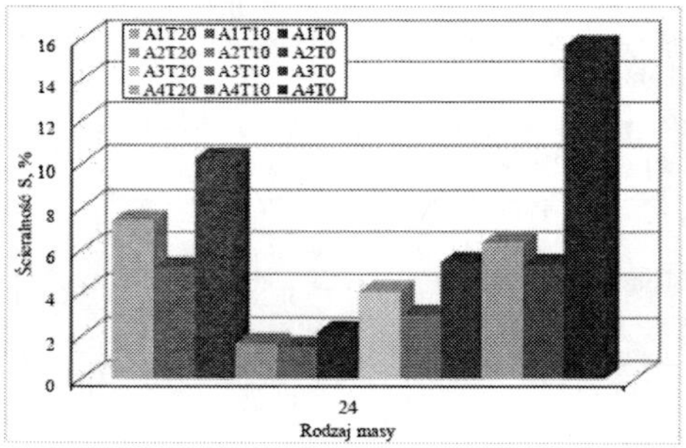

***Fig. 3.** Abrasibility S, versus ambient temperature*

In the Figure 1 the results of the tensile strength R_m^u for the prepared sand hardened under various temperature conditions have been presented. It is visible that for all examined compositions hardening time affects the strength of the sand. Regarding the temperature in most cases, a decrease of the sand preparation temperature causes the lowering sand tensile R_m^u.

For the binders being tested the best results in the matter of R_m^u under the low temperature exhibits the sand mixed on the base of binder A2, then with the binder A3 and A4. The worst from this point of view was resin labelled A1. In the figure 2 the bending strength results R_g^u for the examined moulding sands under various temperature conditions have been presented. Similarly, the sand quality evaluated on this property base shown that with the decrease of the temperature of sand preparation and hardening the strength of the sand in the alkaline-phenolic binding process decreases. The best results from the bending strength point of view were achieved for binder A2 then A3, A1 and A4. In the figure 3 the sand abrasibility S results for the examined sands have been presented. The tests have been carried out with use of the apparatus produced by Stalowa Wola Steel Plant (HSW). It is visible for all tested sands that for the lowest temperature during the experiments the best abrasibility values have been achieved. Additionally, it is seen that for the temperature around 10°C the S value for all compositions was the lowest. It has been proven that the higher strength of the moulding sand the lower its abrasibility, what is quite obvious anyway. The lowest mass losses were recorded for the samples prepared with binder A2 then A3. Higher losses were then for A1 and the highest for A4. For the moulding sand with binder A1 and A4 low temperature significantly decreases their abrasibility S.

Acknowledgments
The study was conducted under the project "Demonstration of the technology of manufacturing innovative cast steel castings to be used in extreme operating conditions", co-financed by the National Centre for Research and Development, under Operational Programme Intelligent Development 2014-2020. Action 1.1 R&D projects of enterprises, Sub-action 1.1.2. R&D works related to manufacturing pilot/demo system, contract POIR.01.01.02-00-0148/16-00 dated May 30, 2017

References

[1] Holtzer M., Bobrowski A., Drożyński D., Makselon J., Isendorf B. (2013). Investigations of Properties of Moulding Sands with Resins Applied in the Alphaset Technology. Archives of Foundry Engineering. 13(1), 31-37.

[2] Vasková I., Smolková M., Malik J., Eperješi S. (2008). Experience in forming and core mixtures by Alphaset technology. Archives of Foundry Engineering. 8(2), 141-144.

The 73rd World Foundry Congress
„Creative Foundry"
23 - 27 September 2018, Krakow, Poland

TIME-RESOLVED AND IN-SITU 2D / 3D IMAGING OF SOLIDIFICATION IN DUCTILE CAST IRON

Hideyuki Yasuda[1,*], **Akira Sugiyama**[2], **Kiattisaksri Chatcharit**[1], **Kohei Morishita**[1]

[1] *Kyoto University, Sakyo, Kyoto, 606-8501, Japan*
[2] *Osaka Sangyo University, Osaka, 574-0013, Japan*
**Corresponding address: Yasuda.hideyuki.6s@kyoto-u.ac.jp*

Key words: Ductile Cast Iron, In-situ observation, X-ray tomography, Radiography, Spheroidal graphite

Addition of Ce or Mg to cast iron changed the graphite morphology from a flake shape to a more compacted or spheroidal shape [1,2]. During the solidification, not the coupled eutectic but the divorced eutectic growth is selected in the melt which contains sufficient Mg or Ce, leading to spheroidal or compacted graphite particles in the matrix. There are several explanations for the spheroidal graphite formation; inclusions [3-8] or bubbles [9-11] act as nucleation sites for graphite and contribute to formation of spheroidal shape. In addition, Sulfur (S) solved in the melt can change graphite shape [12]. However, it is still not clear how the spheroidal graphite particles form during solidification.

Time-resolved and in-situ 2D observations using synchrotron radiation X-rays has allowed to observe solidification in cast iron in-situ [13-16]. In addition, time-resolved computed tomography (3D observation, 4D-CT) can be used to observe solidification phenomena. For example, the specimens (3.79C-3.1Si-0.14Mn-0.015S-0.04Mg, in mass%, CE=4.8) were repeatedly melted and solidified. In the first run, the graphite particles with spherical shape grew in the melt and were surrounded by austenite dendrites or the coupled eutectic front of austenite and graphite [14,16]. In the second run and further runs, graphite particles transformed from spheroidal shape to flake-like shape before being engulfed by austenite dendrites or the coupled eutectic front. It was also emphasized that the temperature range in which the graphite particles grew as the primary phase increased with decreasing Mg concentration [15,16]. It indicates that Mg in the melt can change phase equilibrium between liquid, austenite and graphite.

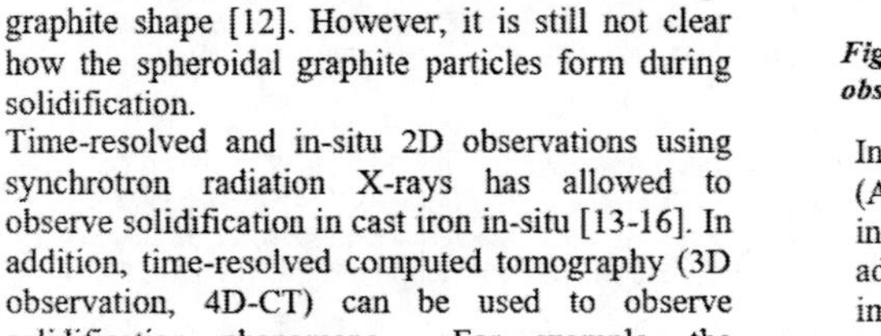

Figure 1. Setup for X-ray transmission imaging [13,16]

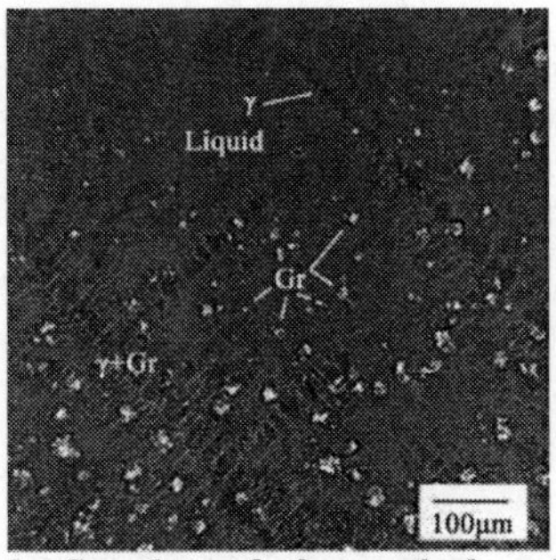

Figure 2. Snapshot of time-resolved and in-situ observation for the spcimen in Al_2O_3 holder [17]

In this presentation, influence of specimen holder (Al_2O_3 and MgO) on the microstructure evolution in the hypereutectic cast iron will be presented. In addition to 2D observation (X-ray transmission images) and the selection of the divorced eutectic and the formation of spheroidal graphite particles.

The chemical compositions of specimen is 3.67C-2.63Si-0.45Mn-0.02Mg-0.003S, in mass% (CE=4.5). Time-resolved and in-situ observations were performed for cast iron (3.67C-2.63Si-0.45Mn-0.02Mg-0.003S, in mass% (CE=4.5)) at a beamline of BL20XU in SPring-8 (Synchrotron radiation facility in Japan). Figure 1 shows a setup for the in-situ observation. A furnace using a graphite heater, a thin specimen (100 μm in thickness) in a holder (Al_2O_3 and MgO) and an X-ray imaging detector were placed along the light axis [13,16]. The pixel size was 0.5μm x 0.5μm. For the observation of solidification, the specimen with the microstructure of ductile cast iron was melted and then the melt was cooled at a cooling rate of 0.5K/s under a vacuum atmosphere (in the order of 1 Pa).

Figure 2 shows a typical microstructure observed in the Al_2O_3 holder [17]. Since the divorced eutectic solidification was selected at the beginning, graphite and austenite grew independently. According to the previous studies [14-16], addition of Mg shifted the eutectic composition to higher carbon concentration. The simultaneous nucleation and growth is explained by considering the

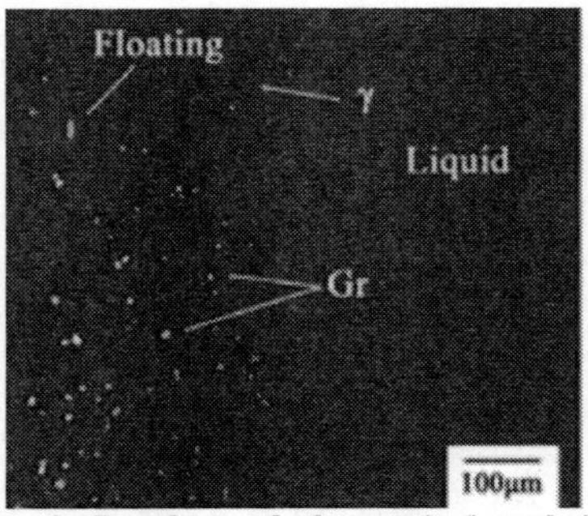

Figure 3 Snapshots of time-resolved and in-situ observation for the specimen in MgO holder [17].

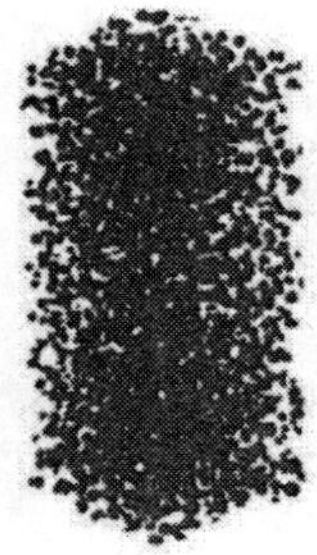

Figure 4 Three-dimensional disteibution of spheroidal graphite particles, obtained by tomography (CT). Domain of reconstruction image is 578μm × 587μm × 1296μm.

shift of eutectic composition. In the bottom part of Fig. 2, the coupled eutectic, in which austenite and lamellar graphite grew side by side, was observed. Thus, the divorced eutectic was followed by the coupled eutectic. In addition, graphite shape deviated from spheroidal shape.

Figure 3 shows microstructure in the MgO holder [17]. Austenite dendrites grew and simultaneously graphite particles nucleated around austenite dendrites. Floating of graphite particles indicated that the nucleation events occurred in the melt. The engulfment of graphite particles into the austenite occurred when the floating graphite particles touched the austenite dendrite arms. Only the divorced eutectic was selected when the specimen was put in the MgO holder. Most of the graphite particles kept spheroidal shape until the end of solidification.

Figure 4 shows a typical distribution of spheroidal graphite particles. Average diameter of graphite particles was 27μm, coordination number was 15 and average distance between graphite particles was 67μm. Evaluation of graphite distribution is useful for considering growth of graphite and apparent volume expansion after engulfment of graphite particles into austenite. In the presentation, time-resolve 3D observation by using CT will be presented and microstructure evolution in ductile cast iron will be discussed.

The results suggested that oxygen potential, which is determined by the holder material (Al_2O_3, MgO), significantly influenced the formation of spheroidal graphite particles. Mg dissolved in the melt and Mg-O-S inclusions can contribute to the formation of spheroidal graphite particles. The results will help further understanding for the microstructure evolution in the ductile cast iron.

Acknowledgments

This study was supported by a Grant-in-Aid for Scientific Research (S) from MEXT. The synchrotron radiation experiments were performed as general projects at beamlines of BL20XU, BL20B2 and BL28B2 at SPring-8 (JASRI).

References

[1] H. Morrogh, J.W. Grant :1947, J. Iron and Steel Inst., **155**, 321.

[2] A.P. Gagnebin, K.D. Millis, N.B. Pilling: 1949, The Iron Age, Feb., p.77.

[3] K. M. Muzumdar, J.F. Wallace: 1972, AFS Transactions, **80**, 317-328.

[4] K. M. Muzumdar, J.F. Wallace: 1973, AFS Transactions, **81**, 412-423.

[5] G.X. Sun, C.R. Loper: 1983, AFS Transactions, **91**, 639-646.

[6] T. Askaland: 1993, Metallurgical Transactions A, **24A**, 2321-2345.

[7] M.H. Jacobs, T J. Law, D.A. Melford, M.J. Stowell: 1974, Metals Technology, **1**, p.490-500.

[8] N. Klorca-Isern, J. Tartera, M. Espanol, M. Marsal, G. Bertran, S. Castel: 2002, Micron, **33**, 357-364.

[9] H. Itofuji: 1996, AFS Trans., **104**, 79-87.

[10] S. Yamamoto, Y. Kawano, Y. Murakami, B. Chang, R. Ozaki: 1978, Metal Science, 1978, **12**, 56-60.

[11] M. J. Lalich, J. R. Hitchings: 1976, AFS Trans., **84**, 652-658.

[12] H. Nakae, S. Jung, H. Inoue, H. Shin: 2004, Proc. 66th World Foundry Congress, Istanbul, p.917-928.

[13] A. Sugiyama, H. Yasuda, T. Nagira, M. Yoshiya, K. Uesugi, K. Umetani, I. Ohnaka: 2011, J. Japan Foundry Eng Soc, **82**, 131-136.

[14] K. Yamane, H. Yasuda, A. Sugiyama, T. Nagira, M. Yoshiya, K. Uesugi, K. Umetani, C. Ushigome, A. Sato: 2013, J. Japan Foundry Eng Soc, **85**, 760-770.

[15] K. Yamane, A. Sugiyama, T. Nagira, M. Yoshiya, H. Yasuda, Y. Tanaka, A, Sato, K. Uesugi, A. Takeuchi, Y. Suzuki, H. Honda, K. Sato: 2014, J. Japan Foundry Eng Soc, **86**, 461-470.

[16] K. Yamane, H. Yasuda, A. Sugiyama, T. Nagira, M. Yoshiya, K. Morishita, K. Uesugi, A. Takeuchi, Y. Suzuki: 2015, Metallurgical and Materials Transactions A, **46A**, 4937-4946.

[17] H. Yasuda, A. Sugiyama, C. Kiattisaksri, K. Morishita, T. Nagira, M. Yoshiya, K. Uesugi, A. Takeuch: 2018, Proc. Int. Symp. on the Science and Processing of Cast Iron, in press

The 73rd World Foundry Congress
„Creative Foundry"
23 - 27 September 2018, Krakow, Poland

FEASIBILITY OF US FOUNDRY SUPPLY CHAIN CONSUMABLES FOR THREE DIMENSIONAL SAND PRINTING

K. Woods[1,*], S. Giese[1], S. Trikha[2]

[1]University of Northern Iowa, Metal Casting Center, Cedar Falls, Iowa, USA
[2]H.A. International, Westmont, Illinois, USA
*Corresponding address: woodskae@uni.edu

Key words: three-dimensional printed sand, silica, furan, tensile strength, scratch hardness, printed density

The purpose of this study was to assess the performance characteristics of commercially available US foundry consumables for use in three-dimensional sand printers. The silica sands selected were a 57-grain fineness number (GFN) and 70 GFN. These sands were a high purity round grain silica and had low trace contaminates, to reduce any possible influence from contaminates. The Furan resin system was selected since it is the most widely utilized resin system for 3D printed sand (3DPS) in the USA. The Furan resin system was from a domestic US foundry supplier. The printed samples were traditional tensile speciemen and rectangle samples for the determination of density. The samples were examined for the following properties: tensile strength, loss on ignition, density, scratch hardness, and permability. The results of this investigation confirmed that the tested domestic resin system and silica sands matched the original equipment manufacturer performance properties.

1. Introduction

Three-dimensional sand printing (3DSP) is the process of selectively bonding the desired part geometry one layer at a time. This process is also known as binder jetting. The sand is first spread evenly and compacted across the build bed. The part geometry is created by a printhead jetting the binder into the required shape. This process is repeated until the final 3D part geometry is created. Since the build box is comprised both of bonded and unbonded sections no support structure is required during printing like other 3D printing processes. The part complexity of a casting can be much greater since the need for draft and parting lines is greatly reduced in sand printing process.

The design freedom provided by 3DSP comes with a higher cost than traditional sand molding processes. Printed sand is generally priced based upon the cubic volume of the mold or core being produced. The biggest factor to the increased cost associated with printed sand is that currently sand printers utilize the consumables provided by the sand printer manufacturer's. The consumables required for printing is the resin system and sand. The purpose of this study is to investigate the printing properties using consumables currently available by the US foundry supply chain.

Previously published printing properties for 3DPS were evaluated to determine the range of acceptable printing performance when the manufacturer resin system and silica sand were utilized. The published tensile strength was 108 PSI[1] to 192 PSI[2]. The published transverse strength was 166 PSI to 392 PSI[3] and 340 PSI[1]. The printed density[1,3,4] ranged from 1.331 to 1.495 g/cm^3. Hackney and Woolridge research showed that compared to traditional hand rammed samples 3DPS process created a mold or core with a much lower density when the same resin and sand was used in both trials[4].

2. Methodology

The print geometry selected for analysis was the American Foundry Society (AFS) tensile specimen defined by AFS 3342-00-3[5]. Additionally, rectangle samples measuring 0.5" x 1" x 1.5" were printed for the determination of the bonded sand density as shown in Figure 1. In total thirty tensile specimens and twelve density samples were printed.

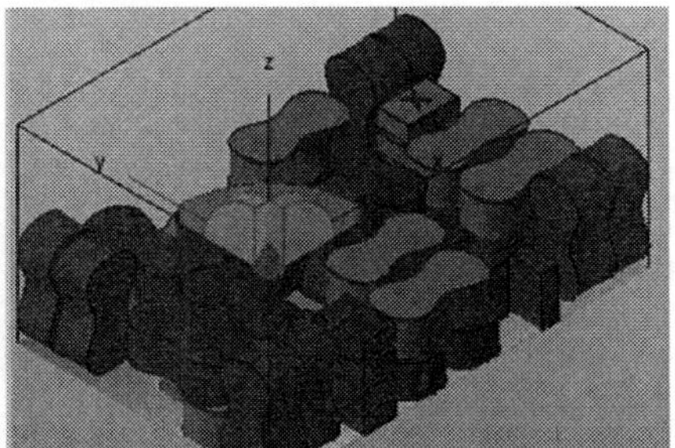

Fig. 1. Build box layout of test specimens per trial.

The resin droplet spacing was adjusted using the X-Resolution machine setting(X_R). The sand layer thickness was adjusted using Z-Resolution(Z_R) machine setting. The combination of X_R and Z_R allowed four different resin contents to be tested. The sand printer utilized for this analysis was a Voxeljet VX200 Sand Printer. The Furan resin system was from

[Wpisz tutaj]

H.A. International. Silica sands were an Illinois round grain, measuring 57 GFN and 70 GFN. The testing was comprised of four trials per sand type with a complete replication of the series for a total of sixteen trials.

3. Discussion of Results

Series 1A-4A and 1B-4B were the trials with the 57 GFN sand. Series 5A-8A and 5B-8B were the trials with the 70 GFN sand. The X_R and Z_R machine settings were the same for trial 1 to 5, 2 to 6, 3 to 7, and 4 to 8. The resin content was evaluated using loss on ignition (LOI) method. The resin content decreased as the sand volume increased. Analysis of variance (ANOVA) confirmed that for the same machine settings the resin content showed good agreement as shown in Figure 2 the orange line plotted on the secondary axis.

The average tensile strength decreased as the sand volume increased per resin droplet. The upper resin content tested of 1.95% to 2.01% showed that the different GFN sands did not change the observed tensile strength range 211 to 233 PSI. Likewise, the lowest resin content tested of 0.91% to 0.99% with tensile strengths of 76 to 107 PSI did not demonstrate strength difference between the two sands types. The only difference between the 57 GFN and 70 GFN sand was observed at the resin content of 1.23% to 1.31%. The 70 GFN sand series 7A/7B retained a tensile strength of 163 to 171 PSI, which was the same as series 6A/6B which had a resin content of 1.50% to 1.55% as shown in Figure 2 the blue bar chart plotted in the primary axis.

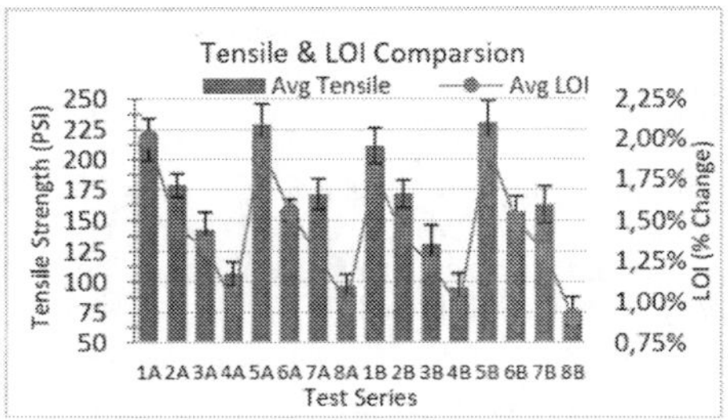

Fig. 2. Sixteen trials tensile and loss on ignition results.

The average printed sand density for all trials was 1.3 g/cm^3. The 57 GFN and 70 GFN sands did not demonstrate a difference in printing density. The printed sand density was consistent regardless of the GFN tested.

The average AFS permeability for the 57 GFN series was 442. The average AFS permeability for the 70 GFN series was 260. The different resin contents tested did not change the permeability for either GFN tested.

The scratch hardness results at the upper and lower tested resin contents mirrored the observed trend from that tensile strength where that GFN did not impact the hardness values. Again, the GFN only impacted the scratch hardness values of series 7A/7B as shown in Figure 3.

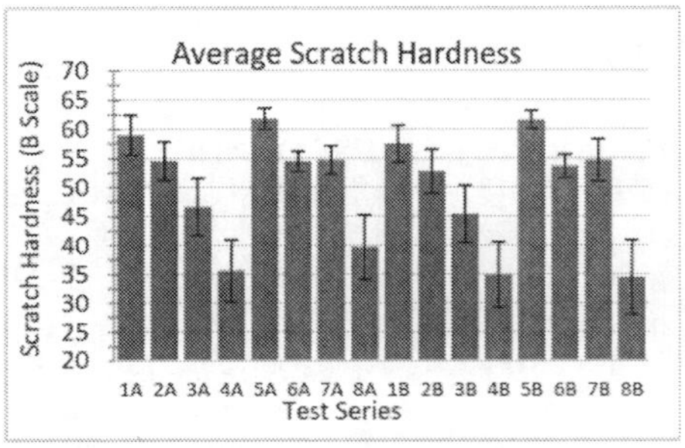

Fig 3. Scratch hardness results for the sixteen trials.

When comparing the previously published tensile strength values of 104 PSI to 192 PSI, the US foundry supply chain consumables achieved similar tensile strengths with the tested machine settings. The test resin and sands were only below the published tensile strengths when the resin content was only 1%. The average printed silica sand density was within the published range from the previous research studies. The GFN did impact printing strengths when the resin value was between 1.23% to 1.31% for the 70 GFN only.

4. Conclusion

The testing completed within this investigation confirmed that US foundry supply chain consumables, do produce sufficiently equivalent performance properties to that of the original equipment manufacturers consumables. Only when the resin content present on the sands was below 1% did the tensile strengths drop below the reported published values.

References

[1] Nyembwe, K., Mashila, M., Van Tonder, P., De Beer, D., & Gonya, E. (2016). Physical properties of sand parts produced using a Voxeljet VX1000 three-dimensional printer. South African Journal of Industrial Engineering, 27(3), 136-142.

[2] Snelling, D., Blount, H., Foram, C., Ramsburg, K., Wentzel, A., Williams, C., & Duschitz, A. (2014) The effects of 3D Printed Molds on Metal Castings, Solid Freedom Fabrication Symposium. Austin, TX.

[3] Coniglio, N., Sivarupan, T., & Mansori, M. (2018). Investigation of process parameter effect on anisotropic properties of 3D printed sand molds. The International Journal of Advanced Manufacturing Technology, 94(5), 2175-2185.

[4] Hackney, P. M., & Wooldridge, R. (2017). Characterisation of direct 3D sand printing process for the production of sand cast mould tools. Rapid Prototyping Journal, 23(1), 7-15

[5] AFS Mold and Core Test Handbook (Vol. 4th Edition). (2015). Schaumburg, IL: American Foundry Society.

The 73rd World Foundry Congress
„Creative Foundry"
23 - 27 September 2018, Krakow, Poland

LIQUID-PHASE BONDING OF CARBON/CARBON AND POROUS CARBON FOR STRUCTURAL AND THERMAL MANAGEMENT APPLICATIONS

Rajiv Asthana[1]*, Mrityunjay Singh[2], Jerzy J. Sobczak[3] and Natalia Sobczak[3,4]

[1]University of Wisconsin-Stout, Menomonie, Wisconsin, USA
[2]Ohio Aerospace Institute, Cleveland, Ohio, USA
[3]Foundry Research Institute, 73 Zakopianska Str., 30-418 Krakow, Poland
[4]Institute of Precision Mechanics, 3 Duchnicka Str., 01-796 Warsaw, Poland
**Corresponding address: AsthanaR@uwstout.edu*

Key words: liquid-phase bonding, wettability, infiltration, reactivity, interfaces, carbon foam, carbon/carbon composites, joint microstructure, joint strength

1. Introduction

Robust liquid-phase bonding techniques such as active brazing play a key role in heat exchangers, turbine engines, fuel cells and thermoelectric systems. While many industrial active braze alloys are available, the effectiveness of such alloys to bond advanced high-temperature ceramics, composites and porous materials must be experimentally evaluated on a case-by-case basis.

In this contribution, some empirical findings based on extensive trials to screen promising brazes to join high-conductivity carbon foam and carbon/carbon composites will be presented. Microstructure and select engineering properties of joints will be presented to identify the most promising material combinations. The role of braze composition and joining conditions on joint microstructure, composition, micro-hardness, residual stress and joint strength will be discussed.

2. Summary on experimental work done

Carbon foams for thermal management and energy storage applications can be designed for density, pore size, and conductivity (e.g., ligament conductivity >1700 W/m.K, bulk conductivity ~175 W/m.K) besides chemical inertness and thermal stability. Mechanical tests on metal-to-foam joints with bonded surfaces either parallel or perpendicular to the foaming direction revealed the effect of foam density and foaming direction [1]. Additionally, the effect of foam surface modification by SiO vapor deposition and reaction to form SiC on the strength of SiC-coated foam-to-Ti joints was investigated (Fig. 1).

The infiltration of foam with SiO vapor was done to produce surfaces comprised of thin layers of reaction-formed SiC on carbon with a view to enhancing the strength without significant detriment to the outstanding thermal characteristics of the foam. In both Ti/foam and Cu-clad-Mo/foam joints, Ti enrichment at the carbon ligament surface (Fig. 2) produced a Cu-Ti phase that is consistent with the wettability enhancement effect of Ti. In carbon foam-to-Ti joints made using a high-conductivity Ag-Cu-Pd braze, no interfacial excess of Pd was noted (Fig. 3).

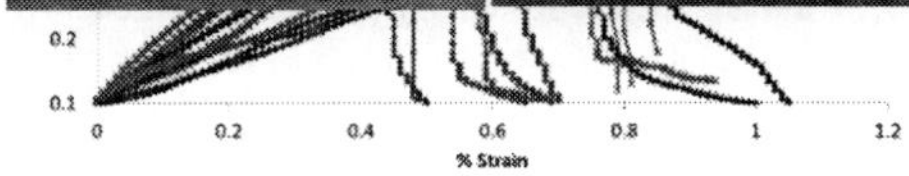

Fig. 1 Mechanical behavior of foam/metal joints (Braze: Cusil-ABA, strain rate: 0.64 mm/min)

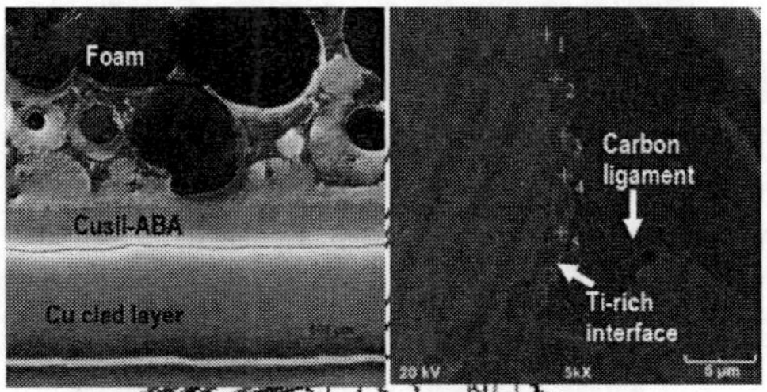

Fig. 2 A carbon foam/Cusil-ABA/Cu-clad-Mo joint, (b) SEM images showing interfacial segregation of Ti

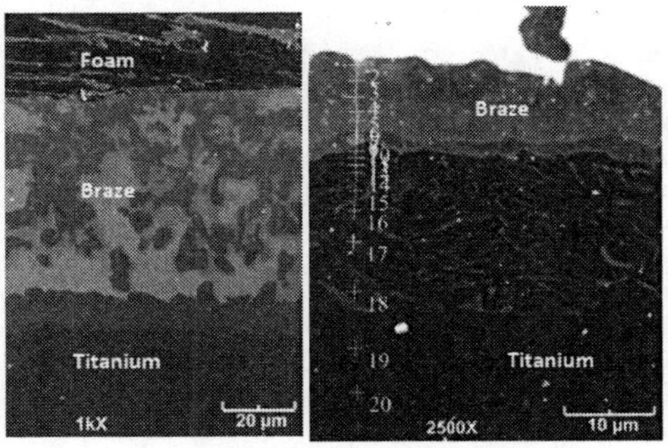

Fig. 3 A Poco foam/Palcusil-5/Ti joint

In SiO-coated foam joints with Ti substrates, a prominent reaction layer containing both Ti and Si developed. Both coated and uncoated carbon foams brazed to Ti were stronger along the foaming ('with-rise') direction than ('against-rise') direction.

Sessile-drop wettability tests with Ag and Pd-base active brazes on C/C composites reveal [2] non-uniform, anisotropic spreading, copious braze infiltration (~100-200 μm in 5 min), and Ti enrichment at interfaces together with dissolution and redistribution of elements in and around the joint, and some tendency toward inter-laminar shear cracking within the composite substrates having low inter-laminar shear strength (Fig. 4).

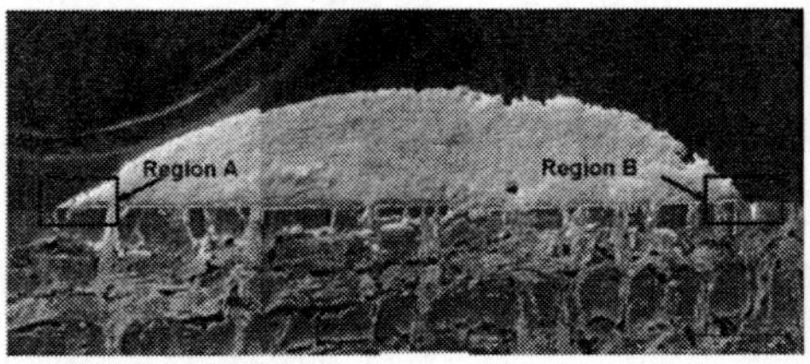

Fig. 4 C/C foam in contact with a Cusil-ABA sessile drop.

Pd-base fillers offer higher use temperatures than Ag-Cu-Ti alloys. Commercial Pd-Co and Pd-Ni braze fillers have very good wettability with C/C [2], and the contact angles attain acute values. In joints made using Ti-bearing fillers, Ti preferentially segregated at the C/C/braze interface. Interdiffusion of alloying elements was noted among the C/C composite braze matrix, and metal substrate.

Carbon/carbon composites have been integrated with titanium tubular subcomponents [3,4] for use in Brayton and Thermoelectric power conversion systems. For light-weight heat exchanger applications, Ti tubes were integrated with high-conductivity C/C using Ag-Cu-Ti braze and flexible, low-density Poco foam that accommodated thermo-elastic strains and prevented cracking. Braze paste exhibited better penetration characteristics than braze foil.

Tensile and shear strength tests were conducted on C/C/Ti joints with and without carbon foam using different brazes and with different fiber orientations at the mating surface. The shear strength of directly bonded C/C/Ti joints, extracted from Butt-Strap Tensile (BST) tests was low (1.5-9.0 MPa, ±0.09-1.62 MPa). Polished C/C samples yielded slightly higher joint strength than unpolished samples. For C/C-to-Ti joints in which Ti tubes were brazed to C/C plates, the average shear strength depended not only on braze type but also on the orientation of the exterior carbon fiber tow at the joint. Joints with fiber tows perpendicular to the Ti tube axis exhibited higher failure loads than joints with parallel fibers. The load-carrying ability of the joint was affected by the bonded area and the fiber-tow orientation. For Ti tube/foam/C/C plate sandwich structure, tension and shear tests showed that shallow grooves led to high stress on the tube-to-foam joint with joint strength in excess of 12 MPa in tension or shear.

Failure always occurred in the foam regardless of the type of C/C composite used and whether the Ti tube was brazed to a curved foam plate to maximize the bonded area or a flat foam surface to maximize stress in the joint.

Both carbon foam and C/C composite were also brazed to Cu-clad-Mo to enhance heat conduction. Joining is beneficial because conductivity and CTE of Cu-clad-Mo can be designed by controlling the clad layer thickness. In the case of Cu-clad-Mo substrate joined to C/C composite using Ag-Cu-Ti active braze alloys the clad layer remains untransformed because the joining temperatures are below the melting point of Cu (1359 K). This provides for stress accommodation by the copper layer, aiding the stress relief provided by the ductile brazes.

Acknowledgments

The research has been done in the frame of US-Poland scientific collaboration with financial support from NASA Glenn Research Centre, the University of Wisconsin-Stout and the Foundry Research Institute.

References

[1] M. Singh, R. Asthana, C.E. Smith, A.L. Gyekenyesi. (2013). Integration of High-Conductivity GraphiteFoam to Metallic Systems using Ag-Cu-Ti and Ag-Cu-Pd Braze Alloys, Int. J. Applied Ceram. Tech., 10(5), 790-800.

[2] R. Asthana, M. Singh, N. Sobczak (2010). Wetting behavior and interfacial microstructure of Pd and Ag-based braze alloys with ceramic matrix composites, J. Mater. Sci., 45(16), 4276-4290.

[3] M. Singh, R. Asthana (2011), Integration of C/C composite to metallic systems for thermal management applications, in M. Singh et al (Eds.), Ceramic Integration and Joining Technologies (163-191). New York: Wiley.

[4] R. Asthana, M. Singh (2013), Active metal brazing of advanced ceramic composites to metallic systems, in D. Sekulic (Ed.), Advances in Brazing (323-360). Cambridge (UK): Woodhead Publishing

The 73rd World Foundry Congress
„Creative Foundry"
23 - 27 September 2018, Krakow, Poland

USE OF HIGH INTENSITY X-RAY ANALYSIS AS TOOL TO CREATE NEW, FUNDAMENTAL MODELS FOR PHASE TRANSFORMATIONS AND RESIDUAL STRESS IN DUCTILE CAST IRON

NS Tiedje[1]*, MK Bjerre[1], T Andriollo[1], M Azeem[2], Y Zhang[1], S Fæster[3], PD Lee[2], JH Hattel[1],

[1]*Technical University of Denmark, Mechanical Engineering, B425-R223, 2800 Kongens Lyngby, Denmark*
[2] *University College London, Mechanical Engineering, Torrington Place, London, WC1E 7JE*
[3]*Technical University of Denmark, Wind Energy, B228-R021, 2800 Kongens Lyngby, Denmark*

**Corresponding address: e-mail: nsti@mek.dtu.dk*

Key words: Ductile Cast Iron, Modelling, Microstructure, X-Ray Tomography, Solidification, Residual Stress

1. introduction

Cast irons have been used for engineering purposes for thousands of years, and still, today they are essential our society. Their mechanical properties are primarily determined by the shape, size and distribution of graphite in the steel matrix. [1]

The mechanisms behind graphite growth and development of shape and their coupling to raw material quality and process conditions remain unknown.[2] Also the mechanisms leading to the formation of residual stresses are also not fully clear.[3;4]

Extensive experimental data has been produced and utilized to generate empirical correlations via statistical methods, showing that e.g. the shape and distribution of graphite affect ductility, yield strength and fatigue properties while carbides influence strength and wear properties.[1] However, little is known on why such correlations hold true, as the individual mechanisms which control the material behavior at the micro-scale have not been clarified.

Recent advances in high resolution X-Ray methods that involve use of synchrotron facilities have made it possible to do high resolution, in-situ experimental studies of phase transformations in engineering materials thus providing detailed and accurate information on the processes that take place during such phase transformations.[5;6]

This paper describes how such facilities can be applied to study solidification of cast iron and formation of residual stress after eutectoid transformation by resolving the processes in 3D and time.

2. 3D studies of solidification and microstructure in ductile cast iron

Solidification was studied in an environmental cell where the sample could be melted and solidified while 3D X-Ray recordings were made as described in.[6] During solidification 62 tomograms consisting of each 360 2D projections were acquired so that in selected sub-volumes it is possible to study in detail formation and growth of nodules. Fig. 1 shows 3 such sub-volumes extracted from the total sample volume. It is possible to follow how the size and shape of individual nodules develop depending during solidification.

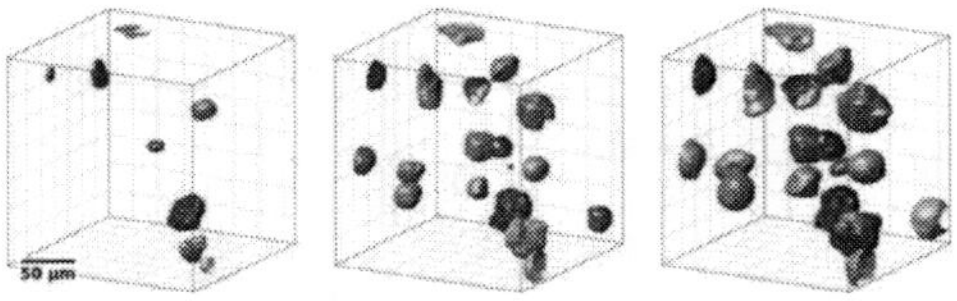

***Fig. 1.** Tomography images from a sub-volume. Images are taken 360, 400 and 480s into the recording. Images show formation of nodules revealing how some nodules remain spherical while others change shape.*

The recordings allow detailed quantitative studies of spheroidal graphite growth and direct comparison to model predictions highlighting how and where models for growth need to be improved. We can show the effect of the local arrangement of graphite, liquid and austenite on the individual graphite nodule growth patterns and while studies of large ensembles illuminates both the average and the extreme nodule growth behavior. At the same time, we can study the development graphite size distribution with time (see figure 2) and of particle shapes during the course of solidification providing new and unexpected insights and addressing unresolved issues regarding the development of particle morphologies.

3. Experimental and numerical analysis of graphite-matrix interaction

It is well known that ductile iron parts, depending on their size, may contain residual stresses developing over distances of a few millimeters or more, which arise due to the presence of constraints that hinder the free thermal contraction of the material during cooling.

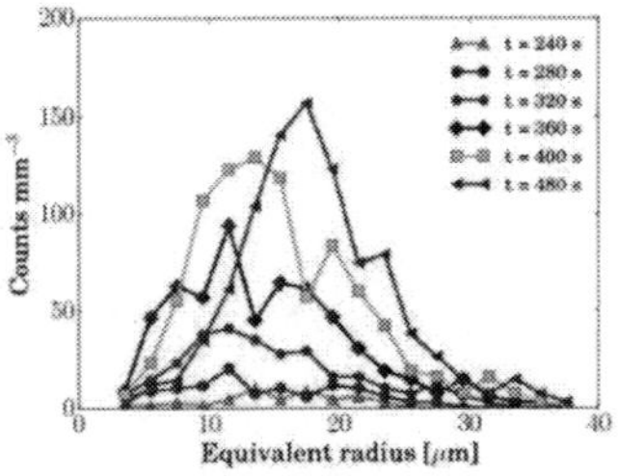

Fig. 2. Nodule size distribution measured in a sub-volume as function of time.

The intrinsic composite nature of ductile iron suggests the possible formation of another type of residual stresses, at much shorter length scales, associated with the thermal contraction mismatch between the two main metallurgical phases forming the material microstructure: the graphite nodules and the metallic matrix.[5] Surprisingly, the subject has not received much consideration in the past, probably due the common belief that the graphite particles are very soft and unable to withstand any kind of loading. However, experimental evidence exists for their mechanical importance, especially at relatively high temperature and under compressive loadings, indicating that ductile iron might not be considered as a merely "voided material" in all situations, as it is illustrated in figure 3a and 3b.

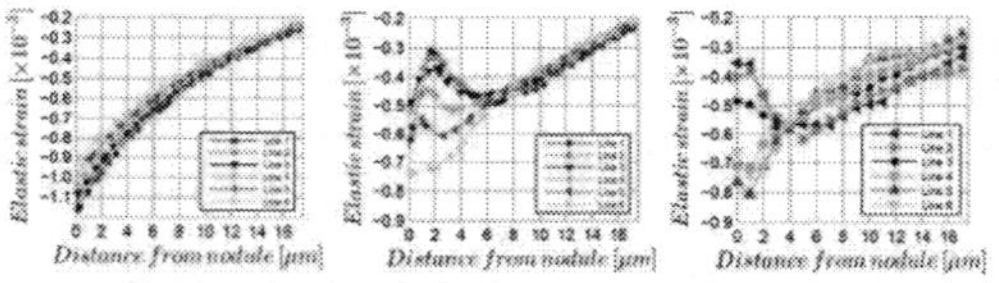

Fig. 3. Residual elastic strain around a graphite nodule assuming that: a) the nodule has no internal microstructure, b) that its internal structure consists of several conical sectors, c) local stress measured with synchrotron X-Ray diffraction

To investigate the existence of residual stresses at the microscale level, the elastic strain in the ferrite matrix surrounding a few nodules lying beneath the surface of a thin sample was measured using a novel 3D X-ray diffraction technique.[5] The measurements revealed that local stresses up to approximately half the macroscopic yield strength of ductile iron may remain in the microstructure after manufacturing. It was also shown that local residual stress varies with orientation relative to crystallographic sectors in the graphite nodule, as shown in fig. 3c.

As a consequence, a theoretical model was developed where the internal subdivision of the nodule into conical sectors was reproduced in detail.[7] In addition, numerical simulations allowed clarifying that the conical sectors making up the nodules create residual stress concentrations in the matrix, which correspond to the elastic strain fluctuations seen in Fig. 3b when the distance from the nodule is less than ≈ 5-6 microns. New studies are being carried out to investigate the impact of this complex residual stress field on the micro-scale deformation of ductile iron upon loading. Fig. 4 shows an example of calculated plastic strain in a ductile iron volume subjected to mechanical load. Colors show local variation in plastic strain based on the distribution of nodules.

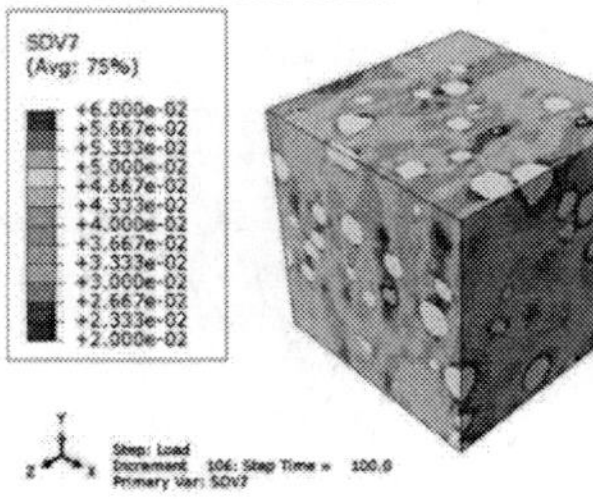

Fig. 4. Calculated local plastic strain around graphite nodules in a volume of ductile cast iron.

4. Conclusion

With the increased use of synchrotron X-rays for in situ experiments on casting-related phenomena we can expect a large improvement in our understanding which provides a much firmer basis for models and predictions of microstructure formation. This paves the way for higher degree of tailoring of processes and alloys for specific purposes and requirements as well as general improved performance of cast components.

5. References

[1] D.M. Stefanescu, ed. (2017), Cast Iron Science and Technology ASM International.

[2] J. Lacaze, N. Valle and K. Theuwissen, (2013), Redistribution and Effect of Various Elements on the Morphology of Primary Graphite in Cast Iron, Advances in Materials Science and Engineering

[3] T. Andriollo, N. Vedel-Smith, J. Thorborg and J. H. Hattel, Internal casting stresses and dimensional stability. In Cast Iron Science and Technology (Edited by D. M. Stefanescu), Chap. 4.11. ASM International, (2017).

[4] N. K. Vedel-Smith, J. Rasmussen and N. S. Tiedje, (2015), Thermal Distortion of Disc-Shaped Ductile Iron Castings in Vertically Parted Moulds, Journal of Materials Processing Technology 217, pp. 262-271

[5] Y. Zhang, T. Andriollo, S. Fæster, W. Liu, J. H. Hattel and R. I. Barabash, (2016), Three-dimensional local residual stress and orientation gradients near graphite nodules in ductile cast iron, Acta Mater. 121, pp. 173-180

[6] M. Azeem, M. K. Bjerre, R. C. Atwood, N. S. Tiedje and P. D. Lee, (2018), Synchrotron quantification of graphite nodule evolution during the solidification of cast iron, Acta Mater. Accepted for publication.

[7] T. Andriollo, K. Hellström, M. R. Sonne, J. Thorborg, N. S. Tiedje and J. H. Hattel, (2018), Uncovering the local inelastic interactions during manufacture of ductile cast iron: How the substructure of the graphite particles can induce residual stress concentrations in the matrix, J. Mech. Phys Solids, Volume 111, pp. 333-357

The 73rd World Foundry Congress
„Creative Foundry"
23 - 27 September 2018, Krakow, Poland

INCREASE PRECISION AND YIELD IN CASTING PRODUCTION BY SIMULATION OF THE SOLIDIFCATION PROCESS BASED ON REALISTIC MATERIAL DATA EVALUATED FROM THERMAL ANALYSIS (USING THE ATAS METSTAR SYSTEM).

Per-Eric Persson [1]*, Zenon Ignaszak [2], Hakan Fransson [1], Valentin Kropotkin [3]

[1] Novacast Systems AB, 372 25 Ronneby, Sweden,
[2] Poznan University of Technology, CAD/CAE of Material Technology and Foundry Laboratory, 60-965Poznan, 3 Piotrowo Street, Poland
[2] Novacast Systems AB, 372 25 Ronneby, Sweden,
[3] MKM Ltd, 24 Ilfata Zakirova St, 426072, Izhevsk, Russia,
**Corresponding address: e-mail: PerEric.Persson@novacast.se*

Key words: ductile iron, simulation and defect prognosis, density calibration, thermal analysis

The experience when a foundryman is comparing the shrinkage defects in castings produced in the foundry and the results from a simulation of the filling and solidification process using a commercial simulation system is that it does not always shows the same result. The practice most frequently used today, is to take parameter values from literature to use as input for the calculations. The simulated result is in many cases close (but not always) to reality in terms of temperatures, cooling speed and solidification time. However, the prognosis for calculation of shrinkage defects (macro and micro) could show a large discrepancy form the real results. The fact is if the same casting is simulated, using the same material data, in another foundry with a similar metallurgical production process, the simulated results and the actual "real" result could be very different.

The material standard of the cast iron is the same, the same target for chemical composition, the same pouring temperatures and filling times but the "real metallurgical production" in a foundry process show variations caused by the metallurgical conditions, chemical composition holding times and metallurgical treatment methods for magnesium addition. The final quality of the raw casting is also influenced by properties in the mold material. The conclusion is that a metallurgical process has a certain process window, there are metallurgical parameters which show variations from one batch to another and the process window is different from one foundry to another. The differences and variation may not be large, but it could have strong influence on the final material quality in the castings.

For example, the Fe-C binary phase diagram, developed for thermodynamic equilibrium conditions, gives important but only general information (guidelines) about how the variation of the carbon content influences the liquidus temperature and allows to estimate (quantitatively/qualitatively) the maximum possible precipitated volume of graphite during the solidification process. The total carbon content and the effect on the precipitation of graphite particles are key parameters for grey- and ductile iron to produce a casting free from macro and micro defects, considering the balance of expansion of graphite and contraction of austenite.

The conducted work shows and confirms how thermal analysis of grey and ductile iron could be used as a source for calculating important metallurgical data to be used as input to increase the precision in simulation of cast iron solidification and to achieve a higher quality in prediction of macro and micro porosities in the castings.

Standard type of sampling cups (M= 0,70 cm) are used for the thermal analysis. The results from thermal analysis elaborated with the ATAS MetStar system are evaluated parallel with the material quality (including tendency to external and internal defects) of the tested ATAS castings (Fig. 1).

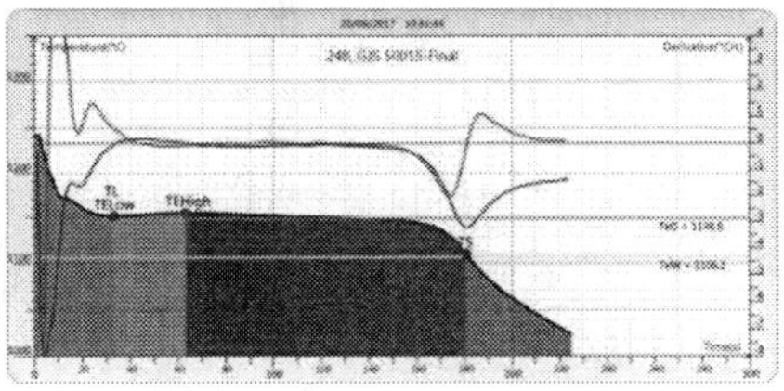

24B, GJS 50015-Final
20/09/2017 10:31:44

MQ = 50 !

ACEL = 4,20

Indicator	Min	Current	Max	Fingerprint
TL	1141,6	1151,7	1156,6	
TES	1143,0	1151,7	1154,0	
dT/dt TES	-2,00	-0,06	0,00	
S1	0,0	0,0	20,0	
TELow	1145,0	1151,7	1152,0	
R	1,0	3,5	5,0	
GRF1	70	91	110	
GRF2	30	42	65	
dT/dt TS	-3,50	-3,25	-2,50	
TS	1080,0	1109,0	1105,0	

Fig.1 Example of DTA curve and corresponding metallurgical quality parameters using ATAS MS system

Significant temperatures and calculated quality parameters are evaluated form the samples and used as input for simulation studies of the same samples using the NovaFlow&Solid (NF&S) simulation system, and in this system - density calibration procedure (Fig. 2).

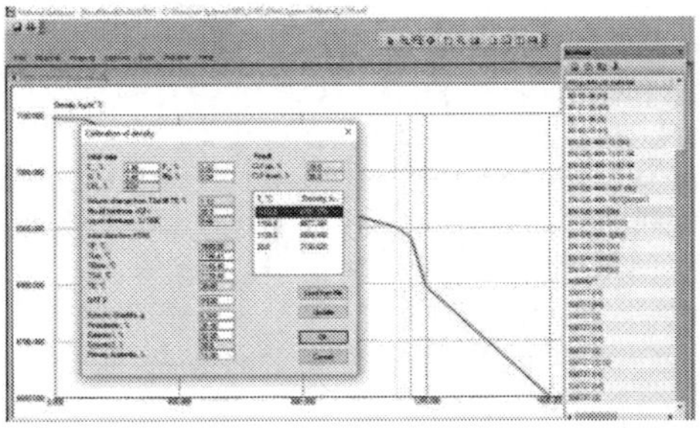

Fig.2 Screenshot of NF&S calibration of density module (possibility of initial data introduction directly from ATAS)

The results are verified with x-ray investigation (Fig.3) and microstructure analysis.

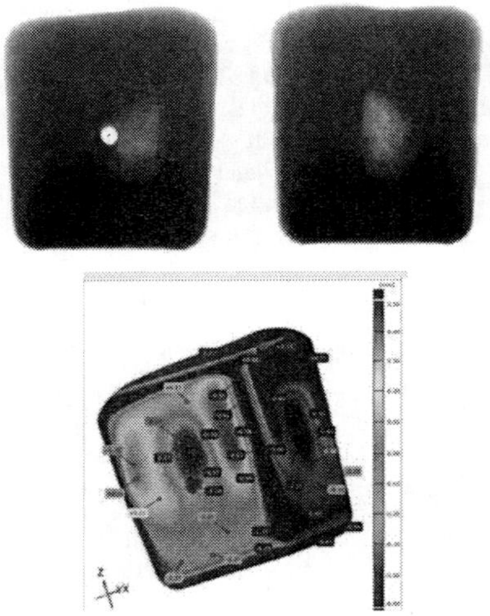

Fig.3 Identification of internal (by X-ray) and external (by 3D scanning) shrinkage defect of ATAS cup. Objective: validation of NF&S simulation results.

The references listed below are materials expanding the thematic area covered by the article
The references listed below ([1] to [10]) are resources expanding the subject area covered by this article.

References

[1] P.E Persson, A.Udriuo, P. Vomacka Wang Xiaojing and T. Sjögren. (2010). ATAS a tool for analyzing, stabilizing the graphite precipitation in grey cast iron. 2010 World Foundry Congress, Hangzhou. Conference Proceedings.

[2] Z.Ignaszak (2004) Examples of model validation for foundry processes virtual prototyping, Advanced Materials Forum II, vol. 455-456, pp. 681-684.

[3] H. Barth, A. Ljung, P.E Persson. (2010). Tight Process Control Essential for Tough Requirements, DIS Heavy Section Ductile Iron Conference 2010. Conference Proceedings.

[4] Z.Ignaszak (2017) Discussion on Usability of the Niyama Criterion for Porosity Predicting in Cast Iron Castings. Archives of Foundry Engineering, vol.17, iss. 3/2017, pp. 196-2014.

[5] Z. Ignaszak, P. Popielarski, J. Hajkowski, E. Codina (2015) Methodology of Comparative Validation of Selected Foundry Simulation Codes, Archives of Foundry Engineering, vol.15, iss. 4/2015, pp. 37-44

[6] G. Bertuzzia, G. Scarpab Comparison of casting simulation results and experimental data in heavy section ductile iron production.

[7] J. Nilsson, Utveckling av produktionssystem för segjärnstillverkning med hjälp avavancerad termisk analys. Rapport Svenska Gjuteriföreningen (Swedish Foundry Association)

[8] H. Barth, A. Ljung, P.E Persson. (2010), Tight Process Control Essential for Tough Requirements, DIS Heavy Section Ductile Iron Conference 2010. Conference Proceedings.

(9) D. M. Stefanescu (2015), Thermal analysis – Theory and applications in metalcasting, International Journal of Metalcasting/Vol. 9, pp. 7-21,

[10] D. Sparkman (2001), Offsetting Shrinkage in Ductile Iron, what Thermal Analysis Shows.

APPLICATION OF DIFFERENTIAL SCANNING CALORIMETRY (DSC) FOR EVALUATION OF ALUMINIUM ALLOYS BILLETS HOMOGENIZATION PARAMETERS

Antoni Woźnicki [1]*, Grzegorz Włoch [1]

[1]AGH University of Science and Technology, Faculty of Non-Ferrous Metals,
Al. Mickiewicza 30, 30-059 Kraków, Poland
**Corresponding address: woznicki@agh.edu.pl*

Key words: aluminium alloys billets, homogenization, DSC, solidus temperature, incipient melting

1. Introduction

Homogenization is an essential part of technological process, strongly influencing deformability of billets as well as structure and properties of obtained aluminium alloys products. This heat treatment is carried out at high temperature range for prolonged time, and in consequence it is very energy-consuming and costs generating process. For numerous aluminium alloys grades parameters of homogenization process can be found in the literature [1, 2]. However, these data need to be treated only as indicative values, which results from two main reasons: (i) ranges of chemical composition defined for particular grades are usually broad [3], what can be reflected in differences of obtained microstructure [4,5], (ii) microstructure of billets depends on casting process conditions. Thus, selection of dedicated homogenization parameters for given (narrowed) alloy composition and billets casting parameters is justifiable from economical and technological point of view.

The goal of present work is to demonstrate of Differential Scanning Calorimetry technique application for evaluation of main structural process which occurs during aluminium alloys billets homogenization, i.e. elimination of alloying additions microsegregation.

2. Experimental

The materials for investigations were received from selected semi-industrial DC-cast billets, which were varying with microstructural effects of unequilibrium solidification. Samples were homogenized in laboratory conditions and examined using mentioned DSC as well as SEM/EDS techniques.

3. Results

In the as cast 5019 alloy, dendritic microsegregation is observed, however no incipient melting peak is found (Fig. 1). During homogenization increase of solidus temperature is observed from 563 °C to about 575 °C (Table 2).

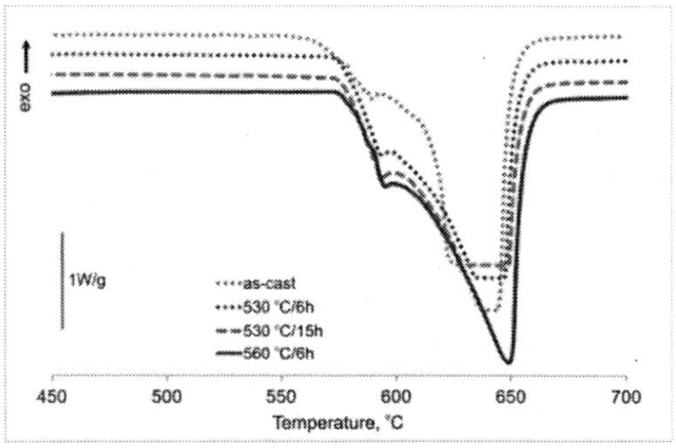

Fig. 1. DSC curves for 5019 alloy billets in as-cast state and after homogenization (soaking time is given in the Figure)[6]

Table 1. Changes of the solidus temperature of the 5019 alloy [6]

State	Solidus temperature, °C
as-cast	563
530 °C/0h	570
530 °C/6h	573
530 °C/15h	575
560 °C/6h	574

Different effect of unequilibrium solidification is found in Al-3.5Cu-1.2Mg alloy, for which significant incipient melting peak is found (Fig. 2).

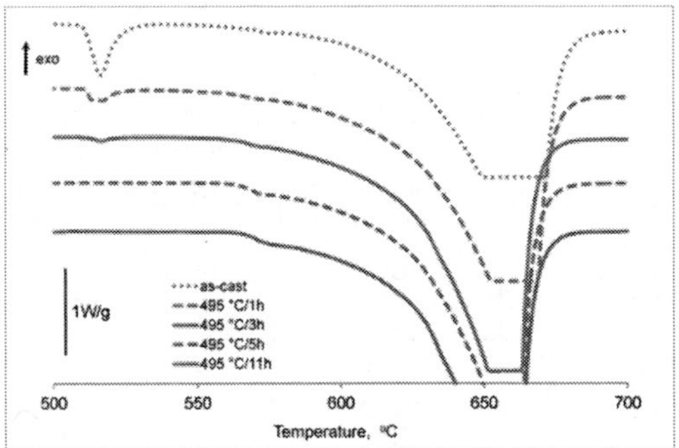

Fig. 2. DSC curves for Al-3.5Cu-1.2Mg alloy billets subjected to homogenization at the temperature of 495 °C (soaking time is given in the Figure) [7]

Table 2. DSC tests results for the Al-3.5Cu-1.2Mg alloy [7]

State	Solidus temperature, °C	Incipient melting heat, J/g
as-cast	508	14,5
495 °C/1h	509	3,2
495 °C/3h	508	0,9
495 °C/5h	561	-
495 °C/11h	565	-

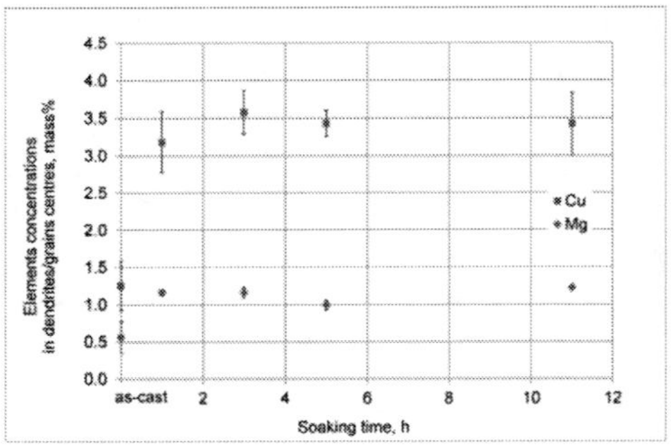

Fig. 3. Elements concentration in grains interiors of Al-3.5Cu-1.2Mg alloy billets [7]

With prolonged homogenization time, incipient melting heat decreases and vanishes after homogenization with 5h soaking time. Simultaneously, significant increase of solidus temperature takes place (Table 2). The decrease of incipient melting heat is accompanied by increment of main alloying elements concentration in grains centres (Fig. 3).

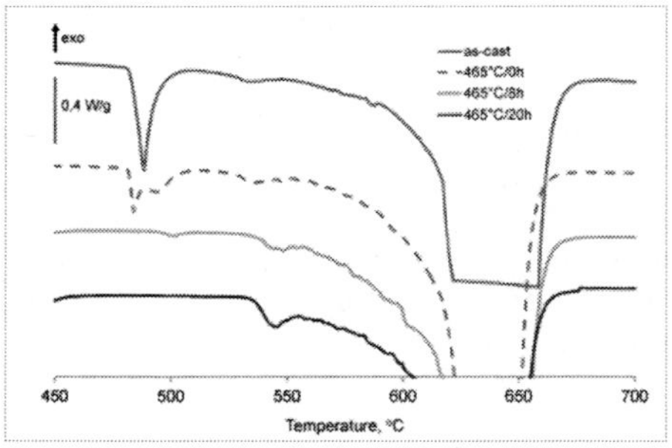

Fig. 4. DSC curves for 7075 alloy billets subjected to homogenization at the temperature of 465 °C

Table 3. DSC tests results for the 7075 alloy billets

State	Solidus temperature, °C	Incipient melting heat, J/g
as-cast	480	13,5
465 °C/0h	481	8,0
465 °C/8h	493	0,5
465 °C/20h	535	-

In the case of 7075 alloy billets the stepwise increment of solidus temperature occurs (Fig. 4, Table 3), resulting from gradual dissolution of low-melting structure components. It creates a possibility of designing the dedicated homogenization courses with increasing soaking temperature, following rise of solidus temperature. The Differential Scanning Calorimetry is a very efficient method for determination of these processes parameters.

4. Conclusions

The obtained results allow to state that microsegregation elimination ending may be determined on the basis of solidus temperature and/or incipient melting heat changes analysis. Moreover, these parameters enable designing the homogenization processes with increasing temperature, taking advantage of solidus temperature rise during the annealing.

Acknowledgments (optional)

Financial support from the State Committee for Scientific Research of Poland under the grant no 11.11.180.960 is kindly acknowledged.

References

[1] P. Saha. (2000). Aluminum Extrusion Technology. ASM International.

[2] B. Dixon. (2000). Extrusion of 2xxx and 7xxx alloys. In Proc. of 7th International Aluminum Extrusion Technology Seminar, 16-19 May 2000, (281-294). Aluminum Association; Aluminum Extruders Council.

[3] PN-EN 573-3 Aluminium and Aluminium Alloys - Chemical composition and forms of wrought products.

[4] D. Leśniak, M. Bronicki, A. Woźnicki. (2010). High-temperature homogenization of AlCuMg alloys for extrusion in T5 temper. Archives of Metallurgy and Materials. 55 (2), 499-513.

[5] A. Woźnicki, D. Leśniak, G. Włoch, B. Leszczyńska-Madej, A. Wojtyna. (2015). The effect of homogenization conditions on the structure and properties of 6082 alloy billets. Archives of Metallurgy and Materials. 60(3a), 1763-1771.

[6] A. Woźnicki, D. Leśniak, G. Włoch, A. Wojtyna, M. Zabrzańska. (2016). Investigation of homogenization process of the 5019 alloy extrusion billets. Key Engineering Materials. 682, 10-17.

[7] A. Woźnicki, G. Włoch, E. Żołnierczyk. (paper in preparation)

The 73rd World Foundry Congress
„Creative Foundry"
23 - 27 September 2018, Krakow, Poland

Analysis of the temperature distribution in the sample during the hot distortion parameter testing

Emilia Wildhirt [1], **Jarosław Jakubski** [1]

[1]*AGH University of Science and Technology, 30-059 Kraków al. Mickiewicza 30, Poland*
**Corresponding address: e-mail: jakubski@agh.edu.pl*

Key words: hot distortion,

Generally applied tests determining the usefulness of molding and core sand for use in production are focused mainly on the analysis of strength properties (flexural strength, stretching or surface hardness). These are very important parameters, but they do not reflect the behavior of cores and molds at elevated temperatures, that is, in real conditions of formation of castings. A number of known methods that are used for their research. The first and best described is the BCIRA method [1-7], in which the parameter determining the behavior of core sands during thermal influence is defined by the English name "hot distortion". The second device for comprehensive study of sand behavior at elevated temperature is the apparatus for studying the hot distortion phenomenon constructed at Western Michigan University. In collaboration with the company MULTISERW Morek apparatus is developed for the study of phenomena at elevated temperatures DMA type, where the halogen lamp is used as a heat source. In the context of this paper the study is to determine the actual temperature of the sample on the results read by the thermocouple disposed at the height of the lower surface of the test sample.

1. Introduction

In order to verify the accuracy of temperature measurement with the aid of a thermocouple used in the DMA device, tests were carried out using thermocouples placed directly in the tested sample.

The study was conducted on moulders made in the technology of cold box. The sand from which the moulders were made had the following composition:

quartz sand from Szczakowa	100 cz. wt.
Isocure resin 300	0.8 parts wt.
Isocure 600 hardener	0.8 parts wt.
Catalyst 702	1.0 cm3 / 250 g by weight

The distribution of thermocouples is shown in Figure 1.

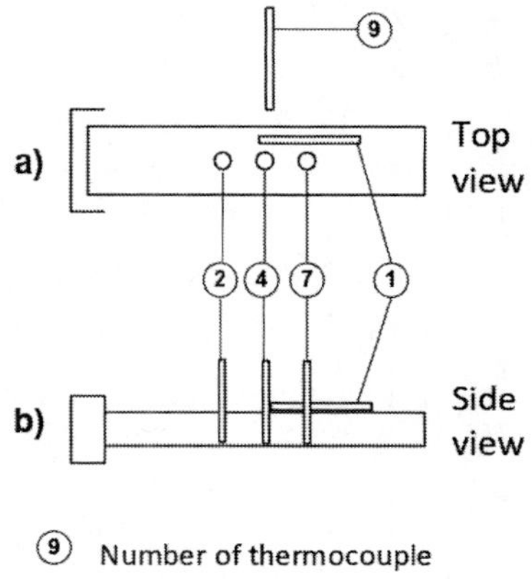

Fig. 1. The distribution of the thermocouples in the test moulder.

The thermocouple number 2, 4 and 7 were placed in the moulder of the hot distortion parameter used for testing. Indication the thermocouple were located inside the moulders at the surface subjected to heat. The thermocouple marked with number 1 was placed on the upper surface of the sample, while the thermocouple number 9 was placed near the thermoelement of the DMA apparatus. Figure 2 presents the results of temperature measurement during heating, which are read using thermocouples marked as 2, 4 and 9.

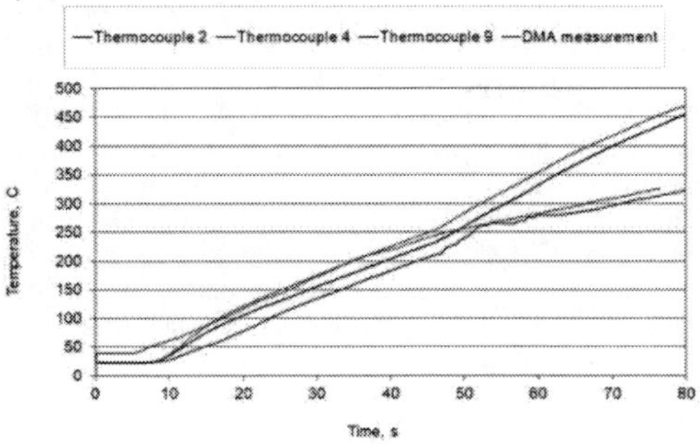

Fig. 2. Temperature changes in the mold during the hot distortion test.

The remaining results, due to large interferences caused by displacement of thermocouples during the measurement, have been omitted.

In the initial stage of heating, no significant changes in the course of the curves plotted for the thermocouple measurement data were found. This stage, however, is short and is probably related to the poor thermal conductivity of the quartz. The time in which this state persists corresponds to the heating of the outer surface of the shapes. For the graph obtained from the DMA apparatus in the initial stage, the difference in the position of the curve is related to the inertia of the apparatus (Fig. 3). During the measurement series, the thermocouple's end is cooled slowly at temperatures close to the ambient temperature. Each subsequent measurement contains an error related to reading the temperature to about 50 ° C.

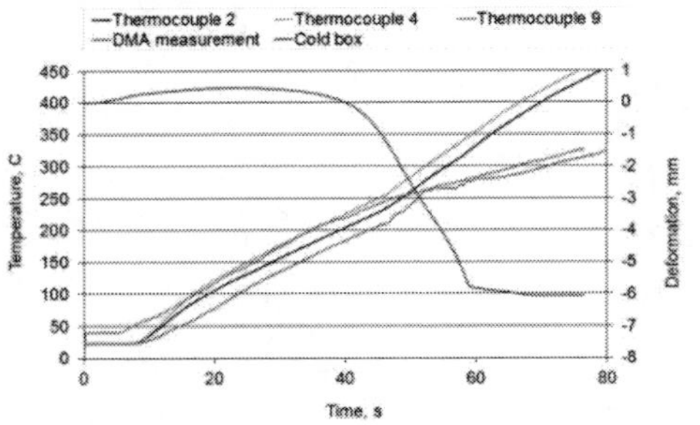

Fig. 3. The hot distortion parameter in relation to the dependence of temperature changes in moulders.

Then, the temperature in the depth of moulders increases as a result of heating, which changes the course of the temperature curves on the graph. It can be seen that the curves obtained for the results collected for thermocouples 2 and 4 have very similar waveforms in the whole range. The graph for data from thermocouples 4 is shifted to higher temperatures by around 20 ° C. The position of thermocouples relative to the heat source may have an impact. The reading curve from the DMA apparatus coincides roughly with the curve from the thermocouple 4 to a temperature of around 220 ° C. After exceeding the temperature of 220 ° C to 250 ° C, this is a small difference, which, however, increases over time. These differences can be explained by observing the graph of thermal deformation of moulders (Figure 3). At a temperature of about 220 ° C, the relative deformation of moulders is 0. Then, due to the many factors and phenomena that occur during heating, the moulders are deformed towards the heat source. Therefore, in contrast to the DMA thermoelement, which has a fixed position, the conditions for heating the moulders and thus the thermocouples in it are changing. Approaching the source of heat increases the heating speed. The graph shows this by increasing the angle of inclination of curves plotted on the basis of measurements from thermocouples 2 and 4. Based on the obtained results, it can be stated that the temperature measurement with the DMA device is correct. However, it gives a good approximation of the real conditions only until the moulders do not begin to change towards the heat source. With a wider range of tests, graphs of strain dependence versus time should be used, which will better reflect actual conditions and facilitate the comparison of results for different masses.

The deformation of moulders consists of two main stages: the deformation upwards from the heat source and downwards. In the further part of the work the deformation from the heat source will be determined as the 1st phase, and the deformation downwards as the 2nd phase of deformation. At the beginning of the second phase, the point where the deformation towards the heat source below the initial position of a moulder takes place.

Conclusion

The presented research was one of many elements that were used to correct the operation of the apparatus. Structural alterations have been made to ensure the best performance and approximation of the actual conditions of exposure of the liquid casting alloy to the core sand.

The changes in the apparatus also forced the correction of the program for processing measurement data.

References

[1] Morgan D., Fashman E. W. (1975). The Bcira Hot Distortion Tester For Quality In Production Of Chemically Bonded Sands. AFS Transaction. Volume 91, 73 – 80.

[2] Hot-Distortion Tester For Bonded Sands – Instruction Manual

[3]Laitar R., Johnson C. K. (1986). High Hot Strength Phenolic Urethane Cold Box System. AFS Transaction. Volume 34, 187 – 194.

[4] Huang N. Y., Su G. X. (1988). Investigation On The Bonding Efficiency Of Chemical Bonded Sand. AFS Transaction. Volume 128, 109 – 114.

[5] Fountaine G. C., Horton K. B. (1990). Hot Distortion Of Cold Box System. AFS Transaction. Volume 70, 9 – 16.

[6] Geoffrey M. M. (1993) Environmental Improvements In Phenolic Urethane Nobake Resins. AFS Transaction. Volume 67, 195 – 197.

[7] Armbruster D. R., Dodd S. F. (1993). New Inorganic Core And Mold Sand Binder System. AFS Transaction, Volume 65, 853 – 856.

THE STUDY ON TENSILE STENGTH OF AlSi21CuNiMg SILUMIN IN THE FINAL STAGE OF SOLIDIFICATION AND THE INITIAL STAGE OF SELF-COOLING

Remigiusz Romankiewicz [1*]

*[1]University of Zielona Góra, Faculty of Mechanical Engineering, Prof. Z. Szafrana str. 4, 64-516 Zielona Góra, Poland, * e-mail: r.romankiewicz@ibem.uz.zgora.pl*

Key words: Silumin, solidification, mechanical properties, tensile strength of solidifying alloys, modification

1. Introduction

A significant effect on the mechanical properties of silumin is their structure, which is mainly formed in the crystallization process [1] and possible heat treatment, the effect of which is also highly dependent on the process of primary crystallization and morphology of the structure being formed [4]. The state of the structure of castings also determines their mechanical properties, especially in the final stage of solidification and the initial stage of self-cooling. Then, the unadvantageous morphology of the leading phase and the uneven distribution of the remaining structural phase components can significantly reduce the ductility and tensile strength of the alloy, causing the tendency of the castings to hot cracking.

2. Description of the tests

The tensile strength tests of solidifying and self-cooling silumin were carried out at the test stand shown in Figure 1 [2, 3]. The construction and operation of the test stand are described below.

In the metal tests form (Fig. 1), the casting of the cylinder with a diameter of Ø40 and a height of 70 mm was subjected to stretching. The test metal was poured from the top into a two-part mold (Fig. 1), the bottom half of which was divided 5 was fixed in the base 2. The split movable upper half 6 was driven by piston rod 14 of the hydraulic cylinder 13. The pump used in the device allowed to obtain pressure up to 160 bar, and the choice of working pressure was carried out with the help of an adjustable pressure valve. The selection of the speed of movement of the hydraulic cylinder rod enabled the knob cooperating with the check-throttle valve placed in the control panel 15. The force with which the cast was stretched was measured by a sensor 17 connecting the movable mold half with the piston rod 14. The elongation of the cast was measured by a displacement sensor 18 by measuring the change in the position of the movable part of the mold. The lower half of the chill mounted on the disc 3 in the base socket was fixed with a retaining ring 4. The metal shield 3, fulfilling the role of the bottom of the mold, had holes enabling the insertion of four mantle thermocouples Ø 1 mm to measure the temperature at the cross-section of the casting.

Tests of tensile strength in the final stage of solidifucation and the initial stage of self-cooling were carried out for AlSi21CuNiMg silumin with chemical composition: 21.2% Si, 1.4% Cu, 1.0% Ni, 0.8% Mg, 0.2% Mn, 0.5% Fe, and the rest of Al. Tensile strength tests for the above-mentioned alloy were carried out at the stand shown in Figure 1. The AlSi21CuNiMg silumin was tested in an unmodified state as well as after the previous modification with phosphorus in form of AlCu19P1.4 master-alloy in the amount of 0.2%.

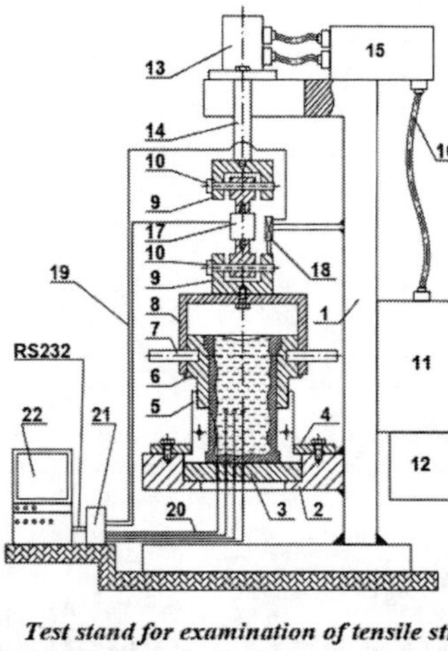

Fig. 1. Test stand for examination of tensile strength of solidifying and self-cooling metals and alloys with testing mould: 1 - frame; 2 - base; 3 - mould bottom; 4 - retainer ring; 5 - stationary bottom part of sectional mould; 6 - upper part of movable sectional mould; 7 - cross-bar; 8 - upper holder of force sensor; 9 - position sensor bracket connected with bottom holder of force sensor, 10 - position sensor, 11 - lever with chain wheel; 12 - chain, 13 - force sensor, 14 — piston rod; 15 – control panel, 16 - electric wire, 17 – force sensor, 18 – position sensor, 19 – conductors, 20 – mantle thermocouples, 21 – microprocessor recorder, 22 – microcomputer.

Metallographic examinations of collected fracture samples were carried out using the NEOPHOT 2 optical microscope.

In order to obtain samples for metallographic examinations, half of the half a disc with high 8 mm was cut off to obtain a quarter-circle sample. On the sample thus obtained, using the SEI detector, along its radius, the photos of the breakthrough surface in the center of the cast were made (Figures 3a and 5a), with a magnification of 250x.

3. Results of the test

Figures 2 and 4 show the diagrams of dependence of tensile stress (σ) and temperature in the center of the casting as a function of relative elongation for the AlSi21CuNiMg silumin. Figure 2 shows that for the unmodified alloy, a tensile stress of 3.5 MPa was obtained with unnoticeable elongation. In the case of alloy after modification with AlCu19P1.4 master-alloy in the amount of 0.2%, the tensile stress increased to 6.6 MPa (Fig. 4) with simultaneous elongation at the level of 4.16%. Figure 3 shows the microfotography of the breakthrough profile in the center of the sample (Figure 3a) and a photograph of the fracture surface (Figure 3b) for the AlSi21CuNiMg alloy in the unmodified state.

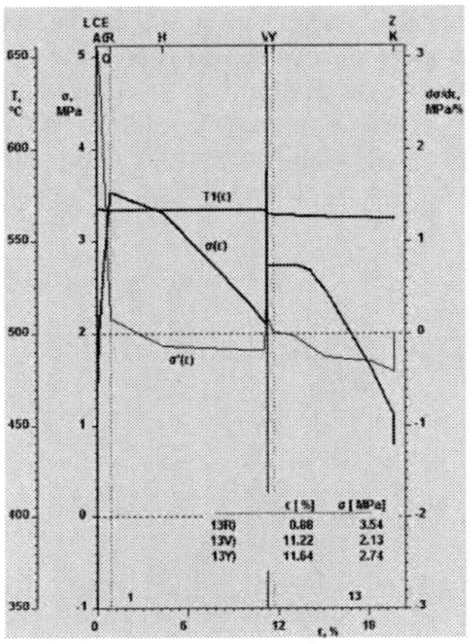

Fig. 2. Silumin AlSi21CuNiMg unmodified - tensile stress (σ) and temperature in the center of the cast as a function of relative elongation (ε)

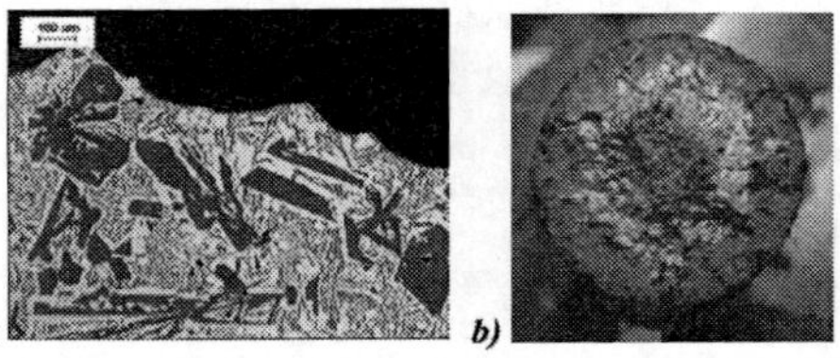

Fig. 3. Silumin AlSi21CuNiMg unmodified: a) microfotography of the breakthrough profile in the center of the sample, b) photograph of the fracture surface

Figure 5 shows a photomicrograph of a breakthrough profile in the center of the sample (Figure 5a) and a photograph of the fracture surface (Figure 5b) for AlSi21CuNiMg silumin modified with AlCu19P1.4 master-alloy in an amount of 0.2%.

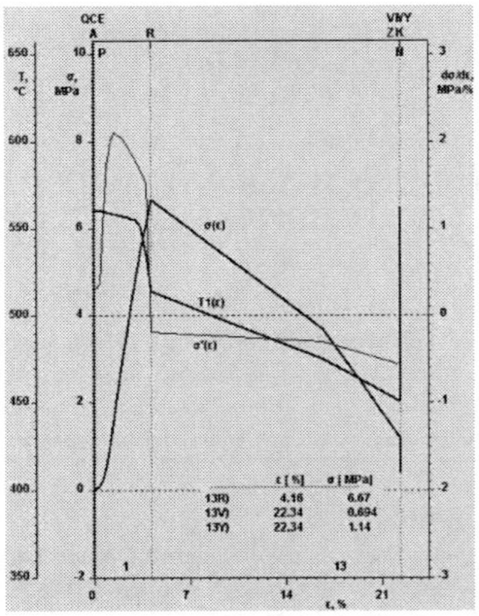

Fig. 4. AlSi21CuNiMg silumin modified with AlCu19P1.4 master alloy - tensile stress (σ) and temperature in the center of the cast as a function of relative elongation (ε)

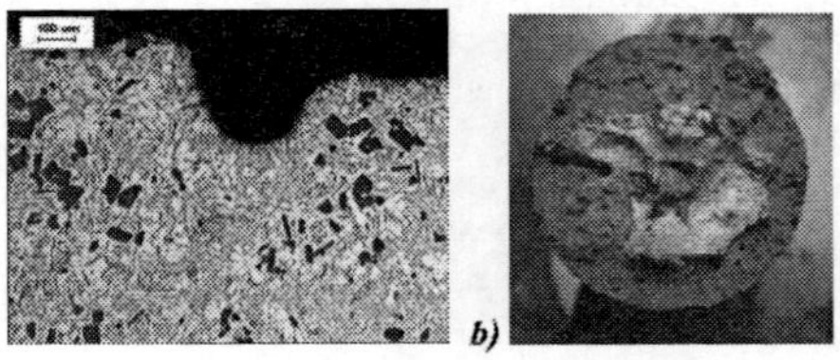

Fig. 5. AlSi21CuNiMg silumin modified with AlCu19P1.4 master alloy: a) microfotography of the breakthrough profile at the center of the sample, b) photograph of the breakthrough surface

4. Conclusions

Modification of the tested silumin resulted in a advantageous refinement of the primary silicon crystals, which as a consequence provided an increase in tensile stress of the silumin samples in the final stage of solidification from 3.5 to 6.6 MPa. As a result of these changes, the resistance of testesd silumin to hot cracks should increase.

References

[1] Höner K.E. (1982). Die thermische Analyse als Kontrollverfahren für die Wirksamkeit einer Schmelzbehandlung von Aluminium – Gusslegierungen durch Veredelung oder Kornfeinung. Giesserei – Forschung. 34 (1), 1-10.

[2] Mutwil J., Romankiewicz R., Kujawa K. (2005). Nowa wersja stanowiska do badania wytrzymałości na rozciąganie krzepnących metali i stopów. Archiwum Odlewnictwa. Nr 17, 183-190.

[3] Romankiewicz R. (2011). Badania nad możliwością szybkiej oceny właściwości technologicznych siluminów w końcowym etapie krzepnięcia i początkowym etapie stygnięcia. Zielona Góra.

[4] Velten B., Rasche C. & Schmidt U H. (1989). Bildung und Ausbreitung von Thermoschockrissen in einer Aluminium-Grusslegierung. Praktische Metallographie Sonderband. Nr 20, 271-280.

The 73rd World Foundry Congress
„Creative Foundry"
23 - 27 September 2018, Krakow, Poland

MICROSTRUCTURAL CHARACTERIZATION OF Fe/TiC COMPOSITE ZONES PRODUCED *IN SITU* USING GLASSY CARBON

Marta Gajewska [1]*, Piotr Natkański [2], Łukasz Szymański[3], Ewa Olejnik[3]

[1]AGH University of Science and Technology, Academic Centre for Materials and Nanotechnology, Mickiewicza 30 Av., 30-059 Kraków, Poland
[2] Jagiellonian University, Faculty of Chemistry, Gronostajowa 2, 30-387 Kraków, Poland
[3]AGH University of Science and Technology, Faculty of Foundry Engineering, Mickiewicza 30 Av., 30-059 Kraków, Poland

**Corresponding address: marta.gajewska@agh.edu.pl*

Key words: Fe/TiC, *in situ* synthesis, MMCs, titanium carbide, composite zone

1. Introduction

One of the main problems of metal matrix composites (MMCs) produced via liquid state in situ methods is their microstructural inhomogeneity. The origin of the problem is twofold: firstly, segregation of reinforcement due to differences in specific gravity between the matrix and the reinforcement (as it is in the case of ex situ cast composites), secondly, difficulties in taking control over reaction of in situ reinforcement formation, especially in large volumes [1,2]. A way to bypass the problem of structural inhomogeneity of cast in situ composite elements is an intentional attempt to produce reinforcement only locally, e.g. in a surface layer directly subjected to abrasion. This approach allows for simultaneous retention of toughness by a casting core [2].

Still, a control of the reaction synthesis is crucial in order to avoid fragmentation of the local composite reinforcement. The literature reports concerning *in situ* Fe/TiC composites indicate that a selection of substrates for the composite synthesis may have a significant impact on obtained composite zone structure [3]. In this study, the effect of different forms of carbon used for TiC synthesis on dimensional stability of local Fe/TiC zones and dispersion of TiC reinforcement within the composite zone were investigated.

2. Experimental

The Fe/TiC composite zones were produced using SHSB process, in which substrates for the SHS synthesis (Ti,C) are introduce into the metal bath in a form of green compacts, and the metal melt heat initiates the reaction. Mixtures of Ti and C (graphite, glassy carbon) powder substrates necessary for the titanium carbide synthesis were prepared in the atomic ratio of 1:1. The powder mixtures were compressed under a pressure of 500 MPa. Then, the formed compacts were placed at the bottom of moulds. Melted base alloy was then poured over the Ti/C compacts.

Phase composition (XRD, electron diffraction), chemical composition (EDS) as well as microstructure (SEM/TEM) of produced ingots were thoroughly characterized.

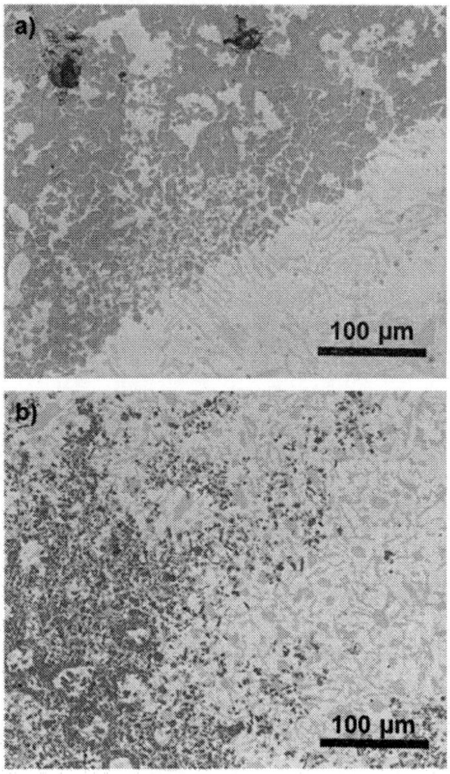

Fig. 1. SEM BSE microstructures of Fe/TiC composite zones produced using a) glassy carbon, b) graphite

3. Results

Performed XRD investigations of both graphite-based and glassy carbon-based materials confirmed presence of TiC in the composite zones. The formation of titanium carbide phases was also confirmed via electron diffraction.

The OM/SEM/TEM observations of the zones revealed however drastic differences in the microstructures of the materials (Fig. 1). The materials produced using amorphous form of carbon exhibited higher dimensional stability of composite zones and more homogeneous distribution of TiC reinforcing particles .

In addition, due to a higher TiC content within the composite zone, the glassy carbon-based materials showed also a higher hardness, both inside and outside the area of the zone.

4. Summary

The performed investigations of produced Fe/TiC composite zones proved that the forms of carbon used for TiC synthesis has a significant effect on their dimensional stability and also on the dispersion of TiC reinforcement within the composite zone.

Acknowledgments

The financial support from the Polish National Centre for Research and Development (Grant No.: LIDER/095/L-6/14/NCBR/2015) is gratefully acknowledged.

References

[1] K. Das, T. K. Bandyopadhyay & S. Das. (2002). A Review on the various synthesis routes of TiC reinforced ferrous based composites. Journal of Materials Science. 37(18), 3881 – 3892.

[2] E. Olejnik, G. Górny, T. Tokarski, B. Grabowska, A. Kmita & G. Sikora. (2013). Composite zones produced in iron castings by in-situ synthesis of TiC carbides. Archives of Metallurgy and Materials. 58 (2), 465-471]

[3] K. Das & T.K. Bandyopadhyay. (2004). Effect of form of carbon on the microstructure of in situ synthesized TiC-reinforced iron-based composite. Materials Letters, 58 (12-13), 1877-1880.

EFFECT OF MODIFICATION ON THE THERMAL ANALYSIS OF GRAIN REFINING IN A319 Al-Si-Cu ALLOYS

Mostafa Mosaad, Waleed Khalifa*

Faculty of Engineering, Cairo University, 12613, Giza, Egypt
** Corresponding address: wkhalifa@eng.cu.edu.eg*

Key words: Si, Cu, thermal analysis, refiner, modifier, cooling curve, cooling rate

The current work investigates the effect of the modification on the thermal analysis of grain refining of A319 alloy. The interactions of modification with the thermal analysis parameters relevant to grain refining were studied using design of experiments (DOE) technique. The factors used for these experiments were 5 factors: Cooling rate, Si content, Cu content, Refiners (Al-5Ti-1B) level and modifier (Al-10Sr) level. The results showed that several time and temperature parameters of grain refining that are obtained from the thermal analysis were obscured at higher levels of modification. This effect is also dependent to some extent on alloys composition and cooling rate.

1. Introduction

The quality and properties of Al-Si cast products; such as the mechanical properties, the hot tearing susceptibility, the surface finish, the machinability, and the pressure tightness of products; are based mainly on certain microstructural characteristics. Therefore, the assessment and prediction of such microstructural characteristics before pouring the melt is of prime interest [1].

Thermal analysis using the cooling curves is called conventional thermal analysis. In this technique, the molten alloy is allowed to cool down slowly (0.1-0.8 Cs-1 [2]), and the data are collected form the thermal analysis curve. Three regions are of significance in this curve: region A indicating the extent of grain refining, region B indicating the level of modification attained, and region C which determines the end of solidification.

The grain refining process provides several advantages to alloy properties and processing [3], including better distribution of porosity and second phase particles, improved surface finish and machinability, greater fatigue strength, better pressure tightness, and reducing tendency of ingot cracking, and provides improved mechanical properties and several other beneficial effects. This is why monitoring of the extent of grain refining is necessary for quality assurance before pouring the melt.

Several research works were carried to study grain refining potential using this technique [7]. The breaking points of cooling curves were calibrated against the growth temperature TAlmax. One the complexities that face this approach is that the growth temperature is a function of the cooling rate. Therefore, constant cooling rate should be experienced to obtain consistent data [5].

When the nucleation undercooling, ΔTAl is large, a relatively large grain size in obtained, and when this value is less than 0.3°C this indicates that there is no barrier for nucleation and the grain size is relatively fine [1]. Hence assessment and comparisons of the grain refining potential of the melts may be done. Grain refinement affects the other thermal analysis parameters such as nucleation and growth temperatures of various phases, e.g., eutectic Si, the solidification range and total solidification time [6].

During solidification of Al-Si alloys, the liquid phase ends with the eutectic reaction resulting in the formation of the eutectic phases. The unmodified alloy exhibits poor ductility leading to brittle fracture of castings, while the properly modified alloy by strontium or sodium shows ductile mode of fracture, with increased elongation and tensile strength [1].

The thermal analysis parameters related to the level of modification are the eutectic temperature depression and the time difference between eutectic nucleation and eutectic finish points. The amount of depression ΔT measured in degrees Celsius is a common indicative parameter for slow cooling rates. The larger the depression the finer and more fibrous is the eutectic structure. For example, well-modified structures occur at depression temperatures of more than 5.5°C in alloys A356.2 [1].

The eutectic temperature is a function of modifier level, and it is lowest at the optimum level of modifier, with over modified alloy showing smaller depression temperatures [7, 8]. The time difference parameter is useful at higher cooling rates [2].

2. Experimental

The thermal analysis experiments of this study were done for various compositions around and beyond that of alloy 319. The specified composition of alloy 319 is Si = 5.5 - 6.5, Cu = 3 – 4, Fe ≤ 1, Zn ≤ 1, and Mg ≤ 0.1, wt%. Three levels of each of Si and Cu were investigated separately. Table 1 lists these targeted compositions. About 450 g of each alloy composition were prepared from commercial pure aluminum alloy AA1050 (containing ~ 0.5% Fe) and commercial pure Si (99% Si) and Cu (99.9% Cu) additions. Small ceramic crucibles and electric resistance furnace were used for this purpose. When the melt temperature was raised at about 800°C, manual stirring with graphite rods (diameter = 15 mm)commenced and continued until dissolving of all alloying additions and homogenous melts obtained. The melt was then degassed for 5 minutes with dry argon using Pyrex tube. After skimming, the grain refiner (Al-5Ti-1B master alloy) was added with the specified level and the melt was stirred for 3 minutes. The melt (at about 750°C) was then poured into a 500°C-heated graphite mold with a K-type thermocouple (diameter = 0.127 mm) inserted from the mold bottom, as shown in Figure 2. The characteristic cooling rate encountered in this first batch of experiments was about 0.8 – 1.1 °C/s. The factors used for these experiments were 5 factors: Cooling rate, Si content, Cu content, Refiners (Al-5Ti-1B) master alloy and modifiers (Al-1-Sr). Each factor mentioned above has three level values except for cooling rate which had only two.

3. Results

Figure 1 shows the effect of grain refiner level on the shape of the cooling curves at the nucleation arrests. This figure shows the cooling curves at three nominal compositions of 6Si-3.5Cu, 7Si-3.5Cu and 6Si-4.5Cu. It is clear that at the same composition, the shape of the curve changed with the level of refiner. At the no-refiner condition, the undercooling (ΔTR-U) is large, and it decreased with refiner addition. In some cases, no undercooling can be identified on the curve.

The values of ΔTR-U, tl and tf of these samples showed that these parameters decreased with increasing the refiner level. The other parameters of No consistent trends with the refiner level were observed. It was also noticeable that both of tl and tf spread over wide ranges. This gives these time parameters a merit for better control of grain refining, even though their values are composition dependent.

However, it is clear that the increase in the depression in eutectic growth temperature and the eutectic solidification time did not scale with the addition level of Sr. The reason is that the values of these parameters do not reflect the amount of Sr existing in the melt; they rather reflect the quality of the melt, i.e. the potency of the added Sr in modifying the size and morphology of the eutectic Si particles. At high levels of Sr, the grain refining parameters were seen unclear and obscured.

At slow cooling rates the thermal analysis parameters precisely reflected the changes occurring in the microstructural parameters. With the A319, the aspect ratio poorly reflected the modification level.

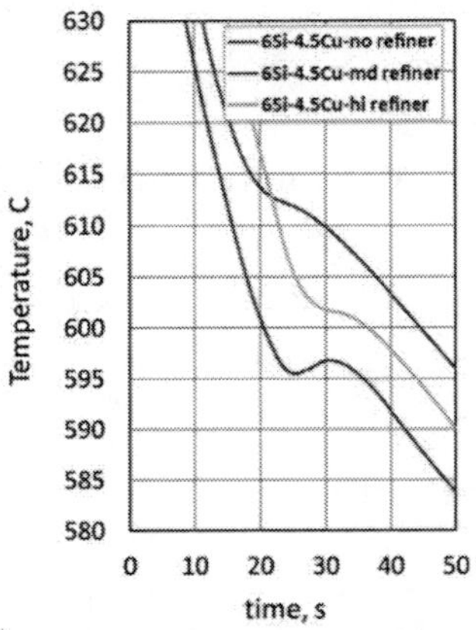

Figure 1 Effect of grain refiner level on the shape of the cooling curves.

4. Conclusions

The results showed that several time and temperature parameters of grain refining that are obtained from the thermal analysis were obscured at higher levels of modification. This effect is also dependent to some extent on alloys composition and cooling rate.

References

[1] D. Apelian, G. Sigworth, K. Whaler: *AFS Trans.* (1984) 297-307.

[2] N. Tenekedjiev, J. Gruzleski: *AFS Trans.* (1991) 1-6.

[3] W. Khalifa, Y. Tsunekawa: *71st World Foundry Congress* 2014.

[4] D. Gloria, J. E. Gruzleski,: *Light Metals* (1999), Met Soc., Ed. by M. Bouchard, A. Faucher.

[5] S. G. Shabestari, M. Malekan: Journal of Alloys and Compounds 492 (2010) 134-142.

[6] R. DasGupta, C. Brown, S. Marek: *AFS Trans.* (1988) 297-310.

[7] H. Beumler, A. Hammerstad, B. Wieting, R. DasGupta: *AFS Trans.* (1988) 1-12.

[8] A. Mitrasinovic, F.C. Robles Hern´andez, M. Djurdjevic, J.H. Sokolowski: Materials Science and Engineering A 428 (2006) 41–46.

The 73rd World Foundry Congress
„Creative Foundry"
23 - 27 September 2018, Krakow, Poland

DESIGN OF COMPETITIVE LIGHT-WEIGHT COMPOSITE MATERIALS: SiC/$TiSi_2$

Donatella Giuranno[1,2]*, Rada Novakovic[2], Rafal Nowak[1], Dmytro Petrin[3], Wojciech Polkolwski[1], Marta Homa[1], Grzegorz Bruzda[1], Artur Kudyba[1], Javier Narciso[3], Natalia Sobczak[1,4]

[1]Foundry Research Institute, Zakopiańska 73 Str., 30-418 Krakow, Poland
[2]National Research Council of Italy-Institute of Condensed Matter Chemistry and Technologies for Energy, Via De Marini 6, 16149 Genova, Italy
[3]Universidad de Alicante-Instituto Universitario de Materiales de Alicante, Apdo. 99, 03080 Alicante, Spain
[4]Institute of Precision Mechanics, Duchnicka 6, 01-796 Warsaw, Poland
**Corresponding address: donatella.giuranno@iod.krakow.pl*

Key words: Si-Ti/C interface; Reactive infiltration; Wetting; MMC-composites; Combined sessile/pendant drop method

1. Introduction

Light, stiff and strong materials capable of bearing load have become increasingly valuable for the design, construction and assembling of lightweight transportation systems such as aircraft, high-speed trains or even for the construction of satellites. In spite of the need, there are few promising new materials in current development that are suited for advanced applications. It is not surprising that R&D is highly focused on advanced ceramics and composites, i.e. MMC-type where the metal phase is an Al, Si or Ti-based alloy. To produce SiC-based composites with desirable properties, the reactive infiltration process has well known advantages over other conventional processing techniques. These processing routes have particular advantages over classical sintering and hot-pressing techniques, i.e. lower processing temperatures, shorter times and near-net shape fabrication capabilities. However, many aspects are still not fully understood mainly related to the interfacial phenomena occuring during infiltration process. Currently, three are the greatest open challenges for scientists: theoretical description of the process by computational models, decrease or replacement of the unreacted Si, and finally, the possibility to "control" the "pore closure" phenomenon. A key contribution could be coming from know-how on wetting characteristics, reactivity, surface properties and thermodynamics concerning reacting phases [1].

Systematic investigations of wetting characteristics, reactivity, surface properties and thermodynamics concerning Si-Ti/C-based substrate couple have been carried out. In addition, infiltration experiments have been performed between Si-Ti/C-preform with different porosity. All the collected findings will be detailed in a next paper. In this work, the more relevant results obtained by wetting tests will be reported and briefly discussed.

2. Materials and Methods

For wetting tests, pure Si, as reference element, and three different Si-Ti alloy compositions have been selected: $TiSi_2$ intermetallic compound, Si-16at%Ti eutectic composition and Si-8at%Ti hypereutectic alloy. The alloys have been produced by mixing high pure Si and Ti (99.998% GoodFellow©) and melted by arc-melting. Each as-produced alloy composition has been checked by SEM/EDS analyses.

As substrates, glassy-C (**GC**-provided by Alfa Aesar©) and Hot Pressed-SiC (**HP-SiC**-provided by GoodFellow©) have been chosen. Prior the experiments, the surface of each substrate has been characterized by 3D-profilometer showing a roughness around 20nm and 1μm for GC and HP-SiC, respectively.

Wettability, spreading kinetics, reactivity, surface and bulk properties (surface tension, density) as a function of composition and operative conditions (time, temperature, total pressure), have been estimated by sessile/pendant drop combined method [2] in the temperature range of (T_m+20)-1500°C (Fig. 1). In particular, wetting behaviour and spreading kinetics have been compared with the results obtained by contact heating sessile drop method [1].

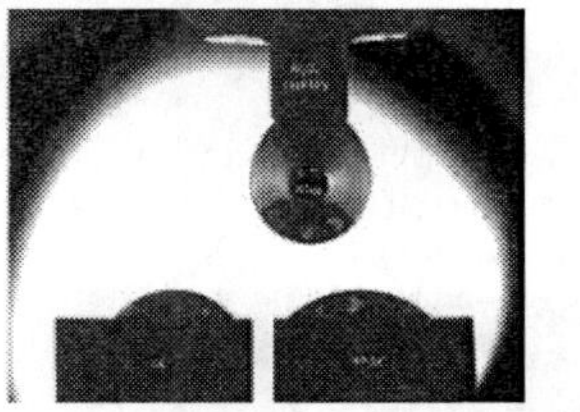

Fig. 1. Surface tension and wetting combined experiments: Si-Ti alloy pendant drop for surface tension measurements combined with dispensed Si-Ti alloy onto GC and HP-SiC for wettability tests.

Moreover, dynamic of infiltration has been evaluated by combining contact heating and dispensed drop methods, as a function of operating conditions, i.e. temperature, average porous radius, C-content in the preform. Interfacial phenomena and resulting microstructure have been carefully analyzed in terms of thickness and composition of the reaction layer, by OM, SEM-EDS and 3D-Tomography techniques. In parallel, surface and bulk properties outcomes have been compared with the predicted values obtained by applying Ideal solution, Quasi-Chemical Approximation (QCA) for regular solution models applied at the base of Si-Ti system as done in [3].

A preliminary thermodynamic study of Si-Ti-C system has been also performed.

3. Results and discussion

In Figure 2, the evolution of wetting behavior of Si-16%Ti eutectics/GC couple within a contact heating experiment (heating rate ≈ 10°/sec) performed at T = 1350°C, is shown. Spreading kinetics is also reported in term of drop radius as a function of time.

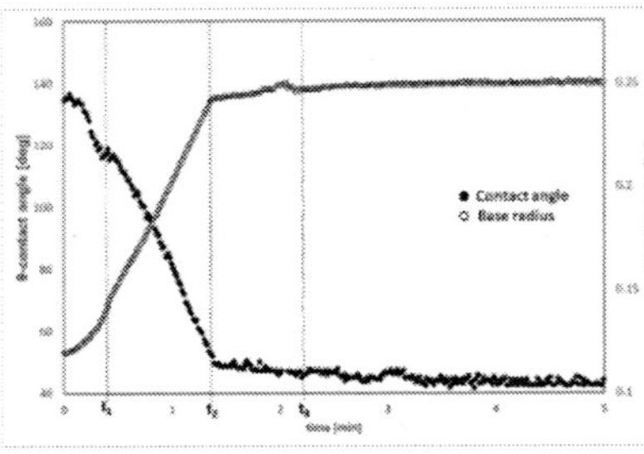

Fig. 2. Contact angle and base drop radius as a function of time of Si-16%Ti /GC couple at T = 1350°C.

As it can be seen, four different spreading stages have been observed. In particular, the stage ($t_o \div t_1$) is typical of wetting experiments performed close to the melting point (T_m Si-16Ti = 1330°C [4]) by contact heating sessile drop method. The presence of SiO_2-primary oxide at the surface affects in delaying the melting of the drop. Moreover, the oxide skin initially prevents the intimate contact between the melt and the substrate. In the second spreading stage (t_1-t_2), the change of the dR/dT, reveals that silicon is dissolving the SiO_2 wetting barrier. As a consequence, the Si-Ti melt is promt to react with C, to form SiC as a reaction product [5]. The linear spreading behaviour observed in this stage suggests that Si-16%Ti/C wetting behaviour is purely governed by the reaction at the triple line, named "linear" reactive wetting by [6]. Within the third stage of spreading (t_2-t_3) a linear dependency is also observed. At the Si-16%Ti/C couple, "linear" reactive wetting is still the dominant mechanism. The decrease of the slope suggests that a "competitive" phenomenon is in progress.

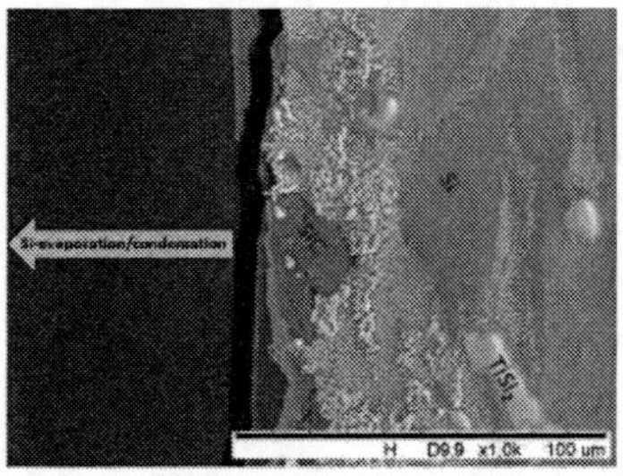

Fig. 3. Contact angle and base drop radius as a function of time of Si-16%Ti eutectic alloy/GC substrate at T = 1350°C.

As evidenced by SEM/EDS analyses (Figure 3), two are the possible competitors for advancing contact angle: the change of the substrate roughness given by the growth/packaging of SiC crystals [7] at Si-16%Ti/GC interface, and evaporation/condensation phenomena close to the triple line.

4. Summary and conclusions

Combined theoretical-experimental approach has been applied to fully investigate the interfacial phenomena occurring at Si-Ti/C-based interfaces.

All the interaction phenomena are strongly dependent on the temperature, as expected. In addition, competitive phenomena, i.e. evaporation/condensation of Si at the triple line, combined with the epitaxial growth/packaging of SiC crystals at Si-Ti/C interfaces, strongly affect the spreading kinetics. Moreover, the advancing contact angle is also influenced by the procedure applied: the presence SiO_2-primary oxide at the alloy surface processed by contact heating sessile drop method, inhibits or delays the first appearance of Si-Ti/C interface. On the contrary, the presence of this oxide skin, mitigates the Si-evaporation, at least at $T \approx T_m$.

In view to relate these evidences with reactive infiltration processes, such knowledge may be helpful to select the optimal operating parameters to obtain tailored SiC-$TiSi_2$ composites.

Acknowledgments

This work was supported by National Science Center of Poland in the framework of "**SIC-WIN**: **R**eactivity of **SI**-alloy/**C**-material system: **W**etting vs **IN**filtration project (G.A. UMO-2016/23/P/ST8/01916), POLONEZ-3. This project received funding under the EU Research and innovation funding programme "Horizon 2020" G.A. no 665778 for the Marie Skłodowska-Curie actions.

References

[1] M. Caccia, S. Amore, D. Giuranno, R. Novakovic, E. Ricci, J. Narciso. (2015). Towards optimization of SiC/CoSi2 composite material manufacture via reactive infiltration: Wetting study of Si–Co alloys on carbon materials. Journal of the European Ceramic Society. 35(15), 4099-4106.

[2] N. Sobczak, R. Nowak, W. Radziwill, J. Budzioch, A. Glenz. (2008). Experimental complex for investigations of high temperature capillarity phenomena. Mater. Sci. Eng. A 495, 43–49.

[3] R. Novakovic, D. Giuranno, M. Caccia, S. Amore, R. Nowak, N. Sobczak, J. Narciso, E. Ricci. (2016). Thermodynamic, surface and structural properties of liquid Co-Si alloys. Journal of Molecular Liquids. 221, 346–353.

[4] T.B. Massalski.. Binary alloy phase diagrams. (1986). vols. 1, 2. Metals Park, OH, ASM.

[5] K. Landry, N.Eustathopoulos. (1996). Dynamics of wetting in reactive metal/ceramic systems: Linear spreading. Acta materialia. 44(10), 3923-3932.

[6] G.W. Liu GW, M.L. Muolo, F. Valenza, A. Passerone. Survey on wetting of SiC by molten metals. (2010). Ceram Int. 36,1177–1188.

[7] R. Voytovych, R. Israel, N. Calderon, F. Hodaj, N. Eustathopoulos. (2012). Reactivity between liquid si or si alloys and graphite. Journal of the European Ceramic Society. 32(14), 3825-3835.

The 73rd World Foundry Congress
„Creative Foundry"
23 - 27 September 2018, Krakow, Poland

STRUCTURE AND PROPERTIES OF HIGH NICKEL AUSTEMPERED DUCTILE IRON

Andrzej Kochański [1]*, Anna Krzyńska [1], Michał Owczarek [2]

[1] *Warsaw University of Technology, Faculty of Production Engineering, 85 Narbutta St. Warsaw, Poland*
[2] *Warsaw University of Technology, Faculty of Production Engineering, 85 Narbutta St. Warsaw, Poland*
**Corresponding address: akochans@wip.pw.edu.pl*

Key words: ADI, Nickel, Structure, Properties, Wear Resistance

Properties of Austempered Ductile Iron depend on chemical composition and heat treatment parameters. Properties of ADI with typical chemical composition are well known. In this study we examined Austempered Ductile Iron with high nickel level. In the present work 2,6% nickel ductile cast iron was treated on different austempering temperatures. Metallurgical structures of high nickel ADI were analyzed. In the present paper results of both hardness test and wear resistance test are presented.

1. Introduction

Austempered Ductile Iron combines high ultimate tensile strength with relatively good elongation or high elongation with very good strength. ADI properties have been described in numerous publications [1-4], and the obtained values of various properties have been collected in databases [5,6]. At the same time, good mechanical and functional properties are achieved at relatively low production costs of castings. This allows the production of responsible machine parts working under high load and wear [7].

2. Experimental procedures

The chemical composition of test samples is shown in Table 1. The specimens for studies were cut from the bottom part of the Y2 type castings. Then the sample for tensile tests, hardness measurements and microstructure observations were prepared.

Table 1. Chemical composition

C [%]	Si [%]	Mn [%]	Mg [%]	Ni [%]	Cu [%]	Mo [%]
3.13	2.20	0.45	0.058	2.60	0.56	0.50

In the experiment, both material as cast and samples after heat treatment were tested. The samples were austenitized at two temperatures and a constant time, and then quenched isothermally at different temperatures. The isothermal quenching was carried out in a tin bath. The detailed heat treatment parameters for individual samples are shown in Table 2.

Table 2. Heat treatment parameters

	austenitization		isothermal quenching	
No of sample	Temp. [°C]	Time [min]	Temp. [°C]	Time [min]
X0	-	-	-	-
X1	950	90	330	90
X2	950	90	300	90
X3	950	90	350	90
X4	900	90	300	90
X5	900	90	350	90
X6	950	90	220	90

3. Results

The work involved metallographic examination using the Olympus optical microscope with the AnaliSys image analysis software. Hardness tests of samples before and after heat treatment as well as wear resistance tests were performed. The wear resistance test was carried out on the T-01M test apparatus using the Ball-on-Disc method.

Metallography

Metallographic examinations were performed on samples etched with Nital using magnification of 200, 500 and 1000 times. Observations of the as cast matrix revealed a structure similar to ausferrite. This material structure is caused by a significant amount of nickel. After the hardening process, the structure of the metallic matrix changed rather quantitatively than qualitatively. In all cases, the structure typical of ADI cast iron was obtained (Figs. 1, 2).

Mechanical properties

The hardness tests were performed using the Brinell method using a 2.5mm diameter ball and load 187.5N for 35 seconds. The hardness of the starting material was 278HB. After heat treatment, the hardness increased to a maximum of 408HB for an austenitized sample at 950°C for 90 minutes and quenched isothermally at 220°C for 90 minutes. The comparison of the hardness is shown in Fig. 3.

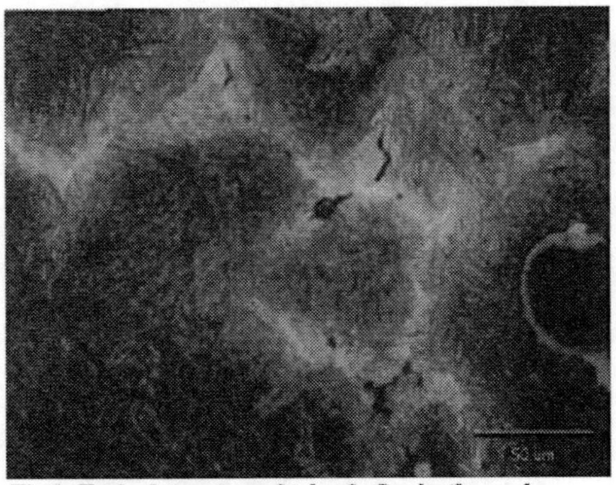

Fig 1. Typical structure obtained after isothermal quenching (mag. x500)

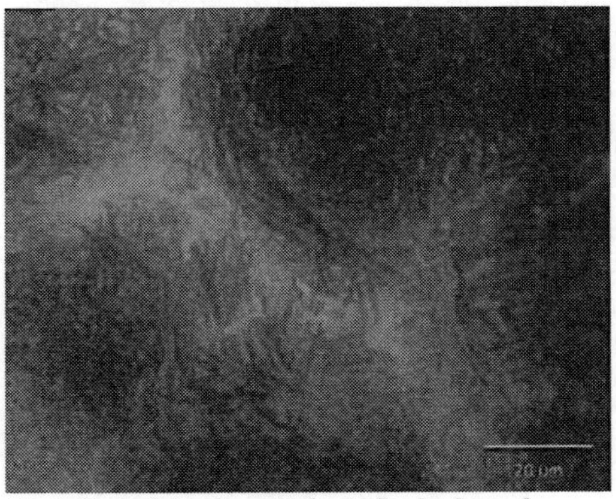

Fig 2. Typical structure obtained after isothermal quenching (mag. x1000)

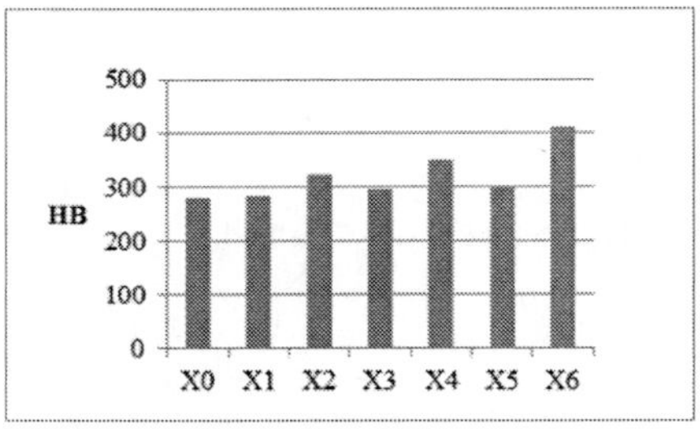

Fig. 3. Hardness of cast iron and after various variants of heat treatment

Wear resistance testing

The tests were carried out for two loads: 20 [N] and 50 [N]. The change in temperature, friction forces and the depth of the ball in the material were recorded. After the wear test, observation under the microscope and the measured width of the resulting trace were made. The main parameter on the basis of which the material abrasion resistance was assessed was the width of the trace created during the test. The average track widths for individual samples are shown in Fig. 4.

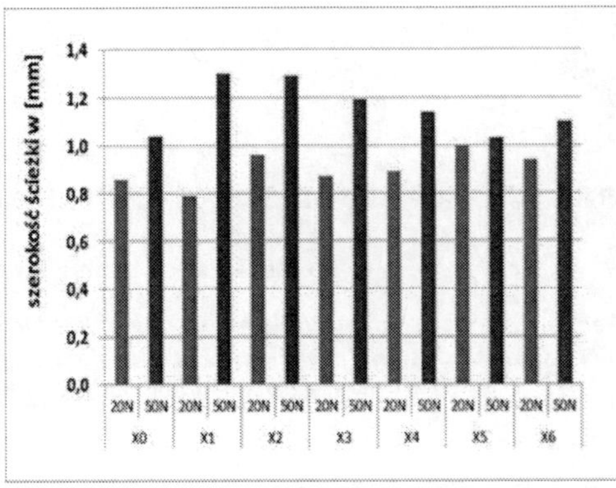

Fig. 4. Track width for cast iron after various heat treatment variants

4. Summary

The study evaluated the effect of isothermal transformation temperature on mechanical properties, including resistance to wear. At 20N load, the best resistance was demonstrated by the X1 sample while at 50N load the X5 sample. Obtained results of hardness and abrasive wear resistance measurements were compared with microstructures in order to evaluate the relationship between volume of austenite and wear resistance. The obtained results were compared with the results obtained for other ADI.

[1] Guzik E., (2006). Some selected problems concerning the formation of microstructure and mechanical properties of ausferritic ductile iron. Archives of Foundry, 6, pp. 33-42 (in Polish).

[2] Fraś, E., Górny, M., (2010),Thin wall ductile and austempered iron castings, Archives of Foundry Engineering, 10, pp. 5-10.

[3] Kochański A., Soroczyński A., Kozłowski J., (2013), Applying Rough Set Theory for the Modelling of Austempered Ductile Iron Properties, Archives of Foundry Engineering, 13/2, pp. 70-73.

[4] Krzyńska A., Kochański A., (2014), Properties and Structure of High-Silicone Austempered Ductile Iron, Archives of Foundry Engineering, 14, pp. 91-94.

[5]Yescas, M.A, Bhadeshia H.K.D.H., MacKay D.J. (2001), Estimation of the amount of retained austenite in austempered ductile irons using neural networks, Materials Science and Engineering, A311, pp. 162–173

[6] Kochański A., Perzyk M., Kłębczyk M., (2012). Knowledge in imperfect data. In ed.: Carlos Ramirez Gutiérrez, Advances in Knowledge Representation (pp. 181-210). Wien: InTech

[7] Wieczorek A.N., (2014), The Role of Operational Factors in Shaping of Wear Properties of Alloyed Austempered Ductile Iron Part I. Experimental Studies Abrasive Wear of Austempered Ductile Iron (Adi) in the Presence of Loose Quartz Abrasive, Archives of Metallurgy and Materials, Vol. 59/Iss. 4, pp. 1665-1674

QUALITY INDEX OF AlSi7Mg0.3 SILUMIN FROM THE PERSPECTIVE OF REFINING METHODS

Edward Czekaj[1*], Z. Kwak[2], A. Garbacz-Klempka[2]

[1] *Foundry Research Institute, Zakopiańska 73 Str., 30-418 Kraków, Poland*
[2] *AGH University of Science and Technology Faculty of Foundry Engineering, Reymonta 23 Str., 30-059 Kraków, Poland*
**Corresponding address: edward.czekaj@iod.krakow.pl*

Key words: Cast aluminium alloys, Refining, Quality Index

1. Introduction

From the middle of 1950s the world can see a constant growth in the production of aluminium, both primary and recycled. The recycled aluminium alloys currently represent over 30% of the global production while the remaining 70% are primary ones. In high developed countries the proportion of recycled alloys is much bigger. In USA for instance, in the first decade of 21st century it exceeded the level of 60% of total aluminium used (Fig. 2) [1, 2].

From the year 1980 to 2007 the production of aluminium originated from recycling had a four times growth (from app. 5 millions to app. 18 millions tonnes/year). In the same time the supply of primary aluminium changed from app. 15 million to app. 38 million tonnes/year – which means 2.5 times increase (Fig. 1).

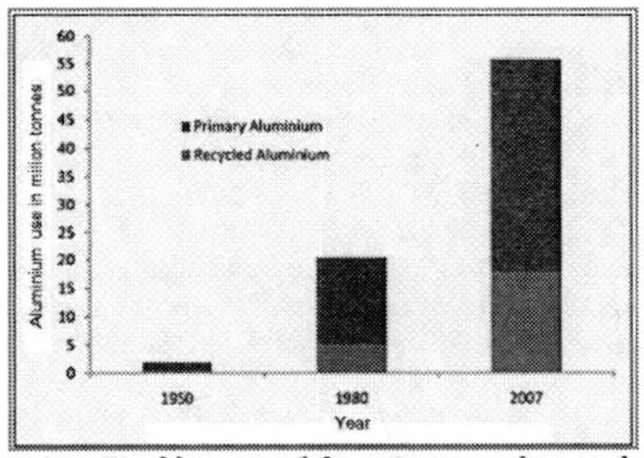

Fig. 1. World usage of the primary and secondary aluminium in years: 1950, 1980 and 2007 [3]

During the last 20 years the primary aluminium production increased by app. 2.5 times. This is mainly caused by the dynamic economical growth in China; from the year 2013 this country produces more than 50% of the total world production of primary aluminium [4].

To 2013 app. 1.1 milliard tonnes of primary aluminium was produced and 75% is still in use [5]. The estimated data given by the International Aluminium Institute (IAI) in 2016 indicate that the world production of primary aluminium was in the year 2015 equal 58.17 million tonnes. After a momentary slump in the market, aluminium plants in China are again operating at full steam. These sources indicate that that the aluminium production grows in the fastest way in Asia [6].

In the total tonnage measure of the products made of aluminium and its alloys using the secondary raw materials is very important. Aluminium originated from the recycling constitutes in last years even additional 20 millions tonnes annually [4]. The fraction of secondary aluminium is constantly increasing in highly industrialised countries.

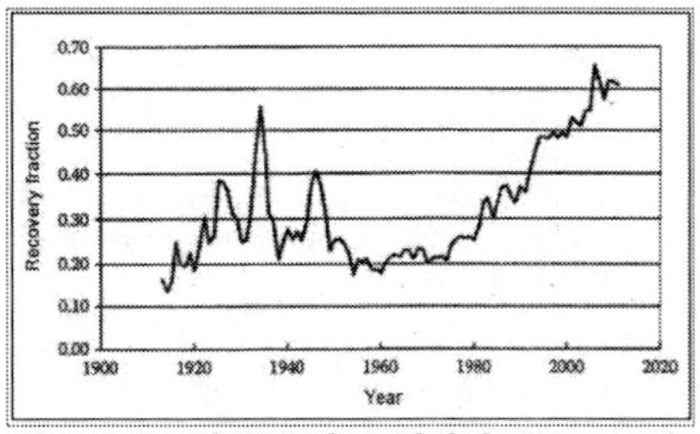

Fig. 2. Usage degree of recycled aluminium in the USA, from 1913 to 2011, in relation to the total usage [7, 8]

The data of March 2012 [2] indicate that app. 150 000 tonnes of secondary aluminium alloys castings were produced in Poland (mainly for motorisation elements). It is simultaneously known, that the production of aluminium castings in 2011 in Poland was equal 256 112 tonnes [9]. Thus it can be easily calculated that the fraction of secondary alloys in the total casting production of Al alloys was app. 58%. It should be assumed that the usage of secondary raw materials will be growing.

The growth in demand for aluminium alloys is generated by the unique material properties of this metal. Aluminium alloys are in demand because they have good technological properties and are easy to recycle. The growth in demand is possible to be met because of large natural reserves, since aluminium is the third most common element in earth's crust and the most common metal [10-12].

2. Methodology and Research

The subject of the study was a standardized aluminium casting alloy EN AC-42100 (EN AC-AlSi7Mg0.3), in accordance with PN EN 1706: 2011 [26] - analogue to American 356 and A356.0 (AA or ASTM) [13, 14], used in low-pressure casting of car rims.

Two experimental (model) AlSi7Mg0.3 alloys were used in the study: a) recycled (secondary alloy) from JN METAL from Ostrowiec Świętokrzyski (Poland); b) primary alloy made in the Institute of Non-Ferrous Metals, Light

Metals Division (IMN OML) in Skawina from technically pure input materials.

The recycling of AlSi7Mg0.3 alloy was carried out at JN METAL in a medium frequency induction furnace with a crucible capacity of approximately 3 tons of aluminium alloy. Experimental secondary alloy was not subjected to additional refining. It was only kept for several minutes at a temperature of 720-760°C. Then the slag was removed from the surface and alloy was poured into metal moulds.

The temperature of liquid metal was measured using 2xNiCr K-2 thermocouple (accuracy of ± 5°C). The temperature was in the range of 720-740°C during casting.

Two gas and two solid refiners were selected for the purpose of the study.

The main criteria for selecting the refiners (fluxes) were ecological considerations. Mainly nitrogen based compounds with minimal quantity of such elements as chlorine or fluorine were used from among the solid refiners.

Gaseous refining was carried out using a lance. The stream of gas was introduced at a height of about 50 [mm] from the bottom of the crucible.

The rods for strength tests were cast in iron mould coated on the inside with a graphite-based colloidal coilite and heated to a temperature of 240-260°C. For each refining variant 5 samples were cast. The results of the measurements were thus statistically averaged.

Static tensile strength tests were carried out at the Centre for Material Testing and Mechatronics of the Motor Transport Institute in Warsaw. An Instron 8802 (250 kN) testing machine was used. Measurements were made at 20°C. The humidity (LB700 hytherograph) ranged from 39% to 45%. The elongation speed was 0.40 mm/min.

HBN Brinell hardness testing was carried out at the Foundry Institute in Cracow, using the Zwick/Roell ZHU 250 universal hardness tester.

Based on the tensile strength and hardness tests, mechanical properties were determined at room temperature. The measured parameters (before and after refining) are listed in Table 1.

Table 1. The measured parameters of mechanical properties of primary and secondary AlSi7Mg0.3 alloys.

Name	Ultimate Tensile Strength	Elastic Limit	Yield Strength	Elongation
Symbol	UTS	EL	YS	A^*
Unit	[MPa]	[MPa]	[MPa]	[%]
Name	Elongation	Necking	Young's Moduls	Brinell Hardness
Symbol	A_5^{**}	N	E	HBN
Unit	[%]	[%]	[GPa]	[HB]

A^* – relative elongation, in relation to displacement of machine jaws;
A_5^{**} relative elongation, in relation to displacement measured with a strain gauge (length 50 mm, sample diameter – Ø10 mm)

3. Conclusions

The conducted studies show that:

- The QI can be a simple, relatively inexpensive and effective way to comprehensively evaluate the quality of casting aluminium alloys, particularly the silumins.
- Flux refiners should be carefully selected not only because of overall quality improvement but also for Quality Index.
- The gaseous, flux as well as the mixed (flux-gas) refining of the recycled casting silumin AlSi7Mg0.3 allow a significant improvement of its basic mechanical properties (UTS, A_5), and as a result improve its quality index.
- Proper refining can bring the quality of the recycled aluminium alloy very close to the parameters of the primary (metallurgical) alloy.
- Refining of primary, metallurgical alloys is not always desirable. The primary aluminium alloy in many cases may not require refining prior to casting, especially if it is melted rapidly (e.g. in induction furnaces).

Acknowledgments

The article was based on the results of the second phase of the project 'Process innovations allowing land filled waste generated during production of aluminium and its alloys to be utilized' under the Operational Program - Intelligent Development 2014-2020, financed by the National Centre for Research and Development (Poland).

References

[1] 46th Census of World Casting Production (2012). *Modern Casting*, December, 25-29, (in English).

[2] Economic Chamber of Non-Ferrous Metals and Recycling. (2012, March). *Recykling metali – niechciane dziecko polskiej gospodarki.* downloaded on 7th April 2017, from http://www.bankier.pl/wiadomosc/Recykling-metali-niechciane-dziecko-polskiej-gospodarki-2504550.html, (in Polish).

[3] http://www.world aluminium.org/media/filer_public/ 2013/01/15/fl0000181.pdf (Global Aluminium Recycling: A Cornerstone of Sustainable Development) (downloaded on 28th March 2017).

[4] http://forumaluminium.pl/pl/forum/6/topic=52 (downloaded on 2nd April 2017).

[5] http://www.aluminum.org/industries/ production/secondary-production (downloaded on 28th March, 2017).

[6] Aluminium: po załamaniu nie ma śladu. Nowa Stal (Rynek), 1 (37), March 2017, pp.13.

[7] Schlesinger M.E.; Aluminum Recycling. Second Edition. CRS press, Taylor & Francis Group, 2014.

[8] Buckinbgham D..A. et al: Aluminum Statistics. U.S. Geological Survey, 3th October 2012.

[9] 46th Census of World Casting Production (2012). *Modern Casting*, December, 25-29, (in English).

[10] Stulgis, G. (2013). Recykling aluminium - ogniwo gospodarki [Recycling of aluminum - the link of economy]. *Recykling odpadów [Waste recycling]*. 6(150), 22-24, (in Polish).

[11] The Aluminum Association (2017). *Secondary production.* Retrieved August 5, 2017, from http://www.aluminium.org/industries/production/secondary-production, (in English).

[12] Brokerage house of BGZ bank. (2015, February). *Report.* Retrieved August 5, 2017, from http://www.bankier.pl/up/rekomendacje/raporty/101000/808 62_23022015_aml_kupuj_57_65.pdf, (in Polish).

[13] *Aluminium: Properties and Physical Metallurgy* (1993) Hatch J.E. (Ed.). Ohio: American Society for Metals (ASM), Metals Park, (in English).

[14] *Metals handbook: properties and Selection: Nonferrous and Special-purpose Materials pol. 2.*(1990) ASM International, Materials Park, OH, 10th, 10th edition, (in English).

The 73rd World Foundry Congress
„Creative Foundry"
23 - 27 September 2018, Krakow, Poland

INFLUENCE OF MAGNESIUM MATRIX COMPOSITION ON PRESSURE INFILTRATION OF GLASSY CARBON FOAM

Anita Olszówka-Myalska[1,*], Marcin Godzierz[1], Patryk Wrześniowski[1]

[1] *Silesian University of Technology, Faculty of Materials Science and Engineering, Krasińskiego 8, Katowice, Poland*
**Corresponding address: anita.olszowka-myalska@polsl.pl*

Key words: carbon open-celled foam, composite fabrication, magnesium matrix composite, pressure infiltration.

1. Introduction

Open-celled porous materials can be applied as a component in composites with different matrices, and in most component consolidation processes, the foam is infiltrated by the liquid matrix. In ceramic-metal systems, two main factors, i.e. the castability and the effects occurring at the interface, affect the final microstructure and properties. In the case of glassy carbon porous material consolidation, the absence of self-infiltration by magnesium is a problem, independently of the alloy composition or foam geometry [1], and that is in good agreement with the literature data reporting poor wettability in carbon-magnesium systems in experiments conducted at up to 700°C [2-4]. Because the further increase of the component consolidation temperature is unfavourable due to reactions at the interface [5], pressure infiltration seems to be an effective solution. In the presented experiment, that method and the same process conditions (temperature, pressure, time) were applied for three magnesium base matrices and one type of carbon material.

2. Materials and methods

In the composite manufacturing process, the open-celled carbon foam (Duocell ppi 100) denoted as C_{of}, of the microstructure presented in Fig.1, was applied, as well as three magnesium base monolithic components i.e. technical purity magnesium and the, commonly used in the technical practice, AZ31 and RZ5 magnesium alloys (Magnesium Elektron, Table 1). The windows in the applied foam, porosity 97 vol.%, were regular and large (351±94 μm) compared with the cell size, and that should facilitate the proper infiltration conditions. In the pressure infiltration process, the Degusa press was employed and the consolidation was conducted in vacuum, for all the foam-metal systems at the temperature of 690°C, according to the patented procedure [6].

The obtained composite samples, 30mm in diameter and 5mm high, were examined, and their open porosity by the Archimedes method in ethanol (POCH BASIC, 99.8%) as well as hardness by the Brinnel technique (2451,66 N, φ=5 mm) were determined (Table 2). Next, the microstructure of the polished non-etched composite samples was characterized by the scanning electron microscopy (Hiatchi 4200) method with the use of the secondary electron mode (SEI).

***Fig. 1.** SEM micrograph of applied Duocell carbon foam (ppi 100)*

Table 1. Chemical composition of applied magnesium alloys (wt%)

Alloy	Al	Zn	Mn	RE	Zr	Mg
AZ31	2.5-3.5	0.7-1.3	0.2	-	-	bal.
RZ5	-	3.5-5.0	-	0.8-1.7	0.4-1.0	bal.

3. Results and discussion

The open porosity and HB hardness determined for the three C_{of}-Mg composites and the reference pure matrix material have been presented in Table 2, and the differences in the changes of both the porosity and hardness compared with the matrix alloy were revealed. Generally, with the application of a magnesium base matrix of higher hardness, the composite material exhibits higher hardness as well, but it is not caused by the presence of the carbon component. The lowest open porosity was registered for the C_{of}-pure magnesium and a slight material hardness increase was obtained. However, in the case of AZ31 and RZ5, where the open porosity was higher, particularly in the C_{of}-RZ5 system, an evident decrease of hardness was obtained. The results of the SEM examinations (Figs. 2a-c) showed good infiltration of C_{of} by pure Mg and the AZ31 alloy as well as the presence of irregular pores in the case of the RZ5 alloy application. That microstructure is in good agreement with the open porosity measurements and explains the dependence of the composite hardness changes on the matrix composition.

Table. 2. Results of porosity and hardness measurements obtained for composites and reference matrix alloys

Material	Open porosity, %	Hardness, HB	Hardness change, %
C_{of} -Mg	1.40±0.05	52.8±6.0	+1.3
Mg	-	52.1±4.7	
C_{of} -AZ31	2.45±0.04	58.7±5.2	-7.8
AZ31 alloy	-	63.7±2.9	
C_{of} -RZ5	9.76±0.05	64.0±9.9	-17.8
RZ5 alloy	-	77.9±2.2	

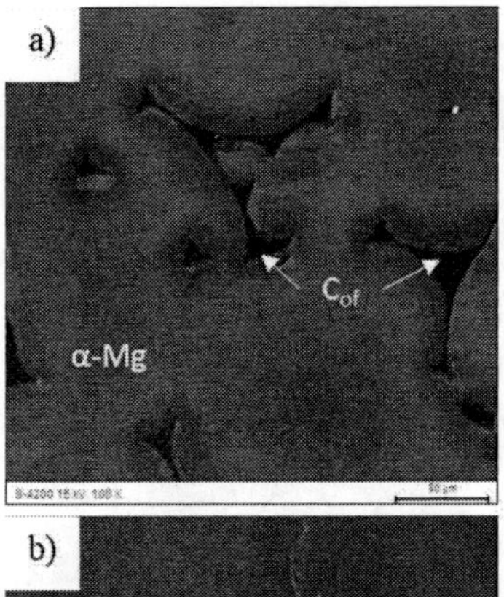

Fig. 2. SEM micrographs of composites fabricated by pressure infiltration method: a) C_{of} -Mg, b) C_{of} -AZ31, c) C_{of} - RZ5

Some effects observed in the obtained materials need to be considered. A small difference in the porosity of the Cof-Mg and Cof-AZ31 (1%) composites and yet a different tendency of the hardness change can be explained by the additional factor that influences the properties, i.e. the interaction of the matrix alloying elements [5]. In the case of the C-AZ31 system, independently of the carbon component shape and size, a thin zone of hydrophilic carbides containing Al mixed with intermetallic phases is formed [7] at the interface, and that is visible in Figure 2b. In the presented experiment, the application of the RZ5 alloy induced not only a zone of interaction products but even a degradation of the carbon foam (Fig. 2c).

The reason for that phenomenon is the presence of rare elements, which has also been observed in the processing of a composite with glassy carbon particles by the stir casting method [5]. Such intense effects of interaction explain the strong decrease of hardness measured in the presented experiment.

4. Conclusions

The proposed method of component consolidation conducted with pressure in vacuum makes it possible to infiltrate successfully the open-celled carbon foam by magnesium matrices. However, the application of pure magnesium ensures better foam infiltration under the same conditions than the commercial magnesium alloys AZ31 and RZ5, selected for the experiment. This effect as well as the magnesium alloying elements influence the hardness of the final composite product.

References

[1] Godzierz M., Olszówka-Myalska A., Wrześniowski P. (2018) *The interaction characteristics of liquid magnesium and selected magnesium alloys with open-celled glassy carbon foams*. Inżynieria Materiałowa/Materials Engineering 222. 61÷68.

[2] Shinozaki N., Morita J., Wasai K.: *Wetting of graphite by molten magnesium*. Journal of the Japan Institute of Light Metals 55 (2005) 310÷314.

[3] Uozumi H., Kobayashi K., Nakanishi K., Matsunaga T., Shinozaki K., Sakamoto H., Tsukada T., Masuda C., Yoshida M.: *Fabrication process of carbon nanotube/light metal matrix composites by squeeze casting*. Materials Science and Engineering A 495 (2008) 282÷287.

[4] Zhang D., Shen P., Shi L., Jiang Q.: *Wetting of B4C, TiC and graphite substrates by molten Mg*. Materials Chemistry and Physics 130 (2011)

[5] Olszówka-Myalska A. (2017) *Magnesium matrix composites. Selected technological issues*. Vol. 677, Gliwice. Wydawnictwo Politechniki Śląskiej

[6] Olszówka-Myalska A., Myalski J. Godzierz M., Hekner B. (2017). Polish patent application No. P.42224

[7] Feldhoff A., Pippel E., Woltersdorf J. (1999). *Carbon-fibre reinforced magnesium alloys: nanostructure and chemistry of interlayers and their effect on mechanical properties*. Journal of Microscopy 196. 185-193.

The 73rd World Foundry Congress
„Creative Foundry"
23 - 27 September 2018, Krakow, Poland

MAIN DIRECTIONS OF RECENT WORKS ON AlZn BASED ALLOYS FOR FOUNDRY ENGINEERING

Witold K. Krajewski [1]*, A. Lindsay Greer [2], Paweł K. Krajewski [1], Katharina Faerber [3,4]

[1] *AGH University of Science and Technology, Faculty of Foundry Engineering, 30 Mickiewicza Av., 30-059 Krakow, Poland*
[2] *University of Cambridge, Department of Materials Science & Metallurgy, 27 Charles Babbage Road, Cambridge CB3 0FS, UK*
[3] *University of Leoben, Department of Metallurgy, Franz-Josef-Strasse 18, A8071 Leoben, Austria*
[4] *Stackpole Powertrain International, Campus-Boulevard 30, 52074 Aachen, Germany*
**Corresponding address: krajwit@agh.edu.pl*

Key words: High-aluminium zinc alloys, High-zinc aluminium alloys, Grain refinement, Mechanical properties, Structural and dimensional stability

Cast alloys with Al- and Zn-based matrices show an increasing range of uses, i.e. the world production of their castings still increases – Fig. 1, [1]. Among the cast alloys based on the Al-Zn system there are high-aluminium zinc (H-Al Zn) alloys containing 15-40 wt% Al and high-zinc aluminium (H-Zn Al) alloys containing 20-40 wt% Zn. These groups have wide range of potential application because of their good strength of 200-400 MPa and simultaneously good tribological and damping properties. Both these groups contain usually 2-5 wt% Cu which increases strength and tribological properties. However, the mentioned alloys show rather low ductility and also structural instability caused by the Cu-containing phase. These problems can be solved by grain refinement aimed at increasing ductility and partial substitution of Cu with other elements, e.g. Ti and/or Mn aimed at improving structural stability with preserved other properties.

The work summarizes development of the H-Al Zn alloys represented here by Zn-(25-26) wt% based ones and the H-Zn Al alloys represented here by Al-(27-30) wt% Al based ones. The results presented below were obtained during joint investigations performed over last years at AGH University of Science and Technology – Faculty of Foundry Engineering and at University of Cambridge, UK – Department of Materials Science and Metallurgy [2-11].

Fig. 2(a) shows coarse dendritic structure of the initial Zn-26Al-2.2Cu alloy while Fig. 2(b) shows refined structure of the same alloy in which some amount of Cu is replaced with Ti. From Fig. 2(c) it can be clearly seen that the alloy with reduced Cu content possesses much lower dimensional changes during natural ageing after homogenization and quenching. This structure is practically stable in a very short time after quenching. At the same time the alloy Zn-26Al-1Cu-1.5Ti shows wear resistance of the same order as the initial Zn-26Al-2.2Cu – Fig. 2(d). The above results show that partial replacement of Cu with Ti gives more stable structure of preserved good wear properties.

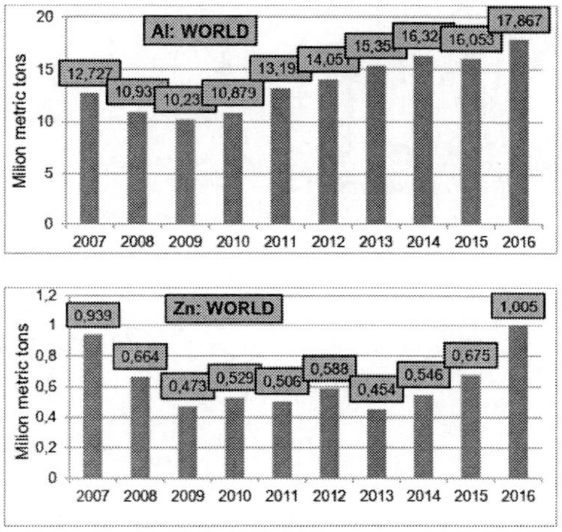

Fig. 1. Structure of world Al and Zn casting production in years 2007-2016 [1]

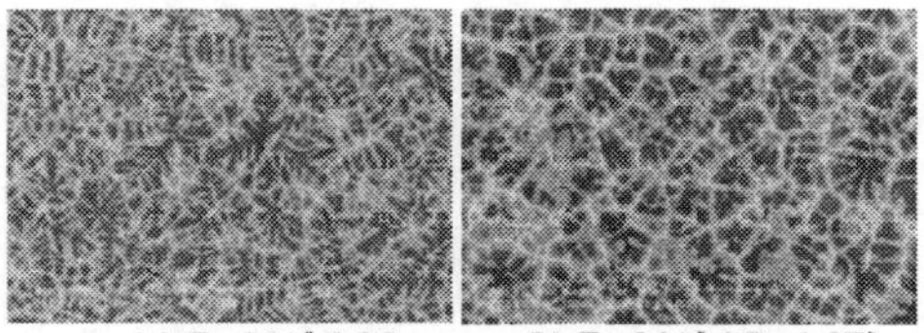

(a) Zn-26Al-2.2Cu **(b)** Zn-26Al-1Cu-1.5Ti

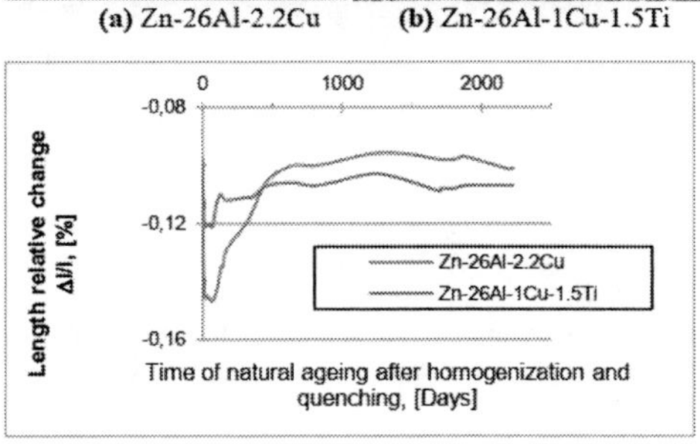

(c)

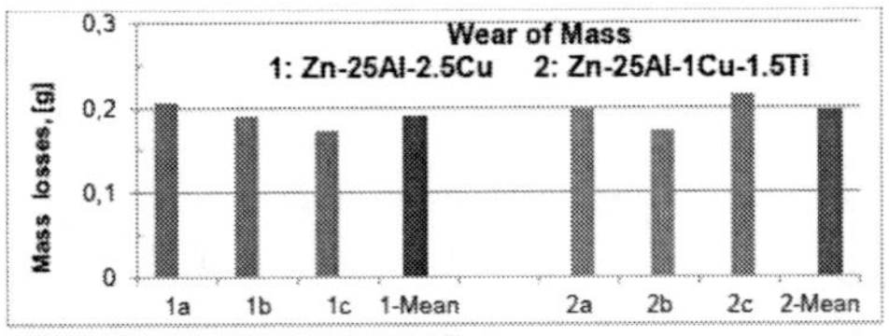

(d)

Fig. 2. ***Microstructure (a) and (b), dimensional changes (c) and wear resistance (d) of the high aluminium Zn-(25-26) wt% Al based alloys [5, 11]***

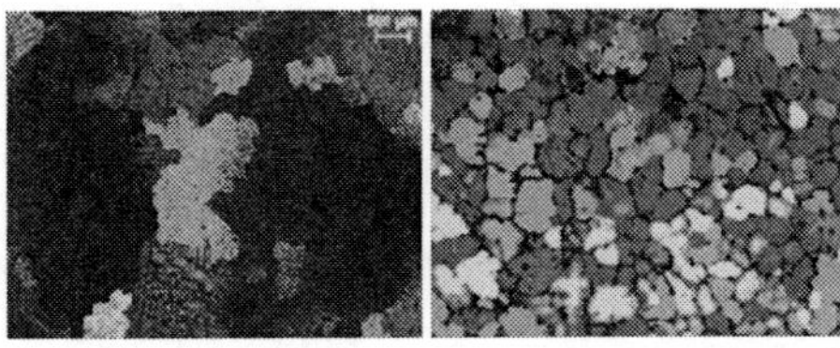

(a) Al-27Zn-2.5Cu (b) Al-27Zn-2.5Cu-Ti

(c) Al-30Zn-2.5Cu-0.5Mn (d) Al-30Zn-2Cu-1Mn-Ti

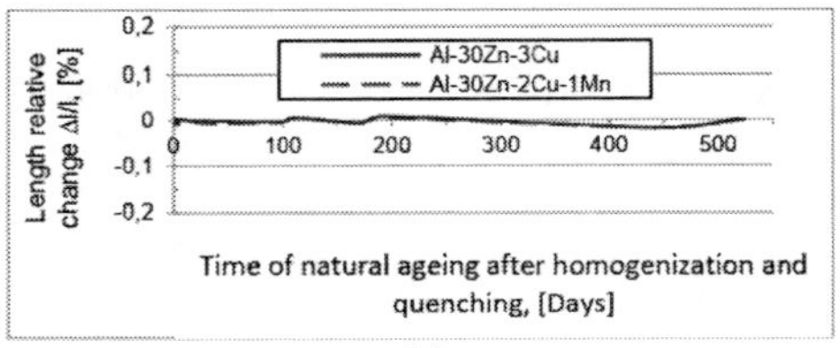

(e)

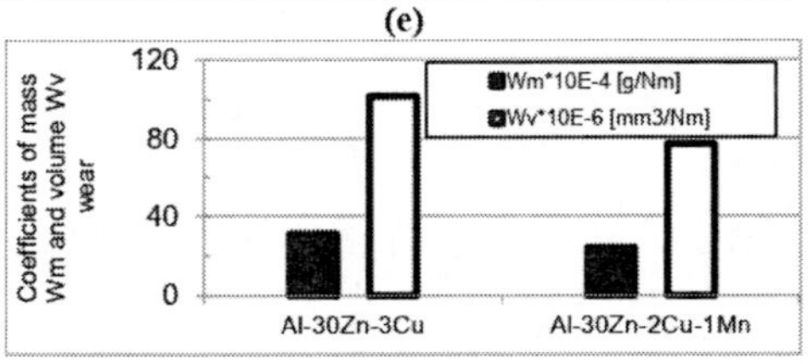

(f)

Fig. 3. Microstructure (a) – (d), dimensional changes (e) and wear resistance (f) of the high zinc Al-(27-30) wt% Zn based alloys [10, 11]

Fig. 3 shows similar results of the investigation performed for high-zinc aluminium alloys, i.e. coarse dendritic structure of the initial Al-27Zn-2.5Cu alloy - Fig. 3(a) and highly refined structure of the same alloy doped with 100 ppm Ti introduced with Al-3Ti-0.15C refiner – Fig. 3(b). Figs. 3(c) and 3(d) show exemplary microstructures of the coarse initial and grain refined Al-30Zn based alloys containing Mn addition. From Figs. 3(e) and 3(f) it can be seen that the alloys of reduced Cu content to 2-3 wt% show practically stable structure during 1.5 year natural ageing after homogenization and quenching. It should be noted that the alloy Al-30Zn-2Cu-1Mn has better wear resistance comparing with the Al-30Zn-3Cu alloy – Fig. 3(f). The mentioned changes are discussed in detail in a full version of this paper which is addressed to the academia and industry workers from area of AlZn-based cast alloys theory and practice.

Acknowledgments

The authors are grateful to the NCN Polish Science National Centre for financial support under grant No. UMO-2017/25/B/ST8/00150. The AGH University of Science and Technology, Krakow, Poland – Faculty of Foundry Engineering and The University of Cambridge, UK – Department of Materials Science and Metallurgy are warmly acknowledged for provision of laboratory facilities.

References

[1] Censuses of World Casting Production, Modern Casting, Issues No. 12 from years 2008 – 2017.

[2] Krajewski W.K., Buras J., Zurakowski M. & Greer A.L. (2009). Structure and Properties of Grain-Refined Al-20 wt.% Zn Sand Cast Alloy. Archives of Metallurgy and Materials. 54 (2), 329-334.

[3] Haberl K., Krajewski W.K. & Schumacher P. (2010). Microstructural features of the grain-refined sand cast AlZn20 alloy. Archives of Metallurgy and Materials 55 (3), 837-841.

[4] Krajewski W.K., Buraś J., Krajewski P.K. & Piwowarski G. (2015). Ultrasound wave attenuation of grain refined high-zinc aluminium sand-cast alloys, Archives of Foundry Engineering 15 (2), 51-54.

[5] Krajewski W.K., Greer A.L. & Krajewski P.K. (2013). Trends in developments of high-aluminium zinc alloys of stable structure and properties. Archives of Metallurgy and Materials 58 (3), 845-847.

[6] Krajewski W.K., Buras J., Krajewski P.K., Greer A.L., Faerber K. & Schumacher P. (2015). New developments of Al-Zn cast alloys. Materials Today: Proceedings 2, Part A, 4978-4983.

[7] Krajewski W.K., Buras J., Krajewski P.K., Greer A.L., Schumacher P. & Haberl K. (2016). New developments on optimizing properties of high-Zn aluminium cast alloys. IOP Conference Series: Materials Science and Engineering 143, Article Number: UNSP.

[8] Krajewski W.K., Greer A.L., Buras J., Piwowarski G., Krajewski P.K. (2017). New developments of high-zinc Al-Zn-Cu-Mn cast alloys. Materials Today: Proceedings, (in printing).

[9] Krajewski W.K., Faerber K., Krajewski P.K. (2018). The influence of grain refinement and feeding quality on damping properties of the Al-20Zn cast alloy. Archives of Foundry Engineering 18, (in printing).

[10] Krajewski P.K., Greer A.L., Faryna M., Krajewski W.K. (2018). Structural stability of novel multicomponent AlZn-based cast alloy. Materials Science Forum (accepted for publication).

[11] Krajewski P.K., Greer A.L., Krajewski W.K. (2018). Towards structural and dimensional stability of selected AlZn-based alloys. Proceedings of 17th International Foundrymen Conference, Opatija, May 16-18, pp. 1-6, University of Zagreb (on CD).

The 73rd World Foundry Congress
„Creative Foundry"
23 - 27 September 2018, Krakow, Poland

EFFECT OF SOLIDIFICATION FACTORS ON CAST STRUCTURE AND 'A' SEGREGATION FOR AN ULTRA LARGE SECTION STRAND OF A VERTICAL SEMI-CONTINUOUS CASTER OF STEEL

Kyung Shik OH[1]*, Joo Dong LEE[2], Changki JEONG[2], Un Kwan CHO[2], Il Hyun CHO[3]

[1] *Head of PosMC Project Team, POSCO, 6261 Donghaeanro Pohang, Gyeongbuk, South Korea*
[2] *PosMC project Team, POSCO, 6261 Donghaeanro Pohang, Gyeongbuk, South Korea*
[3] *Executive Vice President, Investment and Engineering Dept., POSCO, 6261 Donghaeanro Pohang, Gyeongbuk, South Korea*
**Corresponding address: goldsteel@posco.com*

Key words: Large ingot continuous caster, Solidification factor, Ultra large section, 'A' segregation

A vertical semi continuous casting process of an ultra large setion strand such as LICC(a Large Ingot Continuous Caster) [1] has been being developed by the various engineering companies [1~3] due to the high yield, ecofriendly environment, auto controlled process during solidification. For an ultra large section strand, internal soundess like solidification structure and macrosegregation is one of the most important subjects. Therefore, formation of solidifi-cation structure was examined by investigation of macro structure and numerical analysis with and without MEMS. 'A' segregation commonly occur-red in the ultra large section strand, and was investi-gated in view point of solidification factor such as thermal gradient, growth rate, stirring, etc. in this study.

1. Introduction

Ingot casting process of steel has been being mainly used in spite of low yield, not ecofriendly environment, and additional process of forging because it is not easy to adapt a conventional vertical continuous casting process for an ultra large section size steel or high alloy steels due to high construction cost, a small weight order less than about 500ton, and quality problems. Most of steel grades such as carbon structural steels, stainless steels, and high alloy steels have been cast successfully and confidently by LICC process which was developed for both process itself and caster engineering by POSCO.

The cast structures of the 12 tons LICC strand were investigated from the macro-structure of entire longitudinal slices taken from the LICC strand, and was predicted by modelling based on the Hunt [4] and Shibata [5] model. Computer simulation by ANSYS Fluent v.14 were carried out for the casting conditions with and without EMS during casting and solidification. From the measured cast structure and 'A' segregation, prediction of cast structure and formation of 'A' segregation were discussed with the solidification factors.

2. Flow velocity and temperature distribution

Molten steel flow and temperature behaviour during casting with casting speeds and MEMS conditions were investigated through the numerical simulation. We can see wide flow zone by MEMS and it is noticed to be wider for the casting speed of 0.1 m/min than 0.05 m/min. The flow on the molten steel surface was changed to be homogeneous by MEMS, while it varied much with positions by the flow through SEN without MEMS. It can be seen that the solidifying shell at the w/2 and t/2 positions could be melted by the flow through SEN. The temperature distributions in the molten steel with casting speeds and MEMS condition are shown in Fig. 1.

It is noticed that the temperature of both longitudinal section and melt surface was lowered in the wide melt area by MEMS and it is easier for equiaxed crystals to form in the melt in case of MEMS. It is noticed that the temperature of both longitudinal section and melt surface was lowered in the wide melt area by MEMS and it is easier for equiaxed crystals to form in the melt in case of MEMS.

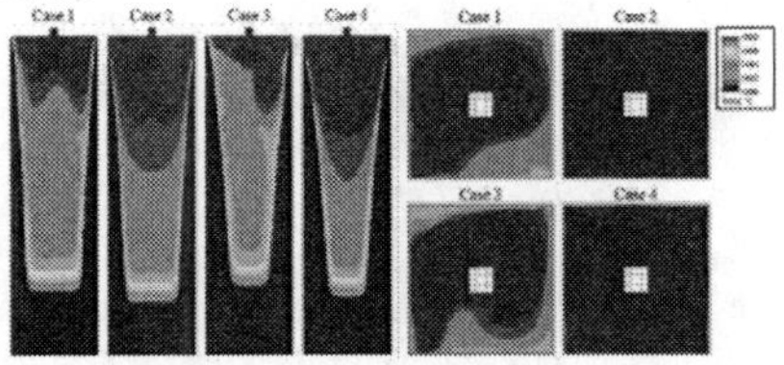

Fig. 1. ***Calculated temperature contour at the end of pouring for the casting speed of 0.1m/min with (Case 1) and without MEMS (Case 2) and for that of 0.05m/min with (Case3) and without MEMS (Case 4).***

3. Cast structure and prediction result

Fig. 2 shows cast structure on the cross-sections of the LICC slabs obtained by real experiments. The zone patterns were similar to the simulated results. Near the columnar zones, dendritic equiaxed grains were developed. The contour plots of $G/V^{1/2}$ (G: temperature gradient, V: growth rate) calculated by computer simulation are shown in Fig. 3. We assumed the heat transfer in the slab during LICC process, as two-dimensional heat conduction with solidification. Fluid flow is approximated with the enhanced thermal conductivity of the liquid. The effective thermal conductivity of the liquid, K_{eff}, is defined as βK_L, where K_L is the thermal conductivity of the liquid. The value of β was 2 for natural convection, and 7~10 for EMS-enhanced convection. The slab system moves during the LICC process. In thus study, the system moved downward with fixed boundary conditions to simulate the process. The low values of $G/V^{1/2}$ mean the possibility of equiaxed grains is high. On the other hand, the high $G/V^{1/2}$ indicates of columnar grains. When no MEMS was applied, the columnar zone along the slab wall was thick and uniform, compared to the case where MEMS applied. MEMS affected not only the equiaxed region but also the columnar region. When MEMS was applied, the columnar zones exist at the wall in the lower slab part, and at the wide region in the upper part. Regardless of MEMS, the equiaxed zone was formed along the centre of the slab. Considering the experimental results, the critical results of $G/V^{1/2}$ is about 0.4. The pipes formed by solidification shrinkage was not predicted with $G/V^{1/2}$. When MEMS was applied, it is noted that white bands were found between columnar and equiaxed zone. The white bands might have been formed by the stirring flow of molten steel by MEMS.

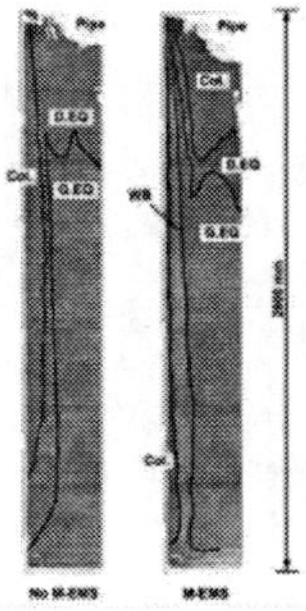

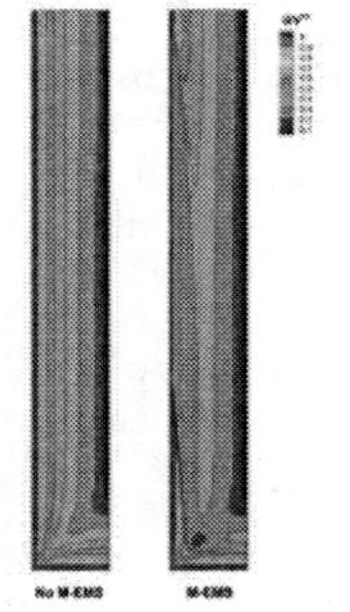

Fig. 2. ***(Left figure) Cast structure on longitudinal section of a LICC strand.***

Fig. 3. ***(Right figure) Predicted cast structure by the simulation of using $G/V^{1/2}$.***

4. Internal soundness and 'A' segregation

It is found that there exist some defects such as shrinkage pipe, centre segregation, 'V' segregation, centre porosity, and 'A' segregation in the LICC strand. Among them, shrinkage pipe causing low yield was much improved by the heating and stirring [1]. Centre segregation and 'V' segregation occurred normally in conventional continuous cast strands, but 'A' segregation did not in them for the case of strand thickness less than 400mm. Location of 'A' segregation occurred in a LICC strand is shown in a sketched figure 4 with and without MEMS together with a growth rate at corresponding position. It was found that growth rate (V) is a vital few of 'A' segregation and it occurs when growth rate decreases rapidly. It was also noticed that 'A' segregation occurs when the temperature gradient at the position decreases rapidly. MEMS is found to be effective to decrease 'A' segregation.

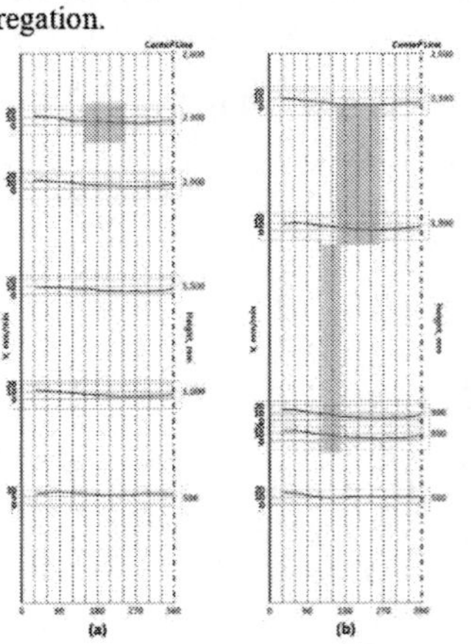

Fig. 4. ***A sketched location of 'A' segregation occurred in a LICC strand with growth rate; (a) with MEMS, (b) without MEMS.***

5. Summary

Solidification structure for an ultra large section size strand of LICC were examined with and without MEMS, and the real zone patterns were similar to the simulated results through the contour plots of $G/V^{1/2}$ calculated by computer simulation. MEMS affected not only the equiaxed region but also the columnar region. Considering the experimental results, the critical results of $G/V^{1/2}$ is about 0.4 for a columnar to equiaxed zone transition. 'A' segregation occur-red in the LICC strand while it did not in conven-tional continuous cast strands for the case of strand thickness less than 400mm. Through the calculated V and G it was found that 'A' segregation occurs when V and G decreases rapidly. V is found to be more critical factor than G.

References

[1] K.S. Oh, J.D. Lee, S.J. Kim and J.Y. Choi, "Development of a large ingot continuous caster", METALLURGICAL Research & Technology (2015, Vol.112, No.2), 203(pp. 1-13)

[2] Feldhaus, S.: "Continuous casting for oversized blooms", Proceedings of 9th ECCC, June 26-29, 2017, Austria Center Vienna, Austria, pp.257~266

[3] Hrazdera, G.: "Semi-continuous casting technology – combining technological advantages of two different casting practices", ibid, 267~276

[4] J.D. Hunt, Mater. Sci. Eng. (1984, Vol.65), p.74

[5] H. Shibata, S. Itoyama, Y. Kishimoto, S. Takeuchi and H.Sekiguchi, Prediction of Equiaxed Crystal Ratio in Continuously Cast Steel Slab by Simplified Columnar-to-Equiaxed Transition Model, ISIJ Int. (2006, Vol.46, No.6), p.921

The 73rd World Foundry Congress
„Creative Foundry"
23 - 27 September 2018, Krakow, Poland

MICROSTRUCTURE QUALITY ASSESSMENT OF ISOTHERMED DUCTILE IRONS THROUGH TENSILE TESTS

Giuliano Angella [1]*, **Franco Zanardi** [2]

[1]*National Research Council of Italy/Institute of Condensed Matter Chemistry and Technologies for Energy (ICMATE), via R. Cozzi 53 – 20125 – Milano, ITALY,*
[2] *Zanardi Fonderie SpA, via Nazionale 3 – 37046 – Minerbe (VR), ITALY,*
**Corresponding author: giuliano.angella@cnr.it*

Key words: Ductile irons, tensile tests, mechanical properties, constitutive equations, quality assessment.

Isothermed Ductile Irons (IDIs - Zanardi Fonderie SpA patented) belong to a new generation of Ductile Irons (DIs) that are very attractive for the good combination of outstanding mechanical properties and low production costs. To produce IDIs, conventional ferritic DIs with no alloying elements are austenitized in the intercritical range, leaving an opportune fraction of pro-eutectoidic ferrite, and then, after quenching in salt bath above the Ms, the austenite transforms into pearlite. The resulting microstructure is called *perferrite* and consists of interconnected ferrite and pearlite, which is different from the conventional bull-eye structure, where ferrite is surrounded by pearlite. An example of perferrite is reported in Fig. 1.

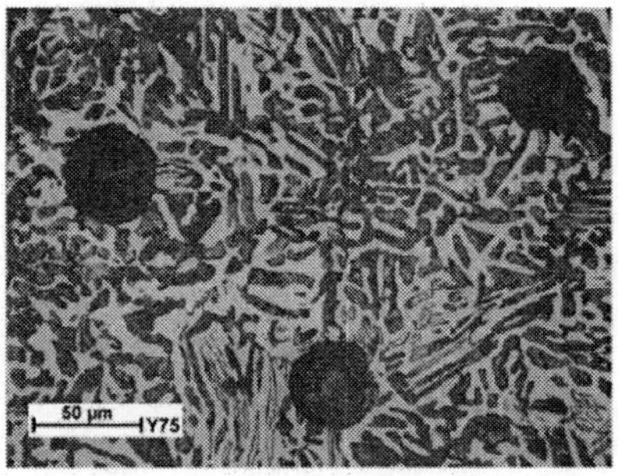

Fig. 1. ***Spheroidal graphite perferritic microstructure from Y-shaped Type IV cast sample etched with 2% Nital.***

A new approach to assess properly these highly performing IDIs is needed [1], since the current international standards are based on the minimal mechanical properties, which cannot give the correct description of the high quality of this new generation of DIs. A mathematical procedure based on the analysis of the tensile plastic behaviour through Voce constitutive equation has been recently proposed [2] to assess the microstructure quality of different ductile irons (DIs). Voce equation is

$$\sigma = \sigma_V + (\sigma_o - \sigma_V) \cdot \exp(-\varepsilon_p/\varepsilon_c) \quad (1)$$

where σ is the true flow stress, ε_p the true plastic strain, σ_V is the saturation stress that is achieved asymptotically with straining, ε_c is the characteristic transient strain that defines the rate with which σ_V is approached, and σ_o is the back-extrapolated stress to $\varepsilon_p = 0$.

The differential form of the Voce equation is used to characterise the strain hardening behaviour of IDIs

$$\frac{d\sigma}{d\varepsilon_p} = \Theta_o - \frac{\sigma}{\varepsilon_c} \quad (2)$$

where Θ_o and ε_c are constants related to the material structure [3,4]. Through fitting Eq. 2 to linear regions at high stresses in the differential experimental data $d\sigma/d\varepsilon_p - \sigma$, the Voce parameters Θ_o and ε_c^{-1} are found, and, as a consequence, σ_V is determined from the condition $d\sigma/d\varepsilon_p = 0$. σ_o is finally found from the best fitting Voce equation to the experimental tensile flow curves.

Microstructure is a generic term that here is meant to come into material perferritic matrix that deforms plastically, and metallurgical defects or discontinuities that can cause premature ruptures, like irregular graphite, shrinkages, slag inclusions or gas holes. The microstructure assessment procedure consists of two diagrams: the first diagram is for the quality assessment of the IDI matrix and is based on the Voce constitutive equation parameters ε_c^{-1} and Θ_o resulting from the analysis with Eq. 2 of the strain hardening data of the experimental tensile flow curves.

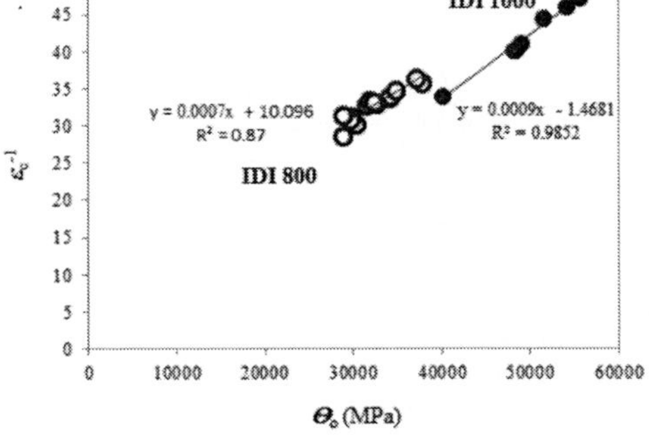

Fig. 2. ***Matrix quality assessment diagram $\varepsilon_c^{-1} - \Theta_o$ for IDI 800 and IDI 1000 produced in Zanardi Fonderie SpA.***

In Fig. 2 the quality diagram $\varepsilon_c^{-1} - \Theta_o$ for IDI 800 and 1000 produced in Zanardi Fonderie SpA are reported. IDI 800 is characterised by a Si content of about 2.5% in weight and coarser perferrite, whilst IDI 1000 by a

Si content of about 3.6% and a finer microstructure. IDI 800 and 1000 have respective quality lines and do not overlap at all, identifying unambiguously the two grades of IDIs. In fact, it has been reported [2] that in a diagram $\varepsilon_c^{-1} - \Theta_0$ different grades of DIs produced in Zanardi Fonderie SpA (austempered DIs, IDIs and ferritic-pearlitic DIs) are well identified by proper quality lines.

A second diagram is for the assessment of the rupture conditions that could be related to possible material defects, and is based on the comparison between the experimental elongations to rupture e_r with the ideal uniform elongations e_u defined according to the Voce formalism. The *uniform plastic strain* ε_u is given according to the relationship

$$\varepsilon_u = \varepsilon_c \cdot ln\left(\frac{\varepsilon_c+1}{\varepsilon_c} \cdot \frac{\sigma_V-\sigma_0}{\sigma_V}\right) \quad (3)$$

ε_u is the true plastic strain that is achieved if the Considére's criterion is fulfilled, i.e. $d\sigma/d\varepsilon_p = \sigma$. The Considére condition identifies the beginning of necking. When a significant amount of perlite is part of the DI structure, necking does typically not occur or only at a little extent, so the Considére condition represents the ultimate tensile stress with the best achievable strain. The uniform elongation e_u, that is

$$e_u(\%) = [\exp(\varepsilon_u) - 1] \cdot 100 \quad (4)$$

Hence, e_u represents the *ideal behaviour* in DIs. However, because of metallurgical defects premature failures can occur with plastic elongations to rupture $e_r < e_u$. In Fig. 3 a second diagram is reported to compare the experimental plastic elongations e_r with the ideal uniform plastic elongations e_u calculated according to Eqs. 3 and 4.

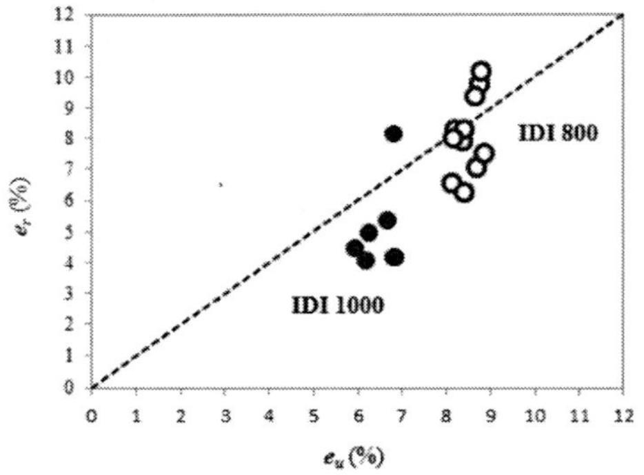

Fig. 3. Defects assessment diagram for IDI 800 and IDI 1000.

In Fig. 3 the data points below the ideal behaviour line ($e_r = e_u$) identify premature rupture because of metallurgical discontinuities, whilst data points lying on it or above (elongations beyond the necking) identify samples with metallurgical integrity.

The quality assessment procedure here proposed is based on the hypothesis that the strain hardening behaviour of DIs matrix is independent on metallurgical defects. This hypothesis is validated by the trends of the experimental differential data $d\sigma/d\varepsilon_p - \sigma$ of IDIs. The linear regions of the experimental strain hardening data on which Eq. 2 has been fitted to find the Voce parameters ε_c^{-1} and Θ_0, are independent on the rupture elongations e_r, and so are independent on defects. Furthermore, the diagram $\varepsilon_c^{-1} - \Theta_0$ identifies specific quality lines for different IDI grades with little lateral scattering regardless e_r. The data point position on the quality line depends on the microstructure refinement, and so the diagram $\varepsilon_c^{-1} - \Theta_0$ identifies specific perferritic matrixes. This is possible because of the little density of metallurgical defects in IDIs (and generally in DIs), in opposition to cast Al alloys where the high density of defects affect significantly their strain hardening behaviour [5].

As a consequence of this hypothesis validation, the defect assessment diagram in Fig. 3 truly represents the presence of metallurgical defects in IDIs. Indeed, the diagram in Fig. 3 is equivalent to the microstructure assessment based on the quality index

$$Q_{Elong} = \frac{e_r}{e_u} \quad (4)$$

that was originally proposed for cast Al alloys [6], where Hollomon constitutive equation was used. Indeed, in Al alloys Q_{Elong} has been criticised [7,8] because of Hollomon equation and the impossibility of discriminating between matrix plastic behaviour and metallurgical defects, so new quality indexes have been proposed [7,8].

In conclusions, the discrimination between matrix strain hardening behaviour and metallurgical defects is possible in IDIs (and also in DIs [2]), and the diagrams in Fig. 2 and Fig. 3 (or Eq. 4) can be used to assess the microstructure quality of IDIs (and DIs).

References

1. F. Zanardi, F. Bonollo, N. Bonora, A. Ruggiero, G. Angella. (2017). A Contribution to New Material Standards to Ductile Irons and Austempered Ductile Irons. Int. J. Metalcasting. Vol. 11, pp. 136-147.

2. R. Donnini, F. Vettore, F. Zanardi, G. Angella. (2017). 10th International Symposium on Science and Processing of Cast Iron – SPCI11 (2017). Evaluation of Microstructure Quality in Ductile Irons Based on Tensile Behaviour Analysis. To be published on Scientific.Net.

3. G. Angella, F. Zanardi, R. Donnini. (2016). On the significance to use dislocation-density-related constitutive equations to correlate strain hardening with microstructure of metallic alloys: the case of conventional and austempered ductile irons. J. Alloys Compd. Vol. 669, pp.262-271.

4. R. Donnini, A. Fabrizi, F. Bonollo, F. Zanardi, G. Angella. (2017). Assessment of Microstructure Evolution of an Austempered Ductile Iron during Austempering Process through Strain Hardening Analysis. Metals and Materials International. Vol. 23, pp. 855-864.

5. M. Tiryakioğlu, J. Campbell, J.T. Staley. (2004). Evaluating Structural Integrity of Al-7%Si-Mg Alloys via Work Hardening Characteristics: II. Concept of Target Properties. Mat. Sci. Eng. A. Vol. 368, pp. 205-211.

6. C.H. Caceres. (1998). A Rationale for the Quality Index of Al-Si-Mg Casting Alloys. Int. J. Cast. Met. Res. Vol. 10, pp. 293-299.

7. M. Tiryakioğlu, J.T. Staley, J. Campbell. (2004). Evaluating Structural Integrity of Al-7%Si-Mg Alloys via Work Hardening Characteristics: II. A New Quality Index. Mat. Sci. Eng. A. Vol. 368, pp. 231-238.

8. M. Tiryakioğlu, N.D. Alexopoulos, J. Campbell. (2004). On The Ductility Potential of Cast Al-Cu-Mg (2009) Alloys. Mat. Sci. Eng. A. Vol. 506, pp. 23-26.

The 73rd World Foundry Congress
„Creative Foundry"
23 - 27 September 2018, Krakow, Poland

NANOMULTILAYERS FOR JOINING APPLICATIONS

Janczak-Rusch Jolanta*, Bastian Rheingans, Claudia Cancellieri, Mirco Chiodi, Vicente Araullo-Peters, Lars P.H. Jeurgens

Empa, Swiss Federal Laboratories for Materials Science and Technology, Joining Technologies & Corrosion, Überlandstrasse 129, 8600 Dübendorf, Switzerland
**Corresponding address: jolanta.janczak@empa.ch*

Key words: nanomultilayers, nanojoining, reactive joining, nanomaterials, brazing

Nanomultilayers (NMLs) offer many unique properties and a lot of design freedom by combining layers of different, nanosized materials. In particular, the fast diffusion along the numerous internal interfaces and the melting point depression of the nanoconfined metals offer new opportunities for the development of advanced joining processes [1,2].
We report the current developments and applications of nanomultilayers for so-called nanojoining processes (joining using nanomaterials and utilising the nanoeffects). This involves i) the application of reactive nanomultilayers for „room temperature" joining of similar and dissimiliar materials, ii) the development of nanomultilayers with special properties for low-temperature joining.

1. Reactive nanomultilayers as a heat source for joining processes

Reactive nanomultilayers (RNMLs) are alternating nanolayers consisting of (at least) two chemical components which can undergo a strongly exothermal reaction upon formation of intermetallic compound phase(s) (Fig.1.). Temperatures well above 1000 °C can be reached locally during this self-propagating reaction [3]. The reaction heat can be used for the execution of a joining process, in particular to melt a brazing filler alloy or solder and to form a bond between components [4]. The resulting joining process is very fast, as the reaction speeds reach up to several dozens of meters per second.

As the RNMLs serve as an internal heat source, there is no need to use a brazing furnace or reflow oven for the joining process. Thanks to localised provision of the heat by the RNML (directly within the joining zone), the joining components stay virtually cold during the process and thus sensitive materials and components can be joined. However, successful application of RNMLs for joining requires precise control of the reaction, which still remains a challenge. In particular, the heat evolution must be controlled by balancing the heat production, temporary heat absorption and heat dissipation, in order to avoid thermal shock or insufficient heating. Furthermore, the mechanical shock-wave, which can be caused by the exothermal reaction, must be mitigated in order to avoid damaging of sensitive devices.

Different examples of joining with RNMLs (e.g. Cu, borosilicate glass joints) are presented. The design of the joint material system (RNMLs, solder, metallisation) will be discussed and the crucial issue of managing the heat flux will be exemplified.

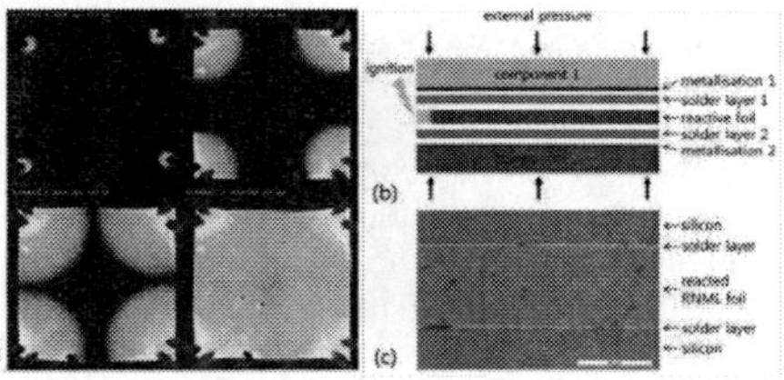

Fig. 1. Joining with reactive nanolayers (a) high-speed video recording of a reacting RNML foil, (b) schematic setup of the joining material system (c) Si-Si joint with Sn-solder and RNML layer

2. Non-reactive nanomultilayers for fast, low-temperature joining processes

Non-reactive multilayers consist of brazing filler metal (alloy) nanolayers separated by nanolayers of a second, inert material, which serve as a functional barrier keeping the nanodimension of the brazing filler upon the process temperature, Fig 2.

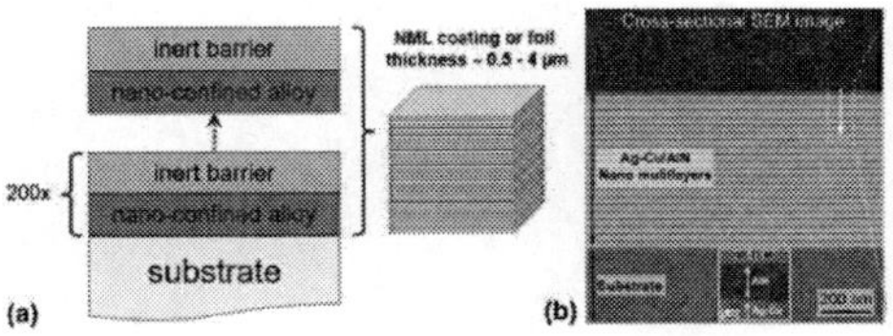

Fig. 2. a) Schematic of the NML-filler, constituted of alternating nanolayers (thickness < 10 nm) of a metal or alloy (e.g. Ag, Cu, Ag-Cu) and a chemically inert barrier material (e.g. carbon, nitride, oxide, refractory metal). (b) Cross-sectional secondary electron image of as-deposited Ag-Cu/AlN NML (light grey: Ag-Cu, dark grey: AlN)

In this way the advantageous properties of the nanostructured filler e.g. melting-point depression or fast (short-circuit) diffusion can be exploit. With the aim to utilise these material characteristics for the development of fast and benign joining processes, we designed and studied various NML configurations (Cu/AlN, Ag/AlN, Ag/Ge/AlN, Ag-Cu/AlN, Cu/W, AlSi/AlN NMLs) [5-12]. In particular the structural evolution of such nanomultilayers upon heating was investigated in detail by ex-situ and in-situ XRD, HR-SEM, TEM, AES methods, Fig. 3.

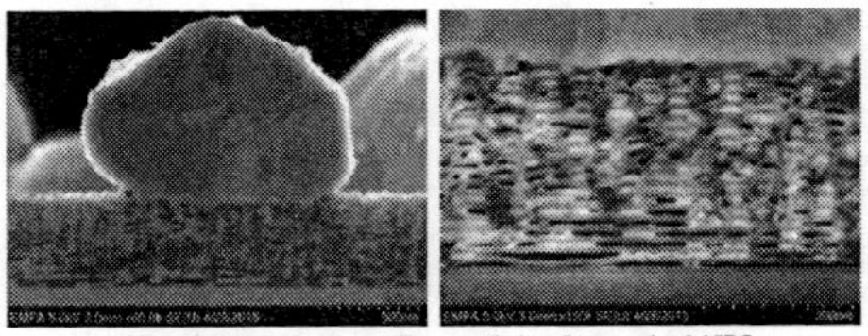

Fig.3 Massive outflow of Ag from Ag/AlN nanomultilayer after heating up to 420°C in air (HR SEM cross-sectional images, light grey: Ag, dark grey: AlN) (a) large Ag particle forming at the top surface of the NML upon heating. (b) voids (dark) in Ag layers forming after Ag migration to the NML surface

Many parameters, such as the NML design (material specification, layer thickness, the number of layer repetitions), the morphology of the layers (as function of the deposition parameters), the structure of the interfaces, the internal stresses and the environmental conditions were found to affect the behaviour upon heating of the investigated nanostructured joining materials. While the complex interplay of these parameters makes the understanding of the multilayer behaviour very difficult, the resulting huge parametrical design space opens new opportunities for the development of nanojoining materials for fast, low-temperature joining processes or localized, selective bonding methods and properties.

Acknowledgments
The authors acknowledge the financial support of Interreg Project ABH028 and FP7-PEOPLE-2013-IRSES Project EXMONAN.

References

[1] D. P. Adams. (2015). Reactive Multilayers Fabricated by Vapor Deposition: A Critical Review. Thin Solid Films. 576, 98-128.

[2] R. Longtin, E. Hack, J. Neuenschwander & J. Janczak -Rusch. (2011). Benign Joining of Ultrafine Grained Aerospace Aluminum Alloys Using Nanotechnology. Advanced Materials. 23, 5812-5816.

[3] G. Kaptay, J. Janczak-Rusch, G. Pigozzi & L.P.H. Jeurgens. (2014). Theoretical Analysis of Melting Point Depression of Pure Metals in Different Initial Configurations. Journal of Materials Engineering and Performance. 23, 1600-1607.

[4] J. Janczak-Rusch, G. Kaptay & L.P.H. Jeurgens. (2014). Interfacial design for joining technologies – An historical perspective. Journal of Materials Engineering and Performance. 23, 1608-1613.

[5] B. Lehmert, J. Janczak-Rusch, G. Pigozzi, P. Zuraw, F. La Mattina, L. Wojarski, W. Tillmann, L.P.H. Jeurgens. (2015). Copper-Based Nanostructured Coatings for Low-Temperature Brazing Applications, Materials Transactions. 56, 1015-1018.

[6] F. Moszner, C. Cancellieri, C. Becker, M. Chiodi, J. Janczak-Rusch & L.P.H. Jeurgens. (2016). Nano-structured Cu/W brazing fillers for advanced joining applications. Journal of Materials Science and Engineering. B 6, 226-230.

[7] C. Cancellieri, F. Moszner, M. Chiodi, S. Yoon, J. Janczak-Rusch & L.P.H. Jeurgens. (2016). The effect of thermal treatment on the stress state and evolving microstructure of Cu/W nano-multilayers. Journal of Applied Physics. 120, 195107.

[8] F. Moszner, C. Cancellieri, M. Chiodi, S. Yoon, D. Ariosa, J. Janczak-Rusch & L.P.H. Jeurgens. (2016). Thermal stability of Cu/W nano-multilayers. Acta Materialia. 107, 345-353.

[9] M. Chiodi, C. Cancellieri, F. Moszner, M. Andrzejczuk, J. Janczak-Rusch & L.P.H. Jeurgens. (2016). Massive Ag migration through metal/ceramic nano-multilayers: interplay between temperature, stress-relaxation and oxygen enhanced mass transport. Journal of Materials Chemistry. C 4, 4927-4938.

[10] D. Ariosa, C. Cancellieri, V. Araullo-Peters, M. Chiodi, E. Klyatskina, J. M. Janczak-Rusch & L.P.H. Jeurgens. (2018). Modelling of interface and internal disorder applied to XRD analysis of Ag based nano-multilayers. ACS Applied Materials & Interfaces. 10(24), 20938-20949.

[11] J. Janczak-Rusch, M. Chiodi, C. Cancellieri, F. Moszner, R. Hauert, G. Pigozzi & L.P.H. Jeurgens. (2015). Structural evolution of Ag-Cu nano-alloys confined between AlN nano-layers upon fast heating, Physical Chemistry Chemical Physics. 17, 28228-28238.

[12] J. Lipecka, J. Janczak-Rusch, M. Lewandowska, M. Andrzejczuk, G. Richter & L.P.H. Jeurgens. (2017). Thermal stability of Al-Si12at.% nano-alloys confined between AlN layers in a nanomultilayer configuration. Scripta Materialia. 130, 210-213.

The 73rd World Foundry Congress
„Creative Foundry"
23 - 27 September 2018, Krakow, Poland

EXPERIMENTAL DEVICE FOR INVESTIGATION OF LOW VAPOUR PRESSURE LIQUID METALS AND THEIR INTERACTION IN CONTACT WITH REFRACTORY MATERIALS

Artur Kudyba[1,*], Natalia Sobczak[1,2], Janusz Budzioch[3], Wojciech Polkowski[1]

[1] *Foundry Research Institute, Zakopianska 73 Str., 30-418 Krakow, Poland*
[2] *Foundry Research Institute, Zakopiańska 73 Str., 30-418 Krakow, Poland*
[3] *MeasLine Sp. z o.o., 37A/308 Wroclawska St., 30-011 Krakow, Poland*

**Corresponding author: artur.kudyba@iod.krakow.pl*

Key words: equipment design, sessile drop, wettability, reactivity, low vapour pressure liquid metal, MMC

1. Introduction

A development of designing and processing of new materials requires a characterization of their properties in various states of matter. While methods for examining solid state materials have been already well developed and corresponding apparatuses are commercially available, a strikingly different situation takes place in the field of techniques and devices for high temperature investigation of liquid state materials.

Examinations of interfacial phenomena has become an emerging and important part of liquid state materials science, while the sessile drop (SD) method is a basic tool applied for the assessment of thermophysical properties of liquid metals and examination of wettability and spreading phenomena. This method allows receiving a quantitative metrics (a contact angle θ, a work of adhesion) describing the wetting behavior of contacting materials. Such determined characteristics are subsequently utilized upon designing and production of new materials or devices, as well as in a liquid-assisted joining of dissimilar materials.

In recent years, both basic and applied research works on experimental determination of thermophysical properties of liquid metals and alloys having low-partial pressure vapors, has gathered a growing attention. This kind of research is especially important for a selection of proper refractories applied for equipment in foundry processes (crucibles, molds, casting spoons etc.) that are in direct contact with liquid metals (magnesium, zinc, etc.).

Beside of applications in the foundry engineering, a determination of thermophysical, technological and utility properties of liquid metals having low vapour pressure, is especially important for an optimization of liquid-assisted processing of metal matrix composites (MMC).

Magnesium matrix lightweight composites, are actually applied in many industry branches (e.g. aviation, automotive, military or medicine), mostly due to a very favorable specific strength, a high stiffness and wear resistance [1].

In order to ensure a high quality of products made of Mg-MMCs, it is necessary to experimentally determine the properties and behavior of liquid Mg in contact with potential reinforcing materials. However, conducting of such experiments is quite problematic, since magnesium exhibits a high chemical affinity to oxygen, what results in a "spontaneous" formation of surface oxide layer. This layer strongly affects the reliability of obtained results, or even make the measurements impossible to be carried out.

Furthermore, an extensive evaporation of magnesium upon high temperature may lead to a severe contamination or damage of laboratory devices. Therefore, only few laboratories struggle with such kind of research, while the number of scientific papers on thermophysical properties of liquid Mg contacting ceramic materials, is strongly limited.

On the other hand, very recently, a new scale laboratory stand for examinations of high temperature properties of liquid metals having low vapour pressure, have been designed, built and initialized by the Centre for High Temperature Studies (CHTS) at the Foundry Research Institute in Krakow [2]. After unique experimental complex reported in [3-7], it represents a new input of CHTS in the development of testing devices for investigation of liquid metals and alloys. Improved methodological and design concepts implemented into the laboratory stand make it unique at the world scale; and at the same time allow for low vapour pressure liquid metals to use experimental procedures that are not available at any other laboratory.

2. Experimental design

A macroview of the new laboratory stand for examinations of high temperature properties of liquid metals having low vapour pressure is shown in Fig. 1a; while various available testing procedures for recording and evaluation of interfacial phenomena, are schematically presented in Fig. 1b.

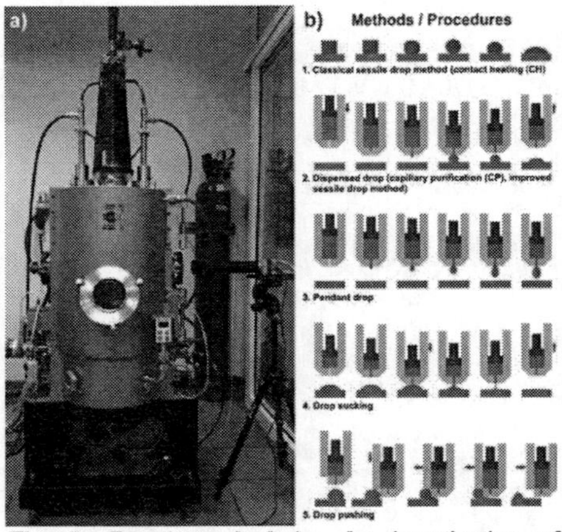

Fig. 1. Experimental device for investigation of properties of low vapour pressure metals in contact with refractory materials: a – front view: b – and possible testing methods and procedures [2]

The main advantages of the new device over actually existed ones, are as follows:

1) an unique ability to simultaneously test up to 10 samples during one single experiment;
2) a wide range of testing temperatures (up to 1000°C);
3) a possibility to use different testing methods and procedures with a particular emphasis on CP procedure (Capillary Purification with up-and-down movement of a capillary with liquid metal), including the proposed ones such as drop sucking and drop pushing, especially useful for *in situ* "opening" of the drop/substrate interface directly at the testing temperature for the characterization of interfacial reaction products,
4) possibility to get high vacuum up to 10^{-7} hPa or to use inert gases atmosphere (a flowing gas with controlled rate at required level of pressure inside vacuum chamber);
5) a controlled contact heating of examined couple of materials to experimental temperature;
6) a specimen table with up-and-down movement;
7) an automatic real-time control of temperature by 4 thermocouples located in selected positions,
8) a fully computerized controlling and recording systems with possibility to work on each unit independently.

Beside of a standard and commonly used classical sessile drop method coupled with a contact heating procedure, the new high temperature device also offers benefits coming from using the capillary purification (CP) procedure. One of the advantage of CP procedure is an elimination of the problem of magnesium oxidation. In the CP procedure a metal and a substrate are separately heated up (without any contact between them) to the testing temperature. Initially, the metal is placed in a capillary located above the surface of substrate; and after reaching the pre-assumed testing temperature it is squeezed through a hole in the capillary. This procedure is very beneficial in terms of:

(I) *non-contact heating of the couple* i.e. an elimination of undesirable effect of heating history, what is especially important in the case of highly reactive systems such Mg/metals or Mg/oxides);
(II) *a purification of the droplet from surface oxide layer* - the native oxide films is mechanically removed during squeezing of liquid magnesium through a hole in capillary, what allows receiving truth contact angles values. A more detailed description of the CP procedure is shown elsewhere [3-7].

Acknowledgments

The studies were realized within the projects financed by the Foundry Research Institute in 2015 – 2017 (projects no.: 5014/00, 6006/00 and 7009/00). Partial financial support from the Ministry of Science and Higher Education of Poland is also acknowledged.

References

[1] A. Olszowka-Myalska, M. Jerzy, M. Godzierz, P. Wrześniowski. (2018). Magnesium Matrix Composite with Open-Celled Carbon Foam obtained by Powder Metallurgy. Archives of Metallurgy and Materials 63(2), 825-831.
[2] A. Kudyba, N. Sobczak, J. Budzioch, G. Bruzda, R. Nowak. (2017). Device for investigation of properties of molten magnesium alloys in contact with refractory substrates. Patent application no P.421216, Poland.
[3] N. Sobczak, R. Nowak, W. Radziwill, J. Budzioch & A. Glenz. (2008). Experimental complex for investigations of high temperature capillarity phenomena. Materials Science Engineering. A495, 43-49.
[4] N. Sobczak, M. Singh & R. Asthana. (2005). High-temperature wettability measurements in metal/ceramic systems – Some methodological issues. Current Opinion in Solid State Materials Science. 9(4), 241–253.
[5] N. Sobczak, J. Sobczak, R. Asthana & R. Purgert. (2010). The mystery of molten metal. China Foundry. 7/4, 425-437.
[6] N. Sobczak Improvement of wetting in metal/ceramic systems. In: Bellosi A, Kosmac T, Tomsia AP editors (1998). Interfacial science in ceramic joining. NATO ASI Series, 58. Kluwer, 27–42.
[7] N. Eustathopoulos, N. Sobczak, A. Passerone, K. Nogi (2005) Measurements of contact angle and work of adhesion at high temperature, Journal of Materials Science. 40, 2271-2280.

ENHANCING THE PERMEABILITY AND PROPERTIES OF CERAMIC SHELL IN INVESTMENT CASTING PROCESS USING ABS POWDER AND NEEDLE COKE

Sanjay Kumar [1]*, D. Benny Karunakar [2]

[1]*Department of Mechanical and Industrial Engineering, Indian Institute of Technology Roorkee, Roorkee - 247667, India,*
[2]*Department of Mechanical and Industrial Engineering, Indian Institute of Technology Roorkee, Roorkee - 247667, India,*
**Corresponding address: bennyfme@iitr.ac.in*

Key words: Investment casting process, Permeability, ABS powder, Needle coke, Mechanical properties

In the investment casting process, initially, a wax pattern is made, around which repeated slurry/stucco coatings are given. The slurry is made up of very fine ceramic ingredients like zircon flour, etc. Once the ceramic slurry is dried out, a ceramic shell is formed around the wax pattern. After drying out of the ceramic shell, the wax is drained out and the shell is fired in an oven for inducing additional strength. Molten metal is then poured into the ceramic shell.
Because of the fine ingredients of the ceramic shell, the investment casting gets a very fine surface roughness, compared to a conventional sand casting. However, the ceramic shell exhibits poor permeability, which often results in porosity defects.
Hence, in the present work, an attempt is made to enhance the permeability of ceramic shell by mixing certain fillers like ABS (Acrylonitrile Butadiene Styrene) powder and needle coke to the ceramic slurry. Fine ABS powder is added to the inner coat slurry in a very small proportion. The ABS powder, upon heating, got burnt and left micro pores inside the inner coating of the shell.
In a similar way, needle coke is added to the outer coat slurry in a very small proportion. During the firing of the shell, the needle coke got burnt and, consequently, pores are created inside the shell, which would enable escape of undesirable gases that may generate inside the shell. This ultimately would reduce the porosity defects in the investment castings.
Tests conducted on the ceramic shell modified with ABS powder and needle coke reveal that the modified shell possesses higher permeability compared to the conventional one. The modified shell also possesses adequate mechanical properties like flexural strength, hot tensile strength, etc., and fair surface roughness.

1. INTRODUCTION

Investment casting process has been broadly utilized for a very long time. This process is well known for manufacturing components with excellent surface finish and close dimensional tolerance. The strength distribution all over the ceramic shell can vary considerably because of the variation in structure and thickness at flat sections and corners. Variation of structure and thickness can cause failure to the ceramic shell under smaller loads [1]. Jones and Yuan [2] have added organic fibers to the binder to improve the strength of shell. Addition of organic fibers also increased the thickness at the flat sections and corners of the ceramic shell. While firing the shell, the fibers were burnt out and left micro pores inside the shell, which has enhanced the permeability of the shell. Amira et al. [3] developed a improve method to evaluate both hot strength and permeability of ceramic shell at high temperatures. Wang et al. [4] added needle coke to the outer coat ceramic slurry and found that after firing, needle coke burnt out and improved the permeability of the shell as compared to the conventional shell.

In the present work, ceramic shells were prepared with some modification to the slurry composition, apart from the conventional shell. In the modified ceramic shell, fine ABS (Acrylonitrile Butadiene Styrene) powder and needle coke were mixed to the primary coat and outer coat respectively. ABS is a thermoplastic polymer having glass transition temperature of 108.98 to 110 °C. ABS thermoplastic polymer has good impact resistance and toughness [5]. Needle coke exhibits high strength, low thermal expansion coefficient and excellent resistance for thermal shock.

2. EXPERIMENTAL PROCEDURE

Ceramic shells were prepared using aluminum silicate as refractory powder and colloidal silica as binder. Modification to the slurry was done by adding fine ABS powder and needle coke to the inner coat and outer coat slurries respectively. Table 1 shows the composition of ingredients for conventional and modified ceramic shell. FE-SEM images of ABS powder and needle coke are shown in Fig. 1.
Permeability of the ceramic shell was calculated using the equation (1):

$$k = \frac{\mu(r_o - r_i)}{\Delta p \, 4 \prod r_o r_i} * f \qquad (1)\ [3]$$

where k is the permeability, μ is the dynamic viscosity of air, r_i and r_o are the inner and outer radii of the specimen respectively, f is the air flow rate and Δp is the air pressure drop.

Table 1. Composition of ingredients for conventional and modified ceramic slurries

Ceramic slurry ingredients	Conventional ceramic slurry	Modified ceramic slurry
Colloidal silica	40 %	40 %
Aluminum silicate	53 %	51-52 %
ABS powder	n/a	0.10-0.50 %
Needle coke	n/a	2 %
Zircon powder	6 %	6 %
Wetting agent	0.20	0.20
Antifoam	0.30	0.30
Deionized water	Few drops	Few drops

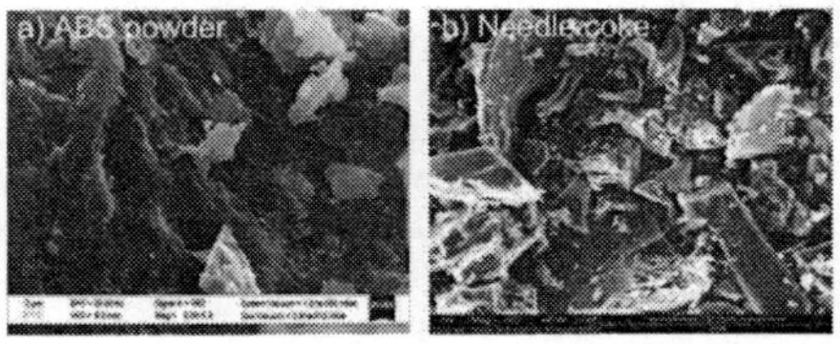

Fig. 1. FE-SEM images of (a) ABS powder (b) Needle coke

Flexural strength (σ_{max}) was calculated using the equation (2).

$$\sigma_{max} = \frac{3P_{max} L}{2WH^2} \qquad (2)\ [9]$$

Where Pmax is the fracture load, W is the width, H is the thickness and L is the span of the test specimen.

3. RESULTS AND DISCUSSION

The consistent permeability of the ceramic shell has a great benefit to the casting because it allows the escape of undesirable gases evolved from the ceramic shell and hence reduces the porosity defects in the cast part.

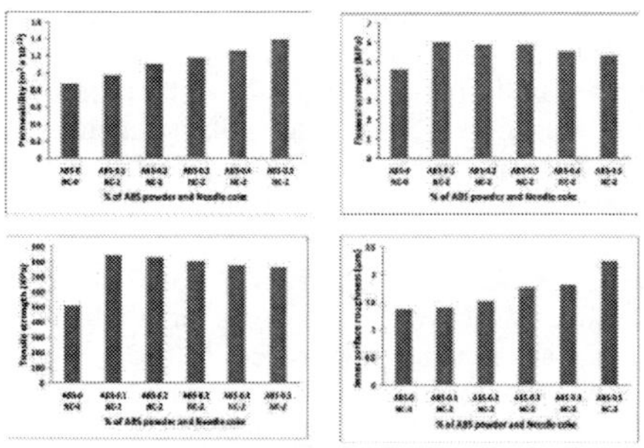

Fig. 2. (a) Permeability (b) Flexural strength (c) Tensile strength (d) Inner surface roughness

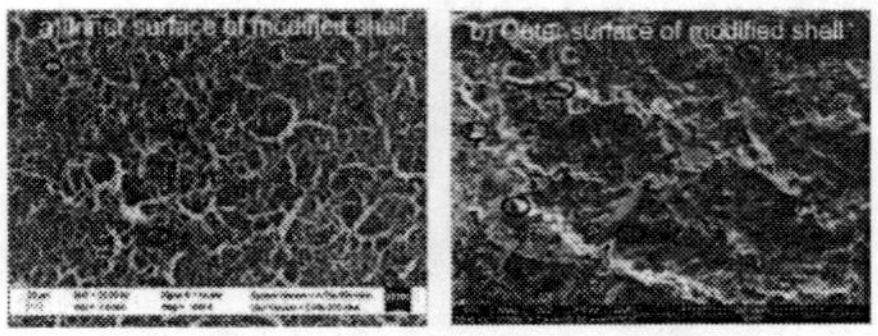

Fig. 3. FE-SEM images showing: (a) Inner surface of modified ceramic shell (b) Outer surface of modified ceramic shell

Fig. 2 (a) indicated that the modified slurry produced a ceramic shell with increased permeability, and permeability of the shell increased as the amount of ABS powder increased. FE-SEM images for the modified ceramic shell are shown in Fig. 3. It can be seen that the fine ABS powder created micro pores at the inner surface while the needle coke created macro pores at the outer surface of the modified ceramic shell. Fig. 2 (b) & (c) indicated that the modified shell has a better flexural strength and tensile strength as compared to the conventional shell. However, strength of the modified ceramic shell decreases as the amount of ABS powder is increased.

Fig. 2 (d) indicated that the conventional ceramic shell has a better surface finish as compared to the modified ceramic shell. As the percentage of fine ABS powder is increased, the surface roughness of the modified ceramic shell also increased.

4. CONCLUSIONS

Modified ceramic shell with fine ABS powder and needle coke has improved permeability and strength as compared to the conventional ceramic shell. The strengths of the modified ceramic shell decreased as the amount of ABS powder increased. The modified ceramic shell has poor inner surface finish because ABS powder got burnt out during firing and left micro pores at the inner surface of the shell. Hence, the addition of ABS powder to the slurry should be optimum.

REFERENCES

[1] S. Pattnaik, D.B. Karunakar & P.K. Jha. (2012). Developments in investment casting process - A review. Journal of Materials Processing Technology. vol. 212, no. 11, pp. 2332–2348.

[2] S. Jones & C. Yuan. (2003). Advances in shell moulding for investment casting. Journal of Materials Processing Technology. vol. 135, pp. 258–265

[3] S. Amira, D. Dubé & R. Tremblay. (2011). Method to determine hot permeability and strength of ceramic shell moulds. Journal of Materials Processing Technology. vol. 211, no. 8, pp. 1336–1340.

[4] F. Wang, F. Li, B. He & B. Sun. (2014). Microstructure and strength of needle coke modified ceramic casting molds. Ceramics International. vol. 40, pp. 479–486.

[5] F.S. Kamelian, E. Saljoughi, P.S. Nasirabadi & S.M. Mousavi. (2017). Modifications and Research Potentials of Acrylonitrile / Butadiene / Styrene (ABS) Membranes : A Review. Polymer Composites, doi:10.1002/pc.24276.

The 73rd World Foundry Congress
„Creative Foundry"
23 - 27 September 2018, Krakow, Poland

COMPARATIVE STUDY OF INTERACTION OF KAOLIN SUBSTRATES WITH LIQUID Ni AND CSMX4 ALLOY

Rafał Nowak [1*], Jerzy J. Sobczak [1], Natalia Sobczak [1,2], Fabrizio Valenza [3], Alberto Passerone [3], Grzegorz Bruzda [1], Piotr Tkaczewski [4], Piotr Kurdziel [4], Gabriele Cacciamani[5]

[1] Foundry Research Institute, 73 Zakopiańska Str., 30-418 Krakow, Poland
[2] Institute of Precision Mechanics, 3 Duchnicka Str., 01-796 Warsaw, Poland
[3] CNR-ICMATE - National Research Council – Institute of Condensed Matter Chemistry and Technologies for Energy, 6 de Marini, 16149, Genoa, Italy
[4] SPECODLEW Sp. z o.o., 3 Witolda Pileckiego Str., 32-050 Skawina, Poland
[5] UNIGE – DCCI – University of Genova – Department of Chemistry and Industrial Chemistry, 31 Dodecaneso, 16146, Genoa, Italy
**Corresponding address: rafal.nowak@iod.krakow.pl*

Key words: Nickel superalloy, wettability, reactivity, kaolin

1. Introduction

Second generation rhenium-bearing nickel-based superalloy CMSX-4 has extraordinary combination of high temperature mechanical properties, good phase stability and oxidation resistance [1]. Consequently, it is widely used for casting many components working at high temperature and in harsh environments, particularly for gas turbine applications. Therefore, information about high temperature interaction between molten CMSX-4 alloy and different refractory materials is of practical importance in a view of proper selection of mold materials and other foundry appliances [2].

This paper is focused on comparative investigation of high temperature wettability and reactivity of kaolin with molten pure Ni and CMSX-4 alloy.

2. Experimental

A sessile drop method coupled with contact heating procedure [3] was applied for real-time observation of high temperature behavior of pure Ni (99.95%) and CMSX-4 alloy (Co-9.6; Cr-6.5; Ta-6.5; W-6.4; Al-5.6; R-3; Ti-1; Mo-0.6; Hf-0.1; Ni– bal., wt%) on polished kaolin substrates produced by HT-HP synthesis. The tests were performed in atmosphere of flowing gas (Ar p~850 mbar) using experimental complex and testing procedure described in details in [3]. Figs. 1 and 2 show temperature profiles of the tests with isothermal heating at a temperature of 1500°C for 15 min.

The images of Ni/kaolin and CSMX4/kaolin couples were recorded with high-speed high-resolution CCD camera and then used for estimation of the contact angle values by ASTRA2 software [4].

After wettability tests, both the surface of the solidified drop/substrate couples as well as their cross-sections were subjected to detailed structural characterization by optical microscopy and scanning electron microscopy (SEM) coupled with energy dispersive X-ray spectroscopy (EDS).

3. Results

The changes of contact angle (θ) vs time for liquid Ni and CSMX4 alloy on kaolin substrates are shown in Fig. 1 and Fig. 2, respectively. Both Ni and CSMX4 alloy do not wet kaolin substrates (θ>90°) but they show a different wetting behavior.

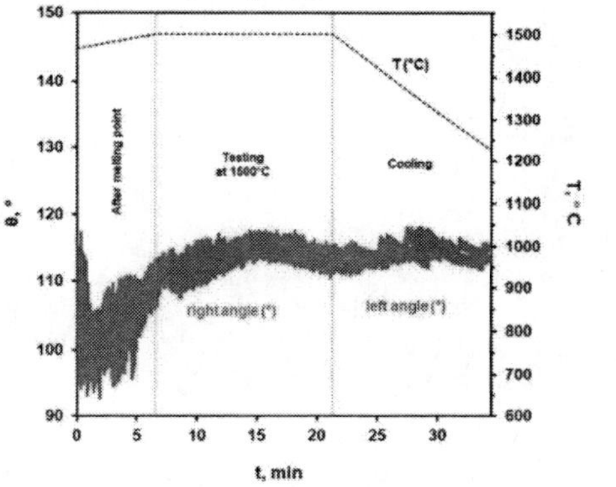

Fig. 1. ***Wetting kinetics of liquid Ni on kaolin substrate.***

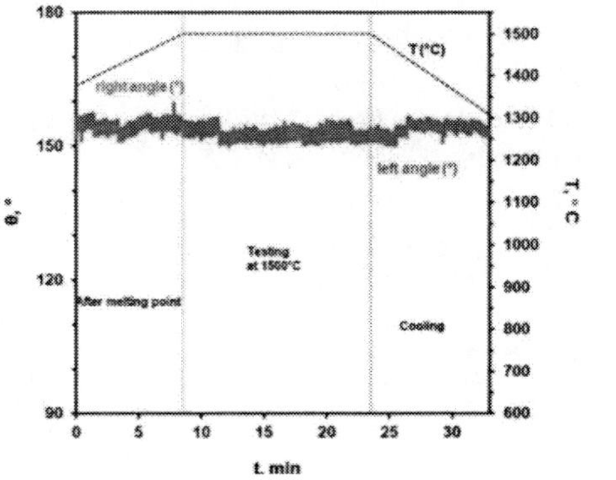

Fig. 2. ***Wetting kinetics of CSMX4 on kaolin substrate.***

Immediately after melting, pure Ni forms the contact angle of ~118° while during further heating to the test temperature, it decreases up to minimum value of ~95° and next increases to the final value of ~ 115°. On the contrary, after formation of spherical shape, corresponding to sample melting, the contact angle of the CSMX4 alloy is very high (153°) and it does not indicate significant change during further heating to and at the test temperature. SEM+EDS analysis of the Ni/kaolin couple showed a clean surface of Ni drop (Fig. 3a) and slight dissolution of kaolin substrate in liquid nickel at high temperature (Fig. 3b).

On the contrary, the surface of CSMX4 drop (Fig. 4a) is covered with regions of different colors. EDS analysis revealed that white color corresponds to chemical composition of the CSMX4 alloy while dark grey regions are rich in oxygen and Al thus suggesting alumina. Most probably, its formation affects wetting behavior of the CSMX4/kaolin couple caused from secondary oxidation of the drop during high temperature test. Moreover, the detailed structural characterization of the substrate surface showed that the area being under the in situ detached CSMX4 drop (Fig. 4b) is covered mainly with alumina (corresponding chemical composition in at.%: 38.8Al, 0.9Si and 60.3O). Occasionally, white regions rich in oxygen and hafnium were also observed hence suggesting reactively formed hafnia.

a)

b)

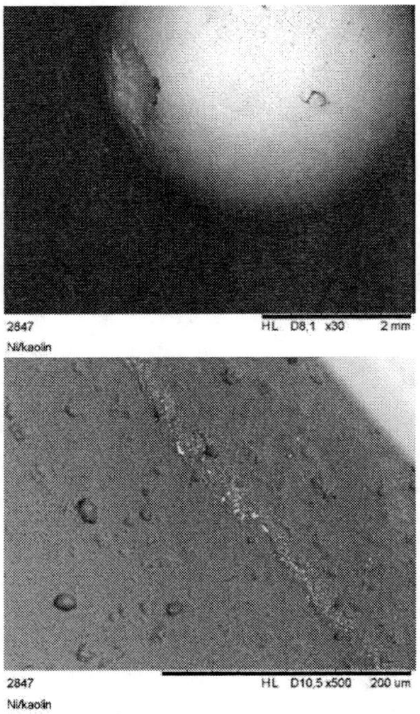

Fig. 3. Top-view SEM images of Ni/couple: a) Ni drop with shrinkage cavity; b) dissolution region of kaolin substrate after moving drop during test at 1500°C.

4. Conclusions

Under conditions of this study, both liquid Ni and CSMX4 alloy do not wet kaolin substrates. A lower contact angle value was observed for the Ni/kaolin couple, in which high temperature interaction is related with slight dissolution of kaolin in liquid Ni. Despite higher contact angle recorded for the CSMX4/kaolin couple, chemical reactions of kaolin with Al and Hf were evidenced by the formation of alumina and hafnia, respectively.

a)

b)

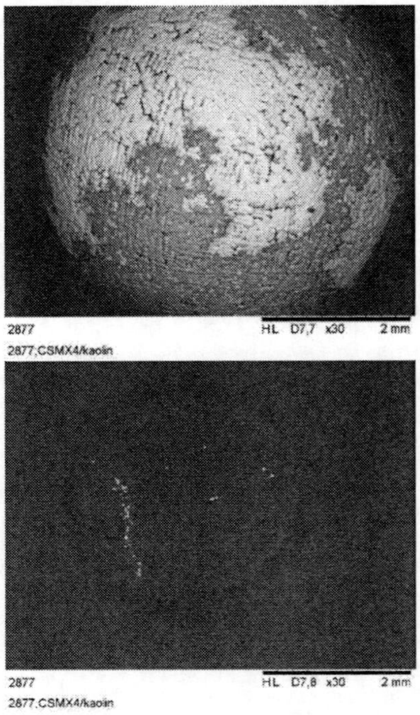

Fig. 4. Top-view SEM images of CSMX4/kaolin couple: a) CSMX4 drop; b) kaolin substrate - the place corresponding to the drop position during wetting test.

5. Acknowledgments

This work is part of the Project INSURFCAST (Innovative Surfaces for Superalloys Casting Processes) financed within the ERA-LEARN 2020 support action funded by EU-H2020, M-ERA.NET Joint Call 2016.

References

[1] Fullagar KPL, et al. (1996) Aero engine test experience with CMSX-4 alloy single crystal turbine blades, J. Eng. Gas Turbines Power 118(2), 380-388

[2] Sobczak N, Sobczak J, Asthana R, Purgert R, The mystery of molten metal, China Foundry (2010) 7, 425–437

[3] Sobczak N, Nowak R, Radziwill W, Budzioch J, Glenz A. (2008) Experimental complex for investigations of high temperature capillarity phenomena. Materials Science and Engineering 495(1-2) 43-49

[4] Liggieri L, Passerone A. (1989) An Automatic Technique for Measuring the Surface-Tension of Liquid-Metals, High Temperature Technology 7(2) 82–86

The 73rd World Foundry Congress
„Creative Foundry"
23 - 27 September 2018, Krakow, Poland

HIGH-TEMPERATURE INTERACTION BETEEN MOLTEN CONVENTIONAL GRAY CAST IRON AND Al_2O_3-ZrO_2 CERAMIC

Mihail Kolev [1]*, L. Drenchev [1], J.J. Sobczak [2], P. Turalska [2], A. Kudyba [2], G. Bruzda [2], N. Sobczak [2,3]

[1]Institute of Metal Science, Equipment and Technologies with Hydroaerodynamics Centre, Sofia 1574 67 Shipchenski prohod Str., Bulgaria
[2]Foundry Research Institute, 73 Zakopianska Str., 30-418 Krakow, Poland
[3] Institute of Precision Mechanics, 3 Duchnicka Str., 01-796 Warsaw, Poland
**Corresponding address: ljudmil.d@ims.bas.bg*

Keywords: MMC, conventional gray cast iron, Al_2O_3-ZrO_2, sessile drop method, wetting kinetics, reactivity

1. Introduction

The contemporary progress in material science and industry without development of advanced materials is inconceivable. The metal matrix composites (MMCs) are immensely desired for various structural and functional applications in every industrial branch. Understanding high-temperature wetting behavior in liquid metal/solid ceramic systems is critical for improving industrial liquid phase-assisted processes and the quality of the final product [1,2].

Therefore, the investigation of the wettability phenomena at elevated temperatures constitutes one of the most important scientific aspects of high-temperature liquid-assisted materials processing stimulated by the demands of modern metallurgy and foundry industry.

Lately, it has been reported that utility properties of iron-based alloys can be improved by introduction of a ceramic phase (e.g. [3-5]).

This study is focused on high temperature interaction between molten conventional gray cast iron and selected ceramic (Al_2O_3-ZrO_2) in order to clarify their wetting behavior and reactivity that affect the structure and properties of MMCs produced from cast iron and reinforced with Al_2O_3+ZrO_2 particles by casting processes (Fig. 1).

Fig. 1. Cross-section of conventional gray cast iron/(Al_2O_3+ZrO_2)$_p$ composite

2. Experimental

The materials used in this study were:

1) conventional gray cast iron containing in wt%: 4.1 C; 1 Si; 0.20 Mn; 0.03 S; 0.04 P, the rest – Fe;

2) Al_2O_3-ZrO_2 substrate containing in wt%: 50 Al_2O_3, 34-40 ZrO_2, 13 SiO_2, and ~1%NaO_2.

The substrate was cut from the Al_2O_3-ZrO_2 ceramic produced by the Institute of Metal Science, Equipment and Technologies with Hydro-aerodynamics Centre.

Wettability study was conducted by a sessile drop method combined with classical contact heating procedure using the experimental complex described in [6]. Testing conditions were as follows: flowing gas atmosphere (Ar, 850–900 hPa), 1450°C/15 min. During wettability test, the drop/substrate images were recorded with a high-speed high-resolution CCD camera. Solidified and cross-sectioned sessile drop couple was investigated by SEM+EDS analysis.

3. Results

The results of wettability test (Figs. 2,3) show that under testing conditions of this study, the selected cast iron alloy does not wet the Al_2O_3-ZrO_2 substrate (the contact angle Θ>90°).

Immediately after melting cast iron sample, Θ~145° and it decreases with time during heating. The first contact angle formed at the test temperature (1450°C, t=0s) is Θ_0~130° while the final value after isothermal heating for 15 min is Θ_f~135°.

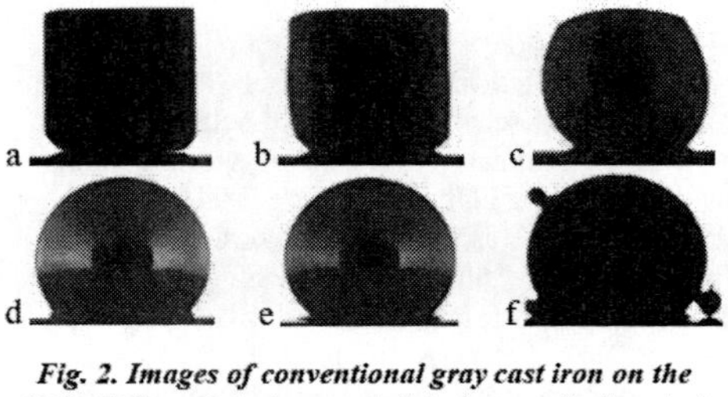

Fig. 2. Images of conventional gray cast iron on the Al_2O_3-ZrO_2 substrate recorded during sessile drop test:
a) the beginning of heating, 21°C;
b) the beginning of melting of cast iron sample, 1130°C;
c) the formation of the drop, 1145°C;
d) the start of wettability study at 1450°C (t=0s);
e) the end of wettability study at 1450°C (t-15 min);
f) the end of the test, 400°C.

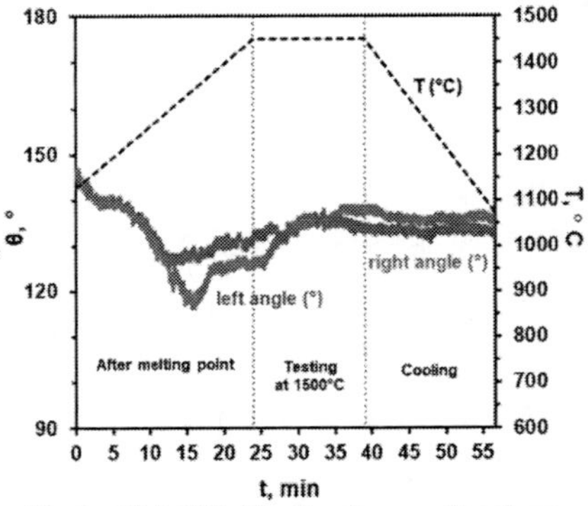

Fig. 3. Wettability kinetics of conventional gray cast iron/(Al_2O_3-ZrO_2) couple

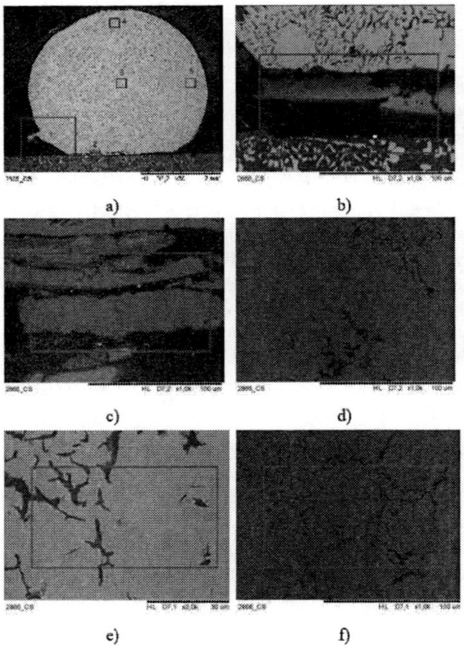

Fig. 4. SEM microstructures of cross-sectioned conventional gray cast iron/(Al_2O_3-ZrO_2) couple after sessile drop test (labelled points correspond to EDS measurements)

It should be noted that after the solidification, the drop was found separated from the substrate (Figs. 4a-c) as experimental evidence of weak bonding established during high temperature test between the cast iron drop and the Al_2O_3-ZrO_2 substrate. Moreover, the mismatch of CTEs contributes to debonding of two materials during cooling. The most critical moment is solidification of cast iron drop accompanied with increase of its volume. This effcet was evidenced by the phenomenon of squeezing smaller droplets from the "mother" drop in the final stage of its cooling (Fig. 2f).

Table 1 and Fig. 4 summarize the most representative results of SEM+EDS structural analysis made in different areas of the cross-sectioned sessile drop couple (in marked points of Fig. 4).

Table 1. Results of SEM+EDS analysis (at.%)

Point	Fe	C	O	Co	Al	Si	Zr	Cr	Na	P	S
1	31.4	48.8	10.4	1.9	3.3	2.4	0.3	0.6	0.3	0.2	-
2	47.7	29.8	8.5	2.6	3.9	4.2	0.6	1.7	0.6	0.3	0.1
3	63.5	25.8	-	4.4	1.2	3.9	-	1.1	-	-	0.1
4	13.7	83.5	0.4	0.8	-	0.8	0.1	0.4	-	-	0.3
5	17.3	80.0	-	1.2	-	1.1	0.1	0.2	-	-	0.1

4. Summary

Under testing conditions of this study, the selected conventional gray cast iron alloy does not wet the Al_2O_3+ZrO_2 ceramic and after melting and holding at T=1450°C for 15 min, it forms a high contact angle of ~135°. Structural characterizations of interfaces by SEM+EDS analysis show negligible reactivity and weak bonding established between two materials during wettability test. These findings suggest that for MMC synthesis by liquid-assisted processes, the improvement of wetting properties of selected alloy/ceramic couple is needed.

Acknowledgments

The authors are grateful for financial support of National Science Centre of Poland within OPUS program (UMO-2015/17/B/ST8/03391).

This work was partially supported by the European Regional Development Fund within the OP "Science and Education for Smart Growth 2014 - 2020", Project CoE "National center for/of mechatronics and clean technologies" (No BG05M2OP001-1.001-0008-C01).

MK wishes to thank for research fellowship from the European Virtual Institute on Knowledge-based Multifunctional Materials AISBL (KMM-VIN).

References

[1] Sobczak N., Singh M., Asthana R. (2005) High-temperature wettability measurements in metal/ceramic systems – Some methodological issues, *Current Opinion in Solid State Materials Science* 9 (4–5): 241–253.

[2] Sobczak N., Sobczak J., Asthana R., Purgert R. (2010) The mystery of molten metal, *China Foundry* 7 (4): 425–437.

[3] Hathaway R.M., Rohatgi P., Sobczak N., Sobczak J. (1998) Ferrous composites: a review. In: N. Eustathopoulos, N. Sobczak (Eds.). Proceedings of 2nd Int. Conf. High Temperature Capillarity, 29 June–2 July 1997, Cracow, Poland, Foundry Research Institute, pp. 267–276.

[4] Dulska N. Studnicki A., Szajnar J. (2017) Reinforcing cast iron with composite insert, *Archives of Metallurgy and Metals* 62 (1): 355–357.

[5] Kan W.H., C. Albino, D. Dias-da-Costa, et al. (2018) Microstructure characterisation and mechanical properties of a functionally-graded NbC/high chromium white cast iron composite, *Materials Characterization* 136: 196–205.

[6] Sobczak N., Nowak R., Radziwill W., Budzioch J., Glenz A. (2008) Experimental complex for investigations of high temperature capillarity phenomena. *Materials Science and Engineering* A 495 (1–2): 43–49.

The 73rd World Foundry Congress
„Creative Foundry"
23 - 27 September 2018, Krakow, Poland

THERMAL CONDUCTIVITY OF SELECTED VERMICULAR CAST IRON ALLOYS

Marta Homa[1], Natalia Sobczak[1,2], Patrycja Turalska[1], Karina Kosiba[3]*

[1] Foundry Research Institute, 73 Zakopiańska Str., 30-418 Krakow, Poland
[2]Institute of Precision Mechanics, 3 Duchnicka Str., 01-796 Warsaw, Poland
[3]Cracow University and Technology, 24 Warszawska Str., 31-155 Krakow, Poland
**Corresponding author: marta.homa@iod.krakow.pl*

Key words: vermicular cast iron, alloying additions, laser flash method, thermal conductivity

1. Introduction

The thermal conductivity of cast iron alloys has a great importance for many practical applications such as brake discs, engine blocks and cylinder heads, since the ability to transfer away heat enhances the resistance to such factors as thermal fatigue and distortion [1]. This thermal parameter is strongly affected by metallic matrix [1,2,3], graphite fraction [1-3], its morphology [2,3], eutectic grains [2], and alloy chemical composition [3].
The introduction of alloying additions may cause changes in scattering of conduction electrons and phonons, which are responsible for heat transfer [1]. For example, alloying cast iron with Si results in strong depressing of thermal conductivity due to its dissolution in the ferritic grains as well as in the ferritic lamellae and pearlite [1]. On the contrary, the addition of Mo increases thermal conductivity of cast iron because it promotes the refinement of graphite and pearlite together with stabilization of ferrite in the matrix [1].

The aim of this work was to experimentally determine thermal conductivity of selected vermicular cast iron (VG) alloys containing small additions of Mo, Cu, V, Sn and Sb.

2. Experimental

2.1 Materials

The samples of selected alloys were prepared in 3 steps process [4]: 1) melting of charge; 2) vermiculitizing at a temperature range of 1440–1450°C with magnesium containing (Mg <6%) master alloy; 3) inoculation with inoculant.
Chemical compositions of the master alloy (S1) and the inoculant (S2) are given in Table 1 while that of produced alloys in as-cast state are presented in Table 2. The contents of the elements in as-cast state were determined by spectrometry analysis.

Table 1. The chemical composition of magnesium master alloy and inoculant (wt %)

Alloy	Si	Mg	Al	Ca	Ba	RE
S1	44.0–48.0	5.6–6.0	Max. 1.0	1.8–2.3	–	5.5–6.5
S2	73.0–78.0	–	0.75–1.25	0.75–1.25	0.75–1.25	–

The sample 0 is a reference VG cast iron used for investigations of the effect of Mo, Cu, V, Sn and Sb additions on thermal conductivity of selected VG cast irons (samples 1-4).

Table 2. Chemical composition of cast alloys (wt %)

Alloy	C	Si	Mn	P	S	Mg	Mo	Cu	V	Sn	Sb
0	3.70	2.30	0.44	0.054	0.015	0.017	-	-	-	-	-
1	3.85	2.07	0.38	0.050	0.020	0.020	0.25	0.83	0.12	0.05	0.032
2	3.60	2.35	0.38	0.045	0.015	0.010	0.25	0.33	0.09	0.045	0.031
3	3.65	2.50	0.55	0.045	0.020	0.025	0.5	0.75	0.28	0.10	0.05
4	3.80	2.45	0.45	0.045	0.020	0.020	0.27	0.40	0.25	0.05	0.010

2.2 Methods

The Laser Flash Analysis (LFA) method was used for measurements of thermal diffusivity. Its main principle is based on the fact that a transient energy pulse, generated by a laser, is focused on the front surface of a sample, and the subsequent temperature increase is measured by IR detector on the far side of the specimen. Thus at adiabatic conditions, thermal diffusivity a, can be determined from the following equation:

$$a = 0.138\frac{l^2}{t_{0.5}} \quad (1)$$

where: l is the sample thickness and $t_{0.5}$ is the time needed for a 50% temperature increase.
Thermal conductivity (TC) was calculated with the consideration of the radiation losses based on the non-linear regression and the Cape-Lehman model [5] using the following equation:

$$TC(T) = a(T)\,c_p(T)\,\frac{L}{L_0}(T) \quad (2)$$

where: $TC(T)$ - thermal conductivity vs. temperature, $a(T)$ - thermal diffusivity vs. temperature, $C_p(T)$ - specific heat vs. temperature, $L/L_0(T)$ - thermal expansion vs. temperature.
In order to determine $TC(T)$, other materials thermophysical properties such as $a(T)$, $C_p(T)$ and $L/L_0(T)$ were measured using the following devices:
1) LFA device Netzsch 427;
2) Differential Scanning Calorimeter (DSC) Netzsch 404C/3/G Pegasus;
3) high-temperature dilatometer (DIL) Netzsch 402C/4/G.

3. Results

Figs. 1-4 summarize the results of thermophysical properties measurements of selected alloys, i.e. $a(T)$ (Fig. 1), $C_p(T)$ (Fig. 2), $L/L_0(T)$ (Fig. 3) and $TC(T)$ (Fig. 4).

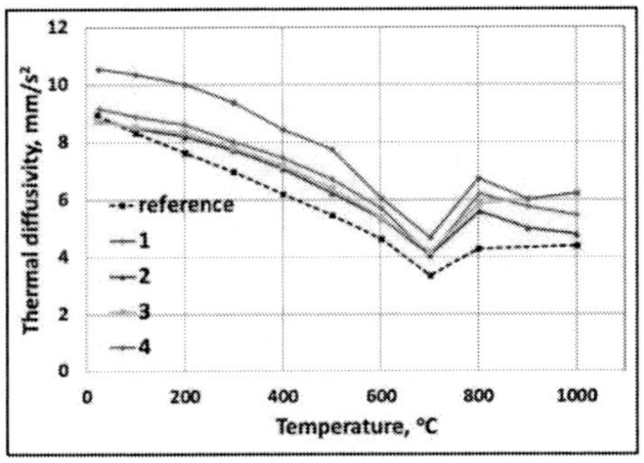

Fig. 1. Thermal diffusivity vs. temperature

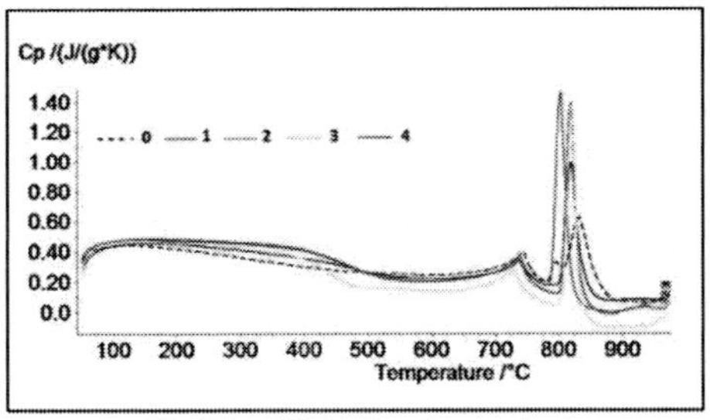

Fig. 2 Specific heat vs. temperature

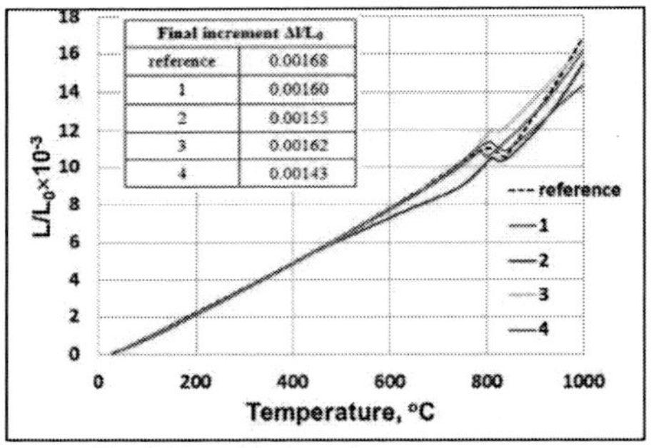

Fig. 3 Thermal expansion vs. temperature

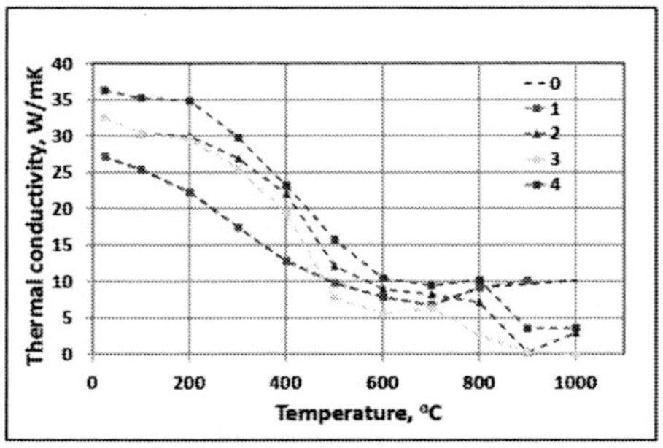

Fig. 4. Thermal conductivity vs. temperature

4. Conclusions

1. Under conditions of this study (atmosphere, temperature range of RT-1000°C), all studied VG cast iron alloys show similar temperature dependences of all examined thermophysical properties.
2. Introduction of Mo, Cu, V, Sn and Sb into selected vermicular cast iron alloys, in the amounts used in this study, does not significantly alter the measured values of specific heat C_p and liner thermal expansion L/L_0, while it slightly increases those of thermal diffusivity a and thermal conductivity TC.
3. For all examined alloys, the values of a and TC decrease with increase in temperature up to about 700°C while at temperature range of 700-800°C, they show noticeable change of the course of the $a(T)$, $C_p(T)$, $L/L_0(T)$ and $TC(T)$ curves related with the *pearlite→austenite* transformation.

Acknowledgments
The authors are grateful for financial support of National Science Centre of Poland within OPUS program (UMO-2015/17/B/ST8/03391).

Reference

[1] D. Holmgren, Review of thermal conductivity of cast iron. 2005. International Journal of Cast Metals Research 18(6) p 331-345
[2] M. Górny, J. Lelito, M. Kawalec. 2015. Thermophysical properties of thin walled compacted graphite cast iron castings. TMS 15-19 March. Springer. p 331-338
[3] J.K. Chen, S.F. Chen. 2011. On thermal conductivity of an in-situ metal matrix composite - cast iron. Materials Science "Composite Materials, Metal, Ceramic and Polymeric Composites for Various Uses. Intech. p 211-224
[4] M. Homa, N. Sobczak, P. Turalska, G. Bruzda, M. Bacior, M. Warmuzek, A. Polkowska. 2017. Thermophysical properties of selected vermicular graphite cast iron alloy, Transaction of Foundry Research Institute, LVII(4) p 345-350
[5] J.A. Cape, G.W. Lehman. 1963. Temperature and finite pulse time effect in the flash method for measuring thermal conductivity, Journal of Applied Physics 34(7) p 1909-1913

The 73rd World Foundry Congress
„Creative Foundry"
23 - 27 September 2018, Krakow, Poland

SIMULTANEOUS TG-DTA THERMAL ANALYSIS OF Si-30B ALLOY

Marta Homa [1], Natalia Sobczak [1,2], Krzysztof Hodor[3], Senol Gezgin[3]*

[1] Foundry Research Institute, 73 Zakopiańska Str., 30-418 Krakow, Poland
[2]Institute of Precision Mechanics, 3 Duchnicka Str., 01-796 Warsaw, Poland
[3]Netzsch GmbH, Germany
**Corresponding address: marta.homa@iod.krakow.pl*

Key words: AMADEUS project, Si-B alloy, graphite, differential thermal analysis, thermogravimetric analysis

1. Introduction

Recently the Si-B alloys have proposed as particularly promising candidates for phase change materials (PCMs) used for ultrahigh temperature latent heat thermal energy storage applications [1] because of their extremely high latent heat (50.02 KJ/mol [2]) and high melting points (e.g. for Si - 1414°C and for B - 2075°C [3]), if compared with conventional PCMs such as salt-based materials. Therefore, information on thermal behaviour of the Si-B alloys is of a great practical importance in view of a lack of such data for Si-B alloys in open literature.

Container-assisted methods are widely used for experimental investigation of factors affecting the *solid-to-liquid* phase change transformations and latent heat measurements. However, their application for high temperature testing, particularly liquid metals, requires a proper selection of a container material that should be non-wettable and non-reactive in contact with liquid metal tested. Following literature data, liquid silicon shows high reactivity with many commercial refractories such as silica, alumina, magnesia [4]; SiC, Si_3N_4, AlN, h-BN [5] and graphite [6,7]. On the other hand, graphite is commonly used as crucible material for melting silicon up to 1450°C taking into account its following advantages: good machinability, high-temperature stability, very low solubility of carbon in solid Si (high solubility of carbon in liquid Si does not affect final purity of solidified Si) and "self-crucible" effect due to the reaction between liquid Si and carbon leading to the formation of continuous interfacial layer of SiC.

In this work, high temperature behavior of commercial graphite container in contact with selected Si-B alloy was examined in a wide temperature rage (25-1800°C) using the combined TG-DTA method of thermal analysis.

2. Experimental

The Si-B alloy with a nominal chemical composition of Si-30B (wt%) corresponding to hypereutectic alloy [3] was used in this study. It was produced from low purity batch materials (Si: 99.99%; B: 99.9%) by the induction melting in a graphite crucible. Prior the DTA-TG measurements, the alloy sample was ultrasonically cleaned with isopropanol and dried.

The following methods were applied for investigation of high temperature behavior of the selected alloy:

1. Simultaneous TG-DTA thermal analysis that involves two methods in one test, i.e. the differential thermal analysis (DTA) and thermogravimetric analysis (TG). The temperature difference and mass gain measurements were performed by means of the thermal microbalance Netzsch STA 449 F3 Jupiter® device equipped with Tungsten furnace and W-Re sample holder.
2. The structural observations by the scanning electron microscopy (SEM) coupled with energy dispersive X-ray analysis (EDS) using TM3000 HITACHI microscope.

The TG-DTA measurements were performed using commercially available graphite container produced by Netzsch company. The TG-DTA test was carried out in a flowing gas atmosphere (He 99.9992%, 100 ml/min) using the following heating/cooling procedure: (1) preheating at 25°C/10 min in a chamber; (2) heating with a rate of 20°C/min from RT to 1820°C; (3) cooling with a rate of 20°C/min from 1820°C to 300°C; (3) slow cooling with a rate of 10°C/min from 300°C to RT.

3. Results

The DTA and TG curves recorded during heating of the Si-30B alloy to 1800°C are shown in Figure 1.

The analysis of the TG curve indicates that during heating of the alloy to the test temperature, its mass is constantly changing. Up to the temperature of 900°C, the mass increases and its total mass gain is +0.3 mg. Above T=900°C, a breakdown is visible on the TG curve and the alloy mass begins to decrease, the loss being the highest above the temperature of 1600°C. At T= 900-1800°C, the total change in sample weight is $\Delta m_2 = -0.33$ mg.

The DTA curve has three peaks: 1378.9°C, 1389.2°C and 1532.1°C. In accordance with the Si-B phase diagram [2], the peak recorded at 1378.9°C corresponds to the melting point of eutectic ($Si+SiB_6$). The shape and course of the other two peaks indicate that a reaction has taken place between the container material and the alloy sample and/or the sensor material that makes difficult a proper interpretation of the DTA results.

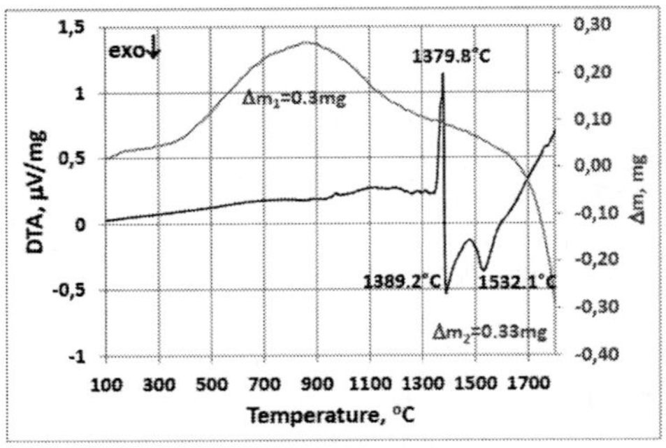

Fig. 1. DTA-TG curves of the Si-30B alloy

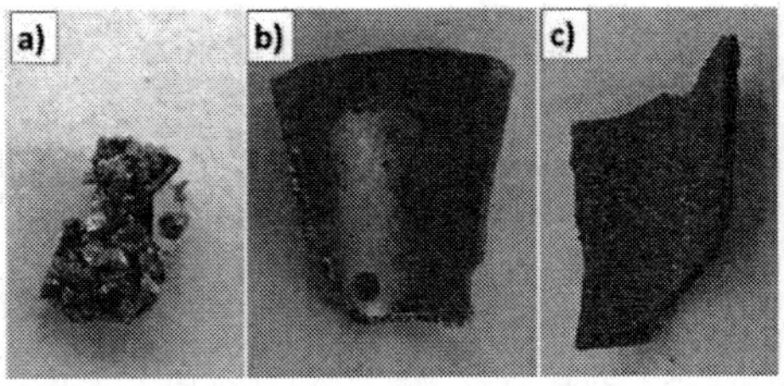

Fig. 2. Photos of the materials after DTA-TG test: a) Si-30B alloy, b,c) fragments of broken graphite container

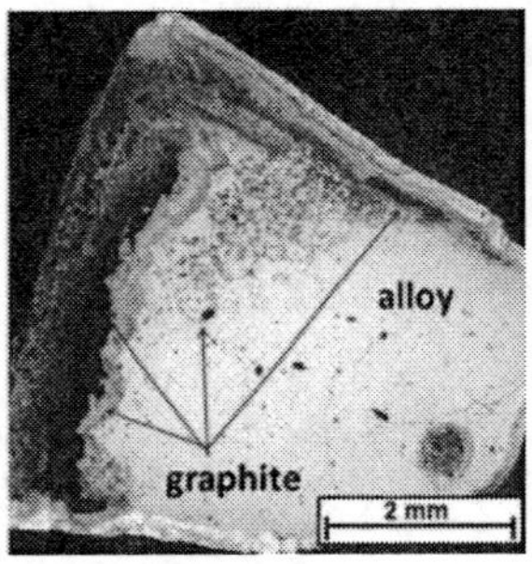

Fig. 3. SEM image of the surface of graphite container after TG-DTA test

After the TG-DTA test, it was found that the container could not be separated from the sensor of the device while the attempt to pull it out resulted in container breakage. Visual observations of the fragments removed from the device (Figs. 2a-c) showed that the fragment corresponding to the Si-30B sample was only partially melted (Fig. 2a). However, fracture surface of some parts of the graphite container had a metallic sheen (Fig. 2c, Fig. 3) due to its infiltration with liquid constituent of the alloy as it was confirmed also by SEM+EDS analysis of cross-sectioned parts of the container. These observations evidence a very good wetting of graphite with molten fraction of the alloy. Moreover, SEM+EDS analysis revealed the formation of carbides SiC and B_4C. It represents experimental proof that during TG-DTA test, graphite reacts with both liquid silicon and boron dissolved in liquid constituent of the alloy to form SiC and B_4C, respectively. Furthermore, chemical reactions taken place during TG-DTA test contribute to the generation of micro-cracks in the container walls and its final failure because of significant mismatch between coefficients of thermal expansion (CTE) of Si, graphite and reactively formed products.

3. Summary

High temperature testing of Si-B alloys by container-assisted methods using graphite containers is accompanied with methodological problems due to high reactivity of these alloys with graphite, especially under conditions when a large amount of liquid fraction is formed (T>1450°C). It might results in unreliable measurements of thermophysical properties of the tested alloys in liquid and semi-solid states caused from chemical reactions of carbon with both Si and B and subsequent change in the alloy chemical composition. Moreover, during high temperature testing by container-assisted methods, micro-cracks generated in the walls of graphite container due to significant CTE mismatch of Si, graphite and reactively formed SiC and B_4C may cause the damage of experimental device since molten fraction of the alloy wets and easily penetrates graphite thus contributing to the degradation of the container.

Acknowledgments

The project AMADEUS has received funds from the European Union's Horizon2020 research and innovation program, FET-OPEN action, under Grant agreement 737054. The authors acknowledge the Department of Materials Science and Engineering of Norwegian University of Science and Technology for manufacture of the Si-30B alloy.

References

1. A. Datas, A.B. Cristobal, C. del Cañizo, et al. AMADEUS: Next Generation Materials and Solid State Devices for Ultra High Temperature Energy Storage and Conversion, Energy Storage Materials, 2017
2. G.V. Belov, V.S. Iorish, V.S. Yungman. 1999. IVTANTHERMO - database on thermodynamic properties and related software. Calphad 23(2), p 173-180
3. A. Mostafa, M. Medraj. 2017. Review binary phase diagrams and thermodynamic properties of silicon and essential doping elements (Al, As, B, Bi, Ga, In, N, P, Sb and Tl). Materials 10 (676) p 2-49
4. Z. Yuan, W.I. Huang, K. Mukai. 2004. Wettability and reactivity of molten silicon with various substrates. Appl. Phys. A-Mater. 78, p 617–622
5. R. Israel, R. Voytovych, P. Protsenko, B. Drevet, D. Camel, N. Eustathopoulos. 2010. Capillary interactions between molten silicon and porous graphite. J. Mater. Sci. 45, p 2210–2217
6. B. Drevet. N. Eustathopoulos. 2012. Wetting of ceramics by molten silicon and silicon alloys: A Review. J. Mater. Sci. 47, p 8247–8260
7. A. Ciftaja, T. Abelengh, M. Tangstad. Wetting properties of molten silicon with graphite materials, Metall. Mat. Trans. A 41A (2010), p 2010-3183

The 73rd World Foundry Congress
„Creative Foundry"
23 - 27 September 2018, Krakow, Poland

WETTABILITY OF TWO-DIMENSIONAL MoS_2 LAYER BY LIQUID TIN

Marta Homa [1], Natalia Sobczak [1,2], Rafał Nowak [1], Grzegorz Bruzda [1], Patrycja Turalska[1], Sudipta Seal [3]*

[1] Foundry Research Institute, 73 Zakopiańska Str., 30-418 Krakow, Poland
[2] Institute of Precision Mechanics, 3 Duchnicka Str., 01-796 Warsaw, Poland
[3] University of Central Florida, P.O. Box 162455 Orlando, FL 32816, USA
**Corresponding address: marta.homa@iod.krakow.pl*

Key words: Wettability, two dimensional MoS_2, liquid Sn, Raman spectroscopy

1. Introduction

Molybdenum disulfide (MoS_2), which contains weakly interacting layers in its crystal structure, arousing a lot of interest as potential substitute of graphene [1]. Advances in recent nanotechnologies permit the fabrication of MoS_2, such as single layer MoS_2 sheets called [2], fullerene-like nanoparticles [2], and nanotubes [2]. Such nanostructures can be used as ultrasensitive electronic devices [1,2] or for the production of innovative composite materials fabricated by liquid-phase methods. In order to suit these purposes, intrinsic surface properties, such wettability and reactivity, are extremely important from both fundamental and application points of view. However, there is no information on wetting behavior of high-temperature liquids (e.g. molten metals) on two dimensional MoS_2 materials while all available literature data refer to water or aqueous solutions [2,3].

This paper is dedicated to experimental study of high temperature wetting behaviour of molten tin on MoS_2-coated silicon substrate.

2. Experimental

The materials were pure Sn (99.95%) and two-dimensional MoS_2 layer deposited on Si⟨100⟩ wafer (2D-MoS_2/Si⟨100⟩).

Real-time melting, wetting, spreading and solidification behaviour of the Sn/2D-MoS_2/Si⟨100⟩ couple was examined by the sessile drop method coupled with non-contact heating and capillary purification procedures [4,5]. During the test, the images of the couple were recorded using Microtron 1310 high resolution digital camera with a rate of 50 frames per second. The values of the contact angles were calculated based on the drop image analysis using the ASTRA2 software [6,7].

The structure and chemistry of the solidified couple were examined using: a) scanning electron microscopy (SEM) coupled with energy dispersive X-ray analysis (EDS), b) versitale automated laser Raman spectrometer (LRS).

3. Results

The obtained representative Raman spectrum of MoS_2/Si⟨100⟩ substrate before wettability test showed a presence of two peaks "finger print" (Fig. 1). The first peak A_{1g} at 407 cm^{-1} is an intralayer mode involving the motion of S atoms along the c axis. The second peak E_{2g} at 383 cm^{-1} is an intralayer vibrational mode involving motion of Mo+S atoms in the basal plane [8].

The images of Sn drop recorded during most representative moments of wettability test are presented in Fig. 2.

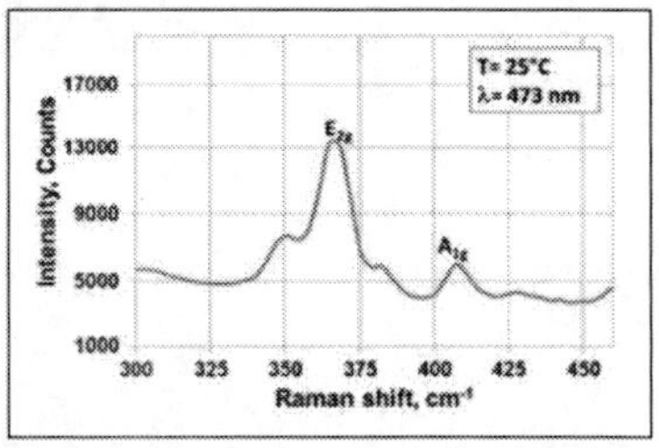

Fig. 1. Representative Raman scattering spectrum of MoS_2/Si⟨100⟩ substrate before wettability test.

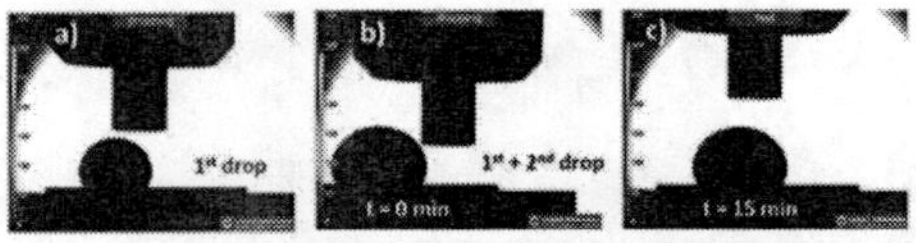

Fig. 2. Images of the most representative moments of the test: a) the 1st drop; b) the 1st +2nd drop, start the test, t=0 min; c) drop under isothermal condition at T = 300°C, t = 15 min, end of the test)

Both the 1st deposited Sn drop and the 2nd squeezed drop that was connected with the 1st one (1st+2nd drop) did not wet the substrate (θ>90°) forming the similar average contact angles of 118° and 121°, respectively. Further isothermal heating for 15 minutes did not affect the contact angle value, showing a final average value of θ_f = 121°.

The image of the solidified Sn/2D-MoS_2/Si⟨100⟩ couple is presented on Fig. 3. Visual observations showed that the Sn drop has almost spherical shape while the drop surface is not smooth. Moreover, the substrate surface is covered by many dark points. The detailed SEM characterization of the Sn drop surface showed the presence of a dendrites-likes structure thus suggesting the alloying Sn with substrate constituents (Fig. 4).

Fig. 3. Photo of solidified Sn/2D-MoS$_2$/Si⟨100⟩ couple

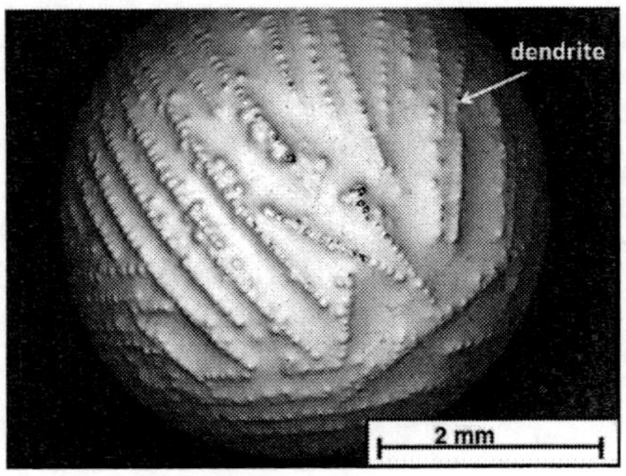

Fig. 4. Top-view SEM image of the solidified Sn drop surface

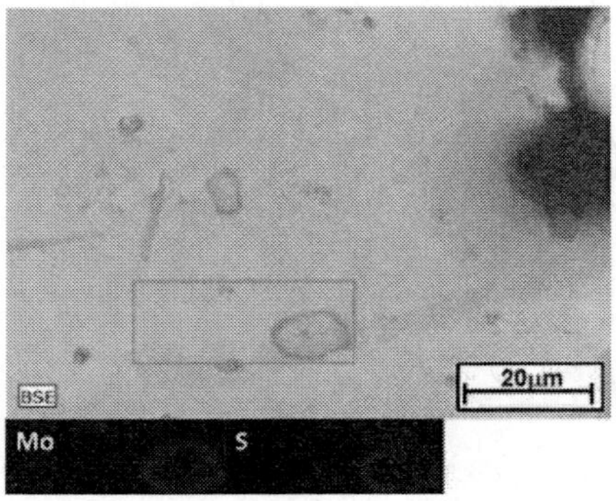

Fig. 5. SEM image of the grey spot with the distribution of Mo and Si in marked area

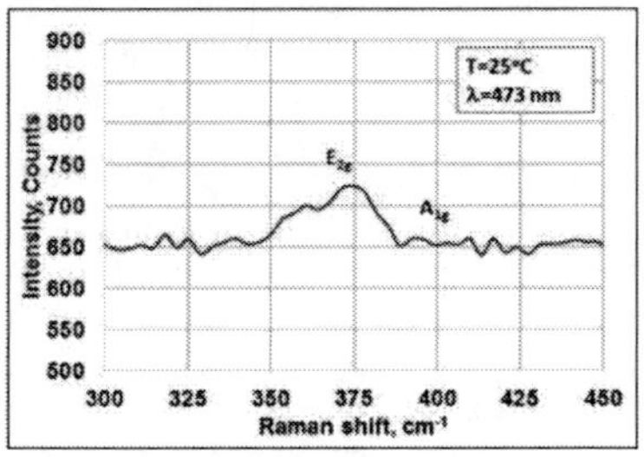

Fig. 7. Representative Raman scattering spectrum of 2D-MoS$_2$/Si⟨100⟩ substrate after wettability test

Under higher magnification of ×1000, many grey spots on the Sn drop surface are well distinguished (Fig. 5). Chemical analysis of these spots revealed the presence of Mo and S. The Raman spectrum taken from the substrate in the vicinity of the Sn drop exhibited a decreased intensity of E_{2g} peak and disappearance of A_{1g} peak (Fig. 7) thus suggesting disappearance of 2D-MoS$_2$ layer.

4. Summary

No-wetting behaviour has been observed for Sn/2D-MoS$_2$/Si⟨100⟩ couple during wetting experiments under conditions used in this study. Nevertheless, the degradation of MoS$_2$ due to its interaction with molten Sn drop took place as evidenced by local EDS analysis and Raman spectroscopy. These findings should be taken into consideration in the design of innovative composite materials and new concept electronic devices using 2D-MoS$_2$ types of materials for liquid-assisted processes.

Acknowledgments

Financial support from the Foundry Research Institute (project No. 7008/00) is acknowledged.

References

[1] A.P.S. Gaur, S. Sahoo, M. Ahmadi, S.P. Dash, M. J-F Guinel, R.S. Katiyar. Surface Energy Engineering for Tunable Wettability through Controlled Synthesis of MoS$_2$. Nano Letters. (2014) 13 14(8), 4314-21

[2] B. Luan, R. Zhou. Wettability and friction of water on a MoS$_2$ sheet. Applied Physics Letters. (2016) 108 131601, doi: 10.1063/1.4944840

[3] A. Kozbial, X. Gonga, H. Liuband, L. Lia. Understanding the intrinsic water wettability of molybdenum disulfide (MoS$_2$). Langmuir. (2015) 31 (30), 8429-8435

[4] N. Sobczak, J. Sobczak, R. Asthana & R. Purgert. (2010). The mystery of molten metal. China Foundry. 7/4, 425-437.

[5] N. Sobczak, R. Nowak, W. Radziwill, J. Budzioch, A. Glenz. (2008). Experimental complex for investigations of high temperature capillarity phenomena. Materials Science Engineering. A495, 43-49.

[6] L. Liggieri & A. Passerone. (1989). An automatic technique for measuring the surface tension of liquid metals. High Temperature Technologies. 7(2), 82–86.

[7] ASTRA Reference Book. (2007). IENRI. Report.

[8] K-G. Zhou, F. Withers,Y. Cao, S. Hu, G. Yu, C. Casiraghi, Raman Modes of MoS$_2$ Used as Fingerprint of van der Waals Interactions in 2-D Crystal-Based Heterostructures. (2014) ASC Nano 8 (10), 9914-9924

The 73rd World Foundry Congress
„Creative Foundry"
23 - 27 September 2018, Krakow, Poland

THE INFLUENCE OF ALLOYING ADDITIONS ON THE INTERACTION BETWEEN RE-MELTED VERMICULAR GRAPHITE CAST IRON ALLOYS AND Al_2O_3 SUBSTRATE

Magdalena Bacior[1], Natalia Sobczak[2,3], Patrycja Turalska[2], Marta Homa[2], Artur Kudyba[2], Grzegorz Bruzda[2*], Adelajda Polkowska[2], Małgorzata Warmuzek[2], Andrzej Pytel[2]

[1] Department of Physics, University of Agriculture in Krakow, Al. Mickiewicza 21, 33-332 Krakow, Poland
[2] Foundry Research Institute, 73 Zakopianska Str., 30-418 Krakow, Poland
[3] Institute of Precision Mechanics, 3 Duchnicka Str., 01-796 Warsaw, Poland

**Corresponding address: grzegorz.bruzda@iod.krakow.pl*

Key words: vermicular graphite cast iron, Al_2O_3, reactivity, sessile drop method, cast iron microstructure

1. Introduction

Vermicular graphite (VG) cast iron has many advantageous mechanical and physical properties, which pay the attention of both scientists and engineers. Compared to grey cast iron, VG cast iron has almost double tensile strength and fatigues strength, 38% higher elastic modulus and 8% smaller thermal conductivity than in grey cast iron [1]. First industrial applications of VG cast iron started in the 1960s and were related with manufacturing of components for trucks, tractors, passenger cars and marine diesel engines [2]. Recently the interest of this material was connected mainly with production of cylinder blocks for diesel engines or railway brake discs [3-5].

Information on wetting and reactivity of molten graphite cast iron alloys with oxides is of practical importance for the development of refractory linings of furnaces and vats used in cast iron production as well as for synthesis of ferrous matrix composites reinforced with oxide particles [7,8].

This study reports the results of high temperature studies of the effect of alloying additions (Cu, Mo, V, Sn and Sb) on the interaction between re-melted perlitic vermicular graphite cast iron alloys with alumina substrates, particularly on wetting behaviour and reactivity of selected alloy/substrate couples.

2. Experimental procedure

A sessile drop method was applied for investigation of high temperature wetting behaviour and reactivity of selected alloy/substrate couples [9,10].

For high temperature tests, three VG cast iron alloys were selected (Table 1), including reference sample 0 and two alloys (sample 1 and 2) containing the additions of Cu, Mo, V, Sn and Sb of a total value of about 1.7 wt% and different additions of Cu. The selected alloys belong to a family of hypereutectic, perlitic vermicular graphite cast iron with a saturation coefficient $Sc > 1$. Melting and casting trials were carried according with procedure described elsewhere [6]. For high temperature tests, the samples of cubic shape and 5x5x5 mm size were cut from the large castings. As a reference material the cast iron without alloying additions was used (sample 0 in Table 1).

Table 1. Chemical composition of VG cast iron before high-temperature measurements (wt%)

VG cast iron	*C*	*Si*	*Mn*	*P*	*S*	*Mg*	*Mo*	*Cu*	*V*	*Sn*	*Sb*
0	3.70	2.30	0.44	0.054	0.015	0.017	-	-	-	-	-
1	3.85	2.07	0.38	0.050	0.020	0.020	0.25	0.83	0.12	0.05	0.032
2	3.60	2.35	0.38	0.045	0.015	0.010	0.25	0.33	0.09	0.045	0.031

For the wettability measurements, polycrystalline Al_2O_3 substrates of 96 % purity (Ceramit, Poland), were used. The examined surface of the Al_2O_3 substrates was polished up to a roughness of about $R_a = 152$ nm.

Wettability tests of VG/alumina couples were performed using experimental complex for high-temperature studies of liquid metals and alloys at the Centre for High Temperature Studies of the Foundry Research Institute [10]. VG cast iron samples and Al_2O_3 were contact heated with a rate of 15°C/min up to experimental temperature of 1520°C in flowing gas (Ar, p=900 mbar).

After 15 min at the test temperature, the couples were cooled to room temperature with a rate of 20°C/min. The images of alloy/substrate couples were recorded using high-speed high-resolution CCD camera MC1310 with a rate of 10 fps (frames per second). Collected images were used to estimate the values of contact angle using the ASTRA2 software [11,12].

After wettability tests, the solidified couples were subjected to detailed structural characterization using optical microscopy (OM) and scanning electron microscopy (SEM) coupled with X-ray energy dispersive spectroscopy (EDS).

3. Results

The analysis of wettability tests evidenced that the all three alloys (re-melted VG cast iron samples) do not wet Al_2O_3 oxide. The average value of contact angle after 15 min at 1450 °C was θ =125° for the sample with higher Cu addition (sample 1) and θ = 130° for the sample 2. For the reference sample without alloying additions (sample 0), the contact angle was θ = 131°.

OM+SEM+EDS analysis of solidified and cross-sectioned drop/substrate couples evidenced that the microstructure of re-melted cast iron drops after high-temperature measurements differs significantly from the microstructure of the initial VG cast iron samples. The regular structure of cast iron with the vermicular graphite after re-melting shows structural heterogeneity in cross section. Both samples containing selected alloying additions before high-temperature measurements were characterized by pearlitic structure (estimated fraction of perlite was above 98%). In the sample with higher Cu addition, 60% of graphite precipitates had vermicular morphology, while the rest of them had spheroidal structure. In the second sample, 80% of graphite precipitates had a vermicular morphology, while the rest of them had spheroidal structure. After high-temperature measurements, it was found that in the solidified cast iron drops the graphite phase took the shape of flakes having a different dispersion degree. The structural characterization of the drop/substrate interfaces revealed in both samples investigated the interfacial layer of about 30 μm thick located at the drop-side interface while at the substrate-side interface, the reaction zone formed in the substrate due to interaction between Al_2O_3 and molten cast iron was well distinguished.

4. Conclusions

Under conditions of this study, the introduction of the additions of Cu, Mo, V, Sn and Sb of a total value of about 1.7 wt% has negligible effect on wetting properties of selected vermicular graphite cast iron in contact with alumina substrates.

Nevertheless, structural characterization of cross-sectioned alloy/substrate couples evidenced that the alloys show notable reactivity resulting in the change of structure and chemistry of interfaces.

Acknowledgment

The authors are grateful for financial support of National Science Centre of Poland within OPUS program (UMO-2015/17/B/ST8/03391, "High temperature investigations of the effect of alloying additions on thermophysical properties of liquid cast iron and its wettability and reactivity in contact with metal oxides").

References

[1] M.A. Suhaimi, K-H. Park, S. Sharif, D.-W. Kim, A.S. Mohruni. (2017). MATEC Web of Conferences 101, 03016, SINCEST

[2] E. Nechtelberger, H. Puhr, J.B. Nesselrode, A. Nakayasu. (1982). Cast iron with vermicular /compacted graphite – state of the art. Development, production, applications. Proc. 49th International Foundry Congress, Chicago. IL. USA, April, CIATF, Paper 1, 1-39

[3] S. Dawson. (2008). Compacted graphite cast iron – a material solution for modern diesel engine cylinder blocks and heads. Proc. Conf. 68th Word Foundry Cong., Chennai, India, February. The Institute of Indian Foundrymen, 93-99

[4] W.L. Guesser, P.V. Duran, W. Krause. (2004). Compacted graphite iron for diesel engine cylinder blocks. Proc. Conf. Congres Le diesel: aujourd'hui et demain, May, Lyon, France, Ecole Centrale Lyon, 1-11.

[5] R.W. Gregorutti, J.E. Gru. (2014). Mechanical properties of compacted graphite cast iron with different microstructures. International Journal of Cast Metals Research. 27(5), 275-281

[6] A. Pytel, A. Gazda. (2014). Evaluation of selected properties in austempered vermicular cast iron (AVCI). Transactions of Foundry Research Institute. LIV(4), 23-29

[7] R.M. Hathaway, P.K. Rohatgi, N. Sobczak, J. Sobczak, Ferrous composites: a review. (1998) In: Proc. 2nd Int. Conf. „High Temperature Capillarity", edited by N. Eustathopoulos and N. Sobczak, 1997, 29 June-2 July, Cracow, Poland, published by Foundry Research Institute, Poland, 1998, 267-276

[8] A. Dulska, A. Studnicki, J. Szajnar, Reinforcing cast iron with composite insert. Archives of Metallurgy and Metals, 2017, 62(1), 355-357

[9] N. Sobczak, M. Singh, R. Asthana. (2005). High-temperature wettability measurements in ceramic-metal systems – some methodological issues. Current Opinion in Solid State Materials Science. 9, 241-253

[10] N. Sobczak et al. (2008). Experimental complex for investigations of high temperature capillarity phenomena. Materials Science Engineering. 495(1-2), 43-49

[11] ASTRA Reference Book, (2007), IENI, Report, Oct.

[12] L. Liggieri, A. Passerone. (1989). An automatic technique for measuring the surface tension of liquid metals. High Temperature Technologies. 7(2), 80-86

THE EFFECT OF BORON CONTENT ON WETTING KINETICS IN Si-B alloy/h-BN SYSTEM

Wojciech Polkowski [1,*], Natalia Sobczak [1,2*], Grzegorz Bruzda[1], Rafał Nowak[1], Donatella Giuranno[1,3], Artur Kudyba[1], Adelajda Polkowska[1], Olga Shuleshova[4], Ivan Kaban[4]

[1] *Foundry Research Institute, Zakopiańska 73 Str., 30-418 Krakow, Poland*
[2] *Institute of Precision Mechanics, Duchnicka 3 Str., 01-796 Warsaw, Poland*
[3] *CNR - Institute of Condensed Matter Chemistry and Technologies for Energy, Via E. De Marini, 6 16149 Genova, Italy*
[4] *IFW Dresden, Institute for Complex Materials, Helmholtzstraße 20, 01-069 Dresden, Germany*

**Corresponding e-mail: grzegorz.bruzda@iod.krakow.pl*

Key words: AMADEUS Project, Si-B alloys, hexagonal boron nitride, wettability, interfaces

1. Introduction

Due to extraordinarily high melting points and latent heat values of silicon and boron, Si-B alloys have been assumed as excellent candidates to be used as phase changes materials (PCMs) for applications in latent heat thermal energy storage (LHTES) and conversion systems working at ultra-high temperatures (up to 2000°C). The main idea of every LHTES based systems is that the latent heat (i.e. that is absorbed during melting and released upon consecutive solidification) can be stored and converted into other energy form e.g. an electricity. The basic principle of AMADEUS Project [1], founded by European Commission within Horizon 2020 Programme, assumes that the utilization of latent heat of Si-B alloys will significantly overcome energy density limitations of *state-of-art* molten salt based LHTES systems. Nevertheless, in order to successfully accomplished the Project goals, some scientific and technological challenges need to be *a priori* faced and solved. Beside of a development of new ultra-high temperature devices for a heat-electricity conversion, a proper ceramic materials able to withstand a long-term heating, holding and cooling in contact with molten Si-B alloys, need to be selected. Thus, an implementation of these materials to real devices must be preceded by careful examinations of their high temperature interaction with refractories which can be used for building PCMs containers.

Firstly, we have assumed that an achievement of long life-times and high reliability of the PCM container requires its inertness toward contacting Si-B alloys under predicted operating conditions. In other words, a proper selection of refractories should be based on two main criteria: (i) an absence of wetting and (ii) a negligible reactivity in contact with Si-B melts.

However, due to pioneering nature of the research, there is a prominent lack of both theoretical and experimental data reported on interaction between molten Si-B alloys and ceramics at any temperature. On the other hand, high temperature capillarity phenomena (i.e. wetting, spreading and infiltration) have been so far widely examined for Si/ceramic systems in terms of a fabrication of photovoltaic grade silicon [2,3]. Although, all reported works deal with temperatures below 1500°C, it is concluded that at temperature near melting point of Si (T_m=1414°C) most of commonly applied in practice ceramics (e.g. oxides, carbides, borides or nitrides) are well wetted by molten Si [2,3]. This behavior originates from a high chemical affinity of Si to oxygen, carbon and nitrogen leading to an easy formation of reaction products at contacting interfaces. In fact, the only one ceramic showing non-wetting behavior with molten Si is hexagonal boron nitride (h-BN). Reported values of the equilibrium contact angle θ measured for Si/h-BN couples at T<1500°C are within the range of 95-145° [e.g. 4,5] (i.e. they correspond to the lack of wetting[1]). In our previous work [6], by using a sessile drop experiments we documented that the non-wetting behavior is maintained up to 1650°C, while at higher temperatures a non-wetting-to-wetting transition takes place. We have also evidenced that the involved reactivity mechanism under static argon atmosphere is mostly based on h-BN dissolution in molten Si (the dissolution rate depends on the temperature) followed by a reprecipitation of h-BN platelets during cooling. Additionally, the interaction was accompanied by a formation of gaseous product (most probably N_2), what was reflected by a lack of completely equilibrated conditions.

[1] The contact angle θ is a metric describing wettability in a considered liquid/solid system. The $\theta<90°$ corresponds to a wetting behavior, while $\theta>90°$ to a lack of wetting.

In this work, we used the sessile drop method experiments to examine the effect of boron content on the high temperature interaction (wetting and spreading) during contact heating of Si-B alloys with hexagonal boron nitride (h-BN) substrates at temperatures up to 1750°C.

2. Materials and methods

The nominal compositions of Si-B alloys (wt%) were selected in accordance to Si-B binary phase diagram [7] as: hypoeutectic (Si-1B), eutectic (Si-3.2B) and hypereutectic (Si-5.7B). Additionally, ultra-highly pure Si (7N) was used as the reference material. The Si-B alloys were fabricated by the electric arc melting technique using properly weight mixtures of pure elements. Commercially available h-BN sinters were used as the substrates. The sessile drop experiments were performed with a dedicated experimental complex [8] working under argon atmosphere, in the temperature range between 1450°C and 1750°C (Fig. 1). More details on experimental procedure are shown elsewhere [6]. During the tests, the images of Si-B/h-BN couples were recorded at 100 fps by using a high speed CDD camera. After the tests, the acquired images were post-processed to compile movies and calculate wetting kinetics.

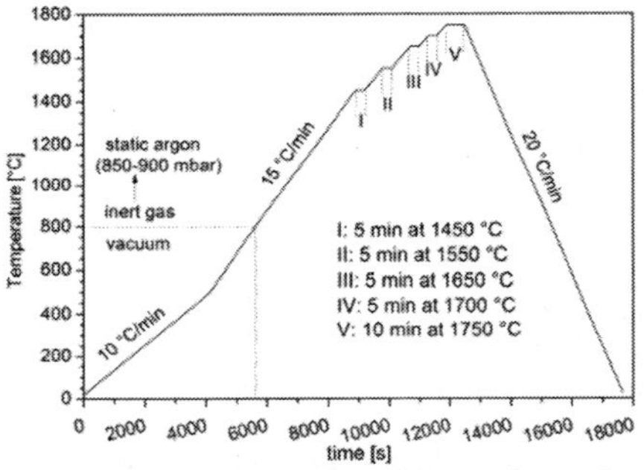

Fig. 1. The heating/cooling scheme used upon the wetting test of Si-B/h-BN couples (more detailed description is shown in [6]).

3. Results and conclusions

The wetting kinetics curves (showing a change of θ vs. time) calculated for pure silicon and Si-B alloys subjected to contact heating with h-BN substrates, are shown in Fig. 2. By comparing the results obtained for Si-B alloys to the behavior of pure silicon on the h-BN substrate (described in details in [6]), it might be concluded that the addition of boron decreases the wettability in the system. For all Si-B alloys, the contact angle values were very high (above 130°) in the whole examined temperature range. The highest value of 150° was received for the eutectic composition of Si-B alloy (Si-3.2B), what beside its lowest melting point and theoretically highest available latent heat [9] allows concluding that this material should be an optimal choice for application as PCM in the ultra-high temperature LHTES system.

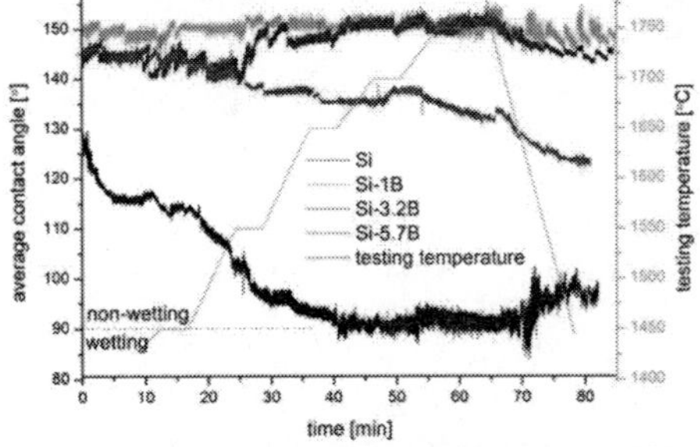

Fig. 2. Wetting kinetics obtained during testing of Si-B/h-BN couples.

Acknowledgments

The project AMADEUS has received funds from the European Union's Horizon2020 research and innovation program, FET-OPEN action, under grant agreement 737054. The sole responsibility for the content of this publication lies with the authors. It does not necessarily reflect the opinion of the European Union. Neither the REA nor the European Commission are responsible for any use that may be made of the information contained therein.

References

[1] Datas, A. Ramos, A. Marti, C. del Canizo, A. Luque (2016) Ultra high temperature latent heat energy storage and thermophotovoltaic energy conversion. Energy 107, 542-549

[2] B. Drevet, N. Eustathopoulos (2012) Wetting of ceramics by molten silicon and silicon alloys: a review. Journal of Materials Science 47, 8247–8260

[3] Z. Yuan, W.I. Huang, K. Mukai (2004) Wettability and reactivity of molten silicon with various substrates. Applied Physics 78, 617–622

[4] J.A. Champion, B.J. Keene, S. Allen (1973) Wetting of refractory materials by molten metallides, Journal of Materials Science 8, 423-426

[5] B. Drevet, R. Voytovych, R. Israel, N. Eustathopoulos (2009) Wetting and adhesion of Si on Si_3N_4 and BN substrates, Journal of European Ceramic Society 29, 2363–2367

[6] W. Polkowski, N. Sobczak, R. Nowak, et al. (2017). Wetting behavior and reactivity of molten silicon with h-BN substrate at ultrahigh temperatures up to 1750°C. Journal of Materials Engineering and Performance. https://doi.org/10.1007/s11665-017-3114-8

[7] R.W. Olesinski, G.J. Abbaschian (1984) The B-Si (Boron-Silicon) system. Bulletin of Alloys Phase Diagrams 5, 479-484

[8] N. Sobczak, R. Nowak, W. Radziwill, J. Budzioch, A. Glenz (2008) Experimental complex for investigations of high temperature capillarity phenomena. Materials Science and Engineering: A 495, 43–49

[9] A. Datas, A.B. Cristobal, C. del Cañizo et al. (2017) AMADEUS: Next Generation Materials and Solid State Devices for Ultra High Temperature Energy Storage and Conversion. SolarPaces 2017 Conference, Santiago, Chile, 2017. Article submitted to AIP Conference Proceedings

THE EFFECT OF SURFACE CONDITION ON WETTING OF HASTELLOY® X BY BRAZER ALLOY OF Ni-Pd-Cr-B-Si SYSTEM

Michał Baranowski [1]*, Artur Kudyba[2], Grzegorz Bruzda[2], Natalia Sobczak [1,2], Mariusz Bober[3]

[1] *Institute of Precision Mechanics, Duchnicka 3 Str., 01-796 Warsaw, Poland*
[2] *Foundry Research Institute, Zakopiańska 73 Str., 30-418 Krakow, Poland*
[3] *Warsaw University of Technology, Department of Welding Engineering, Narbutta 85 Str., 02-524 Warsaw, Poland*

**Corresponding address: michal.baranowski@imp.edu.pl*

Key words: Hastelloy® X, Ni-Pd-Cr-B-Si alloy, brazing, wettability, interfaces

1. Introduction

HASTELLOY® X is solid solution strengthened nickel-based superalloy of Ni-Cr-Fe-Mo system with outstanding combination of high temperature strength, oxidation resistance, weldability and good fabrication characteristics [1]. The components from HASTELLOY® X alloy can be produced by cold and hot forming as well as by casting techniques.

Continuously increasing cost of such components justifies greater interest in joining techniques for both for manufacture of components with a complex shape as well as for their repair.

This work is focused on scientific aspects of joining HASTELLOY® X alloy by brazing, particularly on the effect of alloy surface finish on wetting behaviour with selected commercial brazing filler metal (BFM).

2. Materials and methods

The HASTELLOY® X (nominal composition, wt%: Cr- 22; Fe-18; Mo-9; Co-1.5; W-0.6; C-0.1; Mn-1.0; Si-1.0; B-0.008; Nb-0.5; Al-0.5; Ti-0.15 Ni-bal.) produced by Hayness International Inc. was used as substrates for wettability tests. Three types of brazing metal (BM) substrates with different surface finish were used in this study: 1) in delivery state - D1; 2) after polishing - P1; 3) after polishing and oxidation at a temperature of 600°C for 1h in air - U1.

As brazing filler metal (BFM), commercial high palladium Ni-based brazing filler metal alloy Palnicro 36M (Morgan Advanced Materials) of nominal composition 50Ni-36Pd-10.5Cr-B-0.5Si (wt%) was used in the form of a foil with a thickness of 0.05 mm. This BFM has good strength properties at elevated temperature and good resistance to oxidation [2,3]. A good wetting of Palnicro 36M alloy on Ni-based superalloys of Inconel family has been reported in [3,4].

Wetting properties of selected BFM/BM couples were estimated by a sessile drop method [5] in a vacuum using contact heating to and at a temperature of 995°C. During the wettability test, the types and amounts of residual gases in the vacuum chamber were monitored in real time using Pfeiffer Prisma QMS200 quadrupole device while the images of the BFM/BM couples were recorded by high resolution CCD camera and next used for estimation of contact angle values.

Solidified BFM/BM couples were subjected to structural observations by optical microscopy and scanning electron microscopy (SEM) coupled with X-ray energy dispersive spectroscopy (EDS) analysis.

3. Results

Immediately after complete melting, BFM wets all examined BM substrates forming the contact angle of θ_0=51±4° on the substrate in delivery state D1, θ_0=65±13° on the polished substrate P1 and θ_0=83±2° on the oxidized substrate U1.

However, there is no significant difference in the values of the final contact angles formed under identical conditions on the same substrates, i.e. θ_f=7.6±0.1°, 6.03±0.07° and 6.2±1.0° for D1, P1 and U1, respectively.

Therefore, one may conclude that the oxidation of HASTELLOY® X (under conditions comparable to those used in this study) worsens its wetting with Palnicro 36M only at the initial stage of brazing process in vacuum. Moreover, the type of surface finish used for BM substrates in this study affects the rate of the decrease in the contact angle to its final value showing the slowest rate on the oxidized substrate.

During the thermal process in a vacuum, a large number of gases were released and recorded by residual gas analysis (e.g. for 995°C, Fig. 1). The examined samples show noticeable differences in the amounts of H, H_2, O, OH, H_2O and CO_2 recorded. The largest amounts of gases were released by the sample in the delivery state. Polishing of BM surface caused a slight reduction of gas emissions during a high-temperature process in a vacuum. However, the smallest amounts of released gases were obtained for the polished substrate oxidized in air at 600°C just before wettability test in vacuum.

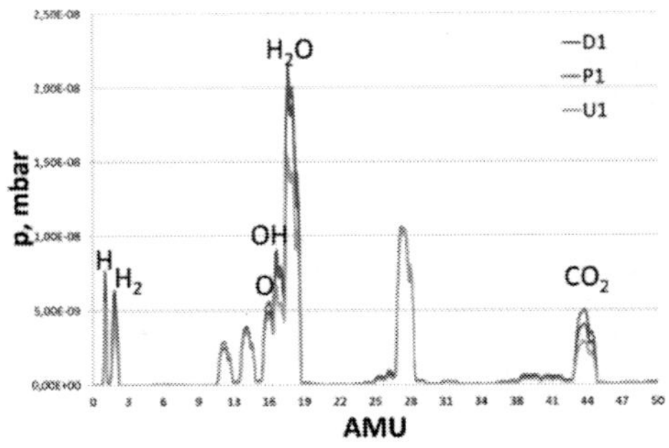

Fig. 1. Partial pressure of residual gases in the test chamber recorded during experiment with Hastelloy X-Palnicro 36M couples, T=995°C, t=0 min

Structural characterization of the cross sections of the solidified sessile drop couples reveals the formation of defect-free bonding between BFM and HASTELLOY® X alloy. A boundary zone is mainly composted of Cr-Mo-B-based precipitations uniformly distributed in the Ni-Cr-Pd-Mo-Fe-Si matrix. It was evidenced that such zone is already formed at 960°C (Fig. 2a). Further increase in temperature and the extension of the holding time at that temperature causes increase in the thickness of the boundary zone (Fig. 2b).

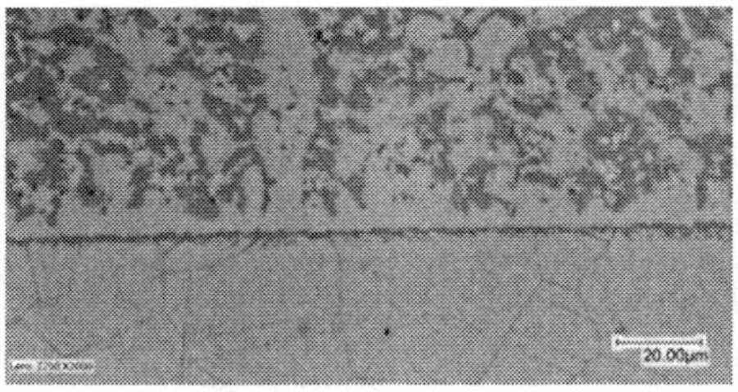

(a)

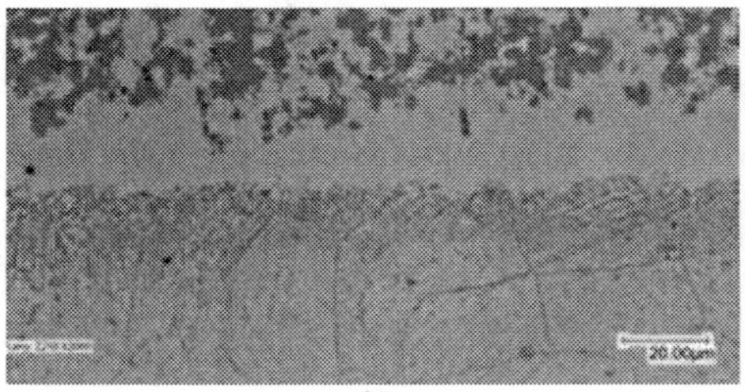

(b)

Fig. 2. SEM microstructure of cross-sectioned BM/BFM couple: a) T=960°C, t=0 min; b) T=995°C, t=10 min

Moreover, the BM surface condition also affects the thickness of the boundary zone (Fig. 3). After holding for 10 minutes at 995°C, the highest average thickness was recorded for the polished substrate oxidized in air at 600°C while the smallest one was recorded for the substrate surface in the delivery state.

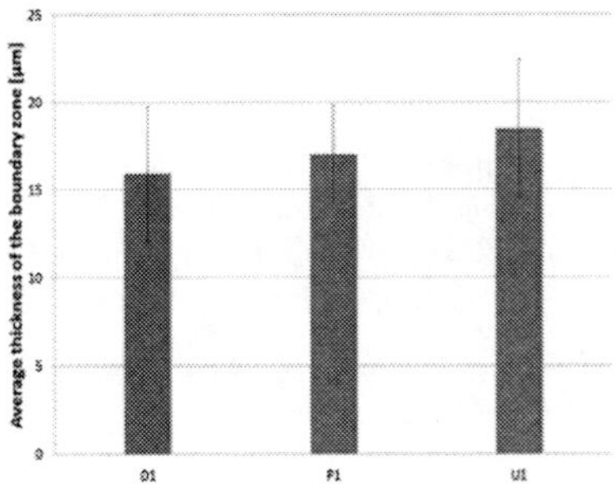

Fig. 3. Average thickness of the boundary zone, T=995°C, t=10 min

4. Conclusions

The results of wettability study coupled with residual gas analysis and measurements of the boundary zone thickness show that independently on the type of substrate surface finish used in this study (i.e. in delivery state, after polishing and polished substrate after oxidation), the HASTELLOY® X substrates display very good wetting with molten Palnicro 36M alloy and during sessile drop tests in a vacuum, they form the boundary zones of similar structure and chemistry. Consequently, the oxidation of HASTELLOY® X alloy in air at 600°C for 1h has insignificant effect on its high temperature interaction with the selected commercial BFM.

References

1. http://www.haynesintl.com/

2. www.morganbrazealloys.com - Brazing Filler Metals Data Sheet, 2013,

3. M.M. Schwartz, Brazing, 2nd Edition, (Materials Park Ohio), ASM International, 2003.

4. I. Dul, J. Senkara, M. Bober, J. Jakubowski, *Przegląd Spaw.*, 2013, **85**(9), p 15–19.

5. N. Sobczak, M. Singh, R. Asthana (2005) High-temperature wettability measurements in metal/ceramic systems – Some methodological issues. *Current Opinion in Solid State and Materials Science*, 9 (4-5) 241-253

The 73rd World Foundry Congress
„Creative Foundry"
23 - 27 September 2018, Krakow, Poland

IMPROVEMENT OF TiC/Fe IN SITU COMPOSITE LAYER FORMATION ON THE SURFACE OF Fe-BASED CASTINGS

Łukasz Szymański[1,2]*, Ewa Olejnik[1,2], Jerzy J. Sobczak[3], Tomasz Tokarski[4]

[1]*AGH University of Science and Technology, Faculty of Foundry Engineering, 23 Reymonta Str., 30-059 Krakow, Poland*
[2] *Innerco Ltd., 43A Jadwigi Majówny Str., 30-298 Krakow, Poland*
[3]*Foundry Research Institute, 73 Zakopiańska Str., 30-418 Krakow, Poland*
[4]*AGH University of Science and Technology, Academic Center for Materials and Nanotechnology, 30 Mickiewicza Av., 30-059 Krakow, Poland*
**Corresponding address: lsz@agh.edu.pl*

Key words: MMC, in situ, TiC, composite layer, casting

1. Introduction

The manufacturing of local layers reinforced with TiC particles has been extensively discussed in many scientific studies. For producing such materials, these methods allow to obtain laminar materials with the application of such techniques as laser plating, thermal spraying, hard facing and welding [1,2]. An interesting solution, applied only rarely, is the application of casting technologies, where an agent initiating the new phase synthesis is a liquid casting alloy. This method specified in the literature as SHSB (Self-propagating High-Temperature Synthesis in Bath) [3]. This is a modification of Self-propagating High-temperature Synthesis (SHS) widely applied in powder metallurgy [4]. In the initial stages of process, the mold cavity is covered with a coating containing substrates for the reaction to form the selected carbide. The synthesis reaction is initiated by the heat supplied by a liquid alloy at a temperature above 1200°C and it proceeds according to the equation: Ti + C = TiC + Q. This reaction is highly exothermic, with the enthalpy of -187 kJ/mol [4]. Intensive energy production in a form of generated heat results in the local increase of temperature of liquid metal, significantly intensifying liquid metal infiltration within composite layer. Thus, TiC particles become separated by a liquid alloy and they move (flow) due to lower density, compared to molten matrix alloy. This phenomenon is called fragmentation of composite layers. Very recently, for confining composite layers fragmentation the application of a moderator, significantly decreasing the heat effect within the area of TiC synthesis reaction, has been proposed [5]. A moderator is to be defined as a mixture of metallic powders with non-metals [2]. The objective of this study was to produce and characterize two different kinds of composite layers based on pure substrates of reaction synthesis of TiC and with the addition of a moderator [5].

2. Experimental

Casting coatings for the in situ fabrication of the composite layers (CL) reinforced by TiC were prepared on the basis of carboxymethylcellulose (CMC) aqueous solutions with 2% concentrations along with additions of pure substrates for the TiC-forming reaction. In order to control the process of reactive infiltration, the addition of 50 wt% of moderator showing the composition of cast iron was introduced into the mixture of titanium and carbon powders. The chemical composition of composite coatings and moderator is given in Table 1. The addition of CMC was introduced as a binding material and stabilizer.

The manufacturing of composite layers consisted two stages. The 1st stage involved covering the casting mold cavity with the coatings. In the 2nd stage, the dried mold was poured with molten steel preheated to a temperature of 1823 K. The samples for investigation were cut out from produced castings.

Table 1. Composition of coatings and moderator

Coating	Mass fraction of substrates for TiC formation [wt%]				Mass fraction of moderator [wt%]	
CL1	100				-	
CL2	50				50	
Chemical composition of moderator [wt%]						
Fe	C	Si	Mn	Cr	Mo	Ni
Bal.	3.6	2.2	0.8	10	0.5	5.5

3. Results and discussion

Figs. 1 and 2 show the most representative results of structural investigations in the form of panoramic macro-views made by putting together a few pictures collected from the cross-sectioned composites layers. They indicate that for CL1, the layer structure is non-homogenous at its whole lengths while the layer thickness varies from 0.5 to 1.5 mm.

For CL2, the modification of chemical composition of composite coatings made possible to obtain its homogeneous structure together with uniform thickness (approx. 1 mm) over the entire length of composite coating. Application of the moderator contributed to decreasing the energy in the area of composite layers during the synthesis reaction of TiC. The residual small porosity observed in the composite layers are related to the water applied as solvent and to the CMC used as a binder.

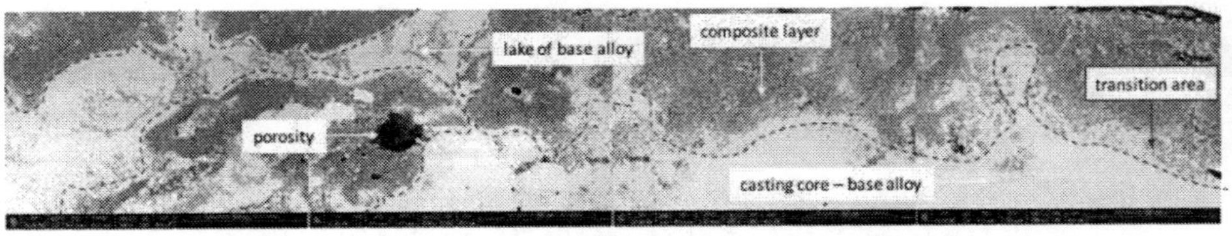

Fig. 1. Macrostructure of composite layer CL1 produced in situ without the moderator

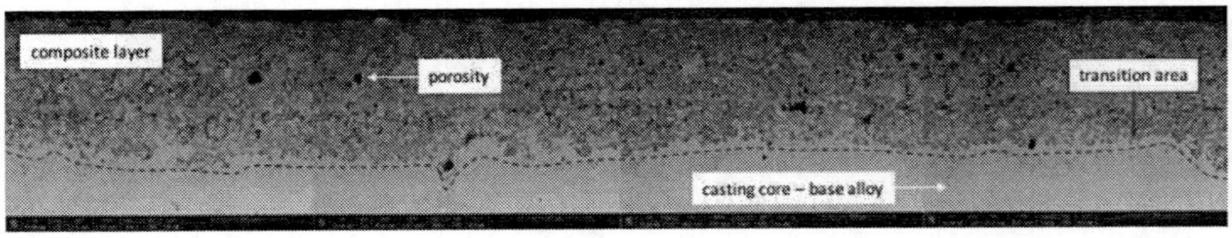

Fig. 2. Macrostructure of composite layer CL2 produced in situ with moderator

The microstructures of composite layers produced are presented in Fig. 3 showing the distribution of dark TiC particles within metal matrix. The in situ fabricated TiC particles are agglomerated in the groups of spherical shapes. The analyzed microstructure is characterized by the locally reinforced region with a relatively dense distribution of TiC particles. The transition area between reinforcing phase and casting core remains discontinuities-free and shows well bonding with the base alloy.

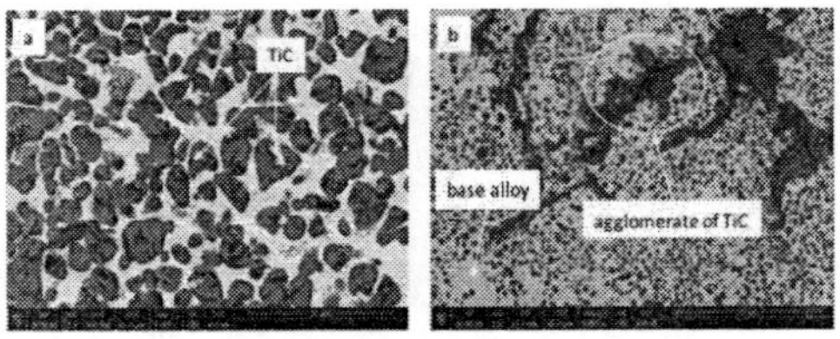

Fig. 3. The microstructure of composite layers: (a) CL1; (b) CL2

The stereological investigations allowed to determine the average size of TiC particles. Their diameter was within 3 μm for CL1 and 1 μm for CL2 (Fig. 4). During the crystallization, the size and number of particles are determined by the degree of undercooling. Relatively large TiC particles were obtained when pure substrates for the TiC-forming reaction were used. Large amounts of heat, released during the TiC-synthesis, caused a lowering of the degree of undercooling. The introduction of the moderator, as the factor stabilizing the reaction of TiC synthesis, significantly decreases the temperature within its area, thus affecting the crystallization of the TiC particles.

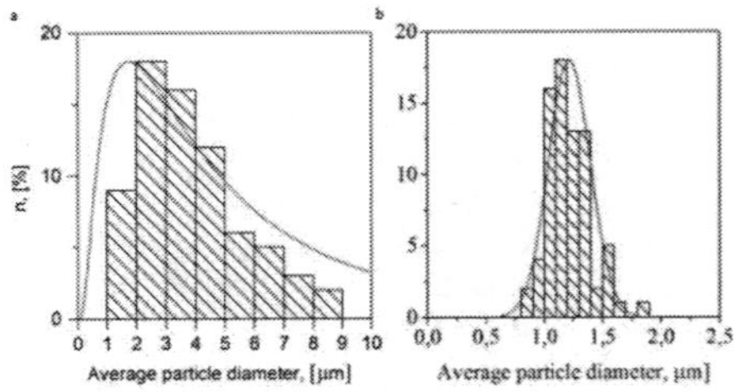

Fig. 4. Average diameter of TiC particles: a) CL1 b) CL2

4. Conclusion

The introduction of a proper moderator in a mold coating is an effective way to confine the in situ formation of a composite layer with uniform thickness and homogeneous structure on the surface of metal castings.

References

[1] L.M. Berger. (2015). Application of hard metals as thermal spray coating. *Int. J. Refract. Met. Hard Mater.* 49, 350-364.

[2] Ł. Szymański, E. Olejnik, T. Tokarski, P. Kurtyka, D. Drożyński, S. Żymankowska-Kumon (2018). Reactive casting coatings for obtaining in situ composite layers based on Fe alloys. *Surf. Coat. Tech.* 350, 346-358

[3] E. Fraś, E. Olejnik, A. Janas, A. Kolbus (2009). FGMs generated method SHSM. Arch. Foun. Eng. 9 (2), 123-128

[4] A.G. Merzhanov (1996). Combustion processes that synthesize materials. *J. Mat. Proces. Tech.* 56 (1-4), 222-241

[5] Patent application No PCT/IB2016/056825, E. Olejnik, A. Jesiołowska. Powder composition for the manufacture of casting inserts, casting insert and method of obtaining local composite zones in castings.

The 73rd World Foundry Congress
„Creative Foundry"
23 - 27 September 2018, Krakow, Poland

LIQUID METAL ENGINEERING FOR CREATIVE FOUNDRY: from Lab to Fab

Natalia Sobczak[1,2]*, Rajiv Asthana[3], Alberto Passerone[4], Robert Purgert[5], Jerzy J. Sobczak[1]

[1] *Foundry Research Institute, 73 Zakopiańska Str., 30-418 Kraków, Poland*
[2] *Institute of Precision Mechanics, 3 Duchnicka Str., 01-796 Warsaw, Poland*
[3] *University of Wisconsin-Stout, Menomonie, Wisconsin, USA*
[4] *CNR-ICMATE - National Research Council – Institute of Condensed Matter Chemistry and Technologies for Energy, 6 de Marini, 16149 Genoa, Italy*
[5] *Energy Industries of Ohio, Park Center Plaza, Suite 200, 6100 Oak Tree Blvd, Independence, OH 44131, USA*

**Corresponding address: natalia.sobczak@iod.krakow.pl*

Key words: materials engineering, liquid metals, high temperature phenomena, thermophysical properties

Historically, metal casting has played a crucial role in the development of human civilization while nowadays casting production is considered a key factor affecting the growth of the world's economies. Further development of the metal casting sector in an era of rapid globalization is facing unprecedented challenges. The last decade significantly impacted the industry's strategy for the future as being a need to focus on:

(1) energy and material savings,
(2) weight reduction of cast components,
(3) substitution of critical raw materials,
(4) minimal machining and net-shape casting,
(5) innovative recycling and reuse solutions,
(6) environmentally-friendly processes,
(7) waste heat recovery,
(8) cost reduction coupled with improved performance.

Therefore, for global industrial progress, the transfer of new knowledge from Lab to Fab and finally to App (Laboratory to Fabrication and finally to Application) [1] is essential for metal casting's next generation that should utilize all achievements of fundamental and applied research in materials science and engineering. Research should focus on understanding structure of high temperature molten alloys and the effects of molten metal treatment on mechanical, physical and utility properties of cast alloys by application of Liquid Metal Engineering, as a new technology for large-scale production and better component performance [2]. In this regard, further research should focus on significant improvements in:

1) computer-aided design,
2) multicomponent alloy synthesis by utilizing the vast amount of information that might be extracted from phase diagrams,
3) testing methods and procedures that mimic real casting processes,
4) interdisciplinary research for utilizing knowledge and experience from other liquid-assisted or high-temperature processes, advanced non-destructive materials characterization, big data analysis to predict materials behaviour as well as for alloy development,
5) theoretical modelling and experimental verification of liquid-metal/ceramic interactions,
6) new design opportunities in cast metal-matrix composites and joining cast alloys (e.g. welding, brazing, soldering).

Advanced computer simulation-aided design is a powerful, fast and economic tool for production of high-quality and defect-free castings with tailored properties. However, reliable numerical simulations require also reliable information on properties of materials involved in casting processes. Particularly for molten alloys of industrial interest, these data quite often are not available, such as thermophysical properties (surface tension, density, viscosity, electrical and thermal conductivity, shrinkage/expansion upon solidification/melting, wetting and reactivity in contact with refractory materials) and thermodynamic properties (specific heat capacity, temperature and enthalpy of phase transformations).

Consequently, progress in measurement methods, validated testing procedures and standards is a prerequisite for successful knowledge-based liquid metal engineering and subsequent fast implementation of creative foundry solutions. It is an essential factor for competitiveness of the metal casting sector. Consequently, scientific efforts in many countries are directed to systematic investigation and collection of data on materials properties. One of the best examples is Materials Genome Initiative of National Institute of Standards and Technology that aims to enable US industry to develop and deploy advanced materials more quickly and efficiently [3].

This contribution is a continuation of authors' previous works [4-6] and it reports new achievements in developing insights into high-temperature liquid-state materials science, solving accompanying methodological problems and creatively implementing scientific findings in industrial practice. Further improvements in the design of an experimental complex [5], built to investigate high-temperature phenomena of molten metals and alloys as well as to measure their thermophysical properties, have been done (Fig. 1). Improved (Fig. 2) and new (Fig. 3) testing procedures for different drop-assisted methods – that are impossible in any other laboratories in the world - have been developed, particularly:

1) testing under non-isothermal conditions (different substrate and metal temperatures),
2) application of electrical current (electro-capillarity phenomenon),
3) tilting a substrate directly in a vacuum chamber at a high temperature,
4) improved residual gas analysis for real-time control of chemical composition of surrounding atmosphere of flowing gas (not only in vacuum).

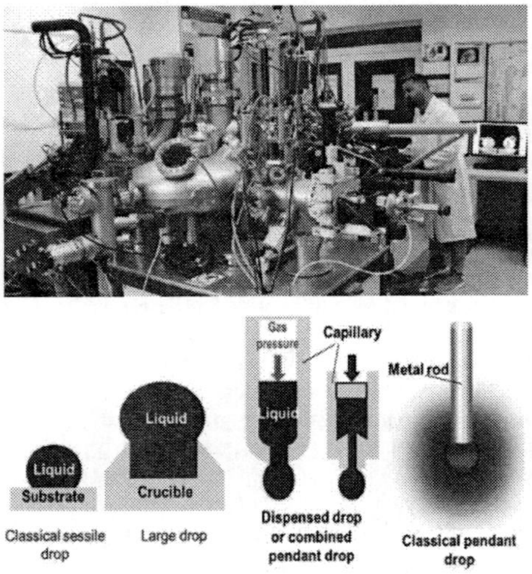

Fig. 1. ***Updated experimental complex for high temperature investigation of liquid metals and alloys by different drop-assisted methods***

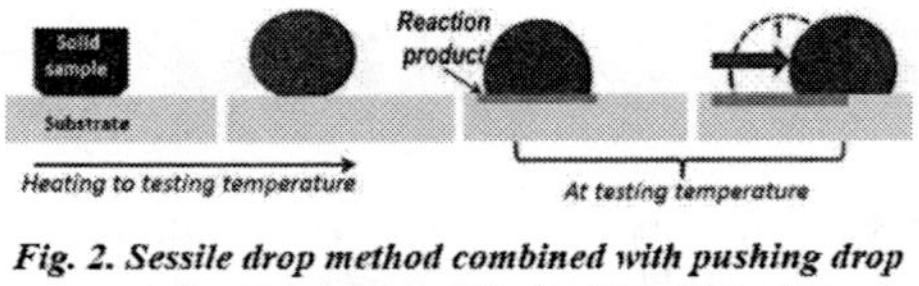

Fig. 2. Sessile drop method combined with pushing drop procedure to expose reactively formed interface

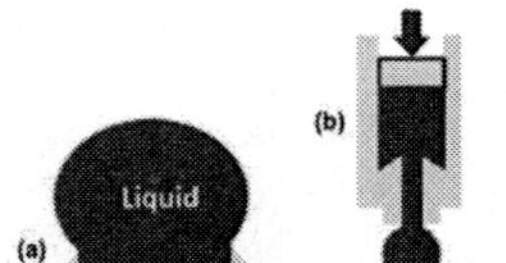

Fig. 3. Three methods combined in one test ("3in1" test): a) large drop; b) pendant drop; c) classical sessile drop

The results obtained using this updated high-temperature facility gave fresh insights into old metal casting problems related to the formation of such casting defects as porosity and non-metallic inclusions. Corresponding findings were used to develop new generation light-weight materials [7] and process improvement [8].

Moreover, the structural and topological heredity of the alloys that experienced external mechanical treatment in a molten state (pressing, extension, mixing) has been confirmed similar to the results reported by other authors for the effects of ultrasound or electric pulse treatment of molten alloys prior to their solidification (e.g. [9,10]). Several examples of application of this effect in foundry practice will be discussed.

Acknowledgments

This contribution contains the results obtained in frame of: 1) Poland-USA Project No. 721/N-NICKEL/2010/0 with financial support from the Ministry of Science and Higher Education of Poland and Energy Industries of Ohio (USA); 2) INSURFCAST project financed within the ERA-LEARN 2020 support action funded by EU-H2020, M-ERA.NET Joint Call 2016; 3) Project UMO-2015/17/B/ST8/03391 financed by National Science Centre of Poland within OPUS program; 4) AMADEUS project financed within the European Union's Horizon2020 research and innovation program, FET-OPEN action, under grant agreement 737054.

References

1. LAB-FAB-APP = Investing in the Europe future we want (2017) EC Report, Brussels; doi:102777/477357
2. F. Czerwinski (2017) Modern aspects of liquid metal engineering, Metal. Mater. Trans. B, 48(1):367-393
3. M. Drosback (2014) Materials Genome Initiative: Advances and Initiatives, JOM 66(3):334-335
4. N. Eustathopoulos, N. Sobczak, A. Passerone, K. Nogi (2005) Measurement of contact angle and work of adhesion at high temperature, J. Mater. Sci. 40(9-10):2271-2280
5. N. Sobczak, J. Sobczak, R. Asthana, R. Purgert (2010) The mystery of molten metal, China Foundry 7:425-37
6. E. Ricci, D. Giuranno, N. Sobczak (2013) Further development of testing procedures for high temperature surface tension measurements: J. Mater. Eng. Perf. 22 (11):3381-3388
7. J. Sobczak, N. Sobczak, L. Drenchev, P. Darlak, P. Długosz, P. Turalska, M. Homa, A. Tchórz, I. Krzak, A. Makaya, L. Pambaguian, Conventional and hybrid gasars: light-weight materials with pressure-temperature managed porosity (2018) – this volume
8. F. Valenza, G. Caccamani, M. Muolo, A. Passerone, R. Nowak, J. Sobczak, N. Sobczak (2018) Interfacial phenomena in Ni alloys-oxides systems of interest for investment casting - this volume
9. I. Brodova, P. Popel, G. Eskin (2002) Liquid metal processing - Applications to aluminum alloy production, Taylor & Francis, London, UK
10. J. Wang, J. Qi, H. Du, Z. Zhang (2007) Heredity of aluminium melt caused by electric pulse modification, J. Iron & Steel Res. 14(4):75-78

TECHNOLOGY

The 73rd World Foundry Congress
„Creative Foundry"
23 - 27 September 2018, Krakow, Poland

USE OF A MATHEMATICAL TREATMENT FOR THE PREDICTION OF STRUCTURAL ZONES LOCALIZATION IN THE CONTINUOUSLY CAST BRASS INGOTS

Waldemar Wołczyński [1]*, Anna A. Ivanova [2]

[1]Institute of Metallurgy and Materials Science, 25 Reymont Str., 30 059 Krakow, Poland
[2]Institute of Applied Mathematics and Mechanics, 74 Rosa Luxemburg Str.,83 114 Donetsk, Ukraine
**Corresponding address: w.wolczynski@imim.pl*

Key words: heat transfer model, continuous casting, brass ingots, structural zones, CET transition

The heat transfer model used in the current predictions is shown schematically in Fig. 1.

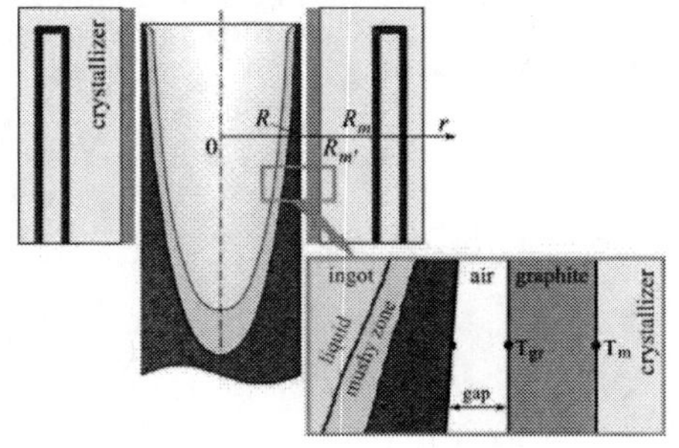

Fig. 1. Outline of the system for the continuous casting of brass ingots as shown in the x,r ***- coo coordinates, [1]***

The governing equation for heat transfer is as follows:

$$c_{ef}(T)\cdot\rho(T)\cdot\frac{\partial T}{\partial t}=\frac{\partial}{\partial r}\left[\lambda(T)\cdot\frac{\partial T}{\partial r}\right]+\frac{\lambda(T)}{r}\cdot\frac{\partial T}{\partial r}$$

(1)

$T=T(t,r)\,,\quad r\in[0,R_m]$

$$c_{ef}(T)=\begin{cases}c_b(T), & T<T_S\\ c_b(T)+\dfrac{L}{T_L-T_S}, & T_S\le T\le T_L\\ c_b(T), & T>T_L\end{cases}$$

where, L is the latent heat, and T - temperature.

The above equations together with the adequate boundary conditions, [2], were applied to the prediction of the structural zones in the continuously solidifying brass ingot. The prediction follows the mathematical method of these zones modelling worked out for the steel static ingot, [3]. Particularly, the columnar into equiaxed structure transformation (CET) was the subject of this mathematical modelling. The mentioned method, [3], takes into account the extreme points (minima and maxima) and also points of inflection of some considered / simulated functions.

In the current model, the functions are the result of the use of Eq. (1) with regard to the current outline, Fig. 1. Mainly, the brass ingots contain the columnar and equiaxed structures, (thus, CET – transition), Fig. 2.

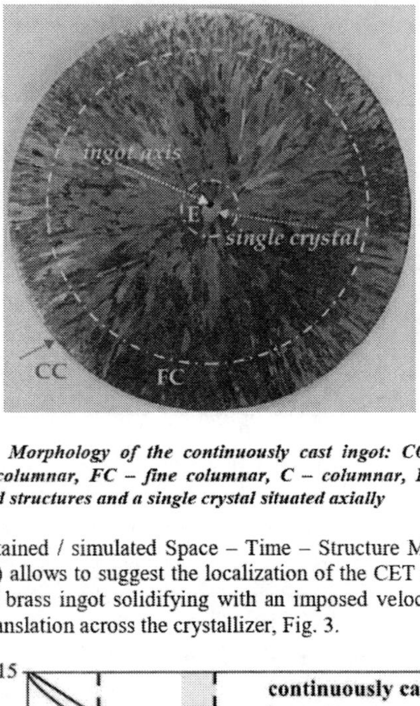

Fig. 2. Morphology of the continuously cast ingot: CC – chilled columnar, FC – fine columnar, C – columnar, E – equiaxed structures and a single crystal situated axially

The obtained / simulated Space – Time – Structure Map (STSM) allows to suggest the localization of the CET for a given brass ingot solidifying with an imposed velocity of its translation across the crystallizer, Fig. 3.

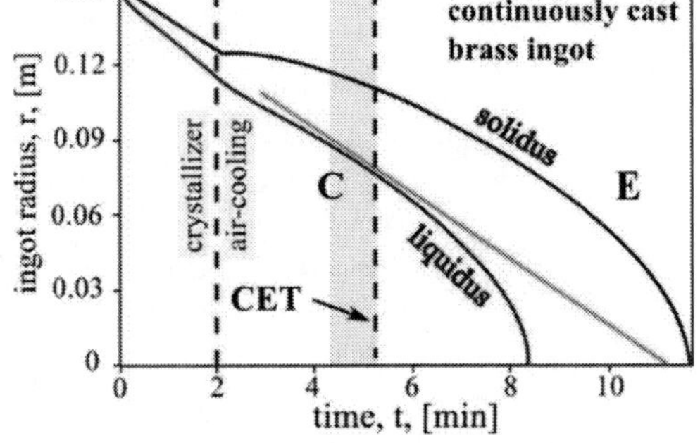

Fig. 3. Prediction of the CET in a brass ingot solidifying with the velocity of translation equal to 0.4 mm/min; localization is in the agreement with the local maximum

Next, the numerically obtained temperature field was transformed into the thermal gradient field. The thermal gradient field and particularly the resultant function, G(t), confirms the formerly predicted CET - localization, Fig. 4.

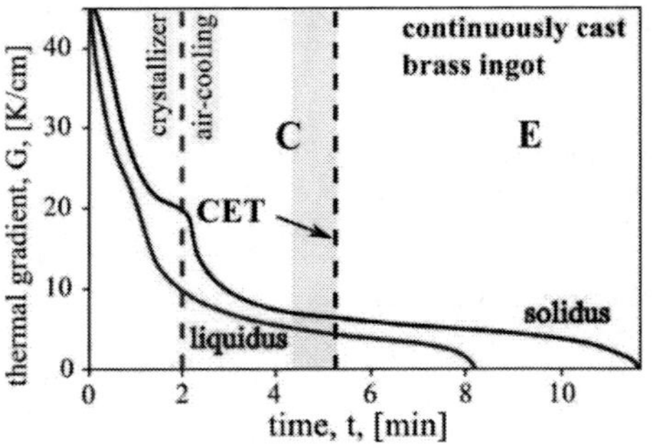

Fig. 4. Localization of the CET as it results from the thermal gradient field; localization is in the agreement with the points of inflection of the analyzed function

A localization of the FCCT is also possible to be distinguished due to the more detailed analysis of the simulated functions, Fig. 5a.

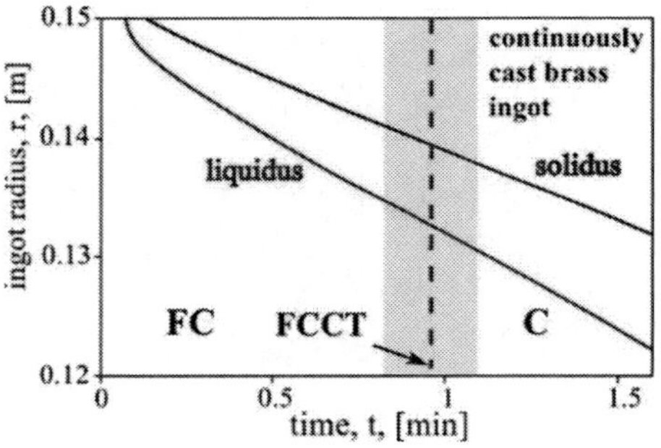

Fig. 5a. Localization of the FCCT being in the accordance with the points of inflection of the simulated STSM

The above FCCT - localization, Fig. 5a, is confirmed by the other simulations, Fig. 5b (local minimum), and Fig. 5c (points of inflection).

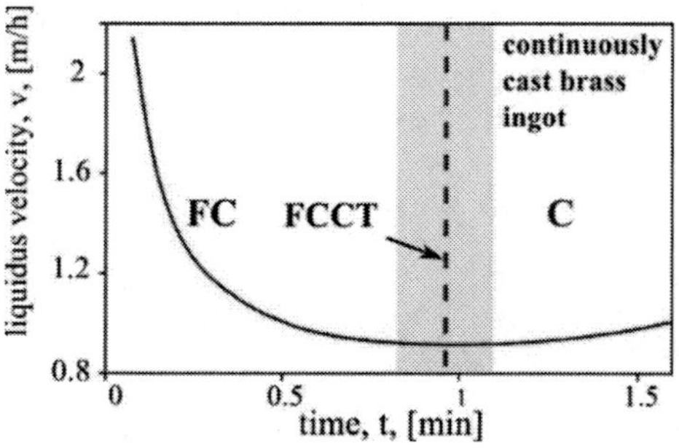

Fig. 5b. Localization of the FCCT being in the accordance with the situation of the local function minimum

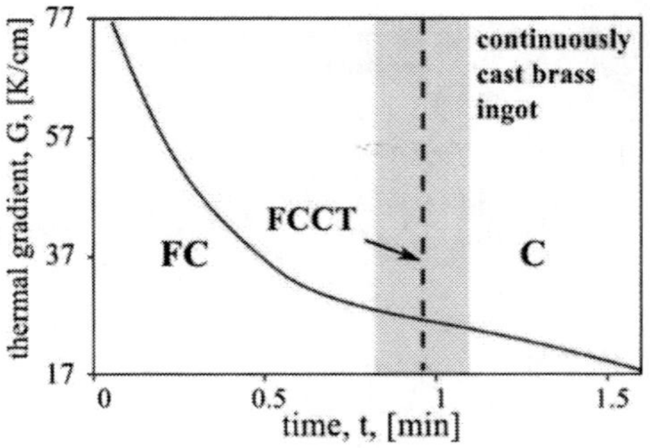

Fig. 5c. Localization of the FCCT being in the accordance with the points of inflection of the simulated G(t) - function

Finally, the current model was used for the prediction of the course of the *liquidus* velocity movement. This function, Fig. 6., shows all the structural zones visible / distinguished within the brass ingot morphology, Fig. 2.

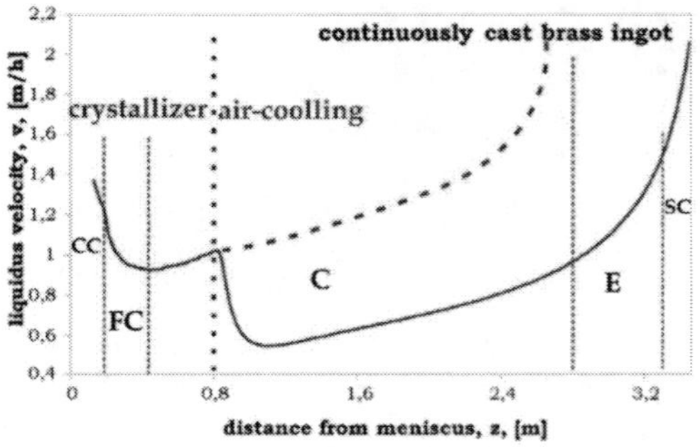

Fig. 6. Localization of the CCFCT, FCCT, CET, and ESCT structural transitions possible in the brass ingot subjected to continuous casting with the crystallizer height equal to 0.8 m

The structural zones of the investigated brass ingot are contained between adequate vertical dashed lines. The crystallizer height is marked by the vertical dotted line.

Acknowledgments

The support was provided by the National Center for Research and Development, Grant No. PBS3/A5/52/2015. The assistance of Mr. P. Kwapisiński with the "KGHM - Polish Copper" is greatly appreciated.

References

[1] P. Kwapisiński, A.A. Ivanova, B. Kania & W. Wołczyński. (2017). Some Similarities / Differences between Steel Static and Virtual Brass Static Casting. Archives of Foundry Engineering. 17(1), 109-114.

[2] W. Wołczyński, A.A. Ivanova, P. Kwapisiński & E. Olejnik. (2017). Control of Structural Zones Formation and Hard Particles Motion in the Brass Ingots. Archives of Metallurgy and Materials. 62(4), 2455-2461.

[3] W. Wołczyński. (2016) Large Steel Ingots: Microstructure Mathematical Modeling. Rafael Colas & George E. Totten, (Eds.), Volume III: Heat Treatment: Special – Molten of the Encyclopedia of Iron, Steel, and Their Alloys (1910-1924), London, New York: CRC Press, Taylor & Francis Group.

The 73rd World Foundry Congress
„Creative Foundry"
23 - 27 September 2018, Krakow, Poland

INCREASING PRODUCTIVE EFFICIENCY OF CASTING OF FERROUS AND NON-FERROUS ALLOYS

Metchislav Sadokha*

BELNIILIT, JSC, Mashinostroiteley 28, 220118 Minsk, Republic of Belarus
**Corresponding address: cadoxa.m@gmail.com*

Key words: castings, alloys, productive efficiency, equipment, production

It is known that castings production is a power- and material-consuming process. It was found [1] that, on the average, while producing casting in Sweden the energy consumption structure is as follows (Table 1):

Table 1. Energy Consumption Structure in Swedish Castings Production

Production step	Proportion in general energy consumption, %	
	Steel and cast-iron castings	Light alloys
Melting, holding, melt pouring	40	44
Moulding, bars production	20	19
Secondary operations	40	37
TOTAL	100	100

According to the data [2] energy consumption structure during the castings production in India is as follows (Table 2):

Table 2. Energy Consumption Structure in Indian Castings Production

Production step	Proportion in general energy consumption, %
Melting, holding, pouring of melt	63
Moulding, production of bars	9
Secondary operations	28
TOTAL	100

Findings of energy consumption research of foundry engineering in Sweden [1] show that for getting melt and its holding chiefly power efficient melting facilities (modern furnaces, melting techniques, lining and heat-insulating materials) are used. A sizeable impact on energy consumption distribution also has the mechanization and automation level of other casting production steps (moulding, production of bars, securing of ecology norms – ventilation, gas purification, etc.). The higher is mechanization level, the higher is energy expenditure on these operations. Thus, the data on energy consumption of foundry engineering of India [2] show the usage of less power efficient melting facilities and usage, in many cases, of manual labour on the stages of moulding and production of bars, as well as relatively low expenditures on secondary operations.

The analysis of production steps of melting and holding of melts shows the following energy sparing reserves:

1. Using efficient right aggregates (medium frequency furnaces for induction melting, etc.).
2. Using energy-sparing aggregates to hold the melt.
3. Shortening the melting storage time prior pouring into the form.
4. Revising details and castings construction for the purpose of their weight decrease.
5. Using casting techniques and activities for decreasing the gating weight.
6. Less castings waste.

In the result of analyzing technical feasibilities of modern melting furnaces, aggregates for melting storage, taking into account possible organizational arrangements during the casting process, transfer to "cold" methods of bars production, decrease in specific weight of gating and so forth, by means of calculation the following data were obtained (Table 3) showing the possible impact of different operations for decreasing energy consumption during the castings production.

Table 3. Impact of Different Operations for Decreasing Energy Consumption during Castings Production

Production step	Energy expenditure variant				
	Average basic variant of energy expenditure	Variant 1 with calc. decrease in energy expend. on melting	Variant 2 with calc. decrease in expend. on moulding & bar production	Variant 3 with calculated decrease in castings waste	Variant 4 with calculated decrease in gating weight
Melting and melt holding	50%	40%	45%	45%	43%
Moulding and bars production	20%	24%	15%	15%	14%
Secondary operations	30%	36%	41%	41%	43%
Total energy expenditure	100 %	100 %	100%	100 %	100 %
Relative energy expenditure	100 %	83%	74%	72%	63%

One of the most efficient ways of decreasing energy consumption while castings production is using the technology which allows to get castings with gating, minimal in size and weight. Let us examine this factor in more detail on the example of castings of aluminum alloys obtained with the gravity die casting method.

The gating weight can be decreased considerably using the self-filling method [3]. A scheme illustrating this method is shown in Figure 1.

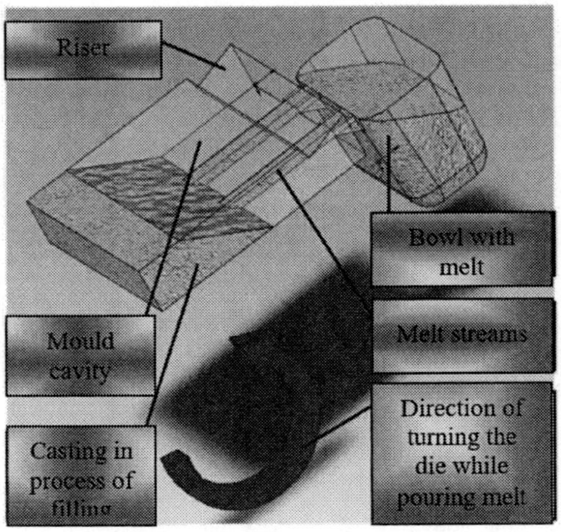

Fig. 1. Self-filling method scheme during top pouring of the alloy into the diecasting die.

Unlike stationary diecasting die, where it is impossible to regulate pressure and the melt stream can only be slowed down with various elements of the gating, while using the self-filling method the length (height) of the casting is not really significant as the height of the casting does not determine pressure at which melt gets into the mould.

Thereby, with self-filling method we can speak about getting castings of considerable height that in stationary die can be obtained only with considerable complications of the gating.

It should be mentioned that even while making high castings the melt starts to get into the die at a low speed, securing laminarity for melt streams. It is conditioned by the fact that first portions of metal get into the diecasting die (into its farther part) at small-angle turn of the die. Pressure (*H*), which in casting process with usage of stationary moulds is constant and is determined mainly with the height of the casting, in this case is a variable quantity and represents the function of the die turn angle (φ):

$$H = f(\varphi) \qquad (1)$$

In its turn, the die turn angle (φ) also is a variable quantity and represents the time function (*t*):

$$\varphi = f(t) \qquad (2)$$

Therefore, pressure (*H*) is, in fact, a time function (*t*):

$$H = f(t) \qquad (3)$$

Thereby, according to the formulas (1-3) using the self-filling method it is possible to control pressure by means of turning the die with controlling the speed of the regime, which is determined by the geometrical peculiarities of castings. In technological layout there is an additional controlled factor of casting process, which substantially expands potentials of influencing the regime of filling the diecasting die with melt and crystallization of the casting.

The self-filling method allows to considerably diminish the gating (up to 2-4 times) while getting high castings in comparison with casting in a stationary die. Thanks to the above-mentioned advantages the self-filling casting method is more and more widely used in industrial production.

References

[1] Rohdin, P., Moshfegh, B. (2007). A Study of Variable Air Volume (VAV) systems in foundries. In Proceedings of Clima 2007 Well Being Indoors.

[2] Energy Efficiency in India, Why and How. Foundry Trade Journal International (2010). Volume 184 No. 3678 (pp. 234-237).

[3] Sadokha, M.A., Melnikov, A.P., Kraev, B.A., etc. (2004). Self-filling Method in Producing Aluminium Castings. In Foundry and Metallurgy, No. 2 (pp. 153-155).

The 73rd World Foundry Congress
„Creative Foundry"
23 - 27 September 2018, Krakow, Poland

ELECTROMAGNETIC METHOD FOR CONTROL THE SOLIDIFICATION OF Al 99.99

Marcela Pokusová[1]*, Alena Brusilová[1], Marta Murgašová[1], Igor Berta[1], Marián Murgaš[1]

[1]Slovak University of Technology, Faculty of Mechanical Engineering, Nám. slobody 17, Bratislava, Slovakia
**Corresponding address: marcela.pokusova@stuba.sk*

Key words: aluminium 99.99%, solidification, as-cast structure, magnetic field, electric current, electromagnetic force

1. Introduction

In metallurgical practice, besides the classic physical-metallurgical methods of modification, inoculation, and microalloying, the electromagnetic methods have been successfully introduced for controlling the process of primary crystallization [1]. These methods offer more functional factors which have different effects on the primary structure formed, than those provided by the physical-metallurgical methods [2].

The paper documents the possibilities for control the solidification process of the pure Al99.99 by applying the electromagnetic methods. It presents the research results from a development of techniques for pouring the flat castings of Al99.99, so called a target intended for depositing the conductive paths on semiconductor chips. The properties of deposits could be only met with the as-cast target's material having either a very fine structure of a grain size below 100 μm, or a very coarse-grain one with the grain size above 10 mm. Because of the essential requirement was keeping the high purity of 99.99%, the castings could not be treated neither by the metal forming technologies nor the physical-metallurgical methods, and thus only electromagnetic methods could be applied.

2. Experiments

To ensure the high purity, the liquid metal was prepared by melting it in the corundum ceramic crucibles using an electric resistance furnace. The samples were poured into the dry-sand moulds; and the surface of the mould cavity was sprayed with a protective coating of Al_2O_3. The samples had the tailored dimensions of targets, namely 10 x 40 x 180 mm, so they could be used directly for sputtering the conductive traces on the chip's substrate.

For experiments the following electromagnetic methods were chosen: treatment using (1) the flow of alternating (AC) current of 50 Hz; and (2) the simultaneous action of a direct magnetic field of 0.035 T or 0.07 T and the alternating current of 50 Hz, when electromagnetic forces are generated. At the same time, the reference samples were poured free of any treatment.

The identical moulds were used for all the experiments. The sketch of the experimental setup with a mould is shown in Fig. 1.

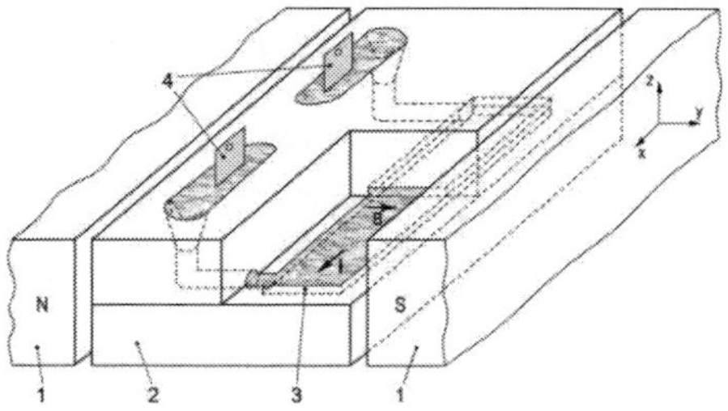

***Fig. 1.** An experimental mould inserted into an operating space of an electromagnet: (1) poles of electromagnet; (2) a dry-sand mould; (3) sample; (4) contacts of the alternating current supply*

The runner system was supplemented with channels which served to supply the electric current (4). After pouring the liquid aluminium into the mould, the electric current *I* was switched on and passed through the sample longitudinally. When examining the effect of the simultaneous action of the direct magnetic field ***B*** and the AC current, the mould (2) was inserted between the poles of electromagnet (1) so that the magnetic field lines were perpendicular to the sample's longitudinal axis (3). Such an arrangement, as in Fig. 1, has ensured the high efficiency of generating the electromagnetic forces. In order to the results were comparable, an alternating current of 500 A was always applied, i.e. the current density was 1.25 $A.mm^{-2}$. Under these parameters, the vibrations have occurred with the amplitude at which contamination of aluminium from the mould surface did not arise.

3. Results and discussion

The character of structure in the cross-section was a criterion for the grain refinement assessment. For studying the macrostructure, specimens were cut out near from the centre of the sample, where the axial heat transfer did not occur. Fig. 2 documents the nature of electromagnetic methods' effects on a macrostructure studied in the 10 x 40 mm cross-sections of samples.

The structure of the reference casting poured without treatment is shown in Fig. 2a. It is formed by the columnar grains which began to grow exogenous on the mould wall, and their growth finished at the joint uneven boundary in the centre. The length of grains is 5 mm, and their width ranges from 1 to 2 mm.

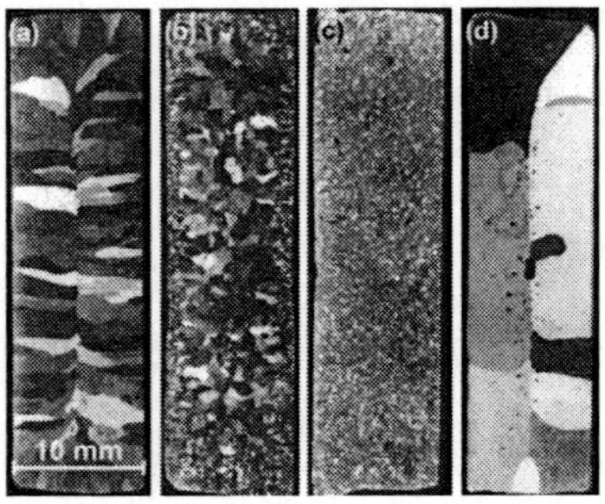

Fig. 2. The macrostructure of the Al 99.99 samples in cross-sections of 10 x 40 mm: (a) a reference; (b) AC current of 500 A and direct magnetic field of 0.035 T; (c) AC current of 500 A and direct magnetic field of 0.07 T; (d) AC current of 500 A.

The macrostructure of the target poured under the simultaneous action of the magnetic field of 0.035 T and the AC current of 500 A is presented in Fig. 2b. At the periphery, the surface layer of the 0.8 mm thickness can be observed, which is made up of the equiaxed grains with the size from 300 to 400 μm. Then, it continues into the central zone formed of the crystallites having the different morphologies and the size from 500 μm to 1.2 mm. The target's macrostructure treated by the simultaneous action of the alternating current of 500 A and the magnetic field of 0.07 T is seen in Fig. 2c. In the whole section, the structure is uniform, consisting of the fine equiaxed grains of 20 – 80 μm.

The macrostructure of Al99.99 influenced only by alternating current of 500 A during solidification is in Fig. 2d. Here, the structure is formed with the coarse crystallites which have arisen on the mould cavity surface, and their solidification was completed on the casting's heat axis, where the plate constituents met at a flat boundary. In a vicinity of the centre, the numerous pores can be observed. Their formation relates to the precipitation of gases dissolved in the liquid metal initially. The different coloration of crystallites can be interpreted as a result of the different orientation of the prevailing crystallographic planes in the section. Moreover, crystallites have the shape of a plate with a thickness of 5 mm and a transverse dimension from 3 to 30 mm. The fine structure presented in Fig. 2c and the very coarse one in Fig. 2d, in spite of their complete differences, were considered to be the successful outcome from the point of view of the targets' needed technological properties.

The theoretical analysis has been carried out on the phenomena occurring when the electromagnetic methods are applied to identify processes involved in influencing the primary crystallization and to partially quantify them. The changes observed in the macrostructure of Al99.99 caused by a passage of AC current can be attributed to the action of the electromagnetic forces generated by an interaction of electric current with a magnetic field due to this current. They always act concentrically to an axis of the liquid metal volume, regardless of the current's polarity, and thus the solidification progressing planarly inwards from the surface advances against the rising pressure, what can explain a flat phase boundary and the observed coarsening of grain.

A distinct refining of macrostructure caused by the simultaneous action of a magnetic field $\boldsymbol{B}$ and the alternating current I can be explained by the action of the electromagnetic forces produced which are acting in a plane perpendicular to vectors of an electric current and a magnetic field and take the form of vibrations. The solidified liquid metal with a crystallization front has represented a complex of the elementary conductors of the different electrical conductivities. As a result of this, there is a non-uniform distribution of current density, as well as of the intensities of the generated forces. Irrespective the electric current oscillation over time, the forces do not act on the metal volume as a whole, but their action is localised at the junction between microvolumes of the different conductivities. Moreover, authors' research has shown a magnetic field itself is actively involved in the control of crystallization process, as well. The macrostructures in Fig. 2b and 2c can be interpreted as the result of the synergic effect of vibrations in the liquid metal, and the magnetic field's effect on the kinetics of phase transformations of primary crystallization.

4. Conclusions

The electromagnetic methods bring about more factors characterized by the diverse effects on the primary structure formed than those provided by the physical-metallurgical ones, or by the influence of mechanically generated forces. Pure aluminium of Al99.99 has provided the best modelling conditions for verifying the character of changes in the as-cast structure, which are caused by the flow of the AC current or by the simultaneous action of a magnetic field and the electric current.

Acknowledgments

This work was supported by the Slovak Research and Development Agency under the contract No. APVV-16-0485.

References

[1] S. Asai. (2012). Electromagnetic Processing of Materials: Materials Processing by Using Electric and Magnetic Functions. Netherlands: Springer.

[2] A.L. Greer, P.S. Cooper et al. (2003). Grain Refinement of Aluminium Alloys by Inoculation. Adv. Eng. Mater. 5 (1-2), 81-91.

The 73rd World Foundry Congress
„Creative Foundry"
23 - 27 September 2018, Krakow, Poland

RECYCLING OF DISPERSED METAL WASTES IN ROTARY FURNACES – A METHOD OF CREATING A NEW SOURCE OF RAW MATERIALS FOR FOUNDRY

S.L. Rovin[1]*, A.S. Kalinichenko[2], L.E. Rovin[3]

[1]UE "Technolit",24, Y. Kolas Str., 220013 Minsk, Belarus
[2]Belarusian National Technical University,65, Nezavisimosti Ave., 220013 Minsk, Belarus
[3]Gomel State Technical University named after P.O. Sukhoi,48, Octyabrya Ave., 220068 Gomel, Belarus
**Corresponding address: technolit@mail.ru*

Key words: recycling, dispersed metal wastes, rotary tilting furnace

The new method proposed of recycling allows processing wastes with almost any composition and state from metal to oxide and multicomponent ones (shavings, fine scrap, mill scale, aspiration dust, sludge, etc.). Moreover, they can be even contaminated with moisture, oil, and organic impurities. The method developed does not require preliminary preparation of the initial materials (cleaning, homogenization, pelletizing, etc.). The finished products are ingots (pigs) for subsequent processing aiming the particular chemical composition or certain cast alloys. Developed process for recycling iron-containing oxide materials based on continuous solid-liquid-phase recovery in one unit – tilting rotary furnace (TRF).

All methods of iron production used today in the steel industry involve the preliminary preparation and pelletizing of raw materials because conventional melting units in foundry are not suitable for melting particulate material. In addition, these technologies are focused on raw materials which are stable and uniform in composition and properties. Processing even the most valuable of the dispersed waste material such as chips without preliminary preparation leads to great losses (the groove) of the metal during melting, a significant increase in energy consumption and reduction in the quality of the alloy.

Cold briquetting is widely used method for preliminary preparation of shavings or chips. But cold briquettes quality does not meet the requirements of melting process. The application of hot briquetting allows to improve their quality including the density of the briquettes – up to $(6.0\text{-}6.5)\cdot 10^3$ kg/m^3, however this leads to increased cost of production to the value of the primary charge materials. But the quality of the briquettes is still not reaching the quality of dense metal.

It is known that the rate of heat transfer intensity and all heterogeneous processes on the boundary "gas – solid body" including solid-phase recovery depend on the specific surface area of the reactants. Study of dispersed iron-containing waste shows their significant heterogeneity due to large number of defects and high porosity at the macro and micro levels. So, one can observe high values of specific reaction surface, which is hundreds of times greater than the specific surface of the pellets and agglomerates. The result is the prerequisite for creating intense and effective recycling process.

The reduction process of disperses porous materials can be adequately described by a "homogeneous" model, in which the reducing agent penetrates the entire depth of the porous particles, which ensures equal conditions and, respectively, the reduction rate throughout the volume of the particle.

The implementation of such a process is possible on the basis of a new type of furnaces – tilting rotary furnaces (TRF), in which the dispersed materials are constantly kept within the dynamic mixing layer. This layer is affected by the speedy high-temperature flow of reducing gases which is characterized with loop-like circulation. Constant mixing layer and its intense purging greatly speed up the processes of mass and heat transfer in the TRF. For example, the volumetric heat transfer coefficient (α_v) reaches the values of 2000-3000 W/(m^{3}K), while in a stationary layer of material it is at the level of 3-4 W/(m^{3}K) [1].

A comprehensive study of motion of gases and particulate material in the TRF, including computer simulation of the movement of non-isothermal gas flows and movement of materials in the furnace, allowed to establish new features and regularities of heat and mass transfer processes, develop ways of controlling these processes and their intensification. This ensured the enhancement of thermal efficiency of the TRF to the level of 50-55%, which is 3-5 times higher than the electric (induction and arc) or stationary fuel furnaces at work on similar materials.

The proposed process of the recycling oxide iron-containing waste is carried out as a two-stage continuous process in one unit of the TRF. The initial material (a metal wastes) and solid reducing agent are loaded into the furnace without any prior treatment and are heated by gas or liquid fuel to a temperature of 1000-1100°C. Further, the air excess factor (α) is reduced to 0.6-0.7 and the process moves to the stage of solid-phase reduction (SPR). The duration of this stage is 1.5-2 h and temperature in a furnace is maintained at 1000-1200°C and reducing atmosphere: $CO/CO_2 = 1.5\text{-}2.5$.

After the recovery level reaches 70-80% the process is moved to the stage of liquid-phase reduction (LPR). Fluxes are loaded into the furnace, if necessary, a reducing agent is added, oxygen is supplied and the temperature rises to 1700-1800°C. Process of transition to the melt is carried out for 5-6 minutes; the rise of temperature is carried out at rate of 2-3 K/s, which eliminates the sintering of the material and the formation of bloom. In the LPR-process the degree of metallization reaches 98-99%. The finish of metal bath boiling indicates the completion of the process. Duration of liquid-phase recovery takes 25-30 minutes. The melt obtained may be poured into ingots and then use as charge in the traditional melting units or transferred in the liquid state in an electric furnace to obtain a fine alloy (duplex-process). In addition, the melt may be treated to the specified composition directly into the TRF by carburizing, refining or alloying.

Testing of the developed technology in production conditions and definition of technological regimes were carried out at the experimental-industrial sample of TRF with capacity 500 kg. Scale, aspiration dust, metallurgical sludge was used as the feedstock in this testing process. Screenings of coke, coal and graphite, as well as lignin was used as a reducing agent. Consumption of reducing agent was in the range of 40 % by weight of oxides (depending on their composition). The specific consumption of gaseous fuel for scale recycling was about 200 m^3/t of the metal produced, and oxygen consumption was approximately 40 m^3/t. Duration and the basic technological parameters of experimental melting of scale and sludge in TRF are shown in Fig.1.

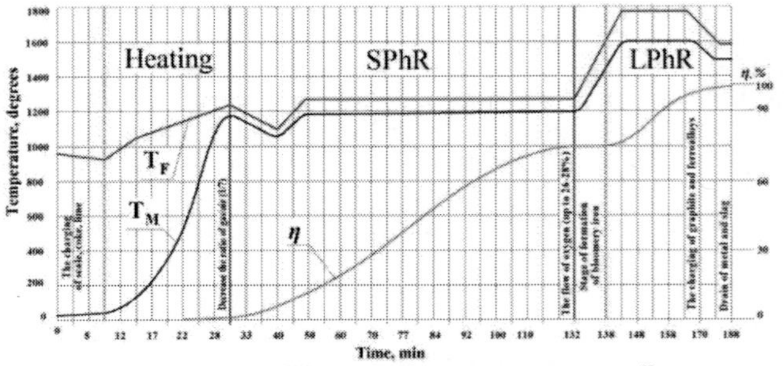

TF – furnace temperature; TM – materials temperature; η – metallization rate

Fig. 1. Mode of scale reduction melting in RTF.

Under optimal technological mode of iron-containing waste recycling in the TRF the metal output (steel or cast iron) reaches 90-92% of the theoretically possible value. The intensity of the process is 5-8 times higher compared to the SPR in the such known processes as "Corex" and "Midrex". The specific energy consumption is 3-5 times less than in liquid-phase processes such as "Hismelt" and "ROMELT". Process developed allows batch processing of materials that enables decentralized recycling including small-scale recycling of dispersed metal waste directly at the enterprises-sources of waste without any pre-treatment. Areas of dispersed iron-containing waste recycling in TRF based on continuous solid-liquid-phase reduction process compared to the known technologies of iron direct production are represented with the diagram Fe-C (Fig. .2). In addition to the recycling of iron containing wastes processing of aluminum particulate, copper and lead-containing waste were developed and introduced into production processes using tilt rotary furnaces. The developed technology and equipment are well adapted to the conditions of production and do not require a large investment due to technological flexibility. The payback period is 3-6 months. The introduction of recycling technology with the use of TRF allows creating a new source of raw materials for foundry production.

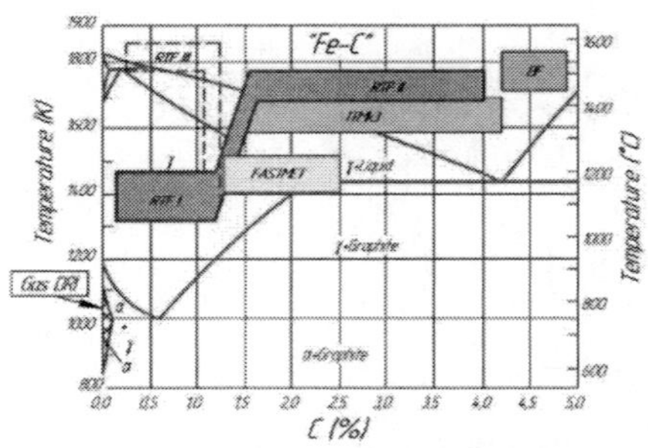

Fig. 2. Technological areas of producing iron carbon alloys in RTF.

Moreover, it is possible to reduce dependence on the primary charge materials, organize a waste-free circulation of metals in industry. In addition, it will eliminate accumulated metal-containing industrial waste, use of solid carbonaceous materials waste, and thus to obtain considerable economic and environmental benefits.

References

[1] Rovin S.L. Recycling of Metal Wastes in Rotary Furnaces. – Minsk: BNTU, 2015. – 382 pp. (in Russian)

The 73rd World Foundry Congress
„Creative Foundry"
23 - 27 September 2018, Krakow, Poland

COMPLETE MASTER OF THE COMPLEX CASTING IN THE TECHNOLOGY OF HIGH PRESSURE DIE-CASTING

Primož Mrvar[1]*, Alojz Križman[2], Grega Gorše [3], Branko Bauer [4], Mitja Petrič[1]*

[1] *University of Ljubljana, Faculty of Natural Sciences and Engineering, Aškerčeva 12, Ljubljana, Slovenia*
[2] *University of Maribor, Faculty of Mechanical Engineering, Smetanova ulica 17, Maribor, Slovenia*
[3] *LTH Castings, Litostrojska cesta 46, Ljubljana, Slovenia*
[4] *University of Zagreb, Faculty of Mechanical Engineering and Naval Architecture, Ivana Lučića 5 Zagreb, Croatia*

**Corresponding address: mitja.petric@omm.ntf.uni-lj.si*

Key words: High pressure die casting, Local squeezing, Pressure analyses, Shrinkage porosity, Calculation of casting technology, CT analyses, Micro and macro structure .

In contemporary high pressure die casting (HPDC) foundries the mastery of each sequence in production cycle is the most important, where the strive to reliable master, as well as planning of composed molten metal, pouring and solidification process, ejection of castings, transport to the cooling place and cutting of gating system and overflows were done.

For castings with a complex geometry and dimensional accuracy, the appropriate planning of pouring and feeding elements according to a heat economy of casting, rapid tooling and prototyping and then reliable manufacturing which includes the mastery of all the edge conditions in the process chain.

In the paper, the example of virtual analysis of casting from Al alloy will be presented with choosing of appropriate foundry technology HPDC, calculation of casting process which includes the filling process of cold chamber, model description of three phases at HPDC, flow of molten metal, solidification with considering the temporary air gap formation between the casting and tool, formation of stress and relaxation of it into deformations in each sequence, when the tool is opening, at ejection, cooling in water or on air and after cutting off the gating system.

Since it is not always possible to produce the castings according to the principle of unidirectional solidification with a traditional approach, for the individual areas, which should be macroscopic dense (without porosity), the local squeezing process is performed in the sequence of the semi-solid state of the region. The location of the impression is marked by a local increase of pressure and a plastic deformation of the already solid part of the casting (solid shell). An example of a very detailed comprehensive treatment of these processes in the case of a real casting will be discussed.

Further, the optimization of the processes of the local squeezing sequence and the overall solidification process will be presented. Comparisons will be made with calculations of volume defects, casting dimensions and deformations with experimentally obtained castings produced from LTH Castings' industrial technology practice.

Proven complete master of high-pressure die-casting have the result of an important financial effect and decreasing of required time to start of serial production of castings.

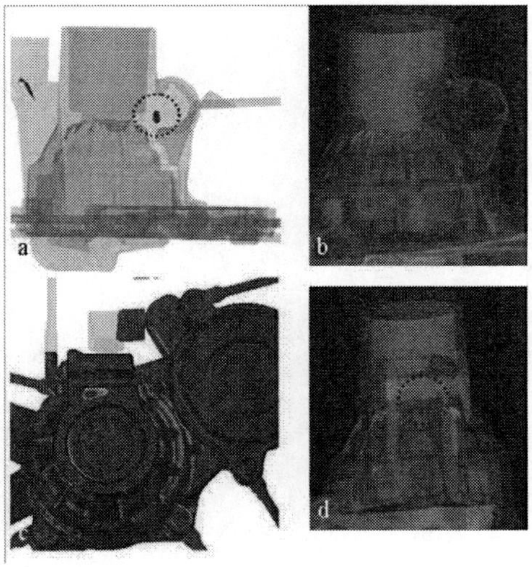

Fig. 1. ***Calculated shrinkage porosity (a), (c) and CT of experimental casting (b), (c).***

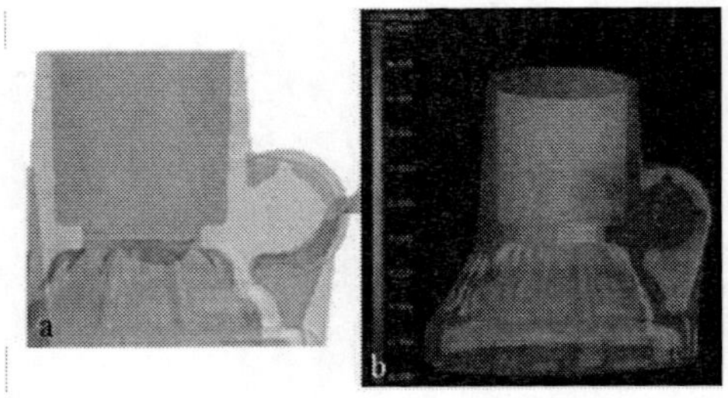

Fig. 2. Calculated shrinkage porosity (a) and CT scan of casting after use of local squeeze technology.

The 73rd World Foundry Congress
„Creative Foundry"
23 - 27 September 2018, Krakow, Poland

STUDIES OF ACCELERATED MICROWAVE DRYING OF CERAMIC MOULDS

Rafał Kaczorowski*, Paweł Just, Tadeusz Pacyniak

[1]*Lodz University of Technology, Stefanowskiego 1/15 Str., 90-924 Lodz Poland*

**Corresponding address: tadeusz.pacyniak@p.lodz.pl*

Key words: casting, ceramic moulds, microwave drying, foamed patterns

The paper presents investigations of accelerated drying of ceramic moulds applied onto patterns made of foamed plastics used in the Ceramic Shell technology.

The studies aimed at determining the microwave radiation parameters (power, time of intervals and uninterrupted work) in order to obtain the shortest possible drying times, not causing model destruction.

The analysis of the test results confirmed that with the increase of the microwave radiation power, the drying time of the particular ceramic mould layers shortens, which is, however, simultaneously connected with an excessive increase of the mould's temperature.

With the microwave power below 1200 W, it is possible to obtain the drying time of one layer at the level of about 30 min, while the mould's temperature reaches the value of 70°C, which does not cause deformation or partial melting of the polystyrene pattern.

The lost wax technology is one of the basic methods of producing complicated precision casts of a low mass [1, 2]. This is connected with the fact that a mould joint is not required and the internal cast surfaces can be obtained without the use of cores. In this technology, various types of (synthetic and natural) waxes are applied for the patterns.

Due to the use of wax (melting point above 40°C), the mould drying temperature usually does not exceed 23°C. This is connected with a prolonged drying time and the fact that, under industrial conditions, each coating is dried at ambient temperature for the time of one day. And so, the preparation of a 7-layer mould lasts about a week.

A technology similar to the lost wax method is the Replicast CS technology [3]. In the replicast CS (ceramic shell) process, wax patterns have been replaced with foamed high-molecular plastics with a high dimension-shape accuracy, made in metal moulds. The patterns were made from extruded polystyrene in the form of closed bushings, presented in Figure 2.

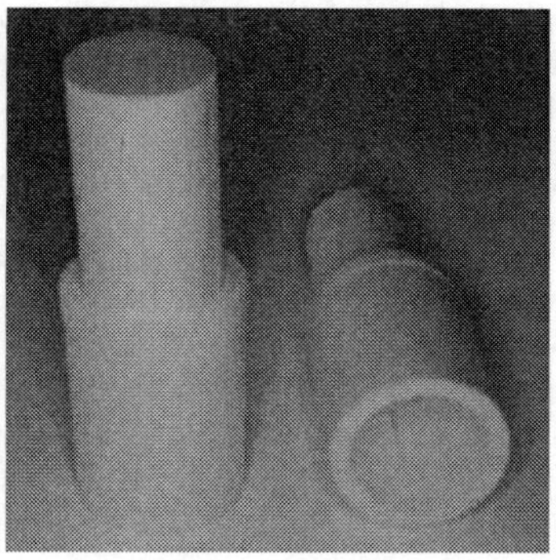

Fig. 1. Pattern of a closed bushing

The models are assembled into sets, onto which layers are applied by being submerged in liquid ceramic mixtures.

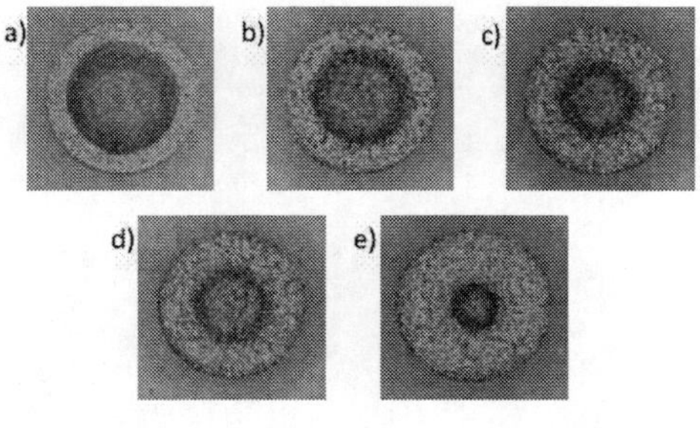

Fig. 2. Bushing patterns with a) 1 layer, b) 2 layers, c) 3 layers, d) 4 layers, e) 6 layers (dried)

After each submersion, the applied coating undergoes a dehydratization-drying process. When models made of plastic are applied, the drying temperature can rise up to about 70°C, which makes it possible to significantly shorten the drying time of the particular layers. Through a multiplied application of the ceramic mixture layers, a ceramic mould is obtained. Next, at 925÷1000°C, the moulds are annealed, during which the foamed pattern is fired and the ceramic mould is hardened.

The replicast CS method is increasingly popular on a global scale as an alternative for the lost wax technology, especially for higher mass patterns. This results from the fact that foamed high-molecular plastic patterns are lighter than melted patterns, while their dimension-shape precision is comparable.

In the presented paper, for the process of drying moulds based on water ceramic coatings, microwave drying was applied – a method more and more frequently used in various global industry branches [4, 5]. Technology of fast microwave dryer for Investment casting was developed by Qbig company, with patent pending. The application of microwave radiation for ceramic mould drying results from its numerous benefits:

- For the ceramic mass, water coatings based on colloidal silica were used (the water absorbs the microwaves and becomes heated very well),
- The plastic – polystyrene – pattern is a dielectric which does not absorb microwaves,
- Low operation costs – it is characteristic of microwaves heating that the microwaves are absorbed only in the thermally treated product, without the necessity to heat the whole device,
- Small dimensions of the microwave installations.

Acknowledgments

Research has been commissioned by Qbig company within project no. POIR.01.01.01-00-0552/15 (competition 1/1.1.1/2015, within the Smart Growth Operation Program, Activities 1.1 „R&D Enterprise Projects", Sub-measure 1.1.1 „Industrial research and development works conducted by enterprises."

Project title: „The development of innovative equipment using a unique high-speed technology to manufacture multilayer ceramic moulds for precision casting."

References

[1] Sarojrani Pattnaik, D. Benny Karunakar, P.K.Jha (2012): Developments in investment casting process - A review. Journal of Materials Processing Technology, Vol. 212, Issue 11, 2332-2348

[2] Z. Żółkiewicz, A. Karwiński (2012): Properties Research of Ceramic Layer. Archives of Foundry Engineering, Volume 12, Special Issue 2/12, 91-94

[3] M.C. Ashton, S.G. Sharman, A.J. Brookes (1984): The replicast CS (Ceramic Shell) process, Materials & Design Vol 5, Issue 2, 66-75

[4] C.M. McLoughlin, W.A.M. McMinn, T.R.A. Magee (2003): Microwave-vacuum drying of pharmaceutical powders, Drying Technology, 21 (9), 1719-1733

[5] A.E. Drouzas, H. Schubert (1996): Microwave application in vacuum drying of fruits, Journal of Food Engineering, 28, 203-209

[6] C. Rapiejko, B. Pisarek, E. Czekaj T. Pacyniak (2014): „Analysis of AM60 and AZ91 alloy crystallization in ceramic moulds by thermal derivative analysis (TDA)", Archives of Metallurgy and Materials. Vol. 59, Issue 4, 1449-1455

[7] C. Rapiejko, B. Pisarek, T. Pacyniak (2014): „Effect of Cr and V alloy additions on the microstructure and mechanical properties of AM60 magnesium alloy", Archives of Metallurgy and Materials. Vol. 59, Issue 2. 762-765

[8] Rapiejko, C., Pisarek, B. & Pacyniak, T. (2017): Effect of intensive cooling of alloy AM60 with chromium and vanadium additions on cast microstructure and mechanical properties. Archives of Metallurgy and Materials., 62 (1), 309-314

[9] B.P. Pisarek, C. Rapiejko, R. Święcik, T. Pacyniak (2015): "Effect of Inhibitor Coating of a Ceramic Mould on the Surface Quality of an AM60 Alloy Cast with Cr and V", Archives of Foundry Engineering, Vol 15, Issue 3, 51-56

[10] S. Pietrowski, C. Rapiejko (2011): Temperature and microstructure characteristics of silumin casting AlSi9 made with investment casting method, Archives of Foundry Engineering, Volume 11, Issue 3, 177-186

The 73rd World Foundry Congress
„Creative Foundry"
23 - 27 September 2018, Krakow, Poland

INFLUENCE OF MN AND CR ON INTERMETALLIC SLUDGE FORMATION IN FE CONTAINING SECONDARY ALSI9CU3 ALLOY WITH AIM OF REDUCING FE LEVEL BY FILTRATION

Björn Dietrich [1*], Hanka Becker [2], Michal Smolka [1], Andreas Keßler [1], Andreas Leineweber [2], Gotthard Wolf [1]

[1]*Technical University Bergakademie Freiberg, Foundry Institute, Bernhard-von-Cotta-Straße 4, 09599 Freiberg, Germany*
[2]*Technical University Bergakademie Freiberg, Institute of Materials Science, Gustav-Zeuner-Straße 5, 09599 Freiberg, Germany*
**Corresponding address: bjoern.dietrich@gi.tu-freiberg.de*

Key words: filtering, aluminium alloy, metallurgy, Fe-removal, intermetallics

1. Introduction

In aluminium cast alloys, iron is mostly considered as an impurity having detrimental effects on mechanical properties. This can be attributed to the formation of plate like β-AlFeSi phase leading to a decrease in fracture toughness, ductility and fatigue resistance. Furthermore, coarse β-plates prevent effective feeding by blocking liquid metal, which results in shrinkage defects. [1]

Unfortunately secondary aluminium alloys usually show quite high iron contents, making them less suitable for high-quality applications like aircraft industry or safety components. Filtration of intentionally generated primary intermetallic sludge particles in the aluminium melt is a promising approach to remove iron from secondary aluminum alloys [2]. The formation of intermetallic sludge particles is promoted in presence of Mn and Cr. These particles bind larger amounts of iron than present in the average in the alloys. The tendency of the formation of sludge can be estimated by the sludge factor [3].

$$SF = 100(w_{Fe} + 2w_{Mn} + 3w_{Cr}) \quad (1)$$

Size and distribution of Fe-enriched intermetallics, depending on different Mn and Cr contents in view of filtration were investigated within the present study.

2. Experimental

A commercial secondary AlSi9Cu3 alloy was used, where the iron level was intentionally andjusted. To increase the level of Mn and Cr, AlMn20 and AlCr20 master alloys were added to the liquid alloy. The obtained chemical compositions are listed in Table 1.

Samples were produced by casting liquid alloy in ceramic crucibles and placed in muffle furnace for 6 hours at 620°C to provoke sludge formation and sedimentation. After dwell time, samples were taken out and cooled in air with a cooling rate of ≈0.15 K/s, representing sand mold conditions.

Metallographically prepared vertical cross sections of the ingots were investigated by light optical micrographs and image analysis to reveal sludge heights, particle size and distribution. For detailed investigations of microstructure and phase identification, EDS mapping and EBSD were carried out.

Table 1. Alloy compositions

Alloy	Fe [wt%]	Mn [wt%]	Cr [wt%]
M1	3.04	0.34	0.03
M2	3.79	0.86	0.04
C1	3.93	0.24	0.42
C2	3.55	0.24	0.51
MC1	3.71	0.41	0.35
MC2	3.39	0.49	0.72

3. Results and discussion

In Figure 1, sludge heights, area density and the number of detected particles is shown. The area density was calculated dividing the total area of detected sludge particles by the area of a predefined region of interest (ROI). The lognormal distribution fitted to observed particle size data are shown in Figure 2.

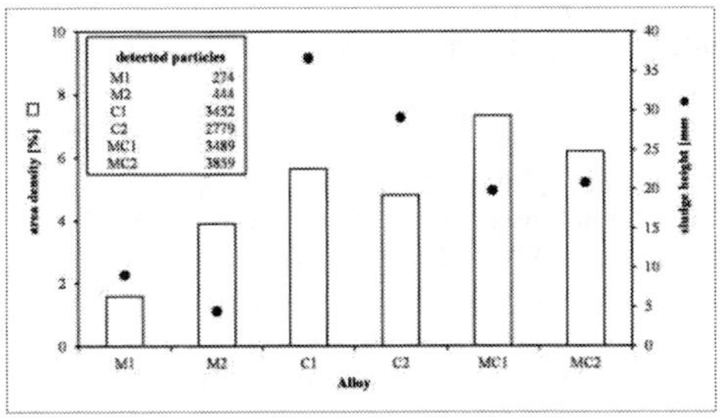

Fig. 1. ***Sludge height and area density of intermetallic particles***

In comparison, Cr addition to the melt has stronger impacts on formation of intermetallics than Mn addition. The number of detected particles strongly increases as well as sludge height, which is in good agreement to equation (1), in which the addition with Cr is rated with factor three and Mn with two. Quite contrary to equation (1), alloys with increased Mn and Cr level sludge height decreases and particle size is reduced. In combination with a further increase in number of detected particles and area density, this can be seen as an indication for a higher volume fraction of sludge in the sludge zone.

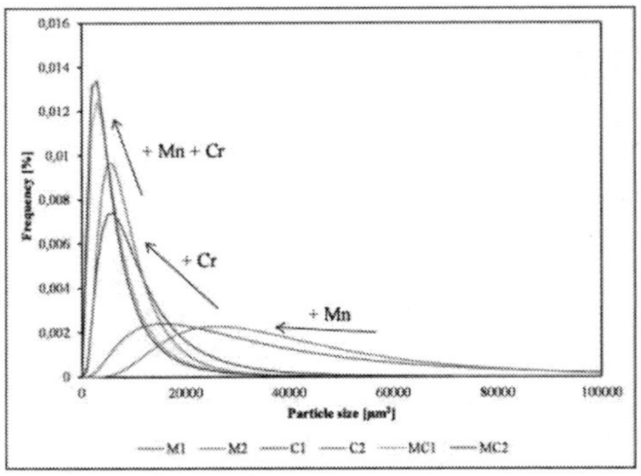

Fig. 2. Particle size distribution of Fe-enriched intermetallic particles

The particle size distribution is illustrated in Figure 2. Adding only Mn (M1 and M2) to the basic Fe-enriched AlSi9Cu3 alloy, the mean particle size exhibits a value of about $4.3{\cdot}10^4$ µm². Adding Cr to the melt, the value shifts to a range of about $1{\cdot}10^4$ µm². Combining Mn and Cr additions, the mean size of particles further decreases to $0.7{\cdot}10^4$ µm².

Besides the size and distribution of Fe-enriched intermetallics, for a successful removal of iron from aluminium alloys by filtration, it is crucial, that the generated primary intermetallic particles are capable to bind large amounts of iron. Hence their removal decreases the iron level in the residual melt. SEM analysis shows that sludge particles consist of various phases, mainly containing the cubic α-phase, which are embedded in the Al-matrix. Adding Mn only to the melt, α-particles bind more than 15 at% Fe. Addition of Cr leads to the formation of the $Al_{13}Cr_4Si_4$-phase [4]. The observed $Al_{13}Cr_4Si_4$-phase is contained in the centers of sludge particles (Figure 3). The phase composition determined by EDS analysis reveals that the $Al_{13}Cr_4Si_4$-phase binds less than 1 at% iron.

Removing the primary intermetallic sludge particles from aluminium melt, leads to an expected Fe reduction of up to 69% for the Mn modified alloy M2.

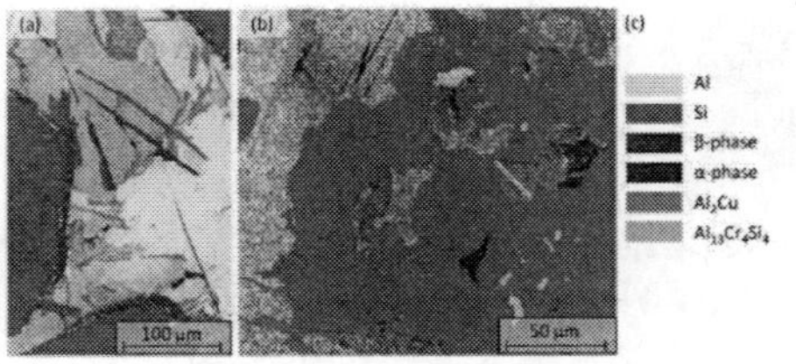

Fig. 3. EBSD phase maps of sample (a) M2 and (b) C2

4. Conclusions

1) The formation of sludge particles in Fe containing secondary AlSi9Cu3 alloys can be influenced by Mn and Cr addition.
2) As compared to Mn addition, Cr addition leads to a larger amount of sludge consisting of a large amount of small particles.
3) The sludge particles consist mainly of the cubic α-phase. Additionally in presence of additional Cr, the $Al_{13}Cr_4Si_4$-phase occurs.
4) The capacity to bind iron from the melt is more than 15 at% at α-phase, but less than 1 at% at $Al_{13}Cr_4Si_4$-phase
5) With aim of filtration, Mn additions are favored due to the formation of larger particles and higher efficiency to bind Fe

Acknowledgments

The authors would like to thank the German Research Foundation (DFG) for supporting the investigations in the Subproject S03 and A07, which are a part of the Collaborative Research Center CRC 920.

References

[1] Mbuya, T., et al. 2003, Influence of iron on castability andproperties of aluminium silicon alloys:literature review, International Journal of Cast Metals Research, Vol. 16 (No. 5), 451-465

[2] Moares, H.L., et al 2006, Removal of Iron from Molten Recycled Aluminum through Intermediate Phase Filtration, Materials Transactions, Vol. 47 (No. 7), 1731-1736

[3] Shabestari, S.G., 2004, The effect of iron and manganese on the formation of intermetalliccompounds in aluminum–silicon, Materials Science and Engineering, A 383, 289–298

[4] Dietrich, B.G., et al, 2017, Intermetallic sludge formation in Fe containing secondary Al-Si alloys influenced by Cr and Mn as preparative tool for metal melt filtration, Advanced Engineering Materials, Vol. 19 (No. 9)

VISUAL INSPECTION OF INVESTMENT CASTING MADE OF NICKEL-BASED SUPERALLOY

Pawel Rokicki*, Rafal Cygan, Kamil Gancarczyk

[1]Rzeszow University of Technology, Research and Development Laboratory for Aerospace Materials, Al. Powstancow Warszawy 12, 35-959 Rzeszów, Poland
**Corresponding address: prokicki@prz.edu.pl*

Key words: investment casting, nickel-based superalloys, visual inspection

The key critical elements of the aircraft engine responsible for operational safety are made of nickel-based superalloys, which are characterized by high fatigue and corrosion resistance as well as good mechanical properties at elevated operating temperature. Elements of this type are most often obtained by investment casting methods, as this is currently the most popular one for obtaining complex geometries such as blades or segments.

The investment casting method is directly combined with the use of multi-layered ceramic moulds. Casting of metal into this type of moulds allows to obtain complex, asymmetrical geometries that do not require further finishing (which affects the size and geometry of the element). Elements casted by investment casting can work under the influence of dynamic stresses at elevated temperature, reaching even in extreme cases 1350 °C. That is why their reliability is so important, as it is determined by durability and stiffness and adequate impact strength - which translates into the properties of the entire engine, and thus the aircraft and its ability to work.

The most important criterion in production of key elements described above is the lack of defects in the mould in meaning of porosity, misruns and metallic inclusions, which occur as a result of various phenomena occurring at the interface of the liquid metal - ceramic mould material. Therefore, the quality requirements regarding the appropriate surface quality of components made in the investment casting method (especially in the aviation industry) are very strict and of great importance.

The most popular currently used ceramic materials in the production for investment casting ceramic shell moulds are aluminosilicates with the addition of a silica-based colloidal ceramic suspension. To improve the metallurgical purity of castings, in the process of pouring, foam filters and single or multi-layered ceramic crucibles are used.

The analysis of various materials and technological factors allows to conclude that the main source of surface defects in castings are: ceramic casting mould, crucible and filters. This is a consequence of their properties and heterogeneity. The occurrence of surface defects directly affects the profitability of the entire production, which is particularly important in multi-series production (as in the case of blades and vane segments).

The main source of defects in the process of pouring and solidification are phenomena occurring in the material of the ceramic mould and the degradation of filters and crucibles in contact with liquid metal (Fig. 1).

Fig. 1. Defects in the internal surface of the melting crucible

The research presented in the paper summarizes the work done for a series of the castings prepared in the CPP Foundry Poland as a real production example. The defect were analyses by means of visual conventional macro inspection, but also by means of scanning electron microscopy. The aim of the work was to determine the statistical chance of different defects types occurrence, depending on the parameters of the casting process.

Acknowledgments

The authors gratefully acknowledge the funding by National Centre for Research and Development, Poland, under grant LIDER/227/L-6/14/NCBR/2015 „ New technology for investment casting manufacturing critical components engine with a ceramic materials new generation".

References

[1] S. Jones, C. Yuan (2003) Advances in shell moulding for investment casting, Journal of Materials Processing Technology, Vol 135, Issue 2-3, 258-265

[2] C.W. Lee, C.K. Chua, C.M. Cheah, L.H. Tan, C. Feng (2004) Rapid investment casting: direct and indirect approaches via fused deposition modelling, The International Journal of Advanced Manufacturing Technology, Vol. 23, Issue 1-2, 93-101

[3] P. Rokicki, G. Budzik, K. Kubiak, J. Bernaczek, T. Dziubek, M. Magniszewski, A. Nowotnik, J. Sieniawski, H. Matysiak, R. Cygan, A. Trojan (2014) Rapid prototyping in manufacturing of core models of aircraft engine blades, Aircraft Engineering and Aerospace Technology: An International Journal, Vol. 86 Issue: 4, 323-327

[4] E. Lassow, P. Johnson, S. Whitaker, M. Guerra Jr. (1986) United States Patent No 4,703,806

The 73rd World Foundry Congress
„Creative Foundry"
23 - 27 September 2018, Krakow, Poland

WETTABILITY MEASUREMENT OF AlSi7Mg ON Al_2O_3, $MgAl_2O_4$, $3Al_2O_3 \cdot 2SiO_2$ and TiO_2 AT 730°C

Claudia Voigt *, Christos Aneziris

Technische Universität Bergakademie Freiberg, Institute of Ceramic, Glass and Construction Materials, Agricolastr. 17,09599 Freiberg, Germany
**Corresponding address: claudia.voigt@ikgb.tu-freiberg.de*

Key words: filtering, sessile drop measurement, aluminum alloy, wettability

1. Introduction

Despite the application of metal melt filtration with ceramic foam filters since the 1960s the process is not fully understood. Outstanding issues are for example the influence of the wetting behavior and filter chemistry on the filtration efficiencies. Regarding the influence of the wetting behavior of a filter material in contact with molten alloy different interpretation can be found in literature.

Syvertsen et al. [1] and Bao et al. [2] assume that a better wetting (lower contact angle) causes an improved convergence of the aluminum melt to the filter wall which increases the probability of a true contact between inclusion and the filter wall. This theory based on the measurement of filtration efficiencies and contact angles of two materials SiC and Al_2O_3. The filtration trials with SiC and Al_2O_3 filters, conducted at SAPA Heat Transfers (Finspång, Sweden) showed a higher but time dependent filtration efficiency for SiC filters [1]. The contact angles were evaluated at 1000°C, 1100°C, 1200°C, 1300°C and extrapolated to a temperature of 700°C, which were for SiC 79°C and for Al_2O_3 97° [3]. Bao et al. [2] conclude that a better wetting would improve the filtration efficiency. On the contrary Uemura et al. [4], who focused at the filtration of steel, assumed an improvement of the filtration efficiency with increasing contact angles between the filter and molten metal. The equilibrium of forces was used for the calculation of the adhesive forces of inclusion at the filter wall and the determination of a relationship between the adhesive forces and the contact angle. The filtration experiments of Uemura et al. [4] with filters made of $ZrO_2 \cdot Al_2O_3$, ZrO_2, $CaO \cdot 6Al_2O_3$ and $3Al_2O_3 \cdot 2SiO_2$ did not show a dependence of the contact angle on the filtration efficiency which was explained by Uemura et al. with the small differences between the contact angles of the tested materials [4].

Fankhänel et al. reported the contact angle measurements of the aluminum alloy AlSi7Mg on four tested ceramic substrates (Al_2O_3, $MgAl_2O_4$, $3Al_2O_3 \cdot 2SiO_2$ and TiO_2) at 950°C [5]. Structural evaluations of the substrate-side interface proved a magnesium impoverishment of $MgAl_2O_4$ and a silicon impoverishment of $3Al_2O_3 \cdot 2SiO_2$ while in the TiO_2/AlSi7Mg couple, the interfacial layer consisting of Al, O, Si and Ti was noted to occur [5].

In this work wetting measurements of AlSi7Mg on four different materials (Al_2O_3, $MgAl_2O_4$, $3Al_2O_3 \cdot 2SiO_2$ and TiO_2) at 730°C were done. The filtration results of the pilot trials at Constellium Technology Center (Voreppe, France) [6] were correlated with the contact angles obtained in this study.

2. Materials and Methods

Four different filter surface chemistries (Al_2O_3, $MgAl_2O_4$, $3Al_2O_3 \cdot 2SiO_2$ and TiO_2) were tested in filtration trials in a filtration pilot line at Constellium Technology Center (Voreppe, France) [6]. Each filter filtered 730 kg of the aluminum alloy AlSi7Mg, which had a temperature of 730°. After the filtration the aluminum alloy AlSi7Mg solidified in the filter. For the investigation of the aluminum alloy filter interfaces samples were cut out and embedded with an epoxy resin. After hardening of the resin the samples were grinded and polished. The prepared samples were investigated with the help of a SEM Philips XL 30 (Philips, Germany). Usually, the substrates for wetting experiments are polished. In order to enable correlation between the wetting test and the filtration trials, comparable conditions regarding temperature, holding time and surface roughness should be assured. Therefore, substrates, used in the sessile drop experiments consisted of pressed alumina substrates coated with the same slurries which were used to coat the filters used for the filtration trials [6] (except the TiO_2 substrate that was produced from TiO_2 disk coated with TiO_2 slurry). The substrates (diameter 12 mm) were uniaxially pressed with 60 MPa and sintered at temperatures of 1600°C (Al_2O_3) and 1300°C (TiO_2). A commercial AlSi7Mg alloy from Trimet Aluminium AG (Essen, Germany) was used as metal.

For the sessile drop measurements at 730°C, a sessile drop apparatus allowing the application of a capillary purification technique were used at the Foundry Research Institute (Krakow, Poland). The test substrate and aluminum alloy (the latter was located in a graphite syringe above the substrate) were non-contact heated (simultaneously but separately) under vacuum with a heating rate of 8 K/min. At the test temperature of 730°C one or two droplets were extruded out of the graphite syringe and oxide-free droplets were dropped on the ceramic substrate already preheated to the test temperature. The extrusion took place at a vacuum smaller than 1×10^{-5} mbar. After dropping, the isothermal heating at the test temperature (730°C) took 30 min. During wetting tests, the drop/substrate images were recorded by high-resolution CCD camera with a speed of 100 images per minute.

3. Results

The cross sections of the filters after the filtration trial were investigated with SEM and EDS mapping. No reaction layers between the aluminum alloy and the filter surface were visible and no areas with decreased magnesium for the spinel filter or decreased silicon for the mullite filter were detectable.

For every wetting experiment two AlSi7Mg droplets were dropped out of the graphite syringe. The contact angles vary during the dropping process due to the fall and impingement of the drops on the substrate. One minute after the dropping the contact angles are very stable, see Fig. 1. Nevertheless a holding time of 30 minutes was chosen to reach similar reaction times in comparison to the filtration trials with durations between 40 min and 76 min. The highest contact angle was measured for $3Al_2O_3 \cdot 2SiO_2$ (136°), followed by 113° ($MgAl_2O_4$) and 108° (Al_2O_3). The lowest contact angle possesses the TiO_2 substrate (81°). These significant differences are caused by the different substrate chemistry and roughness. The roughness of the substrates and of the filters vary. The highest roughness was measured for the $3Al_2O_3 \cdot 2SiO_2$ sample followed by Al_2O_3 and $MgAl_2O_4$. TiO_2 posseses the lowest roughness.

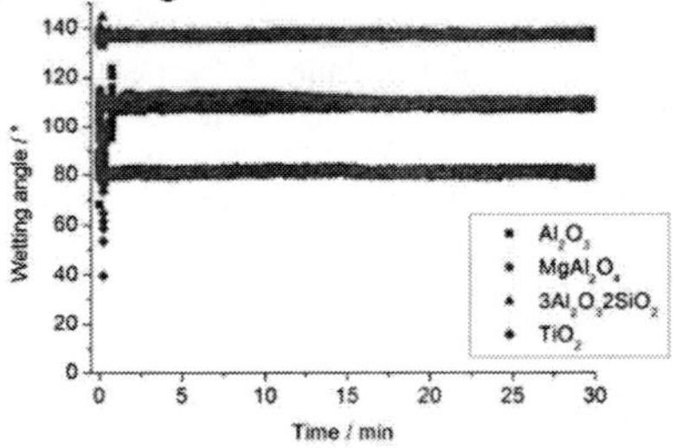

Fig. 1. Time dependence of the contact angle measured by sessile drop measurement with capillary purification

After the wetting measurement, the prepared cross sections of the samples were analyzed by SEM. The interfaces between the substrates and aluminum alloy AlSi7Mg show no reaction zones for Al_2O_3, $MgAl_2O_4$ and TiO_2, whereas the $3Al_2O_3$ $2SiO_2$ possesses changes in the silicon distribution of the filter coating.

3. Discussion

The microstructure of the interface of the wetting test samples and the filters after filtration trial [6] showed no significant difference between selected alloy/ceramic couples of AlSi7Mg/Al_2O_3, AlSi7Mg/$MgAl_2O_4$ and AlSi7Mg/TiO_2. For these couples no impoverishment zones or reaction layers were detected as a result of the comparable temperature and holding time whiles the wetting experiments and the filtration trials. The $3Al_2O_3$ $2SiO_2$ show an impoverishment zone for Si. Due to an absence of a change in the contact angle of the $3Al_2O_3$ $2SiO_2$ in dependence of the time, it is concluded that the contact angle can be used for further considerations.

The measured contact angles were correlated with the evaluated filtration efficiencies of the filtration trial performed at Constellium [6]. The filtration efficiencies are calculated from the N20 – N110 values in the unit K/kg.

The coefficients of determination show a good correlation (R^2=0.87 – 0.92) between the contact angles and the filtration efficiencies for inclusions smaller than 110 µm. The predominant mechanism at the metal melt filtration is called deep-bed filtration which is the result of two successive mechanisms: the transport of the inclusion to the filter wall and the attachment of the inclusion to the filter wall [7]. For small inclusions the determining mechanism is the transport from the metal melt to the filter wall due to the low acting detachment forces to small inclusions. The wetting behavior of filter and metal melt has an influence on the slip length and therefore on the flow velocity whereby it is valid the poorer the wetting (a higher contact angle) the higher is the flow velocity [8]. With increasing melt velocity the probability of contacts between inclusions and filter wall increases.

In order to verify the results and conclusions of these experiments, additional experiments are necessary.

Acknowledgements

The authors would like to thank the German Research Foundation (DFG) for supporting these investigations in terms of the Collaborative Research Centre 920 "Multi-Functional Filters for Metal Melt Filtration – A Contribution towards Zero Defect Materials". Subprojects A02. The authors also would like to acknowledge the support of G. Schmidt, J. Hubálková (S01), E. Werzner (B02) and N. Sobczak, R. Nowak (Foundry Research Institute, Krakow, Poland).

References

[1] M. Syvertsen. An. Kvithyld. S. Bao. A. Nordmark. A. Johansson. (2014). Parallel Laboratory and Industrial Scale Aluminium Filtration Tests with Al_2O_3 and SiC Based CFF Filters. Light Metals. 1041 – 1046.

[2] S. Bao. M. Syvertsen. A. Kvithyld. T. Engh. (2014). Wetting Behavior of Aluminium and Filtration with Al_2O_3 and SiC Ceramic Foam Filters. Transactions of Nonferrous Metals Society of China. 24. 3922 – 3928.

[3] S. Bao. K. Tang. A. Kvithyld. M. Tangstad. T.A. Engh. (2011). Wettability of Aluminum on Alumina. Metallurgical and Materials Transactions B. 42B. 1358 – 1366.

[4] K. Uemura. M. Takahashi. S. Koyama. M. Nitta. (1992). Filtration Mechanism of Non-Metallic Inclusions in Steel by Ceramic Loop Filter. ISIJ International 32. 1. 150 – 156.

[5] B. Fankhänel. M. Stelter. C. Voigt. C.G. Aneziris. (2017) Interaction of AlSi7Mg with Oxide Ceramics. Advanced Engineering Materials. 19. 9. 1700084.

[6] C. Voigt,. E. Jäckel, F. Taina. T. Ziehnert, A. Salomon, G. Wolf, C.G. Aneziris, P. le Brun. (2016). Filtration Efficiency of Functionalized Ceramic Foam Filters for Aluminum Melt Filtration. Metallurgical and Materials Transactions B. 48B. 497-505.

[7] L.N.W. Damoah, L. Zhang. (2010). Removal of Inclusions from Aluminum through Filtration. Metallurgical and Materials Transactions B, 41B. 886-907.

[8] J.-L. Barrat, L.Bocquet. (1999). Large slip effect at a nonwetting fluid-solid interface Physical Review Letters 82. 23. 4671-4674.

The 73rd World Foundry Congress
„Creative Foundry"
23 - 27 September 2018, Krakow, Poland

NUMERICAL SIMULATION OF STEEL MELT FILTRATION

Cornelius Demuth[1*], Eric Werzner[1], Steffen Dudczig[2], Christos G. Aneziris[2], Subhashis Ray [1*]

[1] Institute of Thermal Engineering, Technische Universität Bergakademie Freiberg, Gustav-Zeuner-Str. 7, 09596 Freiberg, Germany
[2] Institute of Ceramic, Glass and Construction Materials, Technische Universität Bergakademie Freiberg, Agricolastr. 17, 09596 Freiberg, Germany
*Corresponding address: ray@iwtt.tu-freiberg.de

Key words: steel filtration, lattice Boltzmann method, large eddy simulation, Lagrangian tracking, filtering

1. Introduction

The desired material characteristics of the cast metal parts require a high purity of the melt during the manufacturing process. Otherwise, the presence of non-metallic inclusions impairs the strength, toughness and fatigue life of the finished products. Melt filtration is an established industrial practice in order to clean the liquid metal, where the ceramic foam filters are employed for collecting the detrimental inclusions.

A profound understanding of the filtration process is imperative in order to estimate the effects of morphological parameters of the filter and experimental conditions on the capture and retention of inclusions. Since the complex geometry of the porous filter is virtually inaccessible to the conventional measurement methods, particularly under the inhospitable conditions of metal casting, the experimental observations are limited only to the integral quantities, e.g., the pressure drop and the filtration efficiency.

Nevertheless, the fundamental phenomena during the liquid metal filtration process can be studied using numerical simulations that facilitate the detailed investigation into the influences of process parameters on the fluid flow and most importantly, the separation or filtration of the inclusions. As a result, this approach allows one to identify specific characteristics of the process that promote better filtration.

In the present investigation, a carefully developed numerical model of the metal melt filtration is applied to the processing of liquid steel. For this purpose, the process conditions inside a steel casting simulator are considered, where the molten steel, poured from a crucible, flows through a ceramic foam filter. The numerical simulation incorporates the detailed filter geometry, obtained from the computed tomography (CT) scanned images of the filter employed in the experiment.

The modelling comprises the liquid metal flow and the motion of microscopic inclusions and gas bubbles during the filtration process. These phenomena are governed by the conservation of mass and momentum, as well as the equation of motion for the particles and bubbles. The fluid flow is computed by the lattice Boltzmann method (LBM), while a Lagrangian tracking is performed for all dispersed phases.

In view of the process conditions encountered during steel filtration, the numerical model also takes the expected turbulent flow regime into account, using the large eddy simulation (LES). In addition to the particulate inclusions, the microscopic gas bubbles, generated during the filtration process and entrained in the flow, are also tracked. Finally, the collision and the agglomeration of inclusions among themselves and with the gas bubbles are simulated.

2. Modelling of steel filtration experiment

The filtration of molten steel in a steel casting simulator facility is considered in the present investigation. In the experiment, 100 kg of steel grade 42CrMo4 (AISI 4142) were molten and held at 1650 °C inside an induction crucible furnace. The melt was further impurified by deliberately adding iron oxides and metallic aluminium in order to generate alumina inclusions. Subsequently, the molten steel was poured through the central opening of a refractory funnel, with a cross section of 50×50 mm^2 covered by a ceramic foam filter, into a tundish. The pouring process took approximately two minutes.

Table 1. Physical properties of molten 42CrMo4 steel and alumina inclusions.

Melt properties	
Density ρ_f	6851 kgm^{-3}
Dynamic viscosity μ	4×10^{-3} Pas
Kinematic viscosity ν	5.8×10^{-7} m^{2}s^{-1}

Inclusion properties	
Density ρ_d	3980 kgm^{-3}
Diameter d_d	2×10^{-5} m

The employed carbon-bonded alumina filter with the dimensions $85\times85\times20$ mm^3 exhibited a nominal pore count of 10 ppi (pores per inch), a porosity of 80 % and an average pore diameter of 6 mm. In addition, the physical properties of the steel melt and the alumina inclusions are presented in Table 1.

Altogether, the experimental conditions correspond to an interstitial melt velocity of 6 cms^{-1}. Consequently, the Reynolds number based on the interstitial velocity and the average pore diameter is in the order of $Re \approx 600$.

As compared to a casting process in an industrial environment, the steel casting simulator is operated under well-controlled conditions, in detail protective gas atmosphere, defined heating and careful insulation. For this reason, an isothermal simulation is performed. The filter geometry incorporated in the numerical simulation is obtained from CT scan images of the ceramic foam used in the experiment. A spatial resolution of 65 µm is employed in the simulation in order to maintain a reasonable computational effort.

Given that the flow inside a porous medium is already in the turbulent regime at the aforementioned Reynolds number, the flow field is solved by the large eddy simulation, which may particularly be used for a coarse discretisation. Nevertheless, the discretisation of the filter section with the dimensions 50×50×20 mm^3 along with both inlet and outlet lengths corresponding to one filter depth by an equidistant Cartesian grid with 65 µm spatial resolution comprises $\approx 5.5\times10^8$ nodes.

3. Fluid flow

The equations, governing the fluid flow during the metal melt filtration, include the conservation of mass and momentum. In the present investigation, the liquid metal is assumed to be an incompressible fluid. Since high flow rates of the melt are present during the steel filtration, the velocity field is expected to be turbulent.

For the LES, a spatial filtering operation is applied to the governing equations, which thus take the form [1,2]:

$$\frac{\partial \bar{v}_j}{\partial x_j} = 0, \tag{1}$$

$$\frac{\partial \bar{v}_i}{\partial t} + \frac{\partial}{\partial x_j}\left(\bar{v}_i\,\bar{v}_j\right) = \frac{\partial}{\partial x_j}\left[\nu\left(\frac{\partial \bar{v}_i}{\partial x_j} + \frac{\partial \bar{v}_j}{\partial x_i}\right) + \frac{1}{\rho}\tau^r_{ij}\right] - \frac{1}{\rho}\frac{\partial \bar{p}^*}{\partial x_i}. \tag{2}$$

Eq. (2) employs the anisotropic residual stress tensor $\tau^r_{ij} = \tau^R_{ij} - \delta_{ij}\tau^R_{kk}/3$ and the modified filtered pressure $\bar{p}^* = \bar{p} - \tau^R_{kk}/3$, where $\tau^R_{ij} = \rho\left[\bar{v}_i\bar{v}_j - \overline{v_i v_j}\right]$ is the residual stress tensor. The unresolved scales in turbulent flow are modelled by an eddy viscosity model [1,2]:

$$\tau^r_{ij} = 2\rho\nu_t\bar{S}_{ij}, \quad \bar{S}_{ij} = \frac{1}{2}\left(\frac{\partial \bar{v}_i}{\partial x_j} + \frac{\partial \bar{v}_j}{\partial x_i}\right). \tag{3}$$

Employing the closure condition, the filtered momentum equation (2) is rewritten as:

$$\frac{\partial \bar{v}_i}{\partial t} + \frac{\partial}{\partial x_j}\left(\bar{v}_i\,\bar{v}_j\right) = -\frac{1}{\rho}\frac{\partial \bar{p}^*}{\partial x_i} + 2\frac{\partial}{\partial x_j}\left[(\nu + \nu_t)\bar{S}_{ij}\right], \tag{4}$$

where ν_t is the eddy viscosity.

4. Transport of dispersed phases

A Lagrangian approach is used in order to model the transport of inclusions and gas bubbles through the filter. Their trajectories are predicted by solving their equation of motion [3]:

$$m_d\frac{d\boldsymbol{v}_d}{dt} = \left(m_d - m_f\right)\boldsymbol{g} + m_f\frac{18\nu_f f_S}{d_d^2}\left(\boldsymbol{v}_f - \boldsymbol{v}_d\right) + m_f\frac{D\boldsymbol{v}_f}{Dt} + \frac{m_f}{2}\left(\frac{D\boldsymbol{v}_f}{Dt} - \frac{d\boldsymbol{v}_d}{dt}\right), \tag{5}$$

where the terms on the right hand side of Eq. (5) represent the buoyancy, the drag force and the forces due to pressure gradient and added mass. The drag force term involves the Stokes correction factor, given as [4]:

$$f_S = 1 + \frac{Re_d}{4\left(1+\sqrt{Re_d}\right)} + \frac{Re_d}{60}, \quad Re_d = \left|\boldsymbol{v}_f - \boldsymbol{v}_d\right|\frac{d_d}{\nu} \tag{6}$$

with Re_d being the dispersed phase Reynolds number. After the velocities $\boldsymbol{v}_d$ are obtained from Eq. (5), the positions $\boldsymbol{x}_d$ are calculated by integrating:

$$\frac{d\boldsymbol{x}_d}{dt} = \boldsymbol{v}_d. \tag{7}$$

5. Numerical simulations

The simulation of liquid metal filtration is performed by an in-house code combining the LES in LBM context and the Lagrangian tracking of dispersed phases. Furthermore, the code has been parallelised for efficient computation using the message passing interface (MPI) protocol.

The collision and agglomeration of inclusions are studied as well as the influence of gas bubbles on the filtration process. From the simulations, the filtration efficiency is determined as the fraction of inclusions captured by the filter. The predicted filtration efficiency is compared with the analysis of steel samples cast with and without the foam filter.

Acknowledgements

The authors gratefully acknowledge the German Research Foundation (DFG) for supporting the Collaborative Research Centre CRC 920, subprojects B02 and C01.

References

[1] Peter A. Davidson. (2015). Turbulence – An Introduction for Scientists and Engineers. 2nd Edition. Oxford: Oxford University Press.

[2] Stephen B. Pope. (2000). Turbulent Flows. Cambridge: Cambridge University Press.

[3] Martin R. Maxey & James J. Riley. (1983). Equation of motion for a small rigid sphere in a nonuniform flow. Physics of Fluids. 26 (4), 883-889.

[4] Frank M. White. (2006). Viscous fluid flow. 3rd Edition. Boston: McGraw-Hill.

The 73rd World Foundry Congress
„Creative Foundry"
23 - 27 September 2018, Krakow, Poland

DETERMINATION OF CRITICAL COOLING RATE FOR MINIMIZATION OF POROSITY IN THE LARGE ALUMINUM CASTING

In-Sung Cho[1*], Si-Young Kwak[1], Yong-Hyun Kim[2], Hee-Kwon Lee[2]

[1]*ICT Manufacturing Process Group, Korea Institute of Industrial Technology, Incheon, 21999, South Korea*
[2]*Daeshin Metals. Co., Ltd. 24-11, Pallyong-dong, Changwon-si, Gyeongsangnam-do, South Korea*
**Corresponding address: ischo@kitech.re.kr*

Key words: porosity, cooling rate, aluminum casting, vacuum chamber, computed tomography, chill

In the present study, effect of cooling rate on the formation of the porosity in the thick aluminum sand casting was investigated. Nowadays large scale thick aluminum casting replaces steel frame for vacuum chamber for semiconductor production, with the consideration of weight and cost reduction. Several thick aluminum castings were manufactured using chill with temperature measurements. The castings were inspected by using 3D computed tomography in order to quantify the porosity defect density in the castings. Effect of the thickness of the chill on the porosity defect density were discussed.

1. Introduction

Traditionally, there have been many studies on the shrinkage cavity of products according to cooling rate in aluminum casting process [1]. Internal defects of casting product were analyzed using industrial CT(Computed Tomography) [2] and a series of studies have been carried out by Kwak [3, 4] to evaluate the mechanical strength of fatigue and impact according to defects which were detected through CT. However, since the cooling rate depends on the location in the product, its characteristics also vary in the distribution of the cooling rate proportionally. Therefore, it is very important to obtain temperature distributions during cooling in the product. In this paper, an investigation was carried out the temperature history of the A356 cast aluminum alloy specimen of 40 mm thickness used for the vacuum chamber by varying the thickness of the chill and the simulation for the temperature history.

2. Experimental Method

In the present study, size of 200x150mm with 40t thickness specimens were prepared, as shown in Fig.1. The chill material was SS400 steel and the thickness was 4 0mm, 60 mm, and 80 mm. The position of the thermocouple was 10mm from the chill, same position with that of porosity defect analysis.

A356 alloy was used for casting materials and compositions were shown in Table 1.

Table 1. Chemical composition of A356 proposed in the present study

	Al	Si	Fe	Cu	Mg	Ti
wt%	92.5	7.0	0.09	0.002	0.35	0.12

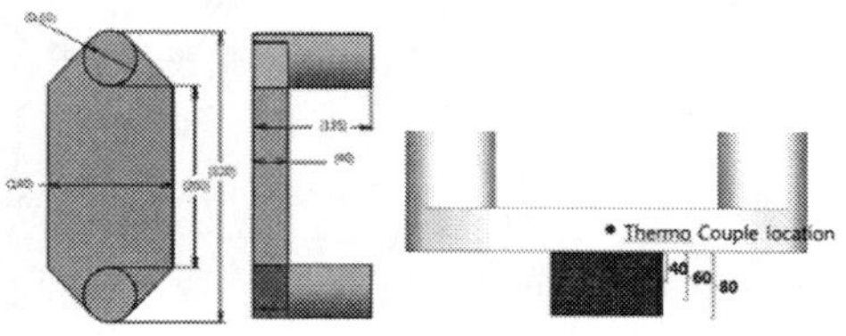

Fig. 1. Schematic diagram of specimen and position of the thermocouple and chill.

3. Result and Discussion

Cooling Rate by Experiment & Numerical Data

The heat transfer and solidification analysis for the specimens was carried out by using the commercial casting process analysis program Z-Cast. Fig. 2 shows the experimental and numerical values at the thermocouple location.

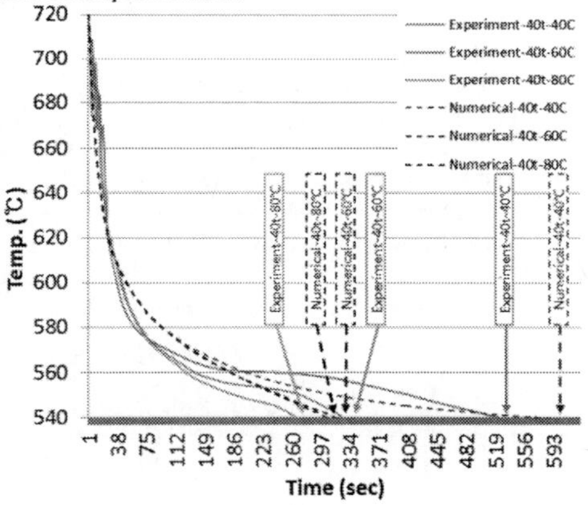

Fig. 2. Temperature histories after pouring at thermocouple location

Since it is difficult to install thermocouples in all areas of the specimen, the numerical conditions for heat transfer analysis those can simulate temperature history correspond to the thermocouple experimental data was set. Then the cooling rates for each area inside the product were obtained through the heat transfer analysis. The cooling rate analysis position in the specimen is 10 mm, 20 mm, and 30 mm in the chill part, and 12.5 mm and 37.5 mm in the center part as Fig. 3.

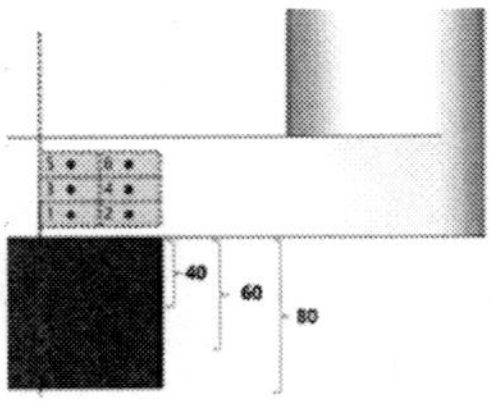

Fig. 3. The location of cooling rate obtained by numerical method

The cooling rate can be evaluated as the temperature interval divided by the cooling time to 542°C from 613°C which are the solidus and liquidus temperatures of A356.

As shown in Fig. 4, the trend of the cooling rates of each specimen was the same for all three specimens as 1 → 3 → 5 → 2 → 4 → 6 orders.

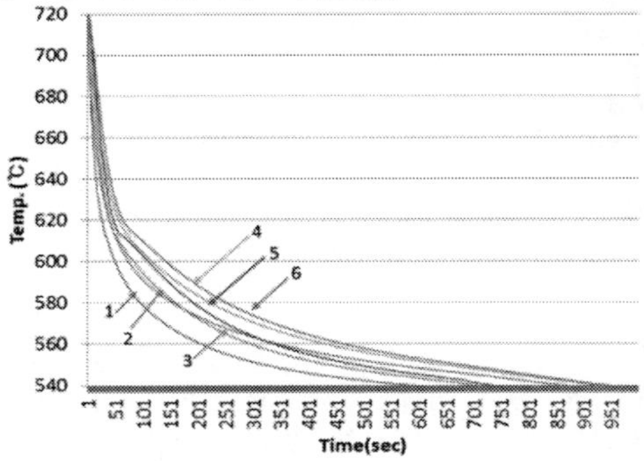

Fig. 4. Cooling curve of 40mm chill

The cooling rates of all specimens are listed in Table 1. The region No.1 of 80C was the fastest at 0.266°C/sec, and the region No.6 of 40C was the slowest at 0.088°C/sec. As shown in Table 2, when comparing the cooling rates according to the thickness, 60C and 80C shows relatively similar cooling rates, but they were different from 40C.

Table 2. Cooling rate according to thickness of chill

	Cooling Rate [°C/sec]		
	40C	60C	80C
1	0.158	0.257	0.266
2	0.098	0.184	0.196
3	0.125	0.210	0.220
4	0.094	0.171	0.182
5	0.118	0.193	0.202
6	0.088	0.168	0.179

4. Conclusions

It was not realistic to measure the temperature histories by installing thermocouples in all areas of the specimen. Therefore, the heat transfer analysis combined with the thermocouple temperature history data was performed to obtain the cooling rate distributions for the whole specimens. In case of 40mm aluminum block, 40 mm chill and 60mm chill were different at cooling rate, but the difference between 60mm and 80 mm was small. The cooling rate distribution inside the specimens could be obtained by using the commercial casting process analysis program Z-Cast. And it can be used to analyze the correlation between the metallographic or defect distribution results and the cooling rate.

.

Acknowledgments

This work was supported by the Ministry of Trade, Industry and Energy (MOTIE, South Korea) [Project Name: Development of large-scale high-quality aluminum sand casting technology.

References

[1] M.J. Oh, S.M. Yoo, I.S. Cho and Y.H. Kim, J. Korea Foundry Society, "A study on the Effect of Insulating sleeve on solidification characteristics of A356 aluminum alloy", 31-4 (2011) 205-211.

[2] I.S. Cho, S.Y. Kwak, M.J. Oh, S.C. Jang, C.H. Lim, 2nd Korea Japan Industrial CT Workshop,"Evaluation of effectiveness of casting defect on physical properties for diecast by using 3D computed tomography"17-23

[3] S.W. Lee H.K. Kim, H.Y. Hwang and S. Y. Kwak, J. Korea Foundry Society, "Fatigue Life Analysis for Al Alloy Casting Containing Internal Defects", 30-5 (2010) 25-28.

[4] S.C. Hwang and S.Y. Kwak, J. Korea Foundry Society, "Effect of Shrinkage Defect on Fracture Impact Energy of A356 Cast Aluminum Alloy", 34-1 (2014) 248-252.

The 73rd World Foundry Congress
„Creative Foundry"
23 - 27 September 2018, Krakow, Poland

OPTIMIZATION OF THE DIE CASTING PROCESS OF THICK-WALL BUSH MADE OF COMPOUND ALUMINUM BRONZE

Bogusław Pisarek [1], Dariusz Kołakowski [1]*

[1]Lodz University of Technology/ Department of Materials Engineering and Production Systems, 1/15 Stefanowskiego Street, 90-924 Lodz, Poland
**Corresponding address: dariusz.kolakowski@p.lodz.pl*

Key words: Solidification Process, Copper Alloy, Die Gravity Casting, Thermal Stress, Optimization

1. Introduction

Compound aluminum bronzes, resistant to tribological wear (abrasive and adhesive) [1], can be used for castings operating in a relatively large temperature range, below 100 °C, for example bush bearings or pistons cold chamber pressure machines operating at about 300 °C. Due to the high casting shrinkage of compound aluminum bronzes [2], incorrect choice of die casting parameters, such as: pouring temperature of cast alloy t_{CA}, preheat temperature of permanent mold t_{PM} and/or temperature removing of casting from mold (die open) t_{DO}, increases the likelihood of hot tears occurring in the casting or cold cracks [3].

In the process of unit and small-lot production of die-casting of a thick-walled bush from studied bronzes, non-uniform distribution of own stresses occurs in the casting - especially thermal stresses [4,5]. The aim of the research, the results of which are described in this paper, was to optimize the die casting parameters of: t_{CA}, t_{PM} and t_{DO}, to minimize hot tears indexes HT or cold crack CCR and equivalent stresses Von Mises's (thermal stresses) VMS.

2. Work methodology

Simulation of the thick-walled bush die casting process (Fig. 1) from CuAl10Fe5Ni5Cr bronze was carried out using the MAGMA5.3 program. In order to minimize the number of experiments, the experiment was developed using the Box-Wilson method (Table 1).

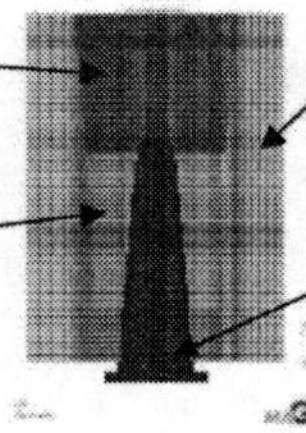

Fig. 1. Model of bush in die

In order to identify one solution in multi-criteria optimization, the weighted sum method was used.

Table 1. Experiment design

x_i i={1,2,3}	x_1	x_2	x_3
	t_{CA}, °C	t_{PM}, °C	t_{DO}, °C
x_i^0	1175	200	625
Δx_i	20	25	25
$x_i^0 - \Delta x_i$	1155	175	600
$x_i^0 + \Delta x_i$	1195	225	650
1	1195	225	650
2	1195	225	600
3	1155	225	600
4	1155	225	650
5	1155	175	650
6	1195	175	650
7	1195	175	600
8	1155	175	600

The functions defining subsequent criteria (reduction of indexes: HT, CCR, VMS) have been combined into one objective function F(k) according to the formula (1):

$$F(k) = \sum_{i=1}^{m} w_i f_i(k) \quad (1)$$

where:

the weight of individual criteria $w_i \epsilon [0,1]$ and $\sum_{i=1}^{m} w_i = 1$.

Assumed in the experiment, that $w_1 = w_2 = w_3$.

In this method, to the purpose of normalization and aggregation of the objective function, all variables used in the evaluation of individual criteria (minimization of: HT, CCR, VMS) were classified as destimulants. Destimulant z_i was determined from formula (2):

$$z_i = \frac{\max f_i(k) - f_i(k)}{\max f_i(k) - \min f_i(k)} \quad (2)$$

where:

$f_i(k)$ - the value of the i-th criterion, (i={1,2,3}).

It was assumed that the final value of the aggregated objective function FC, according to which the optimal solution is sought is in the form (3):

$$FC = \sum(z_{ij} \cdot w_j) \quad (3)$$

for each criterion "j" and "i" of evaluated sets of settings in the process, where w_j is the weight of the given criterion "j". The maximum FC value is sought.

3. Description of achieved results of own researches

3.1. General remarks

The results of the computer simulation of indexes HT, CCR and stresses VMS for the sequence of settings (Table 1) are shown in Table 2.

Table 2. Response $f_i(k)$ (HT, CCR and VMS)

Design	HT	CCR	VMS
x_i^0	21.9652	0.8926	180.98
1	21.7616	0.8971	178.57
2	21.7616	0.8971	182.15
3	23.2900	0.9066	184.13
4	23.2900	0.9066	178.39
5	21.8754	0.8879	186.81
6	22.0300	0.9047	185.8
7	22.0300	0.9047	190.19
8	21.8754	0.8879	192.18

The method of regression analysis was determined the equation of the plane on which lies the gradient vector of the most intense changes in the object's response to the given inputs. Assuming the minimization of the object's response, an example equation describing the change in VMS stresses is described by formula (4):

$$VMS = 310.971 - 0.03 \cdot t_{CA} - 0.1587 \cdot t_{PM} - 0.0954 \cdot t_{DO} \quad (4)$$

The position of the VMS plane in the t_{PM} coordinate system and t_{CA} for the constant value t_{DO} = 625 °C is shown in Figure 2.

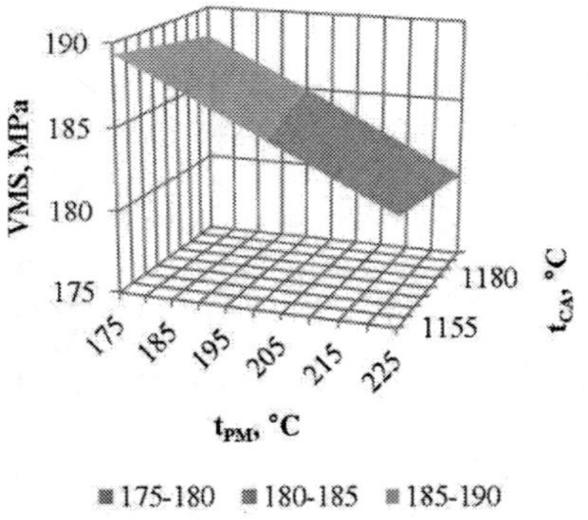

Fig. 2. VMS plane for the constant value t_{DO} = 625 °C

The strength of the influence of particular settings in the process on the value of VMS is shown in Figure 3. The values of the $z_{ij} \cdot w_j$ product for particular criteria after normalization and the value of the FC objective function after sorting are presented in Table 3.

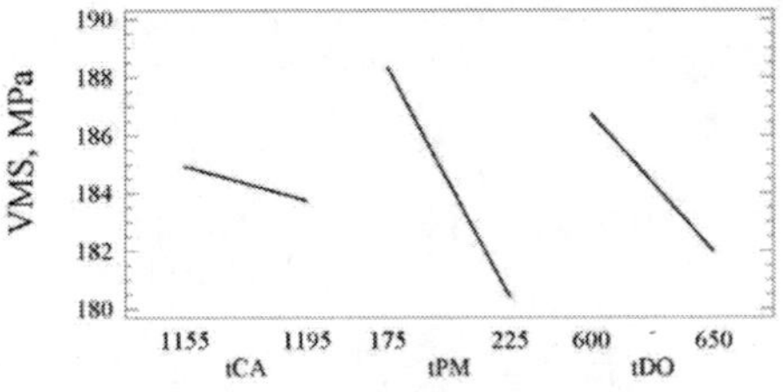

Fig. 3. Main Effects Plot for VMS

Table 3. Value of $z_{ij} \cdot w_j$ product and FC objective function

w_j	1/3	1/3	1/3	
$f_i(k)$	HT	CCR	VMS	
Design	$z_{ij} \cdot w_j$			FC
1	**0.3333**	**0.1152**	**0.2886**	**0.7371**
x_i^0	0.3064	0.1697	0.2375	0.7136
2	0.3333	0.1152	0.2127	0.6612
5	0.3183	0.2267	0.1139	0.6588
8	0.3183	0.2267	0.0000	0.5450
6	0.2979	0.0230	0.1353	0.4562
4	0.1313	0.0000	0.2924	0.4237
7	0.2979	0.0230	0.0422	0.3631
3	0.1313	0.0000	0.1707	0.3020

4. Conclusions

The optimization shows that in order to minimize the values of HT and CCR indexes as well as VMS stresses, in the die casting process of a thick-walled bush made of compound aluminum bronze, the following settings should be used:

- t_{CA} = 1195 °C,
- t_{PM} = 225 °C,
- t_{DO} = 650 °C.

References

[1] Pisarek, B. (2013). Aluminium Bronze Containing Cr, Mo and/or W with High Resistance to Wear (in Polish). Bulletin of the Lodz University of Technology No. 1141, Wydawnictwo Politechniki Łódzkiej, Łódź.

[2] Pisarek, B.P. (2011). Simulation of volumetric shrinkage Sv and surface shrinkage Svp. Pietrowski S. (Eds.), Wysokojakościowe Technologie Odlewnicze. Materiały i Odlewy (167-208). Katowice-Gliwice: PAN.

[3] Pisarek, B.P., Kołakowski, D. & Pacyniak, T. (2016). Analysis of the Causes of Cracks in a Thick-Walled Bush Made of Die-Cast Aluminum Bronze. Archives of Foundry Engineering. Volume 16 (4), 119-124.

[4] Pisarek, B.P., Kołakowski, D. & Pacyniak, T. (2017). Simulation of Stress Distribution in a Thick-Walled Bushing Produced by Die-Casting. Archives of Foundry Engineering. Volume 17 (4), 127-132.

[5] Chojecki, A. Burbiełko, A. & Telejko, I. (2005). Formation of Hot Tears in High Strength Al-Cu Alloys, Archives of Foundry. Volume 5 (15), 53-59.

The 73rd World Foundry Congress
„Creative Foundry"
23 - 27 September 2018, Krakow, Poland

INFLUENCE OF PRINTED CERAMIC FILTERS ON TEMPERATURE FIELD DISTRIBUTION DURING INVESTMENT CASTING OF THIN-WALLED NI-BASED SUPERALLOYS

Mirosław Antosz[1]*, Rafał Cygan [2], and Łukasz Rakoczy[3]

[1] *Consolidated Precision Products, ul. Hetmanska 120, 35-078 Rzeszow, Poland*
[2]*Rzeszow University of Technology Faculty of Mechanical Engineering and Aeronautics Powstancow Warszawy 8 Av., 35-959 Rzeszow, Poland*
[3]*AGH University of Science and Technology, Faculty of Metals Engineering and Industrial Computer Science, av. Mickiewicza 30, 30-059 Krakow, Poland*
**Corresponding address: Miroslaw.Antosz@cppcorp.com*

Key words: investment casting, ceramic filters, 3D printing, nickel-based superalloy, casting defects

1. Introduction

Each manufacturer to become more competitive have to improve position on the market and increase profits. It can be ensured by constant improve technological processes whose effect will be reduced number of poor quality products. This is especially important in the precision casting foundries, where the cost of the single one component is quite high compared to other businesses.

In investment casting foundry the number of scraps for selected products often reach 20-30%, and so for each 10 elements, two or three are rejected. It is caused by the formation of casting defects such as misruns, shrinkage porosity, and non-metallic inclusions. It leads to unable to meet the requirements of grain size, dimensional tolerances, or even mechanical damages. The example of an activity to reduce number of discontinuities is the application of ceramic filters. In the investment casting foundries, the most popular are foam ceramic filters which are brittle and have unrepeatable internal construction [1-4].

During the process, the piece of the filter can be fractured by a stream of metal and run with metal to the mold cavity causing the inclusion. In order to reduce this dangerous, one of the solutions is replacing foam ceramic filters by 3D printed ceramic filters (Fig. 1.). These filters have repeatable construction and much lower tendencies to mechanical damages.

Fig. 1. Printed filters with different level of the filtration: a) fast, b) medium, c) slow

2. Methodology

In order to determine the effects of the application of printed filters in production condition, the four castings with different filters (technological filter 10 PPI, and 3D printed filters: fast, medium and slow) were performed. The experimental set consisted of 3 plates. The plate played the role of a turbine blade in this experiment and it was designed to provide similar conditions of solidification. (Fig. 2). H-type gating system and pouring cup in which ceramic filters are placed before pouring were used. Six thermocouples were placed inside 2 of 3 plates, and also inside H-type gating system (Fig. 3) to register temperature during the process.

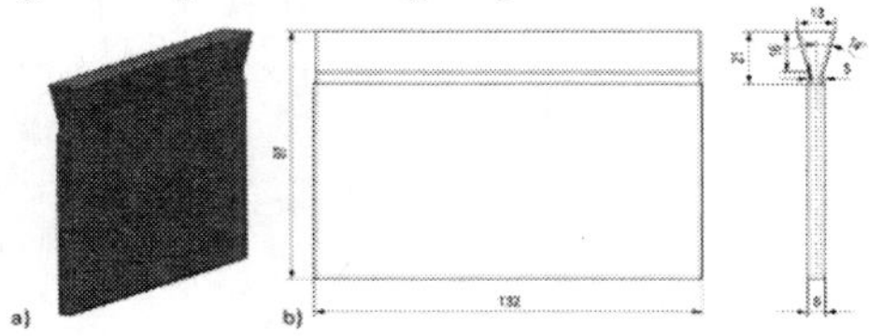

Fig. 2. Plate model a) 3D model, b) geometrical dimensions

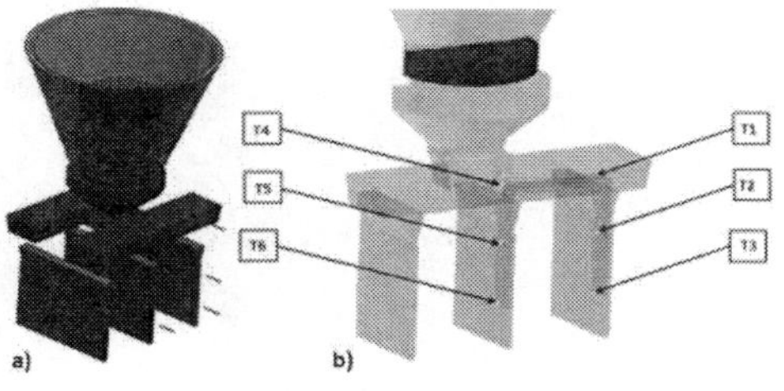

Fig. 3. a) wax assembly with pouring cup and ceramic filter b) location of thermocouples

One of the aims of the experiment was to determine, based on the temperature field distribution, which of the printed filter can be used, instead of a standard foam filter. The pouring process of Ni-based superalloy IN713C casting was performed in Research and Development Laboratory for Aerospace Materials at Rzeszow

University of Technology, in a vacuum furnace VIMIC 2 E – DS/SC (Fig. 4). The charge of 7 kg load was melted, heated up to 1470°C and poured into ceramic mould annealed at 1200°C. Six thermocouples registered metal temperature during pouring and cooling process through 1150 s.

Fig. 4. Mold after pouring

To compare the temperature field distribution, simulations were carried out in Procast software with identical parameters as in the experiment.

3. Results

Based on the data from 6 thermocouples, temperature distribution curves were created. Comparative analyzes were carried out for 3 representative thermocouples: two located on the bottom of the plates T3 (Fig. 5) and T6 (Fig. 6), and one located inside H-type gating system (Fig. 7).

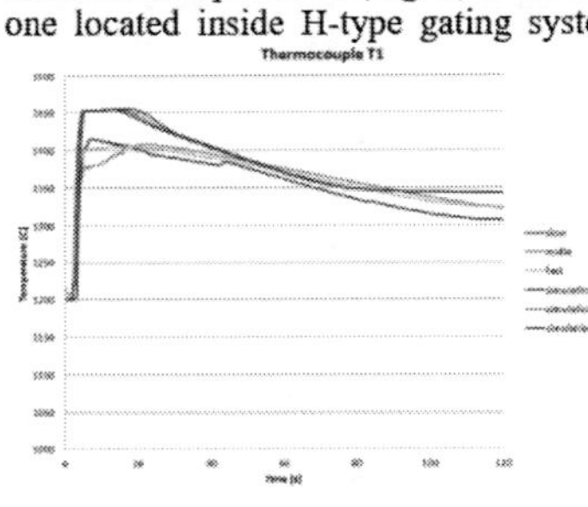

Fig. 5. Cooling curves comparison on location T1

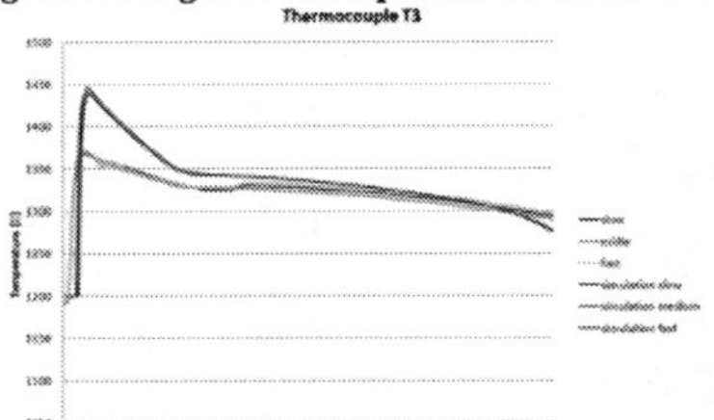

Fig. 6. Cooling curves comparison on location T3

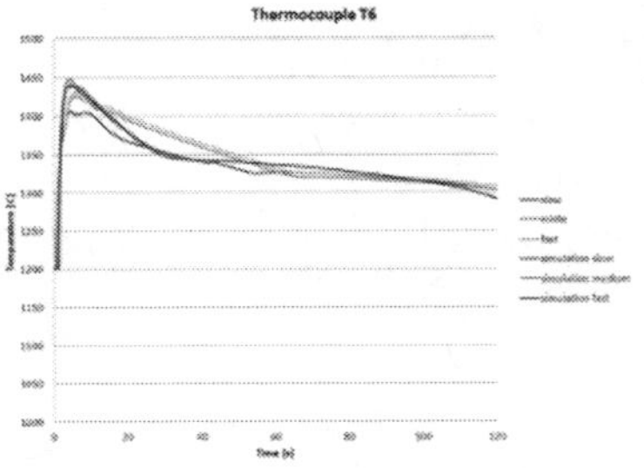

Fig. 7. Cooling curves comparison on location T3

4. Summary

Cost of 3D printed ceramics filters are higher than foam filters. Application of 3D printing filters in serial production is a proposal worth considering and should be preceded by a business analysis. It is certainly a good proposition due to the cost of castings. The non-metallic inclusions originated from filters may give us a lot of problems. The experiment carried out does not give us a definite answer which of the 3 printed filters is the best for production, due to too long pouring time (too low flow rate). To achieve the final answer experiment should be repeated with a reduced pouring time in the laboratory or using only a simulation by Procast software.

Acknowledgments

This research was supported by National Centre for Research and Development, Grant No. LIDER/227/L-6/14/NCBR/2015.

References

[1] M. Gromada, A. Świeca, M. Kostecki, A. Olszyna, R. Cygan.2015. Ceramic cores for turbine blades via injection moulding. Journal of Materials Processing Technology. Vol. 220. 107-112

[2] R. Cygan, P. Zywicki, T. Szczech, M. Kwiatkowski. 2016. Revert alloy impact on mechanical properties of nickel based superalloys used in production of aircraft engine critical components. 72nd World Foundry Congress. 2016. 225-226

[3] D. Szeliga, K. Kubiak, W. Ziaja, R. Cygan. 2014. Influence of silicon carbide chills on solidification process and shrinkage porosity of castings made of nickel based superalloys. International Journal of Cast Metals Research. Vol. 27. I. 3. 146-160

[4] P. Rokicki, G. Budzik, K. Kubiak, J. Bernaczek, T. Dziubek, M. Magniszewski, A. Nowotnik, J. Sieniawski, H. Matysiak, R. Cygan, A. Trojan. 2014. Rapid prototyping in manufacturing of core models of aircraft engine blades. Aircraft Engineering and Aerospace Technology. Vol. 86. I.4. 323-327

The 73rd World Foundry Congress
„Creative Foundry"
23 - 27 September 2018, Krakow, Poland

INFLUENCE OF POWDER ADDITIVES IN TWO COMPONENT ENVIRONMENTALLY FRIENDLY INORGANIC BINDER SYSTEMS ON STRENGTH BEHAVIOR

Iveta Vasková [1]*, Martin Conev [1], Martina Hrubovčáková [1]

[1] *Technical University of Košice, Faculty of Materials, Metallurgy and Recycling, Institute of Metallurgy, Letná 9, 042 00 Košice, Slovakia*
**Corresponding address: iveta.vaskova@tuke.sk*

Key words: Inorganic binder, alkali silicate, additive, bending strength, environment

1. Introduction

Due to the tendency to decrease harmful emissions and odor released during core production and casting when organic binders are used, new two component alkali silicate based binders were developed [1,2]. These inorganic binder systems are environmentally friendly and since their introduction at GIFA 2003 their usage is widely spreading, mainly for serial production of aluminum parts for automotive industry [3,4]. Two component binder systems are based on alkali silicate based solution (liquid part) and solid blend of synthetic and mineral auxiliaries (powder additive). Powder additive composition significantly influences properties of the cores, such as initial strength, de-coring ability, thermal stability or surface of the casting. Curing process takes place in heated tooling (up to 200°C) in combination of hot air purging (over 100°C). During this process, water is expulsed from the mold material via physical and chemical mechanism and Si-O-Si bonds are formed. Due to a loss of water, three dimensional silica network of SiO_4 tetrahedrons is formed. Through a reactive groups of additive, individual binder particles are combined (Fig. 1.) which is also accelerating curing process [5-7].

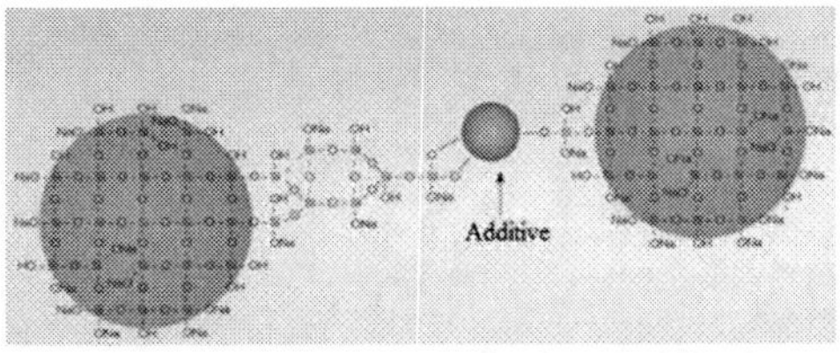

***Fig. 1.** Effect of additive on three-dimensional silicate network creation [7]*

2. Experimental work

Influence of 4 powder additives commercially used for serial production of aluminum castings on strength behavior was examined. As a liquid part commercially available blend of modified alkali silicate binders with specific weigh at 20°C approx. 1.38 g/cm^3 and viscosity at 25°C approx. 25 mPa·s was used. List of tested additives with properties description according to the data sheets of the supplier is presented in Table 1 below:

Table 1. List of tested additives with description

Additive	Properties description
A	designed for filigree geometries, high compaction, improved de-coring properties, organic containing
B	high compaction, improved humidity resistance and de-coring properties, superior dimensional accuracy during pouring, 100% inorganic
C	reduces penetration and sand adhesion, high degree of compaction, high dimensional accuracy during casting, 100% inorganic
D	especially suitable for filigree geometries, improved de-coring, reduces penetration and sand adhesion, high degree of compaction, high dimensional accuracy during casting, improved humidity resistance, 100% inorganic

From selected materials, standard test bars for bending strength measurement were prepared according to the procedure described in [8]. In order to investigate influence of powder additives on strength behavior, following mixtures with silica sand were used:

Table 2. Composition of tested mixtures

Binder	Additive	Additive amount
2,4 %	A	0,4 %; 0,8 %; 1,2 %
2,4 %	B	0,4 %; 0,8 %; 1,2 %
2,4 %	C	1,0 %, 1,5 %, 2,0 %
2,4 %	D	1,0 %; 1,5 %; 2,0 %

Weight of produced test bars was measured to obtain information about compaction when using different additives. Bending strength after 30 seconds (immediate) and after 1 hour (final) was measured. Each result is an average of three measurements.

3. Results

Results were divided for comparison of additives "A" to "B" and additives "C" to "D" because of their different addition amount. Graph in the Fig. 2 shows comparison of strength behavior and Fig. 3 shows compaction with additives "A" and "B". From the results it can be seen that higher amount of additive "A" results in increase of immediate and final strength and also higher compaction of the sample. Higher amount of additive "B" results in higher immediate strength and compaction only up to 0.8 %, final strength is dropping linearly, which can be here considered as an advantage of this additive from the de-coring ability point of view.

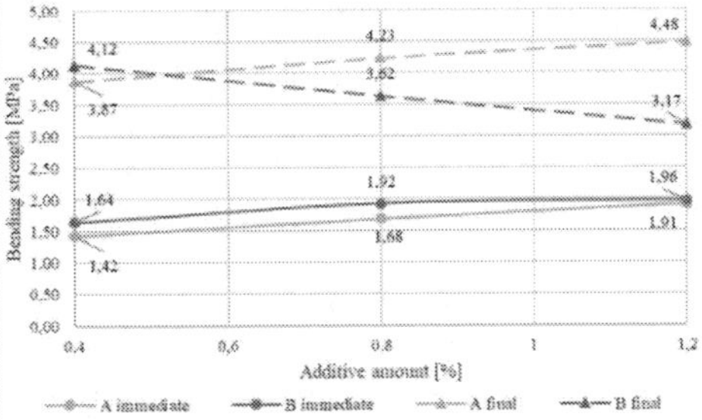

Fig. 2. ***Comparison between strength behavior with the use of additives A and B***

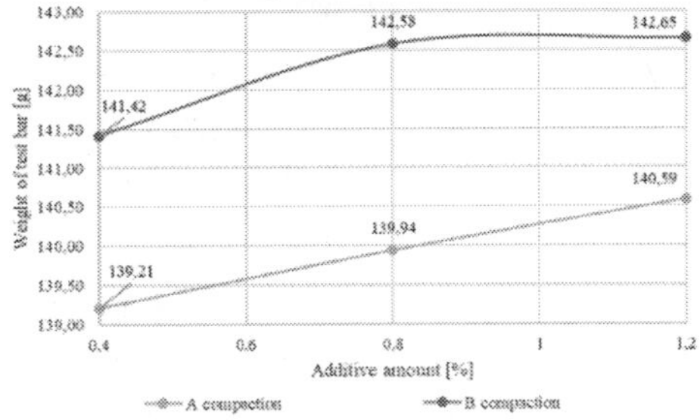

Fig. 3. ***Compaction of samples with additives A and B***

Graphs in the Figs. 4 and 5. present results of strength and compaction measurement for additives "C" and "D". Immediate strength with additive "C" is significantly increasing in interval of addition 1 – 1.5%, opposite effect can be seen when observing compaction. Final strengths are on the same level with different addition amounts of additive "C". Additive "D" amount is not affecting immediate strength, but final strength and also compaction is dropping with increasing amount of this additive.

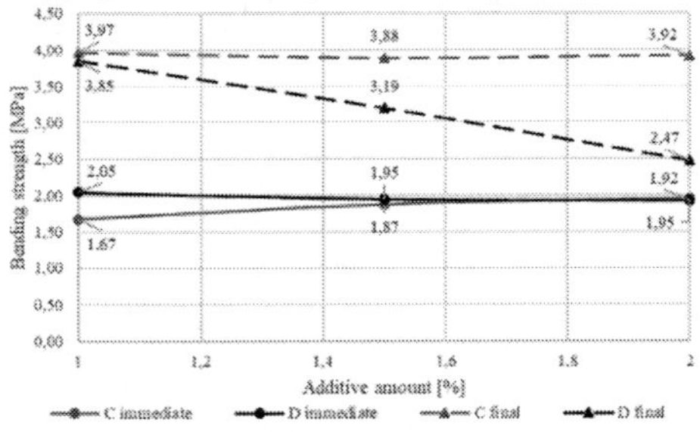

Fig. 4. ***Comparison between strength behavior with the use of additives C and D***

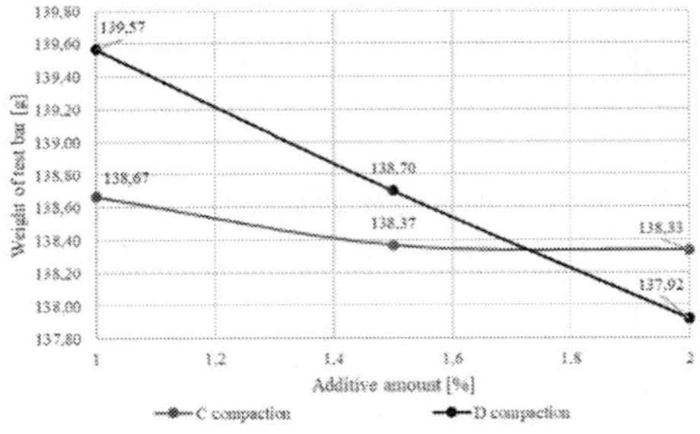

Fig. 5. ***Compaction of samples with additives C and D***

4. Conclusion

Investigation of 4 different powder additives commercially used in new two component alkali silicate based binder systems showed various influence on strength behavior and compaction of produced test bars. Additive "A" is significantly increasing strength properties and compaction. Additive "B" is increasing immediate strength, but decreasing final strength, compaction is significantly increased by addition up to 0,8 %. Additives "C" and "D" contain also compounds for penetration reduction, their standard addition rate is higher than in case of "A" and "B". Increase in amount of "C" is slightly increasing immediate strength while final remains on same level, but with lower addition compaction is higher. High addition rate of "D" results in loss of final strength and lower compaction. Depending on technological needs (e.g. strength, penetration issues) different additive type and amount can be used to adjust core quality.

References

[1] J. Wang, Z. Fan, H. Wang, X. Dong, N. Huang. (2007). An improved sodium silicate binder modified by ultra-fine powder materials. China Foundry. Vol. 4 No.1, 26-30.

[2] K. Loechte, R. Boehm. (2005). Cordis, the inorganic binder system – properties and experience. Casting plant + Technology International. Vol. 3, 6-11.

[3] H. Polzin. (2014). Inorganic binders for mould and core production in the foundry. Berlin: Schiele & Schön GmbH.

[4] L. Zaretskiy. (2016). Modified silicate binders new developments and applications. International Journal of Metalcasting. Vol. 10, Issue 1, 88-99.

[5] Ch. Appelt, S. Voss. (2016). Einfluss der Temperaturleitfähigkeit organisch und anorganisch gebundener Formkerne auf die Gefügeeigenschaften von Aluminiumlegierungen. Giesserei. Issue 8, 22-27.

[6] R.K. Iller. (1979). The chemistry of silica. Chichester: John Wiley & Sons.

[7] C. Wallenhorst. (2010). Grundlagen zum Verständnis der anorganischen Kernfertigung. Giesserei-Praxis.Vol. 6, 181 – 184.

[8] M. Conev, I. Vasková, M. Hrubovčáková, P. Hajdúch. (2016). Impact of Silica Sand Granulometry on Bending Strength of Cores Produced by ASK Inotec Process. Manufacturing Technology. Vol. 16, No. 2. 327-334.

The 73rd World Foundry Congress
„Creative Foundry"
23 - 27 September 2018, Krakow, Poland

IMPROVEMENT OF THE MANUFACTURING TECHNOLOGY OF CAST STEEL CASTINGS TO BE USED IN EXTREME OPERATING CONDITIONS

Bogdan Zdonek [1]*, Artur Żak [1], Ireneusz Szypuła [1], Jan Jezierski [2,3], Mariusz Adamczyk [1], Krzysztof Kostrzewa [2], Włodzimierz Kutera [2]

[1] Institute for Ferrous Metallurgy, ul. Karola Miarki 12 – 14 Gliwice, Poland
[2] PIOMA - Odlewnia Sp. z o.o., ul. Romana Dmowskiego 38, 97 – 300 Piotrków Trybunalski, Poland
[3] Silesian University of Technology, Department of Foundry Engineering, ul. Towarowa 7, 44-100 Gliwice, Poland
**Corresponding address: krzysztof.kostrzewa@pgosa.pl*

Key words: steel casting, steel cleanness, extreme operating conditions, casting quality, AOD process

1. Introduction

Development of Earth regions adjacent to polar circle boosts demand for steel and cast steel (castings) products of low-alloy steels characterized with mechanical properties enabling operations at low (-40°C) and very low temperatures (-60°C), and in particular absorbed energy KV_2 equal to at least 42J. Customers for castings manufactured at PIOMA – Odlewnia sp. z o.o. are world-renowned companies producing mining equipment, machinery, railway engineering equipment etc.

Operational properties of the casting are determined both by the following stages of cast steel production process as well as casting manufacturing, as well as chemical composition and manner of castings heat treatment. A significant challenge for achievement of the required impact resistance at low temperatures is metallurgical purity of castings, measured with the quantity and morphology of non-metallic inclusions.

2. Cast steel manufacturing technology

PIOMA- Odlewnia Sp. z o.o. is equipped with two arc-furnaces with basic lining of the capacities of 3 and 6 tons for low-alloy cast steel melting. Melting technology is a conventional double slag process. Steel is tapped by means of tapping spout to stoppered ladle, where momentary argon treatment of cast steel as well as casting into moulds is executed. Prior to launch of technology modification, ladle was equipped with chamotte lining. So far, steelmaking technology did not allow to produce castings of high purity, as purity level measured with content of total oxygen in castings was 100 ppm and more, while the required levels should range from 10- 20 ppm and be lowest possible.

3. Modification of the technology of cast steel melting and castings manufacturing

Modification of the technology of cast steel manufacturing for castings for oil and gas industry, to be used in operations at low temperatures (-40° C), comprehensively described in study [1], covered introduction of changes in the following areas:

- chemical composition of cast steel G18NiMoCr3-6 by increase to maximum permissible levels the contents of carbon, chromium, molybdenum and nickel, reduction in the content of total oxygen, reduction in the content of nitrogen content in cast steel and introduction of microalloys V, Ti, B to limit growth of austenite size during austenitization prior to hardening and increase in hardenability of cast steel [2];
- cast steel melting technology by application of high alumina lining (andalusite) of the ladle as well as modification of Al_2O_3 inclusions by mans of liquid metal treatment with Ca Si cored wire;
- heat treatment by application of double normalization at temperature of 920°C with still air cooling and subsequent hardening in water following austenitization at temperature of 950°C with application of intensive mixing of cooling bath and tempering at temperature of 620°C with water cooling.

Experimental heats of cast steel according to the modified technology have shown that a majority of non-metallic inclusions in castings was constituted by fine modified calcium aluminates: 12 CAO $7Al_2O_3$ with (Ca, Mn) S border, as well as compound oxide inclusions type CAO Al_2O_3 MgO SiO_2 of equivalent diameter up to 2.6 μm. Apart from non-metallic inclusions, presence of discontinuities in form of small shrinkage voids was found.

Volume fraction of non-metallic inclusions in cast steel samples dropped from 0.50% (without technology modification) to 0,18% in modified technology, and the content of total oxygen decreased from the level of 106 ppm to the level below 70 ppm. This was related directly to improvement of plasticity of castings, measured with (A_5) elongation and area reduction (Z).

The required absorbed energy (KV_2) at temperature of -40° C (min. 42J) was obtained in samples of castings of equivalent wall thickness up to 75mm. At higher equivalent thickness, impact resistance levels were clearly lower.

Improvement in impact resistance at temperature of -40° C in castings of higher equivalent thickness than 75 mm is related to further improvement of castings purity by aiming at further reduction in total oxygen content as a result of application of secondary refining of cast steel in the equipment like ladle furnace, AOD converter, VOD or in VD process [3-6].

In case of castings manufactured both by means of conventional as well as modified technology, defects close in form (porosity and discontinuities), as well as chemical composition to scale ($FeO \cdot MnO \cdot SiO_2 \cdot Al_2O_3$) [7] appeared on their surface. These defects are most probably formed as a result of secondary oxidation of cast steel during casting into moulds. As the most recent research has shown [8], the oxidized surface of liquid metal as a result of turbulences in filling system comes up to the surface in form of slag film, mainly as a result of precipitation and outflow of gas bubbles. Solidification temperature of these inclusions is close to the temperature of cast steel solidification and owing to that as a result of bubbles outflow, capillary tubes of slag film are formed. Films formed as a result of cast steel solidification in casting move along the solidifying surface of casting skin and form "tangles" detected in microscope observation as micro-shrinkages or scale. This may be prevented by protection of the flux of steel cast to mould, reduction in the temperature of non-metallic inclusions solidification by means of introduction of e.g. small quantity of boron, the oxide of which bonds with slag film. Thus, its solidification temperature is reduced which leads to easier separation from gas bubbles.

4. Conclusions

Results of the research conducted so far on manufacturing of steel castings for extreme applications, mainly by way of cast steel purity improvement have shown that:

1. In the present technical conditions of PIOMA – Odlewnia sp. z o.o., it is possible to reduce the content of total oxygen to ca. 70 ppm, and thus increase purity of cast steel. This facilitates obtaining castings of wall thickness equivalent to 75 mm, of the required KV_2 energy of min. 42 J, at temperature of (-40).
2. In order to achieve further improvement of cast steel purity, it is necessary to apply new solutions in scope of ladle treatment of liquid cast steel. At present, works are conducted at PIOMA – Odlewnia Sp. z o.o., under "Demonstrator" project, aimed at launch of innovative technological line for production of cast steel using ladle treatment in AOD converter.

Acknowledgments

The study was conducted under the project "Demonstration of the technology of manufacturing innovative cast steel castings to be used in extreme operating conditions", co-financed by the National Centre for Research and Development, under Operational Programme Intelligent Development 2014-2020. Action 1.1 R&D projects of enterprises, Sub-action 1.1.2. R&D works related to manufacturing pilot/demo system, contract POIR.01.01.02-00-0148/16-00 dated May 30, 2017

References

[1] Żak A., Zdonek B., Adamczyk M., Szypuła I., Kutera W., & Kostrzewa K. (2015). Technologie produkcji wielkogabarytowych odlewów staliwnych przeznaczonych do zastosowania w ekstremalnych warunkach eksploatacji. Prace IMŻ. 2, 18-25.

[2] Garbarz B. & Pickering F.B. (1988). Effect of vanadium and austenitising temperature on hardenability of (0,2-0,3)C-1,6Mn steels with and without additions of titanium, and molybdenum. Materials Science and Technology. 4, 117-126.

[3] Inomoto T., Matsuo M. &Yano M. (2015). Evaluation of the Chemical Vacuum Effect in Decarburization Treatment by Argon Injected Steel under Normal Atmosphere. ISIJ International. 55(4), 723-726.

[4] Kneis L, Trummer B. & Knabl B. (2004). The Hybrid Plug – An Innovative Purging Plug for Steel Ladles. RHI Bulletin. 1, 34-38.

[5] Zdonek B., Szypuła I., Kozłowski J. & Szczęch S. (2012). Secondary Steel Refining for Continuous Sequence Bloom Casting for High Oxide Cleanness Final Products. Archives of Metallurgy and Materials. 57(1), 347-354.

[6] Cotchen J.K., Misra P., Pretorius E. & Marraccini R. (2005). Iron & Steel Technology. May, 216-223.

[7] Zdonek B., Żak A. & Adamczyk M. (2013). Badania i ocena możliwości produkcji w warunkach techniczno – technologicznych PIOMA – Odlewnia Sp. z o.o. nowego asortymentu odlewów ze staliwa do zastosowań w niskich temperaturach. Sprawozdanie IMŻ nr BO-1413. December, unpublished.

[8] Campbell J. (2017). Melting, Remelting and Casting for Clean Steel. Steel Research International. 88(1), 1-13.

The 73rd World Foundry Congress
„Creative Foundry"
23 - 27 September 2018, Krakow, Poland

OPERATIONAL ASSISTANCE SYSTEM WITH DIRECT MANIPULATION OF FLOW RATE AND FALLING POSTION OF OUTFLOW LIQUID IN TILTING-LADLE-TYPE POURING MACHINE

Yuta Sueki [1]*, Yoshiyuki Noda [2]

[1] *University of Yamanashi, Department of Mechanical Engineering, 4-3-11, Takeda, Kofu, 400-8511 Japan*
[2] *University of Yamanashi, Department of Mechanical Engineering, 4-3-11, Takeda, Kofu, 400-8511 Japan*
**Corresponding address: g16tm009@gmail.com*

Key words: pouring, automation, production control, flow rate control

1. Introduction

In the casting industry, the operational pouring machine which can be manipulated by the operator in the remote site is often used for improving the working environment[1,2]. However, it is difficult to pour the molten metal precisely into the pouring basin by operating a tilting-ladle-type pouring machine, because the pouring flow rate is varied by the tilting angle of the ladle. In order to solve this problem, the operational pouring machine with manipulatable pouring flow rate has been developed[3]. With this pouring flow rate manipulation system, the operator can manipulate the pouring flow rate directly. However, the falling position of the outflow molten metal on the upper surface of the pouring basin is varied by varying the pouring flow rate. Then, the spilling of the molten metal can be caused by the variation of the falling position.

Therefore, in this study, we propose the operational assistance system in pouring machine with manipulatable pouring flow rate to pour precisely the liquid into the pouring basin. In this approach, the pouring flow rate estimation method[4] and the falling position control system of the outflow liquid from the ladle[5] are integrated systematically. The pouring flow rate and the falling position of outflow liquid can be estimated by using state estimation algorithm [4] instead of a flowmeter. By transferring the ladle horizontally in accordance with the estimated falling position, the actual falling position of outflow liquid can be held to target position such as the center of the pouring basin. With the proposed operational assistance system, the operator can manipulate the pouring flow rate on his own terms, and the falling position of the outflow liquid can be controlled automatically.

The operational assistance system is applied to laboratory-type operational pouring machine, and the effectiveness of this approach is verified through the experiments.

2. Operational Pouring System

In this study, we use the operational pouring system as shown in Fig. 1. The ladle can be transferred on Y- and Z-axes and rotated on Θ-direction by servomotors.

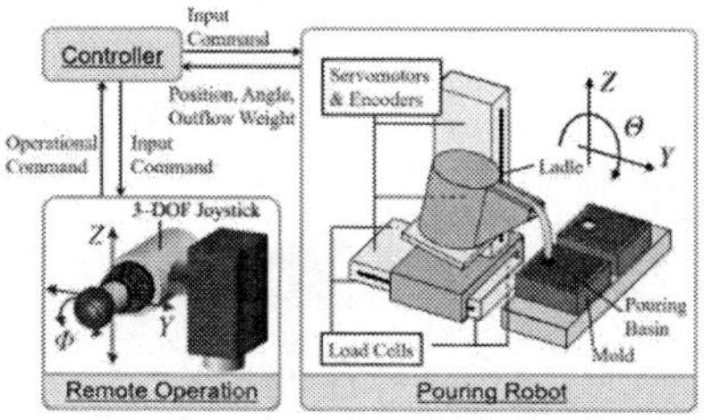

Fig. 1. Control System of Operational Pouring Machine

The driving force of each motor is amplified through a ball-screw mechanism on the Y- and Z-axes. The transfer distance and tilting angle of the ladle can be measured by rotary encoders installed into the motors. The center of the ladle's rotation shaft is placed near the ladle's center of gravity. In case that the ladle is rotated around the center of gravity, the tip of the pouring mouth moves in a circular trajectory. It is difficult to pour the molten metal into the pouring basin, since the pouring mouth is moved by tilting. Therefore, the position of the tip of the pouring mouth is controlled invariably while pouring by means of synchronous control of the Y- and Z-axes for rotational motion around the ladle's θ-direction[6]. The weight of outflow liquid can be measured by the load cells equipped on the base of the pouring machine. The 3-DOF joystick can be operated in horizontal and vertical directions for transferring the ladle on Y- and Z-axes respectively, and rotated for tilting the ladle on θ-direction. The attitude of the joystick can be measured by the rotary encoders installed in each rotation.

3. Operational Assistance System in Automatic Pouring Machine to Pour Precisely

The operational assistance system in the automatic pouring machine is represented by the block diagram as shown in Fig. 2. At the first, the operational angle of the joystick is applied to the pouring flow rate controller and the input command is generated from the controller. The input command is applied to the motor P_t for tilting the ladle. Then, the liquid is poured from the ladle through the pouring process P_f.

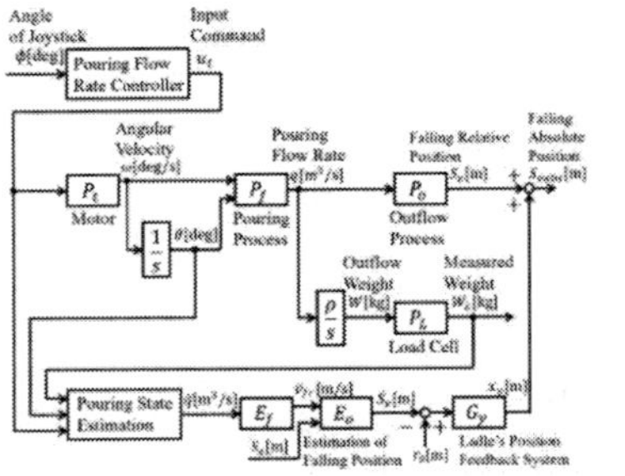

Fig. 2. Block Diagram of Pouring Process

The outflow liquid behaves as parabolic motion in the outflow process P_o and the position of the outflow liquid on the upper surface of the pouring basin can be measured by a camera. The weight of the outflow liquid can be measured by the load cells equipped on the bottom of the pouring machine. While the pouring, the ladle moves automatically to control the falling position of the outflow liquid on the upper surface of the pouring basin. In the falling position control of outflow liquid, the pouring flow rate and the falling position can be estimated by using the extended Kalman filter(EKF) as the state estimation algorithm. In accordance with the estimated falling position, the ladle is transferred horizontally by the ladle's position feedback control system, and the falling position of the outflow liquid can be held to the target position.

4. Experimental Verification

In this study, we implemented the proposed approach into the laboratory-type automatic pouring machine and verified the precision of the pouring flow rate and the falling position of the outflow liquid. Fig. 3 shows the results of the pouring motion, and Fig. 4 shows the results of the falling position of the outflow liquid. In Fig. 3, (a) shows the Angle of Joystick. (b) and (c) are the tilting angular velocity and the tilting angle respectively. (d) and (e) are the liquid height of the pouring mouth and the pouring flow rate respectively. The solid lines are the reference trajectories. The dashed lines show the simulated results obtained from the pouring process model. The chained lines show the estimated results by using EKF. (f) shows the weight of the outflow liquid. The dark solid line is the measured weight by using the load cells. The other lines are shown in the same manner as (d) and (e). According to these results, the operator was able to manipulate the pouring flow rate directly. Fig. 4 shows the results of the falling position of the outflow liquid. The x-indicators show the result with EKF and the dot markers show the result of previous approach without EKF. From these results, it is confirmed that the falling position can be tracked precisely by using the falling position control with the EKF.

5. Conclusion

In this study, we have developed the operational assistance system for improving the precision of

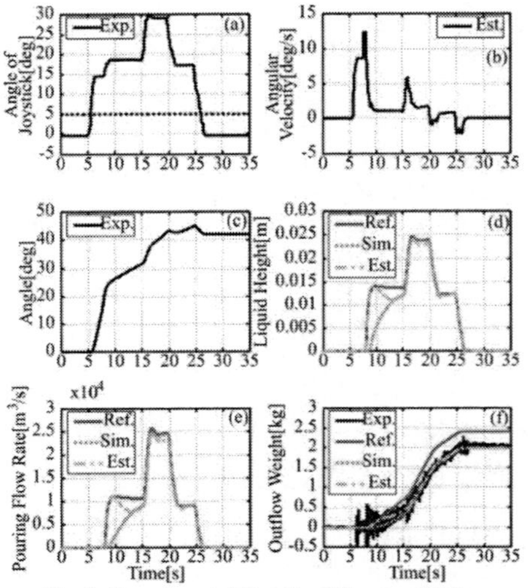

Fig. 3. Experimental Results of Pouring Motion

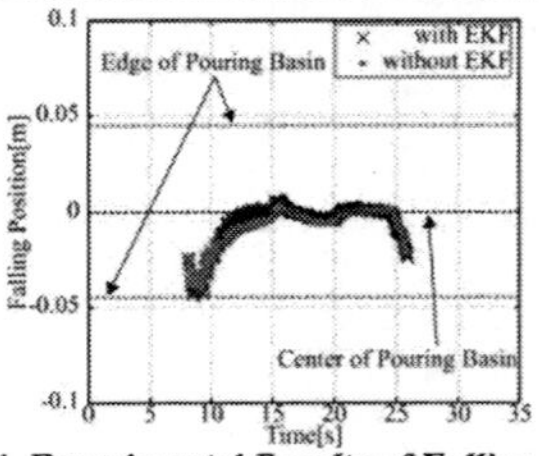

Fig. 4. Experimental Results of Falling Position

pouring. With this approach, the operator can manipulate the pouring flow rate directly and the falling position of the outflow liquid from the ladle can be held to the target position automatically. Through the pouring experiments, we demonstrated that the proposed approach assisted the operator to pour the liquid precisely into the pouring basin.

References

[1] W. Lindsay, (1983), Automatic Pouring and Metal Distribution System, Foundry Trade Journal, pp.151-176

[2] J. Sato & K. Yoshida, (1992), Automatic Pouring Equipment for Casting - Mel Pore System, Industrial Heating (in Japanese), Vol. 29, No. 4, pp. 19-27

[3] T. Yajima & Y. Noda, (2017), Development of Operational Pouring Machine with Direct Manipulation of Pouring Flow Rate(in Japanese), the 169th Japan Foundry Engineering Society Meeting, pp. 31

[4] Y. Noda & K. Terashima, (2009), Estimation of Flow Rate in Automatic Pouring System with Real Ladle Used in Industry, ICROS-SICE International Joint Conference 2009, pp. 354-359

[5] A. Ito, Y, Noda & K. Terashima, (2012), Outflow liquid Falling Position Control by Considering Lower Position and Clash Avoidance with Mold, Preprints of the IFAC Workshop on Automation is the Mining, Mineral and Metal Industries, pp. 240-245

[6] M. Suzuki, Y. Yang, Y. Noda, & K. Terashima, (2008), Sequence Control of Automatic Pouring System, Proceedings of 10th Asian Foundry Congress(AFC), pp. 88-92

The 73rd World Foundry Congress
„Creative Foundry"
23 - 27 September 2018, Krakow, Poland

INFLUENCE OF THERMO-PHYSICAL PROPERTIES OF MOULDING SANDS ON THE SOLIDIFICATION TIME OF DUCTILE CAST IRON

Alena Brusilová[1], Marcela Pokusová[1], Alexander Schrek[1]*

[1]Slovak University of Technology, Faculty of Mechanical Engineering, Nám. slobody 17, Bratislava, Slovakia
*Corresponding address: alena.brusilova@stuba.sk

Key words: spheroidal graphite nucleuses, moulding sand, solidification time, heat conductivity, wall thickness

1. Introduction

For achievement good mechanical properties of ductile cast iron, it is indispensable to obtain the highest quantity of the regular graphite spheroids that should be distributed uniformly in the matrix. The effective wall thickness is also adjusted by the correction coefficient ch expressing the influence of thermo-physical properties of the moulding sand on the solidification time. The graphite takes the shape of nodules due to the solidification conditions produced by the active elements introduced into the melt during metallurgical processing [1,2]. The basic nodulizing element Mg causes the tendency of cast iron solidify metastably and from this reason, the inoculation has to be done simultaneously or as soon as possible. The aim of inoculation is to form the graphite nucleuses of the supercritical size and promote their growth. It is necessary to give increase the regular graphite nodules as much as possible and distribute them uniformly throughout the matrix. This will depend not only on the chemical composition of the melt but also on the cooling gradients given by the casting section thickness and the heat conductivity of the mould. The pouring into the green bentonite sand moulds is the most common method for casting of ductile cast iron. The properties of spheroidal graphite iron (SGI) will depend besides the pouring method also on the casting wall section and the procedure of metallurgical process of the melt. The solidification rate is specified by the thickness of the metal layer χ solidified on the mould wall which will be directly proportional to the square root of solidification time τ expressed by the equation:

$$\chi = k \cdot \sqrt{\tau} \qquad (1)$$

when *k* is the coefficient of solidification allowing for the balance between the quantity of heat liberated from the casting transferred into the mould. The coefficient *k* effects on the solidification time τ which controls solidification course [3]. The values of parameter *k* were determined experimentally. The coefficient *k* can be done by solving the differential equations for heat transfer and conduction in the casting and mould using the finite element method which is often used for the simulation of solidification process. Our experiments have shown the dependence the content of ferrite and pearlite in the slowest and the quickest solidifying parts on the walls thickness [4]. The most of castings made from graphitic irons are poured into green sand moulds where the basic components are silica and bentonite as the binding material. For obtaining higher quality and accuracy it is applied the second generation moulding sands based on synthetic resin binding system which allows easier shaking-out the castings from the mould. However, they introduce the new factor in the solidification process that changes the thermo-physical conditions of solidification. There are exist many relations on the determination of coefficient *k*. The equation for calculating „*k*" according to Přibil is expressed [3]:

$$k = \frac{\sqrt{\pi}.(T_s - T_f).b_f}{2\rho.[c.(T_l - T_s) + l]} \qquad (2)$$

where T_s is temperature of solidus (1423 K for SGI), T_f is starting temperature of mould (K), ρ is density of molten metal (ρ = 7.103 kg m^{-3} for SGI), c is the specific heat of melt (c = 540 J kg^{-1}K^{-1} for SGI), T_l is the temperature of the pour (K), l is the heat of solidification (l = 230.103 J kg^{-1} for SGI), b_f is the coefficient of heat accumulated in the mould (Ws$^{-1/2}$m^{-2}K^{-1}).

2. Experiments

Our experiments have been focused on influence observation of thermo-physical properties of the moulding sands on the solidification time and the results from both measurements were compared. We assumed that the properties of molten metal were constatnt and then the change of moulding technology can only impact the coefficient of heat accumulated in the mould. There were used two types of moulds – 1. the green mould made from bentonite moulding sand containing 3.5% of water and 2. the mould made from resin moulding sand. This combination of moulding sands is very often used because of good reproducibility of results and implementation in practice. The coefficient of solidification *k* has been determined experimentally through thermocouples and compared with the calculated values obtaining by the simulation software WinCast. The coefficient *k* has been determined for bentonite moulding sand and for the casting of cylinder shape or plate. After statistical evaluation, we obtained the value of parameter k = 0.94 $mm.s^{-1/2}$ or 0.0564 $m.h^{-1/2}$. The molten metal was poured under overheating of 200 K. The coefficient of heat accumulation for bentonite $b_{f(b)}$ = 1587$Ws^{-1/2}m^2K^{-1}$ was calculated. The correction factor χ was derived adjusting the thickness of casting wall t against the modulus M:

$$t=2.M_{\chi}=2.M.\frac{b_{f(b)}}{b_{f(m)}}=2.M.\frac{1587}{b_{f(m)}} \qquad (3)$$

Using the moulding sand based on the foundry silica sand S22 bonded with phenol-formaldehyde resin, the cooling conditions have changed substantially. This moulding sand had value of product $\rho_f . c_f$ = 1.4 $MJ.m^{-3}.K^{-1}$, heat conductivity λ_f = 0.8 $W.m^{-1}.K^{-1}$ and the value $b_{f(m)}$ = 1058$Ws^{-1/2}m^2K^{-1}$. The value of correction factor χ = 1.5 and of wall thickness t=3 mm. When the wall thickness is 2 mm, i.e. modulus M = 1 mm, the casting will solidify at an equal rate to casting having thickness 3 mm. When the wall thickness is 2 mm, i.e. modulus M = 1 mm, the casting will solidify at an equal rate to casting having thickness 3 mm. Assuming for equal overheating of molten metal and identical casting when we substituted the bentonite sand for water-free resin sand, only the value of coefficient of heat changed from $b_{f(b)}$ to $b_{f(m)}$. We calculated the corresponding change of the casting solidification time in bentonite sand from $\tau_{0(b)}$ to $\tau_{0(m)}$ and in resin sand:

$$X=\frac{b_{f(m)}}{b_{f(b)}}=\frac{1058}{1587}=\frac{\tau_{0(b)}}{\tau_{0(m)}} \rightarrow \tau_{0(m)}=2.25.\ \tau_{0(b)} \qquad (4)$$

3. Results and discussion

The results from the computer simulation of the casting solidification from ductile cast iron STN 42 2314 are presented in the Fig. 1. The solidification process simulation has been performed for bentonite sand and resin sand. The casting has been formed from conical part with thickness 20 mm and bearing tube with thickness 30 mm. In bentonite mould, the thin part of the casting wall were solidifying during 123 to 400 seconds and cylindrical part within 400 to 680 seconds. The solidification has been completely finished within 700 seconds after pouring when conical and cylindrical parts were joined (Fig. 1a). The solidification time of casting poured into resin mould is presented in the Fig.1b where conical part solidified from 210 to 850 seconds and cylindrical part in the range from 850 to 1300 seconds.The solidification was completed after 1415 seconds. It follows that solidification time of the casting in resin mould was approximately doubly longer than in bentonite mould what corresponds acceptably with simulation results. The experiments confirmed that the casting thinnest sections of 2 mm poured into resin mould solidified in equal time as the section of 3 mm in bentonite mould [1]. The possibility to pour castings having the modulus from 1 to 1.5 was successfully verified in series production.

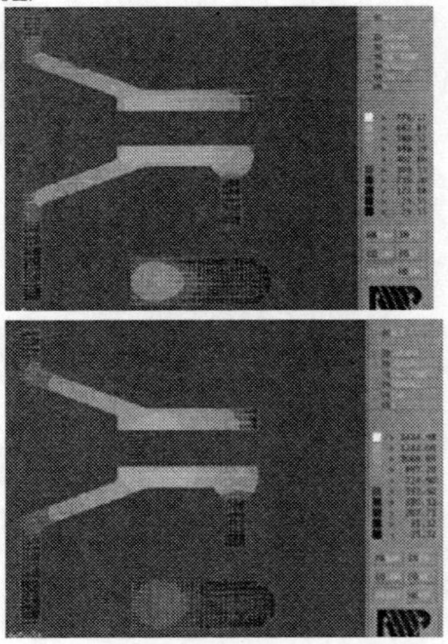

a)

b)

Fig. 1. Solidification time of casting in: a) bentonite mould, b) resin mould

necessary the high quantity of regular spheroidal graphite that are distributed uniformly in ferrite-pearlite matrix. Besides chemical composition of the molten metal, it will depend on the cooling gradients given by the casting section thickness and the heat conductivity of the mould. According to the presented results it can be claimed that solidification time of the casting in resin moulding sand was approximately double than in bentonite sand what corresponded acceptably with simulation results. the casting thinnest sections of 2 mm poured into resin mould solidified in equal time as the section of 3 mm in bentonite mould. The castings pour having the modulus from 1 to 1.5 was successfully verified in series production. In the case of heavy-wall castings (t>50 mm), the considerably extended time of solidification causes inoculation fading and worsening of quality parameters SGI.

Acknowledgments

This work was supported by the SRDA under contract No. APVV-16-0485.

References

[1] Krajewski, P.K., et al. High Temperature Thermal Properties of Bentonite Foundry Sand. Archives of Foundry Engineering, 2015, 15(2), *47–50.*

[2] Lekach, S.N. - Bestužev, N.I. Vnepečnaja obrabotka vysokokačestvennych čugunov v mašinostrojeniji. Minsk: Navuka i technika, 1992.

[3] Pokusová, M. - Murgaš, M. Using the modulus at optimizing of ductile cast iron chemical composition. In: Chvorinovs's colloquium. VŠB - TU Ostrava, 2009.

[4] Pokusová, M. et al.: Computer simulation of solidification for in-mold technology. In Acta Metallurgica Slovaca. 2007, 13, 4, 76-79.

The 73rd World Foundry Congress
„Creative Foundry"
23 - 27 September 2018, Krakow, Poland

CONDUCTIVE STIRRING PROBLEMS OF STEEL IN CONTINUOUS CASTING PROCESS

Alexander Schrek[1]*, Marcela Pokusová[1], Alena Brusilová[1],

[1] *Slovak University of Technology, Faculty of Mechanical Engineering, Nám. slobody 17, Bratislava, Slovakia*
**Corresponding address: alexander.schrek@stuba.sk*

Key words: conductive stirring, steel slab, crystallizer, magnetic field, flow velocity

1. Introduction

This paper is focused on the stress-strain properties study of conductive stirring continuously cast steel ingots. This process is activated through the direct electric current interaction flowing along the slab length with the static field byat least one electromagnet. The conductive stirring method causes the molten metal motion in the crystallizer as well as in the zone of remaining melt in the whole volume of the continuous slab. The steel has in the subsurface layer the thick dendrite structure. In practice, the mixing of melt with linearly engines is used for grain refinement. Their disadvantage is a generated heat over **1MJ.s^{-1}** on 1m^2 of the continuous slab. The surface crust of the continuous slab contracts immediately after solidify, it moves away from the water-cooled walls of crystallizer and the overheated melt melts it downs backwards. The reducing crust is disrupting during exhaust, the melt primary unmixed alloying elements enriched penetrates on the surface through radial cracks and the surface defects create. The inductive strirers enable the temperature equalization of the melt along cross-section and its cooling on the liquidus temperature. The stirrers are parts of crystallizer body conducting generated heat away through the water- cooled walls. The creating of surface defects is limited thanks inductive stirrers. The properties of steel in the slab cross-section are simultaneously changing with a quality of surface.

2. Experiments

In the induction stirring furnaces the intense melt flow with a speed above 1 m.s^{-1} is a factor improving material properties of steel slabs [1]. There is decreasing of alloying and *accompanying* elements in action point of stirrer with characteristic light border or circle according to cross-section form of slab. For this zone it is used the term „white band" [2]. On the Fig.1 the thickness course of stirred area along the length L_A is presented which corresponds to the white band position and velocity flow profile V_A after projection to cross-section of slab. The broken markings line reaching the value 0,18% in the ingot centre presents distribution of C in unstirred slab section and with this related differences of stress-strain characteristics along cross-section of ingot.

The stirring motion activates the columnar and polyhedral grains refinement and the chemical heterogeneity decreasing [3], without transcrystalline solidification [4]. The high energy consumption and technical intensity of inductive stirrers initiated the development of energy more effective stirring methods through magnetostatic field interaction with direct current flowing along continuous casting steel ingot, so-called conductive method [5]. The successful industrial verification of conductive stirrer during continuous casting of crowbar 0.2 x 0.2 m in ŽP Podbrezová, a.s. [6] showed the posibility of stirring in the crystallizer and simultaneously in the part of residual melt zone.

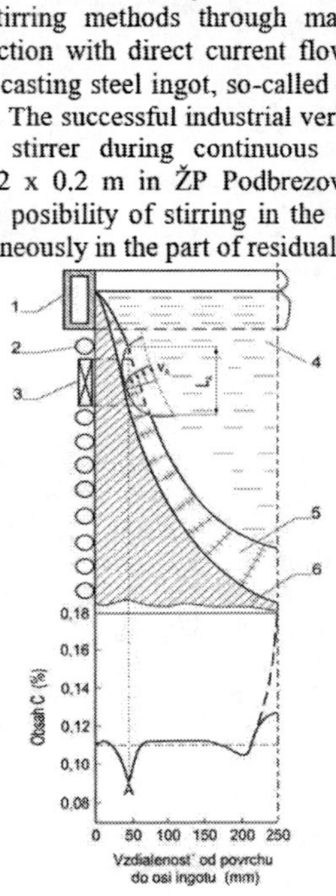

Fig. 1. Distribution of C in the slab section: 1- crystallizer, 2- rolls, 3- stirrer, 4- liquid metal, 5- two phase region, 6- solid phase

3. Results and discussion

The tests have been realized on the pilot production line of continuous casting MANESMANN-DEMAG with the crowbar 200 x 200 mm from the steel with chemical composition 0.1 C – 0.005 S – 0.2 Si – 0.5 Mn and conductive stirring method. The low velocity flow and negligible heat quantity generated in the slab have been allowed to place the electromagnet activated the melt motion about 0.75 m under crystallizer thus significantly nearer than the inductive stirrers allow. The physical centre analysis of stirring motion activation and the other effects developing during conductive stirring method has been realized in conditions according to the Fig. 2.

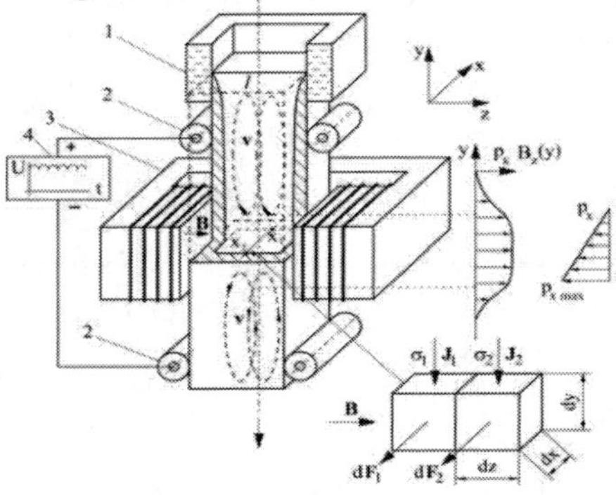

Fig. 2. Arrangement of stirrer: 1- crystallizer, 2 drive rolls , 3- electromagnet, 4- power supply

The melt in the continuous slab is the conductor carrying electrical current. In the places where it exposed to magnetic field activity, the forces and pressure effect are excited. The result of these influences it will be the melt motion.

$$dp_x = \frac{dF_x}{dy.dz} = J_y B_z dx \qquad (1)$$

$$p_m = \int_{x_1}^{x_2} J_y B_z dx = J_y B_z (x_2 - x_1) + p_0 \qquad (2)$$

The pressure increasing in the direction of axis x effects in all directions and causes the melt motion in two circulating flows upwards and downwards from the electromagnet work area. The approximate numerical calculation of generated pressure according to the equations (1) a (2) in the activity conditions of stirrer prototype at 800 A enable the informative idea of action forces. The pressure affecting at the current density J_y in the melt about 14280 A.m^{-2}, induction B_z = 0,072 T and bath dimension $x_2 - x_1 = \Delta x$ = 0,155 m, it will increases from 0 to approximately 160 Pa.

4. Conclusion

Duriing development of stirrer, the purpose was to obtain the information about properties of material solidifying with different stirring intensity. The changes in makrostructure and stress-strain characteristics have been possible to observe already at the velocity of melt motion about 0.05 m.s^{-1}. The optimal speeds have been measured in the range 0.15 to 0.2 m.s^{-1}, i.e. at much lower values than it is necessary in the induction stirrers. Based on the test channel results, the parameters for testing of stirrer have been specified – the flow velocity to 0.2 m.s^{-1} and the average value of induction B = 0.072 T which was observed by measurement on the electromagnet. The graduated current values have been 120 – 300 – 550 – 700 – 800 A which caused metallographic and stress-strain changes in the macrostructure. Stress-strain characteristics have been determined by means of tensile, compressive and hardness tests on the samples made from the cross-section of slabs which have been obtained from the subsurface layer and the cross-section of slabs. These tests showed the marked influence of crystalization way on the properties of slabs which was influenced by the conductive stirring behind crystallizer.

Acknowledgments

This work was supported by the Slovak Research and Development Agency under the contract No. APVV-16-0485.

References

[1] K. AYATA, et al. (1981) Influence of Electro-Magnetic Stirring at Mold on Negative Segregation in Continuously Cast Bloom. Tetsu-To-Hagane, 678, 238-246.

[2] H. TAKEUCHI, et al. (1978) Quality improvement of continuously cast stainless steel blooms through electromagnetic stirring. Transaction ISIJ, Vol. 18, 453-360.

[3] H. TAKEUCHI, et al. (1980) The Effects of Electromagnetic Stirring on Cast Structure of Continuously Cast SUS 430 Stainless Steel Slabs. Tetsu-To-Hagane, 66, 6, 638-646.

[4] M . DISERENS, (1981) et al.Experiences on continuously cast billets influenced by electromagnetic stirring below mould. Scandinavian Journal of Metallurgy, č. 10, 19.

[5] Patent SK 277974.

[6] M. MURGAŠ, et al. (1996) Netradičný spôsob premiešavania zvyškovej taveniny v plynule liatych ingotoch. Hutnické listy 51, 10, s. 11-15.

[7] M. MURGAŠ, M. POKUSOVÁ, (1996) An advanced liqiud-metal stirring method for continuous steel casting. Magnetohydrodynamics, Vol. 32, No. 4, 425-428.

[8] M. MURGAŠ, A. BIACOVSKÁ, (1991), The effect of magnetic field on the kinetics of phase transformations. Chemical papers 45, 6, s. 731-734.

[9] M. MURGAŠ, M. POKUSOVÁ, A. POKUSA (2001), Príspevok ku problematike miešania kontinuálne liatej ocele. Hutnické listy, 56, 1-3, s. 7-12.

[10] J. JOURA, (1994) Netradičný spôsob premiešavania taveniny pri plynulom liatí na ZPO v železiarňach Podbrezová a.s. In CO MAT TECH '94. Bratislava: STU, 310-314.

The 73rd World Foundry Congress
„Creative Foundry"
23 - 27 September 2018, Krakow, Poland

THE APPLICATION OF X-RAY COMPUTED TOMOGRAPHY TO STUDY THE QUALITY OF CERAMIC MOULDS IN PRECISION FOUNDRY

Adam Tchórz[1]*, Izabela Krzak[1], Marzanna Książek[1], Krzysztof Żaba[2], Jarosław Mizera[3], Ryszard Sitek[3], Artur Dydak[4], Piotr Kurdziel[4].

[1]*Foundry Research Institute, 73 Zakopiańska St., 30-418 Cracow, Poland*
[2]*AGH University of Science and Technology, 30 Mickiewicza Ave., 30-059 Cracow,Poland, Department of Non-Ferrous Metals,*
[3] *Warsaw University of Technology, Department of Materials Science, 141 Woloska Str.,02-507 Warsaw, Poland*
[4]*The Specodlew Enterprise of Foundry Innovation Ltd., 3 RotmistrzaWitoldaPileckiego, Str., 32-050 Skawina, Poland*
**Corresponding address: adam.tchorz@iod.krakow.pl*

Key words: X-ray computed tomography, ceramic shell moulds, precision casting

Abstract

The aim of the present work was to detect and visualise specific internal discontinuities in ceramic casting moulds, i.e. porosity, cracks, delaminations or inclusions. Then, analyses were carried out in terms of determining the proportion of the volume porosity in the object, determining the size of the porosity and its location in the object. The tests were carried out for two different technological characteristics of moulding systems made with the method of melted models. The tests were carried out using the X-ray computed tomography (CT) method.

1. Introduction

Precision casting with the use of melted models is widely applied in engineering and aerospace industries to obtain both responsible and complex products. Usually, these are castings with complex shapes as well as high dimensional accuracy and smoothness, which do not require further processing, e.g. machining. Ceramic casting moulds used in precision casting technology have a fundamental impact on the quality of the product and constitute an integral part of this process. Ceramic moulds are made of moulding sands, the basic ingredients of which are ceramic powders, binders and auxiliary materials. The parameters of both moulds and castings depend on their properties [1].

The process of producing multi-layered ceramic moulds consists in immersing the finished model set in a liquid ceramic mass. Next, the coatings produced are sprinkled with highly fireproof material characterised by a defined grain size, starting from the smallest (for the facing layer) and ending with the thickest one in order to strengthen the shell. Depending on the application, a ceramic casting mould consists of several or even a dozen ceramic layers. Hence, the individual layers are often characterised by heterogeneous performance caused by the anisotropy of the materials serving as the matrix of individual layers as well as uneven coverage of the wax model with a liquid ceramic mixture. It is worth noting that the properties of the ceramic mould obtained in the process of melted models determine the dimensional accuracy of castings in more than 50%. Therefore, due to the complexity of the process, castings of the desired quality are not always obtained. The use of the X-ray CT method creates new possibilities of assessing the quality of multi-layer ceramic moulds intended for casting responsible parts, using the method of melted models. Spatial imaging allowes for a more accurate analysis of the internal structure of the material, including the morphology of internal discontinuities and their location in the space of the material. The results obtained with this method can be used to develop new solutions, which in consequence will translate into an increase in the efficiency of the process.

2. Material and methodology of tests

The tested material consisted of ceramic moulds made of seven layers in the technology of melted models. They were characterised by a dense packing of details at the gating system (Fig. 1). Two ceramic mixtures were prepared for the production of moulds. The selection of powders and binders was carried out on the basis of the knowledge of materials used in industry and used commercially. The ceramic moulds were prepared using the methods used at Specodlew company. The composition of mixtures for individual FC layers is summarised in Table 1.

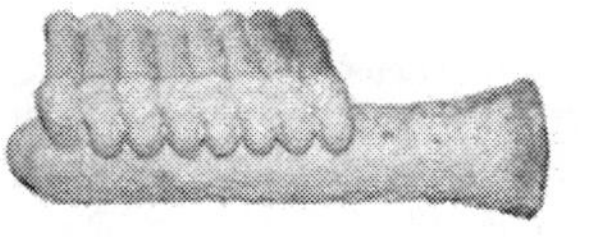

Fig. 1. ***An example of a ceramic mould subjected to X-ray CT testing***

In order to reveal internal discontinuities in the ceramic mould and to understand their character, the method of X-ray computed tomography (CT) was used. This method is a non-destructive technique allowing to obtain cross-sectional (2D) and spatial (3D) images. Not only does it enable a qualitative assessment of the tested material, but it also allows for carrying out advanced analyses [2].

The tests were carried out using the v|tome|x 1-450-type CT station. The test parameters used were: voltage of 250 kV, amperage of 300 μA, and voxel size of 124 μm. The samples subjected to CT tests had maximum dimensions of 300x200x70 mm (Fig. 1).

Table 1. Technological characteristics of moulding systems in application.

Mould type	K	R
1 layer	Matrix quartz flour SiO_2) 75 μm. Ludox binder.	Matrix quartz flour SiO_2) 75 μm. Remasol Premium Plus binder.
Structural layers (2-6)	Matrix fine-grain quartz sand (SiO_2) 0.1-0.3 mm. Hydrated ethyl silicate binder.	Matrix fine-grain quartz sand (SiO_2) 0.1-0.3 mm. Remasol Plus binder.
Structural layer (7)	Matrix coarse sand 0.5-1 mm. Hydrated ethyl silicate binder.	Matrix coarse sand 0.5-1 mm. Remasol Plus binder.

In order to determine the quantitative 3D characteristics of the internal discontinuities in the ceramic moulds, VGStudio MAX 2.0 was used. This programme uses scan data to create a three-dimensional size model that can be displayed in 3D or 2D views. The module of this programme for defect analysis offers several different algorithms to detect internal defects, e.g. discontinuities or inclusions located within the object. In addition, it provides detailed information on specific defects, i.e. their size, volume, area, number and location.

3. Tests results and discussion

The results obtained by means of X-ray tomography were characterised by a satisfactory effect of 3D image reconstruction. Images of individual cross-sections were characterised by sufficient brightness and contrast so that spatial analysis of discontinuities could be carried out. They allowed to determine the nature of the discontinuities, their dimensions and location in the detail area in a significant way. Figure 2a shows the 3D representation of the porosity analysis for K-type and R-type ceramic moulds. In contrast, figure 2b presents a visualisation showing the dependence of the number of porosities in function of their size. On the basis of the quantitative analysis carried out, the total proportion of porosities in the ceramic moulds was determined. Porosity for K-type and R-type moulds is similar and ranges between 2.6 and 2.9%. The pore distribution is uniform over the cross-section of the layers, which guarantees sufficient gas permeability of the moulds. However, it was noticed that in the R-type moulds (Fig. 2) there are larger porosity clusters of considerable size in the areas of "detail" connections with the gating system. However, the location of porosities in these areas should not affect the mechanical durability of the R-type ceramic mould as the thickness of the seven subsequent (structural) layers is greater by approx. 25%. However, it is worth noting that ceramic moulds with a dense packing of details may generate a larger percentage of defective products.

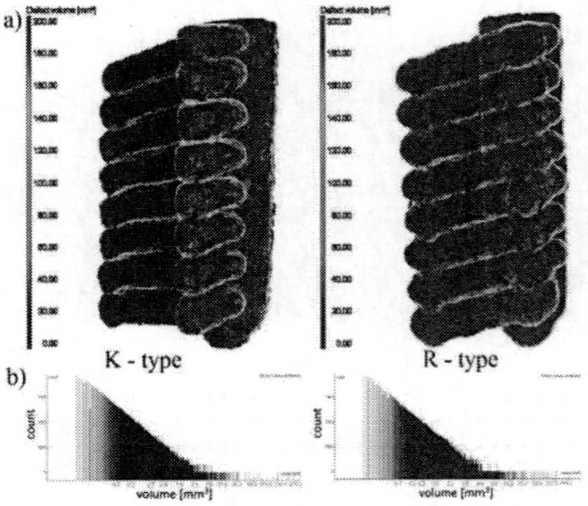

Fig. 2. ***Spatial view of K-type and R-type ceramic moulds with the porosity analysis carried out (a), chart showing the dependence of the number of porosities in function of their size (b)***

4. Conclusions

As a result of the tests, internal and surface discontinuities were revealed. The test results allowed to determine the nature of the discontinuities and their location in the detail area in a significant way. Spatial imaging allowed for a more accurate analysis of the internal structure of the ceramic mould material, including the morphology of internal discontinuities and their location in space. Based on the obtained results, it was demonstrated that the pore distribution is influenced by the binder used. When replacing ethyl silicate with Remasol, the largest porosity clusters occur in the areas of connections between the gating system and the detail. The porosity of the ceramic moulds is comparable for both binders used. In addition, a ceramic mould porosity methodology was developed with the use of X-ray computed tomography method.

Acknowledgments

This work has supported by The National Centre for Research and Development under the project No PBS3/A5/54.

References

[1] Wiśniewski P., Sitek R., Koralnik M. K., Spychalski W. L., Moszczańska D., Mizera J. (2017). Investigations of coolings process of ceramic shell samples by using a thermographic camera. Ceramic Materials (ISSN 1505-1269). 69, 2. 146-153.

[2] Tchorz A., Boroń Ł. (2013). Quantitative characteristics of morphology 3d graphite ductile iron with *x-ray* CT scan. In Pacyna J. (Ed.), Papers of Materials Engineering (p. 392) . Krynica: PAN.

DEVELOPMENT AND APPLICATION OF CAST STEEL NUMERICAL SIMULATION SYSTEM FOR HEAT TREATMENT BASED ON INTECAST

Yin Yajun, Wang Wen, Wang Chongxun, Zhang Jie, Zhou Jianxin*, Shen Xu, Ji Xiaoyuan, Xu Xiaojing

State Key Laboratory of Material Forming and Mold Technology, Huazhong University of Science and Technology, Hubei, China
**Corresponding address: zhoujianxin@hust.edu.cn*

Key words: InteCAST, heat treatment, microstructure and performance prediction, stress-strain

Abstract

Based on InteCAST platform, a numerical simulation system for heat treatment of cast steel was developed using C++ environment. The system mainly includes: heat treatment process inquiry module, initialization and property parameter setting module, automatical import and display module of TTT and TTA curves, heat treatment process design module, Organization, performance and stress-strain prediction module. This paper introduces the mathematical model and numerical solution algorithm in the process of numerical simulation of the heat treatment organization, performance and stress-strain as well as the interface operation process and application examples of the system.

1. Introduction

With the development of computer technology and numerical simulation theory, the application of numerical simulation of casting solidification process has been more and more mature, which plays an irreplaceable role in the design of casting process. In recent years, coupled with the smelting - casting - solidification - heat treatment of the whole casting numerical simulation has become a research hotspot. As an important step during the whole process, numerical simulation of heat treatment is paid more and more attention to by scholars around the world. Many commercial CFD codes such as AVL-FIRE v8.5 and COSMAP as well as many commercial FEA softwares including DANTE, MSC.Marc and ABAQUS have been developed in order to obtain the desired phase changes, to reduce distortion and residual stress, to develop relatively shorter and less expensive thermal cycles [1-6]. FEM and CFD models such as Leidenfrost temperature model and Leblond model are proposed, computational fluid dynamics was also used to predict the distributions of temperature, iron-carbon phases, hardness and residual stress [6-8]. Computational analysis, based on the finite element, finite difference, or finite volume methods, is more and more used to effectively investigate and optimize complex manufacturing processes and thermal treatments, and all the process steps, i.e. heating, holding, and cooling, have been considered, modeling both the austenite formation and decomposition and taking into account nucleation and growth processes, allowing the evaluation of properties not always experimentally measurable [9-10]. Therefore, based on InteCAST platform, this paper developed a numerical simulation system for heat treatment of cast steel using C++ environment.

2. R&D and Application

Considering the design style of InteCAST, heat treatment process and simulation program parameter setting logic, using wxWidgets cross-platform development library, a numerical simulation interface system of heat treatment including cooling medium, TTT and TTA curves, material thermal properties, calculation and analysis, etc. was developed. During the calculation of organization and performance in heat treatment, based on the heat conduction control equation, the influences of nonlinear factors such as interface heat transfer coefficient, latent heat of transformation and thermophysical parameters were considered. Based on the finite difference analysis method and with Scheil superposition principle applied, a simulation program was written and the temperature field of the heat treatment process was obtained.

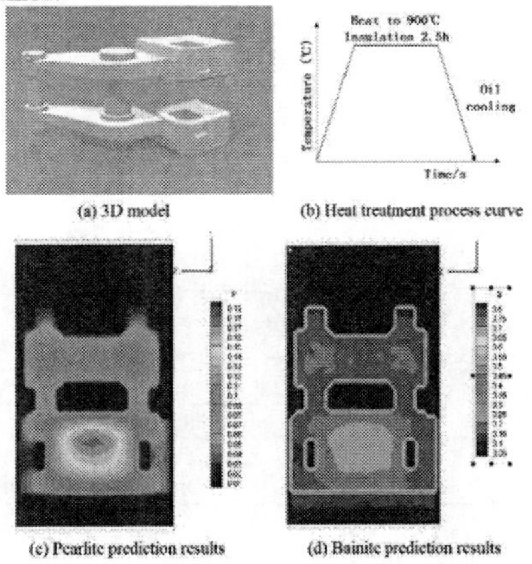

Fig. 1. ***Heat treatment process and microstructure properties prediction of a bending inner pick arm casting***

On the basis of this, in view of the different cases of diffusive and non-diffusive organization transition, several phase volume fraction equations such as Avrami equation and Koistinen-Marburger equation were applied respectively, by which we could accurately simulate the volume fraction of each phase in the organization and obtain the distribution rule of workpiece hardness according to the Meniere hardness model. Figure 1 presents the result of the heat treatment microstructure simulation of a bent inner pick arm casting.

Aiming at the common heat treatment process of cast steel, a corresponding three-dimensional transient coupled thermo-mechanical model was established in the heat treatment stress-strain module. Firstly, the temperature variation inside the workpiece during the heat treatment process of the casting was analyzed by means of thermal analysis. Then use it as the input condition to obtain the stress-strain distribution inside the workpiece. The complex convection-radiation heat exchange among the interior of the cast steel, the steel castings and the environment, the change of thermal physical properties of the material with the temperature and the influence of the latent heat released during the process of solidification were taken into account in the thermal analysis comprehensively. In the analysis of force, the bilinear isotropic strengthened constitutive model was adopted and the thermal elastoplastic method was used to accurately simulate the stress-strain distribution of heat treated cast steel. Figure 2 presents the heat treatment stress simulation results of a casting, and Table 1 presents the process parameters applied in heat treatment.

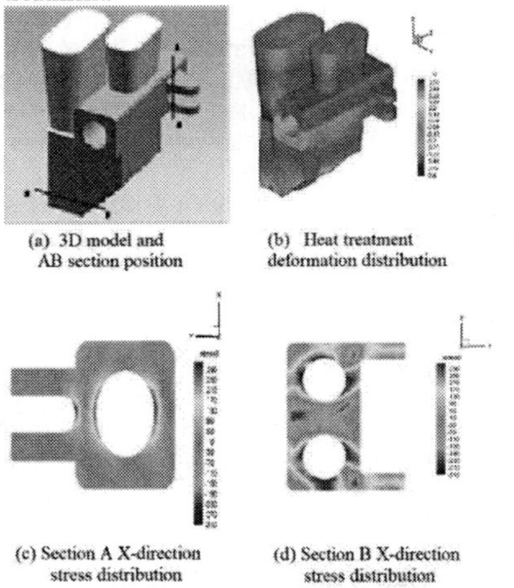

(a) 3D model and AB section position (b) Heat treatment deformation distribution

(c) Section A X-direction stress distribution (d) Section B X-direction stress distribution

Fig. 2. Heat treatment stress simulation results of a casting

Table 1. Heat treatment process parameters

Technology serial number	Heat treatment stage		
	Warming	Insulation	Cooling
1	120℃/h	870℃,5h	Water Cooling

3. Conclusion

In integrated view, based on the actual production requirement of heat treatment of the factory and using the large heat treatment process database of Kunming Guangweitong Machinery equipment co. LTD as the basis, the system, with the forefront and more mature numerical algorithms and mathematical models, sets heat treatment process inquiries, data import, parameter setting to process design, numerical simulation of a variety of functions in one and thus achieves full coverage of the heat treatment production process.

The application example also shows that the numerical simulation system for heat treatment is easy to operate, has a friendly interface, can accurately predict the distribution of organization properties, stress-strain distribution trend and maximum stress distribution position of the workpiece during the heat treatment process, and assist the process staff to optimize the process to reduce the heat treatment defects as well as increase the technological yield thus guide the actual production of heat treatment.

Acknowledgments

This research was financially supported by the National Nature Science Fund Projects, China (no.51605174), and State Key Laboratory of Materials Processing and Die & Mould Technology Research Project (2015-2017).

References

[1] Ferguson B L, Li Z & Freborg A M. (2005). Modeling heat treatment of steel parts[J]. Computational Materials Science. 34(3), 274-281.

[2] Şimşir C & Gür C H. (2008). A FEM based framework for simulation of thermal treatments. Application to steel quenching[J]. Computational Materials Science. 44(2), 588-600.

[3] Srinivasan V, Moon K M, Greif D, Wang D M & Kim M H. (2010). Numerical simulation of immersion quenching process of an engine cylinder head. Applied Mathematical Modelling, 34(8), 2111-2128.

[4] Yaakoubi M, Kchaou M & Dammak F. (2013). Simulation of the thermomechanical and metallurgical behavior of steels by using ABAQUS software[J]. Computational materials science. 68, 297-306.

[5] Deng X, Ju D. (2013). Modeling and simulation of quenching and tempering process in steels[J]. Physics Procedia. 50, 368-374.

[6] Škerget L, Hriberšek M, Zhang D, Stauder B & Greif D. (2014). Numerical simulation of immersion quenching process for cast aluminium part at different pool temperatures. Applied thermal engineering. 65(1-2), 74-84.

[7] Eser A, Broeckmann C & Simsir C. (2016). Multiscale modeling of tempering of AISI H13 hot-work tool steel–Part 1: Prediction of microstructure evolution and coupling with mechanical properties. Computational Materials Science. 113, 280-291.

[8] Wróbel J, Kulawik A & Bokota A. (2017). The numerical analysis of the hardening phenomena of the hot-work tool steel. Procedia Engineering. 177, 33-40.

[9] Berglund D, Alberg H & Runnemalm H. (2003). Simulation of welding and stress relief heat treatment of an aero engine component[J]. Finite Elements in Analysis and Design. 39(9), 865-881.

[10] Carlone P, Palazzo G S & Pasquino R. (2010). Finite element analysis of the steel quenching process: Temperature field and solid–solid phase change[J]. Computers & mathematics with applications. 59(1), 585-594.

The 73rd World Foundry Congress
„Creative Foundry"
23 - 27 September 2018, Krakow, Poland

EFFECTS OF GREEN SAND PARTICLE SIZE DISTRIBUTION ON SQUEEZE COMPACTING BEHAVIOR ANALYZED BY DISCRETE ELEMENT METHOD

Yuuka Ito [1], Shingo Yoshida [1], Yasuhiro Maeda [2]*, Hiroyasu Makino [3]

[1] *Graduate Student, Dept. of Mechanical Engineering, Daido University*
[2] *Dept. of Mechanical Engineering, Daido University / Takiharu-cho 10-3, Minami-ku, Nagoya 457-8530, Japan*
[3] *R&D Center, Sintokogio LTD / Honohara 3-1, Toyokawa, 442-8505, Japan,*
**Corresponding address: y-maeda@daido-it.ac.jp*

Key words: Squeeze Molding, Discrete Element Method, Simulation, Hard-Core/Soft-Shell model, Void Fraction

The green sand mold requires the green sand managed appropriately. Simulation is useful tool for designing and controlling for the sand molding, so the demand for molding simulation is increasing. In the present study, the squeeze molding simulation are carried out considering the particle size distribution. As a result, it was clear that the relationship between the void fraction and the compacting pressure during squeeze molding was clarified. The influence of the particle size distribution on the compacting behavior of the molding sand layers was also clarified. Further, the three-dimensional visualization of the green sand molding was shown in the present study.

1. Introduction

Basic modeling of green sand particle and algorism for molding using the three-dimensional Discrete Element Method (DEM) have been proposed [1,2]. However, development of DEM for sand mold molding has been stopped since Maeda et al. proposed the model [1]. The main reason is that the computer performance to sufficiently demonstrate the function of DEM was insufficient. But, the performance of computers has dramatically increased since then, and this problem is being solved.

Recently there are reports [3,4] using the DEM, simulation using DEM is expected to be a powerful tool of the molding process. Now, it is possible to take into consideration of the size distribution and the cohesion of the green sand particles. In present study, we redeveloped the Discrete Element Method considering the grain size distribution of sand particles and adhesive strength according to the current computer environment and conduct squeeze molding simulation.

2. Discrete Element Method (DEM)

In the present mathematical model of DEM, the green sand particles are assumed to be viscoelastic elements. The motion of each element is obtained by integrating the equations of motion step-by-step:

$$\ddot{r} = \frac{1}{m_e}(f_c + f_d) + g \qquad (1)$$

where r is the position vector, m_e is the element mass, f_c is the summation of the contact force, g is the gravity acceleration, and dot denotes a time derivative. For squeeze molding, the drag force f_d is zero because the air is not used. After a time step Δt, the velocity vector v and the position vector of an element r can be calculated by integration of Eq.(1).

The Hard-Core/Soft-Shell model for the green sand shown in Fig.1 is proposed by Maeda et.al. [1,2]. The Hard-Core is modelled the sand particle such as silica or artificial (ceramic) sand, and the Soft-Shell is also the thin layer of bentonite or oolitics.

In the DEM, it is appropriate to adapt one particle to one discrete element. However, since the actual number of sand particles is too many, it is impossible to handle all the particles. Therefore, we calculate by mapping to large discrete elements while preserving mass and number ratio.

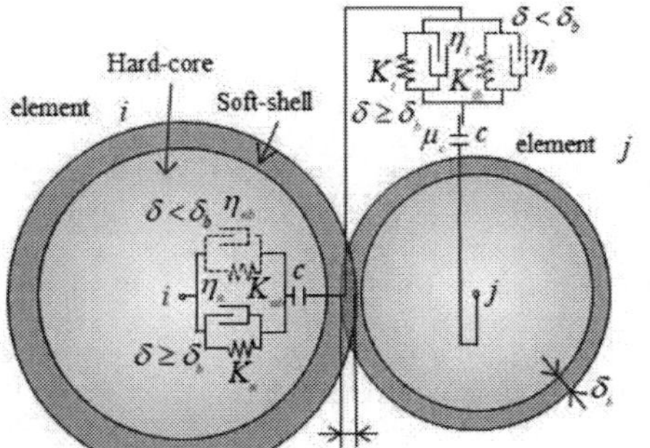

Fig. 1. Hard-Core/Soft-Shell model for green sand.

3. Results and Discussions

The object is the process of molding green sand into a cylindrical flask of φ 50 × h50 and molding from the top with a squeeze pressure of 0.98MPa shown in literature [1].

The total number of elements for the ceramic sand #650 is changed to 942, 2029, 3181,4020. Figure 2 shows the same of the size distribution form. From Fig.3, even if the total number of identified discrete element is different, the squeeze compacting behavior is almost the same tendency.

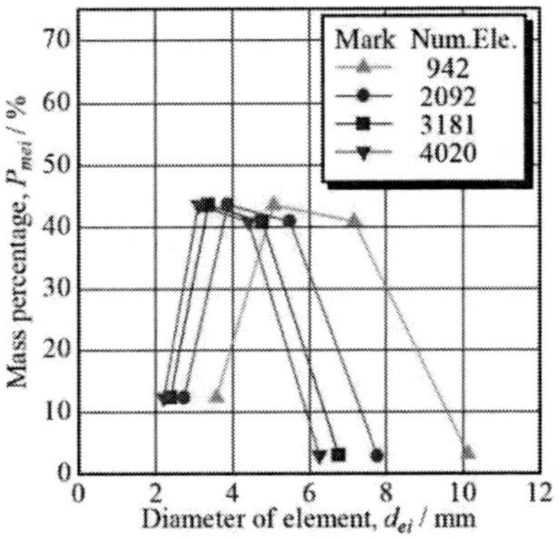

Fig. 2. Size distributions of elements changing with total number of elements for Ceramic sand #650.

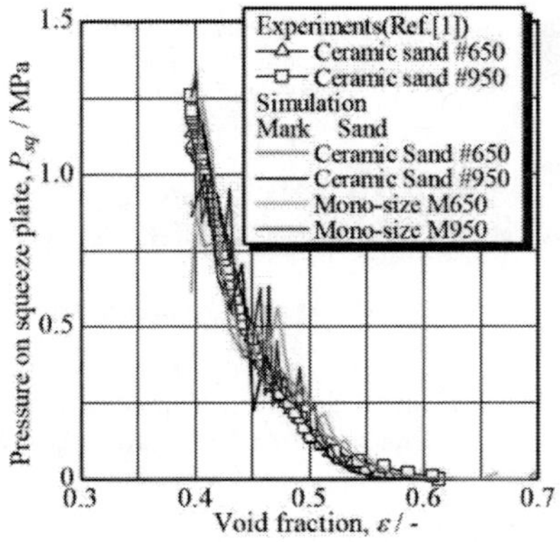

Fig. 4. Relationship between void fraction and pressure on the squeeze plate.

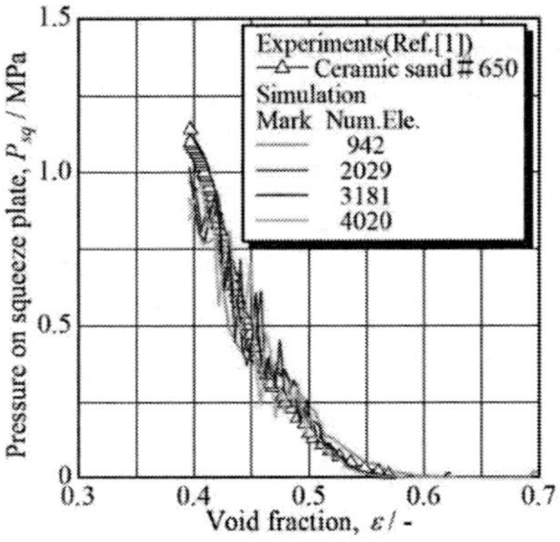

Fig. 3. Compacting pressures on squeeze plate Psq changing with total number of elements.

Fig. 5. Three-dimensional visualization results for Ceramics sand #650 and Mono-size M650 by ParaView.

The ceramic sands without the size distribution are prepared to investigate the influence of the particle size distribution on the compacting behavior. They are the M650 and M950. The relationship between the void fraction and the squeeze pressure shows in Fig.4. Further, the calculated results are visualized three-dimensionally using ParaView [5] as shown in Fig. 5. This makes it possible to track the behavior of the particles in the flask.

4. Summary and conclusion

The characteristics of green sand with particle size distribution and cohesion are modeled in this study. From the simulation with several kinds of sands using proposed model, the identification method from a real particle size distribution to the discrete element is useful for molding simulation using DEM. Three-dimensional display of green sand with particle size distribution is very effective.

References

[1] Y. Maeda, Y. Maruoka, H. Makino and H. Nomura: "Squeeze Molding Simulation Using The Distinct Element Method Considering Green Sand Properties", J. Materials Processing Technology, 135(2003), 172-178.

[2] Y. Maeda, Y. Maruoka, H. Makino and H. Nomura: "Sand Compacting Simulation Using Distinct Element Method Compacting Particle Size Distribution", J. Japan Foundry Engineering Society, 75(2002) 108-114.

[3] E. Havad, P. Larsen, J. H. Walther, J. Thorborg and J. H. Hattel: "Flow Dynamics of green sand in the DISAMATIC moulding process using Discrete element method(DEM)", IOP Conf. Series: Materials Science and Engineering, 84(2015), 012023.

[4] Y. Nakata and M. Yamanoi: "Quantitative Evaluation of Mixed States of Granular Systems with Shannon Entropy", J. Soc. Powder Technology, Japan, 54(2017), 296-304.

[5] Information on https://www.paraview.org/

The 73rd World Foundry Congress
„Creative Foundry"
23 - 27 September 2018, Krakow, Poland

IMPROVED ULTRASONIC DEGASSING OF AlSi10Mg ALLOY AND ITS PERFORMANCE EVALUATION WITH THE REDUCED PRESSURE TEST (RPT) METHOD

Haize Galarraga [1]*, Maider Garcia de Cortazar [1], Eber Arregi [2], Jon Ander Iparraguirre[1], Jose Luis Oncala[1]

[1]Fundación TECNALIA R&I, Mikeletegi Pasealekua 2, Donostia-San Sebastian (Gipuzkoa), Spain
[2] EDERTEK S. Coop, Isasi Kalea, 6, Arrasate-Mondragón, (Gipuzkoa), Spain
**Corresponding address: haize.galarraga@tecnalia.com*

Key words: aluminium, degassing, ultrasonic, oxides, reduced pressure test

Abstract
Porosity is detrimental to the mechanical properties of aluminium alloy castings, thus, in most of the cases, especially in structural parts, reduction of porosity is desired and pursued. Porosity can be caused by gas formation, solidification shrinkage, or non-metallic compound formation. In this work, the reduction of gas porosity formed by the hydrogen dissolved in the molten AlSi10Mg aluminium alloy has been studied employing an ultrasonic degasser in combination with purging gas blowing. Qualitative study of the hydrogen content of different samples has been performed employing the widespread reduced pressure test (RPT). Different factors affecting to RPT test have also been studied.

1. Introduction

The presence of hydrogen in aluminium melts is principally originated from the vapour in the atmosphere. The water vapour reacts with the aluminium, forming aluminium oxide and gaseous hydrogen [1]. Hydrogen comes out of solution during solidification, causing the formation of porosity. Porosity reduces strength and causes leakages in castings, consequently, hydrogen needs to be removed from cast alloy products. Depending on the application of the casting, the minimum acceptable content of hydrogen could differ. Common general target values are around <0.2 ppm [2].

The most effective way to reduce the gas porosity during casting is degassing. Several types of degassing methods are currently in use, but the most common methods in aluminium industry consist of blowing gases like argon, nitrogen or chlorine through the melt via rotary impeller, creating small bubbles and increasing the efficiency of the degassing process.

Several studies have been published showing the effectiveness of the ultrasonic treatment in molten aluminium alloys [3–6]. Ultrasonic degassing uses ultrasonic vibrations to generate oscillating pressures in molten aluminium, producing cavitation phenomena and creating fine bubbles. The bubbles produced during cavitation could provide nuclei for hydrogen bubbles to merge and flow out of the melt. Ultrasonic degassing is considered a promising technology due to its reduced dross formation, absence of brittle moving parts and energy efficiency [7].

In this work, nitrogen gas was blown through the ultrasonic probe, creating a fragmented and dispersed nitrogen micro-bubble stream due to the ultrasonic cavitation. This combined effect generates high capacity degassing system, as it has been demonstrated for the treatment of aluminium for twin roll casting [3].

A method to gauge the hydrogen level in the melt was necessary to study the performance and effectiveness of the ultrasonic degassing. The RPT method is a common tool in foundries which allows the operator to qualitatively assess the hydrogen level of a batch of molten aluminium, allowing corrective action to be followed [8]. In this work, the dependence of the RPT method with several parameters as the relative humidity in the atmosphere, RPT temperature and aluminium temperature was also studied.

2. Experimental

In each test performed in this work, 6 kg of AlSi10Mg alloy were melted in a silicon carbide crucible with an electric furnace. For the comparison between low and high oxide content melts, 100% ingot and 100% rejected parts were employed as feedstock respectively.

RPT samples were obtained solidifying small aluminium quantities under vacuum (-0.9 bar). A 1 mm diameter thermocouple was introduced in the aluminium sample to measure the temperature until vacuum was applied. This temperature was called "RPT temperature". The density of the solidified samples (apparent density) was measured employing the Archimedes′ Principle, where the density is calculated after measuring dry weight and immersed weight of the samples.

Degassing tests were performed with a SCR ultrasonic Ultra-D™ degasser running with 20 kHz frequency. High purity nitrogen with controlled flow rate was employed as a purging gas. RPT densities measured after degassing tests were accomplished with the furnace at 800°C. Vacuum was applied when the aluminium sample cooled down to 650°C (RPT temperature). Same temperatures were applied to evaluate the effect of the humidity.

3. Results

Apparent densities measured at different furnace and RPT temperatures are shown in Fig.1. The influence of these parameters in the RPT density measurements is considerable.

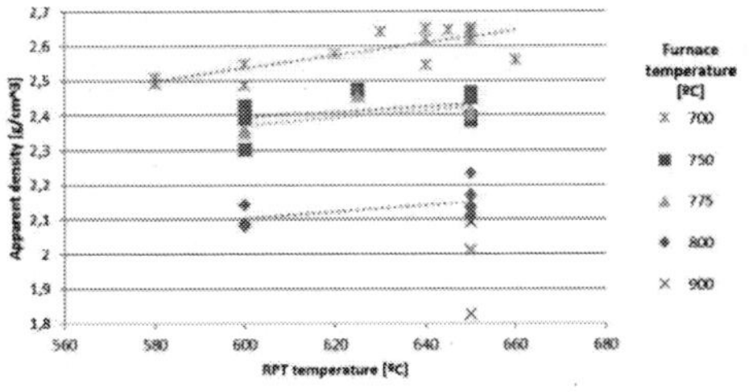

Fig. 1. Apparent density dependence of the RPT and furnace temperatures

Relative humidity also plays an important role in the amount of hydrogen dissolved in the melt as it is shown in the Fig. 2. Although the results look significantly scattered, there is a clear trend showing lower measured densities for higher humidity atmospheres.

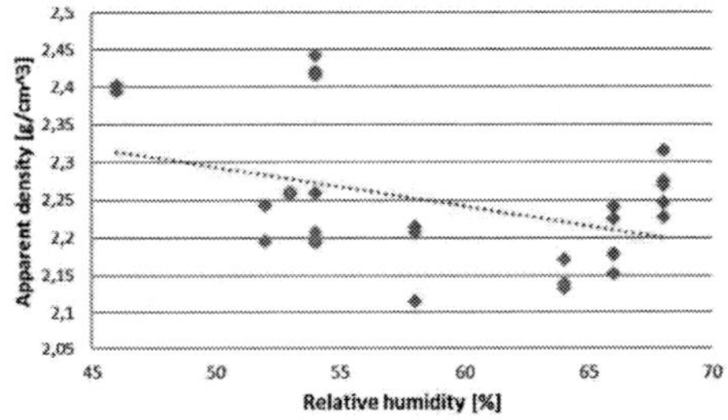

Fig. 2. Effect of the ambient humidity in the measured RPT densities

Fig. 3. Shows measured RPT densities, obtained employing different purging gas flow rates and ultrasonic treatment periods. The results show that density does not increase significantly after 5 minutes long ultrasonic treatment. Besides that, achieved degassing levels are almost identical for the nitrogen flow rate ranges studied in this work.

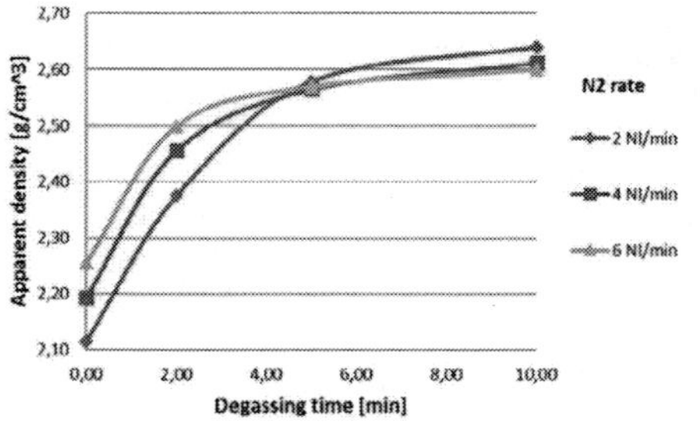

Fig. 3. Density measurements for different degassing times and purging gas flow

A comparison between degasification (2Nl/min N_2) of low and high oxide content melts is shown in Fig. 4. Similar performance can be noticed based on the parallel paths of the density/degassing curves. The offset between two curves is attributed to the different atmosphere humidity values measured during the tests.

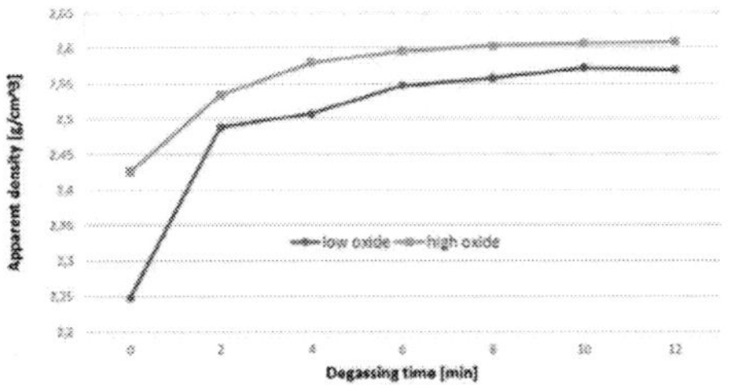

Fig. 4. Ultrasonic degassing (2Nl/min N_2) comparison between low and high oxidized melts

4. Conclusions

- Temperature control of the aluminium sample when vacuum is applied during RPT test is necessary to obtain consistent results.
- The considerable influence of the atmosphere humidity in the RPT measurements advises to consider or keep records of the humidity.
- In the ultrasonic-gas blowing combined degasification method, purging gas flow rate has minor effect in the hydrogen removal.
- Oxide content of the melt has not considerable influence in the ultrasonic-gas blowing combined degasification.

Acknowledgments

This work has been done within the AL-STRUC project funded by the ELKARTEK program of the Basque Government with grant number KK-2017/00020.

References

[1] Robles-Hernandez FC, Herrera JM, Mackay R. (2017). Al-Si Alloys: Automotive, Aeronautical and Aerospace Applications.Springer.

[2] Lumley R. (2010). Fundamentals of aluminium metallurgy: production, processing and applications. Elsevier.

[3] Smith DD, Britt KO. (2016). Theory and Practical Application of Ultrasonic Degassing at JW Aluminum. Light Metals 2016 (pp. 803–808). Springer.

[4] Alba-Baena N, Pabel T, Villa-Sierra N, Eskin DG. (2013). Effect of ultrasonic melt treatment on degassing and structure of aluminium alloys. Materials Science Forum. Vol. 765. pp. 271–275.

[5] Eskin GI. (1995). Cavitation mechanism of ultrasonic melt degassing. Ultrasonics Sonochemistry. Vol. 2(2). pp. S137–S141.

[6] Puga H, Barbosa J, Azevedo T, Ribeiro S, Alves J. (2016). Low pressure sand casting of ultrasonically degassed AlSi7Mg0.3 alloy: modelling and experimental validation of mould filling. Materials & Design. Vol. 94. pp. 384–391.

[7] Eskin DG. (2017). Overview of ultrasonic degassing development. Light Metals 2017 (pp. 1437–1443). Springer.

[8] DJurdjevic MB, Odanovic Z, Pavlovic-Krstic J. (2010). Melt quality control at aluminum casting plants. Metalurgija. Vol. 16. pp. 63–76.

The 73rd World Foundry Congress
„Creative Foundry"
23 - 27 September 2018, Krakow, Poland

DETERMINATION OF THE CHARGE MATERIALS RANGE IN A MULTISTAGE CHARGE BURDEN OPTIMISATION FOR THE FOUNDRY FURNACES

Eugeniusz Ziółkowski[1]*, Kamil Schmalenberg[1,2]

[1] *AGH University of Science and Technology, Faculty of Foundry Engineering, Reymonta 23, 30-059 Kraków, Poland*
[2] *Odlewnie Polskie, Władysława Rogowskiego 22, 27-200 Starachowice, Poland*
**Corresponding address: ez@agh.edu.pl*

Key words: Charge burden, optimization of burden, foundry furnaces, optimization of melting, correction of the chemical composition of liquid metal

Optimization of charge burden for foundry furnaces consists in determining the percentage or mass (kg, t) of the share of each charge material in the burden with the lowest unit cost.
The classic optimization task requires the determination of such values x_j, to make the objective function in the form

$$\sum_{j=1}^{N} c_j x_j \tag{1}$$

has reached the minimum value for restrictions in the form

$$\begin{cases} A_i^{min} \cdot m \leq \sum_{j=1}^{N} A_{ij} x_j \leq A_i^{max} \cdot m \\ 0 \leq x_j^{min} \leq x_j \leq x_j^{max} \leq m \\ \sum_{j=1}^{N} x_j = m \\ i = 1,2,\ldots,M \quad j = 1,2,\ldots,N \end{cases} \tag{2}$$

where:

- M – the number of chemical elements taken into account in the charge burden calculations,
- N – the number of batch materials,
- c_j – unit price of the j-th charge component, e.g. €/kg,
- x_j – mass of the j-th component in the charge to foundry furnace, e.g. kg,
- A_i^{min}, A_i^{max} – lower and upper contents chemical element A_i in the charge, %,
- A_{ij} – content of the chemical element A_i in j-th charge component, %,
- x_j^{min}, x_j^{max} – lower and upper limit of content of the j-th component in the charge material, kg,
- m – mass of the charge, kg.

The x_j values resulting from the solution of the task (1,2) are the optimal starting values of x_j^{opt} in the process of weighing and proportioning charge materials in foundry. Due to the dispersion of individual batch materials and the precision of their weighing or dosing, it is possible to determine acceptable ranges of shares for each charge material to be taken in succession. In further considerations, it was assumed that weighing or dosing batch materials will be carried out in accordance with the order of indices j=1,2,…,N. In order to determine the range of the first material in the burden, the optimization task should be solved:

$$x_{1,min} = x_1 \rightarrow min \tag{3}$$

with restrictions (2) and task in the form

$$x_{1,max} = x_1 \rightarrow max \tag{4}$$

for the same system of restrictions.

The result of weighing or dosing the first charge material, marked for example as x_1^R, must be in the range of $x_{1,min} \div x_{1,max}$. Determining the share of the second charge material requires the introduction of the equation (2) in the system in form of $x_1 = x_1^R$ and solutions for optimization tasks with target functions analogous to (3) and (4) for index 2.

In next stages, optimization tasks are solved, which allow for the previously weighed batch materials to determine the range of the share next component and to quickly determine whether the weighing result of this component is within the acceptable range.

The algorithm presented in this article can find a practical application in multi-step optimization of the charge burden set up from charge materials with different dispersion or with low precision of weighing or dosing [2]. It can be an element of the control system of devices for preparing the burden in the processes of producing liquid metal in various foundry furnaces [3].

References

[1] Ziółkowski Eugeniusz (2009). Methods review for the determination of burden for foundry furnaces. Polish Academy of Science. The Katowice Branch Commission of Foundry Engineering. In *Postępy teorii i praktyki odlewniczej* (pp. 449-460). Katowice-Gliwice.

[2] Schmalenberg Kamil & Ziółkowski Eugeniusz. (2015). Analysis of the quality of the electromagnetic weighting charge materials. Conference of AGH-UST Faculty of Foundry Engineering PhD Students. 07.05.2015. AGH-UST Kraków, Poland.

[3] Ziółkowski Eugeniusz. (2017). Burden optimization of lump charge materials for foundry furnaces. Archives of Metallurgy and Materials. Vol. 62, no. 4, 2217-2221.

The 73rd World Foundry Congress
„Creative Foundry"
23 - 27 September 2018, Krakow, Poland

EVALUATION OF THE EFFECT OF ULTRASONIC DEGASSING ON COMPONENTS PRODUCED BY HIGH PRESSURE DIE CASTING

Manel da Silva[1]*, Attila Bajusz[2], Thomas Pabel[3], Tose Petkov[3] and Xavier Planta[4]

[1]*Eurecat, Centre Tecnològic de Catalunya, Av. Universitat Autonoma 23, Cerdanyola del Vallès, 08290, Spain*
[2]*Certa Kft., Sátoraljaújely, 3980, Hungary*
[3]*Austrian Foundry Research Institute, Leoben, 8700, Austria*
[4]*Ultrasion S.L., Av. Universitat Autonoma 23, Cerdanyola del Vallès E-08290, Spain*
**Corresponding address: manel.dasilva@eurecat.org*

Key words: die casting, ultrasonic, ultrasounds, degassing.

Ultrasonic processing is known to be an efficient means of aluminium melt degassing with additional benefits of being economical and environment friendly. This paper describes the performance of ultrasonic degassing in preparing melt for high pressure die casting. Better efficiency of hydrogen degassing was shown for ultrasonic degassing as compared with conventional $Ar+N_2$ lance bubbling. The results show that the components produced by high pressure die casting after ultrasonic degassing treatment has a similar porosity level as the components degassed with conventional lance bubbling.

1. Introduction

Ultrasonic degassing of liquid metals has long history. As early as in the 1940s Esmarch et al. studied the degassing of Al–Mg alloys by sonic vibrations induced by contactless electromagnetic stirring and vibrations in the crucible [1]. Starting from the 1960s laboratory and pilot-scale trials of ultrasonic degassing for foundry and later wrought alloys have been performed and summarized by G.I. Eskin [2].

Despite successful industrial trials in the 1960–1970s, ultrasonic degassing was not adopted as a mainstream technology due to arrival of Ar-assisted degassing. In recent years, however, the intrinsic features of ultrasonic degassing stipulated comeback interest to this technology that may answer the current environmental challenges. In addition, new level of ultrasonic technology makes its application easier.

This paper reports the results of pilot-scale trials of ultrasonic degassing as applicable to one of major foundry technologies, i.e. high-pressure die casting.

2. Experimental Procedure

The experiments were conducted using a prototype specifically designed to treat large volumes of molten aluminium (Fig. 1a). The used device is described in detail elsewhere [3].

The ultrasonic equipment used in the experiments was composed of an ultrasonic generator, a water-cooled magnetostrictive transducer, a titanium booster, all supplied by Reltec (Russia), and a niobium sonicator. Fig. 1b shows the assembled ultrasonic degassing equipment used in the experiments.

Fig. 1. Images of ultrasonic degassing prototype (left) and ultrasonic equipment used in the trials (right).

An AlSi9Cu3 (EN AC-46000) alloy was used for the trials. The treatment was conducted in holding furnace with a capability of 500 kg fill up to almost its maximum capacity (> 95 %). The degassing treatment was conducted at a metal temperature of 690±10 °C.

During the ultrasonic treatment (US) the sonotrode was moved with the prototype (Fig 1a) inducing circular movements of different diametres at an approximate rotation speed of 1 rpm. Between 4 and 4.5 kW of power in the range of 17-18 kHz were applied in the molten metal, with an approximated vibration amplitude of 25 μm. Alternatively, a 15 min degassing treatment with a porous graphite lance bubbling an $Ar+N_2$ mixture, was introduced in the same amount of metal, with the same temperature and composition.

Measurements of the hydrogen content with a Reduced Pressure Test (RPT) (MK, Germany) were made before and after the degassing treatment in the holding furnace. The values of the resulting

Density Index (DI) were calculated from the extracted samples.
The production of the cast components was made using a High Pressure Die Casting (HPDC) Unit (Weingarten 250 Tn) with a 50 mm plunger diameter, 3 m/s of injection speed and 180-220 bar of compacting pressure. A standard production die of an actual industrial component was used to cast the specimens.
The porosity of one randomly selected part produced without melt treatment, after each of the melt treatments (US and lance) and about 1-1.5 hour after starting the production, were analysed by computed tomography in a v|tome|x.

3. Results and Discussion

In Fig. 2 are presented the results of the RPT measures conducted in the molten aluminium before and after the degassing treatment. In the graph it can be observed a decrement on the DI after the degassing treatments. Nevertheless, it can be observed that the US treatment is much more effective than the lance degassing using an $Ar+N_2$ mixture. This result suggest that ultrasounds are a more effective degassing method to remove H_2 than the standard lance degassing treatment for this AlSi9Cu3 alloy.

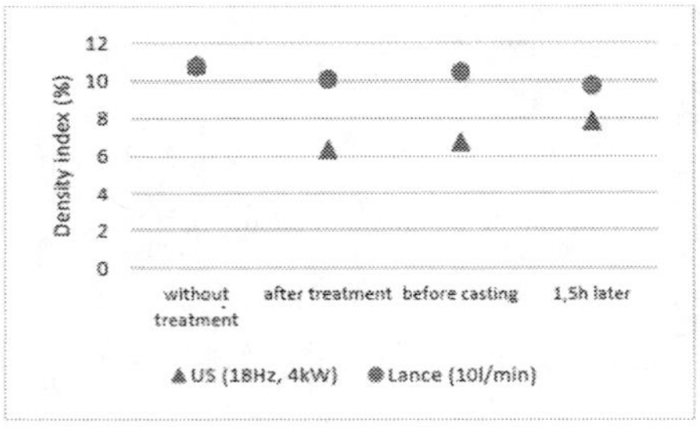

Fig. 2. Density index obtained after ultrasonic treatment (US) and lance degassing treatment (Lance).

In Fig. 3 are presented the 3D reconstruction with the pore distribution obtained by the X-ray tomography for each of the die-casted components analysed, treated with US and with Lance. In the images it can be observed that the defects are concentrated in the lower part of the component, mainly in the junction between the main body and the two lower arms.
In Fig. 4 is presented in form of histogram a comparison of the pores observed by X-ray tomography in the different components inspected. It can be observed that both degassing treatment considerably reduce the number of pores, especially of the small ones, however no difference can be observed between the parts produced by HPDC regarding the degassing treatment conducted, presumably due to the high porosity intrinsically related to this casting process [4].

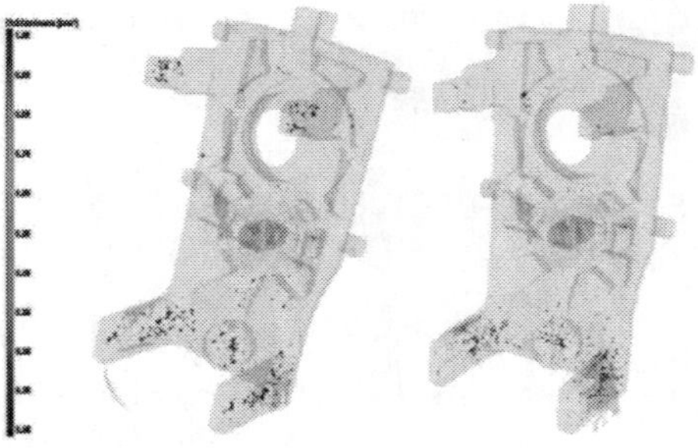

Fig. 3. 3D reconstruction of the tomography images of the part degassed with US (left) and with Lance (right).

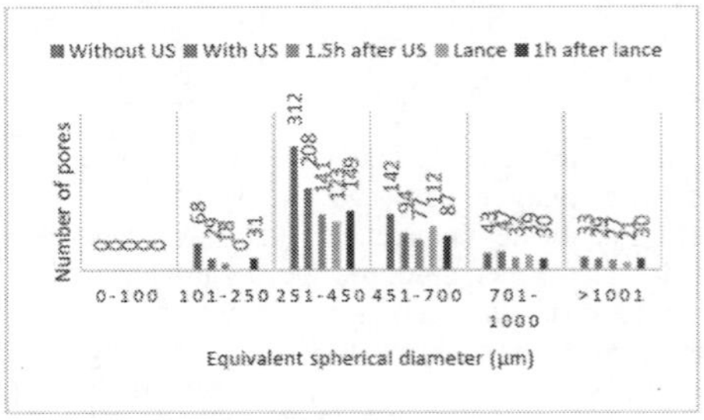

Fig. 4. Porosity results of HPDC parts

4. Conclusions

From the results obtained in the present study the following conclusions can be inferred:
-Ultrasonic degassing using a prototype-level setup is able to achieve lower hydrogen content in the melt than a commercial porous lance for a melt volume of 500 kg of AlSi9Cu3(Fe) alloy.
-Porosity of the castings produced after ultrasonic degassing or lance degassing are similar.

Acknowledgments

The authors would like to acknowledge the financial support provided by the European Union's Seventh Framework Program managed by REA under Grant Agreement number 286344 (www.doshormat.eu). The authors gratefully acknowledge all help provided by the members of the consortium.

References

[1] W. Esmarch, T. Rommel, K. Benther (1940). Werkstoff-Sonderheft. W.V. Siemens Werke, Berlin. 78–87.
[2] G.I. Eskin,D.G. Eskin (2014). Ultrasonic Treatment of Light Alloy Melts, 2nd edn. CRC Press, Boca Raton.
[3] M. da Silva, L. Rebolledo, T. Pabel, T. Petkov, X. Planta, J. Tort, D. Eskin (2015). Evaluation of effect of ultrasonic degassing on components produced by low pressure die casting. Int. J. Cast Metal. Res. 28. 193-200.
[4] A. Niklas, S. Orden, A. Bakedano, M. da Silva, E. Nogués, A.I. Fernández-Calvo (2016). Effect of solution heat treatment on gas porosity and mechanical properties in a die cast step test part manufactured with a new AlSi10MnMg(Fe) secondary alloy. Mater. Sci. & Eng. A, 667. 376-382.

The 73rd World Foundry Congress
„Creative Foundry"
23 - 27 September 2018, Krakow, Poland

EFFECT OF CURING PARAMETERS ON SELECTED TECHNOLOGICAL PROPERTIES OF THE MOULDING SAND WITH INORGANIC CORDIS BINDER USED FOR ABLATION CASTING OF ALUMINIUM ALLOYS

Małgorzata Hosadyna-Kondracka [1]*, Jadwiga Kamińska [1], Katarzyna Major-Gabryś [2], Aleksandra Grabarczyk[2], Michał Angrecki[1]

[1]Foundry Research Institute, Zakopiańska 73, 30-418 Krakow, Poland
[2]AGH University of Science and Technology, Faculty of Foundry Engineering, Department of Moulding Materials, Mould Technology and Casting of Non-Ferrous Metals, Mickiewicza 30, 30-059 Krakow, Poland
**Corresponding address: malgorzata.kondracka@iod.krakow.pl*

Key words: innovative technology, ablation casting, moulding sand, inorganic binder, thermal curing

1. Introduction

Ablation casting is a term taken probably from the Latin word *ablatio* meaning removal or *ablutio* which means washing. The essence of the ablation casting technology consists in pouring castings in single-use moulds made from the mixture of sand and water-soluble binder. After pouring the mould with liquid metal, while the casting is still solidifying, the mould destruction (washing out, erosion) takes place using a stream of cooling medium, which in this case is water.

Ablation casting technology offers a number of advantages, which include, among others, the elimination of an additional operation of casting knocking out from the mould and avoiding in this way the dust pollution in a foundry.
There is no air gap between casting mould and top layer of the cooled casting, which limits the heat flow to the outside. This enables solidification under the conditions of unprecedentedly large temperature gradients and high solidification rates. The mechanical properties of the obtained castings are equal to or higher than the level of analogical characteristics obtained in the pressure die casting process [1]. Another advantage of this innovative technology is controlled directional cooling, which favours the elimination of shrinkage porosity.

China is considered to be the precursor of ablation casting technology [1]. In the clay mould, a material such as a rope was placed. It was fired and then washed out. Thus prepared mould was filled with molten metal. This technique was used to produce dish handles from around 1600 BC.

In 2006, Alotech patented a casting stand using the ablation casting method [2]. The sand mould was located on a conveyor belt and after being poured with liquid metal it moved horizontally to a set of spray nozzles, where under high water pressure it was broken down.

At the Foundry Research Institute in Krakow, a modification was introduced to the casting stand for its subsequent use in the ablation casting (Figure 3) [3]. The device consists of a chamber (1) open from the top, inside which there is a movable work table (2) mounted on a hoist set in vertical and rotary movement. In the side walls of the chamber, nozzles (5) with holes of various diameters are installed permanently or on articulated joints for feeding a liquid cooling medium. Under the chamber, a tank (6) is installed for the flowing liquid cooling medium. It is equipped with a basket (7) to collect sand from the broken down mould.

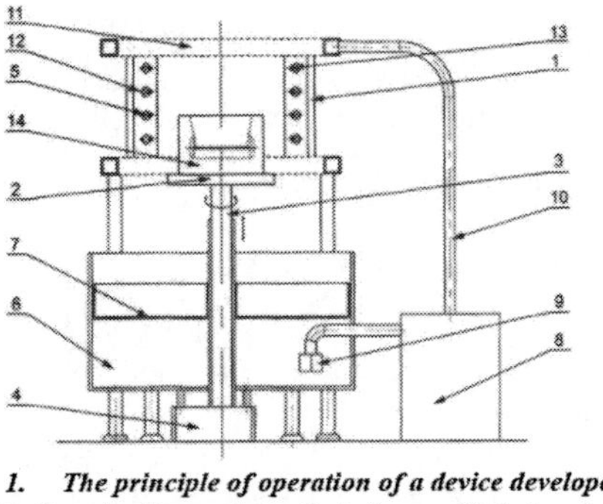

Fig. 1. The principle of operation of a device developed at the Foundry Research Institute for mould destruction and controlled casting cooling

2. Own research

As a part of this work, it was proposed to use moulds made from sands with hydrated sodium silicate thermally hardened by dehydration for ablation casting of aluminium alloys. The advantage of the dehydration process is the achievement of high strength properties, much higher than in the case of chemical curing, and absence of by-products of the chemical curing reaction harmful for the sand mixture quality.

Cordis binder produced by the Hüttenes-Albertus Company was used in the studies [4]. It is a new generation inorganic binder based on hydrated sodium silicate. Its curing takes place under the effect of high temperature. In this technology, the moulding sand also contains the addition of a loose inorganic agent called Anorgit.

Studies involved the preparation of loose moulding sands based on 1K silica sand from the Grudzeń Las mine with a main fraction of 0.20/0.40/0.315 and containing the addition of an inorganic Cordis binder and Anorgit agent. The sands were mixed in a laboratory ribbon mixer, type LMR-2. The base sand was mixed with the Anorgit agent for 60 seconds, then the binder was introduced and the whole mixture was mixed for further 150 seconds. The prepared samples were cured at 180°C for 3, 4 or 5 minutes. The sand mixture properties were examined in hot state, i.e. directly after being taken out from the furnace, and cold state, i.e. after 1 hour of air curing. The compositions of the tested moulding sands are shown in Table 1.

Table 1. Compositions of the tested moulding sands

Sand symbol	Sand composition in parts by weight
SAND 1 T=24°C, W=21%	Silica sand – 100 Cordis binder – 1.5 Anorgit agent – 1.0
SAND 2 T=24°C, W=21%	Silica sand – 100 Cordis binder – 2.0 Anorgit agent – 1.0
SAND 3 T=24°C, W=20%	Silica sand – 100 Cordis binder – 2.5 Anorgit agent – 1.0

The results of the conducted tests are presented in Figures 2-4.

As shown by the tests, SAND 1 had optimal bending strength values after curing at 180°C for 4 minutes. The "hot" bending strength of this sand was 0.80 MPa, and after 1 hour of standing out it increased to 1.80 MPa. SAND 2 had optimal bending strength values after curing at 180°C for 5 minutes. The sand "hot" bending strength was 0.64 MPa, and after 1 hour of standing out it increased to 4.81 MPa. The strength of this sand cured for 4 minutes was 3.8 MPa. SAND 3 had optimal bending strength values after curing at 180°C for 4 minutes. The sand "hot" bending strength was 0.74 MPa, and after 1 hour of standing out it increased to 3.87 MPa.

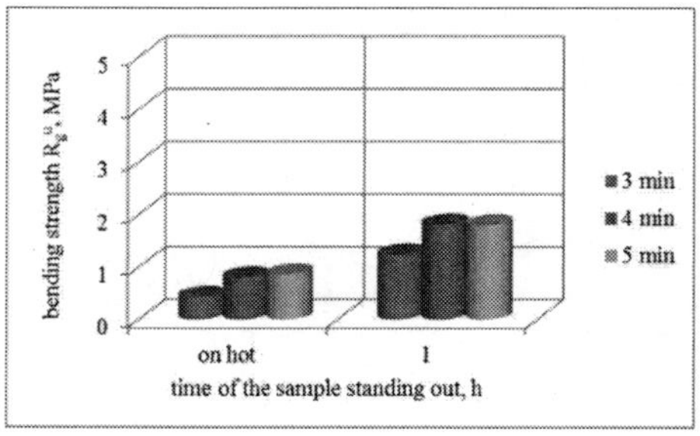

Fig. 2. Effect of curing time on bending strength of SAND 1

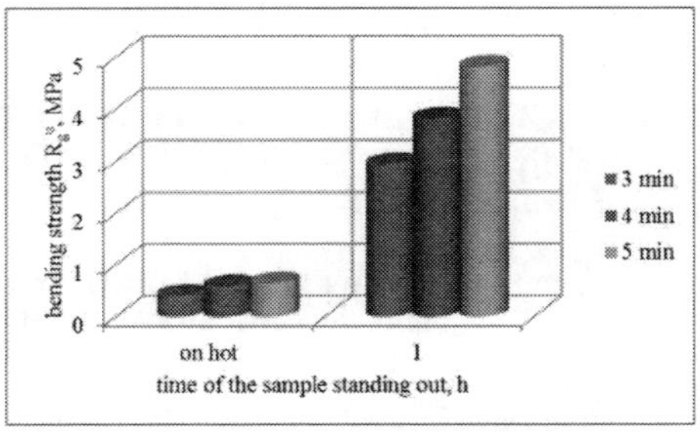

Fig. 3. Effect of curing time on bending strength of SAND 2

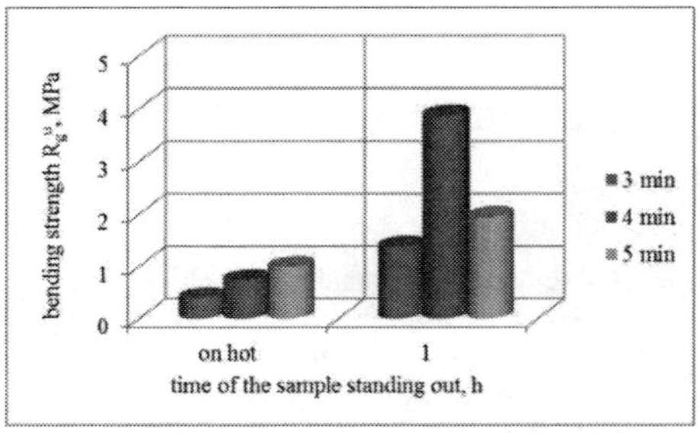

Fig. 4. Effect of curing time on bending strength of SAND 3

Studies have shown that the strength properties of the sands with Cordis binder increase with the increasing amount of this binder. It should be noted, however, that the addition of 1.5 parts by weight of the binder is already sufficient from the required mould strength point of view. What is more the smaller amount of the binder is very advantageous in terms of both ecology and economy.

3. Conclusions

Moulding sands made by the Cordis technology achieve high bending strength with only a small amount of the binder added. Their strength is much higher than the strength of the sands based on the chemically cured hydrated sodium silicate.

References

[1] Dudek P., Fajkiel A., Reguła T., Bochenek J. (2014). Research on ablation casting technology for aluminium alloys. Transactions of Foundry Research Institute 2/2014, 23-35.

[2] John R. Grassi, John Campbell. (2006). Patent US 2008/0041499 A1. Solidification microstructure of aggregate moulded shaped castings.

[3] Dudek P., Fajkiel A., Saja K., Reguła T., Bochenek J. (2013). The device for mould destruction and controlled casting cooling. Patent application no 404518.

[4] Hüttenes-Albertus Poland (2016). The characteristics card of Cordis.

The 73rd World Foundry Congress
„Creative Foundry"
23 - 27 September 2018, Krakow, Poland

GEOMETRIC FORM OF GATING SYSTEM ELEMENTS AND ITS INFLUENCE ON THE INITIAL FILLING PHASE

Rafał Dojka [1,2]*, **Jan Jezierski** [1], **Niels Skat Tiedje** [3]

[1] *Silesian University of Technology, Dept. of Foundry Engineering, ul. Towarowa 7, 44-100 Gliwice, Poland*
[2] *Odlewnia Staliwa Łabędy, ul. Mechaników 9, 44-109 Gliwice, Poland*
[3] *Dept. of Mechanical Engineering, Technical University of Denmark (DTU), DK-2800 Kgs. Lyngby, Denmark*

**Corresponding address: rafal.dojka@polsl.pl*

Key words: flow analysis, gating system, internal quality, simulation, cast steel

1. Introduction

Undoubtedly the geometric form of gating system has crucial influence on the mould filling process which is one of the key components in ensuring the quality of the casting [1-3]. At the time when commonly used principles for gating system design were created (1950s and 1960s) the possibility to simulate the mold filling process was highly limited in comparison to the present. However, since then, the activity in the field of scientific research associated with the gating systems design has clearly decreased and the rules formed back then are still present and the majority of foundries is still using them during manufacturing technology designing. Authors are convinced that it is highly important to take advantage of modern technology to reveal that gating system design can be improved and often used rules are at most good and far from perfect or even optimal [4-6]. Presented research described in this article is a part of a wide experimental plan meant to develop a scientifically optimized and practically useful molding technology. In this article authors want to present the influence of geometric form of the sprue, cross runner and the ingate on the flow kinetics in the initial filling phase of the mold.

2. Methodology

In order to analyze the differences in flow kinetics in each system authors prepared an experimental plan consisting of 24 simulations performed in MAGMASOFT MAGMA5. Two types of naturally pressurized sprues were compared. First sprue was a variation of traditionally used tapered sprue with circular cross-section (circular sprue), however the tapering was not straight, but was changing according to the hyperbolic curve as the falling stream of metal does due to the stream continuity law. The second sprue (transitional sprue) shape was changing from circular cross-section to rectangular cross-section according to special shape transition, the cross-section surface reduction was analogous to one in the first sprue. First sprue was connected with a circular cross runner, the second sprue was connected with a rectangular cross runner. In every variant the cross runner was extended and ended with a spin trap which allowed to successfully trap the first portion of cooled oxidized metal and reduce the metal pressure in the system and thus the velocity of metal entering the mold. Examined ingates had circular and rectangular cross-sections, different height (25mm and 50mm) and different expansion angles (0°,5° and 10°) in order to examine the possibility of reducing the metal injection effect without the detachment of the stream from the mould wall. Each gating system with the same level of ingate expansion had exactly the same volume. All simulated variants were presented in Table 1.

Table 1. Geometric form of gating system elements

Sprue and cross runner	Ingate, expansion angle	Height
Circular sprue, circular cross runner	circular, 0°	25mm
		50mm
	circular, 5°	25mm
		50mm
	circular, 10°	25mm
		50mm
Circular sprue, circular cross runner	rectangular, 0°	25mm
		50mm
	rectangular, 5°	25mm
		50mm
	rectangular, 10°	25mm
		50mm
Transitional sprue (circular to rectangular), rectangular cross runner	circular, 0°	25mm
		50mm
	circular, 5°	25mm
		50mm
	circular, 10°	25mm
		50mm
Transitional sprue (circular to rectangular), rectangular cross runner	rectangular, 0°	25mm
		50mm
	rectangular, 5°	25mm
		50mm
	rectangular, 10°	25mm
		50mm

Simulated alloy was GS-52 cast steel acc. DIN 1681, pouring temperature was set to 1570°C, mould material was alphaset molding sand.

3. Results and discussion

Obtained simulations allowed the assessment of each examined solution. Figure 1 presents the initial mold filling phase for gating system with circular sprue and cross runner with 25mm ingate with 0° (a), 5° (b) and 10° (c) expansion. The jetting effect is present in all 3 variants, the expansion allows to reduce the jetting, however it does not eliminate it.

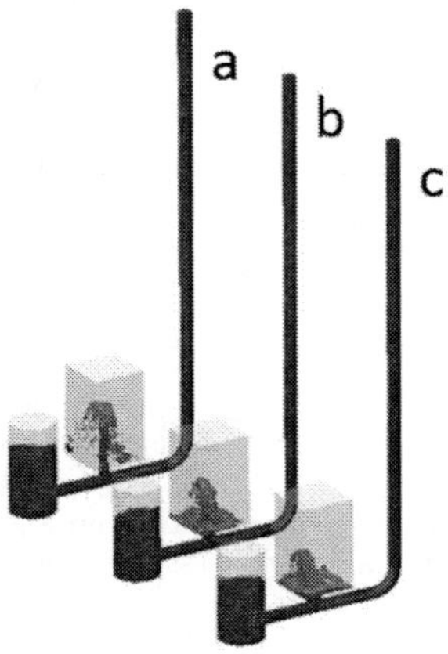

Fig. 1. Initial filling phase, circular sprue, circular cross runner, circular 25mm ingate with 0° (a), 5° (b) and 10° (c) expansion

Figure 2 presents the initial filling phase for gating system with transitional sprue, rectangular cross runner and rectangular 25mm ingate with 0° (a), 5° (b) and 10° (c) expansion.

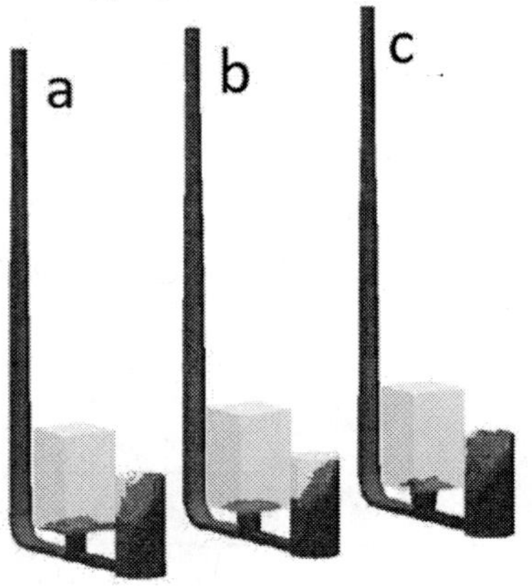

Fig. 2. Initial filling phase, transitional sprue, rectangular cross runner, rectangular 25mm ingate with 0° (a), 5° (b) and 10° (c) expansion

The change of geometric form of the system from circular to rectangular allowed to reduce the velocity of the metal entering the mold thanks to increased friction of metal against the mold. The jetting effect was completely eliminated even with non-expanded ingate. The metal velocity field during the pouring was noticeably more stationary in the system presented on figure 2 in comparison to system presented on figure 1 what with less turbulent flow of the metal which can cause reduced possibility of oxide formation during the flow and thus enhanced quality of the casting. Unfortunately, software used for simulation of casting processes provides nearly no information about oxide bifilm formation during the flow, their negative effect on the castings quality nor the deleterious impact of bubbles [7,8].

4. Summary

The results of the gating system design optimization led to the following conclusions:

- geometric form of gating system element has key role in the initial filing of the mold;
- conducted modeling confirms the effectiveness of presented gating system designs in reduction of velocity of metal entering the mould cavity;
- improper expansion of the ingate may cause the detachment of the stream from the mould wall which would cause reoxidation of the metal and degrade castings quality;
- further development of simulation software would allow to account the influence of oxide and bubble formation on the quality of the casting.

References

[1] Li D., Sun M., Wang P., Kang X., Fu P., Li Y. (2013). In process modelings and simulations of heavy castings and forgings. Proceedings of the 11th International Conference on Numerical Methods in Industrial Forming Processes (NUMIFORM). Shenyang, China. 81-94.

[2] Foglio E., Gelfi M., Pola A., Goffelli S. & Lusuardi D. (2017). Fatigue characterization and optimization of the production process of heavy section ductile iron castings. International Journal of Metalcasting. 11. 33-43.

[3] Yang D., Li S., He F., Sung W., Kao J. & Chen R. (2014). Twin gating system design for typical thin wall stainless steel castings based on fast pouring mechanism. Frontiers of Mechanical Engineering and Materials Engineering II. 457-458, 1657-1660.

[4] Huang P. & Lin C. (2015). Computer-aided modeling and experimental verification of optimal gating system design for investment casting of precision rotor. International Journal of Advanced Manufacturing Technology. 79. 997-1006.

[5] Ducic N., Slavkovic R., Milicevic I., Cojbasic Z., Manasijevic S. & Radisa R. (2017). Optimization of the gating system for sand casting using genetic algorithm. International Journal of Metalcasting. 11. 255-265.

[6] Fourie J., Lelito J., Zak P., Krajewski, P. & Wolczynski W. (2015). Numerical optimization of the gating system for an inlet valve casting made of titanium alloy. Archives of Metallurgy and Materials. 60. 2437-2446.

[7] Campbell J. (2015). Cavitation in liquid and solid metals: Role of bifilms. Materials Science and Technology. 31. 565-572.

[8] Campbell J. (2015). Sixty years of casting research. Metallurgical and Materials Transactions A - Physical Metallurgy and Materials Science. 46A. 4848-4853.

The 73rd World Foundry Congress
„Creative Foundry"
23 - 27 September 2018, Krakow, Poland

EFFECT OF INCREASED TEMPERATURE ON DIMENSIONAL AND SHAPE ACCURACY OF CASTINGS PRODUCED FROM THE EN AC-AlSi11 ALLOY BY PRESSURE DIE CASTING PROCESS

Aleksandra Jarco*

University of Bielsko-Biała, Department of Production Engineering and Automation, ul. Willowa 2, 43- 300 Bielsko-Biała, Poland
**Corresponding address: e-mail: ajarco@ath.bielsko.pl*

Key words: high pressure die casting, Al-Si alloy, blistering, dimensional and shape accuracy, elevated temperature behavior

1. Introduction

Components produced by traditional pressure die casting technology are not, as a standard, heated in elevated temperatures due to presence of numerous gaseous pores contained inside walls of castings [1,2]. Presence of the pores comes directly from run of the pressure die casting process, when liquid metal rapidly fills cavity of the mould. Pouring into mould is turbulent, what disables complete removal of gaseous phase from cavity of the mould in very short time of metal injection [2,3]. Appearance of a surface defects on castings in elevated temperatures depends on temperature effect on mechanical properties of the alloy, depends also on value of pressure of the gas closed inside pores, and on primary shape and location of such defects [4].

Performed investigations were aimed at assessment of effects of increased temperature on dimensional-shape changes of a crowns of connecting member of „110" type, produced from the EN AC-AlSi11 alloy in process of pressure die casting.

2. Experimental Methods

The investigations were performed on castings of fire-fighting fixture (Fig.1). Requirements imposed on this element are specified in the PN-M-51031 standard [5].

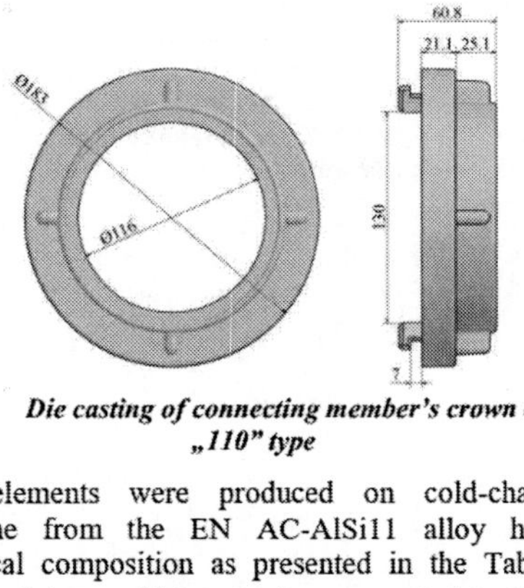

Fig. 1. Die casting of connecting member's crown of the „110" type

The elements were produced on cold-chamber machine from the EN AC-AlSi11 alloy having chemical composition as presented in the Table 1. Pouring into mould temperature amounted to 610°C, while injection pressure of metal into the mould amounted to 94 MPa.

Table 1. Chemical composition of the EN AC-AlSi11 alloy

Si	Fe	Cu	Mn	Mg	Zn	Ti	Al
10.1	0.83	0.38	0.08	0.19	0.19	0.10	rest

The castings were held at temperatures 460, 475, 490°C for 3.5 h. Verification of dimensional and shape behavior of the castings after holding them at increased temperatures was performed with use of the 3D Heavy Duty Optima scanner (accuracy of 0.018 mm) and *GOM Inspect 2017* computer program. Results of the verification were compared with raw casting.

3. Results and discussion

Obtained results (Figs. 2-4), presented in graphical form, depict effects of holding the castings at increased temperature on accuracy of dimensional and shape behavior of the connecting member's crown of "110" type in comparison to initial conditions.

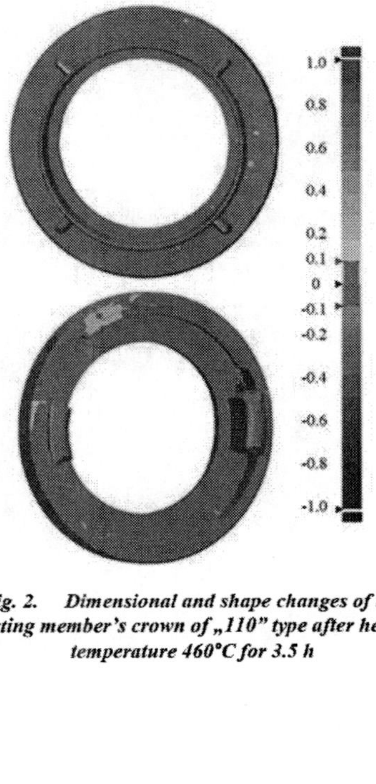

Fig. 2. Dimensional and shape changes of the connecting member's crown of „110" type after heating at temperature 460°C for 3.5 h

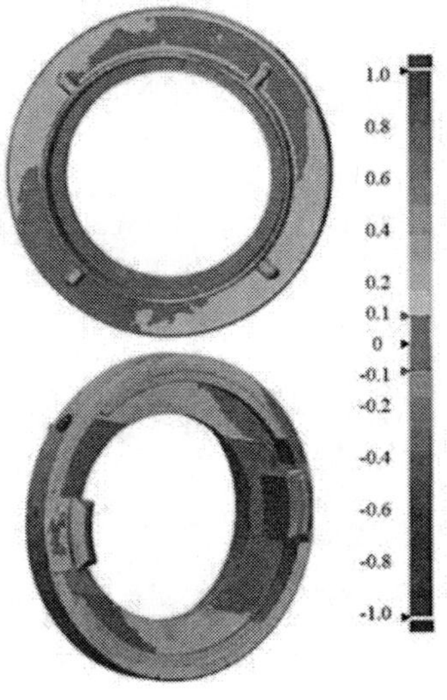

Fig. 3. Dimensional and shape changes of the connecting member's crown of „110" type after heating at temperature 475°C for 3.5 h

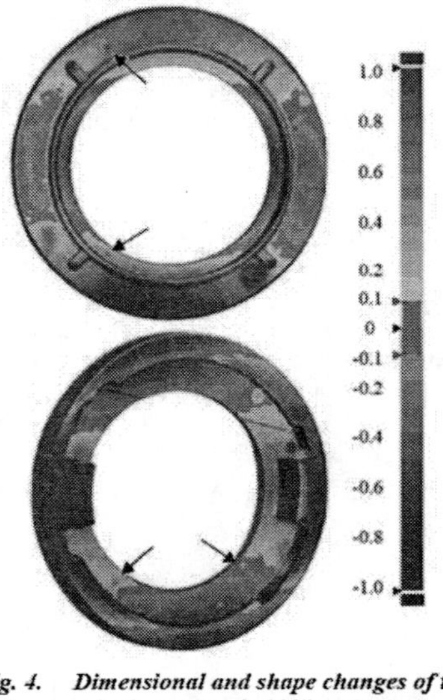

Fig. 4. Dimensional and shape changes of the connecting member's crown of „110" type after heating at temperature 490°C for 3.5 h

Holding at temperature of 460°C didn't result in any distinct changes in dimensions and shape of the castings. Maximal deviations have amounted to ±0.15 mm (Fig.2), comparing to the initial element. For temperature of 475°C the dimensions differ from the nominal ones with ±0.2 (Fig.3), what is also within allowable range. However, further increase of heating temperature results in already visible changes in dimensions and shape of investigated castings. For heating temperature of 490°C, the dimensions exceed allowable limits with about ±1 mm (Fig.4). At such temperature are also visible a surface defects in form of blisters, marked with arrows in the figure 4. The gaseous blisters closed inside walls of the casting became compressed in effect of pressing pressure, what results in decrease of their volume, and the same, results in increased pressure of gas inside the blisters. Their exposure is caused by heating of the castings to temperature causing higher plasticity properties of the alloy and increase of pressure of gas closed inside the pores [1,4]. It leads to deformation of walls of the casting being in direct vicinity of gaseous pores, what results in exposure of surface defects and deformations of the element shape.

4. Conclusion

Increase of heating temperature, in case of fire-fighting fixtures castings produced from EN AC-AlSi11 alloy, in process of pressure die casting leads to considerable changes in dimensions and shape, and to exposure of surface defects as early as at temperature 490°C. It effects in significant operational limitations of such elements at increased temperatures and problems in case of a possible heat treatment. The heat treatment should be performed at a lower solution temperatures (of the order of 460-475°C) to eliminate shape deformation of the element and appearance of defects in form of blisters on surface of the casting.

References

[1] A. Białobrzeski. (1992). Odlewnictwo ciśnieniowe maszyny, urządzenia i technologia. Warszawa: WNT.

[2] H. Toda, P. Qu, S. Ito, K. Shimizu, K. Uesugi, A. Takeuchi, Y. Suzuki & M. Kobayashi. (2014). Formation behavior of blister in cast aluminium alloy. International Journal of Cast Metals Research. 27 (6), 369-377.

[3] X.G. Hu, Q. Zhu, S.P. Midson, H.V. Atkinson, H.B. Dong, F. Zhang & Y.L. Kang. (2017). Blistering in semi-solid die casting of aluminium alloys and its avoidance. Acta Materialia. 124, 446-455.

[4] O. Ozhoga-Maslovskaja, E. Gariboldi & J. N. Lemke. (2016). Conditions for blister formation during thermal cycles of Al–Si–Cu–Fe alloys for high pressure die-casting. Materials and Design. 92, 151-159.

[5] PN-M-51031. (2015). Sprzęt pożarniczy. Łączniki do węży pożarniczych. Warszawa: PKN.

The 73rd World Foundry Congress
„Creative Foundry"
23 - 27 September 2018, Krakow, Poland

APPLICATION OF PATTERNS FABRICATED BY THE FDM TECHNIQUE (FUSED DEPOSITION MODELING) IN PRECISION CASTING

Rafał Kaczorowski, Paweł Just, Tadeusz Pacyniak*

Lodz University of Technology, Stefanowskiego 1/15 Str., 90-924 Lodz Poland
**Corresponding address: tadeusz.pacyniak@p.lodz.pl*

Key words: foundry engineering, ceramic moulds, rapid prototyping techniques, Fused Deposition Modeling (FDM)

The investment casting process is an economical technology in the case of large lot production of elements with complex and complicated geometries. However, it is not suitable for the production of individual parts, due to its high costs (metal matrices for wax pattern fabrication). The paper presents the investigations of the process of obtaining casts with the use of this technology, where the wax pattern was replaced with an HIPS pattern made by the rapid prototyping technique FDM. Owing to this, it became profitable to produce single casts or short test series in ceramic moulds. A comparison was made of the dimensional accuracy of the obtained patterns and test casts, as well as of their surface roughness. Also, a presentation was made of the studies of the strength of ceramic moulds subjected to the process of accelerated drying with microwaves.

Casting with the use of the investment casting process belongs to the group of precision casting methods making it possible to obtain casts of small masses (up to 5 kg), with high dimensional accuracies and complicated shapes [1]. Despite its many advantages – the possibility to obtain casts with thin walls, with difficult to access cavities and openings – it is an expensive method. It requires the use of expensive instrumentation, e.g. metal matrices for the fabrication of wax patterns, and the process itself of producing the ceramic moulds is quite long and labour-consuming. That is why the investment casting process is used in large-lot or mass production of casts. The high costs of instrumentation in this technology do not make it possible to apply it in piece or even small-lot production. The rapid prototyping techniques have recently become standard tools in the product design and the manufacturing industry. Owing to the possibility of a fast fabrication of three-dimensional patterns, the time of designing a new product as well as its development cycle becomes shorter. Also, the rapid prototyping techniques can be applied to manufacture elements of very high complexity, and the geometrical shape is unlimited. For example, it is possible to fabricate hollow elements [2-5]. With the continuous development of the rapid prototyping techniques, the costs of producing a model drop and the time of its preparation shortens [6]. In particular, plastic patterns made by the FDM method (Fused Deposition Modeling) can successfully replace the wax patterns, as they can be melted and fired from the ceramic mould without its damage, and also, they characterize in a higher melting point, owing to which it is possible to significantly accelerate the moulds' drying process.

In the tests carried out, the patterns were fabricated from HIPS polystyrene on the industrial printer 3DFactory 2.0 by Omni3D. The printer had a closed heated chamber with the dimensions: 500×500×500 mm. The print accuracy provided by the producer was +/- 0.2 mm.

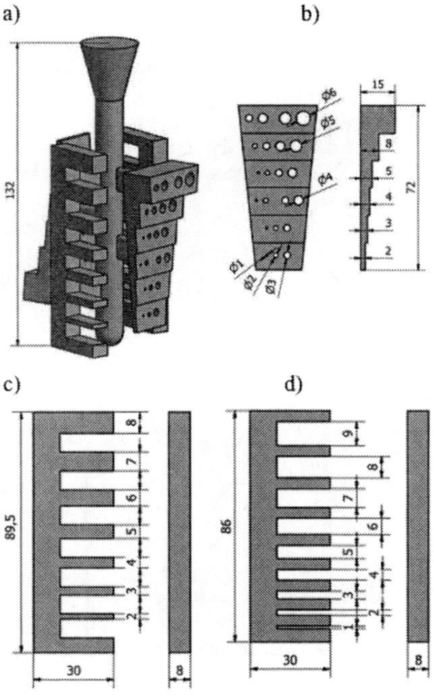

Fig. 1. Study patterns: a) pattern unit, b) a step, c) a "comb" with different rib thicknesses, d) a "comb" with different gap widths

The pattern unit presented in Figure 1 consisted of the following models:

- a stepped pattern (Fig. 1b) with the step height of 2-15 mm, with fabricated ports, 1-6 mm in diameter.

- a "comb" with different rib thicknesses (Fig. 1c), rib thickness 2-8 mm – it enabled an evaluation of the obtained cast (minimal wall thickness, shape representation, surface quality, etc.)
- a "comb" with different gap widths (Fig. 1d), gap width 1-10 mm – it was used to determine the properness of the ceramic mould fabrication and the minimal thickness of the mould's wall.

In order to fabricate the ceramic moulds, the pattern units were submerged in a ceramic slurry based on colloidal silica (LUDOX) with aluminium silicate flour Refracoarse with the addition of a surface-active agent [7]. On the first layer, mixtures of kinematic viscosity 240 cSt were used, and it was covered with fine sand Refracoarse, granularity 0.1÷0.3 mm. On the following layers, mixtures of a lower viscosity were applied (170 cSt), and they were covered with coarse sand Refracoarse, granularity 0.5÷1.0.

The moulds were dried in a prototype microwave drier made available by Qbig. The total drying of each layer was obtained after about 30 minutes for the microwave power of 1200 W. Owing to the use of microwave radiation as well as plastic models, it was possible to dry the moulds at higher temperatures (up to 70°C), which significantly shortened the time of fabricating the ceramic moulds.

Next, the moulds were subjected to thermal treatment according to the annealing cycle presented in Figure 2, in the chamber furnace by Nabertherm.

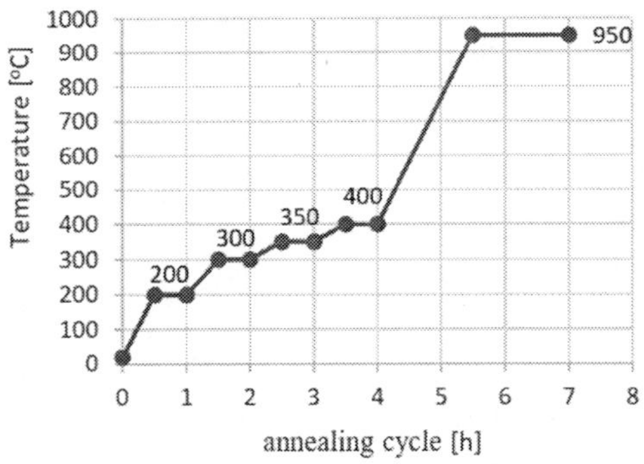

Fig. 2. Ceramic mould annealing cycle

Acknowledgments

Research has been commissioned by Qbig company within project no. POIR.01.01.01-00-0552/15 (competition 1/1.1.1/2015, within the Smart Growth Operation Program, Activities 1.1 „R&D Enterprise Projects", Sub-measure 1.1.1 „Industrial research and development works conducted by enterprises."

Project title: „The development of innovative equipment using a unique high-speed technology to manufacture multilayer ceramic moulds for precision casting."

References

[1] Green-Spikesley E. (1971) Investment Casting. In: Beadle J.D. (eds) Castings. Production Engineering Series. Palgrave, London

[2] C.W. Lee, C.K. Chua, C.M. Cheah, L. H. Tan, C. Feng (2004): Rapid investment casting: direct and indirect approaches via fused deposition modelling,Int J Adv Manuf Technol 23: 93–101

[3] Nannan Guo, Ming C. Leu (2013): Additive manufacturing: technology, applications and research needs: Front. Mech. Eng., 8(3): 215–243

[4] Lalehpour, A., Janeteas, C. &Barari, A. (2017): Int J Adv Manuf Technol. https://doi.org/10.1007/s00170-017-1165-5

[5] S. Wang, A.G. Miranda & C. Shih (2010): A Study of Investment Casting with Plastic Patterns, Materials and Manufacturing Processes, Vol. 25, Issue 12, p. 1482-1488

[6] Matthew Frank, Drs. Sanjay B. Joshi, Richard A. Wysk (2003):Rapid prototyping as an integrated product/process development tool an overview of issues and economics, Journal of the Chinese Institute of Industrial Engineers, 20:3, 240-246

[7] R. Kaczorowski, P. Just, T. Pacyniak (2017): Selected Technological Properties of Liquid Ceramic Slurries Used to Produce Moulds in the Replicastcs Technology. Archives of Foundry Engineering – Vol. 17, Issue 4, 190-194

The 73rd World Foundry Congress
„Creative Foundry"
23 - 27 September 2018, Krakow, Poland

THE INOCULATION EFFECT OF ALUMINUM ADDITION ON SELECTED HIGH-CHROME CAST IRON PROPERTIES

Andrzej Szczęsny[1]*, Dariusz Kopyciński[1], Dorota Siekaniec[1], Edward Guzik[1], Marek Sokolnicki[2], Adam Nowak[2]

[1]AGH University of Science and Technology, Department of Foundry Engineering, Mickiewicza 30 Av., 30-059 Kraków, Poland
[2] Odlewnie Polskie S.A., Inżyniera Władysława Rogowskiego 22 Str., 27-200 Starachowice, Poland
**Corresponding address: ascn@agh.edu.pl*

Key words: High-chromium cast iron, inoculation, hot cracking, wear resistance, toughness, hardness

Abstract

The presented work includes the subject of aluminum addition as inoculant to high-chromium cast iron, especially in the aspect of eliminating hot cracks and its impact on wear resistance, hardness and toughness.

1. Introduction

High-chromium cast iron is widely used as a material with increased abrasion resistance in the steel, energy and mining industry.

One of the significant problems are hot cracks that appear during the foundry production process [1-6]. One of the ways to reduce this phenomenon is to modify the alloy by aluminum.

2. Methodology

The melts were carried out in industrial conditions. Liquid metal overheated to 1600°C in induction furnace. The liquid metal poured from furnace to ladle. Aluminum was carried out on the stream of liquid metal. The addition of aluminum was in the shape of rods with a diameter of 3 mm and a length of 5 mm.The following amount of aluminum have been used; 0.17% Al, 0.33% Al and 0.66% Al. Each time, an Althoff-Radke test (A-F) [7] was performed, fig. 1. and a mold containing, among other things, an impact test.

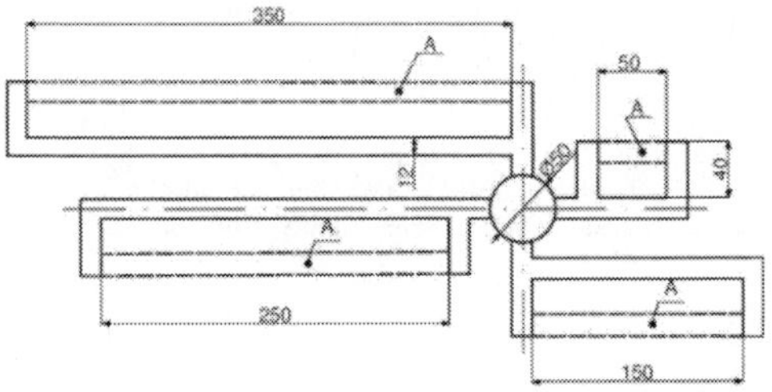

Fig. 1. A-F Hot cracking test, (A) - resistance rod

The research was carried out at the Department of Casting Engineering and Cast Composites at the Faculty of Foundry Engineering. Vickers hardness test, Charpie hammer and abrasion test device MAN company was used.

3. Results and discussion

The results of the Althoff-Radke test are shown in Fig. 2-5, selected microstructures in Fig. 6.

Fig. 2. A-F Hot cracking test for reference cast iron

Fig. 3. A-F Hot cracking test for 0.17% Al addition to cast iron

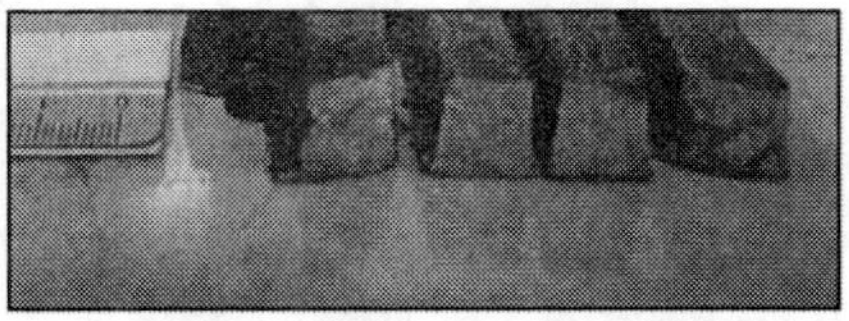

Fig. 4. A-F Hot cracking test for 0.33% Al addition to cast iron

Fig. 5. A-F Hot cracking test for 0.5% Al addition to cast iron

a) b)

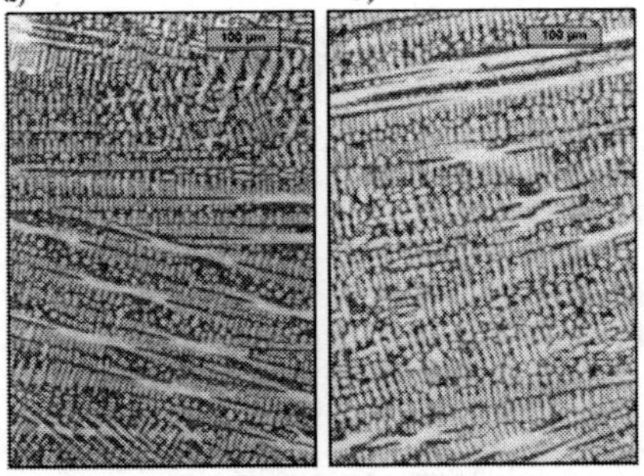

Fig. 6. Microstructure of high-chromium cast iron without - a) and with addition of 0.5% Al

Fig. 7 shows the test results obtained during the research.

a)

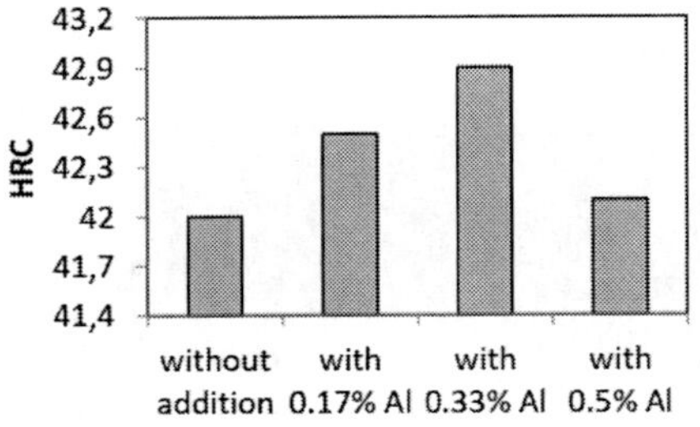

b)

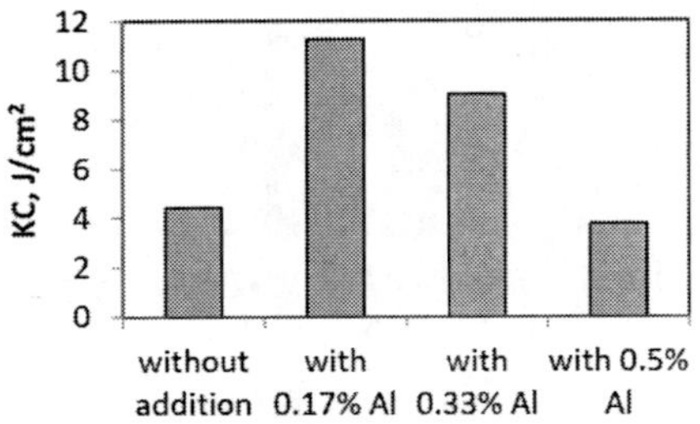

c)

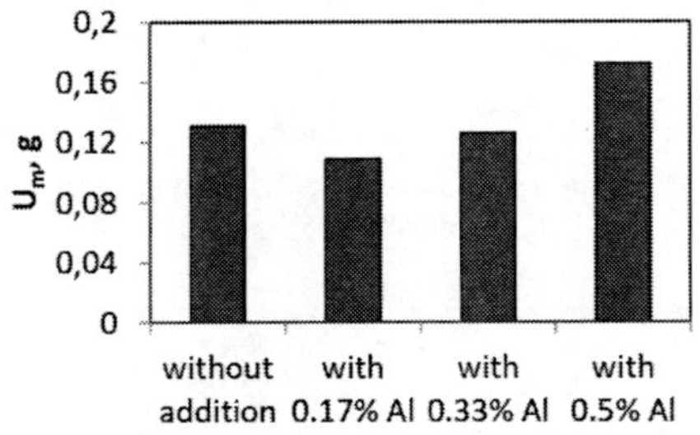

Fig. 7. Test results of high-chromium cast iron: HRC-toughness, impact resistance - b), weight loss during the abrasion resistance test – c)

4. Conclusions

The presented results show that the addition of aluminum has an impact on the reduction (addition of 0.17% and 0.5%) or elimination (0.33%) of hot cracks.

An increase in the hardness of the alloy was observed with each addition of Al. The highest values were obtained at 0.17 and 0.33%.

The impact strength in the case of the 0.5% addition decreased compared to the initial sample, while at 0.17% and 0.33% the impact strength was more than twice as high.

In the case of abrasion resistance, its value at the first two doses (0.17 and 0.33%) of Al decreased to increase at 0.5% Al.

Acknowledgments

The study was co-financed by NCBiR as a targeted project No. PBS/B5/44/20.

References

[1] D. Siekaniec, D. Kopyciński, E. Guzik, E. Tyrała, A. Nowak. (2017). Effect of different inoculants on impact toughness in high chromium cast iron. Archives of Foundry Engineering. 17(2), 85–88.

[2] C. P Tabrett, I. R. Sare, M. R. Ghomashchi. (1996). Microstructure-property relationships in high chromium white iron alloys. International Materials Reviews 41(2), 59-82.

[3] A. Drotlew, M. Garbiak. (1998). Cracking behaviour of carbides in erosion wear. Solidification of Metals and Alloys, 38, 195-200.

[4] D. Kopyciński, E. Guzik, A. Szczęsny, R. Gilewski, D. Siekaniec. (2014). White cast iron with the microstructure in situ composite. Archives of Foundry Engineering. 14, 147–150.

[5] Ö. Doğan, J. Hawk, G. Laird. (1997) Solidification structure and abrasion resistance of high chromium white irons. Metallurgical and Materials Transactions A. 28, 1315–1328.

[6] A. Studnicki, R. Dojka, M. Gromczyk, M. Kondracki (2016) Influence of titanium on crystallization and wear resistance of high chromium cast iron. Archives of Foundry Engineering. 16, 117-123.

[7] D. Kopyciński, D. Siekaniec, A. Szczęsny, M. Sokolnicki, A. Nowak. (2016) The Althoff-Radtke test adapter for high chromium cast iron. Archives of Foundry Engineering. 16, 61–64.

The 73rd World Foundry Congress
„Creative Foundry"
23 - 27 September 2018, Krakow, Poland

AZ91 MAGNESIUM BASED NANOCOMPOSITES OBTANED USING THIXOMOLDING TECHNOLOGY

Łukasz Rogal*, Lidia Lityńska-Dobrzyńska, Bogusław Baran

Institute of Metallurgy and Materials Science, Polish Academy of Sciences
25 Reymonta St., 30-059 Krakow, Poland

**Corresponding address: l.rogal@imim.pl*

Key words: thixomolding, composites, semi-solid state, MgO, SiC

1. Introduction

Magnesium alloys are attractive for the automobile and aircraft engine industries, because they have the lowest density of all technologically-relevant alloys [1-3]. However, the relatively low ductility and strength limit their wide commercialization [4]. Magnesium matrix composites (MgMCs) provide alternative for the improvement of their mechanical properties [5, 6]. For this purpose, we developed new method based on the semi-solid metal (SSM) processing technology - thixomolding, which involves the in-situ formation of nano-scale native MgO by the application of mixture of reactive CO_2 and Ar gases as the inert atmosphere to the partially melted AZ91 magnesium alloy.

An additional advantage is high shearing and mixing in the semi-solid state range by the rotation of screw and the transport of suspension through thin slits, which induce the refinement and homogeneous distribution of particles. The detailed microstructural studies have been conducted, using scanning and transmission electron microscopes to identify the strengthening particles and to explain the mechanisms of their formation in the AZ91 composites.

2. Methodology

Magnesium injection molding technology was used to obtain Mg base composites using ex-situ and in-situ methods. AZ91 chips (8.7% Al, 0.7% Zn, 0.2% Mn, 0.02% Si, Mg balance) were mixed with 5 wt% of β-SiC nanoparticles in solid state and next fed to the cylinder of injection molding machine (ex-situ process). Using screw rotation the granules were transferred to nozzle area at simultaneous intensive mixing of slurry containing reinforcement phases. Injection process was conducted at 595 °C, which correspond about 90% liquid phase, to steel die.

In in-situ technology magnesium matrix nano-composites were produced by reaction of CO_2 with AZ91 alloy at semi-solid temperature range. Reactive carbon dioxide atmosphere was introduced to hot zone of cylinder to conduct controlled oxidation reaction with partially melted Mg alloy.

Detailed characterization of microstructure was carried out using scanning (FEI SEM XL30) and transmission (FEI Tecnai G2 operating at 200 keV) electron microscopes.

3. Results and discussion

Composites consist of α(Mg) globular grains with size of 20-50 μm and volume of 7-10% surrounded by fine eutectic mixture (α(Mg) + β-$Mg_{17}Al_{12}$) containing nano β-SiC (of 20-100 nm) in the case of ex-situ process (Fig. 1a) and MgO particles with size of about 50 nm for in-situ (Fig. 1b).

In both materials, significant increase in mechanical properties have been noted.

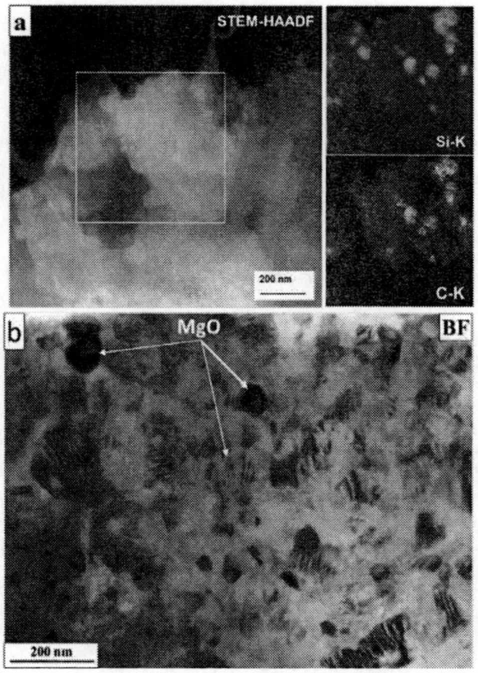

Figure 1. TEM microstructures of AZ91 base composites obtained using thixomolding; a) with addition 5 wt% of β-SiC - STEM-HAADF image and elemental maps, b) under Ar and CO₂ atmosphere - BF image of eutectic area.

The room temperature compression test (Fig. 2) was carried out to determine the deformation behavior of the thixomolded AZ91 alloy with 5 wt% of SiC (dashed curve) as well as obtain in CO_2 reactive gas (continuous curve). Additionally, reference AZ91 after thixomolding (process conducted at Ar gas and without strengthening particles) has been showed.

The AZ91 sample with SiC particles had the yield strength (YS) of 125 MPa at the compression strength (CS) of 235 MPa and hardness 75 ± 3 HV. The plasticity of AZ91 composites was 3.5%.

The obtained results were similar to those of commercial AZ91 processed using thixomolding technology: YS – 100 MPa, CS – 198 MPa. The AZ91 obtained at reactive CO_2 revealed the yield strength of 220 MPa at the compression strength of 460 MPa and hardness 103 ± 2 HV.

The increase of properties was attributed mainly to the in situ formed hard oxide particles in the matrix, and their uniform distribution throughout the matrix. In the MgMCs, when forces acted upon the matrix they were sustained by the reinforcing particles, which offered resistance to the movement of dislocations thereby increasing the hardness. It is consistent with the overall tendency presented in literature.

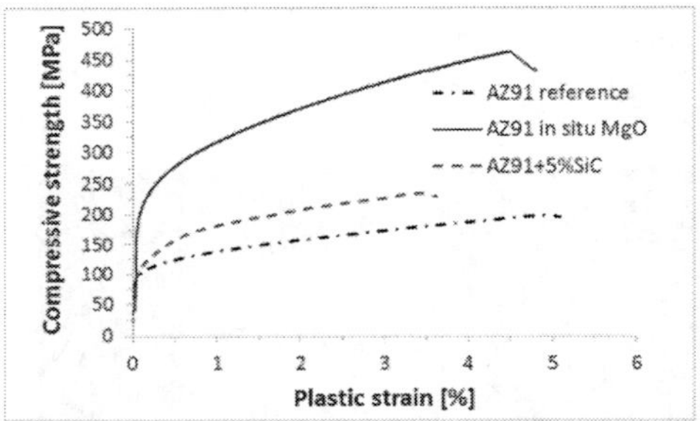

Figure 2. Compression curves of AZ91 thixomolded samples: reference (dashed doted curve), process conducted at in CO_2 (continuous curve) and with addition of 5 wt% SiC (dashed curve).

Acknowledgments

The authors gratefully acknowledge the financial support by the Polish National Centre for Research and Development, Grant No. LIDER/007/151/L-5/13/NCBR/2014.

References

[1] K. Maruyama, M. Suzuki, H. Sato, Creep strength of magnesium-based alloys, Metall. Mater. Trans. A 33 (2002) 875–882.

[2] Z. Wang, W. Zhao, Mg-Based Quasicrystals, in: W. Alfredo Monteiro (Eds.), New Features on Magnesium Alloys, InTech CC BY 3.0 license, 2012, pp. 1–28.

[3] L. Wei, Y. Doudou, Q. Gaofeng, Z. Yingbo, Y. Dandan, Microstructure Evolution of Semisolid Mg-2Zn-0.5Y Alloy during Isothermal Heat Treatment, Rare Metal Materials and Engineering, 45/8 (2016) 1967–1972.

[4] L.L. Rokhlin, Magnesium Alloys Containing Rare Earth Metals: Structure and Properties, J.N. Fridlyander, G.G. Eskin (Eds.), Advances in Metallic Alloys, 2003, pp. 34–68.

[5] H. Yan, Z.X. Huang, H.X. Qiu, Microstructure and Mechanical Properties of CNTs/A356 Nanocomposites Fabricated by High-Intensity Ultrasonic Processing, Metallurgical And Materials Transactions A, 48(2) (2017) 910–918.

[6] Z.Y. Hai, Y.L. Xing, Review of recent studies in magnesium matrix composites, Journal of Materials Science 39(20) (2004) 6153–6171.

The 73rd World Foundry Congress
„Creative Foundry"
23 - 27 September 2018, Krakow, Poland

NONDESTRUCTIVE EVALUATION OF DUCTILE CAST IRON MATRIX VIA CASTING SURFACE BY ELECTROMAGNETIC METHOD

Noritaka HORIKAWA [1]*, Lin WANG [2], Kazuma TANI [3]

[1] *National Institute of Technology, Asahikawa College, 1-6,2-2 Shunkodai, Asahikawa, 072-8142, Japan*
[2] *Hitachi Metals Ltd., Casting Technology Research Laboratory, 11 Kinugaoka, Moka, 321-4367, Japan*
[3] *National institute of technology, Asahikawa college, Advanced Course of Production System Engineering*
**Corresponding address: horikawa@asahikawa-nct.ac.jp*

Key words: Ductile cast iron, Eddy current method, Nondestructive testing, Casting surface, Microstructure

1. Introduction

Mechanical properties of ductile cast iron depend on the matrix and the nodule count of the microstructure. Microstructure affected by the thickness of the casting and the chemical compositions. Mechanical properties of castings are changed by those conditions and nondestructive evaluation method of the properties is required. There are some attempts to apply an eddy current technique to evaluate the microstructure of cast iron [1]. However, the eddy current technique is affected by the surface condition of the target[2]. This study examined the effect of the casting surface on evaluating ductile cast iron microstructure with the eddy current method. Stair-type specimens with three different thickness were prepared. Microstructures of the specimens varied by addition of chemical compositions. Eddy current evaluation was carried out on the casting surface and machined surface. Eddy current signal on the machined surface and pearlite ratio inside of the specimens showed good correlation. With the eddy current signal on the casting surface, pearlite ratio beneath the casting surface indicated good correlation.

2. Materials

Stair-type specimens with the thickness of 40mm, 10mm, 3mm were prepared. Fig. 1 indicates the dimensions of the specimen. Chemical compositions of the specimen were shown in Table 1. Cu, Mn and Cr were added to change microstructure of the specimens. Shot blast was used to remove sand on the surface of the specimens. Under surface of the specimen were cut by milling machine. Four specimens were cast and three of them were annealed to relief residual stress on the surface induced by machining. Fig. 2 shows the microstructure at the subsurface and inside of the specimens. Addition of the alloying element increases the area of pearlite in the matrix. There is the transient layer with 100~200μm thick beneath the casting surface of 10mm and 40mm thickness part. Nodule count beneath the casting surface is higher than that at the inside of the specimens.

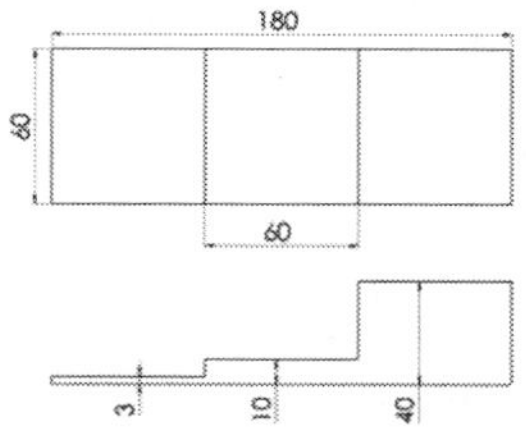

Fig. 1. Dimensions of the specimen

Table 1. Chemical compositions of the specimens

Specimen	C	Si	Mn	Cu	Cr
YD6-9	3.76	2.35	0.31	**0.10**	0.02
YD10-13	3.67	2.32	0.31	**0.23**	0.02
YD14-17	3.70	2.35	0.31	**0.40**	0.02
YD18-21	3.70	2.35	0.31	**0.53**	0.02
YD22-25	3.74	2.33	**0.39**	0.24	0.02
YD26-29	3.7	2.33	**0.49**	0.24	0.02
YD30-33	3.74	2.26	**0.58**	0.24	0.03
YD34-37	3.74	2.25	**0.68**	0.24	0.03
YD38-41	3.74	2.35	0.49	0.24	**0.06**
YD42-45	3.67	2.38	0.49	0.24	**0.11**
YD46-49	3.70	2.38	0.49	0.25	**0.15**
YD50-53	3.69	2.38	0.49	0.24	**0.20**

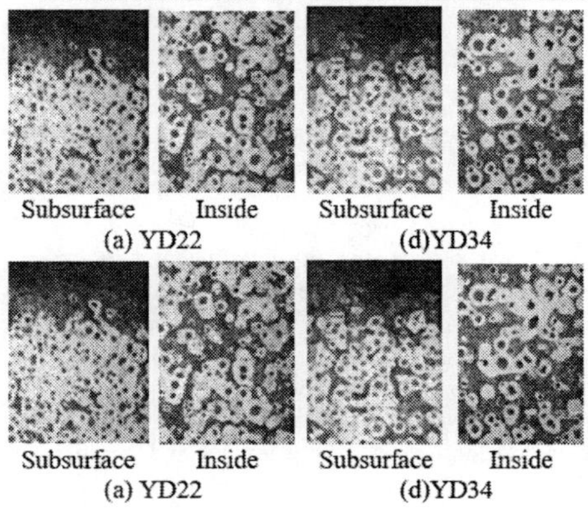

Fig. 2 Microstructure of the specimens *200μm*

3. Equipment

Schematic illustration of an eddy current testing instrument is shown in Fig. 3. Testing probe consists exiting coil (89turn) and detecting coil (360turn). Sinusoidal current was applied to the exiting coil and output signal of the detecting coil was measured by lock-in amplifier. Output of lock-in amplifier Vx and Vy were recorded. Vx related with the output signal from the detecting coil having the same phase of the exciting current. When the signal of the detecting coil have 90 degree phase lag, only Vy can be measured.

Measurements were performed at the center of the machined surface and casting surface in each thickness. Testing frequency are 500Hz and 1kHz.

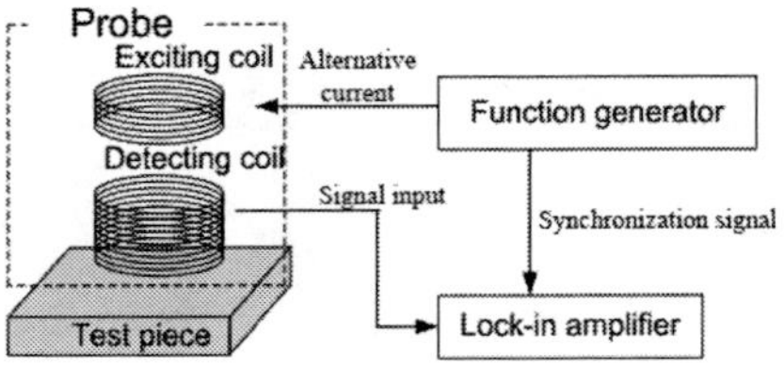

Fig. 3. Schematic illustration of an eddy current testing instrument

4. Results and discussion

Fig.4 shows the eddy current signal of the specimens on the machined surface with the testing frequency of 1kHz. Plots of the eddy current signals of different thickness locate different area in the Fig.4. Plots on the thinner part indicate small in Vx and large in Vy.

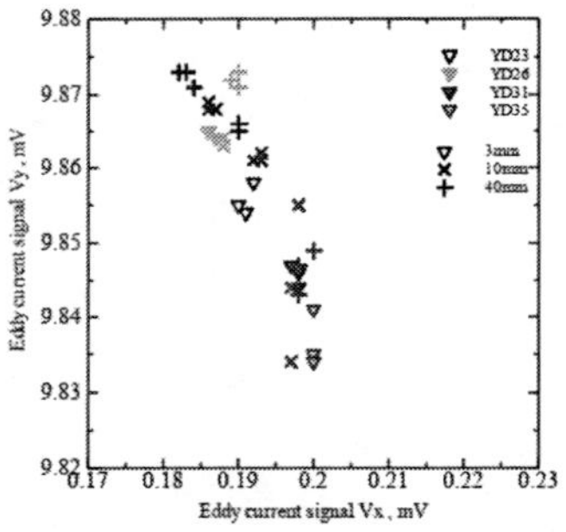

Fig. 4 Eddy current signals of the specimen with different chemical compositions and thickness. (Machined surface)

Relation between the eddy current signal Vy on the machined surface and the pearlite ratio in the matrix of the specimens are shown in Fig. 5. Pearlite ratio and Vy shows significant correlation, and there is no difference on that correlation among the plots of the specimens with different alloying element. On the other hand, on the casting surface, poor correlation obtained between Vy and pearlite ratio. Fig. 6 shows relation between Vx and the pearlite ratio. Good correlation between them can be observed. Better correlation obtained with lower testing frequency 500Hz. It suggests that the effect of casting surface was suppressed with lower frequency.

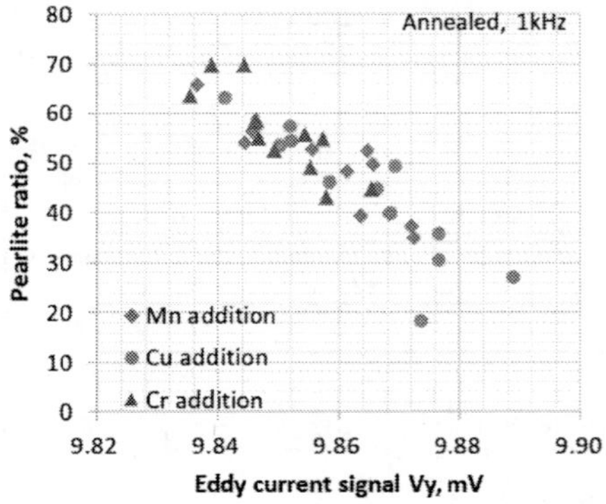

Fig. 5 Relation between the pearlite ratio and eddy current signal Vy measured on the machined surface.

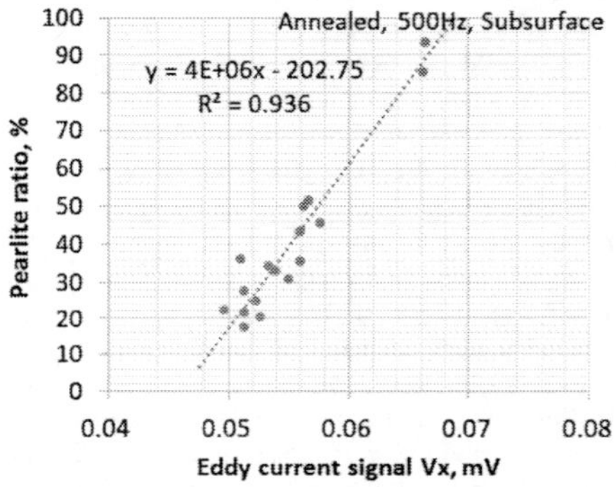

Fig. 6 Relation between the pearlite ratio and eddy current signal Vx measured on the casting surface.

5. Conclusions

(1) Eddy current signal that measured on both the machined surface and casting surface showed correlation with the pearlite ratio of the ductile cast iron.

(2) Eddy current signal components Vx and Vy were suitable for evaluation on the casting surface and the machined surface respectively.

(3) With lower testing frequency, better correlation between the microstructure and the eddy current signal.

References

[1] S. Konoplyuk, T. Abe, T. Uchimoto, T. Takagi, M. Kurosawa, (2005), Characterization of ductile cast iron by eddy current method, NDT&E International, Vol. 38, pp. 623–626

[2] N. HORIKAWA, T. HONMA, (2016) Effect of Surface Condition on Eddy Current Evaluation of Ductile Cast Iron Matrix, Proceedings of the 72nd World Foundry Congress

The 73rd World Foundry Congress
„Creative Foundry"
23 - 27 September 2018, Krakow, Poland

CHALLENGES IN GRAVITY SAND CASTING OF ZE41 Mg ALLOY

Ilhan Aygün[1], Erhan Körpe[1], Murat Akçin[2], Özen Gürsoy[3], Eray Erzi[3], Derya Dispinar [3*]

[1] *VIG Metal, Kütahya, Turkey*
[2] *Magmasoft, Istanbul, Turkey*
[3] *Istanbul University, Faculty of Engineering, 34320 Avcilar-Istanbul, Turkey*
**Corresponding address: e-mail: deryad@istanbul.edu.tr*

Key words: ZE41, Mg casting, Mould design

1. Introduction

Casting of magnesium alloys is always a challenge. Its significantly high affinity to oxygen makes it difficult to control during melting and pouring processes. Magnesium is the lightest metal almost half of density of aluminum [1]. The alloying of magnesium with various elements improves its mechanical properties which may become close to aluminum alloys. Therefore, magnesium alloys are still the attractive choice of metals and research on casting of these alloys have increased over the past decade. It has high ductility and damping properties. Thus, combined with its specific strength, major application areas are automotive and aerospace parts [2].

The current alloys are expensive and have not reached the desired mechanical properties. Therefore, high pressure die casting, particularly hot chamber applications are quite common [3-4]. In this work, Zr grain refined ZE41 alloy was sand cast and its castability and mechanical properties were evaluated.

2. Experimental Study

The chemical composition of ZE41 alloy used in casting trials is given in Table 1. The cylindrical part geometry was selected which is given in Figure 1.

Table 1. Chemical composition of ZE41

Zn	Zr	Rare earth	Mg
3.7	0.7	0.8	Rem.

Fig. 1. Schematic drawing of the cast part

Casting experiments were carried out in an air tight-sealed unit where the melting furnace is on top with a stopper at the bottom of the crucible, and the cast part is placed underneath the stopper connected with the runner system to the sand mould as seen in Figure 2. This original design was proposed by Prof.Ali Kalkanli (METU, Ankara, Turkey).

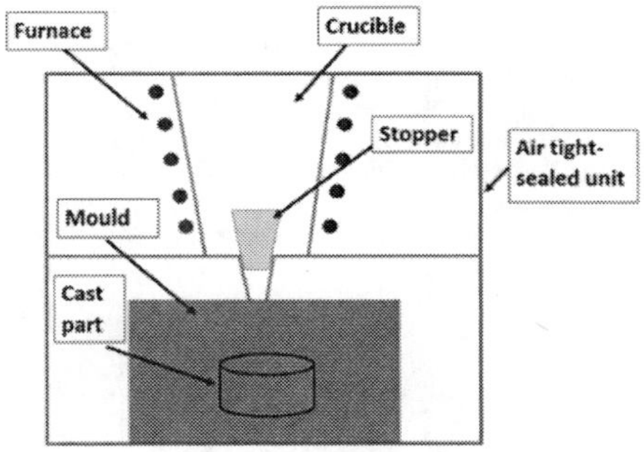

Fig. 2. The casting unit

The casting temperatures were selected as 675 and 710°C. Alphaset resin binder was used with silica sand. Magmasoft Simulation was used for optimization of casting parameters.

3. Results

In the initial mould design, cylindrical down sprue was used with the cast part being filled by 4 ingates from the outside with a C shaped runner that surrounds the cast piece as seen in Figure 3.

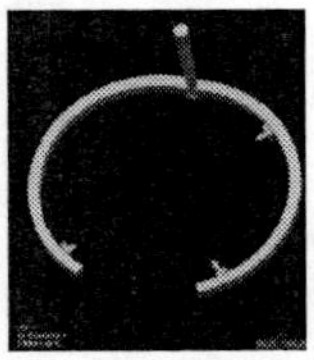

Fig. 3. The initial runner design

However, this design had resulted in misrun, cold shots and porosity. Such examples can be seen in Figure 4 and 5. X-ray results had shown that the cast piece had Grade 8 and 9 porosity, gas hole, shrinkage and foreign materials.

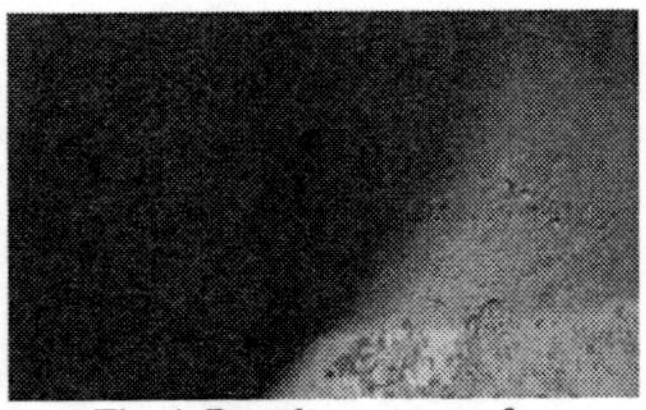

Fig. 4. Porosity on top surface

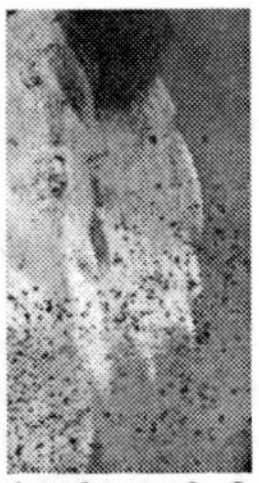

Fig. 5. Porosity observed after machining

A new design was proposed and Campbell's 10 rules of casting [5] were applied. A tapered downsprue with natural fall of the liquid magnesium was measured and applied which was placed at the center of the cylindrical cast piece. Four runners were placed at the bottom of the gates that looked like chandelier and the cast piece was filled from the bottom. AFS 100 sand was dropped down to AFS 40. The binder content was decreased from 3 wt% to 1.0 wt%. Magmasoft optimization tool was used to optimize the casting parameters. These can be seen in Figure 6 and 7.

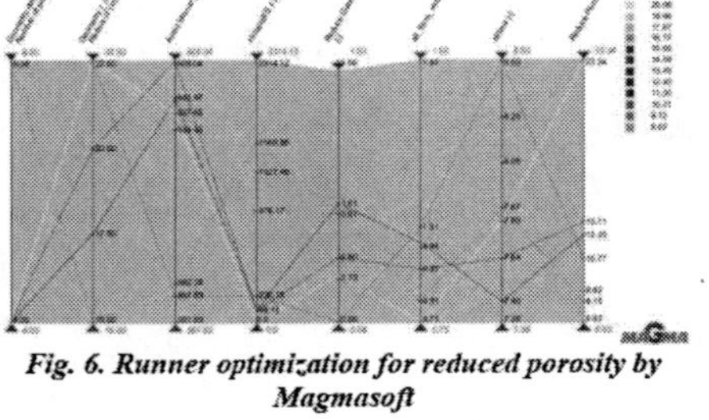

Fig. 6. Runner optimization for reduced porosity by Magmasoft

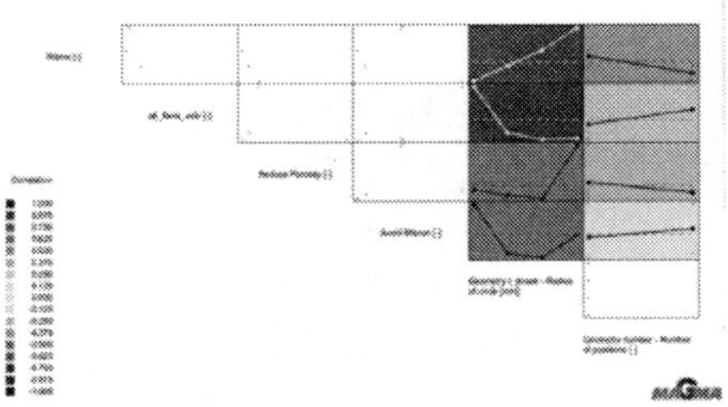

Fig. 7. Correlation Matrix results by Magmasoft

4. Conclusions

- After optimization studies of the runner design, sand grain size and binder content, X-ray studies show zero grade porosity, shrinkage and foreign material.
- ZE41 was successfully sand cast with minimized defects.
- Magmasoft Optimization tool is effective and comprehensive for improving quality and cost of metal casting.

References

[1] Mordike, B. L., & Ebert, T. (2001). Magnesium: properties—applications—potential. Materials Science and Engineering: A, 302(1), 37-45.

[2] Luo, A. A. (2013). Magnesium casting technology for structural applications. Journal of Magnesium and Alloys, 1(1), 2-22.

[3] Bettles, C. J., Forwood, C. T., StJohn, D., Frost, M. T., Jones, D. S., Qian, M., Nie, J. F. (2003). An elevated temperature magnesium alloy suitable for precision sand casting of powertrain components. *Magnesium technology*, 223-226.

[4] Qian, M., & Das, A. (2006). Grain refinement of magnesium alloys by zirconium: Formation of equiaxed grains. *Scripta materialia*, *54*(5), 881-886.

[5] Campbell, J. (2015). *Complete casting handbook: metal casting processes, metallurgy, techniques and design*. Butterworth-Heinemann.

The 73rd World Foundry Congress
„Creative Foundry"
23 - 27 September 2018, Krakow, Poland

LIGHTWEIGHT DIE CASTING TOOLS - A PROMISING OPTION FOR ENHANCING THE HIGH PRESSURE DIE CASTING PROCESS

Müller Sebastian[1]*, Dilger Klaus[1]
Müller Anke[2], Rothe Felix[2], Dröder Klaus[2]

[1]*Institute of Joining and Welding, TU Braunschweig, Braunschweig, Germany*
[2]*Institute of Machine Tools and Production Technology, TU Braunschweig, Braunschweig, Germany*
**Corresponding address: sebastian.mueller@tu-braunschweig.de*

Key words: die casting, process modeling, die deflections, lightweight design, energy reduction

The approach presented in this article focuses on improving the conventional manufacturing concept of die casting tools towards a flexible, modular design, where only a few contour elements of the cavity die have to be replaced for achieving different castings. In order to pave the way for this design method, basic thermal and structural properties of a lightweight design die casting tool were analyzed. The finite element calculations carried out revealed that the modular lightweight die casting tool consumes considerably less energy for preheating and during operation. Due to the material reduction, the calculated deformations and stresses in the tool during the casting process deviate from the properties of conventional die casting tools. Although initial calculation results look very promising, further knowledge must be acquired in this field in order to ensure the future success of modular lightweight die casting tools.

1. Introduction

Most of the components for conventional die casting tools are still designed to order and are manufactured using subtractive processes. There is currently no official standard or even cross company standardization of the individual components within the die. Hence, most of the components for conventional die casting tools are still designed to order and are manufactured using subtractive processes. Promising approaches for the increased use of a modular or standardized design to reduce production costs while maintaining ordinary subtractive manufacturing processes have already been demonstrated [1]. *Queudeville* et al. have analyzed the modularization and standardization of high-pressure die casting tools, aiming at breaking the one-to-one relationship between the cast component and the corresponding die casting tool [2-3]. It was found that modularization and standardization of dies could be a promising approach for foundries.

However, the current production principle suffers from a number of disadvantages. For example, high tool weights and a high resource input during production lead to thermally inert tools. Due to the high tool mass, a considerable amount of energy is still required for the tool temperature control. Moreover, taking into account the usual weights of die casting tools and the characteristic deposition rates of various metallic generative processes [4], it becomes evident that there are currently narrow limits to the increased use of additive manufacturing processes.

2. Design concept of a lightweight die casting tool

Casting tools could be made available with considerably reduced material input and manufacturing efforts as well as improved energetic properties, if the tool design would be based on lightweight design principles. As a result, only the individual tool cavity itself is produced by cutting processes. The other tool components can be designed modular and reusable, e.g. by means of supporting structures with members and nodes, which can be adapted to the geometric requirements by incremental manufacturing methods.

Fig. 1 illustrates the basic design principle of the lightweight tool. As a reference cast component, the existing 4 mm plate of the research institute was selected. The die casting tool consists of a cavity die and a mold base surrounding the cavity die. This principle applies to the fixed and ejector tool halves.

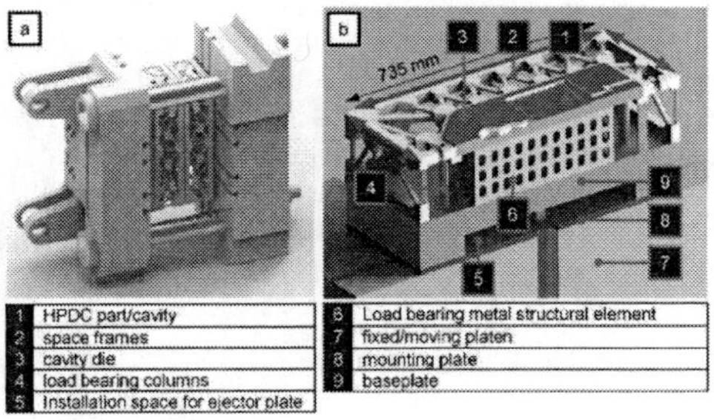

Fig. 1. Principal design approach of the lightweight design tool; cross-sectional (a) and isometric (b) view

However, there are major differences in the design of the mold base. In contrast to the conventional mold bases, which usually consists of a solid steel block, a highly segmented design was used here. The mold base consists of a spaceframe, a rigid, light, truss-like structure composed of interlocking struts in a geometric pattern. Due to the inherent rigidity of a triangle, bending moments are transmitted as tensile and compressive forces over the length of the individual struts. Compared to the conventional die casting tool, a weight saving factor of approx. 2.3 (880 kg) was achieved.

3. Simulation of thermal properties

In order to quantify the differences in energy consumption during preheating and the casting process, finite element models for transient thermal analysis were applied. The investigations take both the preheating and the casting process into account. Fig. 2 shows the time-dependent values for the thermal output and the average cavity temperature during the die casting process. The values displayed were calculated from the 20th casting cycle after it was determined that for this number of casting cycles, the average difference between the individual cycles for both design approaches is less than 1% of the relative change. It is noticeable that the time-temperature curves show a similar pattern for both tool designs.

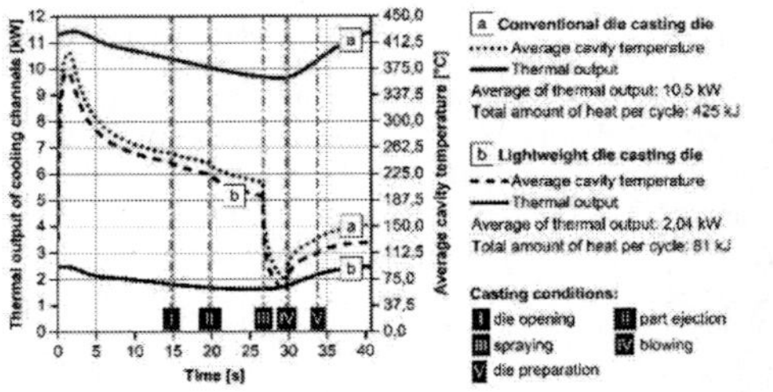

Fig. 2. Calculated transient thermal output and average cavity temperature after 20 casting cycles

With regard to the time-dependent thermal output of the cooling channels during the casting process, it becomes obvious that there is a fivefold difference between both design principles. This surprisingly high factor results from the fact that less heat is extracted from the lightweight tool during the casting process, as there is significantly less surface area exposed to the spray lubricant and thus to energy transfer. Due to its low wall thickness, the lightweight design principle also serves as thermal insulation, which effectively suppresses the heat conduction from the tool to the fixed and movable panel. In addition, the cooling channels of the conventional tool are further away from the cavity surface and thus heat the adjacent areas of the conventional mold more intensely. These areas in turn exchange more energy with the environment.

4. Simulation of die deflection

Ensuring a safe operation over the entire service life and compliance with form and position tolerances are two essential requirements for a die casting tool. Hence, a thorough analysis of the deformations and stresses during the casting process is of great importance in order to ensure sufficient functional properties of the lightweight design tools. A comprehensive simulation model was created which is partly based on the approaches published in [5]. The model takes into account both the thermal and mechanical loads during the HPDC process. As shown in Fig. 3, the die casting tools exhibit a different deflection behavior under identical load boundary conditions. In this case, the deviation in tool stiffness also leads to a considerable difference in the deflection of the machine platen.

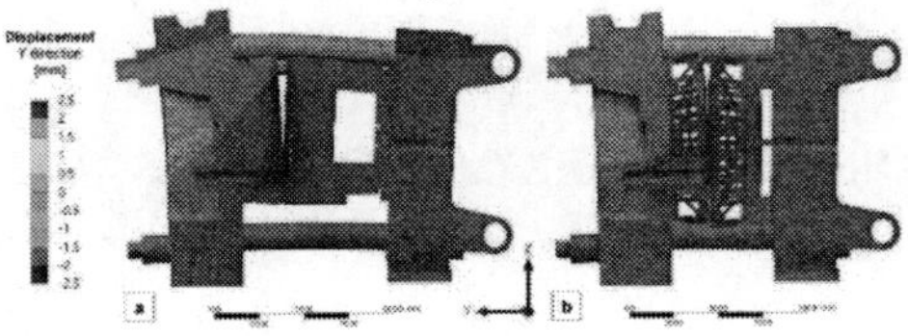

Fig. 3. Calculated displacement in y-direction of the functional unit of a high pressure die casting machine, magnification of 40x

For the evaluation of the wall thickness, the deformation of both die halves was exported from the FE mesh and the deviation from the nominal wall thickness (4 mm) was calculated and displayed interpolated (Fig. 4). A direct comparison of the wall thickness deviation shows that the cavity deformation of the lightweight die casting tool (Fig. 4b) during the casting process is comparable to or even lower than those of the conventional die casting tool (Fig. 4a).

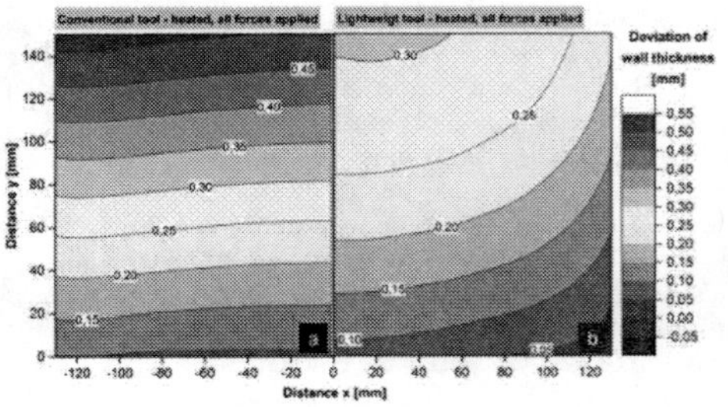

Fig. 4. Graphical representation of the casting component (4 mm plate) wall thickness with applied thermal and mechanical loads during casting process

5. Conclusion and Outlook

From the authors' point of view, the concept represents a promising possibility to further increase the performance of the die casting process in terms of energy consumption and production flexibility. However, it should be clear that this is initially still a concept that serves to evaluate the performance of the design principle. The results obtained must be transferred step by step into a real die casting tool and experimental measurements of the mechanical and thermal properties as well as investigations of the HPDC component quality must be carried out.

References

[1] Brecher C, Özdemir D (eds) (2017) Integrative production technology: Theory and applications. Springer, Switzerland

[2] Queudeville Y, Vroomen U, Bührig-Polaczek A (2014) Modularization methodology for high pressure die casting dies. Int J Adv Manuf Technol 71(9-12): 1677–1686.

[3] Queudeville Y, Vroomen U, Bührig-Polaczek A (2014) Modularization methodology for high pressure die casting dies. Int J Adv Manuf Technol 71(9-12): 1677–1686.

[4] Ding, D. et al. (2015) Wire-feed additive manufacturing of metal components: Technologies, developments and future interests. Int J Adv Manuf Technol 81(1-4): 465–481.

[5] Yamagata H, Shigenaga Y, Tanikawa S et al. (2013) Deflection Simulation of a Die Installed in a High-Pressure Die-Casting Machine. Transactions of the Japan Society of Mechanical Engineers 79(806): 3920–3929

The 73rd World Foundry Congress
„Creative Foundry"
23 - 27 September 2018, Krakow, Poland

RECONSTRUCTION OF THE CASTING TECHNOLOGY OF PREHISTORIC BRONZE ORNAMENTS WORKED WITH THE LOST-WAX TECHNIQUE ON THE BASES OF METAL SCIENCE ANALYSES, COMPUTER MODELLING AND MODEL ALLOYS

Aldona Garbacz-Klempka[1]*, Józef Szczepan Suchy[1], Zofia Kwak[1], Jacek Gackowski[2], Łukasz Kowalski[2], Szymon Rosołowski[3], Janusz Kozana[1], Marcin Piękoś[1]

[1]AGH University of Science and Technology, Faculty of Foundry Engineering, Historical Layers Research Centre, Reymonta 23 Str., 30-059 Kraków, Poland
[2]Nicolaus Copernicus University in Toruń, Institute of Archaeology, Szosa Bydgoska 44/48, 87–100 Toruń, Poland
[3] Biskupin Archaeological Museum, Biskupin 17, 88-410 Gąsawa, Poland

**Corresponding address: agarbacz@agh.edu.pl*

Key words: Archaeometallurgy, Lost-wax casting, Tin bronzes, Computer Modelling, Bronze Age

1. Introduction

This work describes an approach to the reconstruction of the casting technology of the bronze ornaments recovered from the so-called hoard from Brudzyń (Żnin district, Kujawsko-Pomorskie Voivodeship). The hoard was accidentally discovered in 2016 and consists of more than 60 bronzes, and several iron artefacts. Among bronze metalwork deposited in the hoard, there are band ornaments made of wire, sheet, and rod (necklaces and bracelets), socketed axes, fibulae, pins, and phalerae (Fig. 1). The hoard should be dated to the beginning of the Early Iron Age (650BC-550BC).

Analytical tests of the collective find from Brudzyń were carried out by an interdisciplinary team of researchers from the Faculty of Foundry Engineering in cooperation with the Historical Layers Research Centre of the University of Science and Technology in Kraków, Institute of Archaeology of the Nicolaus Copernicus University in Toruń, and Archaeological Museum in Biskupin.

2. Methodology and Research

Band ornaments (Fig. 1) were analyzed with respect to the bronze casting technology of the Late Bronze Age and the Early Iron Age. The investigations involved the employment of the analytical methods (EDXRF, SEM-EDS, OM), which were supported by computer techniques. Model alloys referring in their raw material profiles to the analyzed artefacts were also prepared.

The results of the EDXRF analyses were applied in composing the model alloys. They were analyzed in terms of their composition, microstructure, phase transformations as well as casting and mechanical properties [1-2].

Fig. 1. The selected bronze artefacts deposited in the hoard from Brudzyń (a frame stands for the analyzed objects)

In order to present more complete characteristic of the band ornaments from Brudzyń, the modeling of the pouring process and casting solidification in the clay mould was performed.

Such an analysis required the computer visualization of the artefacts (Fig. 2) and reconstruction of the hypothetical casting moulds (Figs. 3-4), in which they were cast.

The visualization was supported with the SolidWorks software, while the modeling was processed in the environment of the MagmaSoft software. This allowed us to determine the direction of the metal flow in the mould as well as the porosity distribution in the castings.

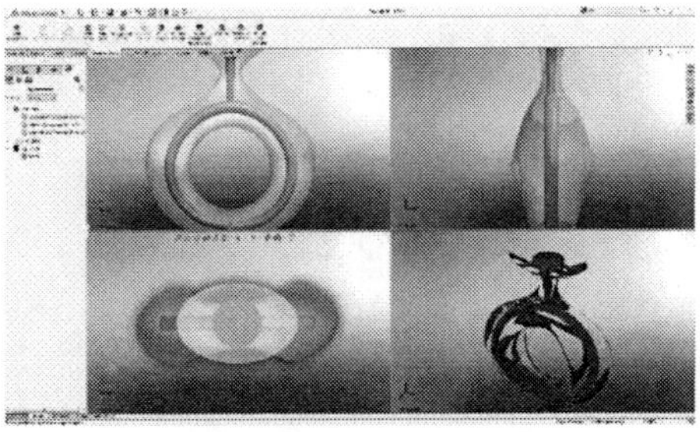

Fig. 2. 3D reconstruction of the artefact from the Brudzyń hoard

Fig. 3. Artefact from the Brudzyń hoard fitted into the 3D reconstruction of the ceramic casting mould

Fig. 4. 3D reconstruction of the ceramic casting mould for the band ornament manufacturing

On the basis of the previous investigations performed on the corresponding artefacts collected from other places, the shape and placement of the gating system were designed. Also, the additional parameters such as the pouring time and temperature, or mould temperature, were indicated. The porosities observed macroscopically on the surfaces of the band ornaments were compared with the corresponding casting defects resulted from the simulation of the metal solidifying process in the mould [3-5].

3. Conclusions

The obtained results allowed the more precise and detailed recognition of some aspects related to the lost-wax technique which was commonly applied during the Late Bronze Age and the Early Iron Age in Poland. Based on the metallographic data, the raw material profiles as well as structures and properties of the alloys, employed for manufacturing the band ornaments deposited in the hoard from Brudzyń, were recognized. Also, the casting moulds and gating systems were reconstructed. Analyses of the casting defects, sparsely distributed in the ornaments, provided an additional information about the level of the bronze casting during the prehistoric times.

Acknowledgments
The financial support of the State Committee for Scientific Research of Poland under the grant numbers 11.11.170.318–11 and 11.11.170.318–1 is acknowledged.

References

[1] A. Garbacz-Klempka, J. S. Suchy, Z. Kwak, T. Tokarski, R. Klempka & T. Stolarczyk (2018) Study of investment casting technology from Bronze Age. Casting workshop in Grzybiany (southwest Poland). Archives of Metallurgy and Matrials. 63(2)

[2] A. Garbacz-Klempka, J. S. Suchy, Z. Kwak, P. Długosz & T. Stolarczyk (2018) Casting technology experiment and computer modeling of ornaments from Bronze Age. Archives of Metallurgy and Matrials. 63(2)

[3] Garbacz-Klempka A., Kozana J., Piękoś M., Kwak Z., Długosz P. & Stolarczyk T. (2015). Ceramic moulds for precision casting in the bronze age and computer reconstruction of the casts. Archives of Foundry Engineering.15 (spec. iss. 1), 21–26

[4] Garbacz-Klempka A., Kowalski Ł., Kozana J., Gackowski J., Perek-Nowak M., Szczepańska G. & Piękoś M. (2016) Archaeometallurgical investigations of the early iron age casting workshop at Kamieniec: a preliminary study. Archives of Foundry Engineering 16(3), 29–34

[5] Garbacz-Klempka A., Kowalski Ł., Gackowski J., Kozana J., Piękoś, M., Kwak, Z., Cieślak, W. (2016). The workshop of the Lusatian culture metalworker in Kamieniec, Toruń district : the research results of casting hoop ornaments using model alloys. In: A. Garbacz-Klempka, J. Kozana, M. Piękoś (Eds.), Nauka i Technologia, Monografia. Kraków: Akapit, 47-70.

The 73rd World Foundry Congress
„Creative Foundry"
23 - 27 September 2018, Krakow, Poland

STUDY OF $(\%TiO_2)/[\%Ti]$ PARTITION COEFFICIENT IN CUPOLA FURNACE

Viridiana Herrera-Vázquez[1], Manuel Castro-Román[1]*, Hector Mancha-Molinar[2], Martín Herrera-Trejo[1], Arturo Juárez-Hernández[3]

[1]*Cinvestav Unidad Saltillo, Department of Metallurgical Engineering, Av. Industria Metalúrgica 1062, Parque Industrial Saltillo-Ramos Arizpe, 25900 Ramos Arizpe, Coahuila, Mexico Country*
[2] *Tupy Saltillo Mexico, Blvd Isidro López Zertuche 4003, Zona Industrial, 25230 Saltillo, Coahuila, Mexico*
[3] *UANL, FIME, Av. Universidad s/n, Ciudad Universitaria, 66451 San Nicolás de los Garza, Nuevo León, Mexico*
**Corresponding address: manuel.castro@gmail.com*

Key words: Titanium, Cupola Furnace, Titanium Nitrides, Slag Viscosity, Slag Basicity.

1. Introduction

This study was conducted to provide information about the deleterious Ti enrichment in iron melts obtained from Cupola. It was divided into two stages.

The first one was focused on determining whether the ratio $(\%TiO_2)/[\%Ti]$ observed in the production was close or not to thermodynamic equilibrium. The second stage corresponded to characterization of the Ti content evolution during two production campaigns of one-week long each one.

To characterize the $(\%TiO_2)/[\%Ti]$ partition coefficient, samples of metal and slag were taken out from the Cupola in relevant production steps of each campaign: start up, normal and end step of week campaign. Also samples of the initial charge materials were collected to identify the possible sources of Ti.

2. Experimental

Value of $(\%TiO_2)/[\%Ti]$ close to thermodynamic equilibrium was estimated with experiments of isothermal holding at 1762 K of industrial samples of metal and slag. This tests were done in a laboratory induction furnace with a graphite susceptor as heating element. Holding time was 2, 4 or 6 hours.

To characterize the $(\%TiO_2)/[\%Ti]$ partition coefficient throughout the typical production steps, samples of metal and slag that corresponds to them were taken out from Cupola. Also, to identify the important sources of Ti, it was performed both the sampling the initial material charge and the materials balance corresponding to Cupola operation. As silicon carbide briquettes were identify as an important source of Ti, the results of conventional chemical analysis technique were complemented by chemical microanalysis of Ti precipitates in these briquettes.

3. Results

The results of the isothermal experiments shown that the equilibrium $(\%TiO_2)/[\% Ti]$ ratio must be close to five (Fig. 1). However in the industrial samples the values of partition coefficient $(\%TiO_2)/[\%Ti]$ were above fifteen (Fig. 2), far away of the thermodynamic equilibrium. That suggests that any aspect that increases the contact time of the liquid metal droplets with the slag or enhance in some way the Ti absorption in metal would imply an increase of Ti in the melt in order to achieve thermodynamic equilibrium.

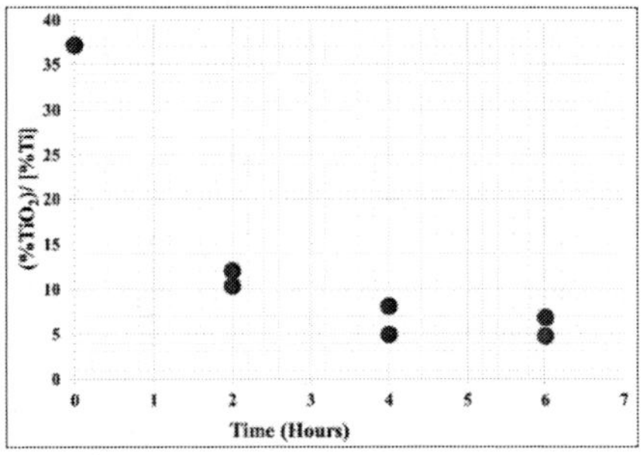

Fig. 1. Evolution of $(\%TiO_2)/[\% Ti]$ during isothermal holding at 1762 K (Laboratory experiments).

Regarding slag properties, an increase in their viscosities must increase the drops-slag contact time if slag volume is keep constant. Using the model proposed by Urbain et al. [1], the effect of both slag composition and temperature on viscosity was estimated. In the range of variability observed in the process studied, the temperatures changes leads to an increase of slag viscosity form 9 Pa.s to around 22 Pa.s for the lower and the higher process temperature, respectively (Fig. 3). The change of viscosity due to variation of slag composition tends to be narrower than observed by temperature effect, exception made of the two points that are placed out of typical tendency (Fig. 4). However, other factors as slag volume or titanium sources must be taken account to explain the whole variation observed in (%TiO_2)/[%Ti] values. Slag volume in start-up operation of Cupola could tends to be higher to one of the normal or the end operations, according to material balance. The lowest (%TiO_2)/[%Ti] ratio observed in an start-up sample (Fig. 2) corresponds to one operation were slag volume is very large.

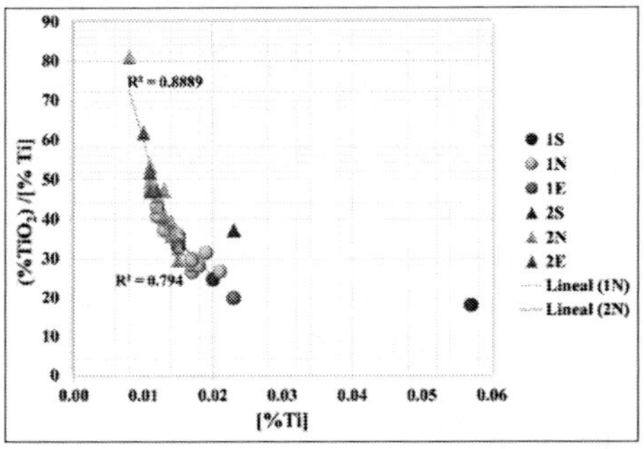

Fig. 2. Relationship of (%TiO₂) / [%Ti] with [%Ti] according to samples take out from Cupola at different production steps: 1S, 1N and 1E means start, normal and end step of week one, while 2S,2N and 2E corresponds to same steps in week two.

Regarding to titanium sources, SiC briquettes that shown titanium presence as nitrides has been identify as potentially important contributors to Ti increase. This kind of briquettes where avoided in campaign of week two, where more suitable (%TiO_2)/[%Ti] values were observed as compared with the corresponding values of ratios of week one (Fig. 2).

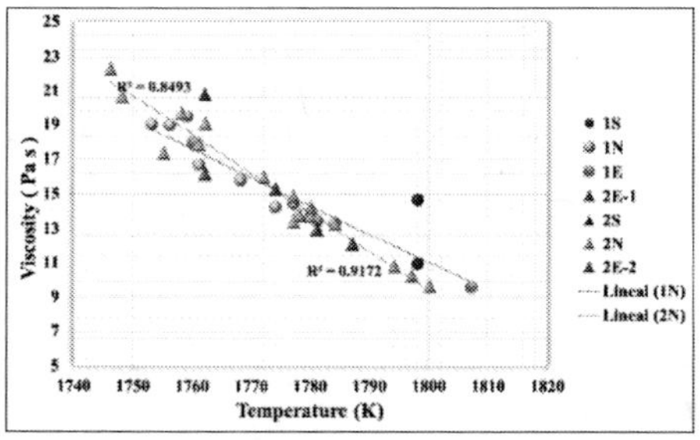

Fig. 3. Viscosity of the system SiO₂-CaO-Al₂O₃-MgO as a function of melting temperature.

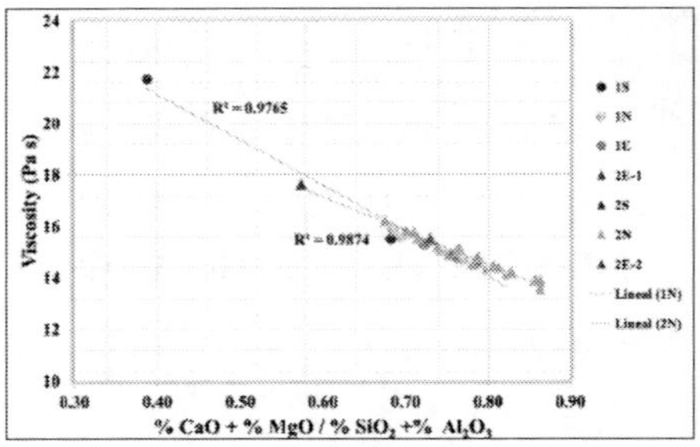

Fig. 4. Viscosity of the system SiO₂-CaO-Al₂O₃-MgO as a function of slag basicity. 1S,1N and 1E means start, normal and end step of week one, while 2S,2N and 2E corresponds to same steps in week two

4. Conclusions

The values of (%TiO_2)/[%Ti] ratio observed in samples from two week plant operation were above fifteen, exceeding the value of five that could be expected in conditions approaching the thermodynamic equilibrium.

The SiC briquettes that shown titanium presence mostly as nitrides instead of oxides has been identify as potentially important contributors to Ti increase in melt.

Other factors as slag viscosity and slag volume shown also non negligible variations in process. These variations must explain part of whole variation observed in (%TiO_2)/[%Ti] values, considering that they must affect the contact time of liquid drops with the slag.

Acknowledgments

The author V. Herrera-Vázquez acknowledges the scholarship provided by the National Council of Science and Technology (CONACyT) for the realization of this work. This work was made as a part of an IBEROEKA project.

Reference

[1] Urbain, G., Bottinga, Y., & Richet, P. (1982). Viscosity of liquid silica, silicates and alumino-silicates. Geochimica et Cosmochimica Acta, 46(6), 1061-1072.

The 73rd World Foundry Congress
„Creative Foundry"
23 - 27 September 2018, Krakow, Poland

20 YEARS OF RESEARCH PROJECTS TARGETED TO ZERO DEFECT MANUFACTURING IN DIECASTING

Franco Bonollo[1]*, Nicola Gramegna[2], Lars Arnberg[3]

[1] Padova University, Dept. of Engineering and Management (DTG), Italy,
[2] Enginsoft SpA, Padova, Italy,
[3] Trondheim University, NTNU, Norway
**Corresponding address: bonollo@gest.unipd.it*

Key words: Aluminium alloys, High Pressure Die casting, Microstructure, Defects & Imperfections, Mechanical behaviour

1. Introduction

Metal manufacturing is a relevant part of whole GDP of industry sector in Europe. In particular, the field of non-ferrous foundry, led by Aluminium casting technology, is constituted by about 2700 European companies, with more than 4 million tons of castings produced in 2016. The last survey carried out by CAEF individuates Germany and Italy as producers of 60% of Aluminium alloys castings in Europe.

The key-process, in this field, is High Pressure Die Casting (HPDC), which had an enormous evolution in last 20 years, and is facing relevant challenges in next years. One of the main driving forces for this evolution has been the effort for achieving the Zero Defect Manufacturing (ZDM) target.

Various R&D projects strongly contributed to generate innovation, thanks to well-balanced public-private funding. This paper reviews the most relevant EU-funded research and education projects carried out in the field in the last 20 years, showing the approach that created the best synergy among all strategic elements needed for an intelligent, efficient and innovative Aluminium foundry.

2. The innovation path

Innovation is the results of several and concerted actions, as summarised in Fig.1. A key-issue is certainly training and education, which was targeted since 1999, with the Leonardo Pilot Project Coprofound, aimed at improving the use of process simulation as a powerful tool to achieve a better knowledge on foundry processes.

The FP5-IDEAL research project (2002-2005) was devoted to Integrated Development Routes for Optimised Cast Aluminium Components, in the context of vehicle weight reduction through sustainable use of light alloys.

The FP6-NADIA research project (2006-2010) contributed to the development of New Automotive components Designed for and manufactured by Intelligent processing of light Alloys, in the context of :

- nanotechnologies and nanosciences,
- knowledge based multifunctional materials,
- new production processes and devices.

The FP7-StaCast Coordinated Action (2012-2014) allowed the elaboration of two new EN Standards:

- classification of Defects and Imperfections in Aluminium castings,
- evaluation of Mechanical Potential of Aluminium casting alloys.

The FP7-MUSIC research project (2012-2016) was addressed to the digital transformation in HPDC process. In detail, MUSIC was referred to MUlti-layers control&cognitive System to drive metal and plastic production line for Injected Components in the context of Smart Factories: Energy-aware, agile manufacturing and customization.

The challenge of MUSIC was to transform a production-rate-dominated manufacturing field into a quality/efficiency-driven and integration-oriented one, to exploit the enormous (and still underestimated) potential of HPDC through collaborative research and technological development, along the value chain with research groups, design, engineering and manufacturing companies and through advances in manufacturing, ICT and model process technologies.

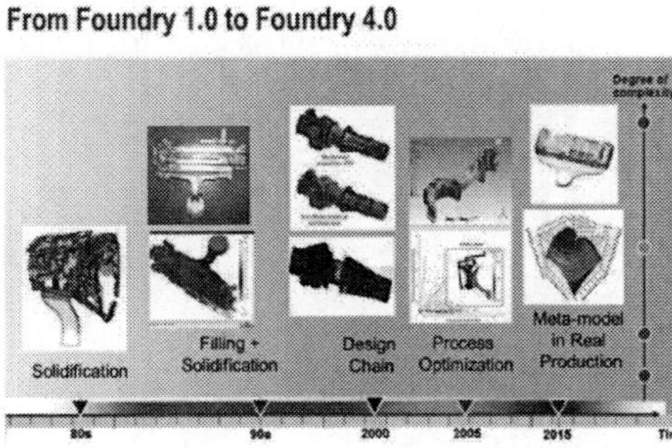

Fig. 1. Evolution and Innovation in Aluminium foundry processes: from Foundry 1.0 to Foundry 4.0

These projects collect a budget of about 28 millions of €, with a EU contribution of about 16 millions of €. More than 50 Partners have been involved, with an effort of almost 3000 person month.

3. The new challenges

Several challenges are still open in the field, with some common elements:

- further knowledge must be achieved for properly linking multi-scale material properties and transformation processes (solidification, cooling, heat treatments);
- material modelling, virtualization of processes and factory digitalization need to increase their use and penetration level in foundries;
- skills of all the persons involved in foundry processes have to be improved and developed, paying strong attention to individuation of new profiles and competences.

The intelligent management of these elements in next years will generate competiveness in the global market and in the supply chain, taking into account also eco-sustainability issues, based on circular economy and adoption of decision support system including energy and cost models.

Acknowledgments
All the projects mentioned in the paper have been supported by EU, in the frame of FP5, FP6 and FP7 Programs.

References

[1] Bonollo, F., Gramegna, N., Timelli, G. (2015), High-pressure die-casting: Contradictions and challenges, JOM, 67 (5), pp. 901-908.
[2] F. Bonollo & G. Timelli editors (2014): Aluminium alloy castings: the EU StaCast guide to defects classification, mechanical potential and design issues, Assomet Servizi srl, Milano
[3] F. Bonollo & N. Gramegna editors (2014), The MUSIC guide to key-parameters in High Pressure Die Casting, Assomet Servizi srl, Milano
[4] N. Gramegna & F. Bonollo editors (2016), Smart Control and Cognitive System applied to the HPDC Foundry 4.0, Assomet Servizi srl, Milano

The 73rd World Foundry Congress
„Creative Foundry"
23 - 27 September 2018, Krakow, Poland

INVESTIGATIONS OF THE THICKNESS OF PROTECTIVE COATINGS DEPOSITED ON MOULDS AND CORES

Łukasz Jamrozowicz[1]*, Joanna Kolczyk-Tylka[1], Anna Siatko

[1]*AGH University of Science and Technology, Al. Mickiewicza 30, 30-059 Kraków, Poland*

**Corresponding address: ljamroz@agh.edu.pl*

Key words: protective coatings, coating thickness, coating viscosity, core, mold,

1. Introduction

The casting production is a multistage and complex process. The main and superior aim is obtaining the ready product of the determined parameters and without defects. The casting quality in a large measure depends on the mould or core preparation. One of the preparation stages is covering the mould (core) surface by the protective coating. The application of such coating improves the smoothness of the mould surface and due to that the casting surface becomes more smooth and less rough. In addition, the protective coating covering the mould surface layers minimizes reactions between the mould and liquid metal. The application of these coatings protects the casting against certain casting defects such as burn-on, veins, sand buckles and others. In addition, some coatings can favourably influence the casting structure.

The efficiency of the protective coating influence depends significantly on its thickness on the mould (core) during covering. Too thin coating will not provide the satisfactory protection. The thickness of the deposited coating depends on several factors: coating viscosity, way of deposition, number of deposited layers, grain size of the matrix out of which the core or moulds are made [1-13].

2. Investigations aim and methodology

The aim of the hereby study was the determination in which way such parameters as the coating viscosity, the way of its deposition and the grain size of the matrix of the moulding sand, influence the thickness of the deposited coating.

The alcoholic coating and water coating of three apparent densities 10s, 20s and 30s, were used in investigations. The viscosity measurements were done by means of the 'Ford's cup' of a clearance of 4 mm. Coatings were deposited on cores made of furan moulding sand. High-silica sand of three various matrix sizes: 'fine' sand $d_L = 0.22$ mm, 'medium' sand $d_L = 0.32$ mm , 'coarse' sand $d_L = 0.46$ mm, were used for making cores.

Deposition of coatings was done by immersing the core in the coating and also by painting.

The thickness of the coating in wet condition was measured directly after its deposition, by the device called the 'comb'. In addition, after drying the coating, its thickness on the core as well as its penetration into intergranular spaces were measured. In order to perform this last measurement the sample was cut out and the measurement was done by means of the microscope.

The measurement results are presented as diagrams.

3. Results of investigations

Figure 1 presents examples of images of cores cross-sections with the deposited alcoholic coating of the arbitrary viscosity being 20s. It can be noticed that at larger matrix grains ('coarse' sand) the coating penetrates deeper into intergranular spaces. However, the coating thickness on the core is at a similar level. These observations are confirmed by measurements presented in Figure 2.

The thickness of the coating covering the core equals 0.3 mm regardless of the matrix grain size, the coating penetration into intergranular spaces increases with the grain size increase, from approximately 1.2 mm to 2.1 mm.

***Fig. 1.** Penetration of the coating into the intergranular spaces depending on the matrix grain size – coating with arbitrary viscosity 20s. magnification 10×.*

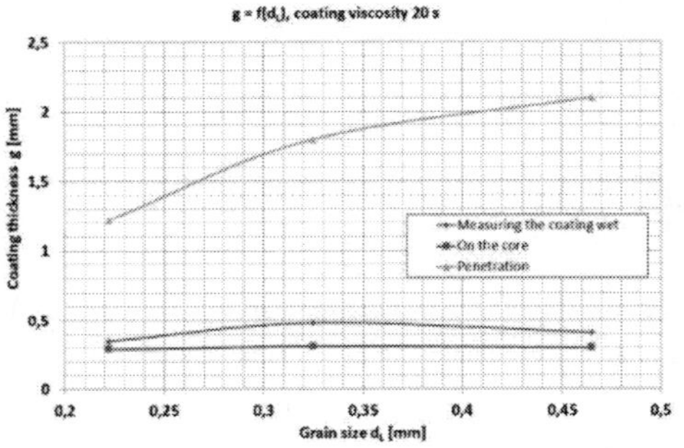

Fig. 2. Thickness of the protective coating depending on the core grain size. Coating with arbitrary viscosity 20s

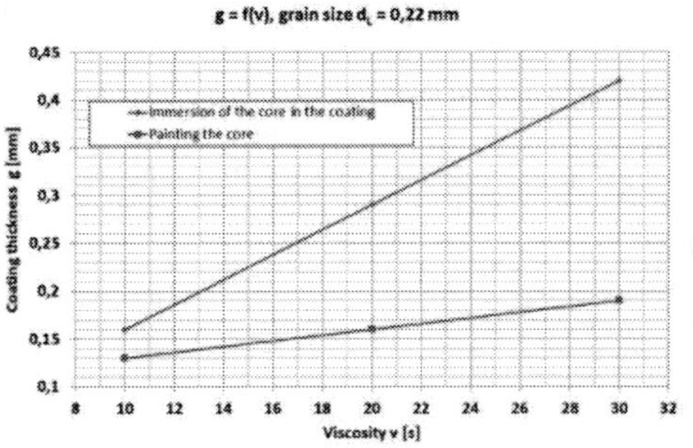

Fig. 3. Influence of viscosity on the thickness of the coating for different methods of its application to the core. The matrix grain size $d_L = 0,22$ [mm]

The influence of the coating viscosity on the coating thickness on the core, at various methods of the deposition, is presented in Figure 3. It results from tests, that when the coating viscosity increases its thickness on the core also increases.

The largest changes could be seen when the coating was deposited by immersing, the change was the most intensive: from 0.15 mm (at the coating viscosity 10s) to 0.45 mm (at the viscosity 30s), while the smallest changes were in case of painting: increase from 0.12 mm to 0.2 mm.

4. Conclusions

The performed investigations indicate that the thickness of the coating formed on the core depends on two factors, on the coating viscosity and on the way of this coating deposition. In turn, the coating penetration into intergranular spaces depends on the grain size of the core matrix and on the coating viscosity.

Acknowledgments

The research received founding from NCBIR no. POIR 01.01.01-00-0042/17-00.

References

[1] Lewandowski L.J. (1997). Tworzywa na formy odlewnicze. Kraków. Wydawnictwo „Akapit".

[2] Seeger K.: (2012). Zastosowanie pokryć wodnych przy formowaniu ręcznym. Przegląd Odlewnictwa. 7-8 s. 322 – 326.

[3] Zych J., Snopkiewicz T. (2011). Drying and strengthening of protective coatings after deposition on casting moulds and sand cores – new investigation methods. Przegląd Odlewnictwa. 9–10 s. 506–512.

[4] Zych J. (2006). Application of the novel ultrasonic method in the on line research of the setting, and hardening process of ceramic materials. Inżynieria Materiałowa. R. 27 nr 3 s. 680–683

[5] Holtzer M., Bobrowski A., Drożyński D., Mocek J. (2013). Investigations of protective coatings for castings of high-manganese cast steels. Archives of foundry engineering. Volume 13 Issue 1/2013 s.39 – 44

[6] Jakubski J., Dobosz St., M Jelinek P. (2005). The influence of the protective coating type on thermal deformation of casting cores. Archives Of Foundry Engineering. Volume 5 Issue 15

[7] Jakubski J., Dobosz St. (2006). Wpływ powłoki ochronnej na zjawiska cieplne w rdzeniach odlewniczych. Archives Of Foundry Engineering. Volume 6 Issue 18

[8] Kubicki J., Kochmańska A. (2001). Wieloskładnikowe powłoki ochronne na staliwie żarowytrzymałym otrzymywane metodą pasty. Archives Of Foundry Engineering. Volume 1, Issue 1

[9] Kochmańska A., Kubicki J. (2009). Efficiency of protective coatings on high creep resistant cast steel. Archives Of Foundry Engineering. Volume 9, Issue 2/2009, s, 129-132

[10] Wróbel J. (2014). Badanie Jakości cyrkonowych powłok ochronnych w aspekcie ich przydatności na formy i rdzenie piaskowe dla staliwa. Przegląd odlewnictwa. Volume 64, issue 9-10.

[11] Jamrozowicz Ł. Zych J. Kolczyk J. (2015). The drying kinetics of protective coatings used on sand molds. Metalurgija. vol. 54, nr 1, 23÷26

[12] Jamrozowicz Ł., Zych J., Kolczyk J., Wróblewski D. (2014). Rola kształtu powierzchni formy w procesie wysychania wybranych powłok ochronnych. Archives of Foundry Engineering. vol. 14, nr 2

The 73rd World Foundry Congress
„Creative Foundry"
23 - 27 September 2018, Krakow, Poland

EFFECT OF CaSiAl MODIFICATION ON THE MICROSTRUCTURE AND MECHANICAL PROPERTIES OF LOW-CARBON MICROALLOYED CAST STEEL WITH 0.04% Nb and 0.07% V

Barbara Kalandyk [1*], **Renata Zapała**[1*], **Sebastian Sobula**[1*], **Grzegorz Tęcza**[1*],

[1] *AGH University of Science And Technology, Faculty of Foundry Engineering, Mickiewicza 30 Av., 30-059 Kraków, Poland*

**Corresponding address: e-mail: bk@agh.edu.pl*

Key words: low-carbon microalloyed cast steel, microstructure, mechanical property, nonmetallic inclusions

The calcium treatment in steelmaking and casting processes lead to modification of Al_2O_3 non-metallic inclusions and oxygen decreasing. In consequence the dendritic oxides transform into more rounded a shape. Additional effect of this treatment is sulphide inclusion content decreasion which improves toughness, ductility and fatigue strength [1-3].

The addition of microalloying elements (V, Nb, Ti) to carbon and low alloyed cast steel influence on tensile strength and impact toughness [4-6].

The effect of CaSiAl modification (43-49% Ca, 43-48% Si, 2% Al) on the microstructure and mechanical properties of low-carbon cast steel is discussed. Tests were carried out on the cast steel with 0.2% C and micro-additions of V and Nb. This grade of cast steel is used, among others, for heavy castings like slag ladles. Melting was carried out in an electric induction furnace under laboratory conditions. Armco iron, low-alloy steel scrap and ferro-alloys of Fe-Si and Fe-Mn were used as a charge.

The basic chemical composition of the tested cast steel is shown in Table 1. The powdered modifying mixture was introduced in the form of a cored wire into the previously Al sediment deoxidized metal bath. Three Y-type ingots were cast from the tested steel, i.e. a reference sample and two samples after modification with a mixture of CaSiAl added in the amount of 1.5 and 3 kg/Mg.

Table 1. Chemical composition of investigated low-carbon microalloyed cast steel

C	Si	Mn	Cr	Ni	Al	other
wt. %						
0.18	0.18	1.5	0.35	0.2	0.05	0.04%Nb 0.07%V

Then the samples were heat treated, i.e. normalized at 920°C which is the type of heat treatment used for massive (large-size) castings made of this steel.

The metallographic examinations by light microscopy and scanning microscopy as well as the mechanical tests have confirmed that the use of CaSiAl mixture during steel melting changes not only the shape and chemical composition of non-metallic inclusions, but also the mechanical properties, mainly impact strength. The examinations by scanning microscopy have revealed that the unmodified cast steel contained non-metallic inclusions such as MnS sulphides, oxides and oxide sulphides of various shapes (Fig. 1). The use of the CaSiAl mixture changed the shape (the predominance of globular inclusions) and the chemical composition of inclusions, especially of oxide sulphides due to their enrichment with calcium (Fig. 2).

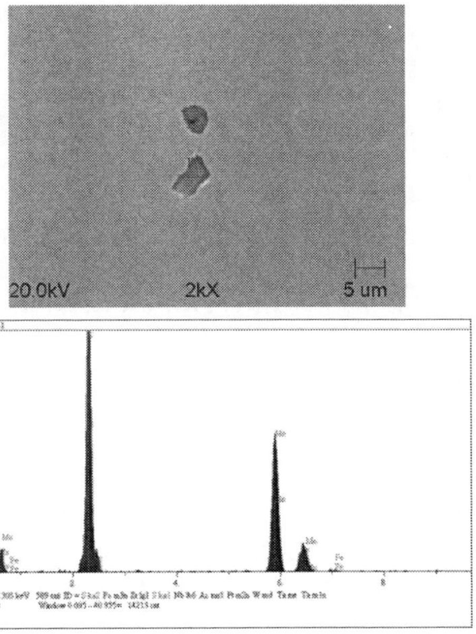

Fig. 1. Example of the X-ray spectrum of a complex inclusion (SEM / EDS) - cast steel before modification

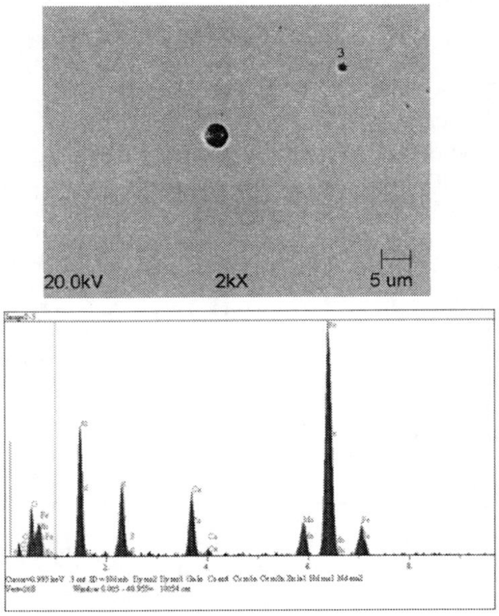

Fig. 2. Example of the X-ray spectrum of a complex inclusion (SEM / EDS) - cast steel after modification with CaSiAl mixture

Compared with reference melt, the change in the morphology of non-metallic inclusions due to the use of the CaSiAl modifying mixture has increased, among others, the impact strength of the tested cast steel (Fig. 3).

On the other hand, modification was found to have no significant effect on the values of Rm (about 715-720 MPa), $R_{p0.2}$ (417-440 MPa) and A_5 (17.3-18%).

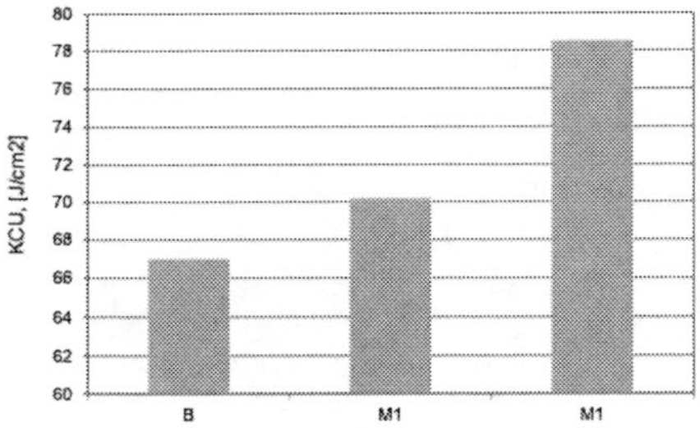

Fig. 3. Impact strength: B - basic chemical composition of the tested cast steel; M1- with 1.5 kg/Mg CaSiAl; M2 - with 3 kg/Mg CaSiAl

Conclusions

Based on the results obtained in the conducted tests, it was found that introducing a CaSiAl mixture during melting of cast low-carbon steel after deoxidation of the melt mainly leads to the modification of non-metallic inclusions (changes in the shape and composition of inclusions). The change also results in an increase of the cast steel.

References

[1] HEN G.J., et al.. (2015). Optimization of Calcium Addition to High-strength Low-alloy Steels.Journal of Iron and Steel Research. vol. (22), 7, pp. 590-597.

[2] Holappa L.E.K., Helle A.S.(1995). Inclusion control in high – performance steels. Journal of Materials Processing Technology. vol. (53), pp. 177-186.

[3] Han Z.J., Liu L. Lind M., Holappa L. (2006). Mechanism and kinetics of transformation of alumina inclusions by calcium treatment. Acta Metallurgica Sinica (English Letters). vol. (19), 1, pp. 1-8.

[4] Kalandyk B., Sierant Z., Sobula S. (2009) Improvement of the microstructure, yield stress and impact toughness of medium carbon steel by vanadium additions. Przegląd Odlewnictwa (Foundry Journal of the Polish Foundrymen'ts Association. vol. (58), 3, pp. 108-113.

[5] Najafi, Hamidreza, Rassizadehghani, Jafar, Asgari, Siroos. (2008). As-cast mechanical properties of vanadium/niobium microalloyed steel. Materials Science and Engineering A. vol. (486), 1-2, pp. 1-7.

[6] Głownia J., Kalandyk B. (2008) Effect of precipitation strengthening in low alloyed Mn-Ni cast steels. Journal of Materials Processing Technology. vol. (207), pp. 147-153. DOI:10.1016

Acknowledgments

This work has been partially executed under a Research Project P1.1.1-PG-09-001.

The 73rd World Foundry Congress
„Creative Foundry"
23 - 27 September 2018, Krakow, Poland

ANALISIS OF RAPID DRYING PROCESS BY SET ON FIRE OF ALCOHOL-BASED PROTECTIVE COATINGS APPLIED ON SAND CORES AND MOULDS

Joanna Kolczyk-Tylka[1*], Natalia Kaźnica[1], Jerzy Zych[1], Łukasz Jamrozowicz[1]

[1]*AGH University of Science and Technology, Faculty of Foundry Engineering, Mickiewicza 30 Av., 30-059 Kraków, Poland*

**Corresponding address: kolczyk@agh.edu.pl*

Key words: protective coatings, drying process, organic binders, inorganic binders

1. Introduction

Castings preparation is associated with the selection of appropriate technology and materials for sand molds such as matrix, binder, protective coating. Both, technologies and materials, have a decisive influence on the quality of castings, mainly on their surface condition.

The task of the coating is primarily protection against penetration of liquid metal into the sand mold, and reduction of the tendency of the molding sand to stick to the surface of the cast [6].

The thickness of the coating layers is adjusted depending on the type of alloy used and the thickness of the casting wall. Usually, one coating is applied, but with a large thermal load on the molds, several layers can be used. When applying only one layer of the coating, it is necessary to reduce the tendency of the surface to crack, that is why the viscosity of the coating has to decrease, resulting in a lower layer thickness [4-5].

Nowadays, for the protection of the environment, the use of water-based coatings is recommended in foundries. However, alcohol-based coatings are still used in large quantities. Coated molds and cores should be dried before the pouring process to prevent any surface imperfections in castings. In the case of water-based coatings, drying occurs most often in specially prepared driers, while in the case of coatings, in which the diluent is alcohol, very often drying is accelerated by setting the evaporating alcohol on fire. The process of drying the coatings involves evaporation of the thinner. Literature distinguishes two basic types of drying processes - natural and artificial. This article analyzes the courses of artificial drying - by setting fire to the protective coating applied to the sand cores [1-3].

2. Aim and methodology

Research of the drying processes of protective coatings using the method of burning included the preparation of flat and unilaterally closed cylindrical core samples made from moulding sands with chemical binders: organic (self-hardening molding sand with furfuryl resin) and inorganic (self-hardening molding sands with hydrated sodium silicate and liquid hardeners).

The alcohol-based zirconium coating were applied to the prepared cores on one and both sides by pouring (Fig. 1).

The gravimetric analysis was used to determine the drying speed. The measurement consisted of continuous recording of the weight loss of the previously coated sample during drying. Different versions of diluent evaporation were compared under conditions of natural evaporation of alcohol from the cores as well as evaporation realize by burning.

One of the assumptions of the work was to determine both the amount of diluent to be removed as well as the drying time of the applied coatings. Another aim was to determine a period when it is still possible to set fire to alcohol vapors. If the concentration of alcohol vapors is too low in the air surrounding the core surface, drying by burning will no longer be possible.

Attempts to ignite the coating applied to the cores were taken immediately after application (maximum after 30 seconds) and gradually lengthened the time by 10 - minute cycles, counting from the application of the coating.

a) b) c)

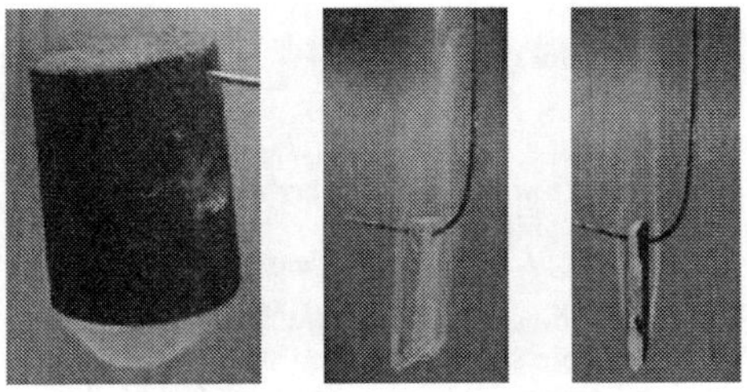

Fig. 1. The core with an alcohol-based protective coating, set on fire during the drying process: a) a one-sidedly closed cylinder with coating applied from the inside, b) a flat core sample coated on both sides, c) a flat sample coated on one side only

This article presents only selected examples of drying processes, the whole research covers a much wider range.

3. Results

Figures 2-4 show an example of the drying process of the cores after previous application of the protective coating. Combination of two types of drying processes are shown, i.e. artificial, is visible - by burning the sample after 10 minutes and 30 seconds from applying the coating and natural - drying in the open in laboratory conditions (atmospheric relative humidity W = 20 - 22%, ambient temperature T = 24 - 25 ° C).

Based on the analysis of the obtained results, it is possible to confirm that an increase occurs in the rate of evaporation in the case burning of the coatings after 30 seconds and after 10 minutes. Shorter drying time compared to naturally drying coatings, is also well visible.

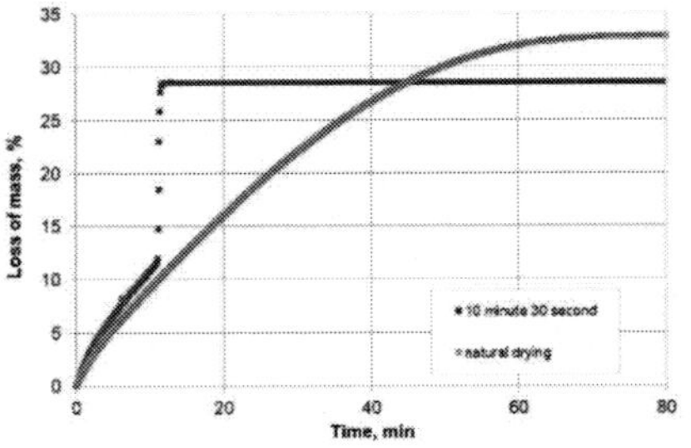

Fig. 2. Influence of time on weight loss. Flat core samples coated on both sides a) ignited after 10 minutes and 30 seconds, b) drying of the core sample in laboratory conditions

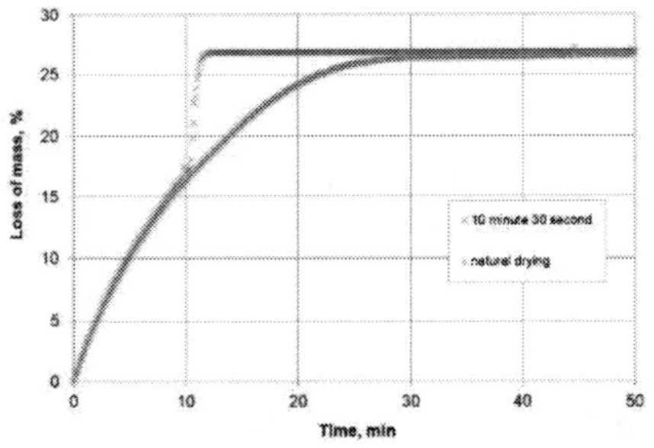

Fig. 3. Influence of time on weight loss. Flat core samples coated on one side a) ignited after 10 minutes and 30 seconds, b) drying of the core sample in laboratory conditions

It is worth noting, however, that in the case of thin-walled core samples covered with a protective coating on both sides (Fig. 2) more moisture is lost from the sample during natural drying than artificial drying by burning. The authors hypothesize that in the case of drying by burning, moisture is blocked inside the mold (core). This is the result of closing of the pores in the molding sand caused by burning of the protective coating and its rapid shrinkage. As a result of this process, the evaporation of moisture from the core to the environment is prevented, and the drying is partially inhibited.

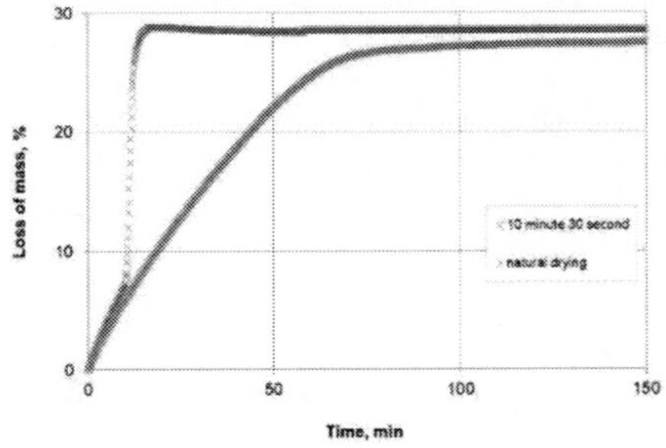

Fig. 4. Influence of time on weight loss. Cylindrical core samples coated on both sides a) ignited after 10 minutes and 30 seconds, b) drying of the core sample in laboratory conditions

4. Summary

Burning of alcohol-based coatings shortens the time of evaporation of the coating components in an irregular manner, however, it leads to incomplete drying. Within very developed surfaces of molds and cores, the presence of thinners and water in the top layer of the molding sand, just under the protective coating, may result in an increased tendency to create defects of gaseous origin.

Acknowledgements

This research received funding from NCBiR No. POIR 01.01.01-00-0042/17-00.

References

[1] J. Kolczyk, N. Kaźnica, S. Żymankowska-Kumon, Ł. Jamrozowicz. (2017). Studies of protective coatings used for metal moulds. Transactions of the Foundry Research Institute. Vol. 57 no. 1, s. 39–50.

[2] J. Zych, J. Kolczyk. Influence of the external surface of wax pattern on the kinetics of drying of ceramic molds. (2013). Archives of Foundry Engineering. Vol. 13 spec. iss. 3, s. 197–202.

[3] Zych J. (2006). Application of the novel ultrasonic method in the on line research of the setting, and hardening process of ceramic materials. Inżynieria Materiałowa. R. 27 nr 3 s. 680–683.

[4] Jamrozowicz Ł. Zych J. Kolczyk J. (2015). The drying kinetics of protective coatings used on sand molds. Metalurgija. vol. 54, nr 1, 23÷26.

[5] Jamrozowicz Ł., Zych J., Kolczyk J., Wróblewski D.: (2014). The Role of the surface shape of the mould in the process of drying of selected protective coatings. Archives of Foundry Engineering. vol. 14, nr 2.

[6] N. Kaźnica, J. Zych, J. Kolczyk. (2017). Analysis of the drying process of moistened surface layers of sand moulds and cores on the example of moulding sand with water glass Transactions of the Foundry Research Institute. Vol. 57 no. 1, s. 29–38.

The 73rd World Foundry Congress
„Creative Foundry"
23 - 27 September 2018, Krakow, Poland

POURING PROCESS CONTROL BASED ON HIGH-SPEED IMAGE ANALYSIS OF LIQUID FLOW

Ryosuke Tasaki [1]*, Junsuke Baba [1], Kazuki Taniguchi [1], Kazuhiko Terashima [1]

[1]Toyohashi University of Technology, Department of Mechanical Engineering,
1-1 Hibarigaoka, Tempaku, Toyohashi, Aichi Japan
**Corresponding address: tasaki@me.tut.ac.jp*

Key words: Gravity casting, Pouring process, Image processing, Flowrate control, Level control

Outflow rate control technology for molten metal from ladle has been actively researched and developed to solve several technical problems in the tilting type automatic pouring system. The outflow liquid weight equals to the product value of the time-integrated flow rate and the density. In the case that liquid density is known normally, the measurement accuracy of flow rate directly effects to the estimation error of the outflow weighing.
We have constructed a feedback system of pouring flow rate by using a 2D camera and aim to operate properly the pouring motion of tilting-ladle type automatic robot on real-time. In this paper, an approach of outflow rate estimation based on image data analysis of free fall streakline from liquid container is proposed.
The results of detection and identification analysis calculations of liquid flow streakline by high-speed image processing has been examined for each known input of flow rate and total amount. Finally, liquid pouring experiment using water has been carried out to evaluate the proposed estimation method and perform the high accuracy pouring control.

1. Introduction

Keeping constant level of liquid inside poured cup as shown in Fig. 1 is one of key technical subjects in the cast pouring process and leads to prevent the overflow from the top and to stabilize the filling/inflow rate into the lower mold cavity. Furtheremore, short weight of poured liquid occurs casting product shape defect or unfilled cavity inside the mold.
On the other hand, extra weight causes overflow from the top of the poured mouse inside the casting mold. The liquid level in the poured mouse during pouring must be stabilized for keeping on higher level to achieve high-quality and quick casting process. So that, it is important to control the outflow rate from liquid container by feedback system with high accuracy of molten metal flow measurement.
We have constructed a visual feedback system shown in Fig. 2 and aim to operate properly the pouring motion of tilting-ladle type automatic machine.

Fig. 1. Pouring process in metal casting factory

2. High-speed Image Analysis

Automatic control pouring machine[1-4] developed by our research group has been used for the confirmation of a proposed measurement system. It is possible to validate the image-based flowrate measurement accuracy by comparing with the exact known flowrate as reference input condition. In the case of water experiment instead of the metal pouring, image processing to detect the water flow streakline is more difficult than the case of metal flow. Captured image of water often displays the light reflection, the low-highlight of flow tube with background, the turbulence flow and the air entrainment due to its characteristics of transparency, low-viscosity and low-surface tension. Therefore, the measurement accuracy by our proposed method will be increased for the metal pouring in the actual casting factory.
An example of dynamic image shown in Fig. 3 displays continuous falling liquid tube from the container during pouring. The time series images and the derived flow rate value for each as real-time measurement information gives outflow weight change and the weight of poured liquid.

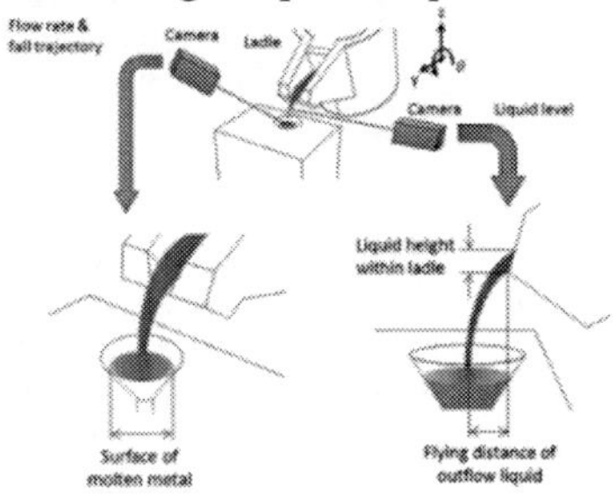

Fig. 2 Visual feedback system for pouring motion

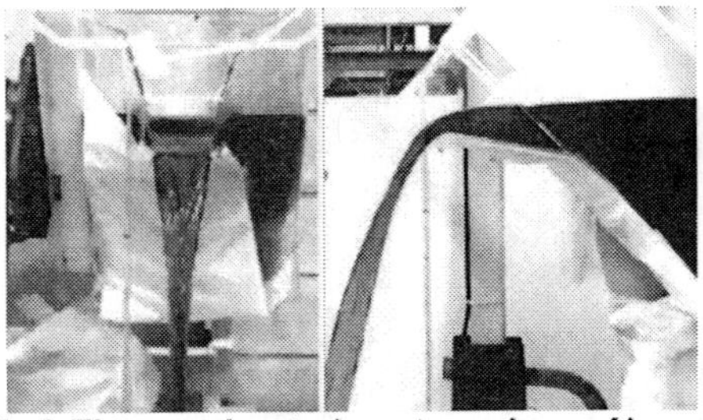

Fig. 3 Water experiment using auto-pouring machine

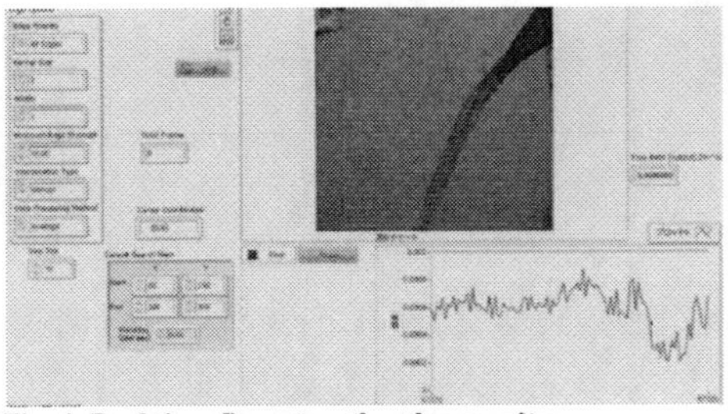

Fig. 4 Real-time flowrate estimation monitor

The results of detection and identifical calculations of liquid streakline by high-speed image processing has been examined for each known input of flow rate. The time series of image are processed in gray scale, binarization, morphology, thinning processing and curve fitting to a parabolic trajectory. Here the coefficients of the quadratic form can be identified for the representative flow streakline to obtain flow velocity at an arbitrary observation point.

The flow rate can be calculated by estimating the cross-sectional area normal to the free-falling flow velocity vector at the same point. Furthermore, the total outflow volume of liquid is obtained by the time integration of flow rate, and the weight is simply derived by multiplying the volume by the density.

3. Water Pouring Experiments

Liquid pouring experiments using colored water has been carried out to evaluate the proposed estimation method of flowrate and perform the high accuracy liquid weighing. As can be seen from Fig. 4, flowrate Q_{est}, related information and the error $\varepsilon = 100(Q_{est} - Q_{ref})/Q_{ref}$ are simultaneously observed in real time. There are three experimental conditions of varying Q_{ref}, shown in Fig. 5. The upper graph shows the tilting angle of the ladle driven by the automatic pouring machine. Here the tiling movement axis is set at the tip of ladle. Then the steady flowrate input as the bell-shaped reference shown in Fig. 6 lower can be generated by using our motion control technology[4]. The accuracy confirmation of the proposed estimation method is confirmed for the steady peaks for each reference flowrate curve. The Estimation result based on streakline analysis makes much difference over +40 % with the exact value. We consider that this huge error is due to the assumption of the flow tube form as the ideal ellipse shape with the tangential width and the depth. The cross-sectional area in perpendicular to flow tube has continuously changing its form in fact. Therefore, the cross-sectional area A_p is remodeled. The modified equation which has terms depending on the flow speed v_m. The newly estimated flowrate is perfectly fitted less than 1 % error, with the reference flowrate for all each condition. Adjustable parameters of c_1 and c_2 are briefly identified for matching to the all exact flowrates. The measurement accuracy with the modified estimation equation and other parameters are shown in Table 1.

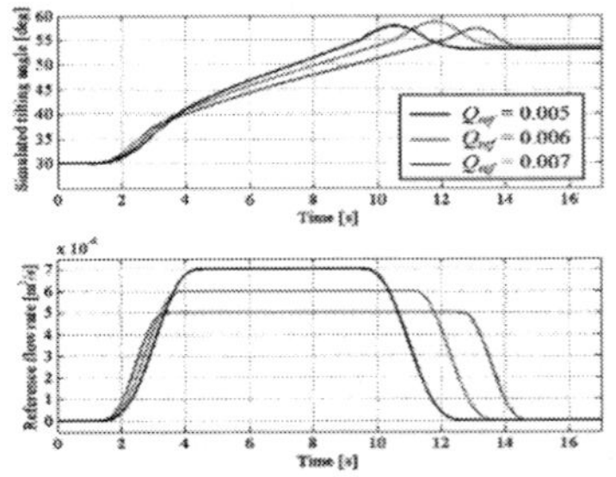

Fig. 5 Experimental conditions of each flowrate reference

Table 1 Parameters and measurement accuracy

Q_{ref} [m³/s]	θ [deg]	A_p [m²]	v_t [m/s]	Q_{est} [m³/s]	Error ε [%]
0.0005	70.63	0.000383	1.528	0.000500	-0.009
0.0006	69.92	0.000507	1.647	0.000595	-0.768
0.0007	69.03	0.000575	1.651	0.000704	0.591

4. Conclusions

We constructed a visual feedback system of pouring motion by using 2D camera. The approach of outflow rate estimation based on image data analysis of free fall streakline from liquid container is newly proposed. The detection and identification analysis calculations of liquid flow streakline is verified with successful performance result within ±1 % error rate. As the near future works, high accuracy measurement for the flow tube cross-section must be suggested, and the poured volume/weight can be measured consequently by calculating the time-integrated flow rate, for the main target: cast metal flow which was shown in Fig. 1.

References

[1] Y. Noda, K. Yano, and K. Terashima, "Pouring and Sloshing Suppression Control of Cylindrical Ladle in Self-Transfer-Type Automatic Pouring Robot with Complex Motion of Pouring and Transfer," Journal of the Robotics Society of Japan, Vol.24, No.1, pp. 94-103, 2006.

[2] Y. Noda and K. Terashima: Modeling and Feedforward Flow Rate Control of Automatic Pouring System with Real Ladle,Journal of Robotics and Mechatronics, Vol.19, No.2, pp.205-211, 2007.

[3] K. Yano and K. Terashima: Supervisory control of automatic pouring machine, Control Engineering Practice, Elsevier, No.18(3), pp.230-241, 2010.

[4] A. Ito, Y. Noda, R. Tasaki and K. Terashima: Outflow Liquid Falling Position Control Considering Lower Pouring Mouth Position with Collision Avoidance for Tilting-Type Automatic Pouring Machine, Material Transactions, Vol. 58, No. 3, pp.485-493, 2017.

The 73rd World Foundry Congress
„Creative Foundry"
23 - 27 September 2018, Krakow, Poland

VERIFICATION AND OPTIMIZATION OF INVESTMENT CASTING TECHNOLOGY FOR PRODUCTION OF CAST METAL SPONGES

Ivana Kroupová [1]*, Václav Merta [1], Petr Lichý [1]

VSB – Technical University of Ostrava, Faculty of Metallurgy and Material Engineering, 17. listopadu 15/2172, 708 33, Ostrava - Poruba, Czech Republic

Corresponding address: ivana.kroupova@vsb.cz

Key words: metal sponge, open cell structure, aluminum alloy, investment casting, evaporable pattern,

The paper deals with an optimization of procedure for production of cast metal sponges with use of evaporable pattern for investment casting technology. The experimental part stresses the solution of metal castability into complicated cavity of the mould, which is an exact negative of the initial pattern of polymeric foam.

There are several ways how to influence (increase) castability of molten metal, e.g. changing the thickness of the polymeric foam fibres, heating of the mould to a higher temperature, increasing of pouring temperature, transformation of the gating system or using of pressure casting.

1. The principle of the production of metal sponges by investment casting technology

Investment casting with use of pattern made of polymer foam is used for production of metal foam with open cell pores (metal sponges), which copies the shape of the polymer foam. A polyurethane foam cavity is first filled with sufficient refractory material, such as plaster (or a mixture of mullite, phenolic resins and calcium carbonate). The assembly is then heated to fire the plaster and remove the polyurethane foam. Molten metal is then poured into the mould - again, combinations of vacuum and high pressure casting can be used to ensure full infiltration. The plaster is then dissolved, to give a net-shape metal foam with an identical structure to the original polymer foam [1-3]. The principle of this technology is shown in the Fig. 1.

In our case a special types of plaster (Satin Cast 20, Kerr Lab; Eurovest, SRS Ltd.) as moulding material was chosen. As evaporable pattern the polyurethane foam Bulpren S 32450 of the Czech manufacturer (Eurofoam) with parametres (below) was used:

- porosity = 10 PPI;
- specific weight = 27 – 31 kg/m^3
- cell diameter = 3800 – 5200 μm.

The casting material was an aluminum alloy (AlSi10MgMn), copper alloy (CuSn12) and Cast Iron. The casting was carried out on a laboratory device MC15 (Co. Indutherm).

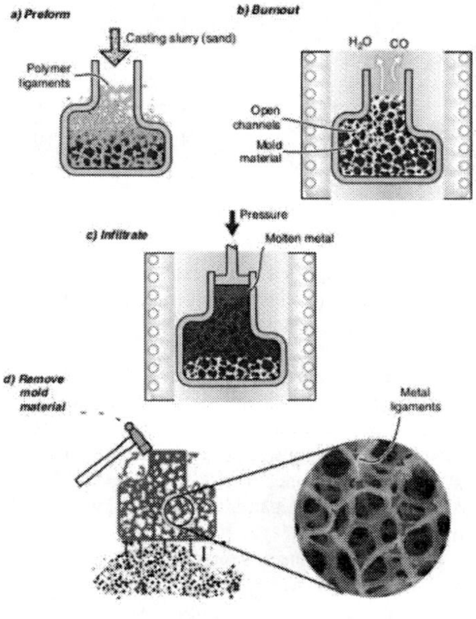

Fig. 1. Investment casting method used to manufacture of open cell metal sponges [4].

1.1. Polymer foam

For manufacture of patterns for investment casting, which are then used to create the metal sponge, polymer foams with a fully open structure are used, i.e. that they do not contain any closed cells. Such types of foams are produced by thermal reticulation process, in which all of the cell membranes, which remained after the foaming, are broken and they coalesce with cell ribs.

The actual ratio of foam ribs makes only 3-5% of the volume of foam, depending on the type of foam. It is, however, impossible to guarantee the fibre thickness of the polymeric foam at the reticulation process. Small thickness of the fibres in the polymer pattern may then cause for example a non-

penetration of metal into narrow cavities of the plaster mould. It is therefore necessary to ensure a bigger thickness of the fibres in the polymeric foam, and thus in the resulting metallic foam.
Two procedures for increasing the thickness of the foam pattern fibres were verified – soaking of the pattern into wax and spraying the pattern with one or two layers of acrylic paint.

1.2. Mould temperature

For casting of various alloys (Cu alloys, Al alloys) it is necessary to define different annealing cycles (see Tab. 1). For casting the Cu and Fe alloys with higher melting temperature (higher pouring temperature) it is necessary to anneal the plaster moulds to higher casting temperatures – to eliminate the thermal shock during casting, and to increase the melt fluidity into the complex mould cavity.

Table 1. Annealing cycles of plaster moulds

	Temperature, increase, soak at temp.	Temperature, increase, soak at temp.	Temperature, increase, soak at temp.
1	120 °C 8 °C/ min; 8 h	320 °C 10 °C/ min; 8 h	**800 °C** 20 °C/ min; 10 h
2	120 °C 8°C/ min; 8 h	550 °C 10 °C/min; 8 h	**1100 °C** 20 °C/ min; 10 h
3	120 °C 8 °C/ min; 8 h	550 °C 10 °C/ min; 8 h	**1000 °C** 20 °C/ min; 10 h

The most commonly used is the annealing cycle No 1 which is suitable for the subsequent casting of Al alloys (low melting temperature or low pouring temperature).
But for casting of Cu and Fe alloys it is necessary to increase the casting temperature of the mould. Therefore the annealing cycle No 2 was recommended. This annealing cycle proved to be unsuitable – the moulds annealed to such high temperatures show impaired collapsibility after metal casting.
A plaster sample has been subjected to differential thermal analysis in which it was found that at temperatures of 1100 °C the $CaSO_4$ is disintegrated to CaO and SO_3 (degradation of the mould).
After the evaluation of the plaster DTA the annealing cycle No 3 was designed. If casting temperature is reduced to the max. 1000 °C the $CaSO_4$ doesn't disintegrate and the plaster (plaster mould respectively) maintains its optimum properties. The moulds annealed in such a way have good collapsibility after casting but for casting the Cu and Fe alloys the mould temperature is still too low. When casting these alloys the metal misruns into the mould cavity due to high temperature jump. Therefore, in other experiments, attention was focused only on aluminum alloys. For casting the Al alloys the mould prepared in such a way is "overheated".

1.3. Pouring temperature

Other way to increase the castability is to increase the pouring temperature of the liquid metal. In this procedure, consideration should be given on the effect of the pouring temperature on the resulting microstructure of the final casting.

1.4. Gating system

Due to the highly complex shape of the evaporable pattern plays a fundamental role also design of gating system. Several types of gating systems have been verified in the experiment, but only the "dot contact" with the pattern was always encountered. Therefore, filling of the mould cavity through the solid wall has been proposed (see Fig. 2).

Fig. 2. Final design of gating system

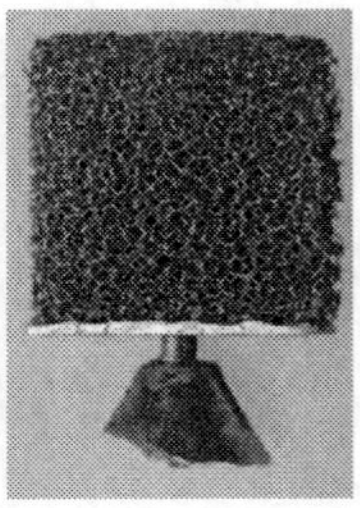

Fig. 3. Casting of metal sponge

2. Conclusion

Thanks adjusting the individual parameters of the investment casting technology, the production process of cast metal sponges from aluminum alloy has been optimized. Final casting is shown in the Fig. 3.

Acknowledgments
This work was elaborated within the research project TH02020668 supported by the Technology Agency of the Czech Republic and projects of "Student Grant Competition" numbers SP2018/77 and SP2018/60.

References

[1] Lichý, P., Bednářová, V. (2012) Casting routes for porous metals production, *Archives of foundry engineering*, vol. 12, no. 1, pp. 71-74.
[2] Banhart, J. (2001) Manufacture, characterisation and application of cellular metals and metal foams. *Progress in Materials Science*, vol. 46, no. 6, pp. 559–632.
[3] Kroupová, I., Bednářová, V., Elbel, T., Radkovský, F. 2014). Proposal of method of removal of mould material from the fine structure of metallic foams used as filters, *Archives of Metallurgy and Materials*, vol. 59, No. 2, pp. 727-730. ISSN 1733-3490.
[4] Ashby, M.F., et al. (2000) *Metal Foams: A Design Guide*. Massachusetts: Butterworth-Heinemann, 251 p., ISBN: 9781493302918.

The 73rd World Foundry Congress
„Creative Foundry"
23 - 27 September 2018, Krakow, Poland

BIOGRAPHY OF BRONZE. ARCHAEOMETALLURGICAL STUDY ON THE CASTING TECHNOLOGY OF THE LUSATIAN CULTURE COMMUNITIES IN GREATER POLAND. PRESESNTATION OF THE PROJECT

Aldona Garbacz-Klempka [1]*, Patrycja Silska [2], Janusz Kozana [1], Michał Piękoś [1], Mateusz Stróżyk [2]

[1] *AGH University of Science and Technology, Faculty of Foundry Engineering, Historical Layers Research Centre, Reymonta 23 Str., 30-059 Kraków, Poland*
[2] *Archeological Museum in Poznań, Wodna 27 Str.- Pałac Górków, 61-781 Poznań, Poland*
**Corresponding address: agarbacz@agh.edu.pl*

Key words: Archaeometallurgy, X-Ray Spectrometry, Bronze Age, Foundry Engineering, Copper Alloys

The paper presents an outline, main objectives, and methods of the research project prepared by AGH University of Science and Technology in cooperation with the Archaeological Museum in Poznań. The project, relates to the field of archaeometallurgy, and it is aimed at the identification and interpretation of bronze–working practices and strategies adopted by prehistoric communities, the Late Bronze Age and the Early Iron Age (13th–7th cent. BC). The chorological scope of the project covers the area of the historical territories of Greater Poland.

In prehistory, metals, especially the knowledge how to work them, played an extremely important role in the process of developing of complex social hierarchies, economic models and systems of symbolic culture.

Archaeometallurgy is a discipline connecting empirical data with their cultural interpretation and scientific experiments through close cooperation between archaeologists and metal scientists.
The application of analytical scientific methods is of crucial importance in the postulated attempt to determine the origins of traditions, identify the networks of exchange of ideas related to metal–working, dynamics of changes and technology transfer, as well as durability and innovativeness of solutions used in metallurgical and casting production. [1, 2]

The beginnings of reflection on the past bronze metallurgy in the Polish lands date back to 1934 when the excavations at Biskupin had started (Fig.1). This site became a real training ground for interdisciplinary research for several generations of scholars specialised in prehistoric bronze–working [3]. Their investigations became a starting point for a complex analysis of metallurgical workshops and practices recorded at the Bronze Age settlement sites.

Fig. 1. Fortified settlement at Biskupin during excavations in the years 1934-1939. Within the settlement site, casters workshops were discovered (photo from the collection of the Archaeological Museum in Poznań)

The increasing availability of analytical methods, which can be observed in recent years, has led to the dynamic rise of the analyses of single finds and small assemblages of artefacts. They have been carried out according to various standards and still not fully formulated methods. Such a situation considerably hinders and often even makes impossible the comparison of the obtained results, particularly due to a rapid increase in their number and a large dispersion of the published data. The source material, including cast artefacts ranging from simple tools to sophisticated ornaments, and the remains of technological processes such as casting moulds (Fig. 2) and crucibles, is rarely

examined in terms of material and technology. Therefore, multifaceted and trans–disciplinary archaeometallurgical studies carried out at a regional level have become a necessity.
Today, a wide range of analytical techniques and material studies based on the material obtained through experimental casting enables not only the detailed reconstruction of a technological process itself, but also the analysis of its variability through the prism of skills, strategies and bronze–working traditions. [4-5]
The use of empirical data in connection with the archaeological interpretation is aimed to provide the fullest possible picture of prehistoric metallurgy in the social and cultural contexts.

Fig. 2. A valve of a stone mould for casting bronze axes, from a caster's grave discovered in Karzec.

Methodology of investigations

Within the project, it is planned to carry out the following analyses:
– the analysis of the development of casting in the context of the product quality; it will consist in examining metal artefacts as items made of novel, innovative materials, along with determining their potential utilitarian properties, their chemical composition and function;
– the analysis of variability and state of advancement of casting technologies observed through the prism of applied metalworking treatments, which are visible in the construction and the degree of refinement of casting moulds, including the construction of the gating system;
– the analysis of the possibility of reconstructing bronze-working practices by examining semi-finished products, slag, and model alloys recreated in laboratory conditions similar to those used at the time;
– the analysis of the possibility of reconstructing bronze-working techniques with the use of tools and digital methods, including computer aided methods of technological processes.
The proposed trans-disciplinary approach to the studies on prehistoric metallurgical and casting practices involves the investigation of metalworking through a multifaceted analysis of its remains: ores and semi–finished metal products, tools and equipment including furnaces, crucibles, casting spoons and moulds, charcoal fuel, production waste such as slag, remains of gating systems, and defective items intended for re–melting.
The source material will undergo quantitative and qualitative chemical and microstructural analyses. The examined objects will be characterised in terms of their construction, structure, and chemical composition. The investigations will include, among others: metallographic macro– and microscopic analyses using optical microscopy (OM) and scanning electron microscopy (SEM), digital X–ray imaging (RT), the analysis of chemical composition with the methods of energy dispersive X–ray spectroscopy (EDS), X–ray fluorescence (XRF), and X–ray diffraction (XRD). In addition, for selected categories of artefacts experimental castings will be made and modern digital methods will be applied to reconstruct metal manufacturing methods used in the Late Bronze Age.
A detailed analysis will involve the assemblages of finds from selected settlement contexts, including open settlements and strongholds of the Biskupin type for which local metallurgical production was confirmed, as well as sepulchral finds corresponding with them chronologically and chorologically, with particular regard to graves containing the elements of equipment of the coppersmith's workshop, and deposits of bronze artefacts and raw material. Only such analytical approach permits further interpretation of the obtained results within individual functional or typological groups of bronze artefacts.

Acknowledgments

The financial support of the National Science Centre, Poland under the grant numbers 2017/26/E/HS3/00656 is acknowledged.

References

[1] Ottaway B.S. (2001). Innovation, production, and specialization in early prehistoric copper metallurgy. European Journal of Archaeology. 4, 87–112.
[2] Pearce M. (2016). Archaeology and archaeometallurgy: some unresolved areas in the interpretation of analytical data, STAR: Science&Technology of Archaeological Research. 2 (1), 46–53.
[3] Piaskowski J. (1957) Technika odlewnictwa w grodzie kultury łużyckiej w Biskupinie, Kraków.
[4] Garbacz-Klempka, A., Suchy, J. S. , Kwak, Z., Długosz P. & Stolarczyk T. (2018) Casting technology experiment and computer modeling of ornaments from Bronze Age. Archives of Metallurgy and Materials. 63(2).

The 73rd World Foundry Congress
„Creative Foundry"
23 - 27 September 2018, Krakow, Poland

THE USE OF SYNTHETIC SLAGS IN POLISH FOUNDRY METALODLEW IN CRACOW

Michał Balicki[1]*, Barbara Kalandyk[2], Sebastian Sobula[2]

[1] *Metalodlew SA, Ujastek 1 Str., 31-752 Kraków, Poland*
[2]*AGH-University of Since and Technology, Faculty of Foundry Engineering, Reymonta 23 Str., 30-059 Kraków, Poland*
* *Corresponding address: balickimich@gmail.com*

Key words: synthetic slag, secondary metallurgy, clean steel

1. Introduction

In Polish steel foundries, for increasing the purity are mainly used processes as: steel argon refining in the casting ladle, the treatment with a core wire containing deoxidizers and modifiers, e.g. calcium, treatment with synthetic slags [1,2]. The advantage of these methods is the ease of adaptation of existing equipment and positive results achieved at low economic costs.

In this research improvement of the purity of steel by means of argon stirring and the use of synthetic slags, is described.

2. Basics of secondary metallurgy

Steel refining with synthetic slags is carried out mainly for its:

- desulphurisation,
- removal of non-metallic inclusions,
- reduction of oxygen content,
- steel insulation from the atmosphere,
- improvement of mechanical and plastic properties of steel.

The basic components of the slag are typically Al_2O_3 and CaO.

Typically, materials for the formation of refining slag are added during the tapping, at the secondary metallurgy station. Large development of the metal slag division is supported by the process of argon steeling [3].

The effectiveness of steel refining with synthetic slags depends primarily on the physicochemical properties of slags, such as: chemical composition, viscosity, density, melting temperature and surface tension [4].

Low sulphur content in liquid steel is possible to obtain only with low oxygen activity. Desulfurization is therefore a process that improves the purity of steel by reducing the content of both sulfur and oxygen in it.

3. Results

In this research, steel was melted in 8,5Mg **EAF** and after tapping the argon refinement in the ladle were carried out. During argon treatment, synthetic slags presented in Table 1 were used.

In Figs. 1 and 2, the chemical composition and morphology of non-metallic inclusions is presented.

Table 1. Chemical composition of the commercial synthetic slags used in Polish foundry, wt%

Slag	Al_2O_3	CaO	SiO_2	MgO	Fe_2O_3	TiO_2 / Al
LDSF LT[1]	**46-50**	**47-53**	**<1.5**	**<2**	**<0.8**	**<0.2**
OPTIMET C[1]	**31-39**	**35-40**	**<8**	**11-15**	**<4**	**<4**
UZRA KP[2]	**41.7**	**42.3**	**3.9**	**4.6**	**1.3**	-/-
LDSF RG[1]	**>39.0**	**>48.0**	**<5.5**	**<3.0**	**<3.0**	**<4.0/-**
UZRA KP +Al[2]	**41.7**	**42.3**	**3.9**	**4.6**	**1.3**	**-/1.5**

-[1]The chemical composition according to Kerneos certificate
-[2] The chemical composition according to ZM Ropczyce certificate

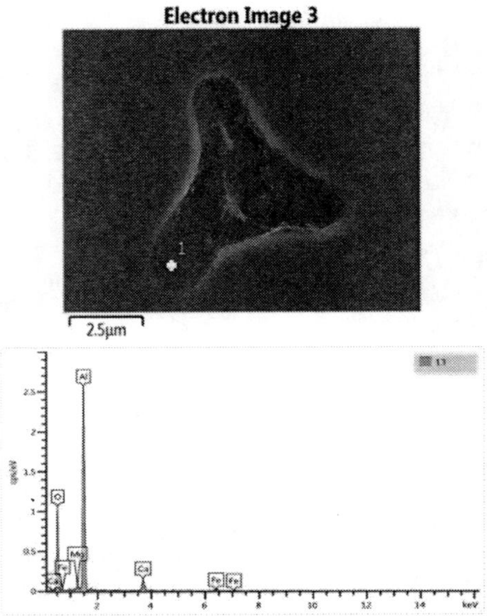

Fig. 1. SEM image of nonmetallic inclusion, and its EDS spectrum

It is shown that main component of this inclusion is aluminum oxide witch also contains calcium.

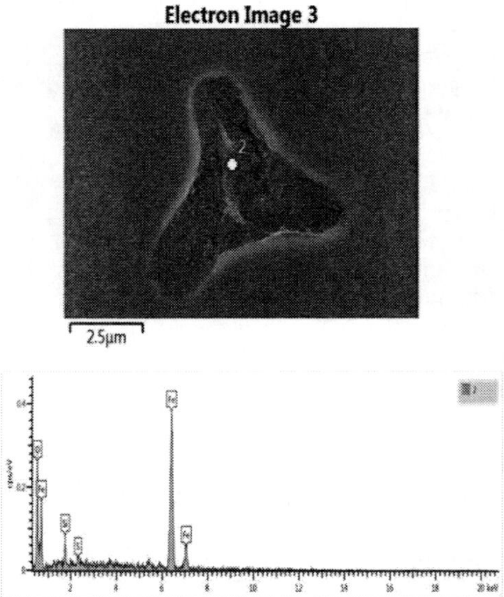

Fig. 2. SEM image of nonmetallic inclusion, and its EDS spectrum

It is shown that main component of this inclusion is silicon oxide which also contains sulphur.

3. Conclusions

-After the tests, we can see the beneficial effect of synthetic slags on lowering the S content in molten cast steel just before the pouring process.
-The best slag used in the tests provide good desulphurization rate even in case of low sulphur content in steel before ledle treatment.
-Secondary treatment reduces the number of non-metallic inclusions by several percent.
- Small differences in the N content during the smelting process of cast steel in the Metalodlew SA conditions confirms the fact that N and H can be effectively eliminated using vacuum;

References

[1] T. Lis, 2009 – Metalurgia stali o wysokiej czystości; Wydawnictwo Politechniki Śląskiej, Gliwice.
[2] W.Terda, 2009, Wapń I jego związki w metalurgii żelaza.
[3] K. Janiszewski, 2012, Rafinacja ciekłej stali z wtrąceń niemetalicznych metodą filtracji, Katowice.
[4] M. Balicki, S. Sobula, B. Kalandyk 2015 Wpływ żużli syntetycznych na proces odsiarczania niskostopowego staliwa Cr-Ni-Mo; Archives of Foundry Engineering Volume 15, 11-14.

The 73rd World Foundry Congress
„Creative Foundry"
23 - 27 September 2018, Krakow, Poland

TWO INOCULATION METHODS FOR REFINING AS-CAST GRAIN STRUCTURE IN AUSTENITIC 316L STEEL

Dustin A. Arvola, Simon N. Lekakh*, Ronald J. O'Malley, Laura N. Bartlett

Missouri University of Science and Technology
1400 N Bishop, Rolla, MO, U.S.A., 65409
**Corresponding address: lekakhs@mst.edu*

Key words: solidification, stainless steel, grain refinement, heterogeneous nucleation, titanium nitride

1. Introduction

Grain refinement of the as-cast structure results in a casting that has higher strength, more isotropic properties, less segregation and porosity, better feeding, and a higher resistance to hot tearing. Inoculation techniques to refine the solidification structure are crucial for austenitic steel castings because significant cast structure modification cannot be achieved by heat treatment or mechanical working of near net shaped components. Inoculation introduces or promotes the formation of "foreign" heterogeneous nuclei prior to the beginning of solidification. These heterogeneous nuclei must: (i) be stable at steelmaking temperatures, (ii) be well dispersed throughout the melt, (iii) be active during austenite solidification (i.e. have lattice similarity, minimum interfaces energy), and (iv) have an appropriate size and shape that promotes nucleation. *TiN* precipitates are effective nuclei for *δ-Fe* grain refinement [1]; however, they are less potent at nucleating *γ-Fe* [2]. A larger lattice disregistry implies an increased difficulty to nucleate austenite phase in comparison to ferrite phase [3].

In this article, two methods grain for refinement of cast austenitic 316L stainless steel are investigated. The first method is based on the in-situ formation of *TiN* by co-precipitation onto preexisting spinel inclusions in the melt [4]. The second method is based on a newly developed master alloy addition that contains preformed *TiN* nuclei.

2. Methodology

The first method (Heat #1) was designed to target the *in-situ* formation of *TiN* precipitates in the melt as potential austenite nucleation sites. *Mg-Al* spinel is formed within the melt first, followed by the co-precipitation of *TiN* onto these spinel inclusions. The advantages of this co-precipitation method include initial deoxidation of the melt when forming spinel which reduces the potential to form undesired titanium oxides and the growth of *TiN* is accelerated in the presence of spinel surfaces thus minimizing the required chemical supersaturation of *Ti-N* to form *TiN* nuclei [4]. To design this complex melt treatment of 316L stainless steel (17%*Cr*, 12%*Ni*), thermodynamic calculations were performed using FactSage 7.0 software (Fig. 1).

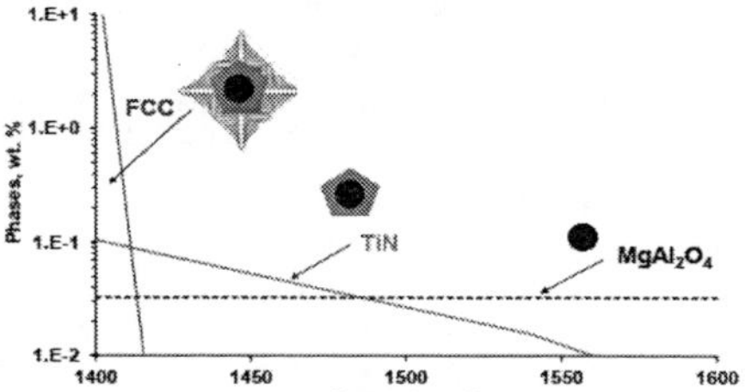

Fig. 1. Simulated in-situ formation of complex heterogeneous nuclei by Mg-Al additions into the melt followed by Ti addition (Heat #1).

The second method is based on a newly developed *Fe-Cr-Ni-Al-Ti-N* master alloy addition that contains preformed *TiN* nuclei. Results from a FactSage simulation of *TiN* stability after the addition of master alloy to the ladle were used to design Heat #2 (Fig. 2).

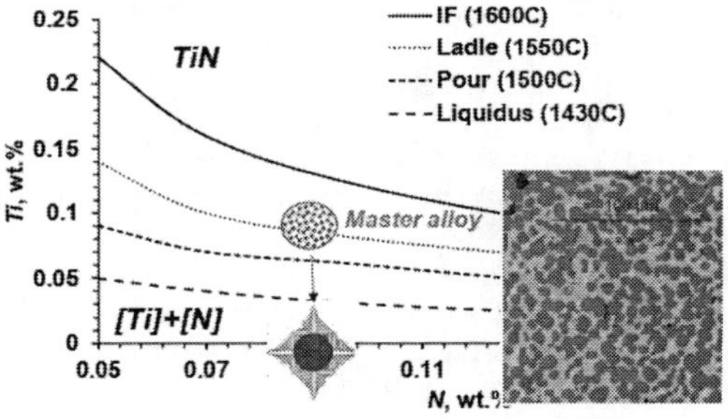

Fig. 2. TiN stability diagram in the melt at different temperatures was used to predict the survivability of nuclei introduced by the master alloy in Heat #2. TiN (grey) in master alloy (shown in insert).

A general outline of the casting processes developed for Heat #1 and Heat #2 are shown in Fig. 3.

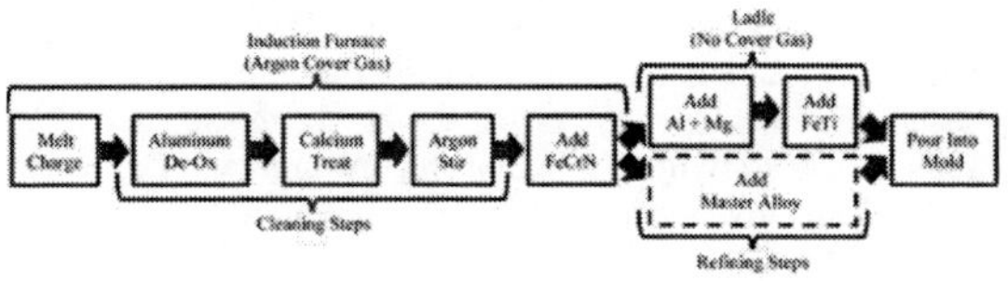

Fig. 3. Designed melt practices for in-situ (Heat #1) and master alloy (Heat #2) grain refining.

Trials were performed using a 100 lb. induction furnace and cylindrical castings (100 mm diameter and 200 mm tall) with a top riser. Heats were poured into no-bake sand molds with a low thermal conductivity side sleeve and a bottom water-cooled *Cu* chill to promote a directional columnar structure (Fig. 4).

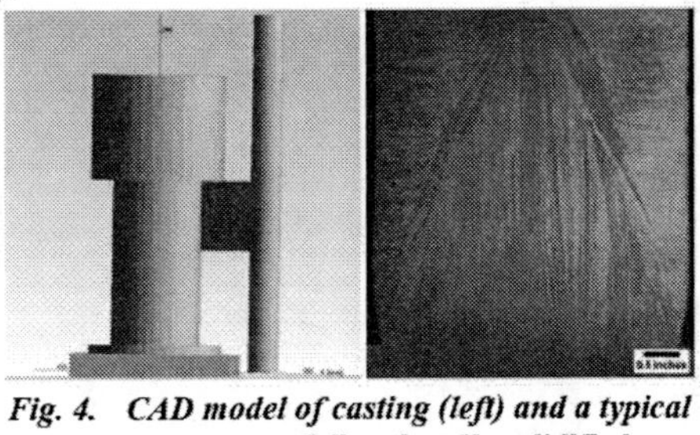

Fig. 4. CAD model of casting (left) and a typical macrostructure of directionally solidified, unmodified 316L austenitic steel (right).

3. Experimental results and discussion

Both inoculation methods achieved refinement of the structure when compared to the base melt that used no grain refining additions. In the *in-situ* method (Heat #1), the columnar zone length extended >25 mm from the chill plate but the equiaxed grain size was small (Fig. 5, left) compared to Heat #2. The master alloy method (Heat #2) also yielded a grain refined structure (Fig. 5, right), but it had a coarser grain size and a shorter columnar zone length (12 mm) than Heat #1.

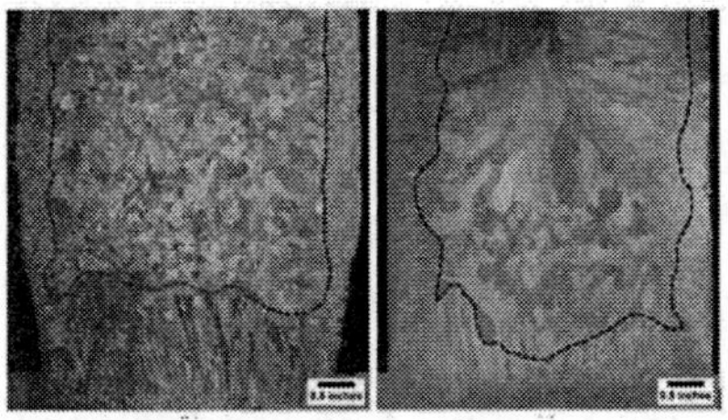

Fig. 5 Optical image of etched macrostructure in vertical section for Heat #1 (left) and Heat #2 (right).

The most common type of precipitate morphology observed in Heat #1 was a *TiN* layer co-precipitated on a spinel core (Fig. 6, left). These complex precipitates had a tendency to form clusters. A variety of *TiN* morphologies were observed in the casting from Heat #2: (i) round mono-phase *TiN* precipitates originating from the master alloy, (ii) fragmented faceted mono-phase *TiN*, and (iii) newly developed, faceted *TiN* with alumina cores (Fig. 6, right). An automated SEM/EDX analysis [5] was used to classify and measure precipitates which could serve as the heterogeneous nucleation sites (Fig. 7). For the *in-situ* method, the main inclusion classes are shown on the ternary diagram along the *Ti-N* binary line for *TiN* and within the *Ti-Al-Mg* ternary for spinel. There were no spinel inclusions in Heat #2 using the master alloy method and the main precipitates were mono-phase *TiN* particles.

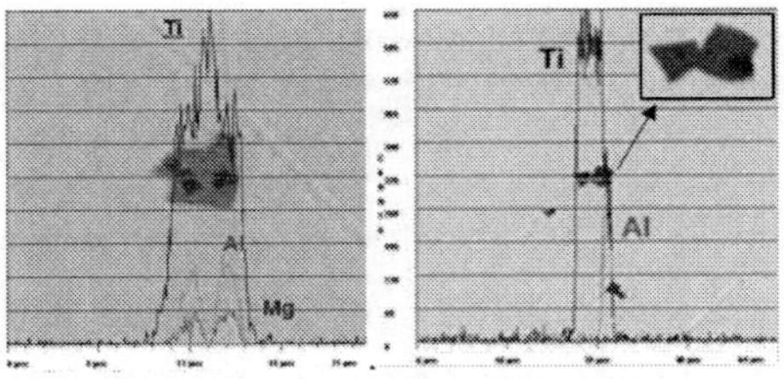

Fig. 6. Line scan of complex TiN inclusions: co-precipitated on Mg-Al-spinel core (left, Heat #1) and newly developed, faceted TiN with alumina cores (right, Heat #2).

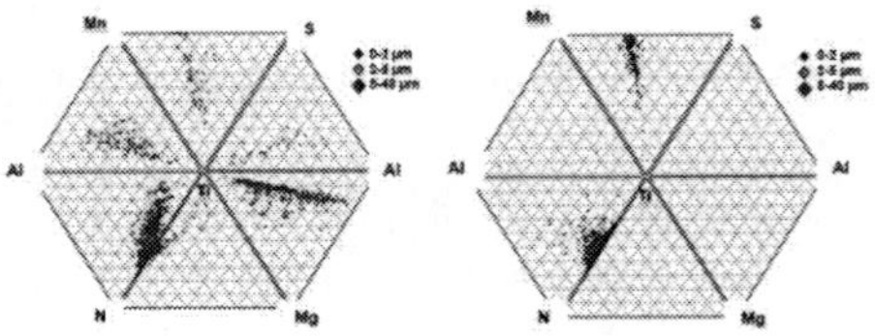

Fig. 7. Ternary diagram for precipitates in Heat #1 (left) and Heat #2 (Right).

Electrolytic extraction was utilized to observe the 3D structure of the *TiN* inclusions. The effect of the *TiN* surface topology on nucleation activity is discussed (Fig. 8).

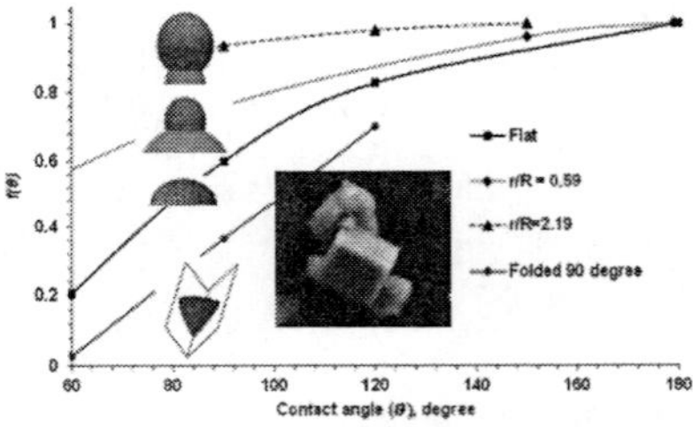

Fig. 8. Shape function for several heterogeneous nuclei shapes.

4. Conclusion

Two grain refining methods that introduced titanium nitride (*TiN*) into 316L melt by *in-situ* formation and by master alloy addition were designed, performed, compared and discussed. The master alloy method extended the equiaxed zone and improved the distribution of *TiN* in the casting (less clustering). However, this method showed less effective grain size refinement when compared to the *in-situ* method.

Acknowledgements

The authors would like to thank the industrial members of the PSMRC for their support for this research.

References

[1] J.S. Park et al, Steel Res. Int, 2014, 85(8), 1303.
[2] S.N. Lekakh et al, ISIJ Int., 2017, 57 (5), 824 .
[3] Bramfitt, Metall. Trans., 1970, 1(7), 1987.
[4] S. Lekakh et al, Metall. Mater. Trans. B, 2017, V 38, 406.
[5] M. Harris et al, AISTech Proceedings, 2015, 3315.

The 73rd World Foundry Congress
„Creative Foundry"
23 - 27 September 2018, Krakow, Poland

Si-Co AND Si-Zr ALLOYS/C-MATERIAL INTERFACES: WETTING VERSUS INFILTRATION

Donatella Giuranno[1,2*], Rada Novakovic[2], Natalia Sobczak[1,3], Rafal Nowak[1], Grzegorz Bruzda[1], Javier Narciso[4]

[1] Foundry Research Institute, Zakopiańska 73 Str., 30-418 Krakow, Poland
[2] National Research Council of Italy-Institute of Condensed Matter Chemistry and Technologies for Energy, Via De Marini 6, 16149 Genova, Italy
[3] Institute of Precision Mechanics, Duchnicka 6, 01-796 Warsaw, Poland
[4] Universidad de Alicante-Instituto Universitario de Materiales de Alicante, Apdo. 99, 03080 Alicante, Spain
**Corresponding address: donatella.giuranno@ge.icmate.cnr.it*

Key words: Si-Co/C; Si-Zr/C; Reactive infiltration; Wetting characteristics; Combined sessile/pendant drop

1. Introduction

In the manufacture of competitive composite materials through reactive infiltration, a great controversy related on the controlling stage of the infiltration process still exists. This controversy is based on the real driven force controlling the infiltration, is it a governed by Darcy's law or by the chemical reaction? To clarify this open issue, a key contribution could be coming from know-how on wetting characteristics, reactivity, surface properties and thermodynamics concerning the reacting phases. The works carried out jointly between N. Eustathopoulos and J. Narciso on different infiltrating systems, i.e. Ni-Si and Al-Si, in C-porous based materials, confirmed that reactivity at the interfaces, is the real governing phenomenon [1].

To further support this conclusion, systematic investigations of wetting characteristics, reactivity, surface properties and thermodynamics concerning Si-Me/C-based substrate couples (Me = Co, Zr) have been carried out and the more relevant results obtained are reported and in this short paper.

The outcomes here presented are doubly profitable: to study reactive infiltration process per se and to make progress in materials design of SiC-based composites.

2. Materials and methods

For wetting and infiltration tests, pure Si, as reference element, and three different Si-Me alloy compositions (Si-rich side) have been selected: Si-Me eutectics and two hypereutectic alloys. The alloys have been produced by mixing high pure Si and Me (3-4N GoodFellow©) and melted by arc-melting. Each as-produced alloy microstructure/composition have been checked by SEM/EDS analyses. As wetting substrate, glassy-C (**C_g**-Alfa Aesar©), with a roughness ≈20nm, has been chosen. A C-porous graphite preform (**G_p**: porosity ≈90%; average porous radius ≈3.0μ, purity 99.7 %) has been selected as substrate for infiltration experiments.

All the aforementioned properties and characteristics have been carefully investigated as a function of operating conditions and composition. Contact heating sessile [2] and dispensed drop procedures [3] have been applied and experiments performed in the temperature range of (T_m+20)-1500°C. Interfacial phenomena and resulting microstructures have been carefully analyzed in terms of thickness and composition of the reaction layer, by OM, SEM-EDS and 3D-tomography techniques. Moreover, to detect the presence of ZrC as additional reaction product, Raman and XRD techniques have been used. In parallel, surface and bulk properties outcomes, i.e. surface tension and density data, have been compared with the predicted values obtained by applying Ideal solution, Quasi-Chemical Approximation for regular solution models [4].

3. Results and discussion

In Fig. 1, the evolution of wetting behavior of Si-27wt%Zr eutectics/C_g couple, within a contact heating experiment performed at T = 1350°C under an Ar (N60) static atmosphere, is shown. Spreading kinetics is also reported in terms of drop radius. As can be seen, four different spreading stages have been observed. In particular, the stage (t_o-t_1) is typical of wetting experiments performed close to the melting point (T_m Si-27Zr ≈ 1370°C [5]) by contact heating sessile drop method. After the melting (detected at T ≈ 1350°C), the presence of SiO_2/ZrO_2-primary oxides at the surface delays the complete melting of the drop.

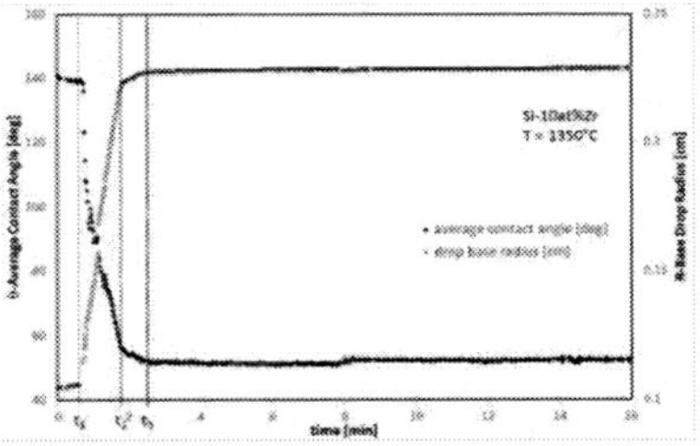

Fig. 1. Contact angle (θ) and drop base radius (R) as a function of time of Si-27%Zr/ C_g couple at T ≈ 1350°C.

Moreover, the oxide skin initially prevents the intimate contact between the melt phase and the substrate. In the 2nd spreading stage (t_1-t_2), several changes in the dR/dt are evident: two possible reasons are 1) the progressive dissolution of the SiO_2 wetting barrier by liquid Si [1] and 2) inhomogeneus melting of the alloy drop. The general behaviour observed in this stage suggests that Si-27Zr/C_g wettability is governed by the reaction at the triple line [1]. Within the 3rd stage of spreading (t_2-t_3), a linear dependency is again observed: reactivity is still the dominant mechanism. The drastic change of the slope suggests that "competitive" phenomena are evolving, strongly influencing the advancing contact angle. In Table 1, the equilibrium contact angle values measured at the end of wetting tests performed by contact heating method on Si-Me eutectics/C_g couples are reported.

Table 1. Final contact angle values (θ_f) as a function of temperature for Si-Me eutectic alloys (Me = Co, Zr)/C_g

Alloy [wt%]/C_g	T [°C]	θ_f	Ref.
Si-37.5Co	1320	36	[2]
Si-37.5Co	1390	38	[2]
Si-37.5Co	1450	<15	[2]
Si-27Zr	1350	52	This work
Si-27Zr	1450	45	This work
Si-27Zr	1500	43	This work

A strong T-dependency of contact angle value has been observed, as expected. The higher values of contact angle of Si-Zr/C_g system might be correlated to the higher surface tension value of Si-Zr liquid metal phase. As main outcome of the theoretical analysis on surface tension behavior as a function of Si-content reported in [4], the Si-Me eutectic alloys in the Si-rich phase are well described by ideal/regular solution models. In addition, even if not detected by XRD and EDS analyses, the presence of ZrC, thermodynamically favoured as reaction product, might be the reason of a different resulting roughness/microstructure at the Si-Zr/C_g interface.

In Fig. 2, the kinetics of infiltration of Si-37.5Co alloy/G_p as a function of temperature is shown.

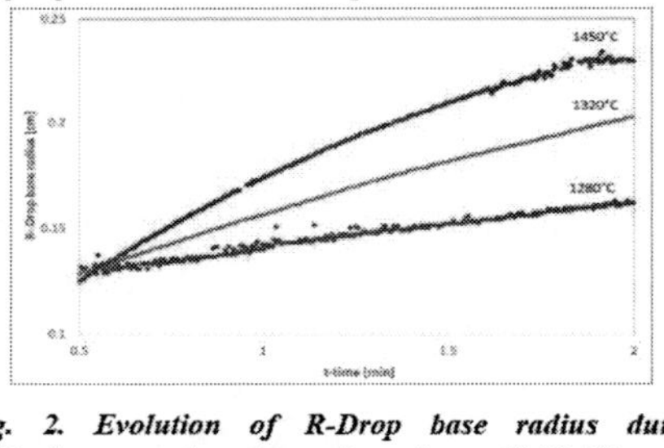

Fig. 2. Evolution of R-Drop base radius during infiltration experiments performed on Si-37.5Co as a function of temperature under a vacuum of $P_{tot} = 10^{-2}$ Pa.

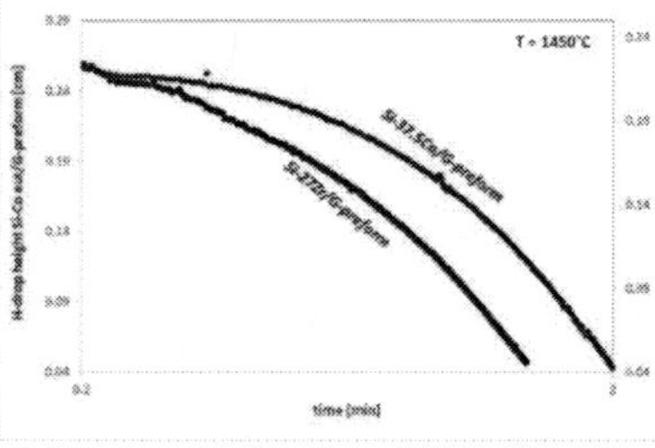

Fig. 3. Kinetics of infiltration of Si-Me/G_p at T = 1450°C and under a vacuum of $P_{tot} = 10^{-2}$ Pa.

A strong T-dependency of the R(t) and dR/dt has been observed, typically related to reactive mechanisms.

In Fig. 3, the evolution of Si-Me infiltrating eutectic melt phases (H-height of the drop vs time), i.e. Si-37.5Co and Si-27Zr, into porous graphite (T=1450°C) is shown. With different infiltration kinetics, both Si-Me alloys spontaneously infiltrate porous graphite. Depending on the operative conditions, the alloying element (Me) might give competitive equilibria making kinetically favoured both SiC and Me-carbides (ZrC). The alloying element may actively influence the capillarity and fluido-dynamics during infiltration owing to its own thermophysical properties (i.e. surface tension, viscosity, etc.).

4. Conclusions and works in progress

Wetting and infiltration experiments have been performed as a function of operating conditions (T, P_{tot}, etc.) and composition by contact heating.

The results obtained further confirm that reactivity at Si-Me/C-preform interfaces is the main driven force controlling infiltration process. In particular, both wettability and infiltration processes are strongly dependent on temperature, Si-content and alloying element. In order to analyze properly the wetting and infiltration kinetics under well-defined operating conditions, experiments will be performed also by dispensed drop method. This method will allow to produce fresh melt surfaces under oxides-free conditions. In addition, thermodynamic studies on Si-Me-C and Raman technique will be used to well distinguish among the produced carbides phases (i.e. SiC and Me-carbides).

Acknowledgments

This work was supported by *National Science Center* of Poland in the framework of "**SIC-WIN**: **Reactivity of SI-alloy/C-material system: Wetting vs Infiltration**" project (G.A. UMO-2016/23/P/ST8/01916), POLONEZ-3, financed under the EU Research and innovation funding programme "Horizon 2020" G.A. no 665778 for the Marie Skłodowska-Curie actions. DG wishes to thank *CNR* for funding part of the research here presented by STM 2016 call (prot. n. AMMCNT-0059260 25/01/2017).

References

[1] B. Drevet, N. Eustathopouls. (2012). Wetting of ceramics by molten silicon and silicon alloys: a review. J Mater Sci. DOI 10.1007/s10853-012-6663-0.

[2] M. Caccia, S. Amore, D. Giuranno, R. Novakovic, E. Ricci, J. Narciso. (2015). Towards optimization of SiC/$CoSi_2$ composite material manufacture via reactive infiltration: Wetting study of Si–Co alloys on carbon materials. J Eur Ceram Soc. 35 (15), 4099-4106.

[3] N. Sobczak, R. Nowak, W. Radziwill, J. Budzioch, A. Glenz. (2008). Experimental complex for investigations of high temperature capillarity phenomena. Mater. Sci. Eng. A 495, 43–49.

[4] R. Novakovic, D. Giuranno, M. Caccia, S. Amore, R. Nowak, N. Sobczak, J. Narciso, E. Ricci. (2016). Thermodynamic, surface and structural properties of liquid Co-Si alloys. J Mol Liq. 221, 346–353.

[5] T.B. Massalski.. Binary alloy phase diagrams. (1986). vols. 1, 2. Metals Park, OH, ASM.

The 73rd World Foundry Congress
„Creative Foundry"
23 - 27 September 2018, Krakow, Poland

The role of recycled ceramic material obtained from the ceramic layered moulds used in the investment casting

A. Soroczyński [1]*, R. Haratym [2], R. Biernacki [1]

Department of Plastic Forming and Foundry Engineering, Warsaw University of Technology, ul. Narbutta 85, 02-524 Warszawa, Poland
[b] University of Ecology and Management, ul. Olszewska 12, 00-792 Warszawa, Poland
**Corresponding author. E-mail address: asoroczy@wip.pw.edu.pl*

Key words: recycling, ceramic mould, investment casting, energy consumption, (use style Key words)

1. Introduction

The article presents the role of the ceramic layered moulds used in the investment casting method with new (certified) and recycled material from ceramic moulds after casting process. The materials that were obtained are mainly aluminosilicates and SiO2. The investigation of changes in the quality of ceramic moulds (including the recycled ceramic material) includes the chemical composition of the ceramics as recovered ceramic material, changes in the particle size of the layered covering material, the gas permeability (CM) during the pouring of liquid metal, and the creation of the porosity are presented. Than the thermophysical parameters and dimensional accuracy of the casting manufactured in the new ceramic layered shell moulds were analysed. Additionally the global cost savings and improved ecological conditions in the foundry and its surroundings was estimated.

The ceramic layered shell mould in the investment casting process have a major influence on the quality of manufactured castings [1]. The dimensional accuracy and the surface quality are the primary objective in this manufacturing technology. Ceramic moulds dimensional distortions are the reason of approximately 60% of overall errors of the casting manufactured mainly with the use of aluminosilicates [2,3]. Based on the quality of the castings the amount of the ceramic moulds with aluminosilicates has increased at least 40% of the global investment casting production. At the same time we need to pay attention to the SiO2. In the sustainable production very important are the environmental aspects affecting human health.

2. Research methodology

The laboratory trials include the special dilatometric analysis of the test samples to achieve the thermal expansion parameter of the ceramic mould samples with the dimensions of ⌀20x60 [4]. The Olympus stereo microscope was used for the evaluation of the ceramic grain size. Additionally the laser measurement of the changes in the particle size and the porosity using mercury porosimetry was checked at random ceramic moulds from the WSK Rzeszow company.

The thermal conductivity was evaluated based on the cylindrical ceramic samples.

3. Verification of the chemical composition of the samples of ceramic moulds

The changes in the amount of SiO_2 in the aluminosilicates of recycled material were found. The amount of SiO_2 increased from 22% to 25,7% due the transition of the certain amount of the binder to the crystalline state. The total content of the Al2O3 + SiO2 has slightly changed and in practice that has no effect on the thermal expansion of the ceramic mould.

4. The evaluation of the ceramic mould porosity

The use of the mercury porosimetry based on the capillarity effect shows very low porosity in the range of 26-28%. The samples of recycled ceramic mould material shows the porosity at similar range 26,5 – 29%.

5. Verification of the particle size of the clean and recycled ceramic moulds material

A new material with the grain diameter of ~ 0,9 mm and a recycled material recovered in the laboratory device using the hammer crushing method, there was a significant fragmentation of the material to the average of 0,5mm grain size. When the recycling process of the ceramic mould was conducted in the laboratory (150mm diameter rollers, the gap between the rolls of 1.2 mm), there was no difference between the grains of a new and recycled material. The rolling mill was used in the Warsaw University of Technology.

For the determination of the particle shape the method commonly used in the powder sintered technology was partially adopted [5]. The shape of the grains was presented by the value of Wk later called the size ratio. Wk value is defined as the ratio of the circumference of grain to the circle inscribed in the outline of the grain. For the evaluation of the Wk value a special software available from the Olympus microscope for the image analysis of each particle was.

6. Gas permeability of ceramic molds

The evaluation of the possibility of multiple recycling of the ceramic material needs a set of further laboratory trials regarding the permeability (gas permeability). The verification was carried out on the laboratory stand presented on the fig. 1. The test was commenced in the 20°C temperature.

7. The impact of the recycled ceramic moulds based on the aluminosilicate on the economic and ecologic aspects

The dorm production delivers approximately 800 Mg of precision casting, which results the amount of over 1000 Mg used ceramic and the amount of used aluminosilicate may be up to 400 Mg. By using the recycled ceramic materials the reduction of the energy consumption used in manufacturing 1kg of ceramic mould by approximately 5MJ (research of the Warsaw University of Technology). Based on the assumption that to produce 1 kWh, the thermal contribution is about 9MJ [9,10]. With the estimated value of 40% used recycled aluminosilicate which is about 160 Mg that could save approximately 88000 kWh of energy. We need to remember that with the primary energy carrier which is the bituminous and lignite coal, the creation of 1kWh falls the emissions of approximately 800g of CO2, SO2 and approximately 7g 3g NOx [9]. That avoids a significant environmental pollution. That ecological damage have influence on the health. The production of 1kWh is giving the 4USD/1kWh of penalty and for the 88000kWh is near 350000USD. Additionally coal energy have great impact on the global warming which is counted to be even 88000 KWh x 12$USD_{avg.}$ [10].

8. Conclusions

- Recycling aluminosilicate can significantly reduce the cost of the production in the investment casting foundry,
- The thermophysical properties of ceramic moulds manufactured using recycled materials are comparable to the moulds from the new ceramic materials
- Dimensional accuracy and the surface quality acquired by using new and recycled materials is similar and ensures the production of high quality castings

References

[1] Haratym R., Biernacki R., Myszka D. (2009). *Ecological production of accurate castings in ceramic molds.* Warszawa: Oficyna Wydawnicza Politechniki Warszawskiej.

[2] Snow, D.H. Scott: (2001). *Comparing Fused Silica and Alumino – Silicate Investment Refractories.* Incast

[3] Karwiński A.: Technological parameters of the process of making molds of ceramic with the participation of "water" silicate binder, monograph "Polska Metalurgia w latach 2011 – 2014", s. 529 – 541, ISBN 978-83-63663-47-6, Wydawnictwo Naukowe „AKAPIT", Kraków, 2014,

[4] Haratym R., Karwiński A.: The influence of the properties of the ceramic mould on its thermal deformations- X Międzynarodowa Konferencja „Spolupraca 2004" 28 – 30.04.2004 Tatrzańska Łomnica, Słowacja, ISBN 80-8070-231-4

[5] Nowacki J., (2005). *Sintered metals and metal matrix composites.* Wyd. Naukowo-Techniczne, Warszawa

[6] Haratym. R., (2006). *The impact of recycled materials from used ceramic moulds for some of the thermo-physical properties in the process of investment casting* Archives of Foundry 6.20

[7] Karwiński A., Żółkiewicz Z.: The Research of Properties of Experimental Ceramic Layers; Archives of Metallurgy and Materials, Vol. 59, Issue 2/2014, s. 703 – 705

[8] Karwiński A., Leśniewski W.: Reasearch on Investment Casting Liquid Ceramic Slurry – Foundry Trade Journal, Volume 179 No.3625, June 2005, s. 158-159,

[9] Central Statistical Office, Fuel and Energy Economy in 2011-2012, Warsaw 2013

[10] Lewandowski W.M., (2007). Environmentally friendly renewable energy source WNT Warszawa

The 73rd World Foundry Congress
„Creative Foundry"
23 - 27 September 2018, Krakow, Poland

PREDICTION OF SHRINKAGE POROSITY IN DUCTILE CAST IRON TEST CASTINGS

Jakub Hajkowski[1]*, Paweł Popielarski[1], Zenon Ignaszak[1]

[1]*Poznan University of Technology, Institute of Materials Technology, 3 Piotrowo St. 60-965 Poznan, Poland*
**Corresponding address: jakub.hajkowski@put.poznan.pl*

Key words: cast iron, density curve, shrinkage porosity, DTA test, computer simulation

1. Introduction

Casting processes are widely used to produce metal components, wherein the cast iron castings represent more than 72% of world production of castings [1,2]. Designing a new casting technology requires incurring large costs associated with the preparation of instrumentation necessary to perform casting moulds. Therefore, before making the mould it is necessary to optimize the procedure of casting technology in order to obtain castings with a high quality at low production costs. Rational procedure of design optimization of casting technology involves the use of computer systems to support this process - simulation systems (codes) .

The effective application of the simulation codes requires the identification and knowledge of adapted to the reality the physical parameters from database (thermal) of the casting - mould system [3,4]. In the most databases of simulation codes, regarding even the cast iron, the density variation character is preserved. However, it is known that the graphite crystallize as a graphite eutectic component is accompanied by the so-called graphite expansion (graphite volume growth), which must causes density decrease in defined temperature interval for case of the eutectic cast iron. Feeding of the solidifying eutectic in cast irons, in a period preceding crystallization of an eutectic phase, is actually done in the same way as in the case of alloys, that do not create this expanding phase at all (cast steel and non-ferrous alloys). Analysis of a process of solidification of the cast irons, on a further stage, requires consideration of compensative eutectic graphite expansion and actual local flows in the balance of need for the liquid metal [5].

2. Methodology

The series of experimental studies were performed in two foundries, in Poland and France. Experimental studies were carried out using simple shape castings i.e. cylindrical and plate made of the same grade ductile cast iron - GJS-400-15. Therefore, it was good way to test repeatability of the shrinkage-expansion phenomenon according to metallurgical alloy quality and specificity of particular foundry. The cast iron GJS-400-15 was selected as a difficult material in respect to compensation of matrix shrinkage by the mechanisms of graphite expansion.

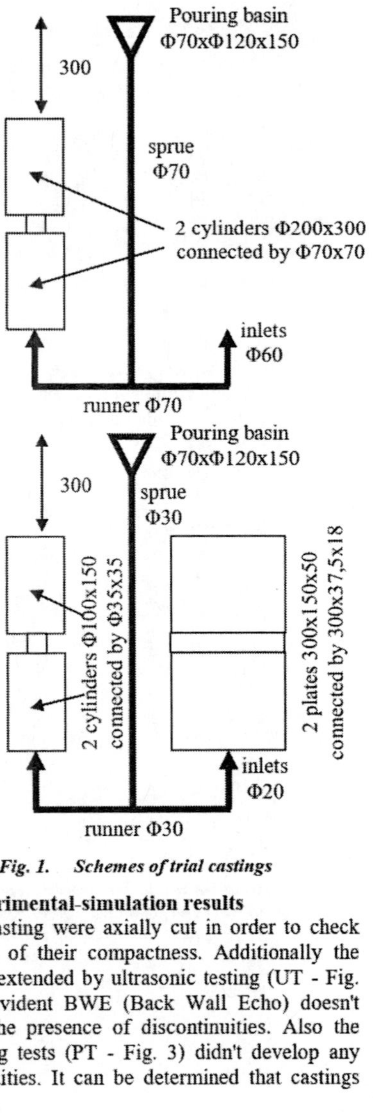

Fig. 1. Schemes of trial castings

3. Experimental-simulation results

The casting were axially cut in order to check the status of their compactness. Additionally the test was extended by ultrasonic testing (UT - Fig. 2). The evident BWE (Back Wall Echo) doesn't indicate the presence of discontinuities. Also the penetrating tests (PT - Fig. 3) didn't develop any discontinuities. It can be determined that castings are sound.

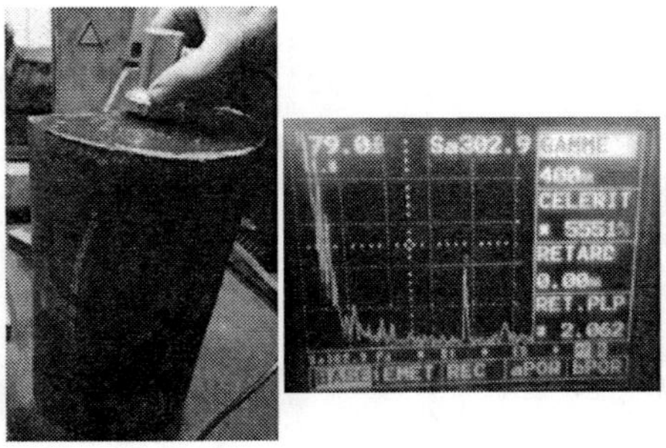

Fig. 2. Ultrasonic testing of upper cylindrical test casting (UT probe 2 MHz, apparatus USN 35)

Fig. 3. Preliminary tests. Results of penetrating testing (PT) [5]

The experimental tests were the starting point to numerical part. The test were elaborated using : NovaFlow&Solid simulation codes. The preliminary results were shown in Fig. 4.

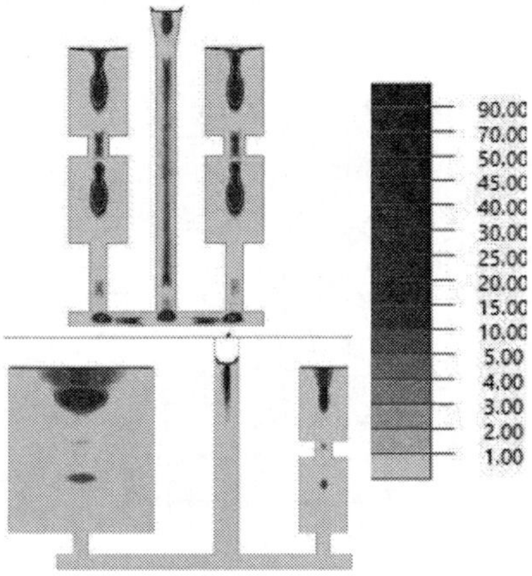

Fig. 4. Simulation results for cylindrical casting (Φ200×300 mm) (top) and combined cylindrical casting (Φ100×150 mm) and plates (300×150×50 mm connected by 300×37.5×18 mm neck) (bottom)

The location and intensity of the shrinkage defects depends not only on variability of alloy density as a function of temperature.

Also Critical Fraction of the Liquid phase (CLF) are condition of flows inside the casting (shrinkage needs and expansion).

Summary

The executed preliminary test let to formulate the following conclusions:

- validation should comprise identification of parameters in a material database used by a given module, especially including critical fractions of mass and capillary feeding, but also, earlier, the rest of thermophysical parameters influencing dynamics of solidification and heat transfer processes in the casting-mould system,
- among these thermophysical parameters, the most important one (especially for cast irons) is variability of alloy density with temperature $\rho=f(T)$, because of a compensating interaction on shrinkage of the graphite matrix expansion.

References

[1] Sobczak J., Balcer E., Kryczek A. (2012). Situation of the foundry engineering in Poland and in the world - the current state and predictions, Przegląd Odlewnictwa 2012, no 1-2.

[2] 48th Census of World Casting Production. (2014) Modern Casting Dec/2014.

[3] Ravi B. (2009). Computer-aided Casting Design and Simulation, STTP, V.N.I.T. Nagpur, July 21, 2009

[4] Jolly M. (2002). Casting simulation: How well do realistic and virtual casting match? State of art. Review. Int. J. Cast Metals Res., 14, 303-313.

[5] Z. Ignaszak (2017), Discussion on Usability of the Niyama Criterion for Porosity Predicting in Cast Iron Castings, Archives of Foundry Engineering, Vol.17(3), pp. 196 – 204.

The 73rd World Foundry Congress
„Creative Foundry"
23 - 27 September 2018, Krakow, Poland

CONVENTIONAL AND HYBRID GASARS: LIGHT-WEIGHT MATERIALS WITH PRESSURE-TEMPERATURE MANAGED POROSITY

Jerzy J. Sobczak[1,*], Natalia Sobczak[1,2], Ludmil Drenchev[3], Pawel Darlak[1], Piotr Długosz[1], Patrycja Turalska[1], Marta Homa[1], Adam Tchórz[1], Izabella Krzak[1], Advenit Makaya[4], Laurent Pambaguian[4]

[1]Foundry Research Institute, 73 Zakopianska Str., 30-418 Krakow, Poland
[2]Institute of Precision Mechanics, Duchnicka 3, Warsaw, Poland
[3]Institute of Metal Science, Bulgarian Academy of Sciences, 1574, Sofia, Shipchenski Prohod Str., 67, Bulgaria
[4]European Space Agency, ESTEC, Keplerlaan 1, 2201 AZ Noordwijk, The Netherlands
**Corresponding address: jerzy.sobczak@iod.krakow.pl*

Key words: high-porosity media, magnesium gasars, gas-eutectic transformation, lotus-like structures, hybrid gasars, pressure – thermal management, computed tomography, mathematical modeling, computer simulation

1. Introduction

The gasars are unique functionally graded materials based on metal-gas system with high porosity, which can be directly oriented. Due to acknowledged and accepted nature of gas pores creation on the front of metal solidification this kind of materials are named either "gasars" (as effect of thermodynamic gas-eutectic transformation [1]) or "lotus-like structures" (as effect of physical gas evacuation from liquid metal being solidified [2]). Effective control of gas porosity is important from both theoretical and practical points of view. Gasars in which an additional solid phase (reinforcement or insert) has been intentionally applied for pressure-temperature management of gas porosity formation were called hybrid gasars.

The paper summarizes the results of trials on the synthesis of conventional and hybrid magnesium gasars fabricated at the Foundry Research Institute using the original set-up [3]. Attempts have been made to control the process of porosity formation by introducing a foreign solid phase into the liquid metal. Mathematical modeling and initial computer simulation were implemented to clarify the phenomenon of porous structure formation. Non-destructive computed tomography method was extensively used for the complex evaluation of the structure of Mg gasars ingots produced.

2. Experimental details

Special experimental unit with improved design of crystallizer (Fig. 1) was used for Mg gasar trials [3].

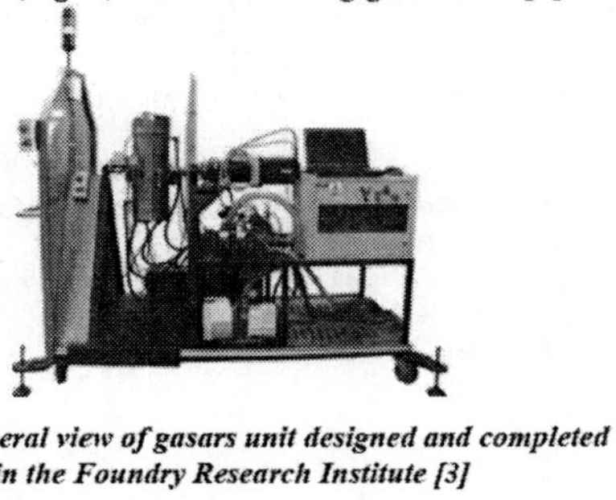

Fig. 1. General view of gasars unit designed and completed in the Foundry Research Institute [3]

Contrary to previous reports (e.g. [4]), it was decided to eliminate the operationally hazardous fact of using pure hydrogen and in experiments, mixtures of Ar with 35 and 50 vol.% H_2 were used. Pure magnesium as a metal matrix and steel wool as an insert were selected. Computed X-ray tomograph NANOTOM (resolution 0.1 μm) was used for non-destructive structural characterization by 2D- and 3D- imaging as well as qualitative and quantitative analysis.

3. Results and discussion

Computer-assisted process management was applied for gasar synthesis and numerical experiments were performed using the software elaborated. For conventional gasars, the analysis of kinetics of gas bubbles formation on the front of liquid/solid interfaces based on the diffusion equations shows an evident relations between porosity degree and applied gas pressure during saturation of magnesium ingot (Fig. 2). Estimated relations served the choice of process parameters and, as a result, a number of conventional gasars ingots of various porosity structure were produced (Fig. 3).

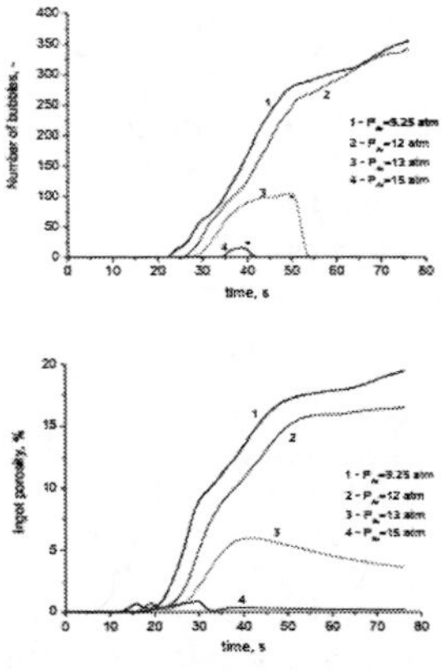

Fig. 2. Number of floating gas bubbles nucleated at S/L interface (a) and changes of total porosity in Mg ingot during solidification (b). Mg melt was saturated in Ar-38 vol.%H₂ mixture at 0.925 MPa

Fig. 3. CT-images of conventional Mg gasars with axial pores produced with Ar-50 vol.% H_2 of 10 MPa using mold side heating (650 °C) with 11 vol.% (a) and 15 vol.% (b) porosity). Mg gasars with radial (left) and axial pores (c)

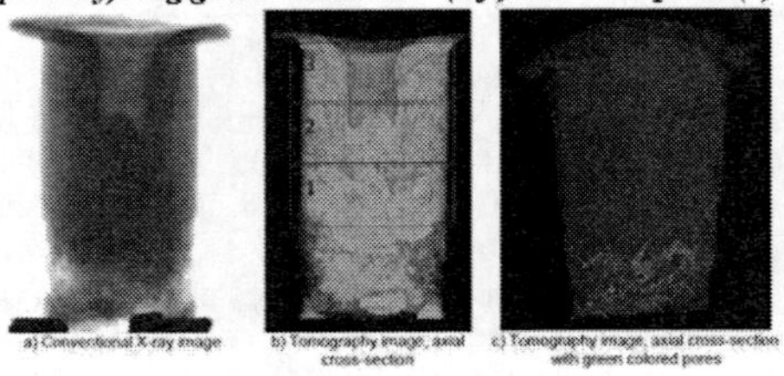

Fig. 4. X-ray images of Mg hybrid gasar produced with steel wool

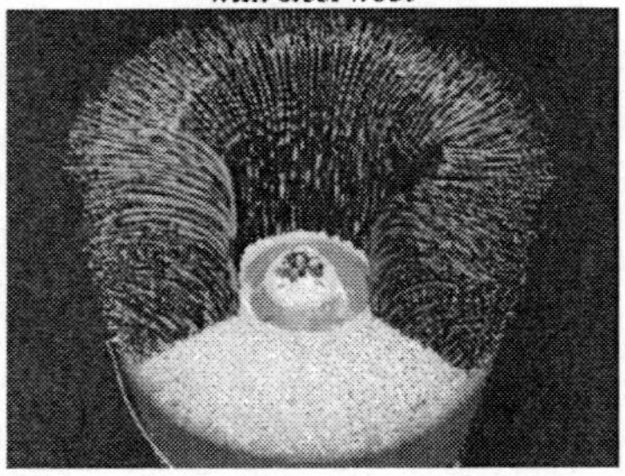

Fig. 5. CT-image of third zone of hybrid Mg gasar produced with steel wool with centrally located shrinkage cavity (estimated average porosity more than 25%)

Hybrid Mg gasars ingots were produced by placing a two-centimeter layer of steel wool on the bottom of the mold. Fig. 4 shows the cross sections of hybrid gasar ingot, produced with Ar-35%H_2 mixture. Its lower part represents steel wool partially infiltrated by molten magnesium. Above this part, three zones of different thermal conditions can be identified. In the first zone, the porosity is not ordered most likely due to volumetric solidification. In the second zone, the pores are axially distributed what is an effect of nearly axial oriented thermal gradient (identical with the direction of heat removal). In the third zone, the thermal gradient is the resultant of two competing directions: radial and axial - the pores are ordered and directed at nearly 45° (Fig. 5). The middle part of the ingot in this zone is characterized by a deep shrinkage cavity, starting from the second zone.

The results of the computer simulation of ingot solidification using Magma Soft are summarized in Fig. 6. Before melt pouring, the upper grey part is an empty mold while the lower green part is filled with steel wool (Fig. 6a). After pouring, the solidification is started and small empty region is formed at the upper part (Fig. 6b). During cooling, the shrinkage cavity becomes deeper (Figs. 6c-d), reaching the appropriate dimension and location with the end of solidification of the ingot. Since the heat removal from the melt was axially symmetric and more intensive at upper part of casting-mould system, the nearly cylindrical cavity was formed inside the gasar ingot.

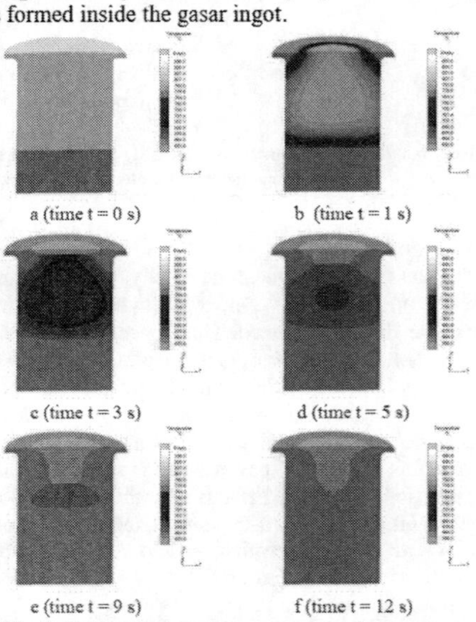

Fig. 6. Volume/temperature sequences of computer simulation for Mg hybrid gasar produced with steel wool

4. Conclusions

Application of reinforcing phase and/or foreign inserts is an useful route for proper pressure-thermal management upon the gasar process with reduced amount of hydrogen in Ar-H_2 mixture (35-50 vol.%) that allows receiving materials with up to 25 vol.% porosity and controlled ordered, directional, arranged and functional structure.

Acknowledgments

This work has been done in the frame of ESA project (Contract No: 4000109468/13/NL/Cbi). Discussions and consultancy with prof. V. Shapovalov and dr. L. Boyko are highly appreciated.

References

[1] Shapovalov V.I., Serdiuk N.P.(1980), Izv. VUZov, Chernaya Metallurgiya, 90-93

[2] Nakajima H. et al. (2006), Materials Science Forum, 512, 295-300

[3] Patent No 14732942.9-1373/2925469, The pressure reactor for producing materials having directed porosity (J.J. Sobczak, N. Sobczak, P. Dlugosz, P. Darlak, L. Boyko, V. Shapovalov) European Patent Office, 30.05.16

[4] Shapovalov V.I. (2013) Legirovanie vodorodom. Zurfond. ISBN 9788-966-1696-92-0

DIGITALIZATION

The 73rd World Foundry Congress
„Creative Foundry"
23 - 27 September 2018, Krakow, Poland

PHOTOGRAPHY OF ATOMIZATION FLOW AND LES-VOF SIMULATION OF DIE INTERIOR FLOW BEHAVIOR UNDER HIGH PRESSURE DIECASTING

Eitaro Koya [1]*, Masahiko Nakagawa[1] Shinya Kitagawa [1]
Jun Ishimoto [2], Yoshikatsu Nakano [2], Naoya Ochiai [2]

[1]Honda R&D Co., Ltd., Motor Cycle R&D Center, 3-15-1 Senzui, Asaka-Shi, Saitama, Japan
[2] Tohoku University, Institute of Fluid Science, 2-1-1 Katahira, Aoba-Ku, Sendai, Japan
*Corresponding address: eitaro.koya@mail.a.rd.honda.co.jp

Key words: HPDC, Atomization, LES-VOF method, Multi-phase flow, Photography

1. Introduction

High-pressure diecasting (HPDC), in which molten metal is injected into a die at a high speed, is notable for its short cycle times and ability to form thin sections. One issue, however, is gas entrapment within the metal during filling, a situation that can produce porosity in the casting and impair its quality. Research conducted in recent years has shown that casting quality can be improved by increasing the "J value," an indicator of the degree of melt atomization after it passes through the ingate [1].

By statistical study of the impact of HPDC defects on casting strength, the authors found that high-strength sections can be obtained by finely dispersing gas porosity in the products [2]. Fine dispersions of gas porosity are observed only in castings formed when the J value exceeds a threshold for atomized flow. In our research, we seek to establish the technology for producing high-strength castings with the HPDC process by assuring fine dispersion of gas porosity. Here, in this paper, we discuss (a) photography of the metal atomization as it passes through the ingate and (b) simulation of how atomized the metal flows into the die.

2. Photography of the metal atomization

We used a method developed by Aida et al. to photograph the metal as it is injected into an open space [3]. Fig. 1 shows a horizontal planar view of the die arrangement we used to take such photographs. The area past the ingate is open. Metal inflow was photographed through an acrylic plate affixed to one side of the die. To observe atomization, photographs were taken under illumination provided by a metal hydride lamp. Next, by means of a transmitted pulsed laser directed through a 30-mm circular window above the ingate, we photographed the metal micro boundaries at that position.

Injection speed curve are shown in Fig. 1. Here, the J value is 1,400, which exceeds a 1,064 threshold value presented for atomized flow [4]. The metal is JIS ADC12 aluminum alloy for die casting.

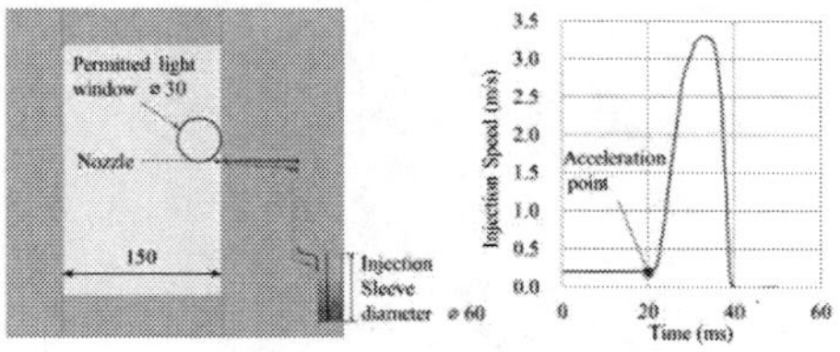

Fig. 1. Die arrangement for atomization experiments and Injection speed curve

3. Simulation of the metal atomization

Currently, many analyses of casting flow are conducted with the volume of fluid (VOF) or a similar method applied to the filling process on a macro scale over a comparatively large mesh size. No attempt is made to directly treat turbulent the metal flow. Here, the authors developed a system to simulate the process of the metal atomization during injection. Such a simulation requires fine-mesh calculations of the metal behavior under surface turbulence. The system we thus developed entails numerical analysis of turbulence by means of large eddy simulation (LES). This said, the VOF method is well suited to simulate changes in surface condition of the metal within the atomization process. We thus coupled LES and the VOF method to create a simulation system (LES- VOF simulation) based on the finite volume method. This offers a high degree of resolution. Fig. 2 shows our computational model and its boundary conditions.

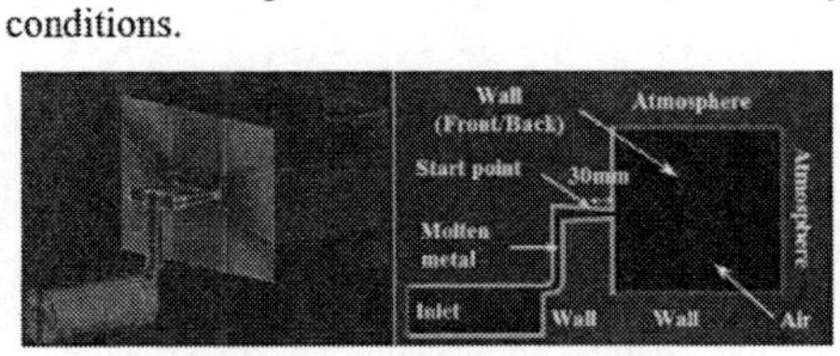

Fig. 2. Computational model

We adopted the Smagorinski model to treat turbulence on a subgrid scale within LES. As clearly seen in Fig. 1, the HPDC method entails somewhat slow filling of the metal up to the ingate, followed by a quick acceleration beyond the ingate. To reproduce this acceleration of the metal, we applied a moving boundary model at the injection sleeve.

4. Discussion of simulation results

Fig. 3 compares (a) macro-level photographs of the metal injection into the open space taken under metal hydride illumination with (b) results of numerical simulation aluminum iso-surfaces.

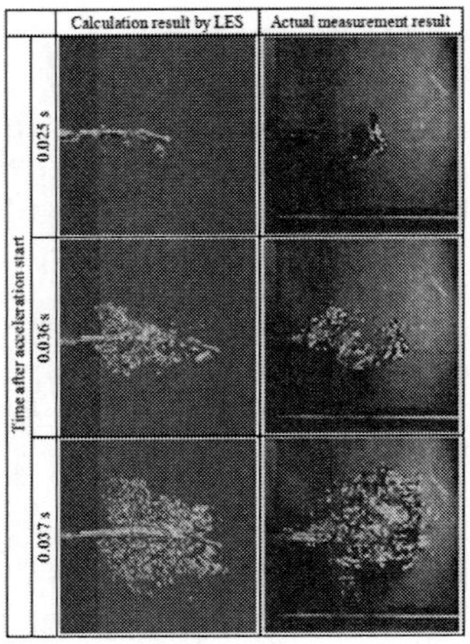

Fig. 3. Comparison between calculated (LES) and measured (photographic) results

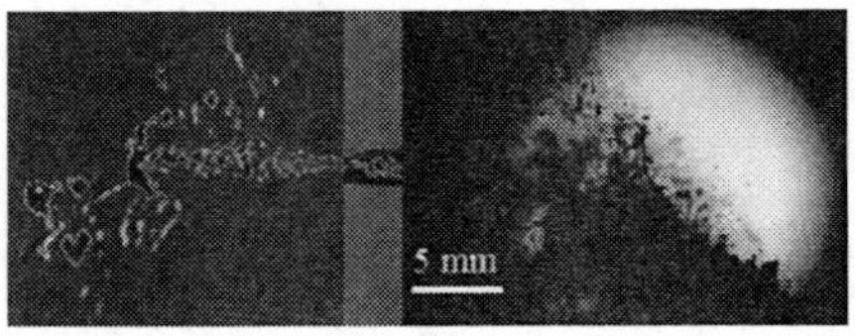

Fig. 4. Comparison between calculated results and photographic results

Successfully reproduced here is the vertical splattering that occurs when accelerated the metal splashes against melt that has passed through the ingate at a lower speed.

Next, Fig. 4 compares (a) a transmitted laser light photograph of the micro condition of the metal near the ingate and (b) results of numerical simulation.

Simulation results suggest that molten aluminum will atomize as ribbon-shaped liquid columns emanating from the surface of the injected flow. This accords to what we observe in the photograph.

5. Simulation of atomized flow within the die

We next applied our system to a simulation of the flow of atomized aluminum from the nozzle into the interior of the die. For this calculation, we chose to model a casting shape in which a thick section is located far from the ingate and could thus present a challenge in terms of HPDC quality [5]. Dimensions for the calculation are shown in Fig. 5. The shape from the injector up to the gate is the same as that with the die used for photography. Calculation results are shown in Fig. 6.

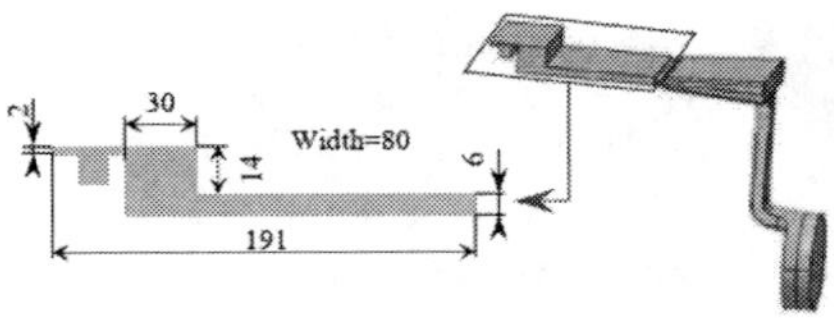

Fig. 5. Model dimensions [mm]

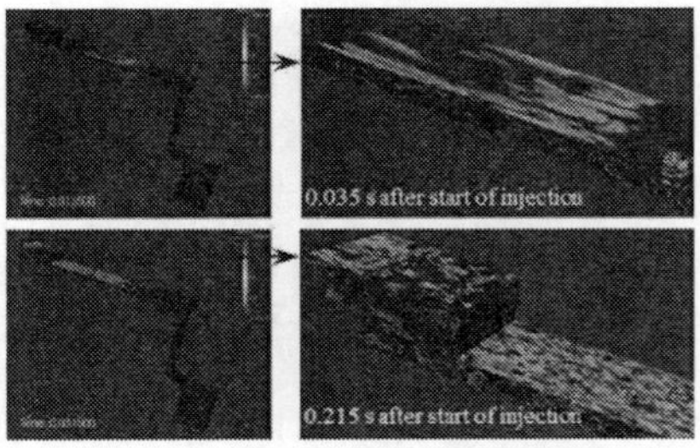

Fig. 6. Fluid volume under filling (simple-shape model)

With this newly developed system, we see that molten aluminum is atomized at the ingate and remains in a multiphase state with air as fills into the die. This shows that by atomizing the metal at the ingate, we can effectively disperse entrapped air as fine bubbles.

6. Conclusions

By photographing atomization under HPDC and applying the LES-VOF method under turbulent flow conditions, we successfully simulated the process of molten aluminum atomization during injection. With this system, we showed that atomized flow as conventionally indicated with the J value is a multiphase flow of aluminum and fine air bubbles.

References

[1] Kaneuchi, T. et al, "Effect of J-Value for Internal Quality of HPDC Products at Various Casting Alloys," Japan Die Casting Congress Transactions 131-136, 2008.

[2] Koya, E. and Nakagawa, M., "Deterioration of mechanical properties of JIS ADC12 alloy by defects and trial for the prediction of casting strength by casting CAE Aluminum," Journal of Japan Foundry Engineering Society (J. JFS). Vol. 89, no.12, 781-786, 2017.

[3] Aida, S., Kuboki, I., Hayashi, Y., Fujimoto, M., et al., "Inference of gate plan/casting conditions on melt spurting," Japan Die Casting Congress Transactions, 111-117, 2012.

[4] North America Die Casting Association (NADCA), "GATING MANUAL", 35-36, Publication #5.

[5] Iwata, Y., Sugiyama, Y., Iwahori, H., et al., "Transferred Pressure and Shrinkage Cavity of Aluminum Alloys on Squeeze Casting," J. JFS, Vol.72, No.4, 263-267, 2000

The 73rd World Foundry Congress
„Creative Foundry"
23 - 27 September 2018, Krakow, Poland

APPLICATION OF EXPLICIT AND IMPLICIT SMOOTHED PARTICLE HYDRODYNAMICS SIMULATION TO CASTING PROCESSES

Tamon Suwa [1]*, Masaki Kazama [1], Yasuhiro Maeda [2]

[1] *Fujitsu Limited, Next Generation Technical Computing Unit, 4-1-1, Kamikodanaka, Nakahara-ku, Kawasaki, Japan*
[2] *Daido University, Department of Mechanical Engineering, 10-3 Takiharu-cho, Minami-ku, Nagoya, Japan*
** Corresponding address: suwa.tamon@jp.fujitsu.com*

Key words: Particle-based method, Simulation, Transfer process, Pouring process, Implicit scheme

1. Introduction

The particle-based fluid analysis methods, such as smoothed particle hydrodynamics (SPH) method, are expected to be applicable and useful to transfer process and pouring process of casting because of advantage of the treatment of moving boundary and free boundary.

In order to simulate fluid behavior with particle-based method, the pressure of the particles should be evaluated to solve equation of motion. The pressure evaluation schemes are roughly divided into an explicit scheme that regards the fluid as weakly compressible and an implicit scheme that satisfies incompressibility.

The explicit scheme is easy to set calculation, but it requires artificial viscosity to stabilize calculation. Therefore, energy dissipation is likely to occur. Since the implicit scheme does not require artificial viscosity, energy dissipation is unlikely to occur, but it is difficult to handle calculation setting, especially boundary conditions.

There are advantages to both scheme of particle-based simulation of casting, however, few studies have validated the accuracy of each scheme compared to experiments of the casting processes. Therefore we have applied the SPH method of both explicit and implicit scheme to the transfer processes and pouring processes of casting and validated the results with experimental data.

2. Numerical method

The Navier-Stokes equation with imcompressible condition can be written as

$$\rho \frac{D\mathbf{v}}{Dt} = -\nabla p + \eta \nabla^2 \mathbf{v} + \rho \mathbf{g}, \quad (1)$$

$$\nabla \cdot \mathbf{v} = 0, \quad (2)$$

where ρ, t, $\mathbf{v}$, p, η and $\mathbf{g}$ are the density, the time, the velocity, the pressure, the viscous tensor, and the gravitational acceleration, respectively.

We employ the explicit weakly compressible SPH scheme [1,2] to discretize the governing equations.

$$\frac{d\rho_a}{dt} = \sum_b \frac{m_b}{\rho_b}(\mathbf{v}_a - \mathbf{v}_b) \cdot \frac{\partial W(|\mathbf{x}_a - \mathbf{x}_b|, h)}{\partial \mathbf{x}_a}, \quad (3)$$

$$\frac{d\mathbf{v}_a}{dt} = -\sum_b m_b \left(\frac{p_a + p_b}{\rho_a \rho_b} - \tau_{ab} \right) \frac{\partial W(|\mathbf{x}_a - \mathbf{x}_b|, h)}{\partial \mathbf{x}_a} + \mathbf{g}. \quad (4)$$

where m, $\mathbf{x}$, τ and h are the mass, position, viscous tensor of a particle and the smoothing length. As a kernel function $W(|\mathbf{x}_a - \mathbf{x}_b|, h)$, we employ the quintic spline. The pressure is given by the equation of state $p = c_s^2(\rho - \rho_0)$, where c_s and ρ_0 are numerical sound speed and initial density.

The stabilized incompressible SPH method by [3] is also used as implicit scheme. With implicit scheme, the pressure is evaluated by the pressure Poison equation,

$$\nabla^2 p_a = \frac{\rho_0}{\Delta t} \nabla \cdot \mathbf{v}_a + \alpha \frac{\rho_0 - \rho_a}{\Delta t^2} \quad (5)$$

The pressure laplacian is discretized as

$$\nabla^2 p_a = \frac{2}{\rho_0} \sum_b \left\{ m_b \left(\frac{r_{ab} \nabla W(r_{ab}, h)}{r_{ab}^2 + (0.01h)^2} \right) (p_a - p_b) \right\}, \quad (6)$$

and the equation (5) is solved with linear solver such as conjugate gradient scheme.

Furthermore, in order to reproduce fluid behavior of pouring of aluminum alloy, we developed the surface model of the effect of the oxide film [4]. The viscosity of the particles positioned surface are set large value.

3. Numerical results

The transportation of liquid container were analyzed as well as the experiment [5]. Figure 1 shows model of the experiment. Figure 2 shows time series of the level of liquid under the transportation condition. The result of the experiment is indicated as the red line. The results of the SPH simulations are indicated as the yellow (explicit) and the blue (implicit) lines, respectively. It is shown that the shape of the first wave is corresponding to the experiment very well. The wave attenuation occurs in the explicit scheme, while implicit scheme shows good agreement with second and after waves.

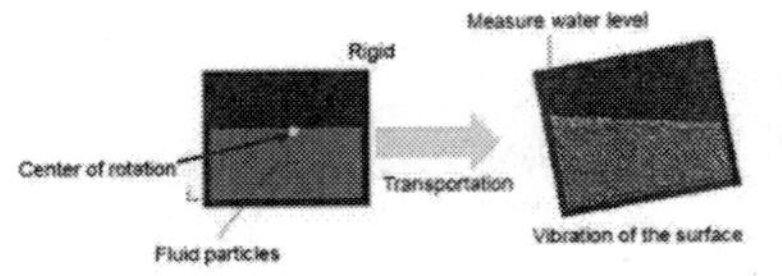

Fig. 1. Model of transfer experiment

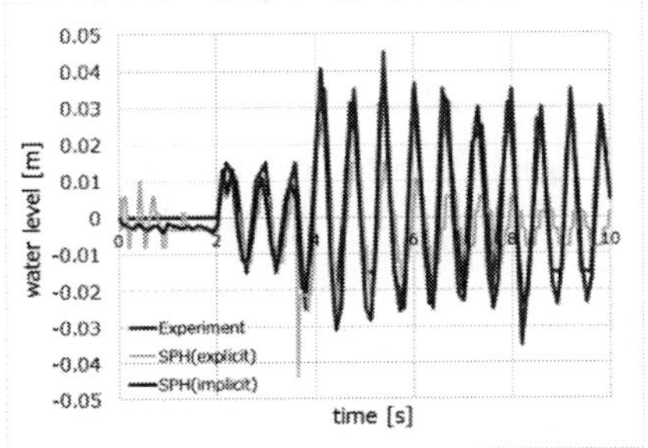

Fig. 2. Level of the liquid surface

In Figure 3 and 4, the results of experiment and simulations of pouring to sleeve with water and molten aluminum alloy are shown [4]. In the case of aluminum alloy, the height of the wave is low and the traveling speed of the wave is slow, in comparison with the case of water. The numerical results of our surface model can properly express the experimental results with both of explicit and implicit scheme.

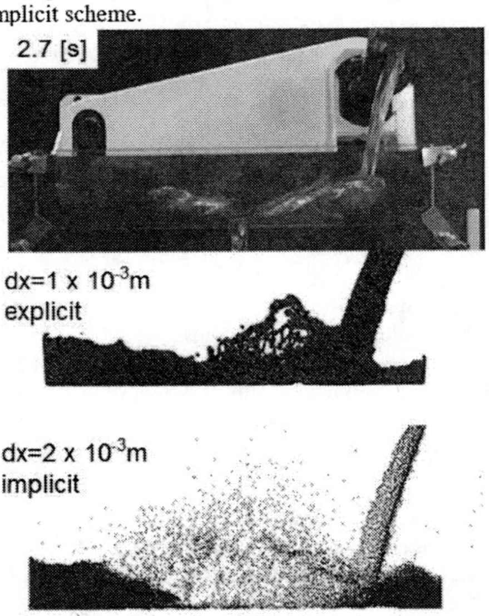

Fig. 3. Results of pouring experiment and simulation with water

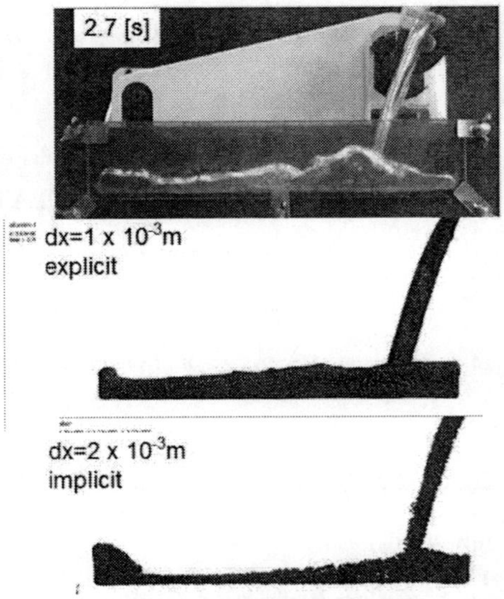

Fig. 4. Results of pouring experiment and simulation with molten aluminum alloy (ADC12)

4. Summary and conclusions

We simulated the casting processes using SPH method and compared them with the experiments. The results reproduce the experiments well, which suggests advantage of the features of SPH method. It was also found that the implicit method gives better results about oscillation dissipation.

References

[1] T. Suwa, T. Nakagawa & K. Murakami. (2012). A Study of the Wave Transformation Passing over an Artificial Reef using SPH Method. Journal of computational science and technology. 7 (2), p.126-133.

[2] M. Kazama, T. Suwa, K. Ogasawara, Y. Maeda & Ito. (2017). Particle Simulation of the Transfer and the Pouring Processes of the Molten Metal. J. JFS. Vol.89 (7), p.389-395.

[3] M. Asai, A. M. Aly, Y. Sonoda & Y. Sakai. (2012). A Stabilized Incompressible SPH Method by Relaxing the Density Invariance Condition. Journal of Applied Mathematics, Vol.2012.

[4] M. Kazama, T. Suwa & Y. Maeda. (2017). Modeling and Computation of the Molten Aluminum Alloy Flow with Oxide Film by Smoothed Particle Hydrodynamics. J. JFS. (in press).

[5] R. Shibuya, H. Okatsuka, Y. Noda & K. Terashima. (2013). Sloshing Suppression Control with Designed Transfer and Tilt Input by Using Generalized Predictive Method. Transactions of the Society of Instrument and Control Engineers. 49, p.134-141.

The 73rd World Foundry Congress
„Creative Foundry"
23 - 27 September 2018, Krakow, Poland

MODELING AND COMPUTATION OF CASTING PROCESS BY PARTICLE METHOD

Masaki Kazama[1]*, Tamon Suwa[1], Yasuhiro Maeda[2]

[1]*Application development div., Next generation technical computing unit, FUJITSU Limited, Japan*
[2]*Department of Mechanical Engineering, Daido University, Japan*
**Corresponding address: kazama.masaki@jp.fujitsu.com*

Key words: Particle method, Simulation, Casting process, Comparing with Experiment, Solidification

1. Introduction

Particle-based computational methods, such as Smoothed Particle Hydrodynamics (SPH) method, do not require numerical mesh. Because of the mesh-less characteristics, these methods are suitable for the numerical simulation including the free surface or the moving boundary. In the casting processes, it is expected that the particle-based methods make the simulation of the flow of the molten metal and the shrinkage cavity more accurate. However, the researches that quantitatively compare the computational results of the particle-based simulation and the experimental results and that evaluate the validity of the technique are few. Then we have applied the SPH method to the casting processes to validate the results with experimental data, as shown in Fig. 1.

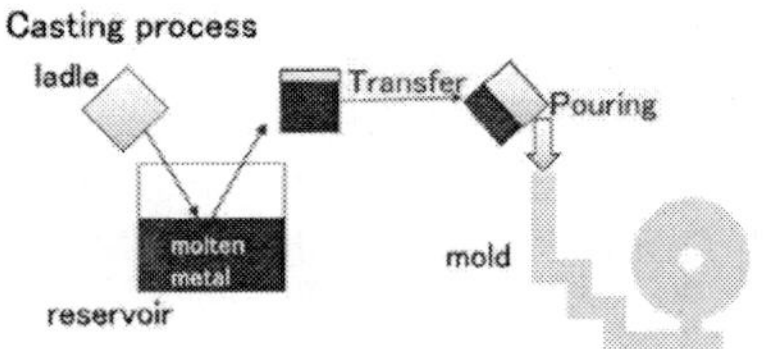

Fig. 1. The diagram of casting process

2. Numerical method

We apply SPH method to the casting processes. In SPH method we divide fluid into particles and treat particles with following equations [1].

$$\frac{d\rho_a}{dt} = \sum_b \frac{m_b}{\rho_b}(\mathbf{v}_a - \mathbf{v}_b)\cdot\frac{\partial W(|\mathbf{x}_a - \mathbf{x}_b|, h)}{\partial \mathbf{x}_a}, \quad (1)$$

$$\frac{d\mathbf{v}_a}{dt} = -\sum_b m_b\left(\frac{p_a + p_b}{\rho_a \rho_b} - \tau_{ab}\right)\frac{\partial W(|\mathbf{x}_a - \mathbf{x}_b|, h)}{\partial \mathbf{x}_a} + \mathbf{g}, \quad (2)$$

$$p_a = c^2(\rho_a - \rho_{a,0}). \quad (3)$$

Where, $\mathbf{x}, \mathbf{v}, m, \rho, p, c$, and $\mathbf{g}$ are position, velocity, mass, density, pressure, sound speed, and gravitational acceleration, respectively. The subscript a means the value of particle a. $\rho_{a,0}$ is the standard density of a-th particle, which is the density when pressure is zero.

W is the following kernel function,

$$W(r,h) = \begin{cases} \alpha(1-q)^4(1+4q) & (0 \le q \le 1) \\ 0 & (1 < q) \end{cases}, \quad (4)$$

where, α is fixed so that the integral of W of whole space equal to 1, τ_{ab} is the viscosity effect between partice a and b. When η_a is the viscosity of a-th particle, τ_{ab} is calculated as follows.

$$\tau_{ab} = \frac{4}{\rho_a \rho_b}\left(\frac{\eta_a \eta_b}{\eta_a + \eta_b}\right)\left(\frac{(\mathbf{v}_a - \mathbf{v}_b)\cdot(\mathbf{x}_a - \mathbf{x}_b)}{|\mathbf{x}_a - \mathbf{x}_b|^2}\right). \quad (5)$$

In addition to the above equations, we introduce the surface resistance model in order to realize the effect of the oxide flim in thin layer of the surface of molten metal[2]. Furthermore, we apply Cleary's numerical model of temperature dependent viscosity and density as the model of solidification and shirinkage [3].

We calculate the above numerical models by explicit method [4]. In addition, we add artificial viscosity and perform shifting method when calculation [5].

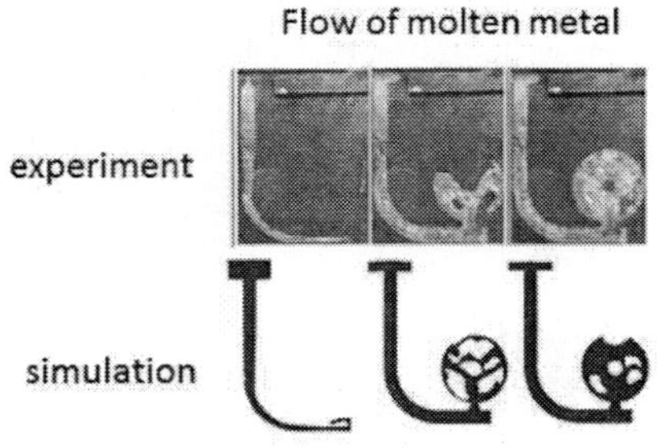

Fig. 2. Numerical results of mold filling behavior of molten aluminum alloy (Curved model)

3. Numerical results

3.1 Mold filling process

Fig. 2 and Fig. 3 are the experiments and simulations of mold filling process. We apply our SPH method to the following two models, and then compare the flow pattern of respective runner shape. As the tendency between the models, the velocity of fluid into the cavity of Curved model (Fig. 2) is faster than that of

Stair like sprue model (Fig. 3). In Stair like model the energy loss has happened when filling the stair like sprue. The qualitative tendency is sufficiently expressed by SPH results.

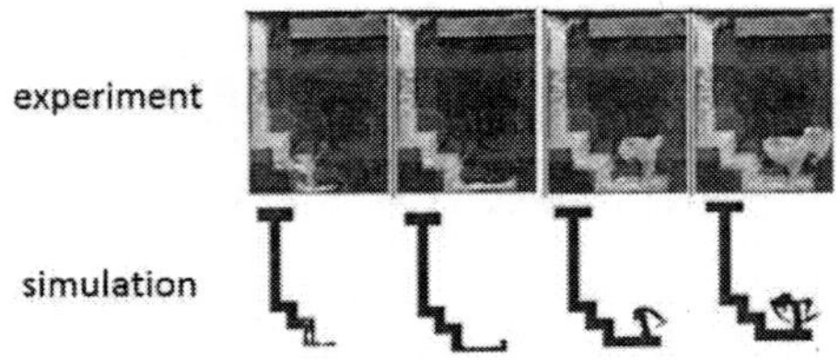

Fig. 3. Mold filling process of Star like model

3.2 Pouring process

Fig. 4 shows pouring process of molten aluminum alloy. In this experiment, molten aluminum alloy is poured into sleeve from ladle. We also performed the numerical analysis with our SPH method on the same condition as the experiment.

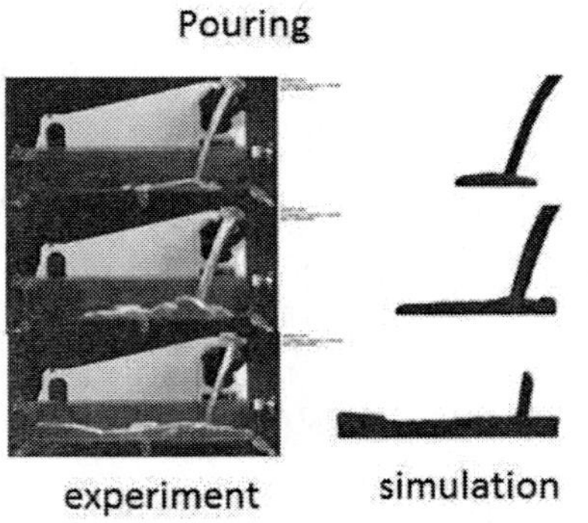

Fig. 4. Numerical results of pouring process by ladle for molten aluminum alloy

Fig. 5 shows the time series of wave front of pouring molten aluminum alloy. Our numerical results can properly express the experimental results from the viewpoint of the traveling speed of wave front.

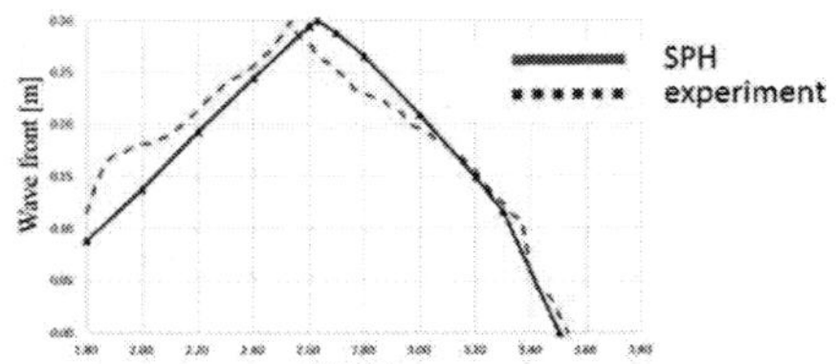

Fig. 5. Time series of wave front. Solid line is SPH result and dotted line is experimental result.

3.3 Cooling and solidification process

Now we are going to compare the shape of shrinkage cavity of numerical results and the experimental results.

For example, Fig. 6 is Conical Mold [6], which is an experiment to confirm the shape of solidification shrinkage by pouring aluminum alloy into conical sand mold and solidifying it. We calculated this process by SPH method.

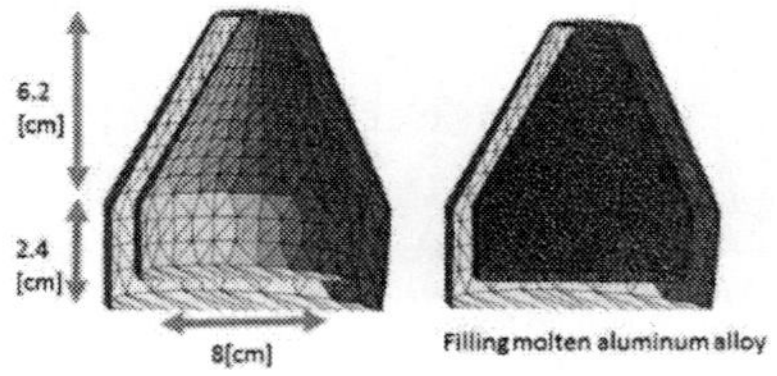

Fig. 6. Diagram of conical mold

Calculation is performed from filling to solidification completion. We adjust the heat transfer coefficient of sand mold and molten metal so that solidification time corresponding with the experimental results.

Fig. 7. Numerical results of cooling process in the conical mold

Fig. 7 shows cooling process calculated by our SPH method. We will show detail of this results in congress.

4. Summary and conclusion

In this work, the SPH method was applied to the casting, then it is shown that the qualitative tendency is sufficiently expressed by SPH results.

In the congress, we will report the detail of these results and numerical method, because we saw the experimental data and our numerical results are show good agreement. Then we would like to show the utility of this simulation by comparing them with the experimental results.

References

[1] J.J. Monaghan. (1992).Smoothed particle hydrodynamics. Annual review of astronomy and astrophysics. Vol. 30 (A93-25826 09-90), p.543-574.

[2] M. Kazama, T. Suwa and Y. Maeda. (2017). Modeling and Computation of the Molten Aluminum Alloy Flow with Oxide Film by Smoothed Particle Hydrodynamics. J. JFS. (in press).

[3] P.W. Cleary. (2010). Extension of SPH to predict feeding freezing and defect creation in low pressure die casting. Applied Mathematical Modelling, 34, p.3189-3201.

[4] T. Suwa, T. Nakagawa and K. Murakami. (2013). A Study of the Wave Transformation Passing over an Artificial Reef using SPH Method, Journal of computational science and technology 7, No. 2, p.126-133.

[5] E. Jahanbakhsh, O. Pacot, F. Avellan and P. Maruzewski. (2011). 6th international SPHERIC workshop, June (08-10). Hamburg, Germany.

[6] Y. Awano, K. Morimoto, Y. Shimizu and H. Takamiya. (1992). Shrinkage Morphology in Al-Si System Casting Alloys, R&D Review of Toyota CRDL, 27(1), p.51-62.

INTELLIGENT DATA ANALYTICS FOR FOUNDRY INDUSTRY 4.0

Marcin Perzyk *, Jacek Kozlowski, Andrzej Kochanski

Warsaw University of Technology, Faculty of Production Engineering, Narbutta 85, 02-524 Warszawa, Poland
**Corresponding address: M.Perzyk@wip.pw.edu.pl*

Key words: foundry industry 4.0, data analytics, data-driven models, advanced statistics, learning systems

The 4th revolution in industry, being a combination of modern technologies to create a 'smart factory' is coming, also in the foundry industry [1]. In a broad sense, the Industry 4.0 is all about data and analytics [2]. Big data analytics and machine learning technologies are now being applied to almost all industries. Data-orientated technologies are doubtless driving greater efficiency, higher production and faster fulfillment, which equal new opportunities [3]. One of the main challenges is to develop the methodologies aimed at extracting useful knowledge from the huge amounts of data now available in most of manufacturing companies and further to utilize it in control of the processes, predicting process behavior or equipment failures and diagnosing of the process and products faults. Compared to most of manufacturing processes metal casting is particularly complex due to its two features. First, material of the final product is not just processed but in most cases must be produced within the foundry (in the melting shop). Second, expendable tools (i.e. molds and cores) are often used for shaping the products, which results in necessity of a parallel production of elements made of completely different material (also prepared within the plant) and using different technologies. All this makes the data acquisition, processing and utilizing particularly difficult. Currently many foundries are able to acquire all relevant data using automated systems, however, making good use of these data is still very limited. In this paper we present some of the possibilities and examples of advanced data analytics in foundries.

1. Types of control and fault diagnosis of foundry processes

Generally, control of manufacturing processes has one goal: limiting the process variability. This can be achieved using two basic approaches: Engineering Process Control (EPC) and Statistical Process Control (SPC). EPC is oriented at process regulation which means that it counteracts the process disturbances by making adjustments to process variables in order to keep the output quality parameter on target. The most commonly used type of EPC is probably the feedback control in which deviations of the output from the target are used to calculate the amount of adjustment. A typical foundry process in which this type of control is applied is melting where the results of chemical composition or information obtained from DTA are utilized for the melt adjustment. Similarly, molding sand preparation systems utilize the current properties of the sand such as moldability and moisture content to adjust the amount of added water and bentonite.

SPC assumes that the process output can be described by statistically independent observations fluctuating around a constant mean and is intended to detect signals which represent the special (assignable) causes of external disturbances and subsequently remove them. There are two main types of SPC and the corresponding types of control charts: for continuous variables (e.g. product dimensions) and for attributes, e.g. defining product as acceptable or non-acceptable (e.g. defective). Typically, samples consisting of several products are checked in regular time intervals. Both types of SPC are used in foundries, however, it seems that, unlike in other manufacturing industries, single measurements are common. The extreme case is that all products are checked which is typical for some automated systems, e.g. modern die casting foundries.

It should be emphasized that diagnosis of process faults and causes of casting defects may be a difficult task regardless of whether they are detected as abnormal pattern on SPC charts or are defined as increased fraction of defective castings. The process disturbances and product defects can be influenced by a large number of parameters of very diverse nature which is a consequence of the complexity of foundry processes mentioned in the previous section.

2. Examples of data-driven modeling for control of foundry processes

The type of information which is primarily important for establishing proper control procedures and finding the most probable causes of process and product faults is the relative significance of process input from the standpoint of the process results. This kind of information can be obtained by different methods, however the relative simple statistical methods appeared to be sufficiently reliable and practical [4].

The most advanced data-driven models are certainly learning systems, often associated with the discipline named Computational Intelligence. They have proved to be very successful in many areas and have found various and valuable applications also in control and fault diagnosis of complex foundry processes. Good examples from a US iron foundry were presented in [4]: based on three neural models control system for an arc furnace and two neural controllers used for green sand processing in a mulling machine. The latter include updating the controller parameters using the data from the plant PLC system and uses the fuzzy-type variables. Another example is a neural regression model used for control of the heat treatment of ADI castings [5]. To obtain a desired grade of the alloy the optimal process temperatures and times along with the iron chemical composition can be determined. The model was build using a large number of data records collected from the world literature and subsequently cleaned and filtered.

An often observed phenomenon related to applicability of SPC is occurrence of autocorrelations in the data having the forms of general trends and periodicities. One of their consequences may be appearance of false out-of-control warning signals. To avoid such misleading situations, the Special Cause Control (SCC) charts can be used, being the Shewhart charts for residual data obtained by substracting from the original data the non-random components found by the time-series analysis. Successful applications of SCC for green sand processing system were presented in [6,7]. However, as indicated in [7], the autocorrelations appearing in data can be also considered as a kind of the process fault or at least its particularity and must be analyzed by the foundry staff. A positive consequence of the autocorrelations is that some useful forecasts of the coming values can be made using the time-series analysis and then used in a feedforward (anticipating) type of process control. Examples of such applications for some foundry processes are presented in [8].

3. Examples of advanced data analytics in fault diagnosis in foundry production

Identification of the causes of out-of-control signals appearing on SPC charts can be facilitated by comparison them with those obtained from the process model predictions. It was demonstrated [9] that a neural regression model of SG iron melting process (with chemical composition as inputs and tensile strength as the output) imitates the real output sequences of points fairly well. By simulated changes of selected elements contents one can conclude what was the 'wrongdoer'. Somewhat similar problem was presented in [10] where the process result was defined as the SG iron grade. The model was used to identify the causes of missed melts, i.e. those in which the combination of tensile strength and elongation was out of specification for the produced grade. In this case a classification type task was encountered in which a classification tree model appeared to be the best, giving only 2% wrong predictions. Again, by simulated changes of the input variables for the missed melts it was possible to find their most probable causes.

A successful application of a neural model for finding root causes of gas porosity, being one of the most difficult to diagnose casting defects, is described in [11]. Simulations using the model allowed to find that the direct cause was too high water vapor pressure in the mold, being a result of low sand permeability and actual high moisture content, affected also by the atmospheric conditions.

4. Knowledge rules from data

Digging in data (Data Mining) can also provide engineers with a knowledge which can be formulated in terms of verbally expressed rules. The appropriate rule systems can be semi-automatically obtained by different methods such as classification trees and the rough sets theory. Thus obtained sets of rules can be utilized in design and adjustment of process parameters. Some examples concerning foundry and related processes can be found in [5,12].

5. Perspectives of data analytics in foundries

The availability of Big Data seems to have two important consequences: first it makes an excellent basis for analyses and advantageous utilization of the data, second it presents a challenge to analysts and engineers. Without good understanding the nature of the data-driven models engineers will not be able to recognize their possibilities and successfully apply them in industrial practice. The role of updating all forms of education is therefore essential.

References

[1] M. Lewis/Foundry Trade Journal (2016). Industry 4.0 and what it means to the FOUNDRY INDUSTRY. Retrieved 11.12.2017, from http://www.foundrytradejournal.com/features/industry-40-and-what-it-means-to-the-foundry-industry-

[2] i-SCOOP (2016). Industry 4.0 – is it all about industrial data and analytics. Retrieved 11.12.2017, from https://www.i-scoop.eu/industrial-data-analytics/

[3] J. Zavalishina/CFE Media (2016). Retrieved 11.12.2017, from https://internetofbusiness.com/manufacturings-4th-revolution/

[4] M. Perzyk, R. Biernacki & J. Kozlowski. (2008). Data mining in manufacturing: significance analysis of process parameters. Journal of Engineering Manufacture. 222(11), 1503-1516.

[4] P.F. Bartelt & J. Moberly. (1996). Applying artificial intelligence to the modern foundry. Modern Casting. 86(2), 52-55.

[5] M. A. Perzyk, A. Kochanski, J. Kozlowski & D. Myszka. (2014). Control of ductile iron austempering process by advanced data driven modeling. In 71st WFC, 19-21 May 2014. Bilbao (Spain): WFO.

[6] Green M. E. (2009). Critical Assessment of Current Metalcasting Green Sand System Control and Monitoring Processes. The Pennsylvania State University, USA.

[7] M. Perzyk & A. Rodziewicz. (2015). Application of Special Cause Control Charts to Green Sand Process. Archives of Foundry Engineering. 15(4), 55-60.

[8] M. Perzyk & A. Rodziewicz. (2012). Application of Time-series Analysis in Control of Chemical Composition of Grey Cast Iron. Archives of Foundry Engineering. 12(4), 171-175.

[9] M. Perzyk & J. Kozlowski. (2016). Methodology of Fault Diagnosis in Ductile Iron Melting Process. Archives of Foundry Engineering. 16(4), 101-108.

[10] M. Perzyk & M. Werlaty. (2017). Diagnosis of missed ductile iron melts with process modelling. Archives of Foundry Engineering. (accepted for print).

[11] M. Perzyk & A Kochanski. (2003). Detection of causes of casting defects assisted by artificial neural networks. J. Eng. Manuf. 217(9), 1279-1284.

[12] M. Perzyk & A. Soroczyński. (2008). Comparison of selected tools for generation of knowledge for foundry production. Archives of Foundry Engineering. 8(4), 163-166.

The 73rd World Foundry Congress
„Creative Foundry"
23 - 27 September 2018, Krakow, Poland

INTRODUCING INDUSTRY 4.0 IN A DIE CASTING FOUNDRY

Bartlomiej Dybowski [1], Marcin Perzyk [2]*, Jacek Kozlowski [2]

[1]*NEMAK POLAND Sp. z o.o., Komorowicka 53, Bielsko-Biala, Poland*
[2] *Warsaw University of Technology, Faculty of Production Engineering, Narbutta 85, Warszawa, Poland*
**Corresponding address: M.Perzyk@wip.pw.edu.pl*

Key words: die casting, Industry 4.0, process control, data analytics, learning systems

The 4th industrial revolution, usually referred to as Industry 4.0, became already a reality in many branches of industry. After the steam machine, production line, application of computers, the time of full automation has come. The 4th industrial revolution is focused on creating self-teaching, intelligent factory, which will be able to predict the quality of the product and, in the future, even improve it by itself. The lifeblood of the system is an internet of things – the process parameters gathered in a cloud or on reliable servers. As the product's quality is a function of process parameters, the on-line collected data may be used in the prediction of final product's quality.

Among all types of foundries the high pressure die casting foundries are probably the most advanced in the process automation and gathering complete production data, making the basis for Industry 4.0. On the other hand, due to the high number of process variables involved and to the non-synchronization of all process parameters in a unique and integrated process control unit, die casting is perceived as one of the most defect-generating process showing less flexibility to any changes in products and in process evolution [1]. As indicated in an in-depth analysis presented in [2], monitoring and correlating all the main process variables and making the process set up and optimization a knowledge-based issue are the key issue in the quality assurance of die castings production.

Nemak Poland, as a leading manufacturer of the aluminum castings for the automotive industry, has to pay special attention to the quality of the products. Currently the company database gathers majority of the process parameters from all high pressure die casting machines resulting in more than 15 000 sets of new data per week. The data cover the most important process parameters, including liquid alloy temperature, injection parameters, die thermo-balance parameters etc. The systems for liquid alloy tracking and more accurate description of cooling circuits flow are being developed. The Nemak project includes two main tasks. One aims at the development of complete systems for recording the process and product parameters. The other task is a development of algorithms capable of predicting the casting quality basing on the historical data, which can be further used for optimal control and adjustments of the process parameters. In the present paper the data analytics system, being the second part of the Nemak project, is presented.

1. Process dependent properties of die castings

The critical properties of aluminum die castings are certainly their mechanical properties such as ultimate tensile and yield strengths, elongation and hardness. In the case of engine blocks, the customer demands increase with increasing engine performance. Nemak has started production of the vehicle structure parts such as frame rails, shock towers or tank cover frames. These components are crucial for the passengers safety, thus the proper production of these components guaranteeing high mechanical properties is essential.

In order to obtain reliable data, the material properties are tested on samples taken from the castings. In engine blocks the bearing regions are chosen as the most massive but also high loaded sections with a thickness reaching 20 mm, making it a challenge to achieve the customer specifications. As the mechanical testing is a destructive investigation, it is conducted mainly during the product development phase.

2. General characteristics of data analytics for casting process control

There are many process and material parameters, i.e. potential input variables, which may affect the mechanical properties of the cast alloy. The first selection has been made by the foundry engineering staff and it covers a wide range of almost all recorded variables related to mechanical and thermal parameters of the casting process, the alloy chemical composition and its heat treatment parameters. Total number of these variables is about 60.

The development of the data analytics system has been divided into three main stages. The first one is the preliminary analysis of the data aimed at detection of the correlations between the input variables; strong correlations would make it possible to reduce the number of the variables worked in the next steps. The goal of the second stage is to identify those variables which are most significant for obtaining the desired mechanical properties of the cast alloy. This kind of analysis provides very important information. At first information is generated about process parameters that are relevant to control of the casting quality. Secondly, it supports the selection of the variables to be used in the next stage of the project. In the third stage an advanced multivariate data-driven model will be constructed linking the mechanical properties of the cast alloy with process parameters. It can be used as an important aid for anticipated setting optimal process parameters and also process fault diagnostics.

3. Correlations between input variables

Two types of correlation coefficients are computed for all possible pairs of input variables: the linear Pearson and the non-parametric Spearman. The number of highly intercorrelated variables with at least one other variable (absolute values of Pearson or Spearman coefficients above 0.9) was about 25, i.e. almost half of the total number of input variables.

The following three kinds of the correlation sources have been identified:

1) Physical (natural) correlation, for example between two temperatures measured in adjacent areas of the die or between water flow and its temperature in the same cooling channel. This kind of correlated variables can be easily replaced by one variable, more apparent from the viewpoint of the product quality.

2) Intentional correlations being a result of the operators' or engineering staff actions. They are observed mainly for new developed products and shall be eliminated in the following stages of the product development.

3) Coincidental correlations, i.e. resulting from simultaneous occurrences of some values in some time periods. In further project stages they shall be avoided but some additional tests may be necessary.

4. Significance analysis of process parameters

Relative significance of an input variable are defined in different ways and calculated by various methods. In the present project the one-way ANOVA was adopted and implemented in a form of a dedicated MS Excel based software. The main features of this tool include automated computations of ANOVA parameters for up to 60 variables, their relative significances and the estimation of achievable output changes due to the given input changes. Fig. 1 shows an exemplary graph in which characteristics of statistically significant variables are presented. The character and direction of the influence of a selected input variable on the output can be also displayed in the form of linear graph including also the appropriate confidence intervals.

It is important that the input variables in ANOVA have to be of discrete type, representing their levels or simply nominal values (e.g. shift ID). The data preparation phase appeared to be extremely important as it can highly influence the ANOVA results. It includes a careful discretization of continuous type variables and representativeness of the data records for each of the input variable category or level.

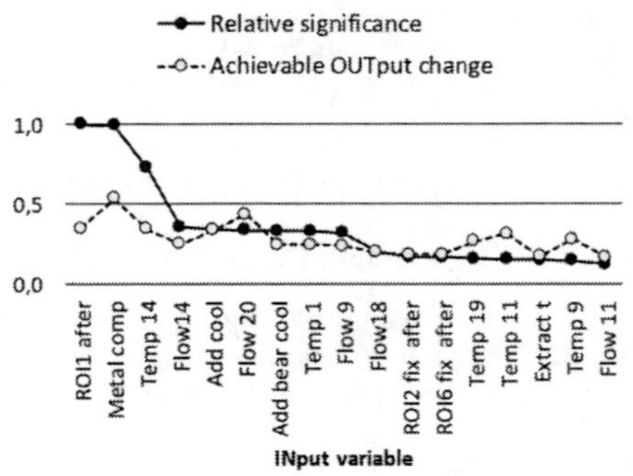

Example of ANOVA results for statistically significant variables for ultimate tensile strength

5. Big Data and 'small data' based models

A data-driven model, linking the mechanical properties of the cast alloy with process parameters selected as indicated in the previous sections, will be a final task of the project. It is expected that it will be possible to detect complex hidden interrelations between variables, including synergies. There are many model types within the Computational Intelligence area (discussed in [3]). The crucial problem in selecting the appropriate type are the small sizes of data sets containing measured values of mechanical properties, though the values of process parameters are available in the form of automatically recorded, large databases. It appears to be a more general observation that in the era of Industry 4.0 some important limitations due to the physical nature of products and processes may play a significant role.

6. Conclusions

The data-driven advanced model developed within the project can serve as a key element of 'Virtual Operator' and/or 'Analytical Operator' defined in [4]. It is expected that the results of Nemak 4.0 project will remarkably improve the stability and predictability of the die castings production process. It will also become a significant part of the Industry 4.0 framework in the foundry production area.

References

[1] N. Gramegna & F. Bonollo. (2016). HPDC foundry competitiveness based on smart Control and Cognitive system in Al-alloy products. La Metallurgia Italiana. No 6, 21-24.

[2] F. Bonollo. N. Gramegna & G. Timelli. (2015). High-Pressure Die-Casting: Contradictions and Challenges. JOM. 67(5), 901-908.

[3] M. Perzyk, J. Kozlowski & A. Kochanski. (2018). Intelligent data analytics for foundry industry 4.0. In 73th WFC, 23-27 September 2018. Krakow, Poland: Polish Foundrymen's Association (STOP).

[4] D. Romero, J. Stahre, T. Wuest, O. Noran, P. Bernus, Å. Fast-Berglund & D. Gorecky. (2016). Towards an operator 4.0 typology: a human-centric perspective on the fourth industrial revolution technologies. In CIE46 Proceedings, 29-31 October 2016 (1-11). Tianjin, China.

The 73rd World Foundry Congress
„Creative Foundry"
23 - 27 September 2018, Krakow, Poland

ANALYSIS OF COMPUTER SIMULATION DATA APPLICATION FOR STEEL CASTING DESING IN ORDER TO REDUCE ITS WEIGHT

Paweł Leszek Żak*, Michał Szucki, Józef Szczepan Suchy

[1]AGH University of Science and Technology, Faculty of Foundry Engineering,
23 Reymonta Str, 30-059 Krakow, Poland
*Corresponding address: pawelzak@agh.edu.pl

Key words: cooling rate, large-size castings, numerical simulation, steel, optimization

1. Introduction

Constant competition between casting companies requires them to fully control the product quality and reduce the expenses as much as possible. Especially in the case of large-size castings, the even small difference in its shape can lead to significant reduction of product weight what may have an impact on income. One of the most important parameter that effects on the casting microstructure is its solidification rate which is connected to heat transfer kinetics [1-3]. Reducing the wall thickness, in a reasonable limit, can have an influence on the casting microstructure and strength [4].

The aim of the research is to analyze the possibility of using MAGMA software to suggest changes in casting geometry. For this purpose, MAGMA optimization module [5] was used. Authors prepared the virtual geometry, which consists of segments of wall thickness that can be matched. Within presented work Authors developed some criterion which allows introducing the rank of assumed simulation variations.

2. Experimental system

The numerical experiments were performed with MAGMA 5.3 software. The analyzed casting geometry consists of five cuboids of given heights (*h*) and depths (*d*). Their widths were controlled by MAGMA's optimization module. The module allows analyzing defined set of simulations under declared by user rank criterions. Using this model requires from user to set group of variable parameters and their range of changes. In discussed case, this parameter was the width of the cuboid, as shown in Fig. 1. The remaining material data was assumed to be constant as can be found in Table 1.

Table 1. Parameters for numerical simulation [6]

Parameter	Value
Casting Alloy	Steel (GS18CrMo9_10)
Mould material	Furan
Initial temperature (metal)	1600 [°C]
Initial temperature (mould)	20 [°C]
Heat Transfer Coefficient	Steel-Sand HTC (from MAGMA Database)

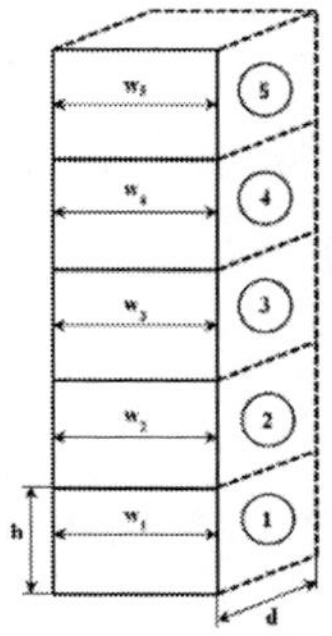

Fig. 1. Experimental system scheme

3. Optimization objectives and criteria

Main optimization objectives were total porosity in the casting, weighted porosity in each cuboidal segment, solidification time (maximal solidification time in the whole casting), solidification time in each cuboidal segment. The weight of porosity summation was the volume of the segment. It was assumed to treat the 5th cuboidal segment as a casting feeder, so the weight for it was zero. The future examination should approach to the specific solidification time value. For this value desired microstructure and casting strength can be obtained. The following criterion was declared to rank the simulations for the solidification time:

$$\sum_{i=1}^{4} v_i \, (t_{sol,i} - t^*_{sol})^2 \rightarrow MIN, \qquad (1)$$

where: v_i denotes volume of ith cuboid, $t_{sol,i}$ solidification time of ith cuboid and t^*_{sol} is desired solidification time, that assure strength of the casting. Criterion described with equation (1) is standard least squares approach and takes into account differences symmetrically equal. This value is multiplied by volume of the section, to prefer the smaller volumes, which production costs are less.

4. Results of numerical analysis

Several numerical investigations were performed with included and excluded some objectives type, criterion and size limitation for cuboidal segments. Size was set constant to $h = 300$ [mm],

$d = 400$ [mm] and w_i could only change in one direction.

During first numerical experiment, solidification time was not taken into account and w_i size change within limit 100 [mm] to 800 [mm] with the step 100 [mm].

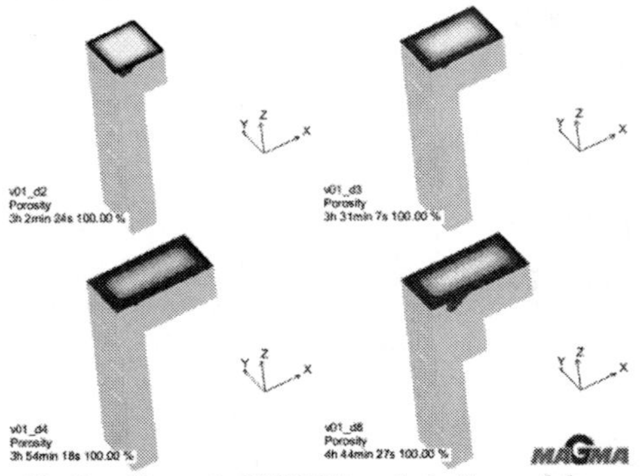

Fig. 2. Chosen by MAGMA's optimization module results that are characterized by the smallest weighted porosity

Results shows, Fig. 2, that each following cuboid from the top of the casting to bottom should have smaller width. The lowest porosity was noticed in the castings where all four segments has width 100 [mm] and has a significantly larger feeder.

Further examinations took into account the limits for chosen segments: their size was set to be larger than 100 [mm], which can affect the directional solidification (Fig. 3). Another attempt was to assume the width of the first cuboid to be larger than fifth "feeder" cuboid (Fig. 4). During those numerical experiments criterion described with Eq. (1) was taken into account with solidification time, t^{*}_{sol}, equal 8 000 [s].

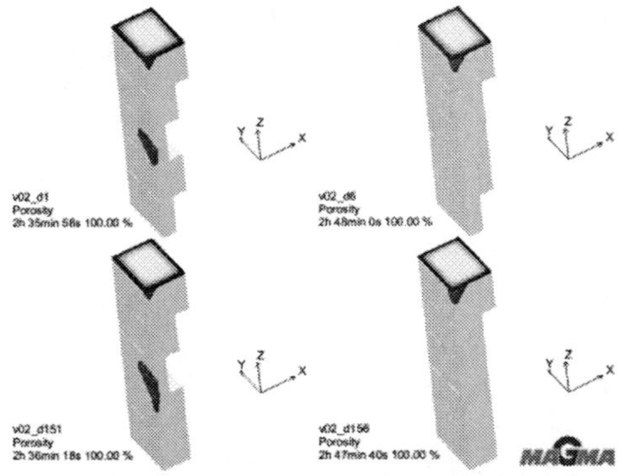

Fig. 3. Optimization results for the assumed size of the second cuboid with solidification time criterion

It was also analyzed how the limits for second segment size will effect on the porosity and properties of the casting. It was assumed that second segment width is set to be 200 [mm]. Obtained data shows that additional limits for geometry can result in the porosity increase. General volume of the casting is greater what has an impact on total porosity volume.

Next analysis took into account case when segment no.1 is large. Again in this case it is shown that porosity can be limited by the special shape of casting segmented geometry, however, it is very difficult to remove porosity completely.

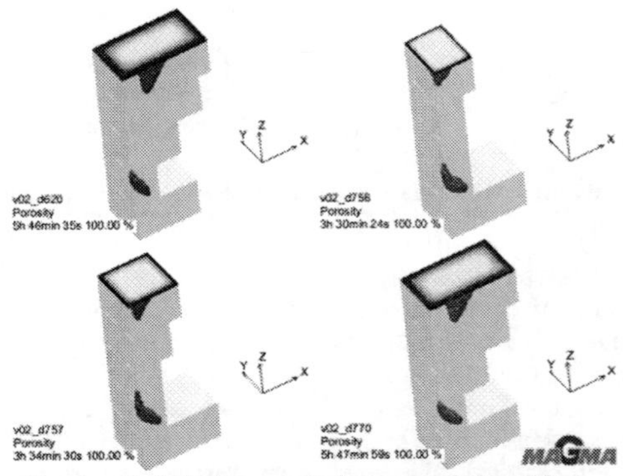

Fig. 4. Optimization results for assumed size of first cuboidal segment to be larger than fifth

5. Conclusions

Presented examination shows that it is possible to use MAGMA optimization module to automatization of casting shape design. Results show that it is possible to use the described procedure for steel casting in order to reduce its wall thickness. Some technological demands can highly complicate the re-designing process thus it is important to analyse designs chosen by the software for the factors that ranked then. Still presented tool can be highly helpful during casting geometry improvement for economic and ecological reasons.

Acknowledgements

The Authors kindly acknowledge the financial support under AGH Project no. 11.11.170.318 task 1.

References

[1] Gradowski A., Suchy J.S., Zak P. L., Lomnicki R., Lelito J. & Gracz B. (2012) Mathematical model of the tensile strength – on the basis of the casting solidification kinetics. Przeglad Odlewnictwa 62 (1-2), 38-41.

[2] Kenawy M.A., Abdel – Fattah A.M., Oksha N. & El-Gazery M. (2001). Mechanical and Structural Properties of Ductile Cast Iron. J. Sol. 24, 151-159.

[3] Havlicek F., Eldbel T. (2011) Geometrical modulus of a casting and its influence on solidification process. Archives of Foundry Engineering 11(4), 170-176.

[4] Adnan Calik (2009). Effect of cooling rate on hardness and microstructure of AISI 1020, AISI 1040 and AISI 1060 Steels. International Journal of Physical Sciences 4 (9), 514-518.

[5] MAGMA 5.3 Manual; build-in the software

[6] MAGMA 5.3 Database; build-in the software

The 73rd World Foundry Congress
„Creative Foundry"
23 - 27 September 2018, Krakow, Poland

NUMERICAL AND EXPERIMENTAL STUDIES ON THE COOLING CONDITIONS OF CAST ALUMINIUM SEMI-FINISHED PRODUCT FOR FORGING PROCESS

Michał Szucki [1]*, Janusz Buraś [1], Anna Dziubińska [2], Paweł Leszek Żak [1]

[1] AGH University of Science and Technology, Faculty of Foundry Engineering, 23 Reymonta Street, 30-059 Krakow, Poland
[2] Lublin University of Technology, Faculty of Mechanical Engineering, 36 Nadbystrzycka Street, 20-618 Lublin, Poland
**Corresponding address: mszucki@agh.edu.pl*

Key words: aluminium alloys, numerical simulations, heat transfer, casting, forging

1. Introduction

Forging is a technology that allows obtaining metal products with high mechanical properties and a homogeneous structure [1]. On the other hand, its disadvantage is the higher production costs and less versatility in comparison to the casting. Therefore, nowadays attempts are made to combine these two manufacturing techniques in one process [2,3]. In this case, the semi-finished products are made by casting technology, and then the final shape and properties are obtained by forging. Unfortunately, one of the disadvantages of such approach is the problem with the right choice of material - metals alloy. As is well known the metal alloys used in forging and casting differ in composition and properties. Previous research [4] indicates that the use of pre-cast aluminium alloys of the 2xxx and 7xxx series for plastic forming can result in a tendency to crack during forging. At the same time, it was observed [4] that these problems can be overcome by providing adequate cooling rates during casting resulting in finer microstructure and improved plastic properties. Therefore, the aim of this work was to determine the proper cooling conditions based on the results of experimental studies and numerical simulations.

2. Experimental studies

Typical aluminium alloy 7075 (Table 1) for forging process was selected as the testing material.

Table 1. Composition of 7075 aluminium alloy [5]

Element	wt %
Al	Balance
Zn	5.1-6.1
Mg	2.1-2.9
Cu	1.2-2.0
Fe	0.5
Si	0.4
Mn	0.3
Cr	0.18-0.28
Ti	0.2

First, experimental studies were carried out including casting cylindrical samples of dimensions ϕ32×75 mm.

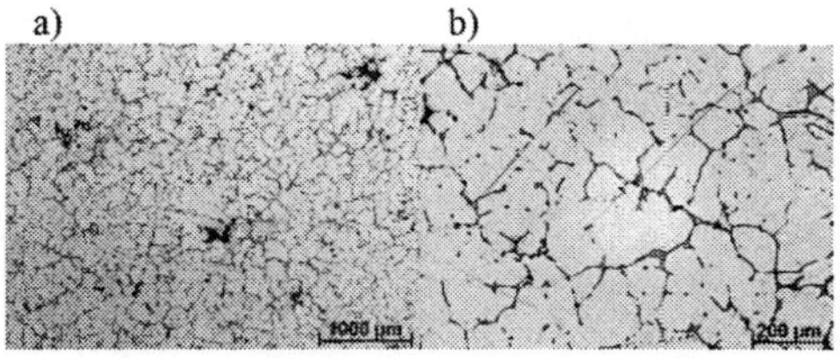

Fig. 1. ***Microstructure of 7075 alloy cast to the sand mould; magnification 2.5× (a) and 10× (b)***

Melted alloy was overheated to 750 [°C] and cast into the: green sand mould, cold (in environmental temperature) permanent (steel) mould and permanent (steel) mould heated up to 300 [°C] respectively. Thus it was possible to obtain a different cooling rate for each casting . In addition, a thermocouple (in the geometric centre of the samples) was placed in each of the mould. The specimens obtained in this way were then subjected to microscopic (Fig. 1) examination and upsetting in the temperature range between 420÷460 [°C] (Fig. 2).

Fig. 2. ***Surface of 7075 alloy (sand mould) samples after upsetting***

Data from thermocouples indicate a clear difference in the cooling rate of 7075 alloy in individual moulds, which translates directly to the different microstructure of the castings obtained. It was observed that castings from the sand mould are characterized by a much more coarse microstructure. Thus, during upsetting, these samples have a tendency to crack. This problem was avoided in the case of permanent moulds.

3. Numerical simulations

The next stage of the research included reflection of the above experiments in numerical simulations in the COMSOL Multiphysics software. This allowed to determine the local cooling rates depending on the type of mould used.

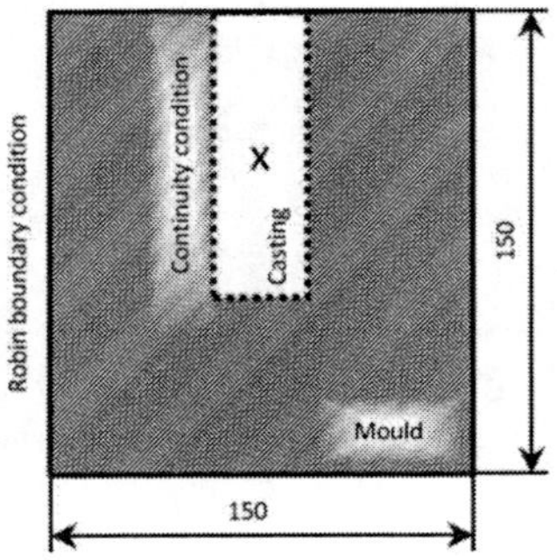

Fig. 3. Computational domain scheme; mark "x" shows a location of thermocouple

The two-dimensional system was used in the simulations with continuity and Robin boundary condition on the internal and external boundary respectively (Fig. 3). The solidification process was included by the substitute thermal capacity method.

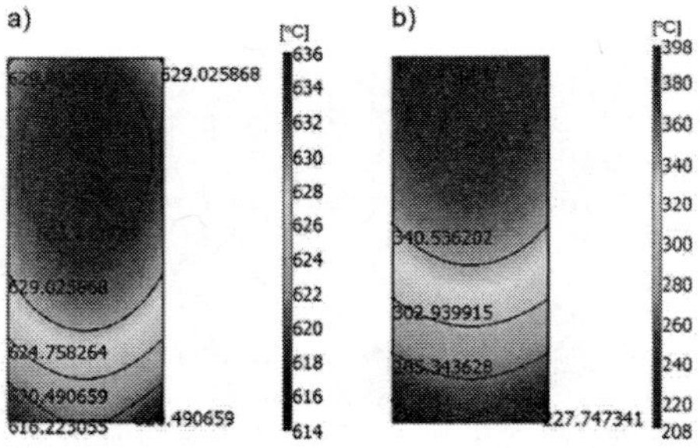

Fig. 4. Sample temperature fields at t=20 [s] obtained in numerical simulations: sand mould (a), cold permanent mould (b)

The results obtained (Fig. 4) were characterized by a good agreement with the experimental data. The simulations also provide the link between local cooling rate/solidification time and the microstructure.

This part of the study also includes additional numerical experiment where mould was omitted and heat exchange between casting and environment was set directly by proper boundary condition (assumed heat flux value) on the casting boundary.

As can be seen in Figure 5, the heat dissipation rate directly affects the casting solidification time.

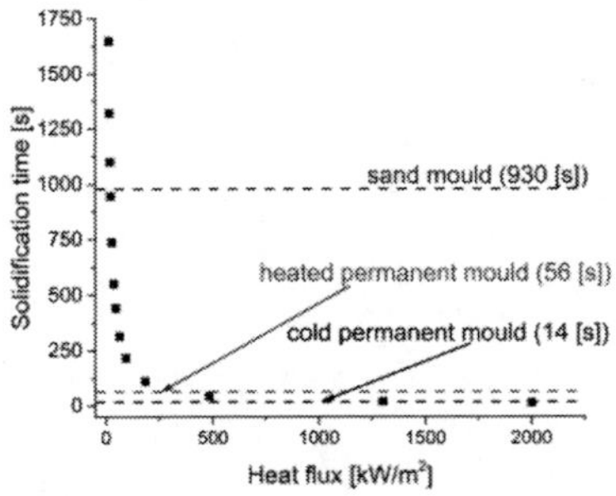

Fig. 5. Solidification time as a function of the heat flux at the boundary

4. Conclusions

In order to use semi-finished cast products in the forging process, it is necessary to provide an appropriate casting microstructure. If the chemical composition of the alloy cannot be changed (e.g. through the grain refinement), the main factor determining the obtaining of a finer grain size is the cooling rate and indirectly solidification time. The presented results showed that it is possible to determine the optimal cooling condition in order to obtain proper casting microstructure.

Acknowledgements

The National Centre for Research and Development

LIDER

Project "The Design and Implementation of an Innovative Low-Waste Metal Forming Technology for High-Strength Light Metal Alloys Used in the Aircraft and Automotive Industry"; No. LIDER/10/0058/L-8/16/NCBR/2017.

The project is financed by the National Centre for Research and Development under the 8th edition of the LIDER Programme.

References

[1] Kuhlman G.W. (2005). Forging of Aluminum Alloys. ASM Handbook. Vol. 14, pp. 299-312.

[2] Chang F.-C., Hwang W.-S., Lee C.-H., Wu C.-F., Yang J.-B. (2004). Forging Condition for Removing Porosities in the Hybrid Casting and Forging Process of 7075 Aluminum Alloy Casting. Materials Transactions. Vol. 54, pp. 1886-1890.

[3] Zhang Q., Cao M., Zhang D., Zhang S., Sun J. (2015). Research on Integrated Casting and Forging Process of Aluminum Automobile Wheel. Advances in Mechanical Engineering. Vol. 2014, pp. 1-12.

[4] Winiarski G., Dziubińska A., Majerski K., Szucki M., Drozdowski K. (2018). Investigation of the deformability of aluminium-copper casting alloys. Advances in Science and Technology Research Journal (in print).

[5] Rajan K., Wallace W., Beddoes J.C. (1982). Microstructural study of a high-strength stress-corrosion resistant 7075 aluminum alloy. Journal of Material Science. Vol. 17, pp. 2817-2824.

The 73rd World Foundry Congress
„Creative Foundry"
23 - 27 September 2018, Krakow, Poland

CONCEPT OF THE SMART FOUNDRY PLATFORM INTEGRATING INDUSTRY 4.0 TECHNOLOGIES

Pawel Malinowski [1*]

[1]*AGH University of Science and Technology, Faculty of Foundry Engineers, Mickiewicza 30 Av., 30-059 Krakow, Poland*
**Corresponding address: pamalino@agh.edu.pl*

Key words: Smart Foundry, Industry 4.0, knowledge base, system integration

1. Introduction

Industry 4.0 is primarily the integration of various technologies, devices, sensors, and computer systems into one. Such an ecosystem can communicate, exchange information between nodes, process real-time data from different stages of production, use complex artificial intelligence algorithms, and as a result significantly increases the added value of the integrated ecosystem. The integration of systems is a combination of them into one unit, enabling efficient communication, data exchange between subsystems.

2. Smart Foundry system as a core platform

The Smart Foundry platform as an open hybrid system that integrates various INDUSTRY 4.0 technologies.

Foundry 4.0 - using the Smart Foundry platform - an ecosystem of technologies, machines, IoT (Internet of Things) systems, devices, measurement sensors, computer systems, robots, computer image analysis algorithms, databases, CAx systems (Computer Aided Technologies) dedicated to the foundry industry , data mining systems (Data Mining), advanced artificial intelligence algorithms, VR systems (Virtual Reality) cooperating with each other, communicating with each other generates great added value for each enterprise.

The integration of systems is becoming more and more important. Thanks to this, a synergy effect is achieved - communication, collaboration, exchange and advanced data analysis. The advantage of implementing the Smart Foundry concept flowing from system integration is much greater than the sum of the benefits of individual systems operating separately.

The value of information, the possibility of sharing it by various subsystems is becoming more and more important in the era of computerization, automation of production processes and gives an advantage over the competition.

The construction of the Smart Foundry system, which combines selected areas of Industry 4.0, enables better optimization of production while maintaining the current quality. By archiving and analyzing information from various stages of production, using artificial intelligence algorithms, neural networks, image analysis algorithms, data mining methods (data mining), we are able to obtain information on the causes of casting defects, predict their formation, etc.

Listing the INDUSTRY 4.0 ecosystem in one Smart Foundry platform is a big challenge, but it brings many benefits.

3. Idea of Smart Foundry platform

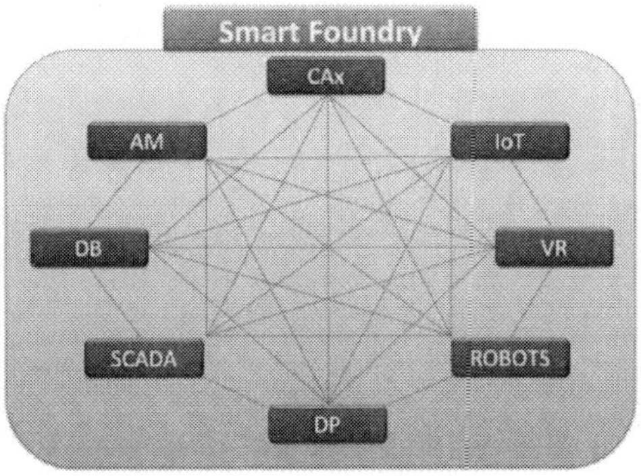

Fig. 1. Schematic of Smart Foundry.

4. Example of using industry 4.0 technology

The article describes the application of industry 4.0 in the foundry industry using a simulation database and a computer image analysis system.

References

[1] P. Malinowski, J.S. Suchy, J. Jakubski. 2013. Technological knowledge management system for foundry industry. Archives of Metallurgy and Materials / Polish Academy of Sciences. Committee of Metallurgy. Institute of Metallurgy and Materials Science. ISSN 1733-3490. 58, 965–968.

[2] Decker PT. Presidential address: False choices, policy framing, and the promise of "Big Data". Journal of Policy Analysis and Management 2014; 33(2): 252-62.
[3] Smart Foundry-die Industrie 4.0 - Gießerei-stahl-online.de.
[4] Thareja P. Competitive foundry through integration of TIPS (Technology, Innovation, Product Design, and Process Systems), (Key note talk at BRICS Foundry Forum, Brazil) [also in Indian Foundry Journal 2016; 58(1): Jan 2012.
[5] Löffler M, Tschiesner A. The Internet of Things and the future of manufacturing. McKinsey's Interview of Siegfried Dais June 2013. Available from: http://www.mckinsey.com/business-functions/digitalmckinsey/our-insights/the-internet-ofthings-and-the-future-of-manufacturing.

APPLICATION OF SELECTED ARTIFICIAL INTELLIGENCE METHODS IN THE SYSTEM PREDICTING THE MICROSTRUCTURE OF COMPACTED GRAPHITE IRON

Barbara Mrzygłód [1]*, Grzegorz Gumienny [2], Dorota Wilk-Kołodziejczyk [1,3]

[1]*AGH University of Science and Technology, Faculty of Metals Engineering and Industrial Computer Science, Mickiewicza 30 Av., 30-059 Cracow, Poland*
[2] *Lodz University of Technology, Faculty of Mechanical Engineering, 116 Żeromskiego Str. 90-924 Lodz, Poland*
[3] *Foundry Research Institute, Zakopianska 73 Str., 30-418 Cracow, Poland*

**Corresponding address: mrzyglod@agh.edu.pl*

Key words: application of Information Technology to the Foundry Industry, compacted graphite iron, data mining, algorithm ANFIS, artificial neural networks (ANN)

CGI (Compacted Graphite Iron), having a specific graphite form with large matrix contact surface, is a unique casting material. It is characterized by a minimum tensile strength of 300 to 500 MPa with an elongation from 0.5 to 2% respectively. Compared to grey cast iron, this material is characterized by higher strength properties and better ductility. In comparison to spheroidal graphite iron, vermicular graphite iron has a lower coefficient of thermal expansion, higher thermal conductivity, higher resistance to dynamic temperature changes, higher vibration damping capacity and better casting properties. All these advantages predispose CGI to a variety of applications.

Technical literature describes numerous studies of the properties of compacted graphite iron [1-8], while there are relatively few publications on the influence of alloy additives on the microstructure and properties of this cast iron [9-10]. By changing the chemical composition, it is possible to modify to a large extent the cast iron microstructure and, as a consequence, its properties.

As part of many research works conducted at the Lodz University of Technology, a number of material experiments were carried out to examine the effect of the content of selected alloying elements on the microstructure of compacted graphite iron. Many years of studies have finally resulted in the development of a database that integrates the results of these studies. The database contains the results of examinations and measurements of the content of individual constituents of the microstructure of compacted graphite iron (ferrite, pearlite, carbides, martensite, ausferrite, austenite) examined as a function of the content of individual alloy additives within the studied range of the chemical composition, including:

- Cu content [wt%],
- Ni content [wt%] ,
- Mo content [wt%],

for different casting wall thicknesses [mm]. The developed database contains 170 records; its fragment is presented in Figure 1.

INPUTS OUTPUTS

Cu	Ni	Mo	[illegible]	[illegible]	[illegible]	[illegible]	[illegible]	[illegible]	[illegible]
0	9,26	0	24	0	0	0	95	0	5
0	14,89	0	3	0	0	0	15	0	85
0	14,89	0	6	0	0	0	21	0	79
0	14,86	0	12	0	0	0	30	0	70
0	14,86	0	24	0	0	0	37	0	63
0	16,53	0	3	0	0	0	7	0	93
0	16,53	0	6	0	0	0	8	0	92
0	16,53	0	12	0	0	0	10	0	90
0	16,53	0	24	0	0	0	11	0	89
0	23,04	0	3	0	0	0	0	0	100
0	23,04	0	6	0	0	0	0	0	100
0	23,04	0	12	0	0	0	0	0	100
0	23,04	0	24	0	0	0	0	0	100
0	0	1,98	3	8	49	43	0	0	0
0	0	1,98	6	18	57	18	0	7	0
0	0	1,96	12	31	26	3	0	40	0
0	0	1,96	24	30	62	3	0	5	0
0,49	0	2,42	3	0	0	3	0	97	0
0,49	0	2,42	6	0	0	3	0	97	0
0,49	0	2,42	12	1	0	4	0	95	0
0,49	0	2,42	24	1	1	5	0	93	0
0,98	0	2,98	3	0	0	2	0	98	0
0,98	0	2,98	6	0	0	1	0	99	0
0	0	0	24	27	73	0	0	0	0
0	0	0	3	0	72	28	0	0	0
...	...	...	...	...	...	...	...	...	...

Fig. 1. Fragment of the developed database

With the integrated knowledge of selected process available in the database, one can attempt to construct algorithms and inference systems that enable automatic processing of this knowledge. As a consequence, it is possible to build computer systems that give the opportunity to create new knowledge without the need to perform additional material experiments.

In this study, model learning cases were used to determine the parameters of neural networks and the parameters of adaptive neural-fuzzy inference system (ANFIS). Intelligent computing tools, such as artificial neural network (ANN) [11-12] and fuzzy logic (FL) [13-14], can be used as modelling tools supporting the prediction of selected process parameters. They are useful mainly when it is necessary to model phenomena of a strongly non-linear nature and multidimensional functional dependencies that are difficult to determine in a purely analytical form.

The developed models are a form of knowledge representation in the developed computer system, whose main purpose is to determine the content of individual phases in the microstructure as a function of the content of individual alloy additives.

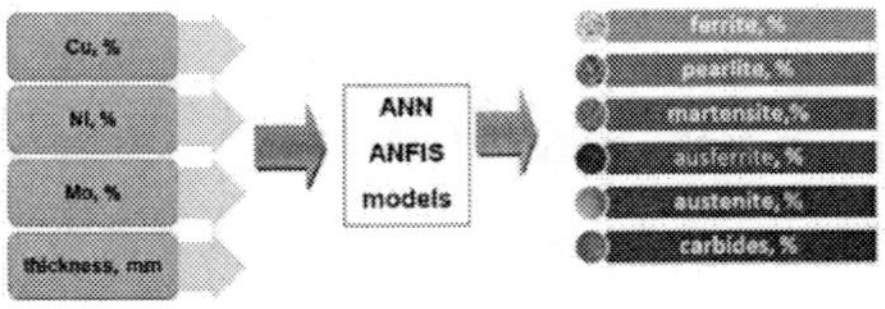

Fig. 2. List of variables in the system predicting the Microstructure of Compacted Graphite Iron

Acknowledgments

Financial support of The National Centre for Research and Development LIDER/028/593/L-4/12/NCBR/2013

References

[1] Pietrowski, S. (2000). Compendium of knowledge about vermicular cast iron. *Solidification of Metals and Alloys*. *2*(44), 279-292. (in Polish).

[2] Abele, E., Sahm, A. & Schulz, H. (2002). Wear Mechanism when Machining Compacted Graphite Iron. *CIRP Annals*. *51*(1), 53-56.

[3] Slatter, T., Lewis, R & Jones, A.H. (2011). The influence of induction hardening on the impact wear resistance of compacted graphite iron (CGI). *Wear*, *270*(3-4), 302.

[4] Qiu, Y., Pang, J.C., Yang, E.N., Li, S.X. & Zhang, Z.F. (2016). Transition of tensile strength and damaging mechanisms of compacted graphite iron with temperature. *Materials Science and Engineering: A*. *677*, 290-301.

[5] Górny, M. & Kawalec, M. (2013). Role of Titanium in Thin Wall Vermicular Graphite Iron Castings Production. *Archives of Foundry Engineering*. *13*(2), 25-28.

[6] Pina, J.C., Shafqat, S., Kouznetsova, V.G. Hoefnagels, J.P.M. & Geers, M.G.D. (2016). Microstructural study of the mechanical response of compacted graphite iron: An experimental and numerical approach. *Materials Science and Engineering: A*. *658*, 439-449.

[7] Kopyciński, D., Guzik, E., Nowak, A., Ronduda, M. & Sokolnicki, M. (2012). Preparation Vermicular Graphite in Thin and Thick Wall Iron Castings. *Archives of Foundry Engineering*. *12*(2), 41-44.

[8] Pietrowski, S. (1998). A mechanism of the vermicular graphite crystallization in cast iron. *Solidification of Metals and Alloys*. *37*, 97-104. (in Polish).

[9] Soiński, M.S. & Mierzwa, P. (2011). Effectiveness of cast iron vermicularization including 'conditioning' of the alloy. *Archives of Foundry Engineering*. *11*(2). 133-138.

[10] Guo, Y., Stalbaum, T., Mann, J., Yeung, H. & Chandrasekar, S. (2016). Modulation-assisted high speed machining of compacted graphite iron (CGI). *Journal of Manufacturing Processes*. *15*(4), 426-431.

[11] Piegat A. (1998). Modelowanie i sterowanie rozmyte. Akademicka Oficyna Wydawnicza EXIT, Warszawa1998.

[12] Tadeusiewicz R. Neural networks in mining sciences – general overview and some representative examples, Arch. Min. Scien. 2015; 60 (4): 971–984.

[13] Nayak PC, Sudheer KP, Rangan DM, et al. (2004). A neuro-fuzzy computing technique for modeling hydrological time series, JOURNAL OF HYDROLOGY, 291(1-2), 52-66

[14] Altug S., Chow M.Y., Trussell H.J., (1999). Fuzzy inference systems implemented on neural architectures for motor fault detection and diagnosis, IEEE Trans Ind Electron, 46 (6), 1069–1079

COMPUTER SIMULATION OF SOLIDIFICATION OF CASTING WITH COMPOSITE ZONE BASED ON TiC REINFORCEMENT

Sebastian Sobula[1], Andriy Burbelko[1], Daniel Gurgul[1], Ewa Olejnik[2],

[1]AGH – University of Science and Technology, Faculty of Foundry Engineering, ul. Reymonta 23, 30-059 Kraków, Polska
[2] Innerco Sp z o.o., ul. Jadwigi Majówny 43A, 30-298 Kraków, Polska
**Corresponding address: e-mail: sobula@agh.edu.pl*

Key words: composite casting, solidification, numerical simulation

1. Abstract

The most used materials in crushers, mills, etc., for abrasive resistant parts are iron alloys. These materials are characterized by high wear resistance due to their hardness. Presently crusher manufacturers are searching for materials which extent working life of these parts. The solution of their expectations are composite materials, for example, cast steel or cast iron reinforced with ceramic particles. These materials offer higher hardness and significantly improve wear resistance of equipment applied in wear environment [1-2]. In Fig.1. the summary weight loss of low carbon cast steel in comparison to the same steel reinforced with titanium carbide particles is presented, [3].

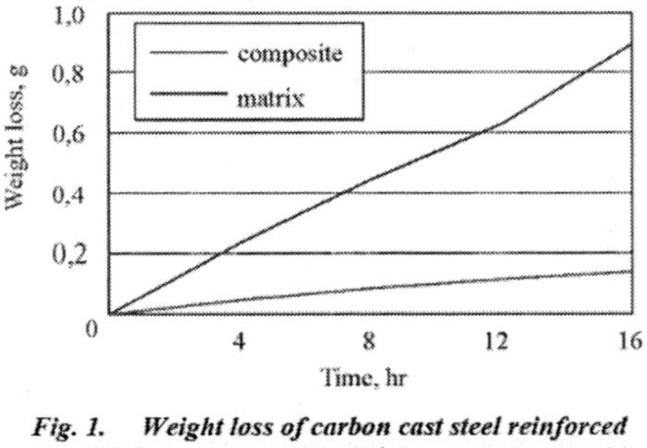

Fig. 1. Weight loss of carbon cast steel reinforced with TiC (marked as composite) in comparison to this same steel without reinforcement obtained in Miller slurry test, [3]

Manufacturing of composite casting is difficult due to many factors which influence the solidification process and cause occurrence of porosities in composite zones.

In this paper in-situ reaction of the self-propagating high temperature synthesis (SHS), [4-5] of TiC carbides synthesis was utilized for the formation of the composite area in casting. Compacts were prepared from the powders materials of the SHS reaction by cold pressing. Next they were placed in the foundry mold cavity before filling. In the initial stage of the mold filling the partial solidification of the poured alloy is possible on the surface of these compacts. Over time after liquid metal is poured into the mold, after the heating of the powder compacts over the temperature of the SHS reaction start, synthesis of TiC in the compacts is started. During synthesis a significant amount of heat is released, and frozen layer is melted. This phenomenon helps in good infiltration of alloy into carbide particle agglomerates. If the weight of compact is too large in relation to the weight of liquid alloy, the synthesis may not occur due to low temperature of the raw materials of the SHS reaction. To analyze the risk of this situation computer modeling of the temperature field in the casting and raw material inserts during mold cavity filling, computer simulations were used with ProCAST software, ESI Group®.

Among of the main factors, which is of great importance for this process accuracy, is the temperature of initiation of SHS reaction of the carbide particles synthesis which are main reinforced constituent. Very important are also the parameters such as reaction time, rate of the TiC formation and the heat released during this transformation.

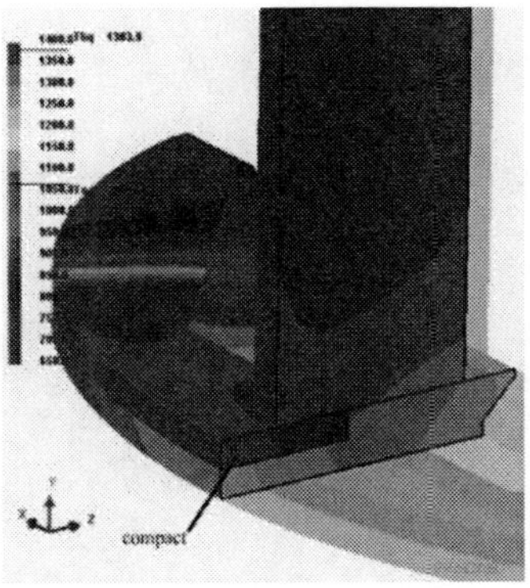

Fig. 2. Temperature field in the experimental casting

To verify the results of the simulation an experimental casting made from high chromium cast iron with composite zone contained TiC reinforcement was produced. The field of temperature in experimental casting with composite zone is shown in Fig. 2. It is shown that temperature in compact is higher than in the casting and even in the raiser. Finally that phenomenon may cause hot spot and porosity in the composite zone and in the region near them.

During the experiment temperature in selected measuring points of casting was measured by thermocouples and recorded by computer system of data measurement and acquisition. Set of the experimental data was compared with data obtained by simulation with ProCAST software.

Acknowledgments
Project is funded by European Funds as part of the National Centre of Research and Development project nr: POIR.04.01.04-00-0067/15, entitled: Development of technology for the production ultra w ear resistant castings with composite zones type TiC/Fe base alloys.

References

[1] Rohatgi P. & Asthana R. (2001). Solidification Science in Cast MMCs: The influence of Merton Flemings. JOM, Volume 53 (9), pp. 9-13

[2] Głownia J., Tęcza G., Asłanowicz M. & Ościłowski A. (2013). Tools cast from the steel of composite structure. Archives of Metallurgy and Materials. Volume 58, (3), pp. 803-808

[3] Sobula S., Olejnik E. & Tokarski T. (2017). Wear resistance of TiC reinforced cast steel matrix composite. Archives of Foundry Engineering. Volume 17 (1), pp. 143-146

[4] Merzhanov A.G. (1969). The theory of stable homogeneous combustion of condensed substances. Combustion and Flame. Volume 13, (2), pp. 143-156

[5] Fraś E. Janas A., Wierzbiński S. & Kolbus A. Odlewane aluminiowe kompozyty „insitu" umacniane węglikami tytanu. (2000) Solidification of Metals and Alloys, Volume 2 (43) pp. 167-174

ECOLOGY

The 73rd World Foundry Congress
„Creative Foundry"
23 - 27 September 2018, Krakow, Poland

FROM WASTE FOUNDRY SAND TO A NEW BIODEGRADED RAW MATERIAL AN ECOLOGICAL SOLUTION FOR FOUNDRIES

Patricia Caballero [1]*, Ester Villanueva [1], Iñigo Crespo [1], Sara Tapola [2]

[1]*Tecnalia Research and Innovation, Mikeletegi 2-20009 San Sebastian, Spain*
[2] *Meehanite Technology Oy, Kuokkamaantie 4, 33800 Tampere, Finland*
**e-mail: patricia.caballero@tecnalia.com*

Key words: Creative-foundry, Ecology, Foundry-sand, Bio-compost, Circular-economy-foundry

1. Introduction

Today, the majority of spent moulding sand from foundries (around 18M tons per year in Europe) is disposed of in landfill sites. This method of disposal represents a waste of material and carries environmental and industrial costs. Additionally, the formaldehyde present in used foundry sand has recently been classified as carcinogenic. With this backdrop, the European Life project "Foundry Sand" described herein, set out to investigate the viability of transforming waste sand into inert compost via biodegradation. This material would potentially have a commercial use as fertilizer or as added-value filler for geothermal energy applications, for example.

The principle objective was to investigate the potential of a natural process (biodegradation) as a method to neutralize contaminants resulting from an industrial process. Other aims included reducing the volume of waste sent to landfill sites by foundries, and, if found to be viable, encouraging the uptake of the method through appropriate dissemination channels Europe wide; around 3,000 foundries and around 1,000 potential customers for the sand.

2. Materials and Methods

Using the principle of biodegradation as the transformation process, all types of waste foundry sand (containing added binding agents) were mixed with organic material in open air "composting" heaps, weighing from 12 to 30 tons. Over a period of three years in two climatically different European test sites (located inland in Finland, and on the coast of northern Spain), the evolution of the heaps was periodically monitored and analyzed for the following: temperature, humidity, gas emissions, DOCs, TOCs, BTEXs, phenols, conductivity and chemical and metallic components.

Temperature, an indicator of the different stages of biodegradation, was monitored via Wi-Fi connected thermocouples inserted in the heaps. Humidity was monitored using similarly placed digital hygrometers.

Gas emissions were captured by overhead intake and analyzed using portable equipment. Liquid seepage and solid matter were collected periodically for laboratory analysis.

In order to take into account, the effects of seasonal climate conditions, the heaps were assembled at different times throughout the year. The total weight of sand processed was 600 tons, and the heaps occupied a combined total area of 2,000m^2. In order to accelerate biodegradation, the heaps were aerated every fifteen days using forklift trucks or mechanical grabs. To maintain humidity levels when RH < 60% in year one (2015) threatened to compromise biodegradation, irrigation water was applied to the heaps in Spain.

As a complementary test in Finland, odor emissions in some of the heaps were measured using specific bio-filtration equipment. The composition of each heap reflected the variety of sand and moulding processes used in the wide range of foundries in operation, e.g. silica sand used in green sand moulding was combined with a standard mix of organic material (forest waste and horse manure). Similarly, olivine sand used in the silicate moulding process was combined with the organic mix to constitute a different trial heap. See full Table 1. The forest waste combined with the sand constituted the bulk of the final material, while the horse manure primarily acted as an accelerant of biodegradation. Prior to the trials, horse manure was found to be more effective than that of other ruminants due to its higher concentration of elements such as ammonia and phenol. The optimum proportion of forest waste to manure had been researched over a period of one year at Helsinki University.

Laboratory tests were made using silicate sand (formaldehyde binder). In field tests, after three months on average, the material was considered ready for shredding to create a compost-like texture. This was then analyzed as compost, fertilizer and filler, in order to evaluate compliance with the corresponding national and European regulations on these materials.

Table 1.

MOULDING PROCESS	SAND TYPE	BINDER TYPE
Green sand	Silica	Bentonite
Silicate	Silica	Silicate and ester
Silicate	Olivine	Silicate and ester
Furanic	Silica	Furanic resins
Furanic	Chromite	Furanic resins
Phenol	Silica	Alkaline phenolic
Alfa-set	Silica	Phenolic no bake
Pep-set	Silica	Phenolic urethane

During the project over 6,000 chemical and biological analyses and tests were carried out in Finland and over 2,900 analyses in Spain from surplus foundry sand samples, composting materials and waste, including water runoff.

3. Results

Initial concentrations of hazardous components in the sand, such as fluoride, phenols, PAH and BTEX, diminished sufficiently during the process to fall within the required limit values: 10 mg/kg for fluoride, PAH 55mg/kg, BTEX 6mg/kg.

In the majority of cases, analysis of the biodegraded material gave results that met the norms and pertinent regulations for compost, fertilizer and filler at national and European levels. Following the biodegradation process, BTEX and PAH found in all foundry sand types were reduced to levels considered safe and non-carcinogenic.

Other hazardous substances (chloride, fluoride and phenol etc.) were also found to diminish sufficiently to meet limit values (Fig. 1).

Hazardous parameters	Green sand (stage 3)	Phenolic sand (stage 3)	Silicate sand (Stage 3)	Limit value for inert waste
Chlorides (mg/kg)	89,5	79	768,75	800
Fluorides (mg/kg)	2,3	<5	5,3	10
Suphates (mg/kg)	<50	<50	153,25	1000
Phenol (mg/kg)	<0,5	0,5	<0,5	1
DOC mg/kg)	846	736	2987,5	500
BTEX (mg/kg)	0	<0,040	0,031	6
PCB (mg/kg)	<0,1	<0,1	<0,1	1
Mineral oil (mg/kg)	<20	<20	<20	500
PAH (mg/kg)	0,302	<0,160	0,15	55
pH (UpH)	8,75	8,7	7,35	>6

Fig. 1. ***Relation sands and hazardous components***

In the case of Furanic sand used in valve casting (duplex steel) which represents around 3.5% of total foundry sand, DOC and TOC levels were reduced, but remained above thresholds. As well silicate sand. See Fig. 2.

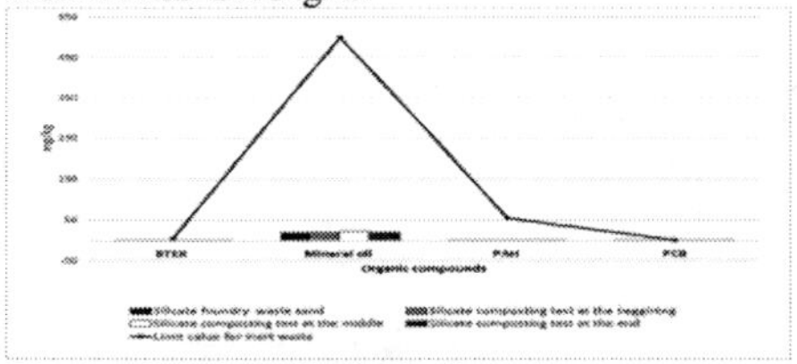

Fig. 2. ***Relations of reducing hazardous components by biodegradation compos***

The irregular results in phenol content were originally thought to be due to binder type (organic or non-organic). However, tests show that the principle source of phenol was pasture grass, traces of which are found in horse manure, and, to a lesser extent, the type of binder used to harden the mold sand, and elements of the feeder system such as sleeves.

Microbiological and phytotoxicity tests were carried out and gave results which were not considered toxic. In addition, the germination index was 82%, above the threshold for toxicity set at 75%. The results achieved (Fig.3) are in line with the principles of a circular economy in which material and energy usage are optimized in a regenerative loop.

Fig. 3. ***Circular economy applied to foundry sand***

4. Conclusions and future actions

The experiments carried out between 2015 and 2017 demonstrate that the proposed method is an innovative, viable and ecological way to recycle 96.5% of waste foundry sand destined for landfill. The final product complied with the pertinent limit values for compost, fertilizer and construction filler in Europe. Foundries adopting this solution would make an industrial cost saving in landfill disposal costs (transport, space occupied and admin) and could even obtain a return on the sand. Further investigations for geo-thermal applications include pilot trials (scheduled for the near future) into conductivity and resistivity.

Acknowledgments

The authors are grateful for the support received from the EU LIFE programme (LIFE13 ENV/FI/285).

References

[1] Recycling of Waste in Foundries', International Committee of Foundry Technical Associations, Sweden, August 1997.

[2] Kniese, B. 1992. Reduction and/or reuse of foundry waste materials. pp. 3.1–3.14.59th World Foundry Congress, St. Paulo, Brazil, September

[3] Felipó, M.T., Soliva, M. 2003. Organic wastes as a resource for Mediterranean soils.

[4] Langenkamp, H., Marmo, M. 2002. Biological Treatment of Biodegradable Wastes. Technical Aspects. Workshop in Brussels.

[5] Huerta, O., López, M., Soliva, M., Zaloña, M. 2010. Compostaje de residuos municipales: control del proceso, rendimiento y calidad del producto obtenido.

[6]http://upcommons.upc.edu/e-prints/handle/2117/9086

[7] López, M., Huerta-Pujol, O., Martínez-Farré, F.X., Soliva, M. 2010. Approaching compost stability from Klason lignin modified method: Chemical stability degree for OM and N quality assessment. Resources, Conservation and Recycling, 55: 171-181.

[8] Martin, D.L. Gershuny, G. 1992. The Rodale book of composting. Rodale Press, Emmaus,

[9] LIFE13 ENV/FI/285 "Foundry sand" 2014: Re-use of surplus foundry Sand by Composting. http://ec.europa.eu/environment/life/project/Projects/index.cfm?fuseaction=search.dspPage&n_proj_id=4995&docType=pdf. [Accessed on 10 June 2015].

The 73rd World Foundry Congress
„Creative Foundry"
23 - 27 September 2018, Krakow, Poland

CORRELATION OF THERMAL ANALYSIS TO BINDER EMISSIONS

Scott R. Giese [1]

[1*]*University of Northern Iowa, Department of Technology, Cedar Falls, IA, USA*
**Corresponding address: scott.giese@uni.edu*

Key words: Differential Scanning Calorimetry (DSC), Mass Spectroscopy (MS), Hazardous Air Pollutants (HAPs), Volatile Organic Compounds (VOCs), phenolic urethane.

The objective of the research work was to investigate the feasibility of replicating thermal and mold atmosphere conditions in a differential scanning calorimeter (DSC) coupled with a mass spectrometer (MS) to develop standard procedures in measuring HAPs and VOCs. Initial research focused on the development of appropriate DSC parameters for an A356 aluminum casting. Using results from the actual casting, the DSC was programmed with a 5 segment profile adjusting the heating/cooling rate and gas atmosphere to represent the conditions for aluminium cast in a phenolic urethane no-bake mold. The emissions were recorded and compared to emissions collected in a hood stack test. Using a scaling factor to compare certain emission compounds between the DSC and hood stack test, satisfactory results were obtained showing an average scaling factor of around 5 for most HAPs and VOCs. Results demonstrated an emission testing protocol using the DSC was feasible.

1. Introduction

Emissions of hazardous air pollutants and organic volatile components have been a major concern for the foundry industry in providing a safe work environment for their employees. The major concern in conducting emission studies is the research expense to pour actual castings and measure the emissions during the pouring, cooling, and shakeout segments of the metal casting process. Variability in emission components is introduced when different casting alloys are used to represent various mold heating rates. Additionally, continuous sampling over the casting regime only provides the total emission amount over that particular time reference and fails to illuminate the peak production of certain emission products. From an applied research approach, it is unclear if the components of emission production are a product of complex chemical reactions occurring between the coated sand grains in a soup of VOCs, HAPs, and light gases as the casting cools or are from the rapid exposure of the heated molding sand and core binder to an oxygenated environment during the shakeout of the mold.

The author presented a paper at the 2014 World Foundry Conference discussing the emission characteristics when experimental atmospheres are changed from a neutral type atmosphere to a more representative mold atmosphere. The thermal analysis technique employed coupled differential scanning calorimetry (DSC), thermo-gravimetric analysis (TGA), and mass spectroscopy (MS) to understand how a binder decomposes during a constant heating rate. As an example, xylene emission from a pure phenolic urethane binder sample produced two large emission peaks around 300°C and 700°C in a pure nitrogen atmosphere. When the atmosphere was changed to a reducing atmosphere, a decrease in emission rate and a shift of the high temperature emission was observed. For an oxidizing atmosphere, strong emission were observed at 300°C and absent at 700°C.

The changes in emission rates and compounds at which hazardous air pollutants (HAPs) and Volatile organic compounds (VOCs) lead to an American Foundry Society investigation in exploring if DSC-TGA-MS technique can lead to a standardized procedure that would be representative of hood stack emissions. Long term research benefits would include a more defined quantification understanding for emission models required for casting simulation software.

The preliminary study investigated if the DSC-TGA-MS technique could be adaptable for an extensive evaluation of binder emission and correlated to actual measured hood stack emissions. Initially, the main objective of the investigation was to assess the feasibility of using small sand coated samples, performing a heating and cooling rate representing the actual thermal profiles developed in a mold, and replicating a representative dynamic mold atmosphere. The research challenge was assessing the correlation of the accuracy and reliability of the DSC-TGA-MS methodology with actual hood stack measurements.

2. Design of experiments

The investigation was divided into four components. Initially, a standardized test casting geometry was assessed to identify strategic locations for MS instrumentation based on mold heating rates capable for the DSC. Once the instrumentation was determined, several test castings were poured in grey iron and aluminum to collect mold atmosphere data to replicate in the DSC furnace. Using the fixed gas mold atmosphere data, the DSC was programed with the correlated thermal and fixed gas emission profile to replicate the conditions in the mold to the DSC furnace chamber. Finally, castings were poured again and the emissions recorded with the DSC-MS were correlated to hood stack emission data.

3. Experimental results

The emission measurement from the initial casting trials indicated a rapid decrease in nitrogen and oxygen as result of the phenolic urethane emission and gas expansion of the heating of the mold. Because phenolic urethane has a carbonyl linkage, rapid increase in carbon dioxide was observed for the first 200 seconds. As the mold came to thermal equilibrium, emission of methane and carbon monoxide with a decrease in carbon dioxide was measured. Carbon monoxide stabilized for the duration of 2000 seconds and methane gradually decreased to around 1200 seconds. Carbon dioxide rapidly decreased to around 600 seconds indicating the phenolic urethane binder linkage completed it's initial decomposition.

Using the casting simulation data, embedded thermocouples, and MS data, the following DSC experimental parameters were designed to replicate the thermal and mold atmosphere conditions of A356 aluminum poured in a phenolic urethane no-bake mold:

Segment 1 – 150 mL/min CO_2 and 20 mL/min N_2 at a heating rate of 50°C/min from 25°C to 275°C

Segment 2 – 120 mL/min CO_2 and 20 mL/min N_2 at a heating rate of 25°C/min from 275°C to 400°C

Segment 3 – 80 mL/min CO_2 and 20 mL/min N_2 at a heating rate of 10°C/min from 400°C to 500°C

Segment 4 – Isothermal hold at 500°C with 80 mL/min CO_2 and 20 mL/min N_2 for 10 minutes

Segment 5 – 100 mL/min N_2 at a cooling rate of 4°C/min from 500°C to 320°C

A sampling collection system was designed to collect the gases, including VOCs and HAPs, using 6L SUMMA canisters. The experimental apparatus is shown in figure 1. Tests performed were: Fixed Gas Analysis using ASTM D1946-90 methodology, EPA Method TO-15 to determine VOC's and HAPs, and GC/MS Tentative Identified Compound Scan to determine if other decomposition species were present.

A casting experiment was performed using the SUMMA canister approach employed in the DSC-MS experiment connected to a hood stack. An aluminum step casting was poured in an enclosed hood connected to a sampling duct conforming to U.S. Environmental Protection Agency (EPA) standards. The casting was poured and allowed to cool for 60 minutes followed by shakeout. During this time, both sampling techniques were used simultaneously during the 60 minutes cooling phase and 10 minute shakeout sampling.

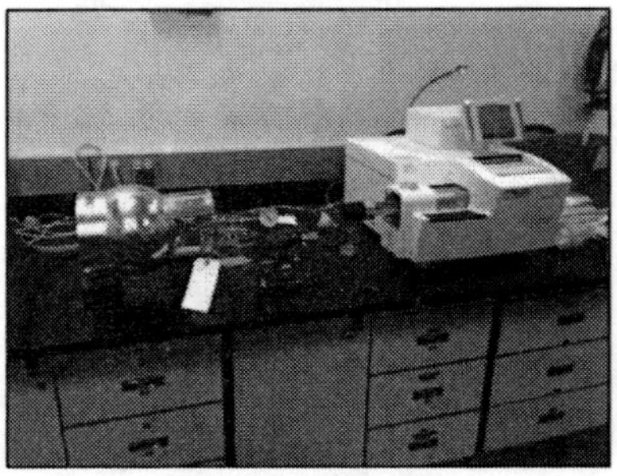

Fig. 1. Experimental configuration used to measure the light gases, VOCs, and HAPs using the DSC and TGA. On the left is the 6L SUMMA canister used to collect the furnace atmosphere.

Table 1 shows the comparison between the key emissions emitted in the DSC and hood stack emissions. A DSC to Hood Stack emission ratio was determined, indicating higher emissions in the DSC. This observation was anticipated because the mass of the emitted compound per volumetric size of the DSC furnace chamber is significantly higher whereas for the hood stack test, the air flow per emitted compound is substantially greater.

Table 1. Emission comparison between the DSC technique and hood stack emission.

Compound	DSC ($\mu g/m^3$)	Hood Stack ($\mu g/m^3$)	DSC to Hood Stack Ratio
Benzene	22.8	2.90	7.86
Toluene	59.9	25.0	2.40
Propylene	24.5	4.87	5.03
1,2,4 Tri-methyl-benzene	43.8	5.60	7.82
Naphthalene	926	17.9	51.7

4. Conclusions

The research project established the preliminary testing protocols for conducting a DSC-MS test to assess the emissions of sand binder systems using either natural or chemical binders. Using the phenolic urethane no-bake binder system as the initial test case, the investigation demonstrated good correlation between DSC results and hood stack measurements. However, reproducibility needs to be addressed in future investigations.

The 73rd World Foundry Congress
„Creative Foundry"
23 - 27 September 2018, Krakow, Poland

CUPOLA FURNACE SLAG – ITS ORIGIN, PROPERTIES AND UTILIZATION

Alena Pribulova[1]*, Peter Futas[1], Dana Baricova[1], Stefan Eperjesi[1]

[1]*Technical university in Kosice, Faculty of Materials, Metallurgy and Recycling, Letna 9, Kosice, Slovakia*
**Corresponding address: alena.pribulova@tuke.sk*

Key words: slag, cupola furnace, chemical composition, granulated slag, slag gravel

1. Origin of slag from cupola furnace and its properties

Cupola furnace is the most widely used furnace for cast iron melting. Typical, and main secondary product from cast iron production is a cupola furnace slag. Its production is approximately from 40 to 80 kg per one tone of the cast iron. The sources of the formation of the cupola slag are the slag forming additives, the coke ash, the metal batch impurities, the molten parts of the refractory lining. The basic, major components of the cupola slag are CaO, SiO_2 and Al_2O_3, with minor components being MgO, FeO, MnO and P_2O_5.

Cupola furnace slag tends to be dense, solid, vitrified material that varies in colour from cream to black, but the predominant colouration would be green to brown [1]. Light coloured slag is sometimes associated with high basicity materials and darker slag with more oxidizing conditions in furnace. If the iron content is too high and the lime level too low, the slag converts to Aero-chocolate appearance instead of being solid. Some slags will exhibit light coloured grains in their makeup which are usually refractory particles eroded from the furnace lining.

Cupola slag solidifies in relatively large pieces subsequently it may break down into various sized lumps. Fig. 1 shows the slowly cooled piece of cupola furnace Alternatively, some large furnaces operate with a wet slag granulation system where the slag exiting the furnace tapping/ slagging box is broken up by a water steam. The slag resulting from this type of operation is relatively fine, Fig. 2, compared with slag pieces produced by conventional slagging practice.

Fig. 1. Pieces of slowly cooled cupola slag

The physical properties of the slag produced during cupola operation are follows: specific gravity – 3.5, impact value – 6.1%, crushing strength – 35.5% and water absorption – 0.4% [2].

Fig. 2. Pieces of granulated cupola slag

When the slag is allowed to cool slowly, it solidifies into a grey, crystalline, stony material, know as air cooled, or dense slag. When the molten slag is chilled very rapidly either by pouring into a large excess of water, or by subjecting the slag stream to jets of water. The quenching breaks up the material into small particles, solidifies as a glass. The product is called granulated slag and is used as cement, or as sand for concrete [3]. The granulated slag can replaced the natural sand in concrete, the same can be said about the crystallized slag, which after crushing constitutes excellent normalized gravels.

Chemical composition of the cupola slag is directly dependent on the composition of input materials and technology of melting in cupola furnace. The chemical composition directly influences the slag viscosity and crystallization rate during cooling, thereby affecting the porosity, nature and size of crystals in the solid slag. This is the most important factor affecting its potential use. The resulting basicity of cupola slags ranges from acidic to slightly alkaline slags.

Table 1 shows the range of chemical composition of slags from different foundries in Slovakia and in India.

The slags from cupola furnaces are mainly composed of silicates with occurrence of a large number of other types of complex minerals. The most common minerals are wollastonite ($CaO.SiO_2$), fayalite ($2FeO.SiO_2$) and other known components, in particular the combination of SiO_2-Al_2O_3-CaO.

Table 1. **Chemical composition of cupola furnace slags from Slovakia and from India**

Country of origin/source	Chemical composition, %											
	SiO_2	Al_2O_3	CaO	FeO	MgO	MnO	K_2O	TiO_2	Na_2O	P_2O	Fekov	B
Slovakia [4]	55.19	7.29	33.68	1.15	0.49	-	-	-	-	-	1.68	0.62
Slovakia [5]	42.07	13.26	30.92	4.31	5.58	0.33	0.33	0.38	0.35	0.03		0.87
Slovakia [5]	49.84	18.51	17.24	5.06	3.69	2.12	0.52	0.71	0.63	0.69		0.42
India [1]	34.97	6.06	19.36	29.95	6.02	3.26	0.74	-	-	-		0.73
India [1]	47.99	16.58	24.33	1.56	7.79	1.19	0.56	-	-	-		0.67
India [1]	42.65	21.84	21.84	15.19	6.85	3.69	0.92	-	-	-		0.67

2. Use of cupola furnace slag

Metallurgical slags represent a very valuable secondary raw material for their chemical composition.That is why new possibilities for their utilization can be found.In general, we can divide the possibilities of utilizing metallurgical slags into three areas:

- road and construction industry (road construction material, cement production material, etc.)
- agriculture (neutralization of acidic soils in the case of basic slags)
- special methods of utilization (material for the production of silicate fibers)

Slag is a real binder, during the hydration of Portland cement the $Ca(OH)_2$ enters in reaction with slag components, forming the CSH, which playing the role of filling empty spaces in binding medium thus reinforcing the concrete structure.
The authors of the article have dealt with the various possibilities of utilizing metallurgical slags arising in the process of production of pig iron and steel. Very positive results have been obtained in the preparation of cement concrete mixtures.For the production of concrete were used different types of metallurgical slag gravel with fractions0-4 mm, 4-8 mm and 8-16 mm. Concrete mixture consists also of cement, water, and aeration and plastificator additives. Plastification additives reduce the water surface tension, thereby increasing the wetting power of mixing water. Aeration additives cause the controlled dispersion of air in the form of very small, stable and reciprocally separated bubbles in fresh concrete, which remain in the concrete even after hardening. Concrete mixtures from blast-furnace slag, or from an electric arc furnace slag, met the strict standards for road concrete STN 73 6123, which is used for demanding applications in roads of first and second classes.

For these reasons, the possibilities of utilization of cupola slags in the production of cement concrete mixtures were analyzed in further experiments. Amount of cement in all of types of concrete mixtures was 7,93 kg. The concrete mixture, which consisted exclusively of cupola slags had an absolutely unsatisfactory strength parameters. In following experiments, a combination of blast-furnace and cupola slags was therefore used. This concrete mixture does not comply with STN 73 6123 for road concretes, however mixtures B, C and D complies with the standard for concretes with lower strength properties, so called common concretes, STN ENV 206 –C 8/10 and STN ENV 206 –C 12/15.

Another analysed possibility was the use of cupola slag for the preparation of slag-alkaline concretes, i.e. cementless concretes. The basis of the preparation is the grinding of a suitable slag to the required fineness and the subsequent alkaline activation of the hydration reactions, consisting in mixing the slag with the solution of one of the dissolved alkali metal compounds (Na or K).
In our case, it was a hardening of the mixtures with water glass.This is the so-called selfsolidifying hardening.The hardening runs with the help of dikalcium silicate from slag and partly also CO_2 from air .

Acknowledgments
This work was supported by the Scientific Grant Agency of The Ministry of Education of the Slovak republic No. VEGA 1/0703/16, VEGA1/0073/17 and APVV-16-0485.

References

[1] Singh M. (2014). Utilization of cupola furnace slag as partial replacement to coarse aggregates in concrete mix. Patiala, Punjab.
[2] Behera G.B., Mohanty M. &Bagh S.R. (2011). A Slag as coarse aggregate and its effect on mechanical properties of concrete. International Journal of Erath Science and Engineering. Vol.4, pp. 899 – 902.
[3] Zeghichi L. (2006). The effect of replacement of naturals aggregates by slag products on the strength of concrete. Asian Journal of Civil engineering (Building and Housing". Vol.7, No.1, pp. 301 – 3014.
[4] Baricova D., Pribulova,A. & Demeter P. (2010). Comparison of possibilities the blast furnace and cupola slag utilization by concrete production. Archives of Foundry Engineering. Vol.10. Spec. Issue 2/2010, pp.15 – 18.
[5] Ladomersky J., Nosal E., Hroncova E. (2009). The research of possibilities of utilization of slags from cupola furnace. Waste forum. Vol.2009. Issue 2, pp. 125 – 132.

The 73rd World Foundry Congress
„Creative Foundry"
23 - 27 September 2018, Krakow, Poland

INNOVATION IN THE PRODUCTION OF CAST PARTS IN THE ERA OF DIGITALIZATION AND TRANZITION TOWARDS CIRCULAR ECONOMY ACCORDING TO THE "SHARING"-TYPE ECONOMIC MODELS

Vescan Mihail-Marius[2], Soporan Vasile-Filip[1*], Pădureţu Sanda[3], Samuilă Viorica[2], Tibori-Szabo Zoltan[4], Vasâlie Ioan[5]

[1]Technical University of Cluj-Napoca, CPADDD, Cluj-Napoca, B-dul Muncii, nr.103-105, Romania
[2]Technical University of Cluj-Napoca, DIMADD, Cluj-Napoca, B-dul Muncii, nr.103-105, Romania
[3]Technical University of Cluj-Napoca, DLMC, Cluj-Napoca, B-dul Muncii, nr.103-105, Romania
[4] Babeş-Bolyai University, FSPAC, Cluj-Napoca, str. Traian Moşoiu, nr.71, Romania
Saturn S.A. Alba Iulia, Alba Iulia, Str. Cabanei, nr. 57, Romania

**Corresponding address: vfsoporan@gmail.com*

Key words: circular economy, digitalization, casting production, virtual space communication, business "sharing" model

1. Motivation of the paper

Given the importance of the "production of cast parts" on the global economy, the paper aims to analyze the determinations that the new ways of communication, specific to virtual spaces, have on a new business model, the "sharing" model, in the sense of influencing the transition to the circular economy and, hence, the foundry activity.

2. Objective of the paper

The objective of the research is to determine a procedure for assessing the capability of castings production to adapt to the requirements and functionalities of a new business model, the "sharing" model, as a way of ensuring the requirements of the circular economy.

3. Problem formulation

The problem set forth for achieving the proposed objective is to determine the conditions and modalities by which the "sharing" collaborative communication platforms can ensure the adaptation - modeling of the production functions of the cast parts to the new **"business model"**, through the facilities provided by the **"virtual communication"** and institutional adaptation to the principles of **"circular economy"**.

4. Research method

At methodological level, the analyzes followed the objectives of institutionalizing the obtained results, which is why the studies had the following sequence of deployments: critical analysis of the existing situation; the interpretation of public policies and their consequences; assessing the regulatory framework and quantifying its effects; presenting the technical solutions and the framework for optimizing the choices made within them; understanding the financial mechanisms and financial instruments operating in the field and on the topic adopted.

The specificity of the research was an interdisciplinary one, by trying to study the specific elements of four domains: "production of castings", "virtual space communication", "circular economy" and "construction and operation of business models". The novelty of the research consists in structuring the elements acquired at the level of studies in the particular domains in order to determine the constructive elements of the "collaborative sharing platform", capable of responding to the imposed conditions.

5. Achievements of research activity

One of the achievements of the studies made consists in structuring the symbiosis that exists between the analyzed domains in order to structure collaborative platforms. In this context the "conception and production of cast parts" is correlated with the structuring of the following stages: "the developments at the manufacturing level", "the principles of the circular economy" and "the virtual communication". They determine the basis of the "new business model" so that at the higher level the "new value created" or "the consolidation of the existing one" can be valued at the higher level through the "functional value", "material value", "energy value", "environmental value" and "social value".

At the level of analysis of the production of castings from the perspective of "circular economy" and "virtual communication", a phasing and stepping-up of the process and the inclusion of the specific activities were realized (2). In the analysis, 7 phases of the manufacturing process were established: analysis of the domain specific information; analysis of manufacturing preparation from the point of view of the available technologies

and technological infrastructure; analyzing the manufacturing process, analyzing the results of the manufacturing process; end-of-life exploitation and disposal analysis; analysis of other phases contributing to the production of castings; process analysis at the level of the whole from the perspective of respecting eco-responsibilities.
The European document, which focuses on the innovative components of the circular economy (1), has been analyzed from the perspective of adapting its requirements to the specifics of the castings industry. Thus, the following are considered: technological innovation (material and process design, product design, resource management - raw materials, water, energy, resources); organizational innovation (integrated solutions and systems, logistics, business models, policy support tools); social innovation (new patterns of production and consumption, product service models, citizens' involvement, new patterns of production and consumption). In view of the measures taken to strengthen the circular economy, it is appreciated that virtual sharing platforms provide multiple roles and functionalities, generating social networks that generate new business models. Within them, based on the new structures, the following specific elements are generated: networking, information, connection, professional orientation, collaboration and support relations. Therefore, two of the major roles of the "sharing business model", ie dissemination and co-use, can be distinguished. Concluding, it is clear from the preliminary analysis that the specifics of the "production of castings", "circular economy", "virtual platforms" and "business model" have been identified, which will be used in the appreciation of their use in the provision of good practices in ensuring the transition of castings production to the circular economy.

6. Major contribution of research

The major contribution of the research is the determination of how the casting industry can migrate to the new business model based on informational access and distribution through sharing platforms. The specificity of the ITP sector in terms of the new model involves two essential aspects: large production capacities, large investments, a long product lifecycle, a complex value chain, high energy consumption, increased environmental impact etc; the resulting IPT products are not necessarily for final use, but generally parts and subassemblies and components of other end-use products. The major elements that determine the "sharing" business model, specific to the virtual environment, as well as the mutations it can make in the production of castings in the circular economy, are interconnection and integration.

7. Conclusions of the research activity

For IPT, the effects of using sharing collaboration platforms within the new economic model can be multiple: effective asset use, net profit growth, increased business development capability, product/service design for repeated use, less pressure on natural resources, respecting the principles of the circular economy, therefore creating and assigning new types of value. This is only possible by using complex computer systems capable of integrating and coordinating the multitude of information (preferences, reservations, consumer profiles etc.) and especially communicating with users. Here are two basic conditionings for the IPT mutation to the sharing model to be possible: 1. Creating the relational infrastructure in IPT through informatization, information, communication, interconnection as the basic elements of this business model. 2. Modeling of IPT activities, products and services that are suitable for distribution in view of co-use or repeated use by multiple consumers.
Consequently, identifying aspects and IPT components that can be migrated to the sharing model can be structured according to the roles described above.
Dissemination or sharing of experiences, abilities, information, knowledge, contacts, data etc. can be done in various sectors of the IPT: market analysis, regulations, processes, technological infrastructure, constructive conception, technological design, production preparation, flows, deliveries, recycling, resources, maintenance, storage, transport, pollution, consumption, technical control, energy, waste management, regulation etc.
Co-use, or booking and accessing various resources, services or products for their exploitation can be done in the following sectors: innovation (centers), design (institutes), storage (buildings, warehouses), logistics (parks), quality control (laboratories), training (schools, laboratories), sales (specialized platforms), marketing (specialized tools) etc.

References

1. Comunicarea Comisiei către Parlamentul European, Consiliu, Comitetul economic şi Social european şi Comitetul Regiunilor „Spre o economie circulară: un program deşeuri zero pentru Europa", Bruxelles, 25.09.2014, COM(2014) 398 final 2.
2. M.M. Vescan, V.F. Soporan1, D.M. Crişan1, T.R. Lehene1, S. Pădureţu and V. Samuila, Engineering of the institutionalization of the circular economy at the level of casting production, International Conference on Innovative Research — ICIR EUROINVENT 2017 IOP Publishing, IOP Conf. Series: Materials Science and Engineering 209 (2017) 012096 doi:10.1088/1757-899X/209/1/012096.

MANAGEMENT

The 73rd World Foundry Congress
„Creative Foundry"
23 - 27 September 2018, Krakow, Poland

THE APPLICATION OF RESOURCE AND PROCESS CONSUMPTION ACCOUNTING (RPCA/RCA) IN A FOUNDRY ENTITY

Marcin Latallo-Anulewicz[1*], Tomasz M. Zielinski[2]

[1]Foundry Research Institute, 73 Zakopianska Street, 30-418 Krakow, Poland
[2] ABC Akademia Ltd Company, 35/403 Towarowa Street, 61-896 Poznan, Poland
**Corresponding address: marcin.latallo@iod.krakow.pl*

Key words: RPCA/RCA, Resource Consumption Accounting, cost optimization in foundry, modern cost accounting, Process Consumption Accounting

Today's global competition drives manufacturing companies to make technologies and production methods more flexible to meet the constantly increasing requirements of their customes. In that circumstances it is necessary to incorporate the micro-marketing rules in each company, and the relevant information regarding the costs related to production, service, logistics, customers, and other [6]. Most of production companies uses traditional cost accounting concepts (in Poland – about 75%) [1], which use arbitrary volume based drivers (i.e. direct labor hours, tons) for allocating indirect costs to products and services [2]. The methods of measuring costs results in a detached information about the actual costs of processes and operations, such as manufacturing, final treatment, services, logistic, internal transport, quality control, etc. Traditional costing systems are still functioning in companies, although they are no longer able to explain the phenomena occurring on today's market.

Accounting and corporate finance are mostly the same, as they were 100 years ago, when manufacturing was booming, and the growth seemed to be limitless. But for over 30 year the industry has been devoid of such growth. The transformation of accounting analysis and toolbox must be in line with marketing success and a contemporary profit making. It is very important for every CEO to control the costs and understand the contribution of each process and resource to profitability [6].

In response to these challenges the modern concept of cost accounting was introduced, which combine German Grenzplankostenrechung (GPK) [4] and American Activity Based Costing (ABC), turning to Resource Consumption Accounting (RCA) and Resource Process Consumption Accounting (RPCA). The GPK method is concentrated on resource cost and capacity, while ABC is concentrated on processes, activities, and their effectiveness. RPCA/RCA is in general an integration of those two methods resulting in a comprehensive, systematic and management cost accounting that allows to provide a reliable information of the cost of resources, products, services and customers [3]. The implementation of RPCA/RCA brings crucial benefits for the company. Such integrated cost management system allows for optimization of production techniques, technologies, increase the allocation of production resources, decrease the number of unused resources, optimization of costs related to customers, it helps to define any manufacturing processes at any stage with direct cost allocation, and the contribution of each process and resources into the profit making.

The aim of this article is to show how RPCA/RCA can be implemented in a foundry entity. Various casting techniques, manufacturing operations, like molds and cores making, preparation of batch material, melting, pouring molds, drying-out the castings, mechanical and heat treatment, cleaning and finishing, represents the processes and activities that exist in each foundry company. The identification of such processes both in direct and indirect operations (i.e. sales, marketing, administration, logistic, IT) and resources (primary and secondary), like furnaces, mold and core manufacturing facilities, melting department, cleaning-finishing department, sand delivery facilities, core and metal removal, labor, rooms, warehouses, buildings, computers and IT systems, media – remains crucial for RPCA/RCA implementation.

Allocating the costs of these resources in a cause and effect manner to products, services and customers results in much more relevant information about planned and actual costs at each management tiers of a company [7]. Furthermore, the information about marginal and unused capacity costs gives deep insight into usage of resources and enable avoiding negative effect of fixed costs downward spiral.

The relevant benefits for the foundry company, that lies upon the implementation of RPCA/RCA are: increased competitiveness and profitability through the access to key information about profitability of products, services and customers – based on cost-

and-effect cost accounting; improving the use of resources and to avoid occurring costs of unused capacity; relevant information of marginal cost, support for the process of pricing, bidding and negotiating with customers, by providing a reliable information on the costs and profitability of products, services, and customers costs.

RPCA/RCA will help to capture complexities of the contemporary casting manufacturing activities, by distinguishing the complex of inter-relationships between resource pools and cost objects (incl. processes) and relies on the integration with IT production and management systems (like ERP). It is the more suitable the more multiple heterogeneous resources are employed to operate the production.

Implementation of RPCA/RCA in a foundry company will also help to enhance the investment and innovation policy. Comprehensive construction of RPCA/RCA in a foundry will help to compare actual cost of any operating facility to potential investment or upgrade, and its influence for profitability. Coherent RPCA/RCA will help to drive the innovation policy. Number of innovations lies upon the operating costs of actual facility. Traditional financial statements usually do not provide a sufficient information for decision making in the scope of real benefits that can be brought via innovation.

The innovations in the scope of management accounting usually gets lost in the crowd of technological, IT and marketing innovations. Although they can be considered as key factor of decision making, that influence the profitability, and competitive position of any company. RPCA/RCA is then recognized as the most advanced tool for support this process.

References

[1] Zielinski T.M. (2014). Resource and Process Consumption Accounting in an enterprise. University of Economics, Poznan, Poland

[2] Cooper R., Kaplan R.S. (1999). The Design of Cost Management Systems, Prentice Hall, Upper Saddle River, New Jersey, USA.

[3] Keys D.E., Van der Merwe A. (2001). The case for RCA: Excess and Idle Capacity, Journal of Cost Management, July-August, Vol. 15, No. 5, p. 21-32.

[4] Kliger W., Pampel J., Vikas R. (2012). Flexible Plankostenrechung und Deckungsbeitragsrechnung, 13 Aufl., Springer Gabler, Wiesbaden.

[5] Latallo-Anulewicz M. (2006). The models of costs accounting and their application in the foundry company – Part 2 – Activity Based Costing, Foundry – Science and Practice, Vol. 5/2006. Foundry Research Institute, Cracow, Poland.

[6] Marcus D. (2014). I Take It Back, Modern Casting, December 2014, p. 41.

[7] Tse M., Gong M.Z. (2009). Recognition of Idle Resources in Time Driven Activity Based Costing and Resource Consumption Accounting Models. Journal of Applied Management Accounting Research. Vol. 7, No. 2, p. 41-54.

CONSIDERATIONS ON THE CORRELATION OF ENGINEERING TRAINING WITH THE DEVELOPMENTS IN THE FIELD OF THE CREATIVE PRODUCTION OF CAST PARTS

Soporan Vasile-Filip[1*], Samuilă Viorica[2], Lehene Tiberiu[2], Pădureţu Sanda[3] and Vescan Mihail[2]

[1]*Technical University of Cluj-Napoca, CPADDD, Cluj-Napoca, B-dul Muncii, nr.103-105, Romania*
[2]*Technical University of Cluj-Napoca, DIMADD, Cluj-Napoca, B-dul Muncii, nr.103-105, Romania*
[3]*Technical University of Cluj-Napoca, DLMC, Cluj-Napoca, B-dul Muncii, nr.103-105, Romania*

**Corresponding address: vfsoporan@gmail.com*

Keywords: cast parts production, major developments, engineering training, critical thinking, circular economy

1. Motivation of the paper

The theme of the paper is motivated by the fact that the forms of engineering training must take into account the developments in the complex field of casting production. The justification for this choice is due to the fact that the latest global professional events (71st WFC 2014 - Bilbao, Spain; Technical Forum 2015; Technical Forum 2017 – Metal Casting Conference World Cast in Africa – Innovate for Sustainability) allocated an important space to this issue. At the same time, at the Technical University of Cluj-Napoca, within CPADDD, there is an important concern in this direction, materialized by the publication of some studies (Vescan, 2017), the existence of an annual report on the conditionality of research activities regarding the production of cast parts and some works in the field of vocational training, virtual communication (Vescan, 2016) and taking into account casting production from the perspective of the circular economy (Soporan, Crişan, Lehene and Pop, 2016). In relation to the general theme of the conference, the "creative foundry" and the areas covered (materials, technologies, digitization, management and greening), it is appreciated that engineering training systems are particularly important for ensuring the creativity of those who will act in the casting industry.

2. Objective of the paper

The main objective of the paper is to determine a methodology for identifying the major evolutions at the level of the production of castings and their incorporation in the engineering training programs.

3. Problem formulation

The problem to become the objective of the research starts from recognizing the fact that there are important multidisciplinary developments in the production of cast pieces (basic fundamental engineering, specialization engineering and complementary domain engineering), which is why a number of questions are still awaiting the institutional response involving the professionalism of many participating categories: industrial and banking employers, creators and producers of cast parts, users of cast products, suppliers of products and services used in the field, regulators and executives at the level of ecoresponsibilities which operate.

4. Research method

In order to solve the problem, we adapted the methodology developed in the spirit of the analyzes carried out at the "Foundation for Critical Thinking" at the engineering level, by Richard Paul (Paul, 2008).

Using the elements of engineering behavior (the characteristics of the reasoning, the attributions of the engineering activities, the engineer's behavior, the ways of thinking, the structuring of judgments and the establishment of their structure), the following phases of the reasoning adapted to the analysis of the engineering training systems in the casting industry were structured: (1) Establishing the scope of critical engineering analysis used; (2) Establishing the key problems faced by the production of cast parts (the system under consideration) and their evolution in the perspective of the functioning of the whole system; (3) Identification and enumeration of the most important documents and characterization information of the analyzed system; (4) Presentation of key notions, concepts and ideas to be acquired in order to understand the functioning and possibilities of reforming the engineering training systems in the production of castings; (5) Analysis of the problems faced by the engineering training systems at the production stage of cast parts; (6) Specification of key adopted hypotheses in solving system problems; (7) Logical determination of the structuring of judgments necessary to solve the problem set in the analysis; (8) Determining how key hypotheses influence the results of the analysis; (9) Determining the consequences of judgments made in accordance with the hypotheses adopted in the structure used;

(10) Evaluating the results obtained and presenting the solutions offered by the analysis using the principles of engineering critical thinking.

5. Achievements of research activity

By going through the proposed methodology, it is appreciated that the production of cast parts requires a permanent connection to the conditionings of the socio-economic activities.

The major needs of innovation in this field can be grouped into two levels: the level of technical and technological innovation and the level of changing the approach of the castings production markets.

From a pragmatic point of view, regarding the actors involved in the production of castings, the rules of survival and development make for a substantial collaborative dynamic between the information and the actions carried out.

Unfortunately, from those working in engineering training, there doesn't exist an approach based on the principles of interconnecting vessels between the evolution of the analyzed field (the production of the cast parts) and the engineering solutions provided by the university structures. It is acknowledged that the appreciation is not valid in all cases.

As with other productive areas, the speed of market change, the development of certain products and technologies, the mutations that impose "intelligent materials", "interactive products", "3D printing production", "plant of the future", require a new area of approach to training programs, that of increasing the competitiveness, agility and reactivity of engineering training systems at the manufacturing stage of cast parts. In this context, it is necessary to integrate new solutions at the level of the programs, aimed at: the evolution of human-machine interfaces, the efficient integration of digitized and computerized productive chains, the efficient use of new manufacturing systems accompanying productive activities.

Therefore, it is important that the topics of engineering training be anchored in the problems of technical solutions and not be detached from the current issues, referring to circular economy, eco-design, hybrid design, rapid prototyping, hybrid parts, 3D printing with technological solutions, to ease and increase the safety of designing of traditional parts and manufacturing technologies, to generate intelligent parts with new technology-oriented geometries and much more.

The context described by the fact that the system of conditioning, which aims to ensure ecoresponsibility, will impose major themes that, given the material and energy consumptions specific to castings, will require energy efficiency and climate change, optimization of metallurgical processes, recycling of metallic materials and of the auxiliary ones, reducing the impact of the entire process on the environment (water, air, soil). This will require a rethinking of technological processes and infrastructure to better respond to the demands of environmental protection and sustainable development. These themes are therefore important modules needed to be included in the training programs in the castings industry.

6. Major contribution of research

In view of the analysis performed, we consider that at present there are some major action areas that need to be included in the engineering training programs of the casting industry: the development of the traditional research specific to the field and its adaptation to the evolution of the material investigative means, of algorithmic nature and of structure of computer tools and applications; finding ways to adapt to the new requirements of ecoresponsibility imposed by new realities that develop specific actions to protect the environment and ensure sustainable development; establishing the optimal solutions that respond to opening up and imposing new conditions in the areas of use and competition imposed by other technological processes; adaptation of technological solutions to the evolution of the other fields that accompany and influence the production of cast parts; the integration of the new technological means of the digital age into the totality of the activities carried out in the casting industry.

References

[1] V. Samuilă, V.F. Soporan, G. Conțiu1, S. Pădurețu, T. R. Lehene and M.M. Vescan (2017). Hierarchical modeling of professional skills in the field of casting manufacture engineering. International Conference on Innovative Research — ICIR EUROINVENT 2017 IOP Publishing, IOP Conf. Series: Materials Science and Engineering 209 (2017) 012095 doi:10.1088/1757-899X/209/1/012095.

[2] M.M. Vescan, V.F. Soporan, D.M. Crișan, T.R. Lehene, S. Pădurețu1 and V. Samuila (2017). Engineering of the institutionalization of the circular econo at the level of casting production. International Conference on Innovative Research — ICIR EUROINVENT 2017 IOP Publishing, IOP Conf. Series: Materials Science and Engineering 209 (2017) 012096 doi:10.1088/1757-899X/209/1/012096.

[3] V.F. Soporan, M. Crișan, T. Lehene and A.L. Pop. (2016). Methodology for appreciation the manufacturing castings from perspective of circular economy. International Conference on Innovative Research 2016 - ICIR Euroinvent 2016 IOP Publishing IOP Conf. Series: Materials Science and Engineering 133 (2016) 012063 doi:10.1088/1757-899X/133/1/012063.

[4] M.M. Vescan and V.F. Soporan. (2016). Castings manufacture from the perspective of digital communication in virtual space. International Conference on Innovative Technologies, IN-TECH 2016, Prague, 6. - 8. 9. 2016, p. 143-146.

[5] R. Paul and L. Elder. (2008). Mini-guide de la pensée critique. Foundation for Critical Thinking.

Technical

MATERIALS

The 73rd World Foundry Congress
„Creative Foundry"
23 - 27 September 2018, Krakow, Poland

THE FREEDOM OF CREATIVITY: COATINGS AND ADDITIVES CONCEPTS ENABLING ENHANCED CONSTRUCTION AND CASTING PROPERTIES

Reinhard STÖTZEL[1*], Christian KOCH [2], Shengyin LIU [3], Lee HORVATH [4]

[1*] *ASK Chemicals GmbH, Reisholzstrasse 16 – 18, 40721 Hilden, Germany ,*
[2] *ASK Chemicals GmbH, Reisholzstrasse 16 – 18, 40721 Hilden, Germany,*
[3] *ASK (Zhenjiang) New Materials Technology Company Limited, Room 702 - 703 Block A, No 596 Middle Longhua Road, Xuhui District, Shanghai 200032, China,*
[4] *ASK Chemicals LLC, 495 Metro Place South Suite 250, Dublin OH 43017 USA,*
**Corresponding address: e-mail: Reinhard.stoetzel@ask-chemicals.com*

Key words: Mould and core coating, sand additives, Lost Foam & 3D printing coating, Residue free casting, casting defect

The capability and casting quality is the focus of enterprises due to the great demand of good and high-quality casting.

To have new concepts that can quickly match customer KPI (key performance indicators) using sustainable solutions by optimization of the casting processes, in particular by using core/mould coatings or sand additives, is crucial.

Increased demands, such as thin-walled castings, the use of advanced alloys and changes in sand quality, are also causing the technical demands on the applied coating technology to rise, making the use of high-grade coatings unavoidable. The guideline VDA 19 (ISO 16232) formulates a very clear requirement for automotive series castings. The residuals in the components are limited in order to prolong the maintenance intervals for the motor vehicles. Residuals in the cavities of the engine blocks cause the filters to clog early and are detrimental to the longer maintenance intervals required.

Lost Foam castings or 3D printing which require high-tech performance of the coatings can cover the freedom of design. Last but not least there is a trend towards steel castings in the automotive market that requires special features too.

1. Residue free castings with Novel TS Coating

OEM's are forcing the foundries to deliver castings with a minimal amount of residue in their castings. For motor blocks these limits could be down to 300mg per castings. The fact that the water jackets or oil galleries are almost impossible to clean via shot blasting leaves a serious demand on the coatings to provide flawless castings but also zero adherence of the coating on the casting surface.

A special coating based on good anti-veining and anti-penetration properties has been developed, which reduce the coating residue after pouring to a minimum. SEM investigations show, that the coating has self-releasing properties after pouring.

2. Lost Foam and 3D printing coatings for complex casting design

The trend in Lost Foam and 3D castings is to produce more complex and intricate castings, which lead to following demands on coating properties:

- Controlled insulation characteristics
- Tailored permeability
- Perfect rheology
- Short drying time

Examples of complex castings using these high tech coatings have proven to provide benefits to the foundries and their customers by allowing more integration of components.

The 73rd World Foundry Congress
„Creative Foundry"
23 - 27 September 2018, Krakow, Poland

NEW TECHNOLOGY PLATFORM ECOCURE BLUE: REDUCTION OF EMISSIONS IN FOUNDRY PROCESSES – FIRST PRACTICAL EXPERIENCES

Lenzen, Frank [1*]

[1]ASK Chemicals GmbH, Reisholzstrasse 16 – 18, 40721 Hilden/ Germany

Key words: emissions, BTX, Cold Box, hazard-free, binder reduction

Nowadays, foundries are facing new major challenges. On the one site, higher productivity and cost effectiveness is expected, on the other side environmental protection and occupational health and safety are becoming more and more important.

Ever stricter demand of statutory regulations and compliance with stronger emission limit guidelines, for example for phenol, formaldehyde, benzole, toluene, xylene, are creating high investment pressure for the foundries. As secondary measures only remain quite often solutions as additional enclosure, suction removals respectively thermal incineration.

For years, foundries are constantly seeking to contribute to reduce emissions in the foundry process, both from the exhaust point of view as well as when it comes to the disposal of used sand.

In the past, binder generations have been established on the market with the primary goal to reduce the BTX-emissions. One of the possible answer to this problem is represented by the new generation of inorganic binders.

Within the segment of organic binders, especially in the field of Cold Box, a reduction of emission has mainly and so far been solved by the substitution of solvents. Thus, classic aromatic solvents have been replaced at first by biodiesel and later on by tetraethylorthosilicate.

In 2007, ASK Chemicals started to face the problem of emissions and their reductions in a completely new way. However, compared to other developments mentioned before, this approach seems to be more holistic. Not only solvents were considered, but also and specially the resin components, which in total, as studies indicated, have a far greater influence on emissions as solvents.

The primary goal was to maximize the productivity, which means for the foundries to produce cores with the smallest possible quantity of amine. The result of the development work was the first generation of the so-called HE (High Efficiency) systems. Considerable reductions of emissions were already achieved in this way.

After successful implementation of the HE system, the focus of development in the following years was mainly to minimize further hazardous substances that are typically part of binders as unreacted substances (monomers). This is mostly phenol and formaldehyde.

In 2015, for the first time in history of binder development, a cold box part 1 component free of hazardous substances that have to be declared according GHS / GLP was developed and was ready for serious production.

Practical examples showed that with the use of this generation of binders a reduction of BTX emissions can be achieved. Thus, the company Düker for example, has achieved in their processes a 66 % reduction of BTX emissions by a 22% reduction of binders as well as reduction of BTX in the used sand of 34 %.

The 73rd World Foundry Congress
„Creative Foundry"
23 - 27 September 2018, Krakow, Poland

EFFECT OF MAGNESIUM INOCULATION ON THE MICROSTRUCTURE AND MECHANICAL PROPERTIES OF A SPHEROIDAL CAST IRON KNUCKLE: A FOCUS ON THE STEERING ARM.

Steven Mavhungu[1]*, Kasongo Nyembwe [2], Morella Sekotlong[3]

[1] *Metal Casting Technology Station, Faculty of Engineering and the Built Environment, University of Johannesburg,*

[2]*Department of Metallurgy, Faculty of Engineering and the Built Environment, University of Johannesburg, P.O. Box 17011, Doornfontein, 2024, Johannesburg, South Africa*

Key words: Inoculation, Magnesium cored wire, Spheroidal graphite, Steering knuckle.

The steering knuckle is integral component of the Suspension and stability control system of modern vehicles. Good mechanical properties with an emphasis on the fatigue properties are essential for this component as it is subjected to a cyclical load of significant magnitude during service. These properties are the function of the microstructure achieved in the component during the various manufacturing processes including forging and casting. The strut mount of the knuckle is required to meet specified microstructure and mechanical properties. However, in line with the recent trend of stringent quality requirements of cast components, Original Equipment Manufacturers (OEMs) have had to extend the specifications to other sections of the knuckle. This paper evaluates the effect of cored wire inoculation on the microstructure and mechanical properties of the steering arm of a typical spheroidal cast iron component. The investigation shows that the use of a cored wire having higher rare earth content formulation could lead to a homogeneous matrix containing consistent graphite nodule morphology. However, this was found not to be the condition for better mechanical properties along the knuckle arm in line with required specifications. The findings in this paper contribute to a better understanding of steering knuckle properties to allow its production for safer automobile applications.

Introduction

The steering knuckle is one of the essential components of the suspension and stability control of modern vehicles [1]. This component is typically connected to the tie rod, stub axle and axle housing with the aid of bearings. The function of the steering knuckle is to convert the linear motion of the tie rod into angular motion of the stub axle [2]. Figure 1 shows the parts of a typical automotive steering knuckle [3].

Steering knuckle components for automobile applications are manufactured by forging or metal casting processes. Effects of manufacturing alternatives and costs, material behaviour, and processing conditions such as surface finish and residual stresses are discussed in the literature [4], [5]. Forging produces steering components that are free of blowholes defects thus providing higher durability and fatigue performance of steering knuckles compared to the ones produced by metal casting [1]. On the other hand, the metal casting is cheaper for mass production of steering knuckles in steel, cast iron or aluminium alloys.

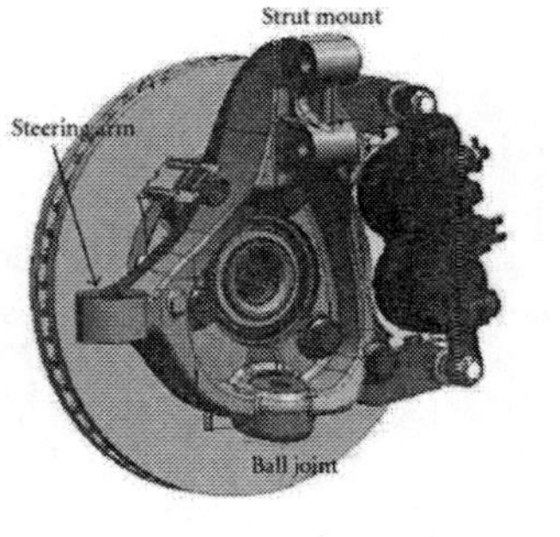

Fig. 1. Steering knuckle system [3].

Experimental Procedure

Steering knuckle and area of interest

Figure 2 shows the steering knuckle used in the study as well as the area of interest which was the focus of the metallurgical investigation.

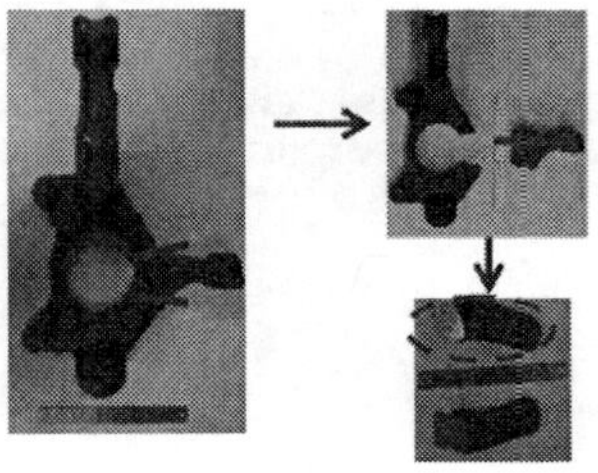

Fig. 2. Automotive steering knuckle used in the study. Sectioned steering arm is shown on the right.

Magnesium cored wires

Two types of commercial cored wires of significant different chemical compositions in terms of Si, Ca, Mg and Rare earth content. were used in for the production of steering knuckle castings.

Metallurgical characterization of steering arm

The steering knuckle castings produced with the two different Mg cored wires were characterized for chemical composition, microstructure and mechanical properties including hardness, tensile strenght and elongation.

Initial Results

Table 1 shows the chemical compositions along the steering arm of the knuckle using the two cored wires.

Table 1. Chemical analysis results of steering wheel knuckles produced from the different wires

Element	Position 1		Position 2		Position 3		Position 4		Position 5	
	Wire 1	Wire 2	Wire 1	Wire 2	Wire 1	Wire 2	Wire 1	Wire 2	Wire 1	Wire 2
C	4.31	3.87	4.31	3.87	4.31	3.67	4.31	3.67	4.31	3.87
S	0.0102	0.0063	0.0102	0.0063	0.0102	0.0063	0.0102	0.0063	0.0102	0.0063
Si	1.978	2.303	2.086	2.331	1.744	2.384	1.87	2.272	1.802	2.289
Mn	0.282	0.165	0.288	0.16	0.243	0.17	0.265	0.177	0.243	0.175
P	0.0036	0.02	0.0038	0.016	0.0011	0.04	0.0033	0.017	0.0011	0.018
Cr	0.098	0.07	0.042	0.025	0.044	0.049	0.035	0.027	0.026	0.026
Mg	0.029	0.068	0.034	0.08	0.037	0.083	0.044	0.069	0.041	0.076
Pb	0	0.134	0	0.105	0.024	0.201	0	0.107	0.034	0.11
Ti	0.016	0.016	0.017	0.016	0.013	0.017	0.015	0.015	0.013	0.017
Al	0.011	0.052	0.0092	0.021	0.011	0.043	0.014	0.013	0.013	0.014

Table 2 shows the metallurgical matrices of steering obtained in the steering arms produced with the two cored wires.

Table 2. Steering microstructure in the steering arms

Position	Wire 1	Wire 2
Position 3	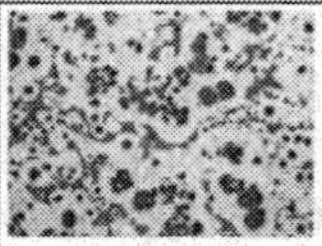 % Ferrite: 74 - 81%	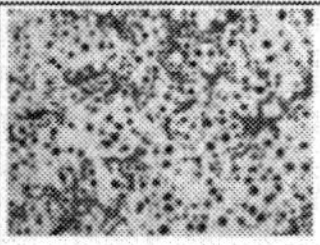 % Ferrite: 74 - 81%
Position 4	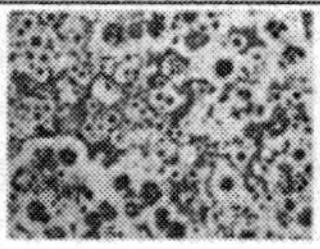 % Ferrite: 74 - 81%	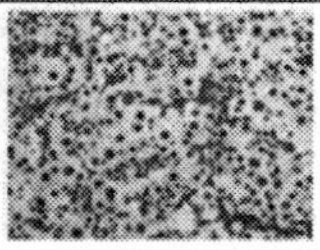 % Ferrite: 74 - 81%

Table 3 illustrates the tensile test results obtained for the steering arms produced with the two different Mg cored wires.

Table 3. Tensile test results and associated fracture microstructure

	Tensile Test	Microstructure at fracture - 100x Magnification	UTS (Mpa)	Elongation (%)
Wire 1	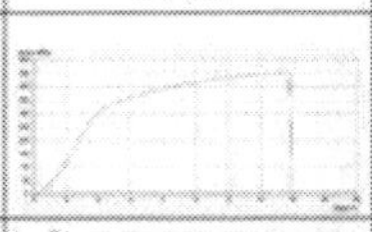	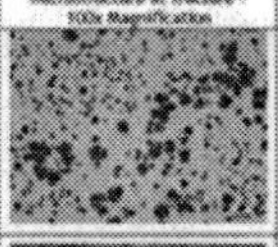	451.1	15.31
Wire 2	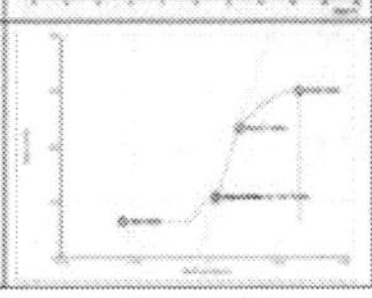	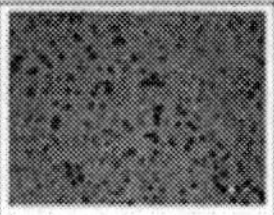	502.68	10.542

Conclusion

The study assessed the metallurgical properties of a sand cast knuckle in ductile iron produced with two types of cored wires. Unlike previous studies, the focus of the investigation was the steering arm of the knuckle. Differences in mechanical and metallographic properties between the two Mg cored wires were observed suggesting a complex interaction of chemical composition, matrix, and graphite nodule characteristics at work to produce the final structure of the steering arm. These initial findings on the steering arm contribute to a better understanding of steering knuckle properties to allow its production for safer automobile applications.

Acknowledgments

The Authors acknowledge the funding support of National Foundry Technology Network and the Metal Casting Technology Station at the University of Johannesburg.

References

[1] V.Manoj Kumar, M.Ayyappan, D.Sudarsan. 2017. Design and Analysis of Steering Knuckle for Fatigue Loading. International Journal of Advanced Research in Mechanical Engineering & Technology.

[2] V.Sivananth, S.Vijayarangan, and R.Aswathaman2015. Fatigue and Impact Analysis of Automotive Steering Knuckle under Operating Load Cases. Altair Technology Conference.

[3] E. A. Azrulhisham, Y. M. Asri, A. W. Dzuraidah, N. M. Nik Abdullah, A. Shahram, and C. H. Che Hassan. 2010. "Evaluation of Fatigue Life Reliability of Steering Knuckle Using Pearson Parametric Distribution Model," International Journal of Quality, Statistics, and Reliability, vol. 2010, Article ID 816407, eight pages,

[4] A. Fatemi, M. Zoroufi, P. Shenoy, and A. Afzal. 2005. "Comparative durability study of competing manufacturing process technologies". Proceedings of the 26th Forging Industry Technical conference.

[5] S.K Chandrakar, D.L Soni and S Gardia. 2013. "FEA of A Steering Knuckle for Life Prediction," International Journal of Engineering Research and Technology. ISSN 0974-3154 Volume 6, Number 5,

The 73rd World Foundry Congress
„Creative Foundry"
23 - 27 September 2018, Krakow, Poland

Study on properties of Hyper Duplex Stainless Steel 7A of ASTM A-890 (CD3MWN)

B. Raha[1*] ; Nithin Raj P[2] ; K Sekar[3]

[1]*Peekaysteel Castings(P) Ltd., Calicut , Kerala, 673027, India*
[2,3] *National Institute of Technology., Calicut, Kerala / Ph.D research Scholar & Assitt. Professor.*
**Corresponding address: e-mail: b.raha@peekaysteel.in*

Key words: Hyper Duplex Stainless Steel, Sigma Phase, Ageing, CPT

Abstract

Casting grade Hyper Duplex Stainless Steel ,7A of ASTM A-890, is a high alloyed duplex stainless steel having higher Cr, Mo, Nitrogen , W and other elements like B, Ba, Ce+La. They are added to make the steel stronger and better for use in extreme operational conditions. Due to increase in alloy content, hyper duplex steel has higher PREN(>45) than other Super Duplex Stainless Steel grades like 5A & 6A & so exhibits better corrosion performance. With good combination of high corrosion resistance and strength, hyper duplex stainless steels can be successfully used in the extreme environmental condition in ultra-deep formations or deep water wells, particularly for oil and gas sector. In this paper, studies conducted on effect of heat treatment on mechanical properties, corrosion characteristics and test values compared with super duplex grade 6A. Study also conducted on sigma phase formation by ageing the sample in temperature range from 700-1000°C for different times. Electrochemical corrosion characteristics obtained using tafel fit in various conditions and found to be better than 6A. Pitting corrosion & Critical pitting temperature determined (CPT)as per ASTM G-48 method E .Mechanical properties like UTS,YS & BHN found to have improved as compared to 6A. However impact values at -50°C is found to be inferior to grade 6A. Sigma phase formation found to have effect on holding time at both 900 & 1000°C with corresponding increase in BHN .Metallographic study carried out by optical (OM)and scanning electron microscopy(SEM). Finally overall performance of grade 7A of ASTM A-890 was compared with Hyper Duplex Stainless Steel developed by SANDVIK SAF 2707 HD which does not contain Tungsten as in case of casting grade 7A of ASTM 890

1. Experimentation & Discussion

Two experimental heats of grade 7A were made in a Induction Furnace as per composition given in Table 1. Specified chemistry mentioned in italics. Test blocks from 1st heat was heat @ 110°C/Hr (Max) up to 1130°C and soak at that temperature as per 1 inch per Hour of section thickness + 1 Hour Extra. After that temperature brought down to 1105°C & next test blocks were water quenched with good agitation of bath. Mechanical testing of test pieces carried out as per ASTM standard.Test result is shown in Fig2. Sample preparation for metallographic work carried out by electro Polishing & etching with 40% NaOH solution (6- 9V, 1.5-2 Amps, 7-9 Sec).

Table 1 Chemical Composition of Two Heats of 7A [1stHeat /2nd Heat]

C	Si	Mn	P	S	
0.30 max	***1.0 max***	***1.0-3.0***	***0.30 max***	***0.20 max***	
0.030	0.798	1.11	0.027	0.007	
0.0296	0.887	2.808	0.005	0.006	
Cr	Mo	Ni	Cu	W	
26.0-28.0	***2.0-3.5***	***6.0-8.0***	***1.0 max***	***3.0-4.0***	
26.59	2.99	7.74	0.39	3.45	
26.36	2.446	7.74	0.512	3.924	
B	N	Ba	Ce+La	PREN	%F
0.001-0.01	***0.30-0.40***	***0.0002-0.01***	***0.005-0.030***		
0.0039	0.313	0.001	0.0055	47.18	48
0.007	0.385	0.001	0.013	47.06	47

Table 2 Mechanical Properties(after 1st HT)

	UTS, MPa 690 Min	0.2% YS MPa 515Min	% EL 20 Min	BHN	ISO-V At -50°C Not Specified
7A	859	684	29.8	269	Av. 20J
	868	591	32	255	Av. 25J
6A	690	450	25	234	Av. 120J

From this test we find that strength and ductility values of grade 7A are much superior to grade 6A. However toughness values at -50°C are lower than grade 6A.

Micro structure of the sample of grade 7A is shown in Fig 1. Ferrite & Austenite is found to be in equal proportion.

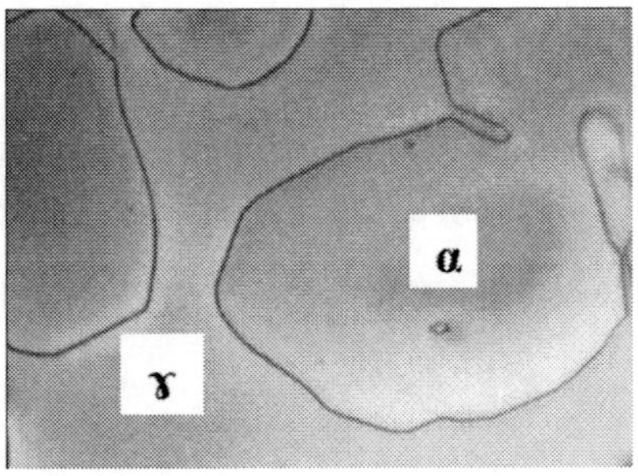

Fig 1 shows α+γ in equal proportion. NaOH electro etched. Magnification 500X

1st Experimentation:

In order to find out effect of heat treatment on toughness ,we carried out heat treatment of samples from second heat in muffle furnace as shown in Table 3.

Table 3 Heat Treatment carried out on samples

Cycle	Heat Treatment done
A	Heat to 1160°C and soak for 4Hr, Next furnace cool to 1060°C and water quenched
B	Heat to 1130°C and soak for 4Hr, Next furnace cool to 1060°C and water quenched
C	Heat to 1130°C and soak for 4Hr, Next furnace cool to 1030°C and water quenched
D	Heat to 1130°C and soak for 4Hr, Next furnace cool to 1080°C and soak for 2Hrs & next water quenched

Mechanical test results after above cycles are shown in Table 4.

Table 4 Mechanical Testing Results

	UTS MPa	YS MPa	%El	ISO-V J @ -50°C	%F
A	880	644	31	Av.43	44
B	868	629	26.7	Av. 56	40
C	Only Impact test carried out.			Av.16	36-40
D	883	628	28.4	Av. 36J	44

With all combination of heat treatment cycles ,we find that , all properties improved considerably as compared to 6A. This is shown in Fig2. However , toughness values at -50°C was much lower than 6A.

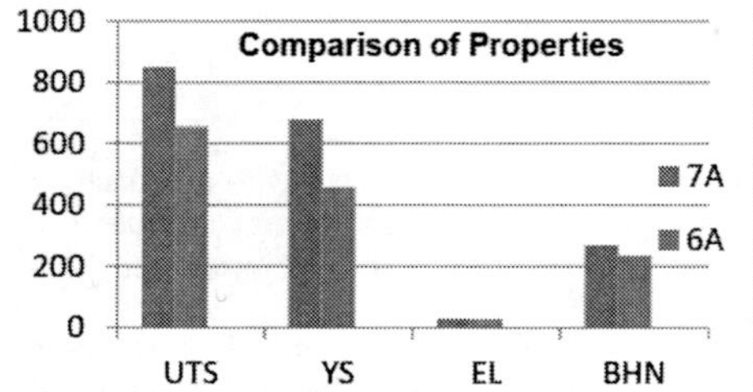

Fig 2 Comparison of Tensile Properties

2nd Experimentation:

For finding deleterious phase precipitaion in grade 7A, we carried out thermal ageing treatment on 30*30*50mm samples in laboratory muffle furnace. Sample was quenched in water after soaking.
This is shown in Table 5

Table 5 Ageing treatment

Temp.° C	Time ,Min	%Ferrite	% Sigma	BHN
1000	10	0.95	38	415
900	20	2.2	34	388
800	30	10	28	363
700	60	32	18	302

It shows that at 1000°C for 10Min ageing , almost whole ferrite was consumed and 38% sigma formed. Fig 3 shows micro structure of this sample , where sigma is revealed as red particles .

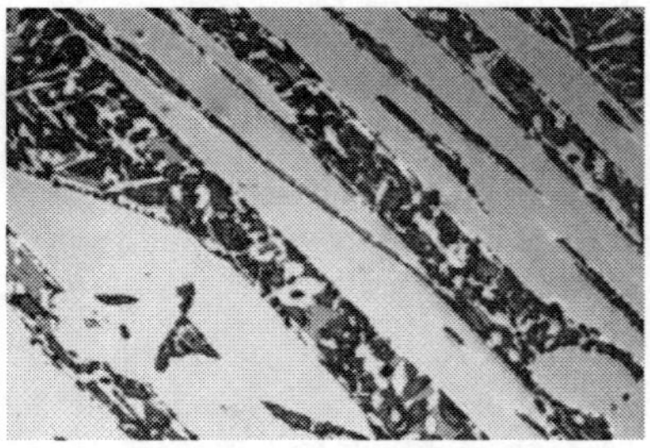

Fig 3 Sample shows sigma as red particles (200X).

3rd Experimentation.

Pitting Corrosion test was carried out as per ASTM G 48 method A and found to have no weight loss in ferric chloride solution at +40°C for 24 hours and there was no pitting observed. CPT determined as per ASTM G48 method E. Electro chemical pitting resistance was carried out using potentiostat (CHI Instruments) in a voltage range of -0.6V to 1.8V. Scan rate used was 0.01V/s. . Tafel Fit obtained for pitting corrosion resistance & it shows better performance of grade 7A than 6A

Conclusion

Grade 7A showed higher strength after solution annealing treatment when compared with grade 6A. Low temperature toughness for grade 6A is better than grade 7A. 38%Sigma formed in 10min at 1000°C Electro chemical pitting test shows better pitting corrosion resistance for 7A when compared with 6A.

Further Studies required for characterization of 7A.

Acknowledgement.
The authors are grateful to management of Peekay Steel and Staff members of Materials Science Department of NIT Calicut for carrying out this research work.

The 73rd World Foundry Congress
„Creative Foundry"
23 - 27 September 2018, Krakow, Poland

ADDITIVE MANUFACTURED HYBRID SEGMENTS FOR DIE CASTIG DIES EQUIPPED WITH CONFORMAL COOLING CHANNELS

Dr. Waldemar Sokolowski

Oskar Frech GmbH + Co. KG
Schorndorfer Street 32
73614 Schorndorf
Germany

Key words: additive manufacturing, additive manufactured hybrid parts, conformal cooling, die casting, die casting dies

1. Conformal heating & cooling

The conformal heating & cooling of segments (as e.g. cores or slides) for die casting dies is a very efficient method to adopt the appropriate temperature to the whole casting process. It enables e.g. the reduction of hot spots, it decreases the cycle time and as a consequence it increases the productivity, it improves the quality of the die casting products and increases the life time of the components and of the whole die [1]. It can also improve the ejection of casting products e.g. in such cases when they have a minimal ejection slope [2]. The conformal channels of slides or cores have usually very complex geometries and cannot be produced conventionally. The only method which can be employed is additive manufacturing, in this particular case the Selective Laser Melting (SLM).

2. Hybrid segments for die casting dies

Hotspots occur generally only in critical areas of casting dies. It is sufficient to use conformal heating & cooling only for these areas. In such cases slides and cores can be manufactured much more economically as so called hybrid segments. It means that only top parts of them equipped with the complex conformal tempering channels for elimination of hotspots are additive manufactured. Their base bodies can be produced conventionally at lower costs. In the die casting technology martensitic steel 1.2343 is widely used for die segments, but this steel can be processed in 3D printing technology only with the preheating above 450°C. Therefore other hot work steel, the maraging tool steel 1.2709 is used in the additive processes. The top parts and the base bodies of hybrid segments can of course be made of the same material, e.g. of the maraging tool steel 1.2709 (Fig. 1). However, this steel is relatively expensive as a row material. The combination of the cheaper and widely used tool steel 1.2343 as a base body and 1.2709 as a top part is a more economical solution. The fusion of these two steels in the additive process was examined concerning their usage for die casting segments. Tensile test as well SEM (Scanning Electron Microscopy) analysis was used for this purpose.

3. Tensile tests of hybrid segments

The geometry of the test samples is shown in Fig. 2. The tensile tests were conducted at the ambient temperature of about 20°C, at 400°C and 600°C. The last two temperature regions correspond to the temperatures in die casting dies when zinc and aluminum are processed, i.e. to their melting temperatures.

Fig. 1. An example of hybrid segments in the powder bed before removing them from the 3D printers. The base body is a metal block made of the maraging tool steel 1.2709. The additive manufactured top parts are made also of the same material.

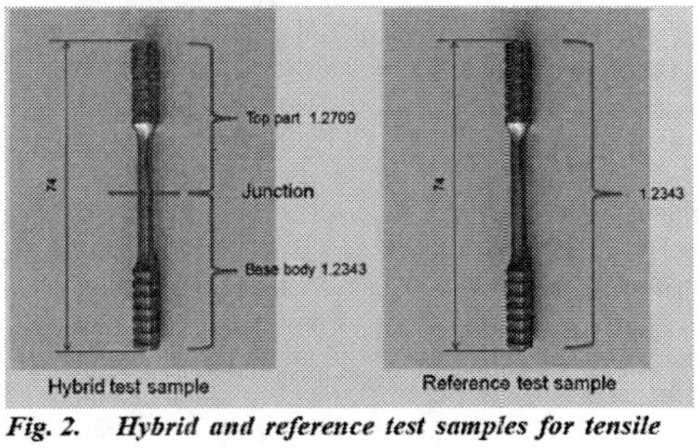

Fig. 2. Hybrid and reference test samples for tensile tests.

4. Experimental Results of Tensile Tests

The conducted tensile experiments show that the fractures of the hybrid samples occurred only above the junction area between 1.2343 and 1.2709. This behavior is the first indication that the junction area has a very good quality. The diagrams of tensile strength versus material stretching are shown in Fig. 3 and Fig. 4. The ultimate tensile strength of hybrid samples at the ambient temperature is between 1.459 and 1.493 N/mm^2, whereas for the reference samples it is ≈1.600N/mm^2. At higher temperatures 400°C and 600°C, the ultimate tensile strength of hybrid samples is about 1200N/mm^2 and 820N/mm^2 respectively. These results show that the hybrid segments made of these tools steels 1.2709 and 1.2343 are suitable for die casting segments.

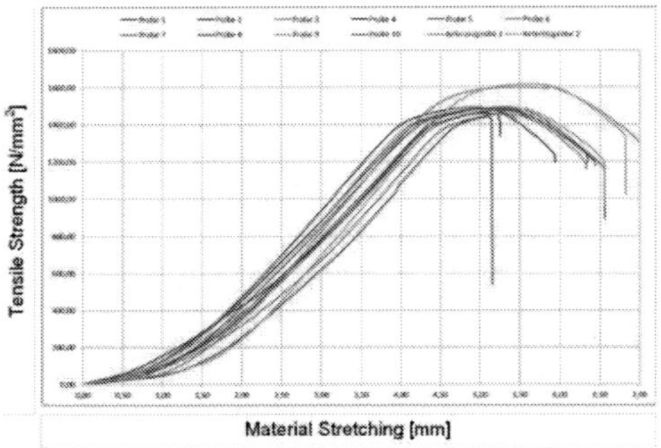

Fig. 3. Tensile strength of hybrid probes (no. 1-10) and reference probes measured at ambient temperature.

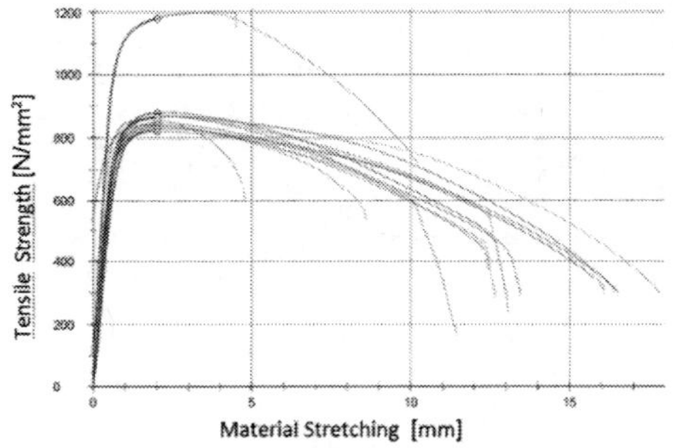

Fig. 4. Tensile strength of hybrid probes (no. 1-10) and reference probes measured at the 400°C (upper curves) and 600°C temperature (lower curves).

5. Experimental Results of SEM

As an additional method for analyzing the quality of the junction area between the steel 1.2709 and 1.2343 the Scanning Electron Microscopy (SEM) was used. The SEM images with 12x and 666x magnifications presented in Fig.6 show that after simple mechanical post processing of the hybrid samples the junction area cannot be recognized even with the help of this method. Therefore the junction area was examined with EDX-Mapping (Energy-Dispersive X-ray Spectroscopy). The results are shown in Fig. 7. They indicate that there are no cracks and no porosity in the fusion zone. The EDX-Scan along a line over the junction enables to measure the width of the fusion by estimating the content of the elements which belong either to 1.2709 or 1.2343 steel (see Tab.1). The width of the fusion zone is very narrow and is approximately 30µm. These results confirm that the junction between 1.2709 and 1.2343 has an excellent quality.

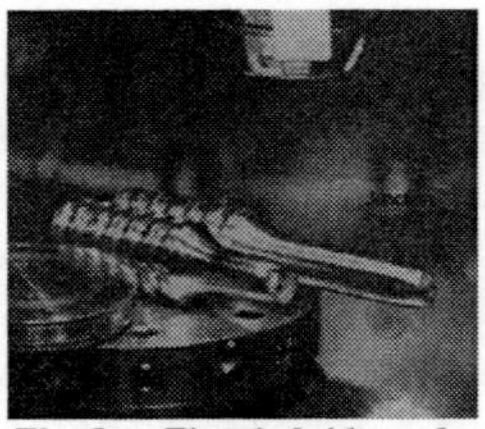

Fig. 5. The hybrid probes for Scanning Electron Microscopy (SEM) and for analyzing of the junction area.

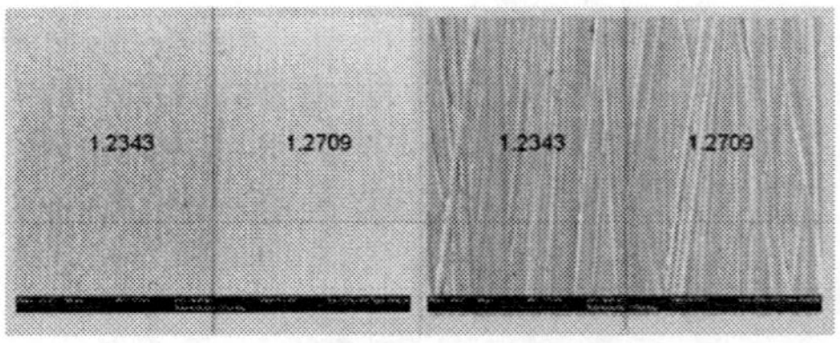

Fig. 6. The mechanically post processed junction area cannot be recognized in der SEM images (magnifications 121x and 666x respectively).

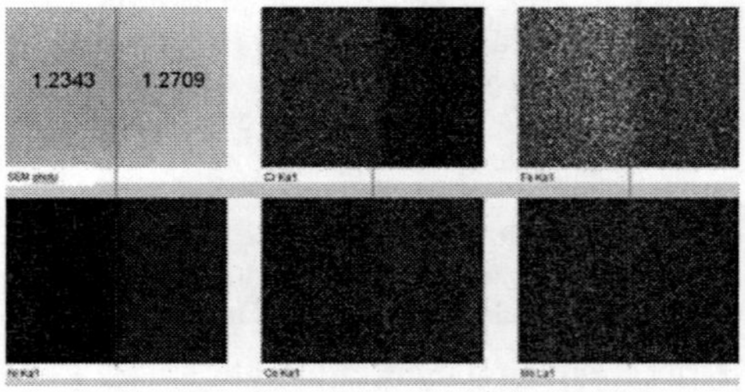

Fig. 7. The junction area was examined with the EDX-Mapping. The results show a very good fusion between the 1.2343 and 1.2709 without any cracks and porosity.

Table 1. EDX-Mapping shows the transition between the 1.2343 and 1.2709 at the junction area.

Material	Fe	Cr	Mo	Co
1.2343 (X37CrMoV5-1)	91,6	5,3	11	0
1.2709 (X3NiCoMoTi_18_9_5)	66	18	5	10

References

[1] Gerd Bühner, Dr. Waldemar Sokolowski. (2016). Prozessoptimierung durch beheizte Gießkanäle und konturnahe Temperierung in Druckgießwerkzeugen. Ranshofener Leichtmetalltage, November 9-10, 2016. Tagungsband der 9. Ranshofener Leichtmetalltage 2016. ISBN-13:978-3-902092-09-0.

[2] Oskar Frech GmbH + Co. KG, Success Story: FEIX Druckguss – Optimization of the production process by additive made cooling canals (2018), EUROGUSS 2018, Germany.

The 73rd World Foundry Congress
„Creative Foundry"
23 - 27 September 2018, Krakow, Poland

MICROSTRUCTURE CHARACTERISATION OF GRADED ALLOY CAST IRON FOR FLASHING/GRINDING PLATES USED IN GRINDING OF BALL BEARINGS

Prayut Bhamawat [1]*, J N Baad [2]

[1]M/s Mangalam Steelcast Pvt Ltd, Plot No. 1104-1105, GIDC Waghodia - 391760, Vadodara, Gujarat, India,
[2] M/s Reiki Metallica, 701, Sanket Heights, Nr. Akshar Chowk, Vadodara-390020, Gujarat, India,
*Corresponding address: e-mail: prayut.mangalam@gmail.com

Key words: Grinding-Plate, Flashing-Plate, Ball-Bearings, Alloy Cast Iron, Self Lubrication Graphite

Alloy cast iron has been widely studied and used for numerous applications. It has characteristics like strength, weight, price, ability to cast complex geometries etc, and can be modified by varying its chemical composition and microstructure, hence it is widely used in engineering application. In recent years study of alloy cast iron having tribological properties has been of great interest as there is great potential in manufacturing industries by reducing the wastage of resources resulting from high friction and wear. There have been various studies to investigate the tribological behavior of alloy cast iron with different graphite morphology and microstructures, which would influence wear rate and the lifetime of the parts [1].

This paper covers mainly about development of grinding plate and flashing plate castings for the application of grinding steel bearing balls. In order to grind & polish the balls efficiently, casting plate should have uniform structure, high strength and wear-resist matrix [2,3]. The aim of this paper is to study the factors like matrix microstructure, graphite flakes morphology, carbides quantum & its distribution on which wear resistance of cast iron is dependent [4] & evaluate the performance (cut-rate of the bearing balls) with change in these characteristics.

1. Requirement for Grinding/Flashing Plate

Cold-working for surface finish operation is a challenging job. For such application, have to consider the flowing set of properties to acquire the desired results.

1. Wear Resistant – to withstand the high hardness of material to be ground.
2. Uniform Hardness through-out the thickness – for uniform cut rate and improve the life of the plate.
3. Freedom from foundry defects like blow holes, pin holes, sand inclusions etc. – Even the minute defect disturbs the quality and surface finish of the bearing ball.
4. Close grain structure – Required to impart smoothness after grinding
5. Self lubrication – Presence of graphite in microstructure in optimum amount is necessary to have lubrication in micro-level
6. Machinability – Plates should be machinable at the hardness around 50 HRC
7. Heat absorption – The plates should have sufficient thermal conductivity to promote the heat dissipation during grinding operation.
8. Damping Capacity – Plates should have sufficient damping capacity to absorb the vibration during service.

2. Manufacturing Process

In this study, alloy cast iron Flashing plates were manufactured in a medium frequency induction furnace of 750 Kg capacity. Main raw material used for manufacturing the same are pig iron, cast iron scrap & alloying elements. Healthy inoculation practice is sine-qua-non for favorable graphite distribution and its size.

Wear resistant is imparted by having high carbon content in range of cast iron along with the alloying addition to promote the optimum amount of alloying carbides. Machinability is improved by addition of non-carbides forming elements which also enhances the hardenability of the casting. The subsequent transformation in their stoichiometry is done by the way of heat treatment. Separate cycle of hardening is carried out so as to achieve the desired hardness range (between 48 – 53 HRC). Combination of both heat treatment cycles is decided in a way to get the uniform hardness across the thickness of 100 mm.

Meticulous foundry practice is adopted to ensure the defect free & consistent castings. Defined practice of inoculation is followed so as to achieve close grain structure. Final Heat Treatment operation is carried out to obtain uniformly distributed fine graphite structure so as to have self-lubrication property as well as to improve damping capacity.

3. Observations

Three different variations of the plates (named as Plate No. 1, 2 & 3) were manufactured and observations mainly hardness, microstructure and cut rate of bearing balls – to check performance (conducted trials on the ball grinding machines with 13mm 420C balls) were noted and compared with standard plates (i.e., imported plates that is already developed and is in regular use).

Table 1. Comparison of the observations of various plates

Plate No.	Microstructure	Hardness	Cut Rate

			(μ/hr)
1	Graphite Flakes with Carbides	60 HRC	118
2	Smaller size graphite Flakes w/ dis-continuous elongated primary carbides	51 HRC	44.9
3	Finely distributed smaller size graphite flakes, w/ scattered distribution of primary carbides in tempered maternsite	52 HRC	98.7
Standard Plate	Finely distributed longer size graphite flake with network of primary carbides in tempered martensite	48 HRC	114.9

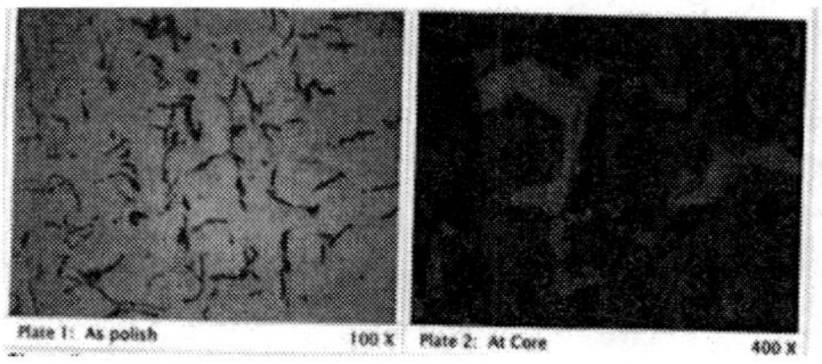

Fig. 1. Microstructure observation of Plate No. 1 (Microstructure shows graphite flakes with presence of carbide in matrix of martensite)

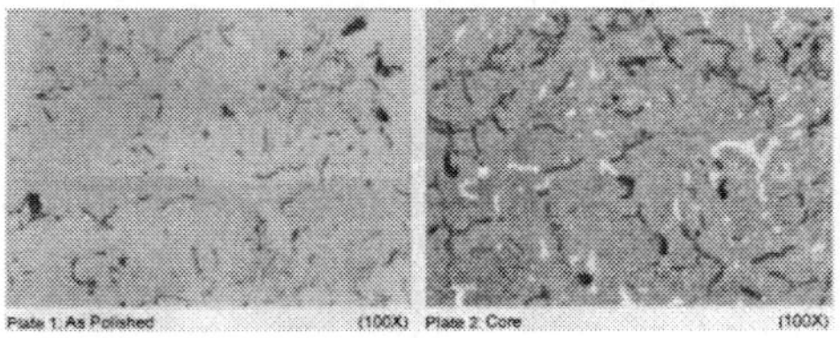

Fig. 2. Microstructure observation of Plate No. 2 (Small graphite flakes along with discontinuous spots of elongated primary carbides in tempered martensite matrix)

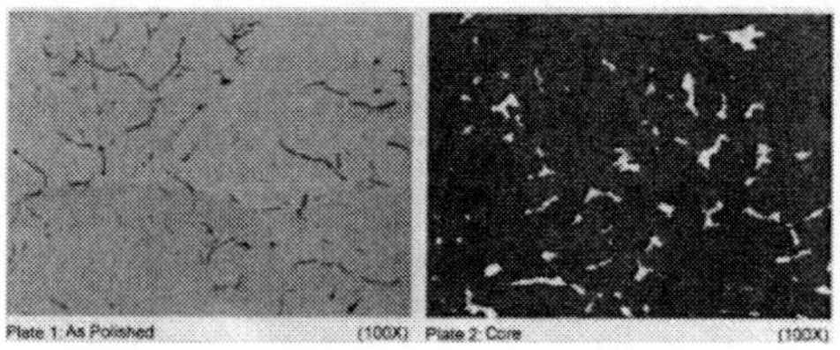

Fig. 3. Microstructure observation of Plate No. 3 (Scattered Graphite flakes with size around 3-6 along with scattered primary carbides in tempered martensite matrix)

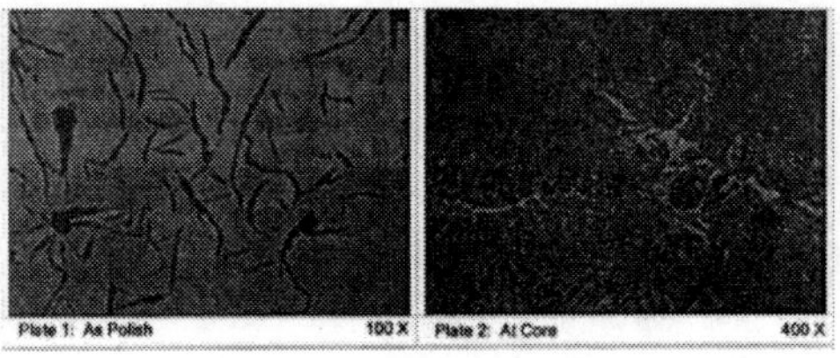

Fig. 4. Microstructure observation of Standard Plate (Larger graphite flakes with elongated primary carbide in martensite matrix)

4. Discussion

Plate No. 1: Hardness too high and excellent cut rate but observed to have such plates with higher tendency of cracking with slight variation in thermal stress. Moreover during ball grinding operation, because of higher carbides, the balls were getting polished instead of grinding that is not acceptable in flashing operation.

Plate No. 2: Hardness within desired range but cut rate observed to be very less. This might be due to graphite quantum is more (distribution is OK) and Carbide percentage and distribution is lean. So not performed well with respect to standard plate.

Plant No. 3: Received hardness within desired range. Graphite flake size is smaller, finer, more uniform and slightly vermicular. Carbide % and distribution is normal. Performance was better & close to the standard plate.

5. Conclusion & Future Plan

Comparing to the standard plate, it can be concluded that for the better performance of the plates, it is important to maintain optimum carbide quantum, uniformly distributed in tempered martensite matrix.

For next trial (Plate No. 4), we will be doing micro-alloying to have the more uniform distribution of graphite flakes & carbides. Both these changes should enhance plate performance further. An alternative of having longer & dense graphite flakes as in standard plate but with an altered heat treatment of austempering is also in the pipeline.

References

[1] You Wang, Zhaoyi Pan, Zheng Wang, Xiaoguang Sun, & Liang Wang. (2011). Sliding wear behavior of Cr-Mo-Cu alloy cast irons with and without nano-additives. Wear. 271, 2953-2962.

[2] Junbai Xing, Qiang Wang & Jinpeng Chang. (2007). Microstructure of martensite/bainite dual-phase grey cast iron and its strengthening mechanism. ISIJ International. 47(12),1776-1780.

[3] C. Scandian, C. Boher, J.D.B. de Mello, & F. Rexai-Aria. (2009). Effect of molybdenum and chromium contents in sliding wear of high-chromium white cast iron: The relationship between microstructure and wear. Wear. 267, 401-408.

[4] Aigbodion, V. S., Isah, L. A., Achiv, F. M., Ezema, I. C., Daniel-Mkpume, C. C., & Asuke, F. (2016). Anti-Wear Enhancement of Silicon-Grey Cast Iron for the Production of Grinding Plates: Effect of Snail Shell Powder as Carbide Forming Material. Silicon, 1-10.

The 73rd World Foundry Congress
„Creative Foundry"
23 - 27 September 2018, Krakow, Poland

ADSORPTIVE NATURE OF GASES PRESENT ON HIGH TEMPERATURE TREATED CARBONACEOUS MATERIALS

Andreas Hahn[1], Rich Nelson [2], Rijo Jacob Robin [2]*, Carsten Wehling [2]

[1]*ZetA Partikelanalytik GmbH, Bischheimer Weg 1, Mainz, Germany*
[2] *Superior Graphite, 10 South Riverside Plaza, Chicago, USA*
**Corresponding address: e-mail: rjacobrobin@superiorgraphite.com*

Key words: Gas content, Recarburizers, Desulco®, TPD, Adsorption

1. Introduction

Gas porosity is one of the most serious problems in the casting of cast iron, aluminum and copper. It is generally caused by the evolution of gases during the casting and solidification process [1]. The melting method and charge materials influence the nitrogen content of the iron. For example, in cupola melting, the higher the steel content used, the higher the final nitrogen level. In electric melting practice, steel charge levels can be as high as 80 percent and 3 percent carbon may be added in the form of a carburizer. The carburizer quality is important, since carburizers can contain variable levels of gases. For example, some carburizers can contain high levels of nitrogen, as shown in (Table 1) below. A carburizer containing 3,000 ppm nitrogen can contribute 30 ppm of the gas for every 1% carbon addition. Therefore, a 3% addition would certainly result in an iron having a high tendency for nitrogen fissure defects to occur. If hydrogen is present, cumulative effect of both gases then results in the formation of fissure or pinhole type defects. [2] [3]

Recarburizer material	H_2 (ppm)	N_2 (ppm)	O_2 (ppm)
H.T.Thermally Purified Carbon	10	40	120
Graphitized Petroleum Coke	32	16	115
Electrically Calcined Anthracite	3100	2790	14650
Calcined Acetylene Coke	3300	382	1025
Pitch Coke	2000	12700	1056

Table 1. Gas content in recarburizers

In order to ensure low transfer of gases into the melt, it is important to quantify the gas content in the carburizer material. Industry is relying on several techniques to quantify different types of carburizers, even though they might be fundamentally different type of carbonaceous materials. To date, there are no known standard methods available to gas content in manufactured graphitic carbons. Standards like ASTM D5373 are available for coke and coal to quantify hydrogen and nitrogen but not oxygen.[4] Most of the instrumental techniques based on inert gas fusion method rely on the standards that are meant for non-graphitic carbons and this tends to give inconsistent results on HT (high temperature) thermally purified carbons (Desulco®), which are partially graphitic. Different sample preparation methods contribute to further errors especially in the oxygen content values.

It is the objective of this paper to verify the adsorptive nature of gases present on high temperature treated carbon, which would help clarify whether the gases measured using the gas analyzers are really present inside the material or physisorbed (Fig. 1) on the surface. In addition, study the effect of humidity on gas content.

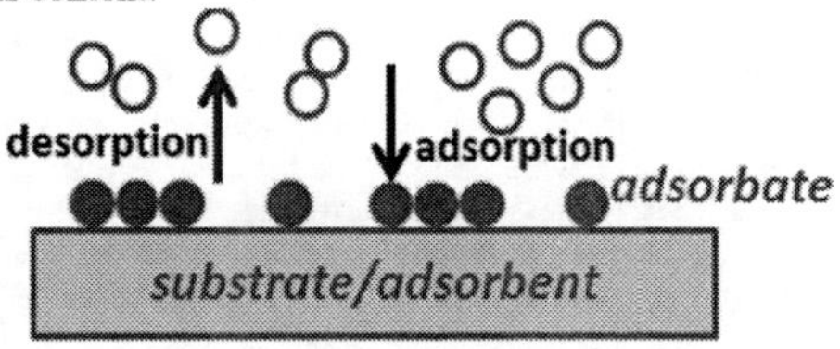

Fig. 1. Schematic illustration of adsorption onto, desorption from, a substrate [5]

Standard gas analyzers lack the ability to differentiate between physisorbed and chemisorbed species. Adsoprtion in case of a non-porous media is a surface phenomenon of the substrate, where preferential accumulation of adsorbate molecules takes place. Molecules at the surface of a material experience imbalanced forces of intermolecular interaction, which contribute to the surface energy. Pure physical adsorption or physisorption is based on weak interactions like the van der Waals and electrostatic forces between the solid surface (adsorbent) and gas molecules (adsorbate), while chemisorption is based on the chemical molecular bonds resulting in much stronger interactions. Thus, the strength by which adsorbate molecules are attached to the adsorbent molecules determines the nature of adsorption. [6] It is common to regard the upper limit of bond strength in physisorption as around 0.6 eV per atom or molecule or 58 kJ/mol (1 eV/molecule = 96.5 kJ/mol) [7] [8]

Temperature programmed desorption (TPD) technique offers the ability to investigate the temperature dependence of specific adsorption or desorption process in a well-defined gas atmosphere. It offers insight to the kinetic and thermodynamic parameters of the sorption reactions and evaluates the amount of desorbing species. [9] Individual steps including sample pretreatment, adsorption and desorption can be monitored by a thermal conductivity detector (TCD) in the exhaust (Fig. 2) gas stream with variation of the parameters: temperature, gas flow, gas species or gas composition.

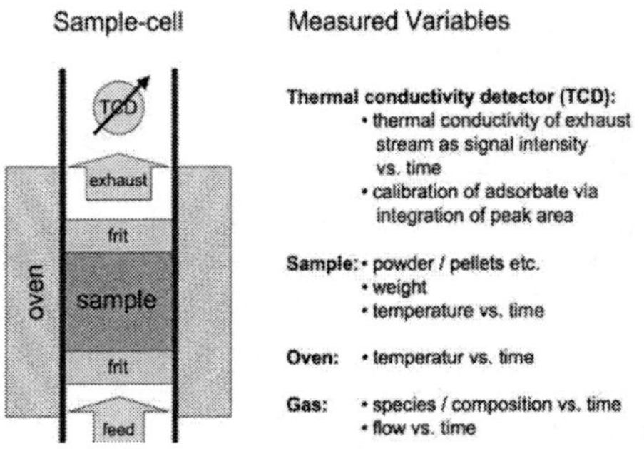

Fig. 2. Sketch of TPD instrumentation and measured variables

2. Experiments and Results

A series of temperature programmed desorption (TPD) experiments were performed on a 'fresh' HT-thermally purified carbon (Desulco®) sample and on samples, which were exposed to different levels of humidity before measurement. Loading of the samples was performed via storage in desiccators with defined humidity levels of 45% rH and 95% rH for at least 48 hours before measurement. Effect of humidity on desorption was compared against recarburizers listed in (Table 1), at similar conditions.

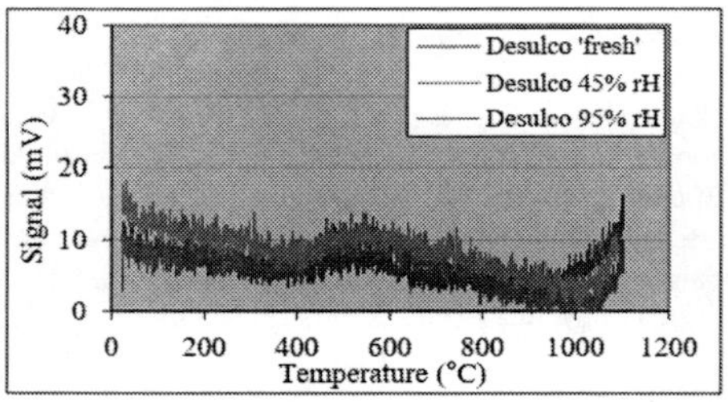

Fig. 3. Humidity dependence of desorbed gas from Desulco®

Although a distinct increase in sample weight was observed during pretreatment, similar desorption profiles obtained in (Fig. 3) hint that the water is only weakly adsorbed and was desorbed easily during starting up of the measurement, while Helium is flowing through the sample.

The rate constant for the desorption process may be expressed in an Arrhenius form, [10]

$$k_{des} = A\, e^{-Ea_{des}/RT} \qquad (1)$$

Where Ea_{des} is the activation energy for desorption, and A is the pre-exponential factor; this can also be considered to be the 'attempt frequency', ν, at overcoming the barrier to desorption. The TPD experiments were performed at different heating rates and energy of desoprtion was obtained from the slope of an Arrhenius plot.

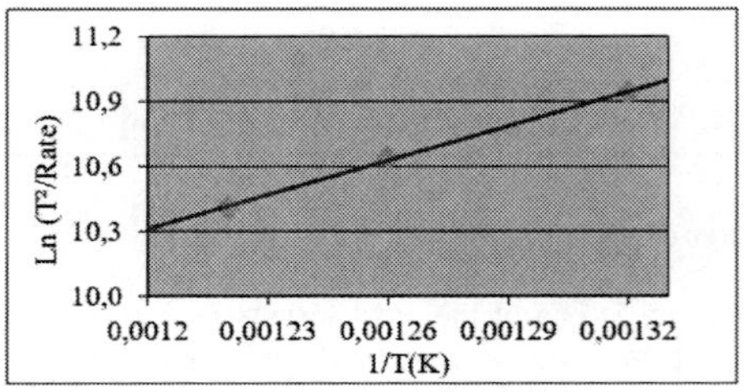

Fig. 4. Desorption energy calculation from varying heating rates: 10 K/min, 15 K/min, 20 K/min

The value of 42.8 kJ/mol for the energy of desorption was determined (Fig. 4) for Desulco® sample with good correlation. This value is below the level of chemical binding energy.

3. Conclusion

This TPD humidity study indicated that there is virtually no dependancy of humidity on the gas desorbed from Desulco®, however it did seem to have an effect on the other recarburizers tested. Energy of desoprtion obtained on Desulco®, is below the level of chemical binding energy, clearly showing that gas present or being measured by gas analyzers are present only as loosely bound on the surface of Desulco® but not present 'inside' the material.

4. References

[1] Stefanescu D. M. (1988) Volume 15 Casting Metals Handbook 9th Edition (page 82), ASM International

[2] Greenhill J.M. (1989) AFS Elemental Effects on Gray Iron Conference, Sept 1989 (page 4-7) BCIRA England

[3] Greenhill J.M & Reynolds N.M. (1981) Foundry Trade Journal, AFS Elemental Effects on Gray Iron Conference, Sept 1989 (page 111)

[4] ASTM International (2017). Petroleum standards. Dec 2017 (www.astm.org/Standards/petroleum-standards.html)

[5] Zhang Y.-Q (2017). Surface Analysis with TPD and Low-Energy Electron Diffraction. Physik-Dept, TUM

[6] Tembe B.L Adsorption, module 8 surface chemistry, Engineering Chemistry, NPTEL, IIT Bombay

[7] Woodruff D.P (1994) Modern Techniques of Surface Science 2nd Edition (page 11), Cambridge University Press

[8] Davis M.E. (2003) Fundamentals of Chemical Reaction Engineering, McGraw-Hill Higher Education

[9] Schroeder S.L.M., Gottfried M. (2002) TPD and Thermal Desorption Spectroscopy. Advanced Physical Chemistry Lab, FU Berlin

[10] QMUoL (2005) Adsorption of Molecules on Surfaces, Dec 2017 (www.chem.qmul.ac.uk/surfaces/scc/scat2_6.htm)

The 73rd World Foundry Congress
„Creative Foundry"
23 - 27 September 2018, Krakow, Poland

PROBLEMS AND IMPROVEMENTS ON THE PRODUCTION OF LARGE CASTING WITH HI-SI DUCTILE IRON

Toshitake Kanno[1*]

[1]*Kimura Foundry Co., Ltd, R & D, 1157 Nagasawa Shimizu-cho, Sunto-gun, Shizuoka-ken 411-0905 Japan*
**Corresponding address: e-mail: kanno@kimuragrp.co.jp*

Key words: high silicon, ferritic spheroidal graphite cast iron, mechanical properties, nodule count, embrittlement, P, Mg

We studied on the problems and improvement on producing large casting with solid solution strengthened ferritic spheroidal graphite cast irons (hearafter, Hi-Si).

As the matrix of Hi-Si material becomes hardened with silico-ferrite, its hardness, tensle strength and yield strength increase but its elongation and impact value decrease conversely. If Si amount increases or if the cooling rate decreases due to its heavy section, as Si segregation progresses, the elongation and impact value are significantly reduced. When increasing the graphite nodule count with setting optimum inoculation or chiller, the segregation of Si or other impurity elements decreases and then the mechanical properties are recovered. In addition, when Hi-Si reaches around 400 Celsius (hereafter, C), its elongation decreases largely, which is called 400C embrittlement. This high temp. embrittlement can be prevented by setting Mg/P less than 1.5, because the grain boundary fracture occurs due to the free Mg not fixed as Mg_3P_2 at the grain boundry.

1. Introduction

In the past, solid solution strengthened ferrite spheroidal graphite cast iron (hereafter Hi-Si material) has been used as a heat-resistant component for small casting such as manifolds. However, because of its high yield strength, the researches for applying it to middle or large size castings as a substitute material for cast steel have been carried out. However, in large castings, since the cooling rate is slow, abnormal graphite forms due to segregation of Si or various impurity elements at grain boundaries. As the result, the elongation and the impact value are remarkably lowered.

In this research, we investigated the cause of the reduction of the mechanical properties in this Hi - Si material and its prevention method.

2. Experimental method

4.3% of high silicon molten metal was prepared with 3000Hz induction furnace. (Basic composition: C3.24%- Si4.3%- Mn0.3%- P0.034%- (base melt) S0.012%- Mg0.044%) In the Si changing test (Si2.5%~5.0%), carbon amount was adjusted not to become hypereutectic composition. In high temp. embrittlement test, the Mg/P ratio was adjusted from 1.3 to 2.2. After spheroidizing treatment at 1380C, 25mmY block (7kg, hereafter t25YB) and D200xH200mm test piece (50kg, hereafter D200TP) were cast. In the high temp. embrittlement test, the tensile tests were conducted at from room temp. to 550C, and the change of the mechanical properties (especially elongation) were investigated.

As shown in ***Fig.1***, in order to investigate micro-structure and mechanical properties of large casting, we cast as follows; (a) step test pieces (t=130, 200, 270mm, 420kg, hereafter Step TP), (b) attached test pieces of ENt70mm (12kg, hereafter t70ENTP), and (c) concave type test pieces of thickness 500mm (900kg, hereafter t500 TP). On the surface of drag side of t500 TP, chiller is set as like (c) (hereafter, t500Chill).

3. Experimental result

3.1 Influence of Si amount and heavy section on mechanical properties

As shown in ***Fig.2.***, as Si amount increases, the hardness, the tensile strength and the yield strength increase, because the matrix is hardened by silico-ferrite. However, the elongation and the impact energy decrease conversely. Since Hi-Si has ferritic matrix and pearlite transformation does not happen, the mechanical properties difference between 25mm Y-block TP (rapid cooling) and D200TP (slow cooling) is small. And Hi-Si shows high yield strength, relatively.

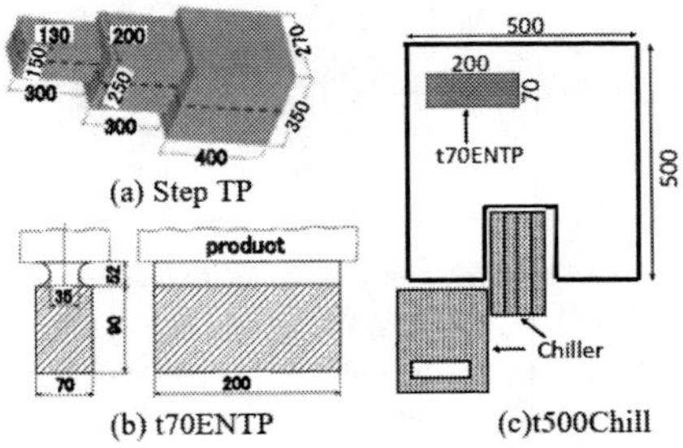

Fig.1 Size of the test pieces cast in this study

Fig.3 shows the mechanical properties of Si 4.3% Hi-Si. As the product becomes heavy section and cooling rate slows down, the mechanical properties decreases due to Si segregation. The reduction of the tensile strength (a) or the yield strength (b) is small, but the reduction of the elongation (c) is remarkable. However with setting chiller, the mechanical properties are recovered.

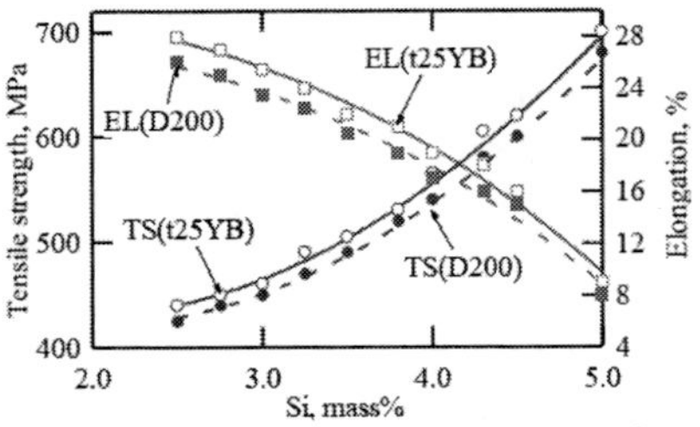

Fig.2 Influence of Si on the tensile strength and the elongation (t25YB and D200TP)

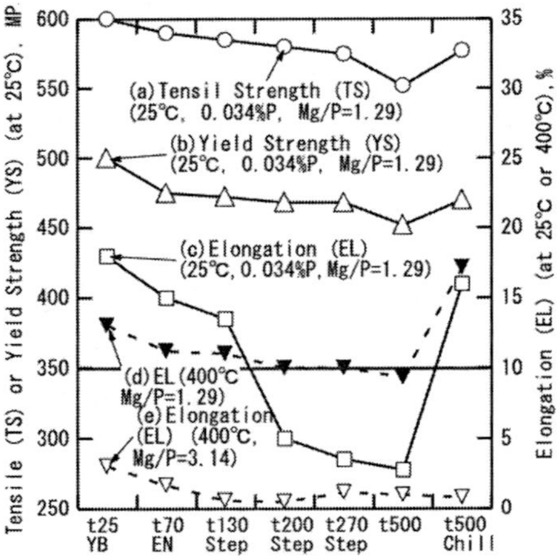

Fig.3 TP thickness vs mechanical properties (Si:4.3%, Mg:0.044%)

3.2 Improvement of mechanical properties with setting chiller

As shown in *Fig.4.*, with setting chiller to t500TP, the graphite nodule count increases markedly. Therefore, it seems that the reduction of Si segregation or other impurity elements causes the recovery of the mechanical properties.

(a) Without chiller	(b) With chiller
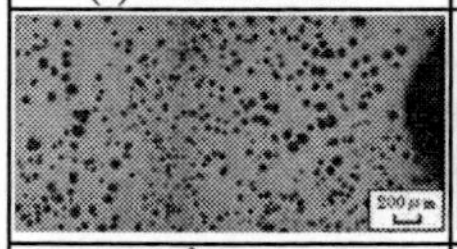	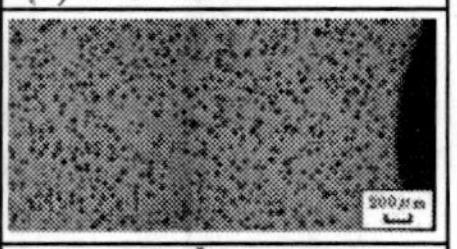
NC:96/mm^2,TS: 552MPa, YS:452MPa,EL:3.8%	NC:225/mm^2,TS:577MPa, YS:470MPa,EL:16%

NC:Graphite nodule count

Fig.4 Increasing graphite nodule count with setting chiller (t500TP)

3.3 Cause of high temp. (400C) embrittlement and its prevention method

Fig.3.(d) and ***Fig.3 (e)*** show an example of the 400C embrittlement of Si4.3% Hi-Si. When Mg/P ratio is 1.29, the elongation at 400C exceeds 9%, but when Mg/P ratio is 3.14, the elongation is as low as 1 to 3%.

Fig.5. shows the relationship between tensile test temp. and elongation when the Mg/P is 3.14. If Si amount is high, the elongation reduction near 400C becomes large. In case of pearlite type material such as GGG70 with 2.1% Si, the embrittlement at 400C is not remarkable.

Fig.6. shows the relationship between Mg/P ratio and elongation at 400C of 4.3%Si Hi-Si. Elongation increases, when Mg/P becomes under 1.5. Yanagisawa suggests that the high temp. embrittlement occurs due to the grainboundary fracture caused by free Mg not fixed as Mg_3P_2 at the grain boundry[1)].

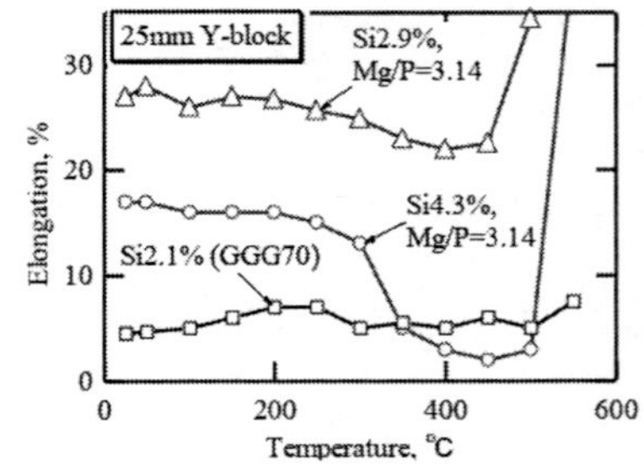

Fig.5. Test temp. vs elongation (P:0.014%, Mg:0.044%, Mg/P=3.14)

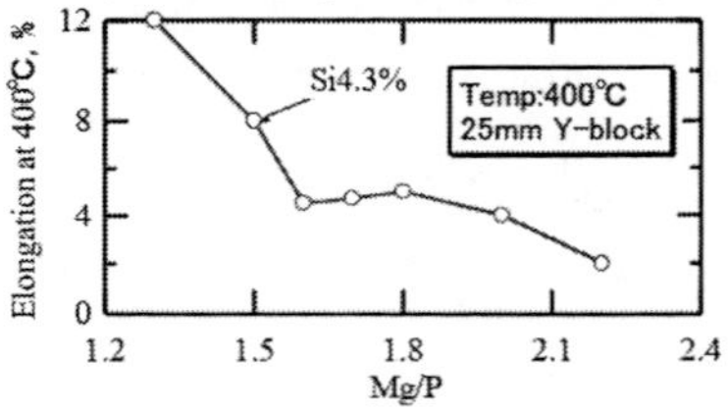

Fig.6. Mg/P vs 400C *elongation (Mg:0.044%)*

4. Conclusion

(1) As for Hi-Si material, as Si amount increases, the hardness, the tensile strength and the yield strength increase, because the matrix is hardened by silico-ferrite. However, the elongation and the impact energy decrease conversely. Especially, the yield strength shows higher value than that of cast steel.

(2) Since Hi-Si is ferritic matrix material, the tensile strength or the yield strength does not decreases largely, even if the product becomes heavy section, but the elongation decreases largely.

(3) In orde to prevent the elongation reduction, it is important to increase the graphite nodule count with setting optimum inoculation or chiller.

(4) The elongation of Hi-Si decreases near 400C largely. This high temp. embrittlement can be prevented with adjusting Mg/P ratio under 1.5.

References

[1] O.Yanagisawa: J.JFS, 72(2000) pp604-609.

The 73rd World Foundry Congress
„Creative Foundry"
23 - 27 September 2018, Krakow, Poland

USE OF DILATOMETRY TO EVALUTE THE HIGH TEMPERATURE CHARACTERISTICS OF SILICA IN CHROMITE SAND

S. Ravi [1]*, J. Thiel [1], N. Bryant [1], S. Giese[1], T. J. Schneiter[2]

[1]*University of Northern Iowa, Metal Casting Center, Cedar Falls, Iowa, USA*
[2]*Sivyer Steel Corporation, Bettendorf, Iowa, USA*
**Corresponding address: e-mail: ravis@uni.edu*

Key words: Steel Casting, Double Skin, Glazing, Chromite, Penetration

1. Abstract

The steel casting industry incorporates chromite sand in their core rooms, mainly due to the desirable properties exhibited by the aggregate. Chromite sand provides a cheaper alternative to specialty aggregates such as Zircon, while reducing or eliminating casting defects associated with silica sand. However, chromite sand is associated with surface defects in heavy section steel castings, known as the double skin defect. One of the causes for this defect is the presence of impurities, such as silica or silicates, in the sand. The University of Northern Iowa Metal Casting Center conducted tests using dilatometry to analyze the high temperature characteristics of chromite at varying silica content. Additionally, sinter temperatures were calculated for the samples. The testing and results detailed may be useful to monitor chromite sand for their high temperature characteristics. The sinter temperature results can be compared with sand temperature curves from simulation software to evaluate casting applications for defects related to Chromite sand.

2. Introduction

Chromite sand is widely used in the steel casting industry, primarily due to its high thermal conductivity, low thermal expansion and wettability resistance to molten steel. Sand cores made with Chromite usually show good chilling characteristics, hence reducing the need for metal chills. Additionally, the lower thermal expansion reduces or eliminates expansion related defects, such as veining or finning.

Impurities in Chromite sand may cause an adverse effect on casting quality, especially for heavy section castings. These impurities may consist of silica, and silicates, calcium and other elements. Research conducted by Biel et al.[1] determined that the iron reduced from chromite sand combines with the silicates within the sand, under oxidizing conditions and high temperature, to form fayalite. Fayalite is a low melting point iron silicate and is capable of softening or fluxing the chromite aggregate. This results in the chromite sand being more easily wetted by molten steel and forming a surface defect known as the double skin defect or glazing. This defect is usually observed when the sand temperature at the mold-metal interface is at or above 1250 °C (2282 °F).

Trials were conducted at the University of Northern Iowa Metal Casting Center to evaluate the effect of silica content on the high temperature characteristics of chromite sand. Test samples of Chromite sand were received from a local foundry and characterized using an X-Ray Fluorescence Spectrometer (XRF) for their composition, due to observed defects with this sand in heavy section steel castings. These samples were taken from different batches. Due to the high variability observed in the silica content in the five samples, a control sample, with a known silica content of <1.0%, was used. The test samples were doped with additional silica and tested. Thermal expansion tests were conducted to evaluate the samples for their expansion characteristics. Surface viscosity and sinter temperature for the samples were calculated from the thermal expansion results.

3. Methodology

The chemical composition of the samples was determined using a Benchtop XRF Spectrometer. The samples were first crushed to a fine powder using a mortar and pestle and then placed in a sample holder, which contains a 3 micron gauge polypropylene film. Three point calibration curves were conducted for the elements of interest.

The University's high temperature aggregate dilatometer was used to run thermal expansion and sinter temperature tests on the chromite sand samples. The unit is capable of testing samples up to a temperature of 1650 °C (3002 °F). The samples were cylindrical and had a height of approximately 1.6 inches (4.06 cm) and a diameter of 1.1 inches (2.8 cm). A heating rate of 10 °C per minute was used for the tests. The resulting deformation was recorded. Surface viscosity[2] and sinter temperature was calculated from the deformation recorded from the dilatometer and is useful to describe the sintering characteristics of an aggregate. A Phenolic Ester binder system was used to prepare all sand samples, with a resin content of 1.5%, based on sand weight.

4. Discussion of Results

Table 1 shows the silica content measured in the control and test samples. The control sample was measured to have a silica content of 0.126%. The silica content in the test sample was observed to vary widely from batch to batch, ranging from 0.18% to 1.81%. Based on the XRF results, chromite samples with 1.8% silica was tested with dilatometry and compared to the control sample. Additionally, to evaluate higher silica contents, chromite samples with 2.2% silica and 3% silica were tested.

Table 1. X-Ray Spectrometry Results for Control and Test Samples.

Sample ID	*Cr_2O_3*	*Fe_2O_3*	*SiO_2*
Control	47.92	29.96	0.126
Test #1	56.84	41.81	0.365
Test #2	56.19	41.05	1.810
Test #3	57.62	40.91	0.183
Test #4	57.28	41.03	0.563
Test #5	56.97	40.85	1.03

The thermal expansion results for the four samples are shown in figure 1. The control sample was measured to have a steady expansion upon heating from room temperature to approximately 1380 °C (2516 °F). A sharp contraction was seen after this point. In comparison, increasing the silica content in chromite sand can be observed to decrease the temperature at which the onset of contraction occurs. The three chromite with silica samples were measured to have significantly lower contraction temperatures. Further, the total expansion before the onset of contraction was observed to increase from ~0.002 in/in for the control sample to 0.0058 in/in for chromite with 3% silica.

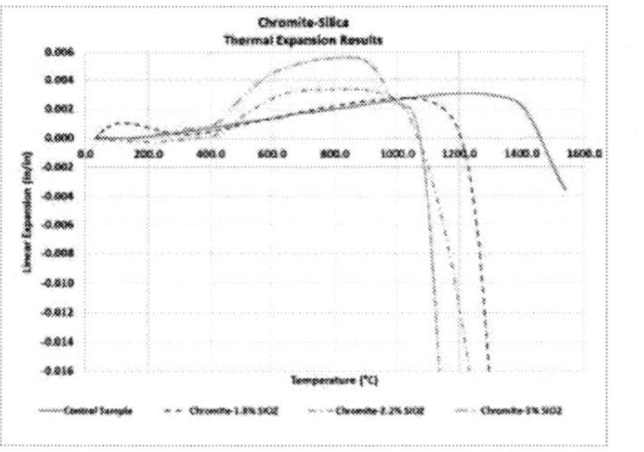

Fig. 1. ***Linear Expansion Results***

The sinter temperatures calculated for the samples are shown in table 2. A large decrease in sinter temperature can be observed from the control sample, which contained 0.126% silica, to the chromite with 1.8% silica sample. The sinter temperature was measured to be ~253 °C (487.4 °F) lower. The 2.2% and 3% silica samples were observed to have a lower sinter temperature when compared to the control and 1.8% silica sample.

Table 2. Sinter Temperature Results

Sample ID	*Sinter Temperature (°C)*
Control	1460.1
Chromite-1.8% SiO_2	1206.3
Chromite-2.2% SiO_2	1066.7
Chromite-3.0% SiO_2	1072.6

A simulation was conducted by the local foundry to evaluate the temperature profile in the chromite sand core. Figure 2 shows the results. It can be observed that at a distance of 0.5 inches (1.27 cm) from the mold-metal interface, the core stays at temperature at or above 1330 °C (2426 °F) for approximately 240 minutes (~4 hours). This temperature is significantly higher than the sinter temperature measured for the three high-silica chromite samples. Hence, these samples can be expected to soften or sinter, due to the high heat input from the casting. Further, from the temperature profile, it can be observed that the temperature of core 1 inch (2.54 cm) away from the mold-metal interface reaches 1200 °C (2192 °F), which may create issues for chromite with higher impurity content.

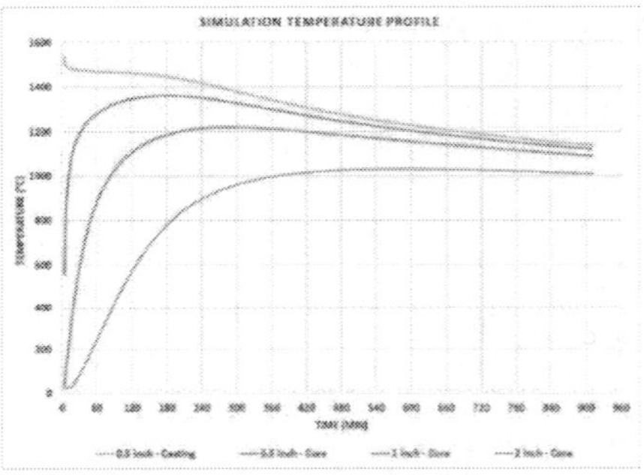

Fig. 2. ***Simulation Temperature Profile***

5. Conclusion

As observed, double skin defects in steel castings with chromite molds or cores can be attributed to high silica content in Chromite sand. While the industry specifies a maximum amount of silica in chromite, that is 3.0%, the section thickness of the casting will also play a role and a lower amount of silica may still lead to this defect for heavy section steel castings. The methodology and results detailed, in conjunction with simulation temperature curves, present quick testing options for a steel casting foundry to evaluate and identify potential problems with the use of chromite sand.

Acknowledgments

The authors would like to acknowledge the efforts of the student employees of the University of Northern Iowa Metal Casting Center for their assistance in the experiments.

References

[1] Biel, J., Petro, A., Flinn, R. A., "Variables Affecting Chromite Sand Performance in Molds", AFS Transactions, paper 80-111 (1980).

[2] Gabriel Tardos et al. from the Department of Chemical Engineering, City College of New York.

The 73rd World Foundry Congress
„Creative Foundry"
23 - 27 September 2018, Krakow, Poland

Effect of Nd Content on Structures and Mechanical Properties of Mg-Gd Alloys Cast into Sand Molds

Noboru Sunayama[1], Kota Takahashi [2], Susumu Ikeno [2], Seiji Saikawa[2], Koichi Komai [1]

[1]TANIDA LTD, 28-2, Higashikagatsume-machi Kanazawa-shi, Ishikawa 920-0209 Japan
[2] Graduate School of Science and Engineering for Research, University of Toyama, Toyama-shi, Toyama 930-8555, Japan

Key words: sand casting, heat resistant magnesium alloy, Mg-Nd-Gd, structures, mechanical property

1. Introduction

The widely used commercial magnesium alloys are Mg-Al-Zn type (AZ type) and Mg-Al-Mn type (AM type) alloys. Because the melting point of the eutectic phase is low, the use of these alloys are limited by relatively poor mechanical properties and creep resistance at elevated temperature. Research is being focused on the development of alloys able to withstand high stresses at elevated temperature.

In the 1980's, magnesium alloys containing additions of rare-earth elements, WE 54 and WE 43, which are excellent in strength and corrosion resistance, has been introduced in the aerospace and speciality applications. However, the high cost and low castability problem is still a problem to be concerned.

In recent years, Mg - Nd - Gd alloys have become more widely used as heat-resistant magnesium alloys for sand casting. In the 1990's, phase diagrams of magnesium-rich portion, aging characteristics and tensile properties of Mg - 3 to 9 mass% Gd - 1 to 7 mass% Nd alloys with a high rare-earth concentration were investigated[1)].

In addition, the Mg-3mass%Nd- 1mass%Gd alloy (hereafter, mass% is omitted) with a lower rare-earth concentration is a new magnesium sand casting alloy for used to elevated temperature. This alloy, which has high strength, good corrosion resistance and excellent castability is being used in civilian, military aircraft, automobile and motorsport industry in Europe.

However, recent studies on this alloy system are mostly reported in practical composition, and few reports have investigated the influence on the low concentration side of the main solute Nd and Gd [2)3)4)]. In this study, the influence of Nd content in Mg-Nd-Gd alloy on the solidification structure and the mechanical properties was investigated.

2. Experimental method

The chemical composition of the foundry cast in this study is shown in Table 1 (hereafter, composition where 0%, 1.5%, 3.0% and 6.0% of Nd element were added with Mg - 1.5% Gd - 0.3% Zn - 0.55% Zr alloys referred to as 0 Nd, 1.5 Nd, 3.0 Nd, 6.0 Nd, respectively). The experimental alloy was melted under the mixed atmosphere of CO2 and SF6. The melt was poured into a sand mold standard tensile test bar. the specimens were sand-casted bars with a gauge length of 50 mm and a diameter of 14 mm, which can simultaneously cast two pieces. The sand mold was prepared by mixed silica sand and phenolic resin.

Appearance of casting is shown in Fig. 1. The cooling curve was measured by the thermocouple at the center of riser on the pouring side, due to investigated solidificate sequence. The as-cast specimens were solution treated at 520 °C for 8h under the mixed atmosphere of CO2 and SF6 and cooled in forced air. Aging treatments were performed at 205°C with cooled in air.

The aging behavior was investigated using by Vickers hardness testing with at a load of 0.98 N for a dwelling time of 20 s. The mechanical properties at room temperature were tested with Instron - 100KN material test machine at an initial strain rate $3.0 \times 10^{-4}\ s^{-1}$. The results of ultimate tensile strength (UTS), tensile yield strength (YS, offset=0.2%) and elongation to fracture (El) were obtained.

For microstructure observations, the specimen which cut tensile bar were polished and etched by a solution of 1 vol.% nitric acid ethanol. The microstructures were analyzed with optical microscope (OM), field emission scanning electron microscope (FE-SEM). The phases identified by X-ray diffraction (XRD).

Table 1 Chemical composition of Mg-Nd-Gd experimental alloys

Alloys	code	Alloying elements(mass%)				
		Nd	Gd	Zn	Zr	Mg
Mg-0Nd-1.5Gd-0.3Zn-0.55Zr	0Nd	0	1.48	0.307	0.453	Bal.
Mg-1.5Nd-1.5Gd-0.3Zn-0.55Zr	1.5Nd	1.71	1.51	0.286	0.529	Bal.
Mg-3.0Nd-1.5Gd-0.3Zn-0.55Zr	3.0Nd	3.32	1.57	0.329	0.569	Bal.
Mg-6.0Nd-1.5Gd-0.3Zn-0.55Zr	6.0Nd	5.98	1.68	0.316	0.608	Bal.

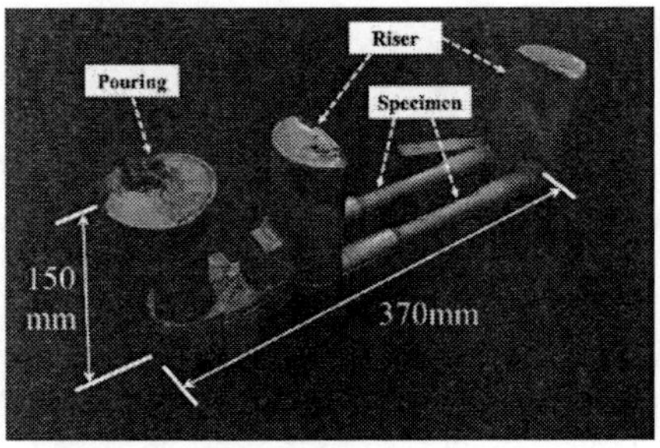

Fig.1 Appearance of castings

3. Experimental Results and Discussion

The microstructure of as-cast conditions, as-quenched conditions and peak-aged conditions in 0Nd, 1.5Nd, 3.0Nd and 6.0Nd experimental alloy are shown in Fig.2.

In the case of as-cast conditions, 0Nd experimental alloy was mainly composed of a white α(Mg) grain. 1.5Nd, 3.0Nd and 6.0Nd experimental alloy with addition of Nd were composed of a white α (Mg) grain and a grayish Mg - RE eutectic phases at grain boundaries. From these observation results, the amount of eutectic phase along the grain boundaries becomes greater with the increase in Nd content. Furthermore, the addition of Nd affects to decrease the grain size.

In the case of as-quenched conditions, 0Nd experimental alloy and 1.5Nd experimental alloy, which were composed of only the α - Mg matrix, and the gray eutectic phases were almost dissolved into α (Mg) matrix after solution treatment. However, for 3.0 Nd experimental alloy and 6.0 Nd experimental alloy, undissolved eutectic phases formed along the grain boundary after the solution treatment.

Crystallized and precipitated phase constituents of as-cast conditions, as-quenched conditions and peak-aged conditions of each experimental alloys identified by XRD and OM are shown in Table 2. According to the XRD results, the main constituent phases of the 0Nd experimental alloy is α - Mg and Mg3Gd phase, and the main constituent phases of the 1.5Nd, 3.0Nd and 6.0Nd experimental alloys with addition of Nd are α-Mg, Mg12Nd and Mg_3Nd. Estimate precipitation processes of Mg-Nd base alloys have been SSSS(supersaturated solid solution)→G.P. zone(Mg-Nd)→β''(Mg3Nd)hcp、D019→β'(Mg3Nd)fcc→β(Mg12Nd)bct, and the phase contributing to hardness was reported as β'' phase 5). In this study, Mg3Nd phase detected for 1.5 Nd, 3.0 Nd and 6.0 Nd experimental alloys may also become the main strengthening phase.

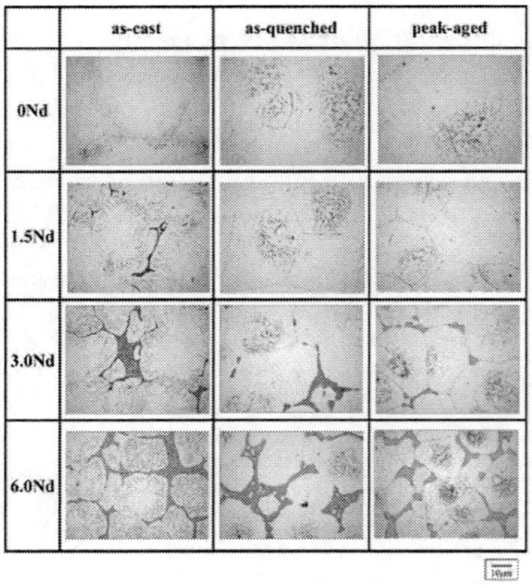

Fig.2 Microstructures of as-cast and as-quenched peak-aged conditions by high magnification (a)0Nd, (b)1.5Nd , (c)3. 0 Nd and (d)6.0Nd ,respectively

Table2 Crystallized and precipitated phases identified by results of OM and XRD

	as-cast	as-quenched	peak-aged
0Nd	α-Mg Mg_3Gd	α-Mg Mg_3Gd	α-Mg Mg_3Gd
1.5Nd	α-Mg $Mg_{12}Nd$ Mg_3Nd	α-Mg	α-Mg Mg_3Nd
3.0Nd	α-Mg $Mg_{12}Nd$ Mg_3Nd Mg_3Gd	α-Mg $Mg_{12}Nd$ Mg_3Nd Mg_3Gd	α-Mg $Mg_{12}Nd$ Mg_3Nd Mg_3Gd
6.0Nd	α-Mg $Mg_{12}Nd$ Mg_3Nd Mg_3Gd Mg_5Gd	α-Mg $Mg_{12}Nd$ Mg_3Nd Mg_3Gd	α-Mg $Mg_{12}Nd$ Mg_3Nd Mg_3Gd

References

[1]Y.Kojima, S. Kamado : J.JILM 45 ,No.2(1995)
[2]Rafael Ferraguta, Fabio Moiaa, Fabrizio Fiori b, Danilo Lussanac, Giuseppe Riontinoc : Journal of Alloys and Compounds 495 (2010) 408–411
[3]G. Riontino, D. Lussana, M. Massazza,G.Barucca, P. Mengucci, R. Ferragut : Journal of Alloys and Compounds 463 (2008) 200–206
[4]Abdel Salam Hamdya, Darryl.P. ButtbaCentral : Electrochimica Acta 108 (2013) 852– 859
[5]B.L. Mordike , Creep-resistant magnesium alloys, Materials Science and Engineering A324 (2002) 103–112

The 73rd World Foundry Congress
„Creative Foundry"
23 - 27 September 2018, Krakow, Poland

MANUFACTURING OF CORROSIVE-RESISTANT Cr-Ni STEELS AND Ni-BASED ALLOYS IN VACUUM FURNACES

Antonin ZADERA [1]*, Vaclav KANA [1], Vladimir KRUTIS [2], Vitezslav PERNICA [1], Martin DULAVA [3]

[1]*Brno University of Technology, Faculty of Mechanical Engineering, Brno- Technicka 2, Czech Republic,*
[2] *MECAS ESI s.r.o., Plzen, Czech Republic,*
[3] *S+C ALFANAMETAL s.r.o., Tršice, Czech Republic,*
**Corresponding address: zadera@fme.vutbr.cz*

Key words: austenitic stainless steel, vacuum metallurgy, induction furnace, decarburization, degassing

Corrosive-resistant austenitic and austenitic-ferritic steels have broad applications in chemical, power and food industries. This paper describes the conditions of manufacturing such steels and some nickel-based alloys in the vacuum induction furnace.

1. Introduction

Austenitic and austenitic-ferritic Cr-Ni steels are used as corrosive-resistant steels. Their resistance to corrosion is achieved by a high content of chromium and also nickel, molybdenum nitrogen, etc. The typical characteristic of such steels is the low content of carbon. The lower corrosive resistance of the steels with higher carbon content could be accounted for especially by a lower Cr content in the metal matrix on the grain boundaries due to the formation of carbides $M_{23}C_6$, or M_7C_3. The highest carbon content in today's corrosive-resistant steels with the austenite or austenite-ferritic structures is usually 0.03% [1, 2]. The same or even lower values of max carbon concentration are allowed in some nickel based alloys e.g. HASTALLOY or ALLOY 59, etc.

There are two principal ways of reaching a low carbon content. The first one is based on using charging materials and ferroalloys with the lowest possible carbon content or by using pure metals when melting in induction furnaces (IF). The disadvantage of such a method is especially the higher price of charging materials [3]. The second possibility is based on melting the charging materials with higher carbon content followed by the melt processing in the AOD (Argon Oxygen Decarburization) convertor or by using vacuum procedures VOD (Vacuum Oxygen Decarburization). The carbon content is lowered by the carbon reaction occurring under the low partial pressure of CO in the vacuum (VOD) or in the inert gas (AOD). Together with the decreasing of carbon content, the oxygen, hydrogen, and nitrogen values are lowered, too [4]. The pouring of the high-alloyed Cr-Ni based alloys is commonly done in the open air (i.e. as the air castings). During the pouring process, we can expect a flow turbulence and formation of oxide films (so called bi-films), resulting in many casting defects [5]. The alloys with the high chrome content are inclined to the bi-films occurrence. Thus, it is necessary to eliminate their formation by modifying the gating system, i.e. by optimizing the filling of the mould cavity. Ni based alloys containing elements with a very high affinity to oxygen (such as Al, Ti, Zr, etc) must be (due to the oxide film formation) not only melted but also poured in a vacuum.

2. Experiments

The main goal of the experiments was to describe theoretically and to verify experimentally the technology of melting and metallurgical processing of the austenitic and austenitic-ferrite steels as well as Ni alloys in a vacuum induction furnace. The melts/heats were carried out in the CONSARC vacuum induction furnace with neutral lining (VEDORIT) based on Al_2O_3 and MgO.

The maximum weight of the melt was 80 kg. The melting took place in the open air, after reaching a temperature of about 1600°C, the furnace was covered with a top cover. Then the vacuum process took about 30-60 minutes, depending on the amount of the oxidized carbon.

The vacuum process was controlled by monitoring and adjusting the chemical composition of the melt, measuring the oxygen activity of the melt and setting the temperature and the pressure of the gases in the furnace. During the vacuum process, argon or nitrogen was blown into the melt through a porous brick in the bottom of the furnace. About 60 heats of austenitic, duplex steels and nickel alloys were carried out using the above technology. The change of the carbon content during the vacuum process varied in the range of 0.03 – 0.1 %.

3. Modelling of the de-gassing kinetic under vacuum and the course of chemical reactions

Reactions taking place in the vacuum furnace during the melting of the alloys based on Fe-Cr-Ni were investigated in terms of thermodynamics and kinetics. The results obtained were then compared and verified with the results of the experiment.

3.1 Modelling of the de-gassing kinetic in a vacuum

De-gassing kinetic in the vacuum furnace was monitored during the de-gassing reaction (shown below for nitrogen). The model of the decreasing of the speed of nitrogen content created by the authors was based on Fick's first law [4].

$$dn = -D * F \frac{dc}{dx} dt \qquad (1)$$

The equation (1) says how many molls (*dn*) would go in time *dt* through the surface unite *F* with the concertation gradient dc/dx and difusion coeficient *D*. The driving force of diffusion is the concentration gradient dc/dx. To keep the conditions for the vacuum process in the vacuum furnace constant, the initial constant parameters were included in the constant K_i. To decrease the nitrogen content in a particular vacuum furnace, equation (1) can be modified into the form (2).

$$ln \frac{N_t - N_{rov}}{N_0 - N_{rov}} = K_i * t \qquad (2)$$

The constant *Ki* is the kinetic constant, N_t is the nitrogen concentration in time *t* and N_{rov} is the equilibrium of nitrogen dissolution under the given pressure and temperature. Based on the modelling results, the kinetic constant and the kinetics of de-gassing and decarburization were determined. This makes it possible to control the vacuum process leading to the required final chemical composition and properties of the metal.

3.2 Measuring of the oxygen activity

Oxygen activity was measured for 18 austenitic steels in the melt in the furnace before vacuuming (**a_{O1}**) and after the vacuum processing (**a_{O3}**) as shown in (Fig.1). The TSO6 combined oxygen probes manufactured by Termosondy Kladno were used for measuring.

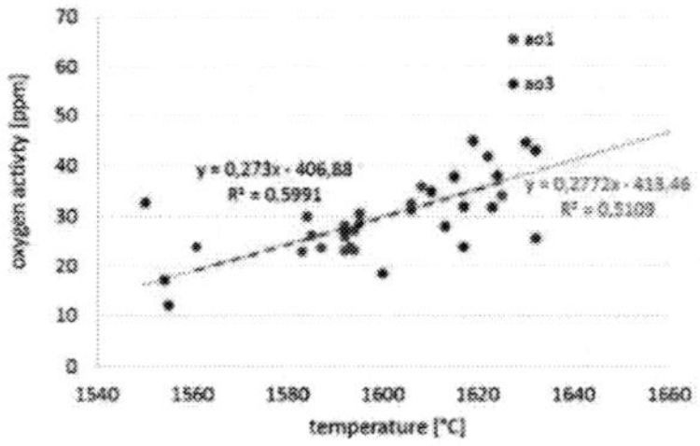

Fig. 1 Oxygen activities in the IF before and after vacuuming depending on temperature

The measured values of oxygen activity before and after vacuuming were put into a single statistical file. The regression coefficient R^2 of oxygen activity dependence on temperature for all measurements of oxygen activity before and after vacuuming is approx. 0.5477. The given dependence is statistically significant. For the given melting process and vacuum processing, the oxygen activity in austenitic steels treated by the given process can be predicted on the basis of equation (3) with substituting the given metal temperature (T [°C]) only.

$$a_O = 0{,}278 \cdot T - 414 \quad [ppm] \qquad (3)$$

3.3 Modelling of the course of chemical reactions during solidification and cooling

The measured oxygen activities for the above melts were compared with thermodynamic calculation. Considered was the calculation of the chemical composition of the melt determined from the melt samples taken from each melt. The calculation was carried out using the thermodynamic model developed by the authors. The oxygen activities in the set up model are based on the calculation using equations (4), (5), and (6).

$$2[Cr] + 3[O] \rightarrow (Cr_2O_3) \qquad (4)$$

$$K_{Cr_2O_3} = \frac{a_{Cr_2O_3}}{a_{Cr}^2 \cdot a_O^3} = \frac{N_{Cr_2O_3} \cdot \gamma_{Cr_2O_3}}{a_{Cr}^2 \cdot a_O^3} \qquad (5)$$

$$a_O = \sqrt[3]{\frac{N_{Cr_2O_3} \cdot \gamma_{Cr_2O_3}}{a_{Cr}^2} \cdot \frac{1}{K_{Cr_2O_3}}} \qquad (6)$$

$N_{Cr_2O_3}$ concentration of Cr oxide in the inclusion [-]

$a_{Cr_2O_3}$ activity of chromite oxide in the inclusion [-]

$\gamma_{Cr_2O_3}$ activity coefficient of Cr oxide in the inclusion [-]

a_O – oxygen activity [-]

a_{Cr} – chromium activity [-]

The value of the activity coefficient of chromic oxide $\gamma_{Cr_2O_3}$ was used as determined by the authors previously when measuring the austenitic steels.

4. Conclusion

This contribution describes only some areas of the research connected with the production of high-alloyed steels and alloys using the vacuum metallurgy in an induction furnace. The core part of this research is the area of decarburization and de-gassing kinetics and the thermodynamic solutions to the oxide-reduction reactions and the formation of inclusions depending on the vacuum treatment and steel de-oxidation.

As a result, we now have technology used in further research as well as in the production of castings at the BUT foundry.

The paper was written with the support of the Technological Agency of the Czech Republic as part of the project, evidence No TH02020076.

References

[1] MARSHALL, P. *Austenitic Stainless Steels – Microstructure and Mechanical Properties*, 1984, p.431, ISBN 0-85334-277-6.

[2] ALVAREZ, A. I.; DEGALLAIX-MOREUIL, S. *Duplex Stainless Steels*. GREAT BRITAIN, 2009, p.437, ISBN 978-1-84821-137-7.

[3] DULAVA, M.; ZÁDĚRA, A.; ŠENBERGER, J. *Dezoxidace žárovzdorných ocelí*. Slévárenství. Vol. 61. Issue 9-10. p. 321-324. 2013. ISSN 0037-6825.

[4] KNÜPPEL, H. Desoxidation und Vakuumbehandlung von Stahlschmelzen, Band 1. Düsseldorf: Verlag Stahleisen. 1970.

[5] CAMPBELL, J.: *Casting*, Butterworth-Heinemann, Oxford, 2000, ISBN 0 7506 1696 2.

Gap in pagination due to unavailable paper.

Pages 453-454

The 73rd World Foundry Congress
„Creative Foundry"
23 - 27 September 2018, Krakow, Poland

Prediction of shrinkage cavities of using casting simulation

Hayato MITSUYA [1*], Kazumichi SHIMIZU [1], Kenta KUSUMOTO [1], Takashi FUNABIKI [1]

[1]Muroran Institute of Technology, 27-1 Mizumoto-cho muroran-shi Hokkaido 050-8585, Japan
e-mail : 14022122@mmm.muroran-it.ac.jp

Key words: Differential case, Casting simulation, Shrinkage cavity, Casting defect, Thin-walled casting

1. Introduction

The foundry industry in Japan is supplying high-quality, low-cost parts to the automobile industry and machinery industry, contributing to the maintenance of international competitiveness [1]. The largest users of the foundry industry are transportation agencies, accounting for over 60% of total demand. Castings are an important part supporting the basic performance of automobiles, and it can be said that the automobile industry will not be available without casting. For automotive parts, thickness reduction and complex shape are required for weight reduction [2, 3]. Thereby, Casting defects are generated and it is difficult to secure the yield rate. The setting of the shape, size and position of the runner, weir, and feeder in the casting method is important, leading to the prevention of casting defects. Casting defects increase as the yield rate improves. As a solution to that problem, by using casting simulation at the time of product manufacture, it is possible to predict where a casting defect occurs on the computer, and it is expected to improve production efficiency. The purpose of this research is to estimate the occurrence of shrinkage cavities by performing molten metal flow and solidification analysis by cast simulation using automobile parts differential case. In addition, casting experiments are carried out to measure the temperature change of the casting to match the temperature change curve, so that the physical properties of castings and molds during the experiment are selected. From this result, we examine the consistency of shrinkage cavity defect position in actual product and analysis.

2. Materials and Experimental Method

Create a model to investigate locations where shrinkage cavity defects may occur, perform melt flow and solidification analysis. For the analysis, use casting simulation software "JSCAST". Casting method is created using 4 differential case. FCD450 was used as the casting material, sand was used as the casting mold, the pouring temperature was 1350 ° C, and the pouring time was 9 seconds. For prediction of casting defects of products, temperature change of molten metal by molten metal flow analysis, speed change, final solidification part of molten metal by solidification analysis, $G/\sqrt{R}$ method which is shrinkage prediction parameter were used. In the $G/\sqrt{R}$ method, shrinkage cavities are more likely to occur as the value is smaller, so the values were empirically set to 0.1 and analyzed.

Fig. 1 (a) shows the basic model created with reference to the product differential case. For the basic model, molten metal flow, solidification analysis was performed. In the basic model, a temperature decrease was observed in the thin part far from the position where the molten metal flowed from the weir to the product part. Furthermore, temperature drop was more noticeable in the lower part of the product. The final solidified part is seen in the product center and in the lower part of the product, and the occurrence of casting defects such as shrinkage cavities is considered at that position. Therefore, in order to move the final solidified part to a position other than the product part, changed model was prepared by adding riser to the center part of the product shown in Fig. 1 (b). In order to set the basic model and the amount of molten metal to be the same, the diameter of the riser in the middle of the runner was reduced by 20 mm. As a result, the temperature change of the molten metal did not change, but the final solidified part could be moved to the riser added to the center of the product. As a result, it was possible to alleviate the shrinkage cavity in the center of the product and in the lower part of the product for analysis.

In order to examine the consistency of the analysis, experiments are carried out in which a model prepared on the CAD is actually cast, and the change in temperature at the time of molten metal flow and solidification is measured at three points shown in Fig. 1 (a). The pouring temperature in the casting experiment was 1470℃, and the pouring time was 9.43s.

FCD450 was used as the casting material, and alkaline phenol self-hard sand mold was used for the sand mold. X-ray CT measurement that can perform nondestructive inspection is used for observing the position of the shrinkage cavity of the product.

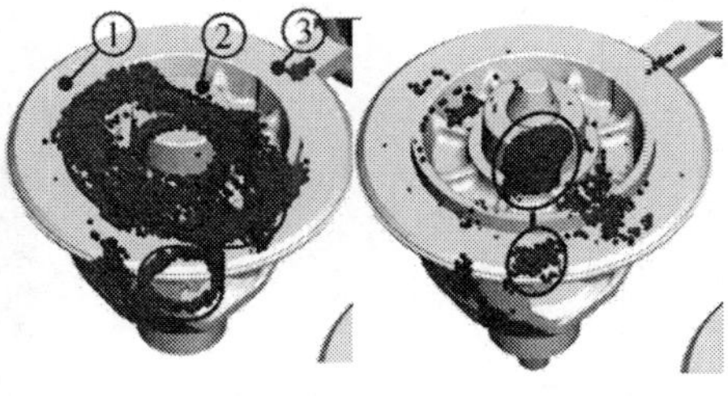

(a) Basic model *(b) Changed model*

Fig. 1 Shrinkage cavity distribution

3. Results and Discussion

Casting simulation software (ADSTEFAN) is suitable for selecting physical properties of castings and molds. Because ADSTEFAN can perform relay analysis within the specified range of physical properties. Adjustment of the temperature change curve obtained from the casting experiment was performed using ADSTEFAN. Fig. 2 shows the result of fitting the temperature change curve of the changed model. For the measurement points 2 and 3, it was possible to obtain a temperature change curve approximated by the experiment value and the analysis value. However, since the measurement point 1 is a complicated shape and a thin wall position, it was possible to approximate the trend of the experimental value, but a similar curve could not be obtained. Table 1 shows the physical properties of castings and molds obtained by the fitting. Using physical properties obtained from ADSTEFAN, JSCAST conducted molten metal flow simulation and solidification simulation in the basic model in which many shrinkage cavities were observed. Based on the results, we examine the consistency of the analysis result of simulation and the occurrence position of shrinkage cavity in the actually cast product. By the non-destructive possible X-ray CT measurement, the position where the shrinkage cavity of the product was generated was confirmed. Fig. 3 shows the results of comparing shrinkage location in analysis and product. In both analysis and product, it was confirmed that shrinkage occurrence positions coincided with the product center portion which is the final solidified portion and the product lower portion. From this, it is assumed that shrinkage cavity defects can be predicted by casting simulation if physical properties of actual castings and molds are available. In addition, it is considered that more accurate prediction of shrinkage cavity defects is possible by increasing the accuracy of fitting. From the results of X-ray CT measurements, the occurrence of shrinkage cavity in the modified model was relaxed in the product as compared with the basic model.

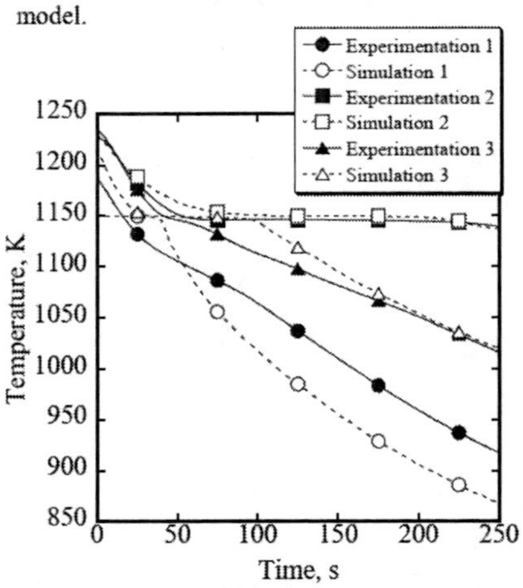

Fig. 2 Comparison of temperature change of experimentation

4. Conclusions

In this study, we investigated the distribution of shrinkage cavity defects in a differential case using casting simulation. The results are shown below.

(1) Temperature change curves were fitted using casting simulation software (ADSTEFAN).. As a result, the temperature change curves of the experimental values and the analysis values could be matched at the measurement points 2 and 3. Measurement point 1 could be brought closer to the experimental value trend, but similar curves could not be obtained because of the large temperature change at the thin-walled part.

(2) As a result of comparing shrinkage occurrence positions in the analysis and products, it was confirmed that it coincided at the product center and the product bottom.

(3) It is assumed that shrinkage cavity defects can be predicted by casting simulation if physical properties of actual castings and molds are available. In addition, it is considered that more accurate prediction of shrinkage cavity defects is possible by increasing the accuracy of fitting.

References

[1] RIETI. (2006). Research on responding to changes in the business environment surrounding the foundry industry. P6

[2,3] Kikuo KATO. (2005). On the efforts of the Japan Casting Association from now on based on the present situation of Japanese foundry industry.

[4] Hiroki SOTOZAKI, Kazumichi SHIMIZU. (2015). Prediction of shrinkage cavities of using casting simulation. Japan Foundry Engineering Society The 167th.

Table 1 Physical properties

	Casting	Template
Density (g / cm^3)	7.2	1.4
Thermal conductivity (cal / cm s deg)	0.06	0.002
Specific heat (cal / g deg)	0.1	0.25
Initial temperature (deg C)	1300	
Liquidus temperature (deg C)	1150	
Solidus temperature (deg C)	1145	
latent heat (cal / g)	50	

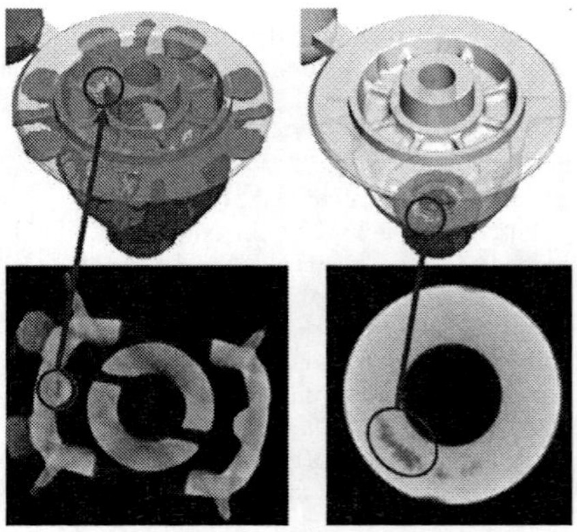

Fig. 3 Installation position of thermocouple

The 73rd World Foundry Congress
„Creative Foundry"
23 - 27 September 2018, Krakow, Poland

Thermal Stress Analysis of TiC Cast-in Insertion Multi-component White Cast Iron

Shigeya Murase [1*], Kazumichi Shimizu [1], Kenta Kusumoto [1]
Hiroya Hara [2], Masato Shirai [3] and Hideto Matsumoto [4]

[1]Muroran Institute of Technology, 27-1, Mizumoto-cho, Muroran-shi, Hokkaido, 050-8585, Japan
[2] Hokkaido special cast steel. Co., Ltd.., 497-3, Yura, Kurisawa-cho, Iwamizawa-shi, Hokkaido, 068-0011, Japan
[3] Iwamizawa Imono. Co., Ltd.., Yura, Kurisawa-cho, Iwamizawa-shi, Hokkaido, 068-0111, Japan
[4] Sankyo Gokin foundry. Co., Ltd.., 5-10-7, Tsukuda, Nishiyodogawa-ku, Osaka-shi, Osaka, 555-0001, Japan
** Corresponding address: e-mail: 14022125@mmm.muroran-it.ac.jp*

Key words: Erosive wear, High Temperature, Inserted TiC, Bondability, Thermal stress

1. Introduction

Erosion is a kind of surface damage that caused by the impact of dispersed particles in fluid [1]. The wear resistant liner is set in inner wall of turning chute of blast furnaces. However, heavy erosive wear has occurred by the impact of iron ores and cokes at elevated temperature for an average from 873 to 973 K. Therefore, the regularly maintenance and exchange for the liner is conducted. Nevertheless, there are the problem about blowing-stop for exchange and increase in running cost. Based on this, it is urgent to develop new wear resistant materials which could be used at elevated temperature.

According to the research of Mr. Shimizu et al., the multi-component white cast irons that crystallize carbides with high hardness in the matrix during solidification have been showed good resistance property to erosive wear under high temperature conditions, by the composite with the Tungsten carbides (WC) of less reduction in hardness at high temperature by the cast-in insertion method [2]. However, there is a possibility of cracking due to thermal stress and residual stress occurred by difference of liner expansion coefficient between base metal and cemented carbides. In addition, there is the problem that the bonding state of boundary surface differs by carbide forms or chemical composition of base metal. Therefore, the Titanium carbides (TiC) that could be expected improving bondability and weight reduction were focused on.

In the present study, the thermal analysis by finite element method (FEM) used software for structure analysis simulation "ANSYS" was conducted, and the possibility of TiC cast-in insertion method was examined.

2. Analytical method

The values of general multi-component white cast iron and TiC were used as the physical property for the analysis [3,4]. The physical property of each material at 298 K and 873 K are as shown in Table 1. Liner expansion coefficient of each material was used Thermomecanical analyzer (TMA). Also, Young's modulus was used resonance method. This analysis was elastic analysis not to set the plastic conditions.

Table 1. Physical property of each material.

Material	White cast iron	TiC
Temperature [K]	298-873	298-873
Young's modulus [GPa]	214	356
Poisson's ratio	0.28	0.21
Liner expansion coefficient [/K]	12.2×10^{-6}	8.61×10^{-6}
Thermal conductivity [W/mK]	34	17.2
Yield stress [MPa]	220-1030	4000
Tensile strength [MPa]	965	—
Compressive strength [MPa]	2060	1350

The dimension of analytical model is a tabular of 50×50×15 mm, and TiC is a cuboid with dimension of □10×50 mm (with/without R2 at the corners) and a cylinder with dimension of φ10×50 mm. Analytical model is connected four 1/4 models to set the constraints. The constraint conditions of analytical model is indicated in Fig. 1. The constraints of each area (X-axis and Z-axis) in connected surface portion of 1/4 models and the omni-directional constraint in the center of the bottom nodes that is the origin of analytical model are set.

The analysis time is 2.7 ks of actual testing time and the time history response analysis of 30 s/step. The heat is loaded to all the nodes as load condition. First, it is heated from 298 K to 873 K of test temperature over 5.4 ks. Second, it is dropped at 853K by the impact of particles after 30 s when it reached to 873 K, and heated to 873 K again after 1.77 ks. This process is repeated every 1.8 ks in a total of 10 times. Finally, it is cooled from 853 K to 298 K over 5.37 ks.

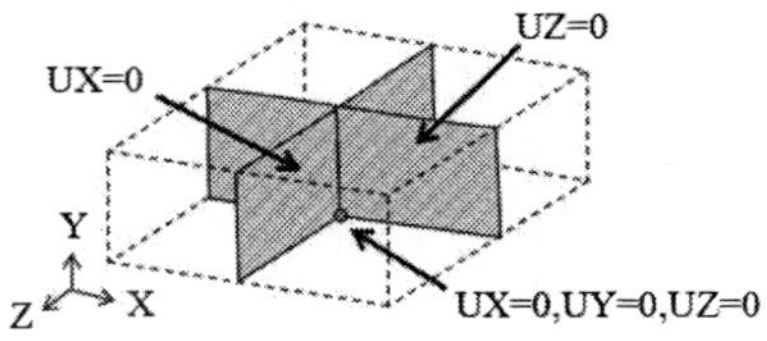

Fig. 1. The constraint conditions of analytical model.

3. Analysis result and discussion

The thermal stress analysis result of four TiC shapes is indicated in Fig. 2 and Fig. 3. Each analytical model was compared by Mises stress and Maximum principal stress which were used in yield criterion. Each stress value was the maximum at 873 K to be elastic analysis in this analysis. According to Fig. 2 and Fig. 3, TiC showed the maximum stress value on the near end surface ragardless of TiC shapes. Base metal showed it on the near boundary surface with TiC. As a result, Trapezoidal model showed especially excellent analysis result in comparison with conventional cuboidal model to be absorbed the stress concentration by adding R2 at the four corners. However, it is difficult to process cemented carbides as TiC with high hardness. On the one hand, thermal strain of each model was on the same extent of 0.026 %.

Next, the strength evaluation of cuboidal model which was specimens for high temperature erosive wear test was conducted. The maximum and minimum principal stress were used to evaluate the maximum tensile/compressive stress to be generated in each material. From these, tensile/compressive stress of TiC was 884/782 MPa and multi-component white cast iron was 460/675 MPa. As a result of discussion by maximum principal stress theory, TiC does not yield not to exceed yield stress. It's considering that each material dose not result in a breaking in comparison with tensile and compressive strength. However, there is a pssibility that base metal exceeds yield point and occurs plastic deformation to reach tensile strength of 48 % and compressive strength of 33 %. As the reason, the trace of deformation toward Y-axis direction was confirmed on the near four corners according to the degree of freedom solution vector for displacement.

WC-Co cast-in insertion multi-component white cast iron that was specimens for erosive test so far showed 1873 MPa (tensile) in maximum principal stress and 1772 Mpa (compressive) in minimum principal stress. TiC specimens showed comparably low thermal stress about from 1000 to 1100 MPa. The difference of liner expansion coefficient between base metal and cemented carbides was small by changing core from WC-Co to TiC. From this, it is sugested that the risk of specimen breaking by thermal stress under high temperature conditions is reduced. In terms of thermal stress, it was clear that TiC suited as core for the cast-in insertion.

4. Conclusion

The thermal analysis of TiC specimens under high temperature erosion test by FEM used software for structure analysis simulation "ANSYS" was conducted in the present study. The conclutions are summarized as follows:

(1) Trapezoidal model showed excellent analysis result to be absorbed the stress concentration by adding R2 at the corners.

(2) As a result of the elastic analysis, this spicimen does not result in a breaking by thermal stress under high temperature eroion test. However, there is a possibility that base metal exceeds yeild point and occurs plastic deformaton to reach its tensile strength of 48 % and its compressive strength of 33 %.

(3) TiC specimens showed low stress values in comparison with WC-Co specimens. Based on this, the result of validity of TiC cast-in insertion method and utilizable possibility as heat and wear resistant materials was obtained.

References

[1] I. Finnie, Erosion of surfaces by solid particles, Wear 3 (1960) 87-103.

[2] K. Shimizu, Y. Xinba, H. Matsumoto, Y. Maeda, Erosive Wear Properties of Hard Carbides Cast-in Insertion Cast Irons, Jpn. Foundary Eng. Soc. 80 (2008) 457-465.

[3] M. Hashimoto, Y. Nishiyama, K. Yamamoto, S. Sasaguri, Y. Matsubara, Effects of Carbon and Tungsten on Mechanical and Hot Wear Properties of Multi-component White Cast Irons, Jpn. Foundary Eng. Soc. 79 (2007) 732.

[4] H. Suzuki, Hard Carbides and Sintering Hard Material, Maruzen. Co. (1986) 2.

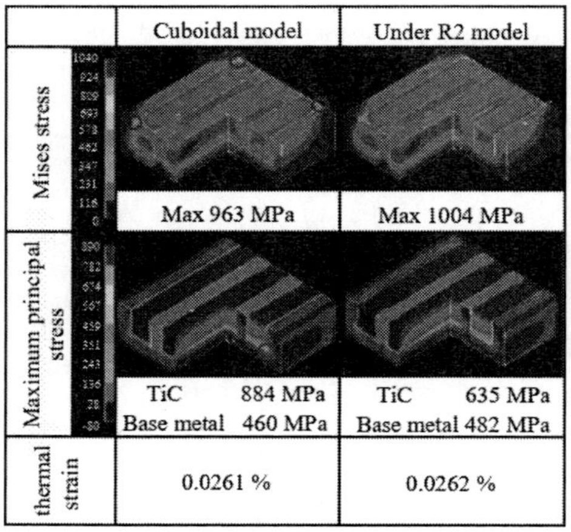

Fig. 2. The result of thermal stress analysis.

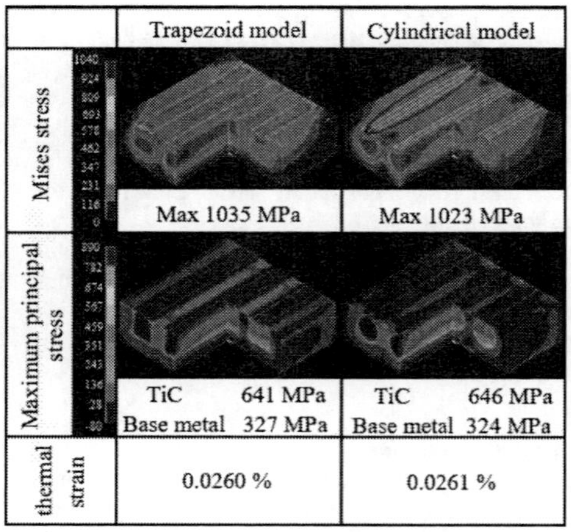

Fig. 3. The result of thermal stress analysis.

The 73rd World Foundry Congress
„Creative Foundry"
23 - 27 September 2018, Krakow, Poland

Effect of Alloying Element on the Mechanical Properties of High Silicon Ferritic Ductile Cast Iron

Tae-Woo Kim[1], Ki-Ho Park[2], Hae-Wook Kwon[1]
[1]*Yeungnam University, 280 Daehak-ro, Gyeongsan-si, Gyeongbuk, Korea.*
[2]*Namyang Metals Co., 20 Nongong-ro 53-gil, Nongong-eub, Dalsung-gun, Daegu, Korea.*
**Corresponding address: e-mail: hwkwon@ynu.ac.kr*

Key words: cast-iron, ductile cast-iron, high silicon ductile cast iron, mechanical properties, solid solution strengthening

Abstract
The effect of alloying element on the mechanical properties of high silicon ferritic ductile cast iron was investigated. The strength and hardness increased and the elongation decreased with the silicon content. The impact energy absorbed increased with the increased silicon content up to 4.0wt% and then decreased drastically. The ductile-to-brittle transition temperature seemed to be increased with it. The strength and hardness increased and the elongation decreased with increased copper content because of the decreased volume fraction of ferrite and the impact energy decreased. The similar trend of mechanical properties were obtained with increased nickel content and nickel appeared to be the more effective solid solution strengthening element then copper.

1. Introduction

The mechanical properties of ductile cast iron is determined by the microstructure, which is composed of graphite nodules and a ferritic and/or pearlitic matrix[1]. That is significantly influenced by the volume fraction of ferrite or pearlite. Generally ductile cast iron with ferritic matrix has lower yield and tensile strengths and higher ductility and toughness than that with pearlitic one.

The morphology of graphite and structure of metal matrix ductile cast iron are greatly influenced by the chemical composition [1, 2]. Silicon promotes the formations of graphite and ferrite matrix. Conventional ductile cast irons contain silicon of 2.0 to 3.0wt%. Moreover, the silicon is an effective element for solid solution strengthening of ferrite matrix[1, 3]. The strengthening effect of silicon progressively increased with the increased silicon content, up to 4.3wt%.

The strength of ductile cast iron under as-cast condition is increased substantially by the addition of copper[4]. The impact energy absorbed of this material is decreased and the ductile–to-brittle transition temperature is increased by that[5]. These are caused by the promoting formation of pearlite because the diffusion of carbon during the decomposition of austenite is inhibited by the addition of copper.

Nickel is soluble in iron at all levels but it reduces the solubility of carbon in iron, lowering the eutectic carbon content. Since nickel raises the stable austenite-graphite eutectic temperature and lowers the metastable austenite-iron carbide eutectic temperature, it reduces chill forming tendency.

The effects of silicon content on the tensile and compact proportion of the high silicon ferritic ductile cast iron were investigated and those of copper and nickel ones were also. It was determined whether copper or nickel was better for solid solution strengthening of the material.

2. Experimental Procedure

The test casting of Y-block of DIN EN 1563 Type II standard with the thickness of 25 mm was poured after melting. Molten metal was obtained using the medium frequency induction furnace of 3,500 kVA capacity. After the charging material has been melted, the carbon content was adjusted. Ferosilicon and silicon carbide were added to adjust the silicon content. The melt was tapped into the ladle of 1 ton capacity. During tapping the melt was treated with 10wt%Mg-Fe-Si by sandwich method. The copper chip and/or metallic nickel were added into the ladle as needed. Then, it was poured into the green sand mold at the temperature range of 1,420~1,440℃. During pouring, it was stream-inoculated.

The small amounts of melt were also poured into the shell cup of C. E. meter and metallic mold of chill specimen. The former was used for the analysis of carbon content and then latter for the analysis of other element busing spectrometer.

The silicon content was varied at 0.5wt%(interval) in 2.5~4.5wt% range under the condition of same carbon equivalent of 4.3%. The copper content was varied at in 0.10~1.30wt% range with the same silicon and nickel contents of 4.0 and 0.10wt%, respectively. The nickel content was varied at in 0.10~1.20wt% range with the same silicon and copper contents of 4.0 and 0.10wt%, respectively.

The tensile specimen of EN ISO 6892-12009 standard for each condition was obtained and tested using the universal test machine of 50 tons capacity at the cross-head speed of 10 ㎜/min. The impact specimens of EN ISO 148-1:2010 standard was also obtained for each condition and tested at 25℃ interval in -50~50℃ range. The cubic specimens with the edge of 25 ㎜ for the observation of microstructure were obtained by cutting the Y-block test casting. These were ground and polished by conventional method. The graphite morphology was observed under the polished condition and the nodularity was measured using image analysis. The matrix structure was observed with etched specimen and the volume fraction of ferrite was measured.

3. Results and discussion

3.1 Effect of silicon

The silicon content was varied in the range of 2.5~4.5wt% under the condition of same carbon equivalent. The strength and hardness increased and elongation rather decreased with the increased silicon content. Silicon is a strong ferrite stabilizer and the volume fraction of ferrite increased from 75 to 100% as the silicon content increased from 2.50 to 4.0wt%. The former was 100% also when the latter was 4.5wt%. The silicon is a strong solid solution strengthening element for ferrite

The impact energy absorbed decreased with the increased silicon content at the same test temperature, even though it was hard to conclude like this at the test temperature of 50℃ and decreased drastically when the silicon content was high as 4.5wt%. The ductile-to-brittle transition temperature seemed to be increased with the increased silicon content. That was 25℃ below zero when the silicon content was 3.50wt% and increased to 0℃ when 4.0wt%. Silicon acted as embrittling element and the long range ordered crystal lattice could be formed with high silicon content in the solid solution.

3.2 Effect of copper

The copper content was varied in the range of 0.10~1.20wt% with the same silicon content of 4.0wt%. The strength and hardness increased and the elongation decreased with the increased amount of copper added. The copper was dissolved in the austenite matrix and inhibited the diffusion of carbon during the transformation of austenite. Therefore the copper acted as promoting element of pearlite formation. The volume fraction of ferrite in the matrix decreased with the increased amount of copper added, even though the silicon content is as high as 4.0wt%.

The impact energy absorbed decreased with the increased amount of copper added. The impact energy would be absorbed largely as plastic strain energy during the test. The interface area between the ferrite and cementite increased with the copper content. The crack during the impact test could be nucleated at this interface area and also by the fracture of cementite itself in the pearlite.

3.3 Effect of nickel

The nickel content was varied in the range of 0.10~1.20wt% with the same silicon content of 4.0wt%. The strength and hardness increased and the elongation decreased slightly with the increased amount of nickel added. Nickel also acts as a solid solution strengthening element of ferrite because the solubility limit of nickel in the ferrite is as high as 10wt%. The impact energy absorbed decreased with the increased amount of nickel added because of solid solution strengthening effect of this element.

4. Conclusion

The strength and hardness increased and the elongation decreased with the silicon content. The impact energy absorbed increased with the silicon content up to 4.0wt% and then decreased drastically. The ductile-to-brittle transition temperature seemed to be increased with it. The strength and hardness increased and the elongation decreased with increased copper content because of the decreased volume fraction of ferrite and the impact energy decreased. The similar trend of mechanical properties were obtained with increased nickel content and nickel appeared to be the more effective solid solution strengthening element then copper.

References

[1] M. Gagn'e. (2004). The Sorelmetal Book of Ductile Iron, Montreal; Rio Tinto Iron & Titanium. 3-38.

[2] J. Sertucha, J. Lacaze, J. Serrallach, R. Suarez, F. Osuna. (2012). Materials Science and Technology. 28(2). 184-191.

[3] W. Stets, H. Loblich, G. Gassner, P. Schumacher. (2014). International Journal of Metalcasting. 8(2). 35-40.

[4] A. Alagarsamy et. al. (1993). Ductile Iron Handbook. AFS. 93.

[5] G. M. Goodrich ed. (2003). Iron Castings Engineering Handbook. AFS. 275.

[6] Z. Galvas, A. Strkalj, A. Stojakovic. (2016). Metallurgia. 55(3). 293-296.

[7] F. Huyan, R. Larker, P. Rubin, P. Hedstròm. (2014). ISIJ International. 54(1). 248-250.

The 73rd World Foundry Congress
„Creative Foundry"
23 - 27 September 2018, Krakow, Poland

TECHNOLOGICAL ASPECTS OF PRODUCING CERTIFIED REFERENCE MATERIAL (CRM) FOR ZINC ALLOYS

Beata Cwolek[1], Łukasz Jan Wierzbicki[1], Barbara Bolibrzuch[2], Joanna Kulasa[1], Joanna Kostrzewa[2], Barbara Juszczyk[1]

[1] Metal Processing Department,[2] The Department of Analytical Chemistry
Institute of Non-Ferrous Metals, Sowinskiego Str.5, 44-100 Gliwice, Poland
**Corresponding address: e-mail: lukasz.wierzbicki@imn.gliwice.pl*

Key words: zinc, zinc alloys, melting, casting, certified reference materials

1. Introduction

Spectral analysis, especially in the case of non-ferrous metals and their alloys, guarantees in the industrial conditions a quick and accurate assessment of the chemical composition of manufactured alloys and products. The accuracy of the results is determined not only by the method of analysis but also by the quality of the reference materials. One should emphasize, that the requirements for castings for certified reference materials (CRM) are very high. These requirements concern the chemical composition, which should be consistent with the designed composition and material homogeneity in terms of structure and distribution of elements in the volume of the melt. The chemical composition of series of certified reference materials for zinc alloys, intended for the calibration of spectral equipment, has been developed based on the requirements of customers who are producers of these alloys. The elements of the metal batch are the metallic components – direct materials, namely metals and their alloys for the castings production as well as non-metallic components – indirect materials (i.e. fluxes and protective coverings) used for technological operations but not included in the composition of the casting. The components of the metal batch can be characterized by the chemical composition, state (primary, master alloy), origin (fire, electrolytic) and the form (ingots, cathodes, granulate, powder). Therefore, before starting castings production, the chemical composition of the metal batch should be determined taking into account changes in the content of particular chemical components during melting. The decisive factor on the order in which alloy components are introduced is the quantity and properties of individual components in the liquid state, such as: heat of melting, weight and specific volume, melting and boiling point, vapor pressure, activity in the environment and solubility in other components [1÷7].

2. Results

In the article, the processes of producing materials for a series of certified reference materials for ZA8 and ZA12 zinc alloys are presented. The chemical compositions of these alloys are shown in Table 1.

Table 1. Chemical compositions of ZA8 and ZA12 zinc alloys grades, wt. %

Element	Grade of alloy	
	ZA8	ZA12
Al	8,2÷8,8	10,8÷11,5
Cu	0,9÷1,3	0,5÷1,2
Mg	0,02÷0,03	0,02÷0,03
Pb	max.0,005	max.0,005
Fe	max.0,035	max.0,050
Ni	max.0,001	max.0,001
Cd	max.0,005	max.0,005
Si	max.0,035	max.0,05
Sn	max.0,002	max.0,002
Zn	balance	

Zinc alloys are foundry alloys used mainly for pressure castings of bodies, housings and covers of various devices in the automotive and electro-technical industries. These alloys are used also for manufacturing for example parts of carburettors, slide bearings and sleeves. According to the present standards for chemical compositions and requirements of the producers of aforementioned alloys, the preliminary chemical compositions of the reference materials for the ZA8 and ZA12 alloys were designed. The chemical compositions of designed alloys are collected in the Table 2.

Table 2. Chemical compositions of the CRM materials, wt. %

Element	No of standard				
	ZK1	ZK2	ZK3	ZK4	ZK5
Al	11,5	10,5	9,5	8,0	6,5
Cu	1,5	2,2	3,0	5,0	6,5
Mg	0,0005	0,003	0,025	0,05	0,035
Pb	0,001	0,008	0,02	0,03	0,04
Cd	0,0005	0,001	0,005	0,01	0,02
Sn	0,0005	0,001	0,0025	0,0045	0,006
Bi	0,15	0,13	0,1	0,05	0,002
Sb	0,15	0,12	0,07	0,03	0,002
Zn	balance				

Materials for certified reference materials were made by melting and casting in an induction crucible furnace with a capacity of 10 kg Cu. Due to different concentrations, different melting points and volatility, copper was introduced in the form of a master alloy, aluminium as an alloying component and impurities such as Pb, Sn, Mg, Sb, Cd and Bi were introduced in pure form. After elaboration of the batch recipes and preparing the batch material weighing, the melt for CRM was made. The production process consisted of melting in the graphite crucible of zinc, overheating the bath to a temperature of 680°C and fusing individual elements in a strictly defined order. Copper was first melted in the form of ZnCu15 master alloy. Then after thorough mixing of the bath by a graphite rod for about 5 minutes, further components were added in the following order: Al, Mg, Sb. After reducing the liquid bath temperature to about 500°C, Pb, Cd, Sn and Bi were melted while stirring the liquid metal inductively. After addition of all components, the liquid metal was mixed by a graphite rod for about 15 minutes, reducing the temperature of the liquid metal to about 460°C, and then casted into a cold cast iron mould. Obtained in this way rods with dimensions close to ϕ40 mm x 350 mm, were cut on cylinders of height about 25 mm and subjected to initial testing of homogeneity of all the ingredients melted in zinc and checking the accordance of the chemical composition with the designed one. For initial homogeneity testing, samples were taken alternately from the bottom and from the top of casting bars which is one reference material (standard). The homogeneity assessment of all standard components was carried out by comparing the intensities of the characteristic lines of the elements included in the reference materials (standards), inducing four times in the low-voltage spark on both sides, the active surfaces of each cylinder. The chemical composition was determined based on chemical markings made on a sequential emission spectrometer equipped with an inductively coupled plasma.

3. Summary

The material homogeneity tests carried out on CRM for zinc alloys produced by melting and casting showed full homogeneity of Al, Cu, Pb, Mg, Cd, Bi, Sb and Sn in the whole mass of ingots for individual melts in the series, and the results of determinations of these elements do not differ in a significant way from the designed chemical composition. In the Table 3, the results of the chemical composition analysis for the ZK series are presented.

Table 3. Results of the analysis of the chemical composition of ZK series, wt. %

	No of standard				
Element	ZK1	ZK2	ZK3	ZK4	ZK5
Al	11,76	10,49	9,70	8,35	6,45
Cu	1,11	2,48	3,16	4,40	6,54
Mg	0,00045	0,0034	0,0027	0,055	0,038
Pb	0,00065	0,0065	0,015	0,029	0,050
Cd	0,0008	0,0014	0,0052	0,011	0,019
Sn	0,00061	0,0012	0,0027	0,0050	0,0056
Bi	0,16	0,17	0,10	0,019	0,0011
Sb	0,16	0,11	0,070	0,014	0,0029
Zn	balance				

Positive results of homogeneity and chemical composition tests allowed the statistical evaluation of homogeneity and attestation based on the results of analysis of all reference components. The tests were made in three specialized research laboratories. Statistical homogeneity assessment was based on the Snedecor F examination using 30% of the cast material. The reference material produced according to the presented scheme is now a commercial product in the form of a certified reference materials (CRM) designed for the spectral determination of Cu, Al, Pb, Sn, Mg, Sb, Cd and Bi in zinc.

References

[1] Namieśnik J., Konieczka P., Zygmunt B. (2009): „Ocena i kontrola jakości wyników pomiarów analitycznych", Wydawnictwo Naukowo-Techniczne,

[2] Górny Z., Sobczak J.(2005): „Nowoczesne tworzywa odlewnicze na bazie metali nieżelaznych", ZA-PIS, Kraków,

[3] Korzec R., Fajkel A., Czekaj E. (2002): „Instrukcja topienia stopów metali nieżelaznych", Instytut Odlewnictwa, Kraków,

[4] Müller E., Bolibrzuch B. (2002).: „Spektralne certyfikowane materiały odniesienia", Analityka, nr 4, str. 23-27,

[5] Konieczka P. (2007): „Materiały odniesienia – kryteria wyboru i zasady stosowania", Politechnika Gdańska,

[6] Baker H.: (ed.), (1992) „Alloy Phase Diagrams", ASM Handbook, vol. 3,

[7] Cwolek B.; Wierzbicki Ł.J.; Gałeczka J.; Juszczyk B. (2014) „Opracowanie serii certyfikowanych materiałów odniesienia dla dwóch gatunków stopów cynku ZA8 i ZA12 przeznaczonej do kalibracji aparatury spektralnej"; Sprawozdanie wewnętrzne IMN nr 7333/14 z realizacji pracy statutowej, Gliwice

The 73rd World Foundry Congress
„Creative Foundry"
23 - 27 September 2018, Krakow, Poland

Development of Mg-Al-Sr-Ca system Heat Resistant Alloy for Die-casting Excellent in Castability and Recyclability

Seiji Saikawa [1]*, Manabe Mizutani[2], Katsuhito Yoshida[2], Nozomu Kawabe[2]

[1] University of Toyama,Science and Engineering for Research, 3190, Gofuku, Toyama, 930-8555, Japan
[2] Sumitomo Electric Industries, ltd, 1-1-1, Koyakita, Itami-shi, Hyogo 664-0016, Japan
**Corresponding e-mail: saikawa@eng.u-toyama.ac.jp*

Key words: Magnesium alloy, Heat resistance, Mg-Al-Sr-Ca, Die Casting, Castability, Recyclability

1. Introduction

Magnesium alloys have the characteristic with high specific strength and lightweight property, it is widely used for auto mobile industry with required the weight saving of structural material [1-3]. Mg-Al system alloys as typified by AZ91 and AM60 magnesium alloys are the most widely used for automobile components, such as seat frame, steering core and cover parts. However, the poor heat resistance of those conventional Mg-Al alloys has prevented their use in heat-resistant part of the powertrain, such as the engine and its structured parts, where the operating temperature can be as high as about 130-150°C. In order to solve this problem, significant effort has been made to develop heat-resistant magnesium alloys and these efforts have resulted in a number of new alloys[1-3].

In recent year, one of the most used heat-resistant alloy is Mg-3%Al-1%Si system AS31 alloy, which is applied as a gearbox case of the automobile in a European car company[3], because of the good die castability, hot-tear resistance and cheaper material cost . However, the heat resistance of the AS31 alloy was lower than a recent heat-resistant alloys, this is limited to use for more high-temperature area[4]. On the other hand, other heat-resistant magnesium alloys except the AS31 alloy, such as Mg-Al-RE (AE system), Mg-Al-Ca(AX system) and Mg-Al-Ca-RE(AXE system) alloys had a different industrial problem to be high in a material cost to contain the expensive rare-earth element and/or castability was inferior. In order to solve the problem, we pushed forward the development of the new composition magnesium alloy for HPDC. In late years we succeeded in the development of a new heat-resistant alloy excellent in castability and recyclability..

In the present study, characteristics of Mg-9mass % Al-3mass % Sr with containing small amount of Ca alloy AJX931 , which was developed by us was investigated.

2. Experiment procedure

The chemical compositions of the alloys are shown in Table 1. Six kinds of heat resistant alloys and one kind of conventional alloy AZ91 were cast in this study. Using a cold-chamber die cast machine with a clamping force of 5000-6500kN, the power-train component shown in Fig.1 was cast at molten alloys temperature of apporoximately 650-690°C, a die temperature of 200-250°C, an injection speed of 1.5-4m/s and a casting pressure of 50-70MPa. The component is 240mm in length, 150mm in height and 720 g in weight.

Table 1 Chemical composition (mass%)

alloy	Al	Ca	Sr	RE	Si	Zn	Mn	Fe	Be	Cu	Mg
AS31	3.7				0.9		0.26	8	9	<10	bal.
MACS	6.9	2.8	0.25		0.0		0.20	59	14	10	bal.
ACM522	4.9	1.7		1.9	0.0		0.21	29	34	10	bal.
AE44	3.9			4.0	0.0	0.0	0.27	6	15	10	bal.
MRI153M	7.7	1.0	0.25		0.0	0.0	0.23	11	26	<10	bal.
AJX931	9.0	1.0	3.1				0.31				bal.
AZ91	9.0				0.0	0.7	0.19	10	9	<10	bal.

(The value of Fe, Be and the Cu is ppm indication)

Fig. 1 Appearance of prototype component produced by die-cast process .

After casting, tensile, impact and fatigue test pieces were cut out from the column shapes part of the upper side of casting. Due to the evaluation of heat resistance, ring shaped portions of lower side of casting clamped to the aluminum alloy A380 block by the M10 steel bolt with an axis force of 10kN, and then those assembled castings held at 150°C for 300hours. During this test, bolt-load retention (BLR,%) was measured by the strain gauge attached to inside of the bolt.

On the other hand, due to the evaluation of recyclability of alloys shown in Table1, each alloy with a weight about 4kg was melted in Ni free stainless steel crucible using an electrical furnace with CO2-2%SF6 protective cover gas flow(1). After held at 650-690°C for 3hours(2), molten alloy cast into permanent mold, and then it was degassing by Ar gas bubbling after the remelted at 670°C(3). By each process of (1)-(3), small columnar sample(44mm in length,35mm in diameter) was cast due to analisys of quality of molten metal in each alloy. Using the columnar samples, observation of inclusions by OM and measuring of hydrogen gas content were carried out for evaluation of quality in castings.

3. Results and Discussion

Because of the most of motor parts are made up from the fastening structure by bolt. Therefore, new heat resistant magnesium alloys are being considered to replace aluminum alloy(e.g., 380 alloy, etc.,,) in automobile component(e.g., transmission cases, etc,.) at high-temperature use, BLR under the compression stress is the most important characteristic than a creep examination in the normal tensile stress.

Fig.2 shows result of a measurement of the BLR holding at 150°C for 300 hours. The conventional alloy AZ91, BLR value is lowest in all alloy. On the other hand, the heat resistance properties are improved dramatically by adding Ca and/or RE over about 1.5mass%, and the BLR level improves remarkably from 28% of AS31(Ca and RE free alloy) to 66~75% of Ca, RE containing alloys such as AE44, ACM522 and MACS. However, AE44 alloy has industrial problem to be expensive to contain a large amount of RE(rare-earth). ACM522 and MACS are easy to produce hot-tearing in as-cast components, both alloys are inferior in a castability. In the case of our developed alloy, BLR value of AJX931 is 64% of superior value, without containing RE and large amount of Ca.

Fig.3 shows evaluation result of recyclability on heat resistant alloys. Number and area furaction of inculusion in casting component are lower in the AE44 and AJX931 alloys. Hydrogen gas content is lowest in the AJX931 alloy. It is cleary that AJX931 is the best in recyclability.

4. Conclusion

We have successfully developed a superior alloy with castability, heat-resistability and recyclability..

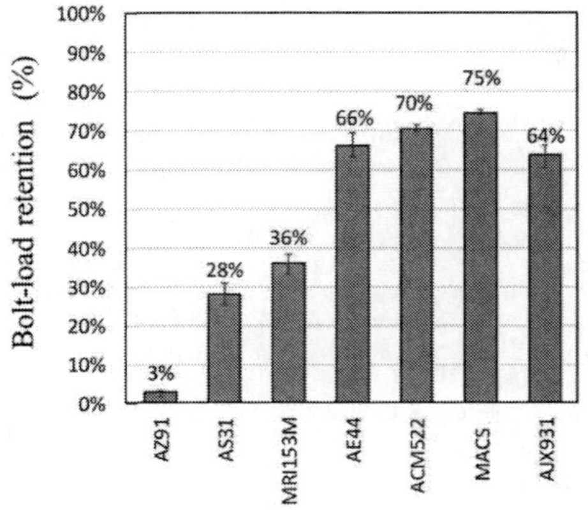

Fig. 2 Result of a measurement of the bolt load retention (BLR) held at 150 °C for 300hours.

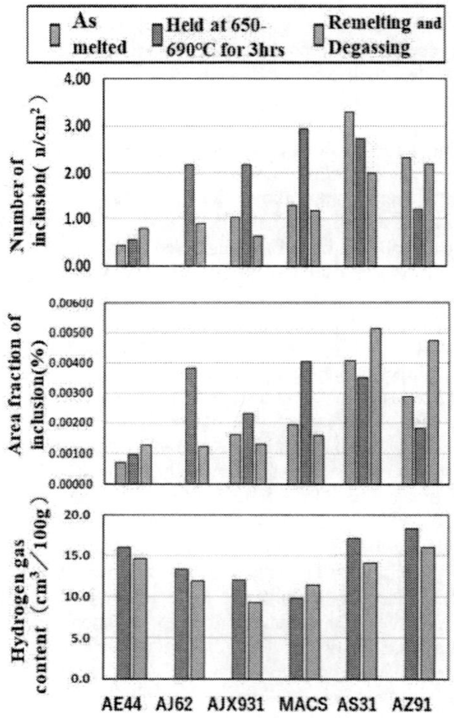

Fig. 3 Evaluation result of recyclability on heat resistant alloys

References

[1]S.Kamado, J. Jpn Inst. Light Met, 59(2009), 216-224.

[2]S.Saikawa, Japan Institute of Light Metals., 60(2010), 571-577.

[3]Alan A. Luo, Journal of Magnesium and Alloys, 1(2013),2-22.

[4]S.Saikawa, Japan Die Casting Congress, JD16-26(2016),147-155.

The 73rd World Foundry Congress
„Creative Foundry"
23 - 27 September 2018, Krakow, Poland

OXIDE CERAMIC AND REFRACTORY MATERIALS FOR METALLURGICAL PROCESSES AND INDUSTRIES

Fiodar Pantsialeyenka[1*], Andrzej Kuczumow[1], Valerij Szmuradko[2], Katsiaryna Pantsialeyenka[2]

[1]*The John Paul II Catholic University of Lublin, Faculty of Law and Social Sciences, Institute of Environmental Engineering, 37 – 450 Stalowa Wola, ul. Kwiatkowskiego 3A, POLAND*
[2]*Belarusian National Technical University, mechanical-technological faculty, 220013 Minsk, Nezawisimosti Ave. 65, BELARUS*
[*]*Corresponding address: e-mail: panteleyenkofi@tut.by*

Key words: ceramics, refractories, metallurgy, batching cup

1. Introduction

Assessing the role of various materials in the development of modern civilization in the historical context **[Błąd! Nie można odnaleźć źródła odwołania.]**, it should be noted that the role of ceramics after the 2nd world war has increased significantly.
This is explained by the unique set of properties of technical ceramics and, above all, be the high hardness, wear resistance and corrosion resistance, melting point and stability of mechanical properties at elevated temperatures, small electrical and thermal conductivity, optimal mass-to-volume ratio and some other special properties [2].

At the Byelorussian Steel Works it is used a wide range of refractories, including batching cups of continuous casting machines and wear-resistant supporting ceramic rollers with high tribotechnical characteristics. We develop such ceramic materials. It's required properties depend on the number of factors [3,4]:

1)material science (chemical composition of initial ingredients; structure, types of crystal lattice and modification of phases at all stages of the technology, the nature of the interaction of phases in ceramics at different temperatures);

2) technological (granulometric composition and degree of dispersion of initial powders, the degree of activation of powders and the compacting method , temperature and time parameters of preparation, processing, mechanical activation of components and binders, mixing, pre-and final sintering, final machining modes).

2. Research methods

Structure and chemical composition research was realized by scanning electronic microscopy method. Other characteristics were discovered by common methods.

3. Results and discussion

For continuous casting machine the most effective is construction of double-layered batching cup which consists of zirconia insert covered by high-alumina refractory concrete body.

Required properties of two-layered batching cup are: high refractoriness (more 1770°C), wear and corrosion-erosion resistance to melts of steels during continuous 11-30 hours steel casting (melt temperature 1600-1650°C).

In connection with more stringent working conditions, the inserts must have higher properties: melting temperature mote than 2715°C, heat resistance (17-21 thermocykle, 1200°C – water), mechanical strength (compressive strength 500-900 MPa, bending strength 150-330 MPa), low ferrodynamic wear (erosion of the insertion channel-for example, for a ø 14 mm not more than 1 mm). In addition, the insert material must have a low thermal conductivity, which excludes metal deposition on the channel walls and its narrowing.

For production of batching cup corundum and zirconium dioxide powders were used. The choice of zirconia as a material for batching cup's inserts production is based on a combination of its properties: density (depends on the crystal lattice type) 5,56-6,1-6,27 g/cm^3; refractoriness -2300-2500°C; temperature of start of deformation under load 0,2 MPa - 2200°C; the energy of chemical bonding of atoms Zr – O(757 ± 32 kJ / mol) in comparion with Al – O (480 ± 60 kJ / mol) и Mg – O (418 ± 17 kJ / mol); melting point 2715°C; boiling point 4300°C [5]

Body of batching cup produced using ceramic beton's technologies [6]. Refractory properties are provided by corundum fillers and binders derived by tribochemical (mechanical) grinding of corundum (100 μm) with oxide additives in the ceramic attritors (colloid mills) for 3 to 18 hours. As a result this mixture transforms into an activated highly concentrated ceramic binder suspension.

Depending on the content of introduced (no more than 2% by weight) CaO ($CaO-Al_2O^3$, $CaO-2Al_2O_3$) the intermediate strength in hydration hardening and the final strength in the process of the thermal preparation and exploitation of batching cup's concrete body are provided. The gradient strength of the batching cup is formed both during the thermal preparation for use (1000°C) and during subsequent operation during casting of blanks in the continuous casting machines at 1600-1650°C. The linear shrinkage does not exceed - 1%, ensuring the integrity and tightness of the system.

The most effective raw material for the manufacture of batching cup's inserts were zirconia, partially stabilized zirconia powders (3-9% by weight of CaO or MgO or Y_2O_3) or their compositions [4]. For batching cup's inserts, compositions based on polyfraction powders from partially stabilized zirconia (3-9% by weight of CaO) with the oxide-doping and structure-modifying additives were used. ZrO_2 has a low thermal resistance due to the high thermal coefficient of linear expansion (at 1500 °C $\alpha = 11 - 11.5 \times 10$-6) in combination with low thermal conductivity. The addition of CaO, MgO, Y2O3 oxides stabilizes the zirconium dioxide (baddeleyite) matrix, creating heat-resistant structures and the insert becomes heat-resistant. It excludes metal deposition on the channel walls and its narrowing and matrial of insert has high deformation temperature (about 2200°C at 0,2 MPa).

Molding of the inserts was carried out at 100 - 150 MPa on the installations of its own design for dry isostatic pressing. Sintering was carried out at 1700 - 1800 °C in the oven "TermConzept" (Germany) for 60-120 minutes.

The fixing of the sintered insert (Figure 1, on the right) in the body of the batching cup (Figure 1, on the left) was carried out by two methods: pressing in a mechanized mold on a hydraulic press, and also by vibroforming in a special form with the geometric dimensions of the housing of the batching cup with frequency and amplitude vibrations t = 60 - 180 Hz and Δ a = 0.3 - 1 mm, respectively.

Fig. 1. Batching cup with insert

Also, supporting rollers for production of metal cord (figure 2,a,b) at Byelorussian Steel Works were produced from the developed oxide-ceramic materials, which showed a 2.5 times higher resistance than the ceramic ones previously used.

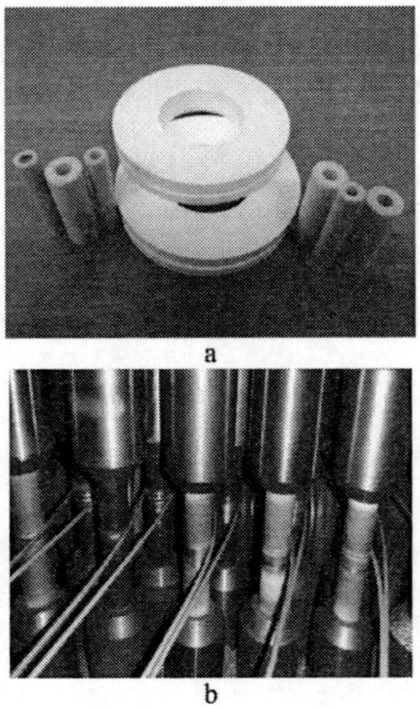

Fig. 2. Supporting rollers for production of metal cord

The developed effective materials and products have been tested at Byelorussian Steel Works. Now the development and testing of materials and constructions of batching cup working on the principle of a slide gate with automatic replacement during the casting process are carried out.

Acknowledgments (optional)

We express our gratitude to the management of BMZ for assistance in the implementation of research and testing

References

[1] AshbyM., JohnsonK. 2003. Materials and Design. The Art and Science of Materials Selection in Product Design. Amsterdam...Tokio.

[2] Gnesin G.G., Skorohod V.V. 2008. Inorganic Materials Science. Materials and technologies. Encyclopedic Edition. Kiev: Naukova dumka.

[3]Roman O.V., Pantsialeyenka F.I., Reut O.P., Shmuradko V.T., Kirshina N.V., Zhilevich A.V. 2010. Scientific and practical approaches to the creation of ceramic - refractory materials and technologies design. New refractories. №9, p.17-27.

[4]Pantsialeyenka F.I., Reut O.P., Shmuradko V.T., Grigorjev S.V., Stepkin M.O. 2016. Features of creating ceramic materials-products for various technical purposes. In the 19th international symposium "Technologies. Equipment. Quality". P.90-93. Minsk.

[5] Strelov K.K. 1985. Theoretical foundations of refractory materials technology. Moscow. Metallurgia.

[6] Zamyatin S.R. et.all. 1982. Refractory concretes. Moscow. Metallurgia.

The 73rd World Foundry Congress
„Creative Foundry"
23 - 27 September 2018, Krakow, Poland

CAST COMPONENTS IN SUPER DUPLEX ALLOYS INTERCOMPARISON BETWEEN BENCH MOLDING AND THREE DIMENSIONAL PRINTING

Yusuke Tomita [1]*, Hidetoshi Fujii [2], Julián Izaga [3], Ibai Gallastegi [3],

[1]*KIMURA FOUNDRY CO., LTD., 1798-17 Kitaema Izunokuni Shizuoka 410-2223, Japan.*
[2] *Osaka University, Joining and Welding Research Institute, 11-1 Mihogaoka Ibaraki Osaka 567-0047, Japan*
[3] *IK4-AZTERLAN, Metallurgical Technical Centre, Aliendalde Auzunea N°6*
E-48200 Durango (Bizkaia), Spain.
**Corresponding address: e-mail: tomita-y@kimuragrp.co.jp*

Key words: 3D printing, sand mold casting, binder systems, synthetic sands, super duplex stainless steel

1. Introduction

Cast Super Duplex Stainless Steel (SDSS) combines high strength and corrosion resistance [1] and it is mostly applied in complex parts that are produced in small batches. This combination of short batches and high complexity is the perfect frame for the application of Additive Manufacturing (AM) technologies to the sand molds, avoiding the cost and lead time of conventional sand molding patterns. This study compares the use of AM sand molding against conventional self-hardening binder system sand molds for SDSS castings.

On the other hand, it is known that the corrosion resistance becomes inferior at near the surface of the stainless cast steel products due to carburization by the decomposition gas of the resin [2]. Especially in the case of sand molds by AM, since the amount of resin addition is large, the carburized structure tends to be deep. However, AM molding is advantageous for building complicated sand molds. In this study, it is examined whether the depth of carburization can be controlled by the change of wall thickness of AM sand mold. The depth of carburization, using various organic binder system sand molds, has been also examined and compared.

2. Experimental procedure

As shown in Fig. 1, six kinds of cores with Φ70 mm and length of 150 mm, with different size of gas vent holes, and the main mold were fabricated by AM. The 3D printer S-MAX (ExOne) applying the furan self-hardening system is used. The amount of furan resin addition is about 1.8%, and the amount of acid hardener added is about 0.28% with respect to sand. Material of ASTM A890M Grade 6A was cast in the casting system as shown in Fig. 2. The solution heat treatment was done with water-cooling after holding at 1100 ° C for 1 hour. The obtained test pieces were cut and polished, electrolytically etched with 10% oxalic acid aqueous solution under the conditions of 1 A/cm^2 for 90 s, and the structure near the casting surface in contact with the core was observed.

Impeller test pieces were cast for evaluation, using both conventional phenol urethane self-hardening system and furan AM system. The casting material was ASTM A890M grade 5A.

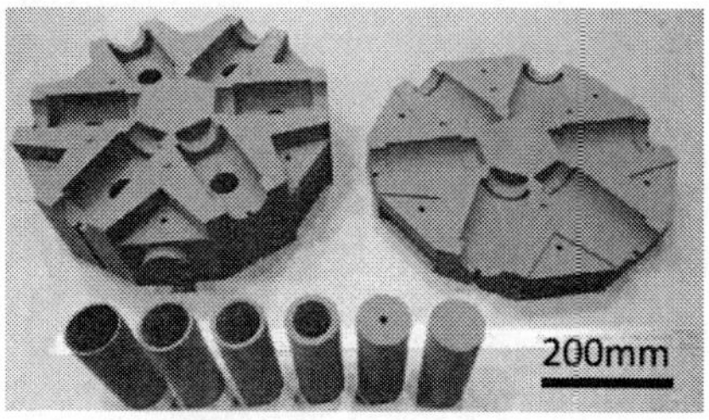

Fig. 1. AM sand molds and cores with the wall thickness of 3 mm, 5 mm, 10 mm, 15 mm, 30 mm, and without vent hole, from the left

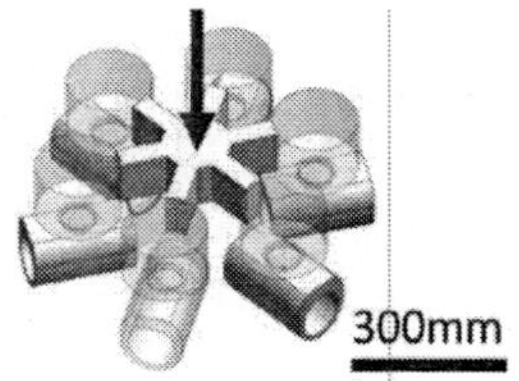

Fig. 2. Schematic drawing of casting system (poured from the arrow direction)

3. Result and discussions

The 3 mm thick core was broken and the molten metal invaded into the core. Cores having a thickness of 5 mm or more were sound, and the castings had a shape as designed. The microstructures of the area near the core (inner surface) are shown in Fig. 3. The average depth of carburization by changing core thickness is shown in Fig. 4. The result of using 30 mm thick core (Fig. 3 (b)), with a small vent hole as used in general sand molds, shows no improvement compared to that of without a vent hole core (Fig. 3 (c)). On the other hand, it can be seen that the carburization is remarkably reduced with the result of using 5 mm thick core (Fig. 3 (a)).

A model of gas pressure distribution in sand molds is considered with the following conditions:

- The resin decomposition gas generated only from the area near the boundary surface of sand molds and molten metal.
- The gas generation rate is constant.
- Steady State.

The diffusion velocity V of the generated gas, which is moving toward the gas vent / outside is defined by the following equation. In this equation, p is gas permeability of the sand molds, ΔP is the gas pressure difference of the molten metal side and the vent side, and t is the thickness of the sand mold.

$$V = p\frac{\Delta P}{t}$$

In a steady condition, V is equal to the gas generation rate. p depends only on the properties of the sand mold. Therefore, the thickness of the sand mold has no effect on V and p. As the result, the gradient of gas pressure ($\Delta P/t$) is not related with the thickness of the sand mold, as shown in Fig. 5.

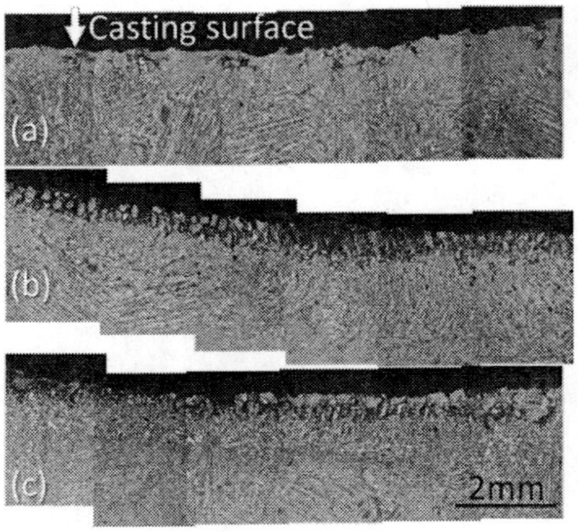

Fig. 3. Microstructures in the vicinity of SDSS casting surface by using (a) 5 mm thick core, (b) 30 mm thick core, and (c) without vent hole core

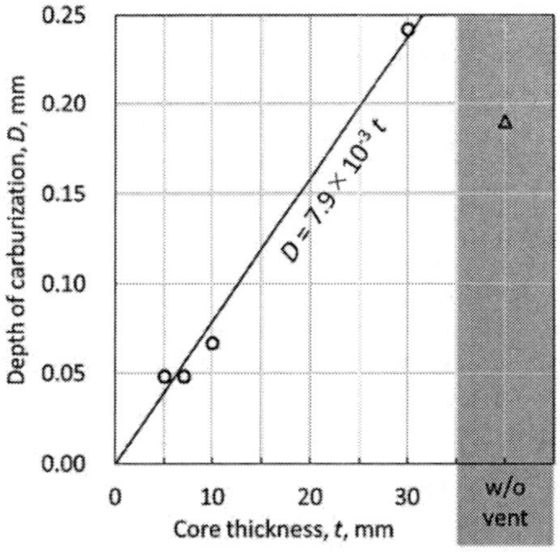

Fig. 4. Average depth of carburization near the casting surface cast with 3D printed various thickness core and w/o vent hole core

In cases when the solved gas concentration in the melt is lower enough than equilibrium concentration, the gas solving rate is proportional to the partial pressure of the gas. Therefore, by reducing the wall thickness of the sand mold, it is considered that the partial pressure of the gas in the vicinity of the casting surface decreased, and the depth of carburization is reduced. Under the condition of no vent, less carburization is found. There is a possibility that the carburized melt layer was stirred due to the bubbling from the decomposed gas, and the appearance of carburization decreases.

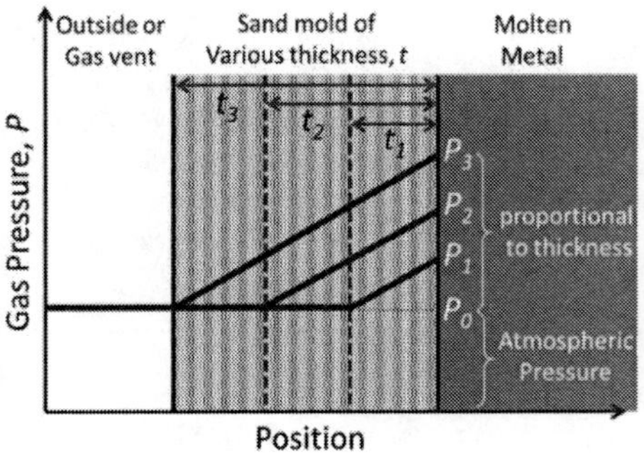

Fig. 5. The partial pressure distribution of decomposed gas in the sand mold, when conditions are steady

The cast impeller test piece is shown in Fig. 6. There were no gas porosity or shrinkage defects and the dimensions were very accurate.

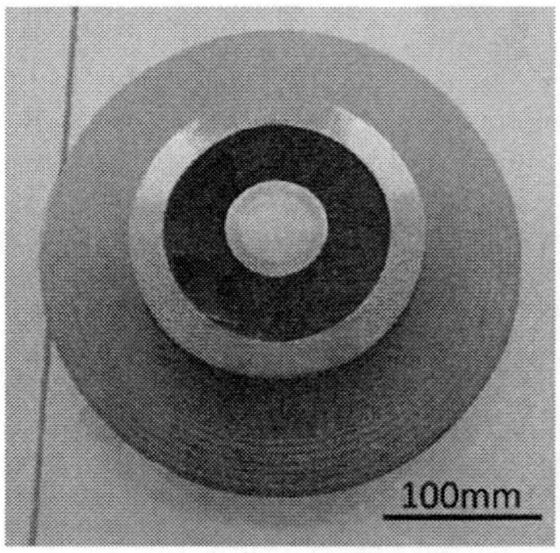

Fig. 6. Cast Impeller specimen of ASTM A890M 5A

4. Conclusion

The effect of furan self-hardening AM sand mold wall thickness on the depth of carburization of cast SDSS has been studied. Although the improvement of the carburization depth is hardly observed in the normal wall thickness of sand mold, it becomes clear that carburization can be reduced by making the thickness of the sand mold extremely thin.

References

[1] M. Kowaka. (1978). The Corrosion Behavior of Duplex Stainless Steels. Bulletin of the Japan Institute of Metals. 17 (8), 657-665.

[2] M. Tsuda, T. Kobayashi, K. Tamada and H. Ito. (1981). On carbides and Sulfides Formed on the Surface of Stainless Cast Steel in Furan Resin Molds. Imono. 53 (5). 227-232.

INFLUENCE OF NICKEL CONTENT ON EROSIVE WEAR AND HEAT TREATMENT CONDITONS BEHAVIOUR OF MULTI COMPONENT WHITE CAST IRON

Kenta Kusumoto [1]*, Kazumichi Shimizu [1], Hiroya Hara [2], Masato Shirai [3] and Jun Ito [2]

[1] *Muroran Institute of Technology, 27-1, Mizumoto-cho, Muroran-shi, Hokkaido, 050-8585, Japan,*
[2] *Hokkaido special cast steel Co., Ltd., 497-3, Yura, Kurisawa-cho, Iwamizawa-shi, Hokkaido, 068-0111, Japan*
[3] *Iwamizawa Imono. Co., Ltd., Yura, Kurisawa-cho, Iwamizawa-shi, Hokkaido, 068-0111, Japan,*
**Corresponding address: e-mail: kusumoto@mmm.muroran-it.ac.jp*

Key words: Erosive wear, Multi-component cast iron, Ni content, Destabilization heat treatment, Secondary carbides

1. Introduction

In recent years, many researchers have paid attention to study various alloys for the purpose to develop wear resistant materials. In particular, the multi-component white cast iron has been proved to exhibit good resistance to erosion and abrasive wear characteristics due to the large amount of carbides with high hardness in the matrix [1]. Some researchers were also focus on the elements effect in the materials [2], such as tungsten, niobium, vanadium, etc. Nickle (Ni) is known as having an effect of improving the hardenability and toughness of the heat-treated steel. In addition, Ni will be preferentially solid solution into the matrix instead of forming carbides compared to the other carbide-forming elements such as Cr and Mo. When Ni is added into the material, it can be expected that the amount of solid solution of carbide-forming elements will be suppressed, and on contrast, these carbide-forming elements will be encouraged to form eutectic carbides during crystallization.

Therefore, in the present study, 0, 3, 5 mass % (Hereinafter referred to as %) of Ni were added into the traditional multi-component white cast irons with 2% of carbon (C) and 5% of chromium (Cr), molybdenum (Mo), tungsten (W), niobium (Nb), and the improvement of the wear resistance due to the increase of the carbides amount can be expected.

2. Experimental procedure

2.1 Materials

Three kinds of multi-component white cast irons with about 2 % of C, 5 % of Cr, Mo, W, Nb, 0, 3 and 5 % of Ni were used as specimens (abbrev. 0Ni, 3Ni and 5Ni). The chemical compositions of experimental materials are as shown in Table 1. In order to strengthen the matrix by secondary hardening with secondary carbides, experimental materials were conducted heat treatment after the specimens were prepared.

Table 1. Chemical compositions of specimens. (mass%)
P:0.04, S:0.02

	C	Cr	W	Nb	Mo	Ni	Fe
0Ni	1.87	4.82	5.72	5.16	5.44	0.00	Bal.
3Ni	1.89	4.65	5.46	5.21	4.28	3.07	Bal.
5Ni	1.76	4.55	5.25	4.88	4.16	5.08	Bal.

The optimum heat treatment conditions vary depending on the type and amount of contained elements. Therefore, the optimum heat treatment conditions to fully demonstrate the performance of this cast iron were investigated. A high performance muffle furnace was used for the heat treatment, and the test piece size was 10 × 10 × 15 mm. As the heat treatment condition, the quenching temperature was set at 1123 K, 1373 K and 1423 K and held for 1～3 hour, then forced air cooling was performed. After that, it was kept at 673 K to 873 K for 1 hour and then tempered by air cooling. The tempering temperature was investigated at 20 K intervals.

Optical microscopy, Energy dispersive X-ray analysis spectroscopy (EDS) and hardness tester were used to observe the microstructure and hardness measurement of the experimental materials.

2.2 Experimental method

The test specimens for erosive test were mechanically machined to be sized 50×50×10 mm. Specimens were tested using a blast machine. Irregularly shaped steel grits with average diameter 770 μm, Vickers hardness of 810 HV were used in this study. A schematic view of the testing machine and steel grits are shown in Fig. 1. The examined particle speed was 20 m/s, and the particle feed rate was measured with about 20.0 g/s, and the particle input 2 kg with changing their impingement angles respectively 30, 60 and 90 deg. All the erosion tests were conducted at room temperature in 3600 sec.

Erosion rate was used to the evaluation of wear resistant. It refers to a value calculated from the volumetric loss divided by total feed of erodent.

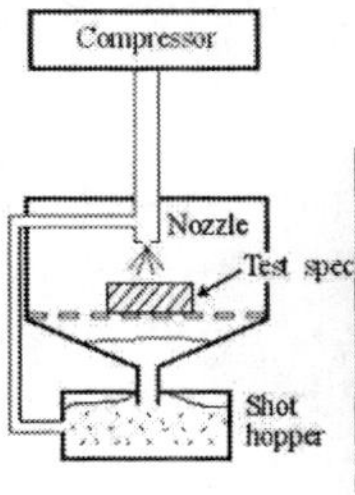

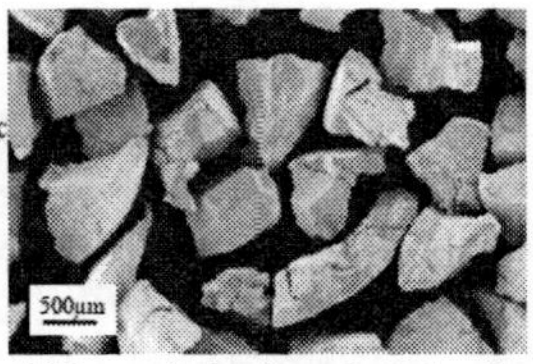

(a) Testing machine (b) Irregularly shaped steel grits
Fig. 1 Schematic view of the testing machine, steel grits.

3. Result and Discussion

3.1 Examination of heat treatment condition

Microstructure observation results of experimental materials after heat treatment are as shown in Table 2.

The matrix of all specimens changed into martensite after heat treatment. And the precipitation of secondary carbides can be confirmed. And the amounts of carbides were increased. With the addition of Ni increased in the matrix, there is a slight increase in the amount of eutectic carbides from 15 % to 20 % and a sharp increase in the amount of secondary carbides from 4.3 % to 15.3 % for material 0Ni and 5Ni.

From the above, the heat treatment conditions of 1123K_forced air cooling is determined to be appropriate.

The hardening holding time at the temperature of 1123 K was examined. Metallographic observation and Vickers hardness with the hardening holding time changed are shown in Fig. 2. As the holding time increased, the secondary carbide showed a tendency to precipitate each particle slightly larger in both test materials. Also, as the holding time increased, the hardness of the specimen and the area fraction of the secondary carbide increased.

From the result of the tempering condition, the highest hardness was 3 Ni and the tempering temperature was 733 K, and 5 Ni was 773 K. However, the hardness value was 729 HV for 3Ni and 728 HV for 5Ni, which was equal to or less than that of the specimen with only hardening. From these heat treatment test results, optimum heat treatment conditions for 3Ni and 5Ni are hardening only, hardening temperature is 1123 K, holding time is 3h.

Table 2. Structural components, hardness of specimens.

Hardening temperature	Cooling method	5Nb-3Ni			5Nb-5Ni		
		HV10	Area ratio of secondary carbides (%)	Matrix	HV10	Area ratio of secondary carbides (%)	Matrix
850°C (1123K)	F.A.C	790	13.3	M, γ	789	15.3	M, γ
	A.C	666	10.7	γ, M	684	8.29	γ, M
	F.C.	653	9.47	γ, M	636	8.67	γ, M
1100°C (1373K)	F.A.C	611	3.18	γ	624	2.11	γ
	A.C	698	2.72	γ	573	1.98	γ
	F.C.	651	6.15	γ	839	6.91	M, γ
1150°C (1423K)	F.A.C	678	-	γ	431	1.14	γ
	A.C	610	1.12	γ	558	0.82	γ
	F.C.	708	1.43	γ	602	0.90	γ

γ : Austenite, M : Martensite

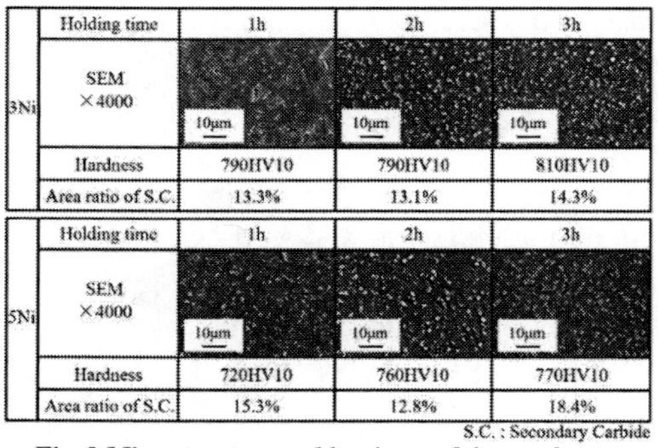

Fig. 2 Microstructure and hardness of the specimens.

3.2 Result of erosive wear test

Erosion rates of specimens and comparable material are as shown in Fig. 3. According to the result, multi-component cast irons which contain 5 % Ni showed the most excellent erosive wear resistance. As reasons, it can be considered that with Ni addition, the area ratio of eutectic carbides and secondary carbides was increased which enhanced the matrix structure and suppressed the plastic deformation of the surface of experimental material. When comparing specimens with hardening retention times of 1 hour and 3 hours, secondary carbides grew and precipitated largely in the specimens kept for 3 hours and it showed better erosive wear performance for specimen with longer hardening retention.

From these results, it is considered that the amount of secondary carbides and the shape to be precipitated are important for improving erosion resistance wear resistance. Therefore, determination of optimum heat treatment conditions to control the secondary carbides amount was clarified that an important factor.

4. Conclusions

In this study, the influence of heat treatment conditions on erosion wear characteristics of multi-component white cast iron with varying Ni content was investigated. The results are summarized below.

(1) As the amount of Ni added increased, the amount of carbide increased and the erosion rate showed a tendency to decrease. From this fact, it was confirmed that the wear resistance was improved by addition of Ni.

(2) As a result of investigating the heat treatment conditions, it was confirmed that the amount of secondary carbide was increased by increasing the hardening retention time, it was clear that grasping the heat treatment condition is an important factor.

(3) 5Ni (3h) showed the most excellent wear resistance.

References

[1]Yasuhiro Matsubara, Nobuya Sasaguri, Kazumichi Shimizu, Sung and Kon Yu, Solidification and abrasion wear of white cast irons alloyed with 20% carbide forming elements, Wear 250 (2001) 502-510.

[2]C.R. Loper Jr., H.K. Baik, Influence of molybdenum and titanium on the microstructures of Fe–C–Cr–Nb white cast irons, AFS Trans. 97 (1989) 1001-1008.

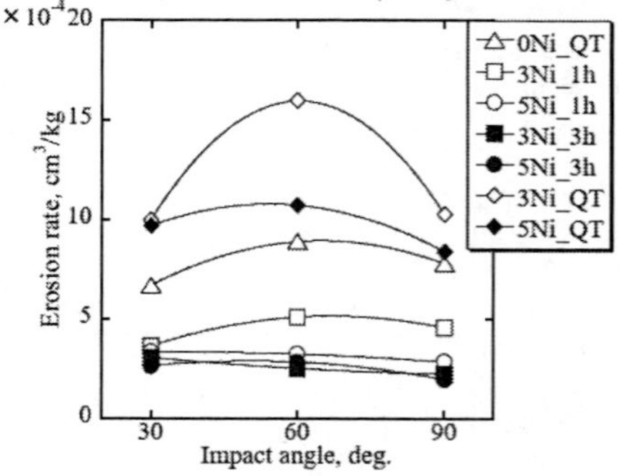

Fig. 3 Erosion rate as a function of impact angle.

The 73rd World Foundry Congress
„Creative Foundry"
23 - 27 September 2018, Krakow, Poland

ABRASIVE WEAR CHARACTERICTICS OF MULTI COMPONENT WHITE CAST IRON

Kazumichi Shimizu [1]*, Kenta Kusumoto [2], Hiroya Hara [3], Masato Shirai [4] and Jun Ito [3]

[1] *Muroran Institute of Technology, 27-1, Mizumoto-cho, Muroran-shi, Hokkaido, 050-8585, Japan,*
[2] *Hokkaido special cast steel Co., Ltd.., 497-3, Yura, Kurisawa-cho, Iwamizawa-shi, Hokkaido, 068-0111, Japan*
[3] *Iwamizawa Imono. Co., Ltd.., Yura, Kurisawa-cho, Iwamizawa-shi, Hokkaido, 068-0111, Japan,*
**Corresponding address: e-mail: shimizu@mmm.muroran-it.ac.jp*

Key words: Abrasive wear, Multi-component white cast iron, Ni content, Hardness, Carbides

1. Introduction

Castings are mainly used for attachment members such as tooth and underbody parts in construction machines such as shredding blades of shredders such as shredders and power shovels and bulldozers in the recycling process. These parts are always exposed to earth and sand and metal in excavation work, dismantling work, etc., and are the most abrasive wearing part. By attachments are worn, reduction in operating rate due to replacement, and increased running costs are concerned.

In order to reach an aim of high safety and low cost, it's important to develop wear resistant material and to estimate of life service during abrasion.

Previous studies, abrasive wear characteristics of various steel and cast iron using Suga type wear tester which reproduces a dual abrasive wear is simply investigated. As a result, it has been clear that high chromium cast irons are good wear resistance materials. And multi-component white cast irons which contain a plenty of carbide forming elements such as Cr, Mo, W, V and Nb exhibit better resistance to abrasive and erosive wear due to the crystallization of MC, M_2C and M_7C_3 carbides with high hardness during solidification and the secondary hardening of matrix during heat treatment [1-3]. The amount of alloy element dissolved in the matrix is changed with the type of carbide and microstructure due to the addition content of alloy. However, when considering adaptation to actual, the actual wear situation similar to that environment reproducible testing machine is important.

Therefore, an evaluation tester capable of reproducing three body type abrasive wear was introduced, and the three body type abrasive wear characteristics of multi-component white cast iron was investigated in this study.

2. Experimental procedure

2.1 Materials

Six kinds of multi-component white cast iron with about 2 % of C, 5 % of Cr, Mo, W, Nb, 0, 3 and 5 % of Ni, 0 and 5%Co were used as specimens (abbrev. Ni and Co content). The chemical compositions of experimental materials are as shown in Table 1. In order to strengthen the matrix by secondary hardening with secondary carbides, specimens were conducted heat treatment after the specimens were prepared.

2.2 Experimental method

The test piece size was a flat plate shape of 50 × 50 × 10 mm, and a rubber wheel type abrasion tester capable of reproducing ternary abrasive wear was used for the abrasion tester. A schematic diagram of the testing machine is shown in Fig. 1. A test piece was pressed against the circumferential portion of the rubber wheel rotating at 100 ± 1 rpm with a load of 196 N. Silica sand (average particle diameter: 300 μm, hardness: 1100 HV) was supplied at a rate of about 4.2 g / s from the popper between them, An abrasion test was conducted for 5 minutes (500 rpm).

The mass of the specimen was measured with an electronic balance (measuring accuracy: 0.1 mg) before and after the abrasion test, and the difference (wear loss) was calculated. The abrasive wear characteristics were evaluated using a value obtained by dividing the abrasion loss by the distance.

Optical microscopy, Energy dispersive X-ray analysis spectroscopy (EDS) and hardness tester were used to observe the microstructure and hardness measurement of the experimental materials.

Table 1 Chemical composition of specimens

(mass%)

	C	Ni	Cr	Mo	W	Nb	Co	Fe
0Ni-0Co	1.87	-	4.82	4.44	5.72	5.19	-	Bal
3Ni-0Co	1.89	3.07	4.65	4.28	5.46	5.06	-	Bal
5Ni-0Co	1.76	5.08	4.55	4.16	5.25	5.04	-	Bal
0Ni-5Co	1.90	-	5.23	5.35	4.86	3.14	5.38	Bal.
3Ni-5Co	1.82	3.43	5.32	4.27	5.51	4.60	4.62	Bal.
5Ni-5Co	1.73	5.03	5.03	5.15	5.24	4.51	4.57	Bal.

P≦0.05, S≦0.04

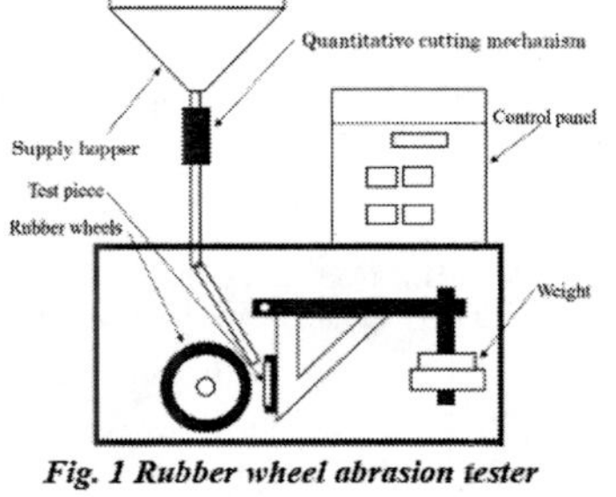

Fig. 1 Rubber wheel abrasion tester

3. Result and Discussion

3.1 Microstructure and hardness of specimens

Microstructure observation and hardness measurement results of specimens after heat treatment are as shown in fig. 2. When comparing the hardness of 0Co type specimens, 690 HV 10 for 0Ni, 740 HV 10 for 3Ni, 620 HV 10 for 5Ni, 3Ni-0Co showed the highest hardness. Comparing the hardness of the 5Co type specimens, 660 HV 10 for 0Ni, 790 HV 10 for 3Ni, 670 HV 10 for 5Ni, and 3Ni-5Co showed the highest hardness.

The area ratio of eutectic carbides in the form of plates and lamellae in each specimen was 11.4 to 16.1 %. Also, as a result of calculation of the area ratio of secondary carbide, it was 4.3 to 5.9 % for 0Ni typr specimens and increased to 10.3 to 13.5 % for specimens with Ni. There was no correlation between the hardness of each specimen and the carbide area fraction. The combined area ratio of eutectic carbide and secondary carbide was 20 to 25 %.

As a result of XRD diffraction, diffraction peaks of α-Fe, NbC, Fe_3C, Fe_3W_3C, Fe_3Mo_3C and M_7C_3 were confirmed in all the specimens regardless of whether Ni or Co was added or not.

In addition, α-Fe is a diffraction peak of the matrix. It is a mixed structure of martensite and austenite acording to the hardness and the microstructure observation result.

Material	0Ni-0Co	3Ni-0Co	5Ni-0Co
Microstructure	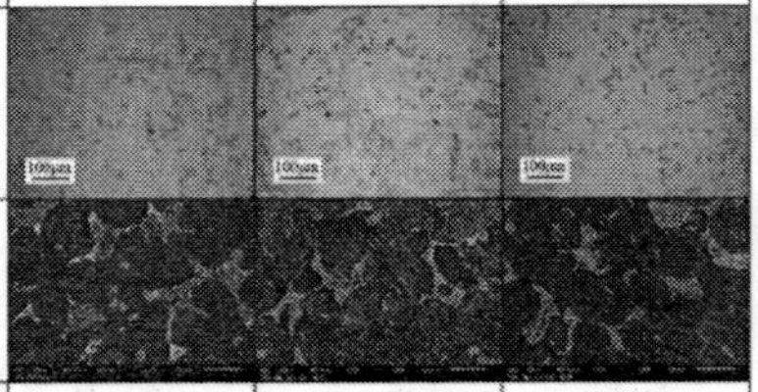		
SEM photograph			
Vickers hardness	690HV10	740HV10	620HV10
Area ratio of E. C.	15.0%	14.3%	11.4%
Area ratio of S. C.	4.3%	11.7%	13.5%
Material	0Ni-5Co	3Ni-5Co	5Ni-5Co
Microstructure	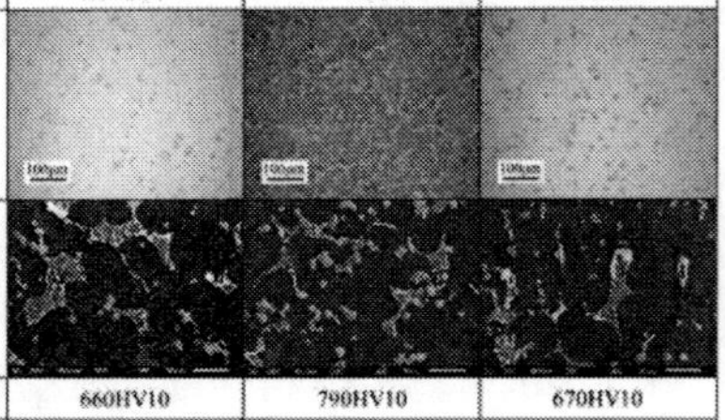		
SEM photograph			
Vickers hardness	660HV10	790HV10	670HV10
Area ratio of E. C.	12.2%	12.0%	16.1%
Area ratio of S. C.	5.9%	13.4%	10.3%

E.C. : Eutectic carbide
S.C. : Secondary carbide

Fig. 2 Microstructure and hardness of the specimens

3.2 Result of erosive wear test

In both 0Co and 5Co type specimens, the specimens added with 3% Ni showed good three body tipe abrasive wear characteristics.

The results of rubber wheel wear test of each specimens together with the hardness mentioned above were shown fig. 3.

In particular, the hardness of 3Ni-5Co showed a high value of 790 HV 10, showing the most excellent three type abrasive wear characteristics.

With or without addition of Co, there was no difference in wear loss for 0Ni and 5Ni type specimens. This is because there was no significant difference in hardness between 620 and 690 HV 10.

From the above, the wear loss tended to decrease with increasing hardness of specimens.

In this experiment, there was a correlation between hardness and abrasion loss.

From this, it can be inferred that ternary abrasive wear depends on the hardness of the test material. This is the same trend as previous research results [1-4].

Therefore, the rubber wheel abrasion tester was used in this study, three type abrasive wear characteristics can be quantitatively evaluated.

4. Conclusions

The results of this research are summarized below.

(1) The rubber wheel type abrasion test introduced in this experiment, the wear loss tended to decrease as the hardness of specimens increased, and there was a correlation between the hardness and the wear loss. This is the same as the previous research result by this testing machine, it is possible to quantitatively evaluate ternary abrasive wear characteristics.

(2) Hardness of 3Ni-5Co showed the highest value and exhibited the excellent resistance to three bod type abrasive wear characteristics compared with other specimens.

References

[1] Kenta Kusumoto, Kazumichi Shimizu, Hiroya Hara, J. JFS 87 (2015) 16

[2] Yasuo Nakajo, Kazumichi Shimizu, Kenta Kusumoto, Hiroya Hara, J. JFS 88 (2016) 234

[3] K. Kusumoto, K. Shimizu, X. Yaer, Y. Zhang, Y. Ota, J. Ito, Wear 376-377 Part A (2017) 22

[4] K. Shimizu, T. Noguchi, Wear 176 (1994) 255

5) Hiroya Hara, Kazumichi Shimizu, Kenta Kusumoto, Jun Ito, J. JFS 86 (2014) 43

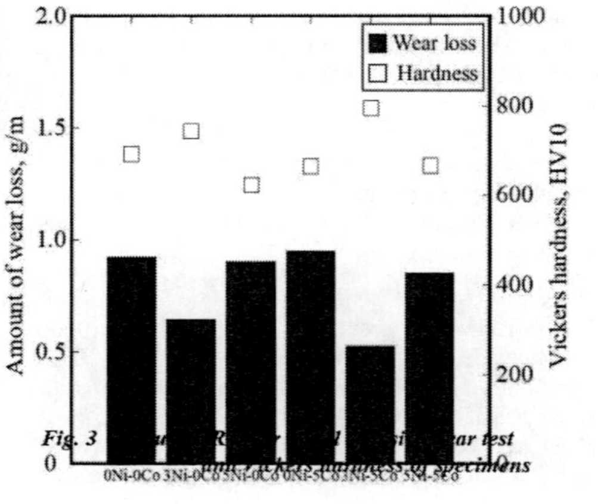

Fig. 3 … Rubber wheel abrasive wear test and Vickers hardness of specimens

The 73rd World Foundry Congress
„Creative Foundry"
23 - 27 September 2018, Krakow, Poland

NANOKARB: ENGINEERED CARBON ADDITIVE FOR GREEN SAND FERROUS FOUNDRIES

Vijay Gurunath [1*]

[1]Refcoat Chemicals Private Limited Company, Managing Director, No.10, Nanjappa Layout, Yelachenhalli, Kanakpura Road, Bangalore 560062, India.
*Corresponding address: e-mail: vijay@refcoat.com

Key words: Coal dust replacement, Nano technology, VOC's, WTS, Surface finish

1. Overview

Surface Finish on Grey Iron castings is always a factor for continuous improvement. Efforts are continuously being made to improve the surface finish by altering various parameters. Majority of work has been to either go in for using finer sand or to change the process. Effort to improve the surface finish in Green sand ferrous castings is the key issue. Use of Coal Dust is the proven economical method but recent studies on the impact on environment discourage its use. Latest developments in Nano Technology have helped the production of suitable material with reduced Volatile Organic Compounds (VOC) and more efficient activity.

2. Problem Statement

The exact same product that works in foundry A does not work in foundry B. Why? This is especially true if we are talking of carbon additives in green sand foundries. VOC is a growing concern to foundries across the world. There is a need to understand the mechanism by which carbon forms a barrier between the molten metal and sand.

3. Solution

Using advanced data analysis algorithms and cutting-edge material science, we have understood the complex interaction at the molten metal and sand interface. With this understanding, we have engineered a nano-ceramic impregnated carbon (Nanokarb) which will eliminate the use to coal dust in green sand.

Ceramic-oxide nanoparticles impregnated into carbon form a barrier between the molten metal and the sand. The presence of ceramic-oxide nanoparticles in quasi graphitic carbon exhibit excellent non-wetting characteristics at extreme temperature, as compared with conventionally used coal-dust in green sand. The high surface area to volume ratio is instrumental in the formation of an effective barrier between molten metal and silica sand.

Nanokarb properties:
- Excellent non-wetting properties;
- 0% Volatile Matter < 400 C
- <10% Volatile Matter @ 925 C;
- Drastically reduced VOC
- Altered thermal conductivity

4. Working

With Nanokarb, the effectiveness of the barrier formed between the molten metal and sand is far superior compared to conventional technology. Fig. 1, is a photo of an engine block made out of green sand, without Nanokarb, but a refractory coating was used on the mould surface to prevent burn-on defects/metal penetration common in castings this size.

Fig. 1. (a) Without Nanokarb (Shot blasting for 6 min)

Fig. 2. (b) With Nanokarb (shot blasting for 5 min)

Fig.2, is a photo of the same engine block made out of green sand with Nanokarb but NO refractory coating was used on the mould surface. The superior barrier formed between the molten metal and sand is quite evident.

The engineered Nanokarb technology alters the thermal conductivity of the sand system slightly just enough to reduce the sinter rate of bentonite and this not only reduces bentonite consumption, but also increases Wet Tensile Strength (WTS). Fig.3 below clearly shows that there is a 11.62% increase in WTS with the introduction of Nanokarb into the system.

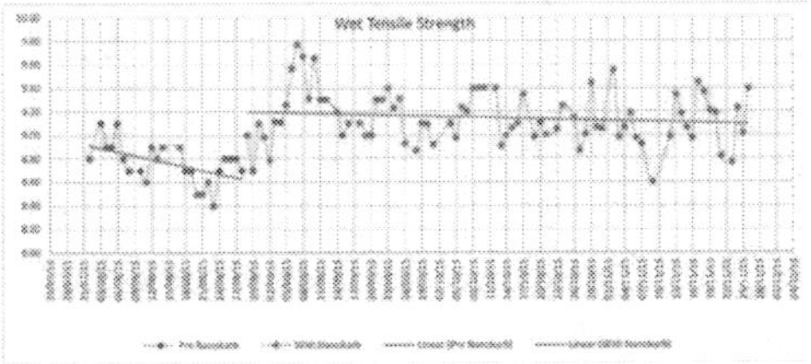

Fig. 3. Increase in WTS with Nanokarb

Fig.4 below shows that the Green Compressive Strength is maintained as desired with much lesser bentonite.

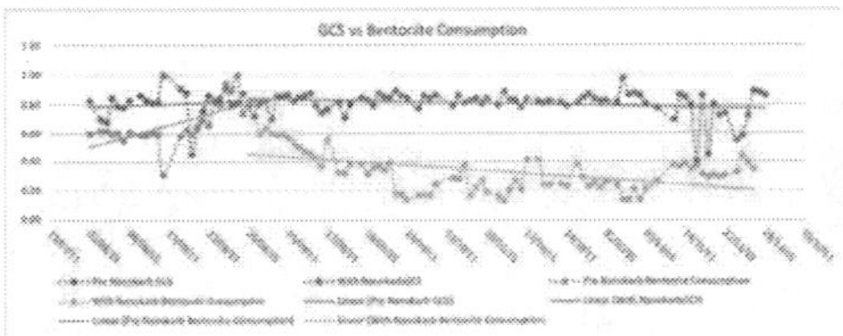

Fig. 4. Reduced Bentonite consumption with Nanokarb

Low temperature volatiles ever present in naturally occurring coal dust can present a peculiar challenge in green sand foundries. By virtue of radiation, mould surfaces reach temperatures of 400C even before molten metal comes in contact with it. At this temperature, any low temperature volatiles present in the green sand will spontaneously result in an exothermic combustion increasing the temperature on the mould surface. It is well known that bentonite loses its strength at such elevated temperatures. This results in reduced mould surface stability leading to Sand Fall, Edge and Cod Disintegration and Sand Inclusions.

This has been completely eliminated in Nanokarb ensuring that there are no volatiles at less than 400 C. This ensures that the bentonite remains activated for a longer time and increases the mould stability at higher temperatures resulting in lesser sand inclusions.

The high surface area to volume ratio of Nanokarb is instrumental in ensuring reduced carbon consumption. With precision control on the barrier formation, much lesser Nanokarb is required as compared with regular coal dust.

Table 1. Reduction in consumption of Carbon with Nanokarb

	Before Nanokarb	With Nanokarb	% Reduction in consumption
Carbon	13 kg per 3000 kg batch mix	5.5 kg per 3000 kg batch mix	**57.69%**
Bentonite	24 kg per 3000 kg batch mix	19.2 kg per 3000 kg batch mix	**20%**

The results shown in Fig.2, is with the above reduced consumption of Carbon. Even with just 0.18% Nanokarb addition into the sand mix, outstanding barrier between the molten metal and sand has been formed which also eliminated the requirement of refractory coating on the mould surface.

Reduction of carbon consumption by this degree has never been possible before. The environmental impact of this is reduced emission is phenomenal.

5. Conclusion

We now have access to nano-materials that not only reduce the impact on the environment by a huge factor but is also engineered keeping in mind the performance required at the molten metal interface.

Benefits of Nanokarb:

- Drastically reduced Carbon consumption and reduced VOC
- Eliminates the possibility of sub-surface porosity arising out of carbon additives
- Drastically reduces metal penetration and burn on defects
- Increases WTS even with reduced bentonite consumption making the sand more resistant to expansion and erosion defects.

With this organic evolution in green sand, the impact on the environment is drastically reduced along with a hope of achieving reduced sand related rejections.

The 73rd World Foundry Congress
„Creative Foundry"
23 - 27 September 2018, Krakow, Poland

MICROSTRUKTURE AND HARDNESS OF HIGH VANADIUM MARTENSITIC CAST STEEL FOR WEAR RESISTANT APPLICATIONS

GRZEGORZ TECZA [1], ŁUKASZ ROGAL [2], JAN GŁOWNIA [1]*, JAN DUTKIEWICZ [2]

[1] *AGH University of Science and Technology, Faculty of Foundry Engineering, 30-059 Cracow, Reymonta 23 Str., Poland,*

[2] *Institute of Metallurgy and Materials Science of the Polish Academy of Science, Reymonta 25 Str., 30-059 Cracow, Poland,*

**Corresponding address: e-mail: 14.1.jang@gmail.com*

Key words: wear resistant, cast steel, microstructure, carbides, hardness,

In the mining industry crushing the walls is done using the set of tools, consisting of hard teeth made of sintered carbides (WC) that are embedded in the forged low alloy steel body. We propose new approach in which application of cast steels instead of WC teeth and hot forged handle decrease manufacturing costs. Additions of carbide-forming elements e.g. vanadium or tungsten to the high carbon cast steel lead to significantly increase of hardness and wear resistance after complex heat treatment consisting of high temperature austenitization and tempering. As a result such parts could widely use in industrial applications e.g. earthmoving, quarrying, mining an dredging. Fig. 1 shows an example of a casting of a cutting tool made of high alloy cast steel for application in the earth-moving industry and abrasive wear conditions.

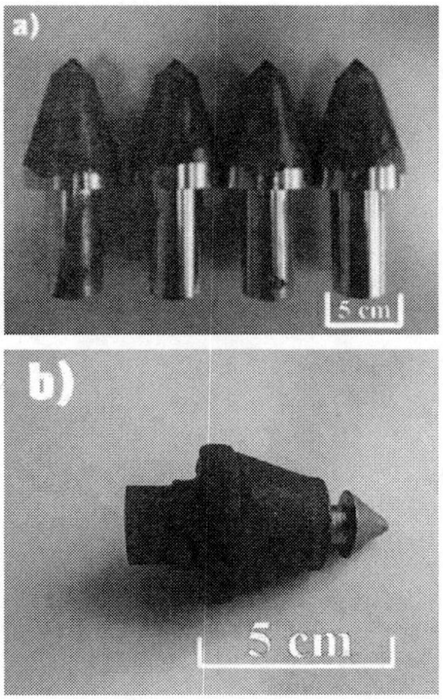

Fig. 1. ***An examples of wear resistant castings used in the coal-moving equipment-a), and in road machine-b)***

Cutting tool sets are made of lost wax casting technology and thixoforming. An example of composition studied in present work are listed in Tab. 1. Depends on chemical composition the significant differences in type and content primary and secondary carbides in the matrix are visible (Fig. 2). The grade No 3 (Tab. 1) with the higher carbon content (3.88%) has the highest hardness after quenching in polymer solution (HI) ≈ 950 HV (Tab. 2). In thixoformed sample tempered at 600°C for 2h hardness 600 HV was obtained.

Table 1. The chemical compositions of investigated materials (1-3 obtain using lost wax casting, 4-5 – thixoforming), content of elements in wt.%

Composition	1	2	3	4	5
C	2.77	3.44	3.88	0.96	2.5
Mn	0.85	0.95	1.06	0.48	8.14
Si	0.64	0.72	0.71	0.49	1.35
Cr	0.55	2.22	0.05	3.6	1.09
V	17.5	17.7	17.9	8.0	5.27
W	0.03	4.61	0.02	-	-
Mo	-	-	-	-	0.3
Ti	0.04	0.03	0.03	-	-
Ni	0.02	0.02	0.02	-	0.19

Fig. 2. ***Microstructures of the grades 1-3 in as cast conditions***

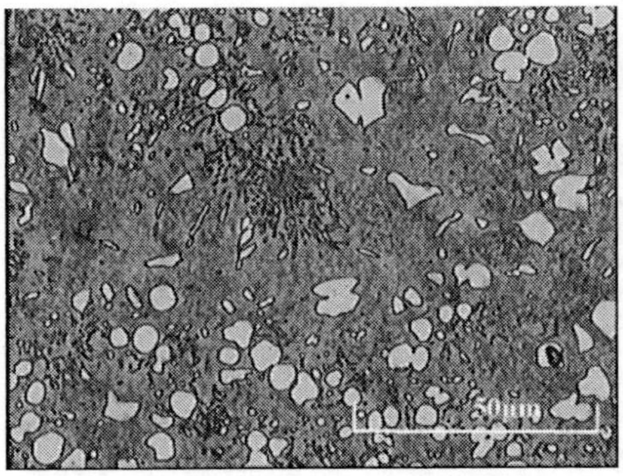

Fig. 3. Microstructures of grade nr.3 after quenching from 920°C

Table 2. List of applied various conditions of quenching as function of hardness

Composition	Austenitization temperature [°C]	Cooling medium	Average hardness [HV]
1	880	oil	274
	880	HI 15%	263
	920	HI 15%	320
2	880	oil	530
	880	HI 15%	590
	920	HI 15%	631
3	880	oil	876
	880	HI 15%	966
	920	HI 15%	883
4, 5	Thixoformed and tempering at 600°C/2h	Cold steel die	580

Conclusions

1.Wear resistant teeth made by thixoforming method have globular microstructure after its modification in solid – liquid conditions,

2 The highest hardness obtained for the quenching he alloy with 3.7%C after modification,

3.Next investigations are provide with more complicated grades for hardness higher than 1000-1200 HV

The 73rd World Foundry Congress
„Creative Foundry"
23 - 27 September 2018, Krakow, Poland

ACTICOTE CG RANGE OF COATINGS TO REDUCE GRAPHITE DEGENERATION AT THE SURFACE ZONE OF COMPACTED GRAPHITE IRON CASTINGS

Dr Ugo Nwaogu, R & D Scientist

Foseco Nederland BV, Pantheon 30, 7521 PR Enschede, The Netherlands
e-mail: Ugo.Nwaogu@foseco.com

Key words: Compacted graphite iron, Compacted graphite degeneration, Foundry coatings, Metallurgy, Mechanical properties

With increasing emission legislation for internal combustion engines, there is an increasing trend for the wider adoption of Compacted Graphite iron (CGI) for automotive components. Emissions legislation and the demand for higher performance from smaller engines have together driven the development of diesel engine technology over the past ten years. One of the most significant of these developments has been the advent of common rail and unit injector fuel management and delivery systems, which allow for higher cylinder pressure in direct injection diesel engines. The higher peak firing pressures provide more efficient combustion, improved performance, reduced emissions and quieter engine operation. Simultaneously, the increased firing pressure placed increased mechanical and thermal loads on the main bearing region of the cylinder block, potentially resulting in premature fatigue failures. The irreversible trend toward higher peak firing pressures have prompted engine designers to seek stronger materials to meet their durability targets without increasing the size or weight of their engines. Given that the new engine programs are typically intended to last for approximately three vehicle generations, the chosen engine materials not only need to satisfy current design criteria but must also provide the potential for future performance improvements without changing the overall block architecture. Therefore, with at least 75% increase in ultimate tensile strength, 35-40% increase in elastic modulus and approximately double that of the fatigue strength of grey cast iron, compacted graphite iron is ideally suited to meet the current and future requirements for diesel engine design.

Compacted Graphite Iron is intermediate to grey and ductile iron in its microstructure, its properties and its production in the foundry. The compacted or "worm-shaped" graphite morphology when the magnesium content of the iron is insufficient to fully stabilize the spheroidal graphite of ductile iron.

Researchers eventually recognized that the "intermediate" graphite shape provided an advantageous combination of the strength and stiffness of ductile iron along with the thermal conductivity, machinability and vibration damping capacity of grey iron. This unique combination of properties makes CGI an excellent candidate for state-of-the-art lightweight engine blocks, diesel cylinder head applications and a variety of other components that simultaneously require strength and thermal conductivity. Initially, the production series were restricted to products with a wide microstructure specification or to those that can tolerate the use of titanium to increase the stable CGI range. Specific examples included manifolds, bedplates, brake components, pump housings, fly wheels and brackets. The application of CGI to the high-volume production of complex components such as cylinder blocks and heads, that require a narrow microstructure specification and cannot tolerate the use of titanium due to machinability requirements followed with the development of improved process control systems, such as the Sintercast process.

Currently, component design in CGI is limited by the reversion or degeneration of the compacted graphite (CG), resulting in a layer of flake graphite (FG) in the rim zone of the compacted graphite iron castings. This defect commonly occurs in ductile iron (DI) and compacted graphite iron (CGI) grades of cast iron. The average thickness of the degenerated layer for thin wall DI castings ranged from 0.15 to 0.45 mm, while for CGI it ranged from 0.7 to 2.5 mm. The presence of this degenerated layer of graphite significantly reduces the mechanical properties (fatigue limit, elongation, etc.) of the component in the area where it occurs.

Controlling the degeneration of CG at the surface rim zone requires an understanding of the effects of melt chemistry, melt temperature, cooling and solidification rates and the interaction between the metal and the mould/core surface. The assumption is that the direct cause of graphite degeneration, resulting in a layer of FG, is the depletion of magnesium (Mg) at the surface layer of the castings. This is due to the reaction of Mg with sulphur and oxygen present in the moulding materials and/or in the mould atmosphere. The use of active coatings provides an insulating barrier between the metal and mould/core surface, thereby preventing the interaction between the metal and mould or core surface. This barrier promotes the reduction/elimination of this reversion thereby creating significant values for the engine designers.

This paper highlights the development of a new foundry coating for CGI application to reduce CG degeneration to FG at the rim zone of compacted graphite iron castings. The new range of coatings are:

- Water-based and environmentally friendly.
- Especially designed for core dipping applications.

The development process first characterized a range of coatings according to foundry application properties and by using advanced analytical technologies. Casting simulation was applied to validate a new test model for the casting trials. The new coatings were subsequently benchmarked alongside other commercial coatings, for the same application, using the customized test model and on commercial CGI engine blocks and cylinder heads, in a foundry known for series production of CGI castings. Other tests included:

- The effect of different coating dry layer thicknesses on the reversion level after casting.
- Investigation of the microstructures at the surface zone of the castings, performed using optical microscopy.
- The effect of the coatings on the mechanical properties of CGI castings.

The results from the trials and investigations show that the new coating showed superior performance in the reduction of the CG reversion to FG at the rim zone of CGI castings. The mechanical property test results show that the coating has little or no effects on the mechanical properties of the castings.

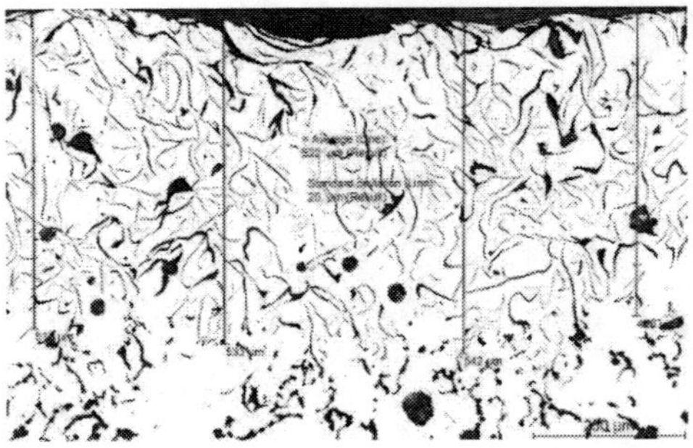

Fig.1: Microstructure of CGI casting produced using a commercial coating showing severe CGI reversion

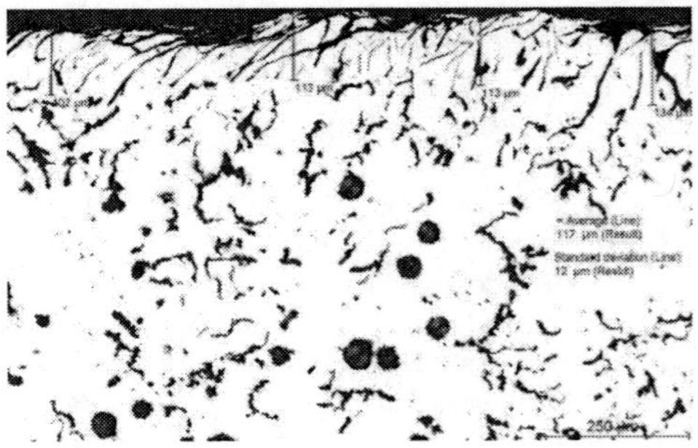

Fig.2: Microstructure of CGI casting produced using ACTICOTE CG 800 showing reduced CGI reversion

The 73rd World Foundry Congress
„Creative Foundry"
23 - 27 September 2018, Krakow, Poland

THE OPTIMISATION OF THE FLOWABILITY OF SAND MIXTURES TO PRODUCE HIGH DENSITY AND DEFECT-FREE INORGANICALLY BONDED SAND CORES

Dr Vincent Haanappel, R & D Manager Binders

Foseco Nederland BV, Pantheon 30, 7521 PR Enschede, The Netherlands
e-mail: Vincent.Haanappel@foseco.com

Key words: inorganic, binder, additive, flowability, cores

Environment-friendly binder systems for moulds and cores are generally based on the use of a 2-component inorganic binder system consisting of a liquid sodium silicate and a powder-like solid consisting of various minerals. As environmental pressures increase, the replacement of organic by inorganic binders is almost inescapable, with the potential to deliver a significant reduction of unpleasant odours and toxic emissions.

The development of the new inorganic binder systems has required significant research activities to ensure the performance of the current organic binder systems have been matched. In this context, with the developed technology to heat cure the inorganic bonded cores, high strength can be achieved, comparable with that obtained with the existing organic binder systems.

The new environment-friendly binder system SOLOSIL TX developed by Foseco, is an inorganic thermosetting core binder, which combines the advantages of the organic binder regarding strength and productivity with environmental friendliness [1].

To produce high quality castings whilst using inorganically bonded cores depends not only on the casting process itself, but also on the quality of these cores. Use of cores with insufficient strength or with locally low compaction results in lower surface smoothness and often in defects on the casting surface. To achieve high quality castings whilst remaining cost competitive some of the requirements of the binder are:

- Use of relatively cheap and environment-friendly materials
- Easy to handle and to mix with various types of sand
- Sufficient bench life of the sand mixture
- Good flowability of the sand mixture
- No blocking of the inlet shooting nozzles
- High quality cores with sufficient compaction and surface smoothness
- Short cycle times = short core production time
- Sufficient breakdown / de-coring properties
- No gas formation during the casting process
- Sufficient storage stability of the produced cores.

This study will focus on how to produce good quality cores (i.e. high compaction, no surface defects and high surface smoothness) using quartz sand and an inorganic binder system, the latter to achieve an environment-friendly working area.

High quality cores can only be produced when flowability of the sand mixture is such that the core box is filled completely. This means that the surface friction between individual sand grains themselves and with the surface of the core box should be sufficiently low. Figure 1 shows a small part of two types of cores, one (top picture) with insufficient compaction due to low flowability, and the other (bottom picture) with high compaction – high flowability of the sand mixture.

Flowability of sand mixtures depends on various parameters: the nature and amount of the liquid binder, the nature and amount of the optional additive(s), and the type of sand. Regarding the additive and sand, flow properties of these depends on several parameters, such as particle size distribution, particle shape, chemical composition, moisture and temperature [2]. Generally, the sand type is fixed because it is most cost effective to source it from local quarries. This means that only the nature and properties of the liquid binder and sand additive are accessible for modification.

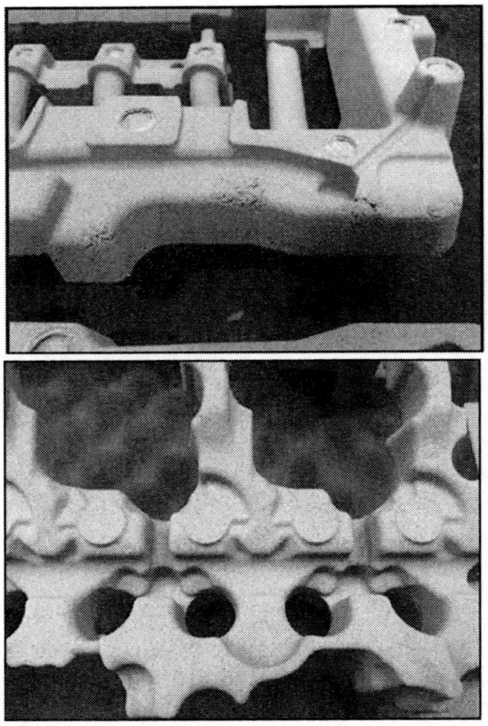

Fig. 1: Top: core with locally insufficient compaction (low flowability of the sand mixture); Bottom: core with high compaction (high flowability of the sand mixture). Both cores were manufactured with inorganic binder systems.

The objective of this study is to highlight some methods to determine, to influence and to optimize the physical and chemical properties of inorganic binder systems, aiming for a high flowability of the sand mixture.

Fig. 2: Attension Optical Tensiometer.

The first part of this contribution is to determine the surface tension of the liquid binder using the Attension Optical Tensiometer. Surface tension is a parameter that affects the flowability of sand mixtures containing a liquid inorganic binder. This system determines the critical micelle concentration of a surface-active agent; higher concentrations (above the micelle concentration) do not influence the surface tension of the liquid. Using also the PFT Powder Flow Tester to quantify flowability of sand mixtures and various types of core boxes to physically observe flowability, it was found that the lower the surface tension of the liquid, the higher the flowability.

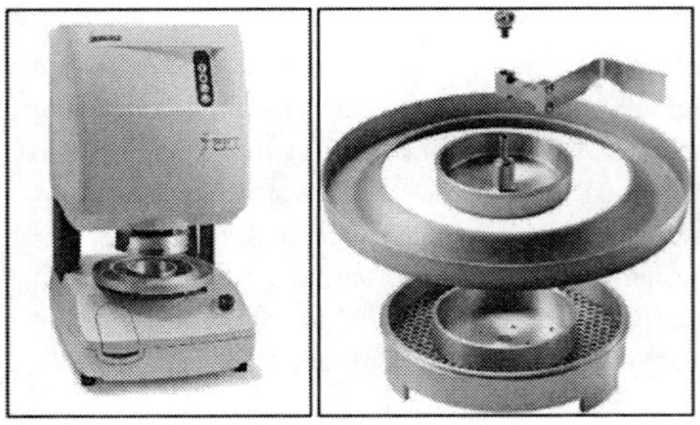

Fig. 3: PFT Powder Flow Tester.

A further improvement of flowability was achieved by adding a small amount of a well-chosen additive, mostly micro-sized spherical particles to the sand mixture. Balancing the powder-like additive and the adapted liquid binder, flowability can be controlled and adapted for various applications; finally achieving high quality cores. Other requirements from the foundry industry, such as strength, wettability, resistance against hot distortion, de-coring properties etc. are considered as well to define the most appropriate inorganic binder system called SOLOSIL TX.

Acknowledgments

The authors wish to thank all colleagues of the R&D Global Research Centre of Vesuvius for their help and support, in particular R. Wattimena, M. Huusken, J. Morsink and M. Goorhuis for all the laboratory work.

References

[1] Schrey, A., The new environment-friendly binder system SOLOSIL TX for core making for the mass production of complex cast components, Foundry Practice 263, The Authoritative Magazine for Foundry Engineers, June, 2015, pp. 15-23.

[2] Schulze, D., Powders and Bulk Solids - Behavior, Characterization, Storage and Flow, Springer (2007).

The 73rd World Foundry Congress
„Creative Foundry"
23 - 27 September 2018, Krakow, Poland

FEEDING TECHNOLOGY

Jürgen Kmetsch, European Product Manager Feeding Systems

Foseco Europe, Gelsenkirchener Straße 10, 46325 Germany
e-mail: Juergen.Kmetsch@foseco.com

Key words: new, kompressor, spot, feeding, development.

The requirement for cast parts with thinner wall sections and a more complex design is a continuous trend in the foundry market. This leads to castings with more isolated sections and as a result the requirement for spot feeding. This paper describes the incremental development and optimization of the ram-up sleeve application technology using a patented collapsible metal breaker core design. The latest development is the result of a consequent implementation of experiences made in the area of spot feeding on high pressure moulding lines.

1. Header

Spot Feeding Technology – a new development for iron castings

2. Introduction

A high exothermic feeder concept with a thick-walled design emerged in the mid 1980's. This was the starting point for a new generation of feeder sleeves. The invention was patented by a German foundry during 1985 [1]. The technical drawing of such a sleeve described in the patent is still similar to the today's spot feeder designs (Fig.1).

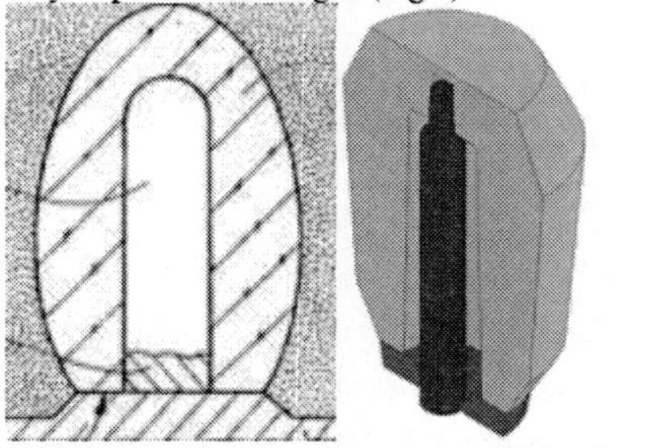

Fig. 1 1985's patent design and current spot feeder shape

The new feeder sleeve concept showed several new features for the foundry industry, providing significant benefits on application, in particular:

- Demand for higher compressive strength for ram-up application since higher moulding pressures were used with modern green sand moulding lines.
- Cost pressure required reduced feed metal volumes with higher moduli, improved casting yield and reduced footprint area on the casting.

In the next 20 years the development of spot feeding concepts was driven by specific technical requirements of the foundries. Process cost pressure and increasingly complex casting designs demanded by the end customers influenced the market for reduced feeder contact and footprint areas. As a result, solutions which gave the lowest fettling and cleaning costs, in combination with optimized yield, were developed.

In 2004 a new generation of spot feeders was launched [2], which addressed the foundry demands (Fig. 2).

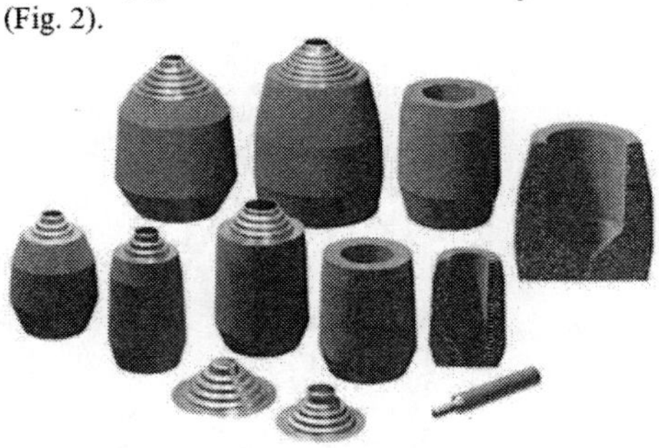

Fig. 2 Product range of spot feeders with collapsible breaker core

The concept of a collapsible metal breaker core provides two key advantages:

- The moulding pressure is partly absorbed by the collapse of the metal breaker core and compaction of the moulding sand beneath it, enabling this concept to withstand higher moulding pressures than other conventional solutions.
- The feeder neck height is reduced during moulding, this helps to ensure that the feeder neck remains open until after the casting solidifies (Fig. 3 – Right)

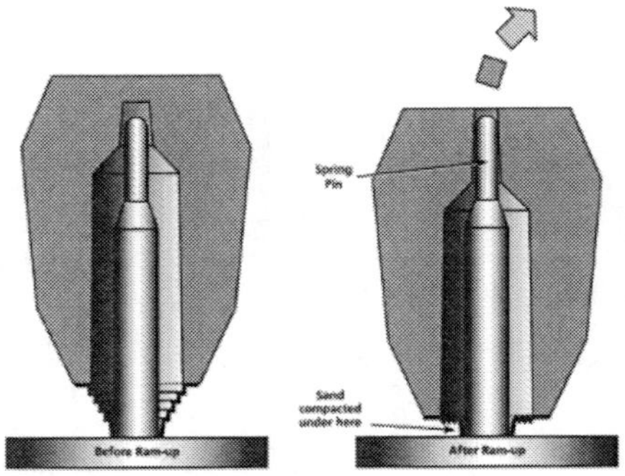

Fig. 3 Sleeve and metal breaker core before and after compaction

This feeder sleeve product line proved to be a robust solution for many foundries.

A further development, which started in 2015, adopted the advantages of spot feeders with collapsible metal breaker cores such as consistent knock-off areas, constant feeder volume and excellent moulding results. The new design allowed the reduction of the breaker core metal wall section whereas these were still suitable for the use in high pressure moulding applications.

The design changed from a stepped to a grooved design (Fig. 4).

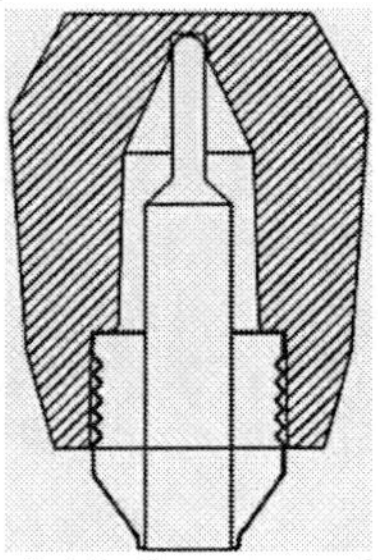

Fig. 4 Principle sketch of new generation of collapsible metal breaker core together with spot feeder

The new sleeve concept allowed the combination of metal breaker core with feeder sleeve without the help of hot glue in a press fit operation. The breaker core itself rests on a small socket in the opening of the feeder sleeve. During moulding the grooved part of the metal core is compacted and the main feeder sleeve body is moving down towards the pattern plate, compacting the mould sand between sleeve and pattern plate (Fig. 5).

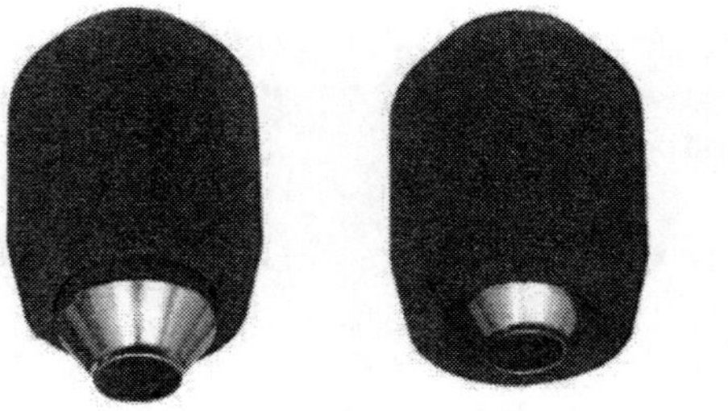

Fig. 5 New sleeve concept before and after compaction – typically 8 to 15mm distance between sleeve main body and casting

After full compaction the major part of the metal breaker core is super-heated by the high exothermic sleeve material. As a result, the metal neck surface in direct contact to the mould sand is reduced by 50% compared to the prior stepped design (Fig. 6).

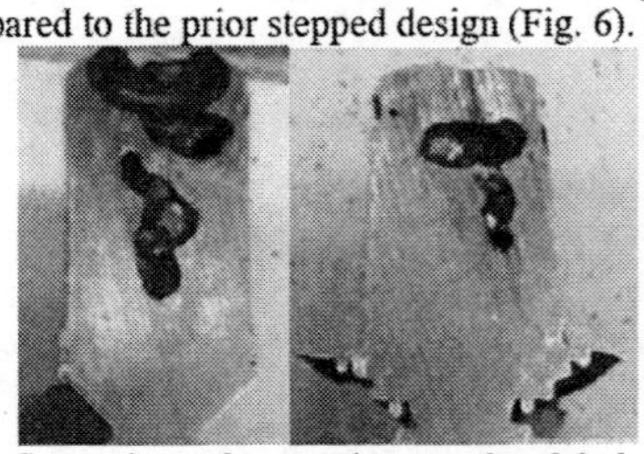

Fig.6 Comparison of new and conventional design after application

Several foundry trials proved an equivalent or better feeder neck pass-through performance compared to the existing concept with similar knock-off results (Fig 7).

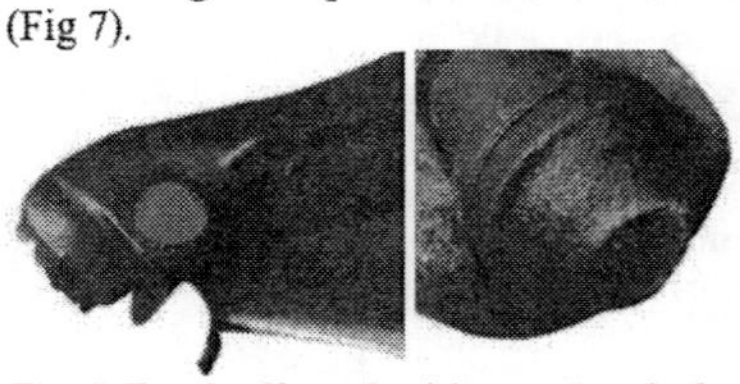

Fig. 7 Knock off result with new spot feeder concept – sound casting with virtually no further fettling work required

Conclusion

The concept of collapsible metal breaker cores provides several key advantages such as consistent moulding results even with highest moulding pressures as well as reduced footprint and neck contact areas. Further developments resulted into an improved design providing excellent feeder neck pass-through characteristics.

References

[1] Scherer, M., German Patent No. 3418137, October 02nd 1986

[2]. Powell, C., Sällström, J., Pehrsson, J. E., European Patent No. 1567294, May 5th 2006

The 73rd World Foundry Congress
„Creative Foundry"
23 - 27 September 2018, Krakow, Poland

AUTOMATED INTELLIGENT COATING CONCEPT FOR FERROUS FOUNDRIES

Christoph Genzler, European Product Manager Coatings

Foseco Europe, Binnenhavenstr. 20 P.O. Box 8. 7550 AA Hengelo (O), The Netherlands
e-mail: Christoph.Genzler@foseco.com

Key words: consistency, layer-thickness, intelligence, coating, density.

Increasing requirements from foundry applications such as thinner wall sections, complex internal geometries and new alloys, is driving the demand for high-performance coatings. However, high performance coatings need to be applied consistently to achieve optimum results on a repeatable basis. Poor application control can undermine coating performance and can in some instances lead to excessive scrap or rectification work.

Controlling coating density, and ultimately wet layer thickness, enables the foundry to optimize coating application. Wet layer thickness can be adjusted to suit the demands of specific applications.

Consequently, a number of benefits are achieved

→ consistent coating layer thickness
→ improved coating application efficiency
→ reduction of coating layer related defects
→ reduced casting scrap
→ fewer scrap moulds / cores
→ optimized drying
→ improved productivity
→ reduced coating dumping costs
→ improved working environment
→ lower casting manufacturing costs
→ improved foundry profitability

Foseco has further developed and optimized the concept of using online measured coating density as the ultimate measure to control, adjust and prepare coating in use at foundries.

This new fully automated and almost maintenance free "Intelligent Coating Unit" still controls coating density in real-time delivering even greater consistency in wet layer thickness applied and is now even more simple to use.

The new intelligent concept offers new options of even further integrating the measurement technique in existing coating related equipment or stand alone solutions.

Thus, coating performance can be further optimized to deliver a range of benefits to the foundryman including reduced coating layer related defects and subsequent casting scrap; lower core scrap; reduced coating consumption and waste disposal, improved productivity by shortest possible drying times and an enhanced working environment.

In addition, the continuous measurement system provides excellent data for improved process control, calibration and auditing. The equipment employs very little moving parts and is therefore the most easy and robust coating automation system in the industry. The ICU measures now density more accurate than any other equipment – up to the 4th digit.

The 73rd World Foundry Congress
„Creative Foundry"
23 - 27 September 2018, Krakow, Poland

INVESTIGATION AND CHARACTERIZATION OF INCLUSIONS IN ALUMINIUM CAST ALLOYS FOR AUTOMOTIVE INDUSTRY

Onur Ozaydin [1]*, Tugce Yagci [2], Alper Kaya[1], Osman Culha[2], Elvan Armakan[1]

[1]Cevher Wheels, R&D Department, İzmir, Turkey,
[2] Manisa Celal Bayar University, Faculty of Engineering, Department of Metallurgical and Material Engineering, Manisa, Turkey,
*Corresponding address: e-mail: oozaydin@cevherwheels.com

Key words: Aluminium Alloys, Inclusion, Microstructural Analyses, Scanning electron microscopy, Elemental Mapping.

1. Introduction

The concepts of sustainability and energy saving are very important for the future of the world. Ecological concerns have led to the manufacturers to research and develop new materials and mass production processes. In automotive industry, one way of reducing the strictly regulated CO_2 emission values is the reduction of the fuel consumption and this can be achieved by increasing the amount of light alloy components in a vehicle. For this reason, aluminum alloys are widely used in the automotive industry.

2. Materials

Compared to other light alloys, aluminum alloys are at the forefront in terms of high specific strength (strength/density) and visual quality. Aluminium cast alloys are the preferred material for critical automotive components such as cylinder heads, support brackets and wheels.

Wheels are not only critical components for safety, but they also visually enhance the vehicle. Since all Original Equipment Manufacturer (OEM) wheels must satisfy certain mechanical and metallurgical requirements, visual characteristics can have a distinct impact on the customer. In the last decades, market share of re-machined wheels (also named as 'diamond cut wheels') is growing dramatically in market. Figure 1 shows diamond cut wheel.

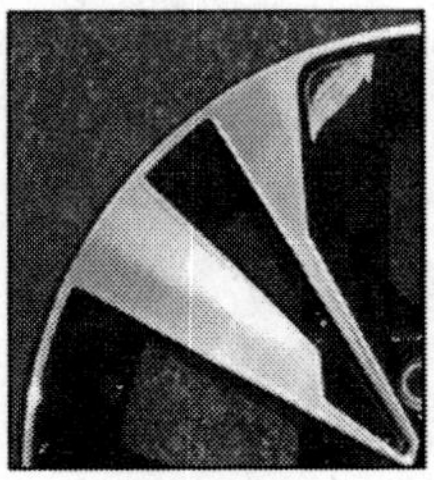

Fig. 1. Diamond cut wheel.

However, a challenge is inherent in re-machined wheels, the machined surface must have a perfect appearance, even a very small defect may be deemed to be unacceptable by current automotive standards.

3. Failure Modes

One of the most encountered failure modes on diamond cut surfaces is the inclusions in aluminum alloys. Inclusions are solid particle embedded in liquid aluminium. Inclusions display different morphologies and chemical compositions. The sources of the inclusions are usually non-metallic and pigments, refractory material particles, ambient conditions and additional alloying elements. They also affect the mechanical properties and machinability of the aluminium alloy parts. The inclusions must be removed to obtain a cleaner melt and satisfy required product quality.

4. Experimental Approaches and Discussion

In this study, inclusions and their root causes have been investigated in detail.

Firstly, a literature research is conducted to determine theoretical root causes. [1-4] It should be noted that, the identification of a root cause is not always a straightforward process. Since there are multiple mechanisms for inclusion formation, occasionally the emergence of inclusions is a result of synergetic conditions.
Secondly, the effects of these root causes are evaluated step by step. Thus, the types of inclusions which may affect the surface quality directly are determined. To identify the effect of inclusions, different samples are taken from melting and holding furnaces at different steps during the wheel production process. Samples taken from molten aluminum alloy before and after degassing were also examined to decide the effects of degassing on the inclusion formation. Figure 2 shows cast samples taken from different steps of the production process aluminium alloy wheels.

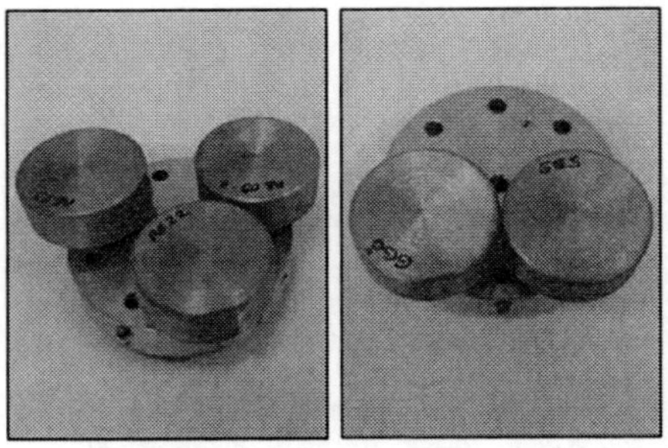

Fig. 2. Cast samples taken from different steps of the production process aluminium alloy wheels.

Thirdly, these inclusions are analyzed morphologically and chemically (as weight %) with scanning electron microscopy (SEM) technique with energy dispersive spectroscopy (EDS). Elemental mapping technique was used for the determination of the spatial distribution of the elements in the region containing inclusions. By means of these microstructure examinations, the structures of inclusions are clarified and linked with potential root causes.

5. Conclusion

Finally, the impurities, oxides and related compounds in the structure are examined the interactions between the impurities and root causes are examined in detail. Figure 3 and Figure 4 show the inclusions with different chemical compositions on the surfaces of the diamond cut wheels.

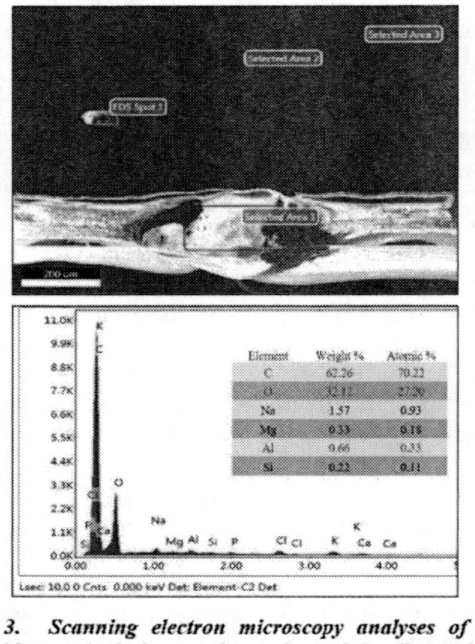

Fig. 3. Scanning electron microscopy analyses of the carbide type inclusion on the surface of the diamond cut wheel.

According to the results of the compositional analyses, the inclusion is containing some oxides and carbides. There is no aluminium alloy in this region. The carbide inclusion affects the mechanical properties negatively and reduces the fatigue life of the wheels significantly.

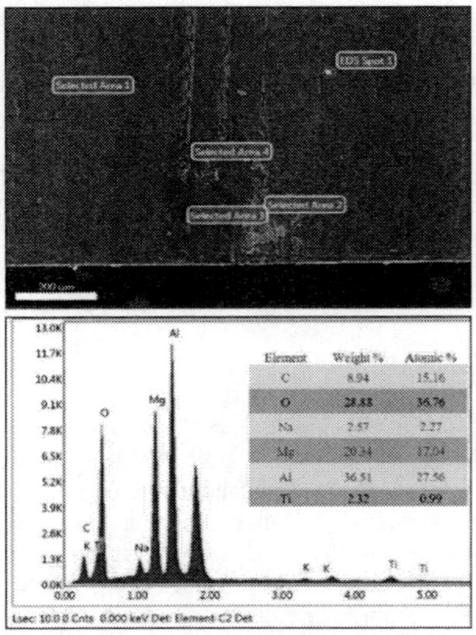

Fig. 4. Scanning electron microscopy analyses of the complex oxide type inclusion on the surface of a diamond cut wheel.

Figure 4 shows the microstructural image and chemical composition of a complex oxide type inclusion. This type of inclusions containing aluminium magnesium oxides (solid solutions of $MgAl_2O_4$ and Al_2O_3) are named as 'spinel' in the literature [3]. As shown in Figure 4, they have an irregular shape and when present in the product, their surface acts as a preferential site for crack nucleation and propagation. Therefore, precautions must be taken during the production process of the aluminium alloy wheels.

Acknowledgments

The authors would thank to all colleagues from Cevher Alloy Wheels Company, especially A. Oktem, and Manisa Celal Bayar University for collaboration and academic support.

References

[1] J. Hirsch, B. Skrotzki, G. Gottstein. (2002). Aluminium Alloys. Germany: Wiley.

[2] D. Doutre, B. Gariepy, J.P. Martin, G. Dube, (1985). Aluminum Cleanliness Monitoring: Methods and Applications in Process Development and Quality Control. Light Metals. (pp 1 1 79-1196) Wiley Online Library.

[3] B. Prillhofer, H. Antrekowitsch, H. Böttcher, P. Enright. (2008). Nonmetallic inclusions in the secondary aluminium industry for the productşon of aerospace alloys. Light metals. Wiley Online Library.

[4] R. Guthrie, D.A Doutre. (1985) "On-Line Measurements of Inclusions in Liquid Metals, " Refining and Alloying of Liquid, Aluminum and Ferro Alloys. (pp 145-164)

The 73rd World Foundry Congress
„Creative Foundry"
23 - 27 September 2018, Krakow, Poland

LOCAL COMPOSITE REINFORCEMENT TYPE TiC/FeCr FABRICATED IN SITU IN BLADE CASTING

Ewa Olejnik[1,2]*, Łukasz Szymański [1,2], Sebastian Sobula[1], Tomasz Tokarski[3], G. Cios[3], Paweł Kurtyka[4], P. Czapla[1,5]

[1] *AGH University of Science and Technology, Faculty of Foundry Engineering, 23 Reymonta St., 30-059 Krakow, Poland*
[2] *Innerco Ltd., 43A Jadwigi Majówny St., 30-298 Krakow, Poland*
[3] *AGH University of Science and Technology, Academic Center for Materials and Nanotechnology, 30 Mickiewicza A., 30-059 Krakow, Poland*
[4] *UP Pedagogical University of Krakow, Faculty of Mathematics, Physics and Technical Science, 2 Podchorążych St., 30-084 Krakow, Poland*
[5] *EUROCAST Ltd., 19 Grabowa St., 30-227 Krakow, Poland*
**Corresponding address: e-mail: ewa.olejnik@innerco.com.pl*

Key words: Metal Matrix Composite, in situ, TiC, wear resistance, casting

1. Introduction

Abrasive wear parts of machines and devices are one of the most popular problems which can be observed in the very field of industry. Ferrous alloys of an increased hardness constitute the largest group of materials used for the production of structural elements resistant to wear abrasive. Nowadays when the working parameters still increasing their properties are insufficient. Metal matrix composites based on Fe alloys reinforced with ceramic particles are characterized by good mechanical properties and high wear resistance. The most popular ceramic phase used for casting reinforcements is TiC. Titanium carbide is characterized by a high hardness, melting temperature, thermodynamic stability and a relatively good wettability in Fe alloys [1]. For producing such materials the methods allowing to obtain laminar materials with the application of such techniques as laser plating, thermal spraying, hardfacing and welding (TIG – tungsten inert gas) [2]. The mentioned methods of obtaining composite reinforcement have several drawbacks, such as: obtaining heterogeneous structures, low process accuracy, a possibility of the occurrence of porosities and inclusions as well as significant limits in designing elements of a complex geometry. An interesting solution, which however applies only rarely, is the application of casting technologies, where an agent initiating the synthesis is a liquid casting alloy – SHSB (Self- propagating High Temperature Synthesis in Bath). SHSB is the modification of the Self- propagating High-Temperature Synthesis (SHS) reaction applies in the powder metallurgy. SHSB is a very complex process, which is strongly limited by reactive infiltration phenomenon. Merzhanov determined the enthalpy of TiC formation (amounting -187 kJ/mol) resulting in intensive energy production in the form of heat within the area of the composited zone manufactured in situ [3]. This phenomenon is unfavorable due to local temperature increases, resulting in composite zone fragmentation. One of the latest methods of confining composite zone fragmentation is the application of a moderator addition, significantly decreasing the heat effect within the area of TiC synthesis reaction [4].
The objective of this research was the manufacturing of a blade reinforced locally with the composite zone containing TiC reinforcing phase and comparison with commercially available castings (blade reinforced by the insert of sintered tungsten carbide).

2. Experiment

The mixture of substrates necessary for TiC in situ formation was prepared with the addition of moderator. The chemical composition of compacts and moderator used for the synthesis reaction of local composite reinforcements is shown in Table 1.

Table 1 Chemical composition: the compacts used for synthesis reaction in situ composite reinforcement and the moderator – cast chromium iron

Lp.	Mass fraction of substrates of TiC formation [wt. %]			Mass fraction of moderator [wt.%]		
1	60			40		
Chemical composition of moderator [% wt.]						
Fe	C	Si	Mn	Cr	Mo	Ni
78.7	2.7	0.4	0.6	14.5	2.3	0.3

The powder of substrates of TiC and moderator were subjected to mixing for 6 hours. Next the so-prepared weight was cold pressed in the the uniaxail hydrualic press under of 500 MPa. The components produced were given a shape corresponding to the mold cavity. The obtained compacts were placed into the mold cavity according to the diagram presented in Figure 1.

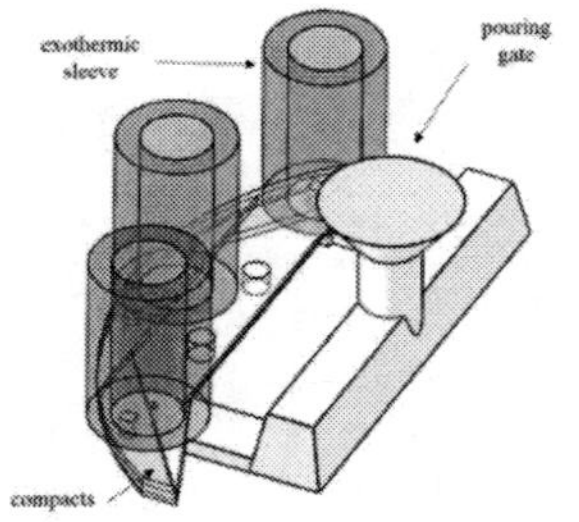

Fig. 1 Diagram of the mold cavity with compacts mounted for the synthesis reaction of composites zone

The mold cavity was poured with cast chromium iron at the temperature of 1673 K; chemical composition of which is given in Table 2.

Table 2 Chemical composition of base alloy

Chemical composition, [% at.]						
Fe	C	Si	Mn	Cr	Mo	Ni
78.7	2.7	0.4	0.6	14.5	2.3	0.3

The final castings were cut in order to take sample from the areas of manufactures composite zones and base alloys. The samples taken from castings were analyzed considering their structure, microstructure, mechanical properties and wear resistance.

3. Results and discussion

Figure 2 demonstrates the macrostructure of casting with the local composites reinforcements (LCR) type TiC-FeCr.

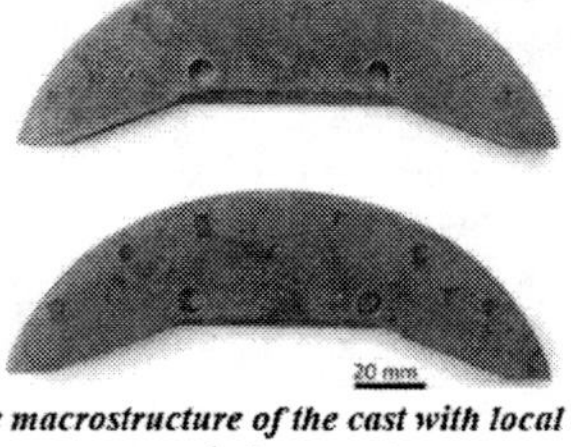

Fig. 2 The macrostructure of the cast with local composite reinforcement

The microstructure of the transition area between the local composites reinforcement and the cast iron core of the casting is presented in Figure 3. The in situ fabricated phase reinforcement – TiC is well bond with the metallic matrix, the transition area is remains free form impurities. This effect is attributed to the infiltration of chain containing the TiC substrates throught liquid metal. The microstructure of a composite zone is characterized by clusters of carbides are observed in coagulated forms.

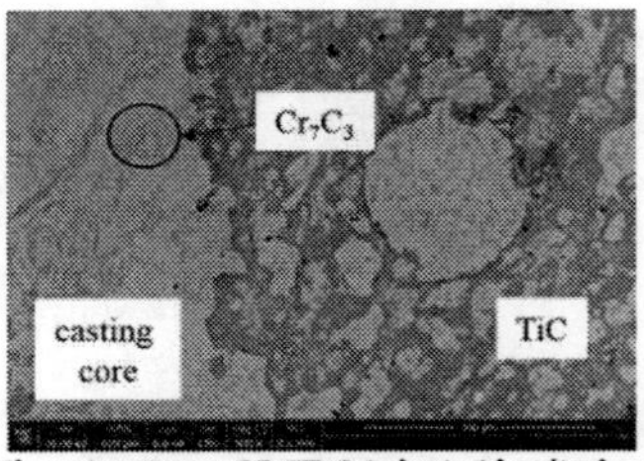

Fig. 3 Microstructure of LCR fabricated in situ in cast iron

Figure 4 shows the graph illustrating hardness in the area of the base alloy and composite reinforcement. It can be observed that there is a distinct increase in hardness in the composite zone containing particles of phase of TiC in comparison to base alloy ranging 500 units.

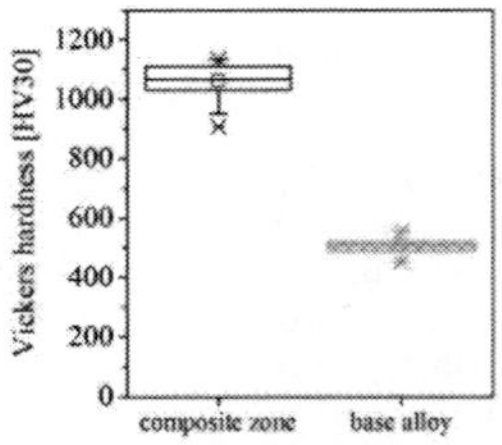

Fig. 4 Hardness Vickers HV30 LCR and base alloy

Acknowledgments

This study was carried out within the scope of "Development of technology for the production ultra-wear resistant castings with composite zones type TiC/Fe base alloys" project No. POIR 04.01.04-00-0067/15. The project financially supported by European Funds Smart Growth, Operational Programme Smart Growth.

Conslusions

Local composite reinforcements (LCR) in steel casting were obtained the in situ reaction of TiC. The carried out mechanical investigation revealed the growth by several hundred times in hardness of the LCR compared to base alloy.

References

[1] K. Feng, Y. Yang, B. Shen & L. Guo. (2005). In situ synthesis of TiC/Fe composites by reaction casting. Materials & design. 26, 37-40.

[2] L.M. Berger. (2015). Application of hard metals as thermal spray coating. Int. J. Refract. Met. Hard Mater. 49, 350-364.

[3] A.G. Merzhanov. (1996). Combustion process that synthesize materials. J. of Mat. Proc. Tech. 56, 222-241.

[4] E. Olejnik, A. Jesiołowska. No PCT/IB2016/056825. Powder composition for the manufacture of casting inserts, casting insert and method of obtaining local composite zones in castings.

The 73rd World Foundry Congress
„Creative Foundry"
23 - 27 September 2018, Krakow, Poland

NEW MATERIALS IN MOULD MAKING ARE REQUIRED TO MEET THE NEW CHALLENGES IN DIE CASTING

Andreas Endemann[1], Thomas Hoehn[1]

[1] Weldstone GmbH, Kunstmühlstrasse 12, 832026 Rosenheim, Germany
*Corresponding address: e-mail: hoehn@weldstone.com

Key words: Anviloy®, Conformal Cooling, 3D-design, thermal conductivity, die making

Modern light metal die casting is facing increasing challenges regarding efficiency, costs and quality of the casts.

Higher cooling rates reduce cycle times which leads to reduced production costs and increased capacities.

Faster cooling also reduces dendrite arm spacing with positive effects regarding mechanical properties and smaller wall thicknesses of the casts.

The solution is Anviloy® Products for Die Casting

Anviloy® Tungsten Heavy Alloy products replace or add on to tool steel solutions. They demonstrate outstanding properties of tungsten combined with some other unique effects.

Anviloy® 1150 and Anviloy® 1350 are characterized by:

- Very high melting point
- High thermal shock resistance
- Good chemical resistance
- Remarkable mechanical strengths and hardness especially under high temperatures
- Excellent thermal conductivity (4-times higher than tool steel)
- Almost no sticking
- No heat treatment necessary

Typical applications:

- Gate components
- Distributors
- core pins
- sprue bushes
- sprue posts
- runners
- cooling inserts
- wear resistance inserts

Additionally, Anviloy® Weld Rod and Wire, a TIG-welding filler metal, is available to repair or armour both tool steel and Anviloy® Tungsten Alloys.

Latest developments:

Anviloy® C – super hard coated Anviloy® products
Anviloy® 3D – additive manufacturing design Anviloy® products

1. Anviloy® Products for Die Casting

Anviloy® Products are made of tungsten based heavy alloys. They are used in die casting processes to increase the performance of moulds in regards to thermal conductivity and wear resistance. To replace the commonly used tool steel in die casting moulds, Anviloy® Products can be used as inserts or core-pins.

Recently more and more die casting moulds are equipped with Anviloy® Products in areas such as

gates, distributors, core pins, sprue bushes, sprue posts, runners and cooling or wear resistance inserts. In addition, much success has been seen by manufacturing the entire die casting mould completely out of Anviloy®.

Additionally, Anviloy® Weld Rod, a TIG-welding filler metal, is available to repair or armour both tool steel and Anviloy® Tungsten Alloys.

Anviloy® Tungsten Heavy Alloy products demonstrate outstanding properties of tungsten combined with some other unique effects.

<u>Anviloy® 1150 and Anviloy® 1350 are characterized by:</u>

- Very high melting point
- High thermal shock resistance
- Good chemical resistance
- Remarkable mechanical strengths and hardness especially under high temperatures
- Excellent thermal conductivity
- Almost no sticking

These properties perfectly meet the requirements for mould materials in the die casting process. The service life in various casting applications is several times higher than it would be with hot work tool steels such as H13 / 1.2343. In many applications, we have seen the life of the Anviloy® components last 25 times longer than the traditional hot work tool steels. Consequently, contamination of the aluminium with iron is almost completely eliminated. Similarly, due to the high thermal conductivity and low thermal expansion, heat checks can be prevented. Anviloy® 1150 has a very low tendency to stick, or solder, to the cast product. This eliminates the need for coating and thus the hydrogen formation due to the coating.

One of the biggest advantages is the up to 4 times higher thermal conductivity compared to hot work tool steel. This feature not only enables shorter cycle times but also improves the finished product's material properties and drastically reduces the formation of bubbles. This allows reducing wall thicknesses and material savings.

Only with higher casting temperatures, which usually are beyond the permitted temperature range of hot work tool steels, a complete mould filling can be achieved. Anviloy is able to withstand this loading. Its advantages even increase with increasing temperature.

In contrast to hot work tool steels (H13), Anviloy® 1150 does not require any heat treatment prior to use. This then add another level of cost savings during the die manufacturing process.

These are the reasons why Anviloy® 1150 and Anviloy® 1350 are widely used for core pins, spreaders, gating areas and inlets and entire dies.

The 73rd World Foundry Congress
„Creative Foundry"
23 - 27 September 2018, Krakow, Poland

THE SHIPBUILDING INDUSTRY IN THE POLISH ECONOMY AS AN IMPORTANT ELEMENT OF THE DEVELOPMENT OF A MODERN MACHINERY INDUSTRY, INCLUDING FOUNDRY

Andrzej Strzeboński

President of Szczecińska Stocznia Sp. z o.o.
Antosiewicza 1, 71-642 Szczecin, astrzebonski@stocznia-szczecinska.pl

Key words: Szczecin, Shipyard, Polish, machinery, Shipbuilding

1. History of Stocznia Szczecińska

The Stocznia Szczecińska [*Szczecin Shipyard*] has been operating for over 70 years in various legal forms. Structural changes in the shipyard have been related to the need to adapt the it to the changing external conditions in the area of shipbuilding. However, the basic task of "Szczecin Shipyard" has always been to build ships.

In 1992-2002, the Szczecin Shipyard Porta Holding S.A. was one of the largest shipyards in the world, the top shipyard in Europe. In the decade 1992-2002, the shipyard would design, build and sell, at **a profit,** about 160 vessels worth over 4 billion USD. In addition, it introduced into production 15 series of prototypes (container ships, tankers, Ro-Ro vessels, products tankers, bulk cargo carriers, paper carriers and other), including very modern chemical tankers.

The shipyard would cooperate with many companies in our country in the field of manufacturing subassemblies and parts necessary to build the final product, which was a ship. In Szczecin, over 11,000 employees worked for the needs of the shipyard, while on the national scale, in the implementation of semi-finished products, in the peak period, several hundred thousand people were involved. For example, at the Engine Makers Zakłady Cegielski in Poznań, 85% of the value of production was marine engines for the Polish shipyards. In 2009, "Szczecin Shipyard" was put into liquidation in accordance with the EC recommendations. Such plants as the Engine Makers Zakłady Cegielski in Poznań were forced to liquidate plants producing semi-finished products for Szczecin and disposed of highly specialised engineering staff.

2. Revitalization of the Szczecin Shipyard as an opportunity for the Polish machinery industry

Resumption of shipbuilding in Szczecin took place after 2015.

In 2016, the Szczecińskie Konsorcjum Okrętowe (SKO) [*Szczecin Shipbuilding Consortium*] was established, whose main mission is the construction of fully equipped ships in Szczecin. The Consortium incorporates companies from Poland, including STOP - a co-organizer of the 73th World Foundry Congress, and companies from abroad. Last year, on the basis of the remaining shipbuilding property, the Stocznia Szczecińska sp. z o.o. was founded.

At present, in the field of shipbuilding, Polish shipyards deal mainly with building specialised ships, which are technically advanced and innovative and tailor-made. In the premises of the Szczecin Shipyard, fishing vessels, fish farm service ships, ships for specialised technical works, passenger and car ferries, special ferries for local traffic service including innovative electric powered ships, tourist ships, mega yachts, small multi-purpose freighters, also those for mixed sea and inland shipping, have been built. The recipients of the vessels built are shipping companies, fishing ship owners from Scandinavian countries, local ferry and cargo carriers.

3. Materials and new technologies in shipbuilding

Materials used in the construction of ships are, firstly, rolled steel in the form of sheets and profiles. Also stainless steels, aluminium alloys, copper alloys and cast parts from iron and non-ferrous alloys are applied. The weight of cast steel castings for the hulls used on a single ship depends on the design adopted and is usually from a few to a dozen or so tonnes / ship. The share of castings in items of equipment for ships is definitely higher - some of these castings in the form of castings of on-board equipment such as: hawse pipes, chocks, anchors are ordered by shipbuilders as components ready-made for installation, whereas the prevalent majority of castings is supplied as component parts of machinery and equipment such as for example parts of: windlasses and mooring winches; casings of pumps and separators; elements of hydraulic cylinders; steering gears; internal combustion engines; turbines; etc. In the case of marine equipment the share of cast components is significant and depends upon the actual type of a ship.

Recently, composite materials have been used in shipbuilding. The paper [1] discusses the layered composites used in the construction of hulls of sea-going vessels, in which between the steel plates there is a rigid core, made of properly selected material. The hulls of such vessels are lighter and fireproof.

At the Instytut Technologii Bezpieczeństwa Moratex w Łodzi [*Institute of Security Technology Moratex in Łódź*] [2], composites have been developed that can replace armoured steel and increase the protection of seafarers on warships.

In the design and construction of ships digital modelling techniques are used [1]. Computer technologies enable the construction of a virtual marine system both in the design, construction, crew training and the ship's operation. The French industrial group Naval Group [4] has been conducting intensive work in the field of virtual reality. The virtual model of a ship can be used in the design and construction processes by cooperating teams involved in the design and physical production of materials, subassemblies and assembly of the ship. We anticipate a renewal of cooperation with companies in the machinery industry in regard to the supply of subassemblies and parts manufactured in the Polish foundries.

References

[1]. Owczarzak W.: Przegląd konstrukcji warstwowych; Autobusy 6/2016; Politechnika Gdańska

[2]. htps://tech.wp.pl/polski-wynalazek-dla-okrętow-wojennych-kompozyty-lepsze-niz-stal-pancerna-6034822403490433a

[3]. Górny Z., Nawarecka S., Połcik H.: *Simulation of the Solidification Process of Composite Castings made by Gaseous Counter-Pressure Process* - Part I. International Committee of Foundry Technical Associations. Commission 8.1. Proc. of II Cast Composites, Conference'98, Polanica-Zdrój, 1998.

[4]. Dura M. : Wirtualna rzeczywistość a budowa i użytkowanie okrętów, Defence 24.

TECHNOLOGY

The 73rd World Foundry Congress
„Creative Foundry"
23 - 27 September 2018, Krakow, Poland

NEW INDUCTION WIRELESS MANUFACTURING EFFICIENT PROCESS FOR ENERGY INTENSIVE INDUSTRIES (NIWE)

Meléndez A.*, Obregón A.

Tecnalia Research & Innovation, C/Geldo. Edificio 700. E-48160. Derio – Bizkaia, Spain
** antton.melendez@tecnalia.com*

Key words: Advanced manufacturing, Foundry, Efficiency, Flexibility, Aluminium, Cast iron, Steel

The increasing global concurrence and episodes like the last financial crisis have leaded manufacturers in general, and foundries in particular, to increase as much as possible the flexibility in their processes due to the high variability of orders.

In other words, manufacturers are being forced to start and stop their production chains. Changes on their production programs with undesirable energy penalties could put companies in risk of collapse or, as minimum, suffering a strong decrease of the cost efficiency of their processes.

Trying to tackle this, NIWE project was born to develop a different production process able to reduce the embodied energy on the entire process of foundry products by over 25%, reducing drastically its carbon footprint. The demonstration has been performed in aluminium, iron and steel processes.

According to project targets the idea to be explained in the future paper is summarized in next picture (Fig. 1).

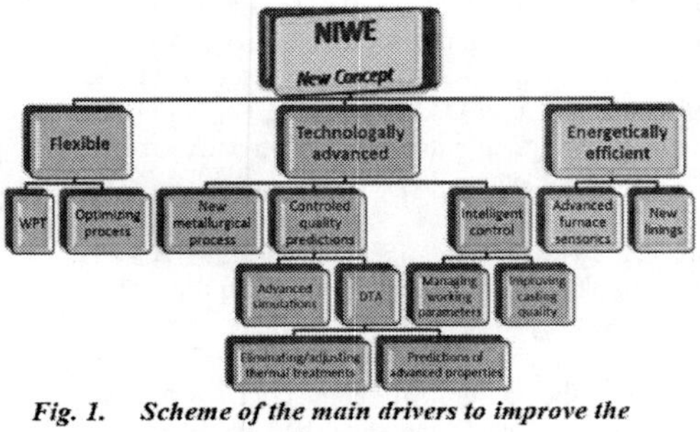

Fig. 1. Scheme of the main drivers to improve the foundry process

The energy efficiency gains were focused along the entire process mainly by acting over a previous deep revision and implementing some innovative equipment. More in detail, a new furnace that will provide wirelessly and fast the necessary power to melt, by means of an inductive coupling as well as a strong metallurgical control of the process which, by treating the metal in the liquid state, will avoid thermal treatments in solid state later.

The power is applied to the heating system depending on the foundry material. So, resistances for melting aluminium or induction for cast iron or steel were used initially on lab scale and later on industrial demonstrators.

Wireless system has been a key part into the technical solution because it is able to make possible highly flexible production through a quick modification of the factory layout based on easy interchangeable positions of furnaces, depending on the demand of pouring lines. NIWE process is focused to reinforce the efficiency of four of the steps of productive processes in foundries: melting, pouring, scrap generation (internal scraps like rejected parts or filling/feeding runners) and, very important, later thermal treatments. These strengths are closely related to which can be seen in Fig. 2:

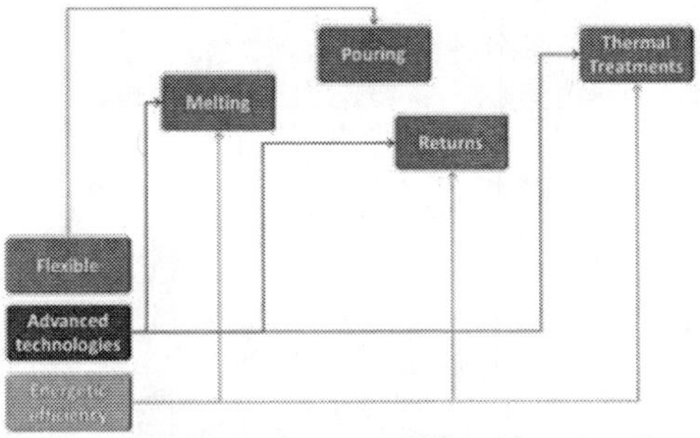

Fig. 2. Relationship between NIWE process strengths and productive steps in foundries

Flexibility

The main challenge for the NIWE proposal was to work with a furnace electrically fed by means of a Wireless System. This singularity should provide the foundries enough flexibility:

- by making easy to change the furnace position or to interchange positions, if possible, among different types and sizes of furnaces.
- by providing the possibility of change furnaces operations (melting ↔ holding) depending on needs, places or production programs in plants.

This flexibility should make achievable speeding up daily activity and modifying easily plants' layout, in order to accomplish narrowly the customers' orders. In case of aluminium HPDC machines the solution required using two interchangeable furnaces, melting and holding, by alternating them when required (one furnace is melting whilst the other one feeds the die machine). In this way the HPDC machine only stops during furnaces' interchanges.

Other important reason to do that could be the necessity of changing alloys.
In cases of iron or steel furnaces, interchanges or changes have been revealed as more difficult.
Once detected the necessity, by working with mobile furnaces, NIWE system provides the possibility to manufacture different sizes of series and so, a production near to *just in time* is possible.

Advanced technologies
NIWE project provides technical advances derivate from a deep study of foundry process, from raw materials till final products.
To obtain this goal it has to be implemented the Wireless Power Transfer (WPT, **Błąd! Nie można odnaleźć źródła odwołania.** and **Błąd! Nie można odnaleźć źródła odwołania.**), a highly disruptive equipment in comparison with the rest of them as well as crucial materials involved (innovative power management solutions by means of smart controls, new refractories and so on…).

Fig. 3. WPT system for aluminium (industrial demo)

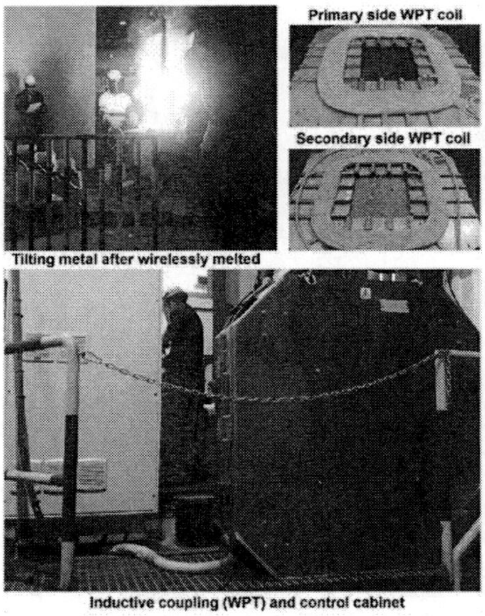

Fig. 4. WPT system for cast iron and steel (scale lab demo)

Related to refractories, latest generation of them have been previously selected, carefully developed and tested in trials.

The third point concerns the use of smart controls, based on sensorics, to manage the energy consumption. As an example of reducing the required energy, iron and steel furnaces will have available a smart system named VIP Melt Manager. It will control the casting process by acting, in intelligent mode, together with the WPT system. The particular characteristics:

- Control of the complete melting process (kWh/t, in order to optimize power usage and improve operational safety.
- Scheduling furnace preheatings, the sintering processes and controlling the melting process to reduce manpower requirements and lengthen lining life.
- Smart detection (with self-diagnostic) of water cooling system failures.

Energetic efficiency
Despite around 5% of extra energy consumption due to the work of WPT system, the rest of benefits of NIWE balances as improves energy consumption and helps users to reach significant reductions. So, besides the new design, DTA techniques (Differential Thermal Analysis) resulted in other interesting, powerful and competitive advantage tool in cases of aluminium or iron. For steel, these techniques have no sense.
Results based on DTA techniques will promote saving thermal treatments in solid state by modifying and narrowly controlling the liquid state, with accurate predictions of casting quality. DTA techniques also permit a proper estimation of alloys properties, reducing lot of casting defects with important associated energy savings.
In cases of unavoidable thermal treatments (need of properties of parts improvement), DTA analyses will allow adjustments on temperature for them. In the same way, for iron DTA is going to help to predict some properties in situ, before other later analysis that can delay the whole process.
All these things mean saving important energy process costs (around 15%), shortening delivery times of parts, an increase of production capabilities and, obviously, providing competitive services to customers.
By having the possibility of connection the furnaces where the user needs, the metal waiting for pouring moulds as well as the corresponding energy to hold it in temperature should decrease drastically, a crucial point of contribution to energetic efficiency.
Finally, referring to previously mentioned refractories it is expected they will be able to reduce heat losses with sensible associated reduction of energy required for melting. In case of aluminium, in the external wall strong temperature reduction is achieved (30÷55%). For cast iron and steel, the spinel refractory was improved by adding an especial external layer, instead of conventional mica.

The 73rd World Foundry Congress
„Creative Foundry"
23 - 27 September 2018, Krakow, Poland

OPTIMIZATION AND CONTROL OF MODERN LADLE POURING PROCESS

Thomas Voss*

OTTO JUNKER GmbH, Jaegerhausstrasse 22, 52152 Simmerath, Germany
**Corresponding address: e-mail: vt@otto-junker.de*

Key words: pouring process, ladle pouring, process flow, pouring control, energy consumption

Abstract

The pouring process is the important key process in the foundry industry. The energy consumption and quality results are mainly defined by the equipment, process flow and pouring control.

It is mandatory to understand the complete process for any improvements and for the decision how to optimize the key process.

In the field of ladle lip pouring process, modern trends of ladle design, control technology and quality documentation will be presented and discussed.

Introduction

Basic types of pouring device

A classification can be made according to the pouring method, distinguishing between stopper-controlled pouring devices and systems emptied by moving the entire pouring ladle.

The first of the above categories can be further subdivided into pressurized and unpressurized stopper-controlled pouring devices. In the second category, metal is poured by tilting or rotating the pouring ladle.

A basic observation to be gained from the above is that with stopper-controlled pouring devices, the output of molten metal can be easily and reproducibly controlled. Moreover, the shape of the pouring stream is clearly defined and restricted by the nozzle diameter and stopper stroke. In addition, the stream of molten metal enters the sprue cup vertically. On ladle pouring systems, on the other hand, the pouring rate varies in a big range with the ladle tilting speed. As a result, the pouring automation requirements are much more complex in the case of ladle pouring devices.

Ladle pouring devices

Characterisation

In a ladle-pouring device, the pouring ladle is tilted over its pouring spout until molten metal is dispensed into the mould sprue cup. In this operation, the pouring spout is fixed in position just above the sprue cup and the ladle is rotated about this point. In selecting the pivot point, it is important to ensure that the pouring spout will not drift. This pouring process corresponds to manual pouring, although the mould is filled in a clearly more reproducible and accurate manner depending on the degree of automation.

The market for ladle pouring devices is split into semi-automatic and fully automatic types.

Fully automatic ladle pouring devices minimize operator intervention through the additional integration of various sensor systems. The main function of this sensor equipment is to monitor the melt level in the sprue cup and the ladle weight, thus providing information on the molten metal output and pouring weight.

Programming of the automatic pouring operation is achieved by diverse methods today. The more basic devices rely on a teach-in routine to provide controlled filling of the sprue cup. The end of the pouring operation is determined essentially via an integrated weighing scale. On advanced pouring devices, the melt level in the sprue cup is sensed by means of a camera or laser system. As the measured reading is processed by a control algorithm, melt output is varied to maintain the predefined level. Automation quality is indicated by the dynamic response of the sprue cup level control function and by the accuracy with which the pouring weight is determined. Process traceability is likewise ensured because all relevant process parameters – including the pouring weight, pouring time, inoculation time, inoculant quantity, pour temperature, ladle number and type of pouring process (automatic, manual, etc.) – are documented automatically on modern ladle pouring machines.

Design types and sizes

The temperature losses occurring in the unheated ladle, in conjunction with the acceptable melt temperature tolerances, impose limits on the dwell time and hence, on the maximum size of the ladle pouring system.

As regards ladle shape, circular cross-section types prevail, with rectangular geometries being used in only two cases. The latter configuration, also referred to as segmental design, is intended to ensure that the pouring rate changes proportionally with the tilting angle speed [1], as mentioned earlier. On pouring systems with a cylindrical ladle body this is achieved via the machine control system.

The melt outlet is frequently in the form of a siphon to ensure slag-free tapping.

The newly developed INDUGA system (Puma Pro) [2] relies above all on full automation of the entire pouring process plus sophisticated machine technology.

One innovation is the two-dimensional pouring stream position control feature. For monitoring the melt level in the sprue cup, the engineers could draw on the extensive experience gained previously with a camera-based pour control system on a pouring furnace. It was also important to optimize the weighing technology and, associated therewith, the accuracy with which the pouring weight is determined. Innovative systems delivering unsurpassed accuracy are used for this purpose.

Stopper-controlled pouring devices

Unpressurized stopper-controlled pouring devices

To explain the basic principle of this equipment type, let us consider the stopper-controlled pouring device with automatic metal dosing of Otto Junker design (see Fig. 1).

The device consists of a trough-shaped ladle, lined with a ceramic castable refractory.

Metal is poured under stopper control through a bottom nozzle in the pouring chamber. A nozzle cleaning system is provided to remove potential accretions.

Heat loss from the pouring device remains very low due to the good insulation and chosen design but cannot, on principle, be avoided altogether. As a result, the temperature of the molten iron will drop, calling for an end of the pouring process and a refill of the ladle after approx. 15 – 20 minutes.

Pressurized stopper-controlled pouring devices

This design differs only marginally from induction-heated pouring furnaces, being virtually identical thereto except for the heating. Our discussion shall therefore be limited to the differences here. A near-identical design means that the device can be easily retrofitted with an inductor without any major revamping work; at least this is possible with Otto Junker equipment. Other manufacturers [3] see the advantage in a favorable total-to-useful-capacity ratio and have therefore opted for a flatter design with the siphon openings merely terminating in a channel in the bottom. This design facilitates alloy changes due to the fairly small heel. In addition, the siphon is easier to clean. On the other hand, the subsequent installation of an inductor is not possible.

It should also be mentioned that the quick-change system for Otto Junker pouring furnaces mentioned in Part 1 can also be used for the unheated stopper-controlled pouring device.

Comparison of the different pouring devices

Ladle pouring devices are characterized by a rapid and easy alloy change and the fact that they provide high application versatility. On the downside, every ladle change takes time and some moulds will invariably go unfilled during this interval, which can be fully compensated by a second machine.

Further, it must be noted that an automatic pouring and melt dosing process based on a ladle pouring device does require a rather sophisticated sensor and control system. Without a doubt it is an advantage that the ladle, before being placed in the pouring device, can be subjected to metallurgical procedures such as inoculation or Mg treatment. Here the treatment ladle is also the pouring ladle. The temperature loss associated with an extra transfer of the molten iron is avoided, and it may be possible to rely on the ladle inoculation method instead of inoculation into the pouring stream.

With smaller to medium pouring weights and longer cycle times, ladle pouring devices provide a good alternative. On high-performance moulding lines set to very short cycle times, a solution may be achieved by installing two ladle pouring devices.

Another benefit is that a pouring device of this type may, in some applications, support the use of differently sized ladles.

The advantages of pressurized stopper-controlled pouring devices lie in the low metal temperature loss, the fact that simultaneous pouring and filling is supported, good automation capabilities and a longer metal dwell time. On the downside, each alloy change is time consuming. Pressurized stopper-controlled pouring devices are ideal for use on high-performance moulding lines with shorter cycles and medium to high pouring weights.

Unpressurized stopper-controlled pouring devices differ from their pressurized counterparts in that the melt level and hence, the metallostatic pressure inside the ladle drops with each pour. In order to ensure constant pouring parameters it is therefore necessary to determine the melt level in the sprue cup very accurately while making use of a sensitive and responsive pouring control system to compensate for pressure changes.

Conclusion

Pouring devices in their various basic designs and configurations constitute a good complement to pouring furnaces and are a good alternative in certain application environments. The development potential resides in improving the automation of the pouring cycle and in optimizing the overall production process, including online interfacing with the melting and moulding processes.

References

[1] Heinrich Wagner Sinto Maschinenfabrik GmbH. Automatic pouring units & pouring machines brochure.

[2] INDUGA GmbH. Ladle Pouring Technology Puma Pro brochure.

[3] Dötsch, E.: Induction melting and holding 2nd ed., Vulkan-Verlag, 2013

[4] ABP Induction Systems GmbH. Pressurized unheated pouring system Pouromat Type OCU brochure.

[5] Otto Junker GmbH. Brochure on unheated pouring devices.

The 73rd World Foundry Congress
„Creative Foundry"
23 - 27 September 2018, Krakow, Poland

THE GREEN SAND FOUNDRY OF TOMORROW

Per Larsen

DISA Industries A/S, Hoejager 8, DK-2630 Taastrup, Denmark,
e-mail: per.larsen@noricangroup.com

Key words: Green Sand, Moulding, Vertical, Sustainability, Industry 4.0

1. Introduction

Even though clay based moulds have been used since the 3rd millennium BC and the clay bond sand moulds as we know them today have been used industrially for many years, the clay bond or green sand moulding technology still shows great potential for further development for the next many years to come.

While previously, the focus in the development of the Green Sand Moulding Technology to some extend was on productivity, a variety of other dimensions are also explored for the further development of the technology. This said, productivity will also be a key driver in the future.

This presentation will describe many of the different dimensions being addressed for preparing the well-known Vertical Green Sand Moulding Technology for the Foundry of Tomorrow. Both hands-on approaches for the short term and measures with a longer horizon will be highlighted.

2. Sustainability

Strong focus is on Sustainability, this including, among other things, using less resources in the form of materials and energy and putting less stress on the manpower operating the foundry.

An example of improving the working conditions is the removal of highly repeated movements. For instance, at the setting of filters at un-cored moulds an operator had earlier to repeat the same movement up to 400 – 500 times per hours, putting high stress on the musculoskeletal system. With new inventions, the highly repeated movements are removed, and the task is reduced to an hourly refilling operation possible to do in combination with other responsibilities in the foundry.

Different possibilities exist for reducing the usage of materials and energy. Two of them are the Fast Pour System and the Double Index System. Both solutions can contribute to increasing the yield, i.e. the ratio between the weight of the castings in the mould and the total weight poured in the mould. An example from one foundry introducing the Fast Pour System shows an average saving of 1.4 kg liquid iron per mould. An example from another foundry introducing the Double Index System in combination with an optimized mould size shows and increase of yield of 25 %.

Another way of reducing the used resources is to secure high and consistent quality avoiding scrap and rework. This can for instance be done by checking the moulds for mismatch and mould gabs just before the pouring operation.

Looking for instance at mould gabs, which can lead to both out of tolerance castings and down time due to mould run out, the following can be achieved: By checking for mould gabs mould by mould just before pouring an alarm stopping the pouring of the actual mould can be set, if the measured value for the mould gab is out of tolerance. In this way both scrap/rework and downtime can be avoided, and hence loss of valuable resources is avoided.

3. Industry 4.0

A long list of possibilities both on the short term and on the long term un-folds when considering the 4.0 discussions with the eyes of a foundry man.

A theme, that becomes apparent at a first glance, is data availability. In most foundries today, data is not available from one central system. Separate systems often exist for sand preparation, melt preparation, moulding, pouring, scrap accounting etc. Some data may even not be in an electronic format but are recorded on paper manually. By getting access to all data from the casting process in one system opens for many new opportunities. As a starting point makes easy access to data monitoring and analyzing much simpler and hence strongly speed up troubleshooting in case of problems.

Apart from securing access to all available process data in one system, is expanding the available data to incorporate all process critical data essential for being able to do efficient process optimization and through that reduce the scrap to an absolute minimum. One critical area is the pouring operation. Many quality controlling parameters originates from the pouring operation, this both being more basic parameters like pouring time and temperature, but also more sophisticated parameters as time to fill the pouring cup and final melt level in the pouring cup. Having full access to the quality controlling parameters originating from the pouring operation and being able to correlate them with the rest of the process data in the foundry and with scrap data leads to a very strong tool for scrap analysis and reduction.

Being able to check the moulds for mismatch and mould gabs just before pouring leads to operational improvements in first hand as described above. Monitoring and recording the data also opens op for predicting problems before they happen by looking on trends of the data.

With the newly developed data handling and processing systems being developed presently, having access to data will open for many new opportunities for optimizing the foundry operation.

4. Productivity

As mentioned in the introduction is the focus in the development of the Green Sand Moulding Technology becoming more diverse and hence opens for more dimensions. This is by no means to abandon the focus on productivity. Improved productivity is still seen as a part of the solutions taken us into the future. Most of the examples mentioned above have also positive impact on the productivity.

Being able to set filters at un-cored moulds without the need for highly repeated manual movements opens for up to 70 moulds extra per hour.

Introducing the Fast Pour System or the Double Index System for reducing the poured weight per mould will also lead to possibilities of increasing the moulds per hour without harming quality control parameters like melt flow rate in the mould. Depending on the actual pattern plate, increases of 20 % in productivity is not uncommon.

Checking the moulds for mismatch and mould gab just before pouring the moulds does not directly increase the moulding rate. Early warnings of problems avoiding unplanned stops and reducing scrap/rework will although increase the total productivity of the foundry.

5. Conclusion

The opportunities described above and a variety of additional opportunities for reducing both the use of materials and energy, and the load on the man power operating the foundry including further improvements of the productivity, all sum up to a positive outlook for the future of the Green Sand Moulding Technology.

The 73rd World Foundry Congress
„Creative Foundry"
23 - 27 September 2018, Krakow, Poland

TEN YEARS OF INDUSTRIAL EXPERIENCE WITH LOW-EMISSION ADDITIVES FOR MOLDING SAND

Engelhardt, Thomas [1]

[1] *Clariant Produkte (Deutschland) GmbH, Ostenriederstr. 15, 85368 Moosburg, Germany*
thomas.engelhardt@clariant.com

Key words: industrial experience, reduction emission, BTEX, molding sand, low emission technology

In order to reduce emissions from molding sand systems, an innovative environmentally friendly foundry additive was introduced. Essential component of these additives are special natural graphites with which at least a part of the conventional lustrous carbon formers like sea coal and resins can be substituted. In the last ten years, the practicability of this system, named LE (low emission) technology, was proved in many foundries worldwide. In some foundries, the emission of volatile aromatics from the molding sand could be reduced by up to 85 %.

In addition to the reduction of the emission, graphite acts as a kind of lubricant within the molding sand. This leads to a more homogeneous compaction of the mold, related casting defects can be minimized.

1. Introduction

Strict rules of the Goteborg protocol stipulate that, from the year 2020 onwards, a drastic country-specific emissions reduction of volatile organic compounds has to be followed. Therefore, the foundry industry has to intensify its efforts to lower its emissions, especially those of volatile aromatics. Thereby, the supplier industry is responsible for offering a technology to the foundry industry with which emissions can be reduced without increasing production costs. This concerns emissions from the lustrous carbon formers of the molding sand system as well as the today commonly used organic cold box binders for cores. Therefore it was the goal of a three year development project to develop an emission-free respectively low-emission alternatives for the lustrous carbon formers like sea coal and resins being in use today in the molding sand.

2. Sources of emission in a grey iron foundry

In a discussion about emissions from a grey iron foundry, it is important to focus on the lustrous carbon formers in the molding sand system as well as on the cold box binders for the cores. During pouring of liquid iron, the lustrous carbon formers in the molding sand are pyrolyzed and form low-boiling aromatics like benzene, toluene, xylene and ethyl benzene (BTXE). At the same time, polycyclic aromatic hydrocarbons (PAH) which have with a high boiling point are formed. Therefore, BTXE aromatics are a topic of gaseous emissions, while PAHs are adsorbed by the molding sand systems. These PAHs can create problems during waste disposal.

The cold box core binders also form low-boiling aromatics as well as PAHs during pouring. In addition, certain amounts of phenol are formed.

3. Technologies to reduce emissions from molding sand

Some different technologies are available to reduce the emissions from molding sand.

One technical solution is the adsorption of the aromatics in an external adsorption unit. Such external adsorption unit offers good efficiency, but additional investments are necessary. This leads to higher total process costs.

Other concepts follow the idea to adsorb the emissions in the molding sand system by adding special adsorber materials like zeolite to the molding sand. Such adsorber materials can offer also good efficiency, but transfer the emission problem from the gas phase to the waste disposal.

4. Low-emission (LE) technology

The most efficient way to reduce the level of emissions from a molding sand system is preventing the formation of toxic aromatics during pouring. The LE technology follows this concept. The challenge of such concept is to find a molding sand additive, which can overtake the function of classic lustrous carbon formers without formation of aromatics during pouring.

Graphite is a natural inorganic carbon-based material, which contains between 70 and 95 % pure carbon. This natural material is available with different morphologies and particle size distributions. All graphite grades are completely emission-free.

A special type of such natural graphite is the key component of the LE technology. This special type of graphite forms – like bentonite – very thin layers that can be separated from each other during the reprocessing of molding sand. By applying shear energy to the graphite agglomerates, the graphite can delaminate in very thin single sheets and can surround the silica grains like sea coal with a thin layer of carbon. So, the wetting characteristics of the silica grain by liquid iron is influenced in the same positive way like with sea coal.

While the morphology of graphite and bentonite is very similar, the surface characteristics are completely different. In contrast to bentonite, graphite has hydrophobic surface properties and therefore can be hardly be wetted by water. While bentonite is wetted very well, bentonite can swell in water very easily and forms well-dispersed bentonite layers which act as binders for the molding sand. Unmodified graphite is not wetted by water and swims on the surface in the form of coarse agglomerates. However, as the coverage of silica grains with graphite has to take place during molding sand processing in the mixer, the wetting of graphite by water has a special importance. When suitable dispersing agents are present, the dispersion will be successful like with bentonite.

The combination of special graphites with suitable dispersing agents is patented in the meantime [1].

5. LE technology in foundry practice

By comparing the emission of benzene as most critical component from much more than 60 different molding sand systems worldwide, the huge differences in the benzene emission level are remarkable. These huge differences reflect the different levels of lustrous carbon formers in the molding sand systems as well as the different level of core binders.

The comparison of the benzene emission of currently 30 European foundries processing LE technology shows very clearly that in most of the cases a reduction of the benzene emission by 50 % and even more can be realized (fig. 1). If a foundry is starting with a very high level of emission, a reduction of up to 85 % is possible.

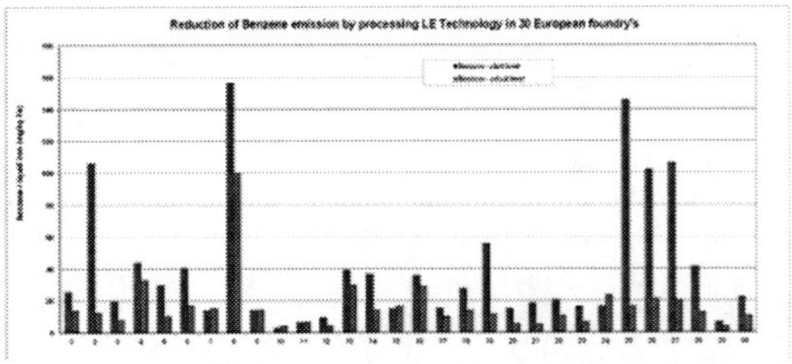

Fig. 1 Reduction of benzene emission by processing LE Technology in 30 European foundries

References

[1] T. Engelhardt et al. (2016) EP 2477766

The 73rd World Foundry Congress
„Creative Foundry"
23 - 27 September 2018, Krakow, Poland

DEVELOPMENT OF MOLTEN METAL TRANSPORT SUPPORT SYSTEM WITH AN OVERHEAD CRANE

Akihiro Kaneshige[1]*, Satoshi Ueki[1], Takanori Miyoshi[2] and Kazuhiko Terashima[2]

[1]National Institute of Technology, Toyota College, Mechanical Engineering Department,
2-1 Eisei-cho Toyota City, Aichi Pref., Japan,
[2] Toyohashi University of Technology, Mechanical Engineering Department,
1-1 Hibarigaoka Toyohashi City, Aichi Pref., Japan,
**Corresponding address: e-mail: kanesige@toyota-ct.ac.jp*

Key words: Ladle Transportation, Overhead Crane, Support System, Molten Metal, Vibration(sloshing) Suppression Control

1. Introduction

In transport of heavy and high temperature molten metal by an overhead crane in the steel or the casting industries, in order to carry safely and efficiently, an experienced operator is carrying out operations that suppress vibration (sloshing) of the molten metal in the ladle under a poor environment such as high temperature, high humidity, many dust. From the viewpoint of lack of skilled operators and improvement of work environment, development of a transport support system that suppresses to vibration(sloshing) of the molten metal and to transports to the destination in a short time is desirable. Research and development concerning a transport of molten metal have been carried out from various viewpoints[1-5].

This paper presents a design of the ladle transport assisted control system with an overhead traveling crane, which is allowed to operate by an inexperienced operator easily. In particular, an overhead crane transportation by the operator, which is allowed to select freely pathway, is needed from the point of freedoms of the operation. A presented support control system is designed as to suppress both the rod vibration and the sloshing without measuring the sloshing. A control system is two DOFs controller, a feed-forward controller is based on two notch filter, and a feedback controller is based on the H_2/H_∞ controller design method using the LMI. An effectiveness of a proposed transport support control system is verified by experiment.

2. Experimental Equipment

General view of an overhead traveling crane experimental equipment is shown in Fig.1. In a present apparatus, a cart moves in the traveling direction, and a command voltage is given from a computer to a motor driver, and a moving speed of cart is controlled. A cart position is calculated by an encoder of a motor built-in. The rod fixing a ladle is assembled to a cart via a rotating shaft, and a swing angle of a rod is measured by a laser sensor. Also, a height of the sloshing for verification of proposed a control system is measured by the ultrasonic sensor. Then, it can be operated with the joystick shown in Fig.2.

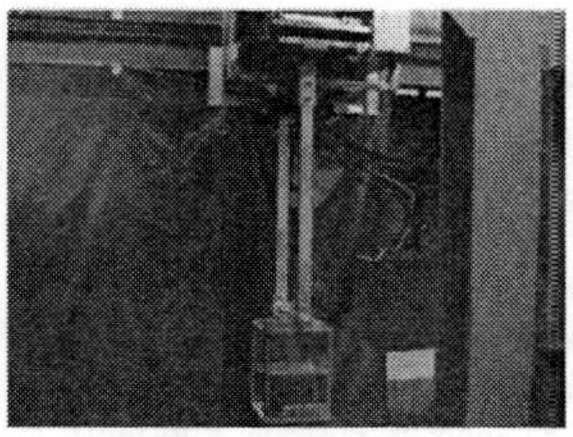

Fig.1. Overhead traveling crane experimental equipment

Fig.3. Input device(Joystick)

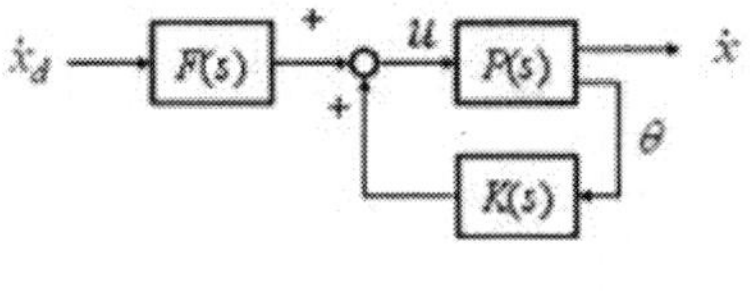

Fig.3. Block diagram of 2DOF controller

3. Transport Support Control System

Figure 3 shows a control system of a transport support system that suppresses both a rod vibration and a sloshing. A control system is two DOFs controller, a feed-forward controller $F(s)$ is based on two notch filter, and a feedback controller $K(s)$ is based on the H_2/H_∞ controller design method using LMI[5].

A feedforward controller $F(s)$ is design using a notch filter which can easily construct a control system even in damping control, A natural frequency of rod and liquid level is excluded. A natural frequency of a rod is obtained by the following equation[4].

$$\omega_{n1} = \sqrt{\frac{g}{l}} \qquad (1)$$

where l is a rod length. Also, A natural frequency of a molten metal is obtained by the following equation[4].

$$\omega_{n2} = \sqrt{g\kappa \tanh(\kappa h_s)} \qquad \kappa = (n\pi)/(2L) \qquad (2)$$

where h_s is level of a molten metal, and L is width of a ladle. A notch filter is constructed from these two natural frequencies by the following equation[4].

$$F(s) = \frac{s^2 + 2\zeta_1\omega_{n1}s + \omega_{n1}^2}{s^2 + \omega_{n1}s + \omega_{n1}^2} \cdot \frac{s^2 + 2\zeta_2\omega_{n2}s + \omega_{n2}^2}{s^2 + \omega_{n2}s + \omega_{n2}^2} \qquad (3)$$

In the H_2 controller, the inertia force of the molten metal is regarded as an external force, and the speed of a cart of the crane is quickly converged to the speed by the external force, so that the design is made due to the purpose of reducing the vibration of a liquid surface in a ladle. Weight function of W_2 is defined by low pass filter of the following equation[5].

$$W_2(s) = \frac{1}{0.3s + 1} \qquad (4)$$

In the H_∞ controller, design is made as a vibration suppression controller considering changes in a rod length, the size of a ladle, and the amount of molten metal. The designed weight function W_∞ is shown in the following equation[5].

$$W_\infty(s) = \frac{s^2 + 8s + 4}{s^2 + 4s + 275.6} \qquad (5)$$

4. Verification of Transport Support System

Verification of effectiveness of the proposed support system was confirmed by control experiments. The experimental results are shown in Fig. 4. For transporting conditions, an operator moved about 1.5 m from the start point and returned to the start point again using a joystick. In addition, in order to confirm the difference in operation by the operator, a transportation experiment was conducted by seven inexperienced persons. Figure 4 shows the results of one control experiment. Proposed control system carried out that a transportation time is shorter, vibration of the rod and molten metal vibration are suppressed compared with without control system. Other subjects showed similar results.

5. Conclusion

In this research, a ladle transport support control system with an overhead traveling crane was proposed which is allowed to operate by an inexperienced operator easily. The proposed system is a robust control system against changes in the amount of a molten metal and a rod length and suppresses both the deflection angle of a rod and a molten metal level vibration, reduces the residual vibration, and is considered to be a practical method.

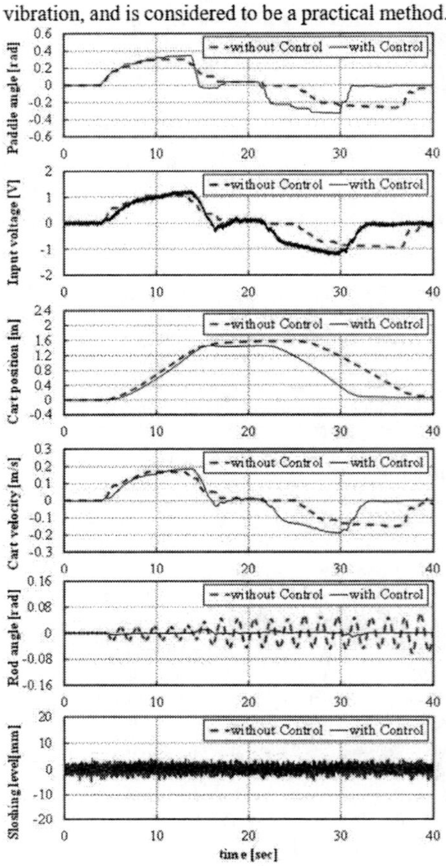

Fig.4. Experimental Results

References

[1] Aravind Samba Murthy, Arto Kivila and William Singhose, Slosh Suppression of Liquid in a suspended Container using Robust Input Shaping, The proceedings of 19th International Congress on Sound and Vibration, (2012), pp. 1163-1170.

[2] Carlos Vazquez and Joaquin Collado, Control of a parametrically excited crane, Proceedings of the 18th World Congress of the International Federation of Automatic Control, (2011), pp. 6703-6708.

[3] Hamaguchi, M. and Taniguchi, T., Damping Control of Sloshing in a Liquid Container on a Cart with an Active Vibration Reducer of 6-DOF Parallel Linkage (The Case of a Straight Path on Horizontal Plane), Transactions of the Society of Instrument and Control Engineers, Vol. 51, No. 2, (2015), pp. 101-109 (in Japanese).

[4] Kaneshige, A., Kaneshige, N., Hasegawa, S., Miyoshi and T., Terashima, K., Model and control system for 3D transfer of liquid tank with overhead crane considering suppression of liquid vibration, International Journal of Cast Metals Research, Vol. 21, (2008), pp. 293-298.

[5] Ueki,S., Yamauchi Y., Kanshige A., Miyosi T.and Treashima K. Design of liquid container transport control system with overhead traveling crane, Transaction of the JSME, Vo.83, No855, (2017) p. 17-00237 (in Japanese).

The 73rd World Foundry Congress
„Creative Foundry"
23 - 27 September 2018, Krakow, Poland

MODERN GREEN SAND MOULDING FOR EVERYONE

Bo Haugbølle[1], Per Larsen[1]*

[1]DISA Industries A/S, Hoejager 8, DK-2630 Taastrup, Denmark,
**e-mail: per.larsen@noricangroup.com*

Key words: Low investment, Less labor, Casting quality, Green sand moulding, Vertical

1. The Challenge

The constant pursuit of innovative ways to boost efficiency, enhance productivity, improve casting quality and increase savings, while battling environmental concerns, seeking to improve overall working conditions and ensuring the employment of sufficient manpower and resources – preferably all at the same time – is universal. Irrespective of foundry size, geography and market, they remain factors that all foundries can relate to and continuously try to optimize.

To make modern, state-of-the-art foundry equipment available and accessible to small and midsized foundries worldwide, DISA's goal has been to offer a compact vertical moulding machine with the power and precision to increase competitive edge foundries seeking to produce higher quality of casting at lower costs per kg.

2. The Solution

In order to reinforce and consolidate the competitive edge of today's small and midsized foundries, DISA has chosen to meet the following key critical pains and challenges in particular:

a) Lack of labor
Even for foundries situated in so-called growth areas, sufficient labor is a scarce resource, and it can prove consequential not just for the areas in questions, but also the world market for industrial goods in general.

b) Lack of resources to invest
Where the investment scenario for larger foundries is a question of continuous innovation, the lack of monetary funding can be a question of mere survival for many small and midsized foundries.

c) Lack of consistency in casting quality
Due to the existing jolt squeeze or manual process used in many small and midsized foundries, the casting quality tend to be inconsistent and the mismatch to be high.

Being renowned for its world-class engineering, DISA has always sought to optimize both efficiency and productivity in its moulding technology and moulding lines, and it is no secret that DISA is traditionally known for making solutions available to foundries requiring both very high capacity and very high speed.

Although DISA's very intensive trend spotting has shown that high efficiency and productivity remain paramount for the foundry business in general, DISA has attempted to counter very specific needs and requirements put forth by small and midsized foundries. In that process, DISA has made a fundamental shift in its way of thinking, entering into new territory in order to innovate and launch products specifically targeted and designed for this market segment.

3. Maximum capacity at minimum investment – from manual to automated processes

It is an evident turning point for small floor moulding foundries in particular to have both affordable and accessible vertical moulding machines at their disposal. It will allow them to convert from existing jolt squeeze or manual process to an automated vertical moulding process, thereby enabling a relatively fast production of high-integrity moulds with low mismatch and minimal finishing costs.

Thus, an entry-level vertical moulding machine is DISA's way of going head-to-head in its way of addressing the challenges of small and midsized foundries mentioned above:

In terms of struggling to obtain sufficient and qualified labor, the consideration has been to base a new moulding machine on very straightforward principles that allow generally skilled labor to handle all operation and maintenance in a safe working environment. Additionally, it has been a top priority to accentuate not less than eight ways to sustainable growth and safe working conditions, including aspects such as reducing physical exertion and risk of accidents, enhancing employee safety and allowing for quieter operation of the machines and, thus, a more comfortable all-round working environment.

As regards the limited resources to invest in new moulding investment, the aim has been to design a machine that offers maximum capacity at minimum investment, delivering a high Return On Investment

(ROI) over a long service life – even in cases where capacity will not be fully utilized. Additionally, a simplified operation with innovative mechanical, pneumatic, hydraulic and electrical systems inherited from other DISA solutions has resulted in high uptime and minimal maintenance requirements.

The fight of small and midsized foundries to offer a consistent casting quality has been sought met through the introduction of an automated vertical moulding process that enhances surface quality and overall mould accuracy. Ultimately, this will also add competitive edge and bring less cost per casting.

4. Much more than a singular machine

The final result builds on DISA's accumulated knowledge from entry markets worldwide in order to thoroughly assess and, ultimately, address and fully meet the needs of small and midsize foundries everywhere. As such, an entry-level vertical moulding machine of this type, including both an automatic mould conveyor system and an automatic core setter as part of its basic equipment, serves to boost both the capacity and efficiency, and make up-to-date affordable and quality casting production available to everyone.

5. Conclusion

The aim has been to understand the needs of small and midsized foundries – both right now and over time – by providing no-fuzz moulding solutions and candid, straightforward advice to the entry-level market that will always seek to fit with potential, future growth plans.

Some of the main advantages of such an entry-level vertical moulding machine are:

1. Simple system that can be implemented with a very small investment

2. Straightforward operation and maintenance

3. Improved casting quality and increased productivity – converting from manual to automated process

4. Perfect entry capacity

5. On-site local support, including support with vertical process technology, to help maximize complete foundry performance at all times.

The 73rd World Foundry Congress
„Creative Foundry"
23 - 27 September 2018, Krakow, Poland

FOCUS ON DEVELOPMENT OF QUALITY, HIGH PRESSURE DIE CASTING PROCESS

Ildiko Peter[1], **Mario Rosso** [2*]

[1, 2] *Politecnico di Torino, Departent of Applied Science and Technology*
Corso Duca degli Abruzzi 24, Torino, Italy
**Corresponding address: e-mail: mario.rosso@polito.it*

Key words: die casting, numerical simulations, development of casting technology, HPDC optimization

Abstract

In contemporary high-pressure die casting (HPDC) foundries, the mastery of each sequence in the production cycle is more and more important. For castings of complex geometry and at high dimensional accuracy, the appropriate planning of pouring and feeding elements is of fundamental importance. According to the heat economy of casting, rapid tooling and prototyping and then reliable manufacturing includes the mastery of all the edge conditions in the process chain.

In the paper, an example of virtual analysis of gearbox casting from Al alloy will be presented. It includes a large variety of parameters, as follows: choosing of appropriate foundry technology, calculation of computer simulation of casting process which takes into account the filling process of cold chamber and filling of cavity, model description of three phases at HPDC, flow of molten metal, solidification, formation of stress and deformations.

Additionally, the optimization of cooling and heating systems will be compared with calculated volume defects, dimensions of castings and their deformations with experimentally obtained values.

The characterization has been performed by computer tomography and 3D measurement. Please do not change the document settings and be wary when pasting elements into this template.

1. Introduction

Nowadays, using numerical simulations in the foundries is a consolidated event. Some foundries that have just started to use numerical simulations have realized that the benefits from using them are not only a competitor's advantage, but they are useful in order to produce quality parts with a reasonable price. To cover all foundry technologies, software developers must be in constant connection with the foundry R&D units to cover all the new features and industry demands.

Software solutions are based on several different methods such as Finite Differences (FD) method, Finite Element Method (FEM), Control Volume Method, etc.. In addition to the basics results like gas and shrinkage porosity, flow analysis, temperature field in the casting and in the die, the software is able to predict deformations and residual stresses. Conventional numerical tools cover all basic foundry technologies like gravity casting in unique and permanent mold, low pressure die casting, high pressure die casting and also more specific processes like investment casting, semi-solid modeling, core blowing, centrifugal casting, lost foam and the continuous casting process. Useful for all castable alloys, the metal casting simulation software helps to address defect detection in some situations, individuation of residual stresses, to evidence if any part distortion occurs, prediction of the microstructure and of the mechanical properties, etc..

The appropriate use of the numerical simulations in casting technology development should be included from the start of the design. To find where the casting's defect will be most problematic, at the beginning of the simulation only the solidification of the casting without gating system has to be considered. As a second step, in case HPDC technology is used, the gating system has been designed including the cooling and/or heating system. When casting defects are in acceptable limits, optimization of the heating and cooling system is carried out to get deformation of the casting in the desirable limits. With optimization of heating and cooling system in the die an influence on the tool fatigue can be obtained.

In this paper, optimization of heating and cooling system for a part that was already in production, but not optimally developed during foundry technology design, is presented and discussed. In the course of the production, it was found that in some areas in the die overheats, porosity in the casting meets the requirements but actually, it could be lower. Calculations and measurements confirmed problems at machining of the casting due to casting deformation. The optimization of casting deformation is the task for designing the next series of the tool.

The 73rd World Foundry Congress
„Creative Foundry"
23 - 27 September 2018, Krakow, Poland

BENEFITS OF USING PROCAST SIMULATION SOFTWARE TO SOLVE CASTING DEFECTS IN ODLEWNIE POLSKIE S.A.

Maciej Królikowski [1], Vlastimil Kolda [2], Krzysztof Bialek [3]

[1]*ODLEWNIE POLSKIE S.A., 27-200, Starachowice, Inż. Władysława Rogowskiego 2, Polska*
[2] *ESI Group/Mecas ESI s.r.o., Technická 15, Brno, Czech Republic*
[3] *VPI POLSKA SP Z O. O., ul. Ciepłownicza 23, 31-574 Krakó, Polska*

Key words: Numerical Simulation, Cast Iron, Casting Technology, Solving Defects, Porosity Prediction, Optimization

1. Abstract

Numerical simulation and its use in the casting process is an increasingly common tool for analyzing the causes of casting defects. Conventional methods are often not sufficient to pinpoint the causes, and moreover they always follow real tests. Only numerical simulation allows analysis of defects already at the design stage of the technology or casting design. These two stages of casting development, i.e. the design of the construction and the production technology, are closely related and at the same time they are going in many cases against each other. A casting design that is designed to meet the needs of the final part functions often does not consider the technological requirements for its production in the foundry. The numerical simulation allows the evaluation of the impact of each design change of the casting with respect to its final quality. From the point of view of technology, the numerical simulation offers many possibilities to ascertain in advance its shortcomings and to optimize the technological solution of the cast before the production in the development phase.

For many foundries, numerical simulation has been incorporated into the casting process chain a long time ago, when the possibilities of numerical simulation were, compared to nowadays, very limited. This is also the case of the company ODLEWNIE POLSKIE S.A. where they started using ProCAST casting simulation suite in 1998. Software has since become an integral part of the technology design process and well used tool to detect problems and its causes.

Nowadays, numerical simulation is not only used in the casting process on Loramendi and HWS production lines, but thanks to a special module also to predict defects associated with the process of cold-box production of sand cores. This fact, in combination with deep expertise and years of experiences of technologists in ODLEWNIE POLSKIE makes the company one of the most successful cast iron foundries on Polish market.

The article aims to show several examples of real castings from the ODLEWNIE POLSKIE portfolio, a method of implementing simulation in the standard

cycle of each casting from the prototype stage to its serial production. It will also demonstrate the cost reduction done by implementation of such virtual tool which allows to avoid the traditional problems associated with the trial-error approach.

Numerical simulation allows user to see different kind of results based on which one can analyze many different metal behaviors. In case of gravity sand casting process starts with casting filling. Following given technological parameters like pouring temperature, pouring velocity or permeability of the mold, user is able to see what is really happening inside the casting mold. Using many different simulation results user of simulation software ProCAST is able to analyze process of filling, address all pain points from the temperature, velocity perspective, or evaluate those exact criteria which shows problematic behavior from the oxides or other kind of undesirable inclusions point of view. User is even able to see mold erosion risk and decide, based on the simulation results, where such released mold particle can end its path through the metal stream.

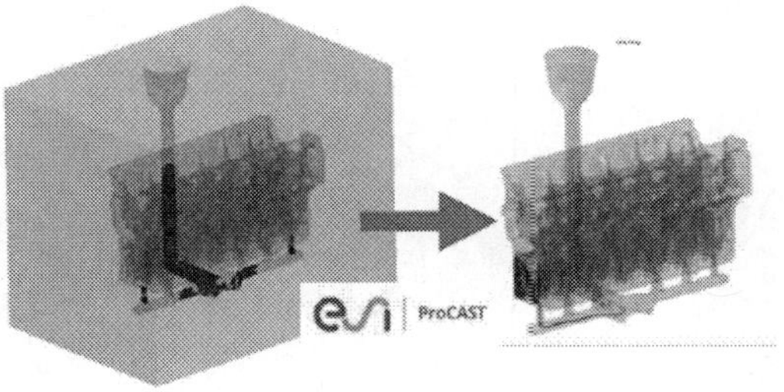

Fig. 1. Mold erosion risk / particle tracing

One of the most important behavior for any casting is its solidification pattern. Casting technologist have to have at least minimum level of experience to be able to design suitable casting technology with respect to wanted solidification pattern. Numerical simulation can help to find biggest hot spots of the casting body and even calculate exact value of thermal modulus. Casting engineer is then simply able to select proper feeder from the catalog. Of course, situation is usually not so easy, especially if the casted material is cast iron or ductile iron, where so many different phenomena entering the casting process. To predict such complex behavior, simulation software must have proper functions. In ProCAST, most important module to handle different aspects of every possible material is a special thermo-dynamic material database Computherm. Based on alloy chemical composition, software is able to create all material parameters which are needed to perform the calculation. Such tool allows also to see effect of various chemical composition on the final solidification/porosity results.

Speaking about porosity prediction. As mentioned above cast iron is not simple material. Many different aspects can have big influence on final metal quality and therefor on shrinkage porosity. Beside of described thermos-dynamic database, software ProCAST implemented special features to handle material quality. Those features can consider level of inoculation and modification, fading effect of such treatment or the way how it was done during the process.

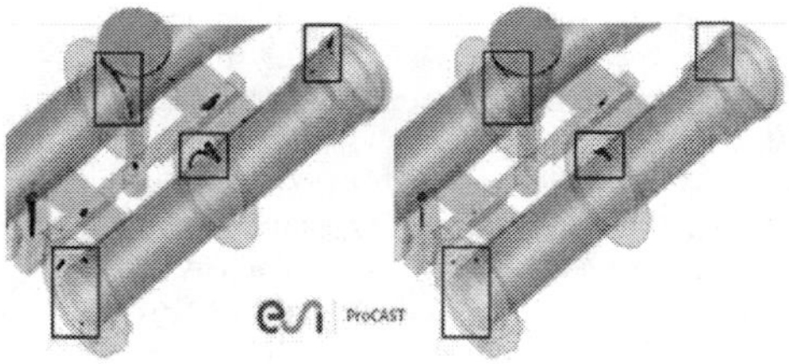

Fig. 2. ***Difference in the level of inoculation (left-average inoculation, right-improved inoculation)***

Finally, concerning cast iron and ductile iron materials, question of huge importance is also final material microstructure and related mechanical properties. Software ProCAST contains special "Microstructure module" which allows user to determine distribution of mechanical properties across the casting body. This module is fully connected with above mentioned thermo-dynamical material database. Every kind of cast iron material can be then investigated.

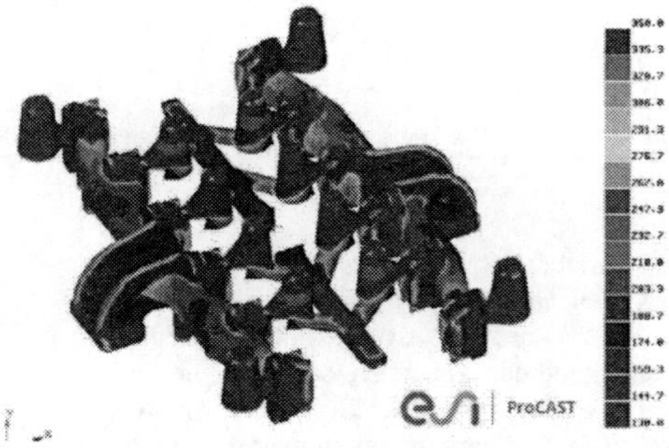

Fig. 3. ***Prediction of casting hardness in ProCAST***

All this virtual steps and features mentioned above give casting engineer great opportunity to know many information about casting process even before real production starts. It allows him to validate casting design and technology from many different perspectives. This is what in Odlewnie Polskie know well and that is why simulation is valid part of the casting production process.

The 73rd World Foundry Congress
„Creative Foundry"
23 - 27 September 2018, Krakow, Poland

INTELLIGENT MANUAL AND AUTOMATED MICROPARTICLE DRY ICE CLEANING SYSTEMS FOR THE FOUNDRY – DIE CASTING INDUSTRY

Dietmar Juchmes, General Director EMEA, *e-mail: djuchmes@coldjet.com,* **Cold Jet BVBA, Zone 1 Researchark 330, B-1731 Zellik, Belgium**

Key words: cleaning solutions, dry ice technology, dry ice cleaning, dry ice production, automated cleaning solutions

Today, proven lean manufacturing management methodologies, such as 5S and TPM, are encouraging organizations to embrace maintenance as one of the critical business function. Innovative foundries are constantly looking for ways to increase the productivity of their equipment and the quality of their parts, and being at the same time more reactive to customer needs.

For foundries very often die cleaning remains a critical component for the production of quality products. Deposits on core box vents or openings and cleaning of die castings can cause a number of problems. The reliable removal of residues of the cold-box binder from the core boxes for the quality of products is of fundamental importance. Therefore, foundries should clean their equipment at regular intervals.

There are few traditional cleaning methods well known on the market like sand/soda/water blasting, chemical cleaning, hand tools which often generate delays in mold cleaning and cause longer downtimes. When chemical or mechanical media are used it is harmful for the employees as well as abrasive for the tools and thus wear critical mold tolerances. To understand this problem foundries need to clean molds more often, faster, and non-abrasively, in a sustainable, environmentally responsible manner.

The other obstacle for foundries is downtime caused when cleaning permanent aluminum molds, shell core molds, semi-solid castings and die casting tooling. Typical manual cleaning methods require cool down, disassembly, unproductive hours of manual cleaning and then reassembly. In addition, most traditional cleaning methods are not fully effective and often result in damage to the part or equipment (more detailed comparison of chosen cleaning methods is available in Appendix 1).

Cleaning with dry ice is a more effective technology comparing to traditional cleaning methods. This environmentally friendly cleaning method is the best alternative for sand, soda, chemicals or metallic abrasive media blasting. Instead of abrasive media this technology use dry ice (solid CO_2) accelerated at supersonic speeds through a specialized hose and nozzle. Once the dry ice hit the surface, the combination of kinetic energy and thermal effect of the dry ice blast breaks the bond between the surface and the residue. Upon impact the CO_2 sublimates returning into gas form, leaving no secondary waste to clean up. As an example, cleaning core boxes is often done with dry ice MicroParticle due to the very complex geometries that require a precise and consistent clean that only dry ice can provide. Dry ice cleaning offers significant improvement in cleaning times (up to 60%) as well as reduction of damage to equipment and resulting scrap product. It allows for a non-conductive, in-place method to clean parts.

To achieve better results foundries use dry ice MicroParticle technology which provide the most thorough clean for many applications. Micro Particles clean faster, do not damage surfaces and saving customer 30% on dry ice and air compare to 3 mm dry ice pellets. By resizing dry ice media into the finest MicroParticles, more particles enter the air stream and strike the surface per second than traditional methods, quickly removing the contaminant and providing the fastest clean compared to 3 mm pellets. Due to the particle size, less compressed air is needed to propel the MicroParticles to peak velocities, resulting in a much quieter clean.

Nowadays technology goes even further by providing robotic automation handling for a wide range of applications including cleaning equipment, parts and surface preparation. With the integration of an automated dry ice MicroParticle blasting system within the existing production line, manufacturing operations are seeing further reductions in downtime, improvements in part quality, as well as a quicker ROI. Integrated cleaning systems are cost effective solutions with superior cleaning performance and minimal space requirements. They provide continuous, uninterrupted and completely automated operations.

Manufacturers also provides systems which combines dry ice MicroPellets production and dry ice blasting into one system (Combi-System), able to integrate with a robot. High-density dry ice is produced on demand and is always available on site and in the exact quantity required. The pelletizer unit automatically adapts the production of dry ice pellets to the dry ice consumption, predefined on the blasting unit. The system can also be used to produce dry ice pellets for other applications, not only cleaning. It is fitted with an extra pellet outlet from where pellets can be produced directly into an external dry ice container.

Customize integrated Combi-System helps to improve quality and lower operating costs. It enables customer to focus on their production needs on a 24/7 base and saves them a lot of time, trouble and money. Intelligent dry ice cleaning & production system communicate via Remote Systems with the customer and track all information and settings which are required for a modern foundry production facility.

Appendix 1.Comparison of cleaning methods

Cleaning method	Operation cost	Equipment investment	Repeatable cleaning & trackable	Damage & increased mold wear	Secondary & residual waste	Maintenance cost	Simple & safe operation	Cleaning time
Power-Wash	High	High	Medium	Low	Yes High cost	Low	No	Long
Chemical Cleaning	Medium	Medium	Medium	Low	Yes High cost	High cost	No	Long
Grinding	Medium to low	Medium to low	Medium	Yes & very high	Yes Low cost	No	No	Medium
Brust	Medium to low	Medium to low	Medium	Yes & high	Yes Low cost	No	No	Medium
Sand-Blasting	Medium	Medium	Medium	Yes & very high	Yes High cost	No	No	Fast
Dry Ice 3mm Pellets	Medium	Medium	Good	Very low	No & 100% environ-mental friendly	Low	Yes	Fast
Dry Ice Micro-Particles	Medium to low	Medium	**Standard & Excellent**	**No damage or mold wear**	No &100% environ-mental friendly	Low	**Yes**	**Very Fast**

The 73rd World Foundry Congress
„Creative Foundry"
23 - 27 September 2018, Krakow, Poland

NEW INVESTIGATION OF MATERIAL-DEPENDENT-CONTROL OF FLOWABILITY IN GREEN SAND MOLDING PROCESS

Jürgen Bast[1*], Dheya Abdulamer[2], Abdulkader Kadauw[1], Bertram Hentschel[1]

[1]TU Bergakademie Freiberg IMKF, Agricolastr. 1 09599 Freiberg, Germany,
[2]TU Bergakademie Freiberg IMKF, Agricolastr.1 09599 Freiberg, Germany, [2]ERETC-UOT-Baghdad, Iraq,
**Corresponding address: bast@imkf.tu-freiberg.de*

Key words: Molding process, Green molding mixture, Compactibility, Material-dependent flowability control, Bluetooth technique

Quality management and control are a basic and important activity needed along the production process. In order to realize the aspects of "Industry 4.0" and, based on this, an integrated digitalization of the casting production processes, the foundry industry must solved a lot of considerable problems. These problems result from the insufficient amount of useful data sets and missing of sensors to detect effects of different process steps. This applies in particular to the compaction process of the bentonite-bonded molding material in the green sand molding process. The currently offered time-, pressure- or displacement-controls for compaction of molding mixture do not consider its compactibility potential, which is technologically possible as a result of its composition and preparation. These controls merely realize the pre-setting control factors of the molding machine, which often leads to over- or under-compacted molds, with all problems of insufficient gas permeability or too low dimensional stability or broken mold parts during handling and pouring of the molds. To solve this problem, a compactibility sensor (Fig. 1) has been developed in the research work [1].

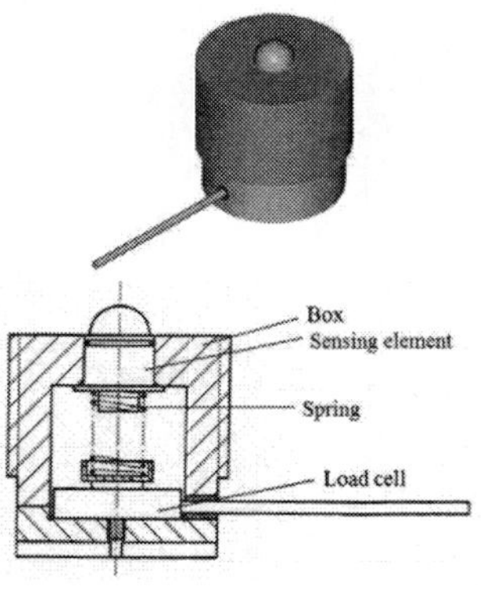

Fig. 1 Compactibility sensor

This makes it possible to measure the compactibility potential of the molding sand mixture and thus to supply exact signals for controlling the compaction process. The further development of this sensor has led to a material-dependent-flowability measurement method for molding sand materials [2, 3], which can be used both in the laboratory and directly in the molding process of the molding machine (Fig. 2). For the first time in the molding material testing this makes it possible to compare directly the laboratory values with those from the mold production.

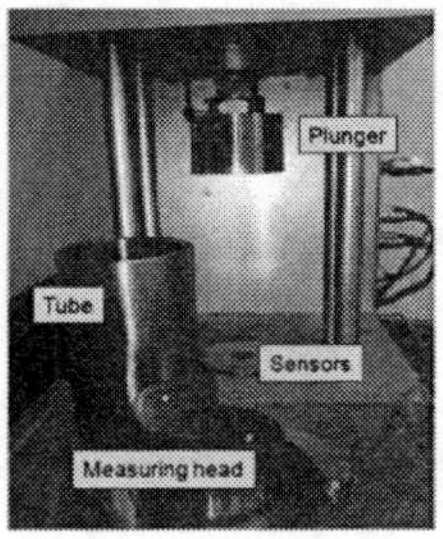

Fig. 2 Material-dependent-flowability measurement lab device

As a result of the incipient molding sand compaction, the upper sensor generates an electrical signal. After a certain time depending on the molding material properties, the lower sensor starts to display the changes in the state of stress (density of molding sand).

In the lower side of the measuring head magnets are installed so that it can be free positioned on the pattern plate (Fig. 3 and 4).

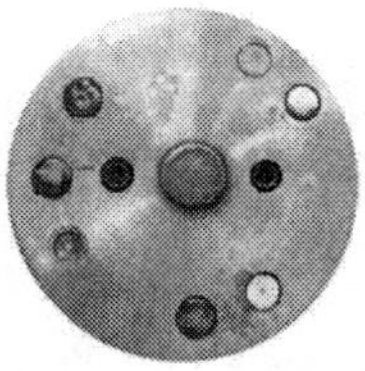

Fig. 3 Installed magnets in the lower side of the measuring head

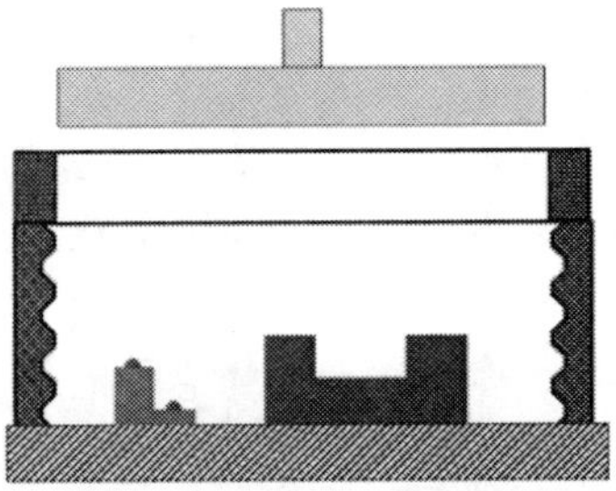

Fig.4 Free position of the material-dependent-flowability measurement head on the pattern plate

In accordance with the state of art for the increasing of the flexibility of the measuring head, a wireless transmission of the measured values was realized. For this purpose, the Bluetooth technology known from computer technology was used.

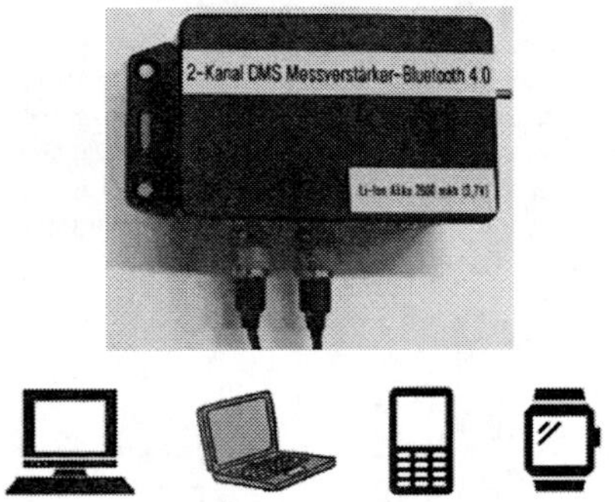

Fig.5 Data amplifier for the transmission of measurement readings

This makes it possible to send the data both to the machine control as well as to computers, laptops, smart phones or operator watches (Fig. 5). Since the data acquisition takes place in real time, the machine operator has the opportunity to regulate in the cases of mistakes. Since the introduced compaction energy is transferred directly and exclusively to the molding material, the energy consumption per mold can be reduced. It also significantly reduces scrap risk and increases production.

References

[1] Malaschkin, A., (2002). Entwicklung einer Methode zur kontinuierlichen Qualitätskontrolle beim Nassgussformverfahren mit automatischen Formanlagen, DISSERTATION, TU Bergakademie Freiberg

[2] Bast, J., Kadauw, A., Malaschkin, A., (2009). Optimising of moulding parameters for green sand compaction by computer simulation and a new compaction measuring device, International Journal of Metalcasting 3(2009) Nr. 2, S. 55 - 64

[3] Bast, J., (2014). A new method for the measurement of flowability of green moulding sand, Proceedings World Foundry Congress 2014 Bilbao

The 73rd World Foundry Congress
„Creative Foundry"
23 - 27 September 2018, Krakow, Poland

"TOWARDS VISION ZERO" IN THE AUTOMOTIVE INDUSTRY FOUNDRY CHALLENGES AND OPPORTUNITIES

Ata, Mustafa

Continental, Division Chassis & Safety, Guerickestr. 7, 60488 Frankfurt am Main, Germany
e-mail: mustafa.ata@continental-corporation.com

Key words: Resonance Frequency, Noise, NVH, Quality, Automotive Industry, Foundry Process, Zero Noise, Zero Defects

Introduction

Zero Defect, Zero Noise, Zero Accident and Zero Emission are the main driver in the automotive and effecting also the foundry industry. In order to position itself as a supplier to the automotive industry, is it necessary to have a high level of quality and driving of lean principles. Management has to drive actively quality improvement and models with quality oriented behavior.

Zero Defects strategy should change the perspective thinking cost of quality and derived from that doing strongly continuous improvement activities. Application of right quality tools and driving of necessary activities with right investments will ensure of having advanced and sustainable foundry with high quality of performance. Quality has to be first and a core competence. "Zero Failure Culture" along the PLC (Product Life Cycle) requires keeping high quality level.

Zero Noise topic is one of the important topic in the automotive industry and touching more and more the foundry industry. It provides more comfort to the final customer. The challenge is how to control foundry processes. What are the main drivers?

Zero Accident vision requires that nobody should be injured. Accidents can be prevented. There's no avoiding the uncomfortable truth that there are still 1.24 million people killed globally on roads every year.

Tomorrows vehicle will be always connected, user friendly, comfortable and intelligent.

Information management – in the vehicle and beyond is the key for "Zero-Accidents".

Zero Emission has the vision to protect the environment. This can be achieved for example with a good engine management and lightweight components.

Approach

High complexity and quality of casting carts is challenging every day. This paper will be focus on "Zero Noise". To achieve this target requires an appropriate quality management. Quality is to produce and no to check. The generation of quality in production process is one of the key for success.

How can be such a kind of process established in the foundry?

Each sub-component in the car has its own natural frequencies. It is very important to control the critical mode of each part. As example is taken brake caliper. How to ensure similar natural frequencies of the produced casting parts, which process parameter in the foundry are influencing and what are the main drivers.

Tests and Analysis

Influence of Casting Process and Product Variation on Resonance Frequencies.

There are a lot of parameters which are influencing the frequency spectrum of the casting part. The influencing parameter from the point of foundry view are like weight of casting part, production date (batch-code), material structure with their mechanical properties, chemical composition, molding process, position of cavity, casting defects, dimensions (e. g. wear of tool), shape, etc.

In Figure 1 is a principal Pattern-Layout shown for a vertical molding line with eight cavities.

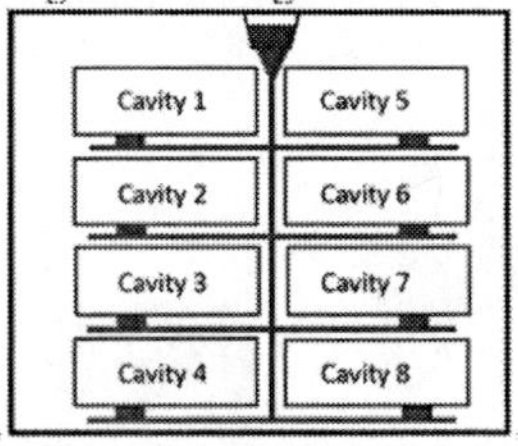

Fig. 1. Principal Pattern-Layout of Vertical Molding Process

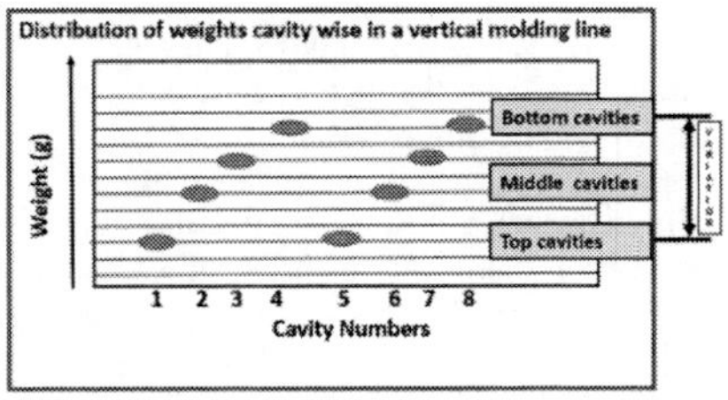

Fig. 2. Weight Distribution of Cavities by Vertical Molding Process without any dimensional adjustment

The frequency (f) is defined as the quotient of sound velocity (c) and the wave length (λ). The sound velocity is depending on E-Modulus (E) and density (ρ).

Some main influencing parameter for testing of casting part can be described as following:

- The part temperature is influencing the E-Modulus and density
- The shape or geometry of the casting part is influencing the wave length (λ).

That means that the part will expand with increasing and contract with decreasing of temperature. Having of casting part with higher temperature means having of lower resonance frequency. The resonance frequency is correlating with the temperature of casting part.

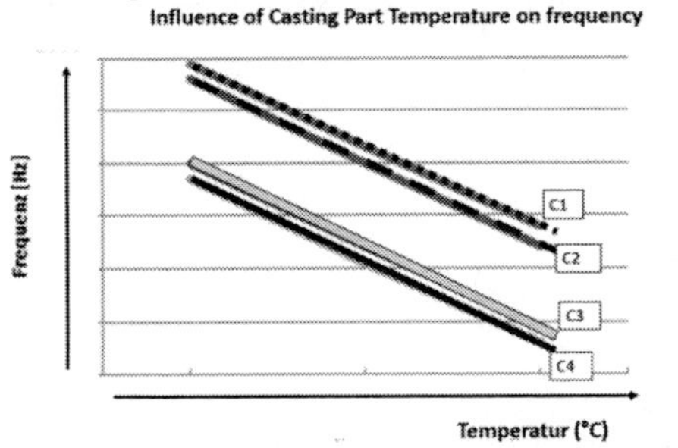

Fig. 3. ***Influence of Casting Part Temperature on Resonance Frequency***

Figure 3 shows the influence of casting part temperature on frequency on a certain frequency mode. Different frequency modes showing different correlations.

Before testing the resonance frequency of casting part it is very important to have the parts on similar temperature level.

Resonance frequency measurement is meanwhile well established in the foundries that are producing for the automotive industry like brake discs and meanwhile also in the foundries which are producing calipers (housing and brackets).

Each component has to be aligned so, that their resonance frequencies (eigenfrequency) cannot be stimulated.

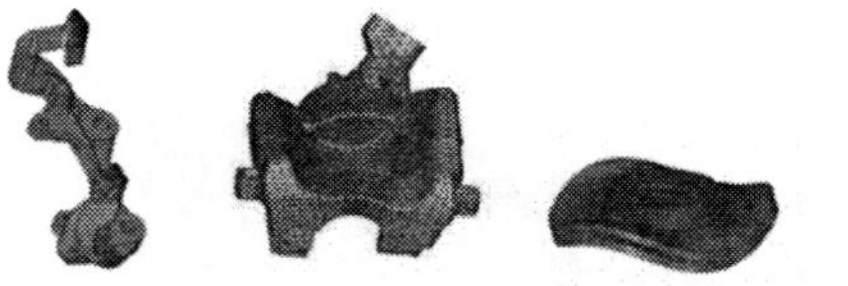

Fig. 4. ***Natural Frequency behavior as example of Bracket, Housing and Brake Disc in certain mode of Frequency***

Following test trials are done with following results:

The aging time has no significant influence on eigenfrequency by ductile iron calipers.

Resonance frequency of machined casing parts are correlating with frequencies of raw casting parts.

The test results are based on acoustic resonance analysis. This method uses the known physical effect that a body having been suitably excited (e. g. by tapping), vibrates in certain characteristic forms and frequencies. That can be measured by a microphone (airborne sound) or by a laser vibrometer (structure-borne sound) and digitally analyzed. The decisive factor for the number and spread of natural resonances of a work piece is, its geometry. A bar has few natural resonances and a complex work piece has multiple natural resonances.

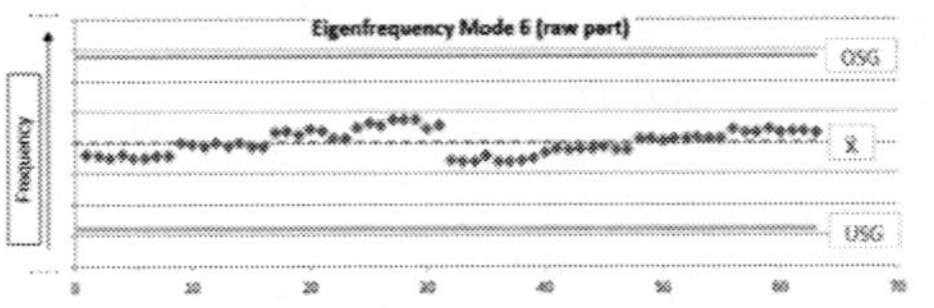

Fig. 5. ***Eigenfrequency of raw casting parts for a certain frequency mode***

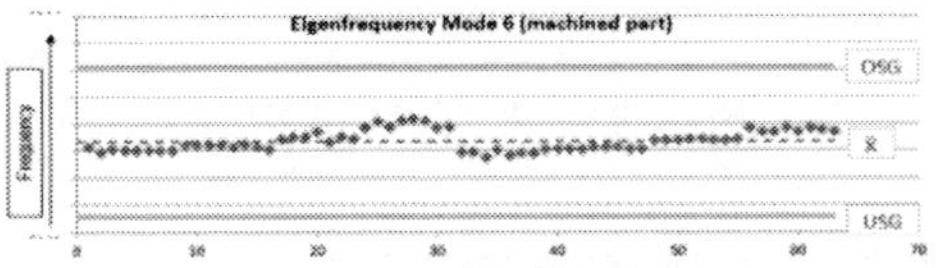

Fig. 6. ***Eigenfrequency of machined parts for a certain frequency mode***

Figure 5 (frequency of raw casting part) and figure 6 (frequency of machined parts) showing the investigation of frequency results.

Significant frequency correlation between raw and machined casting parts are given.

Results

Main influence of different resonant frequencies is related to different weights of cavities (dimensional deviation). It is very important to keep the weight of casting parts on the same level in order to have similar frequencies at different production date and cavity by cavity.

Resonance frequencies of raw and machined casting parts are correlating. Machining operation of casting parts will more equalize the differences in weight between cavities.

Temperature of part is influencing the resonance frequency. Therefore it is important before measuring resonance frequency to have the casting parts in similar temperature range. Summer and winter condition should be taken in consideration.

Tool wear and chemical composition in ductile iron are having minor influences on resonance frequencies.

Aging is meaningless regarding GJS-Caliper production date.

Conclusion

Each construction element has his own fingerprint "Natural frequency". The natural frequencies are characterized by the geometry and by the material of the element.

The main challenge is to keep stable process condition. As long as a foundry will have the material under control, the biggest influence to the frequency will come from dimensional variation (weight).

The foundry has to keep the spread of dimensional variations in a stable condition. This requires stable production process. Stable production requires collecting of production parameters, analyzing them and in case of necessity acting accordingly.

References

Continental internal researches, analysis and investigations.

The 73rd World Foundry Congress
„Creative Foundry"
23 - 27 September 2018, Krakow, Poland

Flexible Manufacturing through in-house conversion of Shell Core Making Machine to Cold Box Process Making Machine with easy reversibility

Deepak Gaur[1]*, Dharmendra Pratap Singh[2]*, K Kiran Kumar [3]*,

[1]*Material Analysis lab1 Quality Assurance Lab Maruti Suzuki India Ltd Gurgaon - 122015*
[2]*Materials Analysis lab1 Quality Assurance Lab Maruti Suzuki India Ltd Gurgaon - 122015*
[3]*Materials Analysis lab1 Quality Assurance Lab Maruti Suzuki India Ltd Gurgaon - 122015*
**** Materials Analysis Lab Quality Assurance Maruti Suzuki India Ltd Gurgaon 122015:***
e-mail: dharmendra.singh1@maruti.co.in

Key words: Flexible Manufacturing, Low cost Automation, Digitization, Cold Box Process, Scrubber Unit, Amine Generation Panel

Maruti Suzuki India Limited, formerly known as Maruti Udyog Limited, is an automobile manufacturer in India. It is a 56.21%-owned subsidiary of Suzuki and motorcycle manufacturer Suzuki Motor Corporation. As of January 2018, it had a market share of 56% of the Indian passenger car market. In February 2012, the company sold its ten millionth vehicle in India.
Maruti Suzuki has adopted several advanced tools like ISO 15001, ISO 14001 & OHSAS as well as various in-house developed tools has resulted in to effective and efficient way of working under Productivity, Quality, Cost, Delivery, and Safety Moral & Environment.

Indian Foundry Industry is growing rapidly and competing with Global Standards. This has given a tremendous pressure on all foundries to be cost efficient and give Best Quality Product with first time Right Approach.

To overcome this Team at QA MA Maruti Suzuki studied the Current Senario of Indian Foundry Industry Core Making Process and implimented a Innovative Flexible Manufacturing setup to Convert the Existing Setup of Shell core Making Equipment to Cold Box Process for Easy Reversiblity.

Along with team took the challenge for innovating Low Cost Automation for Amine Gas Generation Panel and Amine Scrubber Unit.

Current Senario

Shell Core Manufacturing Process Involving High Heat and High Natural Resource Consumption (i.e. 65 Tons of PNG for Barrel Core Making)
World Wide Shell core making process is renowned Process for getting core of all types.
Cold Box Process is a new process with no heat involved as Amine gas is used to activate the resin and Binder at room temperature. Amine Gas act as a catalyst and is neutralised with installed Scrubber Unit.
OEM Across World Supplies the Cold Box Equipment with a CAPEX investment of 4 Cr. **This is not an economical option for Small and Medium scale foundries.**
Cold Box eliminates any heating of the Core (Shell Core box needed to be heated upto 300°C) The resin% is reduced from 3% to 1% in cold box process.

New Innovative frugal Technology

Team has inhouse modified the Shell Core Box Equipment with innovative technology i.e Flexible Manufacturing with a limited budget of 12 Lacs without going for conventional Nex Capex Item i.e. New Cold Box Equipment.
If anything goes wrong we can go back to existing Shell core making process anytime.

Inhouse Offline Machine Modification of the Existing Equipment to Cold Box Process done avoid any production loss as Foundry.

Following Modification :-

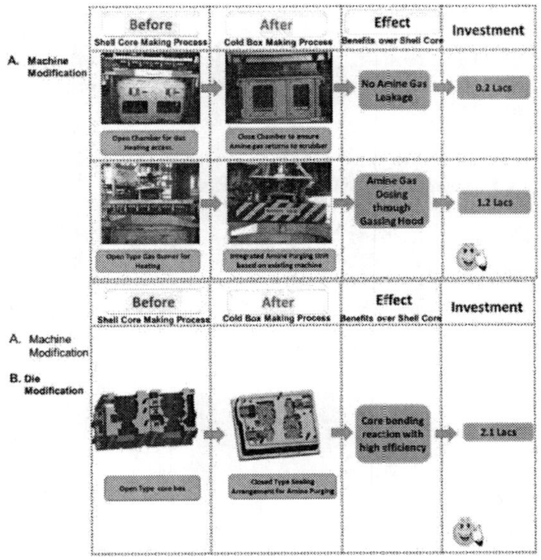

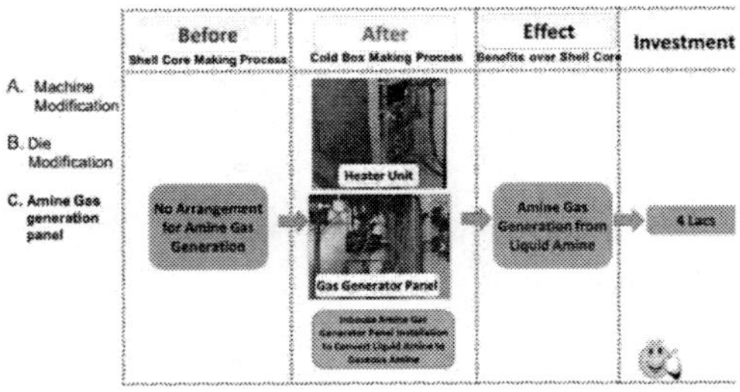

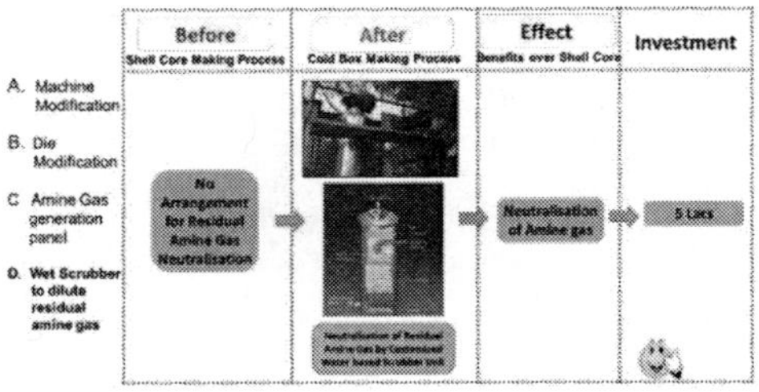

Flexible manufacturing Done so as to revert to Original process for any component to be manufactured through Old Shell Cold Process

This solution is Unique in its kind as this is flexible manufacturing in this Equipment can be modified to Either Shell on Cold Box Within Two Days time for Shell Core Manufacturing and Cold Box core Manufacturing

This is Flexible Manufacturing and all modification is carried out Offline and conversion back to old process can be carried out in two days time. As all the spares are easily Available.

Core making cell productivity has increased 37% (Cycle Time reduction by 37%)

Opportunity Cost Saving of 18 Cr/Annum (New Business Generation)

Conclusion

Team has sucessfully converted the Existing Shell Box Equipment to Cold Box Equipment with a minimal cost of 12 Lacs.

This has given the Foundry a emense productivity improvement with reduction in Operational Cost.

The Modification has been done Inhouse without been dependant on an OEM. This has enabled the Foundry to be able to Service the Equipment Inhouse and Run the Machine without Breakdown.

The 73rd World Foundry Congress
„Creative Foundry"
23 - 27 September 2018, Krakow, Poland

Seeing through the Cloud of Industry 4.0

Mark Lewis

Omega Sinto Foundry Machinery Ltd,
Peterborough, UK
markl@ofml.net

Key words: Industry 4.0, Cloud, Technology, Internet of Things, Control

A look at how cloud connectivity and Industry 4.0 are shaping the chemically bonded foundry industry and what the benefits are for those investing in the future.

1. What is Industry 4.0

To break it down into a very simple form, there are four main components that make up Industry 4.0:

Cyber Physical Systems – The integration of computer control with physical operations creating an interface between the machine and the operator.

Internet of Things (IoT)– A network of smart devices that connect to the internet and communicate their live status.

Cloud Computing – The web-based server that allows remote storage and connection to the smart devices.

Cognitive Computing – Artificial intelligence or a program that uses the above three principles to intelligently manage the plant.

The word "Industry 4.0" is being used a lot in manufacturing at present but its clear there is a lack of understanding of what it can do for businesses. In a lot of industries, it will be some time before we see an implementation of a truly Smart Factory, but that doesn't mean we can't work towards that now and benefit from certain aspects of this revolution in control.

2. How can Industry 4.0 help the foundry industry

The biggest benefit to any industry is the improvement in efficiency and flow of information.

There are a huge range of benefits 4.0 can bring to the foundry industry and below are some of the key ones:

- Efficiency
- Innovation
- Cost reduction
- Improved communication
- Improve customer relations
- Remote connectivity

Each of these points can be achieved through many different paths. The main point to understand is Industry 4.0 isn't a set standard but an accumulation of various technologies joined to create a smart factory.

3. Implementing the next industrial revolution

The first point to understand is implementing a complete smart factory is a mammoth task and would be both time consuming and costly. The future of 4.0 is wide open at present and many people are stating they are offering a 4.0 solution. Its hard to see through all the marketing and technical blurb to truly see how to move a business on the path of 4.0.

The best route is a simple one of small steps meaning reduced down-time and cost spreading. Even if the foundry does not see the need for a full 4.0 smart factory implementation they should not discount the benefits some of the technologies can bring.

Of the four main components that make up 4.0 the simplest one to implement is cloud computing. This can easily be added to existing or new machines and offers huge benefits as a stand along technology.

4. What is Cloud Computing

Cloud computing is all about putting computing services over the internet. The most common examples are remote servers, storage, and software. Most people are using these types of services daily whether it be web based email or business software. More and more providers are moving to cloud based services and with faster broadband speeds the need for local servers is becoming less and less.

The issue within an industrial environment is that we cannot yet rely on the remote servers to run the local machines, as this would bring the plant to a halt if internet connection was lost. With this in mind it's important we find the correct solution to allow the flexibility of remote connection without compromising the local machinery and systems.

5. Using the Cloud within your foundry

These days most machinery within a foundry will be run with a PLC (Programmable Logic Controller) and most likely some form of local interface screen for setting up the plant and controlling it. These systems are tried and tested giving reliable operation in an industrial environment and many are capable of connecting to the internal network of the factory.

The information available on the local control system will vary depending on the machine but a lot of useful production and maintenance data can often be found.

For example, sand mixers will often report sand and chemical usage along with temperature variations. A common task is for a staff member to note down usage data and maybe temperatures throughout the working week. If this information could be sent to the cloud the user can access it from anywhere in the world on any device with an internet connection, it could save huge amounts of wasted time and drastically improve the data flow.

Another example would be a thermal reclamation plant that is run 24/7. Often the plant will run through the night and weekend with no staff on site. If the unit were to shut down due to alarm condition no one would be aware, and the foundry could be short of sand the following day. But if the unit was connected to the cloud, it would allow remote notification to maintenance or production staff allowing them the ability to check the machine status and rectify remotely where possible.

Suddenly by removing the restrictions of local access the possibilities become huge. The machines can report on their own status enabling the maintenance and production teams to become more efficient by knowing the plant is running without leaving their office or home.

As mentioned earlier the implementation of this type of system is relatively easy and low in cost. A simple device is connected to the PLC and some form of internet connection, be it ethernet, wireless, or cellular. Data is then sent from the PLC to the cloud server where the user can securely log in and retrieve the stored information.

6. What are the risks

The main concern many people raise is security. This is an important factor to consider and comes in two parts, machine and end user. Firstly, the system needs to be segregated from the existing network so no conflicts can occur between the existing IT infrastructure and the machinery. It also needs to support strong firewall protection and user management so only the appropriate people can access the data.
From a user's standpoint strong password policies and two factor authentication are critical.

By keeping the system as data acquisition and not affecting the machine operation it also means the risks of machine down-time due to network loss are removed.

7. Conclusion

The consumer technology world is advancing at an incredible rate with industrial technology carefully following in its footsteps. Industry normally takes a much more methodical and careful approach to the new technologies as reliability and performance are key factors to consider. The good news is we are now seeing huge advances in products allowing better connectivity for manufacturers to utilize them in their equipment.

The all encompassing Smart Factory may be out of reach for many businesses but by choosing the elements and technologies that suit your business giving you the immediate benefits there's no reason not to invest.
Cloud connectivity brings industry 4.0 one step closer and is easy to install plus simple to use. Correct implementation is vital, as understanding how to use this new technology and get the best from it is as important as the hardware itself.

Some may see Industry 4.0 as a gimmick or fad but one thing is for sure, whether you invest in a full smart factory or just use parts of the technologies available, it can bring clear benefits to most businesses. It would certainly be unwise to ignore this so-called revolution.

The 73rd World Foundry Congress
„Creative Foundry"
23 - 27 September 2018, Krakow, Poland

Wireless measurement of mold temperature during centrifugal casting and heat transfer analysis

Naomichi Iwata*, Takeshi Suzuki

JFE Steel Corporation, Steel Research Laboratory, 1, Kawasaki-cho 1-chome, Handa, Aichi 475-8611, Japan.
**Corresponding address: e-mail: n-iwata@jfe-steel.co.jp*

Key words: centrifugal casting, heat transfer coefficient, wireless measurement, heat transfer analysis, local solidification time

1. Introduction

In centrifugal casting process, molten metal is poured into high-speed rotating mold. Because of the rotation, it is difficult to measure the temperature of metal and/or the mold. Therefore, heat transfer and solidification phenomena during centrifugal casting is not fully clarified. Since the phenomena is strongly affected by heat transfer at the metal/mold interface, the estimation of heat transfer coefficient at the metal/mold interface (h_M) is very important for the analysis. There is a few reports on the measurement of h_M during centrifugal casting. Vacca et al. carried out measurement of molten metal and mold temperature during centrifugal casting by slip-ring system and determination of h_M by inverse solution method.[1] In this study, mold temperature during centrifugal casting was measured by wireless data collection system as simpler method than the previous one. In addition, h_M was determined by the measured data and numerical heat transfer analysis.

2. Experimental procedure

2.1 Experimental apparatus

Figure 1 shows schematic illustration of the centrifugal casting machine used in this study and the set-up of mold temperature measurement. The mold made of cast steel has the size of 250 mm in inner diameter and 75 mm in axial length. The inner surface of the mold was coated by zircon with 3 mm in thickness. Type K (chromel-alumel) thermo-couples were installed within the mold at 1, 3 and 7 mm from mold inner surface to measure mold temperature. Transmitters were connected to the end of the thermo-couples and fixed on the shaft of motor driving the mold rotation. Thermo-electromotive force generated by thermo-couples was collected by wireless data logging system at a sampling rate of 1-second.

2.2 Casting conditions

Casting conditions are shown in Table 1. A 17 kg charge of high carbon steel was melted at 1460 °C in an induction furnace and poured into the rotating mold. Mold temperature just before the pouring was 200 °C, and the rotating speed was 1140 rpm.

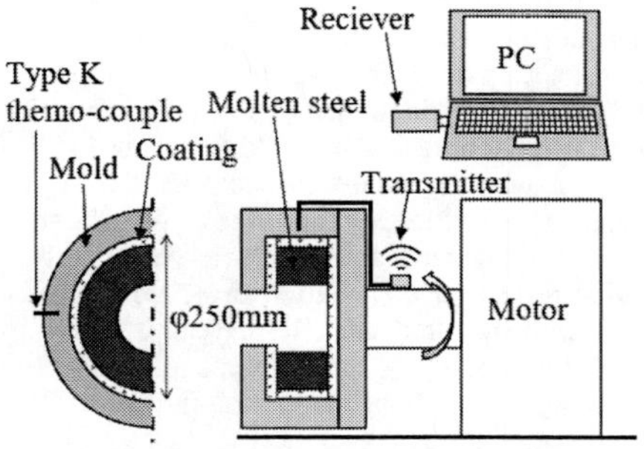

Fig. 1. Schematic illustration of the centrifugal casting machine and the set-up of mold temperature measurement

Table 1. Casting conditions

Casting size	φ244 x 70mm
Pouring temperature	1460 °C
Rotation speed	1140 rpm
Preheating temperature of mold	200 °C

Table 2. Material properties and boundary conditions

Thermal conductivity of metal at 20 °C	16.7W/m/K
Thermal conductivity of mold at 20 °C	51.7 W/m/K
Thermal conductivity of coating in all temperature range	1.3 W/m/K
Liquidus temperature of metal, T_L	1401 °C
Solidus temperature of metal, T_S	1267 °C
Latent heat of solidification	316000 J/kg
Heat transfer coefficient at the mold outer surface	900 W/m²/K
Heat transfer coefficient at the metal inner surface	100 W/m²/K
Emissivity of metal inner surface	0.9
Air temperature	120 °C

2.3 Numerical heat transfer analysis

A two-dimensional heat transfer analysis model was made in order to determine h_M. Abaqus ver.6.14, commercial finite element analysis software, was used for the analysis. Calculation was started in the initial condition of that the mold was filled by the metal with the temperature of 1460 °C, then transient temperature field in the mold was solved. Material properties and boundary conditions used in the calculation are listed in Table 2. h_M was determined by fitting based on the measured mold temperature.

3. Results and Discussion

Figure 2 shows measured mold temperature at 3 mm from the inner surface of the mold during centrifugal casting. Wireless data collection could be successfully carried out even though the transmitters were rotated in extreme high-speed. After pouring, mold temperature increased to about 400 °C, then gradually decreased. Figure 3 shows measured temperature distribution of the mold at 300 s after pouring. Mold temperature decreased linearly from the inner surface to the outer surface. Temperature difference between 1 mm and 7 mm from the inner surface of the mold was about 40 °C. Calculated mold temperature change with time and mold temperature distribution for h_M = 420 W/m^2/K are also shown in Figure 2 and Figure 3, respectively. Calculation results for h_M = 420 W/m^2/K are in good agreement with measured data. The present value of h_M is generally correspond to previous report.[1] On the other hand, in the case of static casting, h_M = 190 W/m^2/K was reported by Okumoto et al.[2] This value is smaller than our value for centrifugal casting.[2] It is considered that centrifugal force acts on metal to press against the mold, and to decrease gap between outer shell of metal and mold's inner surface.

Figure 4 shows calculated metal temperature at 1 and 20 mm from the outer surface of the metal. Based on the results, local solidification time between liquidus temperature (T_L) and solidus temperature (T_S) was about 252 s and 365 s, respectively. Vacca et al. reported that local solidification time of molten metal at 20 mm from outer surface during centrifugal casting was about 258 s.[1] Therefore, present values of local solidification time are in good agreement with previous report.

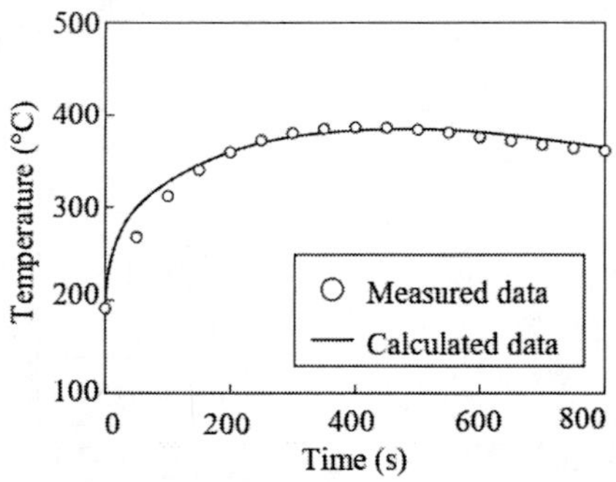

Fig. 2. Mold temperature at 3 mm from the inner surface of the mold after pouring

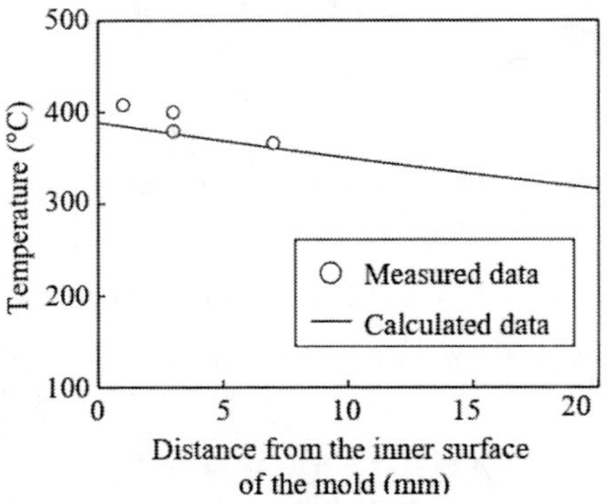

Fig. 3. Mold temperature distribution along radial direction at 300 s after pouring

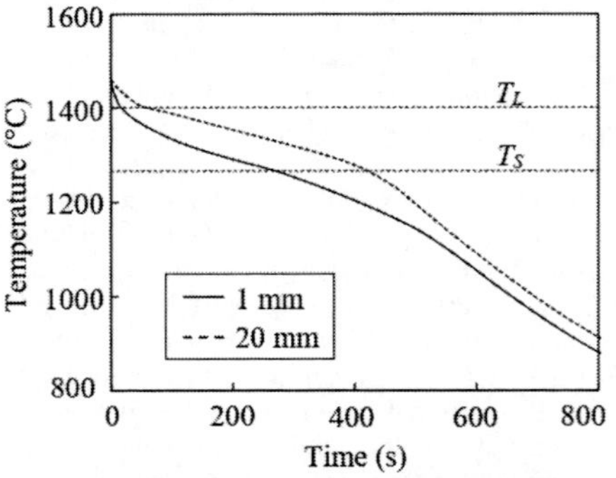

Fig. 4. Calculated metal temperature at 1 and 20 mm from the outer surface of the metal

4. Conclusions

Measurement of mold temperature during centrifugal casting by wireless data collection system and numerical heat transfer analysis were conducted to determine heat transfer coefficient at the metal/mold interface. The following results were obtained.

Wireless data collection could be successfully carried out even though the transmitters were rotated in extreme high-speed.

Estimated heat transfer coefficient at the metal/mold interface was 420 W/m^2/K. This value is generally correspond to previous report. On the other hand, heat transfer coefficient at the metal/mold interface of static casting is smaller than that of centrifugal casting. It is considered that centrifugal force acts on metal to press against the mold, and to decrease gap between outer shell of metal and mold's inner surface.

References

[1] S. Vacca, M.A. Martorano, R. Heringer and M. Boccalini. (2015). Determination of the Heat Transfer Coefficient of at the Metal-Mold Interface During Centrifugal Casting. Metall. Mater. Trans. A. 46, 2238-2248.

[2] T. Okumoto, S. Okada, T. Aizawa and M. Tanigawa. (1974). Study on the Temperature, Coating and Cooling of Mold in Metal Mold Casting Process of Cast Iron (in Japanese). Imono. 46, 609-614.

RAPID DETERMINATION OF NODULARITY INDEX IN DUCTILE CAST IRON PRODUCTION

Ramón Suárez [1,2], Pello Larrañaga [2*], Gorka Zarrabeitia [2], Fernando Vilela [3]

[1] *Veigalán Estudio 2010, Askatasun Etorbidea 16, 2I, E-48200, Durango, Spain,*
[2] *IK4-Azterlan, Aliendalde 6, E-48200, Durango, Spain,*
[3] *Sakthi Portugal, Rua Jorge Ferreirinha 679, 4470-314, Maia, Portugal*
**Corresponding address: plarranaga@azterlan.es*

Key words: Nodularity Index, Thermal Analysis, Cooling Curve, Free Magnesium, Residual Magnesium

Abstract

One of the most important facts in foundry industry when producing ductile iron castings is the rapid determination of the nodularity index. The traditional methods based in the analysis of Mg content by spark spectrometry technique and/or metallographic inspection of samples taken from the mold are time consuming and do not allow fast corrections if the nodularity index is lower than the specified one. In this paper, a new approach to control nodularity index based on thermal analysis technique is presented. It is based on the shape and parameters of the cooling curve resulting from Te coated standard thermal analysis cup where an accurate amount of S and inoculant is added.

Introduction

Graphite is the most crystalline phase of solid carbon but its nucleation and growth during ductile cast iron (SGI) production is a complex process.

Mg is the main element used to modify the shape of the graphite from lamellar to spheroidal. RE (specially Ce) and Ca are also elements that promote nodules but from industrial point of view fully spheroidal graphite cannot be obtained using only RE or Ca. On the contrary, S and O are elements that promote lamellar graphite. Therefore, the effective control of S, Mg and O together with an appropriate inoculation procedure are critical to ensure a correct formation of graphite nodules.

The determination of S and Mg in foundries is usually performed by Optical Emission Spectroscopy (OES) while O content needs techniques that commonly are not available in foundry plants. OES provides the total amount of Mg (also known as "residual Mg"). However, only free Mg (or active Mg), this is, not combined Mg forming MgO, MgS or any other more complex compound, is able to modify the graphite shape. This free Mg can be analyzed by thermal analysis technique [1].

The amount of free Mg may vary for the same residual Mg and for this reason it is common to observe different nodularity rates for the same residual Mg.

A new approach based on evaluating the cooling curve shape of the melt poured into a standard thermal analysis cup containing some Te and an accurate amount of mixture of S and commercial inoculant has been developed.

S is very reactive with Mg and it will counteract the Mg that is not previously combined, this is, the free Mg.

Te additions, even in a very small quantity (10-50 ppm), induce a high undercooling on the eutectic temperature of gray cast iron, causing the alloy to solidify fully following the metastable diagram, i.e., a complete carbidic iron (white iron) is obtained. Notwithstanding, if there is some free Mg (above 0.006%) there is not significant undercooling promoted by the amount of Te added in standard thermal analysis cups and therefore not fully metastable solidification occurs [2].

The present paper shows the relationship between the undercooling and further recalescence of the cooling curves recorded with Te-S-inoculant containing cups and the nodularity index obtained in standard cups with the same inoculation rate that the one used for the castings.

Experimental Procedure

Samples analyzed in the present work were taken from a cold pouring unit of about 2 t capacity. Nodulizing treatments were carried out by sandwich method using commercial ferroalloy. The amount of nodulizer employed varies from 0.80% to 0.90% depending on the product manufactured.

Two thermal analysis samples were obtained from normal production:

- Active Mg control cup that contains 410 mg of mixture composed by 22% of high purity S and 78% of a commercial inoculant
- Plain cup previously inoculated with the same inoculation rate that the castings.

Optical metallographic inspection to evaluate the graphite shape was performed in all plain cup.

Results and Discussion

When a special mixture of Mg neutralizer, based on S and commercial inoculant, is added into a Te coated standard thermal analysis cup, Mg preferably reacts with S. Then, if the amount of S counteracts all free Mg of the melt, Te promotes such undercooling that the solidification occurs according to metastable diagram, i.e., white iron is obtained. If there is some free Mg left in the melt after S reaction, then the undercooling promoted by Te decreases resulting in some graphite precipitation. If free Mg is high enough the undercooling promoted by Te is almost negligible and fully graphitic iron results. The differences between the corresponding cooling curves are easily distinguishable (Fig. 1).

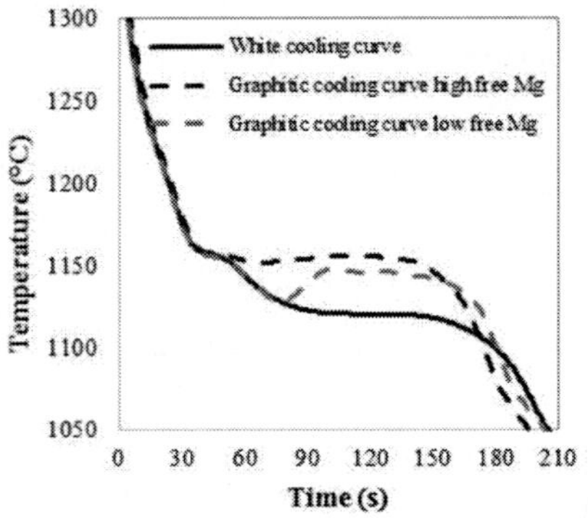

Fig. 1. Cooling curves corresponding to different undercooling promoted by Te.

The cooling curves recorded during this work showed that both undercooling ($\Delta T = T_{stable} - T_{emin}$) and Recalescence ($Recal = T_{emax} - T_{emin}$) decreased as free Mg increased. The same trend was observed with residual Mg increase. This trend is shown in Figure 2 in which only the cooling curves that are not corresponding to fully white iron are represented.

A relationship was found between Temin and Recal [3] determined in these cooling curves (from Fig. 2) and the nodularity index (N.I.) observed in the thermal center of the plain cups poured at the same time (1).

$$N.I. = f(Temin, Recal) \quad (1)$$

The accuracy resulting from the obtained equation and the evaluation of the metallographic inspection is shown in Figure 3.

This approach allowed determining the nodularity index in a fast and reliable way. Therefore, it can be assured every time that the requirements of the customers are completely fulfilled.

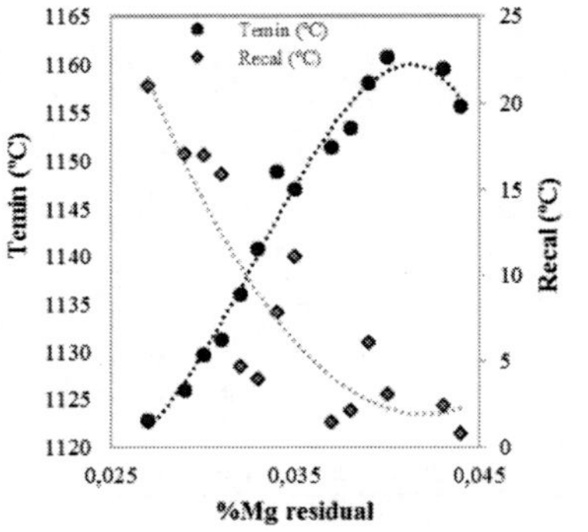

Fig. 2. Temin and Recal evolution with residual %Mg.

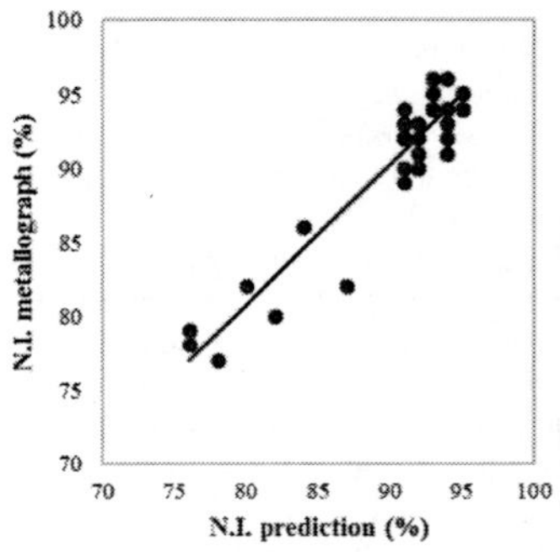

Fig. 3. Relationship between the N.I. prediction and the one observed by metallographic inspection.

Conclusions

Active Mg control in ductile iron production can be performed in a fast and reliable manner using cooling curve recording method. From these cooling curves there is a relationship between the minimum eutectic temperature (Temin) and Recalescence (Recal) with the nodularity index (N.I.) in the corresponding plain cup.

Acknowledgments

Authors acknowledge Sakthi (Portugal).

References

1. Suárez R, Garay J & Larrañaga P, (2014) Patent EP2749657lA1.
2. Suárez R, Sertucha J, Larrañaga P & Lacaze J (2016). Metall. and Mat. Trans. B, 47(5), 2744-2753.
3. Larrañaga P, Gutierrez JM, Loizaga A, Sertucha J & Suárez R, (2008). AFS Trans. (08-008).

The 73rd World Foundry Congress
„Creative Foundry"
23 - 27 September 2018, Krakow, Poland

PUR COLD-BOX SYSTEMS – PAST – PRESENT - FUTURE

Dipl.-Ing. Peter Gröning

Hüttenes-Albertus Chemische Werke GmbH, Wiesenstrasse 23, 40549 Düsseldorf, Germany
PGroening@huettenes-albertus.com

Key words: PUR Cold-Box, Past, Present, Future, 50 Years of Development

1. Introduction

50 years have passed since the introduction of the Cold-Box Process to the Foundry industry. In 1968 it was presented at the AFS Metalcasting Congress and Exposition. The history of the Cold-Box process is a remarkable story of a technology and its further development until today. The development and application was going hand in hand together with automation processes and technology changes in the foundry and automotive industry.

2. Past

The first Cold-Box Systems used solvents based on hydrocarbons. In the early years the supplier and the user had to learn the basic characteristic/needs of the process, like: dry and pure silica sand, dry purging air, characteristics of the catalyst, need of special gas generators, different shooting and venting characteristic. The Cold-Box process started quickly to replace older and slower core making technologies. Catalyst usage was high in the beginning. Used catalysts were mainly DMEA and TEA. Specialized machinery (core shooter, gas generators) have been developed and introduced accordingly. Milestones of the Cold-Box development:

- 80s - Cold-Box is becoming water resistant.

That was an important milestone for the introduction of water based coatings in automotive foundries on a large scale. The introduction of water based coatings improved strongly the foundry environment and the workplace safety. It also was possible to coat the cores now directly after the production. That was boosting the automatization in the core shop and shortened the cycling times and not to forget improved the casting quality.

- 90s – Modification of solvents

Introduction of plant based (RME) solvents for Cold-Box Part 1 and 2. These new solvents improved significantly the productivity in the core shop. Advantages are less sticking, less smell, less pollution and very low catalyst consumption. In 1999, the first Cold Box system with silicium-based solvents (TEOS) was introduced to the foundry industry. Compared to the aliphatic and aromatic

solvents that had been used in the Cold Box process until then, the use of Tetraethyl Oorthosilicate was the first step towards a partly inorganic Cold Box system. With these Silicatic Systems another big step forward was taken to reduce the pollutant emissions in the foundry. Advantages of these systems are especially a reduced carbon content (Figure 1) – reduced gas evolution, smell, pollutant emissions and condensates.

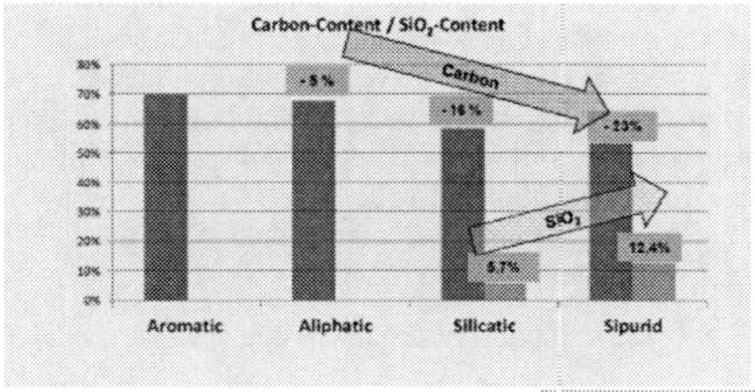

Figure 1 : Carbon content Cold-Box systems

Introduction of different new types of catalyst. DMPA (Dimethylpropylamine) was introduced to the foundry industry in the late 90s. This catalyst has a higher boiling point and a reduced odour characteristic compared to the most common catalyst DMEA. The advantage of DMPA is the lower smell level in the core shop and around the foundry.

3. Present

Organic binder systems and, above all, the Cold-Box process still remains the dominating process for core production, especially in the field of serial casting – a position that this process has attained due to its broad range of applications, its effectiveness and also the technical and technological further advancements. Modern Cold-Box Systems have to fulfil many requirements. For serial foundries the crucial property is high productivity. To reach high productivity modern Cold-Box systems have to achieve a lot of different characteristics like:

- high strength level
- high reactivity, low catalyst demand

- long benchlife
- low sticking
- low gas evolution,
- low pollutants emissions,
- high thermal stability

Newly developed gassing units allow a faster cycling speed and a reduced catalyst demand. Together with the use of catalysts with reduced smell (DMPA) that makes a great difference in the core shop.

4. Future

Today's Foundry Industry faces many challenges. On one side real challenging technical requirements like very high complexity and increased core intensity combined with the need of very low dimensional tolerances. The goal is to reduce the casting weight. That means the casting walls are becoming thinner. The standard casting wall in modern engine castings is now around 2.5 mm. On the other side the environmental legislation has led foundries to shift their focus increasingly to the application of sustainable, environment- and resource-saving technologies. Foundries will face even more and different changes and challenges in the near future. A real big change is expected with the rise of the E mobility. E mobility will sharply reduce the need of casting components and will also shift the focus further towards light metals. The 3 D printing is a relatively new technology but will change the traditional foundry business. From the core processes side we will see a growing share of inorganic applications. In the future of Cold-Box Systems we can see two main ways of offering sustainable solutions to the foundry industry:

- the reduction of organic content
- the reduction of monomer content

Reduction of organic content: We have been reducing the organic content in cold-box systems for many years – firstly by substituting the conventional aromatic or aliphatic solvents of the resin and activator components with inorganic solvents. In further development steps, the base resin was then designed in partly silicated form. The benefits of these systems include not only the reduction of odour, condensate and gas, but also the reduction of pollutant emissions such as BTX, BTEX, TOC and DOC.

Reduction of monomer content: Another option for reducing the pollutant load especially in the waste sand is to cut the monomer content of the cold box resins. Developments in this direction already took place in past decades. In the 1970s, 80s and 90s all industrial partners were heavily involved in the fields of research and application. Cold box resins with reduced monomeric shares were also developed and already brought to market at that time. Landfill costs progressively decreased again over the next few years, greatly reducing the pressure to use such systems. Investments in core sand reclamation equally declined. By reducing the monomer share in the cold-box binder, i.e., the proportion of free phenol + formaldehyde, the landfill suitability of waste sand can be improved. In particular, this allows the phenol index to be significantly reduced. These newly developed cold-box resins are manufactured using a special formulation and production technology. Standard resins have a monomer level of < 5.0%, usually 4.5%. So called low-phenol resins with a monomer content of 2.5% have long time been used in foundries (Figure 2).

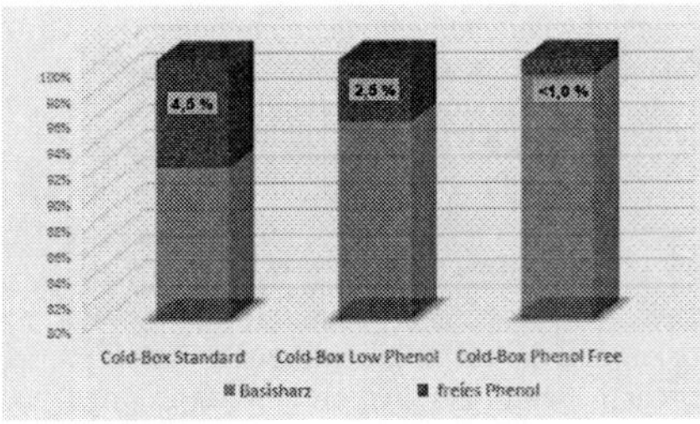

Fig. 2: Typical phenol content of cold-box resins

The phenol index input from the core sand can be significantly reduced with the new resins. Unfortunately, reducing the phenol content alone is not sufficient to lower all of the foundry's key emission parameters. The BTX emissions, TOC, DOC, the amount of gas and condensate as well as the odour load cannot be reduced by just lowering the phenol content.

Bringing two ways together: From our point of view, it is necessary to combine the advantages of the two basic approaches. For this reason, we have developed an organic-reduced resin system with a phenol (monomer) content of < 1.0%. This system enables us to reduce both the foundry's key emission parameters and the phenol index in the waste sand.

OPTIMIZATION AND AUTOMATION OF CHEMICAL CONTROL OF ALLOYS IN SMART FOUNDRY 4.0

Alberto Montenegro

AMV Soluciones, Av. Madrid 40 36214 Vigo, Spain
e-mail: alberto@amvsoluciones.com

Key words: Optimization, chemical control, automation, charge calculation, Big Data.

1. INTRODUCTION

Optimization of the casting process, whether for ferrous or non-ferrous materials, is determined by different factors that, as well as affecting the production costs and the quality of the alloy produced, directly influence the levels of rejected castings and are decisive in process efficiency and company competitiveness.

Chemistry is one of the key factors when it comes to getting high quality castings. That's why the following points are a must for any foundry or melt shop that aspires to be competitive and stay on the market:

1. Improve the chemical control of the alloys, achieving composition increasingly stable, according to the desired physical and mechanical properties.
2. Do it in the most efficient way possible:
   a. Optimizing the use of raw material inventory, obtaining the necessary qualities at the least possible cost.
   b. Minimizing the number of steps necessary to achieve the desired chemistry, with the consequent reduction in energy and processing time.
3. Avoid human errors in the input of the different parameters that will determine the final chemistry of the casting.
4. Have complete traceability of the process and learn from data history to continuously improve results.
5. Combine all these factors in a reliable system that guarantees the customer that their orders will be produced with maximum quality, optimizing costs, minimizing time and eliminating rejections.

How can we have control of such important points? Here are some important measures to be taken:

1. Calculate furnace charges and adjustments using mathematical algorithms that guarantee the results are free of errors, considering chemistry, recovery, melting losses and material costs, ensuring accurate chemical results at minimum cost.
2. Integrating the different measurement and analysis devices (scales, spectrometers, thermal analysis systems, etc), so that their results are automatically incorporated in the system, showing alarms if necessary and carrying out the following calculations based on those results:
   a. Furnace adjustments, if required.
   b. Ladle additions.
3. Applying Linear Regression techniques and, in general, Big Data analysis to the casting history and refining the input parameters from them, so that the actual chemical results are closer and closer to those calculated by the system.

Those techniques, which AMV has recently started to apply in several foundries of different countries, with interesting initial results, will form the basis of a Case Study AMV will be presenting together with United Cast Bar (UCB) - Chesterfield (UK), a world-class producer of Continuous Cast Iron Bars, both in Grey and Ductile Iron.

The study will be conducted based on cumulative production results over 6 months, analyzing at each stage different KPIs: melting time, energy consumption, chemical variability of heats and raw material costs.

2. CASE STUDY: UNITED CAST BAR (Chesterfield)

CURRENT STATE:

The study is being carried out through the implementation of the Linear Programming based system ALEA. The following points describe how the system is being used, as well as the first improvements observed:

1. ALEA system is implemented in a local server of UCB, with PC terminals in different areas of the plant, in particular in the laboratory, crane cabin and furnace platforms. The system is used by the laboratory supervisors to calculate furnace charges and chemical adjustments.

2. The system works integrated with (*Fig. 1*):
    a. Spectrometer: chemical analysis can be imported into the system in real time and stored together with the heat number. If the heat is out of spec, ALEA calculates the optimal adjustment considering the actual chemistry.
    b. Scales: 1 crane scale and 2 manual scales at the corresponding furnace platforms.

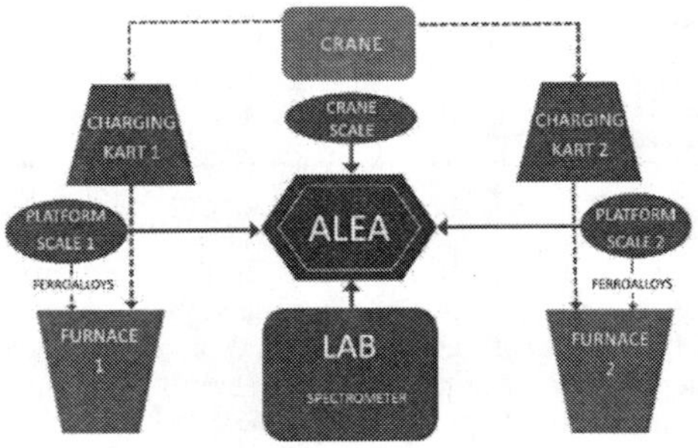

Fig. 1. ALEA system implementation scheme.

3. The operator reads on ALEA screen the different materials to be charged, including amount and charging order. The system reads the charged weight from the scales and warns when it is within the admissible limits. The operator does not type any quantity, but validates the actual values by pressing an industrial button under the screen when the system asks for confirmation.

4. Based on the chemical results of the heats prepared during the first 2 days under the conditions described, a first review of the chemical composition of the raw materials was made, in particular the C, Si, Mn, Cu and Cr content of scrap and returns. Also, the ranges of those elements used by ALEA for charge calculation were slightly narrowed.

5. After 30 days working under the above conditions, a 70% reduction in the number of melts requiring adjustment has been observed, as well as a higher consistency of chemical analysis, with increasingly narrower chemical variation bands.

NEXT STEPS (results to be included in extended paper):

Once a 60-day uninterrupted casting history has been accumulated using the system under the conditions described above, the following steps will be performed:

1. Analysis of deviations between expected chemistry (calculated by ALEA) and the actual one (given by spectrometer). Let p_{ij} be the recovery of the element j for the material i, $[0\% \leq p_{ij} \leq 100\%]$, C_i the amount of material i used for the charge. The expected amount of element j in the molten metal is:

$$E_j = \sum_i p_{ij} \cdot C_i \qquad (1)$$

So that the expected proportion of element j in the molten metal is:

$$P_j = E_j \Big/ \sum_k E_k \qquad (2)$$

If $\hat{P}_j$ represents the actual proportion (read from spectrometer), we will analyse the difference:

$$\hat{\varepsilon}_j = (P_j - \hat{P}_j)/P_j = 1 - \hat{P}_j/P_j \qquad (3)$$

for each alloy, calculating its average (ideally close to 0) and its variance (which we want to minimize).

2. Apply Linear Regression analysis to this 60-day sample, estimating the recoveries of the main elements for the different raw materials, which will provide experimental values $\hat{p}_{ij}$ (recovery of element j in material i) with their corresponding Coefficient of Variation $CV = \sigma(\hat{p}_{ij})/\mu(\hat{p}_{ij})$.

3. Update the chemical composition and recovery of the raw materials in ALEA, considering only the most significant $\hat{p}_{ij}$ (CV not higher than 20%). Also, narrow the charge calculation range of the different elements according to their weighted CV (higher CV, narrower range). These new values will be automatically applied by the system in the following calculations.

4. Repeat the same analysis and recalibration of the system every 60 days. The following KPIs will be calculated at each step:
    a. Heats requiring chemical adjustment (%).
    b. Average melting time per heat (min / heat).
    c. Average energy used per heat (kW h / heat).
    d. For each alloy, chemical variation ranges, overall and per element (%).
    e. Cost in raw materials corrected to market prices (cost / tonne produced).

References

[1] D. Alevras, M. Padberg. (1999). "Linear Optimization and Extensions". Berlin: Springer-Verlag.

[2] A. Gelman, J. Hill. (2006). "Data Analysis Using Regression and Multilevel/Hierarchical Models". New York: Cambridge University Press.

[3] W. Winston. (1995). "Operations Research: Applications and Algorithms". Duxbury MA: Duxbury Press.

[4] Arman Bolourian. (2010). "Evaluation of Energy Efficiency Improvement". Göterborg: Chalmers.

[5] S. Zinn, S. L. Semiatin. (1988). "Elements of Induction Heat". Palo Alto CA: ASM International.

[6] H. Sahai, M. I. Ageel. (2000). "The Analysis of Variance: Fixed, Random and Mixed Models". New York: Springer Science & Business Media.

The 73rd World Foundry Congress
„Creative Foundry"
23 - 27 September 2018, Krakow, Poland

CORDIS® PROCESS – THE NEXT GENERATION

Dipl.-Ing. Klaus Löchte

Hüttenes-Albertus Chemische Werke GmbH, Wiesenstraße 23,40549 Düsseldorf, Germany
Kloechte@huettenes-albertus.com

Key words: Inorganic binders – 3-component-system – climate adaption-

Inorganic binders on the rise

The use of inorganic binding systems in foundries has significantly increased during the last years. In 2010 approximately 600 tons of inorganic binders have been used worldwide, but in 2015 it were already 8,000 tons. Various German automotive manufacturers as well as some globally acting foundries established the inorganic systems successfully for their serial production. Meanwhile, also in other important industrial nations a stronger environmental awareness causes new ways of thinking in the foundry industry. Strict emission standards accelerate for example in Asia product and process innovations. Furthermore, technology innovators such as the German automotive producers VW, Daimler, and BMW contribute to the prevailing of in Europe established technologies as „State-of-the-Art".

In spite of the fact that inorganic binding systems achieved a certain maturity, providing nowadays obvious commercially and qualitatively high-grade results, the development is by far not finished.

Challenges of IOB

Over the last years, HA foundry experts and chemists worked hard to further improve the IOB technology and products. This technology is rising various challenges, in each step of the manufacturing process. One of these challenges is to optimize the storage stability of the cores under difficult climatic conditions (meaning humid-warm surroundings). Such conditions are typical for Asian countries, but also in temperate areas during the summertime. Due to the hydrophilic properties of the cores manufactured with a binding system based on water glass, they tend to absorb moisture from the air and thus become instable.

New 3-component-system

In order to solve the described problems Hüttenes-Albertus developed an efficient concept. It consists of a new, flexible 3-component-system which is adaptable to changing climate conditions: additives to enhance the store stability are not integrated into the Cordis silicate solution in the production stage, but are delivered separately – as a "summer" additive for high humid conditions > 60 % and as a "winter" additive for low humid conditions.

The individual binder components are long-term stable. Depending on the climatic situation the permanent binder, the respective fluid additive, and the powder additive are being mixed with the sand directly. Consequently, a high-performing binding system is achieved, being perfectly adaptable to the individual conditions and showing a low moisture absorbance as well as a high storage stability of the produced cores.

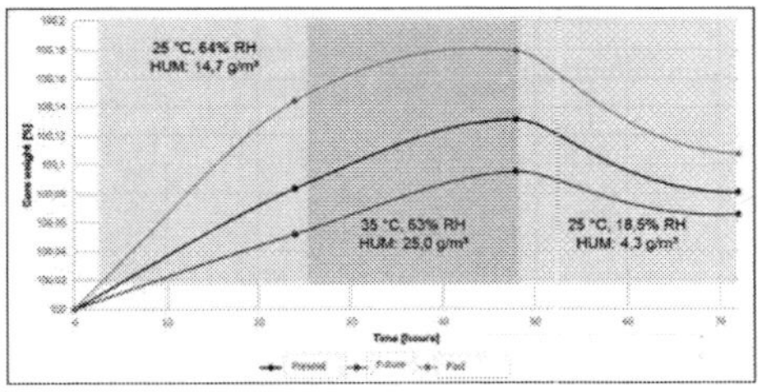

Fig.1. Comparison between different generations of Cordis: water absorption and release under changing air humidity

Successful application in foundries

The 3-C-System is already running on industrial base, for almost one year, in a renowned foundry. Over this period, not only the increase of the storage stability of cores could be established, but also the improvement of the following relevant items:

- Performance during the core production
- Size accuracy of the cores
- Scrap rate of the produced castings
- Size accuracy of the castings and
- Core removal

The 73rd World Foundry Congress
„Creative Foundry"
23 - 27 September 2018, Krakow, Poland

The role of recycled ceramic material obtained from the ceramic layered moulds used in the investment casting

Soroczyński [1]*, R. Haratym [2], R. Biernacki [1]

Department of Plastic Forming and Foundry Engineering, Warsaw University of Technology, ul. Narbutta 85, 02-524 Warszawa, Poland
[b] *University of Ecology and Management, ul. Olszewska 12, 00-792 Warszawa, Poland*
**Corresponding author. E-mail address: asoroczy@wip.pw.edu.pl*

Key words: recycling, ceramic mould, investment casting, energy consumption, (use style Key words)

1. Introduction

The article presents the role of the ceramic layered moulds used in the investment casting method with new (certified) and recycled material from ceramic moulds after casting process. The materials that were obtained are mainly aluminosilicates and SiO2. The investigation of changes in the quality of ceramic moulds (including the recycled ceramic material) includes the chemical composition of the ceramics as recovered ceramic material, changes in the particle size of the layered covering material, the gas permeability (CM) during the pouring of liquid metal, and the creation of the porosity are presented. Than the thermophysical parameters and dimensional accuracy of the casting manufactured in the new ceramic layered shell moulds were analysed. Additionally the global cost savings and improved ecological conditions in the foundry and its surroundings was estimated.
The ceramic layered shell mould in the investment casting process have a major influence on the quality of manufactured castings [1]. The dimensional accuracy and the surface quality are the primary objective in this manufacturing technology. Ceramic moulds dimensional distortions are the reason of approximately 60% of overall errors of the casting manufactured mainly with the use of aluminosilicates [2,3]. Based on the quality of the castings the amount of the ceramic moulds with aluminosilicates has increased at least 40% of the global investment casting production. At the same time we need to pay attention to the SiO2. In the sustainable production very important are the environmental aspects affecting human health.

2. Research methodology

The laboratory trials include the special dilatometric analysis of the test samples to achieve the thermal expansion parameter of the ceramic mould samples with the dimensions of □20x60 [4]. The Olympus stereo microscope was used for the evaluation of the ceramic grain size. Additionally the laser measurement of the changes in the particle size and the porosity using mercury porosimetry was checked at random ceramic moulds from the WSK Rzeszow company.
The thermal conductivity was evaluated based on the cylindrical ceramic samples.

3. Verification of the chemical composition of the samples of ceramic moulds

The changes in the amount of SiO_2 in the aluminosilicates of recycled material were found. The amount of SiO_2 increased from 22% to 25,7% due the transition of the certain amount of the binder to the crystalline state. The total content of the Al2O3 + SiO2 has slightly changed and in practice that has no effect on the thermal expansion of the ceramic mould.

4. The evaluation of the ceramic mould porosity

The use of the mercury porosimetry based on the capillarity effect shows very low porosity in the range of 26-28%. The samples of recycled ceramic mould material shows the porosity at similar range 26,5 – 29%.

5. Verification of the particle size of the clean and recycled ceramic moulds material

A new material with the grain diameter of ~ 0,9 mm and a recycled material recovered in the laboratory device using the hammer crushing method, there was a significant fragmentation of the material to the average of 0,5mm grain size. When the recycling process of the ceramic mould was conducted in the laboratory (150mm diameter rollers, the gap between the rolls of 1.2 mm), there was no difference between the grains of a new and recycled material. The rolling mill was used in the Warsaw University of Technology.
For the determination of the particle shape the method commonly used in the powder sintered technology was partially adopted [5]. The shape of the grains was presented by the value of Wk later called the size ratio. Wk value is defined as the ratio of the circumference of grain to the circle inscribed in the outline of the grain. For the evaluation of the Wk value a special software available from the Olympus microscope for the image analysis of each particle was.

6. Gas permeability of ceramic molds

The evaluation of the possibility of multiple recycling of the ceramic material needs a set of further laboratory trials regarding the permeability (gas permeability). The verification was carried out on the laboratory stand presented on the fig. 1. The test was commenced in the 20°C temperature.

7. The impact of the recycled ceramic moulds based on the aluminosilicate on the economic and ecologic aspects

The dorm production delivers approximately 800 Mg of precision casting, which results the amount of over 1000 Mg used ceramic and the amount of used aluminosilicate may be up to 400 Mg. By using the recycled ceramic materials the reduction of the energy consumption used in manufacturing 1kg of ceramic mould by approximately 5MJ (research of the Warsaw University of Technology). Based on the assumption that to produce 1 kWh, the thermal contribution is about 9MJ [9,10]. With the estimated value of 40% used recycled aluminosilicate which is about 160 Mg that could save approximately 88000 kWh of energy. We need to remember that with the primary energy carrier which is the bituminous and lignite coal, the creation of 1kWh falls the emissions of approximately 800g of CO2, SO2 and approximately 7g 3g NOx [9]. That avoids a significant environmental pollution. That ecological damage have influence on the health. The production of 1kWh is giving the 4USD/1kWh of penalty and for the 88000kWh is near 350000USD. Additionally coal energy have great impact on the global warming which is counted to be even 88000 KWh x $12USD_{avg}$ [10].

8. Conclusions

- Recycling aluminosilicate can significantly reduce the cost of the production in the investment casting foundry,
- The thermophysical properties of ceramic moulds manufactured using recycled materials are comparable to the moulds from the new ceramic materials
- Dimensional accuracy and the surface quality acquired by using new and recycled materials is similar and ensures the production of high quality castings

References

[1] Haratym R., Biernacki R., Myszka D. (2009). *Ecological production of accurate castings in ceramic molds.* Warszawa: Oficyna Wydawnicza Politechniki Warszawskiej.

[2] Snow, D.H. Scott: (2001). *Comparing Fused Silica and Alumino – Silicate Investment Refractories*. Incast

[3] Karwiński A.: Technological parameters of the process of making molds of ceramic with the participation of "water" silicate binder, monograph "Polska Metalurgia w latach 2011 – 2014", s. 529 – 541, ISBN 978-83-63663-47-6, Wydawnictwo Naukowe „AKAPIT", Kraków, 2014,

[4] Haratym R., Karwiński A.: The influence of the properties of the ceramic mould on its thermal deformations- X Międzynarodowa Konferencja „Spolupraca 2004" 28 – 30.04.2004 Tatrzańska Łomnica, Słowacja, ISBN 80-8070-231-4

[5] Nowacki J., (2005). *Sintered metals and metal matrix composites.* Wyd. Naukowo-Techniczne, Warszawa

[6] Haratym. R., (2006). *The impact of recycled materials from used ceramic moulds for some of the thermo-physical properties in the process of investment casting* Archives of Foundry 6.20

[7] Karwiński A., Żółkiewicz Z.: The Research of Properties of Experimental Ceramic Layers; Archives of Metallurgy and Materials, Vol. 59, Issue 2/2014, s. 703 – 705

[8] Karwiński A., Leśniewski W.: Reasearch on Investment Casting Liquid Ceramic Slurry – Foundry Trade Journal, Volume 179 No.3625, June 2005, s. 158-159,

[9] Central Statistical Office, Fuel and Energy Economy in 2011-2012, Warsaw 2013

[10] Lewandowski W.M., (2007). Environmentally friendly renewable energy source WNT Warszawa

The 73rd World Foundry Congress
„Creative Foundry"
23 - 27 September 2018, Krakow, Poland

ENVIRONMENTAL AND APPLICATION METHOD IMPROVEMENTS – AN R & D APPROACH

Presentation/Paper Synopsis

Alan Burrows

Hüttenes-Albertus UK, Unit 3 Marrtree Business Park, Kirkwood close Oxspring, Sheffield S36 8ZP United Kingdom,
Corresponding address: ABurrows@huettenes-albertus.com

Key words: Hot toppings – Fluxes - Emissions reduction – Health & safety – Application method

Introduction

Health, safety and the environment are at the forefront of new product development in today's world. This paper describes the developmental approach of a foundry supplier (HA) to meet those challenges.

This is essentially a two-stage approach, with the initial emphasis on designing a product with emissions reductions based on chemistry and selection of raw materials, followed by consideration of which technique/application method is best suited to the customer requirements.

Raw material hazard reduction commenced in early nineties with 'ecoflux' product range. Initial focus was on reducing/removing/replacing toxic fluorides and organo-halides in aluminium flux range. This principle has been extended into products for other alloy groups, including low hazard hot toppings for the ferrous industry. The physical form of the product itself will have an influence on the choice/s of application methods.

Application of the products via the safest and most efficient manner has led to equipment based solutions such as rotary degassing, flux injection, rotary flux injection, and flux application devices for activities such as furnace wall cleaning.

Examples of using chemistry and application methods to improve the working environment

Chemistry approach

An example given is a company producing corrosion resistant components in GJS ductile iron and other ferrous alloys, with casting weights up to 16 tonnes. In close co-operation between customer and our R&D team, new types of hot topping compounds were developed, with focus on emissions reduction which were measured by an accredited independent laboratory appointed by the customer. Traditional hot topping compounds may contain chemicals that generate toxic gaseous compounds such as fluorides, and oxides of nitrogen, sulphur, and carbon. These new generation hot toppings are free from chemicals that generate these hazardous compounds arising from their decomposition.

Initial laboratory evaluation was made comparing existing products with two new formulations. Initial findings comparing the new formulations with the established customer product is summarised as follows:

1) Burn time for both new formulations were more than double that of established product.
2) Expanding ratio of new products at 50% for one product and 100% for second product, the second product matching the established one.
3) Smoke levels were very low on visual assessment for both new products, moderate smoke was seen on established product with smell of 'plastic' reported.
4) The peak temperatures recorded for the two new products were in one case below and in the other above that of the established product, (-8% & + 15%).
5) The time taken to reach maximum temperature for the two new products were longer (11% & 20%), this is to be expected given the extended burn time seen with the new products.
6) 10 minutes after ignition, one product was hotter on residual temperature and one lower than the established product (-7% & + 15%). 20 minutes after ignition, the new products were 15% and 18% hotter, suggesting better insulation/heat retention properties for the new products.

Following the lab assessment, foundry trials were conducted, assessing performance of both feeding characteristics, and also measurement of in-house foundry emissions. The latter was conducted by an

independent research facility experienced in emissions testing within foundries.
This testing was arranged by the foundry, testing their preferred new product compared with the established one. The results are summarised as follows:

1) Fluoride levels as expected were negligible for the new product, also for the established product (limitations on testing equipment confirms levels below 1.7ppm and possibly zero value for both products tested).
2) Carbon monoxide levels were reduced by 50% with the new product compared with the established one.
3) Phenol levels at very low and equivalent levels for both products, and furfuryl alcohol, formaldehyde, benzene and toluene all at lower recorded levels with the new product. Sulphur dioxide reduction of 74% for the new product.
4) The foundry advised that in all cases during the trials, the feeding performance of the hot toppings was deemed equivalent and satisfactory based on casting results.

In the case of items 3), these emissions may be attributed wholly or in part to the resin system used, however the reductions were noted, and the reasons for the apparent improvement are open to debate.

Application approach
An example of using improvements in application method to reduce hazards is given of a furnace wall cleaning flux appli-cator.
In addition to flux formulation improvements, newly designed flux appli-cators are now available, offering several advantages compared with manual application of flux, and older style flux applicators.

Fig. 1.
New Design Flux Applicator

The author describes the benefits of using a wall cleaning flux in combination with an applicator (maintaining furnace efficiency and capacity, reduction of mechanical damage through scraping resulting in extended life of the furnace lining, controlled application, reduction in flux use, reduction in airborne flux/operator exposure).

In summary, the new devices reduce the operator exposure to flux powder and also the furnace heat during use, and are highly portable and lightweight.The features, advantages and benefits of the newly designed flux application equipment compared with previous designs may be summarised as follows:
1) Reduction of device weight (in use) based on separation of flux hopper from the lance and flux dispensing arrangement.
2) User friendly design and ease of portability.
3) Larger hopper size better suited to larger furnaces.
4) Optional extension pieces on lances for larger furnaces.
5) Vibration device at outlet to improve consistency of powder flow.

Table 1. Emissions Table – Hot Toppings

Topping Compound	Hazardous Substance	8 hour TWA Conc.	Exposure Standard	Units
Huttenes-Albertus UK Ltd Hot topping reference 1	Carbon monoxide	67	**30**	ppm
	Carbon dioxide	0.07	**5000**	ppm
	Sulphur dioxide	0.7	-	ppm
	Hydrogen fluoride	<1.7	**1.8**	ppm
	Furfuryl alcohol #	<0.03	-	mg/m^3
	Phenol #	<0.04	**2**	ppm
	Formaldehyde #	0.08	**2.5**	mg/m^3
	Benzene #	1.4	**1**	ppm
	Toluene #	2.7	**191**	mg/m
Established Customer Product	Carbon monoxide	133	**30**	ppm
	Carbon dioxide	0.07	**5000**	ppm
	Sulphur dioxide	2.7	-	ppm
	Hydrogen fluoride	<1.7	**1.8**	ppm
	Furfuryl alcohol #	<0.08	-	mg/m^3
	Phenol #	<0.04	**2**	ppm
	Formaldehyde #	0.12	**2.5**	mg/m^3
	Benzene #	3.1	**1**	ppm
	Toluene #	6.4	**191**	mg/m^3

Notes: Exposure standards are Workplace Exposure Limits as defined in EH40 unless otherwise stated;
 # = Sub-contract laboratory analysis. Oxides of nitrogen not monitored, but based on the new product formulation, zero value is expected as the product is free from nitrates;
Highlighted red = compounds typically seen arising from hot topping chemistry;
Highlighted blue = compounds typically associated with the resin binder system being used.

Conclusion
In many cases, by careful selection of 'low hazard' chemical compounds, and control of their purity and grain size distribution, it is possible to develop products fit for purpose without the need to use compounds that generate hazardous emissions.
Development of new product formulations which are intrinsically safer than the ones currently in use remains the primary objective of the chemist. However, application method also can have an important role in reducing operator exposure during use.

The 73rd World Foundry Congress
„Creative Foundry"
23 - 27 September 2018, Krakow, Poland

CERAMIC-CARBON FILTERS FOR MOLTEN METAL FILTRATION

Maciej Asłanowicz[1], Andrzej Ościłowski[1], Barbara Lipowska[2]*, Zbigniew Robak[3], Roksana Muzyka[3], Piotr Kowalski[4], Krzysztof Wańczyk[4]

[1]Ferro-Term Sp. z o.o., Przybyszewskiego 176/178, 93-120 Łódź, Poland
[2]Institute of Ceramics and Building Materials, Refractory Materials Division, Toszecka 99, 44-100 Gliwice, Poland
[3]Institute for Chemical Processing of Coal, Zamkowa 1, 41-800 Zabrze, Poland
[4]Foundry Research Institute, Zakopiańska 73, 30-418 Kraków, Poland
**Corresponding address: e-mail: b.lipowska@icimb.pl*

Key words: Foundry, Molten metals' filtration, Foam filter

Ferro-Term Sp. z o.o., the only Polish producer of ceramic filters for the foundry branch, in cooperation with a scientific consortium composed of the Institute of Ceramics and Building Materials, Refractory Materials' Division in Gliwice, the Institute for Chemical Processing of Carbon in Zabrze, Foundry Research Institute in Cracow, within the framework of the projects co-financed by the National Centre for Research and Development from EU resources, implemented as part of programmes INNOTECH and PIOR.01.01.01, developed a technology of manufacturing a new generation of foam filters for the filtration of molten metals, in particular high-grade alloy cast steels.

The reason for undertaking the investigations which resulted in the development of the above mentioned technology were problems reported by metallurgical plants due to the use of ZrO2-based filters. When these filters were applied, the alloy was observed to solidify in the filter canals. This phenomenon can be prevented by increasing the alloy temperature (in the case of melting high-grade iron alloys with a high content of chrome and nickel, the temperature of form filling is nearly 1700oC). However, this process is not economical and, additionally, it has an adverse effect on the thermomechanical properties of filters, causing their cracking or deformation.

One of the main components of the newly developed filters is carbon, the presence of which, compared to ZrO2-based filters, has contributed to reducing the density of the filter material and decreasing its heat capacity, thus limiting the cooling of the alloy in the process of filtration.

The new type of filter was obtained by a well-known method [1], which involves coating the elastic polymer foam with appropriately prepared ceramic suspension in such a way that it covers only the elastic bridges surrounding the foam pores. So obtained foams are subjected to thermal treatment, during which the organic skeleton is fired and the coating ceramic material is sintered.

Carbon is added to new filters in the form of caking coal with an addition of pitch characterized by a low emission of harmful polycyclic aromatic hydrocarbons. In the process of mix heating, the caking coke - a novelty in the production of filters, initially reaches a plastic state level, coats and sticks the ceramic components' grains, and, next, as the temperature increases, it gets sintered, forming a ceramic-carbon material [2].

The newly developed filters have been subjected to casting trials in both laboratory and industrial conditions. They were tested in mould gating systems that allow making 50-70 kg casts. The obtained results [3] confirmed the filters' high resistance to thermal shock and to the erosive effect of molten metals as well as their mechanical strength at high temperatures and high effectiveness. The obtained results confirmed the rationality of starting up a serial production of ceramic-carbon filters. For this purpose, the process line owned by Ferro-Term Sp. z o.o. was modernized – it was among others equipped with a specially designed furnace for firing filters in an oxygen-free atmosphere.

The first batches of filters produced on this line are being subjected to detailed examination by research institutes belonging to the consortium and are being tested in the cast iron and cast steel foundry of Ferro-Term Sp. z o.o. The obtained results confirm that the newly developed filters:

- have low heat capacity, which eliminates the effect of metal initial solidification;
- due to their low wettability coefficient, they permeate molten metal more quickly, and, at the same time, are characterized by high efficiency of non-metallic contaminants filtration;
- have a low specific weight, which allows reducing the filter mass in transport packaging;
- although the ceramic-carbon slurry applied to manufacture filters contains carbon, it does not penetrate into molten metals;
- they are characterized by high versatility of application, i.e. they can be used to filter iron

alloys (GJL, GJS, GS) and non-ferrous metals (Cu and Al).

[1] Schwaltzwalder K. & Sommers A.V. (1963). US3090094A. Method of making porous ceramic articles.
[2] Asłanowicz M., Karwiński A., Lipowska B., Muzyka R., Ościłowski A., Robak R., Ściążko M., Wieliczko P., & Witek J. (2014). Zgłoszenie patentowe nr P.40833. Masa ceramiczno-węglowa do wytwarzania filtrów piankowych.
[3] Lipowska B., Witek J., Wala T., Karwiński, A., Wieliczko P., Asłanowicz M. & Ościłowski A. (2014). Cast steel filtration trials using ceramic-carbon filters. Archives of Foundry Engineering. Vol. 14, 4, 61-66.

The investigations were conducted within the projects co-financed by the National Centre for Research and Development: Innotech I Programme „Innovative carbon-ceramic filters for the filtration of metal alloys" and the Smart Growth Innovative Programme, Activity 1.1, Subactivity 1.1.1 Industrial research and development work implemented by enterprises: „Modernization and adaptation of existing technological lines for the purpose of verification technology and production start-up of innovative carbon- ceramic filters for the filtration of liquid metal alloys".

REFRACTORY COATINGS IN CENTRIFUGAL PROCESS: THE CHANGE FROM RELEASE AGENT TO MOULDING MATERIAL

Dr. Klaus Seeger, Dr. Ekaterina Potaturina

Hüttenes-Albertus Chemische Werke GmbH, Hansastraße 1, 30419 Düsseldorf, Germany
e-mail: KSeeger@huettenes-albertus.com, EPotaturina@huettenes-albertus.com

Key words: refractory coating, centrifugal process, surface, casting defects, insulation

Process Description

Centrifugal Casting Process: rotatory symmetric castings are produced in a high speed rotating metal mould (die). The melt is poured into the die as this one is rotated with a defined speed. Thus, the operating centrifugal forces squeeze the melt on the die wall and a high compact structure is developing with less pores and cavities. The internal geometry of the die defines the external geometry of the casting. The wall thickness is resulting from the poured melt quantity. One can distinguish between horizontal and vertical centrifugal casting. Refractory coatings are used for runners as well as for dies – since the development of the process, their significance has grown along with the requirements. Several advantages may be achieved by the application of modern coatings. This paper is focusing on the changes of coatings along time, with regard to new requirements and solutions, offering practical examples.

Release Agent

Originally, when the process started to be used, powder-coatings were often applied, their role being only to achieve a separating effect between the die and the melt. The dry powder was applied directly into the die. It contained a mixture of Ferro silica, powdered graphite, silicon, bentonite or diatomaceous earth. Later coatings were developed by addition of water, being applied by brushing or spraying. The covering of the die's surface with coating allowed a much easier extraction of the casting and a longer lifetime of the dies.

Defect Prevention

The usage of coatings generated also casting defects like inclusions, gas or reaction defects, which could not be completely avoided by changing only the process parameters. There causes of these defects are multiple and various. By optimizing the composition and structure of the refractory coatings, many casting defects may be avoided. For example, replacing the organic compounds with inorganic ones, or the using of calcined raw materials, results in decreasing the loss of ignition, thus reducing the risk of gas defects. Alternatively, the risk of reactions defects may be avoided by choosing low-reaction raw materials.

Insulating Properties

Each coating is offering certain insulating properties, which are depending on its composition and the layer thickness on the die. This parameters significantly influences the cooling rate of the melt and affect the mechanical properties of the casting, which are a crucial part of customers' requirements. The selection of suitable raw materials, has a decisive influence on the thermal conductivity, but also on the chemical reactivity, refractoriness, hardness and application properties of the coating. Additionally, the grainsize distribution is of crucial importance. A fine grainsize creates plenty grain borders, thus slowing down the thermal conductivity; meanwhile, large grainsize creates cavities between the single grains, having an insulating effect as well. Apart from the metallurgy itself, the properties of the coating also have a limited influence on the microstructure and consequently on the mechanical characteristics. For thin-walled pipes targeted insulating properties must be set to avoid the formation of carbides during solidification.

Surface Roughness

For numerous applications there are almost no requirements regarding roughness for the surface quality because most of the castings will be machined after production. Only for a few and very special applications the form and height of the roughness are defined very clearly. For instance sewer pipes (grey cast iron): their production requires very even surfaces with preferably low roughness values. This requirement has no direct impact regarding the functionality of the casting, but some foundries will see it as a quality criterion for themselves as it shows a visible area. Thereby the foundries guarantee actual customer requirements and expect to gain new orders from customers.

Regarding the cylinder liners there are specific requirements to achieve rough surfaces, with a precisely defined structure - in contrast to the even surfaces. Originally, cylinder liners were pressed into engine blocks. With the further development of methods and coatings it was possible to achieve defined surface structures, which allows the direct pouring of the liners into the engine block. The structures with undercuts can only be processed in high-pressure-die-casting. Whereas the liners with a hedgehog- or mushroom-structure can be processed also in the conventional gravity-die-casting. Through the external structure, the cylinder liners interlocks with the engine block and cannot be removed this easily. The coating is significantly involved in the design of those external structures – beside some other machine parameters.

Fig.1. Casting Surface with Undercuts

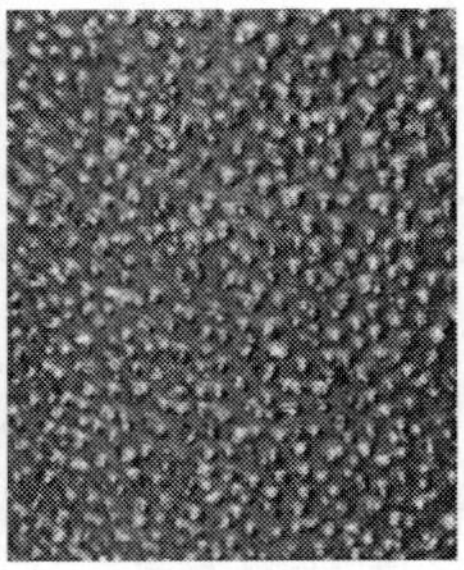

Fig.2. Casting Surface with hedgehog structure

Uniformity

To get a competitive advantage many foundries trim their processes to higher productivity or higher casting quality. This leads to smaller and smaller processing windows, with the risk of high damage at even the smallest deviations from process borders. These requirements do not exclude coatings. It is more and more important to produce high stable coatings with constant characteristics, such as viscosity, composition and application behaviour. Coatings mainly consist of natural raw materials; according to quality, certain variations can occur. Extensive quality checks of raw materials after receiving can detect deviations from the prescribed characteristics, which can be sorted out at an early stage before being used in production. Additionally, a frequent control of the process should be implemented with the help of suitable measuring methods.

Conclusion

The importance of the coating is growing in line with its tasks. At first, the coating was only a release agent between die and casting. With constantly growing requirements, the tasks of the coating increased. Today the coating is used to avoid casting defects, influence mechanical parameters and as a forming medium for an even external structure. To meet the future challenges HA is cooperating with its customers on a "Centrifugal coating for tomorrow".

The 73rd World Foundry Congress
„Creative Foundry"
23 - 27 September 2018, Krakow, Poland

USING OF 3D PRINTED PERMANENT PATTERNS FOR MASS PRODUCTION OF CASTINGS ON „GREEN SAND" MOULDING LINES

Milan HORACEK [1]*, Jaroslav LAUERMAN [2], Vojtech VASEK [3]

[1]Brno University of Technology, Faculty of Mechanical Engineering, Brno- Technicka 2, Czech Republic,
[2] UNEX, a. s., Olomouc, Czech Republic,
[3] ALW, a. s., Olomouc, Czech Republic,
**Corresponding address: horacek@fme.vutbr.cz*

Key words: 3D printing, automatic moulding line, green sand casting, permanent pattern

1. Introduction

To shorten the time required to make first product after demand is a crucial factor in every field of industry, not excluding the foundry. The main aim of this work was to compare the behaviour of different materials used for permanent patterns on an automatic moulding line and especially to concentrate to the possibilities of using patterns made by 3D FDM printer. The material choice for making permanent patterns is quite broad. The final material choice will depend on requirements to the patterns behaviour such as their lifetime, dimensional accuracy and also the speed and the economy of their manufacturing.

Woods - The one of the most used material for permanent patterns

Metals - The metal materials are used especially for mass production of castings, i.e. mostly on the moulding lines

Resins - These materials could be divided into polyurethane pouring resins, epoxy resins or the fast cured poured masses.

Artificial woods - Wood-plastic composites (WPC) are mixtures of wooden flour and polymer. This material is supposed to eliminate the negative properties of the wood itself.

2. 3D printing technologies for foundries

2.1 Basic 3D printing RP methods

Most of these methods could be divided into 3 basic groups: Based on the liquids, powders and solids

2.2 Surface finish of the 3D printed patterns

For improving the surface finish of the printed patterns, i.e. removing so called "stair-stepping effect", the either mechanical or chemical final treatment can be applied. The basic mechanical methods are grinding or shot blasting, the chemical ones could be for example dipping the patterns into a solutions which could smooth the surface by a chemical reaction.

2.3 Rapid Prototyping (RP) for making moulds

There are in principal three areas of using 3D techniques for making castings in foundries (Fig.1):

Direct mould making- By using one of the RP methods the mould is being manufactured, either the permanent metal mould or more often the non-permanent sand mould (Fig.1 – upper line)

Direct mould making via "master pattern"- By using one of the RP methods the permanent pattern (called master pattern) is being made which is then used for making sand moulds on moulding lines (Fig.1 –middle)

Indirect mould making - The sand mould is being made in more steps, i.e. firstly the "master pattern" is made by RP, which is then used for making several needed resin based/urethane by using a silicone/rubber mould (Fig.1- bottom line)

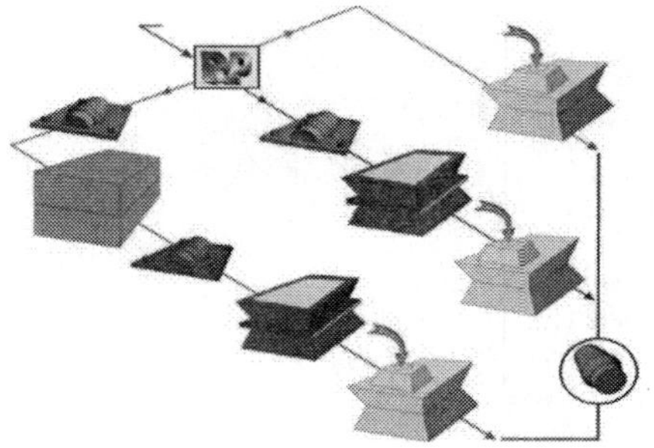

Fig.1 The ways of making quick castings by using RP

3. Experiments – main goals

The main goal of experiments was to evaluate behaviours of different materials used for permanent patterns on the automatic moulding line. All experiments have been carried out simultaneously with the full production in the UNEX foundry in Olomouc. There have been done 3000 moulds in total and the quality (mainly dimensional stability) of tested 8 different patterns always evaluated after defined number of moulding cycles, i.e. 50, 100, 200, 300, 500, 1000, 2000 a 3000.

3.1 Experimental conditions

Design of the tested patterns

The shape of the tested patterns ("comb shape") have been designed in accordance to the similar shapes of majority castings made in the foundry UNEX and also taking into account the easiest possibility of measuring changes in pattern dimensions after certain number of moulding cycles.

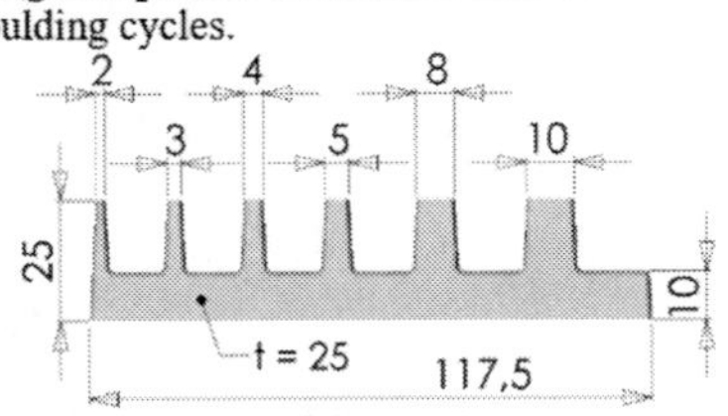

Fig.2 Design of the tested patterns

Tested materials of patterns

A/ Patterns from 3D printing machine

A1/ MakerBot PLA	***A2/ ABSplus-P430***
Cost per pattern: 2.7 Euro	Cost per pattern: 27 Euro
Printed layer: 0, 2 mm	Printed layer: 0,167 mm

B/ Artificial woods

B1/ Ebab 60-1	***B2/ Ebab PW 920***	***B3/ Ebab 140***
Price: 5 Euro/kg	Price: 6.5 Euro/kg	Price: 8.5 Euro/kg
Pattern. 0.25 Euro	Pattern. 0.75 Euro	Pattern. 1.1 Euro

C/ Metallic materials

C1/ Al - EN AW 5083	***C2/ Brass EN CW614N***
Price: 3.8 Euro/kg	Price: 8.5 Euro/kg
Costs/ pattern: 0.9 Euro	*Costs/ pattern: 6.5 Euro*

D/ Pouring mass

D1/ SG 130/PUR 11

Price: 5.5 Euro/kg *Costs/ pattern: 0.56 Euro*

Fig.3 Patterns placement in the flask

Experimental conditions

Air temperature in the foundry 12 – 16 °C

2 moulding lines Wülfel Graue: pressure 4,5 – 4,8 MPa, mould cycle 35 – 50 sec, Sand temperature 35 °C, 6t induction furnace, Pouring temperature: 1365 – 1380 °C

Chemical composition of the poured iron:

C	Mn	Si	P	S	Cu	Cr	Ni
3,23 %	0,45 %	1,95 %	0,05 %	0,055 %	0,30 %	0,04 %	0,03 %

3.2 Evaluation of the results – recommendations

Metal materials

Both nearly the same behaviour. No damages from the moulding technique. More experiments (i.e. longer term – more cycles) to be done

Artificial Woods

Ebaboard 60-1 not suitable for moulding lines (but the cheapest one → for individual casting production, self-hardening sand mixtures technology)

Ebaboard PW 920 neu and Ebaboard 140 – comparable results with Al materials

Ebaboard 140 – the most expensive from the artificial woods, worse results compared with PW920

Generally these artificial woods – for smaller batches (approx. Up to 10 000 moulds) + gating systems

Pouring Mass (resin SG 130 and hardener PUR 11) Broader dimensional deviations, less resistance to mechanical stresses. Useful (i.e. quick) when pouring into 3D made mother dies. Problems with porosity caused by pouring technique (vibration could help)

3D printed patterns

MakerBot and ABSplus P430 similar behaviours and results (without any bigger surface finish defects)

ABSplus – 10 times more expensive compared with PLA but with PLA more work with surface finish operations – deeper grinding etc.

PLA – less expensive, relatively good resistance – using for the first testing castings

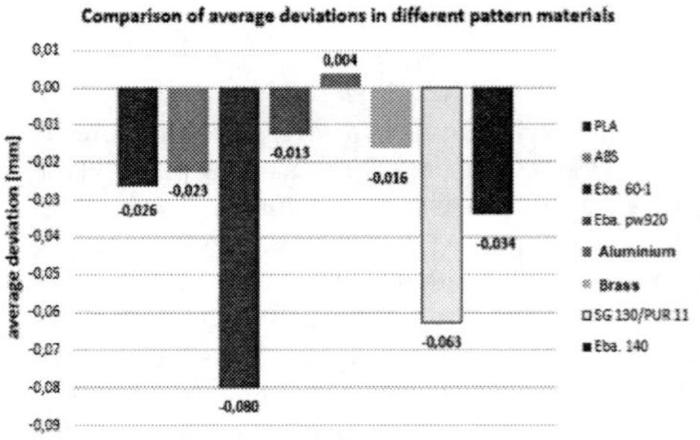

Fig.4 Comparison of different pattern materials

4. Conclusions

3D printed technology could be used for making the permanent patterns. Artificial woods (with exception of Ebaboard 60-1) could be recommended as suitable materials for replacing metal materials of patterns for smaller casting batches. Pouring masses (resins) are suitable for using in combination with 3D printing for permanent patterns (but it is necessary to reduce porosity of such made patterns).Metal patterns did not shown any changes during the experiments.

References

/1/ JAIN, P.L. *Principles of foundry technology*. 5th ed. New Delhi: Tata McGraw-Hill, 2009. ISBN 978-007-0151-291.

/2/ Wood-plastic composite. In: Wikipedia: the free encyclopedia [online]. San Francisco (CA): Wikimedia Foundation, 2001-2017 [cit. 2017-04-08]. Dostupné z: https://cs.wikipedia

/3/ VAŠEK, Vojtěch. *Možnosti uplatnění moderních metod při výrobě prototypových odlitků*. Brno, 2017. Diploma Thesis, Brno University of technology,Faculty of Mechanical Engineering

/4/ Voxeljet [online]. Munich (BRD), 2017 (cit. 2017-0)

Acknowledgements

The contribution was worked out with the support of the specific research project in BUT, Faculty of Mechanical Engineering, evidence No FSI-S-16-3686 Research in the field of vacuum metallurgy and production of castings of iron and nickel alloys

Perspective on Use of green sand additives as essential contaminants

Mr. Deepak Chowdhary

[1]*Managing Director & CEO, MPM Pvt. Ltd., India,*
**Corresponding address: e-mail: deepak@mpmindia.com*

Key words: Green Sand, Additives, optimization, process control

The most refractory material in the sand mold is the raw sand itself. Bentonite, is the most cost effective high temperature, sustainable binder for the raw sand. Water is needed to activate the gelling/binding properties of the Bentonite. And since the mold cavity contains oxygen, which easily diffuses into the casting surface, traditionally, coal dusts and hydrocarbons are used as de-oxidants which create the reducing atmosphere to help prevent sand-metal fusion. Therefore, what is a simple mix of silica sand, bentonite, water and carbonaceous additives, becomes very complex when exposed to the dynamics of the circulating system sand of a foundry where there are varying: (a) sand to metal ratios, (b) core sand infiltration, (c) return sand temperatures, (d) quality of inputs; amongst the many other variables in the sand preparation process.
However, no matter what the process variables or operational dynamics which differ from foundry to foundry, the use of any additive beyond an empirically set optimum point should ideally be determined by experiential and analytical data and events. The over/under use of any one additive can impact several sand properties simultaneously; many a times leading to repetitive casting rejections. Some of the effects of more than necessary use of additives are, for example: (a) increase in total surface area (TSA) of the total sand,(b) increase in moisture demand/Temper water,(c) lower flow-ability of the sand, (d) lowering the fusion point of the molding sand/reduced sand refractoriness, (e) reduced permeability, (f) uneven compaction of the mold, (g) Increase in Inert Fines. These are analyzed in this paper.

The Need for the perspective on additive control:
The need to do so is to control the molding process as consistently as possible, thereby reducing the incidence of casting rejections and reclaim lost profits for the foundry.

1. Effects and Impacts

Higher than necessary additive infusion can result in:

Increase in surface area of the total sand
Arguably one of the least regarded but one of the most important impactor of variation in green sand preparation, is the total surface area (TSA) of the molding sand. Each component of the molding sand, viz: raw sand, bentonite and carbonaceous and other molding additives like cellulose, dextrin, core sand returns, dust collector fines reflux, etc., have a surface area index of their own depending on their sieve distribution/particle size. While it is true that raw sand is the major component of the system sand (almost 80% of the total molding sand) if the incoming sand is not consistent in its sieve distribution curve, variations in the Grain Fineness Number (GFN) of the raw sand itself creates a constant 'noise' in the surface area characteristics of the final prepared sand.
The GFN of the Bentonite impacts the surface area by the addition quantities and or the 'burn-out' percent which may vary from job to job in a foundry.
Carbonaceous additives are rarely quantified for their distribution curve. most often their use is specified as 90% passing through 212 microns or such like depending on individual foundry control documents.
In highly cored patterns, the return of core sand whose GFN is often different from the raw sand due to considerations of resin consumption, coating application, etc; is also significantly relevant in total GFN and surface area control.
Dust Collector Fines (DCF) when refluxed back into the system sand can cause a larger surface area distribution due to its fine particle size. Hence caution is advised in the amounts that can be re-introduced into the system sand.

Increase in moisture demand & Temper water Point
A direct impact of the increase/decrease in the surface area of the system sand is the corresponding increase/decrease in the moisture demand and temper water point.

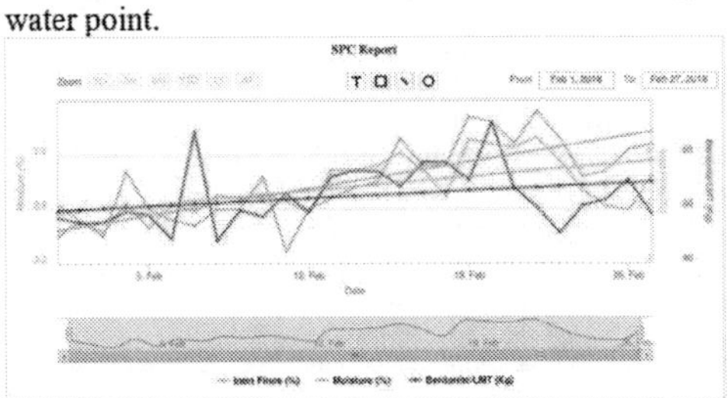

Lowers flow-ability of the system sand
Raw sand is the most freely flow-able component of the system sand. All other additives, while imparting essential properties to prepare this raw natural resource into a moldable format, only serve to reduce its flow-ability. Flow-ability is essential to the molding sand as it enables, or on the other side; inhibits the delivery of

the sand to the all the edges, corners and contours of the cavity.

Lowering the fusion point of the molding sand/reduced sand refractoriness

None of the additives added to the molding sand have the fusion point of the silica sand (fusion point between 1650-1750 °C). Additives like Bentonite (thermal degradation points of Indian Bentonites are between 500-550 °C and a sintering point of 925- 1125°C) lead to lowering the sintering point of the sand as they build up on the sand surface during the re-circulation of the system sand.

Increased sand sticking to the castings/ sand fusion

The cumulative effect of beyond optimum use of additives as a direct consequence (example: lower sintering point of the bentonite) or the indirect consequence (increase in dead clay, moisture demand) most often than not leads to sand sticking and even sand fusion/explosive penetration to the casting surface.

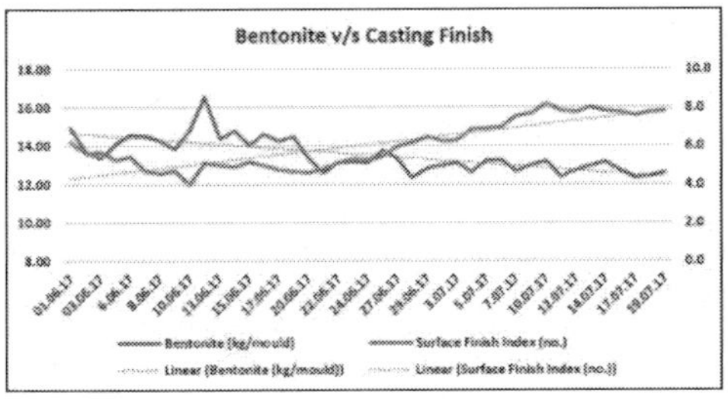

Reduced permeability

higher additions of additives, specially clay and starch lead to reduced permeability with possible related risk of blow hole defects.

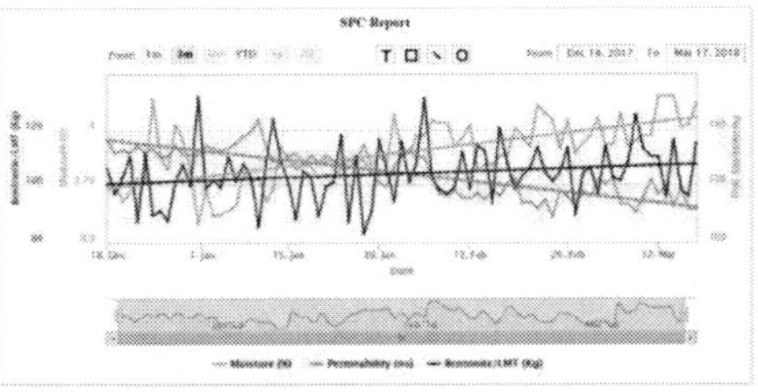

Uneven compaction of the mold

Inert fines in excess of empirical limits of 2.5% to 4%, high quantities of bentonite required due to inefficient development capabilities; can also result in uneven compaction of the molding sand.

2. Outlook on Ash or Inert fines build up

It is empirically accepted that inert fines of maximum 4% levels in the system sand are even beneficial to the molding process because they help in smoothening out the free moisture, enabling thereby, the development of the adjoining active bentonite particles as the mold moves towards the pouring station.

Higher additions of Bentonite and Coal Dusts due to reasons of inefficient or poor-quality; leads to increased burn outs (Ash) and buildup of Inert Fines, the management of which needs extra dust collection efficiencies or compensatory injection of raw sand which can be very costly. The dust collector extraction is not limited to the dead fines generated due to burn outs of the additives. The fines of active bentonite and un-burnt carbonaceous dusts which are also light and dry enough are sucked up in the extraction vacuum.

The result is that significant amounts of Lustrous Carbon additive as well as bentonite are sucked out with the inert fines. (Typically, the LOI of Dust Collector Fines (DCF) ranges between 15-20% and the active Clay around 20%).

These are normally never fully refluxed back into the circulating sand due to reasons of their GFN which can upset the total surface area balance further compounded by handling issues like air pollution, from the POC (Point of Collection) to the POU (Point of Use).

Low ash carbonaceous additives or lower performance bentonites due to non-availability of better quality or for commercial reasons, without taking a view of the addition levels during actual use, can result in losses which are far greater than the commercial considerations of price discounts/savings:

The effects of the perspective of overdosing of additives so gained can be very profound. The same effects in reverse applies to under-dosing of additives.

Goal of essential additions

The goal of optimum additive consumption should be to reduce/increase the: (a) generation of fines (b) dosing of the active clay/Lustrous carbon/water/etc.; rather than controlling the imbalance.

3. What is the optimum additive consumption

A general statement would be that the levels of addition of the additive should be based on strict performance metric of its intended use, achieved by empirical evidence on the casting performance metric or index (example: Surface roughness comparator, etc.) at the lowest/most optimum possible addition levels.

There is no fixed benchmark since additive properties and quality differ from supplier to supplier, across geographic regions and so on.

Acknowledgments

We would also like to gratefully acknowledge NELCAST Foundry and SANDMAN for the data showing the co-relation of additives.

References

1. D. Chowdhary, *""Sand Control – The Next Dimension"*, Foundry Sand Control Seminar, Western Michigan University, 25th August 2016, USA.

The 73rd World Foundry Congress
„Creative Foundry"
23 - 27 September 2018, Krakow, Poland

COMBINING MOLD MILLING AND 3D SAND PRINTING TO OPTIMIZE CASTING DESIGN

Leah Dunlay, Craig Ertl, Wil Tinker

University of Northern Iowa, Metal Casting Center, 360 Westfield Ave. Waterloo IA 50701, USA
University of Northern Iowa, Metal Casting Center, 360 Westfield Ave. Waterloo IA 50701, USA
Tinker Omega Manufacturing LLC, 2424 Columbus Avenue Springfield, OH 45503 USA
ldunlay@uni.edu
ertlc@uni.edu
wtinker@tinkeromega.com
***Key words**: Sand casting, mold design, mold milling, robotic milling, 3D printing*

1. Introduction

Building low quantity production or prototype sand molds and cores is traditionally costly with long lead times. The introduction of 3D sand printing has reduced this cost and lead-time for smaller components, but remains unsuitable for large casting components. Milling sand blocks with a robot to create custom molds reduces these two constraints. The combination of 3D sand printing cores and robotic milling of molds creates the opportunity for producing sand casting tooling for large and complex casting designs efficiently. This technology maintains utilization of mold making equipment in foundries to produce blank sand blocks and builds on the skillsets of staff with expertise in milling and industrial robotics. This paper will present a business case example.

2. Experimental Method and Materials

Robotic milling of the sand is conducted using an industrial 6-axis robot. The system used, below in Fig. 1., integrates the robot with an external system allowing the user to run multiple toolpaths and to change tools without user interuption. The tool changer is allotted with 12 tools.

Fig. 1. Robotic milling cell with tool changer and sand base

The cutting toolpaths are created using an industrial milling software, integrated with an industrial robot software package. The sand block is a phenolic urathane no bake mold. In this example, they are approximately 80 cm x 80 cm x 25 cm and can be seen setup in Fig. 2. The system used is capable of milling sand blocks up to 180 cm x 180 cm x120 cm.

Fig. 2. Sand block milling in progress

Tooling designed for use in metal milling applications is used and includes a face mill, hobb mill, end mills, bull mills, and ball mills. Solid carbide tooling and carbide tooling inserts are used.

3D printing of the sand core and calculations for the business case analysis are conducted with an industrial sand printer. The build enveleope of this printer is 1800 x 1000 x 700mm (x, y, z). The build speed is approximately 25 mm per hour in the z-direction. The sand core is printed in silica sand with a furan binder.

3. Business Case Analysis

The geometry in Fig. 3 represents the mold geometry that is machined and the core geometry that is printed:

Fig. 3. Mold (top) and core (bottom) geometry

The following variables are considered when reviewing the cost of production:

$mc = material\ cost\ (cost\ unit \times mm^3)$
$\%xy = percentage\ of\ xy\ printing\ area\ (mm^2)$
$t = machine\ run\ time\ (hours)$
$V = enveloping\ volume\ size\ (mm^3)$
$D = machine\ depreciation\ (\frac{cost\ unit}{hour})$
$tw = expendable\ tooling\ cost$ (cost unit)

Material cost includes the silica sand and binder. When 3D sand printing, machine depreciation is calculated from a geometric area perspective, as more than one geometry may be printed at one time. The %xy variable represents this factor.

Equations 1 and 2 below represent how the cost of robotic milling of molds compares to 3D printing of molds. The factors in the brackets represents the cost of machine operation.

$$3D\ print\ cost = (mc \times V) + [D \times t \times \%xy] \quad (1)$$

$$Robot\ mill\ cost = (mc \times V) + [(D \times t) + tw\ ] \quad (2)$$

For this case study, the following is known:

1. The 3D sand printer currently costs 5 times that of the robotic milling machine
2. The time to produce 3D printed molds is 3 times that of mold milling
3. The %xy is 70% to produce both the cope and drag

The following is assumed:

1. The volume of the molds are the same
2. The material cost is the same
3. Tool wear is insignificant compared to the cost of machine depretiation times time
4. Differences in design time and setup time is not considered

Therefore the following is to be true:

$$3D\ print\ cost = MC + [5D \times 3t \times 70\%] = MC + [10.5Dt]$$

$$Robot\ mill\ cost = MC + [Dt]$$

This provides an overview for comparative cost; however, a valuable data point is the size at which it is economical to move from 3D printing to mold milling. In this example, this would occur when the XY% is 6%.

4. Results and Discussion

The business case analysis shows that utilizing the robotic sand milling for large molds is cheaper than 3D printing them. While the breakeven point is shown to be when the XY% is 6%, in actuallity this point is approximately 15% due to the following.. The 3D printed volume of the mold is lower than the sand block being milled, as the 3D printed geometry can be optimized based on the casting geometry and the milled sand blocks are standard flask sizes. The cost of tool wear increases the cost of mold milling and the setup time for each block of sand to be milled is higher and that for 3D printing setup.

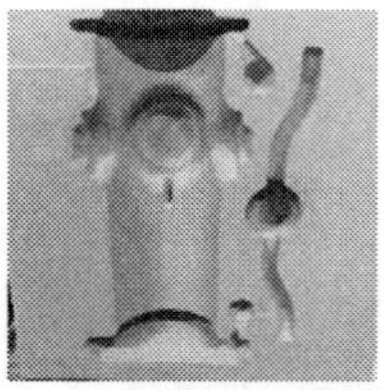

Fig. 4. Robotically milled mold

5. Conclusion

This research demonstrates the successful use of robotic sand mold milling with 3D sand printing of cores. Figure 4 provides an image of a completed, robotically milled mold. The business case analysis shows that mold milling molds of this size is faster and more cost effective than 3D printing the molds. The assumptions made provide a way for the manufacturer to determine when to mold mill molds rather than 3D print molds. This example shows that the technologies are not to operate independently, but complement each other to produce prototype or low quantity casting molds and cores. As the capital investment for a robotic milling is currently a fifth of that for an industrial 3D sand printer, it provides an opportunity for small to medium sized foundries and pattern shops to enter the market to produce custom sand molds competively in house. These companies often have expertise in tool path generation, milling, and industrial robotics, so their learning curve for will be lower for robotic sand milling than that for 3D sand printing. Additionally, by using mold making equipment to create the blank sand blocks, foundries can increase their utilization of that equipment.

6. Future Work

Complexity of casting mold and core geometry is not considered in this work, other than being a consideration in the amount of time to robotically mill the mold. Further work to quantify the effect of complexity in mold milling is needed.
Further research to quantify the cost of tool wear and to identify optimal tool materials, speeds and feeds, and tool shapes is needed. As silica regulations in the USA are becoming more stringint, this work will also be required for alternative aggregates.
Lastly, with the introduction of new 3D sand printers with faster print speeds to the North American market, review of the business case analysis will be required.

The 73rd World Foundry Congress
„Creative Foundry"
23 - 27 September 2018, Krakow, Poland

CASTING SIMULATION: AN AID TO GREEN MANUFACTURING

Amarnath Bhat[1]*, Stephen Barnett[2]

[1] *SoftCAST Technologies Pvt Ltd, Amarnath Bhat, 43/1 Richmond Rd, Bangalore, India*
[2] *INCAST Consultancy, Stephen Barnett, La Niche, Hayes End, South Petherton, Somerset, TA13 5AG, England*
* *email: amar.bhat@softcast.co.in*

Key words: Energy, Carbon Emission, Pollution, Optimization, Simulation

The Foundry industry is an energy-intensive industry. Besides, it has the dubious distinction of being the 2nd most polluting industry, the first being mining. Its societal characteristics are significant energy consumption and pollution generation. Manual systems and processes do not hold a promise of appreciable mitigation of the situation. The need of the hour is elimination of shop trials and quality & yield improvement of production methods using computer-aided simulation systems that enable optimization.

Of the various energy-consuming processes in a Foundry, melting is the A class item, accounting for around 70% of the total energy consumed. A significant fraction of the harmful emissions is also directly connected with melting. We therefore focus on melting in this discussion. A comparison of the theoretical requirement and the actual consumption clearly shows up the operational inefficiency that exists today in melting. Various methods can be used by a Foundry to reduce the energy consumption in melting. We focus here on improvement of rigging, *i.e.* Methoding, using computer-aided tools such as simulation systems. Benefits are assessed in terms of energy savings and reduced pollution.

Use of a Simulation System helps the foundry in improving and assuring quality, improving yield, reducing /eliminating shop trials and making the method robust – outcomes that directly contribute to reduction in energy consumption, and reduction in pollution and harmful emissions (collectively referred to as "carbon emissions").

We present Four Case Studies that illustrate the positive impact on energy consumption and carbon emissions, that result from quality and yield improvement, and elimination of shop trials with a simulation system, SoftCAST. Energy savings and carbon emission reduction, per cast product, ranged from 0.4 kwhs and 0.081 kg resp. for a 0.2 kg Aluminum Die Casting, to over 8000 kwhs and over 1.8 Tonnes resp. for a 2.4 Tonne sand cast steel product. A 5th Case Study illustrates the benefits of virtual manufacturing through simulation of a new product before shop floor implementation. A summary of four of these Case Studies is described here.

1. Developing a New Component using Casting Simulation as an aid

This Case Study is about development of a new large component in the investment casting process, with no tooling supplied, with very little drawing details specified and no CAD files at all. The component (refer Fig. 1) was required to be manufactured in Hastelloy® according to aerospace quality standards.

The customer supplied only a finished cast component. With the help of reverse engineering and casting simulation we were able to create a soft structure. With this information it was possible to obtain a working procedure for the manufacture of the component. It was essential to carry out a casting simulation as otherwise the high number of metallurgical defects and shop trials would have increased the overall manufacturing cost. Further, the loss of time due to possible high reject rate would have jeopardised the project's delivery time. The component's finished weight was 40kg. Seventeen components were required as replacement spares for an aero engine.

Simulation of the casting enabled proper methoding design. The Tree was also simulated to check the method efficacy. This process ensured 1st Time Right. This is a good example of savings on material, labour and time. The elimination of shop trials and rejections also meant reduction in the extent of air emissions and other pollutants.

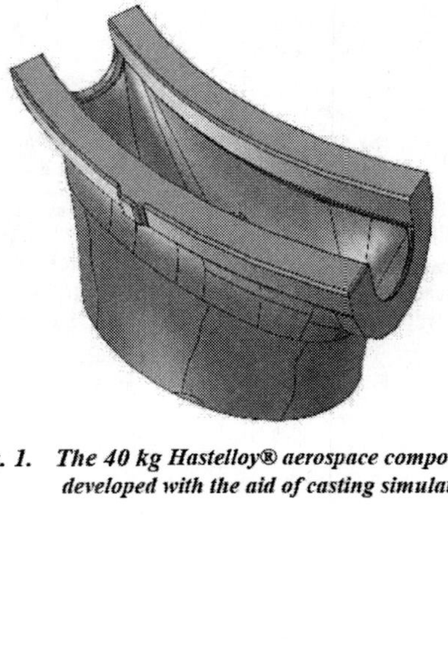

Fig. 1. The 40 kg Hastelloy® aerospace component developed with the aid of casting simulation

2. Yield Improvement with Casting Simulation

This project at one of India's largest integrated manufacturer of Pistons, Piston Rings, Gudgeon Pins and Engine Valves, was aimed at yield improvement of a 0.2 kg Aluminum Die Cast Piston (refer Fig. 2). Yield increased from 45% to 55%. The new method, designed from simulation, was implemented without any shop trial.

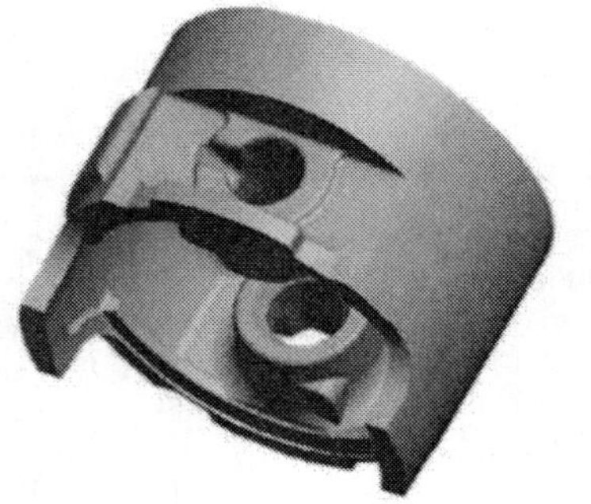

Fig. 2. The 0.2 kg Aluminum GDC component that was optimized for yield through casting simulation

The energy saving with 10% yield improvement was 0.4 kwhs per cast product, and 4.03 kwhs per shop-floor trial reduced. The reduction in emissions was 0.081 kgs per cast product and 0.813 kgs per shop-floor trial reduced. It is worth noting that this component was being produced in 100s of thouands per year.

3. Quality Improvement using Casting Simulation

This project for a reputed steel foundry in South India was aimed at Quality Improvement. The existing method did not assure quality at all, leading to 100% rejection of the 112 kgs Ring castings, with L3 & L4 defects. The new method, designed with simulation, delivered 100% acceptable castings, and was implemented with nil shop floor trials.

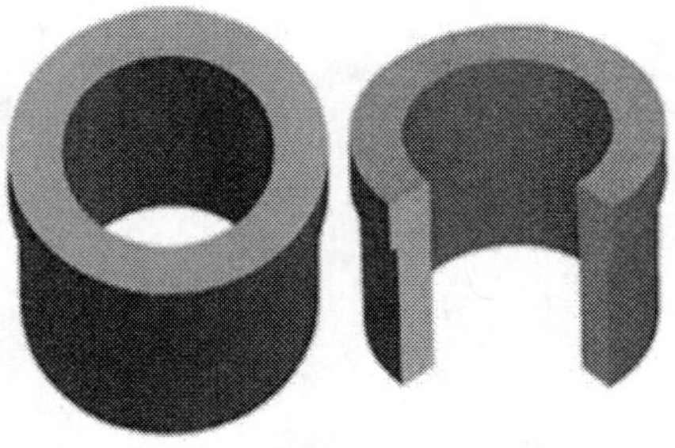

Fig. 3. The 112 kg Steel Component whose quality was improved with casting simulation

The energy saving with 100% quality improvement (and hence, yield improvement also) was 827.16 kwhs per cast product, and 1179.09 kwhs per shop-floor trial reduced. The reduction in emissions was 187 kgs per cast product and 266.5 kgs per shop-floor trial reduced.

4. Quality & Yield Improvement using Casting Simulation

This project was undertaken for one of India's largest engineering corporations. Primarily a yield improvement project, it also addressed quality improvement, as the existing quality was inadequate; further, the method was not robust. Yield improved from 32% (old method) to 65% (new method, designed from simulation). The new method for this 2360 kgs Drum Gear steel casting was implemented after one shop floor trial.

Fig. 4. The 2.36 Tonne Steel casting whose quality and yield were improved with simulation

The energy saving with 33% yield improvement was 8216.74 kwhs per cast product, and 24845.09 kwhs per shop-floor trial reduced. The reduction in emissions was 1.86 tonnes per cast product and 5.62 tonnes per shop-floor trial reduced.

The 73rd World Foundry Congress
„Creative Foundry"
23 - 27 September 2018, Krakow, Poland

OPTIMIZATION OF HEAVY STEEL CASTING MANUFACTURING TECHNOLOGY

Rafał Dojka [1,2]*, **Wojciech Kaczkowski** [1], **Michał Jaromin** [1]

[1] *Odlewnia Staliwa Łabędy, ul. Mechaników 9 44-109 Gliwice, Poland,*
[2] *Silesian University of Technology, Faculty of Mechanical Engineering, Department of Foundry Engineering, ul. Towarowa 7 44-100 Gliwice, Poland,*
**Corresponding address: e-mail: rdojka@odlewstal.pl*

Key words: solidification analysis, manufacturing technology, internal quality, simulation, cast steel

1. Introduction

The ability to control the crystallization process is a powerful tool in the hands of an efficient technologist. Alongside with the quality of molten metal, the possibility to change the crystallization kinetics is without a doubt one of the most important factors affecting the properties of manufactured casting [1,2]. There are numerous ways to achieve this goal, however the authors have focused on the use of external chills. Proper application of chills is an exemplary way to increase metal yield, which especially in case of steel casting is a vital issue, as it reduces the heat modulus and allows to decrease the size of the feeders or even eliminate them [3,4]. Chills are responsible for the creation of feeding end zones. Use of chills allows to avoid casting defects formation and they can be used to relocate the defects into less responsible zones of the casting if their elimination is not necessary, it also refines the grain positively affecting mechanical properties of the casting [5]. However when improperly applied chills can be the source of cracks and can cause an incorrect course of crystallization [6]. This is why the optimization of chill application, especially in manufacturing of heavy steel casting, is a key issue in modern foundry engineering and authors decided to investigate this topic.

2. Methodology

The subject of the optimization process was manufacturing technology of supporting roller used in cement processing industry which shape and overall dimensions were shown in figure 1.

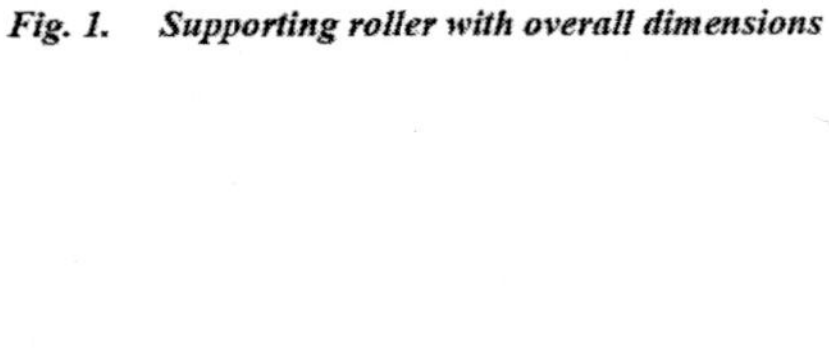

Fig. 1. Supporting roller with overall dimensions

Optimized technology meant to have increased metal yield and accelerate the crystallization at side wall of the casting. In order to analyse the differences in crystallization kinetics and shrinkage of each tested variant authors prepared numerous simulations in MAGMASOFT MAGMA5 software from which they have decided to present selected 5 outcomes. Calculations of steel chills sizes were based on heat modulus reduction method [7,8]. Each variant had different amount of chills representing different modulus reduction, both side and bottom chills were used. Simulated alloy was GS-42CrMo4 cast steel acc. To DIN 17205 which chemical composition is presented in table 1, pouring temperature was 1550°C, filling time was set to 500s.

Table 1. Chemical composition of GS-42CrMo4 cast steel

Element	Amount, % wt.
C	0,38 – 0,45
Si	≤ 0,60
Mn	0,60 – 1,00
Cr	0,80 – 1,20
Mo	0,20 – 0,30
P	≤ 0,020
S	≤ 0,015

Initial temperature of chills was set to 20°C. Whole mold was set as alphaset molding sand, in reality used cores are prepared from synthetic sand consisting of Al_2O_3 and SiO_2 to avoid burn-on sand because of its high sintering temperature (>1800°C). The weight of rough casting in initial technology was 18000kg, the gross weight of the casting with gating system and feeders was 29500kg.

3. Results and discussion

Analysis of the mold filling process allowed to state the temperature of the chill at the end of pouring, temperature of contact chill was 600°C, temperature of chills separated with a layer of synthetic sand was 150°C. All examined variants had the same gating system, so this information allowed authors to shorten the calculation time by elimination of the filling criterion and focusing on solidification by assignment of proper temperatures to the mold and chills at the star of crystallization process. Scheme of initial technology without chills was presented on figure 2.

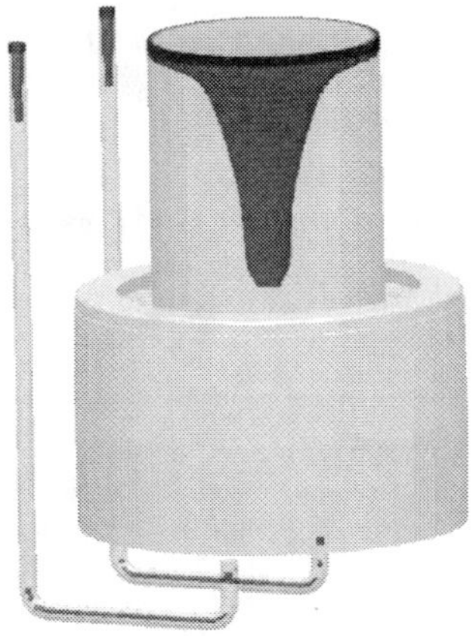

Fig. 2. Initial technology, shrinkage

During crystallization the side part of the casting crystallizes simultaneously from the top and bottom, the middle part starts to crystallize at time of 6h and 42 min. The radius formed by the crystallizing liquid metal is small, what may result in formation of hot cracks. Crystallization takes 38h and 30min. Optimized technology was presented on figure 3.

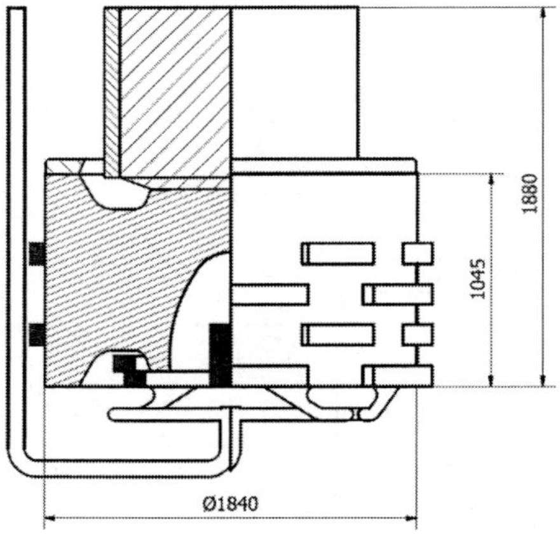

Fig. 3. Optimized technology

Application of 4 rows of sectional contact chills at the sides with two ring chills separated with synthetic sand core and one cylindrical chill in the centre at the bottom allowed to decrease the weight of the main feeder by 3500kg and the weight of rough casting by 500kg. Side part of the casting is solid after 3h and 55min, which would result in enhancement of mechanical properties of side surface layer of the casting. The radius formed by the crystallizing liquid metal is larger than in technology shown on figure 2, as the crystallization in horizontal axis is noticeably faster. Crystallization takes 30h.

4. Summary

Results of experimental optimization of heavy steel casting manufacturing technology led to the following conclusions:

- properly calculated and applied chills are useful tool in controlling the crystallization kinetics of solidifying casting;
- technology optimization allowed to increase the metal yield by reduction the weight of rough casting by 500kg and the weight of main feeder by 3500kg, which summed up to over 13% of gross weight reduction;
- decreasing the radius of liquid zone during solidification allows to avoid hot cracks and results in obtainment of more even distribution of mechanical properties;
- the side wall of the casting in initial technology crystallizes in 6h and 42 min, in optimized technology this time is reduced to 3h and 55 min, which will enhance the mechanical properties of castings side, which is the working surface;
- in optimised technology absolute crystallization time is faster by over 20% in comparison to the initial technology;
- improper use of chills may degrade the quality of the casting by causing hot cracks or disrupting proper course of crystallization.

References

[1] Author(s). (Year of publication). Book title. Place of publication: Publisher.[1] Li D., Sun M., Wang P., Kang X., Fu P., Li Y. (2013). In process modelings and simulations of heavy castings and forgings. Proceedings of the 11th International Conference on Numerical Methods in Industrial Forming Processes (NUMIFORM). Shenyang, China. 81-94.

[2] Camek L., Lichy P., Kroupova I., Duda J., Beno J., Korbas M., Radkovsky F., Bliznyukov S. (2016). Effect of cast steel production metallurgy on the emergence of casting defects. Metalurgija. 55 (4). 701-704.

[3] Ignaszak Z. (2000). Identyfikacja oddziaływania ochładzalników na krzepnięcie odlewów staliwnych . Krzepnięcie metali i stopów. 44. 137-142.

[4] Szczygiol N., Gawrońska E., Grosser A. (2002). Symulacja numeryczna krzepnięcia kierowanego ochładzalnikami zewnętrznymi i wewnętrznymi. Archiwum Odlewnictwa. 4. 239-244.

[5] Suchoń J. (2003). Zastosowanie ochładzalnika w celu rozdrobnienia struktury w odlewie bimetalicznym. Archiwum Odlewnictwa. 10. 214-220.

[6] Campbell J. (2016) Crack populations in metals. Aims Mat Sci. 3(4). 1436-1442.

[7] Gawroński J., Krajczy B., Mochnacki B., Jura S. (1980). Nowoczesne metody przygotowania technologii produkcji odlewów staliwnych gwarantujących uzyskanie wymaganej jakości. Gliwice: Instytut Odlewnictwa Politechniki Śląskiej, STOP Gliwice.

[8] Skarbiński M. (1974). Uruchomienie produkcji w odlewni. Warszawa: WNT.

The 73rd World Foundry Congress
„Creative Foundry"
23 - 27 September 2018, Krakow, Poland

Directional Solidification Casting Technology of Heavy-duty Gas Turbine Blade with LMC Process

Liu Xiaofu[1,2],Lou Yanchun[1,2], Yu Bo[1,2], Su Guiqiao[1,2], Li Changchun[1,2], Guo Xinli[1,2], Li Biao[1,2],Shui Guoyan[1,2]

[1] *Shenyang Research Institute of Foundry, No17 south yunfeng street tiexi district shenyang, China,*
[2] *State Key Laboratory of Light Alloy Foundry Technology for High-end Equipment y, No17 south yunfeng street tiexidistrict shenyang, China,*
**Corresponding address: e-mail: liuxf-01@163.com*

Key words: Liquid Metal Cooling,Heavy-duty Gas Turbine,Large-sized blade,Directional Solidification,Microstructure

In this research, some important factors such as the parameters of the mold shell strength, the heating temperature, the standing time and the withdrawing rates that influence the formation of directionally solidified large-sized heavy-duty gas turbine blades have been studied through the method of microstructure analysis combining with liquid metal cooling process.

1.Effect of ceramic shell strength on formability of blade

the influence of different shell strengths (Table 1) on the formability of the large-size hollow turbine blade is studied.

Tab 1 Strength properties of different ceramic shells

Ceramic shell type	High temperature flexural strength at 1550°C/MPa	Flexural strength after thermal shock resistance/MPa	Residual flexural strength/ MPa
High strength ceramic shell	10	15	30
Low strength ceramic shell	6	8	14
Medium strength ceramic shell	8	12	20

High strength ceramic shell produces high contraction stress in the directional solidification of the blade, which makes the directional hollow blade crack, resultes in incomplete formation of the blade.

The cracking morphology showed a herringbone pattern in both transverse and longitudinal fracture through SEM indicating a brittle property.The pattern suggests that the crack occurs during a shape contraction after the solidification. It is actually a cold crack caused by the poor collapsibility and the high remained strength of the ceramic shell.

A lower strength shell is also tested in the directional solidification process. The result shows that the shell is not able to withstand the dynamic and static pressure of the liquid superalloys because of the poor strength. As a result, alloy melt is easily to leak from the exhaust side, which causes a runout.

To avoid the problems mentioned above, a medium strength ceramic shell is tried.This ceramic shell shows a good deformability.The result shows that the ceramic shell with this level of strength is suitable for the formation of directionally solidified turbine blade under LMC conditions. Fig.1 shows the complete hollow turbine blade.

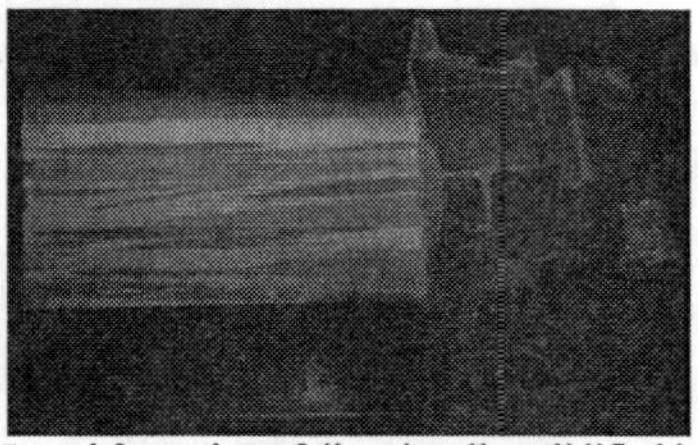

Fig.1 Sound formation of directionally solidified hollow turbine blade

2.Effect of heat preservation temperature on formability of blade

Heat preservation temperature of the mold heating furnace plays an important role in controlling the thermal gradient of the solid-liquid interface during directional solidification process. Three conditions are tested to investigate how the heat preservation temperature influence the formability of the blade.Tab.2 shows the temperature parameters of the heating furnace.

Tab.2 Temperature parameters of the heating furnace

group number	upper zone temperature/℃	lower zone temperature /℃
1	1460	1480
2	1500	1520
3	1540	1560

Lower heat preservation temperature (group 1)produces lower degree of superheating in the superalloy melt and lower thermal gradient at the interface of solid and liquid,which induces higher angle of grainboundry and deviation of crystal grain in the blade.Higher heat preservation temperature (group 3) makes the ceramic shell soften and crack,resulting in some leakage of melt . Sound blade can be achieved in the condition with medium heat preservation temperature (group 2).It proves that an appropriate heat preservation temperature is very important for preparing sound blade.

3.Effect of standing time on formability of blade

When there is no standing, the initial equiaxed crystal grain grow in the direction of spontaneous nucleation without screening through competition. Therefore, the growing direction of crystal are not parallel with that of the heat flow, resulting in a narrow shape columnar grain and a large deviation of orientation from <001> in the formation blade.

When the standing time periods are 3min and 5min,the initial equiaxed crystal grains undergo a competitive screening during the growth. During this process, the crystal grains with orientations that are deviated from the direction of heat flow will be eliminated. As a result, the remaining grains are in good quality with fine and straight columnar shape. These grains grow in a steady state.

When the standing time is 8 min,due to the long incubation time, transversely grown grains occur and cause repeated melting of the top ends of the dendritics in mold heating furnace. The broken dendrites without being melt will become new nuclei that later on grow in their original directions, resulting in broader shape grain columns with serious crystal deviation. Therefore, standing time should be 3 or 5 mins. Fig.2 showes the original structures of the blades with various standing times.

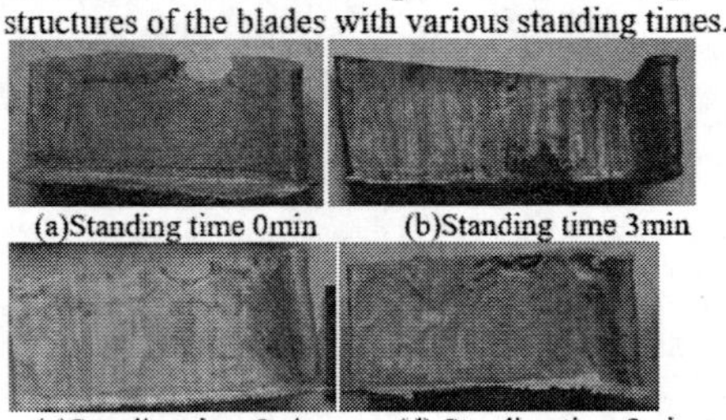

(a)Standing time 0min (b)Standing time 3min

(c)Standing time 5min (d) Standing time 8min

Fig.2 Original structures of the blades with various standing times

4.Effect of withdrawal rate on formability of blade

Withdrawal rate and crystal grain growth rate are required to be as consistent as possible and well match the formation of a blade with ideal structure. Four different withdrawal rates (3, 5, 7, 10mm/min) are studied for their influence on the formability of the blades.

Blades with smooth surface quality can be obtained when the withdrawal rates are 3mm/min, 5mm/min and 7mm/min. Figure 3 shows a large-sized hollow blade with smooth surface. However, when the withdrawal rate is as high as 10mm/min, surface quality reduces.Meanwhile, wrinkles with different depths occur despite the complete formation of blade is finished(Fig.4) because of the excessive withdrawal rate.

Fig.3 Blade with smooth surface

Fig.4 Blade with wrinkle surface

5.Microstructures of directionally solidified blade

To study the microstructures and wall thicknesses of the blade, samples from five different locations of the directionally solidified superalloy blade are taken and observed.Both transverse and longitudinal structures are observed. Transverse structures of all samples present a pattern like "+",but with great difference in dendrite space. The longitudinal structures of samples all present a pattern like "sugar gourd string" with clear structures of the primary dendritic.

The space between dendrites decreases as the blade solidification rate rises,but increases as the distance between the solidification position and the graphite chill plate enlarges. Also, larger space between dendrites is always accompanied with coarser dendritic structure, no matter how thick the wall of blade is.

References

[1] Yutaka KOIZUMI,Toshiharu KOBAYASHI, ZhangJIANXIN,etc.Development of a Next-Generation Ni-baseSingle Crystal Superalloy[C]. Proceedings of the International Gas Turbine Congress 2003 TokyoNovember 2-7,2003:1-6.

[2] A.S.Lebedev and S.V.Kostennikov.Trends in Increasing Gas-Turbine Units Efficiency[J].Thermal Engineering,2008,55(6):461-468.

[3] Giamei A F,Tschinkel J G.Liquid Metal Cooling:A New Solidification Technique. Metall. Trans.,1976,(7A):1427～1434.

[4] John Fernihough , Maxim Konter.Method and apparatus for casting directionally solidified article [P].USP.6311760,2001.6.

[5] Michael Francis Xavier Gigliotti,Jr , Shyh-Chin Huang , Roger John Petterson, Ji-Cheng Zhao. Liquid metal cooled directional solidification process[P].United States Patent, 6276433B1, 2001.

[6] Lou Yanchun.Preparation Technology of Ceramic Shell and Core for Directionally Solidified Gas Turbine Blade in Liquid Metal Cooling Process.High-end Formation Manufacturing Technology [R].Engineering Frontiers Symposium on Advanced Material Processing Technology and through Process Modeling & Simulation , Beijing, 2012.9.

[7] Gerhard Müller,MartinValk , GregorHosse. Large Gas Turbines - the Insurance Aspects: 1-22.

The 73rd World Foundry Congress
„Creative Foundry"
23 - 27 September 2018, Krakow, Poland

INFLUENCE OF QUALITY OF CHARGE MATERIALS ON CHEMICAL COMPOSITION, STRUCTURE AND PROPERTIES OF COPPER PRODUCTS FOR ELECTRICAL APPLICATION

Barbara Juszczyk[1]*, Joanna Kulasa[1], Beata Cwolek[1], Łukasz Wierzbicki[1], Witold Malec[1], Wojciech Burian[1], Jerzy Ilasz [2]

[1]*Institute of Non-Ferrous Metals, Metal Processing Department, Sowińskiego 5 Gliwice, Poland*
[2]*Power Cable Company JSC, Sielecka 1, Będzin*
Corresponding address: e-mail: barbaraj@imn.gliwice.pl

Key words: copper scrap, melting, casting, microstructure, conductivity

In industrial practice, oxygen containing copper Cu-ETP and oxygen free copper Cu-OFE are most commonly used for production of products intended for conducting purposes. For more advanced electrical applications, varieties of OFHC oxygen-free copper with properly shaped grains are also used. The total of the impurities content in high purity copper is clearly defined. In particular, there are three groups of impurities [1, 2], whose total content should not exceed:

- As + Cd + Cr +Mn + P + Sb = max. 0.0015 % - (15 ppm),
- Bi + Se + Te = max. 0.0003 % (3 ppm) → Se + Te = max. 3 ppm,
- Co + Fe + Ni + Si + Sn + Zn = max. 0.002 % (20 ppm).

The influence of impurities on the physical properties of copper depends on the fact whether they form oxides or solutions in the solid state. Elements that form solid solutions (e.g. Fe, Ni, Sn, Zn, Cd) have little effect on the plastic properties of copper, while they increase hardness and strength, and lower the electrical and thermal conductivity. The presence of even small amounts of oxygen and sulfur results in formation of oxides and sulphides in the form of eutectic mixtures located mainly at the grain boundaries. This leads to decrease in plasticity and increase in strength properties. It is assumed that copper oxide inclusions may facilitate formation of a homogeneous and fine-grained structure during recrystallization under both static and dynamic conditions. The presence of oxygen in copper is necessary in most plastic working technologies in which Cu-ETP wire is produced, due to the potential brittleness of the material which may appear during the rolling of a strand. On the other hand, in casting technologies it is possible to almost completely eliminate oxygen. The presence of oxygen in copper and its negative effect on physical properties is a complex problem. Oxygen gets into copper during the melting process mainly and hardly dissolves there [1÷3].

Up to the temperature of 380°C a CuO compound is formed while above that temperature Cu_2O is formed. Its content in copper is usually from 200 to 600 ppm. To summarize the role of oxygen in copper, it should be stated that although it fulfills a number of beneficial functions, its presence in the highest purity copper used in production of specialized conducting wire structures is not recommended due to the risk of electrical conductivity reduction and wire deformability and susceptibility to annealing in the drawing process.

The research in this study focuses on the assessment of the possibility of application of post-production waste for production of high-quality copper products used in electrical applications.

The objective of the study was to determine the relationship between the structure of charge material and the chemical composition, microstructure and properties of the produced copper rods.

In this study comparative analysis of influence of scrap quality and metallic charge structure on chemical purity as well as homogeneity of chemical composition was done. Materials for investigation were manufactured by casting in moulds as well as in continuous casting process. The influence of above mentioned parameters on changes of structure and physical and mechanical properties was examined

1. Experimental

The charge material for the melting and casting tests was in a form of copper production waste from the cable industry. It mainly consisted of lump scrap produced in the process of extrusion of copper wires as well as in the form of wires and coils representing waste from production of control cables. Due to the presence of a certain amount of impurities in the waste material, chemical scrap cleaning operations were considered an important component of the conducted studies as well as preparation of the melting and casting charge, which included attempts to melt out some of the impurities and to transform them into an easily removable form. Ultrasonic method with the simultaneous use of surfactants was used to remove impurities from the waste. The melting tests were carried out using induction furnaces with a capacity of 10 and 25 kg Cu. The purified copper scrap, supplemented with copper cathodes, was used as the input for further melting

and continuous casting, which was carried out in a continuous casting laboratory line. Rods were cast in two strand configuration of rod diameter 12 mm.
The material was melted in a graphite crucible with a capacity of 60 kg Cu under a tight cover of a thick layer of flake graphite and overheated to a temperature in the range of 1150÷1180°C. Because of the specificity of the continuous casting process proper selection of its relevant parameters was required. Therefore, casting tests were carried out using variable casting rates: 0.5 ÷ 0.7 m/min and cooling water flow rate at 1.2 l/min. The stroke was at the level of 5 ÷ 8mm, while the stop time was 0.3÷0.5 s without the use of reverse movement.
The microstructure was characterised with a use of light microscope Olympus GX71F and scanning electron microscope Leo Gemini 1525 equipped with energy dispersive X-ray microanalysis Roentec. Assessment of mechanical properties was done based on tensile test carried in room temperature whereas measurements of conductivity made with the Wheatston - Thomsona Zelap Wh – Th 1 bridge. This article also presents assessment of the susceptibility of obtained materials to mechanical working with regard to the relationship between chemical impurities level and plasticity of examined materials.

2. Results and discussion

Due to the fact that the charge material for the melting process was in a form of technological scrap, in the first stage of the work an attempt was made to melt out some of the impurities from the scrap surface and to transform them into an easily removable form. It was assumed that pre-melted scrap can be used as a charge for a continuous casting process. Analysis of the chemical composition of the produced material showed presence of a large amount of oxygen and other impurities, which directly affect the level of electrical conductivity (Table 1). The obtained test results indicate a high inconsistency of the obtained material when compared to the standard materials for electrical purposes. Therefore, in the next stage of the research, an attempt was made to develop an effective method of removing impurities from the post-production waste which then was used as a charge material for melting and continuous casting process. The material produced in this way was characterized by high chemical purity (O_2 content <10 ppm). On the basis of electrical conductivity tests, it has been established that the material is characterized by conductivity at the level of min. 100% IACS..
The results of the microstructure observations showed existence of some differences depending on the method of material casting. It was found out that in the case of continuous casting a fine-grained structure is observed due to intensive cooling of the material during solidification. The structure of copper cast in a gravitational manner is dominated by large columnar crystals. Additionally, in this material presence of Cu- Cu_2O eutectics is observed (Fig. 1 a) in the form of a continuous lattice occurring over the grain boundaries and in interdendritic spaces.

Table 1. The results of electrical conductivity and oxygen concentration depending on purity of charge

Casting method/ purity of charge	Gravity casting/	Continuous casting
Oxygen concentration, ppm	211	10
Electrical conductivity, %IACS	98,43	101,46

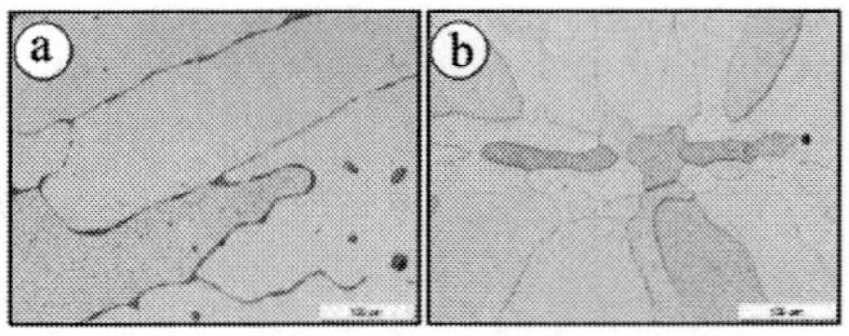

Fig. 1. ***The microstructure of gravity cast copper and continuous cast copper***

3. Conclusions

The quality of final products dedicated to electrical applications depends primarily on the charge materials which are required to present very high chemical purity. Therefore, the manufacturing process should be carried out under conditions which provide protection of liquid metal from contact with the atmosphere, because of the tendency of copper to bond with oxygen. The structure of semi-finished cast products is primarily influenced by the rate and method of heat dissipation during the solidification process. These factors determine the morphology and size of grains.

Acknowledgments The study was conducted as part of Generator of Ecological Concepts Programme's Project GEKON2/O2/266586/2/2015 entitled "Studies into development of innovative proecological technology for production of copper conductors with application of post-production waste". The Project is co-financed from NCBR and NFOŚiGW.

References

1. Camurri C., Carrasco C., Leite R., Mangalaraja R., Dille J. 2012. Influence of Impurities in Cathodic Copper on the Ductility of Copper Wires. Journal of Materials Engineering and Performance. Volume 21. 1474-1478.
2. Knych T., Mamala A., Smyrak B. 2007. Charakteryzacja miedzi beztlenowej, Rudy i Metale R52, nr 11, s. 797-806.
3. Worrell E., Reuter M. 2014. Handbook of Recycling. State-of-the-Art for Practitioners. Analysts and scientists, Elsevier Inc. ISBN: 978-0-12-396459-5

The 73rd World Foundry Congress
„Creative Foundry"
23 - 27 September 2018, Krakow, Poland

The use of Multivariate Control Charts to assess the quality of aluminum alloys melted in crucible furnaces

Marcin Brzeziński[1]

[1] AGH UNIVERSITY OF SCIENCE AND TECHNOLOGY, FACULTY OF FOUNDRY ENGINEERING, 23 REYMONTA STR., 30-059 KRAKÓW, POLAND,
**Corresponding address: e-mail: mbr@agh.edu.pl*

Key words: aluminum cast, quality management of castings, Statistic Process Control, advanced control charts

1. Abstract

The article is the result of research conducted in a aluminum alloy foundry specializing in castings for the automotive industry. The foundry is equipped with crucible furnaces heated with gas, with full installation of control and regulation of the process. The research covered the smelting of alloys in terms of the compliance of the actual content of chemical elements in the metal with the composition assumed in the technological instructions. The actual content of elements is determined by the method of static spectral analysis and recorded in the computer documentation. In everyday practice, the records are subject to a one-time assessment only in terms of compliance with the technological instructions, which does not translate into an improvement in the quality of the melts as a function of the duration of the production process [1].

The introduction of Multivariate Control Charts based on the time analysis of the composition of elements in the melting range makes it possible to take into account the variability of a number of factors affecting the actual content of elements. The content of elements is one of the key information about the final quality of castings [1,2]. Multivariate Control Charts are one of the most modern tools that allow conducting a time analysis of the composition for many alloying elements on one control chart diagram [4,6]. Writing on cards of this type eliminates single control cards [1], depending on the amount of monitored alloying elements. Statistical analysis and the use of multivariate charts allows the user to conduct continuous observation of the melting process with simultaneous detection of the variability of elements as a function of smelting time [3,5,7]. This applies mainly to those elements, which at a given time differ from the assumed values. The method developed in the article is an example of optimizing the quality of processes and is part of the Lean Manufacturing methodology.

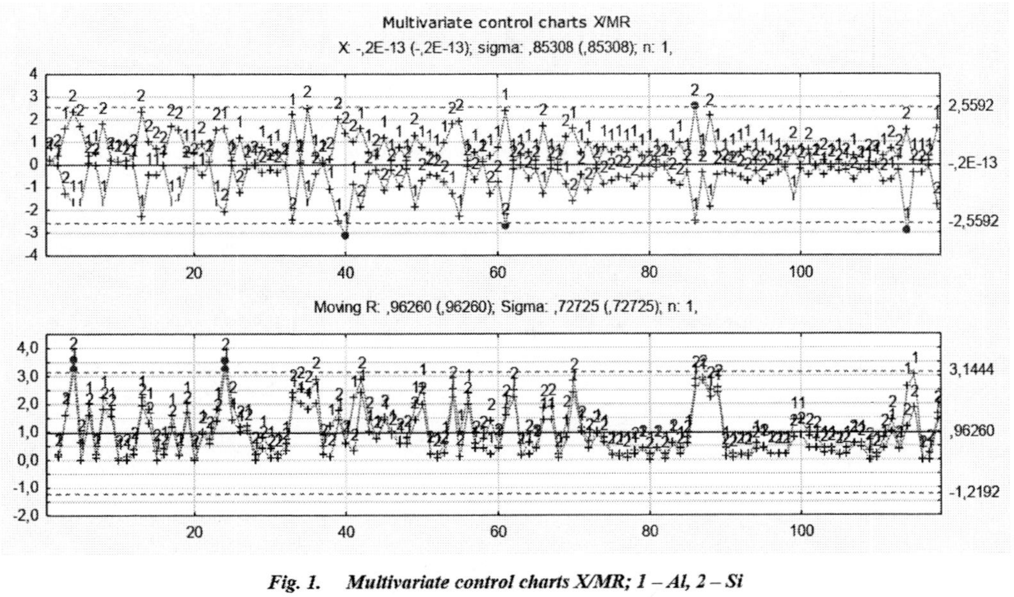

Fig. 1. Multivariate control charts X/MR; 1 – Al, 2 – Si

2. Research methodology and results

The aluminum foundry in which an attempt was made to use multivariate control charts currently occupies a high position among manufacturers of aluminum castings. In connection with the growing competition on the foundry market whose recipient is the automotive industry, the foundry carries out wide investments, among others, equipping the production department with the latest casting machines and the quality control department with advanced software monitoring the production process. This is, among others, the process of monitoring the chemical composition of foundry alloys. The adopted methodology according to which control charts will be used covers those parts of the technological process that have an impact on the quality of the smelted metal. The cards are based on the chemical composition of castings obtained from static spectral analysis. The measurement results have been subjected to standardization, which makes it possible to change the variables into one comparable statistical measure. In the quality control program, multivariate CUSUM charts, multivariate X / MR charts are used, as shown in figures 1 and 2.

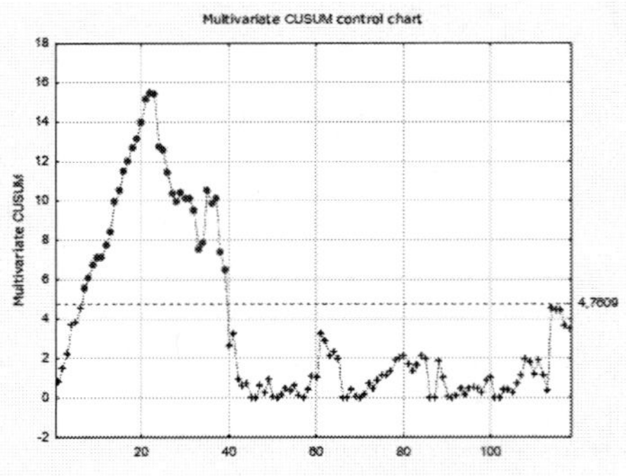

Fig. 2. Multivariate CUSUM control chart (Al, Si)

The program also included ingots supplied by three producers. The obtained results in the form of a multivariate control charts X-bar and s are shown in Figure 3.

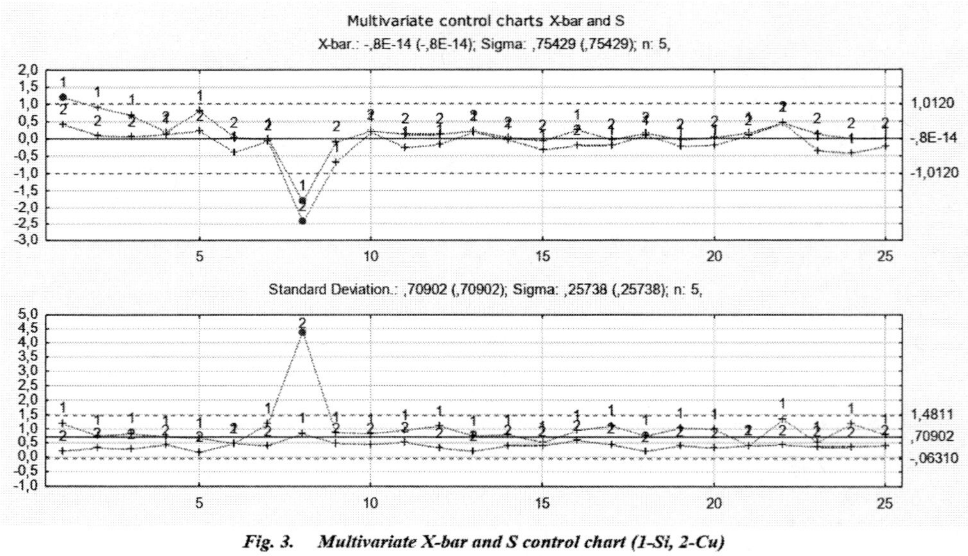

Fig. 3. Multivariate X-bar and S control chart (1-Si, 2-Cu)

References

[1] Brzezinski, M.; DOI: 10.24425/118914

[2] Fedoryszyn A., Brzeziński M.: Assessment of reliability of measurement data and adequacy of the measurement systems in improving quality of casting products. Archives of Foundry Engineering, Volume 15, Special Issue 3/2015

[3] Hamrol A.: Quality management with examples. PWN, Warszawa 2012

[4] Montgomery D.C.: Introduction to Statistical Quality Control. Sixth Edition. Arizona State University

[5] Regulski K., Jakubski J., Opalinski A., Brzezinski M., Glowacki M.: DOI: 10.1515/amm-2016-027

[6] Szymszal J., Lis T., Maliński M., Nowacki K.: Optimisation of foundry production using discrete event simulation (in Polish). PTZP, 2013.

[7] Szymszal J., Gajdzik B., Kaczmarczyk G., DOI: https://doi.org/10.1515/afe-2016-0061

The 73rd World Foundry Congress
„Creative Foundry"
23 - 27 September 2018, Krakow, Poland

HYBRID PATTERNLESS FORMING METHOD AND EQUIPMENT OF MULTI-MATERIAL SAND MOLD

Zhongde Shan [1,3*], **Zhi Guo** [2,3], **Shuai Zhang** [1,3], **Feng Liu** [1,3], **Limin Liu** [1,3]

1. China Academy of Machinery Science & Technology Group Co., Ltd., Beijing 100044, China
2. Department of Mechanical Engineering, Tsinghua University, Beijing 100084, China
3. State Key Laboratory of Advanced Forming Technology and Equipment, Beijing 100083, China
**Corresponding address: e-mail: shanzd@cam.com.cn*

Key words: patternless, multi-material, hybrid forming

1. Introduction

Foundry is one kind of heat processes for metals grasped earlier by human. So far, it has been about 6,000 years of development history. Casting provides raw parts for most of mechanical products. Metal castings have a large proportion in the fields of automobiles, machine tools, heavy machinery, agricultural machinery, fans and aerospace. There are many metal casting technologies. Sand mold casting process is widely used in foundry industry for the advantages of wide raw material source, simple production process and many kinds of alloy. In the traditional sand mold casting process, the design and manufacture of casting mold and core box is a complicated process. The traditional sand mold casting process is characterized by long cycle length, high cost, large waste of materials and low precision. And these shortcomings are highlighted in the manufacture of single and small batch products. Patternless casting technology is a novel digital forming technology. Firstly rudimentary sand mold CAD model is built through reverse calculating the casting CAD model [1, 2]. Secondly accurate sand mold CAD model is built by completing the rudimentary model according to the sand mold casting process database and simulation technology. Finally the sand mold is digital manufactured. Digital sand mold patternless casting technology contains lots of patternless forming methods such as milling, extruding and 3D printing. Many scholars have carried out related research. Zhongde Shan, Xiaoli Dong et al. (2009) investigated the rules of sand mold strength and curing time under different matches of molding sand [3]. And the tool wear pattern was proposed. PCD tool has excellent wear resistance. A. Rodriguez et al. (2012) compared the wear resistance of three kinds of tools such as high speed steel (HSS) tool, carbon steel tool and polycrystalline diamond (PCD) tool [4]. PCD tools are the most suitable for resist abrasion, but its high price is the major drawback. Carbide tools with PVD could give attention to the performance and cost. Feng Liu et al. (2013) used the patternless casting technology to milling the sand mold of binding machine frame, and finally achieving rapid manufacturing of castings [5]. Naisheng et al. (2010) proposed a novel noncontact vibratory coating device to improve the quality of the powder layer by avoiding a collision between the coating device and formed parts [6]. Mindt et al. (2016) established a numerical model to investigate the coating process of Ti-6Al-4V powder and qualitatively analyzed the influence of the powder size on the quality of the powder layer [7].

In addition, there are energy saving and emission reduction, quality and efficient manufacturing requirements in the high-end equipment manufacturing [8-10]. And, key parts of equipment are with integrated, precise, lightweight, green development trend [11]. Because metal parts become more complex and large-scale, technical difficulty of metal parts manufacturing is increased. In particularly, some engine castings, with complex shape and quite uneven walled thickness, whose sand-molds are difficult to form, require a dozen or even dozens of patterns [12, 13]. It is difficult to guarantee the dimensional accuracy. Even, some good design cannot be formed. High quality castings depend on high-quality sand-molds. And, the quality and delivery cycle of castings are influenced by manufacturing cycle, quality, dimensional accuracy and performance of sand-molds, directly [14]. How to form high quality sand-molds, with high speed and efficiency, intelligent and green manufacturing technology, has been the domestic and international research hotspot and difficulty. To address this issue, a hybrid forming method of complex metal parts sand-molds manufacturing, integrated with flexible extruding technology, cutting technology and 3D printing technology, has been put forward, shown as Fig. 1.

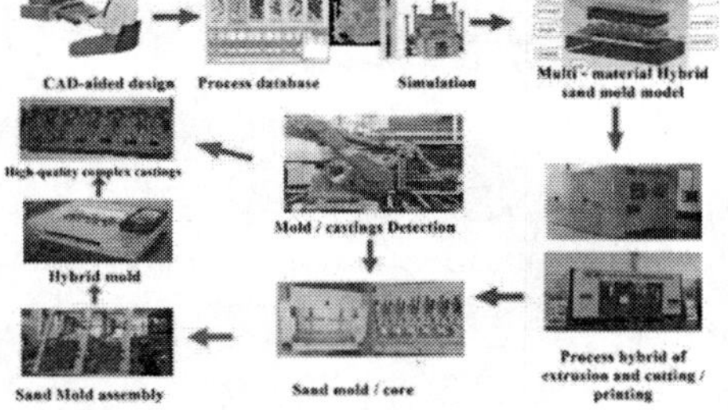

Fig. 1 Hybrid Patternless Forming

Currently, aiming to improve the poor comprehensive abilities such as high performance sand mold manufacturing, high manufacturing efficiency and strong flexibility manufacturing, the paper proposes a novel hybrid patternless forming method and equipment of multi-material sand mold. It contains sand mold flexible extrusion/cutting hybrid forming and multi-material sand mold 3D printing hybrid forming.

2. Hybrid patternless forming method

2.1 Hybrid forming method with sand mold cutting technology and flexible extruding technology

Hybrid forming method with sand-mold cutting technology and flexible extruding technology is a method for sand-mold forming. the near-net forming sand-mold is formed based on flexible extruding technology with sand-mold CAD model. And the precise sand-mold is formed based on sand-mold cutting technology with near-net forming sand-mold. As shown in Fig. 2.

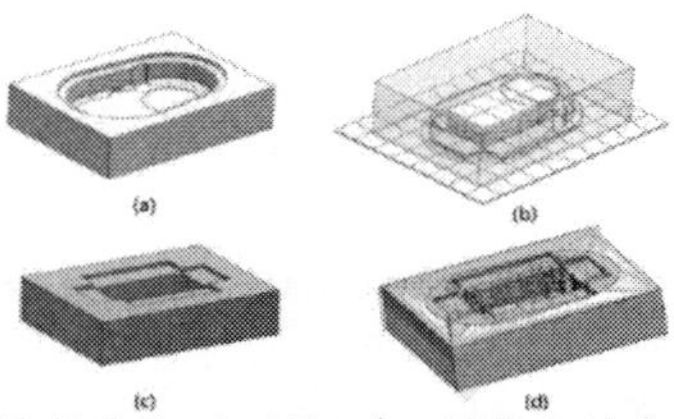

Fig. 2 Hybrid Patternless Forming Method, a) Sand-mold CAD Model; b) Sand-mold Near-net Forming Simulation; c) Near-net Sand-mold CAD Model; d) Sand-mold Precise Forming

To make the hybrid forming method come true, sand-mold hybrid forming algorithm has been studied, shown as Fig. 3.

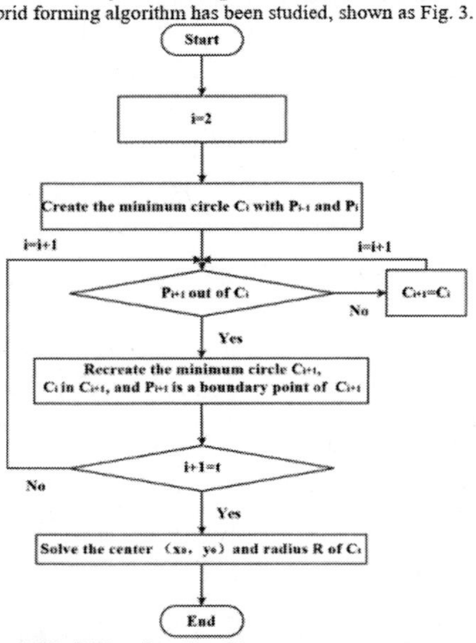

Fig. 3 Hybrid Forming Algorithm

Based on the study of sand-mold hybrid forming algorithm, the hybrid forming control software system has been developed, shown as Fig. 4.

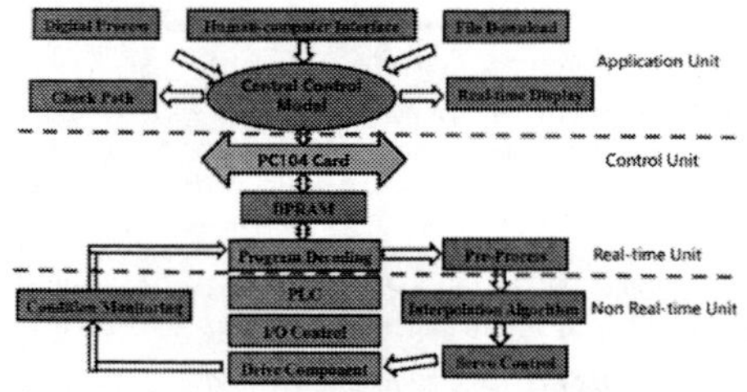

Fig. 4 Hybrid Forming Control Software System

2.2 Hybrid forming method of multi-material sand mold 3D printing

Currently, sand mold 3D printing technologies only fabricate a single material sand mold. Hybrid forming method of multi-material sand mold 3D printing can be used to multi-material sand mold which can control the temperature field of metallic part during the pouring process according to the plan of casting. The multi-material sand mold 3D printing coating process is the key forming procedure. An novel multi-material sand mold 3D printing coating device has been designed. The coating device employed a kind of combinational design of scraper style and travelling hopper style. In order to achieve multi-material sand layer coating, the coating device has many channels with an independent open-close function, as shown in Fig. 5. The removable plate, driven by the air cylinder, is used to open or close the channel. According to the need of foundry technology, the coating device can use different kinds of sand particles on the same layer under the control of the computer. During the coating procedure of the multi-material sand layer, two of the same coating devices were used. One was filled with silica sand particles, the other was filled with zircon sand particles. The multi-material sand layer is obtained by opening or closing the same channel in the two coating devices in complementarity at the same time or the same place.

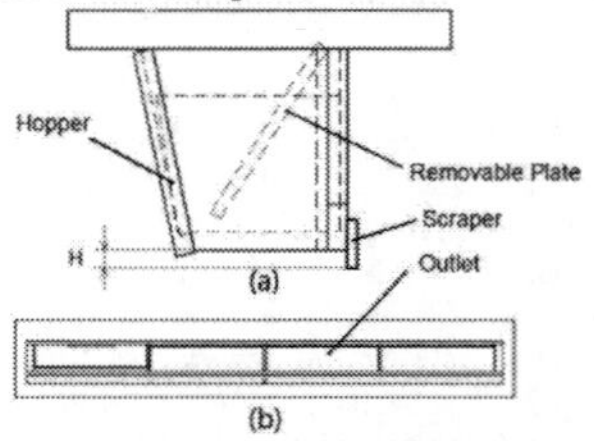

Fig. 5: Coating device of multi-material sand mold print

3. Hybrid Patternless Forming Process Study

3.1 Hybrid forming method with sand mold cutting technology and flexible extruding technology

To make sand mold form without patterns, hybrid forming process with sand-mold cutting technology and flexible extruding technology has been studied. The typical process tests have been carried out, including molding sand mesh effect on cutting-tool wear, sand-mold cutting process performance, cutting mechanism and properties, cutting parameters effect on sand-mold quality and self-adaptive processing technology of high speed and high precision.

In addition, multi-material hybrid mold manufacturing process has been studied, including thermal parameters and casting process performance of special molding sand, shown as Fig. 6.

Fig. 6 Multi-material Hybrid Mold

Based on the method and process study of hybrid forming with sand-mold cutting technology and flexible extruding technology, digital extrusion machine with flexible pattern and digitized precision forming machine without pattern casting have been developed, shown as Fig. 7.

Result shows that, this method can improve the efficiency of sand-molds forming and save sand, effectively. Sand-molds are digitally designed, driven and manufactured, without pattern. The production cycle of sand-molds is shortened. The sand-molds manufacturing costs is reduced. The sand resource is saved. And, the level of digital manufacturing is further improved. Even if castings' design is changed, the pattern does not need to redevelop. The method can meet the complex metal parts manufacturing requirements of high precision, high quality and short cycle. The hybrid forming technology has a good application prospect.

Fig. 7 Hybrid Forming Machines ,a) Digital Extrusion Machine b) Digitized Precision Forming Machine

The sand-mold hybrid forming test has been done with digitized precision forming machine without pattern casting and digital extrusion machine with flexible pattern, shown as Fig. 8.

Fig. 8 Sand-mold Hybrid Forming

3.2 Hybrid forming technology and equipment of multi-material sand mold 3D printing

The coating procedure of multi-material sand particles is the key to accomplish the manufacture of hybrid sand mold. The coating procedure of simultaneous coating silica sand and zircon sand particles has been studied using the above mentioned multi-material coating device. The main content of this research is to control the size of the transition zone between the two kinds of sand. The coating device consists of four independent sand outlets with a size of 150 mm. The silica sand test sample with 0.3% curing agent is added to a single sand hole to form a sand band. The width of the sand band is 177 mm, while the band width of the zircon sand test sample with 0.2% curing agent is 180 mm.

The coating of multi-material hybird sand layer is divided into two parts: the connection of different materials in different sand bands and the connection of different materials in the same sand band. The coating procedure for the multi-material sand layer is made up of two coating devices with the same structure and different sand particles. The outlets of two coating devices are complementary to open and close at the same time to complete the coating of the multi-material composite sand layer. The test results are shown in Fig. 9. Fig. 9 (a) is the ideal composite sand layer, and Fig.12 (b) is the real composite sand layer.

The transition zone is connected by different sands in the same band, and needs two self-adaption coating devices to open and close the same outlet at the same place to complete the coating of multi-material composite sand layers. The test result is shown in Fig. 10.

In addition, digital precision sand mold 3D printing forming machine has been developed shown as fig. 11.

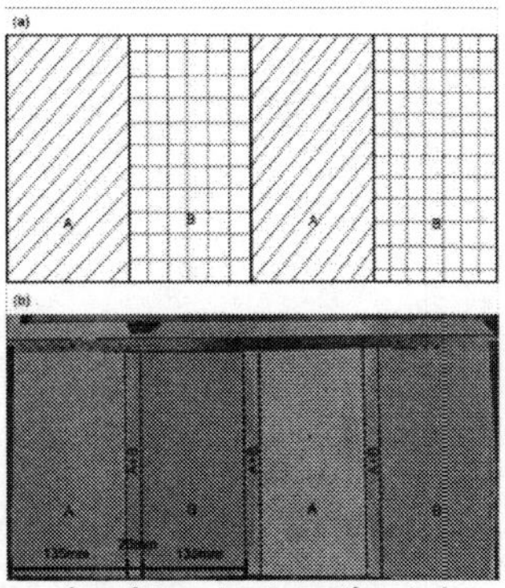

Fig.9 Connection interval of multi-material sand layer between different sand bands

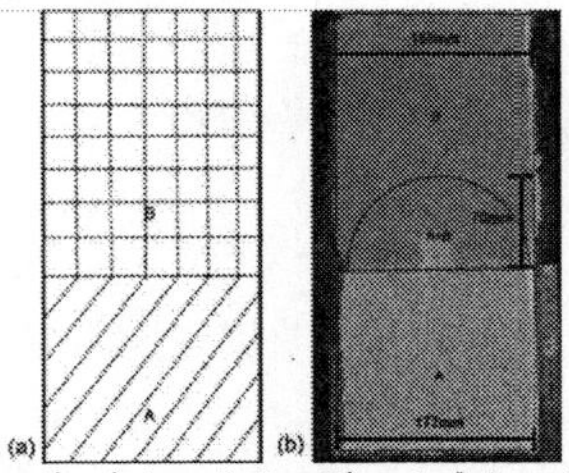

Fig. 10 Connection interval of multi-material sand layer in one sand band

Fig.11 Digital precision sand mold 3D printing forming machine

Conclusions

(1) Hybrid forming method with sand mold cutting technology and flexible extruding technology has been researched. This method can improve the efficiency of sand-molds forming and save sand, effectively. There is a shortened cycle of sand-molds production. And the costs of sand-molds manufacturing is reduced. The multi-material hybrid mold manufacturing process can adapt to the requirements of casting solidification to produce high quality castings.

(2) The multi-material sand particle coating procedure has been researched. There are two kinds of sand particles mixed in the transition zone, the semicircle edge shape will be formed after the coating of silica sand particles, and the sizes of transition zone are determined by the design parameter H and the opening style of the coating device. The size of the transition zone could be reduced by optimizing the open and closing style of the coating device, and high quality multi-material sand layers could be obtained.

Acknowledgements

This work is supported by National Science Fund for Distinguished Young Scholars of China (51525503).

The 73rd World Foundry Congress
„Creative Foundry"
23 - 27 September 2018, Krakow, Poland

Optimum Design of Foam Residue Traps to Manufacture Large-sized Castings for Full Mold Casting

Yuto Takagi [1]*, Masahiro Inagaki [1], Ken'ichi Kanazawa [1], Ken'ichi Yano [1], Shuhei Baba [2], Norimasa Terada [3], Seiichi Iida [3]

[1]*Mie University, Faculty of Eng., Dept. of Mech. Eng., 1577 Kurimamachiya-cho, Tsu, Mie, Japan,*
[2] *Flow Science Japan, Inc., Taito, Tokyo, Japan,*
[3] *Toyochutetsu Co., Ltd., Yokkaichi, Mie, Japan,*
**Corresponding address: e-mail: yanolab@robot.mach.mie-u.ac.jp*

Key words: Full mold casting, CFD simulation, Foam residue, Shape optimization, Casting design

1. INTRODUCTION

One advantage of full-mold casting is that near net shape manufacturing can be achieved more easily than with general sand mold casting that uses a wooden mold because the shape of the cast product does not need a draft. Near net shape manufacturing requires no post-processing, which means fewer man-hours and therefore reduced costs and shorter delivery times. However, in the case of molding for large-sized castings by full-mold casting, a foam residue that is a specific defect of this casting method becomes a serious problem. This defect arises when an undissolved foam pattern enters the product, where it becomes a serious defect that severely compromises the quality of the product.

Studies that pertain to full-mold casting and lost-foam casting are currently being actively pursued. Maruyama obtained and analyzed the thermal decomposition products generated during the filling of the molten metal in lost-foam casting[1], and Koroyashu reports the insulating characteristics of a coating mold on lost-foam casting[2]. Karimian reports the influence of coating mold film thickness on product quality, verified by an experiment in which the residue left by burning was reduced because the amount of combustion gas from the foam pattern discharged to the outside is large if the coating mold film is thin[3]. There have also been studies involving the analysis of fluid behavior using a CFD simulation in lost-foam casting[4,5]. In the present study, we propose a new casting design method for manufacturing large-sized castings for full-mold casting. This proposed method makes use of the optimum design of foam residue traps determined by CFD simulation and a shape optimization method. The placement of these foam residue traps on a product make it possible to avoid the creation of residue defects. The effectiveness of the optimized shape is demonstrated through experiments with actual full-mold casting. For the purposes of the present study, we used the *FLOW-3D* (FlowScience Inc.) fluid analysis software[6], because this software has an effective model for analysis of the full mold casting process[4].

2. FOAM RESIDUE TRAP PROPOSAL AND OPTIMUM DESIGN

We tried to optimize the design of the foam residue traps for capstan drums (Fig. 1). In this optimization, the dimensions defined as design variables are changed within a defined range to determine the optimum combination. The number of possible combinations increases as the number of design variables increases. This requires enormous computational time and makes it necessary to limit the number of design variables to the minimum. The basic shape of the foam residue traps and the set positions of the design variables are shown in Fig. 1. The design variables are defined as x_1 to x_5. Six foam residue traps were set at equal intervals on the side of the rim. All six traps had the same shape, which was determined by these design variables. Also, *FLOW-3D* employs a finite difference method as a space discretization procedure, and subdivides an analytical region into a mesh of rectangular cells. The size of the mesh is 5 mm in all directions.

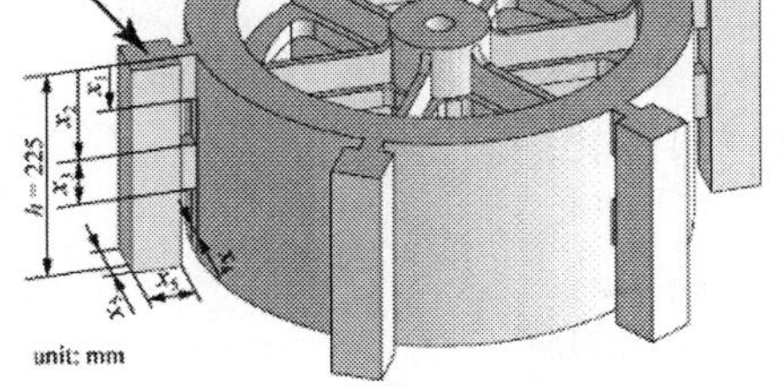

Fig. 1. Design variables for the foam residue traps of capstan drum and mesh block for the CFD simulation

The following three objective functions were identified for evaluation in this optimization. The first objective function J_1 is an index representing the approximate amount of residue in the rim. The second objective function J_2 is defined as the contact area between the rim and the foam residue traps. This is an index that indicates the ease of removal of the foam residue traps from the product after casting is completed. The third objective function J_3 is defined as the volume of the foam residue traps. This definition also takes geometric constraints into consideration and finally formulates the optimization problem as the following equation:

$$\begin{aligned} &\text{Minimize } J_1(x_1, \ldots, x_5), J_2(x_1, x_3, x_4), J_3(x_5) \\ &\text{Subject to } \quad 0 \leq x_1 \leq x_2 \leq 225 \\ &\qquad\qquad\quad 0 \leq x_2 + x_3 \leq 225 \qquad (1) \\ &\qquad\qquad\quad 10 \leq x_4 \leq 30 \\ &\qquad\qquad\quad 30 \leq x_5 \leq 120 \end{aligned}$$

For this optimization problem, optimization was carried out by applying NSGA-II, which is a multi-objective genetic algorithm.

The total number of individuals generated was 775. The design variables of the solution obtained by this optimization were x_1=11.58, x_2=76.50, x_3=108.8, x_4=27.05, and x_5=57.57. Additionally, the value of the objective functions of this optimal solution were J_1=2.250×10^4 and J_2=3.257×10^{-3}m; J_3=7.456×10^{-4}m. Also, the time required for this optimization was about 16 hours by analyzing the CFD simulation in 6 parallels.

3. VALIDATION OF EFFECTIVENESS BY FULL-MOLD CASTING EXPERIMENTS

In order to verify the effectiveness of optimally designed foam residue traps, a casting experiment was conducted. In this experiment, casting was carried out for two kinds of capstan drums according to the conventional casting design without foam residue traps and according to our residue trapping technique using traps of the optimum shape. The experimental results based on the conventional casting design and the corresponding simulation results are shown in Fig. N. Residue defects were observed where indicated in the figure. This is thought to have been caused by residue from the foam pattern generated at the time of filling the molten iron remaining in the rim of the drum.

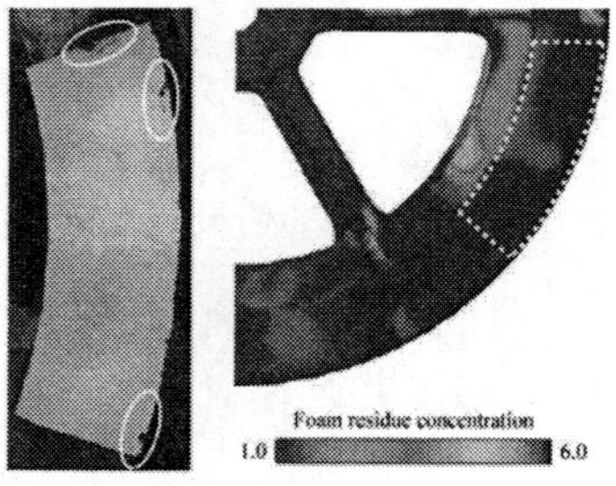

(a) Experimental result (b) Flow analysis result

Fig. 2. Specimen under the previous gating system

Our experimental results for the residue trapping technique using the optimum shape and corresponding simulation results are shown in Fig. N. A slight shrinkage cavity was observed in the test piece, but no residue defects were found. Therefore, the foam residue traps were cut in the same manner as the test pieces, and a color check was carried out inside. Residue defects were observed in the foam residue traps, indicating that residues generated within the product were successfully collected by these traps. Therefore, it was confirmed by this experiment that our residue trapping technique using the optimum shape is effective in reducing residue defects.

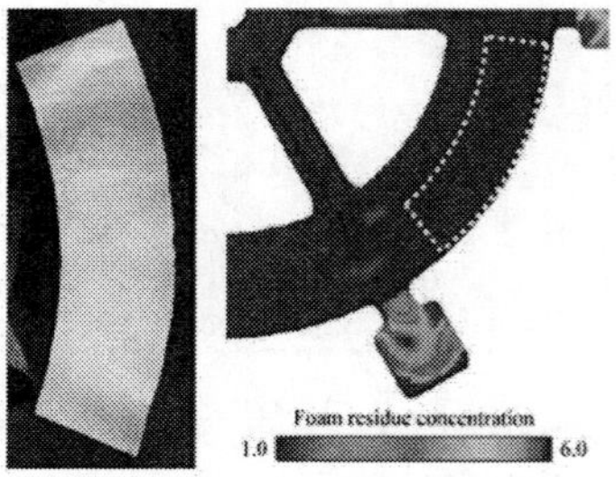

(a) Experimental result (b) Flow analysis result

Fig. 3. Specimen under the gating system using optimum foam residue traps

4. CONCLUSION

In this study, we propose a new casting design with foam residue traps to avoid residue defects. Such defects are a serious problem in manufacturing large-sized castings in full-mold casting. In order to maximize the performance of the foam residue traps, shape optimization was carried out using CFD simulation. The effectiveness of the optimized residue traps was verified by a casting experiment that showed that residue defects can be reduced satisfactorily compared with the case in which no residue traps are used.

References

[1] Maruyama, T., et al., 2009. "Thermal decomposition products of expanded polystyrene pattern in evaporative pattern casting of cast iron". Journal of Japan Foundry Engineering Society, 81(3), pp. 117–122.

[2] Koroyasu, S., and Matsuda, M., 2004. "Effects of high insulating coat on cooling rate of aluminum alloy in evaporative pattern casting process". Journal of Japan Foundry Engineering Society, 76(8), pp. 679–686.

[3] Karimian, M., et al., 2012. "Effect of pattern coating thickness on characteristics of lost foam Al-Si-Cu alloy casting".Transactions of Nonferrous Metals Society of China, 22(9), pp. 2092–2097.

[4] Hirt, C. W., and Barkhudarov, M. R., 1997. Modeling the Lost Foam Process with Defect Prediction. Tech. rep., Flow Science, Inc., Technical Note, FSI-97-TN45.

[5] Hirt, C. W., and Barkhudarov, M. R., 2002. "Predicting Defects in Lost Foam Castings". MODERN CASTING, pp. 31–33.

[6] Flow Science, Inc., http://www.flow3d.com.

The 73rd World Foundry Congress
„Creative Foundry"
23 - 27 September 2018, Krakow, Poland

TRIAD – A MODERN TECHNOLOGY OF NON-CEMENT CONCRETE IN CAST IRON FOUNDRY

Bogdan Cygan [1]*, Józef Dorula [2], Jan Jezierski [3,*]

[1,3] Teksid Iron Poland Sp. z o.o., ul. Ciężarowa, 49, 43-430 Skoczów, Poland
[2] Vesuvius Poland Sp. z o.o. Zakład Foseco w Gliwicach, ul. Leonarda da Vinci 5, 44-100 Gliwice, Poland
[3] Silesian University of Technology, Department of Foundry Engineering
ul. Towarowa 7, 44-100 Gliwice, Poland

**Corresponding address: e-mail: jan.jezierski@polsl.pl*

Key words: refractory material, TRIAD , non-cement concrete, pouring device, cast iron

1. Introduction

The history of the refractory concrete based on the high aluminum cement has 80 years now. The original solution was hi-cement material of traditional application methods. The next huge development in the furnace and ladle lining was use if the low-cement refractory concrete. It was possible thanks to flexibility of the high aluminum cement which is a result of its monolithic properties [1,2].

The use of the monolithic materials based on CAC (Calcium Aluminate Cement) requires the specific technology and following of the procedure like mixing, pouring and then curing process as well as drying and pre-heating. Every step is related to the high aluminum cement hydration. Because the monolithic refractory material must be pre-heated up to working temperature, water necessary for previous liquefaction and hydrates formation is removed. The strength at elevated temperature is achieved thanks to the synthesis and ceramic bonds. The first drying step must be carried out very carefully to not change the properties of the applied concrete and to avoid cracks and the risk of explosion. [2]. To minimize such problems in the refractory materials industry during the last twenty years the works has been conducted to obtain the concrete with the lowest possible cement content. Due to cement content decrease and water addition lowering the physical properties of the refractory materials has been improved. But the cost of such development was more difficult application method and drying process of such lining.

Most of these problems has been sorted out with the new TRIAD products development by Vesuvius company which has no cement content. The bonding mechanism with no cement in the TRIAD materials eliminates calcium oxide CaO present in low- and ultra-low-cement concrete thus its strength in the elevated temperature increases. The phenomena occurring during cement setting process are difficult to control under industrial conditions. Even slight changes of parameters during preparation, mixing and drying process of the monolithic refractory lining based on the high aluminum cement have resulted in instable lining behavior. The technology of the no-cement TRIAD concrete aims to eliminate the lining working parameters variation [3].

One of the unique feature of the TRIAD concrete is its porous structure. During the setting process the venting channels appear. These channels help to rapidly evacuate the steam out of the lining without the influence on the refractory lining properties. The pores diameter is so small that the slag and liquid alloy cannot penetrate the lining what lower its erosion and makes the cleaning out of the build-up much easier. The comparison of the TRIAD concrete porous structure and traditional refractory material of low-cement content was presented in figure 1.

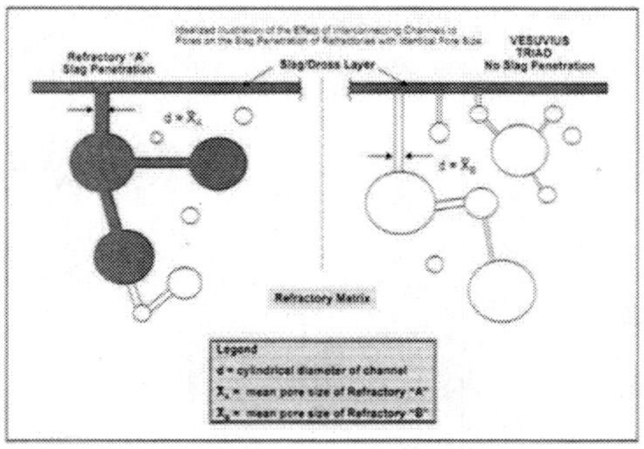

***Fig. 1.** The comparison of the TRIAD concrete porous structure and traditional refractory material of low-cement content [3]*

The above mentioned technological advantages caused that the new monolithic no-cement TRIAD concrete lining has been applied on the cold pouring device on the molding line casting the ductile iron castings in Teksid Iron Poland in Skoczów, Poland. The chemical analysis of the slag taken from the pouring device has shown the very aggressive environment SI = 0,60 (SI – Slag Index calculated according to the formula:

$$([MgO]+[CaO]+[Fe_2O_3]+[MnO])/([Al_2O_3]+[SiO_2])$$

2. Industrial trial in Teksid Iron Poland

The frequent changes of the cast iron grade causing the necessity of the pouring device emptying, various inoculation processes (Sandwich, Inmold) result in the continuous attempts to find the best refractory material to minimize the downtime due to pouring unit maintenance. This is a big challenge both for the producers of these materials and for the foundry plants which are strongly interested in the higher durability of the refractories. It is worth to mention that the very important issue during the exploitation is careful liquid alloy surface cleaning i.e. careful slag removal at the de-slagging stand and inside the pouring device, too. The one being used in TEKSID Iron Poland Sp. z o.o. (Fig. 2.) is a typical gravity type machine.

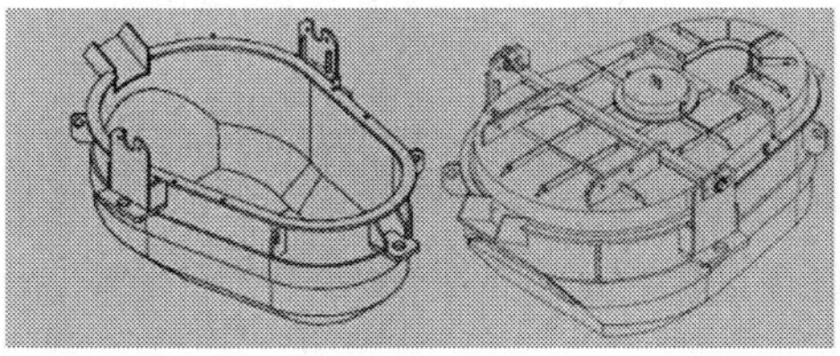

Fig. 2. The pouring device overview

The device being used for the trial was the one working under variable and heavy-duty conditions. It is used both for the ductile (Sandwich-type inoculation) and grey iron (Inmold-type inoculation). It works in 3-shift system for 6 days and then the routine maintenance is going on. The characteristic feature of such unit is variable level of the liquid alloy during the molds pouring process. It is caused by the cyclic topping-out approximately every 5-12 minutes. This operation creates something like 'natural slag aggressive reaction zone' causing the refractory erosion what has been shown in figure 3.

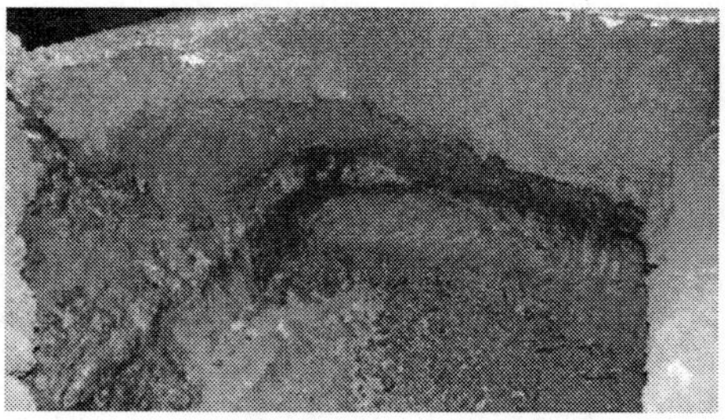

Fig. 3. The view of 'natural slag aggressive reaction zone'

The first step was the installation of the new refractory material supplied by FOSECO inside the selected pouring device. Then the drying and pre-heating process under the supervision of the FOSECO engineer has been conducted and the unit was released for production. The week after the device was analyzed regarding the refractory erosion rate especially in the most sensitive area. Every step was documented on the pictures showing the lining condition during the exploitation period.

Finally, after 4 weeks of continuous exploitation the refractory was knocked out. The experiments have proven that the use of such concrete for the cold gravity pouring device is a clever idea. The significant increase of the refractory durability has been achieved what resulted in significant economic benefits. The observations made during the trails indicated that it is necessary to further improve the lining resistance for the slag especially in the sensitive zones. Therefor it is still a big challenge to look for the better and better chemical composition of the refractory material to make it more slag-resistant.

Summing up, the industrial trials under the conditions of TEKSID Iron Poland Sp. z o.o. have proven the new concrete industrial usability showing that such material is an excellent choice for the cold gravity pouring devices.

Acknowledgments

Financed by the Ministry of Science and Higher Education under the agreement No. 12/DW/2017/01/1 dated on 07.11.2017 r. on the amount of 4 433 520,00 Polish zloty.

References

[1] Drevin J. (2014). Triad – nowa gama przyjaznych dla użytkownika, betonów ogniotrwałych o wysokiej wytrzymałości. Przegląd Odlewnictwa. 9-10, 390, in Polish.

[2] Rybak M. (2011). Wpływ warunków hydratacji cementu glinowego na własności betonu. Piece przemysłowe & kotły. 1, 21, in Polish.

[3] Drevin J. (2011). Triad – Triad high performance castable linings. Foundry Practice. 253 (6), 16.

EVALUATION OF GREEN SAND PREMIXES FOR EMISSION CHARACTERISTICS

S. Ravi [1]*, J. Thiel [1], V. LaFay[2]

[1]University of Northern Iowa, Metal Casting Center, Cedar Falls, Iowa, USA
[2]Imerys Metalcastings, IT Center, Suwanee, GA, USA
*Corresponding address: e-mail: ravis@uni.edu

Key words: Green sand, premix, emissions, VOC, Casting analysis, Carbon Monoxide

1. Abstract

For many years, researchers in the foundry industry and universities have been developing testing protocols to evaluate the emission characteristics of prepared green sand during pouring, cooling and shakeout. This paper will review the testing protocol and methods utilized in the collection and evaluation of these emissions that is the result of the entire metal casting process. An important criteria also is the quality of casting produced (defect free) when the emission characteristics are reduced. This paper will also include the laser imaging evaluation of the castings that is the result of the green sand foundry emissions.

2. Introduction

Green sand is widely used in the foundry industry and accounts for a majority of the castings produced, primarily due to its cost-effectiveness and abundance of the raw materials required. A green sand mixture for ferrous castings primarily consists of sand, bentonite clay, moisture and carbonaceous additives such as sea coal. Despite its wide use in the metal casting industry, little research has been conducted on the emissions characteristics of the green sand process when compared to other processes, such as chemically bonded sand.

Trials were conducted at the University of Northern Iowa Metal Casting Center to evaluate three different green sand premixes for their emission characteristics over pouring, cooling and shakeout (PCS) operations. The three premixes had a bentonite clay content of 78%, with the rest being sea coal and other additives. Two different types of bentonites and sea coals were tested in the three premixes. Green sand mixtures were produced using a 60 GFN silica sand and a target active clay content of 8%. Step-block castings were poured and evaluated for each green sand mixture. Additionally, typical green sand properties were tested for the samples. The samples were tested for a total of four cycles. Emission tests were conducted on the first cycle and fourth cycle. The emission tests consisted of total Volatile Organic Compounds (VOCs), Hazardous Air Pollutants (HAPs) and Carbon Monoxide.

3. Methodology

A Simpson F 1-1/2 Muller was used to produce the green sand mixtures. 350 lbs. batches were produced for each sample. Step-block test castings poured using class 30 grey iron to evaluate the performance of the three premixes over four cycles. The step-block castings weighed approximately 70 lbs. each, with the molds weighing 330 lbs., providing a 4.7:1 sand: metal ratio.

The emission tests were conducted using a custom capture hood and 12 in. x 12 in. steel ducts, as shown in figure 1. The molds were placed inside the hood on the shakeout table prior to pouring. A volumetric flow rate of 1150 – 1200 cfm was measured in the ducts. After pouring, the hood was closed. The molds were then allowed to cool for 100 minutes, prior to shakeout. Shakeout operations were conducted for 5 minutes. The tests ended at 120 minutes.

Fig. 1. Emission collection pouring hood and steel duct.

EPA protocols were followed to determine sampling port locations, velocity traverses and volumetric flow rates. Table 1 shows the EPA tests conducted and corresponding test constituents. Emission tests were conducted for cycle 1 and cycle 4 castings. Total VOCs were measured as propane using a Flame Ionization Detector (FID) unit, shown in figure 2. The FID was calibrated using a three point curve. Additionally, gases were collected using canisters under vacuum over the three operation. These canisters were sent to an external laboratory for detection and analysis of 63 HAPs. A CO/CO_2 analyzer was used to measure carbon monoxide. The total VOC emissions from FID are discussed in this paper.

Table 1. EPA Methodologies used for Emissions Testing.

EPA Method	*Emission Compounds*
Method 25A	Total VOCs
Method TO-15	HAPs
Method 10	Carbon Monoxide

Fig. 2. Flame Ionization Detector (FID) unit.

4. Discussion of Results of the Emission Measurements

The raw FID results from the trials are shown in figure 3. A baseline graphical comparison of the Wyoming natural Sodium Bentonite at cycle 1 and cycle 4 can be observed to follow a trend with an initial increase in hydrocarbons after pouring. After approximately 5 minutes, the total VOCs can be observed to decrease steadily over cooling operations. A peak increase in VOCs can be observed at shakeout (100 minutes). After shakeout, the VOCs were observed to decrease to baseline values again. Both cycles showed similar emission results over the pouring operations. However, cycle 4 can be observed to have lower emissions over shakeout operations, measured to have a peak VOC of approximately 61 ppm. In comparison, the same sample at cycle 1 was measured to have a peak VOC of 102 ppm at shakeout.

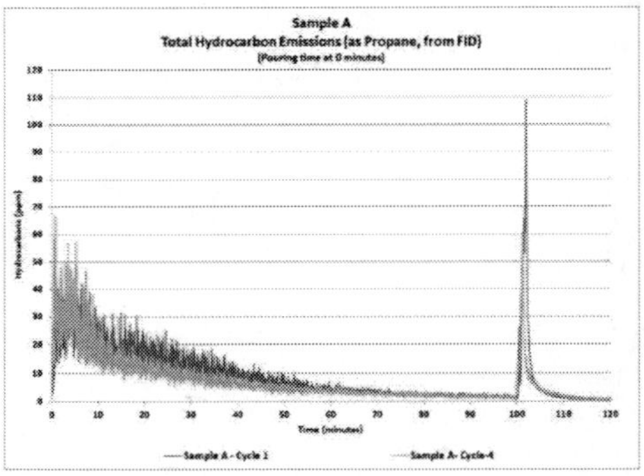

Fig. 3. Cycle 1 VOC Emissions over PCS Operations.

The area under the curves for the Wyoming natural Sodium Bentonite sample, shown in figure 3, were calculated at both cycles using numerical approximation. From the calculated area, and using volumetric flow rate, casting weights and the molecular weight of propane, the total VOC emissions in pounds per ton metal poured were calculated. The results are shown in figure 5. To gain a better understanding of emissions over the various operations, the results were charted as overall, 0 -5 minutes which denotes pouring, 5 – 100 minutes for cooling, 100 – 105 minutes for shakeout and 105 – 120 minutes for post-shakeout operations. Cycle 4 can be observed to have lower overall emissions when compared to cycle 1, mainly due to lower emissions in cooling and shakeout operations. However, cycle 4 was calculated to have slightly higher emissions at pouring when compared to cycle 1.

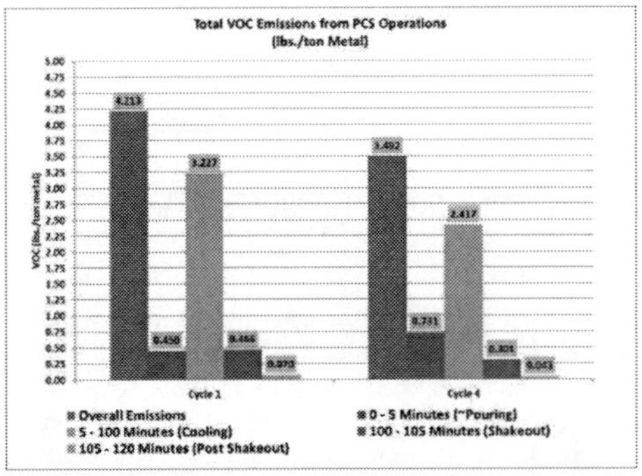

Fig. 4. Total VOC Emissions in Lbs./ton Metal Poured.

5. Evaluation of the Castings

As previously stated the step-block castings weighed approximately 70 lbs. each, with the molds weighing 330 lbs., providing a 4.7:1 sand: metal ratio.

Once the castings are produced and cleaned, they are scanned using a Romer Arm laser scanner and compared to the baseline, that is, the original pattern with shrink allowances for iron. The resulting change is computed and related as a percentage of points that is outside a tolerance of ±0.5 mm. This methodology of comparing and evaluating castings allows for a quantitative approach, as opposed to the qualitative approach usually used based on observations.

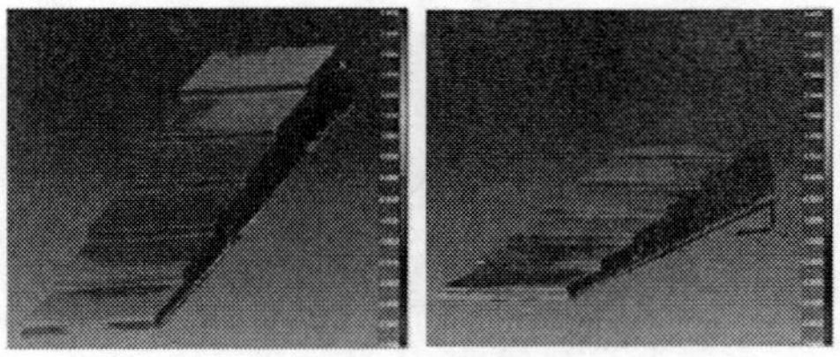

Fig. 5. Color map obtained from laser scanning of castings.

Acknowledgments

The authors would like to acknowledge Imerys Metalcasting Solutions for technical information and materials provided. The authors would also like to acknowledge the efforts of the student employees of the University of Northern Iowa Metal Casting Center for their assistance in the experiments.

The 73rd World Foundry Congress
„Creative Foundry"
23 - 27 September 2018, Krakow, Poland

CONTROL VOLUME SIMULATION OF THE TILT CASTING PROCESS

Dr. Daniel MOLNAR

University of Miskolc, Faculty of Materials Science and Engineering, 3515, Miskolc-Egyetemvaros, Hungary 3515, Miskolc-Egyetemvaros, Hungary, daniel.molnar@uni-miskolc.hu

Key words: Automotive casting, Control Volume, simulation, tilt casting

The casting process itself is the point in manufacture when most of the defects are introduced into the cast part. It is true that the melt might be the wrong composition, and the mould might be the wrong shape, but these misfortunes are unusual, and they are usually modest compared to the damage that is inflicted by most filling system designs.

Tilt casting solutions for gravity pouring could only be described as damage limitation exercises. Only tilt, level transfer and counter-gravity can produce ideal transfer of metal into the mold, thus manufacturing the best quality of casting.

In this paper a Control Volume model is used to simulate the tilt casting process of an automotive aluminum cast part. [1] The target of the experimental process is to identify the critical process parameters and build a simulation protocol, by the help of with, several casting processes and cast parts can be modelled and examined.

1. Tilt casting process

In Figure 1. The tilt casting process can be followed.

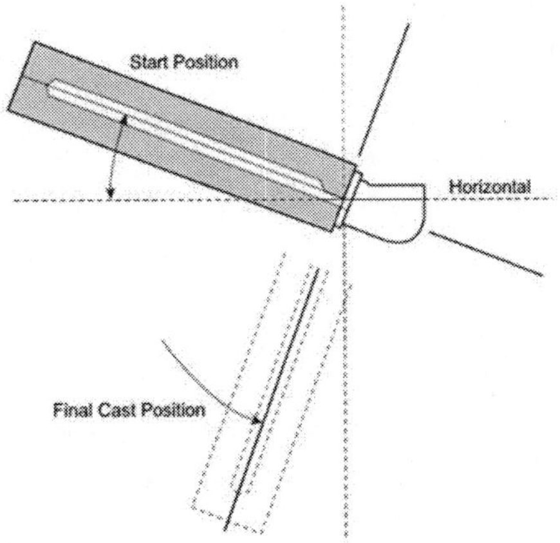

Fig. 1. Tilt casting process

If tilt casting is initiated from a tilt orientation at, or below the horizontal, during the priming of the runner the liquid metal accelerates downhill at a rate out of the control of the operator. The metal runs as a narrow jet, forming a persistent oxide flow tube. In addition, the velocity of the liquid at the far end of the runner is almost certain to exceed the critical condition for surface turbulence. Once the mold is initially inclined by more than 10° below the horizontal at the initiation of flow, it is no longer possible to produce reliable castings.

Tilt casting operations benefit from using a sufficiently positive starting angle that the melt advances into an upward sloping runner. In this way its advance is stable and controlled. This mode of filling is characterized by horizontal liquid metal transfer, promoting a mold filling condition free from surface turbulence.

Tilt filling is preferably slow at the early stages of filling to avoid the high velocities at the far end of the running system. However, after the running system is primed, speeding up the rate of rotation of the mould greatly helps to prevent any consequential non-filling of the castings. [2]

2. The simulation process

In our experiments the NovaFlow&Solid Control Volume simulation code is applied. The mathematical model of a technological problem always is an approximation of the original problem no matter the solution method. Thus, an analytical method, although giving the exact solution to the mathematical problem, results in an approximate solution to the physical problem. It therefore only makes sense to talk all the approximations into account if the value or applicability of a solution to the technological problem should be evaluated. Exact consistency between the mathematical solution and the mathematical model is thus not enough to ensure a good solution to the original problem.

Once the mathematical problem has been solved, the results should be interpreted. Especially in manufacturing processes the misinterpretation of otherwise correct results could lead to totally wrong conclusions. [3]

3. Virtual experiments

The examined geometry is a single cavity turbo charger. Standard EN AC-46000 aluminum alloy (1.66 kg) poured into a steel gravity die. The examined die geometry and the technological process steps can be seen on Figure 2. Process parameters can be seen in Table 1.

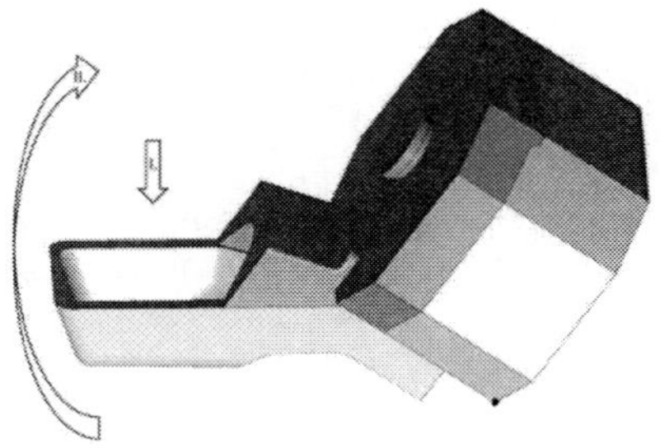

Fig. 2. Die geometry and the process steps

Table 1. Process parameters

Metal stream diameter	Ø 80 mm	
Draining time	0,05 s	
Tilt angle	55°	
	minimum	maximum
Pouring temperature	730°C	750°C
Die temperature	400°C	450°C
Metallostatic height	initial	modified
Flow rate	1,61 kg/s	3,32 kg/s
Tilting time	3 s	5 s
Pouring position	horizontal	non-horizontal
Venting	non-defined	defined

The melt is poured into the basin in the initial die position, using lip-pouring ladle: flow rate, stream diameter and metallostatic pressure are defined. After a given draining time the die tilt to the solidification position with a defined tilting time which determines the tilting velocity. During the whole process the gravity vector is: 0, 0, -1.

10 simulation experiments are implemented and all new simulation is parametrized after the analysis of the previous one:

- Experiments 1-2. are calculated using incorrect flow rate so the metal solidifies in the gating system. Flow rate of the metal is modified during filling.
- In experiment 3. incorrect cavity filling occurs with several unfilled parts. To accelerate the filling velocity pouring method is changed from lip-pouring to bottom pouring ladle and the position of the casting in the cavity is modified.
- In experiment 4. the velocity of the melt in the gating system is too high and the melt enter into the cavity too early. Bottom pouring ladle is not applicable, lip-pouring must be applied.
- In experiment 5. a significant air entrapment developed in the lower part of the geometry. Tilt time is modified.
- In experiment 6-7. air entrapment decreased, but only with the modification of the tilt time the problem cannot be solved. Venting is defined in the area of the riser.
- In experiment 8-10. tilt time is modified to achieve complete filling of the cavity. By the help of venting and tilt time optimization total filling of the cavity accomplished.

Filling process can be seen in figure 3.

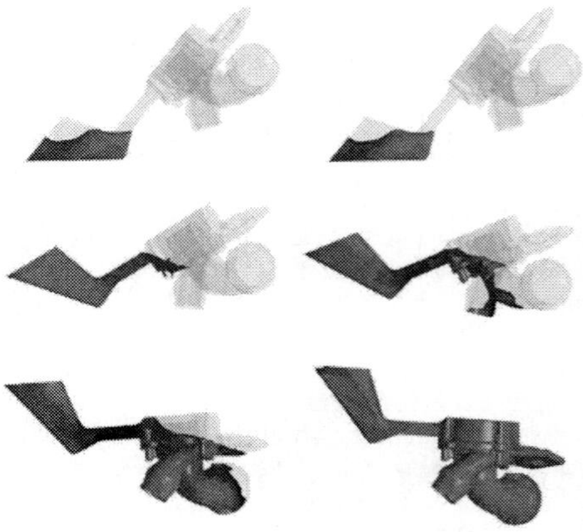

Fig. 3. Filling process

4. Results

The following critical parameters are identified during experiments, by the help of which, tilt casting process can be simulated with adequate accuracy. [4]

Geometrical adequacy: During filling the height of the basin, the cross-section of the basin nose and the geometry of the flow channel between the basin and the gating system are basic parameters.

Pouring conditions: Liquid contraction must be considered, and accurate definition of pressure height is important to avoid splashing of the melt.

Material parameters: Definition of the die material and application of refractory coatings are important to avoid melt cooling.

Tilt conditions: Filling stop criteria of the basin and tilting options such as time and angle are key parameters.

References

[1] Bohdan Mochnacki et al. (1995), Numerical Methods in Computations of Foundry Processes, Krakow: Polish Foundrymen's Technical Association.

[2] John Campbell (2011), Complete Casting Handbook. Oxford: Butterworth-Heinemann.

[3] Jesper Hattel (2005), Fundamentals of Numerical Modelling of Casting Processes. Lyngby: Polyteknisk Forlag.

[4] David Halapi (2015). Die Casting Development with Control Volume Simulation. Thesis, University of Miskolc.

The 73rd World Foundry Congress
„Creative Foundry"
23 - 27 September 2018, Krakow, Poland

SHRINKAGE DEFECT ELIMINATION SUPPORTED BY THERMAL ANALYSIS APPLICATION FOR DUCTILE IRON KNUCKLE PRODUCTION

P. Rodriguez[1], **B. Sagastume**[2], **J. Altuna**[2,] **A. Mura**[3], **N. Segreto**[3]

[1] *EDERTEK Technology Center of FAGOR EDERLAN GROUP, Isasi 6, 20500-Mondragón, Spain*
[2] *FAGOR DERLAN S. Coop., Torrebaso 7, 20540 Eskoriatza, Spain*
[3] *PROSERVICE s.r.l Via Marco Polo, 3 - 35010 Borgorico (Padova) – Italy*
* *e-mail:* p.rodriguez@edertek.es

Key words: Thermal analysis, ductile iron, eutectic behavior, steering knuckle, ItacaX™

1. Introduction

Light weight design is a challenging scenario for the iron steering Knuckles where Aluminum with a much lower density (2,7 kg/dm^3) is becoming a major threat on the iron applications (7,1 kg/dm^3) [1].

Fagor Ederlan´s smart design methodology develops lightweight ductile iron knuckles using material only where is needed reducing to the maximum the final weight of the component.

In this work, the real solidification morphology of 2 "types" (hypoeutectic & real-eutectic) of nodular graphite cast iron were measured using an innovative process control based in ITACAX™ (thermal analysis tool) to establish a supported parallelism between the morphology of the cooling curve and the real solidification morphology of the melt in a ductile iron GJS-400 steering knuckle.

2. Experimental results and discussion

To perform this study, it was taken a prototype EN GJS-400 steering knuckle casted in a LORAMENDI-e VMM 5070B electric molding machine to observe the microstructure and soundness switching from a current hypoeuthectic iron to a real eutectic iron driven by thermal analysis ITACAX™.

It was intended to observe what phases were present immediately after the liquidus temperature, before the eutectic low temperature and during the eutectic and correlate them with the final soundness of the part related to microshrinkage. For this observation, it was necessary to obtain parts and TA curves with the hypoeutectic iron and the same for eutectic iron.

For the execution of this work, a two stations ITACAX™ thermal analysis system was used for sampling, one in the melting area and another in the pouring area. To control the solidification behavior of the melt the TL (Liquidus temperature), TeMin (low eutectic temperature) and HEH (hypo-eutectic-hyper) parameters were been monitored.

Fig1. Simplified Fe-C diagram [3],

3. Foundry plant trials

The representative curve, for the hypo-solidification, of the melts tested (Figure 2). shows two main plateaus. A primary one, without any recalescence, defined as TL (liquidus temperature) and a second one, bigger than the first, with some recalescence, considered to be the eutectic transformation, as indicated in Figure 2.

The HEH (hypo-eutectic-hyper gives a detailed positioning in the Fe-C State Diagram) index values were < 40 giving us an evidence of a not real eutectic solidification.

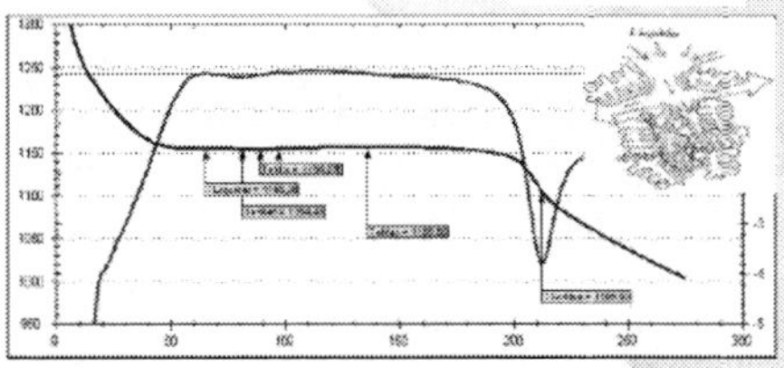

Fig2. Cooling curve of the hipo-iron [3]

As a result of the combination of the part and feeding design and iron a shrinkage defect appeared in the neck of the riser (Fig. 3).

Fig 3. Riser neck shrinkage

In the case study, the hole in the neck was not acceptable and regarding the thermal solidification curve it was necessary to move the iron to the eutectic region (lower contraction area). To drive the solidification behavior and maintain the iron in the safety Fe-C diagram eutectic region, on-line thermal analysis control, C, Si and melt conditioners were balanced.

As a result of the iron conditioning, a low contraction iron was obtained with a HEH between 50 +/- 5 (Fig. 4).
This eutectic curve shows only one plateau of big dimension, with a recalescence of 2°C, considered to the eutectic transformation.

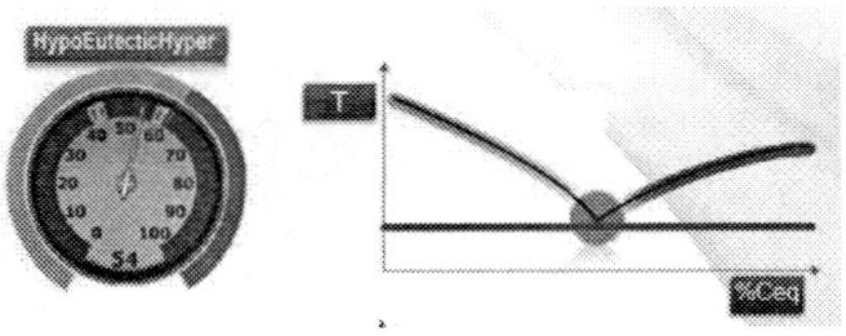

Fig 4. HEH parameter gauge detail [3].

Because of the use of this real eutectic iron in the castings, the shrinkage defective was drastically reduced to a residual <1%. (Fig 5). The neck geometry and size was reduce and nowadays FEG is running a R&D program called EFICAST (efficiency in casting) that is working in riserless & self-feeding castings.

To validate the control procedure (TL, Temin, HEH in melting material treatment and pouring) five batch of 1000 casting parts each, were produced with the defective percentage shown below. This good results, confirms the success of the activity.

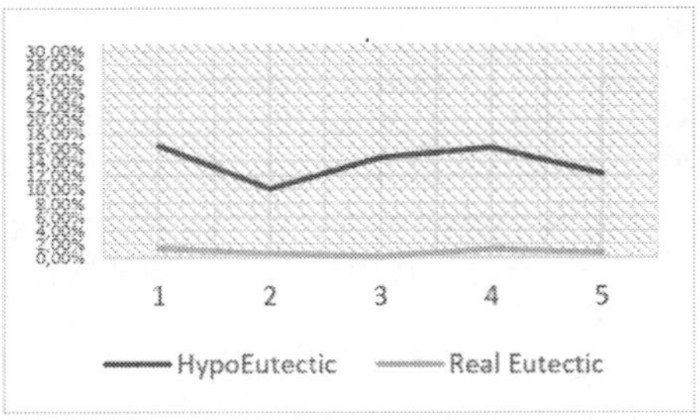

Fig 5.Casting shrinkage defective rate

4. Conclusions

The evaluation of TL, Temin and HEH by ITACAX™ allows the control of the real solidification morphology of a melt that promotes a scrap reduction from 15% to <1% in the case study.

The Metallurgical Process Control ITACAX™ completes the picture of the process and provides advanced integration for knowledge transfer among departments in the foundry.

Controlling the solidification parameters in melting, treatment, and pouring area to promote a 100% real eutectic solidification maximizes the expansion properties of the nodular cast iron melt and therefore reduces the feeding requirements and shrinkage defects.

Combining the part design, the feeding system design and iron eutectic solidification is possible to obtain sound, riserless self-feeding ductile iron knuckles (EFICAST project Fagor Ederlan Eskoriatza).

5. Acknowledgment

The authors would like to thank the great job done by Fagor Ederlan Eskoriatza S. Coop staff and the support given by Proservice s.r.l team.

6. References

[1] P. Rodriguez (2014) "Light weight in iron, new design, process and material strategies developments of nodular graphite iron in suspension components for the automotive industry). WFC 2014 Bilbao

[2] Nicola Segreto, http://www.proservicetech.it/

[3] Pierre Marie Cabanne. (2006) Sorel metal "thermal analysis :a useful technique for the foundryman". www.sorelmetal.com

The 73rd World Foundry Congress
„Creative Foundry"
23 - 27 September 2018, Krakow, Poland

USE OF SUPPLIER QUALITY INDEX FOR ASSESSING PROVIDERS QUALITY IN ALUMINIUM CASTINGS

Eray Erzi[1], Çağlar Yüksel[2], Özen Gürsoy[1], Derya Dispinar [1*]

[1]Istanbul University, Faculty of Engineering, 34320 Avcilar-Istanbul, Turkey
[1]Ataturk University, Faculty of Engineering, 25240, Erzurum, Turkey
*Corresponding address: e-mail: deryad@istanbul.edu.tr

Key words: 356, casting quality, bifilm, index

1. Introduction

In many of the casting operations, whether an ERP (Enterprise Resource Planning) is installed or not, there is a good control of process parameters. The melting temperature is measured, degassing operation is always repeated at the same parameters, runner and sprue design is made and pouring is carried out in the same controlled manner. Although each and every parameter is orderly taken into consideration, yet, rejected part percent can vary from time to time. Operations are checked over and over. One of the most important disregarded parameter in such operations is the measure of the provider's ingot quality. Typically, the chemical analysis of the charge is checked and the density index of the melt is measured. Dispinar [1-2] had shown that density index is misleading. Instead, a numerical indication of melt quality was quantified with bifilm index. In this work, a new quality index was proposed and the assessment of supplier's quality was quantified with the following equation:

$$SQI = (UTS + YS + e + F) - (BI) \qquad (1)$$

Where SQI is Supplier Quality Index, UTS is Ultimate Tensile Strength, YS is Yield Stress, e is elongation at fracture (x100) and BI is Bifilm Index.

Mechanical properties and fluidity need to be high, therefore these values are in positive side. On the other hand, bifilm index is subtracted from the equation, because higher the bifilm index, lower the melt quality; such that bifilm index of zero would indicate a perfectly clean melt.

2. Experimental Study

In this study, A356 alloy was provided from four different suppliers. For commercial reasons, the providers were coded as A, B, C and D.

10 kg of ingots were melted in a resistance furnace in SiC crucible at 745°C. For melt quality assessment, reduced pressure test samples were collected which were solidified under 100 mbar vacuum. The bifilm index was measured from the cross section of the samples [3]. Standard spiral mould was used to measure the fluidity of the alloys. Finally, tensile test bars were produced from the mould given in Figure 1 where the design belongs to John Campbell. In all the experiments, die moulds were used which were heated to 150°C.

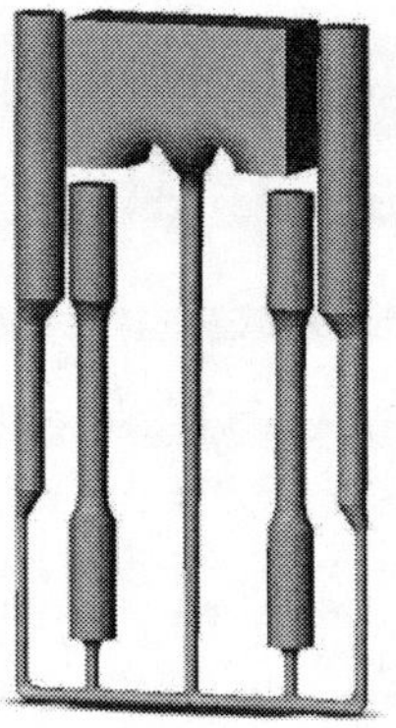

Fig. 1. Tensile test bar

3. Results

The bifilm index measurement had shown that Provider B had the best melt quality and Provider D had the worst as seen in Figure 2.

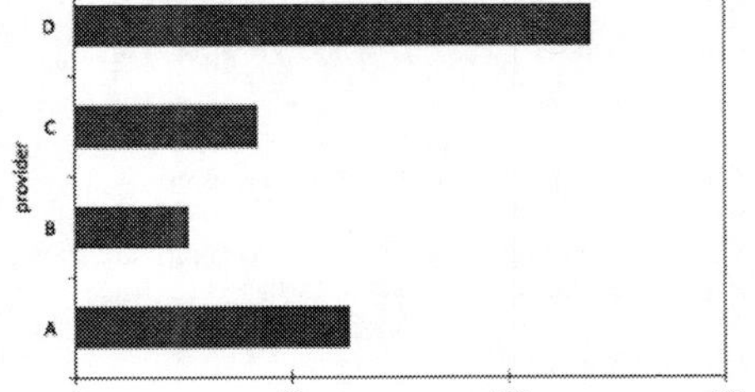

Fig. 2. Melt cleanliness comparison of different providers

The spiral fluidity test results revealed that Provider D was the lowest with 70 mm and Provider B had the highest of 160 mm.

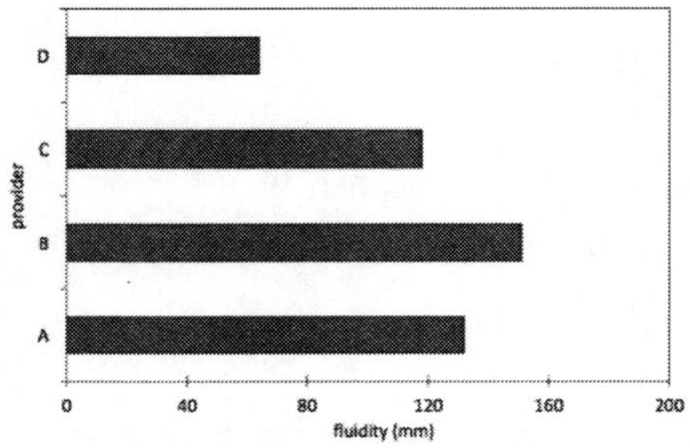

Fig. 3. Fluidity of different providers

After inclusion of mechanical properties in the Supplier Quality Index (Eq.1), Figure 4 was obtained. It can be seem that Provider B had the highest index whereas Provider D had the lowest value.

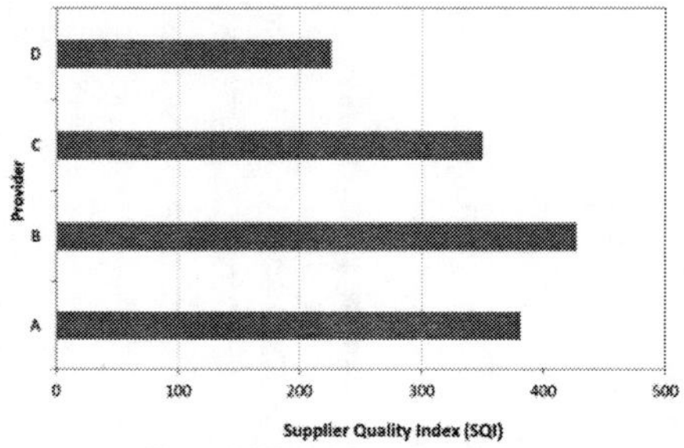

Fig. 4. SQI of different providers

In order to compare the SQI with rejection rates, Figure 5 was plotted.

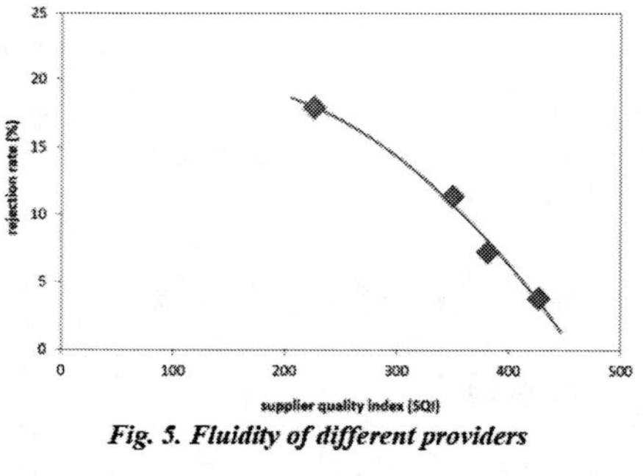

Fig. 5. Fluidity of different providers

It can be seen in Figure 5 that there was a strong correlation between the SQI and the rejection rates.

4. Conclusions

- Supplier quality assessment plays a significant role in determining the quality of the cast parts.
- A simple index was proposed in this work which consisted of tensile values, fluidity and bifilm index.
- There was good correlation between the SQI and rejection rates of cast parts.

Acknowledgment

Authors would like to thank Maxion Wheels, Manisa (Turkey) for their support.

References

[1] Dispinar, D., & Campbell, J. (2004). Critical assessment of reduced pressure test. Part 1: Porosity phenomena. International Journal of Cast Metals Research, 17(5), 280-286.

[2] Dispinar, D., & Campbell, J. (2004). Critical assessment of reduced pressure test. Part 2: Quantification. International Journal of Cast Metals Research, 17(5), 287-294.

[3] Dispinar, D., & Campbell, J. (2014). Reduced Pressure Test (RPT) for Bifilm Assessment. In Shape Casting: 5th International Symposium 2014 (pp. 243-251). Springer, Cham.

The 73rd World Foundry Congress
„Creative Foundry"
23 - 27 September 2018, Krakow, Poland

EVALUATING A CERAMIC RESIN COATED SAND FOR ALUMINUM AND IRON CASTINGS

Sam Ramrattan

Western Michigan University, Engineering Design, Manufacturing and Management Systems, 1903 W Michigan Avenue, Kalamazoo MI 49008-5336, USA
e-mail: sam.ramrattan@wmich.edu

Key words: Chemically Bonded Sand, Disc-shaped Specimen, Hot-Box Shell, Thermal Distortion

There is a lack of information available on shell resin coated sand (RCS) produced with ceramics. This paper relates the physical, mechanical, and thermo-mechanical properties of disc-shaped shell core specimens made from either silica or ceramic sands. Gravity filled disc-shaped shell core specimens were fabricated with silica and ceramic sands. Specimens were laboratory tested and evaluate in aluminum and iron casting trials. Testing included density, abrasion, thermal distortion, hot permeability, impact and retained strength. All tests were accomplished according to newly adopted American Foundry Society (AFS) Standards.

1. Introduction and Objective

The newfangled US "Silica Rule" will constrain conventional foundry silica sand practices. Increasingly foundries are considering alternative molding materials such as synthetic ceramic granular media. Engineers must distinguish difference between silica and ceramic RCS cores and molds that are cured for casting in aluminum and iron foundries [1-3]. The objective was to determine the physical, mechanical, and thermal cured properties and characteristics for gravity fill silica and ceramic RCS at aluminum and cast iron temperatures.

2. Preparation and Testing of Specimens

The shell RCS samples for silica (S) and ceramic (C) sands were prepared to two binder levels (1% & 3%). The shell RCS samples were gravity poured into a disc-shaped cavity of an electronically heated core box (aluminum die tool) manually. The shell RCS disc-shaped specimens were cured at 232°C (450°F) for 3 minutes. The shell RCS disc-shaped specimens were 50 mm in diameter by 8 mm thick and were identified as S1%, S3%, C1% and C3%. The following tests were conducted with a sample size of fifteen specimens: 1) specimen weight, 2) hot permeability, 3) abrasion, 4) impact and 5) TDT [1-3].

3. Casting Trial

Specimens were compared using an experimental doughnut-shaped model (Fig. 1) to perform a casting trials in aluminum and iron. Molds were manually poured with an aluminum (356) alloy and a gray cast iron (Grade 30) alloy. The average pouring time was ~12 sec., temperature at pour ladle was 700 ± 3°C (1292 ± 10°F) and 1430 ± 3°C (2600 ± 10°F); delivered through a pouring sleeve fitted with a ceramic foam filter respectively. All molds were poured to a 10 cm (4") head-height.

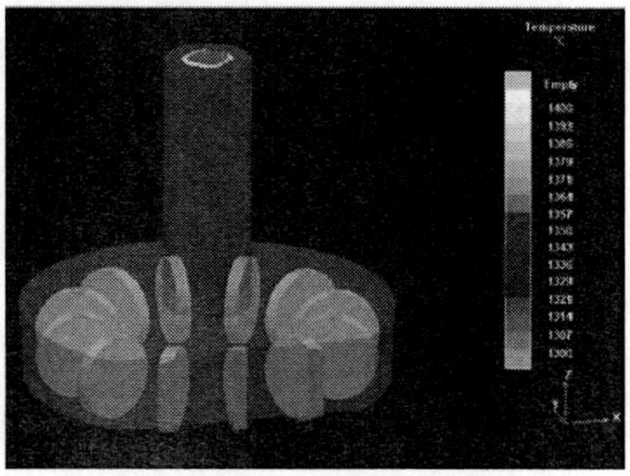

Fig. 1. Doughnut-shaped model with specimens

4. Mechanical and Physical Properties

Table 1 provides a summary of the tested physical and mechanical properties of the shell disc-shaped specimens.

Table 1. Properties of the RCS Specimens

	S1%	S3%	C1%	C3%
Specimen Weight (g)	23.05 (0.21)	24.87 (0.17)	28.80 (0.38)	27.48 (0.22)
Specimen Bulk Density (g/cm3)	1.47	1.58	1.83	1.75
Impact Strength (J)	0.45 (0.09)	0.60 (0.07)	0.51 (0.03)	0.74 (0.08)
Retained Strength (J)	0.41 (0.08)	0.43 (0.07)	0.49 (0.03)	0.58 (0.08)
Abrasion Loss (%)	2.5 (0.78)	0.72 (0.22)	5.68 (1.45)	0.34 (0.10)
Hot Permeability Index (#)	205	200	180	189

5. Thermal Distortion Testing (TDT)

The results from TDT for the various shell RCS showed significant differences. TDT and change in mass results is presented according to shell RCS (Table 2). The thermal distortion curves (TDC) are provided and compared for all systems in Fig. 2 and 3. The thermo-mechanical changes brought forth are in the forms of

TDC, cracks, and percent change in mass on the surface of the test specimens.

For all silica shell RCS specimens tested, the hot surface/specimen interface showed brown to black discolorations due to binder degradation. Cracks were macroscopically evident in both the $S_{1\%}$ and $S_{3\%}$ specimens (Tables 2). The crack propagations on the silica RCS systems were detected as inflection points at ~10 seconds on the radial TDC (Fig. 3). In addition, sand losses were evident at the hot surface/specimen interface where binder bridges pyrolyzed and sand grains broke loose; this was most apparent in the $S_{1\%}$ and $S_{3\%}$ specimens (Table 2).

Table 2. Thermo-mechanical properties of Specimens

	1000°C TDT @ 2.5 N for 90 seconds			Observation During Elevated Temp. Testing	
Specimens	D_L Longitudinal Distortion (mm*sec)	D_R Radial Distortion (mm*sec)	T_D Total Distortion (mm*sec)	% Change in Mass	Cracks and Fractures
$S_{1\%}$	1.64	6.75	8.39	1.60	Present
$S_{3\%}$	2.22	8.65	10.87	1.25	Present
$C_{1\%}$	0.96	1.03	1.99	0.71	None
$C_{3\%}$	2.05	1.39	3.44	0.25	None

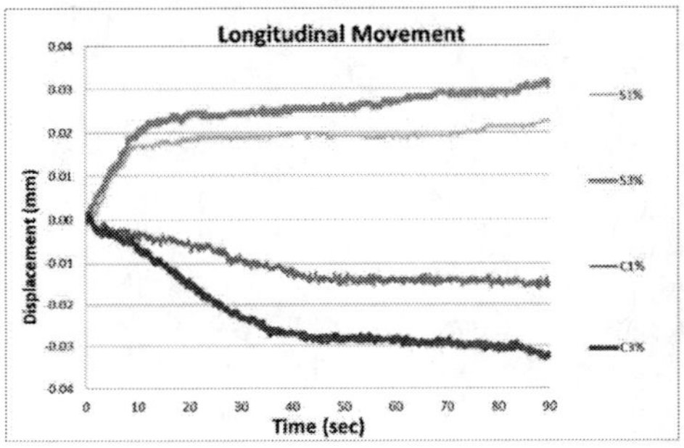

Fig. 2. Longitudinal TDC for Shell RCS specimens

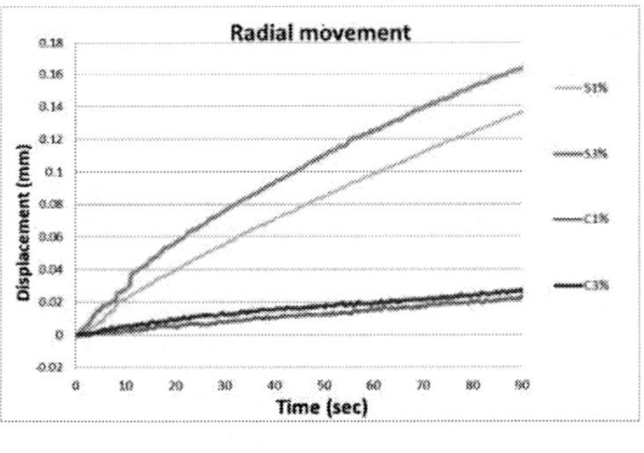

Fig. 3. Radial TDC for Shell RCS specimens

6. Conclusions

Sand cores and molds that are desirable for casting in aluminum and gray cast iron can be produced with ceramic shell RCS technology to a measurable quality Tables (3 and 4). Laboratory testing indicated that the ceramic shell RCS systems were thermo-mechanically superior to the silica shell RCS systems and the casting trial reinforced that fact for both aluminum and gray cast iron.

Table 3. Observations at Specimen/Aluminum Interface after Casting Trial

Specimens	As-Cast Interface: 10 cm head at 700°C	3D Macroscope Images at 12X	Observations within the same casting	Surface Roughness Ra μ-in. (μ-m)
$S_{1\%}$	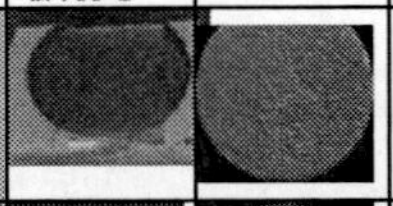		Some sand adhering with discoloration	984 (25)
$S_{3\%}$			Some sand adhering with discoloration	990 (25)
$C_{1\%}$	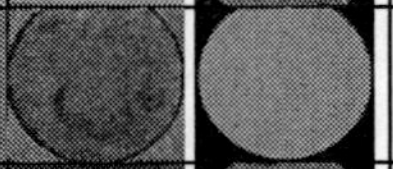		Clean with slight discoloration	375 (9)
$C_{3\%}$	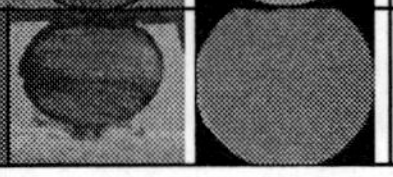		Clean with discoloration	370 (9)

Table 4. Observations at Specimen/Iron Interface after Casting Trial

Specimens	As-Cast Interface: 10 cm head at 1430°C	3D Macroscope Images at 12X	Observations within the same casting	Surface Roughness Ra μ-in. (μ-m)
$S_{1\%}$	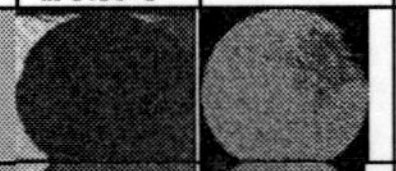		Large penetration	2760 (70)
$S_{3\%}$	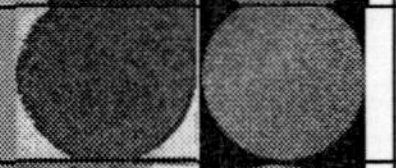		Slight penetration	2660 (68)
$C_{1\%}$			Surface relatively clean	525 (13)
$C_{3\%}$			Surface relatively clean	490 (12)

References

[1] Ramrattan, S. (2015). Non-Standard Tests for Process Control in Chemically Bonded Sands. Proceedings of the WFO International Forum on Moulding Materials and Casting Technologies. Changsha, China. October 2015. https://link.springer.com/article/10.1007/s41230-016-5118-7

[2] Iyer, R., Ramrattan, S., Lannutti, J., Li, W. (2001). Thermo-Mechanical Properties of Chemically Bonded Sands. AFS Transaction. Volume 109, pages 1-9.

[3] Ramrattan, S., Derrick, S., Nagarajan, K, Iyer, R. (2011) "Comparing Casting Evaluation to Thermal Distortion Testing for Various Chemically Bonded Sand Systems Using Image Analysis," AFS Transactions, 11-033.

The 73rd World Foundry Congress
„Creative Foundry"
23 - 27 September 2018, Krakow, Poland

EASY-KNOCK OUT MOULDING AND CORE SANDS – THE FUTURE FOR METAL CASTING

Artur Bobrowski [1]*, Dariusz Drożyński [1], Beata Grabowska[1], Karolina Kaczmarska[1]

[1]*AGH University of Science and Technology, Faculty of Foundry Engineering, Reymonta 23, 30 059 Kraków, Poland*
**Corresponding address: e-mail: arturb@agh.edu.pl*

Keywords: moulding sand, inorganic binder, water glass, geopolymer, knock out

In recent years, interest in moulding and core sand with inorganic binders is on the increase. Until recently, the most widespread inorganic binder was hydrated sodium silicate (water glass). To meet the expectations of customers and increasing requirements in terms of reducing emissions harmful substances to the environment [1], enterprises operating on the market of binders for foundry offered several new inorganic binding systems. Cordis technology (Hüttenes-Albertus) appeared and INOTEC (ASK). In 2003 on the Trade Fair GIFA, an inorganic binder system Hydrobond from MEG was presented for the first time. SAND-TEAM proposes replacing the water glass with geopolymer binders Rudal A and Geopol® [2-5].

The main disadvantage of all moulding sands with inorganic binders is their strengthening caused by high temperature of liquid metal (high final strength) and associated with its difficult knocking-out and poor susceptibility to mechanical reclamation. These are the difficulties determine their limited use in foundry. Various types of activities are undertaken to reduce them, including physical and chemical modification [6-12].

1. The proposed solution – additives for inorganic moulding sands

Modification of the composition of moulding sands with inorganic binders consisted in introducing a relaxation additive. Two additives, designated as S1 and S2, are proposed.

2. Investigation results – improvement of knocking-out

Additives S1 and S2 are introduced into the moulding sand during the mixing process. As shown in Figure 1, their introduction has no negative effect on the mechanical properties of the sands. The sands with additives (IMS + S1 and IMS + S2) have similar values of tensile strength (R_m) to sand without additives (IMS). Figure 2 presents the results of the final compressive strength (R_c^{tk}) determination for the moulding sand without additive and moulding sands with the proposed relaxation additives. Standard samples after 24 hours of curing, were heated at a temperature range of 100-1000°C. Both the moulding sand with addition of S1 and S2 are characterized by several times lower final strength.

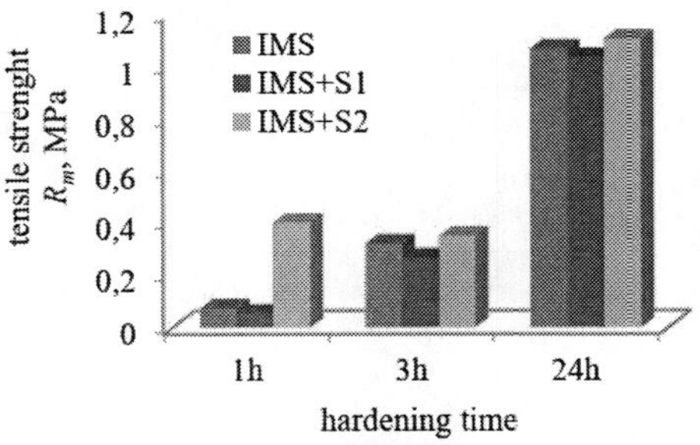

***Fig. 1.** Tensile strength of moulding sands with inorganic binder (IMS – without additive, IMS+S1, IMS+S2 – with relaxation additive)*

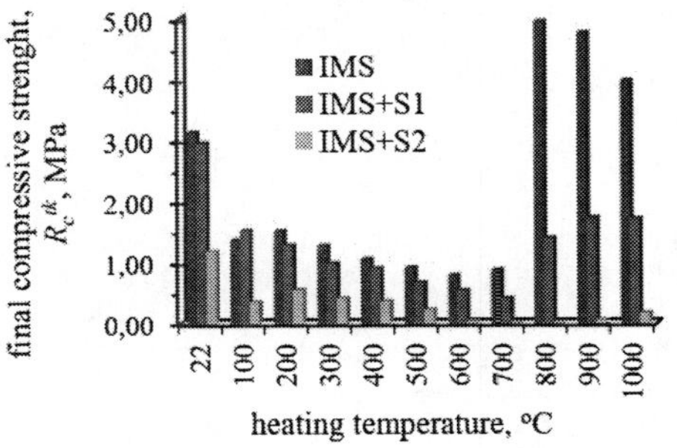

***Fig. 2.** Final compressive strength of moulding sands with inorganic binder (IMS – without additive, IMS+S1, IMS+S2 – with relaxation additive)*

In order to confirm the results of the final strength tests, a knock-out test according to PN-85/11005 was also made (Fig. 3). Standard cylinder-shaped fittings Ø50x50 mm placed in a mould cavity (b) made by using the model (a). The mould is poured with liquid metal, and then knocking-out process is conducted by using a cone-shaped mandrel (c). The measure of the sands susceptibility to knocking-out is the work needed to remove of the core (standard cylinder-shaped) from casting, described as (1) [12]:

$$L_w = 1.63 \cdot n, \text{J} \quad (1)$$

where:
1.63 – work of one stroke weight of 3.3 kg falling from a height of 50 mm above the core, J;
n – the number of strokes of the sinker at the time of the removal of core from the cast.

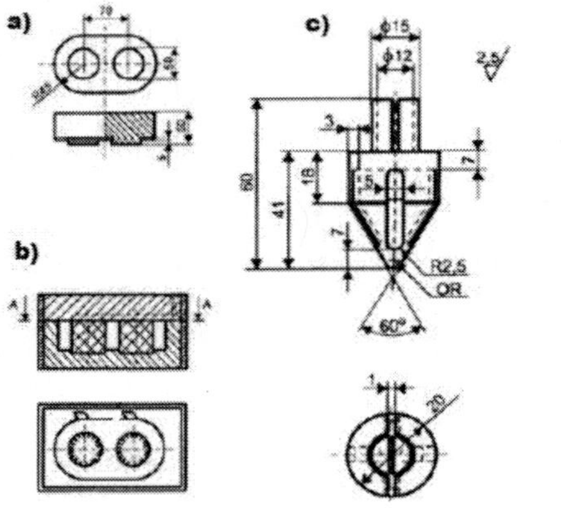

Fig. 3. Model diagram (a), form (b), conical stem (c) to the knocking-out test by PN-85/11005 [12]

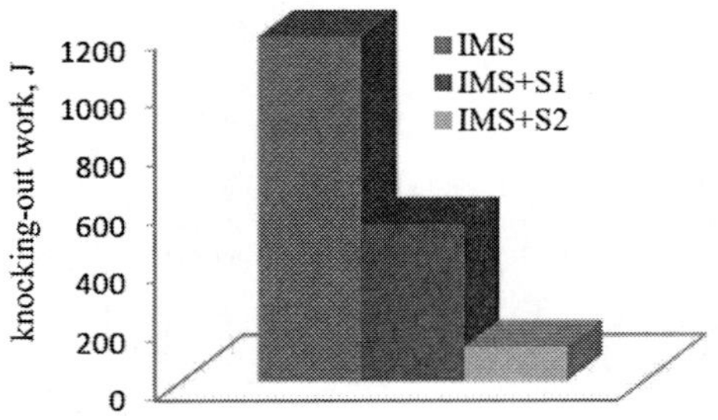

Fig. 4. Results of knocking-out test according to PN-85/11005

Introduction of relaxation addition S1 and S2 significantly improves the knocking-out of moulding sands with inorganic binders - all types (Fig. 4), in relation to moulding sand without additives (IMS). The proposed additives do not cause the emission of harmful gaseous compounds during the casting process, cooling or knocking and can have a positive effect on the mechanical reclamation process of used sands.

Acknowledgments
The research was partially financed in the "Incubator of Innovation +" project is carried out by the consortium of the AGH University of Science and Technology in Krakow AGH Center for Technology Transfer and the Krakow Center for Innovative Technologies INNOAGH Sp. z o.o. in the non-competitive project "Support for managing research and commercialization of Research and Development results in scientific units and enterprises" implemented within POIR 2014-2020, Measure 4.4 - Increase of human potential R+D.

References

[1] Bobrowski A., Holtzer M., Żymankowska-Kumon S., Dańko R. (2015). Harmfulness assessment of moulding sands with a geopolymer binder and a new hardener, in an aspect of the emission of substances from the BTEX group. Archives of Metallurgy and Materials 60(1), 341–344.
[2] Burian A., Novotný J. (2008). Production of molds and cores with an ecological binder system. XI Conference materials from the Technical Foundry Conference 2008. 81-91.
[3] Pezarski F., Smoluchowska E., Izdebska-Szanda I. (2008). The use of geopolymer binder for the production of iron alloy castings. The Transactions of the Foundry Research Institute. XLVIII (2), 19-34.
[4] Polzin H. Anorganische Binder zur Form- und Kernherstellung in der Gießerei, SCHIELE&SCHÖN.
[5] Izdebska-Szanda I., Baliński A. (2011). New generation of ecological silicate binders. Procedia Engineering. 10, 887–893.
[6] Baliński A. (2015). Effect of the WB structure index of hydrated sodium silicate on the final strength of the moulding sand. The Transactions of the Foundry Research Institute. LV (1), 3-9.
[7] Bobrowski A., Stypuła B., Hutera B., Kmita A., Drożyński D., Starowicz M. (2012). FTIR spectroscopy of water glass – the binder moulding modified by ZnO nanoparticles. Metalurgija. 51 (4), 477-480.
[8] Dobosz S.M., Major-Gabryś K. (2004). Glassex - a new addition improves knocking of moulding sands with water glass, Archives of Foundry. 13, 63-68.
[9] Major-Gabryś K, S.M. Dobosz. (2011). The influence of the Glassex additive on technological and knock-out properties of the moulding sands with hydrated sodium silicate and new ester hardeners. Metallurgy and Foundry Engineering. 37 (1), 33-40.
[10] Major-Gabryś K., Dobosz S.M., Zych J. (2005). Criteria for the assessment of the knocking-out of moulding sands with water glass. Archives of Foundry. 17, 163-168.
[11] Chun-xi Z. (2007). Recent advances in waterglass sand technologies. China Foundry. 4 (1), 013-017.
[12] Lewandowski J.L. (1997). Materials for foundry molds. Cracow. Akapit Publisher.

The 73rd World Foundry Congress
„Creative Foundry"
23 - 27 September 2018, Krakow, Poland

NEW & RECLAIMED CHROMITE VS ALUMINO-SILICATE REFRACTORY SAND ADVANTAGES AND DRAWBACKS

Philippe Diaz [1*], Florian Ahouanto [2]

[1] *IMERYS ALUMINATES – Foundry Technical Manager – FRANCE*
[2] *IMERYS ALUMINATES – Andalusite Product Manager – FRANCE*
**Corresponding address: e-mail: philippe.diaz@imerys.com*

Key words: Defects, Refractoriness, Special Sand, Kersand, Chromite, Zircon, Steel, Iron,

Nowadays, global competitive environment and severe economic conditions have led metalcasters to improve their casting quality and production efficiency.
The use of special sands has been widely spread to reduce sand fusion, metal penetration and veining defects in foundries.

However, the first reflex to solve defects is to straightly go to consumables like, new coating, new Sand Additives or a "Special Resin". Foundry management don't want to spend time on defects because "time is money". For technical people, these is the easiest and faster way … at lower cost:
Is it True or Not ?
Most of the time, it's not so simple.
Whatever Sands used, it represented more than 95% of Cores & Mold in volume & weight. Defects reductions required more expertise and deeper investigation on consumable and specially on sands.

Despite the highest price of "special Sands", these types of solutions can generate significant cost reduction in the foundry process, particularly referring to the lower scrap rate as well as the shorter finishing time. On top of the traditional solutions, based on Zircon and Chromite, new type of special sands have been developed over the last 10 years, in order to improve the performance and compensate the high price volatility associated to the traditional sources of those molding sand.

Market trend of all those sands are following the global demand and their main drivers.
Zircon sand is directly liked too Titanium extraction (90% lead by pigment & paper industries) and Chromite driven by ferrochrome alloys and stainless steel demand.

Technically speaking, chromite provide some drawback as high bulk density (generating a low Flowability for shooting process) or some reclaimability risk with silica sand pollution in steel molding (for facing sand).
Bulk Density and high AFS number of Zircon sand drastically reduce the field of applications in foundries.
Regarding Silica Sand and despite its low cost, some issues, in some countries are rising up like the silicosis concern via cristobalite and Respirable Crystalline silica (RCS):
Those indirect topics are becoming more & more relevant in the "General Cost Calculation"

KERSAND™ is an Andalusite based aluminosilicate (58-60% Al_2O_3) special sand designed for Steel, Iron & Copper alloys foundries, for:

- Cores production
- Facing sand application

It has been developed to provide high refractoriness, low thermal expansion, high thermal conductivity and total compatibility with silica sand & all organic/inorganic binders.
This is achieved through a complex purification process taking place in Britany, France where this natural mineral is mined and processed. The Andalusite ore undergoes several process steps, such as milling and classification, heavy media separation, magnetic and electrostatic separation as well as floatation.

This product, available since 2017, has shown significant potential to improve casting quality. This paper provides a global picture of the technical and economic advantages associated to the use of KERSAND™. It is based on feedbacks from foundries using KERSAND™ as an alternative to other types of sand, such as chromite & Zircon.

The 73rd World Foundry Congress
„Creative Foundry"
23 - 27 September 2018, Krakow, Poland

HYBRID RESINS: A GREAT ADVANTAGE FOR QUALITY, ECOLOGY AND COSTS

Dr Amedeo Mazzon *, Luca Gonzo
[1]*F.LLI MAZZON S.p.A. Via Vicenza, 72 - 36015 Schio (Vi) – Italy*

**Corresponding address: e-mail: sales@mazzon.eu*

Key words: hybrid resins, furfuryl alcohol, synthesis, research and development, ecology

After a sudden increase in furfuryl alcohol costs, F.lli Mazzon immediately found an alternative, thus developing a new generation of furan resins. Such resins, besides granting excellent quality and avoiding price increase for customers, also improved working conditions - being harmful and not toxic - and reduced emissions and smells. F.lli Mazzon succeeded in transforming a problem into an opportunity.

1. Market Needs and Trends.

Due to the huge increase in the price of furfuryl alcohol, furan resins almost doubled their prices. Such an increase cyclically happens every 5 -7 years for a number of reasons. As a consequence, foundries found an immediate increase in estimated costs of around 2-3 cents per kg of casting, which final customers didn't want to pay. So the market, i.e. hand-molding foundries, asked for new products with good technical characteristics and more competitive prices.

2. New polymers: Hybrid Resin synthesis.

For a long time F.lli Mazzon has been producing furan-phenolic resins, i.e. mixtures of phenolic resins and furfuryl alcohol to grant more reactivity and mechanical strength. But also these formulations were not competitive because even a low percentage of furfuryl alcohol caused prices to go sky-high.
Our R&D laboratory started to process different synthesis and chemical reactions in order to obtain a resin with a behavior similar to furan resins, but whose costs are similar to phenolic resins.
We found that by making phenolic resins reacting with furfuryl alcohol and other molecules in the reactor for rather long times at given pressure and temperature conditions, we obtain some effective polymers with the following characteristics.

3. Bench life

Furan resins are characterized by a long bench life. Phenolic resins seem to have a bench life, but they actually begin to react immediately with longer stripping times. Our new polymers have similar characteristics to classic furan resins, i.e. good bench life that allows to work within the required times.

4. Stripping time

Furan resins have a very good long bench life / short stripping time rate. For example 10 minutes bench life and 40 minutes stripping time. On the contrary the rate for phenolic resins will be 10 minutes bench life and some hours of stripping time. Our new polymers of Hybrid Resins have good stripping times and enough bench life.

5. Mechanical strengths

The mechanical strengths as well as the thermal resistances of the new polymers of Hybrid Resins are up to the performance of the well-known furan resins. In most cases foundries keep the same percentage of use or sometimes specific catalysts are used to accelerate or slow down the reaction or to obtain maximum resistance.

6. Viscosity

This is definitely the most critical parameter, since hybrid resins are naturally more viscous than pure furan resins. In any way foundries can manage viscosity even at low temperatures. There is no problem if the mixers are equipped with systems that keep the resins e.g. at around 20 ° C. There is no problem too with geared pumps. But it is essential that the mixing system must work efficiently to homogeneously mix the resin with the sand.

7. Emissions and smells

As for the emissions and smells of Hybrid Resins, the working environment is improved. Hybrid Resins are in fact condensed polymers, i.e. high weight molecular chains or interlocked molecules, therefore they can not evaporate in the environment. The content of free molecules is limited so that also smells are fewer. This is definitely an advantage for the working environment.

8. Labelling and Safety Data Sheet

Hybrid Resins are made up of molecules linked to each other and so there is a very low percentage of single free components. The share of free furfuryl alcohol is less than 40% and therefore the labelling symbol is only "harmful" and not "toxic".

In addition, interlocked molecules of furfuryl alcohol do not evaporate and do not increase TLV value that is the allowed concentration in the working environment. That's why Hybrid Resins are eco-and worker-friendly.

9. Formaldehyde

These resins are produced using a formaldehyde content lower than 0,1%, as required by law. Often by using specific reagents we manage to lower its content below this threshold, benefiting the workers and the working environment.

10. Price

Last but not least let's focus on the price. All the tailor-made formulations of these hybrid resins have anyway a cost of raw materials lower than those of traditional resins. For sure the long synthesis process for the polymerization of the resins in the reactor and energy used to remove condensed water are increasing the costs. However these hybrid resins give value for money.

In conclusion, a problem has turned into an opportunity for both F.lli Mazzon and his customers. Surely these resins must meet the needs of each foundry but Hybrid Resins are having an unexpected success both in Italy and abroad.

The 73rd World Foundry Congress
„Creative Foundry"
23 - 27 September 2018, Krakow, Poland

HEAT RECOVERY FROM COMPRESSED AIR STATION – FOUNDRY VOLKSWAGEN POZNAŃ

Tomasz Kamiński

Manager of Planning Department at Foundry Volkswagen Poznań
Volkswagen Poznań sp. z o.o., ul. Warszawska 349, 61-060 Poznań, Poland
Foundry Volkswagen Poznań sp. z o.o., ul. 28 Czerwca 1956r. 240, 61-485 Poznań, Poland
e-mail: tomasz.kaminski@vw-poznan.pl

Key words: Heat recovery, ecology, foundry, innovation, compressed air delivery.

Almost every manufacturer uses compressed air in production processes. Unlikely production of compressed air is one of the lest efficient processes. More than 90% of input energy is changed into waste heat and delivered into air. That situation causes also high costs of compressed air production. One of the solutions is installing of heat recovery system. Energy (in this case heat) recovery is one of the best way to reduce costs of compressed air and minimalize waste.

Fig. 2 Heat recovery piping at compressed air station VW Poznań

Heat recovery is a system of heat exchangers, that recover heat from mechanical parts of compressor. Heat is transferred to secondary water circuit and then for example to sanitary water preheating, heating of social spaces, technology, or as in case of Foundry Volkswagen Poznań to municipal grid. That solution allows to recycle about 75% of total compressor energy consumption.

Compressors station at Foundry Volkswagen Poznań was rebuilt in 2012. During purchasing process one of the main points was Total cost of ownership. It was analyzed and compared compressors with and without heat recovery systems. The analyze shown - that by such installation - heat recovery is viable only by constant heat consumption. There are installed 4 compressors - 3 of them are with heat recovery system. It occurred, that Volkswagen Poznań do not need such amount of heat all the time. To find the solution of the situation Planning Department started to consider another heat receiver. Parallel started meetings with local heat provider for city of Poznań (Veolia). After 2 years of negotiations and technical consulting VW Poznań got technical terms from Veolia for connecting industrial heat recovery to municipal grid. Heat provider had high requirements and the only supplier, that guaranteed all the parameters was Atlas Copco.

Fig. 1 Machinery of heat recovery at VW Poznań

The installation is designed for constant distribution of 1,8 MW of heat. Recovery system of Volkswagen Poznań guarantees constant hot water delivery with temperature over 85°C.

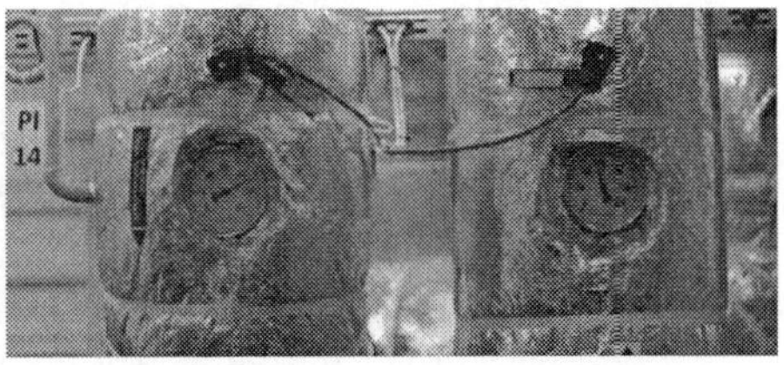

Fig. 3 Temperatures of inlet and outlet

Heat from compressors is transferred to heat distribution center owned by Volkswagen Poznań. It is fully automated station equipped with pumps, heat exchangers, sensors, valves etc. In case of failure on the part of Veolia automatics switches cooling of compressors over cooling towers. From the VW heat distribution center, hot water is transferred to center of Veolia and redistributed to final consumers like Hospital, University, Industry and local citizens. It is more than 30 buildings and 2000 flats.

Profits are significant. Not only economically, but also environmentally. As result of this solution Volkswagen Poznań saves energy, reduces emission of greenhouse gasses and minimalize water consumption. Compressors are cooled with backflow water at temperature of 45°C instead of cool water from cooling towers. That causes reduction of evaporation for almost 20.000m^3 of water each year.

Fig. 4 *Cooling towers used for cooling of compressors*

The 73rd World Foundry Congress
„Creative Foundry"
23 - 27 September 2018, Krakow, Poland

DUCTILE IRON AND COMPACTED GRAPHITE IRON TREATMENT WITH HYBRID MAGNESIUM CORED WIRE PROCESS

Affival sas [1]*, Olivier Bahuon [2]

[1]Affival sas, 70 Rue de l'Abbaye, Solesmes, France,

**Corresponding address: Olivier.bahuon@affival.com*

Key words: Ductile Iron, Compacted graphite, Cored Wire, Magnesium, Nodulizer

The popularity of the family of Ductile Cast Iron comes from their diversity and their interesting cost price. Practical observations and adaptations by thousands of foundries have made the production of Ductile cast iron a routine.

A minimal but well-proportioned addition of magnesium is the basis of the process for changing the shape of graphite particles from lamellar to spheroidal. This treatment is generally carried out with a ferrosilicon-magnesium alloy. The inoculation of the cast iron treated by addition of ferrosilicon in the ladle is also very important.
The control of the chemical composition of the GS cast iron is carried out on about twenty elements. Routine determination of the content of each of them is not necessary, but in such a case it requires the use of reliable raw materials. Nevertheless, the monitoring of the carbon and silicon contents in the furnace is routinely required. Successful foundry has implemented reliable manufacturing and quality control methods.
The development's target of Affival on the nodularization treatment was initially focused on the definition of a means for incorporating some magnesium in an economical way, and very reproducible effect in a bath of molten iron.
The most economical route is the use of pure magnesium, a material that is easily found in sufficient volume to obtain it at a controlled market price.
But to introduce magnesium in liquid iron is difficult because of:

- Magnesium's boiling temperature at 1091°C
- Limited solubility of magnesium in the liquid iron

And more than that, because the magnesium is a strong deoxidizing agent, its using needs also to use a strong inoculant step to balance its wrong impact on nucleation.

1. Good practices for using Cored Wire are to be applied

The use of Cored Wire allows to provide exactly the amount of magnesium required by the liquid metal after the ladle has been filled with known metallurgical and physical characteristics (weight, % S, temperature, ...)

<u>The Cored Wire process offers the following advantages:</u>

- Process can be fully automated
- Better capability than other processes (means lower Magnesium concentration in average required)
- Easy and exact control of wire addition
- Easy to manage the level of final Magnesium
- Possible to use the option of treating high sulfur base iron directly
- Better traceability of data
- Easy to reduce fumes by exhausting system
- Clean iron, less slag, avoid dross formation (longer ladle refractory life)
- Treatment can be done near the pouring station
- Treatment weights from 250 to 40.000 Kg

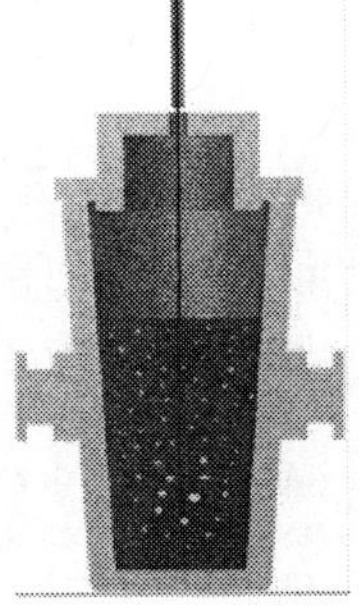

Fig. 1. The Wire should be fed straight and deep on a proper treatment ladle with adapted cover and exhausting system

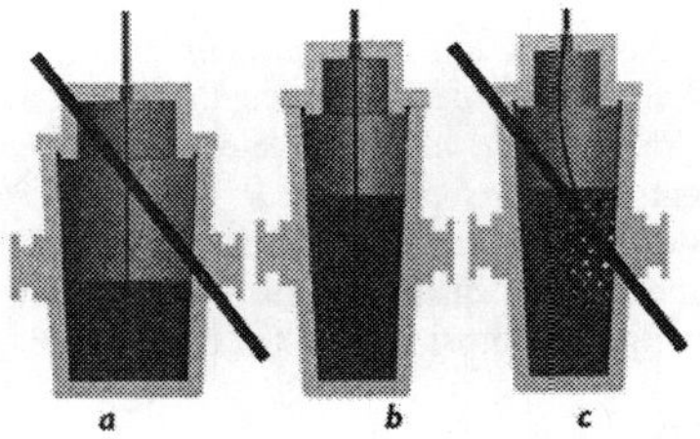

Fig. 2. Examples of Cored wire feeding configurations

2. New generation of Cored Wire for the Ductile Iron treatment: "Hybrid Cored Wire"

The initial principle is to control very precisely the metric weight of magnesium in the cored wire. Two different axis can be taken:

a) The use of two wires: one magnesium and the other inoculant

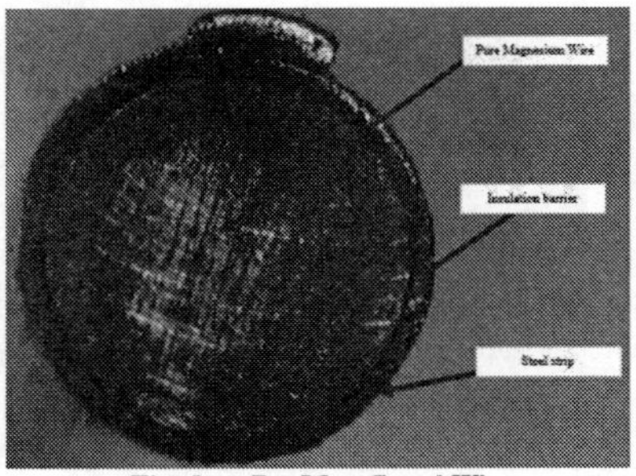

Fig. 3. PapMag Cored Wire

b) The use of a single "hybrid cored wire" included inoculant powder

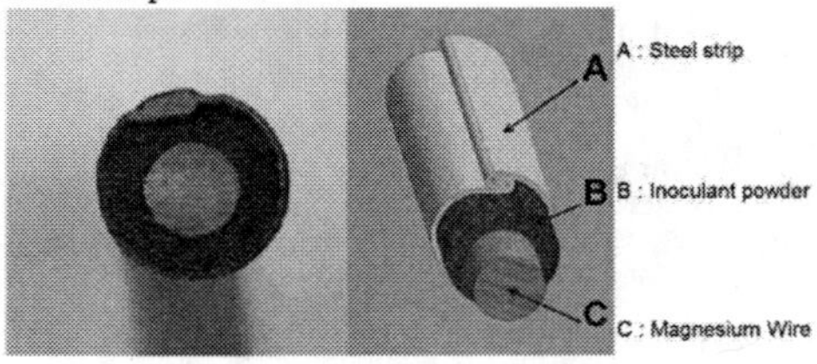

Fig. 4. " Hybrid Cored Wire"

Product Description :

- Round strip (A): steel or copper strapping 0.20 to 0.6 mm thick
- Extruded core (C): extruded magnesium wire diameter 2 to 12 mm
- Powder (B): alloy or mixture of alloys

3. Interest of this concept "Hybrid Cored Wire":

Regularity of the mass of magnesium introduced because of the use of magnesium in the form of solid wire (as opposed to conventional products which uses magnesium powders whose densities and compaction, difficult to control, generate dispersions of metric mass)

The low mass variation of magnesium per meter (+/- 2%) makes it possible to reduce the dispersion of residual magnesium in the treated cast iron and consequently to lower the target level of magnesium, hence a decrease in product consumption as well as an improvement in the quality of castings with this cast iron (reduction of porosities and / or oxide sails)

Positioning the powder mixture around the extruded magnesium wire protects the extruded magnesium wire. When injecting the wire into the liquid bath, the rise in temperature of the magnesium is delayed. The wire can therefore be introduced deeper into the liquid iron column, this optimizes the contact time of the magnesium gas with the liquid and thus improves the magnesium yield.

In addition, reducing the specific surface area of the magnesium metal slows down its vaporization and calms the reactivity of the reaction.

4. Intermediate inoculant layer

The location of the intermediate layer around the extruded magnesium protects the latter. When the flux-cored wire is injected into the molten cast-iron bath, the rise in temperature of the magnesium is delayed by a slow heat transfer. The cored wire can therefore be introduced deeper into the molten iron column. This increases the contact time of the magnesium gas with the liquid iron and thus improves the magnesium addition efficiency.

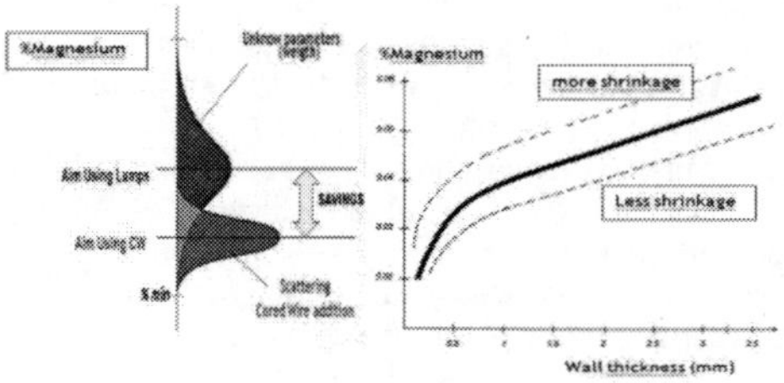

Fig. 5. Thanks to the reduction of standard deviation of %magnesium results we can decrease the average level of magnesium, so reduce the shrinkage.

The inoculant intermediate layer also has a very important metallurgical role. The layer comprises a second active substance in the metallurgical treatment, namely a Barium or Zirconium doped inoculant. Inoculation, as it is known, regenerates the graphitic germination potential after the magnesium treatment. This makes it possible to avoid the formation of cementite and to contribute to obtaining the desired level of spheroidal graphite.

For ferritic cast iron grades, this also promotes the formation of ferrite by increasing the density of spherules.

References

For patents:

[6] A. Markwitz, O. Bahuon, S. Gérardin. (2012). Patent number WO2014072456A1.

The 73rd World Foundry Congress
„Creative Foundry"
23 - 27 September 2018, Krakow, Poland

Development of Additive Manufacturing Sand Molding system Aimed for Mass Production Application

Toshimitsu Okane [1]*, Yoshikazu Ooba [2], Yasuhiro Nagai [3], Syuji Koiwai [4], Takahiro Kaneyoshi [5], Atsuya Toba [6], Hiroshi Kambe [7], Makoto Yoshida [8]

[1] National Institute of Advanced Industrial Science and Technology (AIST), Japan,
[2] CMET Inc., Japan, [3] Gun Ei Chemical Industry Co., Ltd., Japan, [4] KOIWAI Co., Ltd., Japan,
[5] Hyogo Prefectural Institute of Technology, Japan,
[6] Hokkaido Research Organization, Industrial Research Institute, Japan,
[7] NISSAN MOTOR Co., Ltd., Japan, [8] WASEDA University, Japan,
**Corresponding address: e-mail: okane-t@aist.go.jp*

Key words: Additive manufacturing, Sand molding, Mass production, Furan resin self-hardening sand molding, ceramic sand

1. Introduction

Recently, additive manufacturing(AM) casting mold is used to the casting of prototypes and low-volume production. In Japan, additive manufacturing molding and related casting technology development had promoted with the aim of future mass production application. By realization of more complex structures, such as thin lightweight with improved cavities accuracy, the potential improvement in overall casting are expected. This paper reports the outline of the development of additive manufacturing molding apparatus.

Additive manufacturing sand molding for the next generation casting, not only apparatus, but also materials and other development through the casting process is required. It had promoted following development themes.

1) Speeding up of the additive manufacturing molding technology

With the aim of mass production application of high performance casting with complex shape, development of additive manufacturing sand molding system with faster molding speed is required.

2) Materials development for high melting point metal casting by using additive manufacturing sand mold

Performing binder and materials developed to mold material corresponding to the highspeed molding apparatus. In particular, it is targeting the high melting point metal casting by the development of casting technology associated therewith and improving the heat resistance of the mold.

3) Development of casting process connecting casting design, CAE and additive manufacturing molding system

Improves the quality and reliability of castings due to the complexity and thinning of the mold, CAE improvement also required. In parallel with the technical development of apparatus and materials, development of evaluation technology of mold characteristics such as heat resistance, permeability and wettability, and casting materials characteristics such as high temperature deformation is proceeding.

2. Development of materials

Binder jetting(BJT) and selective laser sintering(SLS) have been used for additive manufacturing sand molding system. Particularly, the BJT method has features that the molding size can be increased and the molding speed is high. In the conventional method of BJT for casting mold, the hardened system by using two kinds of liquids, a catalyst and a binder, has been adopted. In this system, wet sand mixed with sand and liquid catalyst is supplied to make the sand layer. In the selected area, hardening to the forming sand occurs by the ink jet binder. In this method, since wet sand is used,

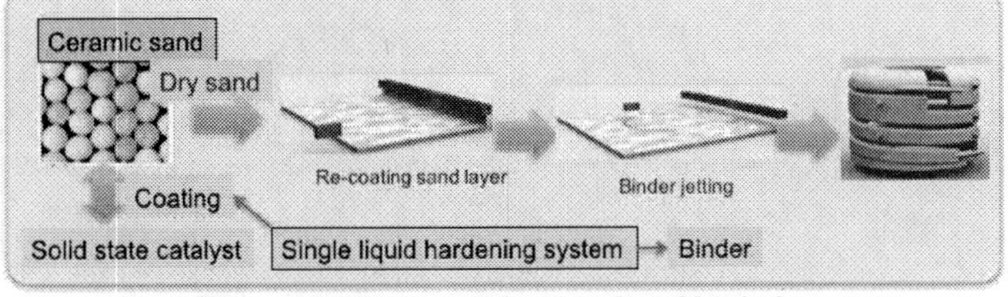

Fig. 1. Developed dry sand - catalyst coated sand hardening system

cohesion tends to occur, which makes it difficult to use ceramic sand.

In order to be able to use ceramic sand, catalyst coated sand hardening system, as shown in Fig. 1, is developed using dry sand coated with solid catalyst.

This hardening system has various features.

- Heat resistance suitable for high melting point materials
- High fluidity of sand and high packing density of mold
- Easy to take out
- Reusable of unhardened sand

This hardening system has additional feature as shown Fig. 2. This figure shows the relationship between compressive strength and elapsed time after molding. The catalyst coated sand hardening system shows that the initial hardening rate is high, therefore, it is possible to make a mold with high precision.

Fig. 3 shows the change in expansion displacement due to temperature. The AM molding ceramic sand shows low thermal expansion and suppression of the veining defect is expected.

3. Development of AM sand molding system

Fig. 4 shows AM sand molding system adapted to the developed materials. Fig. 5 shows printing sequence of this system. The system has an inkjet head and a re-coating system that can operate in both directions for high-speed molding. The size of the molding box is 1800mm x 1000mm x 750mm and molding speed reached about 100,000cc/h.

Acknowledgments (optional)
This report includes the results of the Ministry of Economy, Trade and Industry's Industrial Technology Research and Development "Ultra Precision Three-dimensional Modeling System Technology Development Project" started in July of 2013.

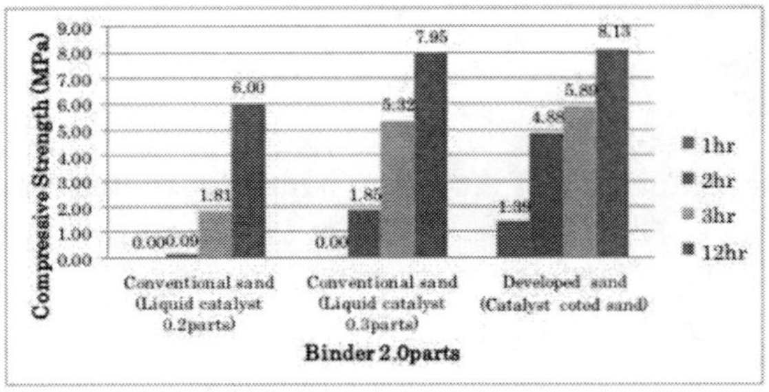

Fig. 2. Compressive strength of catalyst coated sand and conventional liquid catalyst mixed sands

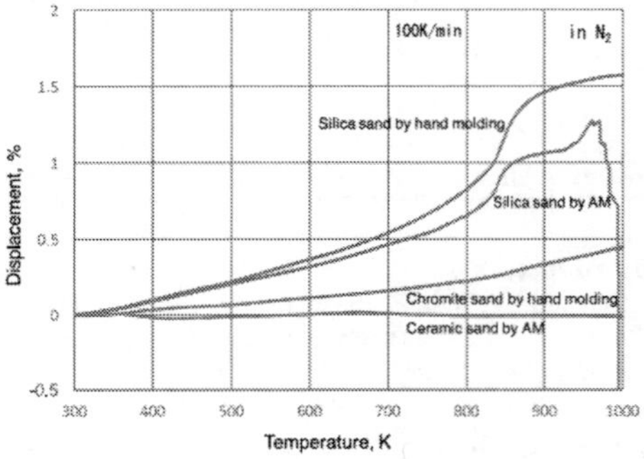

Fig. 3. Change in expansion displacement due to temperature

Fig. 4. Developed AM sand molding system

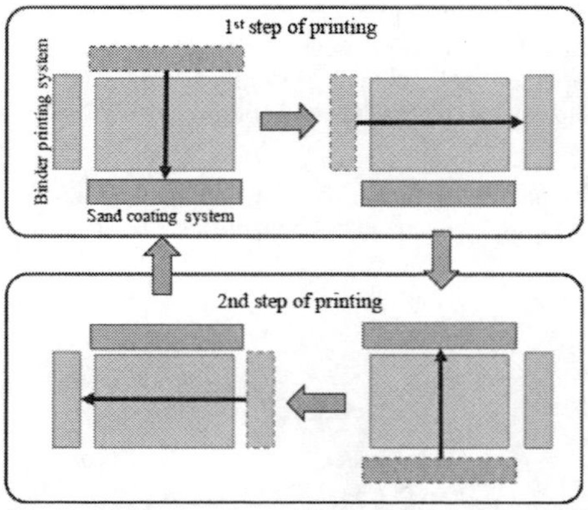

Fig. 5. Printing sequence of developed AM sand molding system

MODERN PERMANENT LINING CONCEPTS FOR CC TUNDISHES – THEORY AND PRACTICE

Paweł Czaja[1,2], Paweł Mirecki [1]*, Pamela Diaz Garcia[2], Jose Prieto Moneo[1]

[1]*Pasek Europe, 1 Piłsudskiego Sq., 00-078 Warsaw, Poland*
[2] *Pasek Espana, Pasek Technical Centre, Calle Dr. Carreño, 1, 33405 Salinas, Asturias, Spain*
**Corresponding address: e-mail: ptc.europe@pasek.es*

Key words: refractory ceramics, refractory concrete, wear lining, tundish, waste management.

Modern steelmaking process requires a number of functional elements capable of sustaining harsh working environments and simultaneously allowing for on the one hand (*i*) time and cost-effective operations and on the other (*ii*) ever more sustainable waste management policy [1]. In continuous casting (CC) technology one of such elements is permanent lining of a tundish. Primarily it is a safety layer, which provides the base for wear lining. Uncompromised wear resistance and sufficient degree of stability expected from permanent lining, including among other requirements high chemical resistance to ladle slag and steel melts as well as high mechanical and thermal shock resistance, have thus critical bearing on a cycle lifetime and efficient performance of the overall CC process. Typically, permanent lining can be deposited either by (*i*) laying bricks, which is disadvantageous in terms of skilled man power and considerable time lengths out of operation it necessitates, or by (*ii*) casting (monolithic permanent lining). An example image of monolithic *vs*. brick permanent lining is shown for illustration in Fig. 1. The foremost advantage of monolithic permanent lining - most commonly executed with an aid of a template, suitable gunning mix and a dedicated drying program – is the short time required for lining work and easy repair. Other benefits include: high reduction of perforation risk, better adaptation than the bricks to mechanical deformation of a tundish skull, lower than with bricks risk of waste material during deskulling, higher lifespan than brick work (~1200 heats) etc. Importantly there is no metallurgical difference between the brick lined and monolithic ladles.

Recently an innovative permanent lining technology has been developed and established under the commercial name "FASTDRY" (FD) [2]. It ventures into the realm of nanoworld, since it utilizes as a binder a solution of silica nanoparticles instead of the Calcium Aluminate Cement (CAC). The primary gain of employing silica nanoparticles over CAC is the faster and more effective disposal of water, circa 95% of water is safely removed with FD at temperatures already below <150°C, which then considerably improves the drying time and reduces the risk of spalling. Other advantages include undisturbed compressive strength of the material,

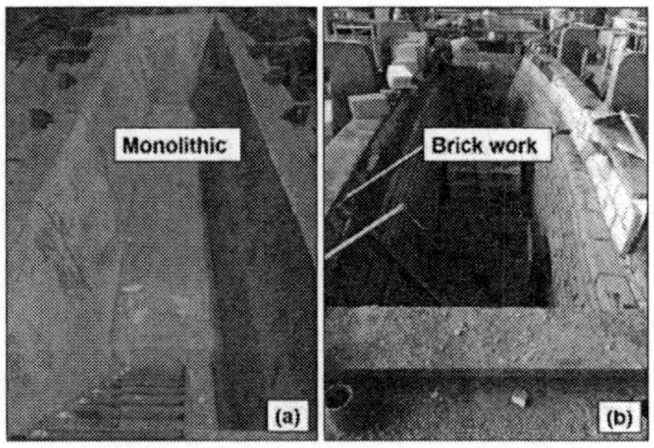

Fig. 1. Example monolithic permanent lining with Pasek FASTDRY (a) vs. brick lining (b).

which is found to progressively increase with temperature unlike with CAC, where the mechanical resistance falls until the temperature range of 600-700 °C is attained. Experimental results have further demonstrated that FD allows for a complete removal of the dry mix following CC without any damage and wear to the permanent lining. This is evidenced in Fig. 2 showing that with FD there is no sintering occurring between the dry mix and the permanent lining and that the dry mix can be easily removed. Overall the FD shows superior properties over more conventional permanent lining concepts including: improved availability of tundish and drying burners, cost reduction in the drying process and energy savings (up to 60%), reduction in spalling risk, extended lining durability and improved deskulling. Altogether this becomes attractive for modern steelmaking.

Fig. 2. FASTDRY concrete, dry mix and steel following 5 h cycles at 1550 °C.

References:
[1] Ebikapade Amasuomo, Jim Baird. (2016). The concept of Waste and Waste Management. Journal of Management and Sustainability. Volume 6, 88-96.
[2] http://www.pasek.es

The 73rd World Foundry Congress
„Creative Foundry"
23 - 27 September 2018, Krakow, Poland

AUTOMATIC BOTTOM POURING OF IRON ALLOYS

pour-tech AB - Industrivägen 39 - 43361 Sävedalen – Sweden

Oliver Schmitz - e-mail: oliver.schmitz@pour-tech.com

Key words: Automatic Bottom Pouring, Laser Control, Inoculation Technology, Temperature control, Data recording, Quality

In today's modern foundry it is increasingly important to improve the quality and reduce the production cost. In general, these two goals are contraire, but with high sophisticated control system there is a way. Over the last 25 years, pour-tech AB has concentrated on developing automatic bottom pouring systems and solutions for iron foundries and has lifted this technology to another level.

At pour-tech AB, we focus on analysing and understanding our customer's needs, drawing from decades of experience from foundry and other industrial areas. This enables us to provide the best and most cost-effective solution for each application, giving our customers the competitive edge.

We have succeeded in building a strong relationship with our customers, exchanging new ideas, concepts, and always improving our technologies and ourselves. Thus, allowing our customers to stay well ahead of their competition.

Keeping up with the latest technological advances, new generation pourTECH™ Systems are currently being installed around the world, designed with the latest Sensor technologies, I/O standards and system features.

1. Automatic Pouring

Automatic Pouring is more than just the control of the iron flow from a pouring furnace into the mold. It also includes a variation of different sensors and software to be able to reduce the operator involvements and to stabilize and improve the whole process. Before starting with a project for the installation of an automatic pouring system, it is important to understand the process and to understand the difficulties to be able to supply the best solution.

The following topics are describing the possibilities:

2. Iron Flow Control

The pourTECH™ automatic pouring control system from pour-tech AB is a real-time, high-speed, closed loop system designed to automatically and precisely regulate the level of molten metal in launders and pouring cups.

The pourTECH™ Line Laser System features a Line Laser and a Line Laser camera. The Line Laser, attached to one side of the pouring vessel, projects a line across the pour cup when the cup is located directly below the pouring nozzle. The Laser camera is mounted on the other side of the vessel, also aimed at the pouring cup.

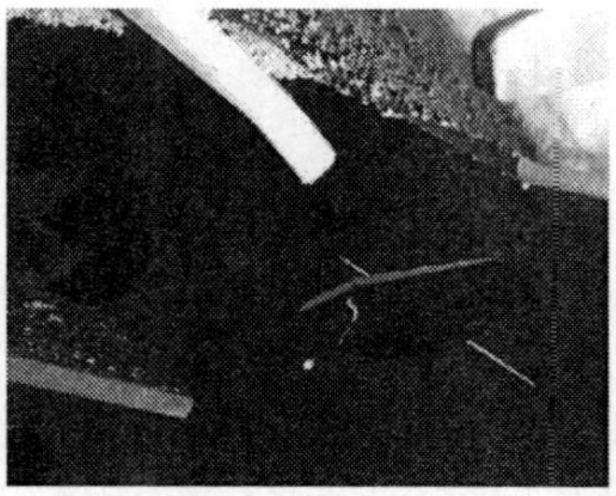

Fig. 1. Laser line across the pouring cup with spot laser for automatic

By viewing the reflection of the line on the metal, the camera uses a triangulation method to determine the iron level in the cup. In order to have a perfect pouring result, it is mandatory to have the nozzle in the correct position over the pouring cup. Therefore, the system comes with a spot laser which enables the system to detect the pouring cup and to automatically reposition the pouring vessel to the correct position. Another important aspect is the level in the launder. In order to keep the level constant, it is measured and regulated with a separate laser.

3. Inoculation Technology

The pourTECH™ control system is designed to interface with in-stream inoculation units. By integrating the inoculation controls with the pouring system, the user has full control of the inoculation process. By selecting an inoculation unit with a variable, controllable feed rate, the rate can be made part of the pattern data for each job. Whenever there is a job change, the feed rate is automatically changed, eliminating the risk of pouring with the wrong inoculation amount.

inoTECH™ is a system developed by pour-tech AB for detecting in-stream inoculation, using a laser and a camera. The inoTECH™ camera counts the granules heading towards the stream, allowing the system to calculate a hit-rate. The hit-rate is displayed after each pour and can trigger alarms and even abort the pour if the rate is too low.

Due to the fact that the inoculation is integrated into the pouring control system a dynamic inoculation feeding is possible. Pouring times, iron qualities and inoculation amounts are synchronized in order to have the correct amount of inoculant in every mold.

Fig. 2. inoTECH™ laser system laser line across the pouring cup with spot laser for automatic

4. Temperature Control

By using a pyrometer to measure the iron stream temperature, the operator receives real-time feedback for every single pour. It also makes it possible to run pouring furnaces' power controls in automatic mode.

Fig. 3. pourTECH System on a pouring furnace

5. Data Recording

All data for each single mold are stored. The result can be displayed in a web-browser for later analysis and process improvements. It is also possible to connect the system with other sources then the pouring system.

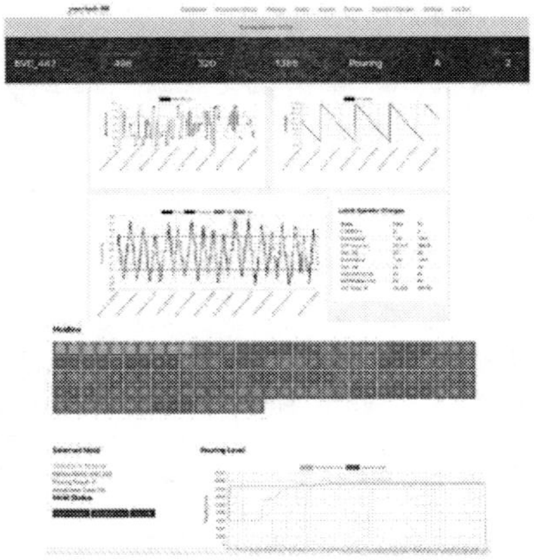

Fig. 4. Database with production overview

It is possible to create production reports to see the production results and react on possible downtimes created by the different systems. Operator changes in the system are recorded for trouble shooting. This database allows the foundry to react on possible problems in all aspects.

6. Data Analysis (Industry 4.0)

By collection data from all sources in the foundry, pour-tech AB is able to analyse them and optimize the whole production cycle with OMNI. This is true artificial intelligence and it is a very strong tool for all foundries. OMNI improves your profit (proven).

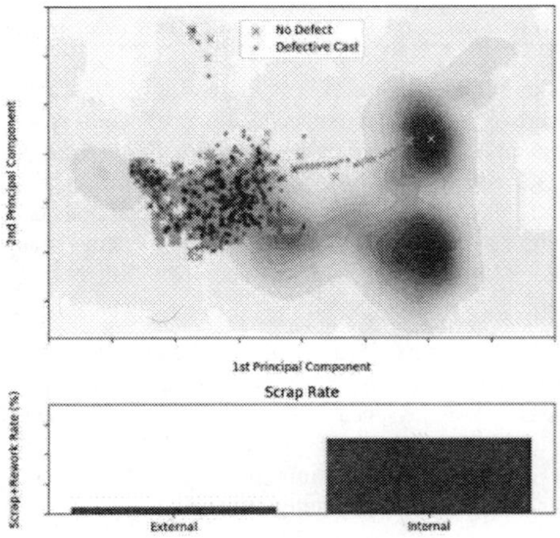

Fig. 5. OMNI system for yield improvement

7. Seamless Integration

Today, the foundries are using different systems for different processes in their production. In order to produce high quality castings it is important to have a seamless integration of these different processes. pour-tech AB has taken this to another level. The integration of pouring furnace, mold line, inoculation technology and laboratory results is already done and leads to optimized production.

Fig. 6. Perfect production result on a vertical mold line

Oliver Schmitz, pour-tech AB Sweden

The 73rd World Foundry Congress
„Creative Foundry"
23 - 27 September 2018, Krakow, Poland

LAC – producer of furnaces and dryers

Szekielda Piotr

LAC, s.r.o., Topolová 933, 667 01 Židlochovice, Czech Republic
**Corresponding address: e-mail: polednova@lac.cz*

Key words: holding furnaces, annealing furnaces, melting furnaces, crucible furnaces, energy saving

In the first part of the presentation, introduction of the company LAC, s.r.o. a Czech producer of furnaces used for heat treatment in foundries will be done. There will be the company history, mission and vision explained.

The main time of the presentation will be focused on standard production line introduction. There will be the most successful standardized technical solutions: electrical and gas furnaces, crucible and retort furnaces and furnaces with recuperation commonly used for melting, holding, annealing provided to foundries introduced highlighting theirs benefits and energy savings options.

Last part of the presentation will be devoted to LAC experience with customs projects and successful installations done at customers.

Company was established by two Czech friends and foundrymen Mr. Ledl and Mr. Crhák in 1992. Since its establishment the company has manufactured more than 13 000 furnaces. We already have supplied our equipment into more than 35 countries around the world. The most distant countries to which we have sold our furnaces and dryers are: Mexico, Vietnam, Kazakhstan, Canada, Malaysia and Hong Kong. Besides, we usually supply our products to European countries such as Poland, Germany, Great Britain, Finland and also Sweden.

The LAC manufacturing program includes the manufacture of a complete standard range of furnace and dryer lines, and also accommodates the individual requirements of the customer through the design and manufacture of atypical furnaces tailor-made to meet customer specifications.

Since 2012 LAC is saving the money of the customers with newly developed energy-efficient PTE Mk. II holding furnaces. With this product, we helped and are still helping our customers reducing costs at a time when everyone is looking for optional savings. Moreover, in 2013, on MSV trade fair which is annually held in Brno, Czech Republic, LAC received an honorable mention for energy efficient product supposing the solution of this type of holding furnace PTE Mk. II.

In 2015 we obtained another valuable award, as we became one of the 100 best Czech companies.

The LAC development and design office works in tandem with a team of service technicians to ensure quality service to customers and pave the way for future company growth.

LAC also supplies heating elements, refractory and insulation materials, regulating elements, and reconstruction of furnaces, heating systems and switchboards to its customers.

LAC commonly manufactures furnaces and dryers which meet the most demanding standards of the aerospace, military and automotive industry. We comply with the standards ČSN EN ISO 17663, AMS 2750 and DIN 17052-1.

In 2008 LAC opened a sales branch in China. Since 2012 we have our own business unit including both sales, after sales and service activities.

The rapid growth of the company is illustrated by its present 150 employees, its transfer to a new production plan in Židlochovice, Czech Republic in April, 2018 with total area of 6,800 m^2 for production, warehousing and company administration facilities.

Second division of LAC which focuses on production of refractory castable shapes with location in Hrušovany nad Jevišovkou, Czech Republic is presently one of the largest manufacturers of refractory concrete shaped blocks in Europe.

The 73rd World Foundry Congress
„Creative Foundry"
23 - 27 September 2018, Krakow, Poland

INDUSTRY 4.0 PROCESS CONTROL AND TRACEABILITY FOR THE FOUNDRY INDUSTRY

SinterCast AB

Kungsgatan 2, 64130, Katrineholm, Sweden

e-mail: arnaud.denis@sintercast.com

Key words: Tracking, Traceability, RFID, Ladle, Core

Increasing demands for process efficiency, product traceability and foundry profitability require the ability to measure every step of the foundry process, and to use these measurements to determine and implement corrective actions that resolve the root of the problem. In support of this demand, SinterCast has developed a suite of tracking technologies that provide new insight for foundry supervisors and managers. These technologies include SinterCast Ladle Tracker® and Cast Tracker™.

1. SinterCast Ladle Tracker®

The SinterCast Ladle Tracker® technology is based on the placement of an identification 'tag' on each ladle, Figure 1, and the installation of 'antennas', Figure 2, at key locations in the foundry to track the time and location of every ladle as it moves through the process.

Fig. 1. RFID tags affixed to ladles

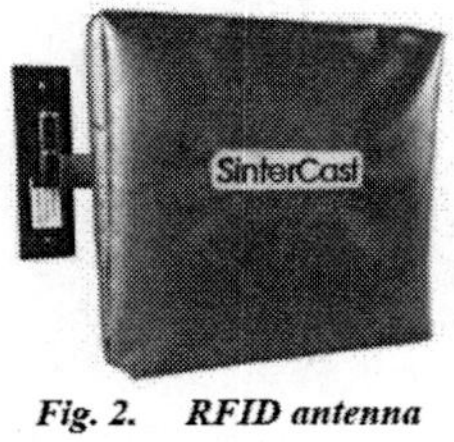

Fig. 2. RFID antenna

The technology automatically ensures that every ladle has successfully passed each step (temperature is correct, weight is correct, treatment was successful, time is within limits), as shown on Figure 3. If any step has not been successfully completed, the pouring car can be automatically locked, without relying on behaviour of the operator. Ladle Tracker enables foundry managers to identify where (and why) every ladle falls out of the process, allowing corrective actions to be implemented to eliminate bottlenecks and to optimise process efficiency. The results provided by the Ladle Tracker can also be used to set quantitative efficiency targets and to track the progress of improvements, month by month.

Fig. 3. Ladle tracking at semi-automatic pouring line

The results can be viewed as a spreadsheet, with one row for each ladle (identifying the individual ladles that fall out of the process), or as a data summary showing average results and highlighting the locations where ladles are falling out, from melting through treatment, transfer and casting.

2. SinterCast Cast Tracker™

The tracking capability has been extended to also track cores, flasks and castings. The SinterCast Cast Tracker™ can provide complete traceability of each casting from the date of manufacture of the cores (inception), shelf storage time, pouring (birth event) and shake out. Together with the Ladle Tracker capability, Cast Tracker provides traceability of the moulding history and the liquid metal history. For the end-user, this novel capability provides complete traceability of each casting. For the foundry, Cast Tracker provides the detailed information (such as cast sequence) needed to determine robust correlations between defects and the process parameters.

All results are available for user-defined downloading, streaming to the foundry quality or

ERP system, or viewing in real time, as shown on Figue 4, on any device that is connected to the foundry network, such as a mobile telephone.

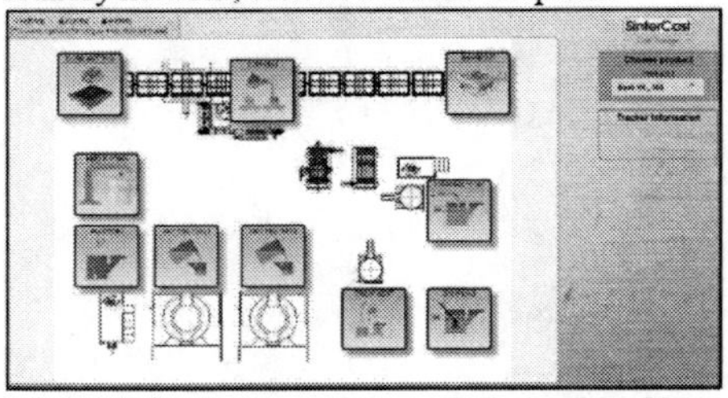

Fig. 4. Real-time position monitoring and tracking

3. SinterCast Tracking Technologies

The benefits of the Ladle Tracker technology have been proven at the Tupy foundry in Saltillo, Mexico. In operation since January 2015, with 20 ladles in service and a typical production pace of 18 ladles/hour, the system ensures that every step in the process is successfully completed and prevents out-of-spec ladles from being poured.

The SinterCast tracking technologies displace operator behaviour with direct measurements and automation, bringing the foundry industry another step closer to Industry 4.0.

The 73rd World Foundry Congress
„Creative Foundry"
23 - 27 September 2018, Krakow, Poland

THE TECHNOLOGY OF CONTINUOUS CASTING OF ALUMINIUM ALLOY INGOTS

Dawid Kapinos [1*], Bogusław Augustyn [1], Marcin Szymanek [1],
Paweł Kumor [1], Artur Woźniczka [1], Zbyszek Zamkotowicz [1]

[1]Instytute of Non Ferrous Metals in Gliwice, Light Metals Division in Skawina, Pilsudskiego Street 19; 32-050 Skawina, Poland,
*Corresponding address: e-mail: dkapinos@imn.skawina.pl

Key words: aluminium, aluminium alloys, ingots, horizontal continuous casting (HDC)

Abstract

The paper comprises information about the continuous casting system of aluminium alloy ingots. The system includes a casting machine which in this study is an example of the device designed and made by the Skawina Light Metals Division of the Institute of Non-Ferrous Metals in Gliwice (IMN OML). The device for horizontal casting of ingots is characterized by high versatility due to the use of, among others, independent drive systems, which control the speed of casting, cooling and lubrication, and a system for recording of casting parameters. A series of trials was carried out on the designed and manufactured device, casting ø70-160mm ingots from various aluminium alloys.

1. Introduction

The global development trends in casting of aluminium and its alloys include, among others, vertical casting (VDC) and horizontal casting (HDC) systems.

Horizontal casting of ingots has economic and technological justification. Cast ingots are used as an intermediate product for plastic processing, mainly extrusion, rolling and forging. Nowadays, also alloys intended for remelting, which so far have been cast in the form of pigs, are more and more commonly cast into ingots.

The process of casting small diameter ingots takes place in a very narrow range of parameters, which for ingots with a diameter of approx. 70mm involves very high casting speeds of up to 400mm/min. With the casting speeds so high, the formation of skin on the ingot surface in a crystallizer of a few-centimetre working length is very difficult and requires very precise setting of all parameters for each alloy individually, and especially for multi-component alloys with a wide range of solidification temperatures[1-4].

Having many years of experience in ingot casting and operating a vertical casting stand, IMN OML has attempted to design and manufacture a multi-purpose work stand to cast ingots in a horizontal arrangement, which is expected to further expand the research capabilities. The authors of the project, taking into account the complex nature of the casting process, accepted the idea of a wide control of the casting parameters, including data recording option. An innovative solution was the use of two independent systems, which enabled setting different cooling and lubrication speeds and intensities during one casting cycle. This allowed parallel casting of several ingots from a single melt applying different casting parameters. The developed solution enabled optimization of casting parameters with the number of tests reduced to minimum.

2. Methodology

The work was carried out on two parallel levels - the design and the technology.

As part of the design work, it was planned to design and manufacture a well-instrumented horizontal casting stand, to enable testing various aluminium alloys, optimization of casting parameters and development of technology for aluminium alloys never so far cast in such a system.

At IMN OML, the device was operated with two crucible resistance furnaces, in which 300 kg of Al was melted, refined and filtered during transport to the casting machine (HDC). The casting machine (Fig.1,2) together with the feeding system has a compact structure that allows its relocation. Additionally, the machine can be equipped with a saw for cutting ingots to a specified dimension still during casting process.

The line configuration allows:

- stepless speed control in the range from 0 - 650 mm / min,
- stepless cooling water flow control in the range of 0 - 200 l / min,
- stepless lubrication control,
- archiving of casting parameters,
- creating technological instructions for the casting process.

Using the designed and built work stand, trials were carried out, casting ingots from alloys included in the 2xxx, 4xxx, 5xxx, 6xxx and 7xxx series. The ingots were cast in the ø70, 100 and 160mm crystallizers designed and made by the IMN OML.

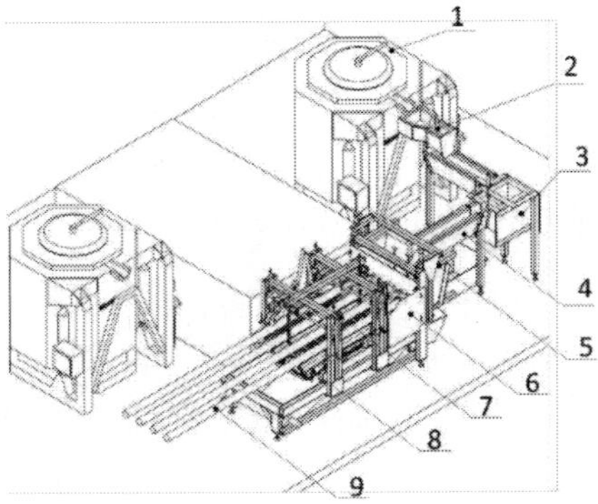

Fig. 1. Diagram of semi-continuous ingot casting stand:
1- furnace, 2- furnace tapping spout, 3- filtration box, 4- tapping spout, 5- launder, 6- casting machine, 7- ingot clamping unit, 8- drive conveyor, 9 – ingots

Fig. 2. Stand of semi-continuous casting in IMN OML

As a result of the studies, new solutions and improvements were introduced to the system design, raising its efficiency and enabling casting high-quality ingots.

To develop guidelines for the casting process, thermophysical parameters of alloys were determined during heating and cooling using a UMSA instrument (Universal Metallurgical Simulator and Analyzer). The tested alloys were additionally subjected to refining and upgrading through modification and grain refinement.

During tests, numerous additional improvements were introduced to the design of the stand and also to the construction of crystallizers and ceramic feeding elements.

Sample photo of ingots cast from aluminium alloy is presented on Fig 3.

Fig. 3. Sample photo of ø70mm ingots cast from EN AW 6082 alloy

3. Conclusions

As part of the work, a universal stand for horizontal casting of Al alloy ingots was designed and its operation was tested. The use of the device enabled casting on a semi-technical scale with strictly defined parameters of ingots from various aluminium alloys included in the 2xxx, 4xxx, 5xxx, 6xxx and 7xxx series.

Positively checked design solutions in the form of a specific device are a tool for the continuation of studies to determine casting parameters for selected alloys and to test the crystallizers, process support systems, materials and feeding system configuration.

References

[1] Zamkotowicz Z., Augustyn B., Kumor P., Kapinos D., Żelechowski J., Bigaj M. (2014). Casting Small-Diameter Ingots from Multicomponent Silumins. Metallurgy and Foundry Engineering. Volume 40 (3), pp.175-190.

[2] Müller H. R. (2006). Continuous Casting. Weinheim: WILEY-VCH.

[3] Sengupta J., Thomas B.G., Wells M.A. (2005). The Use of Water Cooling during the Continuous Casting of Steel and Aluminum Alloys. Metallurgical and Materials Transations A. Volume 36A, pp.187-204.

[4] Zamkotowicz Z., Augustyn B., Kumor P., Kapinos D., Żelechowski J., Bigaj M. (2014). Odlewanie wlewków z siluminów wieloskładnikowych. XVII Międzynarodowa Konferencja Naukowo-Techniczna Odlewnictwa Metali Nieżelaznych - Starachowice, 2014 (pp. 123-132). Kraków: Wydawnictwo Naukowe Akapit.

The 73rd World Foundry Congress
„Creative Foundry"
23 - 27 September 2018, Krakow, Poland

VOX PATRIS - THE BIGGEST SWINGING BELL IN THE WORLD

Piotr Olszewski*

Pracownia Ludwisarska Jana Felczyńskiego, ul. Słowackiego 46/1, 37-700 Przemyśl, Poland
**Corresponding address: biuro@janfelczynski.com*

Key words: biggest swinging bell, bell foundry process, largest bronze casting, church bell

1. Introduction

The biggest cast swinging bell in the world - „Vox Patris" – will be heard from the tower in Sanctuary Basilica of the Divino Pai Eterno in Trindade-Go. The bell was created by Jan Felczynski's Bell Foundry, company Rduch Bells & Clocks and cast in the premises of company Metalodlew S.A. in Krakow. The bell after casting weighs 55 tons and with diameter of 450 cm and more than 4 meters of height is the biggest bell casting ever made in Europe. The initial works on „Vox Patris" started back in 2013 when the first consultations took place. Jan Felczynski's Bell Foundry from Przemyśl was then looking for the foundry capable in terms of technical equipment to handle the project of that size as the premises in Przemyśl were not sufficient to cast such a giant. The requirements were not only the necessary space, modern and powerful furnace but also such things as cranes, appropriate location and professional stuff. Hence after weeks of discussions the co--operation with Metalodlew in Krakow was initiated.

2. First challenges

The initial challenge was to prepare a casting hole (Fig. 1). Magnificent in its size of 7.5m × 7.5m and a depth of almost 6 meters.

Fig. 1. Casting hole

The concerns regarding this construction were mainly relating to the resistance of the cranes' poles as well as the strength of the casting hole itself so it won't collapse. Therefore, it was necessary to prepare a whole technical project and obtain building permit to start any works. The casting hole was prepared using sophisticated pilling technology.
Next start-up challenge related to the necessary construction equipment to build the mould. Where in typical bell casting process all the mould's auxiliary elements are easy to operate, here with the extreme size it was necessary to analyse in detail every element. As a result of many consultations a circle track was created and a steel profile was mounted to this track enabling works in various areas of mould's creation process. Additionally, a channel was created below the ground level of works to enable powerful blowtorch to heat the mould from inside. All the auxiliary elements were created by company Rduch Bell & Clocks that is responsible for mounting and automating this bell. Strong co-operation with the Department of Foundry Engineering of the Silesian Technical University was also a foundation of successful preparations during all phases of works.

3. Mould creation

First the core must be created (Fig. 2). However in this project, even the core was a big challenge. 10 000 bricks were used to prepare a base for further works. The core was empty inside allowing appropriate heating process with the use also of deflector system. The core was then covered with mould material called HS, used in bell foundry manufacturing process, to reflect ideally the inner side of the mould.

Fig. 2. Core

The second part of the mould is the so called „false bell" – the exact shape of the bell created temporary to allow further decorating process in lost-wax technology. The false bell is also initially made of HS material and then covered by beef tallow to ideally reflect the line of the bell and then decorated with the ornaments, reliefs and inscriptions made in wax. To decorate „VOX PATRIS" more than 100 kg of specialized casting wax was used, which corresponds to 2 years demand in normal bells's preparation process. The preparation of the inscriptions and especially of the reliefs was also a challenge. Jan Felczynski's Bell Foundry obtained only a drawings of particular scenes that should be on the bell. Then with the use of sophisticated computer software we were able to prepare a full visualization of the bell. Next step was to prepare a massive amount of reliefs and place them on the false bell (Fig. 3). All the scenes show the history of the Basilica in Trindade-Go together with the beauty of the wildlife of Brazil.

Fig. 3. Placing ornaments

The placing of above mentioned artwork finished the second part of mould creation. The last part was the coat. Coat is also made of HS but in comparison to typical bells manufacturing process must be incomparably more reinforced to assure two areas:

- the process of coat lifting (Fig. 4) which is extremely important and must be done extremely carefully considering that the sole coat that needs to be lifted weighs around 30 tons
- the process of casting itself as the pressure that impacts on the mould's first lock is around 850000N.

The coat was therefore reinforced with the use of steel elements, steel net all based on thick steel circle base and further circle steel elements.

Fig. 4. Coat lifting

4. Heating process & casting

One of the most important parts of the process of production was the heating process both from inside (using the previously mentioned blowtorch located in the channel) as well as from outside using the system of heaters located in the corners of the casting hole. The whole process was preciously monitored by the system of 32 thermocouples placed in various locations of core and coat. Both the core and the coat had to be dry but not overheated because with high temperatures the strength of HS material suddenly goes down.

The successful casting took place on 1st August 2017 (Fig. 5). As the temperatures of that summer were very high the decision was taken to pour the material and cast the bell over night.

Fig. 5. Casting

The furnace used melting for the casting was Otto Junker belonging to Metalodlew S.A. with the capacity of around 15 tons in one pot (two melting pots were available for pouring). Therefore, it was extremely important to co-ordinate the whole casting process in a way to obtain more than 60 tons of bronze (78% of copper and 22% of tin) in one go. To cumulate 60 tons of bronze special ladle was hired and the bell was casted in around 10 minutes with the initial speed of 150 kg/s. After casting the bell was left for three months to cool down. Once it was dug out the additional annealing process was carried to reduce any residual casting stresses.

Now, in August 2018 the bell is being finalized by artists from Jan Felczynski's Bell Foundry and is waiting for long journey to Brazil.

The 73rd World Foundry Congress
„Creative Foundry"
23 - 27 September 2018, Krakow, Poland

DEVELOPMENT OF THE ELECTRIC VEHICLES MARKET BY 2030 IN GERMANY, EUROPE, US AND CHINA

Presentation/Paper Synopsis

Dr. Carsten Kuhlgatz

Hüttenes-Albertus Chemische Werke GmbH, Hansastraße 1, 30419 Hannover, Germany
Corresponding address: CKuhlgatz@huettenes-albertus.com

Key words: E-Mobility – Scenario Analysis – electric vehicle market – Germany - regulation

Introduction

How will the market for electric vehicles develop over the next few years and decades? This is a question on many people's minds. For automotive suppliers, including foundries, much of their future business will depend on the further development of electromobility. The Center of Automotive Management (CAM), an independent scientific institute at the University of Applied Sciences (FHDW) in Bergisch Gladbach, has carried out a comprehensive study on this topic on behalf of the German Industrial Association of the Foundry Chemistry (Industrieverband Gießerei-Chemie e. V). The results of this study are presented by Dr. Carsten Kuhlgatz, President and CEO of Hüttenes-Albertus GmbH and President of the said Association.

Comprehensive analysis

The aim of the study was to analyse developments in drive technology, with a particular focus on the growth of electromobility in the German, EU, US and Chinese target markets by 2030. Based on a review of sales and production trends, three different scenarios were developed to plausibly chart potential future developments based on transparent assumptions. The decisive factors for each scenario were different assessments of how quickly and to what extent important factors such as battery costs, range, the density of (fast) charging sites, charging times and regulatory conditions will develop. Accordingly, the study has modelled a pro-EV, a medium and contra-EV scenario.

Global perspective

A global perspective also provides important insights. The study forecasts sustained growth for the global automotive industry until 2030. The total number of vehicles produced will continue to rise to approximately 120 million vehicles in 2030, compared to 92 million in 2016. In particular, the markets in China and India offer high growth potential.

Scenarios for the development of the electric vehicle market up to 2030

The scenario analysis involves modelling conceivable alternative development paths and visions of the future and considering the driving forces behind them. The future market development of electric vehicles will depend on the parameters described below:

Figure 1: Influencing factors and development paths

1. The **relative costs of electric vehicles** (pure electric vehicles and plug-in hybrids) are a key factor. Buyers compare the purchase prices and the operating costs of electric cars with those of vehicles with conventional power train. Costs for batteries, especially battery cells, are a major factor in the price consumers pay for e-vehicles. In principle, there should be significant cost-cutting potential for battery systems in the future. At the same time, it is highly likely that costs related to exhaust emission purification in cars with combustion engines will rise significantly due to increasingly stringent environmental regulations. Based on the study, price parity between electric and internal combustion engine vehicles could be achieved in the first half of the 2020s.

2. The **functionality of electric vehicles and their suitability for everyday use** is an important prerequisite in purchasing decisions. The electric vehicle (EV) range is a key factor. Since 2009, the mean range of purely EV

increased from 150 km to 400 km by 2016. The various scenarios assume an increase in standard range – from 480 km in the contra-EV scenario to 600 km (by 2025) and 800 km (by 2030) in the pro-EV scenario. Range is considered a key factor for the acceptance of e-vehicles.

3. The subject of **"charging infrastructure and charging time"** is another relevant factor. In the medium scenario, a moderate expansion of private and public charging sites is assumed. It is assumed that in addition to the moderate growth of private charging capacity, there will be around 250,000 (semi-)public charging sites and around 25,000 fast charging sites for long-distance traffic by 2030.

4. The **"regulation/incentive"** factor also has a significant impact on the growth of the e-vehicle market. This includes a large number of instruments, e.g. investments, taxes and levies, as well as incentives to buy and use. The study in particular analyses the impact of CO_2 limits for vehicle manufacturers in Europe. If the limit values were to be lowered from 95g/km in 2021 to 65g/km in 2030 (medium scenario), manufacturers would have to increase sales of electric cars to 28% and reduce CO2 emissions from combustion engines to an average of 90g/km (equivalent to 3.9 litres of petrol/100km).

EV sales development in Germany up to 2030

Taking into account the factors described above, the three scenarios suggest that the electromobility market in Germany will not really gain momentum until the 2020s. In 2030, almost 30% of new registrations in Germany will be for electric vehicles, equivalent to annual sales of around 900,000 cars. Conversely, however, more than 70% of all cars would still have internal combustion engines.

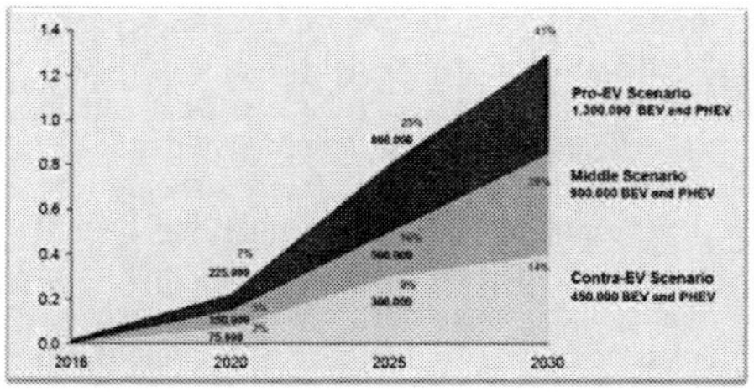

Figure 2: Sales development of electric vehicles according to the three scenarios

Evaluation of the study's findings

Overall, the scenarios show that the supplier market for combustion engine components in Germany will shrink from the 2020s onwards. Companies need to prepare for this by developing and implementing appropriate adaptation strategies.

For HA as a manufacturer of foundry chemicals, however, there are also opportunities for which the company has already developed growth-oriented and forward-looking strategies:

- On the one hand, as the total number of cars produced worldwide grows, so does the automotive industry's demand for castings. As a global company, HA is very well positioned for this.

- Hybrid vehicles will continue to play a major role in the future. Innovative foundry chemistry is needed for the smaller and more complex internal combustion engines used in these hybrid vehicles.

- In terms of pure electric cars, many conventional components outside the powertrain will continue to be cast parts. Structural castings are also becoming increasingly important for car chassis, which are manufactured using the high pressure die casting process. HA will further extend its commitment to die casting solutions.

- Overall, HA will deliver innovative, environmentally friendly binder systems in order to contribute to efficiency gains and emission reductions – both in foundries and in the final product, the automobile.

DIGITALIZATION

The 73rd World Foundry Congress
„Creative Foundry"
23 - 27 September 2018, Krakow, Poland

EXPERIENCES WITH HELIX ATLINE COMPUTER TOMOGRAPHY (CT) AND VGINLINE AT VOLKSWAGEN FOUNDRY HANOVER

CT-inspection of casted Aluminum Cylinder Heads with Volkswagen

Dr. - Ing. Raimund Rösch[1], Dr.-Ing. Ferdinand Hansen[2], Frank Jeltsch[1], Sven Körtge-Sudmann[1]
[1] Volkswagen Foundry Hanover, Germany,
[2] pensioner, formerly Volkswagen Foundry Hanover, Germany,

Key words: worldwide first CT in industrial application, aluminum cylinder heads, measurement of inner wall thickness, detection of pores and shrinkages, dramatically reduced reject rates.

Abstract:
Motivation for fast Atline Computer Tomography was to reduce the scan time from 400 min to less than 5 min, to achieve statistic based part inspection. Manufacturer of this Gantry Computer Tomography, based on modified General Electric Healthcare Technology, is Baker Hughes, a General Electric company.

Details of this Computer Tomography-System are: a 140 kV X-ray source, (53 kW) dual spot, rotating anode tube, a detector wth 16 lines and 912 channels. Measurement volume is 300 x 400 x 800 mm. The voxel resolution is approx. 0, 5 x 0, 5 x 0, 5 mm (typical for casted cylinder heads).

The Healthcare Computer Tomography System was developed for industrial application on light metal castings in cooperation of Volkswagen foundry Hanover and General Electric Sensing & Inspection Technologies. The worldwide first Computer Tomography System was located in Volkswagen Foundry Hanover directly on shop floor area next to production in May 2013. Start of inspection was in June 2013. Until the end of 2017 approx. 115.000 Computer Tomography-Inspections were carried out - most of them automatically with Volume Graphics VGinLine, a software for automated Computer Tomography inspections.

Features for procurement of this Computer Tomography System are rapid testing with automatic results on main deviations, shorter period to production readiness for new parts, quicker process optimization, based on statistics and reduced reject rate in series.

Function modes are rapid scanning in the helix mode in a short sequence time and automatic evaluation for series parts and the axial mode with manual evaluation for new parts.

Important for precision of measurement results for nondestructive testing and metrology application are a detailed work schedule for each operating step, best part position on transport system using specific plates and calibration with specific test phantoms (repeatability test). Inspection scenarios with Atline Computer Tomography System at Volkswagen are pores and shrinkages, remainder from sand cores and measurements of wall thicknesses.

Automated evaluation of deviations and automated evaluation of inner wall thickness with Volume Graphics Inline, latest version, are discussed. Computer Tomography results based on statistics means to scan a number of castings (approx. 30 castings) of a certain quality status automatically in a few hours instead of several days in the past. The Computer Tomography results describe exactly the part quality of this series.

Three examples for process optimization are discussed; a deviation of water jacket, reminder of sand cores and pores could be removed in a very short time.

Internal and external reject rates were reduced dramatically few weeks after start of production readiness with the help of this detailed Computer Tomography analysis. Successful applications for Atline Computer Tomography are extraction of features, wall thickness measurements and detection of deviations like pores or shrinkages.

There are still some themes to optimize. In regions of long penetration length of photons dark stripes and white stripes occur. These scan phenomena are to be removed and some requirements of practical usage are to be fulfilled, for example fully automated part identification and inspection data management. A scan simulation for best part position for scanning will help to achieve the best Computer Tomography scan quality very quickly.

The 73rd World Foundry Congress
„Creative Foundry"
23 - 27 September 2018, Krakow, Poland

Data Analytics: The Next Dimension in Molding Sand Control

Mr. Deepak Chowdhary [1]*, Dr. Himanshu Khandelwal [2]

[1]Managing Director & CEO, MPM Infosoft Pvt. Ltd., India,
[2] Research Scientist, MPM Infosoft Pvt. Ltd., India,
**Corresponding address: e-mail: deepak@mpminfosoft.com*

Key words: Data Analytics, Machine Learning, Green Sand, Optimization, Prediction, Modeling, Smart Foundry

Globally all foundries suffer rejections, some more, some less. The objective of foundries world-wide is to make better castings at the lowest possible cost both in terms of competitiveness as well as the environment. Green sand accounts for almost 50% of casting defects. Management of foundry sand systems has traditionally been considered as an 'art' rather than as a 'science'. Though the physical infrastructure has kept pace with the times through advancements in machine automation, the operational infrastructure is still reliant on the experience of personnel on the foundry floor. The corrective actions typically performed on the sand system are a result of personnel identifying deficiencies or issues in the system through analyses of the sand parameter values and their perceived or estimated effect on the casting rejections. This post-facto process inevitably relies on reactive decision making since this perception or estimation is influenced by the experience of the personnel performing the analyses. Furthermore, the foundry sand system is not only heavily reliant on rapidly depleting resources such as silica sand, carbonaceous materials, water and bentonite but also Energy. With increasing pressure on business sustainability, casting rejections and their control in sand foundries assume a pivotal role in optimization of operational quality and costs.

The power of Predictive and Prescriptive Data Analytics addresses this issue by leveraging the data collected by the foundry from its own, historical data and sand management legacy by means of annotations. The underlying assumption is that the sand parameter, additive, casting rejection and meta-data being recorded by foundries is a repository of cause-effect relationships. These multivariate inter-relationships can be exploited to provide real and tangible value in terms of optimizing molding sand properties as also optimizing the consumption levels of sand additives with component/pattern/part specific granularity (if sufficient data is available) to significantly and consistently reduce casting rejections and also optimize additive consumption.

1. Introduction

The sand system undergoes physical and thermal loadings during its operational cycle and thus loses its actual characteristics, which needs to be recompensed with new additives in the next cycle. In the absence of return sand estimations in most scenarios, compounded by the complexity of varying core sand influx and sand to metal ratios, foundries land up under-dosing or over-dosing of the additives, and this leads to process imbalance [1-2]. Therefore, there is a need to develop a tool, which can help foundries in optimizing their sand system as well additive dosing in optimized levels.

Machine Learning and Analytics modules operate under this assumption, learning the cause-effect relationships hidden in the data [1]. The aim of analytics is to identify similar operational states amongst the data provided. This is achieved through complex algorithms. Once the regions are identified, mathematical models between the various causes i.e. sand parameter, additive data, casting information and their effects i.e. the casting rejections are built.

2. Data Analytics & Cloud Computing for Foundry

SANDMAN which is a patented, cloud-based data analytics driven, decision support application has developed by a project team involving Foundry Domain Experts and Data Scientists to focus on the mentioned un-met need of Foundries.

The fundamental analytics workflow is depicted in Figure 1:

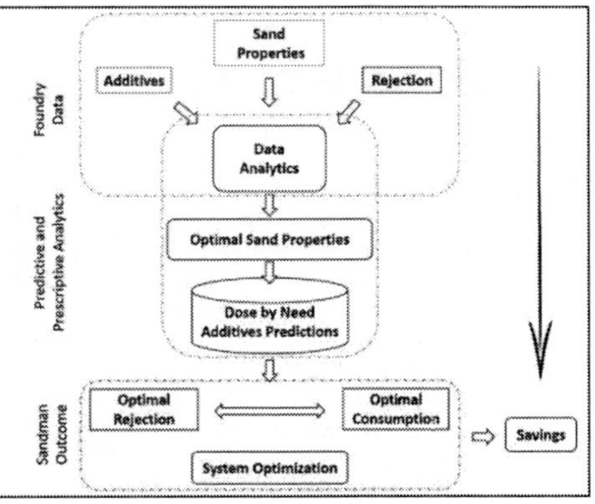

Fig. 1. The building block of machine learning module for green sand foundries

The purpose of this data based machine-learning algorithm is to model the foundry process by underlining it to basic thermo-physical rules of metal casting. The basic methodology used to minimize the occurrence of defects in green sand foundries are shown in Figure 1. The historical foundry data related to the sand properties, additives and casting rejection are collected through various easy-to-use formats and passed through intelligent and comprehensive data-validation engines. The algorithm endorses the correlation between each parameter, and flags the potential outlier data. The resultant clean data is used to build following models:

Predictive Model

This multivariate model depicted as an example in Figure 2 shows the most influencing sand property impacting rejections and the direction the sand property's parameter needs to be modified/tackled for a day/given date (red shows need to reduce the value, green would show a need to increase the value. The mathematical model also throws up a definite figure as a part of the algorithm, but that need not be the absolute value to be achieved. It is more in terms of a guidance), and therefore assist the foundry to take timely remedial measures for maintaining an optimized process balance.

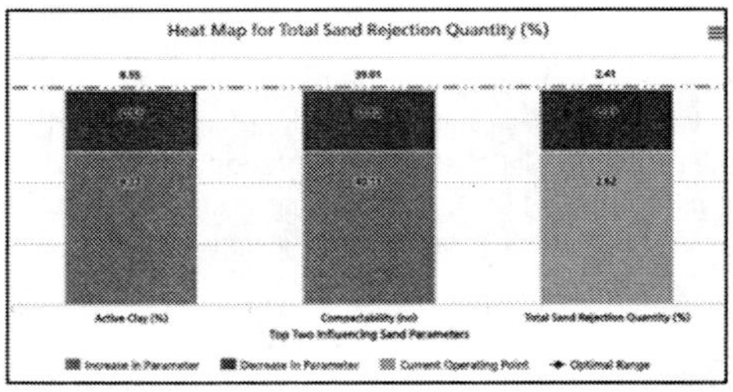

Fig. 2. Top two high influence properties (Active Clay and Compactability), which if maintained around the optimal (8.55 and 39.01), rejections can be reduced from 2.62% to 2.41%

Prescriptive Model

The nonlinear optimization model correlating foundry additive consumption patterns, sand properties data and rejections, provides dose-by-need additives recommendation for each sand mix/batch.

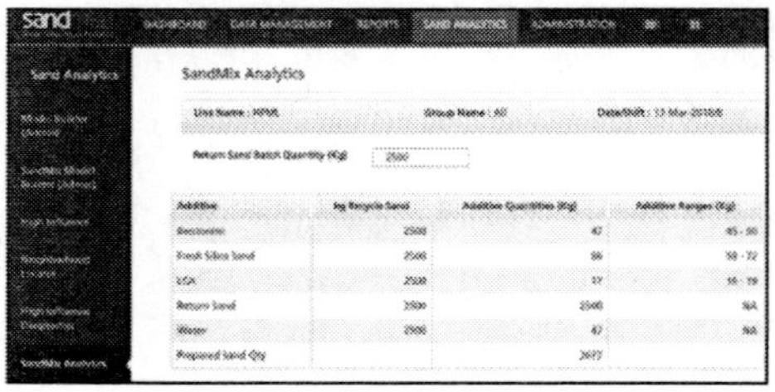

Fig. 3. Predicted Additive (Example: Bentonite 47 kg, new sand 66 kg, lustrous carbon additive 17 kg) for a batch of 2500 kg return sand

3. Industrial Case-study

The software was first tested in NELCAST Foundry, a major foundry group, in India. The advantage in terms of the value proposition is presented in Figure 4 as a proof of concept. After two months of implementation, Foundry's consumption of Bentonite, Lustrous Carbon Additive (LCA) and Fresh Silica Sand (FSS) has dropped by 14.27%, 13.93% and 15.07% respectively. The additive requirement for one ton of liquid metal, before and after analytical support is tinted in the pie chart. The foundry has saved significantly in terms of optimized additive consumption.

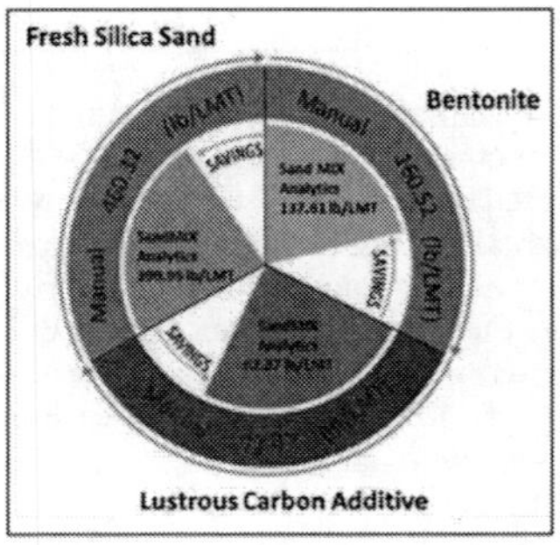

Fig. 4. Additives Savings after initial two month of analytical module operational

4. Conclusion

The paper presented a data analytics driven optimization module for green sand system. This allows foundries to operate their sand system in an optimum and continuously, dynamically balanced manner. A case study is taken up to demonstrate the capability and benefits of this application. This patented methodology keeps the process repeatable, adaptable, dynamically stable for better process capability and consistency, improved productivity, and on-time castings supply and scalable to reducing dependence on the manual interface. The approach has proved to be useful for helping the foundry to reclaim their lost profits corresponding to rejections and consumption.

Acknowledgments

We would also like to gratefully acknowledge NELCAST Foundry, which allowed the first POC at their shop floor along with licensing subscription support and permitted us to highlight the advantages publically.

References

1. S. Ramrattan, P.Patel and D.Chowdhary, *"Evaluation of software systems for green sand control"*,120th AFS Metal Casting Congress 2016, 16th -19th April 2016, Minneapolis, USA.
2. D. Chowdhary, *""Sand Control – The Next Dimension"*, Foundry Sand Control Seminar, Western Michigan University, 25th August 2016, USA..

The 73rd World Foundry Congress
„Creative Foundry"
23 - 27 September 2018, Krakow, Poland

DIGITAL TRANSFORMATION TO FOUNDRY 4.0

Nicola Gramegna[1], Fabrizio Greggio[1], Franco Bonollo[2]

[1] Enginsoft SpA, Padova, Italy
[2]Padova University, Dept. of Engineering and Management (DTG), Italy,
*Corresponding address: e-mail: n.gramegna@enginsoft.com

Key words: Casting process, Cyber-Physical-Systems, Industry 4.0, Intelligent Manufacturing system, OEE, Data mining, Predictive Quality, Cost model

1. Introduction

The competitiveness of foundry sector is driven by innovation oriented to quality and production efficiency. The deeper knowledge of foundry process supports the introduction of optimal solutions to high quality products as requested from all different markets in a global competition. The ability to manage all the stages of foundry production, based on advanced monitoring and cognitive platform, is fundamental to react in real-time with positive impact in terms of energy, environmental and cost.

The digital transformation is essential for innovative foundry (Fig. 1) in agreement with Industry 4.0 paradigm and the application of enabling technologies similarly to previous robotic and automation in assembly. The maturity, transversal usability and fast evolution of ICT technologies is accelerating the application of CPS, IoT and Artificial Intelligence (AI) I in the existing foundry plants or in new production sites. The foundry and metallurgy sector start his digital revolution introducing the advanced process simulation tools in the design and engineering departments of the supply chain. All equipment and process are well design and optimized by casting process simulation.

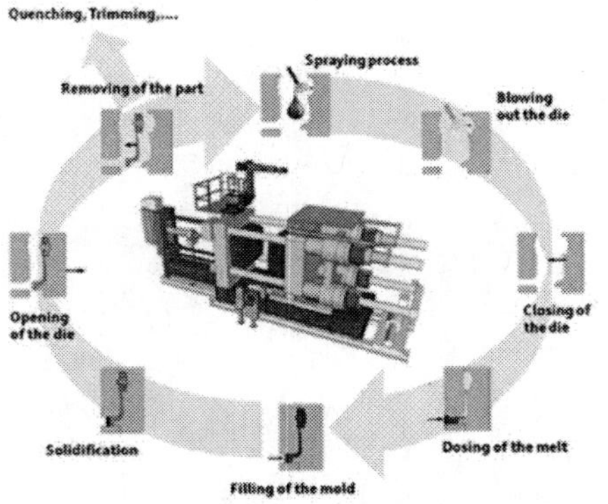

Fig. 1 – Chart of high pressure die casting production line

Nowadays, the challenge is the digitalization of real process introducing intelligent system to control the stability and efficiency of production line as well as developing Data Mining and Artificial Intelligent algorithms to quality assessment and predictive maintenance.

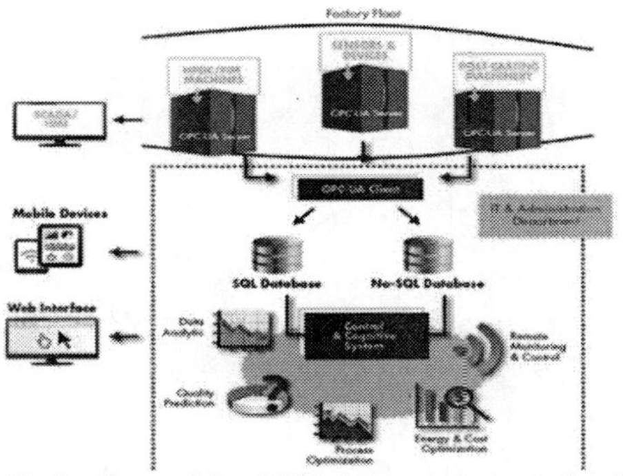

Fig. 2 – Connectivity of ICT platform to for control and cognitive predictive models.

The Smart Manufacturing strategy is oriented to zero defect and system efficiency in foundry with the support of new ICT platform (Fig. 2) with full interoperability of systems, with flexibility for small or large factory, with optimal solution for one size or serial production, improving the human machine interaction as well as the information sharing along the supply chain. All stakeholders in the foundry supply chain (material, implant and equipment suppliers, foundry user and end-user of casting products) are looking for collaborative market place where quality, energy and cost models are key performance indicators (KPIs) of the decision support system (DSS).

Foundry digitalization is facing technologic barriers, limits in people skills and opportunities for new business models. Intelligent control and energy-aware agile manufacturing is giving a new age to the traditional multi-stages production of smart foundry (ref. EU FP7-MUSIC project – [1-6]) in agreement with EU, National and Regional strategies embraced by the Veneto Regional foundry network SINFONET (http://www.retesinfonet.org/.

-

References

[1] F. Bonollo, N.Gramegna : The MUSIC guide to the key-parameters in High Pressure Die Casting – Assomet servizi srl, Enginsoft SpA, ISBN 978-8887786-10-1, 2014

[2] F. Bonollo, N.Gramegna : Smart Control and Cognitive System applied to the HPDC Foundry 4.0 – Assomet servizi srl, Enginsoft SpA, ISBN 978-8887786-11-8, 2016

[3] F. Bonollo, N. Gramegna, Giulio Timelli "High-Pressure Die-Casting: Contradictions and Challenges" JOM: the journal of the Minerals, Metals Society; Materials Society 05/2015

[4] F. Bonollo, N.Gramegna : HPDC Foundry Competitiveness based on smart control and cognitive system in Al-Alloy products – HTDC 22-23 June 2016, Venice (Italy)

[5] C. Raone, N. Gramegna, F. Zuliani: Intelligent management of lubrication phase in High pressure die casting - – HTDC 22-23 June 2016, Venice (Italy)

[6] B. Kujat, N. Gramegna, M. Benvenuti: Innovative control and real-time quality prediction for the casting production of aluminum alloy structural components at AUDI AG – HTDC 22-23 June 2016, Venice (Italy)

The 73rd World Foundry Congress
„Creative Foundry"
23 - 27 September 2018, Krakow, Poland

SINFONET: Smart & Innovative Foundry Network

Franco Bonollo[1]*, Nicola Gramegna[2]

[1] *Padova University, Dept. of Engineering and Management (DTG), Italy,*
[2] *Enginsoft SpA, Padova, Italy,*
**Corresponding address: e-mail: bonollo@gest.unipd.it*

Key words: Innovative network, Foundry, Aluminium alloys, Steel, Cast Iron

1. Introduction

SINFONET (Smart and INovative FOundry NETwork) is the Innovative Network for the Foundry, officially acknowledged by Regione Veneto (L.R. 13/2014) in 2016 (Fig. 1).

Fig. 1. Logo of SINFONET

SINFONET is aimed at developing the "smart foundry of the future", where innovation and integration generate education, jobs, competitiveness. SINFONET collects about 50 partners in the foundry supply chain (Fig. 2) with more than 4500 employees in Regione Veneto, and an overall turnover higher than 1.5 billion of €.

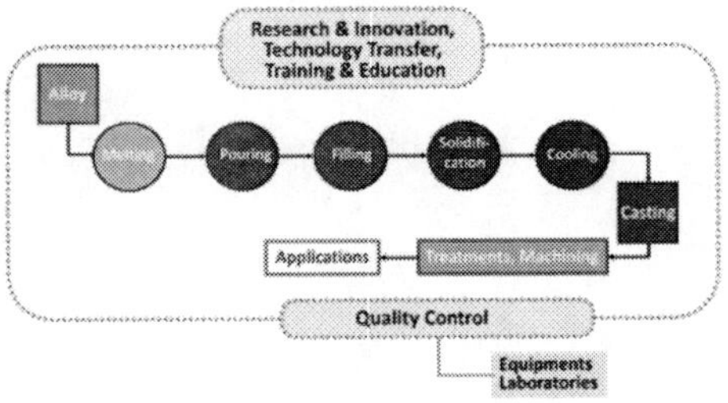

Fig. 2. Supply chain of Foundry

SINFONET is targeted at

- Designing, carrying out, coordinating a wide system of initiatives addressed to Foundry (ferrous and non-ferrous alloys) and Metalworking Companies,
- Building in Regione Veneto an Excellence Centre, collecting all the players in the supply chain: foundries, producers of raw materials and consumables, suppliers of engineering, services, technology, equipments, Research & Innovation Institutions, end-users.

SINFONET promotes multi-disciplinarity and integration, developing methods, products, procedure to support competitiveness (Fig. 3), based on

- Research, development, innovation and technology transfer (materials and processes, control systems, product and process standards, integrated design, energy and environmental management),
- Integration among foundries, suppliers (services, materials, technologies) and other metal-working companies,
- Education and qualification of human resources,
- Participation to Competitive Calls for funding,
- Scouting of new markets and internationalization.

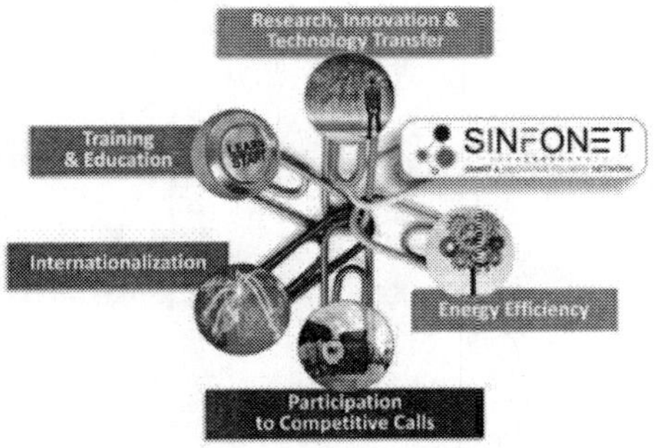

Fig. 3. Mission of SINFONET

2. SINFONET Projects

The activity of SINFONET has officially started during 2017, in the frame of various Research & Innovation and Internationalisation initiatives launched and supported by Regione Veneto (POR-FESR 2014-2020).

The main projects in which SINFONET and its Partners are involved will be presented. They are:

- GAP: Development of innovative foundry alloys and products, both in the field of ferrous (Austempered Ductile Iron, High

Silicon Ferritic Cast Iron) and non-ferrous (Aluminium) alloys;

- FORSAL: Introduction of automation and robot solutions for foundry operation, to be carried out on medium- and big-sized ferrous castings, aimed at improving health and safety of workers;
- SARR: Evaluation of innovative processes for recycling of slags produced by steel foundries;
- PreMANI: Introduction of new techniques for predictive manufacturing and maintenance in foundry
- INFOS: Supporting SMEs in internationalization and market development.

Acknowledgments
All the projects mentioned are supported in the frame of POR-FESR 2014-2020 by Regione Veneto.

The 73rd World Foundry Congress
„Creative Foundry"
23 - 27 September 2018, Krakow, Poland

SIMULATION DRIVEN DESIGN FOR CASTINGS WITH EFFECTIVE CONTROL OF MANUFACTURING CONSTRAINTS

Kunju Ravi[1]*, Solina Martin[2], Afsaridis Kimon[3]

[1]Altair, 1820 E. Big Beaver Rd, Troy, MI 48083, USA,
[2] Altair, Av. Diagonal 682 4th, 08034 – Barcelona – Spain,
[3]Altair, Balkan Center, Building C, 9th km Thessalonikis, Greece
**Corresponding address: e-mail: ravi@altair.com*

Key words: Topology Optimization, Casting Simulation, Design for Manufacture, Simulation Driven Design, Altair Inspire Cast

1. Abstract

Structural optimization methods, namely topology optimization, has paved way for compelling and radical light weight product designs. However, it is prudent to effectively understand performance and manufacturing constraints up-front, for numerical simulation to drive the design and generate shapes that are easier to manufacture. In metal castings, large and thick sections are areas prone to defects such are macro porosity, while thin areas are likely to see micro-porosity and warpage due to uneven cooling. A minimum or a maximum member size to be generated can be defined so that the topology optimization tools like Altair Inspire can generate a design meeting the requirement of the size limitation defined (Fig. 1). Constraints drive the design, and a methodology to understand and mastering them for components that are cast and the ability to evaluate their 'castability' and 'performance' in (Fig. 2) one single comprehensive environment will be presented in the paper. Imposing a draw or split-draw will generate structures devoid of undercuts for conventional castings; and the ability to easily extract machined patterns so they are tapered on the cope and drag side of the mold; or using a free form draw if the mold will be 3D-Printed (Fig. 3). The attempt is to develop guidelines based on the manufacturing methods upfront to generate only those designs that are better suited for casting. Practical exmaples will be demonstrated showcasing the effect of constraints based on desired casting methods illustrating the guidelines required for better castings.

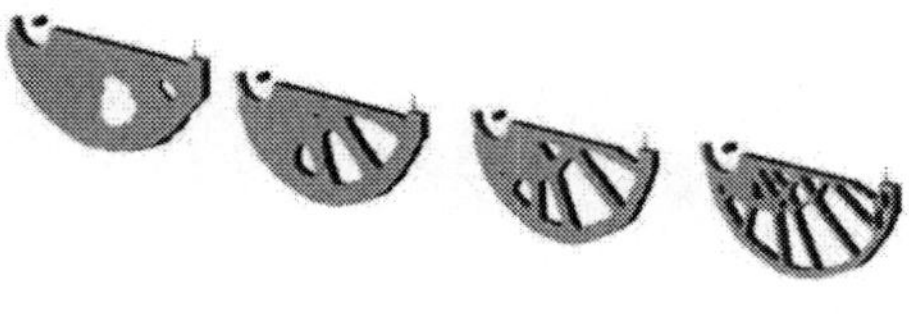

Fig. 1. Figure shows the Topologies generated by changing the of Member size constraints

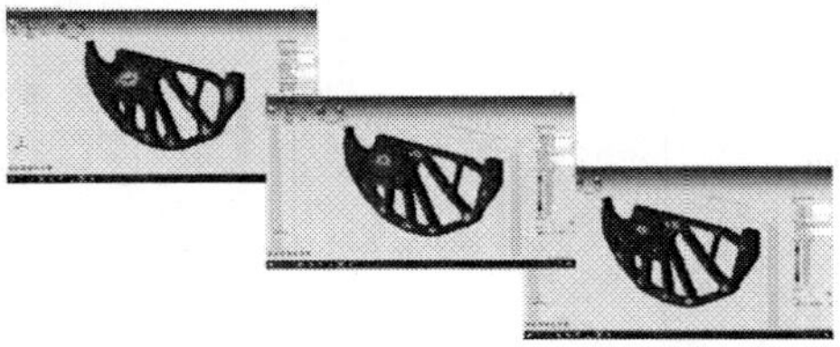

Fig. 2. Casting Porosity Comparison and effect of Member size

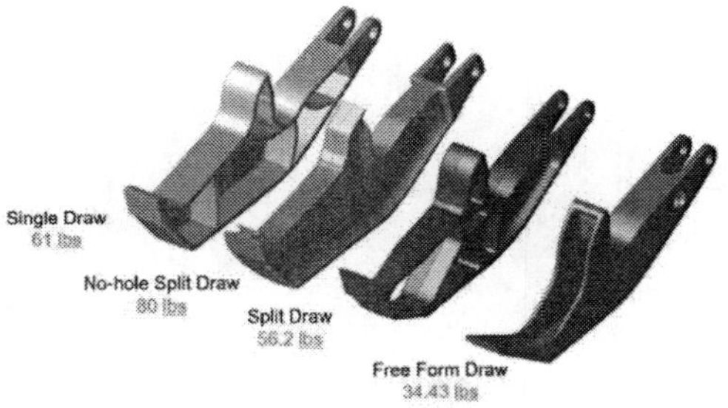

Fig. 3. Example of a Multiple Pole Driver Designs using multiple Casting Draw Constraints (single draw, no-hole split draw, split draw and Free form Casting)

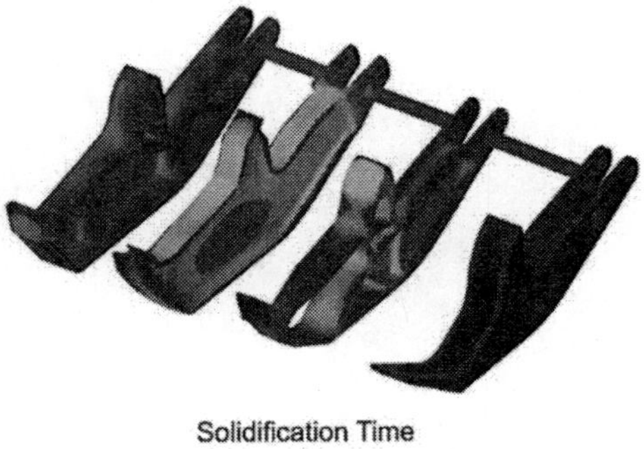

Fig. 4. Comparison of Solidification times for the multiple topological designs

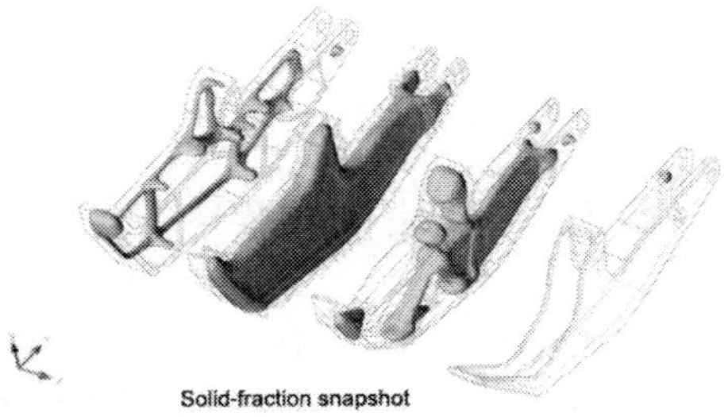

Fig. 5. Comparison of Solidfraction snapshot for the multiple topological designs

ECOLOGY

The 73rd World Foundry Congress
„Creative Foundry"
23 - 27 September 2018, Krakow, Poland

PRO-ECOLOGICAL DIE CASTING FOUNDRY

Wiesław Walczak, Paweł Zawadzki, Igor Zawadzki, Konrad Wrzała

Odlewnia SILUM Sp. z o.o., Opojowice 51, 98-310 Czarnożyły, Poland
w.walczak@silum.com.pl

Key words: Ecology, enviromental protection, pressure die-casting foundry, innovation, modern technology

The dynamic development of industry causes more and more pollution of our world. Therefore, the emphasis on corporate responsibility in environmental protection is increasing. Many companies in their policy take into account pro-ecological activities, the implementation of which guarantees sustainable development.

The aim of the paper is to discuss the issues related to pollution created in die-casting foundries, ways of limiting them and discussing about technologically advanced devices used to minimize emissions and maximize the processing of generated waste.

In particular, the analysis involved the use of ecological shaft aluminum melting furnaces (Figure 1), devices for filtration of dusts and vapors generated in foundry processes (Figures 2a and 2b), as well as devices for processing production wastewater (Figure 3).

Fig. 1. Melting furnace

Fig. 2a. Device for filtration

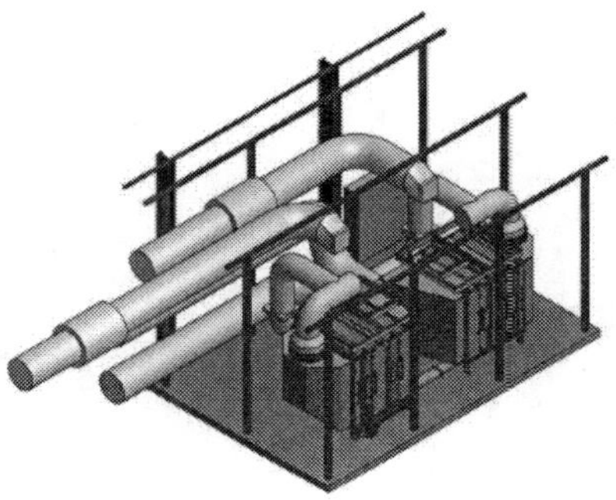

Fig. 2b. Device for filtration

Fig. 3. Device for processing production wastewater

In the paper there is presented the construction and operation principle of individual innovative solutions and their impact on pollution reduction.

The summary of the considerations is to show the legitimacy of using environmentally friendly devices and to indicate the added value obtained through their use. The examples of countermeasures mentioned in the paper shows how to minimize and reduce the impact of the production processes on the environment in a die-casting foundry.

The 73rd World Foundry Congress
„Creative Foundry"
23 - 27 September 2018, Krakow, Poland

SPENT FOUNDRY SAND VALORIZATION IN CONSTRUCTION SECTOR THROUGH THE VALIDATION OF HIGH-PERFORMANCE APPLICATIONS

Erika Garitaonandia [1]*, Clara Delgado [1], Lucia Unamunzaga[1], Antonio Martínez Puche[2], Iñaki García[3], Javier Ríos[4]; Jose Luis Rodríguez [5]

[1] *IK4-AZTERLAN, Aliendalde Auzunea 6, 48200-Durango, Spain*
[2] ACCIONA CONSTRUCIÓN S.A., Valportillo II, nº8, 28108 – Alcobendas, Spain
[3] FUNDICIONES DEL ESTANDA, S.A., Antzitzar 17, 20200 – Beasain, Spain
[4] GAIKER-IK4, Parque Tecnológico Bizkaia, Ed.202, 48170 – Zamudio, Spain
[5] ONDARLAN, S.L., Txirrita-Maleo nº10, 20100 – Renteria, Spain
* *e-mail: egarita@azterlan.es*

Key words: Spent Foundry Sand, reclamation technology, valorization, fine aggregate, construction application.

One of the most pressing problems facing the metal casting industry nowadays is the disposal of Spent Foundry Sand (SFS). Despite having high silica content and good engineering properties, this waste is typically landfilled. Foundry sand could be beneficially reused in the construction sector, if the presence of residual organic compounds from binders and metals from casting are below critical limits.

In the current study, 20 different SFS samples and their leachates have been analyzed. Results indicate that just a few parameters tend to exceed the limit values for inert waste, typically Dissolved Organic Carbon and fluorides in leachate. Regarding physical-chemical specifications for fine aggregates, contents of organic matter, fines and soluble salts in SFS seem to be the most critical.

This paper describes the activities performed in the framework of the ongoing project LIFE ECO-SANDFILL, regarding the integration of a novel mechanical SFS reclamation system in a Spanish foundry, in order to reach the required quality in reclaimed sand for their validation technically and economically in three pilot construction applications.

INTRODUCTION

SFS generation represents more than 80 % of the total solid waste generated in foundries. With more than 3 000 active foundries in Europe [1], this waste amounts to about 9 Mt annually, being less than 40 % of the SFS generated valorized, mainly in cement industries.

The physical and chemical characteristics of SFS depend on the type of casting process and binders used. It consists of a morphologically uniform sand, with an average particle size between 300-500 µm, sub angular-spherical shape and a high content (80-95 %) of valuable silica. Therefore, SFS is potentially recoverable as secondary raw material in diverse applications, in substitution of natural silica sand, provided that applicable environmental and technical requirements are met.

DETERMINATION OF CRITICAL PARAMETERS FOR SFS VALORISATION IN CONSTRUCTION SECTOR

Use of SFS as secondary fine aggregates in a material-intensive sector such as construction offers promising valorization options. The sand must meet relevant technical specifications for each application and, additionally, fulfil several environmental conditions for unbound uses, where the sand is directly in contact with the soil.

Currently, SFS reuse in construction sector is not specifically addressed in any regulation at European or local level. For this reason, regulations existing for valorization of other waste streams have been used as a reference: the quality criteria selected in LIFE ECO-SANDFILL pilot experiences stem from the construction and demolition waste legislation in the Basque Country (Spain) [2], which stipulates that recycled aggregates intended for unbound applications must fulfil leachate limit values for inert waste landfilling [3] and meet contaminant concentration limit values stated in soil pollution regulation [4], whereas no environmental conditions are given to hydraulically bound applications.

SFS (green and chemically bonded sand) generated in 20 ferrous foundries in Spain have been characterized, to identify the parameters in the full set of specifications that are most critical for SFS to comply with limits, from both an environmental and a technical point of view.

Regarding the contaminants content in sand, all the 20 SFS samples studied are below the regulated limits for soil pollution. However, pollutant limits in leachate are exceeded in several samples. 60% of the samples analyzed fail to meet the leaching limits for fluoride and Total Dissolved Solids (TDS) and 70% exceed the content limit for Dissolved Organic Carbon (DOC). The measured values of other leachate parameters such as Zn content and phenol index, in samples from a few foundries, have been found to be around the limits.

Those experimental results suggest that only 5, out of the more than 125 chemical substances of environmental concern for unbound construction applications, may require close monitoring and correction measures in SFS. Table 1 lists the environmental acceptance criteria stablished on them for SFS to be used in LIFE ECO-SANDFILL project.

Table 1. Critical environmental parameters of SFS for use in construction unbound applications

Component in leachate	Limit Value (mg/kg). Inert Waste
Zinc	<4
Total Dissolved Solids (TDS)	4000
Fluoride	<10
DOC	<500
Phenol Index	<1

Considering technical requirements stablished for aggregates, SFS must meet the criteria requested either to the fine aggregates and to the final product, which might vary from one application to another. Table 2 shows the critical parameters defined for sand as an aggregate for bound [5] and unbound [6] applications in Spanish regulations:

Table 2. Critical technical parameters of SFS for use in construction applications

Component in sand	Unbound App.	Bound App.
Organic Matter	< 2%	< 3.5 %
Soluble salts	< 1%	-
Fines	-	< 6%

In this sense, fines and organic matter contents may influence the setting time and mechanical resistance of concrete and mortars, as well as the bearing capacity of embankments. In general, green sand shows values above the limit set for both parameters. Similarly, high values of some soluble salts as sulfates can cause swelling in bound applications and volumetric instability in unbound ones. Furan bonded sand has shown higher values of sulfates than other SFS types.

SFS RECLAMATION TECHNIQUE

The new reclamation equipment integrated and tested in a foundry located in Northern Spain consists on scrubber style reclamation with centrifugal force, in which the sand is rubbed together and against the walls in a closed system, removing effectively impurities from the surface of the silica sand.

An intensive characterization program has been carried out onto sand samples taken from the inflow and outflow of the reclamation equipment, to help define the operation regime that allows achieving the quality required for foundry and construction applications.

THREE PILOT EXPERIENCES

After the adequate reclamation treatment in the equipment, a flow of quality sand, environmentally safe, has been obtained and used in 3 eco-innovative construction applications. Those full-scale case studies have been implemented in the construction works for the High-Speed Train in Spain:

- Embankment: Unbound application. Reclaimed SFS used: ca. 700 t (400 m^3).
- Flowable mortars: Bound application. Reclaimed SFS used: ca. 150 t (100 m^3).
- CLSM (Controlled Low Strength Material): Bound application. Reclaimed SFS used: ca. 10 t (6 m^3).

CONCLUSIONS

Aggregates in hydraulically bound construction applications appears as a promising valorization option for SFS. However, most of the studied SFS samples fail to meet certain environmental limit values for their use in unbound applications and, therefore, need to undergo some conditioning treatment in order to achieve the quality criteria. The trends below have been observed in SFS samples:

- ✓ Regarding environmental compliance, leaching limit values for fluorides, DOC and TDS are exceeded in more than 60% of the samples. These parameters are especially sensitive in unbound applications, since the sand is directly in contact with the soil, but are not critical in bound applications, where the sand is enclosed in a cementitious matrix and the risk of leaching is low.
- ✓ As for technical aspects, parameters such as fines, organic matter and soluble salts contents are considered critical for both bound and unbound applications. The highest percentages of fine and organic matter have been found in green sand samples. In the case of sulfates, the highest values have been measured in furan resin sand.

The characterization carried out over 20 SFS samples from different foundries and the results obtained with the new reclamation process suggest that the new mechanical reclaiming technology enables SFS to reach the acceptance criteria for reusing foundry sand in different construction applications.

ACKNOWLEDGMENT

LIFE ECO-SANDFILL (2016-2019) is co-financed by the LIFE Programme of the European Union (Grant Agreement No. LIFE15 ENV/ES/000612).

REFERENCES

[1] CAEF. Statistics. 2016. Retrieved 20/11/2017 at http://www.caef.eu/downloads/kategorie.asp?kat=9

[2] ORDEN del 12/01/2015 utilización de los áridos reciclados procedentes de la valorización de RCD

[3] Council Decision 2003/33/EC establishing criteria and procedures for the acceptance of waste at landfills

[4] Ley 4/2015, de 25 de junio, para la prevención y corrección de la contaminación del suelo

[5] Instrucción de Hormigón Estructural (EHE-08).

[6] Pliego de Prescripciones Técnicas Generales para Obras de Carreteras y Puentes (PG-3).

The 73rd World Foundry Congress
„Creative Foundry"
23 - 27 September 2018, Krakow, Poland

DATABASE OF MATERIALS FOR THE EVALUATION OF THE IMPACT OF HARMFUL SUBSTANCES IN METALLURGICAL PROCESSES

Szymon Biernat[1]*, Adam Bydałek[2]

[1]Państwowa Wyższa Szkoła Zawodowa w Głogowie, Kierunek Metalurgia, ul. Piotra Skargi 5, 67-200 Głogów, Polska,
[2] Uniwersytet Zielonogórski, Wydział Mechaniczny, ul. Szafrana 4, 65-001 Zielona Góra, Polska,
**Corresponding address: e-mail: szymon_soft@wp.pl*

Key words: environmental protection, substance toxicity, database program, protection

1. Introduction

Continuous technological development entails constant and progressive degradation of the natural environment. In many branches of the economy, there is a tendency to increase the production process at any price without paying attention to ecological aspects. In agriculture itself, the introduction of genetically modified food was to be a panacea for the problem of hunger occurring in the world, while research indicates its very negative impact [1, 2]. Also is the use of fungicides, pesticides and fertilizers, which results in visually nice products in large quantities, however, very heavily contaminated. In addition, the use of nitrates and other preservatives, as well as dyes to an additional extent, poisons products that reach our table, which have a very negative impact on human health [3]. The heavy emitters, in particular of metallurgical industry, are a special emitter of pollutants to the environment. It has a negative impact on the environment in many different areas and in many places, and mining ores and processing them activities may involve even very longtime consequences [4, 5].

2. Problem analysis

The "brass -prop" computer application contains not only an interesting addition describing the determination of process parameters and end properties in copper production [6, 7]. You can find here information on the processes of alloying, refining or the machining process of a given material. There is a fairly detailed description of the purpose of a given material with the analysis of mechanical properties, eg: stretching, hardness, elongation taking into account the dimensions, shapes and geometry of the material as well as a description of the morphology of hard particles that have been tested.

Another element appearing in the computer program is the database of threats related to the substances described in the brass-prop application. It contains a list of all substances present in the database. At this point, the user can find a set of data on specific substances that have been sorted alphabetically in relation to the chemical symbol, but can also provide the symbol of a given compound in a simple search engine, so that without a manual search of the database, to can familiarize yourself with the characteristics of a given compound.

The first and most important information that is obtained by the user from the database is the so-called diamond of fire. This is a standard that was introduced in the United States by the „National Fire Protecion Association". He describes in short the type and degree of threats that the use of a given substance brings. In the case of the database described in the basic form, a table is presented below (Table 1) describing the number of compounds present in the database along with the division into their level of risk in the context of health, flammability and instability.

Table 1. List of substances present in the database in relation to the degree of danger.

Degree of health hazard	The amount of substance	The degree of flammability of the substance	The amount of substance
0	11	0	24
1	3	1	-
2	7	2	-
3	3	3	-
4	-	4	-
Degree of compound instability	The amount of substance	Other threats	Description of danger - Quantity of substance
0	20	W̶	Reacts vigorously with water in an unspecified and dangerous way - 2
1	1		
2	2		
3	1		
4	-		

At this point, one should also pay attention to another very important issue, where compounds can be classified as not posing health hazards in the case of normal contact, meanwhile, in the conditions of high metallurgical temperatures, the employee of both the steelworks and the laboratory may be exposed to inhalation of fumes and dust, which can have a very harmful effect. For example, iron "Fe", which is classified at zero level, for human health can be given, whereas exposure to chronic inhalation of dust of this element in polluted air can lead to so-called non-collagenic pneumoconiosis.

Further information that the database user can read from it concerns the so-called H-phrases and P phrases regarding the substance. This phrases indicate the type of hazard that can be encountered in the case of contact with a given substance. Using appropriate symbols, they can indicate physical, chemical, health and environmental hazards, as well as provide other important information.

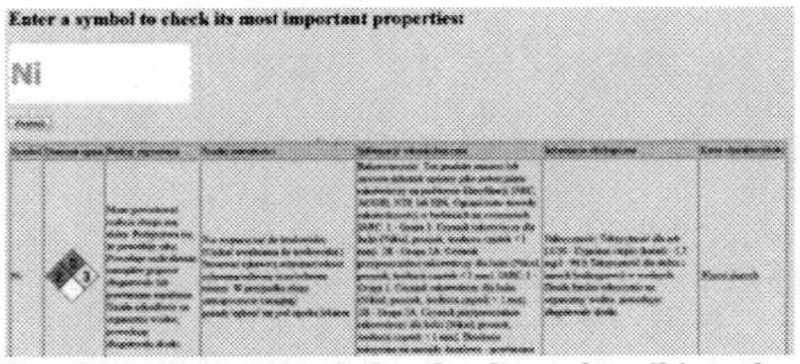

Fig. 1. A fragment of the database describing the types of hazards and precautions for the selected substance.

The next three columns contain supplemental information. They mainly concern data on the toxicological nature of a given material. Let's take a closer look at this information, you can find very rich information about various negative effects on health and even life. The user can learn about an allergic reaction in the case of contact with the skin, the possibility of eye irritation, and even cancer and the emergence of cancer in the event of the substance entering the body regardless of the route. Organs having direct contact with the substance are most exposed to negative effects. It is also important that even substances found to be harmless can lead to serious disorders, and in extreme cases also to death if they get too much into the body.

Information on the substance in the database is much more extensive than described above. However, the remaining advantages of the program and the database are presented in the full article.

In the figure above (Fig. 1) a fragment of the database showing threats, precautions and toxicological information for a specific substance - in this case nickel Ni - is presented.

3. Summary and Conclusions

Summing up the work done, it can be concluded that the next element of the computer application, this time paying special attention to the ecological aspects of materials used in the metallurgical industry, can be invaluable help for people working in this industry. Before proceeding your work, whether for research or other work with a given material, you should always read the information about the substance and be aware of what hazards it may have and which should be particularly used. An essential equipment of every laboratory or industrial plant is appropriate protective clothing, masks, goggles, gloves, aprons, and in special cases also oxygen masks with a closed circuit. It is also worth noting that metallurgical laboratories in which refinery research is conducted at high temperatures should be equipped with fume hoods, that is, laboratory extracts that provide for the discharge of harmful substances from enclosed spaces, which as can be seen from the study, are quite a lot.

References

[1] Jurkiewicz, A. & Bujak, F. (2015). Zagrożenia zdrowotne związane z genetycznie modyfikowaną żywnością (GMF). Medycyna ogólna i nauki o zdrowiu. 21 (1), 62 – 64. DOI: 10.5604/20834543.1142361

[2] Stankiewicz D. Organizmy zmodyfikowane genetycznie. Kancelaria Sejmu, Biuro Studiów i Ekspertyz. Wydział Analiz Ekonomicznych I Społecznych, Warszawa, 2000: 750.

[3] Wasilewska, E. & Małgorzewicz, S. (2015). Niepożądane reakcje pokarmowe na dodatki do żywności. Forum zaburzeń metabolicznych. 6 (1), 8 – 13.

[4] Uberman, R., Pietrzyk – Sokulska, E. & Kulczycka, J. (2014). Ocena wpływu działalności górniczej na środowisko – tendencje zmian. Przyszłość: Świat – Europa – Polska. 2 (30), 87 – 119.

[5] Dudlewski, J., Wtorek, L. (2000). Zagrożenie dla środowiska wypływające z podziemnej eksploatacji surowców mineralnych. In Eksploatacja podziemna a ochrona środowiska w świetle prawa Unii Europejskiej, Ustroń Śl., 29 – 31 maja 2000 (25 – 35). Kraków: Polska Akademia Nauk Instytut Gospodarki Surowcami Mineralnymi i Energią.

[6] Biernat, S. & Bydałek, A.W. (2014). Optimization of the Brass Melting. Archives of Foundry Engineering. 14 (3), 5 - 10

[7] Bydałek, A.W., Najman, K., Kula, A., Biernat, S., Błaż, L. & Wołczyński, W. (2016). Evaluation of Precipitates Type in Brasses as a Function of Charge Material. Archives of Foundry Engineering. 16 (3), 19 – 24

The 73rd World Foundry Congress
„Creative Foundry"
23 - 27 September 2018, Krakow, Poland

PROCESS PERFORMANCE AND ENVIRONMENTAL IMPACT OF THE CaO-Al2O3 SLAG SYSTEM AS ALTERNATIVE TO CALCIUM CARBIDE FOR DESULFURIZATION OF NODULAR CAST IRON

Rodrigo LENCINA [1]*, Emmanuelle HENRY-LANIER [1], Yves ROUX [2]

[1] *Kerneos Aluminate Technologies, 92977 Paris La Defense, France,*
[2] *Procastyr, 74700 Sallanches, France,*
**e-mail: rodrigo.lencina@kerneos.com*

Key words: Calcium aluminate, calcium carbide, cupola, ductile cast iron, desulfurization, LCA, environmental impact

Calcium carbide is an effective reagent for desulfurization of nodular cast iron. For this reason, many foundries operating cupola kilns around the world are still using it, despite the considerable disadvantages, such as the production of acetylene gas when it comes in contact with water.

Moreover, growing environmental standards have raised additional concerns for foundries. Currently, responsible carbide users must be aware of its environmental impact by considering the complete life cycle from manufacturing to waste disposal.

These disadvantages may prompt iron producers to search for valid alternatives to carbide. One of them is lime, which in theory can be almost as effective.

Nevertheless, lime is required to be dissolved in a liquid phase in order to be effective. The creation of a liquid phase or slag can be achieved by the addition of a low melting point metallurgical flux in the ladle.

Fluorspar is one of the most effective fluxes for slag conditioning. However, the $CaO-CaF_2$ slag system still has disadvantages in terms of the lifetime of the ladles' refractory lining.

On the other hand, the $CaO-Al_2O_3$ slag system can be as effective as carbide or fluorspar. Ishikawa *et al.* [1] reported in 2004 the successful use of this slag system in a tea-pot ladle. Also, it can be formed with more environmentally friendly materials, such as lime and pre-fused calcium aluminate.

This work summarizes recent industrial experiences on the use of $CaO-Al_2O_3$ slags. Additionally, the paper presents environmental impact results for two desulfurization system: one based on calcium carbide and the other on calcium aluminate.

1. Description of the $CaO-Al_2O_3$ slag system

A mix made of lime and calcium aluminate that comes in contact with the molten metal will create a slag in the $CaO-Al_2O_3$ system. However, the right slag composition and process conditions must be obtained to ensure an efficient desulfurization. Some of these conditions are described below:

Ladle configuration: Strong interaction between metal and slag is key for efficient desulfurization. Ladle design has to be adapted for an adequate slag flow. The slag evacuation requires special attention because it could slow down the entire process.

Process temperature: At steady state conditions, efficient desulfurization has been recorded at temperatures as low as 1460°C. A sustained operation of the cupola is important for a successful use of the slag. Poor results have been observed when the cupola runs for short periods of time.

Mix of reagents: chemical composition of the C-A mix (C:CaO, A:Al_2O_3) has to be adapted for each plant configuration. The ratio of lime to calcium aluminate in the mix will determine the quantity of liquid phase in the ladle.

2. Performance in industrial operations

Desulfurization efficiency of C-A mix can be equivalent to carbide, provided the right process conditions are attained. The table below compares efficiencies in tea-pot and shaker ladle designs.

Table 1. Performance of C-A Mix and Carbide

Reagent	Carbide	Mix	Carbide	Mix
S initial	0,104%	0,101%	0,096%	0,094%
S target	0,020%	0,020%	0,015%	0,015%
S final	0,017%	0,018%	0,009%	0,011%
Efficiency	84%	82%	91%	88%
Ladle	Shaker		Tea-pot	

The C-A mix will melt and form the slag together with the oxides coming from the molten metal. Table 2 shows the composition of slag samples taken during steady state conditions.

Table 2. Chemical composition of slag samples

Ladle	CaO	MgO	Al_2O_3	SiO_2	Fe_2O_3
Shaker	76,6%	2,0%	5,2%	10,3%	5,6%
Tea-pot	79,5%	1,0%	5,3%	6,4%	7,5%

Figure 1 shows the phases calculated using FactSage

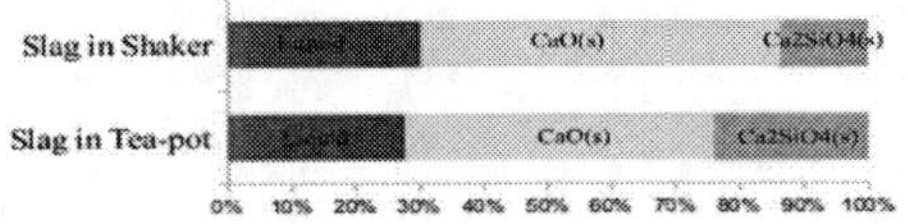

Fig. 1. Phases present in the slags at 1500°C

There is about 30% liquid, which is the most relevant phase because it will absorb the Sulphur when it comes in contact with the molten metal. Since all the

Al_2O_3 content of the slag is found in the liquid phase, it can be concluded that the calcium aluminate fully contributed to its formation

3. Calcium aluminate flux for slag conditioning

A liquid phase must be formed at process temperature in order to promote the desulfurization. Thus, a pre-fused calcium aluminate with a low melting point is key for effective slag formation.

The melting properties of calcium aluminate metallurgical flux depend on the chemical composition and homogeneous mineral phases. The arrow in Figure 2 shows the composition of LDSF®, which is adjusted to the first eutectic of the diagram. The high proportion of the C12A7 mineral phase will ensure a melting point much lower than the process temperature of the metal.

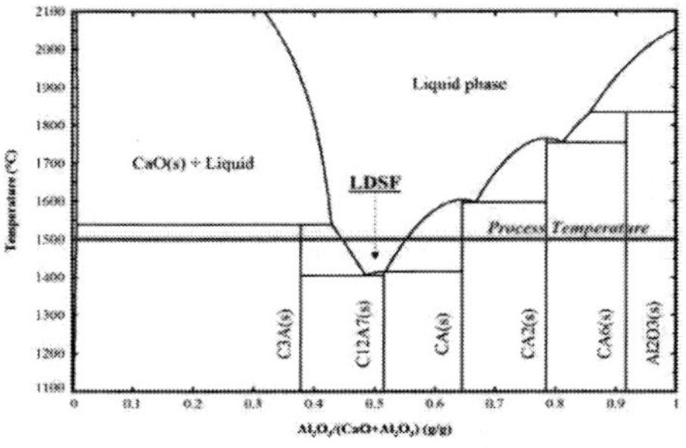

Fig. 2. CaO-Al_2O_3 binary phase diagram

Pre-fusion is an important characteristic for metallurgical fluxes. For example, raw materials such as bauxite can also be sources of Al_2O_3 in a slag. However, due to its high melting point, bauxite will melt poorly at process temperatures, which may cause crust formation and poor desulfurization.

4. Environmental impact

Because it is holistic, systemic and rigorous, Environmental Life Cycle Assessment is the preferred tool when it comes to assessing the potential environmental impacts of products' life cycles. LCA consists of a compilation of energy and material inputs as well as environmental releases. The functional unit, used as a common denominator, provides the basis for the addition of material flows and impacts for any of the life cycle stages for the studied product. See figure below.

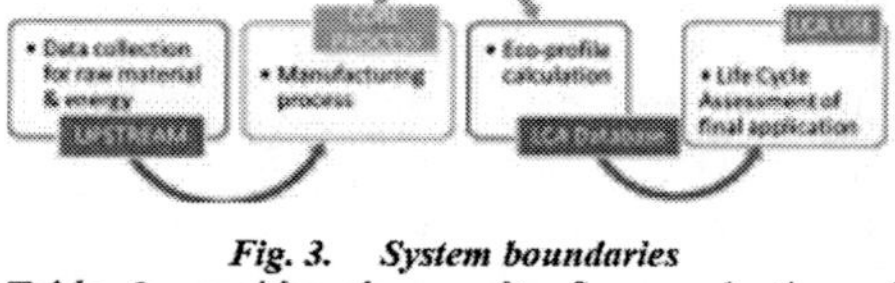

Fig. 3. System boundaries

Table 3 provides the results for a selection of environmental impact indicators also called "eco-profile" for both C-A mix and calcium carbide. Eco-profile results are given for the production of one kilogram of C-A mix or carbide, from "cradle to gate", i.e. including upstream processes (raw material & energy extraction, production and supply) and core processes (production plants). Data used for the modelization of Calcium Aluminate production represents Kerneos' current European technology. Ecoprofile data for lime and calcium carbide comes from the Ecoinvent database [2]. The precision of data used for calcium carbide is relatively low, compared to data used for C-A mix.

Table 3. Environmental impacts from cradle to gate C-A mix and carbide (for 1 ton iron)

For 1 kg		Mix	CaC_2
Climate Change	kg eq. CO2	1,2	3,1
Air acidification	g eq. SO2	2,1	12
Non renew. energy	MJ	6,6	58
Water consumption	litre	1,1	13
Addition in ladle	kg/t iron	6	4

Results per ton are then used to determine the overall impact of final application by considering the quantity of fluxes required to achieve the same desulfurization performance. Results are calculated for 1 ton of cast iron and presented in figure 4.

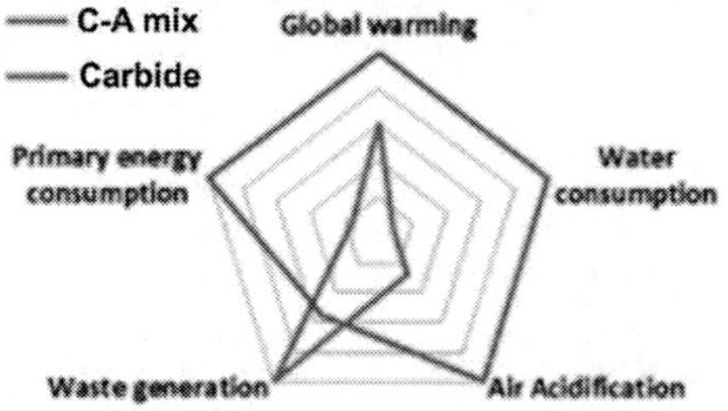

Fig. 4. Environmental impact of the two systems

For all environmental impact categories except waste generation, C-A mix solution has a lower impact than calcium carbide. The waste indicator corresponds to the slag quantity produced. Although the quantity of slag generated with the C-A mix alternative is higher than with carbide, the slag quality is considerably improved and valorization is then easier than with carbide.

5. Conclusions

Industrial experience demonstrates that a slag in the CaO-Al_2O_3 system can efficiently desulfurize in standard process conditions. This slag system can be formed in the ladle by adding a mix of lime and pre-fused calcium aluminate.

LCA shows that the C-A mix has a lower environmental impact than calcium carbide.

Therefore, due to the positive industrial results and eco-friendliness, a mix of lime and calcium aluminate can be considered as a valid alternative to carbide in the desulfurization of nodular cast iron.

References

[1] M. Ishikawa, K. Seino, T. Horii, S. Kawanami. (2004). Recent Improvements and Modifications in Melting Processes. 2nd Cupola Conference. Trier.
[2] ETH Zürich, Ecoinvent database v3.3, (2016)

The 73rd World Foundry Congress
„Creative Foundry"
23 - 27 September 2018, Krakow, Poland

ABSORPTION-BIOCHEMICAL UNITS (ABChU) FOR VENTILATING AIR PURIFICATION IN FOUNDRY

Yury Shapavalau*, Aliaksandr Halibus, Aliaksandr Sudarau

Gazoochistka Engineering LLC, 23-404, Dolgobrodskaya str., Minsk, 220070, Republic of Belarus
**Corresponding address: e-mail: cleangaz.by@gmail.com*

Key words: gas emissions, harmful substances, neutralization, purification plants, water absorption

Clean air is one of the essential conditions for high-efficient labor and human health.

In the core molding and die casting processes (Fig. 1) with subsequent cooling and shaking out a wide range of gaseous substances of the second, third and fourth hazard categories, including phenol, formaldehyde, furfural, methanol, ammonia, acraldehyde, cyanides, benzapyrene, triethylamine, diethylamine, xylene, toluene and a number of hydrocarbons, vents into the atmosphere with ventilating air. This is the cause of an offensive odor and exceedence of the hazardous substance maximum allowable concentration in the atmospheric ground layer of residential areas adjacent to the foundries.

Figure 1 – Mold casting

Volumes of vented emissions from the foundries and shops are from several tens of thousands to several million cubic meters per day and harmful substances emissions are tens and hundreds of tons per year.

The researchers from the Belarusian National Technical University had developed a technical process and a design of units for absorption and biochemical organic substances (Fig. 2).vented air purification from harmful

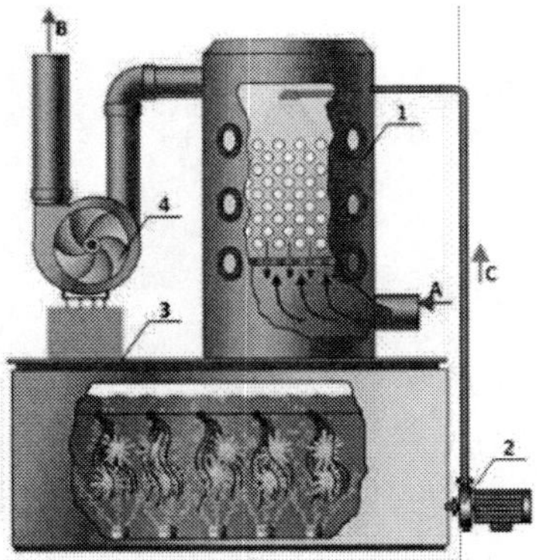

Figure 2 – Diagram of the ABChU:
1 – scrubber with mass exchange grids and packing; 2 – pump; 3 – bioreactor with packing and aerators; 4 – fan; A and B – ventilating air input and output; C – absorption solution

Ventilating air purification with the ABChU is based on natural processes:

- the first principle lies in the fact that the majority of harmful volatile organic compounds are soluble in service water;
- the second principle is based on the ability of specially selected and adapted microorganisms to use dissolved organic and some inorganic substances in water as replenishment source. In the process of these compounds consumption by microorganisms their complete mineralization occurs with formation of water and carbon dioxide.

Harmful substances are trapped with the help of water absorbent in the scrubber with a movable ball-type packing and are neutralized in the bioreactor. Microorganisms in the form of concentrated biomass are added into the bioreactor once before the unit operation commencement. The solution circulation in a closed "scrubber–bioreactor" cycle prevents industrial drains.

The Gazoochistka Engineering LLC develops and manufactures ABChU [1] with a ventilating capacity of 5, 10, 15, 20, 25, 30 thousand m^3 per hour (Fig. 3).

Figure 3 – General layout of the ABChU with a ventilating capacity of 20,000 m^3 per hour.
1 – fan; 2 – scrubber; 3 – sludge chamber; 4 – bioreactor; 5 – drop separator; A and B – ventilating air input and output relevantly.

Condensation of aerosols and vapors of high-boiling substances as well as surface-active substances (surfactants), which are produced under microorganisms' life activity, contributes to the increase in efficiency of water absorption of hardly soluble substances. The surfactants presence is indicated by the foaming in the system of water absorbent regeneration [2].
The nominal characteristics of the harmful impurities separation efficiency remain throughout the whole unit service life and are, as a rule, 90-99%.
Twenty-year experience of the ABChU operation at 52 enterprises of Russia, Belarus and Ukraine showed that upon aggregate index: functionality, efficiency, environmental compatibility, reliability and operation simplicity – the absorption and biochemical technology is the optimal and other times the only technical solution for purification of the air-gas complex compounds (multicomponent, high temperature, presence of suspended, resinous and condensation substances).
Thus, for example, 1) on application of the "dry" methods (coals, catalysts, ion exchange fibers, ozonizers) for the abovementioned purposes, the purification functionality is sharply reduced due to filter working surfaces overgrowing by deposits; 2) in the course of core molding under Cold-box-amin process an increasing number of enterprises prefer to neutralize amines not in "acid" scrubbers but apply the ABChU with significant economic and environmental benefits.
Several ABChU are applied with a total ventilating capacity of 25-30 thousand m^3 per hour in case of necessity to purify large volumes of vened air (Fig. 4).

Figure 4 – The gas cleaning complex of four ABChUs with a total ventilating capacity of 100,000 m^3 per hour

Summary:

1. The absorption and biochemical technology for harmful organic substances neutralization compare favourably by capital and operating costs with other methods of cleaning. This technology is tested under various conditions and could be recommended for usage in foundries.
2. The ABChU installations are especially effective in vented air purification at casting, cooling and knockout of molds.
3. Vented air purification with the ABChU reduce harmful substances emission into the atmosphere and also diminish odor effect significantly in the area of foundry location.

References

[1] Yury P. Shapovalov, Alexandr S. Galibus, Sergey S. Deshchits. (2008). Patent number 010270. Moscow. The Eurasian Patent Organization, the Eurasian Patent Office.
[2] Ganitkevich Y.V. (1988). Surface-active microbially-derived substances. Biotechnology. Moscow.

The 73rd World Foundry Congress
„Creative Foundry"
23 - 27 September 2018, Krakow, Poland

FATA ALUMINUM INORGANIC SAND REGENERATION PROCESS

Bartolomeo Tosco [1], **Paolo Bocca** [2]

[1] *Fata Aluminum S.r.l., Via Stresa 5 – 10044 Pianezza (TO), Italy, bartolomeo.tosco@fataaluminum.com*
[2] *Fata Aluminum S.r.l., Via Stresa 5 – 10044 Pianezza (TO), Italy, paolo.bocca@fataaluminum.com*

Key words: inorganic, sand, regeneration, furnace, emissions

1. Fata Aluminum milestones in sand regeneration

Fata Aluminum experience on thermal regeneration of waste foundry sand started in 1982, both for sand bonded with organic resins and for green sand bonded with bentonite.
At that time Fata Aluminum developed its own first thermal sand regeneration fluid bed furnace.
Since then, Fata supplied more than 60 plants for organic resins all over the world, with capacities from 0,5 t/h to 24 t/h, developing different furnace solutions for different productivities.
For green sand Fata Aluminum supplied 7 plants with mechanical + thermal + mechanical process, with capacities from 1 t/h to 25 t/h, developing different furnace solutions for different productivities.
All the Fata Aluminum furnaces are based on the fluid bed technology.
The mechanical sand surface attrition machine is a Fata Aluminum product, developed for the green sand regeneration.
Our systems include different furnaces, for different productivities and a dedicated mechanical attrition equipment to eliminate the bentonite.

2. Inorganic sand regeneration: a necessity for the "Foundry of the Future"

Due to ever increasing attention to environmental problems related to the traditional foundry activity, mainly linked to the organic resin bonded cores and the consequent solvents emissions, the potential for inorganic resins was very promising due to the zero emissions of this process. As an important benefit this could lead to more "reshoring" of the foundry activities back to the mature-economy countries. This process could be achieved only with the economic advantage given by the regeneration of the used inorganic sand, so as to close the loop and generate a circular economy.

Based on the above assumptions, in 2014 Fata launched his R&D project and the first key step was to invest in a new pilot plant [Fig.1], composed by:

- One mechanical attrition machine, single cell, complete with its own de-dusting unit
- One thermal treatment unit, Easy Rec type.
- An enhanced laboratory to test AFS, LOI, ADV, clay and cores strength.

Fig. 1. ***Fata Aluminum sand regeneration pilot plant***

The system is designed to allow the treatment of multiple batches of mechanical attrition (M) and thermal treatment (T), with the possibility to measure for each treatment the related sand yield.
Starting from 2015 we can regenerate sand with M, T, M+T, T+M, M+T+M treatments.

This pilot plant is to this date utilized on a daily basis for activities tied to:
- Internal R&D tests
- Customer sand tests
- Process qualification

3. Inorganic sand regeneration tests

Since in 2014 the existing literature on this subject was not considered to be fully exhaustive and particularly convincing, we decided to run our plant with all the possible treatment sequences, using both the furnace and the mechanical attrition machine. The first results told us the following:

- The Thermal treatment itself doesn't regenerate the sand
- The Thermal + Mechanical treatment doesn't regenerate the sand
- The Mechanical + Thermal treatment regenerates the sand. The treatment yield of the M+T treatment was close to 85-90%, so we decided not to run test with M+T+M, so as to reduce the % of new sand

addition needed to recover the mechanical treatment losses.

The subsequent tests focused in the optimization of the M+T treatment. We regenerated sand at different temperatures and at different mechanical treatment settings, and at the end of these test we found that:

- Sticking problems during fluidization increase with the increasing the treatment temperature.
- Mechanical treatment has to be set to find the best compromise between sand yield and sand quality.
- Core strength increases with the increasing thermal sand regeneration temperature.

Based on the test results, we decided to develop a new concept of furnace to regenerate the sand at the highest possible temperature, keeping in mind the sticking problems and the fact that none of our existing furnaces could cope with this kind of resin.

This new product, called **Ino Rec** [Fig.2], has a capacity of 1 to 5 t/h and is based on the fluid bed technology. The cooler has a water tube bundle inside the fluid bed and the sand flows by gravity from the furnace to the cooler and then to the final screen.

- The furnace has a shape that allows to work with sand sticking inside too, particularly in the furnace overhead hood.
- The furnace design is patented.

Fig. 2. Ino Rec 1.5 t/h sand regeneration plant

In parallel we kept on running tests with our pilot plant, regenerating inorganic sand for 5 subsequent times, so as to verify the cores strength values after a certain number of recycles. For every cycle me made cores, crashed them, regenerated the sand and made cores again with it.

Being the treatment yield between 85% and 90% of the incoming material, we added at each subsequent cycle 10% of new sand to the previously regenerated sand. The results confirmed that the inorganic sand regeneration system developed by Fata Aluminum can be used in an industrial plant to recycle the inorganic cores used sand since:

- The cores produced with the regenerated sand have strength characteristics higher than the ones manufactured with new sand
- The process yield is close to 90%
- The new sand addition is very limited

The first industrial Ino Rec furnace, having a capacity of 1,5 t/h, has been installed in an Italian foundry at the beginning of 2017. The plant started to be used in full production in May 2017and it is currently supplying the core shop for the manufacture of automotive products.

The customer laboratory tests on this industrial Fata Aluminum plant confirmed that the regenerated sand quality is optimal, even after several re-cycles.

The 73rd World Foundry Congress
„Creative Foundry"
23 - 27 September 2018, Krakow, Poland

Economic solutions for avoiding emissions in foundries

Antonio Cavotta [1*]

[1]xpuris GmbH, Zollhof 2a (Grand Bateau), 40221 Düsseldorf / Germany
*Corresponding address: e-mail: ac@xpuris.com

Key words: emission reduction, exhaust air purification, environmental protection, sustainability, air cleaning system

The requirements of the foundry industry of the 21st century demand a high degree of innovation. The increase in the degree of automation and the use of new digital possibilities are becoming more and more important. The growing complexity of casting geometries, the demand for material efficiency as well as increasingly important applications, such as electro-mobility, are placing new and changed demands on the foundries and the supplier industry. The issue of emissions is already one of the greatest challenges in the foundry industry and will continue to gain in importance in the future.

Regulatory requirements for exhaust air and workplace pollution as well as complaints from the surrounding environment, such as the neighbors, are the main drivers for the increasing calls for ever more emission-reduced production processes. For years, the foundry chemical industry has been researching and developing new products in order to reduce emissions in the various sectors. In addition to these primary measures, secondary exhaust air purification systems are used, which ensure a further reduction in emissions.

1. Exhaust air purification redesigned - the ASK Chemicals start-up xpuris

In addition to the steady optimization of its foundry chemical products, ASK Chemical has decided to take another path which would help foundries tackle the challenges of the emissions and workplace exposure limits.

At the University of Copenhagen and the local Green-Tech Start-up Infuser, new cleaning processes and technologies have been developed that can make an innovative contribution to resolving these challenges. The new company xpuris was founded in the ASK Chemicals Group with the aim of offering these technologies to the foundry industry exclusively and competently.

xpuris pursues the goal of "rethinking exhaust air purification". Thus, in the first step, the emission problems in the foundry process are examined in detail, this is followed by recording the composition of the exhaust air at the different sources in order to develop effective and economical overall concepts. After the technical conception and adaptation to the conditions in the foundries on site, xpuris offers the implementation of "everything from a single source".

2. From theory to practice - the concrete implementation in a core production

The exhaust air purification in the core production Cold Box of a large series foundry can serve as a first practical example. Currently, amine scrubbers and combustion systems are used in this production area to comply with current legal limits.

However, the requirements of the foundries - due to stricter environmental/ work safety requirements or location factors - for sustainable air purification systems are growing steadily:

- Purification capacity: A higher degree of purification is required and will continue to be required in order to comply with limit values (amines, odor and TOC limits)
- Operating safety: Production losses and the risk of fire from condensate deposits and dust sensitivity is reduced, the use of hazardous chemicals can be avoided and the purification system can be monitored.
- Future reliability: Future limits must be complied with, an expansion of the purification system should be possible.
- Efficiency: The investment and operating costs should be as low as possible and adapted to the cleaning requirements for different production capacity utilizations.

In order to meet the requirements described above, the exhaust air strands of various core productions were analyzed and a holistic cleaning concept was developed based on the results. The patented overall concept uses two main systems with different secondary systems and the necessary automation technology:

The xpuris UVP system replicates and enhances the natural self-cleaning mechanism of the atmosphere by more than a factor of 1000. OH radicals generated by UV light attack the pollutants in the exhaust air and form harmless reaction products that can be easily eliminated from the exhaust air stream. The amount of energy used depends on the pollutant and the desired purification result.

The xpuris ACP system uses high-performance catalysts to eliminate pollutants - much like a combustion plant - but operates at significantly reduced temperatures. In these systems, the exhaust air stream is heated to convert the noxious components on the catalyst into harmless gas components.

xpuris systems are modular systems and can be intelligently supplemented by ancillary systems for particle filtration or odor filtering in order to exploit the full potential of the technology. Sensors and state-of-the-art control technology make "on-demand" energy-efficient exhaust air purification possible.

3. Much has already been achieved - but what else is to be expected?

Further research and development will be carried out in order to be able to offer economic solutions for compliance with emissions and workplace exposure limits for other application areas in the foundry industry in the future.

ASK Chemicals and xpuris are partnering with the foundries with their solutions in order to pave the way for a zero-emission future.

The 73rd World Foundry Congress
„Creative Foundry"
23 - 27 September 2018, Krakow, Poland

INCREASE OF OUTPUT - RETROFITS - MODIFICATION

Friedrich Schaefer*

Heinrich Wagner Sinto Maschinenfabrik GmbH, Bahnhofstrasse 101, 57334 Bad Laasphe, Germany,
**Corresponding address: e-mail: Friedrich.Schaefer@wagner-sinto.de*

Key words: Replacement of moulding machine, cooling time extension, add units for automation, preparing for Industry 4.0, sand reclamation

Overview

This presentation focuses on the possibilities of improving, retrofitting as well as replacing of existing equipment. Topics such as an improvement of the mould quality, high repeat accuracy, consistent mould stability by pattern-related parameters as well as sensor technology and automation will play a role for many changeable variables in the moulding process.

SEIATSU moulding process

The extension of the pattern variety thanks to the “SEIATSU“ moulding process with press and combined air flow particularly for complex patterns with deep and narrow pockets, thin-walled partitions, low mould drafts and other challenges of the current modern pattern technology.

The complementary moulding processes "SEIATSU.plus", which accommodate the surface-optimized pattern plates, are also being intensified. Maximized utilization of the pattern surface, particularly in case of assembly up to the edge of the pattern plate, requires an increased mould stability in the area that is close to the moulding box edge. Pattern-side squeezing at the end of the compaction process is the answer of HWS to these requirements of the pattern makers.

These many possible combinations using the parameterized pattern variables for the moulding machine result in “Perfection in every single mould“.

Product portfolio HWS

Besides the introduction of the HWS portfolio, other possibilities of retrofit with the target of increasing productivity are part of this presentation.

Industry 4.0

Not least against the background of the term Industry 4.0, the integration of the "condition monitoring" available in the HWS portfolio with interfaces to the customer's ERP or visualization is presented. See also figure 1.

Practical examples

With the support of various completed projects, the productive, qualitative, ecological and, last but not least, emission-reducing advantages for the foundry are demonstrated exemplarily.

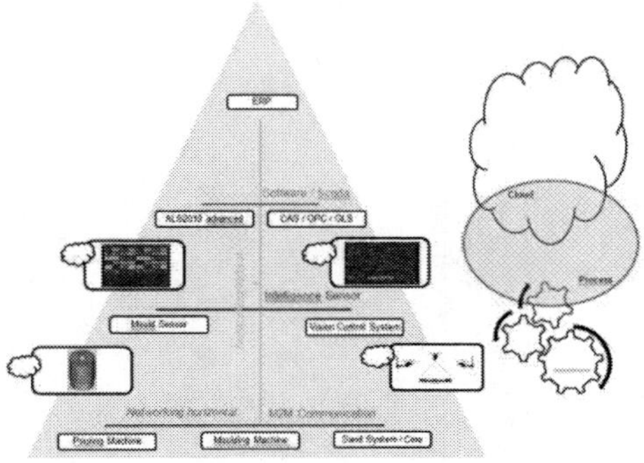

Fig. 1. Perfection in every single mould

The 73rd World Foundry Congress
„Creative Foundry"
23 - 27 September 2018, Krakow, Poland

AUTOMATION OF SAND CORES PRODUCTION – FOUNDRY VOLKSWAGEN POZNAŃ

Tomasz Kamiński

Manager of Planning Department at Foundry Volkswagen Poznań
Volkswagen Poznań sp. z o.o., ul. Warszawska 349, 61-060 Poznań, Poland
Foundry Volkswagen Poznań sp. z o.o., ul. 28 Czerwca 1956r. 240, 61-485 Poznań, Poland
e-mail: tomasz.kaminski@vw-poznan.pl

Key words: Cores production, automation, foundry, innovation, ergonomics.

Volkswagen Poznań consists of four plants dealing with production of cars, special vehicle bodies and castings.

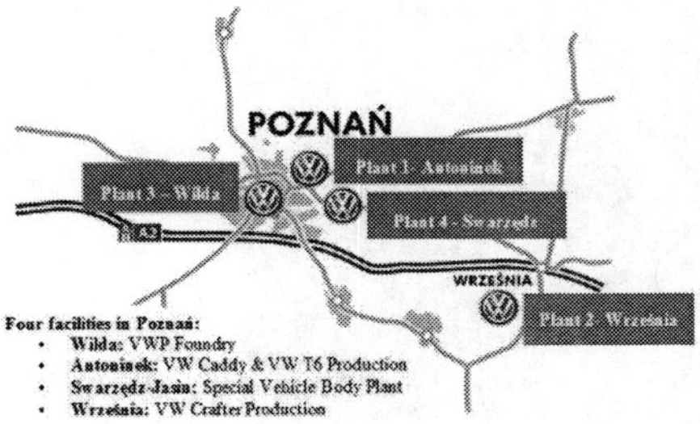

Fig. 2 Location of VW Poznań plants

The Volkswagen Poznan Foundry produces cylinder heads and suspension elements, founded using gravitational technology as well as steering column gearbox housings and clutch housings founded using high-pressure technology. Volkswagen Poznań Foundry provides components for the majority of Volkswagen brands.

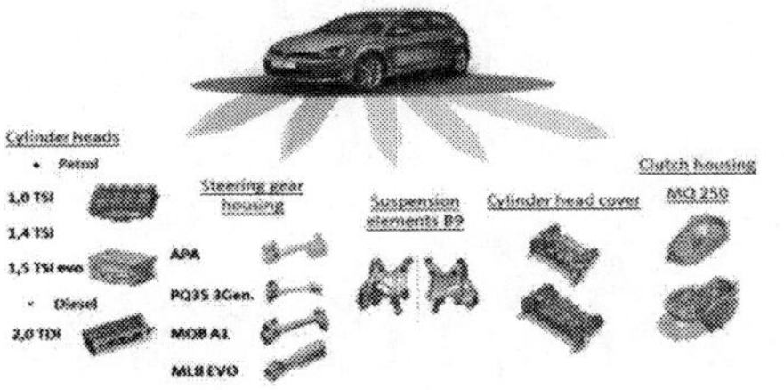

Fig. 3 Foundry VW Poznań - products portfolio

Automation of sand cores production

Sand cores production is the first stage of cylinder heads production process. Different cores are collected in order to model the inner shapes of cylinder head during aluminum casting.

In order to increase the production output, as well as improve the ergonomics of the workplace, Foundry Volkswagen Poznań started to introduce

Fig. 1 Sand core

fully automatic production of sand cores.
Automatic production cell for core DK 1.5 TSI EVO (production capacity 240.000 per annum) consists of core-shooting machine, transportation robot, two deburring robots, deburring table and gluing station.

Cores are picked up from the machine by the transportation robot and put on deburring table. Core shooting machine starts the next cycle, while the cores are being deburred by two deburring robots. Next step is gluing of the cores – deburred cores are picked from deburring table and the glue is dosed by the gluing device. Last step is putting the glued cores onto conveyor belt and transporting them outside of the production cell. This setup allows to extract the cores from the machine and perform all other processes automatically, hence minimizing burdensome labor of the machine operator, whose role gets limited to picking finished cores form conveyor belt and storing them on a trolley.

Fig. 4 Manual deburring

Fig. 5 *Automatic deburring*

Currently three core-shooting machines LHL50 and one machine LCB25 are fully automated and another LCB25 will be automated in Q3 2018.

MANAGEMENT

The 73rd World Foundry Congress
„Creative Foundry"
23 - 27 September 2018, Krakow, Poland

INNOVATION IN KNOWLEDGE TRANSFER FROM ACADEMIA TO THE FOUNDRY INDUSTRY - AN ADVANCED CASE

József Tamás Svidró [1]*, Salem Seifeddine [1], Ehsan Ghassemali [1]

[1]Jönköping University School of Engineering, Department of Materials and Manufacturing, Gjuterigatan 5 Box 1026, SE-55111 Jönköping, Sweden
**Corresponding e-mail: Jozsef.Svidro@ju.se*

Key words: blended learning, casting, education, foundry technology, online

Engineering education seems to lag behind other fields in adopting online methodologies as the special needs of this education have not been perceived to be well served by online education. Despite this, there are increasing evidence of use of blended and online learning in engineering education. For example, online self-study environment to supplement the classroom instructions in engineering courses involving graphics, virtual laboratories and simulation environments, and online learning platforms for developing learning networks for global engineering. Studies on the impact of teaching in blended face-to-face and online formats showed that engineers are interested in using online and blended learning even if campus based teaching is mostly preferred.

The fundamentals of education in cast materials and processes are mainly based on using physics and mathematics, hands-on practical and/or laboratory work. Laboratories in particular are hard to implement online due to the need of direct operation of instruments. Likewise, course materials using mathematics have traditionally not been easy to implement as topics that require only text-based instructions.

This presentation introduces an advanced case of achieving distance learning in the field of cast materials and processes. The case is an educational development project that aimed to plan, develop and create a one-year online master program in materials and manufacturing through collaboration between academia and industry in higher education.

In Sweden, Swedish Foundry Association, Swerea SWECAST and Jönköping University represent the Casting Innovation Centre (CIC) that focuses on development of education and research within cast metals and processes. With the aim of sustaining industrial competitiveness through excellence and innovation and ensuring access to strategic important competences through knowledge transfer from state of the art research to industrial environment, Jönköping University, whithin CIC, developed a globally attractive one-year online master programme in cooperation with the industry.
The most significant opportunity with such online master programme is to reach international students and professionals working in foundries. Some of the major challenges of creating an online programme in general are the lack of previous experience of online education and time to learn new pedagogical approaches and distance tools. A strong campus based tradition in teaching considers that laboratory experiments on campus are necessary. This notion has been more and more questioned. In the online programme, laborations and experiments are demonstrated virtually enabling a more secured, environmentally and cost effective learning approach. Opportunities to apply advanced knowledge on industrial practices become available by incorporating relevant industrial cases. By selecting and producing open educational resources, learning materials such as recorded lectures and laborations can be reused and made available for informal learning in the industry. The development of the online courses in this program has been conducted by multi-professional teams from different disciplines such as media, marketing, pedagogy, web designer, teachers and researchers.

The programme was launched successfully during the autumn of 2017 and has already generated remarkable collaborations among different academic disciplines, as well as with industrial partners involved. A number of professionals were already admitted to the pilot run and attention has already reached national and international universities, following the transition towards more flexible forms of education. From the autumn 2018 the programmewill be open for particpants from all over the world.

By describing one of the courses from the programme in details, this presentation also introduces the examples of distance learning tools applied in the program, the teacher and professional/student roles, the industrial collaboration methods and the best practices for course development and delivery.

The 73rd World Foundry Congress
„Creative Foundry"
23 - 27 September 2018, Krakow, Poland

A Strategy Road Map towards World Class Safety through Innovative Proactive and Unique Daily Routine Activities to achieve Zero Injury in Indian Foundry Industry

Deepak Gaur[1]*, Dharmendra Pratap Singh[2]*, K Kiran Kumar [3]*,

[1]Material Analysis lab1 Quality Assurance Lab Maruti Suzuki India Ltd Gurgaon - 122015
[2]Materials Analysis lab1 Quality Assurance Lab Maruti Suzuki India Ltd Gurgaon - 122015
[3]Materials Analysis lab1 Quality Assurance Lab Maruti Suzuki India Ltd Gurgaon - 122015
**** Materials Analysis Lab Quality Assurance Maruti Suzuki India Ltd Gurgaon 122015:***
e-mail: dharmendra.singh1@maruti.co.in

Key words: Poka Yoke, Benchmarking, Autonomous Maintenance, HIRA, LOTO, Visual dexterity

Maruti Suzuki India Limited, formerly known as Maruti Udyog Limited, is an automobile manufacturer in India. It is a 56.21%-owned subsidiary of Suzuki and motorcycle manufacturer Suzuki Motor Corporation. As of January 2018, it had a market share of 56% of the Indian passenger car market. In February 2012, the company sold its ten millionth vehicle in India.
Maruti Suzuki has adopted several advanced tools like ISO 15001, ISO 14001 & OHSAS as well as various in-house developed tools has resulted in to effective and efficient way of working under Productivity, Quality, Cost, Delivery, and Safety Moral & Environment.

Foundry Industry plays major role in automobile industry. It caters 32% of automobile industry need alone. Foundry by nature of its process considered to be most hazardous and unsafe workplace, consuming large amount of natural resources. In short it has been perceived as 3D process i.e. Dirty, Dangerous and Disliked.

To overcome this Team at QA MA Maruti Suzuki studied the Current Senario of Indian Foundry Industry and implimented various projects to inhance Safety and move towards Zero Injury Plant

Along with team took the challenge for innovating Proactive Measures at many areas in foundry to enhance Safety and Morale of Work force.

1. Unique Routine Activities

- Production meeting should start with SAFETY Oath first.
- Jishu hozen(Autonoums maintaince) activity on machine by involvement of Foundry Head on daily basis
- Behaviorial Based Safety observations by each officer and Senior Mangers
- Pickup rounds leaded by PU Manager on weekly basis at each sub unit of the Plant
- Capturing and analysis of injuries and implementing Countermeasures.
- Hazard identification and Risk Assesment Revisit on Monthly Basis for Each Equipment and Process
- Near miss capturing and discussing on daily basis in Plant Production meeting
- Visual Standard Operating Procedure at Each Work Sation with Do's and Don'ts

Benchmarking with Global Automotive Safety Check Sheet and Preparation of Machine wise Applicability Matrix. Eg :- Benchmarking through JIPM (Japan Institute of Planned Maintenance) and Honda Japan Safety Checksheet for Best Result and Zero Injury Plant.

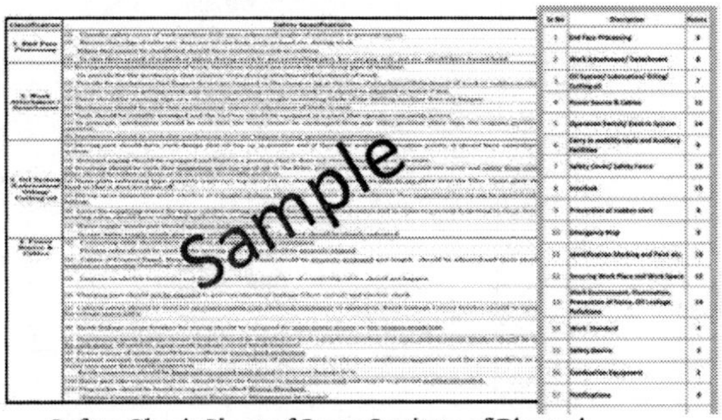

Safety Check Sheet of Japan Institute of Planned Maintenance

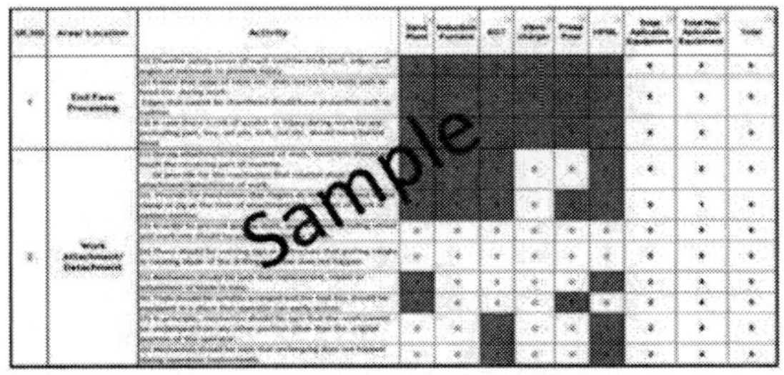

Applicability Matrix for Each Equipment for all safety Measures

Safety Pokayoke at Each Equipment(Min 8 No's Per Equipment) and Process. Checksheet for Checking the Running Condition of Each poka Yoke to Enhace Safety at each Department.

Innovative Unique Activity

- Dedicated Associate for Foundry Process and Safety Training to Associates
- Safety Training Module for Associates
- On Job Training to Associate by Officer
- Visual dexterity (Videography Training Module) for Enhanced Learning Experienc
- Enhancement of associates Morale and Improvement in Associates participation for safety:-
  Associates to be covered under
  A. HIRA and Aspect- Impact Training
  B. Fire Training
  C. LOTO Awareness Training
  D. Atmasurkasha on every second Wednesday of Month
  E. Gemba Safety Quiz
  F. 8 Axis Training
  G. Sweet Distribution for Zero Safety
  H. Recognition for Best Safety Practice

1. Reduction in high/medium risk activities to medium/low risk activity through revisiting of each HIRA proactively and by engineering change or addition of check point added in PM, CLITA and SOP based on degree of difficulties
2. Work Permit system: - Responsibilities & Authorities are distributed as per Hierarchy For extended work dedicated Work Owner officer will be present on work site
3. Training to develop specialized skills
4. Selection of Contractor through Training and Exam(Reputation)
5. Structural Safety Audit and Welding Transformer Check Sheet

Conclution

With Various projects and Methodology under Safety First Approach. Team Maruti has developed a Methedology and Various Tools to eliminate Injury at Indian Foundry Industry. Team has Developed various Unique Routine activity and Innovative Proactive Activities to Enhance Safety and Create Positive Environment.

LIFE LONG EDUCAION OF FOUNDRY EMPLOYEES - A STEP FORWARD

Patrik Svanängen [1]*, *József Tamás Svidró* [2], *Salem Seifeddine* [2], *Conny Gustavsson* [1]

[1]*Swerea SWECAST, Tullportsgatan 3*
Box 2033, SE-55322 Jönköping, Sweden
[2]*Jönköping University School of Engineering, Department of Materials and Manufacturing, Gjuterigatan 5*
Box 1026, SE-55111 Jönköping, Sweden
**Corresponding e-mail: Patrik.Svanangen@swerea.se*

Key words: blended learning, education, foundry technology, online

Access to well-educated and competent employees is commonly seen as the largest future challenge for the global foundry industry. While the global production is showing positive figures and an increasing yearly tonnage of produced components, the European foundry industry is facing increasing difficulties with the recruitment of sufficiently educated and skilled professionals, along with struggles in human capital development of existing staff. A majority of the employees spend most of their career at the same company, where they typically start with a basic introduction which is expected to be enough for 20-40 years of professional work. This in combination with increasing customer demands such as weight reduction, high temperature performance, stricter environmental regulations, increased demands for climate efficient materials and sustainable processes mean that new skills are continuously required. Studies have revealed that for a company to be able to be globally competitive it must undergo a major shift in technology about every 10 years to remain attractive to g demanding costumers. The industry is at a critical point to remedy this situation.

The CLLEFE II project aims to provide a solution to this problem by bringing together European academia, institutes and foundry organizations to collaborate on the development of a new blended learning concept. It will contain a wide variety of short courses where both theoretical and practical elements can be taught remotely. The courses - embedded in a program - will be developed based on industrial needs and quality assured by experts from the foundry industry and end users/customers. A validation system will be introduced as a tool for identifying individual training needs and to individualize education with positive side effects such as cost- and time efficiency. Last but not least, the project will develop a business model so that the training concept creates an income which will be used for continuous development in line with the needs from the society, global customers and the foundry industry.

CLLEFE II is funded by the European Institute of Innovation and Technology, a body of the European Union. Participating countries in the two-year project are initially Sweden, Poland, Italy, Austria, United Kingdom and Spain. The first test-courses will be on the market in the beginning of year 2019.

The 73rd World Foundry Congress
„Creative Foundry"
23 - 27 September 2018, Krakow, Poland

PARTNERSHIP OF INDUSTRY AND KNOWLEDGE CENTER FOR A SUSTAINABLE FOUNDRY INDUSTRY

Lorenzo Sechi [1]*, Vitor Anjos [2]

[1]CLARIANT, 6, rue Louise Michel | 94600 Choisy le Roi, France
[2] CITNM, Rua António da Silva Brinco, 265 3750-122 Águeda, Portugal
**Corresponding address: e-mail: lorenzo.sechi@clariant.com*

Key words: sand preparation efficiency, sand mould quality, green sand technology, sustainable foundry industry, partnership

1. Introduction

Innovation, technology, efficiency and perfection are key drivers of success in the foundry industry.

To reach all of them, many hours of work are needed and different technical expertise are required, many of which the foundry does not have. Therefore, in a world of digitalisation and globalisation, it is important that foundries keep a close contact with partners that can aid in reaching all key driver of success. Suppliers of foundry industry, academia and research institutes are therefore key partners.

2. Cooperation of Industry and Knowledge Center

Long ago, Clariant has realised that, as supplier, they have more to offer to foundries besides Bentonite. They aimed to offer solutions and maintain a constant contact with the customer working in the technical field.

In 2016, Clariant started a cooperation with the Centre for Innovation and Technology N. Mahalingam (CITNM) in Portugal, to join a network of foundry and automotive casting specialists, to explore new opportunities and solutions to foundry industry.

CITNM is a knowledge centre that works closely with all players of the automotive industry supply chain, as a technology transfer centre, to deliver innovation, process improvement, new technologies, new materials and education.

3. Projects handling by cooperation

The best example of cooperation is the work being developed between the partners Clariant, CITNM and Sakthi Portugal. These three institutions are deeply involved in several projects aiming to improve green sand preparation efficiency and mould quality.

The most significant project in development is to increase the overall sand mould quality to eliminate any residual occurrence of sand inclusions in the casting surface. The project involves a multidisciplinary team of 8 technicians, from all partners, to achieve the goals. By using CITNM's network, accesses to other suppliers and technology centres is made available to bring more expertise to the project. Other innovative projects are being planned, such as a permanent connection of Sakthi Portugal's sand preparation system to Clariant offices for monitoring and advise on process control and improvement actions.

This working method based on proximity, cooperation and gathering of different technical expertise, allows Clariant to bring added value to its customers.

References

www.citnm.pt
www.clariant.com/foundry-additives

The 73rd World Foundry Congress
„Creative Foundry"
23 - 27 September 2018, Krakow, Poland

POLISH ENGINEER WITH REGARD TO CHANGES CAUSED BY INDUSTRY 4.0

Katarzyna Liszka [1]*, Józef Szczepan Suchy [2], Paweł Malinowski [3], Katarzyna Klimkiewicz[4]

[1,2,3]AGH University of Science and Technology, Al. Mickiewicza 30, 30-059 Krakow, Poland
Faculty of Foundry Engineering
[4] AGH University of Science and Technology, Al. Mickiewicza 30, 30-059 Krakow, Poland
Faculty of Management
**Corresponding address: e-mail: kasia-liszka@wp.pl*

Key words: competences, engineer, technologist, industry 4.0., training

1. Introduction

The foundry industry is currently facing the fourth industrial revolution called Industry 4.0. This concept is based on information technology and robotizing, which means the total automation of industrial production processes as well as the widespread access to data and machines. Such approach requires changes in applied machines, technologies and employees' competences. The more so, since in the contemporary world, determined by the global and demographic processes, continuous economic changes and the development of new innovative technologies, the competences of employees constitute the element deciding on the company success, aimed at obtaining the competitive advantage. Therefore adjusting the employees' competences to continuously changing reality is so essential.
The aim of the research is to identify the core competencies of employees working in foundry, including the foundry engineers competences. The first part of the research was performed with regard to actual requirements facing foundry engineers and the ways of obtaining the needed competences. On the basis of the analyses of competence requirements related to Industry 4.0. the potential competence range, which will be more and more important in the future, the so-called competence gap, was identified.

2. Competences of the Polish foundry engineer

To recognise the current competences of the Polish foundry engineer the wide analysis of the available professional documentation was carried out (including: education standards of the I-st and II-nd degree in the Faculty of Foundry Engineering of AGH, description of the foundry engineer profession (214607) in: Qualifications of Professions and Specialisations of the Ministry of Family, Work and Social Policy), as well as the survey was performed in 24 Polish foundries concerning the required qualifications of the foundry engineer. Then the findings were verified by comparing them with requirements set by foundries towards candidates for technologists/engineer posts. To this aim, 47 job offers, originated from the internet, published directly by employers or by employment agencies in years 2014-2017, were analysed.

Research indicated that the most important professional tasks of the foundry technologist/engineer post are:
- supervision of the process (55.3%),
- preparation of the technical documentation and its control (48.9%),
- improvement of the process operation and its optimisation with the company resources utilisation (46.8%),
- development of the process parameters and designing of technologies (31.9%),
- initiation and implementation of new technological-metallurgical processes (29.8%),
- analysis of processes and incompatibilities (23.4%).

As a result of the research it is possible to determine the general and widely applied requirements (education, experience, competences) for the foundry technologists/engineer employed in the foundry plant.

Fig. 1. ***Requirements for the foundry technologist/engineer post***

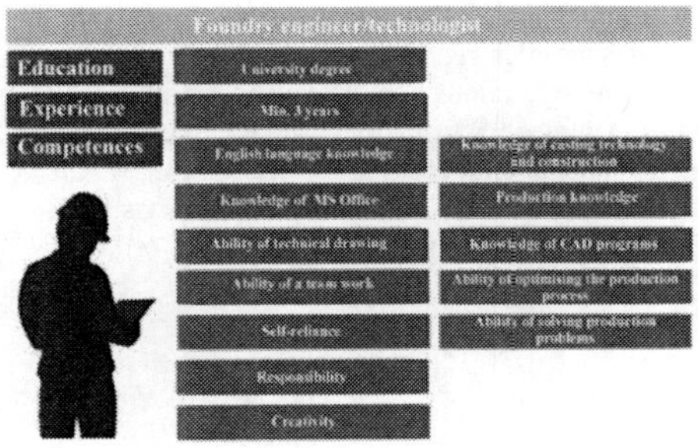

3. Industry 4.0. and the foundry industry

The modern foundry is the automated and robotised plant where processes are coordinated and controlled by intelligent systems allowing for their monitoring from any place in the world. Supervision of machines is possible any time and place due to the control system in the 'cloud'. Orders are deposited by the centralised system and - due to the integrated MRP/ERP system - the casting house automatically controls the deliveries as well as the current and future production needs. Such approach does not mean lowering of the employees engagement but requires another set of skills and competences [1]. Qualifications in such fields as numerical production, robotics and automatics become more important for foundries [2]. Due to adjustment to the concept of Industry 4.0. the foundry plants will improve their production abilities and thanks the system flexibility, allowing for cheaper and shorter production series, will be more reactive to the needs of their customers and more competitive.

4. Foundry engineer of the Industry 4.0. calibre

Industry 4.0. issues high challenges for foundry engineers. The engineer of the fourth industrial revolution calibre will be forced to keep up with the technological progress within his specialisation and specialisations cooperating with it.
In result the contemporary engineer should develop not only new technical competences but also not-engineering soft skills - in much wider scope than nowadays. New technological sectors related to the Industry 4.0. concept i.e. integration of cyber-physical systems, advanced production management systems, robotisation and complex systems of the production data analysis, require the development of new engineering skills.
Thus, the Industry 4.0. engineers are experts having the professional knowledge concerning the special production process as well as wide abilities of the IT range. Data processing and the analysis of large numbers of data originated from several sources, assessing the importance and likelihood of the obtained information as well as drawing the proper conclusions belong to their competencies. Their work is based on electronic media and BigData tools.
Industry 4.0. increase the importance of soft skills. Engineers are required to be more open for changes, flexible in operations, continuously learning and obtaining elements from other fields. Working in virtual - often multicultural - teams causes the necessity of having interpersonal abilities and efficient communication skills [3].

5. Conclusions

Having in view the fact that the Industry 4.0. is the knowledge based operation, Poland - with its high potencial of qualified human resources - should see in it its chances of the development jump.
Entering into the new trend 4.0. demands from the foundry industry the assurance of the availability of the infrastructure providing the production automation and informatisation as well as investments of the managerial and engineering personnel education, since they will be responsible for implementing and maintaining the newest technologies.
The staff management, the new trend Industry 4.0. requires the employee not to apply - as up to now - the previously obtained competences but to be liable to continuous trainings allowing him to improve his skills according to actual requirements and needs of the employer [4].
Educational programmes in universities of technology should be continuously improved and adjusted to the actual needs of the domestic foundry sector and approaching the Industry 4.0. era. This will assure the engineering staff of the highest level allowing for the competitiveness of the Polish foundry plants.

References

[1] M. Lewis, Industry 4.0 and what it means to the foundry industry, Foundry Trade Journal, http://www.foundrytradejournal.com/features/industry-40-and-what-it-means-to-the-foundry-industry-

[2]http://foundrygate.com/en/noticias/ver/4381/industry-40-automated-connectivity, 22/11/2017

[3] Engineers of Industry 4.0. [Not] ready for changes? 2017. ASTOPR Whitepaper

[4] Platform of the Future Industry, Ministry of Development, presentation of July 11-th, 2017, Warsaw.

World Foundry Organization Ltd
Winton House, Lyonshall
Kington, Herefordshire
HR5 3JP, United Kingdom

ISBN 978-1-5108-8041-2

CURRAN ASSOCIATES INC.
proceedings.com